Advances in Intelligent Systems and Computing

Volume 1295

The series "Advances in Intelligent Systems and Computing" contains publications on theory, applications, and design methods of Intelligent Systems and Intelligent Computing. Virtually all disciplines such as engineering, natural sciences, computer and information science, ICT, economics, business, e-commerce, environment, healthcare, life science are covered. The list of topics spans all the areas of modern intelligent systems and computing such as: computational intelligence, soft computing including neural networks, fuzzy systems, evolutionary computing and the fusion of these paradigms, social intelligence, ambient intelligence, computational neuroscience, artificial life, virtual worlds and society, cognitive science and systems, Perception and Vision, DNA and immune based systems, self-organizing and adaptive systems, e-Learning and teaching, human-centered and human-centric computing, recommender systems, intelligent control, robotics and mechatronics including human-machine teaming, knowledge-based paradigms, learning paradigms, machine ethics, intelligent data analysis, knowledge management, intelligent agents, intelligent decision making and support, intelligent network security, trust management, interactive entertainment, Web intelligence and multimedia.

The publications within "Advances in Intelligent Systems and Computing" are primarily proceedings of important conferences, symposia and congresses. They cover significant recent developments in the field, both of a foundational and applicable character. An important characteristic feature of the series is the short publication time and world-wide distribution. This permits a rapid and broad dissemination of research results.

Indexed by SCOPUS, DBLP, EI Compendex, INSPEC, WTI Frankfurt eG, zbMATH, Japanese Science and Technology Agency (JST), SCImago.

All books published in the series are submitted for consideration in Web of Science.

More information about this series at http://www.springer.com/series/11156

Radek Silhavy · Petr Silhavy ·
Zdenka Prokopova
Editors

Software Engineering Perspectives in Intelligent Systems

Proceedings of 4th Computational Methods in Systems and Software 2020, Vol.2

Springer

Editors
Radek Silhavy
Faculty of Applied Informatics
Tomas Bata University in Zlín
Zlín, Czech Republic

Petr Silhavy
Faculty of Applied Informatics
Tomas Bata University in Zlín
Zlín, Czech Republic

Zdenka Prokopova
Faculty of Applied Informatics
Tomas Bata University in Zlín
Zlín, Czech Republic

ISSN 2194-5357 ISSN 2194-5365 (electronic)
Advances in Intelligent Systems and Computing
ISBN 978-3-030-63318-9 ISBN 978-3-030-63319-6 (eBook)
https://doi.org/10.1007/978-3-030-63319-6

This Springer imprint is published by the registered company Springer Nature Switzerland AG
The registered company address is: Gewerbestrasse 11, 6330 Cham, Switzerland

Preface

This book constitutes the refereed proceedings of the Computational Methods in Systems and Software 2020 (CoMeSySo 2020), held in October 2020.

CoMeSySo 2020 conference intends to provide an international forum for the discussion of the latest high-quality research results in all areas related to intelligent systems. The addressed topics are the theoretical aspects and applications of software engineering, computational methods or artificial intelligence. The papers address topics as software engineering, cybernetics and automation control theory, econometrics, mathematical statistics or artificial.

CoMeSySo 2020 has received (all sections) 308 submissions, 184 of them were accepted for publication.

The volume Software Engineering Perspectives in Intelligent Systems brings the discussion of new approaches and methods to real-world problems. Furthermore, the exploratory research that describes novel approaches in the software engineering and informatics in the scope of the intelligent systems is presented.

The editors believe that readers will find the following proceedings interesting and useful for their research work.

September 2020

Radek Silhavy
Petr Silhavy
Zdenka Prokopova

Organization

Program Committee

Program Committee Chairs

Petr Silhavy	Department of Computers and Communication Systems, Faculty of Applied Informatics, Tomas Bata University in Zlin, Czech Republic
Radek Silhavy	Department of Computers and Communication Systems, Faculty of Applied Informatics, Tomas Bata University in Zlin, Czech Republic
Zdenka Prokopova	Department of Computers and Communication Systems, Tomas Bata University in Zlin, Czech Republic
Krzysztof Okarma	Faculty of Electrical Engineering, West Pomeranian University of Technology, Szczecin, Poland
Roman Prokop	Department of Mathematics, Tomas Bata University in Zlin, Czech Republic
Viacheslav Zelentsov	Doctor of Engineering Sciences, Chief Researcher of St. Petersburg Institute for Informatics and Automation of Russian Academy of Sciences (SPIIRAS), Russian Federation
Lipo Wang	School of Electrical and Electronic Engineering, Nanyang Technological University, Singapore
Silvie Belaskova	Head of Biostatistics, St. Anne's University Hospital Brno, International Clinical Research Center, Czech Republic

V. V. Krishna Maddinala	GD Rungta College of Engineering and Technology, Republic of India
Anand N. Khobragade	Scientist, Maharashtra Remote Sensing Applications Centre, Republic of India
Abdallah Handoura	Computer and Communication Laboratory, Telecom Bretagne – France

Organizing Committee Chair

Radek Silhavy — Tomas Bata University in Zlin, Faculty of Applied Informatics
Email: comesyso@openpublish.eu

Conference Organizer (Production)

Silhavy s.r.o.
Web: http://comesyso.openpublish.eu
Email: comesyso@openpublish.eu

Conference website, Call for Papers

http://comesyso.openpublish.eu

Contents

Visualization of Semistructured Information in Organizing Processes of Management of Large Production Systems

Yury Polishchuk(✉)

Orenburg State University, Orenburg, Russia
Youra_Polishuk@bk.ru

Abstract. A method for the automated visualization of semistructured information used in the organization of management processes of large production systems is proposed. This method implements the ability to obtain a schematic representation of expert information in the form of its graphical representation. This allows you to increase the speed of perception of factual data by a group of decision-makers, and, consequently, to reduce the time of synthesis of managerial decisions for large production systems.

Keywords: Semistructured information · Visualization of factual content · Management of large production systems

1 Introduction

The creation and operation of large production systems (LPS), consisting of a set of enterprises and organizations that function as a whole within the framework of one technological process, is one of the directions of industrial development in Russia. Such systems are characterized by a complex interaction between its structural units for the implementation of a single functioning goal. The advantages of such systems are to increase the efficiency of their functioning, increase the scientific potential of its enterprises, and implement their coordination.

The activities of the LPS are related to the solution of managerial decision-making problems, the synthesis of which is carried out by a group of decision-makers (GDM), based on the processing of factual data of exploitative content. Factual data allows you to monitor the current state of the enterprise, determine the strategic, tactical, operational goals and objectives of the enterprise, to develop informed and timely management decisions.

The factual LPS data are typically represented by semistructured information, i.e. information in which a certain structure can be distinguished, however, this structure is completely or partially unknown in advance, or may change over time [1]. It is worth noting that the factual data necessary for the generation

R. Silhavy et al. (Eds.): CoMeSySo 2020, AISC 1295, pp. 1–9, 2020.
https://doi.org/10.1007/978-3-030-63319-6_1

of management decisions of the paperwork can be placed in various information sources: electronic documents, databases, SCADA systems, etc., and in some cases, obtaining them requires a set of studies.

Thus, to obtain expert factual data, it may be necessary to collect and transform it (convolution, aggregation, etc.), followed by presentation in the form of an electronic document with semistructured information content.

To make effective managerial decisions, GDM should receive a final electronic document with a visual representation of semistructured data easy for perception and processing by the experts. In some cases, GDM may require different visualizations of the same factual data to solve decision-making problems.

Thus, a promising area of research is the visualization of semistructured information in the organization of LPS management processes.

2 Methods

Among the studies related to the visualization of document content, the following areas can be distinguished.

Visualization of document collections realizes the possibility of analyzing their factual content demonstrating its connectivity within the collection with the possibility of clustering and grouping documents according to various criteria [2]. Methods of visualization of volume documents, which are aimed at increasing its relevance from the standpoint of the effectiveness of the perception of the material [3]. Development of methods for assessing the effectiveness of visualization of factual information [4].

However, the first two directions were not widely used in the organization of LPS management processes. The third direction is more demanded from the point of view of applicability for the organization of LPS management processes, but it is focused only on increasing the efficiency of perception of visualized factual content by employees included in the composition of GDM. The research direction under consideration complements the approach proposed in the framework of this work to visualize semistructured information.

The large scale of the LPS contributes to increasing the complexity of the process of analyzing its condition, which is due to the large volume of factual data characterizing it. The latter also applies to the structural elements of the LPS, included in its composition. Thus, the task of state analysis is to obtain an array of factual data and its further analysis of GDM. In this regard, to obtain an array of factual data suitable for analysis and visualization, it is necessary to use a model for describing semistructured information that forms this array. Otherwise, the use of the obtained GDM data will be accompanied by difficulties in its processing and visualization.

Thus, the actual tasks for visualizing factual data in the processes of organizing LPS management are: implementing input control for the array of incoming factual data, their structuredness implemented by the model, the possibility of their transformation including convolution and aggregation, as well as the possibility of various visualization options for the same data by GDM requirements.

The process of visualizing semistructured content consists of several stages. Let's consider them in more detail.

Stage 1. At the stage under consideration, GDM forms a task for obtaining factual content, which is necessary for the generate of a management decisions. The formation of the task for the necessary content is implemented on the basis of a semistructured model of the following form [5]:

$$S = \langle root, sObj, LObj, minOccurs, maxOccurs, sMet, Obj\_smet\rangle, \quad (1)$$

where $root$ – root object, $root \in sObj$; $sObj$ – a finite set of objects, each of which contains a fragment of the content of the document (text, drawing, etc.) or acts as a container for one or more objects.

The following meta-properties are available for container objects: $smet_c$ – defines the object as a container; $mixed$ – allows the use of descendant objects in an arbitrary order; $kobj$ – the number of objects in model; $LObj$ – a mapping defined on the set $sObj$, such that $sObj \xrightarrow{LObj} \{Obj_1, \ldots, Obj_{dobj}\}$, where $Obj_{dobj} \in sObj$ – child object; $dobj \in \{1, \ldots, kdobj\}$, $kdobj$ – the number of child objects; $minOccurs$ – a function that determines the minimum possible number of times an object is used in the model; $maxOccurs$ – a function that determines the maximum possible number of times an object is used in the model; $sMet$ – the finite set of available meta-properties of a constraint; Obj_smet – a mapping defined on the set $sObj$, such that $sObj \xrightarrow{Obj\_smet} \{smet_1, \ldots, smet_{mobj}\}$, where $smet_{mobj} \in sMet$ – meta-property of the restriction on the contents of the object; $mobj \in \{1, \ldots, kmobj\}$, $kmobj$ – the number of available meta-properties of the model.

To develop semistructured models, the method proposed by the author [5] can be used.

A semistructured model of content ensures the structuredness of the received data and their input control due to the restrictions defined in the model.

Stage 2. At this stage, an electronic document is formed taking into account the restrictions determined by the model of its content.

Stage 4. The final document with semistructured content is represented by a collection of objects that require visualization. GDM, in accordance with its requirements, forms one or more visualization models $SV_1 \ldots SV_{svn}$, where svn – the total number of visualization models for a given model of semistructured content S.

A mathematical model for the visualization of semistructured content can be written as follows:

$$SV = \langle vObj, vMet, vObj\_vmet\rangle, \quad (2)$$

where $vObj$ – a finite set of visualization objects; $vMet$ – a finite set of available visualization meta-properties; $vObj\_vmet$ – a mapping defined on a set $vObj$ such that $vObj \xrightarrow{vObj\_vmet} \{vmet_1, \ldots, vmet_{vmn}\}$, where vmn – is the number of available visualization meta-properties.

Visualization of the model objects of the semistructured document content is implemented by means of the function of matching the objects of the set of *sObj* to the objects of the set of *vObj*:

$$Fvis\left(sObj_{si}, SV\right) = vObj_{svi}, \tag{3}$$

where $sObj_{si}$ – an object of the set *sObj* of the model *S*, *SV* – a model for visualizing semistructured content; $vObj_{svi}$ – the object of the visualization model of semistructured content.

3 Results

As an example of LPS, we will consider the gas complex of the Orenburg gas condensate field (OGCF) [6].

When operating the OGCF, a large number of automated process control system (APCS) are used. The factual content of the structural diagrams of APCS is used by GDM in the synthesis of various management decisions.

Thus, the creation of documents with semistructured factual content of APCS structural diagrams is relevant for GDM, and visualization of their factual content in the form of a diagram will allow experts to quickly analyze the structure and composition APCS.

For modeling, storage and transformation of semistructured content, we will use the capabilities of the XML language [7].

Stage 1. The model of factographic content of the APCS structural diagram described in the form of an XSD diagram consists of three mandatory segments (Fig. 1):

- ⟨`models`⟩ – a collection of ⟨`model`⟩ segments that describe the devices used in the circuit;
- ⟨`elements`⟩ – a collection of ⟨`element`⟩ segments describing elements (devices, sensors, computers, etc.) of a circuit;
- ⟨`links`⟩ – a collection of ⟨`link`⟩ segments that are used to describe links between elements.

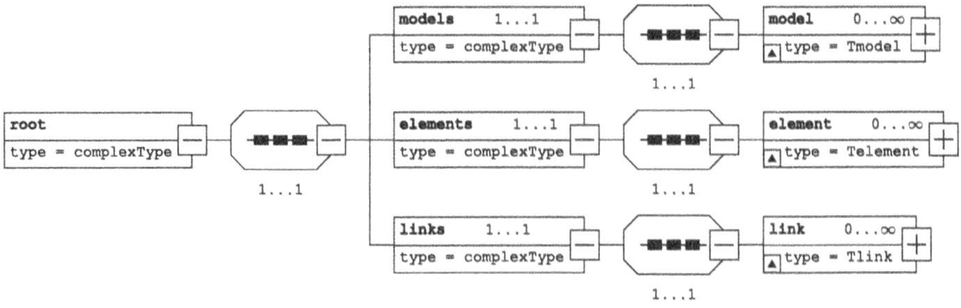

Fig. 1. The structure of the base segment of the model for describing the factographic content of the APCS structural diagram

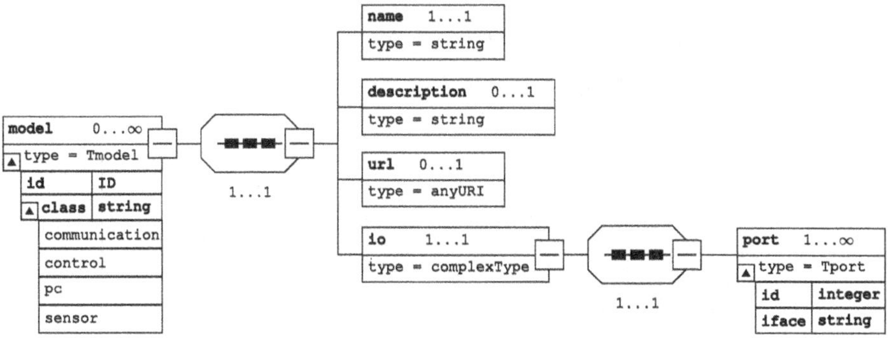

Fig. 2. The structure of the ⟨model⟩ segment of the model for describing the factographic content of the APCS

The repeating ⟨model⟩ segment includes the attributes necessary for a brief description of a specific device model from the structural diagram APCS (Fig. 2).

The following required attributes must be defined for the ⟨model⟩ segment:

- ⟨id⟩ – identifier of type ⟨ID⟩ for further reference to this model in other parts of the document;
- ⟨class⟩ – model class depending on the task it solves (it can have one of the following values: ⟨communication⟩ – communication devices, ⟨control⟩ – control devices, ⟨pc⟩ – control computers, ⟨sensor⟩ – sensors).

The segment under consideration includes the following mandatory objects:

- ⟨name⟩ – model name;
- ⟨description⟩ – a brief description of the model.

The ⟨model⟩ segment also includes an optional object ⟨url⟩ – a link to a detailed description of the device and a mandatory segment ⟨io⟩, which includes one repeating object ⟨port⟩. The object ⟨port⟩ is used to describe the I/O ports on the device. It consists of two attributes:

- ⟨id⟩ – numeric identifier of the port;
- ⟨iface⟩ – port description (for example: RS-232).

The repeating segment ⟨element⟩ is used to describe the devices that make up the control system structure (Fig. 3).

The following required attributes must be defined for the segment ⟨element⟩:

- ⟨model⟩ – link to the description of the model of this device (the same value should be indicated in this attribute as in the ⟨id⟩ attribute of the corresponding object ⟨model⟩);
- ⟨id⟩ – device identifier.

The segment ⟨element⟩ includes the optional object ⟨description⟩, which stores a brief description of the element. The repeating segment ⟨link⟩ is used

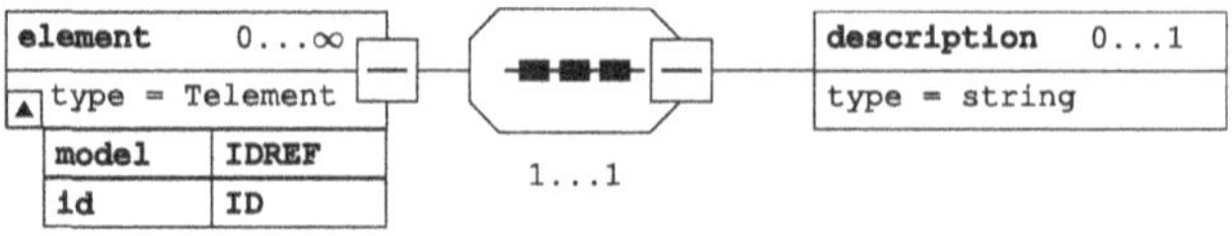

Fig. 3. The structure of the segment ⟨`element`⟩ of the model for describing the factual content of the structural diagram of the APCS

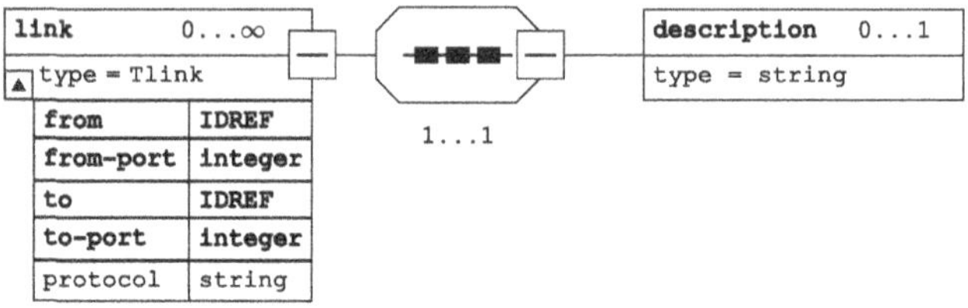

Fig. 4. The structure of the segment ⟨`link`⟩ of the model for describing the factual content of the structural diagram of the APCS

to describe the relationships between the elements of the control system structure (Fig. 4).

The following attributes must be defined for the segment ⟨`link`⟩:

- ⟨`from`⟩ and ⟨`to`⟩ – links to ⟨`element`⟩ objects between which a communication (required attributes);
- ⟨`from-port`⟩ and ⟨`to-port`⟩ – port numbers of related items (required attributes);
- ⟨`protocol`⟩ – an optional attribute that indicates the protocol for transmitting data between two elements.

The segment ⟨`link`⟩ includes an optional object ⟨`description`⟩, which stores a brief description of this connection (for example, information that the communication line is redundant).

Stage 2. Using the proposed model (XSD template), form a electronic document in XML format that stores the factual data of the structural diagram of the APCS for collecting and transporting gas (Fig. 5).

Stage 3. In the considered example, for semistructured content describing the structural diagram of the APCS, the content transformation is not required.

Stage 4. To visualize semistructured content, you can use the capabilities of the XSLT XML document translation language [8]. Using the XSLT template (Fig. 6), we transform the source XML document into an SVG format document, which is a markup language for scalable vector graphics and is a subset of the XML language. This format is intended to describe two-dimensional vector and mixed vector-raster graphics in XML format.

As a result of applying the XSLT template, an SVG format document will be obtained that visualizes the semistructured data describing the structural diagram of the APCS for collecting and transporting gas (Fig. 7).

```
<?xml version="1.0" encoding="utf-8"?>
<root>
  <models>
    <model id="Mzelax" class="communication">
      <name>Zelax M-2B</name>
      <io>
        <port id="0" iface="RJ-45"/>
      </io>
    </model>
    <model id="Mpc" class="pc">
      <name>PC</name>
      <io>
        <port id="0" iface="RJ-45"/>
        <port id="1" iface="RJ-45"/>
      </io>
    </model>
```

Fig. 5. A fragment of an electronic document generated using a model

```
<xsl:template match="element" mode="ElementToNode">
  <xsl:variable name="id" select="@id"/>
  <xsl:variable name="childIds"
  select="//link[@from=$id]/@to"/>
  <xsl:variable name="modelId" select="@model"/>
  <node id="{$id}" model="{$modelId}">
    <xsl:apply-templates select="//element[@id=$childIds]"
    mode="ElementToNode"/>
  </node>
</xsl:template>
<xsl:variable name="coords">
  <xsl:for-each select="$tree//node">
    <xsl:variable name="id" select="@id"/>
    <xsl:variable name="modelId" select="@model"/>
    <xsl:variable name="level"
    select="count(ancestor::*)"/>
    <xsl:variable name="leafChildren"
    select="count(descendant::*[not(child::*)])"/>
    <xsl:variable name="earlierChildren"
    select="count(preceding::*[not(child::*)])"/>
    <xsl:variable name="y" select="70 * $level"/>
    <xsl:variable name="x" select="160 * $earlierChildren
    + 80 * max((0, $leafChildren - 1))"/>
    <coord eid="{$id}" mid="{$modelId}" x="{$x}" y="{$y}"/>
  </xsl:for-each>
</xsl:variable>
```

Fig. 6. A fragment of the XSLT template for converting an XML document to SVG format

4 Discussions

In the search for a way to visualize semistructured content, an approach was proposed to implement the collection of factual content of LPS using models its description. The latter is necessary for GDM in the synthesis of managerial decisions.

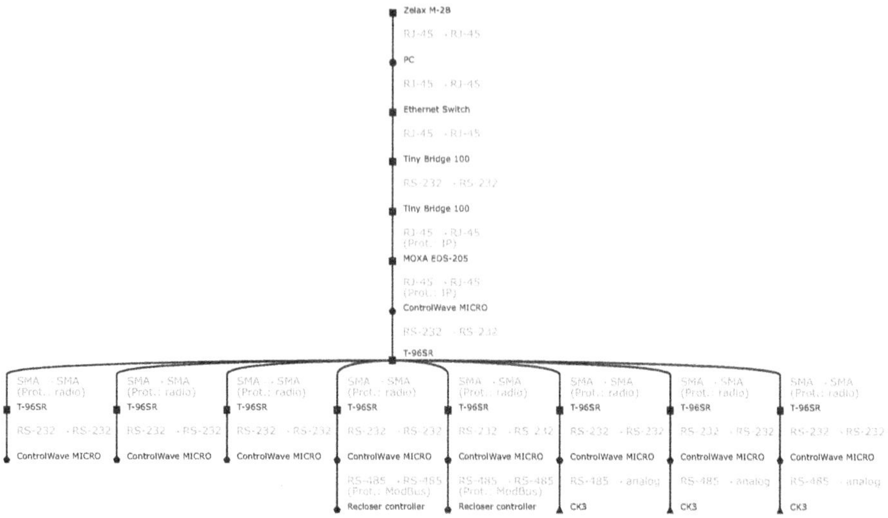

Fig. 7. The result of the visualization of semistructured content

The main advantage of the described method for obtaining data characterizing the state of the LPS is the use of XML format and models that implement the restrictions imposed on the obtained factual data.

All of the above allows us to standardize the process of obtaining factual data and minimize the probability of errors in them.

5 Conclusion

The method of visualization of the semistructured content of LPS proposed in the work allows to automatically obtain its schematic representation, which implements the possibility of analyzing GDM expert information in the form of its graphical representation. The latter allows increasing the speed of perception of GDM factual data, for example, for the considered practical example, in cases of partial modernization of the APCS, its structural scheme will be generated automatically in a minimum time.

References

1. Paley, D.: Modeling of semistructured data. Open Syst. **9**, 57–64 (2002)
2. Apanovich, Z.: Evolution of visualization methods of collections of scientific publications. Electron. Libr. **1**, 2–42 (2018)
3. Zaslavskaya, O., Puchkova, E.: Approaches to the visualization of voluminous text documents. Bull. Peoples Friendship Univ. Russ. **3**, 40–48 (2016)
4. Zakharova, A., Wehter, E., Shklyar, A.: Comprehensive criterion for the applicability of visual analytics in expert systems. Inf. Math. Technol. Sci. Manag. **3**, 37–44 (2018)

5. Polishchuk, Yu.: Synthesis of semistructured models of data content of electronic documents. Bull. Comput. Inf. Technol. **6**, 20–27 (2012)
6. Polishchuk, Yu.: Modeling of a System for Collecting, Processing and Storing Information of a Gas Production Complex. Samgups, Samara (2011)
7. Spencer, P.: Xml Design and Implementation (Professional). John Wiley & Sons, New Jersey (1999)
8. Tidwell, D.: XSLT Mastering XML Transformations. O'Reilly, Sebastopol (2008)

The Development of a Model of the Formation of Cybersecurity Outlines Based on Multi Criteria Optimization and Game Theory

V. A. Lakhno[1], D. Y. Kasatkin[1(✉)], A. I. Blozva[2], Valerii Kozlovskyi[2], Yuriy Balanyuk[2], and Yuliia Boiko[3]

[1] Department of Computer Systems and Networks, National University of Life and Environmental Sciences of Ukraine, Kyiv, Ukraine
{valss21, dm_kasat}@ukr.net

[2] National Aviation University, Kyiv, Ukraine
andriy.blozva@nubip.edu.ua, vvkzeos@gmail.com, rada2606219@gmail.com

[3] Taras Shevchenko National University of Kyiv, Kyiv, Ukraine
julia_boyko2010@ukr.net

Abstract. The models based on game theory which enable to solve the problems of synthesis of information security outlines for information and communication systems (ICS) have been further developed in the article. To the contrary of the existing approaches, the one proposed in the article allows us to consider the situation not only with limited resources (financial, in particular) of the protection party, but also in the context of complex actions by attackers. The proposed additions and the development of game models of the interaction of the protection and attacking parties made it possible to develop a decision making algorithm based on a combination of several methods, in particular, expertise, game theory, etc. The practical value of the developed approach is confirmed by its implementation into the computing core of the decision support system for the selection of rational options of information security means (ISM) for ICS. The use of a decision making support system makes it possible to minimize the time spent on choosing rational sets of ISM according to ICS nodes under the conditions of implementation of complex attack scenarios by intruders. The developed approach is quite flexible in comparison with existing solutions, which allows you to simulate the behavior of intruders of different types and qualifications.

Keywords: Information security · Information and communication systems · Game theory · Protection selection

1 Introduction

The building of an information security system (ISS) to counter the illegal actions of an attacker is a complex procedure that can be carried out on the basis of formal or informal approaches. Nowadays, a number of empirical and formal methods have been developed that solve the task of synthesizing ISS and cybersecurity (CS) [1, 2].

R. Silhavy et al. (Eds.): CoMeSySo 2020, AISC 1295, pp. 10–22, 2020.
https://doi.org/10.1007/978-3-030-63319-6_2

The purpose of the procedure of constructing the ISS outlines is the formation of a set of protection mechanisms that meet the specified requirements of the effectiveness of counteracting intruders. This procedure consists of the analysis stages of the information and communication system (ICS) as an object of protection, the choice of protection mechanisms and the assessment of the protection effectiveness. If at the stage of the assessment of ISS its insufficient effectiveness is detected, the process of selecting protection mechanisms is repeated until the task is solved.

When the interests of a defender and an intruder collide, the parties try to achieve opposite goals having their own certain sets of alternative solutions. Thus, from the point of view of the protection formalized conflict situations are a mathematical model (game), the synthesis of which should be carried out using the apparatus of game theory and other mathematical methods in order to provide the optimal solution according to the degree of security of ICS. The game process is a transition from one state of the game to another as a result of choosing actions among the great number of available alternatives by players.

We admit that the building of ISS involves making decision in conditions of uncertainty considering the conflict relationships of the subjects of the system. However, under the conditions of an ever-increasing number of attacks and the complexity of the scenarios the development of mathematical methods and models that are able to solve optimization problems more efficiently while selecting security mechanisms of ICS simultaneously minimizing the cost of making an effective ISS of the necessary level of security remains an urgent task.

2 Actual Scientific Researches and Issues Analysis

A fairly large number of studies have been devoted to the use of game theory to formalize the procedure of describing the interaction between the protection party of ICS and computer intruders (hackers).

In works [3, 4] the authors mainly consider the existing approaches to the building of ISS based on the game theory. At the same time, the emphasis in the research is placed on two categories – these are the models of pair games and non-cooperative game models. The works do not give specific recommendations on the use of these models in cost minimization procedures by the protection party in the course of the synthesis of information security (IS) outlines of ISS.

In works [5, 6] the authors consider the interaction between attacks and protection mechanisms during the game between an intruder and a defender (system administrator). These publications do not contain specific recommendations for the protection party how to realize the choice of ISS during attacks.

The works [7, 8] contain the results of studies on the appropriateness of using game theory to assess the risks connected with the probability of overcoming the ICS protection outlines by an intruder. The authors, unfortunately, do not provide specific examples of the effectiveness of the models offered in their works.

In [9, 10] the authors propose not only to use the apparatus of game theory to analyze scenarios of the behavior of a computer intruder in the course of attacks on ICS, but also to combine mathematical models obtained on the basis of game theory with elements of control theory. However, this does not consider the possibility of a parallel solution to the problem of minimizing the cost of ISS for ICS. And, in addition, the authors do not take into account the possibility of changing attack scenarios by intruders.

In [11, 12] the authors offer to apply the models and methods of bilinear antagonistic games in the search of ISS investment strategies. However, such models are rather laborious for calculations and require computer support during the search of solutions.

In [13, 14] the possibilities of using computer decision support systems (DSS) and expert systems in the search of rational strategies and alternative options of actions of the protection party of ICS, for example, during various, including complex, attack scenarios by the intruders were considered.

Thus, the existing formal approaches to the modelling and synthesis of IS and CS systems based on the use of the game theory apparatus are mainly devoted to the consideration of static cases of these problems that take into account simple (single-stage) attacks [1, 5, 7, 12], despite the fact that an attack by an intruder is usually complex [1, 2, 4, 14]. Based on the above mentioned, the task of development of the apparatus of game models for the synthesis of information security outlines in ICS remains relevant taking into account the complexity of modern cyber-attacks.

3 The Research Aim

The development of models used in the synthesis of the structure of the information security system based on the use of game theory in order to increase the necessary level of security and minimize the cost of cybersecurity outlines provided that the attacks of the intruders are complex.

4 Methods and Models

The object of the research is distributed ICS with open architecture. ICS consists of many $C = \{c_1, c_2, ..., c_n\}$ interacting components that are involved in the processing and storage of information. Each component is described by a set of characteristics. For example, data transfer rate, bandwidth, functional stability, etc. [15].

The given parameters of the components are valuable for the system which will be denoted by q_c. Considering the peculiarities of the computing environment each of the components is vulnerable to certain types of cyber threats from the set $A = \{a_1, a_2, ..., a_n\}$. We assume that information about the architecture of ICS is open and known to the participants of the game. For example, intruders can obtain this information from an insider or after the implementation of the scanning phase of ICS resources. In addition, the probability of the successful implementation of the threat a_i directed to the component c_i was set. The probability of neutralization of the threat was also set due to the use of information security tools (ISS) from the set $P = \{p_1, p_2, ..., p_m\}$. The effectiveness of the decisions made by an intruder or a defender is affected by random factors which must be taken into account when modelling.

The building of the ISS for the ICS will be considered as an antagonistic game of two players with full information. The peculiarity of the game is that the parties act under the risk [16]. In such game, moves can be deterministic and random. Deterministic moves are the conscious choice according to the strategy of the players among the available alternatives.

The decision of an intruder (hacker – H) determines what threat a_i should be implemented against the component c_i. Then a set of alternatives for a hacker can be represented as a matrix $H = \{h_{ac}\}$. The matrix $H = \{h_{ac}\}$ consists of Boolean elements. Where $h_{ac} = 1$ means making a decision regarding the implementation of the threat a_i against the component c_i. The choice of the strategy of a defender (D) provides the establishment of a protection mechanism (means) $p = 1, 2, ...P$ for the component c_i. A set of alternatives for (D) will be described by a matrix of Boolean elements $D = \{d_{cp}\}$. Where $d_{cp} = 1$ means that a decision is made to establish a protection mechanism p_i in the component c_i.

The random course of the game is a choice made under the influence of random factors but not a specific player. For example, during a hacker's search for passwords, there is a non-zero chance of picking the correct password. Or, on the other hand, during the exploitation by the protection side of the intrusion detection system in the ICS there is a non-zero probability of detecting illegal spy actions by an intruder, for example, a Probe attack aimed at scanning ports. Moreover, for each stochastic move a probability distribution is defined on the set of all alternatives of "nature" [5, 7].

We assume that "nature" affects the decisions made by both the hacker and the ICS defender. We denote by ph_{ac} the probability of success of the hacker during spying or implementation of the threat a_i against the ICS component c_i. Then, we denote by pd_{ap} the probability of detection or neutralization of the threat a_i when using the protection mechanism (means) p_i. The situation in which players find themselves as a result of their moves is called a position. The set of all positions can be divided into the following subsets: positions belonging to the hacker, in each of which he makes a choice from the alternatives available to him; positions belonging to the defender, in each of which he makes a choice among the alternatives available to him.

In game theory [5–7] positions with random moves are singled out separately. However, in the proposed model random moves of "nature" are directly related to the deterministic moves of both players and will be considered together. Thus, applying each of the strategies for players there is a non-zero probability of success (gain) or failure (loss) of the chosen strategy.

The relationship between a defender and a hacker can be formalized using the risk function [8, 16]. A hacker damaging the ICS is trying to maximize the risk. At the same time, a defender countering a hacker and applying ISS tries to minimize the risk or reduce it to zero. Under the limited financial and technical resources [12, 13] according to the given model of an intruder a defender needs to distribute the information security system so that the risk in the ICS would be minimal. In terms of game theory, the risk function is a payment function.

Quantitative value for risk assessment is the damage DA_c.

In general, the ratio for the information security risk function R_{ac} can be described as following [17]:

$$R_{ac} = P_{ac} \cdot DA_c \cdot (1 - NT_{ac}), \tag{1}$$

where R_{ac} – information security risk function for ICS;
R_{ac} – probability of the threat;
DA_c – extent of damage of the component c_i.
NT_{ac} – probability of threat a_i neutralization against the component c_i.

Note that one of the features of the confrontation between a defender and a hacker is the dynamic nature. Typically, an attack is preceded by the monitoring for ICS and reconnaissance. We believe that a hacker attack consists of S stages. It is accepted that damage will only be done in case all steps are completed successfully. After all, if ISS is not overcome at any stage the threat (attack) will not be realized. In this case, the probability of the threat can be represented as follows:

$$R_{ac} = \prod_{s=1}^{S} P^{s}_{ac} \cdot DA_c \cdot (1 - NT_{ac}), \tag{2}$$

where P^{s}_{ac}– probability of the threat at the stage of the attack S.

Expression (2) reflects the dynamic nature of the behavior of the system the state of which changes under the influence of each of the parties. Considering the stages of making an attack, the relationship of a defender and a hacker should be described using the mathematical apparatus of positional games [5, 7, 12, 13]. In such game the participants make moves in turns and try to achieve the maximum benefit for themselves at the next turn. In accordance with the stages of the classical complex attack [18] the following steps of a positional game can be distinguished:

The hacker conducts spying actions (stage 1) in order to analyze, study and search for vulnerabilities in the ICS security system for the subsequent attack. The search of vulnerabilities can be carried out either by passive or active methods [18]. During the active search an attacker runs the risk of being detected. Passive search is less effective but safer for the hacker since the collection of information is done without interfering into the operation of ICS.

The defender's task at this stage is to prevent a possible attack by neutralizing vulnerabilities and the hacker, or at the last stage, if the attack was successful for the hacker. Thus, the a priori probabilities for the implementation of threats ph^{s}_{ac} and their neutralization dp^{s}_{ap} can change over time and that is why they are set for each of the stages S.

As it was mentioned earlier the objective function is expressed through information security risk. Choosing an action strategy, a hacker operates with the probability of threats realization:

$$P^{s}_{ac} = ph^{s}_{ac} \cdot h^{s}_{ac} \tag{3}$$

and potential damage $DA_c = da_c$ (or $DA_c = q_c$) due to technical restrictions on the number of simultaneously realized threats CT. The defender can reduce the risk by installing additional protection mechanisms (means):

$$NT_{ac} = \sum_{p} dp^{s}_{ap} \cdot d^{s}_{cp}. \tag{4}$$

If $NT_{ac} = 1$ then the component (c_i) is completely protected from the threat (a_i). Therefore, a restriction $NT_{ac} \leq 1.$ is introduced in order to prevent the expenditures on excessive ISS. In addition, the defender in most cases has limited financial resources (WD) for the implementation of protection mechanisms (p) the cost of each of which is equal (wd_p). More detailed description of the case when the protection party has limited financial resources (WD) for the construction of information security outlines, and the search of an investment strategy in ISS based on the use of a class of bilinear multi-step games is given in our previous research [11, 13].

Substituting the variables in (2) and taking into account the goal of the defender and the intruder, we can write down the objective function as the following:

$$R = \min_{d} \min_{h} \prod_{s=1}^{S} \sum_{a=1}^{A} \sum_{c=1}^{C} \left\{ ph^{s}_{ac} \cdot h^{s}_{ac} \cdot q_c \cdot \left[1 - \sum_{p}^{P} dp^{s}_{ap} \cdot d^{s}_{cp} \right] \right\}. \tag{5}$$

Under the following limitations:

$$\sum_{a,c,s} h^{s}_{ac} \leq CT, \quad \sum_{c,p,s} wd_p \cdot d^{s}_{cp} \leq WP, \quad \sum_{p} dp^{s}_{ap} \cdot d^{s}_{cp} \leq 1, \quad d^{s}_{cp} = \{0, 1\}, \ h^{s}_{ac} = \{0, 1\}, \tag{6}$$

where $CT-$ the amount of simultaneously of realizable threats CT.

The non-cooperative game of a hacker (H) and a defender (D) of ICS can be written as the following:

$$G\langle I = \{1, 2\}; \quad ST = \{D, H\}; \ R\rangle, \tag{7}$$

where $I-$ great number of players, $ST-$ many acceptable players' strategies, $R-$ winning criterion, objective function.

Provided a finite number of stages, the positional game of a hacker and a defender is considered as a static task. The first step in solving the nonlinear problem (5) is the transition to the dual problem by introducing a variable Ψ and fixing the value of a defender's strategies:

$$R = \min_{\Psi} \prod_{s} \sum_{ac} \Psi^{s}_{ac}, \tag{8}$$

$$\Psi_{ac}^{s} \geq ph_{ac}^{s} \cdot q_c \cdot \left[1 - \sum_{p} dp_{ap}^{s} \cdot d_{cp}^{s}\right], \tag{9}$$

$$\sum_{c,p,s} wd_p \cdot d_{cp}^{s} \leq 1, \quad \sum_{p} dp_{ap}^{s} \cdot d_{cp}^{s} \leq 1, \quad \Psi_{ac}^{s} \geq 0. \tag{10}$$

At the stage of selection of specific ISS for the ICS node we have a task connected with multi-selection (an analogue of the backpack problem). Here the optimization of the placement of "objects" included into the set of ISS must be considered as a modification of the combinatorial problem of a backpack. This approach is distinguished by a fairly simple and rational formalization of the problem statement as well as logical interpretation of the obtained solutions.

If given ISS at the node ICS, $i-$ ISS from the set $P = \{p_1, p_2, \ldots, p_m\}$ provide the level of protection $dpr_i > 0$ and the coefficient of the effectiveness of ISS $\zeta_i > 0$, so it is necessary to choose the set of ISS under which:

$$\Delta DPR = \left[\sum_{i=1}^{N} b_i \cdot dpr_i - DPR\right] \rightarrow \min, \tag{11}$$

$$\overline{Z} = \frac{\sum_{i=1}^{N} b_i \cdot E_i}{\sum_{i=1}^{N} b_i} \rightarrow \max, \tag{12}$$

where $\Delta DPR-$ deviation of the effectiveness of the protection of ICS from the required; $\overline{Z}-$ intermediate coefficient of effectiveness of the set of ISS for the ICS node; $DPR-$ effectiveness (degree of security) which must be provided; $N-$ number of subjects (ISS) at ICS node; $E_i = \sum_{i=1}^{n} u_i \cdot \zeta_i-$ effectiveness of separate ISS, ζ_i- i ISS effectiveness criterion, u_i- weighty coefficient of importance of I –criterion; $W-$ backpack capacity (PBC node), $w = \{w_1, w_2, \ldots, w_N\}-$ the corresponding set of positive integer weights, $dpr = \{dpr_1, dpr_2, \ldots, dpr_N\}-$ the corresponding set of positive integer values. It is necessary to find the set of binary values $B = \{b_1, b_2, \ldots, b_N\}$, where $b_i = 1$, if a subject n_i is included into the set, $b_i = 0$, if the subject n_i is not included.

The further solution to the problem (8)–(12) occurs using classical methods for the solution of multi criteria optimization problems, for example, due to the greedy algorithm or the branch and bound method [19]. As a result of the solution we obtain the optimal set of solutions for a defender d_{cp}^{s*} and an intruder h_{ac}^{s*} which in the theory of games balance Nash [20].

5 Computational Experiment

Let's consider a distributed ICS of a financial institution (for example, a bank or a credit union) that provides transfer, processing and storage of personal data of clients. Suppose that an institution operates in a certain region and has representative offices in all cities. The central office coordinates the work of the entire institution. Thus, the system consists of $c = 1, 2, \ldots, N$ components (N = 25–35 is accepted for our computational experiments) that interact with each other.

The task is to build an information security system for the described ICS. It is assumed that information about information processing technologies and the architecture of ICS is available and may be accessed by cybercriminals. In addition, an important requirement is to take into account the nature of an intruder who can carry out attacks in several stages (S).

To solve this problem it is necessary to formalize the system under protection as well as 1) evaluate potential losses for each segment of the system (q_c); 2) analyze and synthesize the model of the intruder and cyber-attacks taking into account these requirements which will also include possible threats (a) and probability of their implementation $-ph^s_{ac}$; 3) make analysis and synthesis of a protection model for which ISS (p) capable to resist an intruder will be selected.

The software implementation of the models offered in the article, which allow us to find a solution to the problem of searching optimal outlines and sets of ISS for ICS based on multi criteria optimization and game theory, was done in C # language in the Microsoft Visual Studio 2019, Fig. 1.

The DSS module has been conceptually implemented which, based on the available data on the architecture of ICS and the combination of ISS and CS, enables to select the best option for placing available funds among the nodes of the ICS automatically.

Computational experiment results are given in Tables 1, 2 and Fig. 1.

As a result of solution of such problem we obtain the relative value of the risk as well as a set of protection mechanisms that will be optimal when confronting the selected type of intruder.

6 Discussion of the Experiment Results

The upper part of the Table 1 (part 1) shows the ICS components (c) of a financial organization and their values (q_c). Successful implementation of a cyber-threat against one of the ICS components will result in financial and reputational losses equivalent to the value of this component for the functioning of ICS as a whole. In case of the absence of statistical data and financial reports let's suppose that the damage (q_c) is proportional to the number of clients of the ICS branch that was attacked (see Table 1, part 1). In the middle part of the Table 1 (part 2) threats to information security (a) and the probability of their occurrence ph_a. Are shown. Knowing the typical features of a hacker (for example, an internal insider, or a criminal) and his goals you can select typical IS threats using which the hacker can achieve his goal. The developed approach considers the attack as a dynamic process, therefore, it is necessary to divide the threats

Table 1. Computational experiment results

Part 1	Branch ICS (**Component ICS**)	Amount of damage (q_c) (Accepted in proportion to the number of clients of the organization branch
	Component 1	50 000
	…	…
	Component *N*	500 0000

Part 2	**Stages**	*Threat index number* $A = \{a_{1,} a_2, \ldots, a_n\}$	**Cyber threats,** $A = \{a_{1,} a_2, \ldots, a_n\}$	**Probability of implementation,** ph_{ac}^{s}	
	1	**Reconnaissance**			
		1	Network scan	0,55–0,65	
		2	The use of scanners of vulnerability	0,65–0,75	
		3	Protocol Analysis	0,25–0,35	
		4	Others	0,75–0,85	
	2	**Penetration**			
		5	Password matching	0,35–0,45	
		6	Trojan program	0,55–0,65	
		7	Substitution of an object	0,75–0,85	
		8	Others	0,85–0,95	
	3	**Target**			
		9	Data modification	0,85–0,95	
		10	Disclosure of information	0,45–0,55	
		11	Denial of service	0,85–0,95	
		12	Privilege increase	0,75–0,85	

Part 3	**Protection mechanism,** $P = \{p_{1,} p_2, \ldots, p_m\}$	**Cyber Threat Indices for IS** $A = \{a_{1,} a_2, \ldots, a_n\}$												**Cost,** wd_p, *In standard unit*	
		1	2	3	4	5	6	7	8	9	10	11	12		
	1. Access control	0,3	0,3	0,1	0,7	0,2	0,1	0,1	0,1	0,1	0,1	0,1	0,9	15	
	2. Identification and authentication	0,4	0,1	0,1	0,2	0,8	0,1	0,1	0,1	0,1	0,1	0,2	0,1	10	
	3. Cryptographic protection	0,3	0,2	0,1	0,1	0,3	0,2	0,1	0,7	0,7	0,8	0,1	0,1	20	
	4. Integrity	0,1	0,3	0,1	0,2	0,1	0,1	0,2	0,9	0,9	0,1	0,2	0,2	5	
	5. Anti-virus protection	0,1	0,4	0,1	0,1	0,1	0,8	0,1	0,1	0,2	0,1	0,1	0,1	10	
	6. Intrusion detection systems	0,7	0,9	0,3	0,2	0,1	0,3	0,8	0,1	0,3	0,1	0,7	0,1	30	

into stages (S) in the sequence as they will be implemented by the intruder, respectively, in the Table 1, three stages of the threat realization are shown.

It is assumed that in each of the system components (c) the same information processing technologies are used. Consequently, the probability of a threat being realized against each of the components is the same $\forall a, \forall c \Rightarrow ph_{ac} = ph_a$. For the

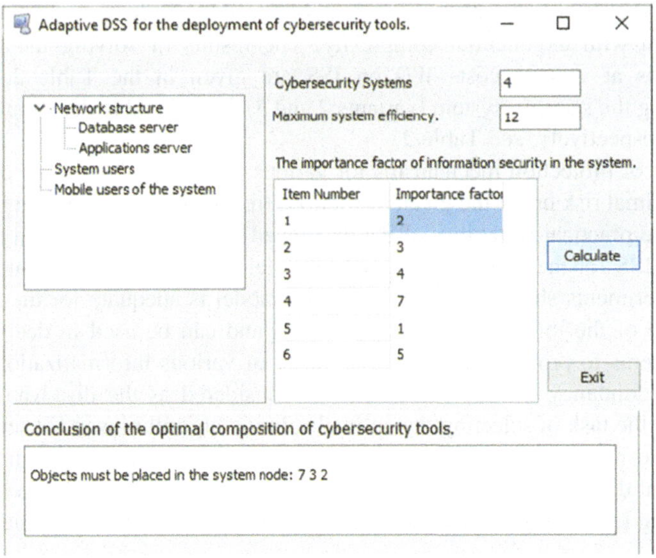

Fig. 1. General view of the DSS interface of the formation of ISS sets for ICS cybersecurity outlines based on multi criteria optimization and game theory

Table 2. Results of calculations

Component ICS	Set of protection mechanisms, $D = \{d_{cp}\}$ (Cost in standard monetary units)	Allocated resources limited by value – *WD*, standard monetary units		
		Variant 1	Variant 2	Variant 3
		500	750	1000
Component 1	10000	100	100	100
Component 2	2000	100	200	500
…	…	…	…	…
Component N	1500	25	25	25
The total estimated risk for IS –*R*		0,187–0,199	0,108–0,119	0,049–0,058

simplicity of calculations we accept that the probability of a threat realization depends directly on the type of threat, however, it does not depend on the characteristics of the system component, at least until the information security system is implemented.

The expert assessment method [22] determines the effectiveness pd_{ap} of each of the protection mechanisms (p) against existing threats (a) in the system, as well as the cost of their implementation wd_p. (see Table 1, part 3). Each of the protection mechanisms (p) provides a certain degree of protection $dpr_i > 0$.

Synthesis of ISS is carried out directly using relation (5) the result of which is a set of protection mechanisms (means) $P = \{p_{1,}p_2, \ldots, p_m\}$ for each component (c_i) of the ICS. The number of possible combinations of the analyzed models equals $3 \cdot 10^5$.

Therefore, the automated mathematical packages Maple and Matlab were used to solve this problem with exponential complexity. The results of solving the problem for several cases at various costs *WD* on ISS are given in the Table 2. In case of strengthening the security system (variants 2 and 3) the risk value decreases to 11–12% and 5–6% respectively, see Table 2.

The sets of protection mechanisms for various initial data have been obtained that provide minimal risk under the given restrictions (resources for the building ISS – *WD*).

Thus, the practical suitability of the developed approach is shown on the example of building ISS for the ICS of a notional financial institution. The conducted computational experiments showed that the proposed model is adequate for the synthesis of the structure of the information security system and can be used in decision making support systems to provide information security of various informatization objects.

Some redundancy of calculations can be considered as the disadvantages of the model since the task of selecting the optimal solution algorithm was not set. Perhaps, it is possible to reduce the number of calculations under setting another task by using more rational decision search algorithms, for example, a modified method of dynamic programming [23] or a genetic algorithm [24, 25]. This will be the continuation of our research.

7 Conclusions

The models based on a combination of several approaches, in particular, expert assessment, game theory, etc. have been further developed in the article. This allows us to quickly and fairly accurately solve the problems on synthesis of outlines of information security for information and communication systems (ICS) taking into account the nature of the attacking party actions. To the contrary of the existing approaches, the one proposed in the article allows us to consider the situation not only with limited resources with the protection party, but also in the context of complex actions by attackers. An algorithm that became the basis of the computing core of a decision making support system (DSS) on the formation of ICS cybersecurity outlines was developed on the base of the proposed models. The key difference of the approach proposed in the article is the consideration of the task in terms of complex attacks by an intruder. That is the variants of actions of computer attackers are simulated changing within the attack time. The usage of the mathematical apparatus of game theory, in particular, previously obtained models based on a new class of multi-step bilinear games, provides the minimum guaranteed risk value for the information resources of the protected object. This approach distinguishes the proposed models from methods based only on expert assessment of the degree of the security of ICS. The practical value of the developed approach is confirmed by its implementation in the DSS computational core. The use of DSS makes it possible to minimize the time spent on the selection of rational sets of ISS for ICS nodes in the conditions of realization of complex attack scenarios by the attackers. The developed approach is flexible in comparison with existing solutions. This allows us to simulate the behavior of intruders of different types and qualifications.

References

1. Ustun, T.S., Hussain, S.S.: A review of cybersecurity issues in smartgrid communication networks. In: 2019 International Conference on Power Electronics, Control and Automation (ICPECA), pp. 1–6 (2019)
2. Srinivas, J., Das, A.K., Kumar, N.: Government regulations in cyber security: framework, standards and recommendations. Fut. Gener. Comput. Syst. **92**, 178–188 (2019)
3. Liang, X., Xiao, Y.: Game theory for network security. IEEE Commun. Surv. Tutor. **15**(1), 472–486 (2012)
4. Luo, Y., et al.: Game theory based network security. J. Inf. Secur. **1**, 41–44 (2010)
5. Shiva, S., Roy, S., Dasgupta, D.: Game theory for cyber security. In: Proceedings of the Sixth Annual Workshop on Cyber Security and Information Intelligence Research, pp. 1–4 (2010)
6. Lye, K., Wing, J.: Game strategies in network security. Int. J. Inf. Secur. **4**(02), 71–86 (2005)
7. Cybenko, G., et al. Overview of control and game theory in adaptive cyber defenses. In: Adversarial and Uncertain Reasoning for Adaptive Cyber Defense, pp. 1–11. Springer, Cham (2019)
8. Ni, Z., Li, Q., Liu, G.: Game-model-based network security risk control. Computer **51**(4), 28–38 (2018)
9. Kim, S.: Game theory for network security. In: Game Theory: Breakthroughs in Research and Practice, pp. 369–382. IGI Global (2018)
10. Zhu, Q., Rass, S.: Game Theory Meets Network Security: A Tutorial at ACM CCS. arXiv preprint arXiv:1808.08066 (2018)
11. Lakhno, V., Malyukov, V., Gerasymchuk, N., Shtuler, I.: Development of the decision making support system to control a procedure of financial investment. East.-Eur. J. Enterp. Technol. **6**(3), 35–41 (2017)
12. Akhmetov, B., et al.: Models and algorithms of vector optimization in selecting security measures for higher education institution's information learning environment. In: Proceedings of the Computational Methods in Systems and Software, pp. 135–142 (2018)
13. Akhmetov, B., et al.: Development of sectoral intellectualized expert systems and decision making support systems in cybersecurity. In: Proceedings of the Computational Methods in Systems and Software, pp. 162–171 (2018)
14. Lakhno, V., Petrov, A., Petrov, A. Development of a support system for managing the cyber security of information and communication environment of transport. In: International Conference on Information Systems Architecture and Technology, pp. 113–127 (2017)
15. Liu, X., Liu, S., Dai, Z., Liang, D., Zhu, Q. Analysis of communication requirements for typical distribution network business. In: 2019 IEEE 3rd Information Technology, Networking, Electronic and Automation Control Conference (ITNEC), pp. 1414–1417. IEEE (2019)
16. Böhme, R., Laube, S., Riek, M.: A fundamental approach to cyber risk analysis. Variance **12** (2), 161–185 (2019)
17. Hlushak, V.V., Novikov, O.M.: Metod proektuvannia system zakhystu informatsii z vykorystanniam determinovanoi hry «zakhysnyk-zlovmysnyk». Naukovi visti NTUU «KPI» **2**, 46–53 (2011)
18. Nichols, W., Hill, Z., Hawrylak, P., Hale, J., Papa, M.: Automatic generation of attack scripts from attack graphs. In: 2018 1st International Conference on Data Intelligence and Security (ICDIS), pp. 267–274 (2018)
19. Quesada, I., Grossmann, I.E.: An LP/NLP based branch and bound algorithm for convex MINLP optimization problems. Comput. Chem. Eng. **16**(10–11), 937–947 (1992)

20. Eicher, T., Osang, T.: Protection for sale: an empirical investigation: comment. Am. Econ. Rev. **92**(5), 1702–1710 (2002)
21. Lakhno, V., Malyukov, V., Parkhuts, L., Buriachok, V., Satzhanov, B., Tabylov, A.: Funding model for port information system cyber security facilities with incomplete hacker information available. J. Theor. Appl. Inf. Technol. **96**(13), 4215–4225 (2018)
22. Cherdantseva, Y., Burnap, P., Blyth, A., Eden, P., Jones, K., Soulsby, H., Stoddart, K.: A review of cyber security risk assessment methods for SCADA systems. Comput. Secur. **56**, 1–27 (2016)
23. Smeraldi, F., Malacaria, P. How to spend it: optimal investment for cyber security, In: Proceedings of the 1st International Workshop on Agents and CyberSecurity, pp. 1–4 (2014)
24. Raman, M.G., Somu, N., Kirthivasan, K., Liscano, R., Sriram, V.S.: An efficient intrusion detection system based on hypergraph-Genetic algorithm for parameter optimization and feature selection in support vector machine. Knowl.-Based Syst. **134**, 1–2 (2017)
25. Demertzis, K., Iliadis, L. A bio-inspired hybrid artificial intelligence framework for cyber security. In: Computation, Cryptography, and Network Security, pp. 161–193 (2015)

Measurement of Energy Efficiency Metrics of Data Centers. Case Study: Higher Education Institution of Barranquilla

Leonel Hernandez[1(✉)], Hugo Hernandez[2], Mario Orozco[3], Gabriel Piñeres[3], and Jesus Garcia-Guiliany[4]

[1] Department of Telematic Engineering, Faculty of Engineering, Institución Universitaria ITSA, Barranquilla, Colombia
lhernandezc@itsa.edu.co

[2] Faculty of Economic Sciences, Corporación Universitaria Reformada CUR, Barranquilla, Colombia
Hugo.hp83@yahoo.com

[3] Department of Electronics and Computer Science, Universidad de La Costa CUC, Barranquilla, Colombia
{morozco5,gpineres1}@cuc.edu.co

[4] Faculty of Administration and Business, Universidad Simon Bolivar, Barranquilla, Colombia
jesus.garcia@unisimonbolivar.edu.co

Abstract. Data centers have become fundamental pillars of the network infrastructures of the various companies or entities regardless of their size. Since they support the processing, analysis, assurance of the data generated in the network, and by the applications in the cloud, which every day increases its volume thanks to diverse and sophisticated technologies. The management and storage of this large volume of information make the data centers consume a lot of energy, generating great concern to owners and administrators. Green Data Center (GDC) is a solution for this problem, reducing the impact produced by the data centers in the environment through the monitoring and control of these and to the application of standards-based on metrics. Although each data center has its particularities and requirements, the metrics are the tools that allow us to measure the energy efficiency of the data center and evaluate if it is friendly to the environment (1.Adv. Intell. Syst. Comput. 574:329–340). The objective of the study is to calculate these metrics in the data centers of a Higher Education Institution in Barranquilla, on both campuses, and the analysis of these will be carried out. It is planned to extend this study by reviewing several metrics to conclude, which is the most efficient and which allows defining the guidelines to update or convert the data center in a friendly environment. The research methodology used for the development of the project is descriptive and no-experimental.

Keywords: Energy · Green data centers · Performance metrics · Environment

R. Silhavy et al. (Eds.): CoMeSySo 2020, AISC 1295, pp. 23–35, 2020.
https://doi.org/10.1007/978-3-030-63319-6_3

1 Introduction

In recent years, the Internet has impacted the way we live, modifying aspects of our daily lives, such as the way we work, entertain or learn, giving great importance to the data generated in all these processes, making them more and more relevant. In the same way, the advent of new technological trends such as the Internet of Things (IoT), Cloud Computing and Virtualization, Sofware Defined Networks, Big Data, and data analytics, AI and Digital Transformation have promoted the existence of large data centers responsible for the secure, reliable and always available storage of data. According to [2], data centers use the network between 5% and 25%, and therefore the energy consumed by inactive devices is wasted, which represents a detriment or impact on the environment. Data centers, composed of many servers, networking devices, and electrical equipment, can absorb as much power as a small city. Numerous studies have shown that average server utilization is usually less than 30% of maximum use [3].

The concept of energy efficiency in data centers was very subjective some years ago since it was often not clear how to measure energy, where it should be estimated, or what units to use. For this reason, energy efficiency metrics were defined to convert Data Centers into "Green Data Centers," friendly to the environment. Some of these metrics, applied to the Data Centers of the University, will be analyzed in this study.

The remainder of this document is structured as follows. A fundamental review of the literature about green data centers is made in the second section. The third section continues with the research method used to make way for calculations of the most relevant metrics in the data centers of the University. Finally, it is mentioned a concise analysis of the results and possible future work that can be developed from this work.

2 Literature Review

Green data centers (GDC) can be defined as that physical space in which the mechanical systems, electrical, lighting, and computation are designed for maximum efficiency and low environmental impact [1, 4]. The core of the internet, or any technological infrastructure of an entity or company, on which all networking services and communications are supported, is the data center. The amount of data they process increases as users' needs demand more storage capacity, speed of transmission, and processing of information. This trend will continue to grow, which is why data centers can generate a high impact on the environment due to the advanced cooling, lighting, and temperature systems they use. Due to this, metrics have been defined that provide guidelines to follow for the design and implementation of green data centers, friendly to the environment.

In recent years, it has become evident the increase in the number of data centers and their size, and with it, the increase in the use of energy [5] in its report on energy use in data centers in the United States, in 2006, data centers consumed 61 kWh (equivalent to the consumption of 5.8 Mio. American households), while in 2008, that figure increased to 69 kWh. According to this report, it is expected that between 2014 and 2020, the increase in the use of energy will be 4%, which is equivalent to 73 billion kWh in 2020, which shows the excessive consumption of energy required for its operation. This extreme energy consumption affects not only the supplier's economy but also becomes a social and environmental problem, to the extent that resources are consumed indiscriminately. On the other hand, [6] shows a power loss model capable of optimally selecting the various devices that may constitute a green data center, presenting its applicability in two case studies. However, there are also some areas in which it is possible to reduce energy consumption [7]. Figure 1 shows an estimate, in which, each one of the parts that generally make up a data center, if they are designed under the parameters stipulated in the regulations of the GDC, can contribute to the reduction of energy consumption:

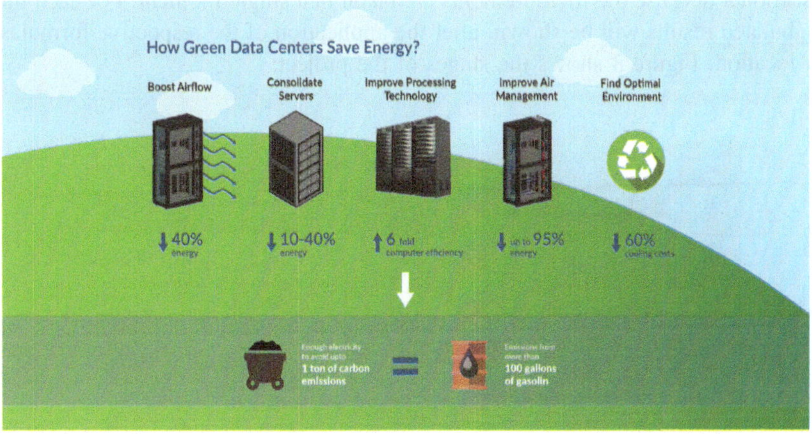

Fig. 1. Reduction of energy consumption with Green Data Centers

Some new technological trends, such as virtualization and cloud and fog computing, have contributed to the design of efficient data centers [8, 9], reducing energy consumption, the use of space, and environmental requirements [10]. Figure 2 shows a traditional energy consumption scheme in data centers:

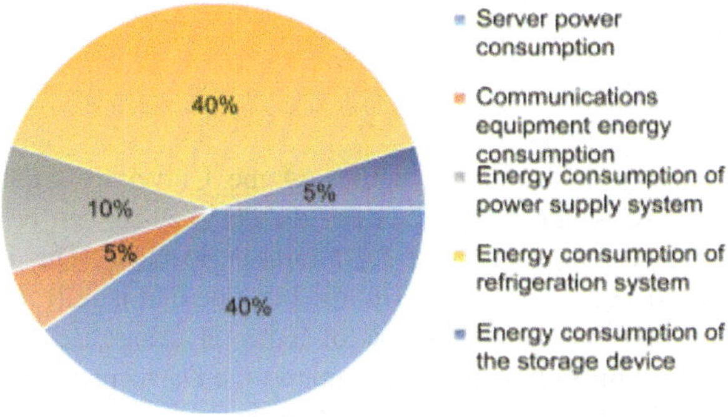

Fig. 2. Traditional energy consumption in data centers

3 Methodology

The research methodologies used to carry out the project are descriptive and non-experimental [11]. Descriptive since all the documentation related to the green data center have been reviewed. Non-experimental because it focuses on the study of the reality of the data centers of the University in its natural dynamics.

The research design to be developed in the project corresponds to a qualitative, transactional design. Qualitative because it is based on a working hypothesis, defined as that the data centers of the Institution are not friendly to the environment, which goes in the opposite direction to the institutional guideline of care for the environment. It is transactional because the measurements are taken in a single moment. For each metric, the tabulated results will be shown, after the application of the respective formulas, for each location. Figure 3 shows the stages of the project:

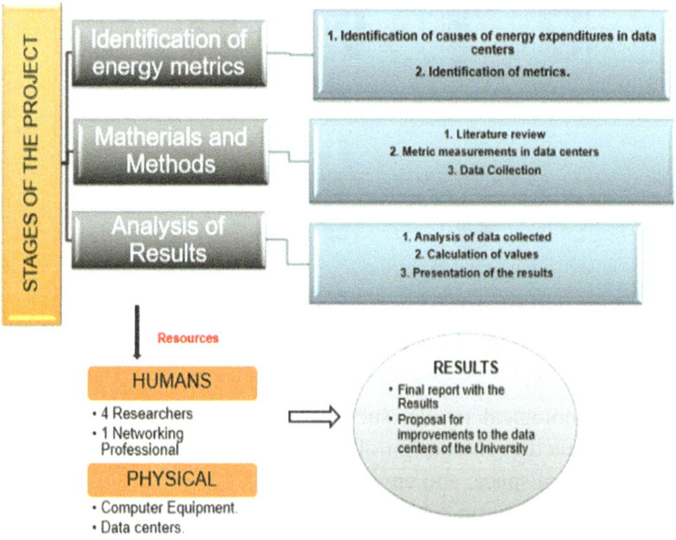

Fig. 3. Stages of the project

4 Results and Discussion

To analyze the energy consumption metrics of the University's data centers, it is advisable first to know which category or tier the data centers can be classified since this allows better decisions to be made for migration to GDC. The TIER of a Data-center is a classification devised by the Uptime Institute that is reflected in the ANSI / TIA-942 standard, and that establishes (as of today) 4 categories, depending on the level of redundancy of the components that support the Datacenter: Tier 1 primary data center without redundancy, Tier 2 is a redundant data center for some topics like the refrigeration system, but they have a single electrical supply path, Tier 3 concurrently maintainable data center (it is Tier 2, and have multiple electrical and cooling distribution lines connected, but only with one active), and Tier 4 fault-tolerant data center, the highest category [12]. For this study, the two data centers analyzed are categorized as Tier II, since they have only one electrical supply path. The data center of campus A has raised floors, auxiliary generators, or UPS, but not campus B that only has UPS. They are connected to a single electrical distribution and cooling line. They are facilities with a certain degree of fault tolerance, and that allows some basic maintenance operations "online".

As noted in the previous sections, the data centers consume a significant amount of energy, according to [13]. The taxonomy of performance metrics of "green computing" is defined, shown in Fig. 4, which is widely used to verify whether or not a data center is "green":

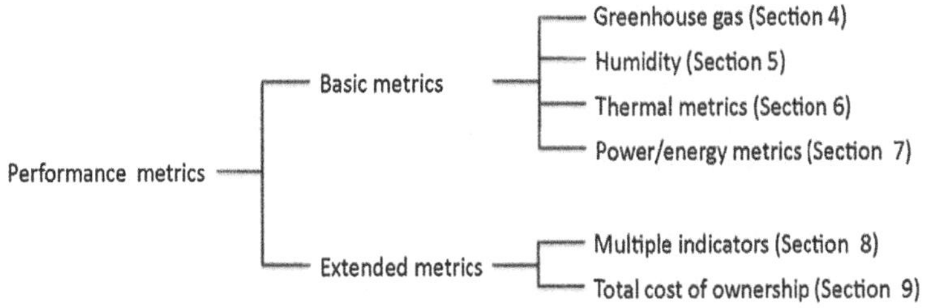

Fig. 4. Taxonomy of "Green Computing" performance metrics

4.1 Power/Energy Metrics

A data center is a particular building, or just a room, which is used to house computer systems, and these are associated with components such as telecommunications and storage systems, backup power sources, redundant communication connections, environmental controls, and security devices. The energy consumption varies greatly, depending on that building or room [14, 15]. For this reason, it is critical to identify the components and devices that make up the data centers of the University.

A. Data Center Infrastructure Efficiency Metric (DCiE)
The DCiE (Data Center Infrastructure Efficiency) is a metric widely accepted by the Green Grid to help IT professionals determine the energy efficiency of data centers and monitor the impact of its efficiency efforts [16], is given by Eq. (1):

$$DCiE = \frac{Power\ Total\ Equipment\ IT}{Power\ Total\ Equipment\ Data\ Center}(watts) \tag{1}$$

A higher DCiE value means that the data center infrastructure is more efficient. The best practice is that this value is above 70% [6] and should never be greater than 100% [17]. The data necessary to calculate the DCiE for Campus B are related in Table 1:

Table 1. DCiE in the data center of campus B.

Quantity	Device	Description	Consumption unit	Total consumption
1	Switch	Huawei s2300	12.8 W–38 W	38 W
2	Switch	Catalyst2960	464 W–870 W (PoE)	1.740 W
1	Switch	Catalyst3560	449 W	449 W
1	Transceiver	Raisecomrc001	36 W–72 W DC	72 W
1	Router	Cisco 3925	85–400 W	400 W
TOTAL CONSUMPTION OF IT				**2.699 W**
1	Conditioned air	ComfortStar	5290W	5.290 W
TOTAL CONSUMPTION OF DATA CENTER				**7.989 W**
DCiE CAMPUS B				**33,78%**

The value of DCiE for campus B is 33.78%, which indicates that it is inefficient, which suggests that actions should be taken to improve this indicator. Table 2 shows the DCiE values for campus A:

Table 2. DCiE in the data center of campus A.

Quantity	Device	Description	Consumption unit	Total consumption
2	Firewall	Sophos WG450	66 W–180 W	360 W
2	Switch	Cisco 2960	464 W–870 W (PoE)	1.740 W
1	Switch	Quidway s3300	100 V–264 V	264 W
1	Switch	TPLINK TL-SG3424	100 V–240 V	240 W
1	Switch	Cisco 2960	464 W–870 W (PoE)	870 W
2	Storage	HP StorageWorks P2000	374 W–432 W	390 W
1	Router	Cisco 2900	80 W–360 W	360 W
3	Servers	HP. Proliant DL 380P GEN8	460 W, 750 W–1200 W	3.600 W
TOTAL CONSUMPTION OF IT				**7.824 W**
1	Conditioned air	ComfortStar	5290 W	5.290 W
1	UPS	Galleon X9B 6K	5400 W	5.400 W
TOTAL CONSUMPTION OF DATA CENTER				**18.514W**
DCiE CAMPUS A				**42,26%**

The value of DCiE for campus A is 42.26%, which indicates that the use of the power in the said campus is average, it is advisable to follow up.

B. PUE

This metric was created by the Green Grid Organization and is calculated using the Eq. (2). The Power Usage Effectiveness-PUE is a metric that allows measuring the efficiency in the use of the energy of a data center:

$$PUE = \frac{Total\ Power\ Equipment\ Data\ Center}{Total\ Power\ IT\ Equipment}(watts) \quad (2)$$

The PUE value indicates the relationship between the energy used by IT equipment and the energy consumed by all data centers [18]. The higher the PUE value, the lower the efficiency of the installation as more "overload" energy is consumed to feed the electrical load. The ideal PUE value is 1, which indicates the maximum achievable efficiency without overload energy [14]. The main disadvantage of this metric is that it only measures the effectiveness of the construction infrastructure that supports a given data center and does not indicate anything about the efficiency of the IT equipment itself [5]. According to Green Grid, the efficiency of the use of the power is measured by levels of efficiency (3 – Very Inefficient; 2.5 – Inefficient; 2 – Average; 1.5 Efficient; 1.2 Very Efficient). Table 3 shows the PUE values for campus B:

Table 3. PUE in the data center of campus B.

Quantity	Device	Description	Consumption unit	Total consumption
1	Switch	Huawei s2300	12.8W–38W	38 W
2	Switch	Catalyst2960	464 W–870 W (PoE)	1.740 W
1	Switch	Catalyst3560	449 W	449 W
1	Transceiver	Raisecomrc001	36 W–72 W DC	72 W
1	Router	Cisco 3925	85–400 W	400 W
TOTAL CONSUMPTION OF IT				**2.699 W**
1	Conditioned air	ComfortStar	5290 W	5.290 W
TOTAL CONSUMPTION OF DATA CENTER				**7.989 W**
PUE CAMPUS B				**2,96**

The value of PUE for campus B is 2.96, which indicates that the use of power on this campus is inefficient, which suggests that actions should be taken to improve this indicator. Table 4 shows the PUE values for campus A:

Table 4. PUE in the data center of campus A.

Quantity	Device	Description	Consumption unit	Total consumption
2	Firewall	Sophos WG450	66 W–180 W	360 W
2	Switch	Cisco 2960	464 W–870 W (PoE)	1.740 W
1	Switch	Quidway s3300	100 V–264 V	264 W
1	Switch	TPLINK TL-SG3424	100 V–240 V	240 W
1	Switch	Cisco 2960	464 W–870 W (PoE)	870 W
2	Storage	HP StorageWorks P2000	374 W–432 W	390 W
1	Router	Cisco 2900	80 W–360 W	360 W
3	Servers	HP. Proliant DL 380P GEN8	460 W,750 W–1200 W	3.600 W
TOTAL CONSUMPTION OF IT				**7.824 W**
1	Conditioned air	ComfortStar	5290 W	5.290 W
1	UPS	Galleon X9B 6K	5400 W	5.400 W
TOTAL CONSUMPTION OF DATA CENTER				**18.514 W**
PUE CAMPUS A				**2,37%**

The value of PUE for campus A is 2.37, which indicates that the use of the power in the said campus is average, it is advisable to follow up.

C. HVAC System Efficiency Metric

The HVAC system (heater, fan, and air conditioning) is the fraction of the data center that is responsible for maintaining appropriate environmental conditions for the proper functioning of the devices included in the data center. The efficiency metric of the HVAC system is calculated by establishing the relationship between the energy consumption of the IT equipment of the data center between the consumption of the HVAC system added to the amount of fuel, steam, and chilled water multiplied by 293, as shown in Eq. (3):

$$HVAC\_Efficiency = \frac{Energy\_IT\_Equipment}{Energy\_HVAC + 293 * (Fuel + Steam + WaterHeat)} \quad (3)$$

If a low HVAC value is obtained, it means that the HVAC system is using a considerable amount of energy, which is necessary to optimize. According to a database of data centers surveyed by Lawrence Berkeley National Laboratory, the effectiveness of the HVAC system can vary from 0.7 (Standard) to 2.5 (Best) [19]. The data necessary to calculate the HVAC efficiency are listed in Tables 5 and 6 for both campuses:

Table 5. HVAC in the data center of campus B.

Quantity	Device	Description	Consumption unit	Total consumption
1	Switch	Huawei s2300	12.8 W–38 W	38 W
2	Switch	Catalyst2960	464 W–870 W (PoE)	1.740 W
1	Switch	Catalyst3560	449 W	449 W
1	Transceiver	Raisecomrc001	36 W–72 W DC	72 W
1	Router	Cisco 3925	85–400 W	400 W
TOTAL CONSUMPTION OF IT				**2.699 W**
1	Conditioned air	ComfortStar	5290 W	5.290 W
TOTAL CONSUMPTION OF DATA CENTER				**7.989 W**
HVAC CAMPUS B				**1,51**

Table 6. HVAC in the data center of campus A.

Quantity	Device	Description	Consumption unit	Total consumption
2	Firewall	Sophos WG450	66 W–180 W	360 W
2	Switch	Cisco 2960	464 W–870 W (PoE)	1.740 W
1	Switch	Quidway s3300	100 V–264 V	264 W
1	Switch	TPLINK TL-SG3424	100 V–240 V	240 W
1	Switch	Cisco 2960	464 W–870 W (PoE)	870 W
2	Storage	HP StorageWorks P2000	374 W–432 W	390 W
1	Router	Cisco 2900	80 W–360 W	360 W
3	Servers	HP. Proliant DL 380P GEN8	460 W,750 W–1200 W	3.600 W
TOTAL CONSUMPTION OF IT				**7.824 W**
1	Conditioned air	ComfortStar	5290 W	5.290 W
1	UPS	Galleon X9B 6K	5400 W	5.400 W
TOTAL CONSUMPTION OF DATA CENTER				**18.514 W**
HVAC CAMPUS A				**3,5**

The value of HVAC for campus B is 1.51, which indicates its effectiveness is better compared to other metrics. Therefore its potential for optimization is minimal. Also, the measure of the efficiency of the HVAC system in the campus A is 3.5, which indicates its effectiveness is good, and that it has the potential to be optimized.

D. Metrics of Space Efficiency, Watts and Performance (SWaP)

The SWaP metric allows measuring the energy efficiency of the reference performance of the server contrasted with the product of the energy consumed by the space used, measured in rack units (RU), as appreciates in the Eq. (4):

$$SWaP = \frac{Perfomance}{Space * EnergyConsumption} \tag{4}$$

Something to highlight about this metric is that it can be applied to any network or storage equipment, as shown in Tables 7 and 8:

Table 7. SWAP in the data center of campus B.

Qty	Dev	Descr	Unit cons	Tot cons	Perf unit (W)	Perf (W)	RU per device	RU Tot	SWAP (%)
1	Switch	Huawei S2300	12.8 W–38 W	38	38	38	1	1	100,00
2	Switch	Catalyst 2960	464 W–870 W (PoE)	1740	370	740	1	2	21,26
1	Switch	Catalyst 3560	449 W	449	60	60	1	1	13,36
1	Transceiver	Raisecom RC001	36 W–72 W DC	72	15	15	2	2	10,42
1	Router	Cisco 3925	85–400 W	400	100	100	1	1	25,00
			Total Cons	**2699**	**Total Perf**	**953**	**RU Tot**	**7**	**5,04**

Table 8. SWAP in the data center of campus A.

Qty	Dev	Descr	Unit cons	Tot cons	Perf unit (W)	Perf (W)	RU per dev	RU Tot	SWAP (%)
2	Firewall	Sophos WG450	66 W–180 W	360	83	166	1	2	23,06
3	Switch	Catalyst 2960	464 W–870 W (PoE)	2610	370	1110	1	3	14,18
1	Switch	Quidway S3300	100 V–264 V	264	92	92	1	1	34,85
1	Switch	TP Link TL-SG3424	100 V–240 V	240	23,3	23,3	1	1	9,71
2	Storage	HP Storage Works P2000	374 W–432 W	390	390	780	2	4	50,00
1	Router	Cisco 2900	80 W–360 W	360	210	210	1	1	58,33
3	Servers	HP. Proliant DL 380P GEN8	460 W,750 W–1200 W	3600	1200	3600	1	3	33,33
			Total Con (W)	**7824**	**Tot Perf**	**5981,3**	**RU Tot**	**15**	**5,10**

4.2 Comparative Analysis of Performance Metrics DCiE and PUE: Standard Value vs. Real Values Obtained

For this comparative analysis, GAPS were calculated, or percentage differences between metrics' standard values and real values obtained in data centers, using the Eq. (5), (where X = DCiE and PUE metrics):

$$GAP(\%) = \frac{X_{(realvalueobtained)} - X_{(standard)}}{X_{(standard)}} \times 100 \quad (5)$$

Table 9 shows the results of the comparative analysis between the standard value of the PUE and DCiE metrics and the actual values obtained in data centers, where GAPS was higher than 100% concerning the high-efficiency standard value established by both metrics, which implies the establishment of priority improvement plans to reduce the PUE and DCIe values and increase the level of efficiency for the Green Grid criterion.

Table 9. Comparative analysis of metrics PUE and DCiE: standard value vs. real values obtained.

Location	PUE real value obtained	PUE value standard	GAP PUE	DCiE real value	DCiE standard value	GAP DCiE
Campus A	2,37	1,2	97,5%	42,26%	100%	57,74%
Campus B	2,96	1,2	146,66%	33,78%	100%	46,66%

5 Conclusions and Future Works

The calculation of each metric allows us to conclude the following. DCiE metric indicates that the efficiency of the infrastructure of the data centers of the University can be improved, since the current performance is inefficient in Campus B, and is average in Campus A. PUE efficiency metric confirms the results obtained with the previous metric, however, it is recommended that this metric is evaluated based on a percentage of measures for periods established by the analyst, to control and monitor the efficiency strategies energy implemented. Concerning HVAC, it indicates that the Campus B data center HVAC system is efficient. Therefore its optimization potential is small, unlike the data center of Campus A, in which the energy efficiency of the HVAC system is regular, so it is possible to optimize it. The data centers of the University have inefficient use of energy, even though the HVAC system installed in both data centers are relatively efficient, so it is recommended to carry out an improvement plan that optimizes energy efficiency. And for SWaP efficiency, we could see which of the installed appliances are being more efficient with the use of energy compared to the others. It could also be observed that Campus B's data center is a little more useful than Campus A's data center concerning the power space ratio. On the other hand, the calculated gap values for the DCiE and PUE metrics allow us to conclude that the two data centers require critical adjustments to be considered as green data centers. Only the PUE gap in Campus A is slightly below 100 (a value equal to or higher indicates that the data center is very inefficient concerning energy consumption). Regarding DCiE, the acceptable values range between 70% and 100%, obtaining for both data centers values well below the minimum allowed.

Among future work that can be done from this study can be mentioned the following: propose a multivariate correspondence analysis (MCA) to evaluate the energy performance of data centers or perform a review of the practices used in different data centers, to improve energy consumption [20, 21] or define an optimization algorithm like [22]. Another future work that can be done is to calculate the gap between the real values obtained from the remaining metrics versus the standard values that should be and validate how far we can be from what is a data center friendly to the environment [23].

References

1. Hernandez, L., Jimenez, G.: Characterization of the current conditions of the itsa data centers according to standards of the green data centers friendly to the environment. Adv. Intell. Syst. Comput. **574**, 329–340 (2017)
2. Baccour, E., Foufou, S., Hamila, R., Tari, Z., Zomaya, A.Y.: PTNet: an efficient and green data center network. J. Parallel Distrib. Comput. **107**, 3–18 (2017)
3. Obaidat, M., Ampalagan, A., Woungang, I., Zhang, Y., Ansari, N.: Handbook of Green Information and Communication Systems. Academic Press, Cambridge (2013)
4. Bin, N., Nor, M., Hasan, M., Selamat, B.: Green data center frameworks and guidelines review **6**, 338–343 (2014)
5. Shehabi, A., et al.: United States Data Center Energy Usage Report. Lawrence Berkeley Natl. Lab. Berkeley, CA, Technical Rep*ort* (2016)
6. Hartmann, B., Farkas, C.: Energy efficient data centre infrastructure—development of a power loss model. Energy Build. **127**, 692–699 (2016)
7. Pierson, J.M.: Large-Scale Distributed Systems and Energy Efficiency: A Holistic View. Wiley, Hoboken (2015)
8. Niles, S.: Virtualization: optimized power and cooling to maximize benefits. Am. Power Convers. White Pap. (2008)
9. Jin, Y., Wen, Y., Chen, Q., Zhu, Z.: An empirical investigation of the impact of server virtualization on energy efficiency for green data center. Comput. J. **56**(8), 977–990 (2013)
10. Uddin, M., Rahman, A.A.: Energy efficiency and low carbon enabler green IT framework for data centers considering green metrics. Renew. Sustain. Energy Rev. **16**(6), 4078–4094 (2012)
11. Hernandez, R., Fernandez, C., Baptista, M.: Metodología de la investigación (2010)
12. Aodbc.: Clasificación TIER en el Datacenter, el estándar ANSI/TIA-942 - Aodbc in the Cloud (2012). https://blog.aodbc.es/2012/07/10/clasificacion-tier-en-el-datacenter-el-estandar-ansitia-942/. Accessed 16 Apr 2020
13. Wang, L., Khan, S.U.: Review of performance metrics for green data centers: a taxonomy study. J. Supercomput. **63**, 656 (2013)

14. Sharma, M., Arunachalam, K., Sharma, D.: Analyzing the data center efficiency by using pue to make data centers more energy efficient by reducing the electrical consumption and exploring new strategies. Procedia Comput. Sci. **42**, 142–148 (2015)
15. Ray, P.P. .: A green initiative to data centers: a review, **1**(4), 333–339 (2012)
16. 42U Data Center Solutions. 42U Data Center Solutions (2018). https://www.42u.com/measurement/pue-dcie.htm.
17. Fernández, P.: Las nuevas métricas de Green Grid para la eficiencia energética. Silicon (2009)
18. Yuventi, J., Mehdizadeh, R.: A critical analysis of Power Usage Effectiveness and its use in communicating data center energy consumption. Energy Build. **64**, 90–94 (2013)
19. Lintner, W., Tschudi, B., VanGeet, O.: Best practices guide for energy-efficient data center design. In: U.S Department Energy, no. March, pp. 1–24 (2011)
20. Dennis, B.: Five ways to reduce data center server power consumption. Green Grid **42**, 12 (2009)
21. Judge, J., Pouchet, J., Ekbote, A., Dixit, S.: Reducing data center energy consumption. ASHRAE J. **50**(November), 14 (2008)
22. Lei, H., Wang, R., Zhang, T., Liu, Y., Zha, Y.: A multi-objective co-evolutionary algorithm for energy-efficient scheduling on a green data center. Comput. Oper. Res. **75**, 103–117 (2016)
23. Hernandez, L., Calderon, Y., Martinez, H., Pranolo, A., Riyanto, I.: Design of a system for detection of environmental variables applied in data centers. In: Proceeding - 2017 3rd International Conference on Science in Information Technology: Theory and Application of IT for Education, Industry and Society in Big Data Era (2017)

Model of Group Pursuit of a Single Target Based on Following Previously Predicted Trajectories

A. A. Dubanov(✉)

Buryat State University, 24 A, Smolin-street, Ulan-Ude, Russian Federation
alandubanov@mail.ru

Abstract. This article describes a geometric model when a group of pursuers pursues a single goal. Movement occurs on a plane, but if necessary, this model can be projected onto an explicitly defined surface. The speed of movement of all participants, both pursuers and targets, is constantly modulated. The goals and strategies of each of the pursuers, despite the difference in trajectories, share one criterion. Their goal is to approach the point of space associated with the object being pursued in a given direction, observing the restrictions on the curvature of the trajectory. The goal and strategy of the target is determined by the behavior of one of the pursuers. #COMESYSO1120

Keywords: Pursuit · Simulation · Algorithm · Target · Pursuer · Trajectory

1 Introduction

Quasi-discrete models in the pursuit problem make it possible to approximate the calculation of dynamic processes with further visualization. In our problem, we introduce a time sampling period, during which the pursuer takes a step and changes the direction of movement. This article discusses the issues of research by the pursuer of a pre-modeled trajectory. It is assumed that the trajectory that meets the specified requirements will be modeled automatically at each time. As an example, we considered a group of four pursuers.

2 Modeling Trajectories

Consider the problem of modeling the path of the pursuer. The path of the pursuer must leave point P with the speed value V_P and reach the target at point T.

Moreover, the trajectory should be constructed in such a way that it goes to the point T in the direction of the vector $V_P^{'}$ (Fig. 1). the Assumed trajectory consists of two parts. From a straight line segment connecting points P and P_{tan}, and from an arc (T, P_{tan}) of radius R_p.

The condition of our simulated trajectory is that the radius of curvature cannot be less than r_p. The angle of the pursuer's entry to point T is determined by the velocity vector $V_P^{'}$. The direction of this vector in our test program is chosen such that the speed

R. Silhavy et al. (Eds.): CoMeSySo 2020, AISC 1295, pp. 36–49, 2020.
https://doi.org/10.1007/978-3-030-63319-6_4

vector of the pursuer V_P' perpendicular to the target speed vector V_T. The center of a circle that meets the specified conditions is calculated as follows:

$$C_T = T + R_P \cdot \left(-\frac{V_T}{|V_T|} \right).$$

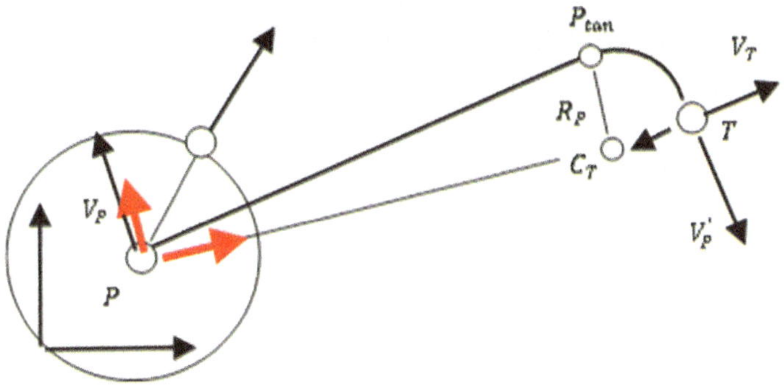

Fig. 1. Estimated trajectory of the pursuer.

To find the point P_{tan}, which is also the interface point of a straight line (P, P_{tan}) and an arc (T, P_{tan}), we implemented a procedure – function that forms a local coordinate system centered at point P and with the basis vectors e_1 and e_2: $e_1 = \frac{C_T - P}{|C_T - P|}$ $e_2 = \begin{bmatrix} -e1_y \\ e1_x \end{bmatrix}$. The basis (e_1, e_2) is orthogonal.

In this coordinate system, the point C_T is converted to the form: $C_{T.n} = \begin{bmatrix} (C_T - P) \cdot e_1 \\ (C_T - P) \cdot e_1 \end{bmatrix}$. Obviously, the coordinates of $C_{T.n}$ will be: $C_{T.n} = \begin{bmatrix} |C_T - P| \\ 0 \end{bmatrix}$. Let the modulus of the vector $|C_T - P|$ will be equal to the number C_x (Fig. 2), then in the local coordinate system (e_1, e_2) centered at point P, the coordinates of the interface point $P_{tan.n}$ will satisfy the system of equations:

$$(P_{tan.n} - C_{T.n}) \cdot (P_{tan.n} - C_{T.n}) = R_P^2$$
$$(P_{tan.n} - C_{T.n}) \cdot P_{tan.n} = 0$$

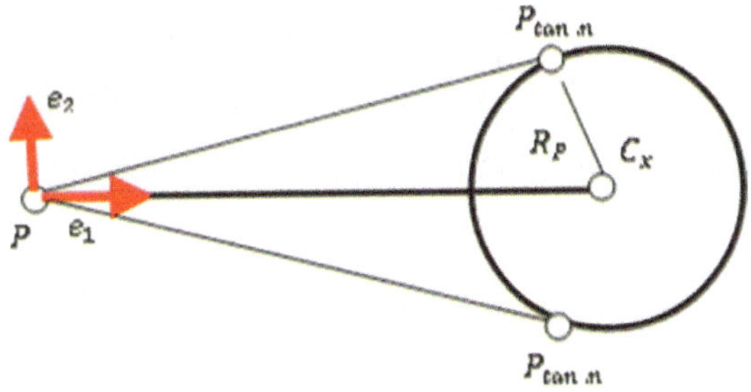

Fig. 2. Determining the interface point in the local coordinate system.

This system of equations has the solution:

$$P_{tan.n} = \begin{bmatrix} \frac{C_x^2 - R_P^2}{C_x} \\ \pm\frac{\left(R_P * \sqrt{(C_x + R_P)*(C_x - R_P)}\right)}{C_x} \end{bmatrix}.$$

This is the solution in the local coordinate system (e_1, e_2) centered at point P. To convert the point $P_{tan.n}$ to the world coordinate system (H_1, H_2), where $H_1 = \begin{bmatrix} 1 \\ 0 \end{bmatrix}, H_2 = \begin{bmatrix} 0 \\ 1 \end{bmatrix}$, you need to get expressions for the basis (H_1, H_2) in the basis (e_1, e_2):

$$h_1 = \begin{bmatrix} H_1 \cdot e_1 \\ H_1 \cdot e_2 \end{bmatrix}, \quad h_2 = \begin{bmatrix} H_2 \cdot e_1 \\ H_2 \cdot e_2 \end{bmatrix}.$$

Then the point $P_{tan.n}$ in the world coordinate system will look like this: $P_{tan} = \begin{bmatrix} P_{tan.n} \cdot h_1 \\ P_{tan.n} \cdot h_2 \end{bmatrix} + P$. In our test program, based on the materials of the Chapter, we chose the option corresponding to the upper position of the point $P_{tan.n}$.

Since we are considering a quasi-discrete model of the pursuit problem, we can introduce a discretization period Δt. Where, within this model, we have that the step of the pursuer over the sampling period has the value $|V_P| \cdot T.t$.

If the minimum radius of curvature of the trajectory of the pursuer is equal, then it is reasonable to assume that the angular rotation frequency of the pursuer P on turns will be equal to $\omega_P = |V_P|/R_P$. The rotation angle during the iteration step cannot exceed the value of $\omega_P \cdot \Delta T$.

3 Analysis of the Coordinates of the Touch Point from the Dynamic Coordinate System of the Pursuer

In our quasi-discrete model of the pursuit problem, the goal of the pursuer is to catch up with the object being pursued, so that at the moment when the coordinates coincide, the pursuer has a given speed vector, while the minimum radius of curvature of the trajectory is not less than the allowed one. Let's form a basis (v_1, v_2) with the origin at point P (Fig. 3): $v_1 = \frac{V_P}{|V_P|}$ $v_2 = \begin{bmatrix} -v_{1_y} \\ v_{1_x} \end{bmatrix}$. In this dynamic coordinate system, which depends on the speed of the pursuer, we translate the coordinates of the point P_{tan}: $P_{tan.v} = \begin{bmatrix} (P_{tan} - P) \cdot v_1 \\ (P_{tan} - P) \cdot v_2 \end{bmatrix}$. The coordinates $P_{i.v}$ in the coordinate system (v_1, v_2) with the beginning at p will be:

$$P_{i,v} = \begin{matrix} \text{if } P_{tan.v_y} \geq 0 \begin{bmatrix} |V_P| \cdot \Delta T \cdot \cos(w_p \cdot \Delta T) \\ |V_P| \cdot \Delta T \cdot \sin(w_p \cdot \Delta T) \end{bmatrix} \\ \text{if } P_{tan.v_y} < 0 \begin{bmatrix} |V_P| \cdot \Delta T \cdot \cos(w_p \cdot \Delta T) \\ -|V_P| \cdot \Delta T \cdot \sin(w_p \cdot \Delta T) \end{bmatrix} \end{matrix}$$

If the angle $\propto$ is less than the angle $\omega\_P \times \Delta T$, then the coordinates of the point $P\_(i.v)$ will look different:

$$P_{i,v} = \begin{matrix} \text{if } P_{tan.v_y} \geq 0 \begin{bmatrix} |V_P| \cdot \Delta T \cdot \cos(\propto) \\ |V_P| \cdot \Delta T \cdot \sin(\propto) \end{bmatrix} \\ \text{if } P_{tan.v_y} < 0 \begin{bmatrix} |V_P| \cdot \Delta T \cdot \cos(\propto) \\ -|V_P| \cdot \Delta T \cdot \sin(\propto) \end{bmatrix} \end{matrix}$$

The angle $\propto$ is the angle between the vector $P_{tan.v}$ and the vector v_1. Next, you should convert the coordinates $P_{i.v}$ from the coordinate system (v_1, v_2) to the world one. To do this, we get expressions for the basis (H_1, H_2) in the basis (v_1, v_2): $h_{1.v} = \begin{bmatrix} H_1 \cdot v_1 \\ H_1 \cdot v_2 \end{bmatrix}$ $h_{2.v} = \begin{bmatrix} H_2 \cdot v_1 \\ H_2 \cdot v_2 \end{bmatrix}$. $H_1 = \begin{bmatrix} 1 \\ 0 \end{bmatrix}$, $H_2 = \begin{bmatrix} 0 \\ 1 \end{bmatrix}$. Where the expression for the point of the pursuer P_i in the next iteration stage will be: $P_i = \begin{bmatrix} P_{i.v} \cdot h_{1.v} \\ P_{i.v} \cdot h_{2.v} \end{bmatrix} + P$. So, the simulated path of the pursuer P_i approaches a straight line (P, P_{tan}). Our simulated trajectory should approach the arc segment (T, P_{tan}) at certain stages of the iterative process (Fig. 1)

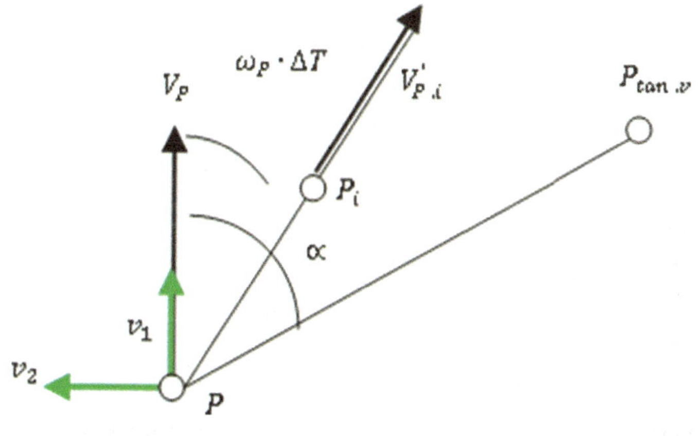

Fig. 3. Choosing the movement direction by the pursuer.

4 Analysis of the Coordinates of the Touch Point from the Dynamic Coordinate System of the Pursuer

In this section, we will consider the situation when the distance between the pursuer P and the center of the circle C_T is less than the minimum radius of curvature of the trajectory R_P (Fig. 4). Figure 4 shows two intersecting circles with radii r_P and R_P, with centers at points P and C_T, respectively, where $r_P = \omega_P \cdot \Delta T$, and R_P - the minimum radius of curvature of the trajectory of the pursuer. The purpose of the problem described in this section is to determine the coordinates of the point P_i in the world coordinate system based on the analysis of the point P_{int} intersection of circles. It is convenient to get the point of intersection of circles in the coordinate system (e_1, e_2) with the center at point P, as before, it is considered: $e_1 = \frac{C_T - P}{|C_T - P|}$ $e_2 = \begin{bmatrix} -e_{1_y} \\ e_{1_x} \end{bmatrix}$. In this coordinate system, expressions for the intersection point of the circles $P_{int.n}$:

$$P_{int.n} = \begin{bmatrix} \frac{C_X^2 - R_P^2 + r_P^2}{2 \cdot C_X} \\ \pm \frac{\sqrt{(C_X + R_P - r_P) \cdot (C_X - R_P + r_P) \cdot (R_P - C_X + r_P) \cdot (R_P + C_X + r_P)}}{2 \cdot C_X} \end{bmatrix}.$$

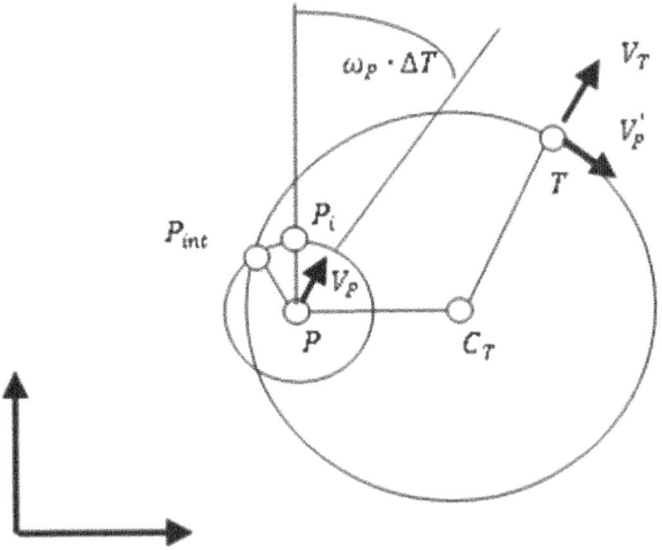

Fig. 4. Analysis of the coordinates of the intersection point of circles.

In our test program, only the upper point (Fig. 4) with a positive sign in the local coordinate system is taken into account. Convert the coordinates $P_{int.n}$ to the world coordinate system: $P_{int} = \begin{bmatrix} P_{int.n} \cdot h_1 \\ P_{int.n} \cdot h_2 \end{bmatrix} + P$. The vectors h_1 and h_2 make sense:

$h_1 = \begin{bmatrix} H_1 \cdot e_1 \\ H_1 \cdot e_2 \end{bmatrix}$ $h_2 = \begin{bmatrix} H_2 \cdot e_1 \\ H_2 \cdot e_2 \end{bmatrix}$, where $H_1 = \begin{bmatrix} 1 \\ 0 \end{bmatrix}$, $H_2 = \begin{bmatrix} 0 \\ 1 \end{bmatrix}$. To analyze the intersection point of the circles, you need to go to the coordinate system (v_1, v_2) with the origin at point P (Fig. 5): $v_1 = \frac{V_p}{|V_p|}$ $v_2 = \begin{bmatrix} -v_{1_y} \\ v_{1_x} \end{bmatrix}$.

In the coordinate system (v_1, v_2), the coordinates Of the P_{int} point look like this: $P_{int.v} = \begin{bmatrix} (P_{int} - P) \cdot v_1 \\ (P_{int} - P) \cdot v_2 \end{bmatrix}$.

The coordinates $P_{i.v}$ in the coordinate system (v_1, v_2) with the beginning at p will be:

$$P_{i,v} = \begin{matrix} if\ P_{int.v_y} \geq 0 \begin{bmatrix} |V_P| \cdot \Delta T \cdot \cos(w_p \cdot \Delta T) \\ |V_P| \cdot \Delta T \cdot \sin(w_p \cdot \Delta T) \end{bmatrix} \\ if\ P_{int.v_y} < 0 \begin{bmatrix} |V_P| \cdot \Delta T \cdot \cos(w_p \cdot \Delta T) \\ -|V_P| \cdot \Delta T \cdot \sin(w_p \cdot \Delta T) \end{bmatrix} \end{matrix}.$$

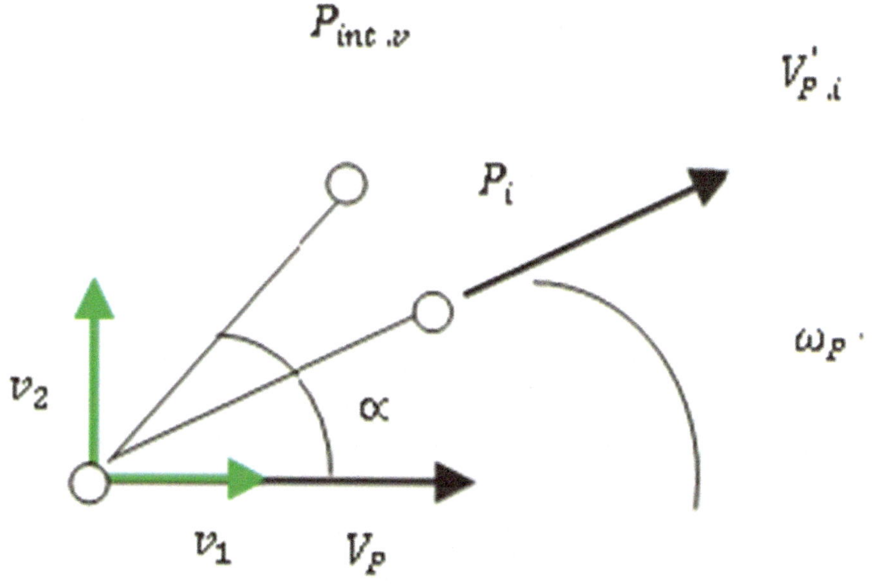

Fig. 5. Analysis from the pursuer's coordinate system.

If the angle $\propto$ is less than the angle $\omega_P \cdot \Delta T$, then the coordinates of the point $P_{i.v}$ will look different:

$$P_{i,v} = \begin{matrix} if\ P_{\text{int}.v_y} \geq 0 \begin{bmatrix} |V_P| \cdot \Delta T \cdot \cos(\propto) \\ |V_P| \cdot \Delta T \cdot \sin(\propto) \end{bmatrix} \\ if\ P_{\text{int}.v_y} < 0 \begin{bmatrix} |V_P| \cdot \Delta T \cdot \cos(\propto) \\ -|V_P| \cdot \Delta T \cdot \sin(\propto) \end{bmatrix} \end{matrix}.$$

The angle $\propto$ is the angle between the vector $P_{int.v}$ and the vector v_1. Next, you should convert the coordinates $P_{i.v}$ from the coordinate system (v_1, v_2) to the world one. To do this, we get expressions for the basis (H_1, H_2) in the basis (v_1, v_2):

$$h_{1.v} = \begin{bmatrix} H_1 \cdot v_1 \\ H_1 \cdot v_2 \end{bmatrix}, \quad h_{2.v} = \begin{bmatrix} H_2 \cdot v_1 \\ H_2 \cdot v_2 \end{bmatrix}.$$

Where $H_1 = \begin{bmatrix} 1 \\ 0 \end{bmatrix}$, $H\_2 = \begin{bmatrix} 0 \\ 1 \end{bmatrix}$. Where the expression for the point of the pursuer P_i in the next iteration stage will be:

$$P_i = \begin{bmatrix} P_{i.v} \cdot h_{1.v} \\ P_{i.v} \cdot h_{2.v} \end{bmatrix} + P.$$

So, in this paragraph, we have analyzed part of the algorithm in which the pursuer seeks to reach the arc (P_{int}, T) (Fig. 4).

5 The Case of Non-intersecting Circles

To avoid unknown situations in our algorithm, consider the case when $|P - C_P| < R_P - r_p$. In this case, we can assign a p_{int} point on the axis (P, C_P). In Fig. 6, to the left of point P. If we consider from the coordinate system (e_1, e_2) with the center at the point C_T, it is the point $P_{int.n} = \begin{bmatrix} -R_P \\ 0 \end{bmatrix}$

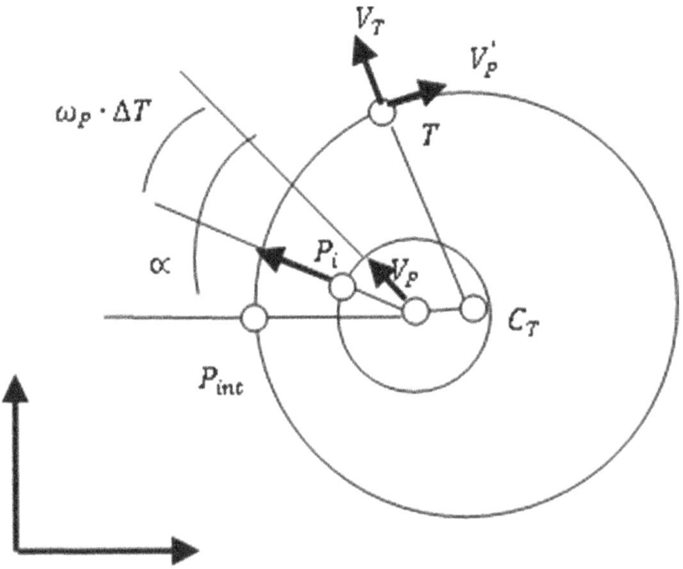

Fig. 6. Analysis of the disjoint circles case.

We need to calculate the angle $\propto$ between the vector V_P applied to the point P and the vector $\overrightarrow{(P, P_{int})}$.

If the angle $\propto$ is less than the angle $\omega_P * \Delta T$, then the coordinates of the point $P_{i.v}$ of the next iteration stage in the coordinate system (v_1, v_2) with the beginning at point P:

$$P_{i.v} = \begin{bmatrix} |V_P| \cdot \Delta T \cdot cos(\propto) \\ |V_P| \cdot \Delta T \cdot sin(\propto) \end{bmatrix}$$

If the point T in the coordinate system (e_1, e_2), with its center at the point C_T, is located in the upper half-plane, then:

$$P_{i.v} = \begin{bmatrix} |V_P| \cdot \Delta T \cdot cos(\propto) \\ |V_P| \cdot \Delta T \cdot sin(\propto) \end{bmatrix}.$$

If in the lower half-plane, then: $P_{i.v} = -\begin{bmatrix} |V_P| \cdot \Delta T \cdot cos(\propto) \\ |V_P| \cdot \Delta T \cdot sin(\propto) \end{bmatrix}$.

If $\propto$ is greater than the angle $\omega_P \cdot \Delta T$, then the coordinates of the point $P_{i.v}$ will be such: $P_{i.v} = \begin{bmatrix} |V_P| \cdot \Delta T \cdot cos(\omega_P \cdot \Delta T) \\ |V_P| \cdot \Delta T \cdot sin(\omega_P \cdot \Delta T) \end{bmatrix}$.

If point T is in the upper half plane, then: $P_{i.v} = \begin{bmatrix} |V_P| \cdot \Delta T \cdot cos(\omega_P \cdot \Delta T) \\ |V_P| \cdot \Delta T \cdot sin(\omega_P \cdot \Delta T) \end{bmatrix}$.

If in the lower half-plane, then: $P_{i.v} = -\begin{bmatrix} |V_P| \cdot \Delta T \cdot \Delta cos(\omega_P \cdot \Delta T) \\ |V_P| \cdot \Delta T \cdot sin(\omega_P \cdot \Delta T) \end{bmatrix}$.

It should be noted that during the simulation, this situation occurred when the speed of the pursuer was much higher than the speed of the target. The low angular speed of the pursuer may also affect it. In other words, a target at high speed with high inertia can get into a situation where $|P - C_P| < R_P - r_p$.

6 Goal and Strategy of the First Pursuer

The pursuer P_1 with the speed V_1 aims simply to catch up with the object T, which means coinciding two points P_1 and T with some degree of accuracy $|P_1 - T| \leq \varepsilon$. As an indicator of accuracy, we can suggest $\varepsilon = |V_1| \cdot \Delta T$, where ΔT is the sampling period in time. In addition, the object P_1 has a maximum angular speed of rotation ω_1, which limits the radius of curvature of the trajectory $R_1 = \frac{|V_1|}{\omega_1}$.

The strategy of the p_1 pursuer is that the coordinates of point T are recalculated to the coordinate system (v_1, v_2) with the origin at point P_1 (Fig. 7):

$$v_1 = \frac{V_P}{|V_P|}$$

$$v_2 = \begin{bmatrix} -v_{1_y} \\ v_{1_x} \end{bmatrix}$$

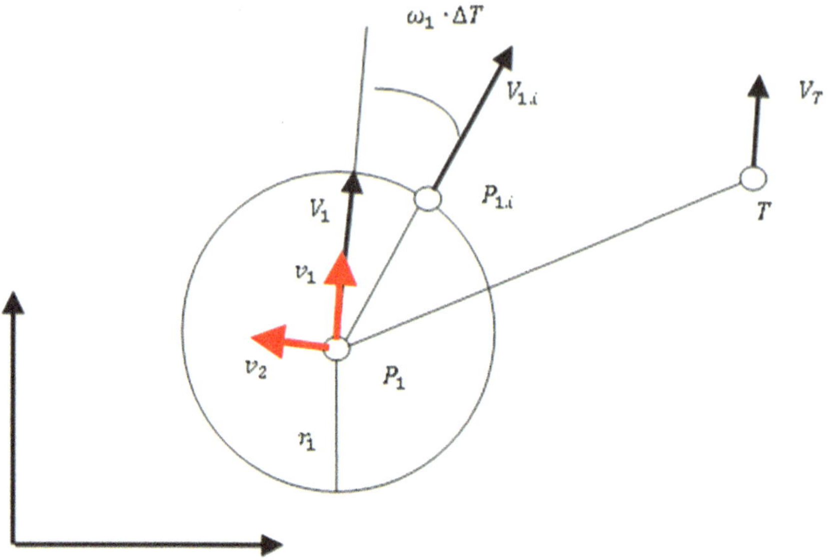

Fig. 7. The first pursuer's strategy.

There, the coordinates of point T will look like this: $T_v = \begin{bmatrix} (T - P_1) \cdot v_1 \\ (T - P_1) \cdot v_2 \end{bmatrix}$.

Next, we analyze the coordinates of point T_v for belonging to the upper or lower half-plane in the coordinate system (v_1, v_2) with the origin at point P_1:

$$P_{i,v} = \begin{matrix} \text{if } T_{v_y} \geq 0 \begin{bmatrix} |V_P| \cdot \Delta T \cdot \cos(w_p \cdot \Delta T) \\ |V_P| \cdot \Delta T \cdot \sin(w_p \cdot \Delta T) \end{bmatrix} \\ \text{if } T_{v_y} < 0 \begin{bmatrix} |V_P| \cdot \Delta T \cdot \cos(w_p \cdot \Delta T) \\ -|V_P| \cdot \Delta T \cdot \sin(w_p \cdot \Delta T) \end{bmatrix} \end{matrix}$$

It is necessary to constantly compare the values of the angles $\omega_1 \cdot \Delta T$ and $\propto$, where $\propto$ is the angle between the vectors $\overrightarrow{(P_1 T)}$ and V_1.

If the angle $\propto$ is less than the angle $\omega_P \times \Delta T$, then the coordinates of the point $P_{1.v}$ will look different:

$$P_{i,v} = \begin{matrix} \text{if } T_{v_y} \geq 0 \begin{bmatrix} |V_P| \cdot \Delta T \cdot \cos(\propto) \\ |V_P| \cdot \Delta T \cdot \sin(\propto) \end{bmatrix} \\ \text{if } T_{v_y} < 0 \begin{bmatrix} |V_P| \cdot \Delta T \cdot \cos(\propto) \\ -|V_P| \cdot \Delta T \cdot \sin(\propto) \end{bmatrix} \end{matrix}$$

7 Goals and Strategies of the Second and Third Pursuers

Pursuers P_2 and P_3 move at speeds V_2 and V_3, respectively. For objects P_2 and P_3, the goal is to combine with a certain degree of accuracy ε not with point T, but with points T_2 and T_3, respectively (Fig. 8). the Coordinates of points T_2 and T_3 are formed as follows: $T_{2,3} = T + n_{2,3}$.

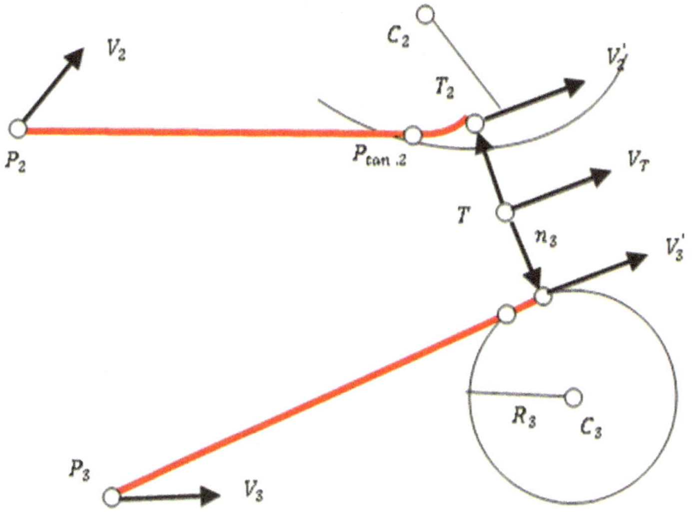

Fig. 8. Strategies of the second and third pursuers.

The normal vector $n_{2,3} = \pm \frac{1}{|V_T|} \begin{bmatrix} -V_{T_Y} \\ V_{T_X} \end{bmatrix} \cdot \Delta S_{2,3}$, where $\Delta S_{2,3}$ is the distance between the points T_2 and T_3 to the point T, respectively. For the object paths P_2 and P_3, the following conditions are selected: they came to the points T_2 and T_3 with the velocity directions V_2' and V_3'.

The radii of curvature of the trajectories must not be less than $R_{2,3} = \frac{|V_{2,3}|}{\omega_{1,2}}$, where $\omega_{2,3}$ are the maximum angular rotation speeds of the pursuers P_2 and P_3. The simulated trajectory at some point in time consists of a straight section $[P_{2,3}, P_{tan.2,3}]$ and an arc segment $(P_{tan.2,3}, T_{2,3})$. At each iteration stage, objects P_2 and P_3 perform discrete rotation and discrete translational movement to reach the simulated paths. In our test program, based on the materials of this paragraph, objects P_2 and P_3, as soon as they enter a course parallel to the course T, begin to move at speeds equal to V_T.

8 Goal and Strategy of the Fourth Pursuer

Consider the fourth member of the group of pursuers. If the behavior of the first participant can be qualified as the main “beater”. The behavior of the second and third pursuers can be qualified as assistants that do not allow the goal to escape, while the role of the fourth pursuer can be interpreted as a player from the “ambush”.

Figure 9 shows two cases of the formation of the fourth pursuer’s trajectory. In the first case, the pursuer’s trajectory enters the position point directly of the target T, perpendicular to its speed V_T. In the second case, the pursuer’s trajectory enters point Q at the speed opposite to the target’s directed speed V_T. The point Q is located on a straight line from the point T with the v_t forming.

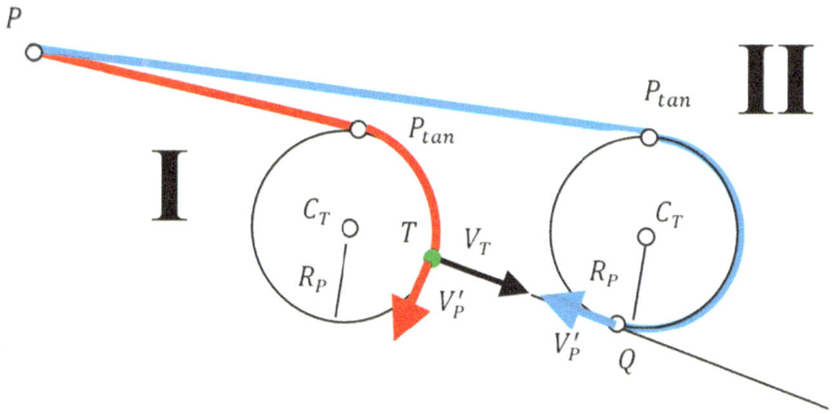

Fig. 9. Ambush pursuer strategy.

The point Q can be placed at any point in the plane, nothing forbids us to do this but the goal may simply not be achieved.

When you reach the Q point, you can change the strategy of the pursuer. Let’s say we slow down to 0 and wait for the target to approach a distance less than ε. You can change the strategy when you reach the Q point to the strategy of the first pursuer.

9 Objective and Strategy the Object of Persecution

Let’s look at the behavior of the object of harassment. In our model, the goal of the object of pursuit is chosen to evade the first pursuer.

Figure 10 illustrates the strategy of the target being pursued. In this figure, the object T with the speed V_T and the angular rotation speed ω_T during the sampling period ΔT rotates by the angle $\omega_T \cdot \Delta T$ and moves the distance $|T_i - T| = |V_T| \cdot \Delta T$.

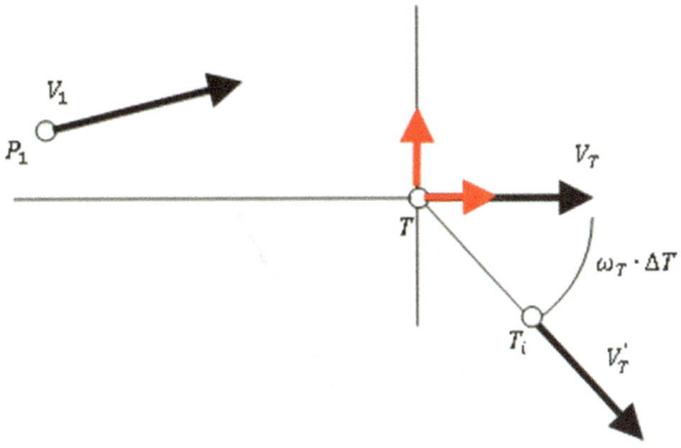

Fig. 10. The target's strategy.

The direction of rotation of the point T depends on the half-plane in which the pursuer P_1 is located. As an alternative strategy, you can suggest a strategy, which is illustrated in Fig. 11.

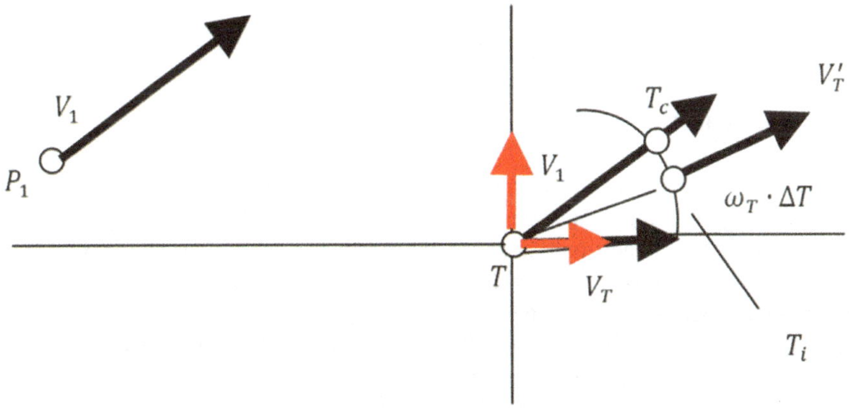

Fig. 11. Additional target's strategy.

Figure 11 shows that the object of pursuit T tends to make its speed V_T parallel to the speed vector of the pursuer V_1. When the pursuer is far away, it is preferable for the target to use a strategy of parallel speeds, as in Fig. 11. When the pursuer approaches a distance of several steps, that is, for the final jump, the goal will benefit from an evasive strategy, as in Fig. 11.

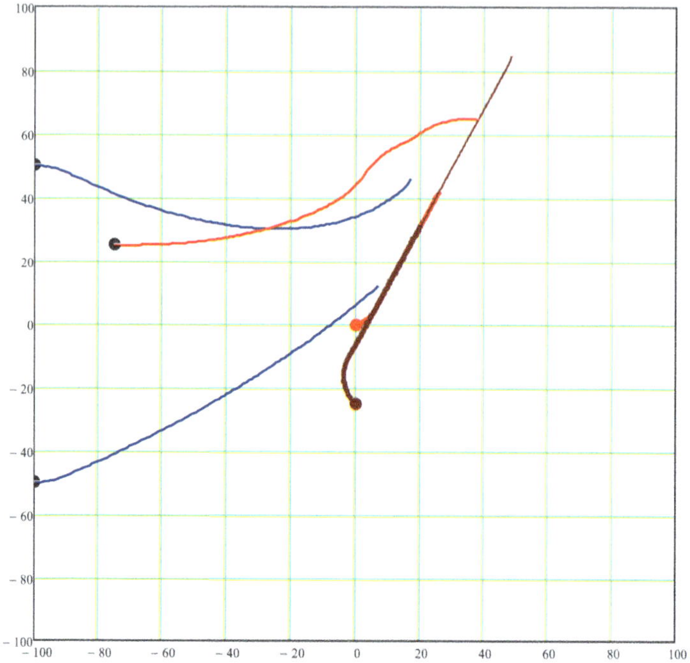

Fig. 12. Result of group pursuit simulation.

10 Conclusion

Based on the materials presented in this Chapter, a test program was written in the MathCAD system, which calculates the trajectories of a group of four pursuers and a target evading them. Each participant in the geometric model has its own goal and strategy. Figure 12 shows a screenshot from the video, where you can see how one pursuer implements a chase on the trail.

Two pursuers take and accompany the target along parallel paths. One pursuer goes perpendicular to the projected trajectory of the target. In the program, we deliberately changed the goal and strategy of the fourth pursuer, to show that within our program, it is quite simple to set the coordinates of entry points and entry vectors to points.

When writing this article, the theoretical provisions set out in [1–6] are used as a basis. The description of the algorithm for following the predicted paths to follow is located on the resource [7]. Figure 12 is supplemented with a link to the resource [8], where the video on the results of the program is posted. The source code of the program is available on the resource [9]. When developing algorithms, they were analyzed and used in writing programs [10–15].

References

1. Isaacs, R.: Differential Games. Mir, Moscow (1967)
2. Pontryagin, L.S.: Linear differential game of evasion. Tr. MIAN SSSR **112**, 30–63 (1971)
3. Krasovsky, N.N., Subbotin, A.I.: Positional Differential Games. Nauka, Moscow (1974)
4. Zhelnin, Y.: Linearized pursuit and evasion problem on the plane. Sci. Notes TSAGI **3**(8), 88–98 (1977)
5. Burdakov, S.V., Sizov, P.A.: Algorithms for motion control by a mobile robot in the pursuit problem. Sci. Tech. Bull. Saint Petersburg State Polytech. Univ. Comput. Sci. Telecommun. Manag. **6**(210), 49–58 (2014)
6. Simakova, E.N.: On a differential game of pursuit. Autom. Telemech. **2**, 5–14 (1967)
7. Algorithm for following predicted paths in the pursuit problem. http://dubanov.exponenta.ru. Accessed 07 May 2020
8. Video, group pursuit of a single target. https://www.youtube.com/watch?v=aC4PuXTgVS0&feature=youtu.be. Accessed 07 May 2020
9. Group pursuit with different strategies for a single goal. http://dubanov.exponenta.ru. Accessed 07 May 2020
10. Vagin, D.A., Petrov, N.N.: The task of chasing tightly coordinated escapees. Izvestiya RAS Theory Control Syst. **5**, 75–79 (2001)
11. Bannikov, A.S.: Some non-stationary problems of group pursuit. Proc. Inst. Math. Comput. Sci. UdSU **1**(41), 3–46 (2013)
12. Bannikov, A.S.: Non-Stationary task of group pursuit. In: Proceedings of the Lobachevsky Mathematical center, vol. 34, pp. 26–28. Publishing house of the Kazan mathematical society, Kazan (2006)
13. Izmestev, I.V., Ukhobotov, V.I.: The problem of chasing small-maneuverable objects with a terminal set in the form of a ring. In: Proceedings of the International Conference "Geometric Methods in Control Theory and Mathematical Physics: Differential Equations, Integrability, Qualitative Theory", vol. 148, pp. 25–31. VINITI RAS, Moscow (2018)
14. Konstantinov, R.V.: On a quasi-linear differential game with simple dynamics in the presence of a phase constraint. Math. Notes **69**(4), 581–590 (2001)
15. Pankratova, Y.B.: A Solution of a cooperative differential game of group pursuit. Discrete Anal. Oper. Res. **17**(2), 57–78 (2010)

The Method "cut cylinder" for Approximation Round and Cylindrical Shape Objects and Its Comparison with Other Methods

A. V. Vasilyev[1,2](✉), G. B. Bolshakova[3], and D. V. Goldstein[4]

[1] Central Research Institute of Dental and Maxillofacial Surgery, Timur Frunze St., 16, Moscow 119021, Russia
vav-stom@yandex.ru

[2] Peoples' Friendship University of Russia, Medical Institute, Miklukho-Maklaya St., 6, Moscow 117198, Russia

[3] Research Institute of Human Morphology, Tsyurupy St., 3, Moscow 117418, Russia

[4] Research Centre for Medical Genetics, Moskvorechye St., 1, Moscow 115478, Russia

Abstract. A method has been developed to approximate the volumes of cylindrical objects by slices. The method is based on an algebraic formula of circle. It allows to determine the thickness by width of the cylindrical object measured on the current and subsequent slices, which enables the morphometry of histological sections made with unequal pitch. The "cut cylinder" method is compared with other popular approximation methods: unilateral rectangles, trapezoids and reconstructions in the Amira software. A theoretical comparison of approximation methods revealed that an error of less than 5% can be achieved for the unilateral rectangle method with more than 11, and for the trapezoids' method more than 6 slices. The "cut cylinder" algorithm allows to reproduce the contour of the cylindrical object in the most accurate way with a smaller number of slices - less than 6. In practical application of methods for round objects reconstructed by 6 histological sections with the equal pitch, the method of approximation by trapezoids gives a more accurate results and is easy to apply. In order to determine the thickness of the slice, the trapezoids' method can be supplemented by a "cut cylinder" method, which contains a formula for determining the thickness of the slice by the width. The developed "cut cylinder" method allows to obtain data comparable to 3D-reconstruction in the Amira software (FEI, USA), however with a large dispersion for sections with equal pitch. #CSOC1120

Keywords: Morphometry · Approximation · "cut cylinder" method · Critical-size defect

R. Silhavy et al. (Eds.): CoMeSySo 2020, AISC 1295, pp. 50–58, 2020.
https://doi.org/10.1007/978-3-030-63319-6_5

1 Introduction

Morphometry in histology often requires the use of additional mathematical techniques for more accurate results [1, 2]. Morphometry is particularly difficult according to thick sections [3]. The step at which they are manufactured is sufficiently large and this can lead to significant distortions in the subsequent conversion of 2D parameters to 3D [2, 4, 5]. That is a commonly used method for multiplying the slice thickness at the object's area can lead to significant errors [3, 6]. Standard protocols for interpolation of volume objects of popular programs such as Amira (FEI, USA), MeVisLab (MeVis Medical Solutions AG, Germany) do not allow to accurately reproduce the contours of round objects [7, 8]. In addition, their algorithms require the same distance between the slices, while during histological processing it is sometimes difficult to obtain serial slices with the same pitch. For example, it is very difficult to achieve a strictly identical cut pitch when producing thick, undecalcified bone slice. In order to comprehensively solve the problem of 3D histomorphometry of objects of cylindrical shape, we developed a method called "cut cylinder," and compared it with mathematical algorithms for accurate reconstruction of objects of cylindrical shape according to histological sections made with different pitch.

2 Materials and Methods

2.1 Theoretical Part

It was performed comparing the theoretical accuracy of developed "cut cylinder" approximation method and other popular approximation methods: unilateral squares and trapezoids (Fig. 1).

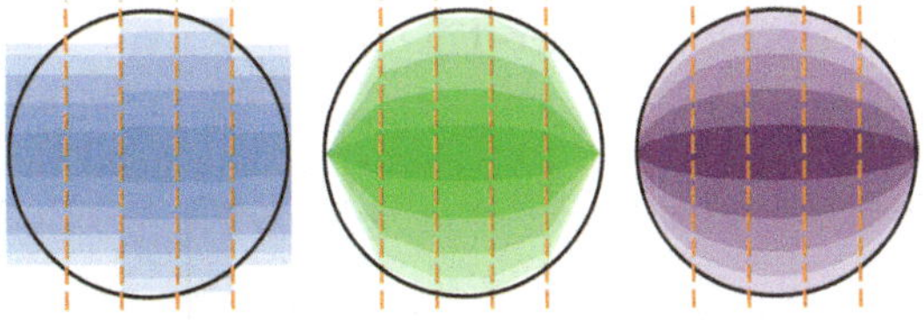

Fig. 1. The reconstruction of the circle's contour and the inner content using the model "cut cylinder" in comparison with the approximation of unilateral squares and trapezoids.

Method of Unilateral Rectangles. Unilateral rectangles method is that the measured area on the flat sections are multiplied by the slice thickness:

$$V = S \cdot s \tag{1}$$

where
V – volume obtained (3D-parameter),
S – measured area (2D-parameter),
s – slice thickness, measured with instruments.

Calculation of the total volume of the object is calculated by adding the received volume from each slice (Fig. 1).

Method of Trapezoids. The trapezoids method is similar to the unilateral rectangles' method, but provides for a bevel going to the next-in-order slice, making the corners of the resulting Fig. less sharp:

$$V = S \cdot \left(\frac{s_n \pm s_{n+1}}{2} \right) \tag{2}$$

where
n – the section number in order.

The total volume is calculated in the same way as the method of unilateral rectangles by summing the resulting areas from each slice (Fig. 1).

Method "cut cylinder". In the basis of the method we are developing, an algebraic circle equation was used for the exact reconstruction of cylindrical objects. In the developed methodology the possibility of automatic determination of histological section location in the space of the examined cylindrical object was laid down.

The simplest application of the technique requires a special approach to the sections manufacture. Thus, the preparation containing the cylindrical defect should be sawed through the center of the defect. Each half should be cut separately, making sections from the edge of the defect to its center (Fig. 2).

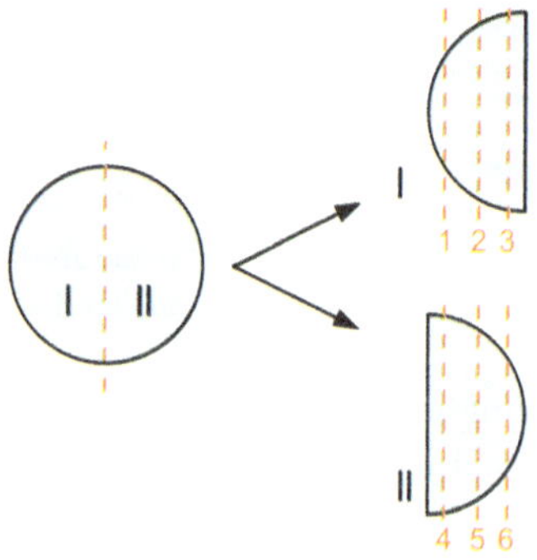

Fig. 2. Methodical reception for unambiguous definition in space of circle center. The studied object of cylindrical shape is cuted in half, after which sections of each half are separately made from the center to the edge.

Preliminary for calculations it is important to know exactly where sections pass in space of our coordinate system (Fig. 3). That is, it is necessary to calculate the x values for the incision at the beginning of the resulting slice – x_n, and for its end, i.e., the beginning of the subsequent slice – x_{n+1}. Within the model, this can be done by the formula:

$$x_n = R \pm \sqrt{R^2 - (\frac{y_n}{2})^2} \tag{3}$$

where
x – distance of the slide from the edge,
y – width,
R – defect radius,
n – section number in order.

The sign "–" should be used for the first half (I) and the sign "+" for the second (II) (Fig. 2).

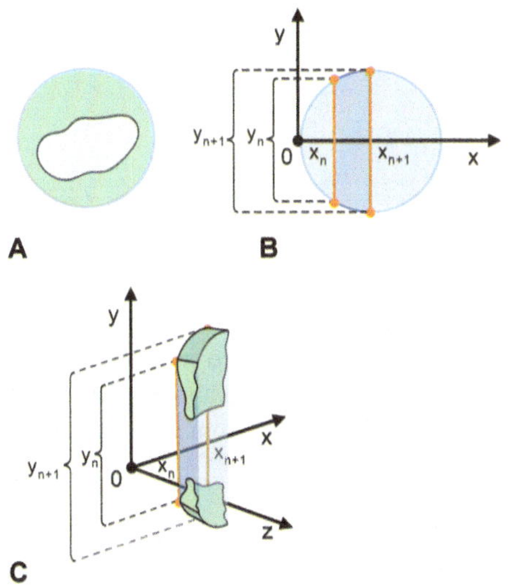

Fig. 3. Calculation of the coefficient for recalculation of 2D to the 3D parameters through the circle equation. A – the space of the circle containing the object under study. B – coordinates that allow to calculate the thickness of the slice. C – example of volume reconstruction of an object using the "cut cylinder" method.

To convert a planar parameter measured on a slice into a volume parameter, it should be multiplied by the following factor:

$$V = S \bullet (x_{n+1} - x_n) \bullet \frac{\int_{x_n}^{x_{n+1}} 2\sqrt{2Rx - x^2}dx}{2(x_{n+1} - x_n)\sqrt{2Rx_n - x_n^2}} \tag{4}$$

where
V – obtained volume (3D-parameter)
S – measured area (2D-parameter)
x – distance of the slide from the edge,
R – defect radius,
n – section number in order.

It should be noted that these calculations are only valid for the calculation of sections up to the middle of the defect. To derive the coefficient after the middle of the circle need not divide the land area on the area of a rectangles, but on the contrary, multiply. That is, if:

$$\frac{\int_{x_n}^{x_{n+1}} 2\sqrt{2Rx - x^2}dx}{2(x_{n+1} + x_n)\sqrt{2Rx_n - x_n^2}} < 1 \tag{5}$$

then use the formula:

$$V = S \bullet (x_{n+1} + x_n) \bullet \frac{2(x_n - x_{n+1})\sqrt{2Rx_n - x_n^2}}{\int_{x_n}^{x_{n+1}} 2\sqrt{2Rx - x^2}dx} \tag{6}$$

Clearly, the first cut will carry no information as it will contain only the edge of the defect. Therefore, it must be presented as zero in order, and next one considered like the first in count (the number n in the formula above). The first slice will determine the distribution of the regenerate in the region from zero to the second slice (x_0 to x_2).

2.2 Practical Part

To verify theoretical data on comparison of accuracy of methods, a popular model was used to study bone regeneration – a model of parietal bones critical-size defect of rats containing tetracycline-like labels [8]. This model illustrates well the problem of 3D-morphometry of cylindrical objects: 1) the object of the study is a defect of cylindrical shape. 2) study material - thick slices of different thickness.

Samples of the necropsy of the skull arch of rats were used from our earlier research of the regeneration of parietal bones critical-size defect. Samples were obtained from male Sprague-Dawley rats with a body weight of 350 g (N = 17). In the first group, samples from non-medicated rats were used; in the second group – with and without exposure together [9]. Under intraperitoneal anesthesia of Zoletil (Virbac, France) at a dose of 125 μg/kg, transverse and lateral-shifted vertical incision of the scalp were produced, forming a triangular flap. Avoiding perforation of the sagittal venous sinus in the middle of the sagittal suture a circular hole was formed in the parietal bones using a C-reamer trepan of 5.5 mm diameter and 1.5 mm height from a set of Neobiotech SLA (Korea). The wound was layer-by-layer taken in. In the postoperative period, the newly formed mineralized bone tissue was labeled with doxycycline and alizarin to visualize the primary neoosteogenesis zone according to a conventional technique [10, 11] On day 28, euthanasia was performed by an overdose of Zoletil anesthesia. A necropsy of the skull arch was taken and fixed in 40% ethyl alcohol for 24 h [12].

Two groups were formed.

In the **first group** (n = 11), reconstruction data was obtained in simulation mode. For this purpose, samples of the skull arch including defect and bone regenerate were subjected to beam-aiming X-ray (n = 11) (Gendex Expert DC). The area of the newly formed mineralized bone tissue within the defect was measured in the ImageJ v.1.51 (NIH, USA). The regenerate was also marked with 6 equidistant parallel lines, after which the length of the sections covering the regenerate was measured (Fig. 4). The resulting numerical data are then used to calculate the original regenerate area using unilateral rectangles, trapezoids, and "cut cylinder" methods in Excel 2019 (Microsoft, USA) and Mathematica v.11 (Volfram, USA) (Fig. 1). This simulation revealed the real errors of each method for measuring circular and cylindrical objects and provided practical guidance on their use.

In the **second group** (n = 6) samples of the skull arch preliminary subjected to a microcomputer tomography (SkyScan1174v2; Bruker-microCT, Belgium) with a size a voxel size 8 μm^3. Further histologic processing includes embedding samples of the skull arch in ethylmethacrylate [8, 12]. Further samples slides with a pitch of 900 μm (6 slides per area of the defect) and was ground to thickness of 20 μm. Using a scanning system based on a fluorescent microscope Carl Zeiss Imager M.1 whole images of all thin sections were obtained. Further, the Amira v5.5 (FEI, USA) software was performed with a three-dimensional reconstruction, followed by measurement of regenerate volumes (Fig. 4). By using ImageJ software marking reclaim zones for each microsection was produced, the defect width is measured to calculate the pitch shear. Next, using the developed model "cut cylinder" the volume of newly formed bone within the defect was calculated.

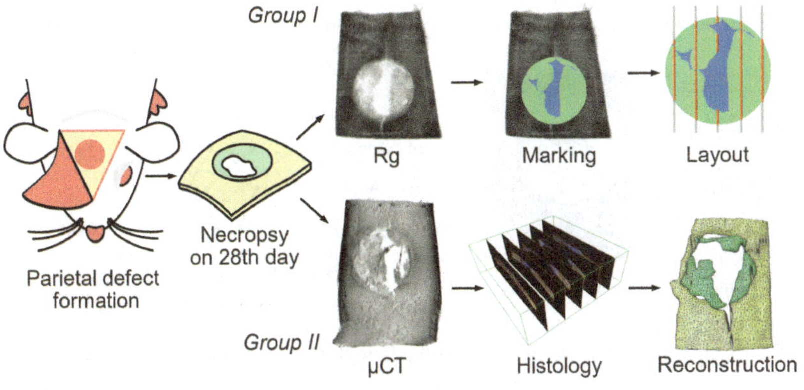

Fig. 4. Scheme of practical part research.

Statistical analyses were performed with Prism v.7 (GraphPad, USA). Intergroup differences were found using Dunn's test for repeated measures.

3 Results

3.1 The Theoretical Advantage of the "cut cylinder" Model

A comparison of the mathematical principles of each approximation method revealed that a theoretical error below 5% can be achieved for a rectangles' method for a number of slices more than 11, and for a trapezoids' method for more than 7 slices (Fig. 5A). The algorithm developed by us, based on the algebraic circle equation, allows to reproduce the contour of a cylindrical object in the most accurate way with a small number of slices - less than 7. The developed algorithm for 3D morphometry of cylindrical objects was called "cut cylinder". The proposed method theoretically provides the most accurate data (Fig. 5A) and has the following advantages:

1) automatically positions the slice in space and finds its thickness;
2) allows to eliminate the use of very expensive equipment, which allows to perform cuts with strictly standardized pitch.
3) more accurately reproduces the regenerate contour and eliminates step-type artifacts.

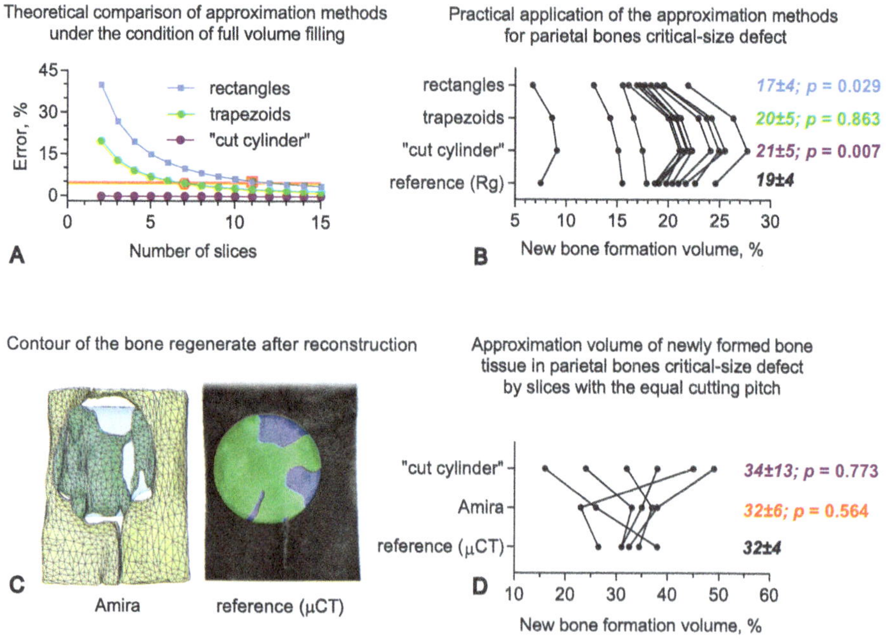

Fig. 5. Comparing the approximation methods with the developed method "cut cylinder".

The developed "cut cylinder" model required testing in practice. For this task, we used histological slides obtained from the study of regeneration of rat's parietal bones critical-size defect.

3.2 Practical Application of the "cut cylinder" Model

As a result of the practical application of approximation techniques for the morphometry of regeneration of rat's parietal bones critical-size defect, the following has been founded. The use of the rectangles method significantly underestimates the actual amount of regenerate volume compared to the real values. At the same time, the developed algorithm – on the contrary, leads to overestimation of results (Fig. 5B). If more than 6 slices are used for reconstruction, the closest to real values are obtained using the trapezoids method. Note that rectangles and trapezoids methods require a preliminary measurement of the thickness of each slice, while the "cut cylinder" method may calculate the thickness by width of the cylindrical object on the histological slice.

A 3D reconstruction performed in Amira (FEI, USA) allows to visualize the data (Fig. 5C). However, this method is inferior to all algorithms due to the fact that it does not allow to arrange in space cuts made with unequal pitch. On the sections obtained with the equal pitch in the Amira software demonstrates comparable results with the developed "cut cylinder" method, but shows a lesser variation of values compared to the reference (Fig. 5D).

4 Conclusion

A clear advantage of the developed method "cut cylinder" is the ability to determine the thickness by width of the cylindrical object measured on the current and subsequent slices. Its combination with the popular and simple method of approximating with trapezoids can show the most optimal results. However, the developed method "cut cylinder" despite of theoretical potential did not showed its appreciable and practical superiority for approximation round and cylindrical shape objects.

Acknowledgments. The mathematical model "cut cylinder" was reviewed and approved by Marchuk Institute of Numerical Mathematics of the Russian Academy of Sciences (IVM RAS). Special gratitude is expressed to the specialist in the field of mathematical segmentation Dr. Alexander A. Danilov for his contribution to the correction and improvement of the method.

References

1. Portero-Muzy, N., Arlot, M., Roux, J.-P., Duboeuf, F., Chavassieux, P., Meunier, P.: Evaluation and development of automatic two-dimensional measurements of histomorphometric parameters reflecting trabecular bone connectivity: correlations with dual-energy X-Ray absorptiometry and quantitative ultrasound in human calcaneum. Calcif. Tissue Int. **77**, 195–204 (2005)

2. Yeh, S.A., Wilk, K., Lin, C.P., Intini, G.: In vivo 3D histomorphometry quantifies bone apposition and skeletal progenitor cell differentiation. Sci Rep. **8**(1), 5580 (2018)
3. Revell P.A.: Quantitative methods in bone biopsy examination. In: Pathology of Bone. Springer, London (1986)
4. Jennane, R., Almhdie, A., Aufort, G., Lespessailles, E.: 3D shape-dependent thinning method for trabecular bone characterization. Med. Phys. **39**(1), 168–178 (2012)
5. Stadlinger, B., Korn, P., Tödtmann, N., Eckelt, U., Range, U., Bürki, A., Ferguson, S., Kramer, I., Kautz, A., Schnabelrauch, M., Kneissel, M., Schlottig, F.: Osseointegration of biochemically modified implants in an osteoporosis rodent model. Eur. Cells Mater. **25**, 326–340 (2013)
6. Dempster, D.W., Compston, J.E., Drezner, M.K., Glorieux, F.H., Kanis, J.A., Malluche, H., Meunier, P.J., Ott, S.M., Recker, R.R., Parfitt, A.M.: Standardized nomenclature, symbols, and units for bone histomorphometry: a 2012 update of the report of the ASBMR histomorphometry nomenclature committee. J. Bone Miner. Res. **28**(1), 2–17 (2013)
7. Cocks, E., Taggart, M., Rind, F.C., White, K.A.: Guide to analysis and reconstruction of serial block face scanning electron microscopy data. J. Microsc. **270**(2), 217–234 (2018)
8. Vasilyev, A.V., Volkov, A.V., Bolshakova, G.B., Goldstein, D.V.: Characteristics of neoosteogenesis in the model of critical defect of rats' parietal bone using traditional and three-dimensional morphometry. Genes Cells **9**(4), 121–127 (2014). in Russian
9. Vasilyev, A.V., Bolshakova, G.B.: The effect of dalarginum on the reparative regeneration of critical size calvarial defect in rats. Morphol. Newsl. **4**, 11–18 (2014). in Russian
10. Allen, M.R., Burr, D.B.: Bone modeling and remodeling. In: Basic and Applied Bone Biology, 1st edn. Academic Press, Cambridge (2014)
11. Tapp, E.: Tetracycline labelling methods of measuring the growth of bones in the rat. J. Bone Joint Surg. Br. **48**(3), 517–525 (1966)
12. Yuehuei, H.A., Kylie, L.M.: Handbook of Histology Methods for Bone and Cartilage, 1st edn. Humana Press, Totowa (2003)

Resource Allocation by the Inverse Function Method

Vitaliy Nikolaevich Tsygichko(✉)

The Federal Research Center "Computer Science and Control" of the Russian Academy of Sciences, Moscow, Russia
vtsygichko@inbox.ru

Abstract. The article presents a method of optimal allocation of resources provided for the synthesis of equilibrium systems. It introduces the concept of 'equilibrium systems' and gives examples of such systems. It provides a meaningful and formal statement of the problem of allocating the resources provided to create equilibrium systems. As a criterion for the resource allocation, the authors propose equal effectiveness of all elements of the system. The article describes the inverse function method of the optimal resource allocation. It describes the analytical and graphical procedures for solving the resource allocation problem by the example of the synthesis of security systems for automated information systems.

Keywords: Equilibrium systems · Optimal allocation · Resource · Efficiency · Inverse function

1 Introduction

There is a very extensive class of objects, the functioning efficiency of which substantially depends on the functioning efficiency of each of their elements. The optimal building of objects of this class implies equal reliability of their elements or equal probability of the elements performing their functions or equal effectiveness of the elements or, lastly, equal strength or equal stability of the elements under the influence of external and internal loads. We call this class of objects the equilibrium systems.

It is often necessary to solve the problem of allocation of the resource provided for such systems to build or to ensure its normal functioning. Moreover, the resource is most often limited, and it must be allocated in such a way as to ensure the equality of key parameters of the system elements.

One such an example of a mechanical system belonging to this class of systems is a supporting structure, for example, a truss which weight is limited and the elements must have equal strength or equal stability, since failure of at least one element results in the destruction of the whole structure.

It is supposed that the structure consists of metal of the same grade. The resource in our example is the structure's weight which is limited and must be allocated between the elements so that the cross sections of the elements provide the condition of equal strength or equal stability of the structure. At the same time, the strength or stability of each element is a nonlinear function of the cross section, and, consequently, weight.

R. Silhavy et al. (Eds.): CoMeSySo 2020, AISC 1295, pp. 59–64, 2020.
https://doi.org/10.1007/978-3-030-63319-6_6

As another example, we consider a transport problem.

Let there be a certain amount of cargo (resource) that must be delivered to its destination in the shortest possible time. The cargo is delivered via several transport routes simultaneously. In general, the transportation time for each route nonlinearly depends on the amount of cargo transported. The minimum time condition is met when transportation in all routes ends simultaneously. It is required to allocate the cargo (resource) so that this condition is met.

2 Inverse function method of resource allocation

We consider, as an example, a security system (SS) [1, 2] of automated information systems (AIS). AIS is a typical equilibrium system. The failure of any element in such systems results in either a sharp decrease in their effectiveness or a complete cessation of their functioning.

SS should ensure equal protection of all AIS elements from all threats to their security [1, 3]. To ensure the security of each AIS element, the SS should spend part of a limited resource, which we define as the cost value of ensuring the security of an AIS element.

We denote the effectiveness of protection of an i^{th}AIS element by certain means of SS through the reliability of its protection p_i against all alleged threats, where $i \in I, I-$is the number of elements that make up the AIS. Protection reliability of an AIS element is understood as the probability of preventing the implementation of threats. Since the security of AIS elements is a number of independent events, the effectiveness of SS of AIS is determined by formula:

$$P_{AIS} = \prod_{i=1}^{i=I} p_i$$

and reaches the maximum with equal values of protection reliability p_i. It is necessary to allocate SS resources between AIS elements in such a way as to obtain the maximum value of the SS efficiency indicator – maxP_{AIS}.

In a similar way, it is possible to define the problems of maximizing the effectiveness of SS at given costs for ensuring the security of AIS or ensuring the minimum cost of SS at a given level of security of AIS elements and a number of other problems.

In general terms, the problem of ensuring the AIS security can be defined as follows.

There is a resource (total costs of SS for AIS security) – S_I and a set of AIS elements – I. Protection reliability of the i-th element of the system is a non-decreasing function of the resource, i.e. funds allocated to this element – S_i

$$p_i = f_i(S_i).$$

It is required for each AIS element $i \in I$ to bring into correspondence such a part S_i of the resource S_I so that the protection reliability p_i of all AIS elements would be maximum and equal to each other.

In accordance with the accepted assumptions, the objective function of the listed problems can be written in the form of equality:

$$p_1 = p_2 = \ldots = p_i = \ldots p_I, \tag{1}$$

where

$$p_i = f_i(S_i), \quad i \in I \tag{2}$$

under conditions

$$\sum_{i=1}^{i=I} S_i = S \leq S_I \tag{3}$$

and

$$A \leq p_i \leq B \tag{4}$$

A and B are some constants that limit the acceptable region p_i.

Unlike most problems of the resource allocation by linear, nonlinear, and dynamic programming methods [3–10], this problem can have an analytical solution, despite the nonlinearity of the objective function (1), (2).

We suppose that the functions $p_i = f_i(S_i)$ are monotonic, nonlinear, and defined on an interval $[0, S_I]$. The functions' monotonicity condition $p_i = f_i(S_i)$ allows to state that there are functions that are inverse to them

$$S_i = f_i^{-1}(p_i). \tag{5}$$

Let us write condition (3) in the form:

$$\sum_{i=1}^{i=I} S_i = S = \sum_{i=1}^{i=I} f_i^{-1}(p_i) \leq S_I \tag{6}$$

It follows from the condition established by the objective function (1) that the sum of the functions in formula (6) can be replaced by one argument p_i function, i.e.

$$\sum S_i = S = F(p_i) \tag{7}$$

Since functions (5) have inverse functions, then function (7) has an inverse function

$$p_i = F^{-1}(S). \tag{8}$$

Now the problem of optimal resource allocation is reduced to determining the maximum (minimum) of function (8) in the interval $[0, S_n]$.

The value $\max(\min)p = p_i$, obtained as a result of examining function (8) for maxima and minima on the interval $[O, S_n]$ and at the ends of this interval, uniquely determines the desired allocation of the resource S_n, i.e. using the function $S_i = f_i^{-1}(p_i)$, each element $n_i \in N$ is assigned a resource value S_i from the found value $\max(\min)p = p_i$.

If the number of objects $n_i \in N$ is large and not all functions $S_i = f_i^{-1}(p_i)$ are defined analytically, then there can be proposed an approximate algorithm for solving this problem that can be easily implemented on a computer.

The solution is a step-by-step approximation procedure with verification of established conditions at each step. We establish an approximation step Δp_i corresponding to the necessary accuracy of the solution (resource allocation error), and the initial value of the parameter $p_i = p_i^{'}$ at which $\sum_{i=1}^{i=I} S_n \prec S_n$ is undoubted.

At the first step, for $p_i = p_i^{'} + \Delta p_i$, we find the corresponding value S_i for all functions $S_i = f_i^{-1}(p_i)$ and check the condition $\sum_{i=1}^{i=I} S_i \prec S_n$.

At the second step, we find the sum $\sum_{i=1}^{i=I} S_i$ for $p_i = p_i^{'} + 2\Delta p_i$, check the condition $\sum_{i=1}^{i=I} S_i \prec S_n$ again, etc. in all subsequent steps.

The procedure ends at the step in which the condition $\sum_{i=1}^{i=I} S_i \prec S_n$ is violated, then p_i and S_i, obtained at the previous step, are taken as the solution to the problem. If condition (4) is established, then it is necessary to introduce an additional check into this algorithm according to this condition.

If the functions $p_i = f_i(S_i)$ and $S_i = f_i^{-1}(p_i)$ can be plotted graphically, then a relatively simple graphic method for solving the problem can be proposed that imitates the step-by-step approximation algorithm considered above. To do this, it is necessary to depict all the functions $p_i = f_i(S_i)$ in one coordinate system and on a single scale on the same graph and place a line next to it, depicting a resource S_n on the same scale, as shown in Fig. 1.

The solution procedure is as follows:

- draw a line parallel to the axis 0, S,
- sequentially transfer the distances from the axis 0, P to the intersection point of the drawn line with the graphs of functions to the resource line.

If the sum $\sum_{i=1}^{i=I} S_i$ thus obtained is less than S_n, draw a second line, etc. Repeat the procedure until condition (1) and (2) are satisfied. The resulting values of p_i and S_i will be the solution to the problem.

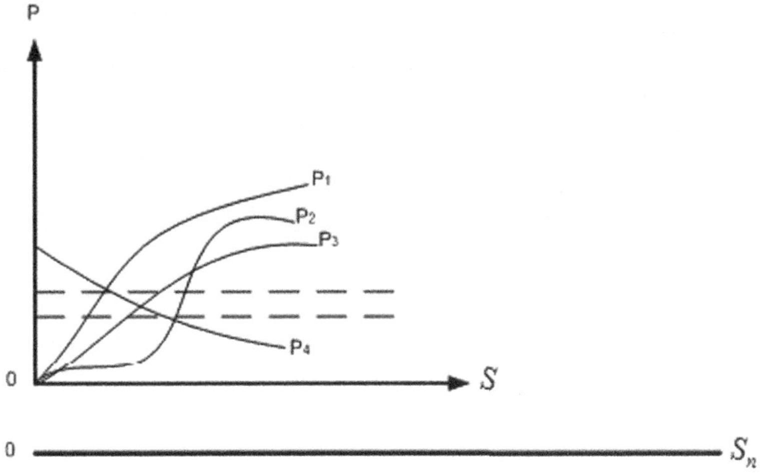

Fig. 1. Graphical interpretation of the inverse function method.

3 Conclusion

In the article, the inverse function method of resource allocation for the synthesis of equilibrium systems was implemented as a software product for solving the problem of synthesis of security systems for a number of information systems, for example, in the transport industry [1, 2]. The Inverse Function Method of Resource Allocation can be extended to a wide class of nonequilibrium systems, i.e. systems in which the equal functioning efficiency of their elements is not required. In this case, the optimality criterion is the established efficiency of each element of the system, which all together constitutes the efficiency of the system as a whole.

Acknowledgement. This work was supported by the Russian Foundation for Basic Research (project 19-07-00522).

References

1. Tsygichko, V.N.: Security risk assessment of critical facilities and critical infrastructures. Risk Anal. Issues **13**(5), 6–10 (2016)
2. Tsygichko, V.N., Chereshkin, D.S.: Security of Critical Transport Facilities. LAP LAMBERT Academic Publishing, Dusseldorf (2014)
3. Tsygichko, V.N.: Forhead About Decision Making, 3rd edn. Krasand, Moscow (2010)
4. Ashmanov, S.A., Timokhov, A.V.: Optimization Theory in Tasks and Exercises. Nauka, Moscow (1991)
5. Karmanov, V.G.: Mathematical Programming, 3rd edn. Nauka, Moscow (1986)
6. Razumovsky, I.V., Kantorovich, L.V.: Reasonable generalization provides more than detailed research. In: Famous Students: Essays on the Graduates from St. Petersburg University. Part 3. Famous Students, St. Petersburg, Russia (2005)

7. Slavyanov, A.S.: Analysis and Practical application of resource allocation models. Sci. Pract. Bull. **4**(9), 228–244 (2018)
8. Cormen, T.H., et al.: Linear programming. Introduction to Algorithms, 2nd ed., Chapter 29. Williams, Moscow (2006)
9. Powell, M.J.D.: A view of algorithms for optimization without derivatives. Department of Applied Mathematics and Theoretical Physics, Cambridge, UK (2007)
10. Verma, S., Ramineni, K., Ian, I.G.: A control-oriented coverage metric and its evaluation for hardware designs. J. Comput. Sci. **5**(4), 302–310 (2010)

Economic Assessment of Investment for the Production of Construction Products, Using the Mathematical Model

D. A. Gercekovich[1], E. Yu. Gorbachevskaya[2(✉)], I. S. Shilnikova[1], O. V. Arkhipkin[1], and Yu. A. Apalchuk[1]

[1] Irkutsk State University, Lenin S., 1, Irkutsk, Russia
[2] Irkutsk National Research Technical University, Lermontov St., 83, Irkutsk, Russia
eugorbachevskaya@mail.ru

Abstract. The sphere of building materials production is one of the most "problematic" as it has a very high share of import dependence. The purpose of this article is to create a mathematical model which may make it possible to evaluate capital investments in the production of certain types of products for construction. Using the obtained model, it is possible to evaluate for investors the most attractive directions for the production of building materials, and those that have negative returns can be considered as potential investors, but require a radical change in the product itself or the production process. Using specific types of investments as an example, the article reveals the investment sense of the "Profitability – Risk" model, based on the theory of portfolio analysis. The synthesized model is used specifically for this industry. The investment attractiveness of certain types of products in the field of production of building materials is assessed. Based on the results of the research, a list of the most profitable sectors for investing money is formed. In general, the sphere of production of building materials is such that it is impossible to abandon the production of economically inefficient at all, therefore this model allows us to determine the areas that require a review of the production and marketing strategy.

Keywords: Risk · Profitability · Portfolio theory · Investment attractiveness estimation · Investment policy · Markowitz model

1 First Section

1.1 Introduction

In today's economy, investment is one of the leading steps in successful business. Only competent and deliberate investment of free capital will allow the company to receive additional profit. However, any investment is associated not only with making a profit, but also both with the possibility of losing not only additional income and with the loss of invested funds. Therefore, any investment is closely related to the concept of risk. To reduce the possibility of losing money, it is necessary to assess the possible degree of

R. Silhavy et al. (Eds.): CoMeSySo 2020, AISC 1295, pp. 65–71, 2020.
https://doi.org/10.1007/978-3-030-63319-6_7

risk when investing in an asset. For example, if the probability of the loss of invested capital is higher than its expected return, then you should rather abandon this investment, however, if the risk is high and the possible profit is high, then it is necessary: a) to analyze the strengths and weaknesses in such form of investments; b) to determine the own attitude to the level of acceptable risk; c) how justified the risk is and, based on the results of the analysis, make a choice whether it is worth investing in this industry.

Industries that are closely related to construction will always be in demand in the investment market. Indeed, new buildings and structures for various purposes are regularly being erected, so the demand for the production of building materials will be at a high level. Every year technologies are improved for the production of certain products for construction, so studies of the effectiveness of investments for the production of certain types of goods for construction will be relevant for a long time.

The purpose of this study is to develop a mathematical model to assess the effectiveness of investing in the production of a particular type of product for the construction industry.

1.2 Methods and Models

The work used statistical data for 2010–2016 on the production of certain types of products for construction in the Russian Federation (Investments in Russia 2017, p. 20). To select the most preferred investment areas in the building materials market, the "Profitability-Risk" model was used. This model is based on the main provisions of the portfolio analysis (Gercekovich and Babushkin 2019). The analysis is based on the obtained values of the expected profitability which for a given period of time is calculated as the arithmetic average of historical data on the dynamics of prices for the products listed below for construction, as well as risk, which implies a root from dispersion. The calculated expected return and risk are rounded to one decimal place. Additionally, for a more complete presentation of the state of the analyzed indicators, the ratio of profitability to risk is calculated (Table 1).

The analysis was carried out in full accordance with the approach proposed by the founder of modern portfolio theory G. Markowitz where the author proposed a new approach to the study of the risk effects of investment distribution, correlation and diversification of expected investment income (Markovitz 1952; Dodie 2007; Gibson 2015; Chekulaev 2002; Sharp 2016) and the methodology for constructing a "winner model" (Damodoran 2007, pp. 168–169; DeBondt and Thaler 1985; Jegadeesh and Titman 1993) and the method of D. O`Shaughnessy (O`Shaughnessy, 2005).

The last two rows of Table 1 show the largest values of the investigated types of products highlighted in the table. As it follows from a cursory analysis of the table the maximum yield on the studied historical data interval showed the production of ceramic facade tiles, and the minimum asbestos-cement corrugated sheets (slate). In terms of risk, the maximum indicator demonstrates another wood paneling parquet, and the minimum indicator shows a sanitary ware made of ceramics. Non-metallic building materials have the best value in relation to profitability to risk, and the worst value for window blocks in assembly.

A point diagram was constructed (Table 1), based on the calculated profitability and risk data (Table 1). The x-axis is the risk, and the y-axis is profitability.

Table 1. Profitability and production risk of certain types of products for construction

№	Name	Expected Return, (%)	Risk, (%)	Return on risk ratio
1.	Non-metallic building materials, mln. m^3	7,4	7,1	**1,05**
2.	including pebbles, gravel, crushed stone, mln. m^3	4,3	9,4	0,46
3.	Non-refractory building brick construction billion of standard unit bricks	0,9	12,2	0,08
4.	Building bricks (including stones) made of cement, concrete or artificial stone, billion standard unit brick	−1,5	16,8	−0,09
5.	Portland cement, alumina cement, slag cement and similar hydraulic cements million tons	1,9	10,2	0,18
6.	Small wall blocks made of cellular concrete, billion conventional bricks	7,9	13,4	0,59
7.	Large wall blocks (including basement wall blocks) made of concrete, million standard unit bricks	−4,2	12,6	−0,33
8.	Prefabricated reinforced concrete structures and parts, mln.m^3	0,1	10,2	0,01
9.	Asbestos-cement corrugated sheets (corrugated) (slate), million standard unit tiles.	**−11,7**	17,9	−0,66
10.	Roofing and waterproofing roll materials from asphalt or similar materials (petroleum bitumen, coal tar pitch, etc.), million m^2	0,2	4,6	0,03
11.	Molded, rolled sheet glass, drawn or blown, but not processed otherwise, mln. m^2	11,3	15,1	0,75
12.	Thermally polished sheet glass and sheet glass with a matte or polished surface, but not otherwise processed, mln. m^2	−0,1	13,0	−0,01
13.	Window unit assembly (complete), million m^2.	−8,3	8,2	**−1,01**
14.	Door blocks units (complete), mln. m^2	0,9	17,9	0,05
15.	Wood panel parquet, mln. m^2	23,4	**50,5**	0,46
16.	Windows and their boxes, polymer window sills, mln. m^2	0,7	10,6	0,07
17.	Doors and their polymeric frames, thousand m^2	1,8	9,5	0,19
18.	Glazed ceramic tiles for interior wall cladding, mln. m^2	3,9	5,4	0,73
19.	Facade ceramic tiles, thousand m^2	**32,4**	37,4	0,87
20.	Ceramic tiles for floors, million m^2	4,9	7,6	0,65
21.	Linoleum on a textile basis, million m^2	2,7	9,1	**0,30**
22.	Ceramic sanitary article, mln units	4,1	**4,2**	0,97
Minimum		−11,7	4,2	−1,01
Maximum		32,4	50,5	1,05

In the process of analyzing the diagrams, points having negative values of expected returns or close to zero indictors were removed from further consideration, since investments in these types of products are more likely not to bring profit, and, therefore, this investment is not of practical interest. These include: building bricks (including stones) made of cement, concrete or artificial stone, billion standard unit bricks −1,5 (№ 4); large wall blocks (including basement wall blocks) made of concrete, million standard unit bricks (−4,2) (№ 7); asbestos-cement corrugated sheets (corrugated) (slate), million standard unit tiles (−11,7) (№ 9); thermally polished sheet glass and sheet glass with a matte or polished surface, but not other-wise processed (−0,1) (№ 12); window unit assembly (complete) (−8,3) (13); non-refractory building brick construction billion of standard unit 0,9 (№ 3); prefabricated reinforced concrete structures and parts (0,1) (№ 8); roofing and waterproofing roll materials from asphalt or similar materials (petroleum bitumen, coal tar pitch, etc.) (0,2) (№ 10); door blocks units (complete) (0,9) (№ 14); windows and their boxes, polymer window sills (0,7) (№ 16).

Further, based on the main points of portfolio analysis, a comparative analysis of pairs of types of production that has an almost identical degree of risk is carried out. Of the two types considered, one that has a lower profitability is removed from further consideration. And, on the contrary, in a pair of two compared types of production, with equal profitability with an accuracy of 0.1%, we delete the one where the risk is higher. In our case, portland cement, alumina cement, slag cement and similar hydraulic cements (1.9) (No. 5) dominate in comparison with polymer doors and frames (1.8) (No. 17). And finally, we select points that have similar values in terms of risk, but different in terms of profitability. Two quantities will be considered equal if the modulus of the difference in their values does not exceed 0.1%. Based on this, we get that the point the doors and their plastic boxes (1.8) (No. 17) has lower profitability than the point pebbles, gravel, gravel with a yield of 4.3 (No. 2), but their risk level is approximately the same 9.5 and 9.4, respectively. In this pair of investment areas under consideration, we give the most preference to the second point, since its profitability is 2.5% higher than that of the first.

2 Results

Based on the conducted analysis as an example, we consider four categories of materials for the construction of those located in the region that are most favorable for investment (above the trend line):

Products sanitary - technical from ceramics;
Non-metallic building metals;
Molded, rolled sheet glass;
Facade ceramic tiles.

Both cautious investors (No. 1, 2, 3) and investors indifferent to risk (No. 4) can “take it into service”. The following conclusions can be drawn from the group of types of products. The most profitable products are facade ceramic tiles and wooden panel board. However, based on the schedule (Table 1), it can be concluded that investments in wood paneling parquet have a large share of risk. In turn, sanitary and technical products made of ceramics and ceramic tiles for walls have the least risk. If we consider the data in relation to profitability to risk, we can conclude that the maximum values in the data category are building non-metallic materials, and the minimum in prefabricated reinforced concrete structures and parts. The manufacturing technology of wooden panel parquet has a rich history; this technology was used in the manufacture of the palace park. The Obninsky factory and the Zarya factory, located in the Kaluga Region are two leading enterprises of parquet production in modern Russia. The production of this type of product requires a lot of attention, especially in the field of technical issues. To solve the problem of which it is necessary to take into account such important factors as uninterrupted supply of electricity, and the location of a forest processing plant near the enterprise. Parquet manufacturing technology requires competent and professional workers. Based on the foregoing, we can conclude that this sphere has a relatively large share of risk, because it is not easy to ensure all of the above factors. However, the demand for this type of product is quite high that means that the industry will develop and this allows us to hope for profit with a high degree of probability.

Ceramic facade tile is one of the oldest facing materials in the world. Its construction characteristics allow it to be used for cladding buildings even in harsh climatic conditions. In Russia, ceramic tiles are in high demand due to their affordability, because most of the tiles on the Russian market belong to the low-price segment of consumers. So, we can conclude that this industry will be continuously improved, and therefore will increase the profitability of this industry. However, the risk of investing in this industry is quite high due to the seasonality of demand for these products. After all, the greatest demand for it relates to the warm season. In addition, the Russian tile market is highly competitive (a high number of manufacturers both domestically and abroad).

A correlation analysis was carried out for the obtained group of leaders. It revealed the presence of several pairs of directions with direct and close correlation. For example, ceramic facade tiles and ceramic tiles for walls correlate at 0.96, and ceramic facade tiles with pebbles, gravel and crushed stone at the level of 0.93. The latter means that this group cannot be considered as a single portfolio. Dividing a group of leaders into several subgroups with to minimize internal correlation relationships will be more effective for the investor (Gercekovich 2017). In the future, the selected subgroups should be combined on the basis of the principle of combining decisions (Lyuis and Raifa 1961; Rastrigin and Erenshtein 1975; Nelson 1972).

Where:	
1. Wooden panel board parquet	6. Pebble, gravel, crushed stone
2. Facade ceramic tiles	7. Sanitary ware from ceramics
3. Molded, rolled sheet glass	8. Linoleum
4. Non-metallic building materials	9. Ceramic tiles for floors
5. Small wall blocks	10. Ceramic tiles for walls

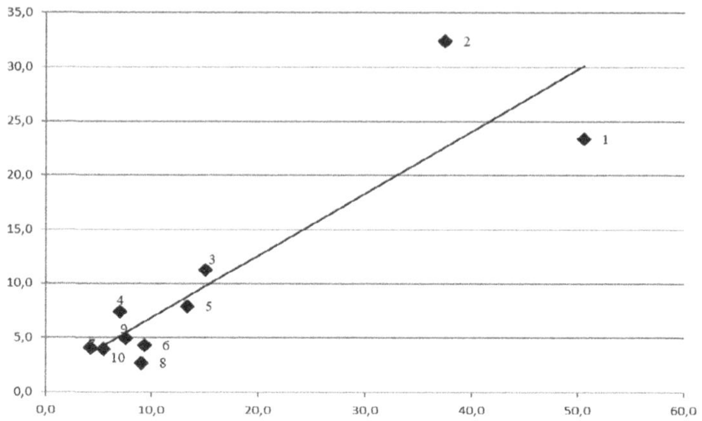

Fig. 1. Graphical display of industry leaders according to the "Profitability- Risk" model

A trend line was drawn for a group of types of leader products (Fig. 1):

$$Dx = 0.57Rs + 1.11; R^2 = 0.81$$

Here Dx is the profitability, Rs is the risk, R^2 is the coefficient of determination. The value of the coefficient of determination indicates the practical suitability of the model "Profitability-risk".

3 Conclusion

The industry for the production of materials for construction is a favorable industry for investment, as this area has great potential in the development and improvement of production technologies which will allow us to hope for profit in case of investing money in this industry. In order to determine the most favorable positions for investment, an analysis of data was carried out according to the "Profitability – Risk" model due to which a list of leaders in this industry was revealed. These are: the production of panelboard wooden parquet, facade ceramic tiles, molded, rolled sheet glass, non-metallic building materials, small wall blocks, pebbles, gravel, crushed stone, ceramic sanitary article, linoleum, glazed ceramic tiles for interior wall cladding and facade ceramic tiles.

References

Chekulaev, M.: Risk Management: Financial Risk Management Based on the Volatility Analysis. Alpina Piblisher, Moscow (2002)

Damodoran, A.: Investment Estimation: Tools and Methods of Valuing and Assets. Alpina, Moscow (2007)

DeBondt, W.F.: Does the stock market overreact. J. Finan. **40**, 793–805 (1985)

Dodie, Z.: Finance. Villiams, Moscow (2017)

Gercekovich, D.A.: Formation of an optimal investment portfolio for a complex of effective portfolios. Bull. Moscow Univ. Ser. Econ. **5**, 86–101 (2017)

Gercekovich, D.A., Babushkin, R.V.: dynamic portfolio analyses of global stock indices. World Econ. Manag. **19**(4), 14–30 (2019)

Gibson, R.: Formation of Investment Portfolio: Management by Financial Risks, 3rd edn. Alpina Piblisher, Moscow (2015)

Investment in Russia. 2017: Statistics. Rosstat, Moscow (2017)

Jegadeesh, N.: Returns to buying winners and selling losers: implications for stock market efficiency. J. Finan. **48**(1), 65–91 (1993)

Lyuis, R.D., Raifa, X.: Plays and Solutions. Publisher of Foreign Literature, Moscow (1961)

Markovitz, H.M.: Portfolio selection. J. Finan. **7**(1), 77–91 (1952)

Nelson, C.R.: The prediction performance of the ERB-MITTPENN model of the U.S. economy. Am. Econ. Rev. **62**(5), 902–917 (1972)

O`Shaughnessy, J.: What Works on Wall Street, pp. 273–295. McGraw-Hill XVI, New York (2005)

Rastrigin, L.A.: Decision making by a team of decision rules in pattern recognition problem. Autom. Telemech. **9**, 133–144 (1975)

Sharp, U.: Investment: INFRA-M. Moscow (2016)

Comparative Analysis of Approaches to Software Identification

K. I. Salakhutdinova[1(✉)], M. E. Sukhoparov[2], I. S. Lebedev[1], and V. V. Semenov[1]

[1] SPIIRAS, 14-Th Linia, VI, 39 St. Petersburg 199178, Russia
kainagr@mail.ru
[2] SPbF AO "NPK "TRISTAN",
47 Nepokorennykh pr., St. Petersburg 195220, Russia

Abstract. The article considers monitoring devices for software installed on automated system users' personal computers. The shortcomings of such software solutions are substantiated and the developed approach to the identification of executable files using the computer-aided learning algorithm is presented, i.e. gradient boosting on decision trees, based on XGBoost, LightGBM, CatBoost libraries. Software identification has been performed using XGBoost and LightGBM. The obtained experimental results have been compared with previous studies by other authors. These results indicate that the developed approach reveals violations of the established security policy when processing information in automated systems.

Keywords: Information security · Software identification · Computer-Aided learning · Gradient boosting on decision trees

1 Introduction

Modern organizations, whether public or private, can no longer be imagined without information technologies. They not only facilitate the automation of previously manual processes but have also become an integral part of the organization's functioning. Therefore, information security is one of the most important components in business. Violation of the established security policies is known to be among the most probable information security incidents. Violation of software policy requirements by users, whether outdated versions or unauthorized installed programs, entails the appearance of a vulnerability, which is further exploited by the attacker. Such software may also be used for personal gain or be subject to copyright from others.

Frequently, the use of organizational protection measures alone is not enough, and it is necessary to back them up with technical measures. For instance, with regard to the task of monitoring software installed on electronic media by automated systems users, software audit tools should be applied as well.

R. Silhavy et al. (Eds.): CoMeSySo 2020, AISC 1295, pp. 72–78, 2020.
https://doi.org/10.1007/978-3-030-63319-6_8

2 Materials and Methods

2.1 Software Monitors

Multiple companies refer to the use of IT asset management (ITAM) systems, which are comprehensive solutions aimed at facilitating physical accounting, financial control and compliance with contractual obligations related to IT assets throughout their entire life cycle. Here, IT assets mean all hardware and software elements of the IT infrastructure that support the business environment activities. In turn, ITAM is divided into Hardware Asset Management (HAM), which covers the management of the IT infrastructure material components: user computers, servers, telephones, etc.; and Software Asset Management (SAM), covering the management of the IT infrastructure non-material components: software, licenses, versions, installation end points, etc.

The service market offers many solutions that identify software assets, manage accounting, monitor their changes and so on. The Microsoft Assessment and Planning Toolkit, Lansweeper, SAManage, AIDA64 Business Edition, and Kaspersky Systems Management can be distinguished from the best-known software products. They are capable of collecting information remotely, while on the server, and with the help of an agent that has been introduced into personal computers. Most of the information provided is collected through the following built-in inventory technologies of the software and hardware environment and operating system: Windows Management Instrumentation, Active Directory, SMS Provider, lshw, dpkg –l, and other technologies.

The shortcomings of this approach are obvious. An inadequate qualification level on the part of the system administrator, the limited availability of computer technology capabilities among users of automated systems, or a low-level approach to recruiting, makes it possible to introduce changes into the configuration data of the installed software. Thus, primitive manipulations with the Windows system registry can enable one to modify the parameters of the installed program by changing its version, name, displayed shortcut, etc. In Linux OS, the same information about programs is contained in configuration files; here it is possible to change the program metadata even before they are installed on the computer, by interfering with the configuration file "control" of the installed .deb package.

Earlier, in [1–3] the authors presented an approach to auditing electronic media that involved generating frequency signatures of a byte or assembly program code and discussed various software identification methods, i.e. enabling the determination of the similarity of the reference signatures of authorized programs with the signatures of the identified files. Thus, the recent publication [4] considered an approach based on the computer-aided learning, namely the gradient boosting decision trees algorithm [5], implemented in the open-source CatBoost Library of Yandex Company [6].

In this article, the authors aim at comparing several devices implementing gradient boosting as applied to the software identification task and correlating the results obtained with the approaches considered by other researchers.

2.2 Gradient Boosting on Decision Tree Libraries

The following most common open libraries were considered:

- XGBoost, an optimized distributed library that is highly efficient, flexible, and portable [7]. It provides parallel boosting of trees (also known as GBDT, GBM) capable of solving many scientific problems quickly and accurately [8]. The written code operates on the majority of the existing platforms, for example, Hadoop, SGE, MPI, and is capable of solving billions of examples. It works with the following programming languages: Python, R, Julia, Scala.

- LightGBM, a Microsoft platform [9] designed to be distributed and efficient with the following advantages over competitive libraries:

- faster workout speed and higher efficiency;
- lower memory usage;
- high accuracy;
- support for parallel and GPU training;
- capability to handle large data arrays.

It works with the following programming languages: Python, R.

- CatBoost, a domestic library, presented in 2017. A special feature of the algorithm is the construction of symmetric trees, the ability to work with category features; in addition, it also allows to be learned on a relatively small amount of heterogeneous data. It works with the following programming languages: Python, R.

2.3 Experiment Set up and Learning Parameter Selection

The experiment was carried out on the same data samples as in [4]. The learning sample is represented by 443 executable files of the Linux operating system (OS) of various versions and bit capacities (32x and 64x) related to 63 different programs. The test sample included 123 files belonging to the same 63 programs, all of them differed from the files used in the learning sample, and had 32x and 64x bit capacities.

The reference program signatures used in the learning sample and the signatures of the identified test sample programs have the same structure and represent the frequency distribution of the feature (one of 10 assembler commands: add, and, call, cmp, je, jmp, lea, mov, pop, push) in each of the 30 received intervals splitting the assembly code of the program.

The following learning parameters were chosen for solving the multi-classification problem [10], using the selected gradient boosting libraries:

For XGBoost:

- booster – a type of boosting;
- eta – the step size used to prevent retraining;
- max_depth – the maximum depth of the tree;
- num_round – the number of boosting iterations;
- objective – an evaluation metric used in model training.

For LightGBM:

- boosting – a type of boosting
- learning_rate – the step size used to prevent retraining;
- max_depth –the maximum depth of the tree;
- num_iterations – the number of boosting iterations;
- objective – an evaluation metric used in model training.

For CatBoost:

- boosting_type – a boosting scheme;
- learning_rate – the learning rate, used to reduce the gradient descent step;
- l2_leaf_reg – L2 regularization coefficient, used to calculate the leaf values;
- depth – the tree depth;
- iterations – the maximum number of trees that will be built when solving the problem of computer-aided learning;
- loss_function – a metric (loss function) used in training.

Table 1 gives the empirically selected values of the selected parameters for solving the identification problem.

Table 1. Learning model parameters

Learning parameters	XGBoost	LightGBM	CatBoost
booster/boosting/boosting_type	gbtree	gbdt	Plain
eta/learning_rate	0.3	0.3	0.7
-/-/l2_leaf_reg	–	–	1
max_depth/depth	2	7	2
num_round/num_iterations/iterations	1000	1000	1000
objective/loss_function	Multi:SoftMax	MultiClass	MultiClass

The remaining values of the model parameters were set by default.

3 Results

The test sample identification results for 10 assembler commands are shown in Fig. 1. Obviously, for almost all assembler commands, the number of correctly identified programs is higher when using the XGBoost library and reaches a maximum value of 95.93% for the jmp command (118 correctly identified executable files of 123 in total).

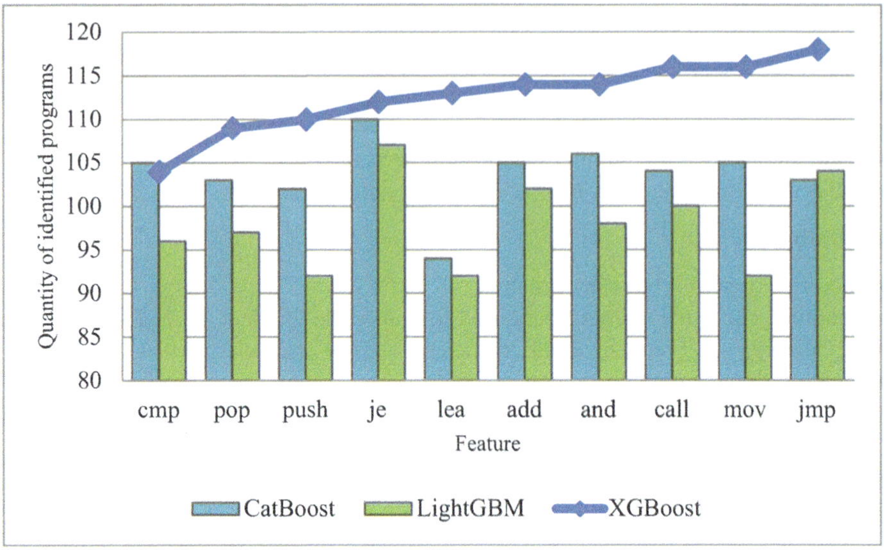

Fig. 1. The number of correctly identified executable files using different Gradient Boosting on Decision Trees Libraries.

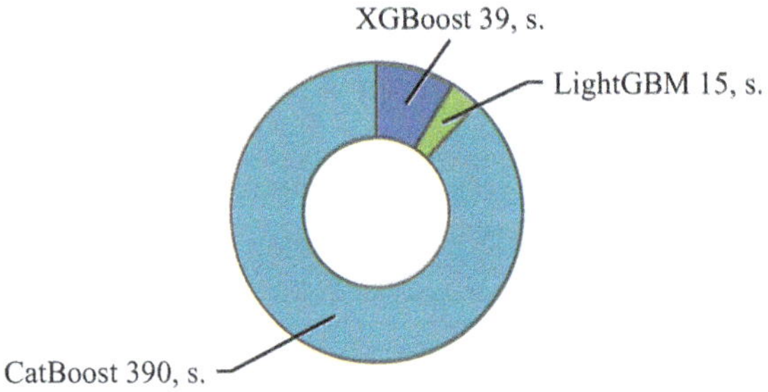

Fig. 2. Training time for the test sample files classification and identification model.

At the same time Fig. 2 shows a significant advantage of the LightGBM in time spent on the model training and identification of 123 executable test sample files.

Various approaches to comparison of executable files were analyzed using the bicubic clustering quality measure [11], the choice of which was determined by the presentation of the results of the experiments performed by the author of [12]; the analysis is given in Table 2.

Table 2. Comparison of approaches to the identification of executable files.

File format identification approach	Bicubic measure maximum	Measurement error
Based on context triggered piecewise hashing [13]	0.29	0.059
Based on the Euclidean distance between vectors for blocks of a fixed dimension [14]	0.55	0.099
Based on the edit distance between vectors for blocks of a variable length	0.69	0.123
LightGBM	0.84	0.071
CatBoost	0.88	0.055
XGBoost	0.96	0.027

In this connection, the results of Table 2 demonstrate that machine learning based on the gradient boosting on decision trees is the most effective method from the viewpoint of calculating a bicubic measure. In turn, XGBoost shows the best result among the reviewed libraries. However, it is worth noting that during the experiment, the authors found that different Gradient Boosting Libraries can be unable to identify a certain number of test sample programs, regardless of the selected feature, while among the three algorithms considered, the names of such programs were different for each algorithm. Thus, the cumulative results of the LightGBM, CatBoost and XGBoost libraries will allow achieving more efficient identification of executable files.

4 Conclusion

Taking into account the above results, the advantage of the software identification approach using computer-aided learning that was developed by the authors is obvious as compared to other approaches using context triggered piecewise hashing, the Euclidean distance between vectors for blocks of a fixed dimension and the edit distance between vectors for blocks of a variable length.

It also should be noted that the best result was shown by the XGBoost Library as compared to other tools for implementing gradient boosting on decision trees in relation to the software identification task.

References

1. Salakhutdinova, K.I., Krivtsova, I.E., Lebedev, I.S., Sukhoparov, M.E.: An approach to selecting an informative feature in software identification. In: Lecture Notes in Computer Science (including subseries Lecture Notes in Artificial Intelligence and Lecture Notes in Bioinformatics), vol. 11118, pp. 318–327 (2018)
2. Krivtsova, I.E., Lebedev, I.S., Salakhutdinova, K.I.: Identification of executable files on the basis of statistical criteria. In: Proceedings of the 20th Conference of Open Innovations Association FRUCT, pp. 202–208 (2017)

3. Salakhutdinova, K.I., Lebedev, I.S., Krivtsova, I.E., Sukhoparov, M.E.: Studying the influence of feature and coefficient (ratio) selection in the formation of a signature in the software identification task. Inf. Secur. Issues Comput. Syst. **1**, 136–141 (2018)
4. Salakhutdinova, K.I., Lebedev, I.S., Krivtsova, I.E.: Gradient boosting trees method in the task of software identification. Sci. Tech. J. Inf. Technol. Mech. Opt. **18**(6/118), 1016–1022 (2018)
5. Druzhkov, P.N., Zolotykh, N.Y., Polovinkin, A.N.: Parallel implementation of prediction algorithm in gradient boosting trees method. Bull. South Ural State University. Ser. Math. Model. Program. **37**(254), 82–89 (2011)
6. CatBoost GitHub Homepage. https://github.com/catboost. Accessed 29 Jan 2019
7. XGBoost GitHub Homepage. https://github.com/dmlc/xgboost. Accessed 29 Jan 2019
8. Kitov, V.V.: Accuracy analysis of the gradient boosting method with random rotations. Stat. Econ. **4**, 22–26 (2016)
9. LightGBM GitHub Homepage. https://github.com/Microsoft/LightGBM. Accessed 2 Feb 2019
10. Kaftannikov, I.L., Parasich, A.V.: Decision tree's features of application in classification problems. Bull. South Ural State University Ser. Comput. Technol. Manag. Electron. **3**(15), 26–32 (2015)
11. Bagga, A., Baldwin, B.: Cross-document event conference: annotations, experiments, and observations. In: Proceedings of the ACL-99 Workshop on Conference and Its Applications, pp. 1–8 (1998). http://aclweb.org/anthology/W99-0201. Accessed 2 Feb 2019
12. Antonov, A.Y., Fedulov, A.S.: File type identification based on structural analysis. Appl. Inf. **2**(44), 68–77 (2013)
13. Kornblum, J.D.: Identifying almost identical files using context triggered piecewise hashing. Digit. Invest. **3**, 91–97 (2006)
14. Ebringer, T., Sun, L., Boztas, S.: A fast randomness test that preserves local detail. In: Proceedings of the 18th Virus Bulletin International Conference — United Kingdom: Virus Bulletin Ltd., pp. 34–42 (2008)

The Digital Random Signal Simulator

Oleg Chernoyarov[1,2,3(✉)], Alexey Glushkov[4,5], Vladimir Litvinenko[5], Yuliya Litvinenko[5], and Kirill Melnikov[1]

[1] National Research University "MPEI", Krasnokazarmennaya str. 14, 111250 Moscow, Russia
chernoyarovov@mpei.ru
[2] National Research Tomsk State University, Lenin Avenue 36, 634050 Tomsk, Russia
[3] Maikop State Technological University, Pervomaiskaya str. 191, 385000 Maikop, Russia
[4] Voronezh Institute of the Ministry of Internal Affairs of the Russian Federation, Patriots Avenue 53, 394065 Voronezh, Russia
[5] Voronezh State Technical University, Moscow Avenue 14, 394026 Voronezh, Russia

Abstract. There is introduced a universal simple digital simulator of the random signal with the specified arbitrary two-dimensional probability distribution of its discrete values based on the Markov model. It allows us to generate an unlimited sequence of samples with the required probabilistic and correlation properties of adjacent random or pseudo-random values of the simulated random process. By applying the simplified version of the simulator, the independent signal samples obeying the specifying arbitrary one-dimensional probability distribution can be formed. The simulator design involves the technique of determining the parameters of the Markov model of the random signal values at the two time moments. The Markov model can be created based on either the specified two-dimensional probability density or the experimental sample of the simulated random process. There is presented the algorithm for signal sample generating that provides a high sampling frequency. Its hardware implementation by means of microprocessor devices or field-programmable gate arrays is also in focus. The computational algorithm procedure does not require complex mathematical transformations of the signal samples and is implemented using an inexpensive circuitry. Changing the statistical properties of the simulated random processes is achieved by either rebooting the storage device with the preformed data array or switching the memory pages in which the necessary arrays are stored. In order to demonstrate the performance of the simulator and its high efficiency, the properties of the simulator, its probabilistic and correlation characteristics are studied. It is shown that a high accuracy of coincidence is ensured between the two-dimensional probability distribution of the selected model and the histogram based on the generated sequence of samples of the random signal. #COMESYSO1120

Keywords: Random signal simulator · Arbitrary two-dimensional probability distribution · Discrete values · Markov model · Matrix of transition probabilities · Histogram · Frequency filtering

R. Silhavy et al. (Eds.): CoMeSySo 2020, AISC 1295, pp. 79–93, 2020.
https://doi.org/10.1007/978-3-030-63319-6_9

1 Introduction

When designing, studying and testing various data-transmission systems, there is a need to apply simulators (generators) of random information or interference signals with the specified statistical properties [1–3]. They are used to simulate the required information signals [4], to jam radio channels [5], to mask the operation of computer devices [6], etc. Applying the acoustic noise sources allows us to protect users from unauthorized eavesdropping [7, 8].

Most often [9–11], in order to design the analog noise generator the thermal noise processes in the electronic elements (Zener diodes, transistors) are used. As a rule, the statistical properties of these processes correspond to the Gaussian probability distribution, while the spectral properties – to white or pink noise. The disadvantages of such generators are both the difficulty of providing the specified probabilistic characteristics with high accuracy and the impossibility of repeating the realization of a random process.

Digital simulators of random processes offer the highest flexibility and provide the required probabilistic characteristics, for instance, by transforming [12] equiprobable random numbers obtained, for example, by means of a generator of a long M-sequence [13]. However, in this case there are computational difficulties with high accuracy implementation of non-linear operations, especially for two-dimensional probability distributions.

The purpose of this paper is to introduce the digital algorithm for generating (simulating) random numbers with the specified two-dimensional probability distribution based on their Markov model [14, 15], which does not have many of the disadvantages that the common methods have mentioned above.

2 The Markov Model

The Markov model [14, 15] of the simulated random discrete process is described by the square transition probability matrix of the form

$$[P_{ij}] = \begin{bmatrix} P_{11} & P_{12} & \dots & P_{1M} \\ P_{21} & P_{22} & \dots & P_{2M} \\ \dots & \dots & \dots & \dots \\ P_{M1} & P_{M2} & \dots & P_{MM} \end{bmatrix},$$

where P_{ij} is the probability of the process transition from the value $z_n = i$, $i = \overline{1, M}$ at the time t_n to the value $z_{n+1} = j$, $j = \overline{1, M}$ at the next time t_{n+1}; $M = 2^m$, m is the length of the binary sample code; n is the number of the sample of the simulated signal, $n = \overline{1, L}$, L is the sample size.

While the two-dimensional probability density $w(x_1, x_2)$ is specified for the values of the continuous random process that are separated by the time interval τ, then these values are quantized, for example, by an analog-to-digital converter (ADC) with the quantization step d and the boundaries of quantization intervals

$$c_k = (k - M/2)\,d + \bar{x}, \quad k = \overline{1, (M-1)}. \tag{1}$$

Here $\bar{x}$ is the mean value (mathematical expectation) of the random process, while $c_0 = -\infty$ and $c_M = \infty$. The quantization step d in (1) is chosen depending on the dispersion (mean square deviation) of the random process [16]. Then for the transition probabilities (1) we get

$$P_{ij} = \int_{c_{i-1}}^{c_i} \int_{c_{j-1}}^{c_j} w(x_1, x_2)\, dx_2 dx_1 \Bigg/ \int_{c_{i-1}}^{c_i} \int_{-\infty}^{\infty} w(x_1, x_2)\, dx_2 dx_1 \tag{2}$$

The Markov model can be built upon the sufficiently long experimental sequence of discrete samples of the random process z_n, $n = \overline{0, L}$ from the ADC output. For this purpose, the numbers l_{ij} of transitions from the value $z_n = i$ to the next value $z_{n+1} = j$ are determined by the observed sampling and then the transition probabilities of the Markov model are estimated as follows:

$$P_{ij} = l_{ij} \Bigg/ \sum_{k=1}^{M} l_{ik}. \tag{3}$$

To eliminate possible zero values in the denominator (3), any small constant (for example, 1) should be added to the values l_{ij}. It should be noted that in this approach no assignment of the probabilistic characteristics of the random process are required to create the Markov model.

To implement the simulator based on the transition probability matrix $\left[P_{ij}\right]$ (2) or (3), the matrix of the two-dimensional probability distribution function is calculated in the following way

$$F_{ij} = \sum_{m=1}^{j} P_{im}. \tag{4}$$

3 The Block Diagram of the Simulator

The block diagram of the simulator is shown in Fig. 1. By the clock generator (GG) signals with the frequency F, the random (or pseudorandom) number generator (RNG) produces the uniformly distributed binary r-bit numbers v_n (where n is the current sample number). To obtain the random numbers, the RNG includes an analog noise generator with an amplifier and r-bit ADC clocked by the CG. If the pseudo-random numbers are required, then they can be obtained by means of the unit including the feedback shifters [13], for example, by means of the M-sequence generator.

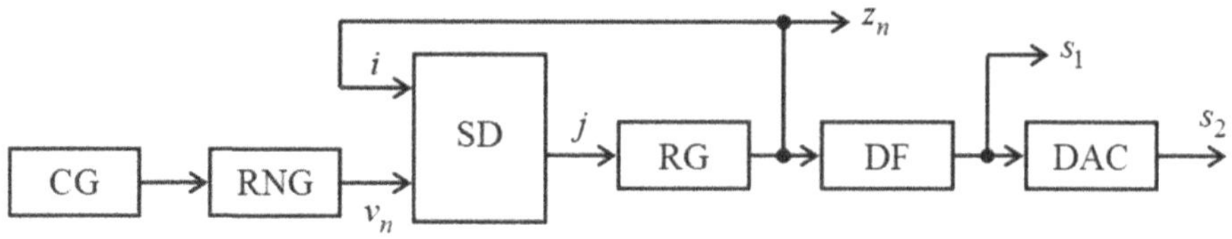

Fig. 1. The block diagram of the simulator of the random signal with the specified two-dimensional probability distribution.

In the storage device (SD), in the memory cells located in the addresses $i2^r + v$, the precomputed minimum binary m-bit codes j are saved, and for them under the specified values of the of binary codes i, $i = \overline{0, (M-1)}$ and v, $v = \overline{0, (2^r - 1)}$ the inequality

$$v < F_{(i+1)(j+1)} \tag{5}$$

is satisfied while the probability distribution function (4) is set.

In the first cycle ($n = 1$), the RNG generates the number v_1 and for an arbitrary initial state i of the register (RG), for example, under $i = 0$, the code of the number j is read from the SD. This number is stored in RG and is the first output sample $z_1 = j$. In the next cycle ($n = 2$), the previously formed sample j becomes the previous one i and the procedure is repeated continuously.

The sequence of the random binary numbers z_n obeying the specified two-dimensional probability distribution is passed to the recipient (Fig. 1) or arrives to the digital filter (DF) to form the digital output signal s_1 with the desired spectrum. Then this signal is fed to the digital-to-analog converter (DAC) which generates the output analog random signal s_2.

A special case of the device presented in Fig. 1 is the digital simulator of discrete random numbers with an arbitrary one-dimensional probability distribution determined by the one-dimensional probability density $w(x)$. The block diagram of such a simulator is shown in Fig. 2.

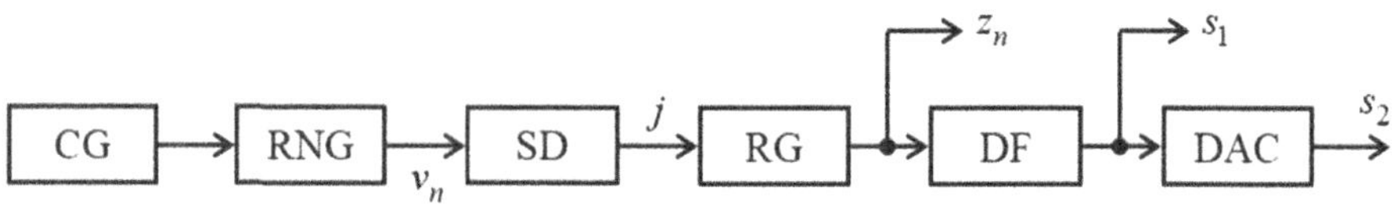

Fig. 2. The block diagram of the simulator of the random signal with the specified one-dimensional probability distribution.

When the thresholds (1) for the specified probability density $w(x)$ are applied, the discrete values of the probability distribution function are determined as follows

$$F_i = \int_{c_0}^{c_i} w(x)\,dx \quad i = \overline{1, (M-1)} \quad \text{and } F_M = 1 \tag{6}$$

Based on (6), for all the values of v, $v = \overline{0, (2^r - 1)}$, the binary codes i are determined, the ones for which the inequality

$$v < F_i \tag{7}$$

is satisfied. The obtained values are saved in the memory cells located in v.

4 Signals with the Specified One-Dimensional Probability Distribution

In radio engineering applications, the models of random processes with various one-dimensional probability distributions are widely used [17–19]. This is primarily a Gaussian (normal) random process with the probability density of the form of

$$w(x) = \exp\left[-(x - \bar{x})^2 / 2\sigma^2\right] / \sigma\sqrt{2\pi}, \tag{8}$$

where $\bar{x}$ is the mean value and σ^2 is the dispersion of the random signal. By means of such processes the noises of electronic equipment are simulated.

As an example, let us present the case when $\bar{x} = 0$, $\sigma^2 = 4$, the ADC width is $m = 6$ (the number of the quantization levels is $M = 64$) and in (1) one chooses $d = 20/M = 0.323$. In Fig. 3a, by dotted line the dependence is shown of the probability p_i that the value of the random process falls within the i-th quantization interval. In Fig. 3b, the corresponding probability distribution function (6) is drawn. Taking into account (7), the values of the samples i are calculated and saved in SD (Fig. 2) at the memory location v, $v = \overline{0, (2^r - 1)}$ to be used then to form the n-th sample of the signal z_n.

In Fig. 4a, there is presented the time realization of the sequence of output discrete samples z_n obtained by simulation of the simulator operation, while in Fig. 4b – the realization of the signal $\tilde{z}_n = (z_n - \bar{z})\, d$ after its transformation into the analog form. In order to test the one-dimensional statistical characteristics of the simulated realization (Fig. 4a), the histogram (probability estimate) of its possible values is plotted by columns in Fig. 3a. As it can be seen, it coincides with the corresponding theoretical probabilities p_i. The samples are uncorrelated and have a uniform spectrum of amplitudes. By applying the digital filter (as it is shown in Fig. 2), the frequency properties of the random output signal can be adjusted.

The random process with the Nakagami-m distribution can adequately describe, for example, the signal amplitude oscillations in a multipath channel [17–19]. Its probability density has the form of

$$w(x) = \frac{2m^m}{\Gamma(m)\,\Omega^m} x^{2m-1} \exp\left(-\frac{m}{\Omega} x^2\right), \quad x \geq 0, \tag{9}$$

where m and Ω are the model parameters and $\Gamma(m)$ is the gamma function. If $m = 1$, then the distribution (9) is transformed into the well-known Raileigh one [17, 18]. In Fig. 5, the Nakagami probability density function is drawn for various model parameters.

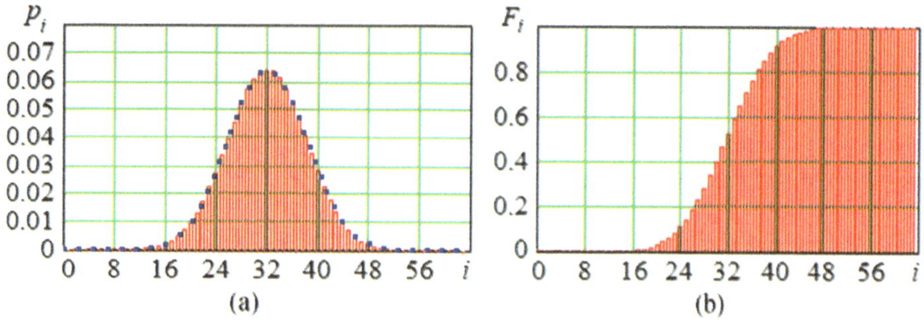

Fig. 3. The estimates of the probability density (a) and the distribution function (b) of the Gaussian random process values.

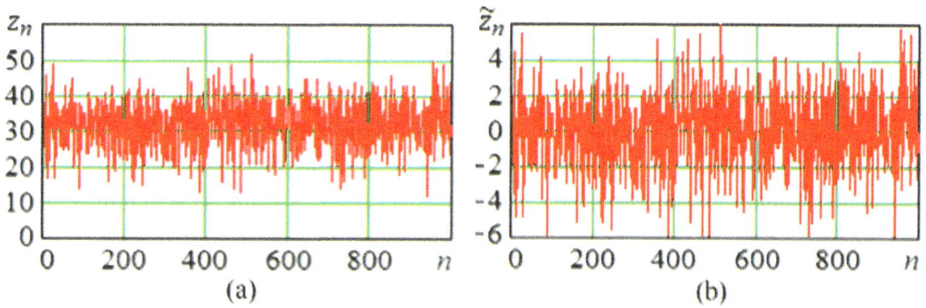

Fig. 4. The digital (a) and analog (b) time realizations of the Gaussian random process.

For example, the simulation of the simulator operation is carried out while $M = 64$, the quantization interval is $d = 4/M = 0.065$ $(x \geq 0)$, $m = 1$ that corresponds to the Rayleigh probability density and $\Omega = 1$.

In Fig. 6a, by dotted line, there is shown the dependence of the probability p_i that the value of the random process is within the i-th quantization interval, while in Fig. 6b – the dependence of the corresponding probability distribution function p_i.

In Fig. 7a, the time diagram is presented of the digital Nakagami signal at the simulator output when $m = 1$ and $\Omega = 1$, that is the simulated random process obeys the Rayleigh random process. In Fig. 6a, by columns, the histogram is shown of the realization drawn in Fig. 7a and we can see that it coincides well with the theoretical values of the corresponding probability density. In Fig. 7b and Fig. 7c, the similar time diagrams are plotted for the other parameters of the model: $m = 4$, $\Omega = 1$ (Fig. 7b) and $m = 4$, $\Omega = 5$ (Fig. 7c). Their histograms coincide with the probabilistic characteristics of the Nakagami distribution shown in Fig. 5.

Thus, the random signal simulator presented in Fig. 2 provides a quite accurate representation of the random process with the specified one-dimensional probabilistic characteristics and uncorrelated samples, while the required spectral characteristics of the simulated process can be formed by the digital filter.

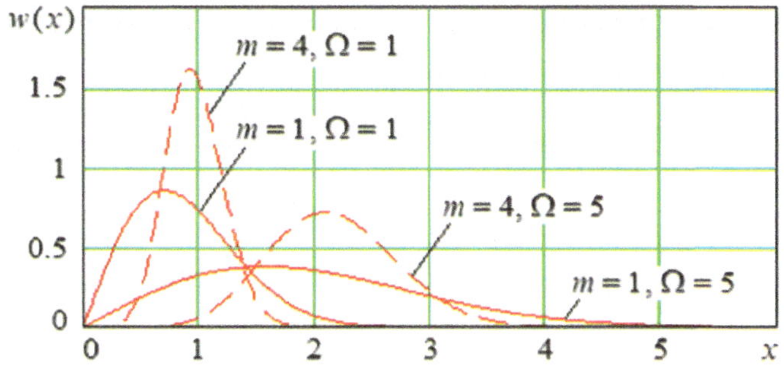

Fig. 5. The Nakagami probability density with various distribution parameters.

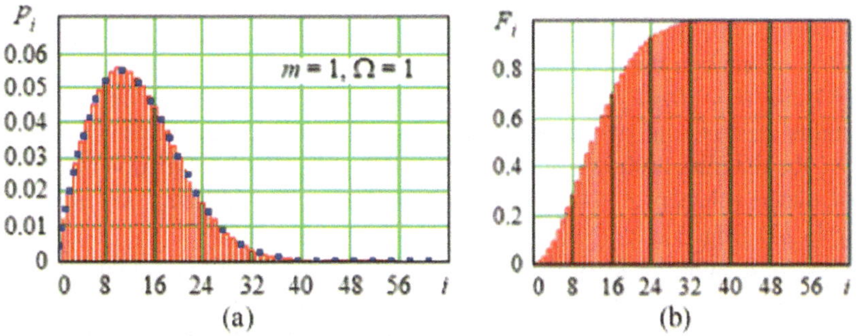

Fig. 6. The estimates of the probability density (a) and the distribution function (b) of the Nakagami random process values.

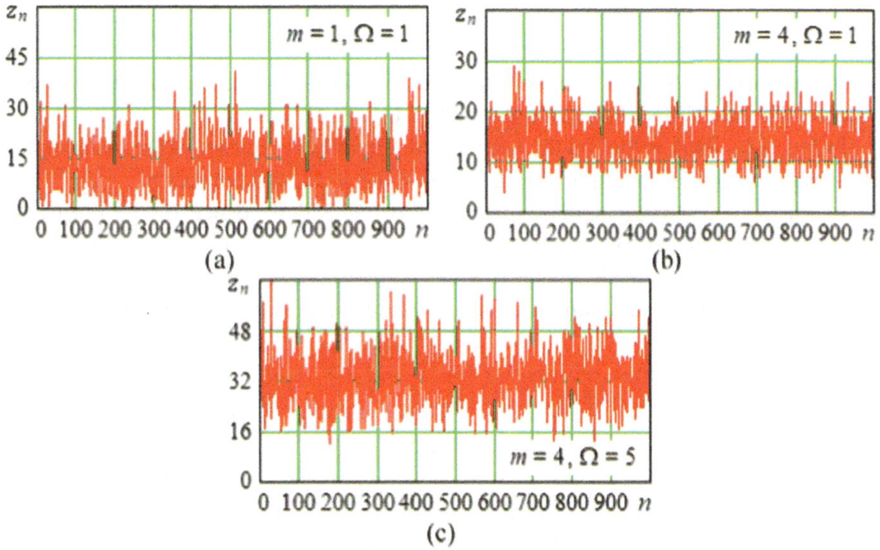

Fig. 7. The time realizations of the Nakagami random signal with various distribution parameters: a) $m = 1$, $\Omega = 1$; b) $m = 4$, $\Omega = 1$; c) $m = 4$, $\Omega = 5$.

5 Pseudorandom Signals with the Specified Two-Dimensional Probabilistic Characteristics

The two-dimensional probability density of the stationary Gaussian random process has the form of [18]

$$w(x_1, x_2) = \frac{1}{2\pi\sigma^2\sqrt{1-\rho^2}} \exp\left[-\frac{(x_1-\bar{x})^2 + 2\rho(x_1-\bar{x})(x_2-\bar{x}) + (x_2-\bar{x})^2}{2\sigma^2(1-\rho^2)}\right], \tag{10}$$

where $\bar{x}$ is the mean value, σ^2 is the dispersion and ρ is the correlation coefficient of the random process. In Fig. 8, an example of the function (10) is presented for the parameters $\bar{x} = 0$, $\sigma = 2$ and $\rho = 0.8$.

When quantizing the random process by applying the quantization levels (1) under $M = 64$ and $d = 20/M = 0.323$, the Markov model is obtained with the transition probability matrix, the two-dimensional diagram of which is shown in Fig. 9a. In Fig. 9b, the diagram is shown of the corresponding distribution function (4). Based on the calculated values of F_{ij} (4) and taking into account (5), the data array is formed for saving in SD (Fig. 1).

With the length $r = 11$ of the binary codes v_n from the RNG output, the SD data array is shown in Fig. 10 where i_k is the value of the sample, k is the SD memory location (its decimal equivalent). In Fig. 10a there are presented the start SD spaces and in Fig. 10b – the final ones, while the total memory content equal to $M \cdot 2^r$ or 128 kilobytes.

In Fig. 11a, the time realization is shown of the simulated Gaussian process with the correlated values. In Fig. 11b, the histogram of this realization and its desired one-dimensional probability density are drawn by vertical and dotted lines, respectively. The obtained diagrams of the two-dimensional probability density of the random process values, the transition probability matrix and the distribution function coincide with the theoretical dependences presented in Fig. 8, Fig. 9.

In Fig. 12a, there is plotted the normalized amplitude spectrum of the realization of 2^{20} signal samples in length as the function of the harmonic number n.

In Fig. 12b, by vertical lines there is drawn the dependence of the correlation coefficient upon the shift k between the numbers of samples, while by dotted line – the corresponding theoretical dependence which is $\rho_k = \rho^k$. As it can be seen, the simulated realization of the random process has the specified correlation properties.

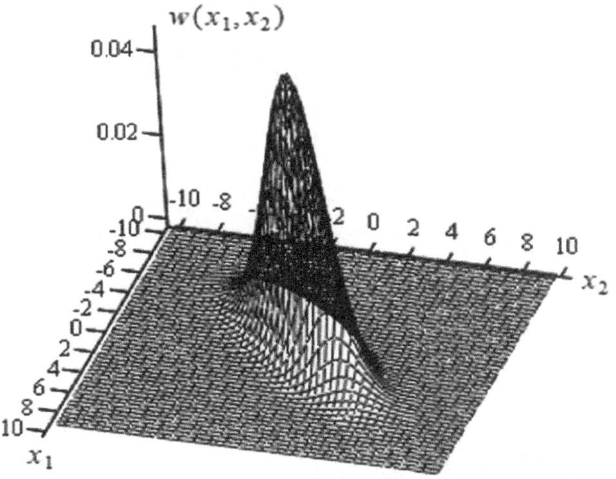

Fig. 8. The two-dimensional probability density of the Gaussian process.

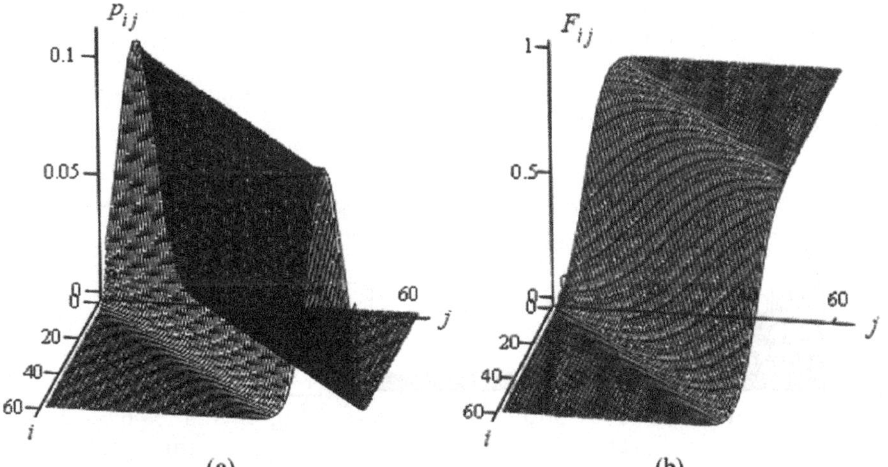

Fig. 9. The transition probability matrix (a) and the two-dimensional distribution function (b) of the Gaussian random process.

Let us now consider the simulation of random processes with the other two-dimensional probability densities of the form of

$$w(x_1, x_2) = \sin(x_1 + x_2)/2, \quad 0 \le x_1, x_2 \le \pi/2, \tag{11}$$

$$w(x_1, x_2) = \sin(x_1)\sin(x_2), \quad 0 \le x_1, x_2 \le \pi/2, \tag{12}$$

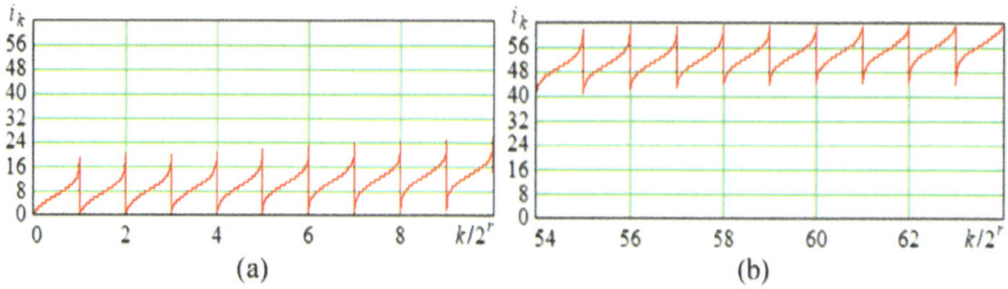

Fig. 10. The data diagrams in SD cells: a) start storage space; b) final storage space.

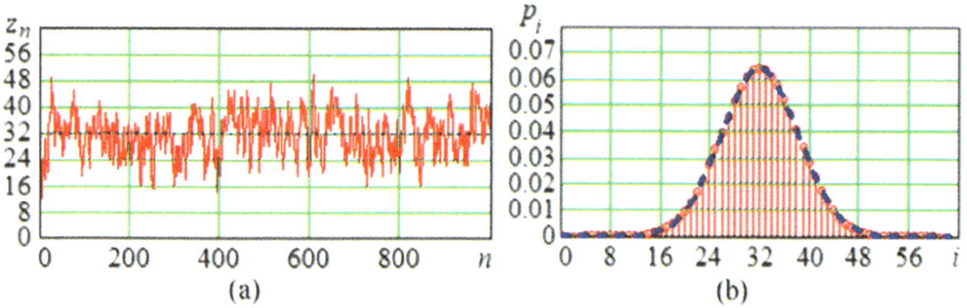

Fig. 11. The simulated realization of the Gaussian random process (a) and its estimated and desired one-dimensional probability density (b).

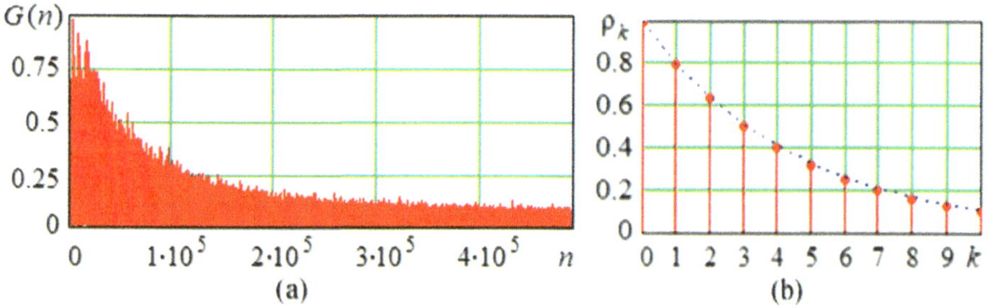

Fig. 12. The normalized amplitude spectrum (a) and the estimated and desired correlation coefficient (b) of the simulated realization of the Gaussian random process.

$$w(x_1, x_2) = \exp[-(x_1 + x_2)], \quad x_1 \geq 0, \quad x_1 \geq 0 \tag{13}$$

The samples of the process (11) are correlated, while the samples of the processes (12) and (13) are not correlated.

In Fig. 13a–c, there are shown the diagrams of the two-dimensional probability densities (11)–(13), respectively, while in Fig. 14a–c – the corresponding Markov models. As it can be seen, the considered random processes have various probabilistic characteristics.

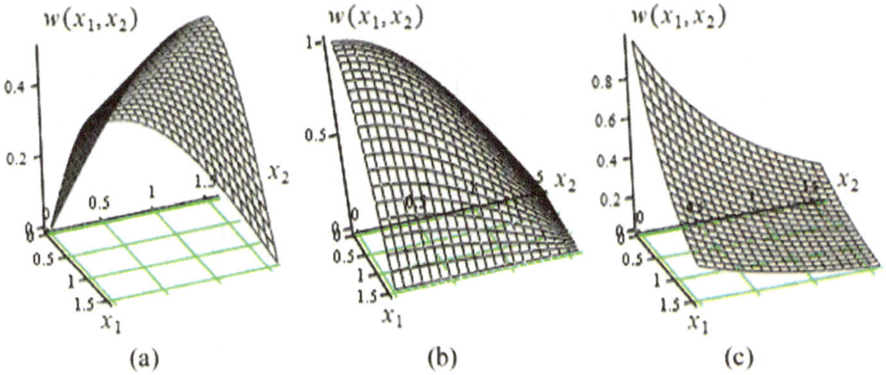

Fig. 13. The two-dimensional probability densities of the simulated non-Gaussian random processes: a) (11); b) (12); c) (13).

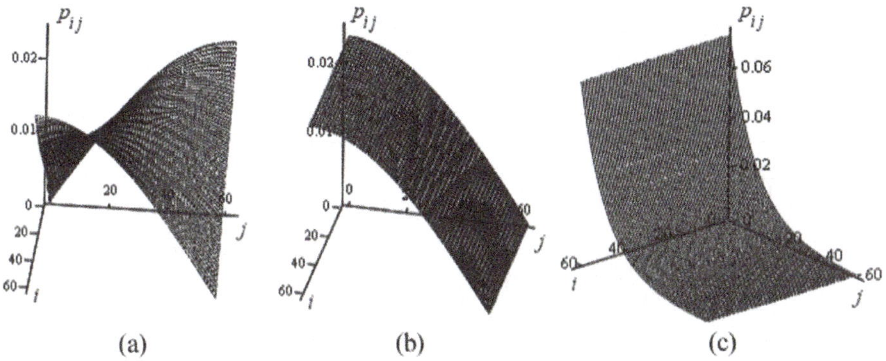

Fig. 14. The two-dimensional diagrams of the transition probability matrices of the simulated non-Gaussian random processes described by the probability densities: a) (11); b) (12); c) (13).

In Fig. 15, the SD data arrays i_k are drawn of the processes (11)–(13), similar to those presented in Fig. 10a. In Fig. 16a–c, the time realizations of discrete samples z_n are plotted of the random processes (11)–(13), respectively, generated by the simulator.

By the samples, the joint probability densities of the discrete values $z_n = i$ and $z_{n+1} = j$ of the processes (11)–(13) are estimated, their two-dimensional diagrams $\tilde{w}(i,j)$ are shown in Fig. 17. From Fig. 17 it follows that the shape of the obtained dependences $\tilde{w}(i,j)$ coincides with the shape of the desired probability densities presented in Fig. 13. Thus, the simulator provides an accurate display of the two-dimensional probabilistic properties of the generated processes.

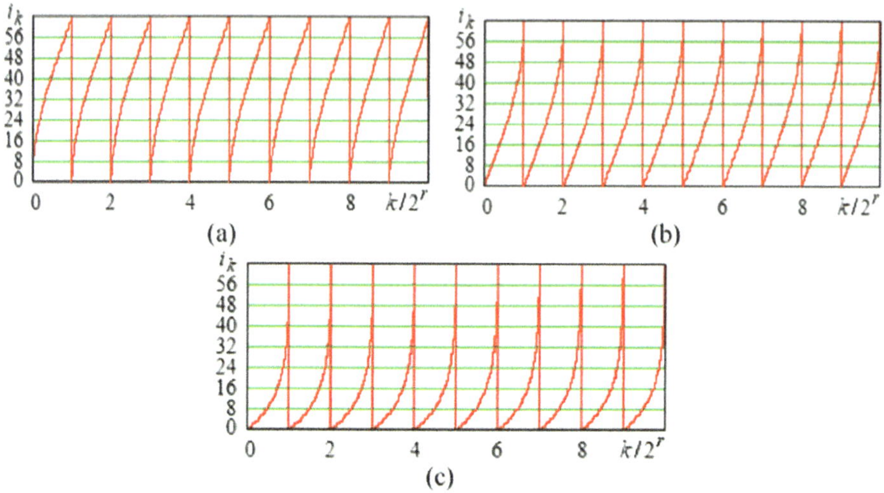

Fig. 15. The SD time diagrams in the simulation of the non-Gaussian processes described by the probability densities: a) (11); b) (12); c) (13).

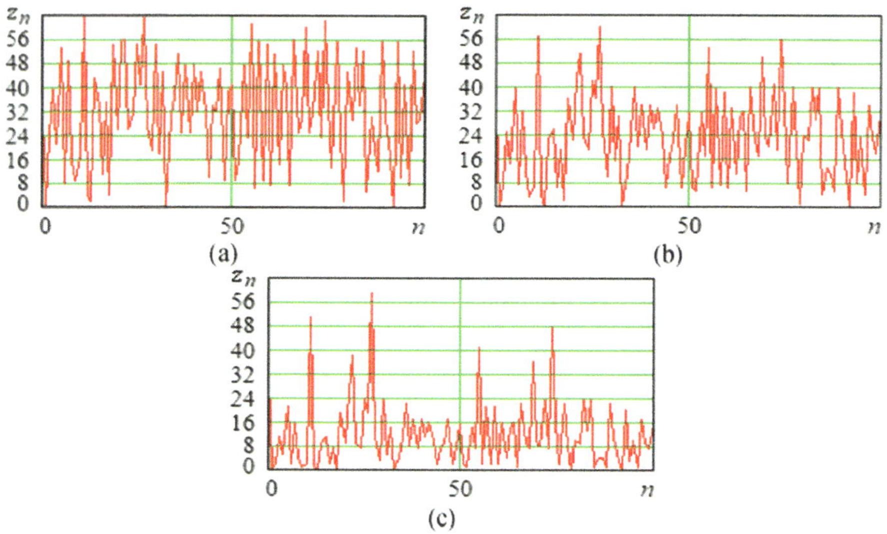

Fig. 16. The realizations of the simulated non-Gaussian processes described by the probability densities: a) (11); b) (12); c) (13).

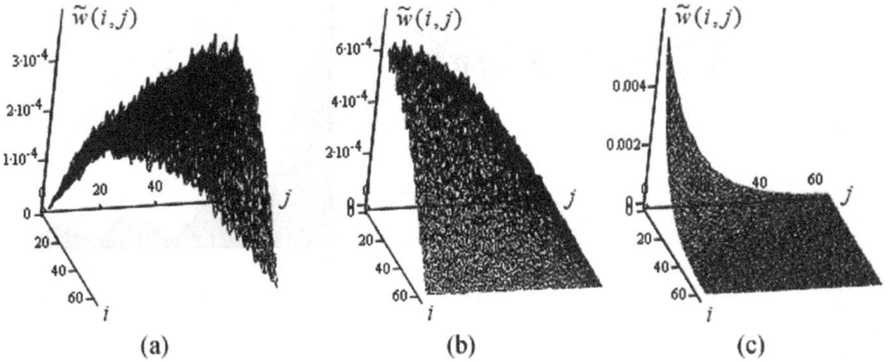

Fig. 17. The two-dimensional diagrams of the estimated probability densities of the simulated non-Gaussian random processes: a) (11); b) (12); c) (13).

6 The Frequency Filtering of the Simulator Signals

The samples z_n of the simulated random processes with the one-dimensional probability distribution are independent and have a uniform amplitude spectrum. When the one-dimensional probability density is the Gaussian one, uncorrelated noise samples with the uniform spectrum are generated. If it is necessary to obtain a non-uniform spectrum of the simulated process (pink noise), then the digital filter DF (Fig. 1, Fig. 2) is applied with the corresponding frequency response.

For the Gaussian random process describing the probability density (8) with $\bar{x} = 0$ and $\sigma^2 = 4$, the time realization of the sequence of the samples is shown in Fig. 4a. As the simplest digital low-pass filter, one considers the moving window of the summation of the K samples (moving average algorithm or linear non-recursive digital filter) whose response is equal to

$$s_{1n} = \frac{1}{K}\sum_{i=0}^{K-1} z_{n-i}. \tag{14}$$

In Fig. 18a, the time realization is shown of the response s_{1n} of the digital filter (14) from the sample number n under $K = 10$. By dotted line, its mean value is drawn. In Fig. 18b, the normalized amplitude spectrum of the filtered process (14) is presented as the function of the harmonic number n. As it can be seen, the DF response changes slower than the signal z_n (Fig. 4a) and its spectrum is substantially non-uniform.

When the DF (14) is exposed to the samples of the Gaussian random process with the mean value $\bar{x}$ and the dispersion σ^2, the mean value $\bar{s}_1$ and the dispersion σ_1^2 of the response are equal to $\bar{s}_1 = \bar{x}$ and $\sigma_1^2 = \sigma^2/K$. In the considered example, these values are $\bar{s}_1 = 32$ and $\sigma_1^2 = 0.4$. In Fig. 19a, by columns, there is shown the histogram of the output samples s_{1n} while by the dotted line – the corresponding desired probability density. In Fig. 19b, the same histogram is presented in the logarithmic scale in order to demonstrate the coincidence of the theoretical and the experimental values of the probability density of the simulated random process in the low-probability area.

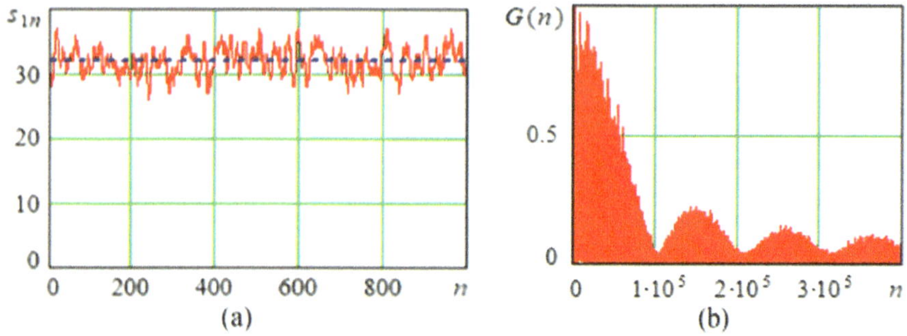

Fig. 18. The time realization of the response of the digital filter on the uncorrelated Gaussian samples (a) and its amplitude spectrum (b).

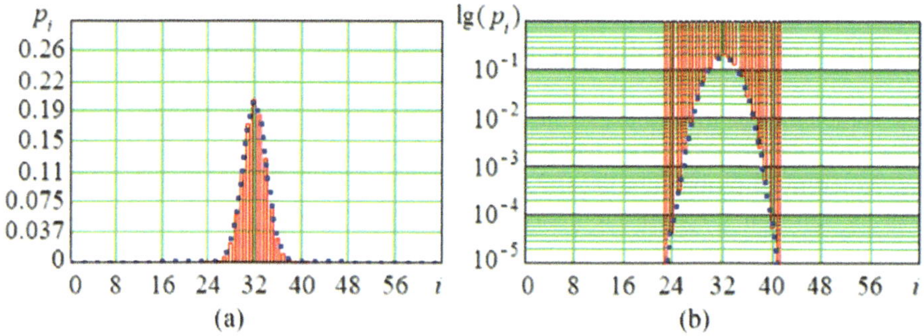

Fig. 19. The desired probability density of the simulated Gaussian process with the non-uniform spectrum and its histogram in the linear (a) and logarithmic (b) scales.

7 Conclusion

There are introduced simple high-speed simulators of the random processes with the specified one-dimensional and two-dimensional probabilistic characteristics. It is shown that they provide high accuracy in generating various random processes with the required statistical and correlation properties including the low-probability areas of the random process values. Switching of the models of the simulated signals is carried out by replacing the contents of the storage device or by switching the pages of the data arrays. Simulators can be implemented in software with low computational costs as well as in hardware by means of the microprocessor systems or field-programmable gate arrays. They allow us to generate high-frequency random signals in real time and can be used while designing, debugging and testing various communication and measuring electronic equipment.

Acknowledgements. This study was financially supported by the Russian Science Foundation (research project No. 20-61-47043).

References

1. Kleijnen, J.P.C.: Statistical Techniques in Simulation, Part II. Marcel Dekker, New York (1975)
2. Knuth, D.E.: The Art of Computer Programming, Vol. 2. Sorting and Searching, 2nd edn. Addison-Wesley Professional, Boston (1998)
3. Law, A.M., Kelton, W.D.: Simulation, Modeling and Analysis, 3rd edn. McGraw-Hill Education, New York (2000)
4. Krasnenko, S.S., Pichkalev A.V., Nedorezov D.A., Lapin A.Y., Nepomnyashcy O.V.: Methods of realizations of satellite radionavigation system simulators. Vestnik Sibirskogo gosudarstvennogo aerokosmicheskogo universiteta imeni akademika M.F. Reshetneva (53), 30–34 (2014)
5. Noori, H., Vilni, S.S.: Jamming and anti-jamming in interference channels: a stochastic game approach. IET Commun. **14**(3), 682–692 (2020)
6. Khan, K.M., Shaheen, M., Wang, Y.: Data confidentiality in cloud-based pervasive system. In: Hamdan, H., Boubiche, D.E., Hidoussi, F. (eds.) Data Confidentiality in Cloud-Based Pervasive System. Second International Conference on Internet of Things, Data and Cloud Computing 2017, ICC, pp. 1–8. ACM, New York (2017)
7. Blintsov, V., Nuzhniy, S., Parkhuts, L., Kasianov, Y.: The objectified procedure and a technology for assessing the state of complex noise speech information protection. East.-Eur. J. Enterp. Technol. **5**(9–95), 26–34 (2018)
8. Lukmanova, O., Horev, A.A., Vorobeyko, E., Volkova, E.A.: Research of the analog and digital noise generators characteristics for protection device. 2020 In: IEEE Conference of Russian Young Researchers in Electrical and Electronic Engineering, ElConRus, pp. 2093–2096. IEEE, Russia (2020)
9. Epstein, M., Hars, L., Krasinski, R., Rosner, M., Zheng, H.: Design and implementation of a true random number generator based on digital circuit artifacts. Lect. Notes Comput. Sci. **2779**, 152–165 (2003)
10. Stipcevic, M.: Fast nondeterministic random bit generator based on weakly correlated physical events. Rev. Sci. Instrum. **75**(10), 4442–4449 (2004)
11. Li, X., Cohen, A.B., Murphy, T.E., Roy, R.: Scalable parallel physical random number generator based on a superluminescent LED. Opt. Lett. **36**(5), 1020–1022 (2011)
12. Devroye, L.: Non-Uniform Random Variate Generation. Springer-Verlag, Berlin (1986)
13. Tam, W.M., Lau, F.C.M., Tse, C.K.: Digital Communications with Chaos: Multiple Access Techniques and Performance. Elsevier Science, Oxford (2006)
14. Dynkin, E.B.: Theory of Markov processes. Dover Publications, New York (2006)
15. Meyn, S.P., Tweedie, R.L.: Markov Chains and Stochastic Stability. Springer-Verlag, Berlin (1993)
16. Gray, R.M., Neuho, D.L.: Quantization. IEEE Trans. Inf. Theory **44**(5), 2325–2383 (1998)
17. Recommendation ITU-R P.1057–1. Probability distributions relevant to radiowave propagation modeling
18. Van Trees, H.L., Bell, K.L., Tian, Z.: Detection, Estimation, and Modulation Theory, Part I. Detection, Estimation, and Filtering Theory, 2nd edn. Wiley, New York (2013)
19. Parsons, J.D.: The Mobile Radio Propagation Channel, 2nd edn. Wiley, New York (1992)

Gradient Descent Method Based on the Multidimensional Voxel Images

Alexey Vyacheslavovich Tolok(✉) and Nataliya Borisovna Tolok

Trapeznikov Institute of Control Sciences, Russian Academy of Sciences, Profsoyusnaya 65, Moscow 117997, Russia
a.tolok@stankin.ru, nat_tolok@mail.ru

Abstract. In the proposed paper one of the approaches to automation of the gradient descent method based on the pre-calculated local geometrical characteristics of space of a function is considered. We represent the local geometrical characteristics by a set of voxel images co-dimensional to this space. Computer geometrical models obtained by means of the Functional Voxel method form the informational basis for the algorithm. The main principles of the gradient motion control along with the peculiarities of the conversion to increase the problem dimensionality are demonstrated. The detected advantages of the algorithm allow to extend the usage of such approach from the mathematical programming problems towards the local optimization on the example of laying the route for the steepest descent.

Keywords: The Functional Voxel method · Gradient descent method · Basic graphical M-image · Mathematical programming · Route planning

1 Introduction

As a rule, gradient descent method is used to calculate extreme points in optimization problems. Multidimensional definitions of such problems induce to search for computer solutions of the gradient descent method which allow to carry out multiparametric evaluation of the chosen motion direction. There are a great number of approaches to computer implementation of the gradient descent among which it is possible to distinguish the basic methods - the steepest descent, the coordinate descent Gauss-Seidel method, nonlinear conjugate gradient method etc. All these methods are based on the sequential calculation of differential characteristics in the current position of the surface point. Such calculation algorithms operate correctly when various additional terms are carried out which depends on the computer representation of a surface model of the investigated function. The complexity of such approaches significantly increases in higher dimensionality of the problem.

To solve these problems, the authors propose to apply the new perspective models created by means of the Functional Voxel Method (FVM). The RFM method, which was thoroughly described in [2], has already proven its effectiveness in problem solving in mathematical modelling [2–7] as it allows to obtain a discrete computer representation of the continuous functional space bounded by the given orthogonal domain.

R. Silhavy et al. (Eds.): CoMeSySo 2020, AISC 1295, pp. 94–103, 2020.
https://doi.org/10.1007/978-3-030-63319-6_10

2 The Functional Voxel Modelling of the Analytical Representation of a Function

The Functional Voxel representation of a computer model is based on the principles of linear approximation of a functional space for computing its local geometric (differential) characteristics. Unlike the existing approaches to linear approximation, the local geometric characteristics of such model are represented by components of the normal of higher dimensionality in comparison with the function space.

In the basis of such representation lies the Assertion:

Assertion: In the space $\mathrm{E}^{\mathrm{m}+1}$ the family of hyperplanes of the form $\mathrm{n_1x_1} + \ldots + \mathrm{n_mx_m} = \mathrm{p}$ is representable as $n_1x_1 + \ldots + n_mx_m + n_{m+1}x_{m+1} = 0$.

This assertion is based on the principle of transformation of the equation to the homogeneous coordinates. For example, the linear approximation in the E^3 space defines a flat neighborhood at the point by the components of the normal $\vec{n}(n_1, n_2, n_3, n_4)$ for the *local* equation $n_1x_1 + n_2x_2 + n_3x_3 + n_4x_4 = 0$.

Hereinafter the components of the normal n_1, n_2, n_3, n_4 will be called the *local geometric characteristics.*

Let us establish some correspondence between the scalar fields n_1, n_2, n_3, n_4 determined on the interval [−1,1] with a color representation C_1, C_2, C_3, C_4 expressing through the intensity gradation of the tone of the monochrome palette, for example, [0, P], where P = 255 is the upper value of the color intensity of the palette.

Such transformation $n_i \xrightarrow{C} C_i$ can be represented as:

$$C_i = \frac{P(1 + n_i)}{2}, \tag{1}$$

where i = 1..4.

Figure 1 illustrates images C_1, C_2, C_3, C_4 obtained by means of the linear approximation of the function space:

$$z = 0.75e^{\frac{-(9x-2)^2 + (9y-2)^2}{4}} + 0.75e^{\frac{(9x+1)^2}{-49} + \frac{9y+1}{10}} - 0.2e^{-(9x-4)^2 - (9y-7)^2} + 0.5e^{\frac{-(9x-7)^2 + (9y-3)^2}{4}}. \tag{2}$$

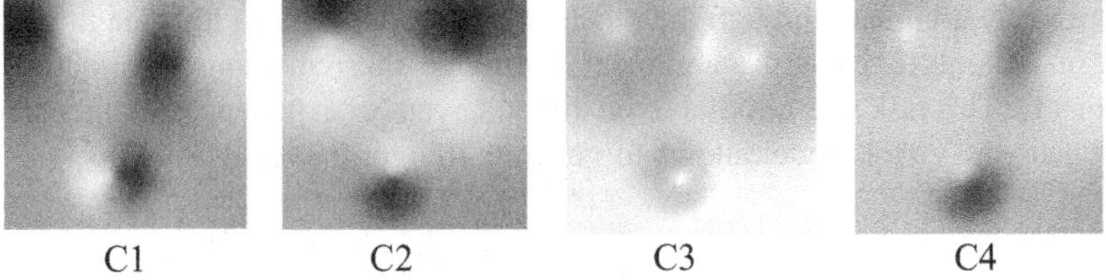

Fig. 1. Images of the normal components n_1, n_2, n_3, n_4.

The graphic image, commensurate with the function space and displaying some single property, will be called the graphic image-model of this function, or the graphical M-image.

Since the created voxel images represent the local geometric characteristics on the bounded function space, each of the voxel images is a graphical M-image.

Since the quantity of the graphical M-images is correlated with the dimensionality of the space as m + 2 which is sufficient for Function Voxel model definition, then such images are allocated to the class of basic M-images.

It is worth mentioned here that at this stage the analytical description of a function of any complexity will be represented by a set of basic M-images, which allow to obtain the local function at each point of the image space by applying the inverse transform $C_i \xrightarrow{N} n_i$ (see Fig. 2):

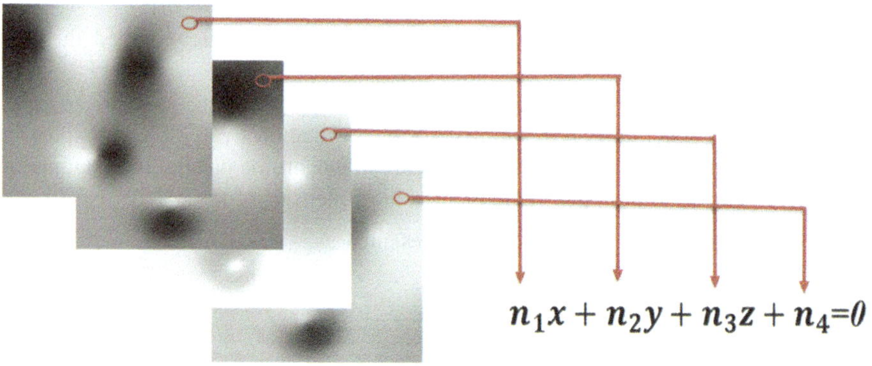

Fig. 2. The Functional Voxel representation of the space of the analytic function (1).

$$n_i = \frac{2\left(C_i - \frac{P}{2}\right)}{P}, i = 1\ldots4. \tag{3}$$

One of the remarkable features of the functional voxel model is that it is convenient to express any argument through the others. It is just sufficient to imagine the problem where one should express the argument x through the remaining y, z in the initial function (1). This task will turn out to be difficult even for experienced mathematicians. The Functional Voxel model allows to generate it algorithmically:

$$x = -\frac{n_2}{n_1}y - \frac{n_3}{n_1}z - \frac{n_4}{n_1}, \quad y = -\frac{n_1}{n_2}x - \frac{n_3}{n_2}z - \frac{n_4}{n_2}. \tag{4}$$

To calculate the value of the function (1) it is sufficient to express the local function z:

$$z = -\frac{n_1}{n_3}x - \frac{n_2}{n_3}y - \frac{n_4}{n_3}. \tag{5}$$

3 The Gradient Descent Algorithm on the Basis of the Functional Voxel Model

The main approaches to solving gradient descent problem are considered in paper [1]. The descent direction is understood as the negative gradient of the function f(x) at a point x_k, leading the algorithm towards the iterative process: $x^{k+1} - x^k - \alpha_k f'(x_k), \alpha_k > 0$, where α_k – the step that defines the main approaches of algorithms.

The most prevalent ways of finding the step size α_k are: constant step size method, gradient method with adaptive step-size, method of the steepest descent etc. The gradient descent algorithm on the basis of the functional voxel model can be referred to the first one, because the motion proceeds on the regular voxel mesh. There are some great advantages over other algorithms with constant step size, for example, the functional voxel model already comprises the gradient in each point of the considered space.

Let's consider the classification: the functional voxel models can be divided into two main categories – *basic M-images*, which describe the geometrical characteristics of the function space, and generated M-images, which are applied as supplementary. *Generated M-images* can be obtained from the basic M-images if required during the solution of the particular problem.

Consider the principle of obtaining generated M-images for gradient descent algorithm implementation. One of the important properties of the normal vector is quite simple transformation of the projections. For example, it is sufficient to carry out the renormalization of the components to transform 4-component vector $\vec{n}(n_1, n_2, n_3, n_4)$ into the 2-component $\vec{n}(n'_1, n'_2)$:

$$n'_1 = \frac{n_1}{\sqrt{n_1^2 + n_2^2}}, \quad n'_2 = \frac{n_2}{\sqrt{n_1^2 + n_2^2}}. \tag{6}$$

This will allow to generate two new images C'_1, C'_2 which indicates for each point of the space *xOy the cosine* of the angle of deviation from the axes *Ox* and *Oy* respectively.

In Fig. 3 two such generated M-images C'_1, C'_2 are shown for the function space (1).

Gradation of the tone of the monochrome palette from white to black characterizes the deviation of the *cosine* from the positive direction of the particular axis to the negative. For example, the white colour in the first image C'_1 indicates that the deviation of an angle α from the axis *Ox* is equal to zero, and it means that the normal at this point completely coincides with the deviation of the positive *Ox* - axis. The black colour indicates the negative direction. In Fig. 4 the diagrams reveal the example of definition of the gradient descent motion towards one of the eight neighboring points.

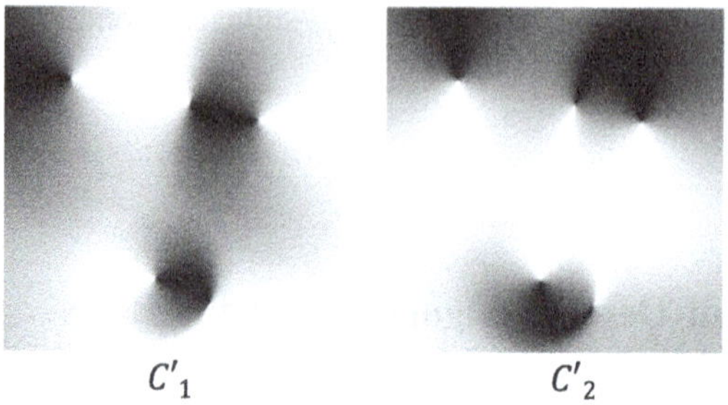

Fig. 3. Generated M-images for gradient descent algorithm implementation.

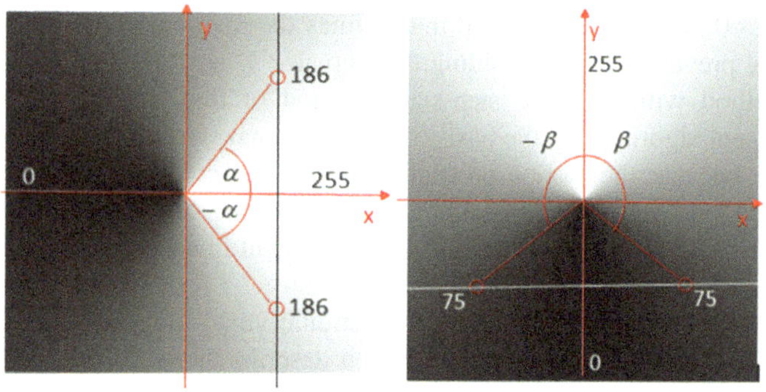

Fig. 4. Colour diagrams of the direction along the axes Ox and Oy.

Let's assume that the current point of the voxel space takes the value 186 from the *M-image* C'_1 and the value 75 – from the *M-image* C'_2.

Diagrams illustrate in a practical way three operations allowed along the axis *Ox* remain at the point, motion along the positive or negative direction of the axis. The value 186, expressed through the intensity gradation of the tone, approaches white colour and that means the motion along the positive direction of the axis *Ox* or to the right. The value 75 taken from the M-image C'_2, on the contrary, demands the motion along the negative direction of the axis *Oy* (or down), which at the junction combines the solution of the motion to the right lower corner.

Formalization of such algorithm can be expressed as:

$$x = C'_1, y = C'_2; \tag{7}$$

$$C_x = \begin{cases} \frac{P}{2} + \frac{P}{8} \leq x \rightarrow C_x = \{C_x(i+1,j+1), C_x(i+1,j), C_x(i+1,j-1)\}; \\ \frac{P}{2} - \frac{P}{8} > x \rightarrow C_x = \{C_x(i-1,j+1), C_x(i-1,j), C_x(i-1,j-1)\}; \\ C_x = \{C_x(i,j+1), C_x(i,j), C_x(i,j-1)\}; \end{cases} \tag{8}$$

$$C_y = \begin{cases} \frac{P}{2} + \frac{P}{8} \leq y \rightarrow C_y = \{C_y(i-1,j+1), C_y(i,j+1), C_y(i+1,j-1)\}; \\ \frac{P}{2} - \frac{P}{8} > y \rightarrow C_y = \{C_y(i-1,j-1), C_y(i,j-1), C_y(i+1,j-1)\}; \\ C_y = \{C_x(i-1,j), C_x(i,j), C_x(i+1,j)\}; \end{cases} \quad (9)$$

where P – is the upper value of the gradation of the colour intensity palette (for example, 255), C_x – the step along the axis Ox; C_y – the step along the axis Oy; C_{res} – the step along the gradient.

The result of the gradient descent algorithm for the function space (1) is shown in Fig. 5. As an image basis we used the traditional colour normalization.

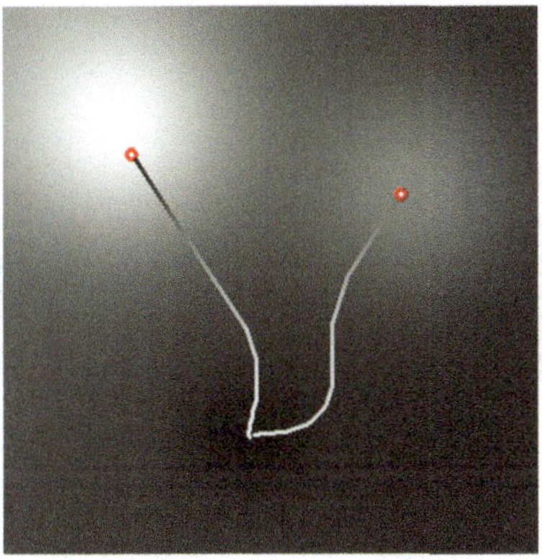

Fig. 5. The gradient descent from two starting points (two red points).

4 Multidimensional Approach to the Gradient Descent Algorithm

Unlike most such problem-solving techniques, the proposed approach can be successfully implemented in multidimensional space.

Let's assume that it is necessary to increase the space of the gradient descent algorithm. In this case we should create and consider the functional voxel model for a function $u = f(x, y, z))$, and to obtain the local geometrical characteristics for the function $n_1x + n_2y + n_3z + n_4u + n_5 = 0$. This process was thoroughly described in [2].

Suppose that such model already exists and it is required to provide a motion from any point of its space along the gradient.

Similarly, to the considered two-dimensional space, here we additionally generate three 3-dimensional M-images C'_1, C'_2, C'_3, which are mapping the components of 3-dimensional normal:

$$n'_1 = \frac{n_1}{\sqrt{n_1^2 + n_2^2 + n_3^2}}, \quad n'_2 = \frac{n_2}{\sqrt{n_1^2 + n_2^2 + n_3^2}}, \quad n'_3 = \frac{n_3}{\sqrt{n_1^2 + n_2^2 + n_3^2}}. \quad (10)$$

Meanwhile, the logic of motion towards the neighbour points remains the same, one more orthogonal direction of evaluation was added. If in two-dimensional case we defined "right-left" and "up - down" motion, then here we analyze the motion "forward - backward" along the axis Oz.

The result of the algorithm in the three-dimensional space on the example of R-functionally modelled cup as a function u = f(x,y,z) is illustrated in Fig. 6. This R-functional model constitutes an intersection and union of such geometrical shapes as the toroid, cylinder and the sphere.

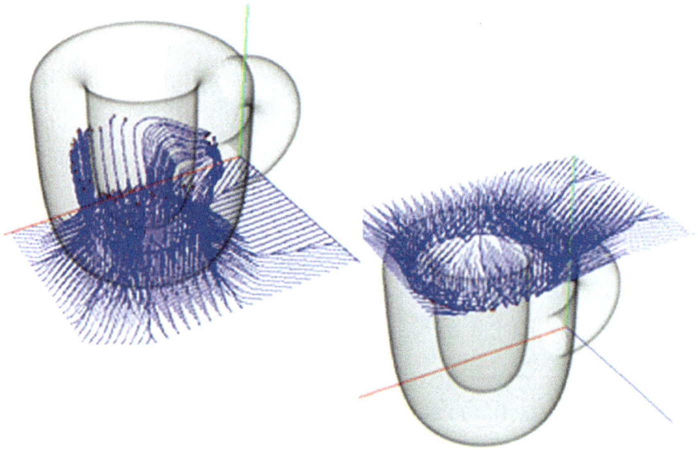

Fig. 6. The algorithm of the gradient descent in 3D.

The initial points of the descent are located correspondingly on the bottom and upper sections of an object.

It is worth mentioned that it is necessary to have the functional voxel model, where $\vec{n}(n_1 \ldots n_{m+2})$, to implement N-dimensional gradient $\vec{n}(n_1, \ldots, n_m)$.

5 Applications for the Designed Functional Voxel Gradient Descent Algorithm

One of the most widely used approaches to solving the mathematical programming problems is the gradient descent method.

Multidimensionality of the FVM approach has provided great opportunities to development of the automatization process, so, we have to pay particular attention to the issue of the functional-voxel modelling of such problems.

The fundamental concept of the R-functional modelling of solving mathematical programming problems was thoroughly described in [3, 4].

In general form the function can be represented as:

$$\begin{cases} F_w = F + w_0(1 + |F|) \rightarrow max; \\ F_w = F + |w_0(1 + |F|)| \rightarrow max, \end{cases} \tag{11}$$

$$w_0 = w - |w|, w = w_1 \bigwedge w_2 \ldots \bigwedge w_n, \tag{12}$$

where $w_1, \ldots w_n$. - predicate functions describing restrictions, F - objective function.

Figure 7 illustrates the example of the algorithm in the following problem statement [3, 4]: Find the global extrema of the function $F = (x-2)^2 + (y-3)^2$ on the set of restrictions: $x + 2y \leq 12, x + y \leq 9, x \geq 0, y \geq 0$.

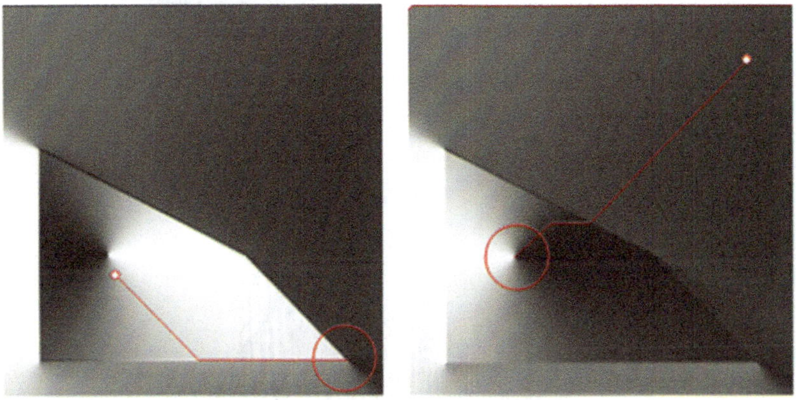

Fig. 7. The result of the algorithm of motion along the gradient and against the gradient respectively on the generated M-Image.

The algorithm was tested and applicated to the problems of modelling motion towards the target point bypassing the unforeseen obstacles [5], using R-functional construction of the surface which geometry allows solving such problem (see Fig. 8).

In Fig. 9 the application of the spacial motion along the gradient for support constructions modelling in additive technologies is demonstrated, the detailed research was presented in [7].

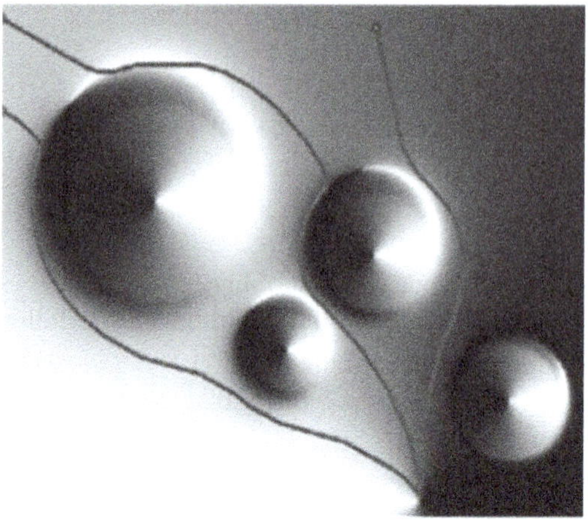

Fig. 8. Modelling the route bypassing the obstacles.

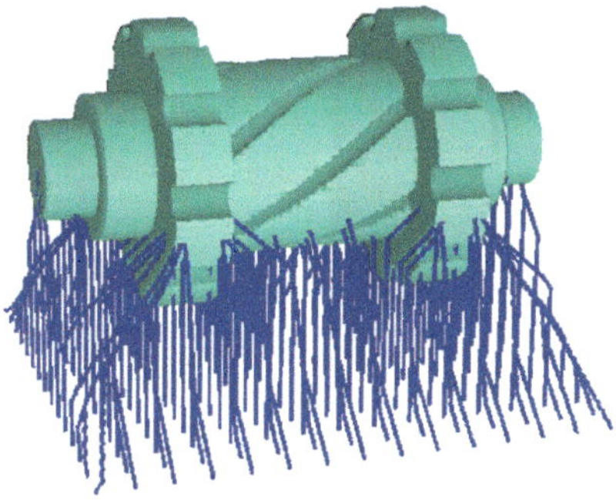

Fig. 9. Application of the gradient descent in support constructions modelling for additive technologies.

6 Conclusions

Applying the Functional Voxel method in various mathematical modelling Tools will allow to significantly expand the class of automated problems as it simplifies the technology of computation and algorithmization, thus, the statement of the problem reduces to the construction of the local geometrical model. The gradient descent algorithm is one of the major tools of the Functional Voxel Modelling which allow to solve not only optimization but control sciences modelling problems as well as the construction of complex structures without much difficulty.

References

1. Glebov, N.I, Kochetov, Y.A., Plyasunov, A.V.: Optimization Methods. Novosibirsk State University, Novosibirsk (2000)
2. Tolok, A.V.: Functional Voxel Method in Computer Modeling. Fizmatlit, Moscow (2016)
3. Tolok, A.V., Tolok, N.B.: Solving mathematical programming problems by means of the functional voxel method. Control Sci. **3**, 37–42 (2017)
4. Tolok, A.V., Tolok, N.B.: Mathematical programming problems solving by functional voxel method. Autom. Remote Control **79**(9), 1703–1712 (2018). https://doi.org/10.1134/S0005117918090138
5. Grigoryev, S.N., Tolok, N.B., Tolok, A.V.: Local search gradient algorithm based on functional voxel modeling. Programm. Comput. Softw. **43**(5), 300–306 (2017)
6. Tolok, A.V., Tolok, N.B., Loktev, M.A.: Modeling function domain for curves constructed based on a linear combination of basis bernstein polynomials. Programm. Comput. Softw. **44**(6), 526–532 (2019)
7. Batuev, E.R., Tolok, A.V., Tolok, N.B.: Voxel modelling of making a prototype control process by means of additive technologies. Autom. Ind. **6**, 42–47 (2019)

Application of the Wald Sequential Procedure in Automatic Network Control with Distributed Generation

Ilyushin Pavel[1], Kulikov Aleksandr[2], and Loskutov Anton[2(✉)]

[1] Petersburg Power Engineering Institute of Professional Development, St. Petersburg, Russia
[2] Nizhny Novgorod State Technical University n.a. R.E. Alekseev, Nizhny Novgorod, Russia
loskutov.nnov@gmail.com

Abstract. Information is presented on the current situation and prospects for integration of distributed generation (DG) and microgeneration facilities into distribution networks. The features of networks with DG objects are considered. The aspects that create significant difficulties for visual recognition and manual control of network modes are given. The necessity of creating automation systems for control of normal and emergency modes, which has the functions of recognizing the membership of the current mode in various mode areas, is substantiated. Also, the need to adapt algorithms and settings to the circuit-mode conditions is shown. It is shown that this problem should be solved at a fundamentally new scientific and technological level, using modern software and hardware systems that implement specialized control algorithms. The advantages of applying the Wald sequential procedure to recognize the current mode under conditions of decreasing frequency, transients, non-sinusoidality of currents and voltages, leading to large errors in the estimation of their parameters, are presented. It is proved that ensuring guaranteed high-speed performance of automatic control devices of modes is possible due to the introduction of a truncation algorithm for sequential analysis. An example of correct decision-making by a frequency unloading automation device in the "island mode" of operation of a distribution network with DG and microgeneration facilities based on the Wald procedure is given. This occurred in conditions of ambiguous frequency measurements and the presence of distorting factors. The authors' recommendations on the application of the Wald sequential procedure for the implementation of relay protection and emergency automation devices in modes accompanied by complex transient processes are presented.

Keywords: Distributed Generation (DG) · Mode control automation · Mode recognition · Wald sequential procedure · Automation speed · Sequential analysis truncation algorithm · Automatic frequency unloading

R. Silhavy et al. (Eds.): CoMeSySo 2020, AISC 1295, pp. 104–120, 2020.
https://doi.org/10.1007/978-3-030-63319-6_11

1 Introduction

In the last decade, there has been a global trend towards the decentralization of the electricity industry. This is accompanied by regular construction and commissioning of new distributed generation facilities (DG).

Large objects of DG, as a rule, are commissioned in Russia by owners of industrial enterprises in the oil and gas, mining, metallurgical, pulp and paper, chemical and other industries. The integration of such DG facilities is carried out in the internal power supply networks of enterprises or distribution networks of medium and high voltage. At distributed generation facilities, gas turbine (GTU), gas piston (GPU), and diesel (DGU) generating units (GU) are usually used.

In the coming years, Russia is planning to commission a significant number of micro-generation facilities that are integrated into low-voltage networks. These include facilities for the production of electricity up to and including 15 kW, operating, including on the basis of renewable energy sources (RES), used by the consumer for their own energy supply (domestic and production needs). With the mass introduction of DG facilities and microgeneration, a variety of circuit-mode conditions may occur, which depend on the modes of generation and consumption of electricity in the nodes of the distribution network. It is not possible to provide reliable power supply to consumers with the specified quality parameters without the development and implementation of automation tools for controlling the modes under the indicated conditions.

It is important to remember that various devices of automation of power systems function in distribution networks. This include devices of automatic frequency unloading, automatic limitation of voltage reduction, automatic switching on of backup power supply, etc., which are used to prevent the occurrence and development of emergency processes, as well as accelerate the restoration of normal modes. The connection of the DG objects to the power system can cause the incorrect functioning of the power systems automation devices, which requires analysis and subsequent changes in the algorithms of their operation and settings. Frequently used automation devices of power systems do not have technical capabilities for recognizing mode areas and adapting work algorithms to the conditions of the current mode [1–7].

Under these conditions, it is necessary to develop a new type of automation for managing normal and emergency modes of distribution networks with DG facilities, which have their own characteristics. Firstly, a significant number of managed devices and active consumers are connected to them. Secondly, the rate of development of normal mode disturbances is much higher here taking into account the small values of the mechanical constant of inertia of the GU. Thirdly, reverse power flows are permissible in networks, which creates difficulties for visual recognition and manual control of modes [8, 9]. Therefore, the task of controlling the modes should be solved at a fundamentally new scientific and technological level, using modern software and hardware systems that implement specialized control algorithms that are adaptable to current circuit-mode conditions.

In the problem of recognizing the belonging of the current mode to different mode areas, the so-called "three-position relays" are used more and more, which are used in various automatic control systems (ACS) [10–12]. In particular, there are known options for the use of such relays in ACS of electric energy storage devices, electric drives, electric water heating, etc. [13]. Automatic control in them is implemented on the basis of discrete controllers, and the regulators of this type are called positional. A feature of the use of three-position relays is the presence of a insensitivity zone ("dead zone"), within which discrete switching of control functions is not performed.

The use of three-position relays in the tasks of constructing automation of control of distribution network modes has not been previously considered. This article is devoted to the principles of recognition of normal and emergency modes using statistical methods.

2 The Statistical Problem of Sequential Decision Making in Recognition of Normal and Emergency Modes

Let there be a mode belonging to one of the two classes $\{H_0, H_1\}$ (0 - normal mode, 1 - emergency mode), characterized by the corresponding hypotheses H_0 and H_1, and let there be a set of sequential measurements of parameters characterizing both normal and emergency mode $\{x_1,. \ldots, x_m\}$, denoted by the generalized variable x. The sequential decision-making strategy corresponds to the set of the decisive function $S = \{S_1,. \ldots, S_m\}$, where S_i is determined by the measurements $\{x_1,. \ldots, x_i\}$ corresponding to the classes $\{H_0, H_1, H_n\}$. Strategy S evaluates each dimension x_i, at the corresponding discrete time instant i, and produces a solution S_i. Hypothesis H_n means "don't know" and assumes "continuation of the solution" by performing one more discrete step. If a decision is made $\widehat{H}_n$, then x_{i+1} is measured and the decision function S_{i+1} is estimated. Otherwise, strategy S implies an unambiguous classification of the mode at step S_i.

Moreover, the following six situations of combining random events of a "decision" $\widehat{H} = \{H_0, H_1, H_n\}$ and a "conditions" $H = \{H_0, H_1\}$ for the current measurement x_i are fundamentally possible:

1. $\widehat{H}_1 H_1$ - correct recognition of emergency mode;
2. $\widehat{H}_0 H_1$ - pass emergency mode;
3. $\widehat{H}_n H_1$ - making the decision "I don't know" under emergency conditions;
4. $\widehat{H}_n H_0$ - making the decision "I don't know" in the conditions of normal operation;
5. $\widehat{H}_1 H_0$ - false decision about emergency mode;
6. $\widehat{H}_0 H_0$ - correct recognition of the normal mode.

Possible indicators of the effectiveness of mode recognition could be considered the probability of combining the above six events. Moreover, any of the probabilities of combining events is reduced to the product of the conditional probability of the solution $P(\widehat{H}_l|H_k)$ and the probability of the condition $P(H_k)$, $l \in \{0, 1, n\}$; $k \in \{0, 1\}$. I.e

$$P(\widehat{H}_l H_k) = P(\widehat{H}_l | H_k) \cdot P(H_k). \tag{1}$$

The probabilities of the presence of $P(H_1)$ or the absence of $P(H_0)$ emergency mode are a priori and usually unknown. The conditional probabilities of solutions $P(\widehat{H}_l|H_k)$ can be estimated based on simulation, experimentally or by calculation methods [9–11]. Accordingly, it is they (conditional probabilities) that are used as indicators of the recognition efficiency of the modes, and not the probability of combining events.

In the problems of pattern recognition, a more general indicator is introduced. This is the average risk of recognition errors (or expected value) [14–21], which is an average payment for decision errors. For a three-alternative mode recognition, it is necessary to take into account the undesirability of making a "don't know" decision. Thus, the average risk is determined by:

$$\overline{C} = C_{01} P(\widehat{H}_0 H_1) + C_{10} P(\widehat{H}_1 H_0) + C_{n1} P(\widehat{H}_n H_1) + C_{n0} P(\widehat{H}_n H_0), \tag{2}$$

where C_{01} and C_{10} - the costs of skipping the emergency mode and making a false decision about the emergency mode, and C_{n1} and C_{n0} - the costs of "don't know" in the presence and absence of the emergency mode.

In addition, we note that multi-stage three-alternative mode recognition refers to sequential procedures. In this case, an important indicator of sequential recognition is the average number of cycles necessary to ensure the specified performance indicators of the recognition mode.

Optimization of the mode recognition procedure in the statistical formulation of the task is achieved by reducing the conditional probabilities of erroneous decisions according to the selected criteria. As a rule, there are no possibilities to lower the conditional probabilities of erroneous decisions, and such requirements are contradictory. For example, you can always achieve the value of the conditional probability of skipping the emergency mode $P(\widehat{H}_0|H_1) = 0$, assuming that the emergency mode corresponds to each realization of currents and voltages. But, then the conditional probability $P(\widehat{H}_1|H_0)$ of the false decision about the emergency mode is almost equal to one. And vice versa, it can be achieved that the conditional probability of a false decision about the emergency mode is $P(\widehat{H}_0|H_1) = 0$, for example, by significantly increasing the set of the protection (automatics) device. In this case, the conditional probability of skipping emergency mode $P(\widehat{H}_1|H_0)$ increases to unity.

In the statistical theory of recognition [14–19] weighted criteria are introduced. The most commonly used criterion is the minimum average risk. In relation to the task of optimizing the three-alternative mode recognition, the equation for minimizing the average risk (2) takes the form:

$$P(\widehat{H}_0|H_1) + \eta_0 P(\widehat{H}_1|H_0) + \eta_1 P(\widehat{H}_n|H_1) + \eta_2 P(\widehat{H}_n|H_0) = \text{minimum}. \tag{3}$$

In Eq. (3), the terms $P(\widehat{H}_n|H_1)$ and $P(\widehat{H}_n|H_1)$ are conditional probabilities of making a "don't know" decision in the presence or absence of an emergency mode.

Moreover, until the last step of the sequential procedure, normalization conditions must be maintained:

$$P(\widehat{H}_1|H_1) + P(\widehat{H}_0|H_1) + P(\widehat{H}_n|H_1) = 1, P(\widehat{H}_1|H_0) + P(\widehat{H}_0|H_0) + P(\widehat{H}_n|H_0) = 1. \tag{4}$$

The weights included in the equation for the minimum average risk (3) are determined by the following equalities:

$$\eta_0 = C_{10}P(H_0)/C_{01}P(H_1), \quad \eta_1 = C_{H1}/C_{01}, \quad \eta_2 = C_{H0}P(H_0)/C_{01}P(H_1). \tag{5}$$

It is believed that the cost of ignorance is less than the cost of errors $C_{n0} < C_{10}$, $C_{n1} < C_{01}$. This condition determines the appropriateness of the transition to a sequential procedure. In addition, we note that $\eta_2 < \eta_0$, $\eta_1 < 1$. This is important for further considerations.

The probability densities of parameter realizations under the condition of normal $p_0(x)$ and emergency $p_1(x)$ modes are assumed to be known (they are obtained, for example, from simulation results). First, we define non-optimal decision functions $\widehat{A}(x)$ и $\widehat{A}_n(x)$. Then, the optimization of the three-alternative mode recognition consists in choosing the most appropriate decision-making rule ("yes", "no" or "don't know"), according to the weight criterion (3), which follows from the equation of average risk (2). With the accepted notation, the conditional probabilities included in the weight criterion correspond to the equalities:

$$P(\widehat{H}_0|H_1) = \int_{(x)} \left[1 - \widehat{A}_n(x)\right] \cdot \left[1 - \widehat{A}(x)\right] \cdot p_1(x)dx, \tag{6}$$

$$P(\widehat{H}_1|H_0) = \int_{(x)} \left[1 - \widehat{A}_n(x)\right] \cdot \widehat{A}(x) \cdot p_0(x)dx, \tag{7}$$

$$P(\widehat{H}_n|H_1) = \int_{(x)} \widehat{A}_n(x) \cdot p_1(x)dx, \tag{8}$$

$$P(\widehat{H}_n|H_1) = \int_{(x)} \widehat{A}_n(x) \cdot p_0(x)dx. \tag{9}$$

Let us note that the numerical value of each equality is determined by the integral of the conditional probability density $p_0(x)$ or $p_1(x)$ over a certain region of the multidimensional space parameters x. The integration regions are determined by the discrete functionals $\widehat{A}_n(x)$, $\widehat{A}(x)$, taking the values {0, 1}. The area of integration for the quantities $P(\widehat{H}_n|H_1)$ and $P(\widehat{H}_n|H_0)$ is $\widehat{A}_n(x) = 1$ the decision-making area "don't know". The integration regions for the quantities $P(\widehat{H}_0|H_1)$ and $P(\widehat{H}_1|H_0)$ are defined as the intersection of the regions:

– "Know" $[1 - \widehat{A}_n(x)] = 1$, that is, $\widehat{A}_n(x) = 0$ (both for $P(\widehat{H}_0|H_1)$ and $P(\widehat{H}_1|H_0)$;

- "Yes" $\widehat{A}(x) = 1$ when calculating $P(\widehat{H}_1|H_0)$ and "no" $[1 - \widehat{A}(x)] = 1$, that is, $\widehat{A}(x) = 0$ when calculating $P(\widehat{H}_0|H_1)$.

Let's put the quantities in Eq. (3), and we will minimize this equality:

$$\int_{(x)} Q\left[\eta(x), \widehat{A}(x), \widehat{A}_n(x)\right] \cdot p_0(x)dx = \text{minimum}, \tag{10}$$

where

$$Q\left[\eta, \widehat{A}, \widehat{A}_n\right] = (1 - \widehat{A}_n) \cdot \left[(1 - \widehat{A}_n)\eta + \widehat{A})\eta_0\right] + \widehat{A}_n)(\eta_1\eta_0 + \eta_2), \tag{11}$$

linear likelihood ratio function:

$$\eta = \eta(x) = p_1(x)/p_0(x). \tag{12}$$

Optimization of the decision rule by minimizing (10) is achieved by minimizing (11) for each fixed $\eta(x)$. To do this, it suffices to compare the values of (11) for different factors $\widehat{A}_n$ и $\widehat{A}$ for each η.

When $\widehat{A}_n = 0$, the decision corresponds to the case of two-alternative recognition (detection) [14–18]. Then

$$Q = (1 - \widehat{A})\eta + \widehat{A}\eta_0, \tag{13}$$

and graphs of the possible values of the function Q (Eq. 13) for the values $\widehat{A} = 1$ and $\widehat{A} = 0$ are presented in Fig. 1 depending on $\eta = \eta(x)$. The optimal (smallest) values of the functions are achieved:

for $\eta\ (x) < \eta_0$ on the line $\widehat{A} = 0$; for $\eta\ (x) > \eta_0$ on the line $\widehat{A} = 1$.

The adoption of such an optimization rule (decision making based on the choice of $\widehat{A}$) is fully consistent with the theory of detection [14]. Similar graphical dependencies presented in Fig. 2, can construct for the function $Q[\eta, \widehat{A}, \widehat{A}_n]$ of η, which corresponds to a three-alternative recognition process with the solutions "yes", "no" and "don't know".

It is important to note that the costs of "don't know" should not exceed the corresponding costs of skipping the emergency mode and making a false decision about the emergency mode, that is, $\eta_2 < \eta_0$, $\eta_1 < 1$. Otherwise, the introduction of a sequential procedure would be pointless. The optimal decision rule corresponds to the minimum values of Q and two-set (threshold) comparison:

when $\eta(x) \geq b$, we take $\widehat{A}(x) = 1$;
when $\eta(x) < a$, we take $\widehat{A}(x) = 0$;
when $a \leq \eta(x) < b$, we take $\widehat{A}_n(x) = 1$,
where $a = \eta_2/(1 - \eta_1)$, $b = (\eta_0 - \eta_2)/\eta_1$.

Thus,

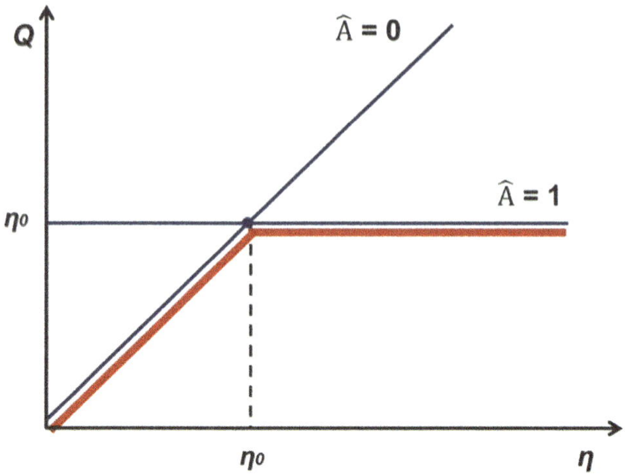

Fig. 1. Explanations for the selection of the smallest values of the linear likelihood ratio function (two-alternative recognition).

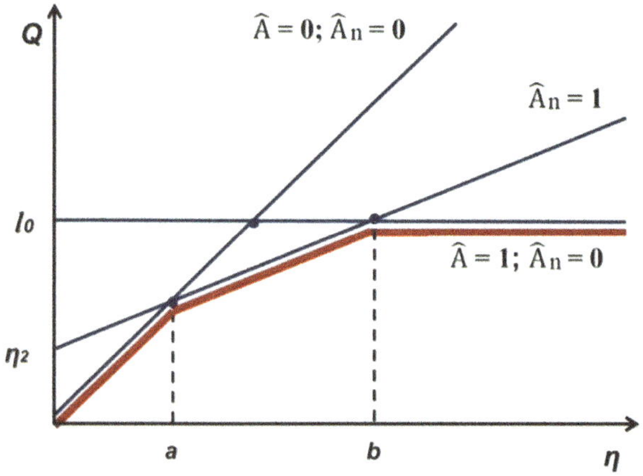

Fig. 2. Explanations for the selection of the smallest values of the linear likelihood ratio function (three-alternative recognition).

if $\eta(x) > b$, then the decision is "yes";
if $\eta(x) < a$, then the decision is "no";
if $a < \eta(x) < b$, then the decision "don't know" is made and the recognition cycle is repeated.

The structural diagram of a mode recognition device using two settings is shown in Fig. 3.

The process of achieving setpoints (a and b) is illustrated in Fig. 4, where k = 1, 2, … When forming the settings a and b, it is not necessary to know the cost indicators of erroneous decisions and a priori probabilities included in the equations for η_0, η_1, η_2.

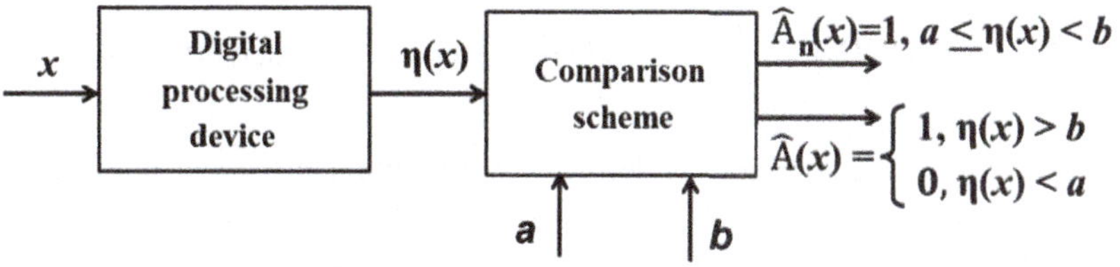

Fig. 3. Block diagram of a mode recognition device.

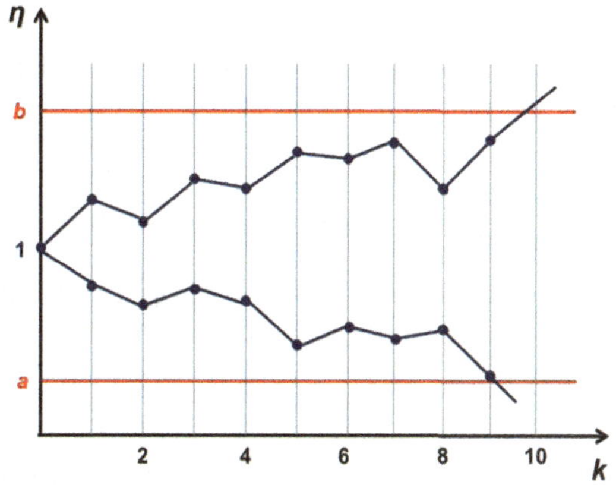

Fig. 4. Explanations for a multi-step decision-making procedure.

The settings for a multistep sequential mode recognition procedure can be determined by the final probabilities of the erroneous solutions $P(\widehat{H}_0\ H_1)$ and $P(\widehat{H}_1\ H_0)$, when $\widehat{\mathrm{A}}_{\mathrm{n}} = 0$.

Due to the ratio between $\eta(x)$ and x from integration over a multidimensional quantity x, we can proceed to integration over a one-dimensional quantity η. For this, we assume that $p(x)dx = p(\eta)d\eta$ and replace the statement of the values 0 or 1 of the function η(x) with the statement of the limits of integration with respect to η.

Then, taking into account Eq. (12), we can obtain:

$$P(\widehat{H}_1|H_1) = \int_{(x)} \widehat{\mathrm{A}}(x)\eta(x)\cdot p_0(x)dx = \int_b^{\infty} \eta\, p_0(\eta)d\eta, \tag{14}$$

$$P(\widehat{H}_1|H_0) = \int_{(x)} \widehat{\mathrm{A}}(\mathrm{x})p_0(x)dx = \int_b^{\infty} p_0(\eta)d\eta, \tag{15}$$

$$P(\widehat{H}_{\mathrm{n}}|H_0) = \int_{(x)} \left[1 - \widehat{\mathrm{A}}(x)\right]\cdot\eta(x)\cdot p_0(x)dx \int_{\infty}^{b} \eta\, p_0(\eta)d\eta, \tag{16}$$

$$P(\widehat{H}_n|H_0) = \int_{(x)} \left[1 - \widehat{A}(x)\right] \cdot p_0(x)dx = \int_{\infty}^{a} p_0(\eta)d\eta. \quad (17)$$

In some cases, a recognition model with an asymptotically larger number of consecutive steps is introduced. For this model, it is valid when the upper threshold is $\eta \approx b$, and when the lower threshold is $\eta \approx a$, but the equations characteristic of the sequential procedure introduced by Wald are also valid [15]:

$$\begin{aligned} &P(\widehat{H}_1|H_1) \approx b\,P(\widehat{H}_1|H_0) \text{ and } P(\widehat{H}_n|H_1) \approx a\,P(\widehat{H}_n|H_0), \\ &\text{or } b \approx P(\widehat{H}_1|H_1)/P(\widehat{H}_1|H_0) \text{ and } a \approx P(\widehat{H}_n|H_1)/P(\widehat{H}_n|H_0). \end{aligned} \quad (18)$$

It is important that the comparison can be made for an arbitrary monotonically increasing (decreasing) function $\mu(\eta)$ of the likelihood ratio. As with two-alternative recognition. The value $\mu(\eta)$ is compared in this case with the settings:

$$\mu(a) = \mu(P(\widehat{H}_n|H_1)/P(\widehat{H}_n|H_0)) \text{ and } \mu(b) = \mu\,(P(\widehat{H}_1|H_1)/P(\widehat{H}_1|H_0)). \quad (19)$$

A sequential theory of decision making was developed by Wald [15], who proved the solution to the optimization problem is achieved by the method of sequential verification of probability ratio.

3 Implementation of Automatic Frequency Unloading Using a Sequential Analysis Procedure

Let us consider an example of the functioning of a fragment of a distribution network with DG facilities in the "island mode" of operation. Under the island mode is meant the mode of operation of a network fragment with one or more DG objects and a load. That is permissible under all conditions of power supply and power consumption. This mode is formed when power lines disconnecting this network fragment from the power system (as a result of short circuit; without short circuit) and exists until it is synchronized with the power system.

Under these conditions, large errors occur in the estimation of currents and voltages with a decrease in the frequency in transients due to non-sinusoidality. Errors in the measurement of frequency can lead to incorrect operation of automatic frequency unloading devices, leading to unnecessary load shedding. Thus, the development of a high-speed algorithm of automatic frequency unloading is relevant, which is capable of making decisions on disconnecting the load with minimal errors under conditions of significant frequency estimation errors. We implement the automatic frequency unloading decision algorithm based on a three-position relay using Wald's sequential analysis procedure.

Let's look at a simplified solution to the recognition task. A sequential test of hypotheses regarding the distribution network mode with DG facilities is performed as follows. For each measured frequency value, one of three hypotheses is accepted:

1. H_0 - the frequency corresponds to the normal mode;
2. H_1 - the frequency corresponds to the emergency mode;
3. H_n - it is not possible to unambiguously determine whether the frequency belongs to the emergency or normal mode, the frequency measurements continue and additional recognition is made based on these measurements.

Verification is carried out sequentially. According to the results of the first observation, one of the three indicated decisions is made. If the first or second decision is made, the verification is completed. The experiment continues if a third decision is made. Further, on the basis of the two observations obtained, in a similar way, one of the three decisions is made. If the third decision is made again, the verification continues, etc.

For the functioning of the algorithm, preliminary simulation with frequency measurements in normal and emergency modes is implemented. Based on the simulation results, the corresponding statistical frequency distributions are formed (Fig. 5). Let us consider an example (Fig. 5), where the red curve on the graph characterizes the distribution for the normal mode (hypothesis H_0), the blue curve for the emergency mode (hypothesis H_1).

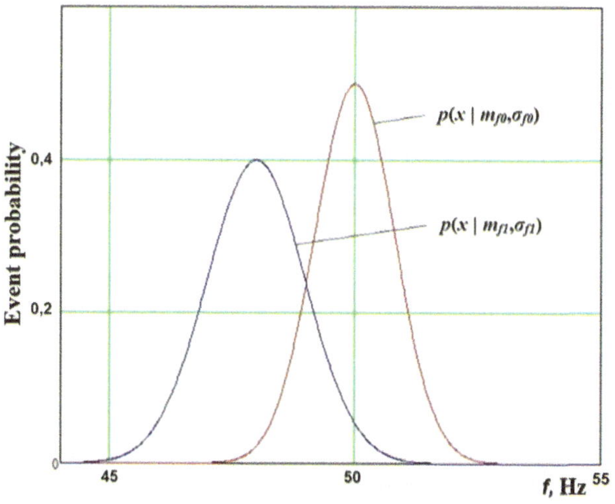

Fig. 5. Statistical frequency distributions for normal and emergency modes.

For the example shown in Fig. 5, we choose that the expected value of the frequency in normal mode is $m_{f0} = 50$ Hz, and in emergency mode, respectively, $m_{f1} = 48.5$ Hz. The laws of frequency distributions (Fig. 5) will be considered Gaussian with standard deviations σ_{f0} and σ_{f1}. The numerical values of σ_{f0} and σ_{f1} are determined by simulation data. Upon receiving the first frequency value, the likelihood ratio is calculated:

$$\begin{aligned}\eta(x1) &= p(x_1|m_{f1},\sigma_{f1})/p(x_1|m_{f0},\sigma_{f0})\\ &= \exp\left[-(x_1-m_{f1})^2/2\sigma_{f1}^2\right]/\exp\left[-(x_1-m_{f0})^2/2\sigma_{f0}^2\right]\\ &= \exp\{1/2\left[-(x_1-m_{f1})^2/\sigma_{f1}^2+(x_1-m_{f0})^2/\sigma_{f0}^2\right]\}.\end{aligned} \tag{20}$$

For k measurements of frequency, the likelihood ratio takes the form:

$$\begin{aligned}\prod_{i=1}^{k}\eta(x_i) = &\left[p(x_1|m_{f1},\sigma_{f1})\ldots p(x_k|m_{f1},\sigma_{f1})\right]/\left[p(x_1|m_{f0},\sigma_{f0})\ldots p(x_k|m_{f0},\sigma_{f0})\right]\\ &= \prod_{i=1}^{k}\exp\left\{1/2\left[-(x_i-m_{f1})_2/\sigma_{f1}^2+(x_i-m_{f0})^2/\sigma_{f0}^2\right]\right\}.\end{aligned} \tag{21}$$

The required number of frequency measurements is a random variable, since it depends on the nature of the transition process and the corresponding errors in the parameter estimates. Recognition of the regime is carried out in relation to the likelihood with the adoption of the following hypotheses:

$$\begin{aligned}&H_1,\ \text{if} \prod_{i=1}^{k}\eta(x_i) > b;\\ &H_0,\ \text{if} \prod_{i=1}^{k}\eta(x_i) \le a;\\ &H_n,\ \text{if}\, a \le \prod_{i=1}^{k}\eta(x_i) < b.\end{aligned} \tag{22}$$

To set the reference values a and b during a sequential analysis, it is necessary to determine the errors of the first α and second β kind. Here, α - the probability of the erroneous choice of hypothesis H_0, and β - the probability of the erroneous choice of hypothesis H_1. The setpoint values a and b for the selection of hypotheses are calculated as follows:

$$a = \alpha/(1-\beta);\, b = (1-\alpha)/\beta. \tag{23}$$

We take the values of errors of the first and second kind equal to $\alpha = 0.01$; $\beta = 0.03$.

Then the settings a and b have the following meanings:

$$\mathrm{a} = 0.01/(1-0.03) = 0.01;$$

$$\mathrm{b} = (1-0.01)/0.03 = 33.$$

Let there be a series of consecutive frequency measurements corresponding to the simulated circuit-mode situation: x_1 = 48.9 Hz; x_2 = 48.8 Hz; x_3 = 48.5 Hz; x_4 = 48.5 Hz.

According to the indicated sequential counts, a decision is made on the existence of a normal or emergency mode.

We calculate the likelihood ratio for the first value x_1 = 48.9 Hz of the frequency according to (20, 21):

$$\eta(x_1) = 1,374; \quad \prod_{i=1}^{1} \eta(x_i) = 1,374.$$

Since the likelihood ratio is in the zone of uncertainty

$$a = 0,01 < \eta(x_1) = 1,374 < b = 33,$$

the H_n hypothesis is accepted and observations continue.

For the second frequency value x_2 = 48.8 Hz, we have:

$$\eta(x_2) = 1,789; \quad \prod_{i=1}^{2} \eta(x_i) = 2,458.$$

However, for the second consecutive measurement, the likelihood ratio is in the zone of uncertainty

$$a = 0,01 < \prod_{i=1}^{2} \eta(x_i) = 2,458 < b = 33,$$

therefore, further observations are required to implement the Wald procedure.

Calculations for the third frequency value x_3 = 48.5 Hz lead to the equalities:

$$\eta(x_3) = 4,098; \quad \prod_{i=1}^{3} \eta(x_i) = 10,074.$$

The results also lead to the need for further calculations, since

$$a = 0,01 < \prod_{i=1}^{3} \eta(x_i) = 10,074 < b = 33.$$

The final decision is formed in the fourth step, corresponding to the measurement of frequency x_4 = 48.5 Hz:

$$\eta(x_4) = 4,098; \quad \prod_{i=1}^{4} \eta(x_i) = 41,267;$$

$$a = 0,01 < \prod_{i=1}^{4} \eta(x_i) = 41,267 > b = 33.$$

Since the result of the product of the likelihood ratio exceeds the response setting $\Pi_{i=1}^{4}\ \eta(x_i)$ = 41,267 > b = 33, a decision is made about the emergency mode in the considered fragment of the distribution network.

The process of sequential decision making by the method of sequential verification by the Wald probability ratio is illustrated in Fig. 6.

An analysis of Fig. 6 shows that the decision required four measurements of frequency and, accordingly, four calculated likelihood ratios. To implement automatic frequency unloading, only simulation data are required, expressed in statistical frequency distributions for normal and emergency modes, as well as current sequential measurements.

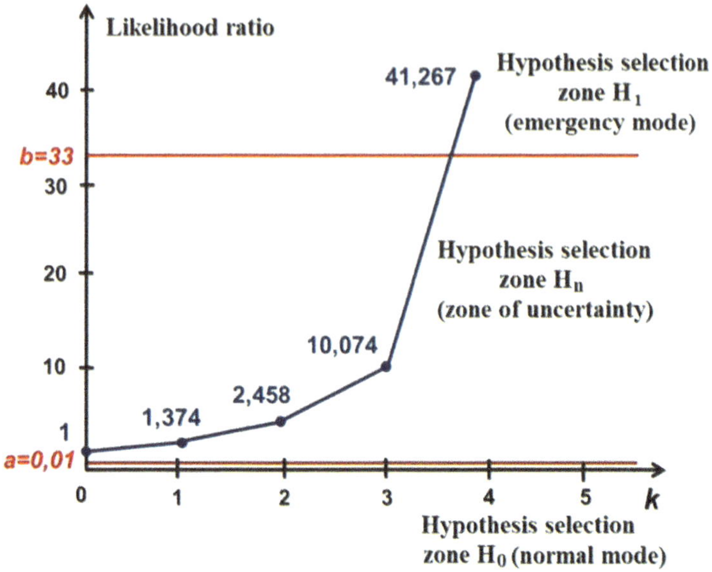

Fig. 6. Wald's sequential decision-making procedure for automatic frequency unloading

In the practice of controlling the modes of distribution networks with DG facilities, it is advisable to introduce several queues of automatic frequency unloading with specified frequency settings. Moreover, each queue corresponds to its normal distribution with a expected value equal to the queue setpoint in frequency and dispersion, calculated according to the results of simulation. The normal mode is also characterized by a Gaussian distribution of frequency values with the expected value of 50 Hz. It is assumed that the mode recognition device has a multi-channel structure corresponding to the automatic frequency unloading queues and normal mode. Each channel implements an independent decision-making procedure according to Wald's method.

To ensure high performance when implementing a multi-channel decision-making scheme, it is advisable to estimate the frequency based on multi-channel filtering. For example, by the maximum likelihood method or the discriminatory method [14]. Thus, when each voltage count arrives at the input of the multi-channel circuit of the automatic frequency unloading device, several parallel calculations are started. Moreover, each new count refines the previous value.

A certain delay in decision making according to the Wald procedure does not practically affect the speed of the emergency automation device at a high sampling frequency of current and voltage signals. In the general case, the delay is determined by the given errors of the first α and second β kind. For example, at sampling frequencies conforming to the IEC 61850, this delay, as a rule, does not exceed 1 ms. An additional reduction in the decision-making time is the introduction of truncation algorithms.

4 Truncation of the Sequential Analysis Procedure

It is known [14, 20] that the sequential analysis procedure with probability equal to one ends at a finite time interval. For relay protection and emergency control devices, this time interval has strict restrictions and should not exceed a certain number of steps k_0 (taking into account the sampling frequency). For this, a special truncation procedure is introduced, according to which, when $k = k_0$ is reached, new decision rules are established.

Wald [15] proposed a simple version, when for $k = k_0$ it is accepted:

- hypothesis H_0, if $b < \Pi_{i=1}^{k0}\ \eta(x_i) \ \leq\ 1$;
- hypothesis H_1, if $1 < \Pi_{i=1}^{k0}\ \eta(x_i) \ \leq\ a$.

Such truncation will change the errors of the first and second kind at the last k_0 step of observation. Obviously, the larger the value of k_0, the less the influence of truncation on the errors of the first (skipping emergency mode) α and second (false decision about the emergency mode) β kind. Denote by $\alpha(k_0)$ and $\beta(k_0)$ the resulting errors at the step $k = k_0$ of truncation. We obtain upper bounds for $\alpha(k_0)$ and $\beta(k_0)$.

To form the upper boundary $\alpha(k_0)$, we investigate the case when truncation leads to a deviation of H_0, while for an un truncated process, H_0 is accepted. Let $p_0(k_0)$ be the probability of obtaining a sample at H_0, which, in a truncated process, leads to a deviation of H_0, while un truncated analysis ensures the adoption of H_0. In this case, the inequality:

$$\alpha(k_0) \leq \alpha + p_0(k_0). \tag{24}$$

The inequality symbol is valid, since samples are probable for which a truncated process leads to the adoption of H_0, and an un truncated process leads to a deviation of H_0. Thus, to form the upper bound of the estimate $\alpha(k_0)$, it is necessary to obtain the upper bound for $p_0(k_0)$. Based on the definition of $p_0(k_0)$, the following conditions are observed for successive observations:

1. $b \ \leq\ \Pi_{i=1}^{k0}\eta(x_i) < a$ for $k = 1, 2,\ldots, k_{0\text{-}1}$;
2. $1 < \Pi_{i=1}^{k0}\eta(x_i) \ \leq\ a$;
3. If the process continues after k_0 tests, then it ends with the adoption of H_0.
4. Denote by $\widehat{p}_0(k_0)$ the probability that condition 2 will hold for H_0:

$$\widehat{p}_0(k_0) = P_0(1 < \prod\nolimits_{i=1}^{k0} \eta(x_i) \leq a). \tag{25}$$

Since the probability of fulfilling condition 2 cannot be less than the probability of simultaneously fulfilling all three conditions, then

$$\widehat{p}_0(k_0) \geq p_0(k_0),$$

therefore

$$\alpha(k_0) \leq \alpha + \widehat{p}_0(k_0). \quad (26)$$

That is, $\alpha + \widehat{p}_0(k_0)$. is the upper bound for the error $\alpha(k_0)$.

Carrying out similar arguments, we can obtain the upper bound for $\beta(k_0)$, which will be determined by the equation:

$$\beta(k_0) \leq \beta + \widehat{p}_1(k_0), \quad (27)$$

where $\widehat{p}_1(k_0) = P_1\ (b\ \leq \Pi_{i=1}^{k0} \eta(x_i) < 1)$.

Equations (26), (27) allow us to estimate the upper probabilistic limits for the implementation of the truncation procedure. It is important that the considered truncation version is not the only one. It is possible to introduce adaptive setpoint values at each step of a sequential analysis [22], as well as truncation algorithms (for example, G. Lorden [23], S. Ayvazyan [24], etc.) based on the use of other motivational principles.

5 Conclusion

The expediency of applying a sequential criterion of Wald's probability ratio for mode recognition in distribution networks with DG and microgeneration facilities is substantiated.

The advantages of applying sequential analysis in the island mode of operation of a distribution network with DG facilities under conditions of decreasing frequency, transient processes, non-sinusoidality of currents and voltages, which contribute to large errors in the estimation of their parameters, are shown.

Due to the presence of a zone of uncertainty, three-position relays implemented on Wald's statistical procedures can lead to a delay in the decision-making process. However, at sampling frequencies of current and voltage signals that comply with the IEC 61850 (for example), this delay does not exceed 1 ms.

To ensure guaranteed high-speed performance of automatic control devices of modes, it is advisable to introduce algorithms for truncating sequential analysis in recognition of modes.

The considered example of the implementation of an automatic frequency unloading device based on the Wald procedure demonstrated the correct decision-making in conditions of ambiguous measurements of the frequency and the presence of distorting factors.

The expediency of applying a sequential analysis in the implementation of relay protection devices and emergency automation in modes accompanied by complex transients is substantiated.

Acknowledgement. The presented research results were obtained with the support of a grant from the President of the Russian Federation for state support of young Russian scientists (MK-3210.2019.8). Agreement No. 075-15-2019-337 of 11.06.2019.

References

1. Buchholz, B.M., Styczynski, Z.: Smart Grids – Fundamentals and Technologies in Electricity Networks. Springer, Heidelberg (2014)
2. Kakran, S., Chanana, S.: Smart operations of smart grids integrated with distributed generation: a review. Renew. Sustain. Energy Rev. **81**(1), 524–535 (2018)
3. Ilyushin, P.V., Sukhanov, O.A.: The structure of emergency-management systems of distribution networks in large cities. Russ. Electr. Eng. **85**(3), 133–137 (2014)
4. Ilyushin, P.V.: Analysis of the specifics of selecting relay protection and automatic (RPA) equipment in distributed networks with auxiliary low-power generating facilities. Power Technol. Eng. **51**(6), 713–718 (2018)
5. Ilyushin, P.V., Kulikov, A.L., Filippov, S.P.: Adaptive algorithm for automated undervoltage protection of industrial power districts with distributed generation facilities. In: International Russian Automation Conference (RusAutoCon), pp. 1–6. IEEE, Sochi (2019)
6. Loskutov, A.A., Mitrovic, M., Pelevin, P.S.: Development of the logical part of the intellectual multi-parameter relay protection. In: Rudenko International Conference on Methodological Problems in Reliability Study of Large Energy Systems, RSES, vol. 139. EDP Sciences, Tashkent (2019)
7. Kulikov, A.L., Loskutov, A.A., Mitrovic, M.: Method of automated synthesis of the logic part of relay protection device which increases its sensitivity. In: IOP Conference Series: Materials Science and Engineering, vol. 643, p. 012124 (2019)
8. Ilyushin, P.V., Suslov, K.V.: Operation of automatic transfer switches in the networks with distributed generation. In: IEEE Milan PowerTech, pp. 1–6. IEEE, Milan (2019)
9. Ilyushin, P.V., Filippov, S.P.: Under-frequency load shedding strategies for power districts with distributed generation. In: International Conference on Industrial Engineering, Applications and Manufacturing, ICIEAM, pp. 1–5. IEEE, Sochi (2019)
10. Yurevich, Ye.N.: Theory of Automatic Control. Energiya, Leningrad, USSR (1975)
11. Sidorenko, Yu.A.: Theory of Automatic Control. BGATU, Minsk, Belarus (2007)
12. Yevsyukov, V.N.: Nonlinear automatic control systems: a textbook for university students. GOU OGU, Orenburg, Russia (2007)
13. Davarifar, M., Rabhi, A., Hajjaji, A., Daneshifar, Z.: Real-time diagnosis of PV system by using the sequential probability ratio test (SPRT). In: 16th International Power Electronics and Motion Control Conference and Exposition, pp. 508–513. IEEE, Antalya (2014)
14. Radio-electronic systems: Fundamentals of construction and theory, 2nd ed. Radiotekhnika, Moscow, Russia (2007)
15. Wald, A.: Sequential analysis. Fizmatlit, Moscow, USSR (1960)
16. Sharygin, M.V., Kulikov, A.L.: Protection and automation of power supply systems with active industrial consumers. NIU RANKhiGS, Nizhny Novgorod, Russia (2017)
17. Sharygin, M.V., Kulikov, A.L.: Statistical methods for recognizing modes in relay protection and automation of power supply networks. Elektricheskiye stantsii **2**, 32–39 (2018)
18. Fukunaga, K.: Introduction to the statistical theory of pattern recognition. Nauka, Moscow, USSR (1979)
19. Fu, K.: Sequential methods in pattern recognition and machine learning. Nauka, Moscow, USSR (1971)
20. Shiryayev, A.N.: Statistical sequential analysis. Optimal stop rules. Nauka, Moscow, USSR (1976)
21. Basharinov, A.Ye., Fleyshman, B.S.: Statistical sequential analysis methods and their radio engineering applications. Sovetskoye radio, Moscow, USSR (1962)

22. Sochman, J., Matas, J.: Waldboost-learning for time constrained sequential detection. In: IEEE Computer Society Conference on Computer Vision and Pattern Recognition (CVPR 2005), vol. 2, pp. 150–156. IEEE, San Diego (2005)
23. Lorden, G.: Structure of sequential tests minimizing an expected sample size. Zeitschrift fur Wahrscheinlichkeits-theorie undverwandte gebiete **51**(3), 291–302 (1980)
24. Ayvazyan, S.A.: Distinguishing close hypotheses about the form of distribution density in the scheme of a generalized sequential criterion. Theory of Probability and its Applications (1965)

Heterostructure Simulation for Optoelectronic Devices Efficiency Improvement

Oleg Rabinovich(✉) and Svetlana Podgornaya

National University of Science and Technology "MISiS", Moscow 119049, Russian Federation
rawork2008@mail.ru

Abstract. In this work nitride heterostructures were simulated for optoelectronic devices such as photodetectors (phototransistors) and solar cells for improving their efficiency. The influence of aluminium atoms and doping as well as temperature on AlGaN/GaN-based heterojunction phototransistors characteristics have been studied. The results suggest that the AlGaN/GaN phototransistor with the Al concentration – 28% and the doping concentration - $N_d = 2 \times 10^{15}$ cm^{-3} and $N_a = 2.1 \times 10^{16}$ cm^{-3}, exhibits a considerable sensitivity and the quantum efficiency approaching about 10%. Solar cells model based on GaN/Si heterostructure was created. The optimum heterostructure design and doping profile were defined. Quite high solar cell efficiencies based on n-GaN–p-Si heterostructures such as 14.35% at 1 · AM 1.5 and 21.10% at 1000 · AM 1.5 were achieved.

Keywords: Simulation · Heterostructure · Optoelectronic devices

1 Introduction

The drastic growth realized in the industry of optoelectronics, demands highly improved materials and systems for detection and reception of optical signals. The outstanding advantages of nitride based heterojunction phototransistors (HPTs) make them attractive for this purpose in short and long wavelength applications. For this reason, a considerable amount of experimental and theoretical works toward understanding how factors such as optical power and radiation wavelength affect nitride based HPTs characteristics have been carried out, yet lack of understanding how heterostructure parameters influence on the performance of HPTs still exists.

The growing demand for generation and detection of optical ultraviolet (UV) signals due to their wide range of applications requires effective optoelectronic devices. At the generation side, UV light emitters are fully used in colored projection displays, space communications, high-density optical storage systems and sterilization of biological and chemical agent. Meanwhile, detection of UV radiations is vital in a number of fields including environmental monitoring, space research, military systems, scientific research and commercial applications.

The high rise in power need will be one of the most serious peculiarities for technical evolution in current century. The greatest opportunity is Sun usage for getting renewable energy source. We need to increase efficiency of solar energy conversion

R. Silhavy et al. (Eds.): CoMeSySo 2020, AISC 1295, pp. 121–133, 2020.
https://doi.org/10.1007/978-3-030-63319-6_12

methods. Now the main goal is to get a high solar energy conversion. We have to create cheap but effective heterostructure for such devices.

Solar cells (SC) connecting combinations in-serious and in-parallel can give possibility to achieve the required parameters in current and voltage, and, therefore power. The current-voltage characteristic of such a device is determined by the expressions:

$$I = I_s(e^{qV/kT} - 1) - I_L, \tag{1}$$

$$J_s = {I_s}/{A} = qN_C N_V \left(\frac{1}{N_A}\sqrt{\frac{D_n}{\tau_n}} + \frac{1}{N_D}\sqrt{\frac{D_p}{\tau_p}} e^{-E_g/kT} \right), \tag{2}$$

solar cells efficiency (η):

$$\eta = \mathrm{P_{EL.}}/\mathrm{P_{OPT.}}, \tag{3}$$

where I_L- a direct current that describes the nonequilibrium carriers excitation by solar radiation, I_s - the diode saturation current, A - the area of the device, τ_n - the electron lifetime, τ_p - the hole lifetime, N_A - the acceptor concentration, N_D - the donor concentration, D_p - the hole diffusion coefficient, D_n - the electron diffusion coefficient, N_c - the concentration of charge carriers in the conduction band, N_V - the concentration of charge carriers in the valence band, $P_{OPT.}$ - the radiation power, $P_{EL.}$ - electric power.

III-nitride semiconductor materials usage in the photodetectors production is very perspective due to their exclusive physical properties such as high absorption coefficient, tunable band gap, high electron mobility and saturation velocity and high thermal stability. These properties have paved a way for developing numerous III-nitride optoelectronic devices including those needed for photodetection and energy converting purposes [1–3].

At the end of last century huge quantity of researches made investigations on multi cascade SC developments. The complication of the multi cascade SC heterostructure is due to the problem of growth the photoactive thinner region with flexible band gap to convert maximum solar energy. New materials for high efficiency solar cells can be direct-gap nitride semiconductors with various bandgaps: InN (E_g ~ 0.65 eV), GaN (E_g ~ 3.4 eV), AlN, (E_g ~ 6.2 eV) and solid solutions based on them. New heterostructures can be synthesized by different growth methods such as MOCVD, MBE and others [4, 5].

It is predicted to reach converting solar energy efficiency into electrical energy about 40–50%. One of possible materials is InGaN heterostructure with an indium atoms concentration about 35–45% with Si substrate. The main part of them is theoretical due to grow high-quality InGaN difficulty with high indium concentration. The main problem is mismatch between the InN and GaN constant lattices (about 11%). Despite difficulties, now it is very promising to create nitride semiconductors for use in solar cells.

In 2009, the American company RoseStreet Energy Labs Inc announced the creation of a silicon nitride solar cell [6–9]. Researchers used a 12-period structure with multiple quantum well (QW) – combination – InGaN (QW) and GaN barriers (3 nm/16 nm) with an In content in the QW about 35%.

The goal of current work is to find possibility for SC efficiency rise based on nitride heterostructure.

Avalanche photodiodes (APDs) demonstrate outstanding inherent advantages such as high gain at high speed over previous devices. However, for APDs to obtain avalanche gain, they need a very high biasing voltage close to a breakdown voltage. Thus, their performance is extremely sensitive to biasing conditions making their use more expensive, due to bias stabilization and temperature compensation, which needs to be monitored for stable operation of the devices. It is thus appropriate to use photodetectors whose internal gain is based on more controllable and well-understandable processes. A nitride-based heterojunction phototransistor (HPT) that amplifies photogenerated carriers collected by the in-built photodiodes is the best choice for this purpose [10–12].

When nitride materials are employed in HPTs, the basic parameters which determine optimum performance in HPTs, such as current-voltage characteristics, quantum efficiency (QE) and sensitivity can be investigated and improved.

Motivated by the inherent advantages of nitride-based HPTs, this work aimed at developing an understanding of how heterostructure parameters and temperature influence on I-V characteristics, quantum efficiency and sensitivity of nitride-based HPTs using Sim Windows simulation software. Thus, the primary objective of this research was to propose a model of the nitride-based HPT whose heterostructure parameters provide optimum performance in short wavelengths applications. The goal for HPT simulation was to find out optimum heterostructure scheme.

2 Methods

HPT with optimum performance requires semiconductor materials that have outstanding properties such as high breakdown field, electron mobility, saturation velocity, thermal conductivity and tunable bandgap. The III-nitride compound semiconductors having these properties are of great interest for fabrication of optoelectronic devices which include phototransistors used for photodetection specifically. These semiconductors are formed by bonding one group III element such as aluminum, gallium, or indium with nitrogen element found in group V of the periodic table [11–15]. III-Nitride group semiconductors AlN, GaN, and InN known as wide-bandgap semiconductors form are an important material system because of their direct bandgaps which span from 1,95 eV to 6,2 eV, covering whole visible region and extending well into the UV range

This research for getting optimum heterostructure for HPT was carried out on $Al_xGa_{1-x}N/GaN/Al_xGa_{1-x}N$ (n-p-n) HPT using the Aluminium mole fraction X = 0.22 – 0.34, doping concentration $10^{16} - 10^{17}$ cm^{-3} for acceptors (N_a) and $10^{15} - 10^{16}$ cm^{-3} for donors (N_d) as well @ an optical power - 1 mWcm^{-2} and energy quanta - 3 eV. During heterostructure simulation the structure consisted: n-$Al_{0.3}Ga_{0.7}N$ collector, n-$Al_{0.3}Ga_{0.7}N$ emitter and a p-GaN base. The emitter and collector thicknesses were

0.875 μm, the base thickness - 0.3 μm. The main HPT parameters are sensitivity and quantum efficiency where calculated by the Eqs. 4 and 5, respectively:

$$Sensitivity(S) = \frac{current\ density(J)}{optical\ power(P)} \tag{4}$$

$$Quantum\ efficiency\ (QE) = \frac{emitting\ recombination}{total\ recombination} \tag{5}$$

The choice for SC heterostructure can be made based on three heterojunctions band diagrams: p-GaN − n-Si, n-GaN − p-Si and n-GaN − n-AlN − p-Si. Diagrams were created during simulation in Sim Windows software [10–15]. The n-GaN layer was 500 nm thick and an electron concentration was 10^{17} cm^{-3}, the p-GaN layer was 500 nm with hole concentration 10^{17} cm^{-3}, the n-AlN layer - 1 nm and the electron concentration - 10^{17} cm^{-3}, n-Si/pSi layers were 500 nm thick and carrier concentrations - 10^{17} cm^{-3}. The creation of SCs based on the p-GaN – n-Si heterojunction, is impossible, since the holes generated by radiation in the n-Si region cannot reach the p-GaN region due to the large gap of the valence bands at the heterojunction boundary. The creation of SCs based on the n-GaN – p-Si heterojunction is possible, since the electrons generated by radiation in the p-Si region can easily fall into the n-GaN region. The creation of SCs based on the n-GaN – n-AlN – p-Si heterojunction is possible, but complicated since the electrons generated by radiation in the p-Si region can reach the n-GaN region only if they can overcome the high energy barrier at the n-AlN – p-Si heterojunction due to tunneling. Tunneling simulations (using the Sim Windows software) at n-AlN layer (1 nm thickness) did not give possibility to create SC. As a result, the n-GaN−p-Si heterostructure was chosen as the main type for the solar cell (see Fig. 1).

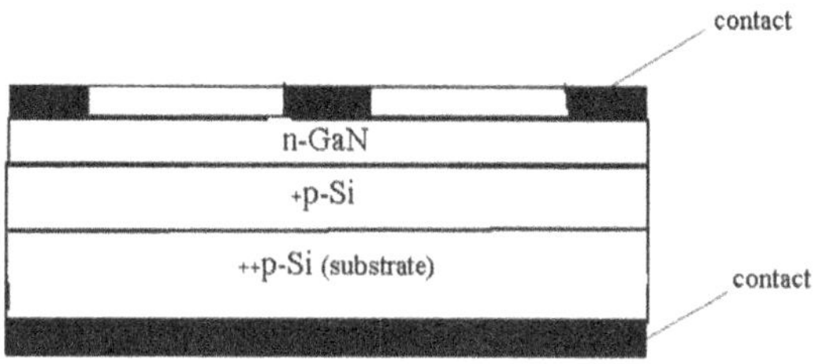

Fig. 1. Design of a solar cell based on a heterostructure n-GaN − p-Si.

For SC simulation such structure n-GaN – p-Si scheme was used and contained the n-GaN layer ($N_d = 10^{17}$ cm^{-3} and a 50 nm thickness), p-Si layer ($N_a = 10^{16}$ cm^{-3} and 20 μm thickness), the Si substrate ($N_a = 10^{18}$ cm^{-3} and 100–500 μm thickness). The SC efficiency is determined by the following expressions:

$$\eta_1 = FF \cdot (U_{OC} \cdot J_{ShC})/(1 \cdot AM1.5), \quad (6)$$

$$\eta_{1000} = FF \cdot (U_{OC} \cdot J_{ShC})/(1000AM1.5), \quad (7)$$

$$FF = (U_m \cdot J_m)/(U_{OC} \cdot I_{ShC}), \quad (8)$$

where U_{OC} - the voltage on the SC in an open circuit, I_{ShC} - the short circuit current through the SC.

These parameters are determined from the current-voltage characteristics (I-V) simulation results at a solar illumination power - 1 · AM - 1.5 = 84.4 mW/cm^2 or 1000 · AM 1.5 = 84.4 W/cm^2. Based on I-V simulation, it is possible to determine the parameters U_m and I_m and by them the maximum value of the electric power P_{el} generated by a solar cell can be calculated.

3 Result and Discussion

3.1 SC Optimum Heterostructure Design Simulation

Figure 2 shows band diagrams without and with irradiation, the optical generation and the nonradiative recombination rate according to the Shockley-Hall-Reed mechanism.

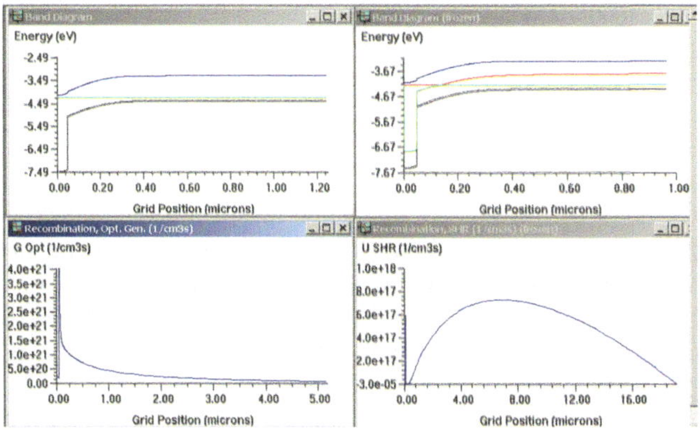

Fig. 2. Short-circuit current mode at conditions of sunlight irradiation with power – 1 · AM - 1.5 = 84.4 mW/cm^2.

It is detected that the recombination rate is lower than the electron – hole pairs generation rate in the p-Si base region. At forward voltage rise U_m, the J_m will reduce to zero, which corresponds to the voltage U_{ShC}. I – V characteristics for a lighting power 1000 · AM 1.5 (see Fig. 3).

It was detected that with radiation 1 · AM 1.5, the maximum power value is achieved at a voltage U_m = 0.45 V and a current J_m = 26.9 mA, at level - 1000 · AM 1.5, the maximum power value is achieved at U_m = 0.60 V and a current J_m = 26.9 A.

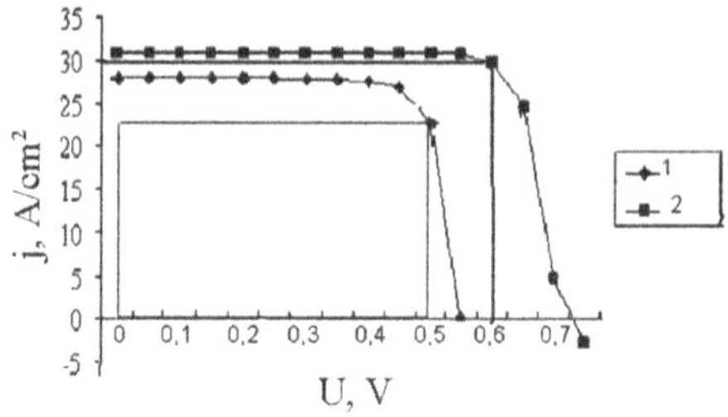

Fig. 3. I – V characteristic for illumination power: 1 - at 1 · AM1.5; 2 - at 1000 · AM1.5.

In Fig. 3 the maximum value of the generated power refers to the areas which are graphically inserted rectangles.

In Table 1 the results of calculating the form factors FF and the efficiency for the solar cells η_1 and η_{1000} according to formulas (6–8) are presented.

Table 1. The data for calculating the form factors FF and the efficiency for the solar cells η1 and η1000

1 · AM 1.5 (S = 1 cm²)					
U_{OC}, V	J_{ShC}, mA	U_m, B	J_m, mA	FF, a. u.	η_1, %
0.54	27.8	0.45	26.9	0.80	14.35
1000 · AM 1.5 (S = 1 cm²)					
U_{OC}, V	J_{ShC}, A	U_m, V	J_m, A	FF, a. u.	η_{1000}, %
0.7	30.9	0.60	29.6	0.81	21.1

The n-GaN – p-Si SC parameters simulation results relate to a heterostructure with the following parameters; n-GaN layer thickness is d_{GaN} = 50 nm; the p- Si layer thickness in which the nonequilibrium charge carriers generation by radiation occurs, d_{Si} = 20 μm; the donor concentration in the n-GaN layer is $N_d = 10^{17}$ cm^{-3}, the acceptor concentration in the p-Si layer is $N_a = 10^{16}$ cm^{-3}, and the lifetime of the injected charge carriers in the p-Si layer is $\tau = 10^{-6}$ s.

3.2 Effects of Varying Al. Concentration on AlGaN HPT Characteristics

It has been observed that altering the aluminium composition at the doping N_d = 1.5 10^{15} cm^{-3} and $N_a = 1.5 \times 10^{16}$ cm^{-3} highly affects the sensitivity, current density as well as the quantum efficiency (QE) of the HPT. The sensitivity of the device increases by increasing the aluminium content in the emitter and collector (n-AlGaN) of the HPT (see Fig. 4).

The sensitivity 12000 A/W was detected at Al. X = 28%. That sensitivity rise was based on increased charge carriers quantity due to the absorption coefficient rise. The sensitivity reduce is based on bandgap energy rise caused by increased electron injection, therefore higher recombination rate is also supported.

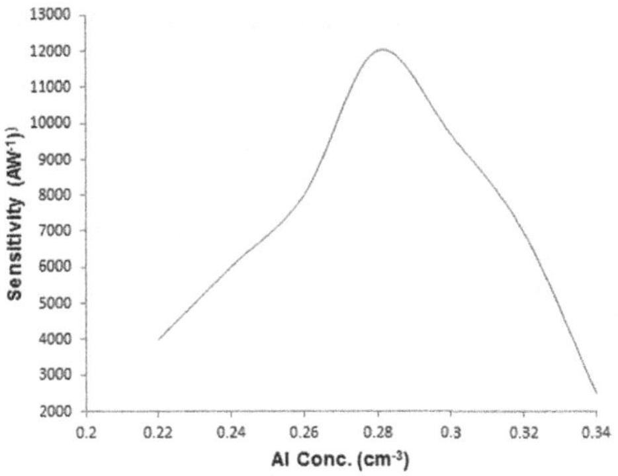

Fig. 4. Sensitivity vs Al conc. @ E = 3 eV, P = 1 mW/cm^2, U = 0.3 V

At low Al. concentration, the lower sensitivity are attributed to few charge carriers that are generated due to the reduced refractive index (absorption coefficient) caused by very low aluminium concentration.

I-V characteristics curve with a maximum current density has been achieved at Al. X = 28% (see Fig. 5). The decrease in current density due to increase in the Al conc. is caused by destructive recombination of the generated charge carriers and the increased bandgap energy which limited the most generated electrons from reaching the conduction band.

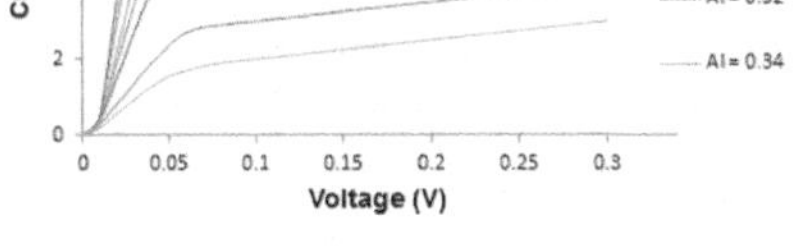

Fig. 5. Al conc. vs I-V for HPT @ $N_d = 1.5 \times 10^{15}$ cm^{-3} and $N_a = 1.5 \times 10^{16}$ cm^{-3}, E = 3 eV, P = 1 mW/cm^2

From another point of view the QE rises by increasing the Al. concentration in the emitter and collector (n-AlGaN) of the HPT (see Fig. 5). The QE about 26–28% was measured at Al. concentration 26%. Such QE rise was due to increased generated charge carriers quantity based on the absorption coefficient rise according to the increased Al. concentration. The QE reduction, which is realized at much higher aluminium concentration (see Fig. 6), is due to photo-generated electrons reduce and absorption coefficient degradation. At a very low Al. concentration, the lower QE are attributed to few charge carriers that are generated due to the increased refractive index in the material caused by very low Al. concentration.

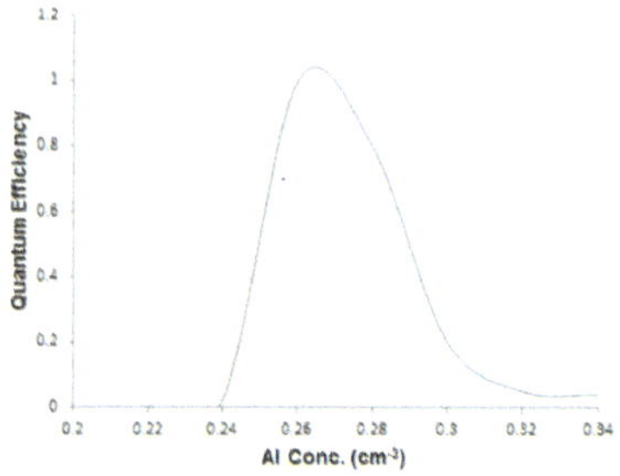

Fig. 6. Quantum efficiency vs Al conc. @ E = 3 eV, P = 1 mW/cm^2

Consequently, aluminium concentration influences highly on the spectral range of the HPT, which could ultimately determine the area of application of the device. An increase in Al conc. reduces the wavelength of the incident radiations. For example, the Al conc. X = 0.28 makes an HPT which can detect light with wavelength 354 nm which corresponds to the bandgap energy 3.5 eV (see Fig. 7). Further results for the dependences of sensitivity and QE for HPT versus different aluminium concentrations are as summarized in Table 2.

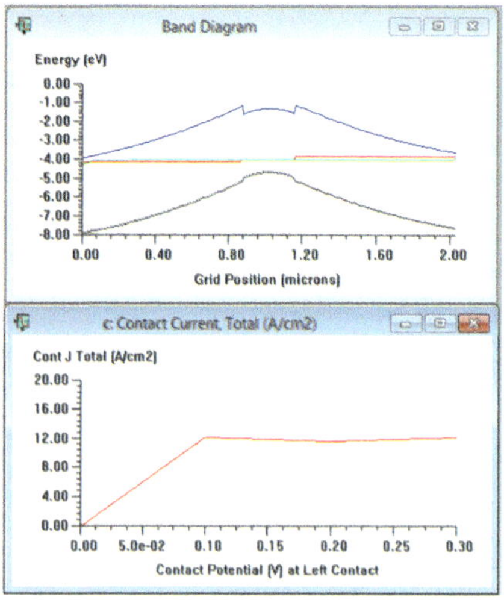

Fig. 7. Phototransistor simulation with illumination @ the Al conc. X = 28%

Table 2. HPT sensitivity and QE dependence vs Al. concentration P = 1 mW/cm^2, E = 3 eV, $N_d = 1.5 \times 10^{15}$ cm^{-3} & $N_a = 1.5 \times 10^{16}$ cm^{-3}

Al conc. X, %	J, A/cm^{-2}	$S = {}^{J}/_{P}$, A/W	$QE = {}^{BR}/_{TR}$
22	4	4000	2.5×10^{-28}
24	6	6000	1.3×10^{-27}
26	8	8000	3.8×10^{-24}
28	12	12000	7.2×10^{-26}
30	10	10000	1.9×10^{-26}
32	7	7000	3.3×10^{-27}
34	3	3000	3.0×10^{-27}

3.2.1 Doping Concentration Influence on AlGaN HPT Characteristics

Changing the doping amount in the HPT structure at the Al conc. 26–28%, seriously affects on the performance characteristics - sensitivity, QE and current density (see Table 2). The maximum sensitivity was detected with N_a in the base - 1.9×10^{16} cm^{-3} and N_d in the emitter and collector - 1.8×10^{15} cm^{-3} at optical power – 1 mW/cm^2 and energy quanta - 3 eV (see Fig. 8).

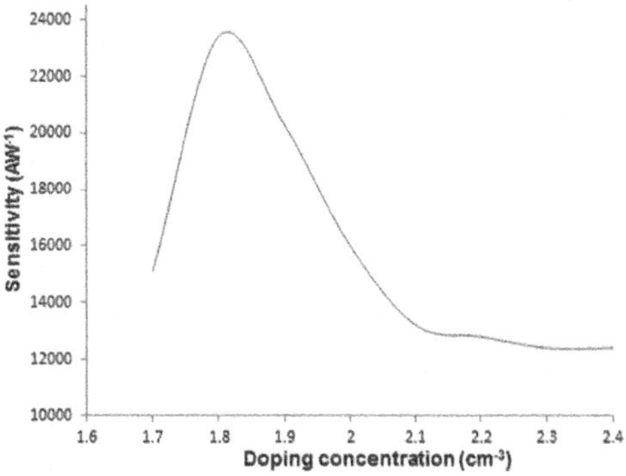

Fig. 8. Sensitivity vs N_a and N_d doping concentration, Al conc. = 0.28 @ E = 3 eV, P = 1 mW/cm^2

This increase in sensitivity was due to increased charge carriers generated due to the doping process and the breakdown voltage (see Fig. 9).

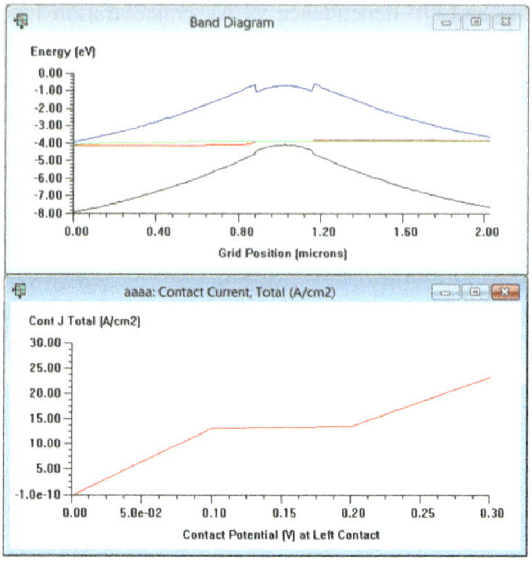

Fig. 9. Phototransistor simulation with illumination @ $N_d = 1.8 \times 10^{15}$ cm^{-3}, $N_a = 1.9 \times 10^{16}$ cm^{-3}, E = 3 eV, P = 1 mW/cm^2, Al X. = 0.28

Further doping reduced sensitivity due to decreased current density generation caused by higher recombination rate because of the defect points caused by excess carriers. At low N_a and N_d in the device, a decrease in sensitivity was due to few charge carriers generated that leads to decrease in the output current (see Table 3).

Table 3. Dependence of HPT sensitivity and QE on doping concentration Al conc. = 0.28, P = 1 mWcm^{-2}, E = 3 eV

N_d, $\times 10^{15}$ cm^{-3}	Na, $\times 10^{16}$ cm^{-3}	J, A/cm^2	$S = {}^{J}/_{P}$, A/W	$QE = {}^{BR}/_{TR}$
1.7	18	15.1	15100	1.8×10^{-27}
1.8	19	23.4	23400	5.8×10^{-28}
1.9	20	20.3	20300	1.7×10^{-27}
2.0	21	16.0	16000	1.7×10^{-27}
2.1	22	13.2	13200	2.8×10^{-27}
2.2	23	13.5	13500	3.0×10^{-27}
2.3	2.4	124	12400	4.3×10^{-27}

As optimum parameters for HPT, the J = 16 A/cm^2 were achieved at N_a in the base - 2.1×10^{16} cm^{-3} and N_d in the emitter and collector – 2×10^{15} cm^{-3} (see Fig. 10). Such current density rise is due to the increase in charge carriers quantity attributed to doping concentration.

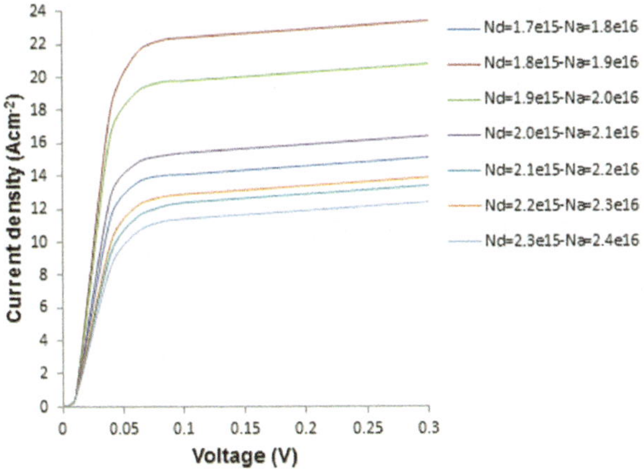

Fig. 10. Doping dependence I-V characteristic curve HPT @ Al X = 0.28, E = 3 eV, P = 1 mW/cm^2

The quantum efficiency HPT dependence versus doping concentration was detected. For example, maximum QE was seen at N_a in the base (p-GaN) - 2.4 × 10^{16} cm^{-3} and N_d in the emitter and collector ($Al_xGa_{1-x}N$) - 2.3 × 10^{15} cm^{-3} @. X = 0.28 (Al conc.) at P = 1 mW/cm^2 and E = 3 eV. The increase in QE was due to the rise in electrons generation due to the doping process (see Fig. 11).

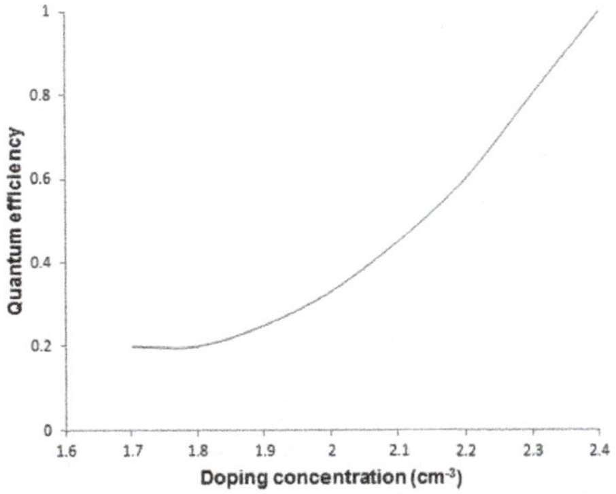

Fig. 11. Quantum efficiency vs N_a and N_d doping concentration @ E = 3 eV, P = 1 mW/cm^2

At the end of simulation HPT sensitivity spectrum was investigated at different doping levels (see Fig. 12).

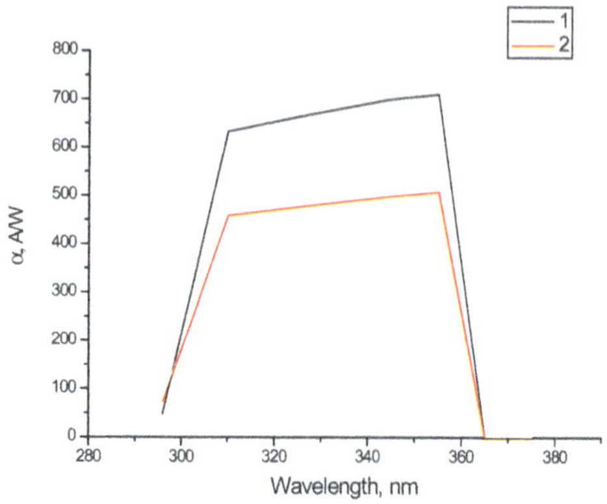

Fig. 12. HPT sensitivity spectrum @ P = 10^{-3} W/cm^2 and U = 9 V, 1 - $N_a = 10^{17}$ cm^{-3}, 2 - $N_a = 10^{18}$ cm^{-3}

4 Conclusions

Detected efficiencies for SC such as 14.35% at 1 · AM 1.5 and 21.10% at 1000 · AM 1.5 are quite high. The advantages of using these elements on silicon versus AlGaAs solar cells are the significantly lower cost of the substrates, significantly higher thermal conductivity and due to this longer device life time.

It is detected that the AlGaN HPT with Al conc. X = 26–28% can achieved maximum efficiency level with maximum sensitivity parameter. The simulation shows that such sensitivity and efficiency rises are based on special doping at the emitter and collector $N_d = 2 \cdot 10^{15}$ cm^{-3} and in the base $N_a = 2 \cdot 10^{16}$ cm^{-3}. The rise in such parameters is due to increased carrier generation in the active region and on the increased device absorption coefficient. Special doping profile in QWs gives the possibility of sensitivity increase for 5–7%.

References

1. Rabinovich, O., Saranin, D., Orlova, M., Yurchuk, S., Panichkin, A., Konovalov, M., Osipov, Yu., Didenko, S., Gostischev, P.: Heterostructure improvements of the solar cells based on perovskite. Procedia Manuf. **37C**, 221–226 (2019)
2. Wu, J., Walukiewicz, W., Yu, K., Shan, W., Ager, E., Haller, E., Hai, L., Schaff, W., Metzger, W., Kurtz, S.: Superior radiation resistance of $In_{1-x}Ga_xN$ alloys: full-solar-spectrum photovoltaic material system. J. Appl. Phys. **94**(10), 6477–6482 (2003)
3. Yamamoto, A., Islam, Md., Kang, T., Hashimoto, A.: Recent advances in InN-based solar cells: status and challenges in InGaN and InAlN solar cells. Physica Status Solidi (c) **7**(5), 1309–1319 (2010)
4. Fedorchenko, I., Kushkov, A., Gaev, A., Rabinovich, O., Marenkin, S., Didenko, S., Legotin, S., Orlova, M., Krasnov, A.: Growth method for $A^{III}B^{V}$ and $A^{IV}B^{VI}$ heterostructures. J. Cryst. Growth **483**, 245–250 (2018)

5. Aleksandrov, S., Zykov, V.: Electric and photoelectric properties of n-$Ga_xIn_{1-x}N$/p-Si anisotypic heterojunctions. Semiconductor **32**(4), 412–416 (1998)
6. Kenichi, S., Hashimoto, A., Yamamoto, A.: InGaN solar cells: present state of the art and important challenges. IEEE J. Photovoltaics **2**(3), 276–293 (2012)
7. Dahal, R., Li, J., Aryal, K., Lin, J., Jiang, H.: InGaN/GaN multiple quantum well concentrator solar cells. Appl. Phys. Lett. **97**, 0731151–0731155 (2010)
8. Routray, S., Lenka, R.: Heterostructures improvement. CSIT **35**, 21–27 (2009)
9. Akmak, H., Engin, A., Rudzin, P., Demirel, H., Unalan, W., Strupin, R., Turan, M.: Effects of filler shape and size on the properties of silver filled epoxy composite for electronic applications. J. Mater. Sci. Mater. Electron. **48**, 56–63 (2011)
10. Rabinovich, O., Legotin, S., Didenko, S.: Impurity influence on nitride LEDs. J. Nano Electron. Phys. **6**(3), 030021–030022 (2014)
11. Rabinovich, O.: InGaN and InGaP heterostructure simulation. J. Alloys Compound **586**(1), S258–S261 (2014)
12. Rabinovich, O., Legotin, S., Didenko, S., Yakimov, E., Osipov, Yu., Fedorchenko, I.: Heterostructure optimization for increasing LED efficiency. Jpn. J. Appl. Phys. **55**, 05FJ131–05FJ134 (2016)
13. Rabinovich, O., Sushkov, V.: The study of specific features of working characteristics of multicomponent heterostructures and AlGaInN – based light-emitting diodes. Semiconductors **43**(4), 524–527 (2009)
14. Urchuk, S., Legotin, S., Osipov, Yu., Rabinovich, O.: Spectral sensitivity characteristics simulation for silicon p-i-n photodiode. J. Phys. Conf. Ser. **643**, 01206811–01206815 (2015)
15. Winston, D.: Physical Simulation of Optoelectronic Semiconductor Devices, 1st edn. Colorado University Press, Colorado (1999)

Multi-well Deconvolution for Well Test Interpretations

Ivan Vladimirovich Afanaskin

Federal State Institution "Scientific Research Institute for System Analysis of the Russian Academy of Science", 36, Nakhimovsky Avenue, Moscow 117218, Russia
ivan@afanaskin.ru

Abstract. An important problem of numerical flow simulation in the frame of an oil field development is the need to improve the quality and quantity of the input data, in particular, the data on cross-well formation properties. Well test is the most informative means of a reservoir borehole-to-borehole survey. When studying the cross-well formation properties based on the interpretation of continuous curves of pressure variations measured by telemetry sensors on down-hole pumps, the impact of adjacent wells and high noise contamination of the obtained data shall be taken into account. In an attempt to solve this issue this article studies multi-well deconvolution as a means to study all components of the pressure variations curve. The multi-well deconvolution can identify the relevant response to a change in the operating mode of a certain well and can process this response in traditional ways. The multi-well deconvolution method makes it possible to assess and account for noise impacts on the pressure curve. This approach also simplifies the curve processing, as the diagnosis of an interpreted formation model is done easier. This work proposes a new approach to building self-influence and influence functions: we propose to set them down as a sum of elementary time functions representing individual filtration modes of the formation. The well bore impact is represented exponentially; the bilinear flow – as a fourth root; the linear flow – as a second root; the radial flow – as a logarithm; the boundary effect – as a linear function. This way the self-influence and influence functions coefficients are set down linearly and Newton method can be used to define them. This approach was tested on two synthetic curves of the bottom-hole pressure obtained by modelling. The simulation and deconvolution curves of the bottom-hole pressure converged closely and this proved that the formation parameters selected for the modelling and retrieved during the self-influence and influence curves processing are close and this, consequently, proves high efficiency of the proposed approach. #COMESYSO1120.

Keywords: Influence function · Self-influence function · Multi-well deconvolution · Well test

R. Silhavy et al. (Eds.): CoMeSySo 2020, AISC 1295, pp. 134–147, 2020.
https://doi.org/10.1007/978-3-030-63319-6_13

1 Introduction

The numerical flow simulation of oil-bearing formations is not only used at the stage of the development planning, but also for the production monitoring and control. It has been proven that the flow simulation is most needed and relevant at mature oil fields because it helps identify unrecovered oil locations and find solutions to extract the remaining oil-in-place. The main problem of the numerical simulation is the lack of credible input data, in particular, on cross-well formation properties.

The most informative methods for a reservoir borehole-to-borehole survey are well tests (by bottom-hole pressure build-up method and well interference method in particular). But unfortunately, the share of such surveys among all survey activities performed in Russia is small, as they require wells shut-in, and the operators tend to avoid this due to economical reasons.

In some cases of insufficient data the analysis of the bottom-hole pressure measured by the telemetry sensors on the down-hole pumps may come in useful. Using such sensors together with electric submersible pumps is rather common nowadays. But in practice, the quality of such input data still causes multiple problems. And thus it is impractical to rely on the telemetry sensors on the down-hole pumps as on sound and stable sources of the data for further analysis. A recent issue with low sensitivity of these sensors has now been solved for the most part, and the sensors now measure borehole environment properties with a good degree of accuracy. But studying cross-well formation properties based on the interpretation of continuous curves of pressure variations, the impact of adjacent wells and high noise contamination of the obtained data shall also be taken into account. In response to the above-mentioned issues, this article studies the multi-well deconvolution as a means to study all components of the pressure variations curve and to extract a useful signal coming from adjacent wells.

The multi-well deconvolution can identify the relevant response to a change in the operating mode of a certain well and can process this response in traditional ways [1–3]. One of the advantages of this approach is that during extraction the response curve is not apparently referenced to any interpretation model, that means no filtration model is needed to extract the response.

Apart from that, the multi-well deconvolution method makes it possible to assess and account for noise impacts on the pressure curve. This approach also significantly simplifies the curve processing, as when influence and self-influence functions are known, the response to operation in any individual interfering well can be calculated separately with a constant equivalent production rate. This ensures a smooth diagnosis curve and, consequently, makes the diagnosis of the formation interpretation model easier and more reliable.

It shall be noted that use of the multi-well deconvolution method for well tests was studied in international [4–7] and Russian works [1, 8–10].

2 Basics of the Deconvolution Method for Flow Dynamics Reverse Problem Solution

Convolution is a mathematical operation on two functions (f and g) that produces a third function that can be viewed as a modified version of any of the initial functions:

$$f * g = w. \tag{1}$$

Usually, w is the recorded signal and f is the signal that needs to be extracted. It is known that w signal was obtained by convolution of f signal with some known g signal. If signal g is unknown, it needs to be estimated. The convolution operation can be defined as a similarity of one function with reversed and shifted copy of the other. In other words, when f function is convoluted with g function, in fact, a multitude of shifted copies of f function weighted by g is summed. In fact, the convolution is a type of integral transformation. It can be expressed as follows:

$$w(x) = f(x) * g(x) = \int_{-\infty}^{+\infty} f(x-y)g(y)dy = \int_{-\infty}^{+\infty} f(y)g(x-y)dy. \tag{2}$$

A discrete view of Formula (2) can be expressed as follows:

$$w(x) = \sum_{y=0}^{\infty} f(x-y)g(y) = \sum_{y=0}^{\infty} f(y)g(x-y). \tag{3}$$

Applying the convolution for well tests, let us write a formula for the fluid inflow (considering unsteady filtration of weakly compressible fluid in elastic formation) during one well operation [4–8]:

$$P_w(t) = P_0 - q(t) * g(t) = P_0 - \int_0^t q(\tau)g(t-\tau)d\tau = P_0 - \int_0^t q(t-\tau)g(\tau)d\tau, \tag{4}$$

where P_w is bottom-hole pressure, P_0 is formation pressure at the time $t = 0$, q is well production rate, g is a self-influence function (i.e. the impact of the well on itself).

The deconvolution is a mathematical operation reverse to the signal convolution. The deconvolution aims to find a solution to a convolution equation. In terms of subsoil flow fluid dynamics, the definition of self-influence function g based on the known measurements of the bottom-hole pressure P_w and production rate q is called deconvolution.

Self-influence function g depends on the formation model and well model. In case of a vertical well in an infinite formation the discrete representation of fluid inflow Eq. (3) is as in [2]:

$$P_w(t) = P_0 - 21,5\frac{B\mu}{kh}\left\{\sum_{j=1}^{N}\left[-(q_j - q_{j-1})\lg(t - t_{j-1})\right] + q_N\left[\lg\left(\frac{k}{\phi\mu c_t r_w^2}\right) - 3,0923 + 0,8686S\right]\right\} \tag{5}$$

where pressure P_w and P_0 are measured in $*10^{-1}$ MPa, and production rate q – in m^3/day; B is a fluid volume factor, m^3/m^3; μ – fluid dynamic viscosity, mPa · sec; k is formation permeability, $*10^{-3}$ μm^2; h is net pay, m; j – a number of well operation mode; N – quantity of well operating modes by the moment of time t; t – time passed from the well operation start, hour; ϕ –porosity, df; c_t – total formation and fluid compressibility, $*10^{-1}$ 1/MPa; r_w – well-bore radius, m; S – well skin factor (the parameter used to assess the bottom hole area conditions), nondimensionalized.

In the case of multiple wells operating simultaneously and impacting each other the inflow equation after convolution will be expressed discretely in the following way:

$$P_{w,i}(t) = P_0 - q_i(t) * g_i(t) - \sum_{l=1}^{M} q_l(t) * g_{l,i}(t),\ l \neq i \tag{6}$$

the integral expression will be:

$$P_{w,i}(t) = P_0 - \int_0^t q_i(\tau)g_i(t-\tau)d\tau - \sum_{l=1}^{M}\int_0^t q_l(\tau)g_{l,i}(t-\tau)d\tau,\ l \neq i, \tag{7}$$

where i index designates the well under test, l index designates the adjacent wells impacting the well under test, M is the number of wells, g_i is a self-influence function of i well, $g_{l,i}$ is a function of l well influence on i well.

In works [4, 6, 7], expression (6) used to determine the self-influence and influence functions is expressed in a matrix-vector form. Yet in this work we propose to express g_i and $g_{l,i}$ functions as a sum of elementary time functions, defining individual filtration modes in a formation. For example, the well bore impact is represented exponentially; the bilinear flow – as a fourth root; the linear flow – as a second root; the radial flow – as a logarithm; the boundary effect – as a linear function.

$$
\begin{aligned}
P_{w,i}(t) = P_0 &+ a_i \sum_{j=1}^{N} \left[-\left(q_{i,j} - q_{i,j-1}\right) \exp\left(t - t_{j-1}\right)\right] + b_i \sum_{j=1}^{N} \left[-\left(q_{i,j} - q_{i,j-1}\right) \sqrt{4t - t_{j-1}}\right] \\
&+ c_i \sum_{j=1}^{N} \left[-\left(q_{i,j} - q_{i,j-1}\right) \sqrt{t - t_{j-1}}\right] + d_i \sum_{j=1}^{N} \left[-\left(q_{i,j} - q_{i,j-1}\right) \lg\left(t - t_{j-1}\right)\right] \\
&+ e_i \sum_{j=1}^{N} \left[-\left(q_{i,j} - q_{i,j-1}\right)\left(t - t_{j-1}\right)\right] + f_i q_{i,N} \\
&+ \sum_{l=1}^{M} \left\{ a_{l,i} \sum_{j=1}^{N} \left[-\left(q_{l,j} - q_{l,j-1}\right) \exp\left(t - t_{j-1}\right)\right] + b_{l,i} \sum_{j=1}^{N} \left[-\left(q_{l,j} - q_{l,j-1}\right) \sqrt{4t - t_{j-1}}\right] \right. \\
&+ c_{l,i} \sum_{j=1}^{N} \left[-\left(q_{l,j} - q_{l,j-1}\right) \sqrt{t - t_{j-1}}\right] + d_{l,i} \sum_{j=1}^{N} \left[-\left(q_{l,j} - q_{l,j-1}\right) \lg\left(t - t_{j-1}\right)\right] \\
&\left. + e_{l,i} \sum_{j=1}^{N} \left[-\left(q_{l,j} - q_{l,j-1}\right)\left(t - t_{j-1}\right)\right] + f_{l,i} q_{l,N} \right\}, \; l \neq i,
\end{aligned}
\tag{8}
$$

where indexes a_i, b_i, c_i, d_i, e_i, f_i, $a_{l,i}$, $b_{l,i}$, $c_{l,i}$, $d_{l,i}$, $e_{l,i}$, $f_{l,i}$ are the model parameters. The bottom-hole pressure of the well under test and production rates of all wells are known. The best-match method can be used to find the above-mentioned indexes, i.e. to deconvolute the bottom-hole pressure curve. When all indexes are known, we can produce the self-influence of i well and the influence of adjacent wells on it. We can further process the produced individual curves of pressure variations using traditional methods and standard software to interpret well test results (for example Topaze и Saphir Kappa Engineering [3]). By processing these curves we can define filtration and volumetric properties of near-wellbore and cross-well environment properties. The above-mentioned coefficients of influence and self-influence functions are expressed linearly in Eq. (8), so we can use Newton method to define them. As production rates are often measured with a rather big inaccuracy, the deconvolution method might require slight change (modification) of the production rate, so that a good match of the calculated and actual curves could be achieved. In this case, the higher priority is given to the minimal production rate modifications. Then the minimised functional can be expressed as follows:

$$
\alpha \sum_{n=1}^{NM} \left(P_{w,i,n}^{m} - P_{w,i,n}^{c}\right)^2 + \beta \sum_{n=1}^{NM} \left(q_{i,n}^{m} - q_{i,n}^{c}\right)^2 + \gamma \sum_{l=1}^{M} \sum_{n=1}^{NM} \left(q_{l,n}^{m} - q_{l,n}^{c}\right)^2 \to 0, \; l \neq i,
\tag{9}
$$

where the lower index n is the ordinal number of the measurement, NM is the quantity of the measurements, superfix m is the measured value, superfix c is the calculated value (for the bottom-hole pressure P_w) and the modified value (for production rate q), α, β and γ are weight indexes.

3 Applying the Multi-well Deconvolution

3.1 Example 1. Infinite Formation

Let us consider the multi-well deconvolution for the bottom-hole pressure curve of an operating production well. For our analysis, we shall take the simulated pressure curve, and the obtained results will further be compared with the formation parameters used in the well flow simulation. The well flow simulation of the pressure curve was executed in Saphir program by Kappa Engineering [3].

For this simulation, we considered 3 vertical producing wells operating without interruptions with a variable production rate in a homogeneous infinite formation. The wells pattern is shown in Fig. 1. The following initial parameters were taken into account: well-bore radii 0.1 m; formation net pay 10 m; porosity 0.1 df.; oil volume factor 1.1 m^3/m^3; oil viscosity 1 mPa • sec; total formation and fluid compressibility $4.3 \cdot 10^{-4}$ 1/MPa; well skin factors 0, nondimensionalized; initial formation pressure 30 MPa; permeability $30 * 10^{-3}$ μm^2. Dynamics of the variable production rates is shown in Fig. 2.

Figure 2 shows the actual bottom-hole pressure curve of the Tested Well (TW). Well 1 and Well 2 are influencing the TW. Figure 2 demonstrates that the deconvolution curve almost completely converges with the actual one. RMS deviation of the curves is $3.85 \cdot 10^{-4}$ MPa.

Figure 2 also shows the response of the TW to its own operation and the response of the TW to a variation in the operation of Well 1 and Well 2. All the three pressure variation curves were processed by the best-match method in Topaze program (pressure decrease curve, Fig. 3) and Saphir program (interference curves, Fig. 4 and 5) by Kappa Engineering [3]. The results of these curves interpretation are shown in Table 1. It demonstrates a good match of the obtained and initial (considered for simulation) parameters.

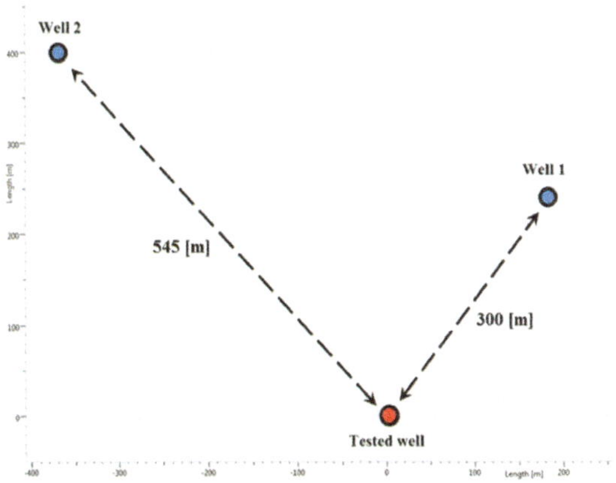

Fig. 1. Wells pattern.

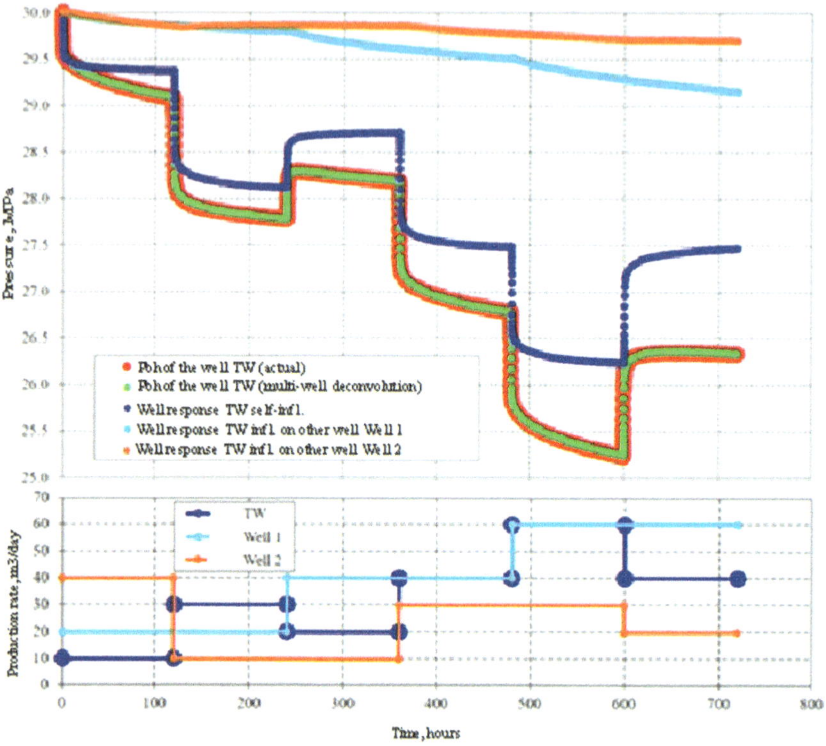

Fig. 2. Pressure, production rate and response curves. Example 1. Infinite Formation.

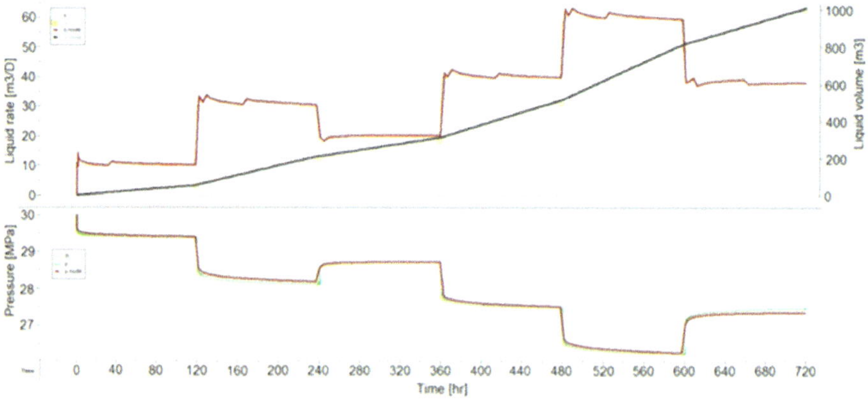

Fig. 3. Processing by the best match method of the TW pressure decrease curve.

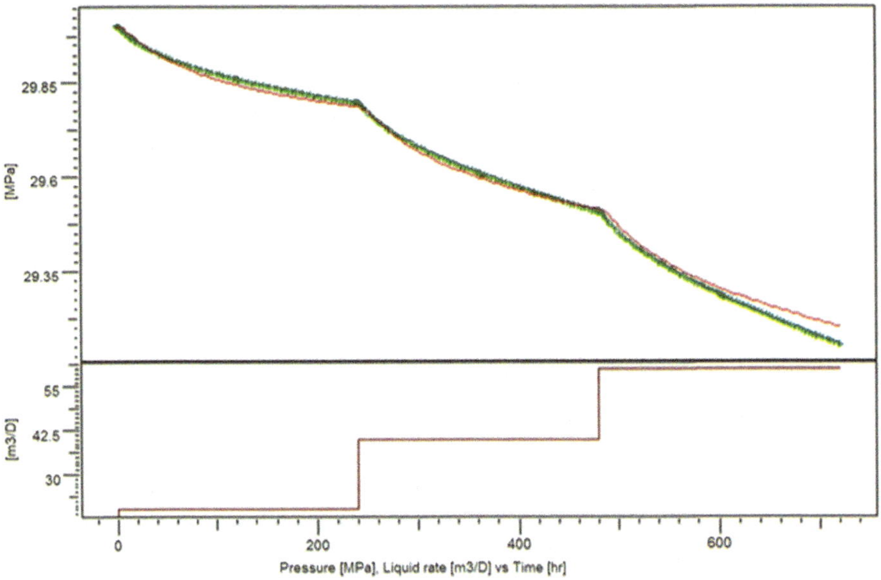

Fig. 4. Processing by the curve best match method of the TW and Well 1 interference test results.

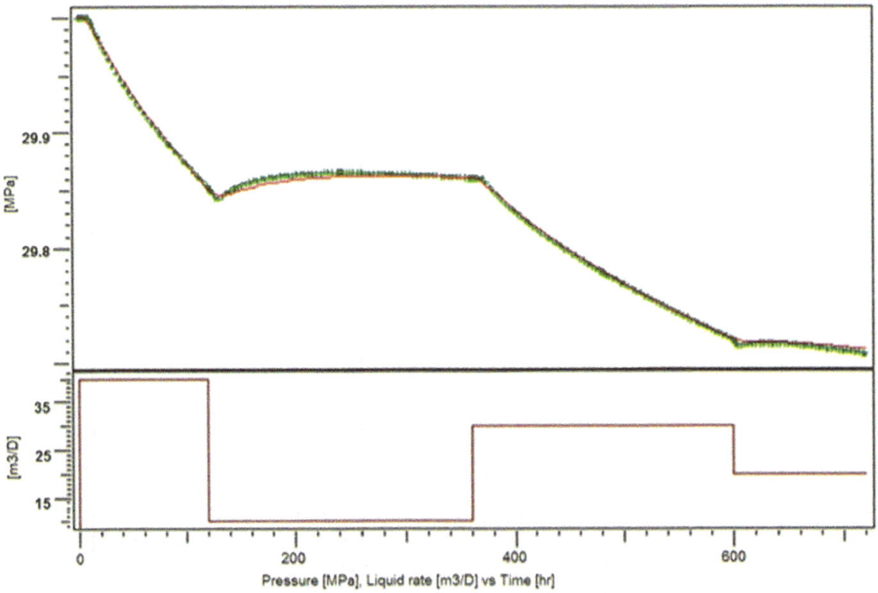

Fig. 5. Processing by the curve best match method of the TW and Well 2 interference test results.

Table 1. Results of the survey interpretation. Example 1. Infinite Formation.

Parameter	Actual value	Pressure decrease curve, TW	Well interference test, Well 1 → TW	Well interference test, Well 2 → TW
Permeability, 10^{-3} μm^2	30	28	22	18
Porosity, df	0.1	–	0.12	0.15
Well skin factors, nondimensionalized	0	−1	–	

3.2 Example 2. Finite Formation

Let us consider the multi-well deconvolution for the analysis of the bottom-hole pressure readings in an operating production well of a finite formation. To assess the results correctly we shall take a simulated pressure curve, as we know the filtration and volumetric properties of the formation that were considered for the model creation. We shall use Saphir program by Kappa Engineering [3] to obtain the simulated curve.

Three wells are considered in the numerical experiment. All the three wells are producing vertical wells operating without interruptions with a variable production rate. Well layouts and the distances between them are shown in Fig. 6. The formation is a homogenous closed system (of a rectangular shape). The following initial parameters were taken into account: well-bores radii 0.1 m; formation net pay 10 m; porosity 0.1 df.; oil volume factor 1.1 m^3/m^3; oil viscosity 1 mPa • sec; total formation and fluid

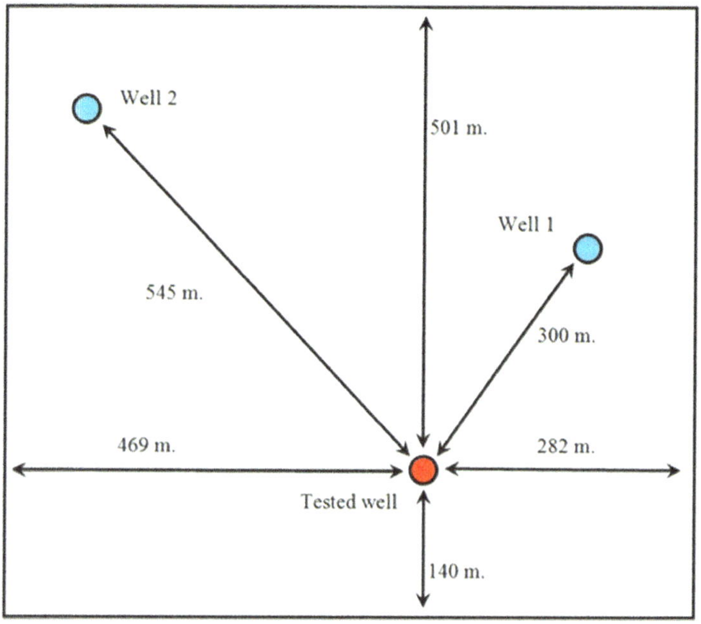

Fig. 6. Map of well layouts.

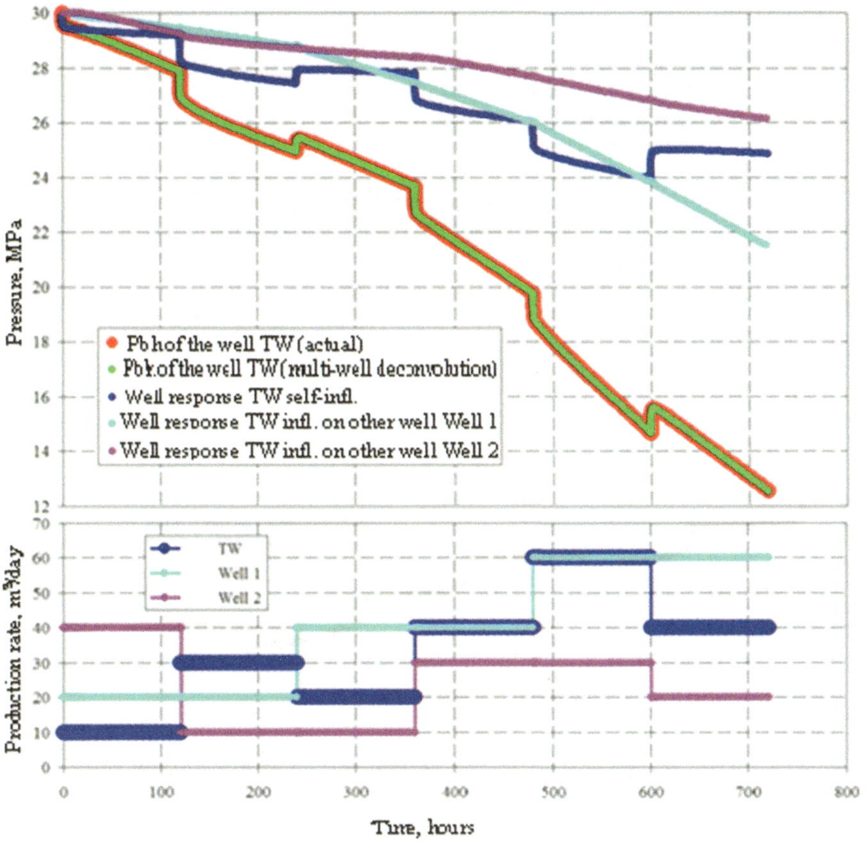

Fig. 7. Pressure, production rate and response curves. Example 2. Finite Formation.

compressibility $4.3 \cdot 10^{-4}$ 1/MPa; well skin factors 0, nondimensionalized; initial formation pressure 30 MPa; permeability $30 * 10^{-3}$ μm^2. Figure 7 shows the production rates of the wells.

For brevity and convenience, the curve obtained by the mathematical model and which we will further study will be called actual. We will study the readings of the bottom-hole pressure in the Tested Well (TW in the Figures). The Tested Well interferes with Well 1 and Well 2. The proposed method of multi-well deconvolution produces a good convergence of actual and calculated curves of the Tested Well bottom-hole pressure (Fig. 7). The RMS deviation of 0.919 MPa for 343 points on the bottom-hole pressure curve was obtained (i.e. less than $2.68 \cdot 10^{-3}$ MPa per point).

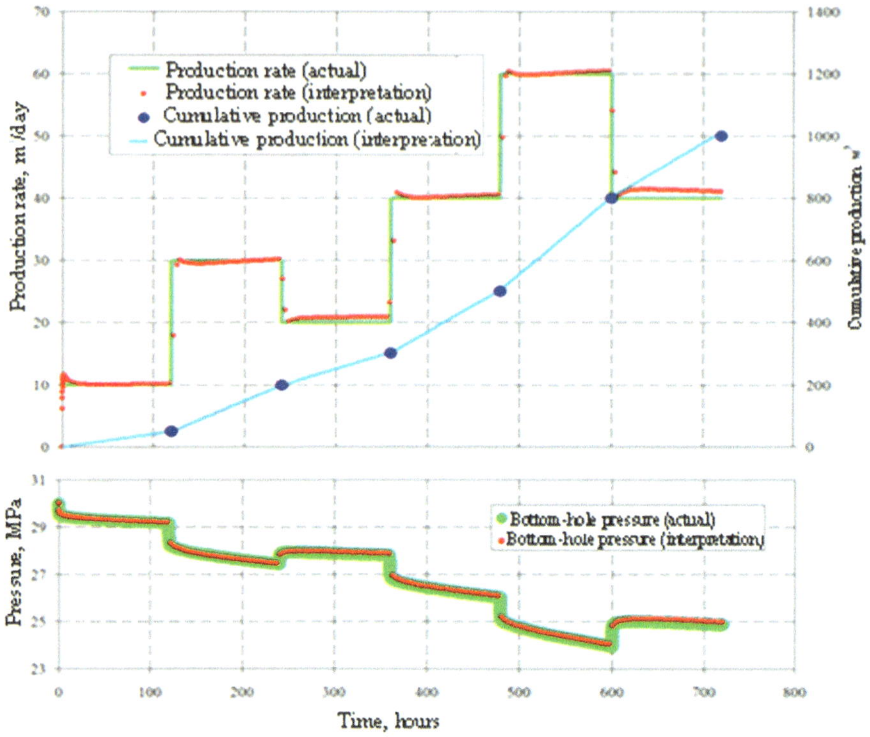

Fig. 8. Processing by the best match method of the Tested Well pressure decrease curve.

Figure 7 also shows the response of the Tested Well to its own operation (the term 'pressure decrease curve' is used in well surveys) and the Tested Well response to the operation of Well 1 and Well 2 (interference curves). These three bottom-hole pressure curves were interpreted by the best-match method in Topaze program (pressure decrease curve, Fig. 8) and Saphir program (interference curves, Fig. 9–10) by Kappa Engineering [3]. The processing results are shown in Table 2. The parameters considered for simulation and obtained by interpretation show good repeatability.

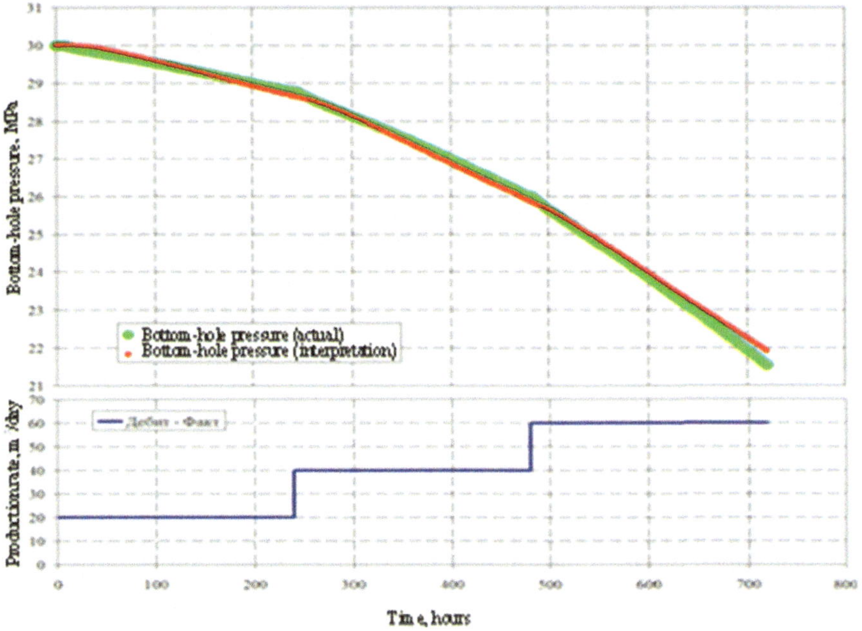

Fig. 9. Processing by the curve best match method of Tested Well and Well 1 inter-well space interference test results.

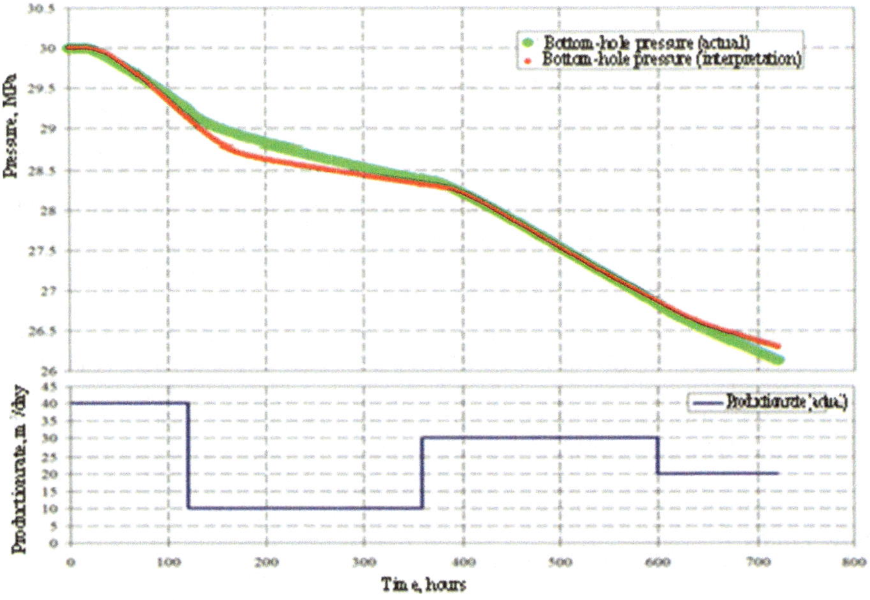

Fig. 10. Processing by the curve best match method of Tested Well and Well 2 inter-well space interference test results.

Table 2. Results of the survey interpretation. Example 2. Finite Formation.

Parameter	Actual value	Pressure decrease curve, TW	Well interference test, Well 1 → TW	Well interference test, Well 2 → TW
Distance to boundaries, m	140; 282; 469; 501	108; 360; 557	85; 110; angle 90°	566; 952; angle 90°
Permeability, 10^{-3} μm^2	30	22	15.7	18.7
Porosity, df	0.1	–	0.18	0.16
Well skin factors, nondimensionalized	0	−1.1	–	–

4 Conclusion

This work analyses the possibility to use the deconvolution for well tests. It proposes a new approach to building self-influence and influence functions: we propose to set them down as a sum of the elementary time functions representing individual filtration modes of the formation (the well bore impact is represented exponentially; the bilinear flow – as a fourth root; the linear flow – as a second root; the radial flow – as a logarithm; the boundary effect – as a linear function). The self-influence and influence functions coefficients are set down linearly and Newton method can be used to define them. This approach was tried out on the analysis of the bottom-hole pressure curves extracted by simulating of two cases: infinite formation and finite rectangular formation. The simulation and deconvolution curves of the bottom-hole pressure converged closely and this proved that the formation parameters selected for modelling and retrieved during the self-influence and influence curves processing are close and this, consequently, proves high efficiency of the proposed approach.

This work was performed as part of the state assignment by SRISA RAS for 2020 in the frame of a project called Elaboration of the Method to Identify Unrecovered Oil Locations In Oilfields and Calculate the Remaining Oil In-Place by Fusion Mathematical Modeling, Development Analysis Based on Well and Formation Surveys.

Acknowledgements. Research conducted with support of state program for SRISA "Fundamental science research (47GP), theme №0065-2019-0019 "Non-developed zones identifications of oil fields and remaining reserves evaluation which is based on complexing of mathematic modelling, field development analysis and reservoir surveillance"" (reg # AAAA-A19-119020190071-7).

References

1. Businov, S.N., Umrikhin, I.D.: Surveys of oil and gas wells and formations. Nedra, Moscow, Russia (1984)

2. Earlougher Jr., R.: Well tests methods. Institute of Computes Sciences, Izhevsk, Russia (2007)
3. Houze, O., Viturat, D., Fjaere, O.S.: Dynamic Data Analysis. Kappa Engineering, Paris, France (2017)
4. Cumming, J.A., Wooff, D.A., Whittle, T., Gringarten, A.C.: Multiwell deconvolution. SPE Reservoir Eval. Eng. **17**(04), 457–465 (2014)
5. Gringarten, A.C.: New Development in Well Test Analysis. Phase 2. Imperial College London, UK (2018)
6. Zheng, S.-Y., Wang, F.: Multi-Well Deconvolution Algorithm for the Diagnostick, Analysis of Transient Pressure With Interference From Permanent Down-hole Gauges. SPE, 121949 (2009)
7. Wang, F.: Processing and analysis of transient pressure from permanent down-hole gauges. Submitted for the Degree of Doctor of Philosophy. Heriot-Watt University, Edinburgh, UK, p. 235 (2010)
8. Guliaev, D.N., Batmanova, O.V.: Pulse-code well interference test and algorithms of multi-well deconvolution – new technologies of cross-well formation properties definition. Vestnik of Russian New University, Series: Complex Systems: Models, Analysis, Control, vol. 4, pp. 26–32 (2017)
9. Krichevsky, V.S.: Well surveys - way to incremental oil production. https://sofoil.com/MRT%20report.pdf. Accessed 08 May 2020
10. SOFOIL: Multiwell well testing. Technology Overview. https://docplayer.ru/79765531-Multiskvazhinnye-gdi-tehnologicheskiy-obzor.html. Accessed 08 May 2020

Collaborative Filtering Recommendation Systems Algorithms, Strengths and Open Issues

Lefats'e Manamolela[1], Tranos Zuva[2], and Martin Appiah[2](✉)

[1] Department of Information and Communication Technology, Vaal University of Technology, Vanderbijlpark, South Africa
lefatse@yahoo.com
[2] Vaal University of Technology, Vanderbijlpark, South Africa
martina@vut.ac.za

Abstract. Recommendation systems recommender systems are a subcategory of information filtering that is utilized to determine the preferences of users towards certain items. These systems emerged in the 1990's and they have since changed the intelligence of both the web and humans. Vast amounts of research papers have been published in various domains. Recommendation systems suggest items to users and their principal purpose is to recommend items that are predicted to be suitable for users. Some of the most popular domains where recommendation systems are used include movies, music, jokes, restaurants, financial services, life insurance, Instagram Facebook and twitter followers. This paper explores different collaborative filtering algorithms. In so doing, the paper looks at the strengths and challenges (open issues) faced by this technique. The open issues give direction of future research work to researchers and also provide information of where to use collaborative filtering recommender systems applications.

Keywords: Recommender system · Collaborative filtering · Matrix factorization

1 Introduction

As one of the most successful approaches to building recommendation systems, Collaborative Filtering (CF) uses the known preferences of a group of users to make recommendations or predictions of the unknown preferences for other users [1]. Such users build a group called neighbourhood. A user gets recommendations to those items that he has not rated before but that were already positively rated by users in his neighbourhood. Recommendations that are produced by CF can be of either prediction or recommendation. Prediction is a numerical value, expressing the predicted score of an item for the user, while Recommendation is a list of top N items that the user will like the most [2, 3].

This paper is arranged as follows: Sect. 2 Collaborative Filtering Algorithms, Sect. 3 Strengths of CF, Sect. 4 Challenges of CF and then finally the conclusion.

R. Silhavy et al. (Eds.): CoMeSySo 2020, AISC 1295, pp. 148–163, 2020.
https://doi.org/10.1007/978-3-030-63319-6_14

2 Collaborative Filtering Systems Algorithms

Based on the method of implementation, recommendation systems generally can be divided into two categories [4, 5], memory-based and model-based as shown by Fig. 1. Memory-based method performs recommendation by accessing the database directly, while model-based method uses the transaction data to create a model that can generate recommendation [6]. By accessing directly to database, memory-based method is adaptive to data changes, but requires large computational time according to the data size. As for model-based method, it has a constant computing time regardless the size of the data but not adaptive to data changes.

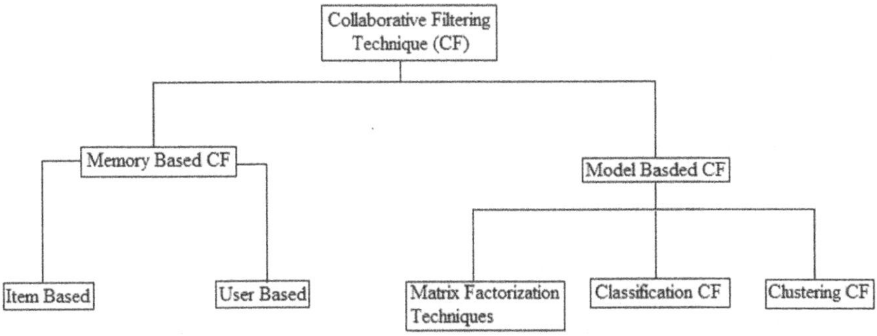

Fig. 1. Collaborative filtering techniques adapted from [7].

2.1 Memory Based Collaborative Filtering

There is no denying that people generally trust recommendations from like-minded friends. Memory-based collaborative filtering applies a neighbor-like pattern to estimate a user's ratings based on the ratings given by like-minded users [8]. The prediction process of memory-based CF typically comprises of three steps: user similarity measurement, neighborhood selection, and estimation generation. This filtering method can be divided into two major categories namely user-item and item-item. User-item approach takes a certain user and searches for other users who have similar rating patterns and suggest items that those similar users have liked. Item-item approach takes an item, search for users who liked that item, and then search for other items that those users or similar users liked. It takes items and output other items as suggestions [9]. For example, in the user based approaches, the value of ratings user u gives to item i is calculated as an aggregation of some similar user's rating of the item. [10] used Eq. 1 for this aggregation.

$$r_{u,i} = aggr_{u' \in U}\, r_{u',i} \tag{1}$$

where U denotes the set of top N users that are most similar to user u who rated item i. Some examples of the aggregation functions include Eqs. 2, 3, 4 and 5 [10, 11].

$$r_{u,i} = \frac{1}{N} \sum_{u' \in U} r_{u',i} \tag{2}$$

$$r_{u,i} = k \sum_{u' \in U} simil(u, u') r_{u',i} \tag{3}$$

where k is a normalizing factor defined as

$$k = 1 / \sum_{u' \in U} |simil(u, u')|, \tag{4}$$

and

$$r_{u,i} = \overline{r_u} + k \sum_{u' \in U} simil(u, u')(r_{u',i} - \overline{r_{u'}}) \tag{5}$$

where $\overline{r_u}$ is the average rating of user u for all the items rated u.

Item-Based CF. Similarity computation in a recommendation system involves identification of the existing users who have the similar tastes to the current user. This is done by using the existing user's reviews. The preferences of the current user are also taken and are compared to the existing user's preferences. Item-based Collaborative Filtering which is also known as Item-Item CF was first introduced by Sarwar and the crew [12]. The technique is based on the assumption that if two items are rated similarly by similar people, those items are similar to each other. Instead of calculating similarities between users' rating behaviour to predict preferences, this technique makes use of the similarities between the rating patterns of items.

The technique was introduced after realizing that although User-based Collaborative Filtering is effective to some extent, it is crippled by the growth of the user base. As an effort to address this issue, the research efforts gave birth to Item-based CF. It was important to extend the Collaborative Filtering technique to cover large user bases so that it can easily be deployed into e-Commerce sites.

Although Item-based Collaborative Filtering may seem inadequate in its raw form, basically because still similarity between items has to be computed (k-NN problem), it complements itself by pre-computing the similarity matrix. User-based CF calculates the neighbourhood when predictions or recommendations are needed. In significantly large User base systems, it is wise to compute similarities based on items because even if one user can decide to add or change the rating, it would not significantly affect or change the similarity between the items especially if the items have many ratings. It is therefore important to pre-compute the similarities between items in an Item-based similarity matrix.

In general Item-based Collaborative Filtering computes the recommendations based on the user's own ratings for other items coalesced with those items' similarity to the target item, instead of other users' ratings and user similarities as in User-based Collaborative Filtering. Nevertheless, the similarities can be calculated by using the same equations mentioned earlier on.

Some of the most popular algorithms used are cosine based similarity, correlation based similarity and adjusted-cosine similarity. The formula for Adjusted-based cosine which is the most popular and believed to be the most accurate [13] is shown by Eq. 6 as used by [14].

$$Itemsim(i,j) = \frac{\sum_{u \in U_{i,j}} \left(R_{u,i} - \overline{R}_u\right)\left(R_{u,j} - \overline{R}_u\right)}{\sqrt{\sum_{u \in U_{i,j}} \left(R_{u,i} - \overline{R}_u\right)^2} \sqrt{\sum_{u \in U_{i,j}} \left(R_{u,j} - \overline{R}_u\right)^2}} \tag{6}$$

where $R_{u,i}$ and $R_{u,j}$ represents the rating of user u on items i and j respectively, $\overline{R}_u$ is the mean of the uth user's ratings and $U_{i,j}$ represents all users who have rated items i and j.

The prediction calculation for item based nearest neighbour algorithm for user u and item j as carried out by [15] is shown by Eq. 7.

$$\underset{item\_based}{P}(u_t, j) = \frac{\sum_{i \in R_{u_t}} Itemsim(i,j) * R_{u_t,j}}{\sum_{i \in R_{u_t}} Itemsim(i,j)} \tag{7}$$

If the predicted rating is high then the system recommends the item to user. The item-based nearest neighbour algorithms are more accurate in predicting ratings than user based nearest neighbour algorithms [13].

User-Based CF. User-based Collaborative Filtering is also known as User-User Collaborative Filtering or k-NN [16] and it is one of the first automated CF methods. It is the one which illustrates the interpretation of the core premise of Collaborative Filtering. This algorithm is based on the assumption that it is highly likely that a user would be interested in what his\her friends showed interest in. So it simply recommends items which were liked by the user's neighbours\friends as illustrated Saptono [17] on Eqs. 8 and 9.

$$sim(u,u') = \frac{\sum_{i \in I(u,u')} R(u,i) \cdot R(u',i)}{\sqrt{\sum_{i \in I(u,u')} R(u,i)^2} \sqrt{\sum_{i \in I(u,u')} R(u',i)^2}} \tag{8}$$

where $I(u,u')$ represents the set of all items rated by both user u' and user u. From this similarity calculation, a set of all neighbours of u is formed $N(u)$. The size of this set can vary depending on the overall expected results. Then $R*(u,i)$ is calculated as the adjusted weighted sum of all known ratings $R(u',i)$, and $u' \in N(u)$ [18].

$$R^*(u,i) = \overline{R(u)} + \frac{\sum_{u \in N(u)} sim(u,u') \cdot \left(R(u',i) - \overline{R(u')}\right)}{\sum_{u \in N(u)} |sim(u,u')|} \tag{9}$$

where $\overline{R(u)}$ represents the average rating of user u.

Figure 2 shows examples of calculating the similarity of two users, u and w, compared with calculating the similarity of two items, i and j, in the process of predicting the value of item i for user u in the user–item matrix.

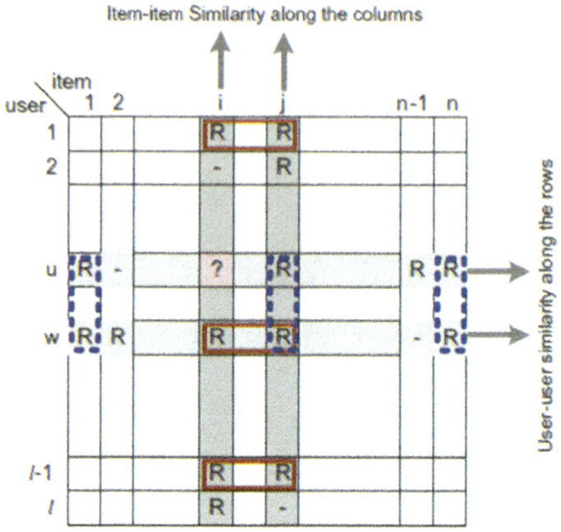

Fig. 2. Similarity computation in a user-based CF and item-based CF [19].

Usually, CF systems take two steps: first, the neighbor group, the users who have a similar preference with the target user (for user-based CF) or the set of items that is similar to the item selected by the target user (for an item-based CF), should be determined by using a variety of similarity computing methods. Based on the group of neighbors, the prediction values of particular items, estimating how the target user is likely to prefer the items, are obtained and then the top-N items with a higher predicted value that will be of interest to the target user are identified.

2.2 Model Based Collaborative Filtering

The main drawback of memory-based technique is the requirement of loading a large amount of in-line memory. The problem is serious when rating matrix becomes so huge in situation that there are extremely many persons using system [20]. Computational resource is consumed much and system performance goes down; so system can't respond user request immediately. Model-based approach intends to solve such problems. In general, latent factor models offer high expressive ability to describe various aspects of the data. Thus, they tend to provide more accurate results than neighborhood models. There are three common approaches for model-based CF such as matrix factorization, classification and clustering [21, 22].

Matrix Factorization. This technique has been adopted from the numerical linear algebra. It is now used widely in recommendation systems due to its capability to improve recommendation accuracy as Adomavicius and A. Tuzhilin, 2012 clearly state. It has been proved to be a better option to address the issues of data sparsity, over-fitting and convergence speed. Most of the best performing algorithms incorporates this technique. It was evident in the earlier mentioned Netflix Prize competition [22] where most of the algorithms presented incorporated technique.

The basic version of the technique relies on the assumption that the user's preference or rating of an item is composed of sum of preferences about various features of that item. The model is stimulated by Singular Value Decomposition (SVD). If users' ratings on items can be represented in the form of matrix M, the SVD of that matrix is the factorization into three component matrices in such a way that $M = U\sum T^T$, Σ represents a diagonal matrix with singular values σ_i of the decomposition. U and T are orthogonal (their determinants are either 1 or -1. U should not be confused with the set of users in this context). This introduces the intermediate vector space represented by Σ. The vectors are transformed from item-space into the intermediate vector space by $\sum T^T$. Classically U is m $\times$ k, Σ is $k \times k$ and M is $m \times n$. M has *rank k* where k is a smaller number representing the reduced dimensional of the rating space. The closest approximation to M is achieved by truncating Σ to $\sum_k$ by retaining only the k largest singular values. The best possible *rank k* approximation can be achieved by using the Frobenius norm to measure the error [23].

There are two more things achieved by the truncation; the decrease of the vector space dimensionality which in turn minimizes the storage and computational requirement of the model, and also the reduction of singular values eliminates the redundant noise and leaves only the strongest effects or trends in the model [16]. This helps to provide high quality recommendations. The computation of the SVD of the ratings matrix yields the following factorization: $R \approx U\sum T^T$ if $m = |U|, n = |I|$, *and* $\sum$ is a $k \times k$ *matrix.*

The feature preference-relevant model is associated with the computation of *rank-k* SVD of $R \approx U\sum T^T$ where by the rows of matrix U are perceived to represent the user's interest in each of the k features and the rows of I are the items relevance for each feature. The singular values in Σ are taken to be preference weights which represent the influence of a particular feature on user-item preferences across the system. A user's preference for a particular item is therefore computed as the weighted sum of the user's interest in each of the features of the item multiplied by the item's relevance to the features.

It is therefore necessary to compute the matrix factorization first in order to use SVD. Singular Value Decomposition can be computed in numerous ways, there are a lot of algorithms like Lanczo's algorithm, the generalized Hebbian algorithm and expectation maximization [24–26].

There must be a dummy data to fill in the missing values of the rating matrix in order to well define the Singular Value Decomposition. This dummy data has to be reasonable, therefore it is computed by taking the item's average rating [12]. Nevertheless there are several methods proposed which can estimate the SVD irrespective of the missing ratings [25]. The most popular method is the gradient descent method. This method trains each feature f in turn using update rules as illustrated Miller, Konstan and Riedl [27] in Eqs. 10 and 11.

$$\Delta u_{j,f} = \lambda(R(u,i) - R*(u,i))i_{k,f} \tag{10}$$

$$\Delta u_{k,f} = \lambda(R(u,i) - R*(u,i))i_{k,f} \tag{11}$$

where λ is the learning rate which is typically 0.001. This method also prevents over fitting by allowing regularization (subtracting constant factor to minimise the variance of predicted regression parameters) [28]. Although the resulting model does not reflect a true SVD due to the fact that the constituent matrices are no longer orthogonal, it tends to be more accurate in predicting latent preferences than the SVD which is not regularized [22]. Miller and the crew [27] added additional term to Eqs. 10 and 11 in an attempt to regularize. This is shown in Eqs. 12 and 13.

$$\Delta u_{j,f} = \lambda((R(u,i) - R*(u,i))i_{k,f} - \lambda u_{k,f}) \tag{12}$$

$$\Delta u_{k,f} = \lambda((R(u,i) - R*(u,i))i_{j,f} - \lambda u_{k,f}) \tag{13}$$

where λ is the regularization factor which is typically 0.1–0.2. The ratings can also be normalized by subtracting the user's average rating or any other baseline predictor before computing the SVD. This may improve the accuracy and induce convergence of iterative methods [29, 30].

After the computation of SVD, it is essential to update it to show new users, items and ratings. This is achieved by employing a commonly known method called *folding-in*. This method practically works well and shows the recommendation of users who were not even considered during the factorization of the rating matrix [31, 32]. It computes a feature relevance vector for the new user or item but it does not re-compute the decomposition itself.

Folding-in computes the feature interest vector u for user u in such a way that $\pi_u \approx u\Sigma T^T$. If r_u is taken to be the rating vector, u is calculated as $u = (\Sigma T^T)^{-1} r_u = T\Sigma^{-1} r_u$. The *folding-in* method ignores the ratings which are zeros. The process is symmetrical; by substituting U for T we get item vectors.

It is necessary to re-compute the complete factorization periodically due to the fact that the accuracy of the SVD deteriorates with time as *folding-in* process updates the user and item vectors. The process can be performed off line in deployed systems when there is less load [32].

Another method for building and maintaining the SVD based on *rank-1*updates was proposed by Brand [33]. This method produced faster real-time updates of the SVD. It bootstraps the SVD with the dense portion of the dataset. Users and items are sorted to make a dense corner in the matrix and this dense portion is extracted from that corner.

The weighted dot product of user-feature preference vector $\boldsymbol{u}$ and the item-feature relevance vector $\boldsymbol{i}$ are perceived to represent the user u's preference for item i as Miller and the company [27] demonstrates by Eq. 14.

$$R*(u,i) = \sum_f u_f \sigma_f i_f \tag{14}$$

Then items can be ranked according to their predicted preference and be recommended to the users.

Classification CF. Classification complications target to find common characteristics that specify the group to which each instance fits. This can be utilized both to recognize the existing data and to predict how new cases will perform. Data mining produces classification models by examining the data that is already classified and inductively discovering a predictive pattern. The existing cases may be derived from ancient databases. They may also be a result of an experiment in which a model of the entire database is tested in the real world and the fallouts used to design a classifier. Sometimes an expert is required to classify a sample of the database, and that sample is used to create the model which will be applied to the entire database [34]. Different classification algorithms are applied and used with different datasets, some of these algorithms are Multi-layer Perceptron, Artificial Neural Network, Logistic regression, JRip and J48.

A Multi-layer Perceptron (MLP) is a class of feedforward artificial neural network. MLP consists of at least three layers of nodes. Except for the input nodes, each node is a neuron that uses a nonlinear activation function. MLP utilizes a supervised learning technique called back-propagation for training. It can distinguish data that is not linearly *separable. Multilayer perceptrons are sometimes colloquially referred to as "vanilla" neural networks, especially when they have a single hidden layer* [35].

Artificial Neural Network (ANN) is a unified group of nodes by means of mathematical approaches to process information. It is a self-adaptive system, which can change its construction based on the internal or external influences. Multiple ANN models have been developed and the most prevalent one is the Multi-Layer Perceptron (MLP) feed forward network [36]. MLP consists of many layers. The most widely used structure was the three-layer structure, due to its capability to solve most image classification problems. The multiple layers include one input layer, one hidden layer, and one output layer as on shown in Fig. 3.

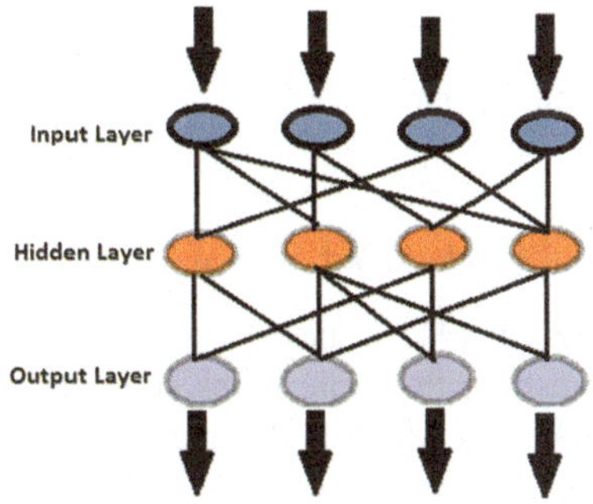

Fig. 3. Illustration of MLP nodes look like adapted [37].

Each layer is composed of artificial neurons. It is visible in Fig. 3 that all the nodes are linked with each other, excluding the nodes in the same layer. The input layer, the hidden layer and the output layer are used for data input, processing, and output, respectively. The downside of this algorithm is that creating a neural network is time consuming.

Logistic regression, by use of a linear combination of independent variables, is a statistical technique used to predict the possibility of occurrence of an event, i.e. its probability [38]. However, it is evident that the algorithm yields low accuracy with high-processing speed, indicating that classification using only a logistic algorithm cannot guarantee the accuracy of the results [38]. For this purpose, logistic algorithms need to be used in conjunction with other algorithms to validate the results.

JRip also known as (RIPPER) is one of the straightforward and most popular algorithms. It examines classes in increasing size and it generates an initial rule set using incremental reduced error. JRip proceeds by treating all the samples of a specific ruling in the training data as a class, thus discovering a set of rules that conceal all the members of the class. This process is iterated until all classes have been covered [34].

J48 is a tree classifier. A tree is moreover a leaf node labeled with a class, or a structure containing a test, then linked to two or more nodes also known as subtrees [39]. To classify some instance, first there is a need to identify its attribute-vector and apply this vector to the tree. The tests are done on the attributes, reaching one or other leaf, to complete the classification process, as illustrated in Fig. 4.

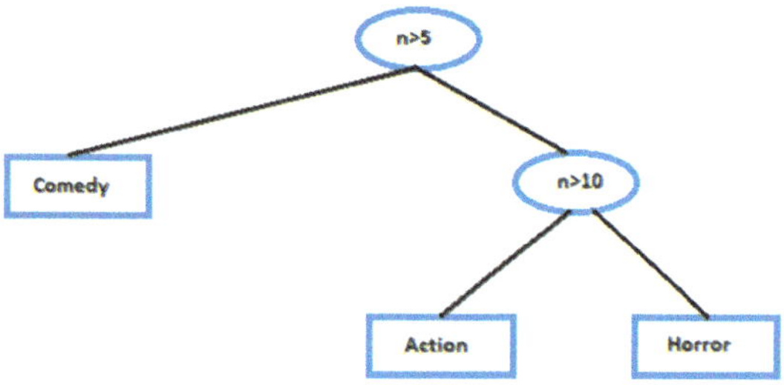

Fig. 4. Simple tree classifier classification process adapted from [40].

From the example on Fig. 4, let n = 5
Then n > 5 = true
n > 10 = true
therefore, the above will be classified as a comedy.

Clustering CF. Clustering CF [41] is based on assumption that users in the same group have the same interest; so they rate items similarly. Therefore users are partitioned into groups called clusters which is defined as a set of similar users.

Suppose each user is represented as rating vector denoted $u_i = (r_{i1}, r_{i2}, \ldots, r_{in})$. The dissimilarity measure between two users is the distance between them. Minkowski distance [42], Euclidian distance [43] or Manhattan distance [44] as shown in Eq. 15, 16 and 17 respectively may be used.

$$distance_{Minkowski}(u_1, u_2) = \sqrt[q]{\sum_j (r_{1j} - r_{2j})^q} \tag{15}$$

$$distance_{Euclidian}(u_1, u_2) = \sqrt{\sum_j (r_{1j} - r_{2j})^2} \tag{16}$$

$$distance_{Manhattan}(u_1, u_2) = \sum_j |r_{1j} - r_{2j}| \tag{17}$$

The less distance u_1, u_2 is, the more similar u_1 and u_2 are. Clustering CF includes two steps:

1. Partitioning users into clusters and each cluster always contains rating values. For example, every cluster resulted from *k*-mean algorithm has a mean which is a rating vector like user vector.
2. The concerned user who needs to be recommended is assigned to concrete cluster and her/his ratings are the same to ratings of such cluster. Of course how to assign a user to right cluster is based on the distance between user and cluster.

So the most important step is how to partition users into clusters. There are many clustering techniques such as k-mean and k-centroid. The most popular clustering algorithm is k-mean algorithm [45] which includes three following steps:

1. It randomly selects k users, each of which initially represents a cluster mean. Of course, we have *k* cluster means. Each mean is considered as the "representative" of one cluster. There are *k* clusters.
2. For each user, the distance between it and *k* cluster means are computed. Such user belongs to the cluster to which it is nearest. In other words, if user u_i belong to cluster c_v, the distance between u_i and mean m_v of cluster c_v, denoted $distance(u_i, m_v)$, is minimal over all clusters.
3. After that, the means of all clusters are re-computed. If stopping condition is met then algorithm is terminated, otherwise returning step 2.

This process is repeated until the stopping condition is met. There are two typical terminating conditions (stopping conditions) for *k*-mean algorithm:

- The *k* means are not changed. In other words, k clusters are not changed. This condition indicates a perfect clustering task.
- Alternatively, error criterion is less than a pre-defined threshold.

If the stopping condition is that the error criterion is less than a pre-defined threshold, the error criterion is defined as shown in Eq. 18 [45]:

$$error = \sum_{v=1}^{k} \sum_{u_i \in c_v} distance(u_i, m_v) \tag{18}$$

where c_v and m_v is cluster v and its mean, respectively. However, clustering CF encounters the problem of sparse rating matrix in which there are many missing values, which cause clustering algorithms to be imprecise.

In order to solve this problem, Ungar and Foster [41] proposed an innovative clustering CF which firstly groups items based on which users rate such items and then uses the item groups to help group users. Their method is a formal statistical model.

3 Strengths of Collaborative Filtering Technique

Collaborative filtering technique tends to produce more serendipitous recommendations [46]. When it comes to recommendations, accuracy isn't always the highest priority. Content-based filtering approaches tend to show users items that are very similar to items they've already liked, which can lead to filter bubble problems. By contrast, most users have interests that span different subsets, which in theory can result in more diverse (and interesting) recommendations. In addition, it is flexible across different domains. Collaborative filtering approaches are well suited to highly diverse sets of items. Where content-based filters rely on metadata, collaborative filtering is based on real-life activity, allowing it to make connections between seemingly disparate items (like say, an outboard motor and a fishing rod) that nonetheless might be relevant to some set of users (in this case, people who like to fish). Moreover, it can capture more nuance around items. Even a highly detailed content-based filtering system will only capture some of the features of a given item. By relying on actual human experience, collaborative filtering can sometimes recommend items that have a greater affinity with one another than a strict comparison of their attributes would suggest. Finally, it benefits from large user bases. Simply put, the more people are using the service, the better your recommendations will become, without doing additional development work or relying on subject area expertise.

4 Open Issues in Collaborative Filtering Recommender Systems

While collaborative filtering is commercially the most successful approach to recommendation generation, it suffers from a number of well-known problems [47]. These problems are highlighted below:

4.1 Data Sparsity

Usage of recommendation system increases very rapidly. Many commercial recommendation systems make use of large datasets. Therefore, the user-item matrix used for filtering could be very large and sparse and because of that performance of recommendation process may get degrade. The cold start problem is caused by the data sparsity. In collaborative filtering method recommendation of item is based on past preferences of users, so that new users will need to rate enough count of items to allow the system to catch their preferences accurately and thus allows for authentic recommendations [48, 49].

4.2 Coldstart

One of the most known problems in RSs is the cold start problem. The cold start problem is related to the sparsity of information (i.e., for users and items) available in the recommendation algorithm. The problem happens in recommendation systems due to the lack of information, on users or items: there is relatively little information about each user, which results in an inability to draw inferences to recommend items to users. The provision of a high QoR in cold start situations is a key challenge in RS [50]. Three types of cold start problems could be identified: (a) recommendations for new users, (b) recommendations for new items, and (c) recommendations on new items for new users [49].

4.3 Scalability

Traditional CF algorithms will suffers from scalability problems as the numbers of users and items increases. For example, consider a ten millions of customers O(M) and millions of items O(N), with that the complexity of algorithm is "n" which is already too large. As recommendation system play an important role in E-commerce application where system must respond to the user requirement immediately and irrespective of users ratings history and purchases system must make recommendations, which requires a higher scalability. Twitter is large web company to scale the recommendations of their millions of users it uses clusters of machines [48, 49].

4.4 Diversity

Recommendation systems are anticipated to increase diversity because they help us to discover new products. Some algorithms, may accidentally do the opposite. Here recommendation system recommend popular and highly rated items which are appreciated by particular user. This lead to lower accuracy in recommendation process. To overcome this problem there is need to develop new hybrid approaches which will enhance the efficiency of recommendation process [48].

4.5 Vulnerability to Attacks

Security is one of major issue in any system which are deployed on web. Recommendation system play an important role in e-commerce applications and because of that recommendation systems are probably targets of harmful attacks trying to promote or inhibit some items. This is one of major challenge faced by the developer of recommendation system [48].

4.6 Synonymy

Synonymy refers to the tendency of a number of the same or very similar items to have different names or entries. Most recommendation systems are unable to discover this latent association and thus treat these products differently. For example, the seemingly different items "children movie" and "children film" are actual the same item, but memory-based CF systems would find no match between them to compute similarity. Indeed, the degree of variability in descriptive term usage is greater than commonly suspected [51].

4.7 Gray Sheep

Gray sheep refers to the users whose opinions do not consistently agree or disagree with any group of people and thus do not benefit from collaborative filtering [51].

4.8 Shilling Attacks

In cases where anyone can provide recommendations, people may give tons of positive recommendations for their own materials and negative recommendations for their competitors. It is desirable for CF systems to introduce precautions that discourage this kind of phenomenon [51].

5 Conclusion

Recommender systems open new opportunities of retrieving personalized information on the Internet. It also helps to alleviate the problem of information overload which is a very common phenomenon with information retrieval systems and enables users to have access to products and services which are not readily available to users on the system. This paper discussed two categories of collaborative filtering technique namely memory-based and model based collaborative filtering. Under memory-based filtering, the paper discussed used-based and item based collaborative filtering while under model-based, the paper discussed Matrix Factorization CF, different Classification algorithms including Multi-Layer Perception (MLP), Logistic Regression, Jrip and J48, Clustering CF. The paper further highlighted the strengths and challenges associated collaborative filtering technique. In the process, challenges and strengths faced by each of the categories of collaborative were discussed. This knowledge will empower researchers and serve as a road map to improve the state of the art recommendation techniques.

References

1. Su, X., Khoshgoftaar, M.T.: A survey of collaborative filtering techniques. Adv. Artif. Intell. 1–20 (2009)
2. Isinkaye, F.O., Folajimi, Y.O., Ojokoh, B.A.: Recommendation systems: principles, methods and evaluation. Egypt. Inform. J. **16**, 261–273 (2015)

3. Breese, J., Heckerma, D., Kadie, C.: Empirical analysis of predictive algorithms for collaborative filtering. In: Proceedings of the 14th Conference on Uncertainty in Artificial Intelligence, San Francisco, CA (1998)
4. Mustafa, N., Osman, A., Ahmed, A., Abdullah, A.: Collaborative filtering: techniques and applications. In: Conference: 2017 International Conference on Communication, Control, Computing and Electronics Engineering (ICCCCEE) (2017)
5. Lee, J., Sun, M., Lebanon, G.: A comparative study of collaborative filtering algorithms. arXiv:1205.3193v1 [cs.IR] (2012)
6. Bobadilla, J., Ortega, F., Hernando, A., Gutiérrez, A.: Recommender systems survey. Knowl.-Based Syst. **46**, 109–132 (2013)
7. Al-Barznji, K., Atanassov, A.: Comparison of memory based filtering techniques for generating recommendations on large data. Eng. Autom. **1**(1), 44–50 (2018)
8. Jannach, D., Zanker, M., Felfernig, A., Friedrich, G.: Recommender Systems: An Introduction. Cambridge University Press, Cambridge (2011)
9. Xiaoyuan, S., Taghi, M.: A survey of collaborative filtering techniques. Adv. Artif. Intell. **2009**, 1–20 (2009)
10. Adomavicius, G., Tuzhilin, A.: Toward the next generation of recommender: a survey of the state-of-the-art and possible extensions. IEEE Trans. Knowl. Data Eng. **17**(6), 734–749 (2005)
11. Breese, J., Heckerman, D., Kadie, C.: Empirical analysis of predictive algorithms for collaborative filtering. Madison, Wisconsin (1998)
12. Sarwar, B., Karypis, G., Konstan, J., Riedl, J.: Item-based collaborative filtering recommendation algorithms. In: ACM 1-58113-348-0/01/0005, Hong Kong (2001)
13. Schafer, B.J., Frankowski, D., Herlocker, J., Sen, S.: Collaborative filtering recommender systems. In: Brusilovsky, P., Kobsa, A., Nejdl, W. (eds.) The Adaptive Web, pp. 291–324. Springer, Heidelberg (2007)
14. Nagpal, D., Kaur, S., Gujral, S., Singh, A.: FR: A Recommender for Finding Faculty Based on CF Technique (2015)
15. Bahadorpour, M., Neysiani, B.S., Shahraki, M.N.: Determining optimal number of neighbors in item-based kNN collaborative filtering algorithm for learning preferences of new users. J. Telecommun. **9**(3), 163–167 (2017)
16. Ekstrand, M.D., Riedl, J.T., Konstan, J.A.: Collaborative Filtering Recommender Systems. Now Publishers Inc., Boston (2011)
17. Saptono, R.: User-Item Based Collaborative Filtering for Improved Recommendation (2010)
18. Nakamura, A., Abe, N.: Collaborative filtering using weighted majority prediction algorithms. In: Proceedings of the Fifteenth International Conference on Machine Learning, San Francisco, CA, USA (1998)
19. Kim, H.-N., Ji, A.-T., Ha, I., Jo, G.-S.: Collaborative filtering based on collaborative tagging for enhancing the quality of recommendation. Electron. Commer. Res. Appl. **9**(1), 73–83 (2010)
20. Al-Bashiri, H., Abdulgabber, M.A., Romli, A., Kahtan, H.: An Improved Memory-Based Collaborative Filtering Method Based on The TOPSIS (2018)
21. Do, T., Phung, M., Nguyen, V.: Model-based approach for collaborative filtering. In: The 6th International Conference on Information Technology for Education, Ho Chi Minh city, Vietnam (2010)
22. Koren, Y., Bell, R., Volinsky, C.: Matrix factorization techniques for recommender systems. Computer **42**(8), 30–37 (2009)
23. Deerwester, S., Dumais, S.T., Furnas, G., Landauer, T.K., Harshman, R.: Indexing by latent semantic analysis. J. Am. Soc. Inf. Sci. **41**(6), 391–407 (1990)

24. Gorrell, G.: Generalized Hebbian algorithm for incremental singular value decomposition in natural language processing. In: EACL, pp. 97–104 (2006)
25. Kurucz, M., Benczúr, A.A., Csalogány, A.: Methods for large scale SVD with missing values. In: KDD Cup and Workshop (2007)
26. Sanger, T.D.: Optimal unsupervised learning in a single-layer linear feedforward neural network. Neural Netw. **2**(6), 459–473 (1989)
27. Miller, B.N., Konstan, J.A., Riedl, J.: PocketLens: toward a personal recommender system. ACM Trans. Inf. Syst. **22**(3), 437–476 (2004)
28. Funk, S. (2006). http://sifter.org/simon/journal/20061211.html
29. Funk, S.: Netflix (2006). http://sifter.org/˜simon/journal/20061211.html
30. Sarwar, B., Karypis, G., Konstan, J.A., Riedl, J.: Application of dimensionality reduction in recommender system, 02 November 2000. Accessed 2019
31. Berry, M.W., Dumais, S.T., O'Brien, G.W.: Using linear algebra for intelligent information retrieval. SIAM Rev. **37**, 573–595 (1995)
32. Sarwar, B., Karypis, G., Konstan, A.J., Riedl, J.: Incremental SVD-based algorithms for highly scalable recommender systems (2002)
33. Brand, M.E.: Incremental Singular Value Decomposition of Incomplete Data (2003)
34. Rajput, A., Aharwal, R.P., Dubey, M., Saxena, S., Raghuvanshi, M.: J48 and JRIP rules for e-governance data. Int. J. Comput. Sci. Secur. (IJCSS) **5**(2), 201 (2011)
35. Hastie, T., Tibshirani, T., Friedman, R.: Unsupervised learning. In: The Elements of Statistical Learning. Springer, New York (2009)
36. Kavzoglu, T., Mather, P.M.: The use of backpropagating artificial neural networks in land cover classification. Int. J. Remote Sens. **24**(23), 4907–4938 (2003)
37. Park, D.C., El-Sharkawi, M.A., Marks, R.J., Atlas, L.E., Damborg, M.J.: Electric load forecasting using artificial neural network. IEEE Trans. Power Syst. **6**(2), 442–449 (1991)
38. Jung, Y.G., Kang, M.S., Heo, J.: Clustering performance comparison using K-means and expectation maximization algorithms. Biotechnol. Biotechnol. Equip. **28**, 44–48 (2014)
39. Shepperd, M., Kadoda, G.: Comparing software prediction techniques using simulation. IEEE Trans. Software Eng. **27**(11), 1014–1022 (2001)
40. Jadhav, S.D., Channe, H.P.: Efficient recommendation system using decision tree classifier and collaborative filtering. Int. Res. J. Eng. Technol. **3**(8), 2114–2118 (2016)
41. Ungar, H.L., Foster, D.P.: Clustering methods for collaborative filtering. In: AAAI Workshop on Recommender Systems (1998)
42. Shrkhorshidi, A.S., Aghabozorgi, S., Wah, T.Y.: A Comparison Study on Similarity and Dissimilarity Measure in Clastering Continuous Data (2015)
43. Jeyasekar, A., Akshay, K., Karan: Collaborative filtering using Euclidean distance in recommendation engine. Indian J. Sci. Technol. **9**(37) (2016)
44. Zheng, M., Min, F., Zhang, H.-R., Chen, W.-B.: Fast Recommendations With the M-Distance (2016)
45. Torres, R.D.: Combining Collaborative and Content-based Filtering to Recommend Research Paper (2004)
46. Keenan, T.: Upwork Global Inc., 28 March 2019. https://www.upwork.com/hiring/data/how-collaborative-filtering-works/
47. Anand, S.S., Mobasher, B.: Intelligent techniques for web personalization. In: IJCAI Workshop on Intelligent Techniques for Web Personalization (2003)
48. Lü, L., Medo, M., Yeung, C.H., Zhang, C.Y., Zhang, Z.K., Zhou, T.: Recommender systems. Phys. Rep. **519**(1), 1–49 (2012)
49. Madhukar, M.: Challenges & limitation in recommender systems. Int. J. Latest Trends Eng. Technol. (IJLTET) **4**(3), 138–142 (2014)

50. Park, S.-T., Chu, W.: Pairwise preference regression for cold-start recommendation. In: Proceedings of the 2009 ACM Conference on Recommender Systems, New York (2009)
51. Shinde, U., Shedge, R.: Comparative analysis of collaborative filtering technique. IOSR J. Comput. Eng. (IOSR-JCE) **10**, 77–82 (2013)

A Computational Simulation of Steady Natural Convection in an H-Form Cavity

Mohamed Loukili[1(✉)], Kamila Kotrasova[2], and Denys Dutykh[3]

[1] Faculty Ben M'sik of Sciences, Hassan II University, Casablanca, Morocco
md.loukili@gmail.com

[2] Faculty of Civil Engineering, Technical University of Kosice, Košice, Slovak Republic

[3] Univ. Grenoble Alpes, Univ. Savoie Mont Blanc, CNRS, LAMA, 73000 Chambéry, France

Abstract. The simulation of natural convection problem based on the Galerkin finite-element method, with the penalty finite-element formulation of the momentum balance equation, is exploited for accurate solutions of equations describing the problem of H-Form cavity differentially heated side walls. The cavity is occupied by the air whose Prandtl number is $Pr = 0.71$, the fluid is assumed to be steady, viscous and incompressible within thermal convection. A numerical investigation has been made for Rayleigh numbers ranging from 10 to 10^6 for three cases of total internal height aspects of H-Form cavity: 0%, 50%, and 85%. Firstly, the goal is to validate the numerical code used to resolve the equations governing the problem of this work. For that, we present a comparison between the profiles at the point (0.5, 0) for the u-component, and u-component obtained in previous work for simple square cavity. Further, a comparison of the averaged Nusselt number with previous works for simple square cavity is realized in order to ensure the numerical accuracy, and the validity of our considered numerical tool. Secondly, the objective is to investigate on the hydrodynamic effects of Rayleigh number for different total internal height aspects of H-Form cavity on the dynamics of natural convection. Shortly after, the ambition is to assess the heat transfer rate for different Rayleigh number for three cases of internal height aspects.

Keywords: Natural convection · H-Form cavity · Heat transfer · Rayleigh number · Nusselt number

1 Introduction

The natural convection flows have been the subject of many investigations due to its basic importance in numerous industrial and natural processes, such as: solar collectors, cooling of electronic equipment, energy storage systems, air conditioned system in buildings, thermal insulation, and fire propensity control in buildings [1–3]. Further, the natural convection mechanism within square cavity has been attracting the interest of many researchers for several decades [4–9], because of its importance and the great number of applications in various fields and industries. In 1983, De Vahl Davis and G. Jones [7] corroborated the accuracy of the benchmark solution and offered a basis

R. Silhavy et al. (Eds.): CoMeSySo 2020, AISC 1295, pp. 164–177, 2020.
https://doi.org/10.1007/978-3-030-63319-6_15

solution to ensure the validity of new contribution in natural convection problem. Next, two sequences of simulations were studied to deal with the impact of buoyancy force on mass transfer rate, and to get the effect of Lewis number on fluid motion [10]. After that, the transient features of natural convection flow in the partitioned cavity were discussed [9], and the authors shed light on three different stages (initial, transitional, and steady) as the time scale was considered. Then, a membrane was introduced to the square cavity to study different aspect of natural convection for different Rayleigh numbers [8]. Two years before, the mechanism of natural convection of air within a square cavity with inferior walls and active side was experimentally and computationally investigated [11]. This last paper showed that the average and maximum velocity, with small values characteristics of natural convection, rise with Rayleigh and even with the angle and attain an extreme value for Ra maximum. More recently, A. Mazgar et al. [12] addressed the impact of gas radiation on laminar natural convection flow inside a square cavity that has an internal heat source, and emphasized the entropy generation. They concluded a significant finding from this investigation is that radiative influence has a key role in the acceleration of the vortices and affording either a homogenizing impact on temperature fields.

Earlier investigators have theoretically and experimentally addressed many aspects of convective heat transfer in cavity enclosures involving conjugate heat transfer effects [13], nanofluids and entropy generation [14], magnetic field effects [15], cavity filled with porous media [16] and the presence of a solid partition [17].

Furthermore, many researchers have studied different geometry aspects of convective heat transfer in simple enclosures, such as the geometry of triangular shape [18], C-shape [19], concentric annulus [20], hemispherical shape [21–23], and parallelogrammic shape [24].

In this work, the objective is to study the natural convection inside H-Form cavity, and to assess the impact of Rayleigh number for three cases of internal height aspects of H-Form cavity: 0%, 50%, and 85% on the fluid flows, and the heat transfers. To realize this work, the paper is illustrated in five sections: after the introduction in the first section, the problem statement and solutions procedures are clearly described in the second section. Shortly after, we will interest to corroborate the accuracy and validate the numerical code used in this work. Next, we shed the light on the hydrodynamic impact of Rayleigh number, and the size of H-Form cavity on both; flow pattern and heat transfer mechanism inside the defined computational domain. Finally, we finish our manuscript by outlining the main conclusions and perspectives of this study.

2 Problem Statement and Solutions Procedures

The studied computational domain is a two-dimensional H-Form cavity, the total internal height aspects of H-Form cavity are stands for *2l* (see Fig. 1). The cavity is occupied by the air whose Prandtl number is $Pr = 0.71$, the fluid considered is assumed to be steady, viscous and incompressible within thermal convection. The upstream and downstream sides are differentially heated with fixed temperatures, whereas the other sides are considered adiabatic and the velocities on every side are zero. The studied configuration is sketched in Fig. 1.

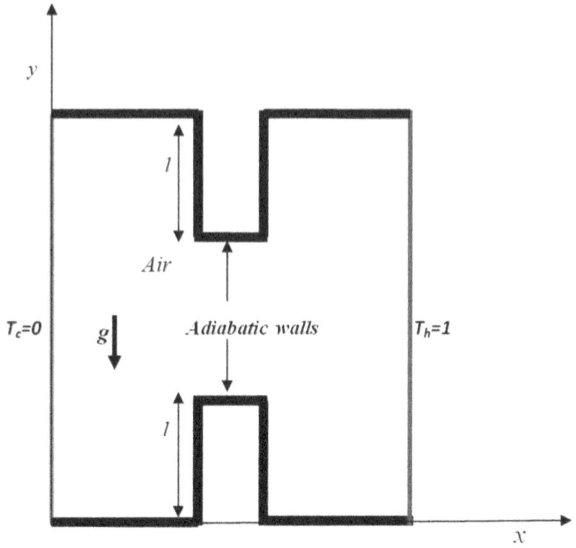

Fig. 1. Physical domain

The non-dimensional equations describing the steady natural convection flow, are expressed as

$$\frac{\partial u}{\partial x} + \frac{\partial v}{\partial y} = 0, \tag{1}$$

$$u\frac{\partial u}{\partial x} + v\frac{\partial u}{\partial y} = -\frac{\partial P}{\partial x} + Pr\left(\frac{\partial^2 u}{\partial x^2} + \frac{\partial^2 u}{\partial y^2}\right), \tag{2}$$

$$u\frac{\partial v}{\partial x} + v\frac{\partial v}{\partial y} = -\frac{\partial P}{\partial y} + Pr\left(\frac{\partial^2 v}{\partial x^2} + \frac{\partial^2 v}{\partial y^2}\right) + Ra\, Pr\, T, \tag{3}$$

$$u\frac{\partial T}{\partial x} + v\frac{\partial T}{\partial y} = \frac{\partial^2 T}{\partial x^2} + \frac{\partial^2 T}{\partial y^2}. \tag{4}$$

and the boundary conditions are expressed as

$$\begin{cases} u = v = 0 \\ T_c = 0 \end{cases} \text{for}\, x = 0, \text{and}\, y \in [0\ 1] \tag{6}$$

$$\begin{cases} u = v = 0 \\ T_h = 1 \end{cases} \text{for}\, x = 1, \text{and}\, y \in [0\ 1] \tag{7}$$

$$\begin{cases} u = v = 0 \\ \frac{\partial T}{\partial y} = 0 \end{cases} \text{for}\, (y = 0\, \text{and}\, y = 1), \text{and}\, x \in [0\ 1] \tag{8}$$

the different parameters are given as follows: u and v denote the non-dimensional velocities, T is the non-dimensional temperature, Ra stands for Rayleigh number, Pr stands for Prandtl number.

The momentum and energy balance equations [Eqs. (2)–(4)] are solved using Galerkin finite element method. The continuity equation [Eq. (1)] will be used as a constraint due to mass conservation and this constraint may be used to obtain the pressure distribution [25–27]. In order to solve Eqs. (2)–(3), we use the penalty finite element method where the pressure P is eliminated by a penalty parameter δ and the incompressibility criteria given by Eq. (1) results in

$$p = -\delta\left(\frac{\partial u}{\partial x} + \frac{\partial v}{\partial y}\right), \tag{9}$$

the continuity Eq. (1) is satisfied for numerous values of δ, the typical value of δ which affords consistent solution is $\delta = 10^7$ [27]. Then, the governing Eqs. (2) and (3) are reduced to

$$u\frac{\partial u}{\partial x} + v\frac{\partial u}{\partial y} = \delta\frac{\partial}{\partial x}\left(\frac{\partial u}{\partial x} + \frac{\partial v}{\partial y}\right) + Pr\left(\frac{\partial^2 u}{\partial x^2} + \frac{\partial^2 u}{\partial y^2}\right), \tag{10}$$

$$u\frac{\partial v}{\partial x} + v\frac{\partial v}{\partial y} = \delta\frac{\partial}{\partial y}\left(\frac{\partial u}{\partial x} + \frac{\partial v}{\partial y}\right) + Pr\left(\frac{\partial^2 v}{\partial x^2} + \frac{\partial^2 v}{\partial y^2}\right) + Ra\,Pr\,T. \tag{11}$$

The penalty finite-element method is a standard technique for solving incompressible viscous flows, the Galerkin finite element method yields the nonlinear residual equations for Eqs. (10)–(11). By adopting biquadratic basis functions with three points Gaussian quadrature in order to evaluate the integrals in the residual equations, the numerical procedure details are depicted in the reference [27].

3 Computational Validation

Firstly, the goal is to ensure the validity and verify the accuracy of the numerical technique adopted to resolve the problem of this work. In Fig. 2, we present a comparison between the profiles at the point (0.5, 0) for the u-component for simple square cavity (a), and u-component obtained in the reference [27] at the mid-width of the cavity (b) for the same boundary conditions. The results of the Fig. 2 are shown a good agreement with the results obtained by Mousa [27] for Rayleigh number Ra ranging from 10 to 10^6.

Secondly, to ensure numerical validity of our numerical code, we illustrate the comparison of the averaged Nusselt number with previous works for square cavity for different values of Rayleigh number, see the Table 1. Thereafter, we calculate the Nusselt's deviation D_v of the present work from the following references (Fusegi et al. [28], De vahl Davis [7], and Barakos et al. [29]).

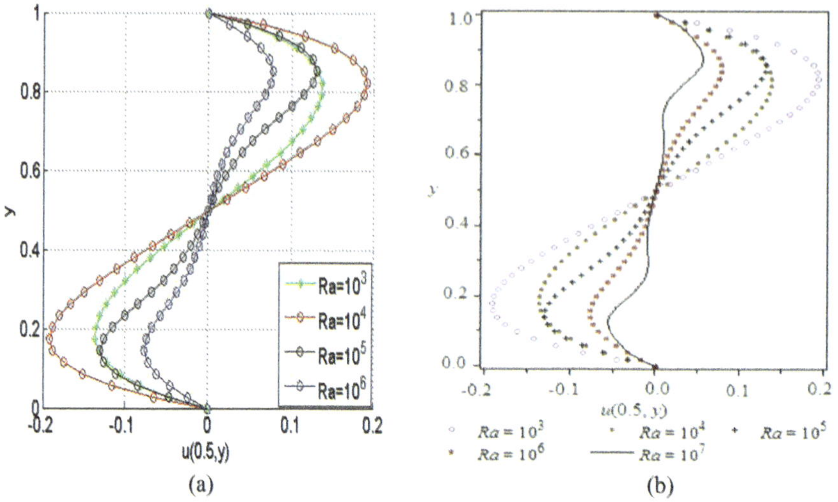

Fig. 2. (a) Distribution of u-velocity component at mid-width of simple square cavity, (b) distribution of u-velocity component at mid-width obtained in [27], for various Ra values.

The Nusselt's deviation D_v is defined as

$$D_v = \frac{|Nu^P - Nu^R|}{Nu^R}$$

Nu^P stands for the Nusselt number of the present work,
Nu^R stands for the Nusselt number of the designed reference.

The Table 1 addresses the comparison of the averaged Nusselt along the hot wall, the results presented show an excellent agreement between the present results and the benchmark solutions [7, 28, 29] for all values of Rayleigh number.

Table 1. The Nusselt's deviation concerning square cavity for different Rayleigh numbers.

Ra	Present work	Fusegi et al. [28]		De vahl Davis [7]		Barakos et al. [29]	
	Nu^P	Nu^R	Dv(%)	Nu^R	Dv(%)	Nu^R	Dv(%)
10^3	1.117	1.105	1,09	1.118	0,09	1.114	0,27
10^4	2.244	2.302	2,52	2.243	0,04	2.245	0,04
10^5	4.520	4.646	2,71	4.519	0,02	4.510	0,22
10^6	8.820	9.01	2,11	8.799	0,24	8.806	0,16

4 Results and Discussion

The present section puts the light on the hydrodynamic impact of Rayleigh number, for three cases of total internal height aspects of H-Form cavity on both; flow pattern and heat transfer mechanism inside the defined computational domain. Firstly, The Fig. 2 highlights streamlines for Rayleigh numbers ranging from 10 to 10^6 for three cases of total internal height aspects: 0%, 50%, and 85%.

Based on the Fig. 3 the outcomes reveal that the Rayleigh number is the major parameter influencing the flow in the case of simple square cavity. In detail, concerning the low values of Rayleigh number (*Ra* from 10 to 10^3) a vortex is created in the center, when Rayleigh number growths to $Ra = 10^4$ the vortex seems to be elliptic, as Rayleigh number increases to $Ra = 10^5$ two vortices are generated at the center giving space to the third vortex to develop at $Ra = 10^6$.

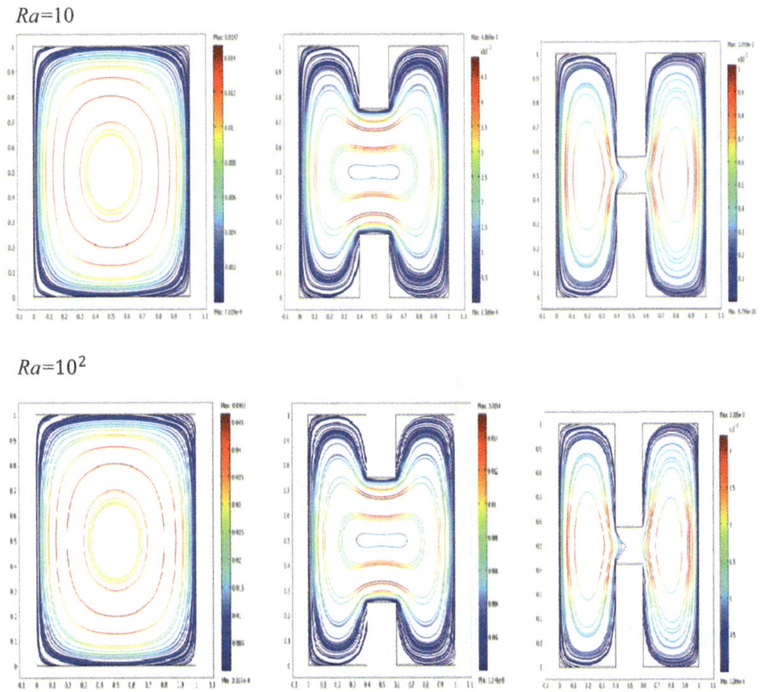

Fig. 3. Streamlines in H form cavity for different total internal height aspects and for Rayleigh number ranging from 10 to 10^6.

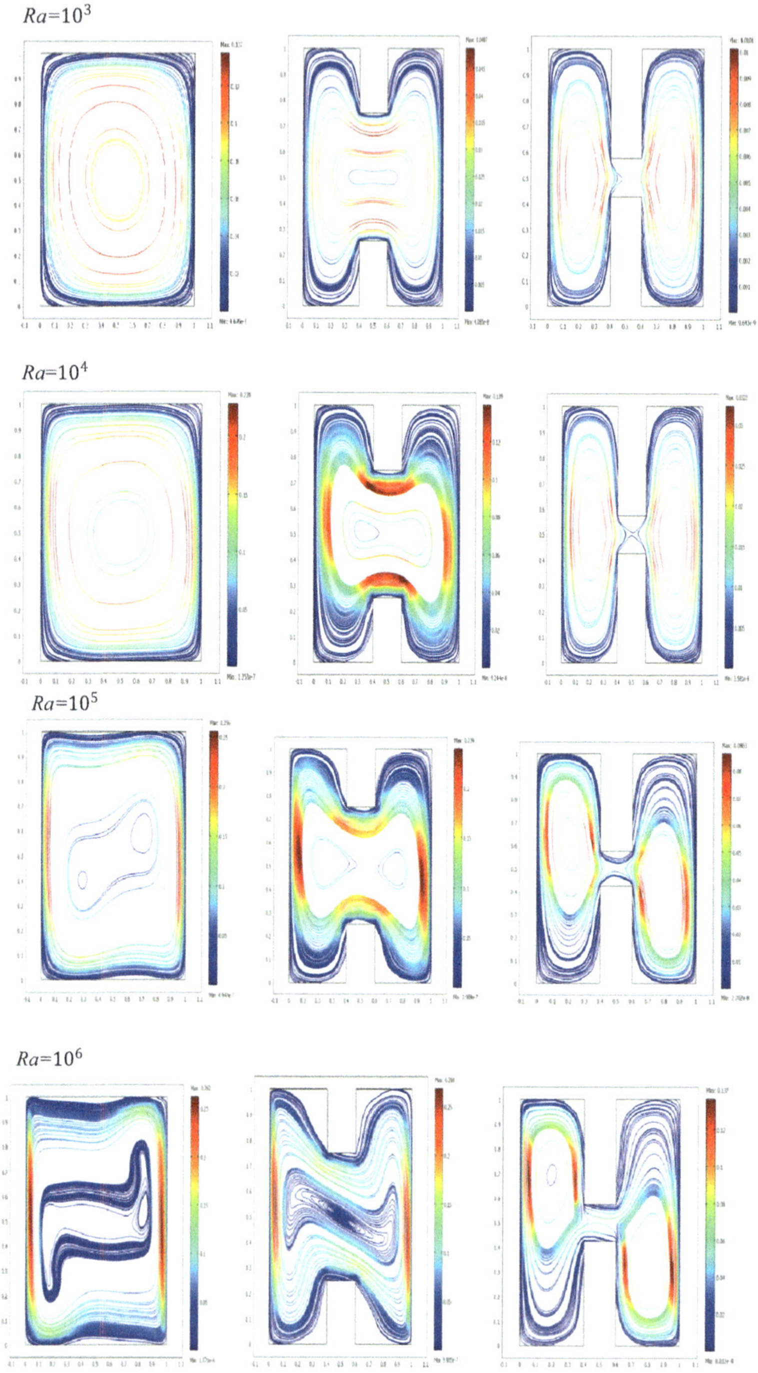

Fig. 3. *(continued)*

Thereafter, for the cavity with 50% of total internal height aspects we remark in general that the velocities are less important than the flow in simple square cavity, which is totally logical regarding the dynamics of the flows-structure interactions. The figures show that Rayleigh number ranging from 10 to 10^3 the streamlines appear elliptic, for $Ra = 10^4$ the velocities are higher at the center creating a vortex and giving space to the second vortex to develop at $Ra = 10^5$, the two vortices meet at center creating a large vortex. By increase of the buoyant force via increase in the Rayleigh number, the flow intensity increases and the streamlines closes to the side walls.

Next, for the cavity with 85% of total internal height aspects we remark that the circulations of fluid flow are blocked in each block of H-form geometry creating vortices at each block separately. Consequently, we notice that the Rayleigh number hasn't a strong effect on the characteristics of the fluid flow, and Rayleigh number effect starts to appear till $Ra = 10^4$ where the two vortices of each block are start to be in contact, once Rayleigh number achieve $Ra = 10^5$ the exchange between two columns becomes important. Lastly, for $\mathrm{Ra} = 10^6$ vortices are more vital, improving stratifications at the left top and downright of each block of H-form cavity.

Secondly, the Fig. 4 presents the isotherms for Rayleigh numbers ranging from 10 to 10^6 for three cases of total internal height aspects: 0%, 50%, and 85%. We remark, in the case of simple square cavity 0% of total internal height aspects, the heat transfer mechanism changes as a function of Rayleigh number. To clear up, we notice that from $Ra = 10$ to $Ra = 10^3$ the isotherms appear vertical, when Rayleigh number increases to $Ra = 10^4$ the heat transfer mechanism changes from conduction to convection, as Rayleigh number raises the isotherms are no longer vertical only inside the very thin boundary layers. Whereas, in the case of cavity with 50% and 85% of total internal height aspects we observe that the circulation of the fluid flow becomes slower compared to the case of simple square cavity, which explain the slow action of the heat transfer mechanism when the total internal height aspects increase.

Next, for the cavity with 85% of total internal height aspects we remark that the Rayleigh number is not dominant till it reaches 10^5 and then the heat transfer mechanism starts to change from conduction to convection, at $Ra = 10^6$ the process of convection is no longer fast due to the blockage of the circulation of the fluid flow inside each block of H-form cavity.

In this subsection, the local Nusselt number along the hot wall of the cavity is presented in Fig. 5 for a wide range of Rayleigh numbers for three cases of total internal height aspects. The goal is to analyze the effects of the increase of total internal height aspects on the local Nusselt number along the hot wall of the cavity at various Rayleigh numbers.

The outcomes of the Fig. 5 present the local Nusselt number along the hot wall of the cavity at various Rayleigh numbers for different total internal height aspects. For all aspect ratios and for different Rayleigh numbers the maximum local Nusselt number occurs at the lower end of the hot wall i.e. $y = 0$. Further, the local Nusselt number decreases as total internal height aspects increases, and then the heat transfer rate decreases with the increase of the total internal height aspects, this is due to the fact that when the total internal height aspects increases the fluid is damped and hence rate of free convection decreases. Furthermore, when the Rayleigh numbers increase the cold fluid moves to hot wall and hence maximum temperature gradient occurs at this region, the cold fluid ascends adjacent to the hot wall, then the fluid temperature increases subsequently the local Nusselt number decreases. For the case of simple square cavity, the local Nusselt profile departs from his vertical position and then the process of the conviction is started at $Ra = 10^4$. Moreover, for the cases of total internal height aspects: 50% and 85% and when Rayleigh numbers ranging from $Ra = 10$ to $Ra = 10^3$

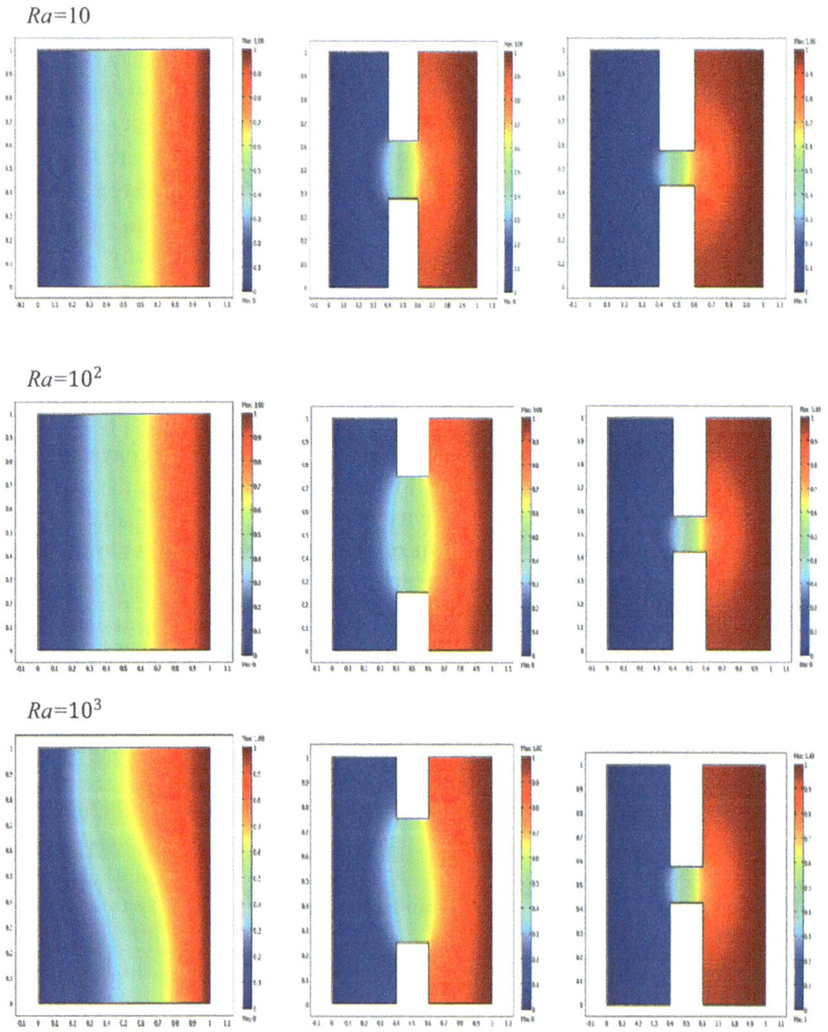

Fig. 4. Isotherms in H form cavity for different total internal height aspects and for Rayleigh number ranging from 10 to 10^6.

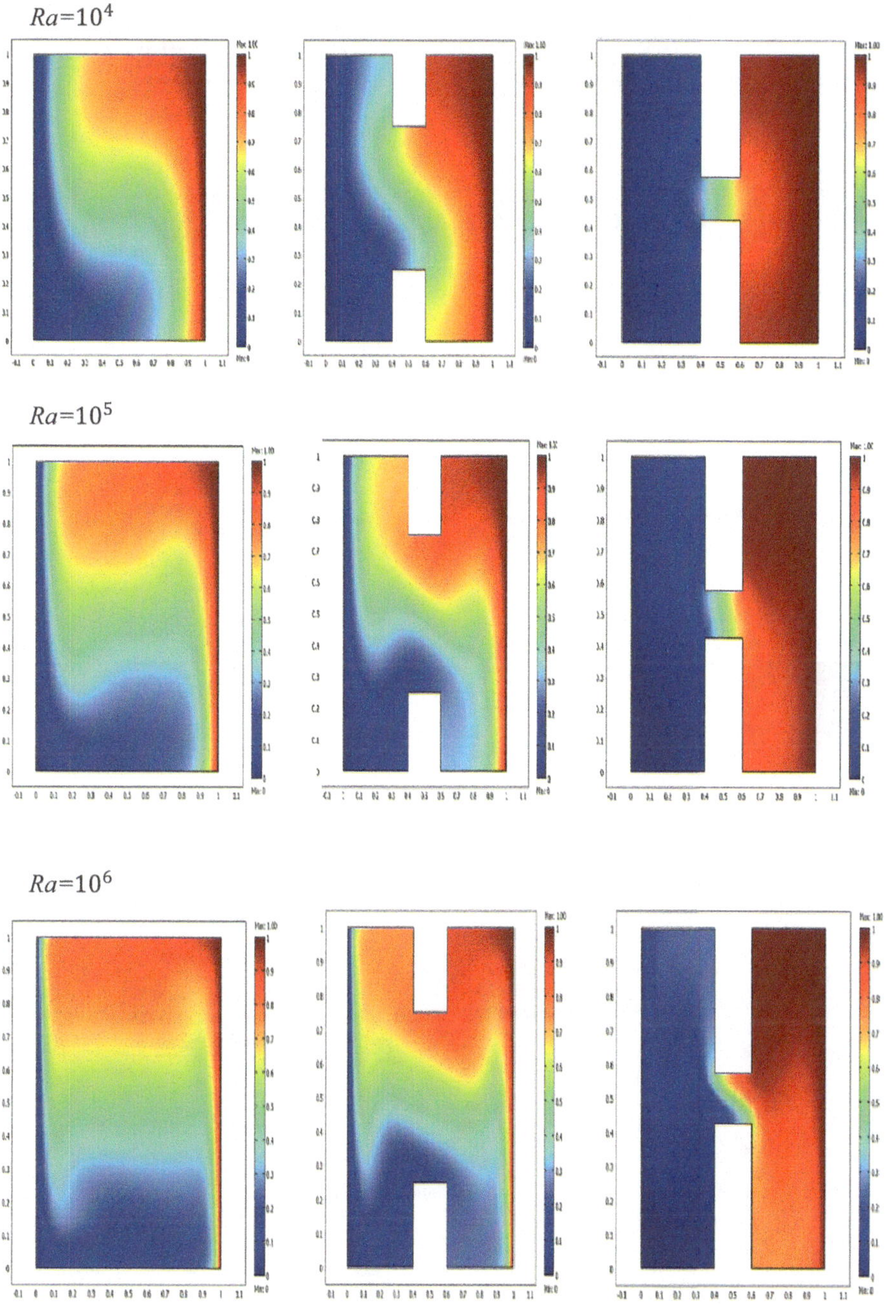

Fig. 4. (*continued*)

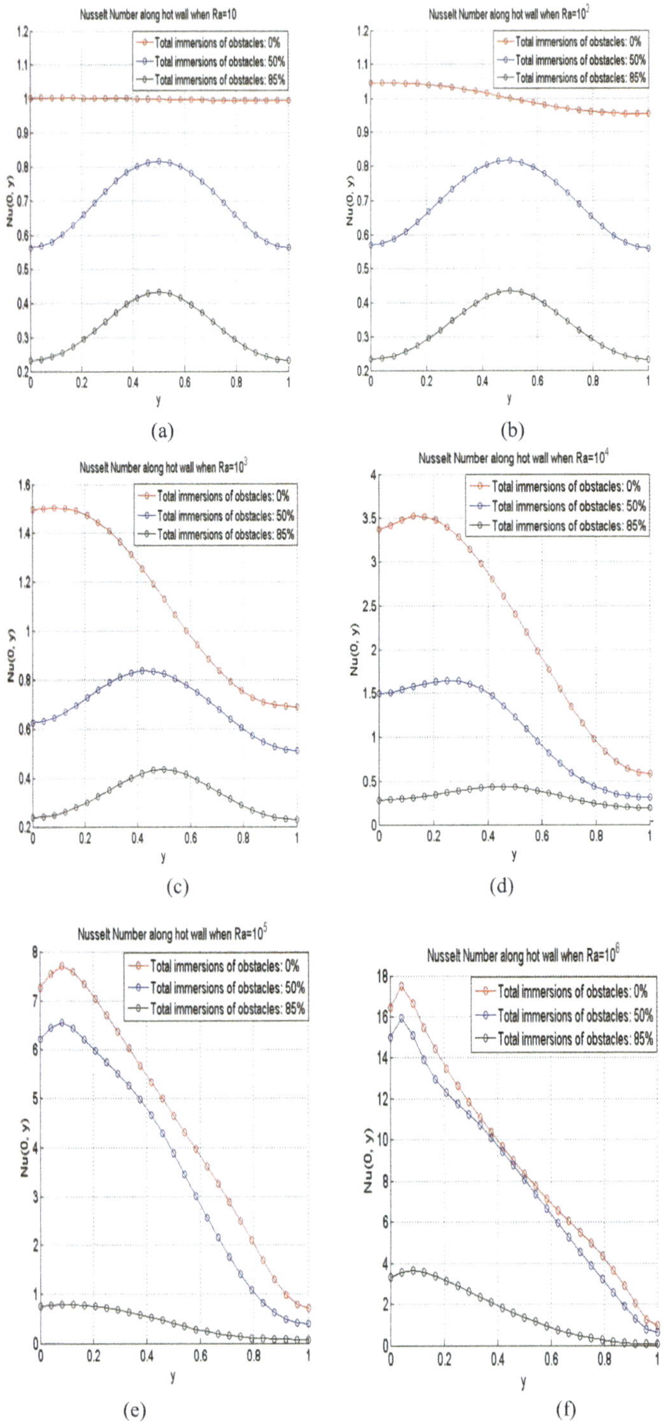

Fig. 5. Variation of the local Nusselt number *Nu* with total internal height aspects at various Rayleigh numbers. **a** Ra =10, **b** $Ra = 10^2$, **c** $Ra = 10^3$ **d** $Ra = 10^4$, **e** $Ra = 10^5$, **f** $Ra = 10^6$.

the location at which the maximum local Nusselt number occurs is about $y = 0.5$. In addition, when Rayleigh numbers ranging from $Ra = 10^5$ to $Ra = 10^6$ for all aspect ratios, maximum rate of heat transfer occurs at about $y < 0.2$ of the hot wall, this position approaches the lower part of the hot wall as Ra increases.

5 Conclusion and Perspectives

In this work, the natural convection in H-form cavity has been studied numerically for different parameters influencing the flow and heat transfer. In details, we have assess the impact of Rayleigh number for three cases of total internal height aspects: 0%, 50%, and 85%. From the foregoing discussion, we remark in the case of simple square cavity (0% of total internal height aspects) that Rayleigh number is the dominant parameter on the heat transfer mechanism and fluid flows. In the case of H-Form cavity with 50% and 85% total internal height aspects, and Rayleigh numbers ranging from $Ra = 10$ to $Ra = 10^3$ the maximum of rate transfers are occurs at the centre. Rayleigh number does not influencing the heat transfer till it reaches 10^4 for 50% of total internal height aspects and 10^5 for 85%. In general, for all aspect ratios, the heat transfer rate increases with Rayleigh number and decreases with the increase of the total internal height aspects. Furthermore, as the total internal height aspects increases the circulation of the fluid flow becomes slower, which causes to decreases effect of heat transfers rate. As perspectives, we endeavor to study the natural convection flows in H-form cavity using meshless methods [30–33], and spectral methods [34] that have been proven a strong efficiency in many nonlinear and engineering fields [30–33].

References

1. Nasrin, R., Alim, M.A., Chamkha, A.J.: Effects of physical parameters on natural convection in a solar collector filled with nanofluid. Heat Transf. Asian Res. **42**(1), 73–88 (2013)
2. Baïri, A., Zarco-Pernia, E., De María, J.M.G.: A review on natural convection in enclosures for engineering applications. The particular case of the parallelogrammic diode cavity. Appl. Therm. Eng. **63**(1), 304–322 (2014)
3. Kaushik, S.C., Kumar, R., Garg, H.P., Prakash, J.: Transient analysis of a triangular built-in-storage solar water heater under winter conditions. Heat Recov. Syst. CHP **14**(4), 337–341 (1994)
4. Raisi, A., Arvin, A.: A numerical study of the effect of fluid-structure interaction on transient natural convection in an air-filled square cavity. Int. J. Therm. Sci. **128**, 1–14 (2018)
5. Dowell, E.H., Hall, K.C.: Modeling of fluid-structure interaction. Ann. Rev. Fluid Mech. **33**, 445–490 (2001)
6. Hou, G., Wang, J., Layton, A.: Numerical methods for fluid-structure interaction. Commun. Comput. Phys. **12**, 337–377 (2012)
7. De Vahl Davis, G., Jones, I.P.: Natural convection in a square cavity: a comparison exercise. Int. J. Numer. Meth. Fluids **3**, 227–248 (1983)
8. Mehryan, S.A.M., Ghalambaz, M., Ismael, M.A., Chamkha, A.J.: Analysis of fluid-solid interaction in MHD natural convection in a square cavity equally partitioned by a vertical flexible membrane. J. Magnetism Magnet. Mater. **424**, 161–173 (2017)

9. Xu, F., Patterson, J.C., Lei, C.W.: Heat transfer through coupled thermal boundary layers induced by a suddenly generated temperature difference. Int. J. Heat Mass Transf. **52**, 4966–4975 (2009)
10. Béghein, C., Haghighat, F., Allard, F.: Numerical study of double-diffusive natural convection in a square cavity. Int. J. Heat Mass Transf. **35**(4), 833–846 (1992)
11. Leporini, M., Corvaro, F., Marchetti, B., Polonara, F., Benucci, M.: Experimental and numerical investigation of natural convection in tilted square cavity filled with air. Exp. Ther. Fluid Sci. **99**, 572–583 (2018). https://doi.org/10.1016/j.expthermflusci.2018.08.023
12. Mazgar, A., Hajji, F., Jarray, K., Ben Nejma, F.: Conjugate non-gray gas radiation combined with natural convection inside a square cavity with internal heat source: entropy generation. In: Conference, 6th International Conference on Green Energy and Environmental Engineering GEEE-2019, 27–29th April 2019, Tabarka, Tunisia (2019)
13. Chamkha, A.J., Ismael, M.A.: Conjugate heat transfer in a porous cavity filled with nanofluids and heated by a triangular thick wall. Int. J. Therm. Sci. **67**, 135–151 (2013)
14. Parvin, S., Chamkha, A.J.: An analysis on free convection flow, heat transfer and entropy generation in an odd-shaped cavity filled with nanofluid. Int. Commun. Heat Mass Transf. **54**, 8–17 (2014)
15. Sathiyamoorthy, M., Chamkha, A.: Effect of magnetic field on natural convection flow in a liquid gallium filled square cavity for linearly heated side wall (s). Int. J. Therm. Sci. **49**(9), 1856–1865 (2010)
16. Tatsuo, N., Mitsuhiro, S., Yuji, K.: Natural convection heat transfer in enclosures with an off-center partition. Int. J. Heat Mass Transf. **30**(8), 1756–1758 (1987)
17. Kaluri, R.S., Anandalakshmi, R., Basak, T.: Bejan's heatline analysis of natural convection in right-angled triangular enclosures: effects of aspect-ratio and thermal boundary conditions. Int. J. Therm. Sci. **49**(9), 1576–1592 (2010)
18. Mansour, M.A., Bakeir, M.A., Chamkha, A.: Natural convection inside a C-shaped nanofluid-filled enclosure with localized heat sources. Int. J. Numer. Meth. Heat Fluid Flow **24**(8), 1954–1978 (2014)
19. Alawi, O.A., Sidik, N.A.C., Dawood, H.K.: Natural convection heat transfer in horizontal concentric annulus between outer cylinder and inner flat tube using nanofluid. Int. Commun. Heat Mass Transfer **57**, 65–71 (2014)
20. Baïri, A., Öztop, H.F.: Free convection in inclined hemispherical cavities with dome faced downwards. Nu-Ra relationships for disk submitted to constant heat flux, Int. J. Heat Mass Transfer **78**, 481–487 (2014)
21. Baïri, A., Monier-Vinard, E., Laraqi, N., Baïri, I., Nguyen, M.N., Dia, C.T.: Natural convection in inclined hemispherical cavities with isothermal disk and dome faced downwards. Exp. Numer. Stud. Appl. Therm. Eng. **73**(1), 1340–1347 (2014)
22. Baïri, A.: Nu–Ra–Fo correlations for thermal control of embarked radars contained in tilted hemispherical cavities and subjected to constant heat flux. Appl. Therm. Eng. **67**(1), 540–544 (2014)
23. Baïri, A., de María, J.G., Laraqi, N.: Transient natural convection in parallelogrammic enclosures with isothermal hot wall. Experimental and numerical study applied to on-board electronics. Appl. Therm. Eng. **30**(10), 1115–1125 (2010)
24. Basak, T., Ayappa, K.G.: Influence of internal convection during microwave thawing of cylinders. AIChE J. **47**, 835–850 (2001)
25. Basaka, T., Royb, S., Thirumalesha, Ch.: Finite element analysis of natural convection in a triangular enclosure: effects of various thermal boundary conditions. Chem. Eng. Sci. **62**, 2623–2640 (2007)

26. Roy, S., Basak, T.: Finite element analysis of natural convection flows in a square cavity with nonuniformly heated wall(s). Int. J. Eng. Sci. **43**, 668–680 (2005)
27. Mousa, M.M.: Modeling of laminar buoyancy convection in a square cavity containing an obstacle. Bull. Malays. Math. Sci. Soc. **39**(2), 483–498 (2015)
28. Fusegi, T., Hyun, J.M., Kuwahara, K., Farouk, B.: A numerical study of three-dimensional natural convection in a differentially heated cubical enclosure. Int. J. Heat Mass Transf. **34**, 1543–1557 (1991)
29. Barakos, G., Mitsoulis, E., Assimacopoulos, D.: Natural convection flow in a square cavity revisited: laminar and turbulent models with wall functions. Int. J. Numer. Methods Fluids **18**, 695–719 (1994)
30. Loukili, M., Mordane, S.: New contribution to stokes equations. Adv. Appl. Fluid Mech. **20** (1), 107–116 (2017)
31. Loukili, M., Mordane, S.: New numerical investigation using meshless methods applied to the linear free surface water waves. Adv. Intell. Syst. Comput. **765**, 37–345 (2019)
32. Loukili, M., Mordane, S.: Numerical analysis of an absorbing boundary condition applied to the free surface water waves using the method of fundamental solutions. In: 8th International Conference on Modeling Simulation and Applied Optimization (ICMSAO), Bahrain (2019)
33. Loukili, M., El Aarabi, L., Mordane, S.: Computation of nonlinear free-surface flows using the method of fundamental solutions. Adv. Intell. Syst. Comput. **763**, 420–430 (2019)
34. Canuto, C., Hussaini, M.Y., Quarteroni, A., Zang, T.: Evolution to Complex Geometries and Applications to Fluid Dynamics. Spectral Methods, p. 596 (2007)

Geometrical Modelling Applied on Particular Constrained Optimization Problems

Lilla Korenova[1], Renata Vagova[2], Tomas Barot[1](✉), and Radek Krpec[1]

[1] Department of Mathematics with Didactics, Faculty of Education, University of Ostrava, Fr. Sramka 3, 709 00 Ostrava, Czech Republic
{Lilla.Korenova, Tomas.Barot, Radek.Krpec}@osu.cz

[2] Department of Mathematics, Faculty of Science, Constantine the Philosopher University in Nitra, Trieda A. Hlinku 1, 949 74 Nitra, Slovakia
Renata.Vagova@ukf.sk

Abstract. In favour of proposals of modified control techniques, geometrical analyses of the theoretical background of parts of the control algorithms should be advantageous. Concretely, the nonlinear optimization has been frequently considered as one of the important parts of the modern control strategies, e.g. the predictive control. Due to improving the control quality and minimization of control errors, the quadratic cost function can be generally included in the control algorithms. Therefore, the applied type of the nonlinear optimization is frequently specified as the quadratic programming problem with constraints. In this paper, the most occurred situations in this quadratic programming are geometrically modelled according to the stereo-metrical approach using the GeoGebra software. Advantages of achieved results can be suitably applied in the further proposals of the modified control methods.

Keywords: Nonlinear optimization · Quadratic programming · Constrained optimization · Geometrical models · GeoGebra

1 Introduction

For purposes of the improving the proposals of the control strategies [1–7], the geometrical analysis of the mathematical principles can be often advantageous. In the stereo-metrical approach displayed in the form of geometrical models, the more difficult principles can be obviously visualized.

The interdisciplinary connection of the pure mathematics with the applied mathematical approaches can be seen in the control system theory [1–7] or also in the applied signal processing, e.g. [8]. The control algorithms are significantly connected with the fulfilling the aims of the improving the control quality [1] together with minimization of the control errors [1]. For these purposes, an obtainment of the required efficient solution is frequently achieved using the optimization problem [3]. According to the control aims, the choice of the quadratic cost function [4] has the significant importance. Many control strategies contain the constraints on the control variables (output variable, manipulated variable or increments of the manipulated variable). As the one

R. Silhavy et al. (Eds.): CoMeSySo 2020, AISC 1295, pp. 178–188, 2020.
https://doi.org/10.1007/978-3-030-63319-6_16

of the modern control strategies, the predictive control [2–3] can be considered. In the predictive control, many modified algorithms have been proposed with connection of improving the optimization parts e.g. in offline case [5] or also in online case, e.g. [6–7]. Also, the similar problems can be seen in a prediction of the time-serialized data, e.g. in [9].

The nonlinear optimization [4] is generally based on minimization of the cost function according to constraint. In particular case of the quadratic programming [4], the quadratic cost function has been applied using the constraints in the control theory. There exists the analytical and numerical approach to solving the quadratic programming [3–4]. The first case is more detailed analyzed in this contribution. [4].

Including the constraints, solving the minimization process [4] is generally computational demanding from the view of the mathematical analysis; however, the quadratic programming has particular properties, which are appropriate for the mathematical expressions, e.g. the positive definitive hessian [10] of the cost function (in case of solving the minimization process).

The mathematical modelling can be divided into several cases e.g. in proposals of simulations [11] or in a visual modelling based on geometry. Geometry has been appeared also in the pure mathematical theory as well as e.g. in using in number theory [12]. The geometrical visualization of the pure mathematical problems is further focused in this paper.

Geometrical visualization of the pure mathematical problems has been widely researched yet. The GeoGebra software [13] with applications [14–19] can be considered as the frequently used geometrical solution, which has been applied by researchers and also in the educational strategies in teaching the mathematics. As can be seen on the main website of GeoGebra [13], this dynamic mathematical software obtained awards across the world in the field of mathematics and educational technologies, e.g. Archimedes 2016: MNU Award in category Mathematics, in Hamburg [13]. The advantageous site of the GeoGebra is the integration of the mathematical core CAS, which can be used for definition of the aimed problems using the analytical geometrical equations [13–19].

The presented geometrical analysis of the constrained quadratic optimization problems considers the following situations: solving the free dimensional extreme [3], bounded extreme with constraints with appearance of their intersections [3] and bounded extreme with constraints without appearance of their intersections [3]. All these three types of constrained optimization can be frequently seen in the technical practice and their geometrical visualization can be helpful to extension of published proposals, e.g. [7] or new further potential proposals of the applied optimization algorithms.

2 Particular Situations in Constrained Quadratic Optimization

The control system theory has the wide spectrum of applications, e.g. [20–22]. In the control algorithms, the nonlinear optimization [4] has been frequently appeared, e.g. in the predictive control [2–3]. For the fulfilling the aims of the minimization of control errors and improving the control quality, the quadratic cost function has been often considered, as can be seen e.g. in [3–4].

The quadratic optimization is a process (1), in which the n-dimensional global optimum $\boldsymbol{x}$ (2) of the quadratic cost function $f(\mathbf{x})$ (3) is determined and obtained with regards to constraints (4). This optimization problem (1)–(4) is the quadratic programming [3–4].

$$\boldsymbol{x} = \operatorname{argmin}(f(\boldsymbol{x}); \boldsymbol{Ax} \leq \boldsymbol{b}) \tag{1}$$

$$\boldsymbol{x} = \begin{bmatrix} x_1 & \cdots & x_n \end{bmatrix}^T \tag{2}$$

$$f(\boldsymbol{x}) = 0.5\boldsymbol{x}^T\boldsymbol{Hx} + \boldsymbol{c}^T\boldsymbol{x} \tag{3}$$

$$\boldsymbol{Ax} \leq \boldsymbol{b} \tag{4}$$

In the process of minimization of the quadratic cost function, matrix $\boldsymbol{H}$ (5) expresses the special property of the positive definiteness (6) and is equal to the Hess matrix [10] of the cost function. Matrix $\boldsymbol{H}$ is consisted of n rows and columns regarding the dimension of the variable $\boldsymbol{x}$. Vector $\boldsymbol{c}$ (7) is included in the linear part of the cost function and has n elements [3–4].

$$\boldsymbol{H} = \begin{bmatrix} H_{11} & \cdots & H_{1n} \\ \vdots & \ddots & \vdots \\ H_{n1} & \cdots & H_{nn} \end{bmatrix} = \begin{bmatrix} \frac{\partial f^2(\boldsymbol{x})}{\partial^2 x_1} & \cdots & \frac{\partial f^2(\boldsymbol{x})}{\partial x_1 \partial x_n} \\ \vdots & \ddots & \vdots \\ \frac{\partial f^2(\boldsymbol{x})}{\partial x_n \partial x_1} & \cdots & \frac{\partial f^2(\boldsymbol{x})}{\partial^2 x_n} \end{bmatrix} \tag{5}$$

$$\forall j \in \langle 1; n \rangle : \left| \frac{\partial f^2(\boldsymbol{x})}{\partial^2 x_1} \right| > 0, \ldots, \begin{vmatrix} \frac{\partial f^2(\boldsymbol{x})}{\partial^2 x_1} & \cdots & \frac{\partial f^2(\boldsymbol{x})}{\partial x_1 \partial x_j} \\ \vdots & \ddots & \vdots \\ \frac{\partial f^2(\boldsymbol{x})}{\partial x_j \partial x_1} & \cdots & \frac{\partial f^2(\boldsymbol{x})}{\partial^2 x_j} \end{vmatrix} > 0 \tag{6}$$

$$\boldsymbol{c}^T = \begin{bmatrix} c_1 & \cdots & c_n \end{bmatrix} \tag{7}$$

Constraints (4) contain the inequalities, which are relations in their geometrical interpretation. Concretely, the matrix-based definition can be specified as (8). A number of rows of matrix $\boldsymbol{A}$ and of vector $\boldsymbol{b}$ correspond with a number of appeared constraints m [3–4].

$$\begin{bmatrix} A_{11} & \cdots & A_{1n} \\ \vdots & \ddots & \vdots \\ A_{m1} & \cdots & A_{mn} \end{bmatrix} \begin{bmatrix} x_1 \\ \vdots \\ x_n \end{bmatrix} \leq \begin{bmatrix} b_1 \\ \vdots \\ b_m \end{bmatrix} \tag{8}$$

According to the [3], situation (9) without constraints can often be occurred also as the situation with parallel constraints and therefore provide the same free dimensional extreme $\boldsymbol{x}$ using rules (10)–(12) [3–4].

$$\boldsymbol{x} = \operatorname{argmin} f(\boldsymbol{x}) = \operatorname{argmin}\ (0.5\boldsymbol{x}^T\boldsymbol{H}x + \boldsymbol{c}^T\boldsymbol{x}) \tag{9}$$

$$\nabla f(\boldsymbol{x}) = [0 \quad \cdots \quad 0]^T \tag{10}$$

$$\frac{\partial f(\boldsymbol{x})}{\partial \boldsymbol{x}} = \boldsymbol{x}^T\boldsymbol{H} + \boldsymbol{c}^T = [0 \quad \cdots \quad 0]^T \tag{11}$$

$$\boldsymbol{x} = -\boldsymbol{H}^{-1}\boldsymbol{c}^T \tag{12}$$

In the situation with constraints, when some of them have an intersection, the solution is a bounded extreme. The inequality constraints are transferred into the form of equalities by including the additional variable $\boldsymbol{y}^2$. The solution of this modified optimization problem is then solved as minimization of the Lagrange function (13). Vector of Lagrange's multipliers $\boldsymbol{\lambda}$ is a column vector with m elements. Using the same structure, the vector of additional variables $\boldsymbol{y}^2$ is built [3–4].

$$\boldsymbol{L}(\boldsymbol{x}, \boldsymbol{\lambda}, \boldsymbol{y}) = f(\boldsymbol{x}) + \boldsymbol{\lambda}(\boldsymbol{A}\boldsymbol{x} - \boldsymbol{b} + \boldsymbol{y}^2) \tag{13}$$

The bounded extreme $\boldsymbol{x}$ (14) can be solved using the matrix Eq. (15) including only the equalities. There are always more possible solutions of this equation; however, only one solution with consideration of the positive Lagrange's multipliers is obtained as final [3–4].

$$[\boldsymbol{x} \quad \boldsymbol{\lambda} \quad \boldsymbol{y}]^T = \operatorname{argmin}(f(\boldsymbol{x}) + \boldsymbol{\lambda}(\boldsymbol{A}\boldsymbol{x} - \boldsymbol{b} + \boldsymbol{y}^2)) \tag{14}$$

$$\nabla \boldsymbol{L}(\boldsymbol{x}, \boldsymbol{\lambda}, \boldsymbol{y}) = [0 \quad \cdots \quad 0]^T \tag{15}$$

3 Proposed Structure of Geometrical Models in GeoGebra

For purposes of proposals of the geometrical modelling, the dimension is further considered as $n = 2$. The definition of the quadratic programming problem respecting this dimension have the form (16)–(19). However, a number of constraints m is not fixed.

In the software GeoGebra [13], the possibilities of the CAS modelling [13] are used in this contribution. The cost function has the algebraic equation without matrices, which were included in the expression (3), respectively (16). The multiplications are applied for an achievement of the required form (17):

$$f(\boldsymbol{x}) = 0.5[x_1 \quad x_2]\begin{bmatrix} H_{11} & H_{12} \\ H_{21} & H_{22} \end{bmatrix}\begin{bmatrix} x_1 \\ x_2 \end{bmatrix} + [c_1 \quad c_2]\begin{bmatrix} x_1 \\ x_2 \end{bmatrix} \tag{16}$$

$$f(\boldsymbol{x}) = 0.5H_{11}x_1^2 + 0.5(H_{21} + H_{12})x_1x_2 + 0.5H_{22}x_2^2 + c_1x_1 + c_2x_2 \tag{17}$$

Frequently, the free dimensional extreme (12) appears e.g. in the predictive control algorithm (e.g. in [3]). This computation can be expressed for the $n = 2$ as (18).

$$\boldsymbol{x} = -\begin{bmatrix} H_{11} & H_{12} \\ H_{21} & H_{22} \end{bmatrix}^{-1} \begin{bmatrix} c_1 & c_2 \end{bmatrix} \tag{18}$$

In case of the appearance of the free dimensional solution (18) in constrained quadratic programming problem, m constraints are parallel. The matrix inequalities (19) display the expression for $n = 2$.

$$\begin{bmatrix} A_{11} & A_{12} \\ \vdots & \vdots \\ A_{m1} & A_{m2} \end{bmatrix} \begin{bmatrix} x_1 \\ x_2 \end{bmatrix} \leq \begin{bmatrix} b_{11} \\ \vdots \\ b_{m1} \end{bmatrix} \tag{19}$$

For case of appearance of constraints with some intersection, the bounded extreme is computed regarding Eqs. (15), concretely as the minimization (20). Constraints can be also written as (19). Only the positive vector $\boldsymbol{\lambda}$ of Lagrange's multipliers is considered. In obtained results, pairs $(\lambda_1, y_1^2), \ldots, (\lambda_m, y_m^2)$ are not both equal to zero. After application of rule (15) on the definition (20), the solution of the system (21) can be obtained [3].

$$\begin{bmatrix} \boldsymbol{x} \\ \boldsymbol{\lambda} \\ \boldsymbol{y} \end{bmatrix} = \operatorname{argmin}\left(f(\boldsymbol{x}) + \begin{bmatrix} \lambda_1 \\ \vdots \\ \lambda_m \end{bmatrix} \left(\left(\begin{bmatrix} A_{11} & A_{12} \\ \vdots & \vdots \\ A_{m1} & A_{m2} \end{bmatrix} \begin{bmatrix} x_1 \\ x_2 \end{bmatrix} - \begin{bmatrix} b_{11} \\ \vdots \\ b_{m1} \end{bmatrix} + \begin{bmatrix} y_1^2 \\ \vdots \\ y_m^2 \end{bmatrix} \right) \right) \right) \tag{20}$$

$$\begin{bmatrix} \boldsymbol{x} \\ \boldsymbol{\lambda} \\ \boldsymbol{y} \end{bmatrix} = \begin{bmatrix} \begin{pmatrix} H_{11} & H_{12} \\ H_{21} & H_{22} \end{pmatrix} & \begin{pmatrix} A_{11} & \cdots & A_{m1} \\ A_{12} & \cdots & A_{m2} \end{pmatrix} & \begin{pmatrix} 0 & 0 & 0 \\ 0 & \ddots & 0 \\ 0 & 0 & 0 \end{pmatrix} \\ \begin{pmatrix} A_{11} & A_{12} \\ \vdots & \vdots \\ A_{m1} & A_{m2} \end{pmatrix} & \begin{pmatrix} 0 & 0 & 0 \\ 0 & \ddots & 0 \\ 0 & 0 & 0 \end{pmatrix} & \begin{pmatrix} 1 & 0 & 0 \\ 0 & \ddots & 0 \\ 0 & 0 & 1 \end{pmatrix} \\ \begin{pmatrix} 0 & 0 \\ \vdots & \vdots \\ 0 & 0 \end{pmatrix} & \begin{pmatrix} 1 & 0 & 0 \\ 0 & \ddots & 0 \\ 0 & 0 & 1 \end{pmatrix} & \begin{pmatrix} 1 & 0 & 0 \\ 0 & \ddots & 0 \\ 0 & 0 & 1 \end{pmatrix} \end{bmatrix}^{-1} \begin{bmatrix} 0 \\ 0 \\ b_1 \\ b_2 \\ 0 \\ 0 \end{bmatrix} \tag{21}$$

For purposes of the modelling in the GeoGebra software, constraints (19) should have the form known in the analytical geometry. These constraints are defined as (22) in their geometrical representation.

$$\left.\begin{array}{c} x_2 \leq \frac{1}{A_{12}}(-A_{11}x_1 + b_1) \\ \vdots \\ x_2 \leq \frac{1}{A_{m2}}(-A_{m1}x_1 + b_m) \end{array}\right\} \tag{22}$$

4 Results

With regards to the possible geometrical modelling, the variable n was set as 2. Then the cost function and constraints can be visualized in the stereo-metrical view in the software GeoGebra. Equations (16)–(22) can be used for the analytical solving the optimization problem, while all combinations of setting of Lagrange's multipliers $\boldsymbol{\lambda}$ and additional variables $\boldsymbol{y}$ are being equal to zero. However, pairs $(\lambda_1, y_1^2), \ldots, (\lambda_m, y_m^2)$ are not both equal to zero. The following examples were experimentally defined.

The first example for the geometrical modelling of the quadratic programming problem is the cost function (23) with the constraints (24). In the software GeoGebra, the cost function was defined in the analytical form (25) and the constraints were defined as (26).

$$f(\boldsymbol{x}) = 0.5[x_1 \quad x_2]\begin{bmatrix} 2 & 1 \\ 1 & 2 \end{bmatrix}\begin{bmatrix} x_1 \\ x_2 \end{bmatrix} \tag{23}$$

$$\begin{bmatrix} -0.75 & 0.8333 \\ -0.2747 & 0.4286 \end{bmatrix}\begin{bmatrix} x_1 \\ x_2 \end{bmatrix} \leq \begin{bmatrix} -0.5 \\ -0.5714 \end{bmatrix} \tag{24}$$

$$f(\boldsymbol{x}) = x_1^2 + x_1x_2 + x_2^2 \tag{25}$$

$$\left.\begin{array}{c} x_2 \leq 0.9x_1 - 0.6 \\ x_2 \leq 0.641x_1 - 1.333 \end{array}\right\} \tag{26}$$

In Fig. 1 obtained in the software GeoGebra for the defined problem (25)–(26), the graphically visualized solution $\boldsymbol{x}$ was achieved in the same form as the analytical expressed solution $\boldsymbol{x}$ in (27) for the assumed particular combination $\lambda_1 = 0, \lambda_2 \neq 0$ and $y_1^2 \neq 0, y_2^2 = 0$. Function value $f(\boldsymbol{x})$ occurred its minimum 0.6498.

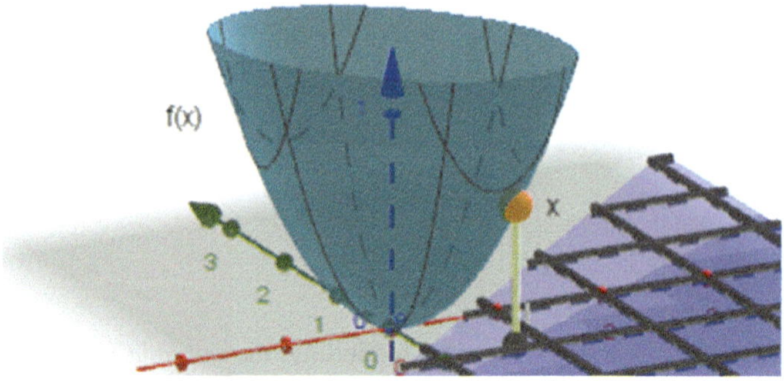

Fig. 1. Solving constrained quadratic programming problem (25)–(26) in GeoGebra

$$\left.\begin{array}{c}\begin{bmatrix} x_1 \\ x_2 \\ \lambda_1 \\ \lambda_2 \\ y_1^2 \\ y_2^2 \end{bmatrix} = \begin{bmatrix} \begin{pmatrix} 2 & 1 \\ 1 & 2 \end{pmatrix} & \begin{pmatrix} 0 & -0.2747 \\ 0 & 0.4286 \end{pmatrix} & \begin{pmatrix} 0 & 0 \\ 0 & 0 \end{pmatrix} \\ \begin{pmatrix} -0.75 & 0.8333 \\ -0.2747 & 0.4286 \end{pmatrix} & \begin{pmatrix} 0 & 0 \\ 0 & 0 \end{pmatrix} & \begin{pmatrix} 1 & 0 \\ 0 & 0 \end{pmatrix} \\ \begin{pmatrix} 0 & 0 \\ 0 & 0 \end{pmatrix} & \begin{pmatrix} 1 & 0 \\ 0 & 0 \end{pmatrix} & \begin{pmatrix} 0 & 0 \\ 0 & 1 \end{pmatrix} \end{bmatrix}^{-1} \begin{bmatrix} 0 \\ 0 \\ -0.5 \\ -0.5714 \\ 0 \\ 0 \end{bmatrix} \\ \begin{bmatrix} x_1 \\ x_2 \\ \lambda_1 \\ \lambda_2 \\ y_1^2 \\ y_2^2 \end{bmatrix} = \begin{bmatrix} 0.7414 \\ -0.8581 \\ 0 \\ 2.2743 \\ 0.7711 \\ 0 \end{bmatrix}\end{array}\right\} \tag{27}$$

In the second example displayed in Fig. 2, modified constraints (28) are considered for the same cost function (23), (25) with the analytical form (29) suitable for GeoGebra.

$$\begin{bmatrix} -0.5 & -0.1274 \\ 0.3167 & -0.3341 \end{bmatrix} \begin{bmatrix} x_1 \\ x_2 \end{bmatrix} \le \begin{bmatrix} 1 \\ 1.4219 \end{bmatrix} \tag{28}$$

$$\left.\begin{array}{c} x_2 \le -3.925x_1 + 7.849 \\ x_2 \le 0.948x_1 + 4.489 \end{array}\right\} \tag{29}$$

Achieved solution $\boldsymbol{x}$ (30) was determined for the assumed particular combination $\lambda_1 = 0, \lambda_2 = 0$ and $y_1^2 \neq 0, y_2^2 \neq 0$. Function value $f(\boldsymbol{x})$ had its minimum 0.

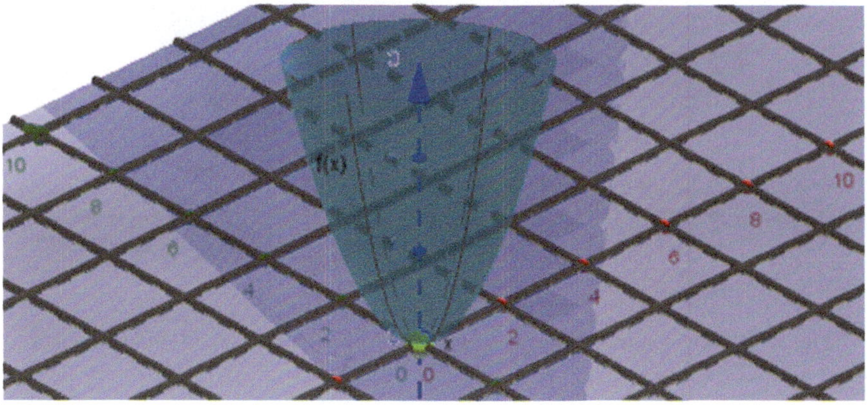

Fig. 2. Solving constrained quadratic programming problem (25), (28) in GeoGebra

$$\left.\begin{array}{l}
\begin{bmatrix} x_1 \\ x_2 \\ \lambda_1 \\ \lambda_2 \\ y_1^2 \\ y_2^2 \end{bmatrix} = \begin{bmatrix} \begin{pmatrix} 2 & 1 \\ 1 & 2 \end{pmatrix} & \begin{pmatrix} 0 & 0 \\ 0 & 0 \end{pmatrix} & \begin{pmatrix} 0 & 0 \\ 0 & 0 \end{pmatrix} \\ \begin{pmatrix} -0.5 & -0.1274 \\ 0.3167 & -0.3341 \end{pmatrix} & \begin{pmatrix} 0 & 0 \\ 0 & 0 \end{pmatrix} & \begin{pmatrix} 1 & 0 \\ 0 & 1 \end{pmatrix} \\ \begin{pmatrix} 0 & 0 \\ 0 & 0 \end{pmatrix} & \begin{pmatrix} 1 & 0 \\ 0 & 1 \end{pmatrix} & \begin{pmatrix} 0 & 0 \\ 0 & 0 \end{pmatrix} \end{bmatrix}^{-1} \begin{bmatrix} 0 \\ 0 \\ 1 \\ 1.4219 \\ 0 \\ 0 \end{bmatrix} \\
\begin{bmatrix} x_1 \\ x_2 \\ \lambda_1 \\ \lambda_2 \\ y_1^2 \\ y_2^2 \end{bmatrix} = \begin{bmatrix} 0 \\ 0 \\ 0 \\ 0 \\ 1 \\ 1.4219 \end{bmatrix}
\end{array}\right\} \tag{30}$$

The third example is based on the mathematical model (25) with constraint (31) realized in GeoGebra by Eq. (32). Achieved solution can be seen also in the geometrical representation in Fig. 3.

$$\begin{bmatrix} -0.75 & 0.8333 \end{bmatrix} \begin{bmatrix} x_1 \\ x_2 \end{bmatrix} \leq -0.5 \tag{31}$$

$$x_2 \leq 0.9x_1 - 0.6 \tag{32}$$

Solution $\boldsymbol{x}$ (33) was obtained and determined for the assumed particular combination $\lambda_1 \neq 0$ and $y_1^2 = 0$. Function value $f(\boldsymbol{x})$ had its minimum 0.0996.

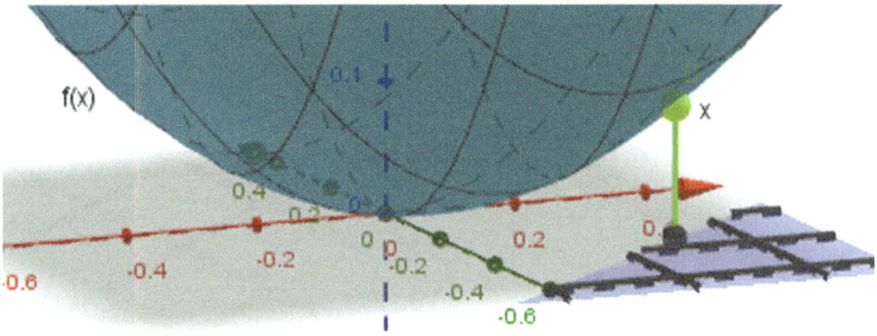

Fig. 3. Solving constrained quadratic programming problem (25), (31) in GeoGebra

$$\left.\begin{array}{c}\begin{bmatrix} x_1 \\ x_2 \\ \lambda_1 \\ y_1^2 \end{bmatrix} = \begin{bmatrix} \begin{pmatrix} 2 & 1 \\ 1 & 2 \end{pmatrix} & \begin{pmatrix} -0.75 \\ 0.8333 \end{pmatrix} & \begin{pmatrix} 0 \\ 0 \end{pmatrix} \\ (-0.75 \quad 0.8333) & (0) & (0) \\ (0 \quad 0) & (0) & (1) \end{bmatrix}^{-1} \begin{bmatrix} 0 \\ 0 \\ -0.5 \\ 0 \end{bmatrix} \\ \begin{bmatrix} x_1 \\ x_2 \\ \lambda_1 \\ y_1^2 \end{bmatrix} = \begin{bmatrix} 0.31 \\ -0.321 \\ 0.3985 \\ 0 \end{bmatrix} \end{array}\right\} \tag{33}$$

5 Conclusion

In this contribution, the geometrical models of particular situations in the quadratic optimization were proposed and practically modelled in the software GeoGebra. With regards to the including the constraints, following specific situations were considered: the free dimensional extreme and the bounded extremes. In case of the bounded extremes, the appearance of the intersection and disappearance of the intersection of the constraints were modelled. As the potential advantage of this presented analysis, the proposals of the modified control strategies can be further improved or extended using the presented types of models. The proposals of the control algorithm can be more efficiently modified and based on the practical visualization in the Euclidian three-dimensional space.

Acknowledgements. This paper was realized with the financial support of the SGS project at University of Ostrava, Faculty of Education: SGS05/PdF/2019–2020.

References

1. Corriou, J.P.: Process Control: Theory and Applications. Springer, Heidelberg (2004)
2. Huang, S.: Applied predictive control. Springer, Heidelberg (2002)

3. Wang, L.: Model Predictive Control System Design and Implementation Using MATLAB. Springer, Heidelberg (2009)
4. Dostal, Z.: Optimal Quadratic Programming Algorithms: With Applications to Variational Inequalities. Springer, Heidelberg (2009)
5. Ingole, D., Holaza, J., Takacs, B., et al.: FPGA-based explicit model predictive control for closed loop control of intravenous anesthesia. In: 20th International Conference on Process Control, pp. 42−47. Institute of Electrical and Electronics Engineers Inc. (2015). https://doi.org/10.1109/PC.2015.7169936
6. Belda, K.: On-line solution of system constraints in generalized predictive control design: convenient way to cope with constraints. In: 20th International Conference on Process Control, pp. 25–30. Institute of Electrical and Electronics Engineers Inc. (2015). https://doi.org/10.1109/PC.2015.7169933
7. Barot, T., Krpec, R., Kubalcik, M.: Applied quadratic programming with principles of statistical paired tests. In: Computational Statistics and Mathematical Modeling Methods in Intelligent Systems, Advances in Intelligent Systems and Computing, vol. 1047, pp. 278−287. Springer, Heidelberg (2019). ISBN 978–3–030–31361–6. https://doi.org/10.1007/978-3-030-31362-3_27
8. Barot, T., Burgsteiner, H., Kolleritsch, W.: Comparison of discrete autocorrelation functions with regards to statistical significance. In: Advances in Intelligent Systems and Computing. Springer, Heidelberg (2020). ISSN 2194–5357 (in Print)
9. Burgsteiner, H., Kroll, M., Leopold, A., et al.: Movement prediction from real-world images using a liquid state machine. Appl. Intell. **26**(2), 99–109 (2007). https://doi.org/10.1007/s10489-006-0007-1
10. Barta, C., Kolar, M., Sustek, J.: On Hessians of composite functions. Appl. Math. Sci. **8** (172), 8601–8609 (2014). https://doi.org/10.12988/ams.2014.410805
11. Vaclavik, M., Sikorova, Z., Barot, T.: Approach of process modeling applied in particular pedagogical research. In: Cybernetics and Automation Control Theory Methods in Intelligent Algorithms, Advances in Intelligent Systems and Computing, vol. 986, pp. 97–106. Springer, Heidelberg (2019). https://doi.org/10.1007/978-3-030-19813-8_11
12. Schmidt, W.M., Summerer, L.: Parametric geometry of numbers and applications. Acta Arithmetica **140**(1), 67–91 (2009). https://doi.org/10.4064/aa140-1-5
13. GeoGebra Homepage. https://www.geogebra.org. Accessed 14 May 2020
14. Vagova, R., Kmetova, M.: The role of visualisation in solid geometry problem solving. In: 17th Conference on Applied Mathematics, pp. 1054–1064. STU Bratislava. (2018)
15. Duris, V., Sumny, T., Pavlovicova, G.: Student solutions of non-traditional geometry tasks. TEM J. **8**(2), 642–647 (2019)
16. Lavicza, Z., Papp-Varga, Z.: Integrating GeoGebra into IWB-equipped teaching environments: preliminary results. Technol. Pedagogy Educ. **19**(2), 245–252 (2010). https://doi.org/10.1080/1475939X.2010.491235
17. Korenova, L.: GeoGebra in teaching of primary school mathematics. Int. J. Technol. Math. Educ. **24**(3), 155–160 (2017)
18. Arzarello, F., Ferrara, F., Robutti, O.: Mathematical modelling with technology: the role of dynamic representations. Teach. Math. Appl. **31**(1), 20–30 (2012). https://doi.org/10.1093/teamat/hrr027
19. Vallo, D., Duris, V.: Geogebra and logarithmic spiral in educational process. In: 17th Conference on Applied Mathematics, pp. 1076–1082. STU Bratislava (2018)
20. Navratil, P., Pekar, L., Matusu, R.: Control of a multivariable system using optimal control pairs: a quadruple-tank process. IEEE Access **8**, 2537–2563 (2020). https://doi.org/10.1109/ACCESS.2019.2962302

21. Pekar, L., Gazdos, F.: A potential use of the balanced tuning method for the control of a class of time-delay systems. In: 22th International Conference on Process Control, pp. 161–166. Institute of Electrical and Electronics Engineers Inc. (2019)
22. Chramcov, B., Marecki, F., Bucki,R.: Heuristic control of the assembly line. Advances in Intelligent Systems and Computing, vol. 348, 189–198. Springer (2015). ISBN 978–3–319–18502–6. https://doi.org/10.1007/978-3-319-18503-3_19

The Mathematical Model for Express Analysis of the Oilfield Development Performance in Waterflooding

Ivan Vladimirovich Afanaskin(✉)

Federal State Institution "Scientific Research Institute for System Analysis of the Russian Academy of Science", 36, Nakhimovsky Avenue, Moscow 117218, Russia
ivan@afanaskin.ru

Abstract. The paper presents an improved mathematical model by V.S. Kovalev and M.L. Surguchev for prompt assessment of the oil reservoir waterflooding. It proposes an original method for the stream line building to transform a well system flow in a curved gallery flow. The model is based on two methods: the stream-tube method and the curved gallery method. A comparison of the values acquired from the proposed model-based calculation with those computed with a commercial hydrodynamic simulator confirms the satisfactory precision of the method from the practical point of view.

Keywords: Waterflooding · Flow paths · Stream tube

1 Introduction

The most common oilfield development method in Russia (and a universally widespread method) is waterflooding, i.e. water injection into oil reservoirs through special injection wells to drive the oil, to push it through to the production wells and to maintain the reservoir pressure. At that, the majority of the oilfields is at the 3rd-4th stage of the development. This indicates a high water content in the well product (up to 90–98%). As the target product of the wells is oil, one of the waterflooding optimization tasks is to minimize the water recovery, at the same time maintaining or even increasing the oil production rate. To fulfil this task, there is a special set of actions referred to as reservoir management employed at the oilfield. At all decision-making levels (from the R&D centre supporting the oilfield development to the oil and gas production directorate), the reservoir management requires many computations including those related to the waterflooding modelling. However, it is not always reasonable to use sophisticated and costly commercial simulation programs. They require a big amount of input data and high qualification of the users. Many reservoir management problems can be solved with analytical, semi-analytical or simple numerical models. One of them is the semi-analytical stream tube model. To a certain extent, it accounts for the non-piston nature of oil displacement (the piston displacement model is used, but the imperfection of water drive is taken into account) and the layout of the actual wells, but does not fully consider the unevenness of the porosity

R. Silhavy et al. (Eds.): CoMeSySo 2020, AISC 1295, pp. 189–203, 2020.
https://doi.org/10.1007/978-3-030-63319-6_17

and permeability properties of the reservoir. The unevenness of the reservoir around the area is indirectly taken into account at the stream line modelling stage. The estimation of the water motion for all the stream tubes requires using one permeability distribution value, describing the waterflooding layer by layer. If necessary, individual well calculations with the permeability distribution adjustment to every well can be carried out. This paper is dedicated to the development of this approach.

2 Stream Tube Model with Account for the Influence of the Phase Mobility on the Uneven Reservoir Waterflooding Indexes

With modern computers, some old calculation methods can be seriously reconsidered. Particularly, one of such methods is using the stream tube model. Different stream tube models [1–6] have been considered in the studies of such famous researchers as N. Dykstra and R. Parsons, J. Arps, I.F. Kuranov, V.S. Kovalev, M.L. Surguchev, Yu. P. Borisov, M.L. Sattarov, B.F. Sazonov, A.M. Pirverdyan and many others.

In the employed oilfield waterflooding calculation methods with account for the unevenness of the reservoir permeability, water is usually assumed to move along the stream tubes starting from the waterflooding commencement to the water breakthrough moment at a constant speed proportionate to the reservoir permeability. However, when the water viscosity is lower than the viscosity of oil in the interlayers of different permeability, the waterflooding frontline displacement proportionality condition is only true at the very beginning and the very end of the reservoir development, while at other waterflooding stages it is broken.

I.F. Kuranov came up with a calculation method with due respect to the fluid motion rate change by layers with the change of their permeability values. But this method is quite complicated to be used in practice. The labour-intensiveness of this method grows with the number of considered layers. The world-famous study by N. Dykstra and R. Parsons was also dedicated to the uneven reservoir waterflooding assessment method with account for the difference in the viscosity of oil and water. A similar method was proposed by V.S. Kovalev and M.L. Surguchev.

Let us take a look at water-oil displacement through the stream tubes. The fluids are incompressible, the solid matrix is non-deformable. The displacement is piston-type displacement. A difference of viscosities and maximum values of the relative oil and water phase permeability is taken into account. Moreover, the imperfection of the water drive is considered. In this situation, it is possible to use the method of V.S. Kovalev and M.L. Surguchev [3–5] for calculating the influence of viscosity and phase permeability on the waterflooding output at the permanent porosity, initial oil saturation and water-oil displacement factor for the gallery fluid recovery situation. When waterflooding unevenly stratified reservoirs with the viscosity values exceeding that of water, the stratum-wise flooding front line displacement causes the increase in the production rate of every layer at the permanent pressure difference between the injection and recovery lines. The production rate of any layer at any time can be calculated with the commonly known formula:

$$q_i = \frac{k_i \cdot A}{\frac{\mu_o}{k_o} \cdot (L - L_i) + \frac{\mu_w}{k_w} \cdot L_i} \tag{1}$$

where A is the coefficient considering the geometric dimensions and the pressure difference between the injection and recovery lines; L is the distance from the initial oil pool outline to the recovery line; L_i is the distance of the outline displacement in the ith layer; k_o/μ_o and k_w/μ_w are oil and water mobility values, respectively; k_i is absolute permeability of the ith layer; k_o and k_w are phase permeability values for oil and water in the oil-saturated and the watered field zone, respectively.

An average production rate of the layer (hereinafter marked with the "AV" index) where the water breakthrough occurred (marked with the "a" index) during the waterflooding operation can be calculated by applying the mean resistance rate at the beginning of waterflooding and after water breakthrough into formula (1):

$$q_{aAV} = \frac{k_a \cdot A}{\frac{L \cdot \left(\frac{\mu_o}{k_o} + \frac{\mu_w}{k_w}\right)}{2}} \tag{2}$$

Similarly, for the ith layer:

$$q_{iAV} = \frac{2 \cdot k_i \cdot A}{L \cdot \left(\frac{\mu_o}{k_o} + \frac{\mu_{AV}}{k_{AV}}\right)} \tag{3}$$

where

$$\frac{\mu_{AV}}{k_{AV}} = \frac{\mu_o}{k_o} + \frac{L_i}{L} \cdot \left(\frac{\mu_w}{k_w} - \frac{\mu_o}{k_o}\right),$$

also

$$\frac{L_i}{L} = \beta_{COV\,i}$$

where $\beta_{COV\,i}$ is the volumetric efficiency for the ith layer.

Therefore, formula (3) can be presented as follows (shifting L coefficient to A coefficient, as for the considered layer element, the distance from the initial oil pool outline to the recovery line is a constant value):

$$q_{iAV} = \frac{2 \cdot k_i \cdot A}{\frac{2 \cdot \mu_o}{k_o} + \beta_{COV\,i} \cdot \left(\frac{\mu_w}{k_w} - \frac{\mu_o}{k_o}\right)} \tag{4}$$

Applying (2) and (4) to the ratio for the volumetric efficiency of waterflooding in the ith layer $\beta_{COV\,i} = q_{iAV}/q_{aAV}$, and indicating the mobility ratio as $E = k_o\ \mu_w/(k_w\ \mu_o)$, after some simple transformations we get:

$$\beta_{COV\,i}^2 \cdot (E-1) \cdot k_a + \beta_{COV\,i} \cdot 2 \cdot k_a - k_i \cdot (E+1) = 0 \tag{5}$$

Having solved the quadratic Eq. (5), we get:

$$\beta_{COV\,i} = \left(-1 \pm \sqrt{1 + \frac{(E^2-1) \cdot k_i}{k_a}}\right) / (E-1) \tag{6}$$

In the formula (6), the second component in the brackets should be a positive value. Considering the entire reservoir, where n is the total number of interlayers and a is the number of the layer waterflooded at the present moment, we get the following equation for the reservoir waterflooding volumetric efficiency:

$$\beta_{COV} = \frac{a}{n} - \frac{n-a}{(E-1) \cdot n} + \frac{1}{(E-1) \cdot n} \cdot \sum_{i=1}^{n-a} \sqrt{1 + \frac{(E^2-1) \cdot k_i}{k_a}} \tag{7}$$

Oil content in the recovered product f_o upon achievement of a certain volumetric efficiency value can be assessed with the following ratio:

$$f_o = \frac{1}{1 + \frac{q_w}{q_o}} \tag{8}$$

where q_w and q_o are the water and oil production rates.

Water and oil production rates at the waterflooding of the layer a can be found with the following formulas:

$$q_w = \sum_{i=n-a+1}^{n} \frac{k_i \cdot A}{\frac{\mu_w}{k_w}} \tag{9}$$

$$q_o = \sum_{i=1}^{n-a} \frac{2 \cdot k_i \cdot A}{\frac{2 \cdot \mu_o}{k_o} + \beta_{COV\,i} \cdot \left(\frac{\mu_w}{k_w} - \frac{\mu_o}{k_o}\right)} \tag{10}$$

Applying formulas (9) and (10) to (8) and having done the required transformations, we get:

$$f_o = \frac{1}{1 + \frac{\sum_{i=n-a+1}^{n} k_i}{E \cdot \sum_{i=1}^{n-a} \frac{k_i}{\sqrt{1 + \frac{(E^2-1) \cdot k_i}{k_a}}}}} \tag{11}$$

To describe the process of water-oil displacement, it is not only the waterflooding volumetric efficiency factor that we need to know but also the amount of the fluid injected through the reservoir by the present moment (dimensionless injected volume) τ:

$$\tau = \frac{q_{oAV} \cdot t_a + q_{w1AV} \cdot t_i + q_{w2AV} \cdot (t_a - t_i)}{Q_o} \tag{12}$$

where Q_o is the mobile oil stock.

The average water and oil production rate for the period from the development commencement to the complete waterflooding of the ith and ath layers equals to:

$$q_{w1AV} = \sum_{i=n-a+1}^{n} \frac{2 \cdot k_i \cdot A}{L \cdot \left(\frac{\mu_o}{k_o} + \frac{\mu_w}{k_w}\right)}, \quad q_{w2AV} = \sum_{i=n-a+1}^{n} \frac{k_i \cdot A}{L \cdot \frac{\mu_w}{k_w}}, \tag{13}$$

$$q_{oAV} = \sum_{i=1}^{n-a} \frac{2 \cdot A \cdot k_i}{\left[2 \cdot \frac{\mu_o}{k_o} + \beta_{COV\,i} \cdot \left(\frac{\mu_w}{k_w} - \frac{\mu_o}{k_o}\right)\right] \cdot L} = \frac{2 \cdot A \cdot k_o}{L \cdot \mu_o} \sum_{i=1}^{n-a} \frac{k_i}{2 + \beta_{COV\,i} \cdot (E-1)} \tag{14}$$

In the formulas (12)–(14), q_{w1AV} is the average production rate of the layers with the permeability from k_a to k_{max} for the water pre-breakthrough period; q_{w2AV} is the production rate of the same layers after the water breakthrough; q_{oAV} is the average oil production rate (mean average production rate of the layers with the permeability from k_{min} to k_a); t_i and t_a are the ith and ath layer waterflooding time, respectively.

If $Q_o = V \cdot m \cdot S_o \cdot \beta_{DISP} = n \cdot q_{aAV} \cdot t_a$ stands for the mobile oil stock where m, S_o and β_{DISP} are the average porosity, reservoir oil saturation and the oil displacement coefficient, V is the reservoir volume, then for the waterflooding of the ith and ath layers it can be assumed that:

$$t_a = \frac{Q_o}{n \cdot q_{aAV}} = \frac{Q_o}{\frac{2 \cdot n \cdot A \cdot k_a}{(1+E) \cdot \frac{\mu_o}{k_o} \cdot L}} \text{ and } t_i = \frac{Q_o}{\frac{2 \cdot n \cdot A \cdot k_i}{(1+E) \cdot \frac{\mu_o}{k_o} \cdot L}}. \tag{15}$$

Consequently, the permeability of the layer waterflooded by the present moment can be recorded as:

$$k_a = \frac{L^2 \cdot m \cdot S_o \cdot \beta_{DISP} \cdot (E+1)}{2 \cdot \Delta p \cdot t_a \cdot \frac{k_o}{\mu_o}}. \tag{16}$$

Applying ratios (13)–(15) to formula (12), we get:

$$\tau = \frac{1+E}{k_a \cdot n} \cdot \left(\sum_{i=n-a+1}^{n} \frac{k_i}{2 \cdot E} + \sum_{i=1}^{n-a} \frac{k_i}{1 + \sqrt{1 + \frac{(E^2+1) \cdot k_i}{k_a}}} \right) + \frac{(E-1) \cdot a}{2 \cdot E \cdot n}. \tag{17}$$

For a reservoir consisting of an unlimited number of layers with the unevenness expressed through the permeability distribution, the volumetric efficiency, oil content in the product and the amount of recovered fluid can be calculated with the formulas:

$$\beta_{COV} = [1 - F(k_a)] - \frac{F(k_a)}{E-1} + \frac{1}{E-1} \cdot \int_0^{k_a} \sqrt{1 + \frac{(E^2-1) \cdot k}{k_a}} dF(k)$$
$$= 1 - F(k_a) \cdot \frac{E}{E-1} + \frac{1}{E-1} \cdot \int_0^{k_a} \sqrt{1 + \frac{(E^2-1) \cdot k}{k_a}} dF(k) \quad (18)$$

$$f_o = \frac{1}{1 + \frac{\int_{k_a}^{k_{MAX}} kdF(k)}{E \cdot \int_0^{k_a} \frac{kdF(k)}{\sqrt{1 + \frac{(E^2-1) \cdot k}{k_a}}}}}, \quad (19)$$

$$\tau = \frac{1+E}{k_a} \cdot \left[\int_{k_a}^{k_{MAX}} \frac{kdF(k)}{2 \cdot E} + \int_0^{k_a} \frac{kdF(k)}{1 + \sqrt{1 + \frac{(E^2-1) \cdot k}{k_a}}} \right] + \frac{(E-1) \cdot (1 - F(k_a))}{2 \cdot E}, \quad (20)$$

where $F(k)$ is the permeability distribution function.

Knowing these dimensionless parameters, the other waterflooding indexes, such as oil, water, fluid recovery and water cut rates can be calculated. Time referencing is done with the permeability of the interlayer waterflooded by the present moment, formula (16).

3 Method of Accounting for the Influence Made by the Well Network Parameters on the Waterflooding Values and the Oil Recovery from Uneven Reservoirs

Let us improve the method proposed in [3–5].

The influence made by the well network parameters on the waterflooding values and the layer oil recovery rates is proposed to be accounted for by transforming the complicated non-linear fluid flow to the well system into a direct flow to a curved gallery (Fig. 1).

The suggested calculation method envisages the following restrictions:

1. porosity, thickness and initial oil saturation values are constant throughout the bed course;
2. for displacement properties calculation, the water-oil piston displacement model for a layered reservoir can be used (unevenness of thickness prevails over the unevenness of area);

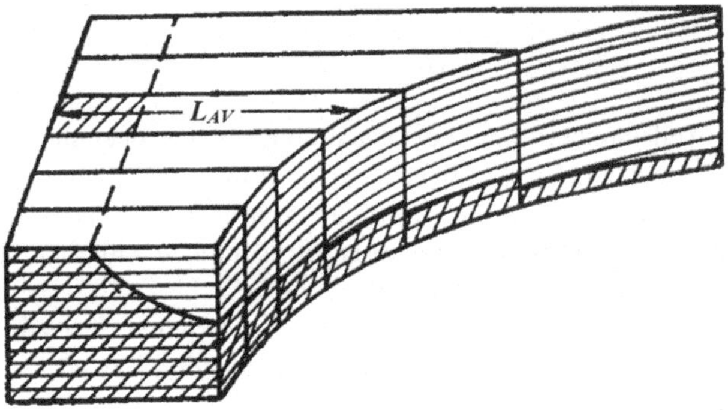

Fig. 1. Curved gallery diagram [4]. Fluid flow from left to right. Water-washed area is shaded.

3. external boundary pressure is constant;
4. bottom-hole pressure is constant and equal at all wells;
5. wells are activated and deactivated at the same time;
6. reservoir shape can be idealized without material loss in the calculation precision;
7. external boundary shape can be idealized without material loss in the calculation precision;
8. the initial water-oil zone of the reservoir is not big and its dimensions can be neglected;
9. for stream line design, the fluid production rate is considered to be constant;
10. bottom-hole pressure of the wells is higher than the oil gas saturation pressure.

Some of these restrictions can be easily bent by introducing minor changes into the model.

For the first approximation, the reservoir is considered even and continuous. The fluid is also considered to be homogenous. At first, an idealized, e.g. rectangular-shaped reservoir is designed with the volumetric method formula for the geological oil reserves calculation:

$$V_{o,0} = a \cdot b \cdot h \cdot m \cdot S_o / B_o, \tag{21}$$

where S_o is the average oil saturation of the reservoir, a and b are the length and width of the reservoir, h is the average thickness of the reservoir, m is open porosity, B_o is the oil volume factor, $V_{o,0}$ are the initial oil-in-place in the volume measurement units. The oil reserves of the idealized figure shall fit the actual site reserves.

Then, using Laplace's equation for the potential Φ (22) and one of the Cauchy-Riemann stream function conditions Ψ (23), the equipotential map and even fluid-stream lines for an even reservoir are made up:

$$\Phi(x, y) = \frac{1}{2 \cdot \pi} \sum_{i=1}^{n} \left[q_i \cdot \ln\left(\sqrt{(x - x_i)^2 + (y - y_i)^2} \right) \right] + const, \tag{22}$$

$$\psi(x,y) = \int \frac{\partial \Phi}{\partial x} dy + \int \frac{\partial \Phi}{\partial y} dx + const, \tag{23}$$

where i is the well number, n is the total number of the operating wells, q_i is the recovered fluid production rate (injected water flow) of the ith well per effective reservoir thickness unit, x and y are Cartesian coordinates, x_i and y_i are the ith well coordinates. The area unevenness of the reservoir is implied with the uneven values of the production (flow) rates of various wells.

If a more precise area unevenness calculation is required, instead of the equipotential map, the isobar map built with the Laplace equation is used for the pressure:

$$\begin{gathered} \frac{\partial}{\partial x}\left[\frac{k(x,y)h(x,y)}{\mu_l}\frac{\partial p}{\partial x}\right] + \frac{\partial}{\partial y}\left[\frac{k(x,y)h(x,y)}{\mu_l}\frac{\partial p}{\partial y}\right] = -q_l^*(x,y)\delta(x,y) \\ \Phi(x,y) = \frac{k(x,y)}{\mu_l} p(x,y), \end{gathered} \tag{24}$$

where $k(x, y)$ is the 2D absolute permeability field of the reservoir, $h(x, y)$ is the effective reservoir thickness, μ_l is the effective fluid viscosity (μl = const), p = p(x, y) ($\mu_l = const$), $p = p(x, y)$ is the pressure field in the reservoir, $q_l^*(x, y)$ is the water source (fluid outlet) density, $\delta(x, y)$ is the Dirac delta function. The value of $q_l^*(x, y)$ is positive for the fluid outlet (production well) and negative for the water source (injection well). It should be noted that modelling filtration processes with the stream line method is quite popular in the oil industry. This technology has been applied at a high level, for example, in [7].

In [3–5], the stream lines were proposed to be built with an electric integrator, i.e. an analogous physical capacitance-resistance model of the oil reservoir based on the electric hydrodynamic analogy method principle. Today, this is quickly and easily done with a computer using the mentioned method and commonly applied mathematical software to be described below.

A stream line in this paper is understood as a line, the direction of the tangent to which at every point coincides with the fluid particle velocity vector at this point, i.e. at every moment of time the fluid particle is moving along the stream line. The stream lines are perpendicular to the equipotentials (equal potential lines) and do not intercross. Equal potential lines are parallel to the isobars.

With a stream line map, the statistic stream line length distribution $F(L)$ can be made up. Then, based on the geophysical well survey data and the core survey, the statistical permeability distribution $F(k)$ can be found.

The second approximation assumes that the reservoir can be regarded as an aggregation of uneven elements (stream tubes) of different length, a function of stream distribution along the length $F(L)$ and with the permeability change in every element described with the permeability distribution function $F(k)$. A stream tube here is understood as a surface limited with two neighbouring stream lines. The changes in the waterflooding indexes of any selected element with the length L can be found as usual for an uneven reservoir with the gallery fluid recovery, for example, using formulas (18)–(20). The values of these indexes for the reservoir as a whole (aggregation of uneven elements) will be expressed with their mean weighed values:

$$\beta_{COV} = \int_{L_{MIN}}^{L_{MAX}} \frac{L}{L_{AV}} \cdot \beta_{COV}(L) dF(L), \tag{25}$$

$$f_o = \int_{L_{MIN}}^{L_{MAX}} \frac{q_l(L)}{q_{lAV}} \cdot f_o(L) dF(L), \tag{26}$$

$$\tau = \int_{L_{MIN}}^{L_{MAX}} \frac{L}{L_{AV}} \cdot \tau(L) dF(L), \tag{27}$$

$$L_{AV} = \int_{L_{MIN}}^{L_{MAX}} L dF(L), \tag{28}$$

$$q_{lAV} = \int_{L_{MIN}}^{L_{MAX}} \frac{L}{L_{AV}} \cdot q_{lAV}(L) dF(L), \tag{29}$$

$$q_l(L) = \int_{k_a(L)}^{k_{MAX}} k dF(k) + E \cdot \int_0^{k_a(L)} \frac{k dF(k)}{\sqrt{1 + \frac{(E^2-1) \cdot k}{k_a(L)}}}, \tag{30}$$

where L_{AV}, L_{MAX} и L_{MIN} are the average, maximum and minimum stream line length in the statistical stream line length distribution $F(L)$ respectively, $q_l(L)$ is the dimensionless fluid production rate of the stream tube with the length L, q_{lAV} is the average dimensionless fluid production rate for all the stream tubes. It should be remarked that in a multi-well system, the dependency of the stream line distribution density $f(L)$ may be quite complex, for example, having several points of extremum.

The calculation can be done both for the reservoir as a whole or for every individual well separately. Separate well calculations make it possible to take account for the non-simultaneous commissioning and decommissioning of the wells.

Such calculations can be easily done with any mathematic software system, such as Maple – Waterloo Maple Inc., Mathematica – Wolfram Research, MATLAB – The MathWorks or Mathcad – PTC [8–11].

With the described method, it is much easier to build a filtration model, compared to a full-scale discrete filtration model in a commercial hydrodynamic simulator using the system of classical mass and impulse conservation equations as a mathematical model, such as [12, 13]. However, that involves some allowances. The promptness of building the stream tubes with the described method is the advantage that allows using it for express analysis of the oilfield development values in case of waterflooding.

4 Example of Applying the Described Mathematical Model

The waterflooding and oil output values with respect to the well network parameters were calculated for an existing oil reservoir of one Stavropol Territory oilfield. Accumulation type: partially lithologically screened sheet oil pool, roof deposit. Reservoir type: terrigenous, porous. Oil productive area constitutes 8,124 thousand square meters, average net oil thickness 2.8 m, net productive volume 23,070 thousand cubic meters, open porosity factor 0.18 u.f., initial oil saturation 0.63 u.f, reservoir oil density 0.806 g/cm^3, initial oil-in-place 1,610 thousand tonnes and gas content 103 m^3/tonne.

Net oil thickness map is presented in Fig. 2 (well numbers modified). Some formation parameters have been modified to keep the commercial secret of the subsoil user company.

At the first approximation, with formula (21) and the idealized reservoir width of 2950 m, the reservoir length of 2813 m was acquired. Then formulas (22) and (23) were used with the well fluid production rate and the open net oil thickness (Table 1) to make up the equipotential and stream line map (Fig. 3).

With the stream line map, the statistical stream line length distribution $F(L)$ was built using the logarithmic-normal distribution law:

$$F(L) = \frac{1}{2} \cdot \left[1 + erf\left(\frac{\ln L - \ln \varepsilon_L}{\sqrt{2} \cdot \sigma_L} \right) \right], \tag{31}$$

distribution parameters σ_L= 0,829 and ε_L= 879, maximum and minimum stream line lengths L_{MAX} = 5860 m, L_{MIN} = 589 m.

Then, based on the geophysical well survey data and the core survey and using M. M. Sattarov's law, the statistical permeability distribution $F(k)$ was found.

$$F(k) = erf\sqrt{\frac{k-a}{k_0}} - \frac{2}{\sqrt{\pi}} \cdot \exp\left(-\frac{k-a}{k_0} \right) \cdot \sqrt{\frac{k-a}{k_0}}, \tag{32}$$

distribution parameters $a = -0{,}0446$ and $k_0 = 0{,}0479$, maximum and minimum permeability $k_{MAX} = 0{,}1709$ μm^2, $k_{MIN} = 0{,}0105$ μm^2.

At the second approximation, the reservoir was regarded as an aggregation of uneven elements (stream tubes) of different length, a function of stream distribution along the length $F(L)$ and with the permeability change in every element described with the permeability distribution function $F(k)$. Relative oil and water phase permeabilities for the piston displacement models were taken from the relative phase permeability graph for the joint fluid stream in the reservoir and assumed the permeabilities equal to the ones at the bound water saturation and residual oil saturation of the reservoir. The respective permeabilities constitute 1.00 and 0.39 u.f. respectively. Oil viscosity is 0.41 mPa•s, water viscosity is 0.36 mPa•s, displacement factor is 0.634 u.f., pressure drawdown is 1.5 mPa. Formulas (16), (18)–(20), (25)–(30) were used to calculate the reservoir waterflooding and oil output values (as a product of the volumetric efficiency coefficient by the displacement factor). Then they were converted into the annual and cumulative production, see Figs. 4 and 5. The forecast period was 23 years. By the end

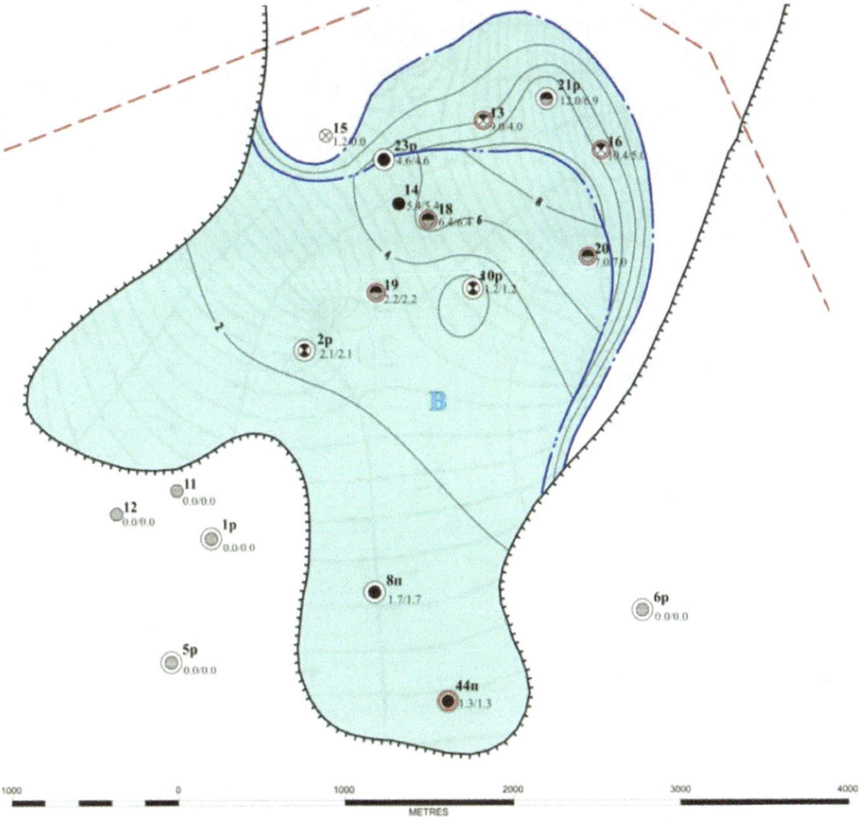

Fig. 2. Net oil thickness map.

Table 1. Equipotential and stream line map building data

Well no.	Well production rate, m^3/day	Net oil thickness, m
18	34.4	6.4
19	34.9	2.2
20	59.5	7.0
21	86.4	6.9
23	35.1	4.6

of the development period, the oil share in the gallery product constituted 1.9% (water cut 98.1%), waterflooding volumetric efficiency 0.453 u.f., injected volume 2.271 units, oil recovery factor 0.287 u.f., cumulative oil production 462 thousand tonnes and cumulative fluid production 3656 thousand tonnes.

To verify the applicability of the stream tube model to the commercial hydrodynamic simulator Rubis Kappa Engineering [14], the entire set of the oilfield properties

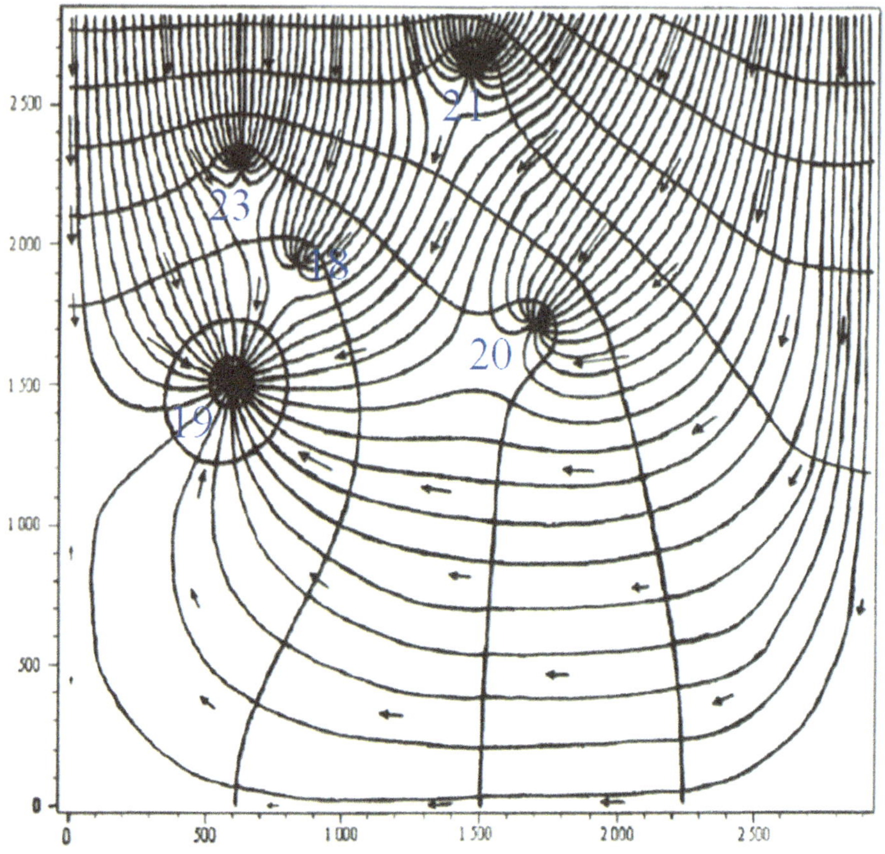

Fig. 3. Equipotential and homogenous fluid stream line field in an even reservoir

was used to build a discreet filtration model. In the reservoir development, wells No. 18, 19, 20, 21, 23 were involved. The well operation modes were determined by the bottom-hole pressure. The initial reservoir pressure was 29.0 MPa, bottom-hole pressure was 27.5 MPa, depression was 1.5 MPa. Well utilisation rate equals 1. The wells were not deactivated until the end of the forecast period, well working modes did not change. The forecast period was the same 23 years. By the end of the development period, the oil share in the gallery product constituted 2.3% (water cut 97.7%), waterflooding volumetric efficiency 0.511 u.f., injected volume 2.070 units, oil recovery factor 0.324 u.f., cumulative oil production 521 thousand tonnes and cumulative fluid production 3333 thousand tonnes. The calculation results are presented in Figs. 4 and 5.

Let us compare the results of the described stream tube model-based calculations with those of the commercial simulator. The annual oil production values change with time, but the change does not exceed 16% (maximum 15.7%, minimum 10.8%). The annual fluid production rates also change with time but do not exceed 13% (maximum

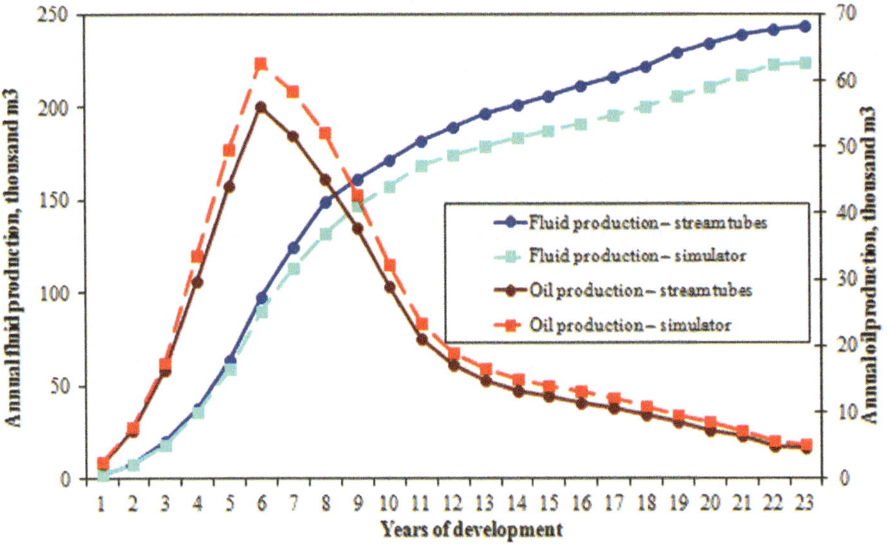

Fig. 4. Comparison of the annual oil and fluid production rates

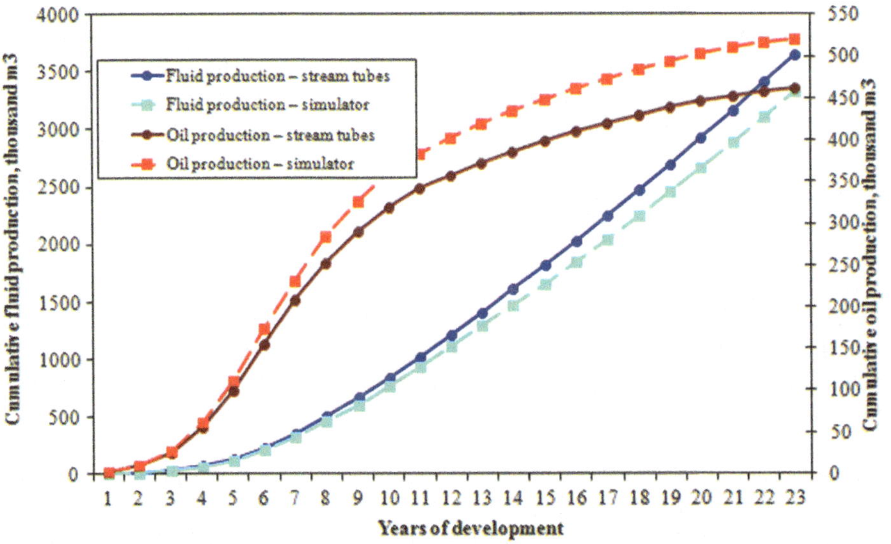

Fig. 5. Comparison of the cumulative annual oil and fluid production rates

12.3%, minimum 7.5%). The change of the cumulative oil and water production grows with time and by the end of the reported period constitute 11.3% and 9.7% respectively. If the simulator calculation results are used as a reference, then the cumulative oil production by the end of the reported period achieved with the stream tube model will be lower than the reference value, and the cumulative fluid production will be higher.

5 Conclusion

The paper proposes a mathematical model for express analysis of the oilfield development performance in waterflooding. The model is based on two methods: the stream-tube method and the curved gallery method.

The model presented herein is different from the similar one (curved gallery model by V.S. Kovalev, M.L. Surguchev) in its proposal to build the equipotential and stream line field with the source and outlet method and the Cauchy-Riemann conditions instead of the electric integrator.

This method takes due regard of the actual well layout but does not account for the unevenness of the permeability and porosity properties of the reservoir to the full extent. The unevenness of the reservoir around the area is indirectly taken into account at the stream line modelling stage. The estimation of the water motion for all the stream tubes requires using one permeability distribution value, describing the waterflooding layer by layer. The fluids are incompressible, the solid matrix is non-deformable. The displacement is piston-type displacement. The difference in viscosities and maximum values of the relative oil and water phase permeability is taken into account. Moreover, the imperfection of the water drive is considered. The calculation can be done both for the reservoir as a whole or every individual well separately. The individual well calculations make it possible to take account for the changes in the operating well stock.

With the described method, it is much easier to build a filtration model, compared to a full-scale discrete filtration model in a commercial hydrodynamic simulator using the system of classical mass and impulse conservation equations as a mathematical model. The promptness of building the stream tubes with the described method is the advantage that allows using it for express analysis of the oilfield development values in case of waterflooding.

The comparison of the values acquired from the proposed model-based calculation with those computed with a commercial hydrodynamic simulator Rubis Kappa Engineering confirms the satisfactory precision of the proposed method from the practical point of view.

Acknowledgement. Research conducted with support of state program for SRISA "Fundamental science research (47GP), theme №0065-2019-0019 "Non-developed zones identifications of oil fields and remaining reserves evaluation which is based on complexing of mathematic modelling, field development analysis and reservoir surveillance" (reg # AAAA-A19-119020190071-7).

References

1. Gimatudinov, Sh.K.: Oilfield Development and Exploitation Guidelines. Development Design. Nedra, Moscow, USSR (1983)

2. Borisov, YuP., Voinov, V.V., Ryabinina, Z.K.: Impact of the Reservoir Unevenness on the Oilfield Development. Nedra, Moscow, USSR (1970)
3. Kovalev, V.S., Zhitomirsky, V.M.: Forecasting Oilfield Development and Waterflooding System Efficiency. Nedra, Moscow, USSR (1976)
4. Surguchev, M.L.: Oilfield Development Process Regulation and Control Methods. Nedra, Moscow, USSR (1968)
5. Kovalev, V.S.: Estimating Oil Pool Waterflooding Process. Nedra, Moscow, USSR (1970)
6. Akulshin, A.I.: Forecasting Oilfield Development. Nedra, Moscow, USSR (1988)
7. Kwong, R.R., Kellogg, R.P., Thiele, M.R., Simmons, D.M. Improving water efficiency in the wilmington field using streamline-based surveillance. In: SPE Western Regional Meeting, 195372-MS. Society of Petroleum Engineers, San Jose, CA, USA
8. Claycomb, J.R.: Mathematical Methods for Physics using Matlab and Maple. Mercury Learning and Information, Massachusetts, US (2018)
9. Pepper, D.W., Heinrich, J.C.: The Finite Element Method: Basic Concepts and Applications with MATLAB, MAPLE, and COMSOL. CRC Press, London (2017)
10. Maxfield, B.: Engineering with Mathcad: Using Mathcad to Create and Organize your Engineering Calculations, Kindle edn. Elsevier, Amsterdam (2017)
11. Lindfield, G., Penny, J.: Numerical Methods: Using MATLAB. Academic Press, London (2019)
12. Ertekin, T., Abou-Kassem, J.H., King, G.R.: Basic Applied Reservoir Simulation. Society of Petroleum Engineers, New York (2001)
13. Fanchi, J.R.: Principles of Applied Reservoir Simulation. Elsevier, Amsterdam (2005)
14. Houze, O., Viturat, D., Fjaere, O.S.: Dynamic Data Analysis. Kappa Engineering, Paris, France (2020)

Numerical Experiments for Stochastic Linear Equations of Higher Order Sobolev Type with White Noise

Jawad Kadhim Tahir(✉)

College of Education, Computer Sciences Department, Mustansiriyah University, Baghdad, Iraq
jawadalisawi@uomustansiriyah.edu.iq

Abstract. In this article presents a numerical experiment for solving high-order Sobolev type linear stochastic equations with white noise. In this study, the method of sequential integration of pseudo-spectrum applied to obtain an approximate solution of the one dimensional pseudoparabolic model. The suggested method established on the mixed between two different methods pseudospectral approach with enhanced finite difference method which used as secondary approach. Obtaining data as a result of processing the problems numerically notice that these data acceptable coinciding with theoretical behavior of solution of the problem. The suggested approach proposed plays an effective role to obtain an approximate solution for stochastic linear equations of high order Sobolev type with white noise model. Numerical experiments are carried out. The enhanced finite difference approach and the energy estimation was implemented, it was shown that a completely discrete scheme is absolutely stable and converges in second-order of space and in first-order of time.

Keywords: El-Gendy approach · Gauss Lobatto points · Pseudospectral sequential integral · Pseudoparabolic model · Sobolev equations

1 Introduction

The finite difference method is used to solve approximately of the semi-linear delay problem in Sobolev type or pseudo-parabolic equations. The approach of integral symmetries constructed the second level of difference proposal scheme. An implicit rule used for temporary integration. The technique which depend on the estimates of energy indicated that a totally discontinuous circuitry absolutely stables and converges in the second order of space and in the first order of time. Computing of estimates of error with norm of the discrete type. Approximate results which obtained affirm that predicted behavior of the suggested approach.

R. Silhavy et al. (Eds.): CoMeSySo 2020, AISC 1295, pp. 204–214, 2020.
https://doi.org/10.1007/978-3-030-63319-6_18

Consider the pseudo-parabolic problem with delay

$$\varpi_t - \frac{\partial}{\partial x}\left(g(x.t)\frac{\partial^3 \varpi}{\partial t \partial x^3}\right) - \frac{\partial}{\partial x}(h(x.t)\,\varpi_x) = \mathfrak{f}(x.t.\varpi(x.t).\varpi(x.t-s)), \quad (1)$$

$$\varpi(x.t) = \varphi(x.t). \quad (x.t) \in \bar{A} \times [-s.0] \quad (2)$$

$$\varpi(0.t) = \varphi(l.t) = 0.t \in [0.\mathrm{T}] \quad (3)$$

A problem (1)–(3) define on the region $\bar{\Omega} = \bar{A} \cup [0.\mathrm{T}].\bar{A} = [0.l].\Omega = \mathrm{A} \cup [0.\mathrm{T}].\mathrm{A} = [0.l]$ and $(x.t) \in \Omega$, where s represents parameter of delay, suppose that $\frac{\mathrm{T}}{s} = m$ integer number; i.e. $\mathrm{T} = sm$ for some integer number m such that $m > 0$, $h(x.t) \geq \alpha > 0$ and $\varphi(x.t).\mathfrak{f}(x.t.\varpi_1.\varpi_2)$ express on functions of smooth which hold a specific regular conditions on the $\bar{\Omega}.\bar{A} \times [-s.0]$, must be indicated respectively in addition to:

$$\left|\frac{\partial \mathfrak{f}}{\partial \varpi_1}\right| \leq c^* < \infty$$

$$\left|\frac{\partial \mathfrak{f}}{\partial \varpi_2}\right| \leq d^* < \infty$$

Application of the Sobolev method for Fredholm nonlinear integral equation was obtained by Kagiwada and Kalaba [7]. The result includes a system of integro-differential equations Sobolev type, this type of models appear in numerous fields of mechanics and physics, these models formulated to represented two phase flows in cancellous circumference, when dynamical impacts are taken into account [9, 10], also used to study the phenomenon of thermal conductivity [13], uniform flow of fluid in rock fissures, shift of liquids of the second kind [8, 11, 15]. The results of the theoretical study of existence and uniqueness of pseudoparabolic models, see [3–5, 7]. Different numerical processing of this type models without delay were considered in [1, 2, 6, 9, 14, 16, 17].

2 Theoretical Study and Solvability of the Stochastic Linear Equations of Higher Order Sobolev Type with White Noise

In this work, the finite difference method used to find the approximate solution for the semilinear delay problem Sobolev type or the pseudo-parabolic problem. the approach of integral symmetries which using basis of piecewise linear functions in the space and the theoretical principles of interpolating quadrature with remainder term in formula of integral constructed of the second level difference approach [6, 12, 17] for singular perturbations without delay. Implicit rules are applied to time integration, shown that discretization with a finite difference is absolutely stable and converges in second and first orders with respect to spacing and timing respectively. The approach of difference presents an analytical aspects of errors of the numerical solution.

Jeffreys proposed a novel approach to get a solution for differential models established on the sequential embedded of Chebyshev extensions, this approach implemented by applying approximation for higher order derivatives and producing another approximations for derivatives with low order by sequentially integrating with higher order derivative, Chebyshev decomposition method introduced for solving boundary value problems using the pseudospectral sequential integral approach, this approach easier than spectral methods in the implementation gives comparable conclusions in precision with less computational requirements. Nasr and El-Hawary used El-Gendi approach with sequential integral approach to solve an equation with boundary value. Elbarbary introduced an enhancement of the sequential integral matrix ElGendi, which gives more accurate results than the results calculated by the usual El-Gendi matrix.

Consider the following problem

$$\begin{cases} \dfrac{\partial \varpi}{\partial t} = a^2 \Delta \varpi + f(P,\ t), \\ \varpi|_{t=0} = \psi(p), \quad P \in \Omega. \end{cases} \tag{4}$$

$$V = \frac{1}{8\pi^{3/2}(t_0 - t)^{3/2}} e^{-\frac{(x-x_0)^2 + (y-y_0)^2 + (z-z_0)^2}{4(t_0-t)}} \tag{5}$$

$$\Delta_{P_0} V - \frac{\partial V}{\partial t_0} = 0, \Delta_P V + \frac{\partial V}{\partial t} = 0, \iiint\limits_{\Omega} V d\Omega = 1. \iiint\limits_{\Omega} V d\Omega = 1.$$

$$\lim_{t_0 \to t+0} V = \begin{cases} 0 & r \neq 0, \\ +\infty & r = 0. \end{cases}$$

$$\lim_{\xi \to +0} \frac{1}{8\pi^{3/2}\xi^{3/2}} \iiint\limits_{\Omega} e^{-\frac{(x-x_0)^2 + (y-y_0)^2 + (z-z_0)^2}{4\xi}} F(x,\ y,\ z) d\Omega = F(x_0,\ y_0,\ z_0).$$

$$= \frac{1}{8\pi^{3/2}\xi^{3/2}} \iiint\limits_{\Omega} e^{-\frac{(x-x_0)^2 + (y-y_0)^2 + (z-z_0)^2}{4\xi}} F(x,\ y,\ z) d\Omega - F(x_0,\ y_0,\ z_0)$$
$$= \frac{1}{8\pi^{3/2}\xi^{3/2}} \iiint\limits_{\Omega} e^{-\frac{(x-x_0)^2 + (y-y_0)^2 + (z-z_0)^2}{4\xi}} (F(x,\ y,\ z) - F(x_0,\ y_0,\ z_0)) d\Omega_0$$

$$\left| \iiint\limits_{\Omega_{P_0,\ \delta}} V(F(x,\ y,\ z) - F(x_0,\ y_0,\ z_0)) d\Omega \right| \leq \frac{\varepsilon}{2} \iiint\limits_{\Omega_{P_0,\ \delta}} V d\Omega \leq \frac{\varepsilon}{2}.$$

$$\left|\frac{1}{8\pi^{3/2}\xi^{3/2}}\iiint\limits_{\Omega/\Omega_{P_0,\,\delta}} e^{-\frac{(x-x_0)^2+(y-y_0)^2+(z-z_0)^2}{4\xi}}(F(x,\ y,\ z)-F(x_0,\ y_0,\ z_0))d\Omega\right| \leq$$

$$= \frac{2M}{8\pi^{3/2}\xi^{3/2}}\iiint\limits_{r\geq\delta} e^{-\frac{r^2}{4\xi}}d\Omega \ldots (\ x = x_0 + \sqrt{\xi}\cdot x_1;\ \ y = y_0 + \sqrt{\xi}\cdot y_1;\ \ z = z_0 + \sqrt{\xi}\cdot z_1)$$

$$= \frac{2M}{8\pi^{3/2}\xi^{3/2}}\iiint\limits_{r_1\geq\frac{\delta}{\sqrt{\xi}}} e^{-\frac{r_1^2}{4\xi}}d\Omega_1 = \frac{M}{\sqrt{\pi}}\int\limits_{\frac{\delta}{\sqrt{\xi}}}^{+\infty} r_1^2 e^{-\frac{r_1^2}{4}}dr_1.$$

$$\iiint\limits_{\omega_R}(\varpi\Delta V - V\Delta\varpi)d\Omega = \iint_s\left(V\frac{\partial\varpi}{\partial n} - \varpi\frac{\partial V}{\partial n}\right)ds.$$

$$\int\limits_0^t\left\{\iiint\limits_{\Omega_R}\left(\varpi\frac{\partial V}{\partial t} + V\frac{\partial V}{\partial t}\right)d\Omega\right\}dt = \iiint\limits_{\Omega_R}\varpi V|_0^{t_0}d\Omega.$$

$$\int\limits_0^{t_0}\left\{\iiint\limits_{\Omega_R}\left(\varpi\left(\Delta V + \frac{\partial V}{\partial t}\right) - V\left(\Delta\varpi - \frac{\partial\varpi}{\partial t}\right)\right)d\Omega\right\}dt = \iiint\limits_{\Omega_R}\varpi V|_0^{t_0}d\Omega + \int\limits_0^{t_0}\left(\iint_s\left(V\frac{\partial\varpi}{\partial n} - \varpi\frac{\partial V}{\partial n}\right)ds\right)dt.$$

$$\int\limits_0^{t_0}\left(\iint_s\left(V\frac{\partial\omega}{\partial n} - \varpi\frac{\partial V}{\partial n}\right)ds\right)dt + \iiint\limits_{\Omega_R}\varpi V|_0^{t_0}d\Omega = -\int\limits_0^{t_0}\left(\iiint\limits_{\Omega_R}Vf(x,\ y,\ z,\ t)d\Omega\right)dt.$$

hence:

$$\varpi(x,\ y_0,\ z_0,\ t_0) = \iiint\limits_{\Omega} f(x,\ y,\ z,\ t)V|_{t=0}d\Omega - \int\limits_0^{t_0}\left(\iiint\limits_{\Omega}Vf(x,\ y,\ z,\ t)d\Omega\right)dt. \tag{6}$$

3 Numerical Study of the Stochastic Linear Equations of Higher Order Sobolev Type with White Noise

Let the set of grid nodes on the discretized Ω is defined as follows:

$$\omega = \omega_N \times \omega_{N_0},$$

moreover

$$\begin{aligned}
\omega_N &= \left\{x_i = ih,\ i = 1,2,\ldots,\ N-1,\ h = \tfrac{l}{N}\right\}\\
\omega_N^+ &= \omega_N \cup \{x_N = l\},\ \overline{\omega}_N = \omega_N \cup \{x_0 = 0,\ x_N = l\}\\
\omega_{N_0} &= \left\{t_j = j\tau,\ j = 1,2,\ldots,\ N_0,\ \tau = \tfrac{T}{N_0} = \tfrac{r}{n_0}\right\}\\
\overline{\omega}_{N_0} &= \omega_{N_0} \cup \{t_0 = 0\}\\
\omega_{n_0^-} &= \left\{t_j = j\tau,\ j = -n_0,\ldots,\ 0\right\}
\end{aligned}$$

Define finite differences for any grid operators as follows, $\eta_i = \eta(x_i)$, defined on $\overline{\omega}_N$, given by the following expressions:

$$\eta_{\bar{x}.i} = \frac{\eta_{i+1} - \eta_i}{h}$$

$$\eta_{x.i} = \frac{\eta_{i+1} - \eta_i}{h}$$

$$\eta_{x.i} = \frac{\eta_{i+1} - \eta_{i-1}}{2h}$$

$$\eta_{\bar{x}.x.i} = \frac{\eta_{x.i} - \eta_{\bar{x}.i}}{h} = \frac{\eta_{i+1} - 2\eta_i + \eta_{i-1}}{h^2}$$

Defined inner product of η_i and ω_i on $\overline{\omega}_N$:

$$(\eta.\omega)_0 = (\eta.\omega)_{\omega_N} := \sum_{i=1}^{N-1} h\eta_i\omega_i$$

$$(\eta.\omega] = (\eta.\omega)_{\omega_N}^{+} := \sum_{i=1}^{N} h\eta_i\omega_i$$

for any grid function $v_i = 0,\ i = 0,\ N$.

Now present the norms

$$\|\eta\|_0^2 = \|\eta\|_{0.\omega_N}^2 := (\eta.\eta)_0$$

$$\|\eta_{\bar{x}}\|_0^2 = \|\eta\|_{0.\omega_N^+}^2 := (\eta_{\bar{x}}.\eta_{\bar{x}}]$$

$$\|\eta\|_1^2 = \|\eta\|_0^2 + \|\eta_{\bar{x}}\|_0^2$$

$$\|\eta\|_\infty = \|\eta\|_{\infty.\omega_N^+} := \max_{1 \le i \le N-1} |\eta_i|$$

and the norm for any $\omega_i(1 \leq i \leq N-1)$

$$\|\omega\|_{-1} := sup \frac{|(\omega.\eta)_0|}{\|\eta_1\|}$$

using the notation:

$$g_{\bar{t},i}^{j} = \frac{g_i^j - g_i^{j-1}}{\tau_j}$$

$$g_{\bar{t},i}^{j} = \frac{g_i^{j+1} - g_i^j}{\tau_{j+1}}$$

next, describe of the difference scheme

$$\tau^{-1}h^{-1}\int_{t_{j-1}}^{t_j}\int_{x_{i-1}}^{x_{i+1}} L\varpi\psi_i(x)dxdt$$

$$= \tau^{-1}h^{-1}\int_{t_{j-1}}^{t_j}\int_{x_{i-1}}^{x_{i+1}} \mathfrak{f}(x.t.\varpi(x.t).\varpi(x.t-s))\psi_i(x)dxdt$$

where the piecewise linear basis functions defined by:

$$\psi_i(\chi) = \begin{cases} h^{-1}(\chi - \chi_{i-1}).\chi_{i-1} < \chi < \chi_i. \\ h^{-1}(\chi_{i+z} - \chi).\chi_i < \chi < \chi_{i+1}. \\ 0.\chi \notin (\chi_{i-1}.\chi_{i+1}).i = 1.2\ldots.N-1 \end{cases}$$

By using the corresponding interpolation quadrature rules and the remainder term in integral form, we obtain the following exact relation:

$$L\varpi := \varpi_{\bar{t}} - (A\varpi_{\bar{t}\bar{x}})_x - (B\varpi_{\bar{x}})_x + \mathcal{R} = \mathfrak{f}(x.t - \tau.\varpi(x.t-\tau).\varpi(x.t-\tau-s))$$

where $(x.t) \in \omega_N \times \omega_{N_0}$

$$A = \tau^{-1}h^{-1}\int_{t_{j-1}}^{t_j}\int_{x_{i-1}}^{x_{i+1}} g(x.t)dxdt$$

$$B = \tau^{-1}h^{-1}\int_{t_{j-1}}^{t_j}\int_{x_{i-1}}^{x_{i+1}} h(x.t)dxdt$$

The remainder has the form:

$$\mathcal{R} = \left(\mathcal{R}^{(0)}\right)_x + \mathcal{R}^{(1)}$$

moreover:

$$\mathcal{R}^{(0)}=\tau^{-1}h^{-1}\int_{t_{j-1}}^{t_j}dt\int_{x_{i-1}}^{x_{i+1}}dx\frac{\partial g}{\partial x}\int_{x_{i-1}}^{x_i}\frac{\partial^3\varpi}{\partial t\partial x^2}(\xi.t)K_1(x.\xi)d\xi-$$

$$\mathcal{R}^{(0)}=\tau^{-1}h^{-1}\int_{t_{j-1}}^{t_j}dt\int_{x_{i-1}}^{x_{i+1}}dx\frac{\partial g}{\partial x}\int_{x_{i-1}}^{x_i}\frac{\partial^3\varpi}{\partial t\partial x^2}(\xi.t)K_1(x.\xi)\,d\xi$$

$$-\tau^{-1}h^{-1}\int_{t_{j-1}}^{t_j}dt\{\int_{x_{i-1}}^{x_{i+1}}\frac{\partial g}{\partial t}dx\}\int_{t_{i-1}}^{t_i}K_1(\mu.t)\,\frac{\partial}{\partial t}\varpi_{\bar{x}.t}\,d\mu \tag{7}$$

$$-\tau^{-1}h^{-1}\int_{t_{j-1}}^{t_j}dt\int_{x_{i-1}}^{x_{i+1}}dx\frac{\partial h}{\partial x}\int_{x_{i-1}}^{x_i}\frac{\partial^2\varpi}{\partial x^2}(\xi.t)\,d\xi$$

$$+\tau^{-1}h^{-1}\int_{t_{j-1}}^{t_j}dt\{\int_{x_{i-1}}^{x_{i+1}}h(x.t)dx\}\int_{t_{i-1}}^{t_i}T_0(t-\xi)\,\frac{\partial}{\partial t}\varpi_{\bar{x}.t}\,d\xi.$$

$$\mathcal{R}^{(1)}=\tau^{-1}h^{-1}\int_{t_{j-1}}^{t_j}dt\int_{x_{i-1}}^{x_{i+1}}dx\varphi_i(x)\int_{x_{i-1}}^{x_{i+1}}\frac{\partial^3\varpi}{\partial t\partial x^2}(\xi.t)k_1^*(x.\xi)d\xi$$

$$+\tau^{-1}h^{-1}\int_{t_{j-1}}^{t_j}dt\int_{x_{i-1}}^{x_{i+1}}\{\textstyle\int dx\varphi_i(x)\}\int_{t_{i-1}}^{t_i}\frac{d^2}{dx^2}\mathfrak{f}(\xi.t.\varpi(\xi.t).\varpi(\xi.t-s))k_1^*(x.\xi)d\xi$$

$$-\tau^{-1}\int_{t_{i-1}}^{t_i}dt\frac{d^2}{dx^2}\mathfrak{f}(\xi.t.\varpi(\xi.t).\varpi(\xi.t-s)). \tag{8}$$

4 Computational Experiments and Results

Now present the results which obtained from the numerical experiments for specific problems.

Example 1.

$$\frac{\partial\varpi}{\partial t}-2\frac{\partial^3\varpi}{\partial t\partial x^2}-\frac{\partial^2\varpi}{\partial x^2}+\varpi(x,t-1)=\mathfrak{f}(x,t).$$

$$\varpi(x,t)=e^{-t}(\sinh(x)-x\sinh(1)$$

$$\varpi(0,t)=\varpi(1,t)=0$$

where $(x,t)\in[0,1]\times[-1,0]$.

Moreover:

$$\mathfrak{f}(x,t) = x\sinh(1)\left(e^{-t} - e^{1-t}\right) + e^{1-t}\sinh(x)$$

The analytical solution of the problem is given by:

$$\iota(x,t) = \mathcal{P}(x)e^{-t}$$

Such that $\mathcal{P}(x) = \sinh(x) - x\sinh(1)$

The computations of numerical experiment of example 1 are presented in Table 1 and Fig. 1.

Table 1. Approximate computations of example 1.

x	t	Exact solution	Approximate solution $h = 0.03$ $\tau = 0.03$	Error of pointwise	Approximate solution $h = 0.03$ $\tau = 0.01$	Error of Pointwise
0.2	0.1	−0.00127	−0.001271	0.01603	−0.001271	0.08060
0.2	0.3	−0.00729	−0.003762	0.02782	−0.003763	0.03908
0.3	0.3	−0.003228	−0.003270	0.0112	−0.003262	0.02779
0.2	0.2	−0.003976	−0.003992	0.0176	−0.003982	0.08827
0.3	0.2	−0.002033	−0.002026	0.03393	−0.002022	0.01123
0.6	0.6	−0.003727	−0.003783	0.03633	−0.00377	0.01332
0.7	0.7	−0.003180	−0.003307	0.03673	−0.003192	0.01323
0.9	0.8	−0.003338	−0.003363	0.03308	−0.00332	0.0116
0.8	0.9	−0.001367	−0.001381	0.01222	−0.001372	0.07361

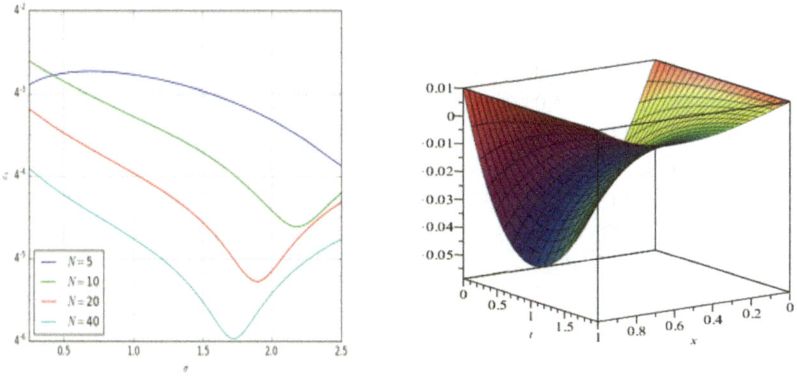

Fig. 1. Graph of numerical solution of example 1.

Example 2.

$$\frac{\partial \varpi}{\partial t} - 3\frac{\partial^3 \varpi}{\partial t \partial x^2} - \frac{\partial^2 \varpi}{\partial x^2} + \tanh(x, t-1) = \mathfrak{f}(x, t).$$

$$\varpi(x, t) = e^{-t}(\cosh(x) - x\cosh(1) + x + 2)$$

$$\varpi(0, t) = \varpi(1, t) = 0$$

where $(x, t) \in [0, 1] \times [0, 2]$.

Moreover:

$$\mathfrak{f}(x, t) = e^{-t}(x\cosh(1) - x + 1) + \tanh(e^{1-t}(\cosh(x) + x\cosh(1) + x + 1))$$

The problem has the following analytical solution:

$$\iota(x, t) = \mathcal{P}(x)e^{-t}$$

Such that $\mathcal{P}(x) = \cosh(x) - x\cosh(1) + x - 1$

The computations of numerical experiment of example 2 are shown in Table 2 and Fig. 2.

It can be observed that the numerical results obtained are essentially consistent with the theoretical analysis of the problem.

Table 2. Approximate computations of example 2.

x	t	Exact solution	Approximate solution $h = 0.03$ $\tau = 0.03$	Error of pointwise	Approximate solution $h = 0.03$ $\tau = 0.01$	Error of pointwise
0.2	2.1	−0.01631	−0.01662	0.002152	−0.01632	0.001101
0.1	1.2	−0.02667	−0.02706	0.002561	−0.02686	0.001550
0.3	1.2	−0.02203	−0.02237	0.003231	−0.0222	0.002627
0.2	1.3	−0.02237	−0.02317	0.006022	−0.02287	0.002025
0.3	1.3	−0.02211	−0.02273	0.006285	−0.02232	0.002138
0.6	1.6	−0.02823	−0.02853	0.006023	−0.02863	0.002023
0.7	1.7	−0.02282	−0.02223	0.003233	−0.02205	0.002622
0.8	1.8	−0.01602	−0.01632	0.002566	−0.01622	0.001551
0.6	1.4	−0.08228	−0.0833	0.002211	−0.08325	0.00111

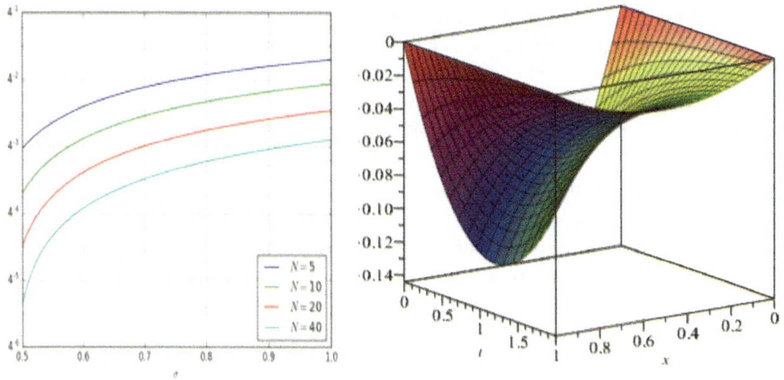

Fig. 2. Graph of numerical solution of example 2.

5 Conclusions

1. Suggested an efficacious approximate approach for solving for stochastic linear equations of higher order Sobolev type with white noise problem.
2. Numerical experiments was implemented by using the enhanced finite difference approach and the energy estimation to solve the problem that shown a completely discrete scheme is absolutely stable and converges in second order of space and in first order of time.
3. Estimations of error were obtained in a norm of the discrete type confirmed that the approximate computations were essentially consistent with the exact results of the problem.

Acknowledgments. I would like to express my thanks to the Mustansiriyah University (www.uomustansiriyah.edu.iq) in Baghdad - Iraq for encouragement and support.

References

1. Amiraliyeva, G.M., Amireliyev, I.G.: Differences schemes for the singularlys perturbed Sobolev equation. In: Proceeding of the Internationals Conference on Differences Equation, Specials Function and Application, Munichs, Germany, pp. 45–50 (2015)
2. Amiraliyev, G.M., Mamedov, Y.D.: Difference schemes on the uniforms meshs for a singularly perturbeds pseudo-parabolics equation. Turk. J. Math. **24**, 220–222 (2015)
3. Barenblsatt, G.I., Zheltovs, I.P., Kochin, I.N.: Basic concept in the theory of seepages of homogeneous liquid in fissured rock. J. Appl. Math. Mech. **45**(6), 186–190 (2019)
4. Chens, P.J., Gurtins, M.E.: On a theory of heats conductions involving two temperatures. J. Appl. Math. Phys. **29**(5), 625–627 (2018)
5. Colemans, B.D., Nolls, W.: An approximation theorems for functional with application in continuums mechanic. Arch. Rational Mech. Anal. **6**(1), 365–370 (2014)
6. Davi, P.L.: A quasilinears parabolics and a related thirds order problems. J. Math. Anal. Appl. **51**, 332–335 (2012)

7. Fans, Y., Posp, I.A.: Class of pseudos-parabolic equations: existences, uniqueness of weak solution, and errors estimate for the Euler-implicit discretization. J. Math. Method Appl. Sci. **38**(28), 229–239 (2011)
8. Fans, Y., Psop, I.S.: Equivalents formulation and numericals scheme for a classes of pseudo-parabolics equation. J. Comput. Appl. Math. **246**, 87–93 (2015)
9. Gabova, S.A., Malyshev, G.Y., Sveshnikova, A.G.: On equations of composite types associated with vibrations of compressible stratified fluids. J. Differ. Equ. **29**(17), 171–180 (2013)
10. Gladkova, A.L.: Uniqueness of solutions of Cauchy problems for some quasi-linear pseudo-parabolic equation. J. Math. Notes **61**(13), 36–38 (2016)
11. Hassanizade, S.M., Grays, W.G.: Thermodynamics basis of capillary pressures in porouss medias. Waters Resour. Res. **29**(11), 389–392 (2003)
12. Huilgols, R.R.: A second orders fluid of the differentials type. Int. J. Non-Linear Mech. **3** (15), 57–58 (2018)
13. Korpusova, M.O.: Destruction of solutions of pseudo-parabolic equations with time derivatives of nonlinear elliptic operators. Comput. Math. Math. Phys. **42**(22), 117–125 (2012)
14. Lightbournes, J.H., Rankins, S.M.: A partials functional differential equations of Sobolev types. J. Math. Anal. Appl. **83**(12), 38–39 (2013)
15. Shishmareva, I.A.: On nonlinear equations of Sobolev type. J. Differ. Equ. **41**(1), 1–13 (2015)
16. Showalters, R.E., Tings, T.W.: Pseudo-parabolics partial differentials equation. SIAM J. Numer. Anal. **1**, 1–27 (2015)
17. Tings, T.W.: Certain non-steady flow of seconds-order fluid. Arch. Rational Mech. Anal. **16**, 1–27 (2013)

Patterns in Navy Systems Using Autonomous Robotic Systems When Running Missions

Gennady P. Vinogradov(✉), Igor A. Konyukhov, and Kirill V. Kupriyanov

CJSC Research Institute "Centerprogramsystem", 50 let Oktyabrya Ave., 3A, 170024 Tver, Russian Federation
wgp272ng@mail.ru

Abstract. Due to the requirement of behavior intellectualization of artificial entities, we have to revise logical and mathematical abstractions in the basis for constructing their onboard control systems. The problem of developing such systems based on the pattern theory is relevant. This provides transfer of effective experience and compatibility of the theological approach and the cause-and-effect approach. The paper considers the problems of pattern identification and modeling. It is proposed to use four positions of information processing for these purposes. There also is the developed method of logical inference on patterns. The approach allows you to transfer the successful experience of a person to the robot control system. The robot's behavior becomes predictable and understandable. To transfer successful experience, a formal model of human behavior pattern is proposed. An approach to pattern model identification is proposed. The approach is focused on mathematical modeling of the observer's subjective judgments about the truth of the model of the pattern of the successful experience carrier object. It is proposed to use four observation positions.

Keywords: Pattern · Agent · Selection model · Autonomous robot

1 Introduction

The relevance of creating and implementing robotic systems in the fleet is associated with the changing concept of warfighting by the Russian Navy. It is assumed that the Russian fleet will be represented by combat reconnaissance groups with submarine and surface-based forces, as strike and detection weapons, command and control of these forces integrated into a unified information management system using network technologies. The elements of such system should be: traditional warfare objects (ships, aircrafts, submarines, etc.) and robotic systems.

Military systems, including units with the robotic system group, should be considered as a cyber-physical object. It is an information-related set of physical components, onboard measuring systems, onboard execution systems, an onboard computing system with implemented control algorithms, as well as a control point with an information-control field. Such object must have self-sufficient behavior that guarantees the fulfillment of a certain mission. Achieving the desired sharp increase in

R. Silhavy et al. (Eds.): CoMeSySo 2020, AISC 1295, pp. 215–229, 2020.
https://doi.org/10.1007/978-3-030-63319-6_19

the effectiveness of such systems and complexes is possible mainly by directing the efforts of designers and scientists to improve the intellectual component of their control system: 1) a set of algorithms for onboard robotic system control systems; 2) algorithms of operators' activities managing robotic system. Together, they form a "corporate intelligence" of a robotic system [1, 2, 8], which allows fulfilling missions according to a robotic system functional purpose in an uncertain and poorly formalized environment.

A robotic system in our opinion is an autonomous intelligent system (hereinafter referred to as an agent) that has human-like behavior and includes [1, 3, 4, 6]:

- onboard measuring devices (or a set of onboard measuring devices), which serve as sensors allowing to obtain information about the environment state and own state;
- onboard execution units (or a set of onboard execution units) that help the system to affect external environment and itself; they serve as effectors;
- means of communication with other systems;
- "cooperative intelligence" that should include onboard computing facilities, their software, operators that are the bearers of a set of algorithms for solving subject domain problems obtained through training.

An agent performs the tasks based on an understanding of its condition, its ideas about the state of the environment and the combat situation development, as well as information received through a communication module. An agent is able to predict environmental changes caused by its actions and to evaluate their usefulness.

The development of the efficiency and functionality of such systems has two directions: a) improvement of a functional and hardware component in Navy advanced samples; b) the development of more advanced algorithms for performing tasks by operators, their transfer through training. One approach to solving this problem is agent-oriented modeling [4–6], which considers an agent as a system that can adequately respond to changes in the external environment that is affected by built-in behavioral mechanisms. However, it should be noted that the vast majority of research in this area remains at the theoretical level. There is a gap between primitive behavior patterns of artificial entities, for example, in swarm robotics, their interaction models and expectations from practice. Improving the agent's capabilities in these areas separately is not enough. Therefore, developers' efforts should be aimed at the system integration of all subsystems of onboard equipment and operators by developing information technologies based on models of artificial intelligence algorithms and network technologies [6–8].

In modern times, when controlling a crewless object, an onboard system mainly solves the following tasks: 1) processing information received from sensors; 2) the use of means of influence on changes in the external environment. The problems of goal-setting (strategies for changing a situation in a favorable direction) and choosing a rational method (tactics) for achieving goals (making decisions) are the prerogative of only a human operator during a remote control of a robotic system [9, 10].

Depending on the degree of operator's inclusion in the robotic system control loop, there can be three classes of intelligent systems [12, 13]:

- Situational awareness system, which provides sufficient information to a control station to: 1) calculate situational assessments and build logical and informational model of an external and internal situation; 2) form a knowledge base; 3) make decisions on modes of action; 4) implement a goal setting stage; 5) adjust the knowledge base based on the results of robotic system actions.
- expert systems presenting to control station operators the real-time modes of action to achieve a mission goal, taking into account their understanding of a situation, with sufficient depth of study for its subsequent implementation after operator's approval [12].
- onboard intelligent control systems combining the two stages listed above, capable of operating without operator intervention and communicating with him to obtain session assessment and adjust a knowledge base.

Intelligent systems use various inference mechanisms to select a rational mode of action in various situations [3]. One of them is an on-line output based on the best patterns of behavior [13].

2 Requirements for the Autonomy and Intelligence of Combat Cyber-Physical Systems

The role of automated systems when performing combat missions should be considered from the standpoint of their influence on a human. They should help a commanding officer to simplify and increase the efficiency of his work. A commanding officer must be a human in the loop control system. Their interaction should ensure the transfer of experience both from a person to a machine and in the opposite direction, thereby ensuring adaptive behavior. This involves the research and development of systems containing the so-called cooperative intelligence. For example, the main difficulty for any autonomous system is recognizing situations in the environment. The complexity and multiplicity of situations arising during a mission make it impossible to identify them according to the results of multiple tests and to form a knowledge base. Therefore, it is necessary to implement an additional monitoring plan for the cyber-physical system to identify classes of situations and successful methods of action for the formation of behavior models (patterns) based on data obtained from tests in real conditions. Such a learning process forms a meta-level, the main link of which is the commander of the control center with its headquarters.

This scheme can be considered as the basic one, guaranteeing a controlled evolution of self-sufficiency, when solving tasks by combat units, which include autonomous work complexes. With this approach, automated robotic systems of varying degrees of "intelligence" and autonomy become an integral part of future developments for the field of combat use by various combat arms.

2.1 Initial Assumptions and Hypotheses

Usually, an autonomous system faces situations that are quite difficult for constructive formalization by traditional formal methods. However, they are well described by the means of the natural language and there is experience how to solve them better. The carrier of such experience is called a leader. The leaders' experience is transmitted through communication in the selected language.

We accept the hypothesis that human experience/behavior should be considered as a function of the interaction of a situation and a person. The purposeful action of a person depends both on situation signs and on his personality traits (motivation degree, a structure of abilities, knowledge, etc.). A situation can be interpreted as a component of the cause that generates a subjective reflection in a person. When choosing a specific behavior based on his subjective representation of a situation, a person affects the situation by changing it. At the same time, the processes occurring in a human mind when performing certain actions lead to the expansion of its structure of abilities (knowledge, experience). The stable human traits, which are manifested in his actions through behavior and experiences, can affect a situation, changing it. And vice versa, a situation can have an opposite effect on human stable traits, when he interprets it, changes his values, norms, modes of action and experience. The agent's behavior model should also take into account this phenomenon of the mutual influence of a robotic system and a situation.

Note that the technology of using a robotic system according to the above involves the development of at least three systems when solving the problem of their "intellectualization" [12]:

- off-board intelligent systems for preparing a robotic system for the current combat mission, which should: a) ensure that all the information necessary for the successful session is transferred to a robotic system (goals, tasks, maps, communication algorithms, methods for obtaining corrections, etc.); b) training a robotic system escort crew;
- onboard intelligent control systems, which ensure combat mission performance in an autonomous mode;
- intelligent systems for analyzing the results of robotic system sessions. The results of their work are the basis for accumulating an effective and relevant base of patterns, as well as forming requirements for a functional and hardware component of a robotic system.

When developing onboard intelligent control systems, the concept of "typical situation" turned out to be constructive. It is a part of a robotic system work that is functionally closed with a clearly defined meaningful purpose; it can be found in various (real) sessions as a whole, concretized in them according to the behavior conditions and the available ways of solving problematic sub-situations that arise in a typical situation [3]. When a robotic system is fully intellectualized, a typical situation and modes of action, as well as a reaction to it, form an individual behavior pattern.

Activities are controlled by certain cognitive models of targeted actions. These models include an idea of a time sequence of performing certain types of actions. By analogy, an agent also purposefully affects environmental objects by implementing modes of action. The basis for developing modes of action should be agent's subjective representations of a purposeful state situation [14]. We will assume that the agent's observed activity is a function of the developed programs, which are the result of its cognitive processes, and which do the assigned tasks or achieve the desired results with varying degree of effectiveness. Thus, the action plan and its implementation are a product of internal programming. Its result is a program (algorithm) of actions using operations available for an agent. It allows the agent to transform the observed situation into the desired (target) one. A person assesses the feasibility of the created algorithm by the linguistic variables "assurance" and "efficiency". The problem of identifying these programs should be solved by analyzing language patterns and non-verbal communication.

A robotic system should be intellectualized through its inclusion in the activity and agreed with an operator (commanding officer). His behavior and actions should be clear and adjustable.

Definition. A pattern is a result of human activity associated with an action, decision-making, behavior, etc., carried out in the past, and considered as a pattern (example) for repeated actions or as a reasoning about actions according to this pattern.

When studying their experience, a person is aimed at aggregating it through creating pattern model. Therefore, a pattern model should be considered as a unit of human experience, for which a person has a certain confidence degree in obtaining the desired states in a situation similar to a typical (cluster) one. Pattern modeling has a limited subset of natural language, including case-based reasoning, which forms a specific part of human experience – meta-experience

3 A Model of Behavior Pattern Fuzzy Description

Purposeful behavior is associated with a choice in a purposeful state situation [8]. We consider a behavioral model as a fuzzy description of the choice situation model. It is proposed to create a possible version of such construction using a "paradigm inoculation" of ideas from other sciences [14], for example [15, 16]. A purposeful state includes the following components.

- A subject making a choice (agent), $k \in K$.
- A choice environment (S), which is many elements and their essential properties, a change in any of which can cause or produce a change in the state of a purposeful choice. Some of these elements may not be system elements and form the external environment for it. The impact of the external environment is described using variables, some of which can remain unchanged over a certain time interval T, and some can change. The first type of variables is called parameters, and the second one is perturbations. Values of both types of variables are generally considered to be independent of an agent.
- Available modes of action $c_j^k \in C^k,\ j = \overline{1,n}$ of the k-th agent, which are known to him and can be used to achieve i-th result (also called alternatives). Each mode of this set is characterized by a set of parameters called control actions.
- Possible results for S environment that are significant for the agent $o_i^k \in O^k,\ i = \overline{1,m}$. The results are estimated using the output parameters of a purposeful state situation.

- The method of estimating the properties of the results after choosing a mode of action. Obviously, result estimates should reflect the value of the result for an agent and thereby reflect its personality.
- Constraints reflecting the requirements imposed by a situation of choice on output variables and control actions.
- A domain model that is a set of relations describing the dependence of control actions, parameters and disturbances with output variables.
- An agent constraint model that is described in detail in [14]. Regardless of the type of description of restrictions, we assume that an agent has a certain degree of confidence about the possibility of changing a part of restrictions by expansion of a set of possible options (alternatives) of choice.

Let us denote U as a set of control parameters, P – a set of parameter vectors, Ω – a set of external disturbance vectors. The model of the purposeful choice state situation is described by F as follows:

$$F : U \times P \times \Omega \rightarrow O. \tag{1}$$

The relation (1) is a domain model, which is the basis of the agent's idea on the control object operation.

Let us introduce measures for the described components that will be used to estimate a purposeful state.

1. We assume that the agent is able to distinguish factors – environmental characteristics $X^k = \left\{ x_i^k,\ i = \overline{1,N} \right\}$. The agent assesses the influence of each factor using a linguistic variable, the influence degree of the factor $\mu_x^k(x_i^k) : x_i^k \rightarrow [0, 1]$. We introduce a parameter for the agent to assess its situational awareness in a purposeful state situation

$$Es^k = \frac{\sum_{i=1}^{N} \mu_x^k(x_i^k) x_i^k}{\sum_{i=1}^{N} \mu_x^k(x_i^k)}.$$

We can define the following restriction:

$$\sigma^k(Es^k) \geq \sigma_0^k,$$

where σ_0^k is some threshold level of agent's awareness from the using its own information sources.

2. We assume that to describe the influence of the selected factors on the results $o_i^k,\ i = \overline{1,m}$, the agent uses approximation in the form of the following production rules:

If x_1 is A^k_{r1} and if x_2 is A^k_{r2} and … and if x_N is A^k_{rN}, then

$$o_i^k = f_{ir}^k(x_1, x_2, \ldots, x_N), \; r = \overline{1, R}, \; i = \overline{1, m} \tag{2}$$

where R is the number of production rules, r is the current production rule number, $o_i^k = f_{ir}^k(x_1, x_2, \ldots, x_N)$ is an explicit function that reflects agent's idea of a causal relationship of input factors with possible results for the r-th rule; A^k_{ri} are fuzzy variables defined on $X^k = \{x_i^k, \; i = \overline{1, N}\}$.

Mathematical models, a verbal description, graphs, tables, algorithms, etc. can be used as $f_{ir}^k(\cdot)$ function.

Since c_j^k is a function of the external environment state parameters, system properties taken into account, a set of assumptions about their possible values forms a scenario of the possible state of the external environment, the system functionality. The implementation of scenarios, for example, using the rules (2) allows forming an idea of the possible results o_i^k. The ambiguity in choosing a mode of action can be described as the degree of certainty of the need for its application to obtain the result o_i^k. This estimate can be described by a linguistic variable $\psi_j^k = \psi_j^k(c_j^k \in C^k | s_i \in S \to o_i^k) \in [0, 1]$.

This is an agent's individual characteristic, which can change after training and gaining experience, as well as after the communication interaction of agents with each other and with an operator. Therefore, $\psi_j^k = \psi_j^k(c_j^k \in C^k | s_i \in S, I^k \to o_i^k) \in [0, 1]$, where I^k is the information that the agent has at the time t_k.

3. When an agent makes a decision in a purposeful state situation to achieve a result o_i^k, the choice of a mode of action c_j^k is associated with constructing a quantitative estimation of the properties of a chosen solution, as shown in [6, 14]. The list of properties and parameters is based on experience, knowledge, intelligence and how well it understands a decision-making situation. The correct description of the properties and parameters of the mode of action is one of the main conditions for the choice c_j^k to lead to the ac result o_i^k. The choice of a list of properties and parameters depends on the agent (its personality). This is the agent's contribution to the decision-making process.

Let us imagine the possible results for agent's choice environment is given in the form $o_i^k \in \left\{o_{ij}^k, j = \overline{1, J}\right\}$, where o_{ij}^k is a set of possible results when choosing the j-th mode of action, $i \in I$ is a set of results taken into account by the k-th agent. It's obvious that $o_{ij}^k = o_{ij}^k(s_i), s_i \in S$.

4. The value of the results o_i^k. Since $o_{ij}^k = o_{ij}^k(s_i)$, and $s_i = S(c_j^k)$, the value of the i-th type of the result can be estimated by the following linguistic variable $\varphi_i^k(o_i^k(c_j^k)) \in [0, 1]$. The function $\varphi_i^k(o_i^k(c_j^k))$ for the result o_i^k will be a monotonic transformation, since $\varphi_i^k(\cdot)$ translates the range of the function $o_i^k(c_j^k)$ into the set of values of the linguistic variable. Since fuzzy variables correspond to the basic value

of the linguistic variable, this transformation transfers the range of the function o_i^k to the range of basic fuzzy variables.

5. The effectiveness of the mode of action in terms of the result is the confidence that this result is obtained by this mode of action at known (or assumed) costs of its implementation. The confidence degree E_{ij}^k that a certain mode of action c_j^k will lead to a result o_i^k in the environment S if the agent chooses it is the following: $E_{ij}^k = E_{ij}^k(o_i^k | A \text{ choses } c_j^k \text{ in } S) \in [0, 1]$.

It is a linguistic variable and expresses an agent's individual assessment of choice consequences in terms of costs.

4 Agent's Choice Model

The three introduced linguistic variables $\mu_i^k(x_i^k)$, ψ_{ij}^k, E_{ij}^k form a model of agent's ideas about a purposeful choice situation.

Since c_j^k can be described in X_i^k terms and the agent has an idea of dependence as a rule base that links c_j^k and the value of the possible i-th result o_i^k, it is possible to determine the value of a purposeful state by the i-th result o_i^k for the k-th agent according to the rule [6, 14]:

$$E\varphi_i^k = \frac{\sum_{j\in J} \varphi_{ij}^k(o_{ij}^k(c_j^k)) \cdot o_{ij}^k(s^k)}{\sum_{j\in J} \varphi_{ij}^k(o_{ij}^k(c_j^k))}.$$

In a similar way, the value of a purposeful state for k-th agent can be evaluated by efficiency for the i-th type of result:

$$EE_i^k = \frac{\sum_{j\in J} EE_{ij}^k(o_i^k(c_j^k)) \cdot \psi_i^k(c_j^k)}{\sum_{j\in J} \psi_i^k(c_j^k)}.$$

The agent evaluates the desirability of a purposeful state according by i-th result and the effectiveness of its achievement in a situation of choice is given in the form of a linguistic variable

$$\chi_{i1}^k = \chi_1^k(E\varphi_i^k) \in [0, 1], \quad \chi_{i2}^k = \chi_2^k(EE_i^k) \in [0, 1]$$

The validity of this statement corresponds to the definition introduced by Zadeh [11] about the types of fuzzy sets. According to that definition, a fuzzy set is a set of type n, $n = 1, 2, 3, \ldots$, if the values of its membership function are fuzzy sets of type $n - 1$.

We can define the following restrictions:

$$\sum_i \chi_{i1}^k (E\varphi_i^k) \geq \chi_1^0 \ u \ \sum_i \chi_{i2}^k (EE_i^k) \geq \chi_2^0$$

where χ_1^0 *and* χ_2^0 are agent's expectations of a mission, which reflect the balance between costs and results o_i^k.

Since s_i is a function of the awareness of k-th purposeful agent $s_k = s_k(I_k)$ and there is an iterative procedure for exchanging views between agents, the following assumption is true:

$$I_{t+1}^k = \omega\ [\sigma_k]\, I_t^k \tag{3}$$

where t is an iteration number during interactive formation of a consistent forecast. This is an assumption about growing awareness of the k-th agent depending on an iteration number. ω is an iterative mapping (generally point-multiple), so that at the initial awareness level I_0^k, any sequence generated by the inclusion $I_t^k \subseteq I_{t+1}^k$ will be bounded and all its limit points are in $M \subset R^n$. This assumption is valid due to the fact that an agent forms a certain stable point of view during communication and analysis. A parameter σ_k shows agent's ability to take new points of view and to review the awareness structure. The introduction of this parameter allows causing the transformation of the agent's choice situation model through producing a change in one or more components or representation parameters during communication or interactive interaction.

Thus, the contribution of a purposeful agent to the situation of choice is shown:

- In evaluating a significance degree of $\varphi_i^k(x_i^k)$ situation factors and through them a situation awareness in the form (3).
- In evaluating the value of results $o_i^k - \varphi_i^k(o_i^k)$.
- In evaluating the admissibility of applying the j-th mode of action to achieve the i-th result $\psi_j^k(c_j^k)$.
- In evaluating the effectiveness $E_{ij}^k(c_j^k)$ of achieving the result o_i^k by the j-th mode of action c_j^k, which help the agent to estimate its own costs of achievement of the result.

The first and the fourth groups of estimates reflect agent's knowledge of a subject domain, the level of its various training types (skills, etc.).

The second and third groups allow describing agent's value system and evaluating a congruence degree of agent's and system values, which largely determine the its performance.

The model of agent's choice situation is a set of structural and functional properties that belong to the choice situation and affect its satisfaction or dissatisfaction with a situation, according to its beliefs.

There is another group of factors that determine the implementation of a result. They include will, risk appetite, self-esteem, motivation. Taking these factors into account, we can talk about such indicator as confidence in obtaining a result o_i^k in a situation of choice $\rho_i^k(o_i^k)$ when using one of the possible mods of action $c_j^k \in C^k$.

According to the hypothesis of rational behavior, the agent forms a decision:

$$
\begin{aligned}
& P_i^k(s \in S) = \underset{c_j^k}{Arg\max}\Big(\sum_{j \in J} E\varphi_i(o_i^k(c_j^k)) - EE_i^k(o_i^k(c_j^k))\Big) \\
& c_j^k \in C^k(I_t^i),\ \ I_t^i \subseteq M,\ \ o_i^k \in O^k \\
& \sum_i \chi_{i1}^k(E\varphi_i^k) \geq \chi_1^0,\ \ \sum_i \chi_{i2}^k(EE_i^k) \geq \chi_2^0 \\
& \sigma^k(Es^k(X)) \geq \sigma_0^k
\end{aligned}
\tag{4}
$$

Since the choice is related to agent's ideas about the choice situation, it is necessary to include the knowledge base (3) in (4).

Relations (4) describe agent's behavior pattern (cyber physical system) when striving to achieve *i*-th result. An agent considers (4) as a pattern, i.e. a way of describing a problem, its solution principle and algorithm. A solution for such problem might be used many times without reinventing anything.

The above characteristics assume that assessing the factor influence degree, the degree of confidence in the need to choose a mode of action, the value of a result, the effectiveness of a mode of action for each result are four personality (individuality) indicators. All other characteristics are derived from them by known methods of the theory of fuzzy sets.

The above indicators: a purposeful state value by the result $E\varphi_i^k$ and a purposeful state value by efficiency EE_i^k – are elements of an integral indicator of a purposeful state value for *k*-th individual $\sum_i E\varphi_i^k \cdot EE_i^k$. Taking into account its degree of confidence in obtaining the result ζ_i^k, an indicator of the expected specific value is the following:

$$
EV_k = \frac{\sum_i (E\varphi_i^k - EE_i^k) \cdot \zeta_i^k}{\sum_i \zeta_i^k}
\tag{5}
$$

This means that if two subjects are in the same situation of choice, then the difference in their behavior should be manifested in the values of specific value estimates by the result and effectiveness and in the degree of confidence in achieving a goal.

The purposeful state of a purposeful agent has the following characteristics:

- the agent is in a selection state: $U(\cdot) > 0$;
- there is at least one potential result o_i^k; if there are other potential results, then their purposeful state values by a result are not equal;
- for an agent, there are at least two potential modes of action c_1^k and c_2^k such that $\psi_{i1}^k > 0$ and $\psi_2^k > 0$;

- the effectiveness of the modes of action c_1^k and c_2^k is such that the sum of the estimates of purposeful state values by the effectiveness of obtaining results o_i^k in these two ways are not equal $\sum_i EE_{i1}^k(o_i^k(c_1^k)) \neq \sum_i EE_{i2}^k(o_i^k(c_2^k))$.

There is at least one potential result o_i^k with a value greater than a certain threshold value φ_i^{k0} for an agent, and its degree of obtaining confidence is also higher than a certain threshold value ζ_i^{k0}.

These rules mean that there is an agent wanting to get some result. For this purpose, it has several alternative ways of achieving it with different effectiveness, and its confidence in obtaining the desired result is significant.

5 Complexity Degrees of Pattern Models for Modeling Skills and Abilities

The pattern modeling process should be aimed at various complexity levels of certain skills and abilities [15, 16].

Simple behavioral patterns include specific, certain actions accessible for direct observation. Their implementation takes a short time period, for example, evading obstacles.

Simple cognitive patterns include specific, easily identifiable and verifiable mental processes that take a short time to complete. This is a memorization of the names of environmental objects (external and internal), formation of a concept ontology, creation of a situation image, assessment, norms, etc. Similar thought patterns are identified through easily observable and measured behavioral results of the carriers of these patterns, as well as through direct feedback.

Simple linguistic patterns include recognizing and using commands containing specific keywords, phrases, the ability to form specific queries, recognizing and responding to commands, reviewing or rejecting key commands, etc. The use of these skills is also available for direct observation and measurement.

Complex behavioral (or interactive) patterns include establishing and coordinating a sequence or combination of simple behavioral actions.

Complex cognitive patterns require a synthesis or sequence of other simple thinking skills. The examples of using complex cognitive patterns are a diagnosis of a problem or current situation, making up a modeling program for situation development or an action program, etc.

Complex linguistic patterns require interactive using of a language in highly dynamic situations. The examples of abilities that require complex linguistic skills are convincing someone of something, negotiating, etc.

6 An Approach to the Identification and Construction of a Pattern Model

There are four basic perception positions for collecting and interpreting information. The first one is personal point of view, the second position is perception of a situation from another person's point of view, the third position is considering the situation from uninterested observer's point of view, the fourth position involves considering the situation from the point of view of the system involved in the situation [15, 16].

Modeling from the first position assumes an attempt to do something by yourself and explore the way it is done. The analysis of the mode of action is performed from researcher's point of view.

It is important to emphasize that in order to obtain a similar description of already performed activities by an agent, the subject in question must leave its previous activity position and move to a new position, which is external both to the actions already completed and to future planned activities. It is called first-level reflection: agent's new position characterized by the previous position will be called a reflective position; the generated knowledge will be reflexive knowledge, since it is taken relative to the knowledge developed in the first position.

The second position implies a possible complete imitation of agent's behavior when a researcher tries to think and act as close as possible to agent's thoughts and actions. Such approach allows understanding significant, but unconscious aspects of the modeled agent's thoughts and actions intuitively. Modeling from the third position consists in observing the modeled agent's behavior as an uninterested observer. This involves constructing a model of a mode of action from the point of view of a specific scientific discipline related to agent's domain area. The fourth position implies an intuitive synthesis of all obtained ideas in order to make a model with maximum values of specific value indicators by the result and effectiveness.

One of the modeling goals is to identify unconscious competence and make it conscious in order to understand it better, to improve it and transfer a skill.

Cognitive and behavioral competence can be modeled either "implicitly" or "explicitly". Implicit modeling assumes taking a second position in relation to a modeling subject in order to achieve intuitive understanding that person's subjective experiences. Explicit modeling consists in moving to the third position in order to describe an explicit structure of the modeled agent's experiences so that it can be transferred to others. Implicit modeling is primarily an inductive process that helps to accept and perceive world structures around us. Explicit modeling is a deductive process for describing and implementing this perception. Both processes are necessary for successful modeling. Without an "implicit" stage, there cannot be an effective intuitive base to create an "explicit" model. On the other hand, without the "explicit" phase, the simulated information will not be able to fulfill itself in methods or means and be transferred to others.

The result should be a model with the following synthesized components: a) an intuitive understanding of agent's abilities, b) direct observation of agent's work, c) researcher's explicit knowledge in agent's domain area.

6.1 The Main Modeling Phases [13, 15, 16]

They reflect the movement from implicit modeling to explicit one.

Preparation. It implies the choice of a person who has the ability to be modeled, and the definition of: a) a modeling context; b) a place and time of accessing a modeled person; c) preferred relations with the modeled person; d) researcher's state in the modeling process.

In addition, this phase includes the creation of favorable conditions that will allow full immersion in a process.

Phase 1. Intuitive Understanding. The first phase of a modeling process includes the involvement of a modeled person in the desired activity or implementation of an ability in an appropriate context. Modeling begins with the adoption of the second position in order to come to an intuitive understanding of a skill that an agent demonstrates. There is no need to define patterns. It is necessary to enter the model's state and identify yourself with it internally, mentally imitating agent's actions. The external behavior of an agent is a surface structure. The second position allows getting information about the underlying structure. This is a reflection phase of the modeled agent's representations or the "unconscious understanding" phase.

Once a sufficient level of intuitive understanding of the studied ability is obtained, it is necessary to create a context for using this ability and to apply it acting in the way the agent acts. Then we must try to achieve the same result while being "ourselves". In this way, we can obtain a "double description" of the simulated skill. The first modeling phase ends when we obtain approximately the same results as the agent gets.

Phase 2. Abstraction. The next stage of the modeling process is separating essential elements of the model's behavior from random, extraneous ones. At this stage, simulated strategies and types of behavior become more explicit. Since there is the ability to achieve results similar to agent's results, it is necessary to build a mode of action based on researcher's views to achieve the same results using the "first position".

The task is to identify specific cognitive and behavioral steps to achieve the desired result in the chosen context (contexts). For this purpose, we use "factoring" of elements of identified strategies or types of behavior in order to assess their significance. If the reactions remain unchanged when some element is missing, then it is not essential for the model. If, the result changes as due to a missing element, it means that something important has been determined. The purpose of such a procedure is to reduce simulated actions to the simplest and most elegant forms, as well as to highlight the essential.

Upon completion of this stage, a "minimal model" of understanding agent's abilities from the "first position" will be obtained and an intuitive understanding of the agent's abilities from the "second position" will be developed. In addition, there will be the "third position", which is an angle showing the difference between the revealed method of reproducing a simulated ability and the way a person displays this ability.

Phase 3. Integration. The final modeling stage involves constructing a context and procedures that allow others to master simulated skills and, therefore, to get the same results as the person who served as the model for this. In order to draw up a plan, it is necessary to synthesize information obtained in all perception positions. In contrast to simple (step-by-step) imitation of the simulated person's actions, the most effective is the creation of appropriate reference experience for students. This experience allows detecting and developing "operation sequences" necessary for the successful implementation of a skill. To acquire this skill, it is not necessary to go through the entire modeling procedure. The main principle at this stage is the "usefulness" of action modes for those agents for which the model is designed.

The described provisions form the basis for the development of a methodology for constructing behavior pattern models of autonomous underwater vehicles during search and rescue missions. The developed modeling complex for testing autonomous underwater vehicle group algorithms is based on patterns. There was a way of constructing an autonomous underwater vehicles architecture with an onboard control system using pattern models and a logical inference system on patterns.

References

1. Bakhanov, L.E., Davydov, A.N., Kornienko, V.N., Slatin, V.V., Fedoseev, E.P., Fedosov, E. A., Fedunov, B.E., Shirokov, L.E.: Fighter weapon control systems: intelligence basics for a multifunctional aircraft. In: Fedosov, E.A. (ed.) Mashinostroenie, Moscow (2005). (in Russian)
2. Fedunov, B.E.: The problems of developing onboard operational-advising expert systems. J. Comput. Syst. Sci. Int. **5**, 147–159 (1996). (in Russian)
3. Fedunov, B.E.: Onboard Tactical Intelligent Systems for Anthropocentric Objects (Examples for Manned Vehicles). De Libri, Moscow (2018). (in Russian)
4. Fedunov, B.E., Prokhorov, M.D.: A conclusion on a precedent in the knowledge bases of onboard intelligent systems. Artif. Intell. Decis. Making **3**, 63–72 (2010). (in Russian)
5. Gorodetsky, V.I., Samoilov, V.V., Trotsky, D.V.: Basic ontology of collective behavior autonomous agents and its extensions. Bull. RAS Theory Control Syst. **5**, 102–121 (2015). (in Russian)
6. Fedunov, B.E.: Constructive semantics for developing onboard intelligence algorithms for anthropocentric objects. J. Comput. Syst. Sci. Int. **5** (1998). (in Russian)
7. Vinogradov, G.P.: Decision modeling by an intelligent agent. Softw. Syst. **3**, 45–51 (2010). (in Russian)
8. Vinogradov, G.P., Kuznetsov, V.N.: Modeling agent behavior taking into account subjective ideas about the situation of choice. Artif. Intell. Decis. Making **3**, 58–72 (2011). (in Russian)
9. Vinogradov, G.P., Kirsanova, N.V., Fomina, E.E.: Game-theoretic model of targeted subjectively rational choice. Neuroinformatics **10**(1), 1–12 (2017). http://www.niisi.ru/iont/ni/Journal/V10/N1/Vinogradov_et_al.pdf. Accessed 10 June 2020
10. Gribov, V.F., Fedunov, B.E.: On-board information intelligent system "crew situational awareness" for combat aircrafts. In: Proceedings of the GosNIIAS, VA Series, vol. 1, no. 18, pp. 5–16 (2010). (in Russian)
11. Fedunov, B.E.: The output mechanisms in the knowledge base of on-board operatively advising expert systems. J. Comput. Syst. Sci. Int. **4** (2002). (in Russian)
12. Zadeh, L.: The Concept of a Linguistic Variable and its Application to Making Approximate Decisions. Radio i svyaz, Moscow (1996). (in Russian)
13. Fedunov, B.E., Shestopalov, E.V.: A shell of an on-board operatively advising expert system for a typical flight situation "group entry into an aerial combat". J. Comput. Syst. Sci. Int. **3** (2010). (in Russian)

14. Vinogradov, G.P.: A subjective rational choice. IOP Conf. Ser. J. Phys. **803**, 012176 (2017). (in Russian)
15. Borisov, P.A., Vinogradov, G.P., Semenov, N.A.: Integration of neural network algorithms, nonlinear dynamics models and fuzzy logic methods in forecasting problems. J. Comput. Syst. Sci. Int. **1**, 78–84 (2008). (in Russian)
16. Dilts, R.: Modeling with NLP. Peter, St. Petersburg (1998). (in Russian)
17. Dilts, R.: Changing Belief Systems with NLP. Class, Moscow (1997). (in Russian)

Renewable Energy and Its Impact on GDP Growth Factors: Spatial Panel Data Analysis of Gross Fixed Capital Formation in Selected EU Countries

Tomáš Formánek(✉)

University of Economics, Prague, Czech Republic
formanek@vse.cz
https://github.com/formanektomas

Abstract. This article evaluates the effects of renewable energy prominence on gross fixed capital formation (GFCF). Energy production from renewable sources tends to be more expensive when compared to fossil-fuel based sources. Hence, there is a legitimate fear that renewable energy sources and the corresponding increase in energy costs may negatively interfere with major GDP growth factors, prominently represented by GFCF. However, multiple beneficial effects of renewable sources exist as well: induced research and development activities, energy security, etc. Given the complex macroeconomic dynamics involved, it may be difficult to adequately quantify the ceteris paribus effects (if any) of renewable energy consumption on GFCF. To address such complex task, this article is based on a stratified macroeconometric spatial panel data model. The analysis is performed at the NUTS2 level, using a contiguous sample of EU countries over the period 2012–2017. Given the analysis performed, there is no statistically significant negative effect of renewables on GFCF.

Keywords: Renewable energy · Gross fixed capital formation · Spatial econometrics · Spatial panel model

1 Introduction

This contribution focuses on the impacts that renewable energy consumption levels may have on a major GDP growth factor – the gross fixed capital formation (GFCF). The motivation for this analysis comes as a direct follow-up to a preceding paper [4] where the relationship between renewables consumption and GDP growth was thoroughly examined. While analysing the effect of renewables on the overall GDP growth is an important endeavour, focusing on GFCF may have prominent benefits over the evaluation of impacts on GDP (which is an all-encompassing aggregated indicator). In particular, focusing on GFCF may provide a more direct information on how investments and thus prospects for future prosperity are affected by the increasing levels of renewable energy.

R. Silhavy et al. (Eds.): CoMeSySo 2020, AISC 1295, pp. 230–242, 2020.
https://doi.org/10.1007/978-3-030-63319-6_20

GFCF is the value of fixed assets acquired by businesses and the government, less the disposed fixed assets (depreciation is ignored here – for detailed and structured description of GFCF, see [7]). GFCF is generally regarded as a major driving factor of present and future GDP growth rates – this variable is often used as an indicator of business confidence, future business activity levels and as a lead indicator for GDP growth.

The energy sector is a known and widely studied strategic factor, necessary for securing favourable background for all economic activities, private consumption, etc. In most developed countries (such as OECD member states), the relationship between energy consumption and economic growth is direct and substantial [1]. In terms of energy production and consumption, renewable sources are a complex phenomenon. On one hand, renewables are just a small and rather costly component of the energy sector, on the other hand, they can be viewed as an incoming and highly disruptive factor that is going to affect life and economy at all levels. To understand and describe the importance of renewable energy sources, one has to start with climate change.

Climate change is a general term describing different long-term shifts in global and/or regional climate patterns. Despite all the past and present academic and political discussions, climate change has transformed into a "grey rhino effect" - a foreseeable, highly probable, high impact yet neglected threat (compare to a "black swan" effect such as the COVID-19 pandemic). To avoid/mitigate the negative effects of climate change, we have to manage a transition from fossil & green house gas emitting energy sources to alternative and renewable sources (see [9] for discussion). In the long-term, such transition is coming as an inevitable and beneficial fact. However, there are potential short-term risks and adaptation costs involved. This contribution focuses on evaluating a specific type of such adaptation costs: the potential inverse relationship between renewable energy consumption and GFCF.

Renewable energy sources tend to be more expensive because they "internalize" the otherwise "external" costs associated with fossil fuels (CO_2 and NO_X emissions, etc.). Increased energy costs induced by the rising share of renewable energy sources may have a negative short-term impact on economic efficiency, thus slowing down planned and realized investment activities – i.e. GFCF. At the same time, most central authorities in developed economies provide incentives and subsidies related to alternative energy sources, aimed at easing the potential stress from increased energy costs. Also, renewable energy sources bring many additional benefits to the economy (besides sustainability of energy production). Research and development investment activities and power-grid upgrades linked to renewables may be quantified relatively easily, while the impact on energy security and climate change may be less straight-forward to asses. Also, many types of alternative energy sources would be available in less developed regions, thus alleviating the demand for traditional infrastructure investments.

Given the above considerations, the actual isolated (ceteris paribus) effect of renewable energy on GFCF may be difficult to quantify and test for significance. To address such a complex task, a spatial panel data model [2,3] is established, covering a representative and diverse sample of 10 adjacent EU countries at the NUTS2 level (annual 2012–2017 data are used). Figure 1 provides a simple insight into the data used: the color-coded choropleth shows relative shares of renewable energy on total energy consumption.

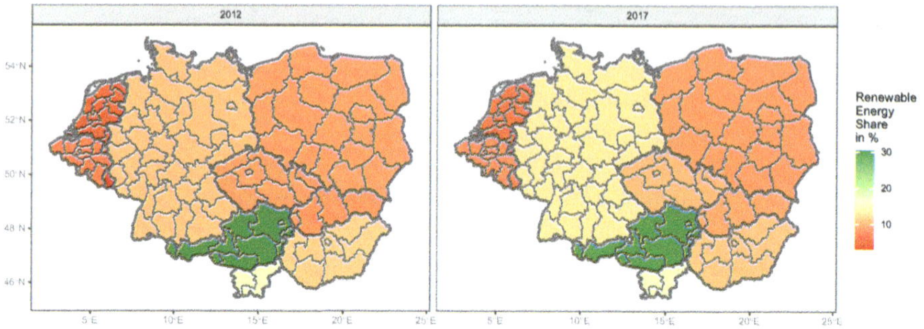

Fig. 1. Renewable energy share on total energy consumption, years 2012 and 2017

As may be seen from Fig. 1, data on renewable energy consumption are only available from Eurostat at the national level (NUTS0). Clearly, a more disaggregated (regional) data would be preferred for this type of analysis. However, the analysis of renewables' impact on GFCF may still be addressed at the NUTS2 level (regions also shown in Fig. 1). Given the state-level differences in legal frameworks, state-wide government incentives and other centralized energy policy actions, renewable energy consumption varies primarily at the state level [4]. Hence, evaluation of the impact of renewables on GFCF may be adequately performed – see Sect. 3 for detailed model and data-set descriptions.

The remainder of this paper is structured as follows: Sect. 2 describes the spatial panel data econometric framework, section three provides the results of empirical analysis (including data-set description, model estimation, testing, robustness evaluation and results discussion). Section 4 and the list of references conclude this paper.

2 Methodology

In this section, motivations for controlling spatial dependencies in cross-sectional and panel data models are explained and the general structure of spatial panel models is outlined.

Modelling spatial dependencies is becoming an increasingly mainstream approach in many fields of economic and non-economic research activities. For example, in most macroeconomic regional analyses based on empirical panel data, the null hypothesis of cross-sectional independence would be violated. If not addressed properly by spatially augmented methods, this situation has important negative effects on model estimation and statistical inference. Neglected cross sectional dependency typically leads to inconsistent estimators (inefficient at best) and flawed test results [2,3].

Spatial econometric models are sometimes described as a generalized form of the autoregressive processes present in time series. In spatial analysis, 2D dependencies are used instead of the one-dimensional time space (sometimes 3D interactions are modelled in environmental studies and bio-statistics). Also, there is one remarkable difference: unlike time series analysis, where the causal direction of autoregressive processes is clear, the case of spatial dependency is much more fluid and complex. Spatial units would typically interact and influence each other (mutual dependency). More importantly, the maximum distances that support spatial interactions are notoriously difficult to assess [8].

In spatial econometrics, spatial structures serve as prior information that is used for estimation of model parameters and marginal effects. As a general rule, neighborhood patterns cannot be estimated along with model coefficients. Unfortunately, a precise neighborhood specification (for a given "map" and observed economic variable) is not generally available/known with precision [6]. Given the large number of neighborhood-definition algorithms and possible options (distance based neighbors, k-nearest neighbors, contiguity patterns, etc.), researches often have to assess multiple spatial settings to evaluate model stability and robustness of their results [5].

To establish a spatial structure used for model estimation, we start by constructing a matrix of distances $\boldsymbol{H}$ $(N \times N)$, where individual elements $h_{ij} = h_{ji}$ depict distances among the N spatial units on a given map. From individual distances, a spatial connectivity matrix $\boldsymbol{C}$ is produced as follows:

$$\boldsymbol{C} = [c_{ij}] = \begin{cases} 0 & \text{if} \quad i = j\,, \\ 0 & \text{if} \quad h_{ij} > \tau\,, \\ 1 & \text{if} \quad h_{ij} \leq \tau\,, \quad i \neq j\,, \end{cases} \tag{1}$$

where $c_{ij} = 1$ denotes two units (i, j) being neighbors and vice versa. Elements on the main diagonal of $\boldsymbol{C}$ equal zero by definition as spatial units are not neighbors to themselves. The parameter τ is a conveniently chosen threshold – the maximum distance allowed between neighbors. While the choice of τ is largely ad-hoc, lower and upper bounds for τ should be observed for estimation and stability of spatial models. For the lower bound, $\tau \geq \max_i \min_j \{h_{ij} | h_{ij} > 0\}$ should be observed. This constraint eliminates "islands" i.e. regions without neighbors and thus zero rows/columns in $\boldsymbol{C}$. The upper bound for τ follows from the weak spatial dependency assumption [3]. As a rule of thumb, τ should not exceed a small fraction of the maximum h_{ij} distance (say, $\frac{1}{4}$).

Following the general approach to spatial panel model specification [3,10], we may cast the model as

$$\begin{aligned} \boldsymbol{y} &= \lambda\,(\boldsymbol{I}_T \otimes \boldsymbol{W})\,\boldsymbol{y} + \boldsymbol{X\beta} + \boldsymbol{u}\,, \\ \boldsymbol{u} &= (\boldsymbol{\iota}_T \otimes \boldsymbol{I}_N)\,\boldsymbol{\mu} + \boldsymbol{\varepsilon}, \end{aligned} \tag{2}$$

where $\boldsymbol{y}$ is a vector of dependent variable observations $(NT \times 1)$ with $i = 1, 2, \ldots, N$ cross-sectional elements and $t = 1, 2, \ldots, T$ time units. The spatial weights matrix $\boldsymbol{W}$ is a row-standardized version of the connectivity matrix (1), such that $\sum_{j=1}^{N} w_{ij} = 1$ and $\boldsymbol{X}$ is a $(NT \times k)$ matrix of exogenous regressors. Elements $\boldsymbol{I}_T$ and $\boldsymbol{I}_N$ denote identity matrices (dimensions given by subscripts) and $\boldsymbol{\iota}_T$ is a $(T \times 1)$ vector of ones. The $\otimes$ operator is the Kronecker product. Vector $\boldsymbol{\beta}$ and the scalar λ are parameters of the model to be estimated by a maximum likelihood (ML) method. The random element $\boldsymbol{u}$ has two constituent parts: unobserved individual effects $\boldsymbol{\mu}$ and potentially heteroskedastic disturbances $\boldsymbol{\varepsilon}$. Vector $\boldsymbol{\mu}$ $(N \times 1)$ contains unobserved individual effects [10].

While model (2) is just one of many possible spatial panel setups, the specification outlined in this section corresponds to the diagnostic Lagrange multiplier tests [10] performed using observed empirical data (see Sect. 3.1). Please note that $\boldsymbol{\beta}$ parameters of model (2) are not the marginal effects. For spatial panel models with spatially dependent endogenous variables, marginal effects are the "direct effects" and "spillovers" (indirect effects) – they can be calculated from an estimated spatial model [3,5].

3 Empirical Analysis

A theoretically well-established spatial panel model is used to describe the dynamics involved in GFCF processes. Specification (3) is based on a relatively common GDP-growth modelling paradigm [4]. Instead of the matrix-based form of outlining model (2), empirical model implementation is conveniently provided in an equation form as

$$\begin{aligned}\log(GFCF_{it}) = {} & \lambda \boldsymbol{w}_i GFCF_t + \beta_0 + \beta_1 \log(GDP_{it}) + \beta_2 RES_{it} + \beta_3 Unem_{it} \\ & + \beta_4 \log(TotEng_{it}) + \beta_5 ActSh15\_64_{it} + \beta_6 \log(R\&D_{it}) \\ & + \beta_7 \log(BaseGDP_i) + \boldsymbol{d}_i'\gamma + \mu_i + \varepsilon_{it}\end{aligned} \tag{3}$$

where $\log(GFCF_{it})$ is the dependent variable: a log-transformed GFCF. Logarithmic transformation is used for simple interpretation of marginal effects (percentage changes). The $\boldsymbol{w}_i GFCF_t$ element corresponds to the spatial lag of $GFCF_{it}$ at the time period t, with $\boldsymbol{w}_i$ being the i-th row of matrix $\boldsymbol{W}$. Variable $\log(GDP_{it})$ is the log-transformed GDP per capita in a given i-th NUTS2 region at time period t, 2015 fixed prices (EUR) are used. RES_{it} is the percentage share of renewable energy consumption on total energy consumption (recorded in percentage points). $Unem_{it}$ is the unemployment rate (also in percentage points). The element $\log(TotEng_{it})$ describes total energy consumption (recorded in thousands of tonnes of oil equivalent) and it provides control for correct interpretation of the main variable of our interest, which is RES_{it} here. Same as RES_{it}, $TotEng_{it}$ is only recorder at the state-wide (NUTS0) level. $ActSh15\_64_{it}$ is the share of economically active individuals in the population (aged 15 to 64) and it proxies overall labor market efficiency. Variable $\log(R\&D_{i,t})$ describes the log-transformed research and development (R&D) expenditures in fixed 2015 prices (EUR). Regressor $\log(BaseGDP_i)$ is the log-transformed GDP per capita from a base year (2011) and it controls for the Sollow-Swan macroeconomic convergence effects [12].

Vector $\boldsymbol{d}_i'$ (1×9) contains nine state-level (NUTS0) binary variables that account for the hierarchical structure of the spatial data. Individual dummies equal one if the i-th NUTS2 region is part of the corresponding state and vice versa. This setup is used to control for country-specific differences in GFCF dynamics. For technical reasons (coefficient identification), there is no dummy variable for Austria in $\boldsymbol{d}_i'$ and Austria is used as a reference level. The last two observable regressors are time-invariant, which simplifies their corresponding subscripts. Model parameters λ, $\boldsymbol{\beta}$ and $\boldsymbol{\gamma}$ are estimated by ML and the interpretation of the remaining elements of the empirical equation (3) follows directly from its general form (2).

3.1 Data

A balanced panel of 10 EU countries (Austria, Belgium, Czech Republic, Germany, Hungary, Luxembourg, the Netherlands, Poland, Slovakia and Slovenia) is used, regional data are observed at the NUTS2 level. Overall, the sample used covers a total of 110 NUTS2 units, belonging to 10 states (NUTS0 regions). In the time dimension, annual 2012–2017 data are used, with the exception of $\log(BaseGDP_i)$ regressor, where 2011-year observations are used to account for the Sollow-Swan type of convergence processes. The choice of economies (and regions) included in the sample is driven by two main factors: first, it is desirable to work with a relatively consistent data-set (e.g. EU member states) and we should aim for data-set with adequate variance in the data. Second, we need a contiguous and unbroken "map" of regions for proper analysis of spatial interactions. In the time dimension, sample choice was mainly driven by data availability, with 2017 annual observations being the latest available time period with a complete data-set, necessary for estimation using spatial panel data. Despite of the limitations discussed, the data used for estimation provide a representative and informative sample, allowing for reasonable model estimation and adequate statistical inference.

For transparency and reproducibility reasons, identification codes of Eurostat's data-sets are provided next. GFCF, our dependent variable, is retrieved from the "nama_10r_2gfcf" dataset. Specifically, total (aggregated) values of GFCF are used for estimation of model (3). While NACE rev.2 dis-aggregation for GFCF is also provided by Eurostat, model (3) dis-aggregated estimation output is omitted from this contribution due to space limitation (and somewhat lesser relevance). However, Fig. 2 provides an overall insight into the nature and spatial distribution of GFCF: the two choropleths in the first row of Fig. 2 show total GFCF for years 2012 and 2017. The second row depicts GFCF in the industrial sector (items B to E of the NACE revision 2 nomenclature), third row displays GFCF in the sector of professional and technical services (sections M and N of NACEr2) and the final row contains GFCF in the information and communication technology (ITC) sector. Different color palettes are used for visual convenience and all data in Fig. 2 are shown on a log-transformed scale.

Data on GDP per capita come from the "nama_10r_2gdp" data-set. Energy consumption is fetched from "nrg_bal_c" (both total and renewable consumptions

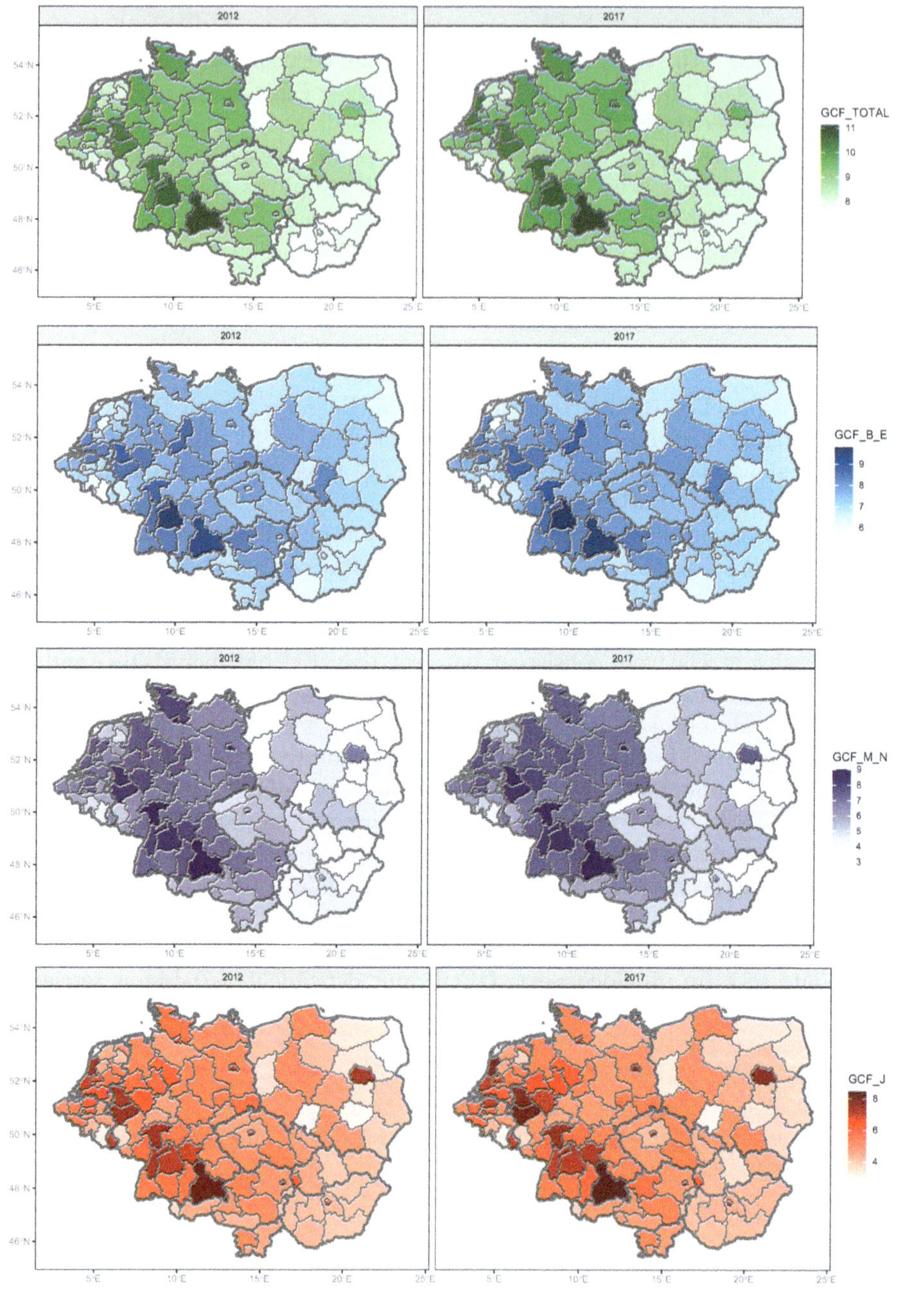

Fig. 2. Gross fixed capital formation: Total, Industry (NACEr2 B to E), Professional and technical services (NACEr2 M & N), ICT (NACEr2 code J); 2015 fixed prices, log-transformed EUR values, years 2012 and 2017 shown

are recorded at the national level only). For labor market data, unemployment is retrieved from "lfst_r_lfu3rt" and economic activity status from "lfst_r_lfp2act". Data-set "rd_e_gerdreg" is used for R&D expenditures. Please note that R&D data are only available on a bi-annual basis from Eurostat, so linear interpolation was used to impute all missing observations for this variable. Finally, 2015 real prices are calculated based on "ei_cphi_m" (simple transformation from monthly to annual data is performed).

3.2 Results and Discussion

Using Moran's I test for panel data [11], null hypothesis of spatial randomness of GFCF is rejected in favour of positive spatial autocorrelation. This result holds under varying spatial definitions (neighborhood structures). To keep this contribution concise, individual Moran's I test are omitted here. However, all relevant estimation & testing outputs, additional figures, R-source codes and data are available from the author upon request.

Estimation of the empirical model (3) is conveniently summarized in Table 1. Three different columns (a) – (c) relate to three alternative setups for model estimation. Columns (a) and (b) refer to alternative spatial structures used for $\boldsymbol{W}$-matrix construction. In column (a), maximum distance threshold between neighbors (τ from the (1) expression) is set to 530 km and this seems to be the optimum value, given the maximized log-likelihood (LL) values as extracted from models estimated using alternative spatial structures. For detailed discussion of robustness of our model to different spatial structures, see Sect. 3.3. Column (b) represents the estimation output from an alternative $\boldsymbol{W}$-matrix, with $\tau = 320$ km. This threshold seems to be the "second best" spatial structure identified by the estimation approach used. Finally, column (c) serves as a reference: this is the output from a pooled, non-spatial version of model (3), where $\lambda = 0$ and $\mu_i = 0$ for $\forall i$. The estimation presented in column (c) are shown as a baseline reference for interpretation results in columns (a) and (b), given the pooled ordinary least squares (OLS) estimation of the model is inconsistent due to violated assumptions for the OLS method (mostly, endogeneity in regressors).

As already discussed in Sect. 2, β_j parameters of model (2) and/or its empirical version (3) do not constitute the marginal effects. Hence, Table 1 contains the calculated direct and spillover effects, along with their bootstrapped standard errors (in parenthesis). By construction, the OLS-estimated pooled model in column (c) has a simpler interpretation, with β_j being the direct effects and with non-existent spillover effects – under zero restrictions imposed on spatial interactions, a change in regressor in a region i can only affects the expected value of the dependent variable in the same region i (in contrast to spatial models, where neighbors are affected as well).

From the estimated marginal effects of individual regressors in Table 1, we may interpret the ceteris paribus influences of different macroeconomic factors on GFCF. From the direct and indirect effects of log(GDP), we may deduce strong positive direct effect of GDP growth on GFCF (conclusion is valid and statistically significant for both spatial structures used in Table 1). Interestingly,

Table 1. Model estimation

Direct/Indirect impacts (DI/II)	(a)	(b)	(c)
DI log(GDP)	0.2699[a]	0.3312[a]	0.3102
(st. error)	(0.1200)	(0.1240)	(0.3001)
II log(GDP)	0.2060	0.1623[b]	—
	(0.1400)	(0.0899)	
DI RES	0.0812	0.4341	−0.4137
	(0.7558)	(0.7336)	(2.4948)
II RES	0.0620	0.2127	—
	(0.6732)	(0.4177)	
DI $Unem$	−0.0048	−0.0047	−0.0073
	(0.0039)	(0.0039)	(0.0056)
II $Unem$	−0.0036	−0.0023	—
	(0.0037)	(0.0024)	
DI log($TotEng$)	−0.4917[a]	−0.4470[a]	−0.6887
	(0.1743)	(0.1699)	(0.5758)
II log($TotEng$)	−0.3754[b]	−0.2190[b]	—
	(0.2219)	(0.1268)	
DI $ActSh$15_64	1.1455[a]	1.0173[a]	0.2680
	(0.3190)	(0.3199)	(0.5494)
II $ActSh$15_64	0.8745[b]	0.4985[a]	—
	(0.5005)	(0.2649)	
DI log($R\&D$)	0.1955[a]	0.1927[a]	0.5064[a]
	(0.0225)	(0.0228)	(0.0148)
II log($R\&D$)	0.1493[a]	0.0944[a]	—
	(0.0678)	(0.0387)	
DI log($BaseGDP$)	0.5663[a]	0.5191[a]	−0.2165
	(0.1784)	(0.1821)	(0.2870)
II log($BaseGDP$)	0.4323[b]	0.2544[b]	—
	(0.2376)	(0.1444)	
λ	0.4360[a]	0.3328[a]	—
	(0.0956)	(0.0814)	
τ (km)	530	320	—
Log likelihood (LL)	447.62	446.00	−103.69

Note: [a] – estimate significant at $\alpha = 5\%$; [b] – significant at $\alpha = 10\%$. Logarithmic transformation of regressors denoted by l.

indirect effects are not statistically significant. The results here may be somewhat affected by the inclusion of the last macroeconomic regressor: log($BaseGDP$). Using the base GDP information, we have to conclude that richer regions would tend to have significantly higher GFCF rates. Significant indirect effects of the base GDP would also imply the formation (historical existence) of high-GFCF clusters. If richer regions invest at higher rates, they may also be expected to grow faster in the future, resulting in a persistent macroeconomic divergence among regions (clusters of regions). Such results support the need for EU's cohesion policies – aimed at mitigating such processes.

From the direct and spillover effects of *RES*, we may see that renewables have no statistically significant influence on GFCF. This conclusion holds for both spatial structures considered. Such result may seem "weak" to many readers – perhaps expecting strong positive or negative influences of renewables on GFCF (upon their personal beliefs and stances). However, I find this results rather satisfactory – the main concern and motivation for this article was to evaluate whether or not renewable energies (their costs) would slow-down GFCF and thus potentially weaken future economic growth. Given the results from our relatively complex model that controls for many different macroeconomic and regional effects as well as spatial interactions, it is possible to conclude that renewables are not a factor that would slow down GFCF significantly.

For accurate ceteris paribus interpretation of renewables and their marginal effects, total energy consumption is also included in model (3). Interestingly, the direct and spillover effects of changes in total energy consumption are all negative, pointing to GFCF flows avoiding regions with predominant energy-intensive sectors. This is a rather interesting finding with potentially significant implications for further research and macroeconomic policy actions.

Model (3) also includes two important control variables describing labor market conditions: the unemployment (*Unem*) and the share of economically active individuals aged 15 to 64 years (*ActSh*15_64). From the marginal effects, we may draw two dramatically different conclusions: while the rate of unemployment is not a statistically significant factor for GFCF, the overall activity level in population has large positive effect on GFCF (both direct and indirect effects are positive and prominent). Again, this has many potentially important implications for future analyses and policy actions – both at the national and state levels.

R&D expenditures seem to have a significant positive effect on GFCF, which seems rather logical and intuitive. Again, both direct and spillover effects are statistically significant and the potential for formation of mutually divergent regional clusters can be expected.

The spatial autocorrelation parameter λ is positive, statistically significant and relatively strong in both (a) and (b) columns of Table 1. This may serve as an additional reassurance of proper model specification. A very similar conclusion may be drawn by comparing the maximized *LL* functions shown in Table 1: changing model specification from a pooled non-spatial model to spatial panel model (3), we observe a dramatic and highly statistically significant increase in model's fit to data.

Finally, state-specific effects (corresponding to the vector $\boldsymbol{d}$) are excluded from Table 1 due to space limitations: while it is important to control for such effects, their interpretation is not very informative with respect to the topics analyzed. Overall, the estimates shown in columns (a) and (b) of Table 1 provide adequate fit to data, explicit control over major relevant factors and important macroeconomic insight with respect of renewables' effects on GFCF.

3.3 Robustness to Changes in Spatial Structure Definition

As pointed out in Sect. 2, spatial structures cannot be estimated along with the parameters of model (2). Rather, the researches has to make a somewhat ad-hoc choice and set the spatial structure (i.e. matrix $\boldsymbol{W}$) used for model estimation. However, the spatial structure is not known, at least not completely. From the discussion corresponding to expression (1), we know there are theoretically determined bounds to the τ maximum neighbor distance threshold. However, such bound are relatively wide. At the same time, we should be aware of the fact that changing spatial structure (e.g. by increasing τ or by applying other approaches to setting spatial structure), we may as well influence/change the estimated marginal effects [8].

A relatively simple solution to the above described problem follows from previous literature [4,5,8] and it may be described as follows: The spatial econometric model under scrutiny is repeatedly estimated with slightly altered spatial structures (e.g. τ is put subject to small increments over the whole plausible range of distances) and the outputs from such estimations are evaluated with respect to stability of marginal effects, fit to data, etc.

Figure 3 provides a concise visual summary of such stability evaluation. Spatial structures were created for $170\,\text{km} \leq \tau \leq 1.000\,\text{km}$ (10-km increments were used to generate alternative $\boldsymbol{W}$ matrices) and maximized LL values, λ-parameters and marginal effects were compared. From the top-left part of Fig. 3, we may see how the LL function changes with increasing τ distances. While the overall optimum of the LL function corresponds to $\tau = 530$ km, there is another prominent local LL-maximum at $\tau = 320$ km: estimations corresponding to both spatial structures are shown in Table 1. Such "bimodal" (multimodal) shape of the LL function is relatively common in spatial analysis. It may be argued that the aggregated GFCF values used for model (3) estimation are constituted by two "main groups" of investment types, each following different spatial (agglomeration) pattern.

From the robustness evaluation of λ (top-right element of Fig. 3), we can see that the estimated parameter remains relatively stable even under changing spatial definitions. The same favourable stability can be identified for the estimated direct effects and spillovers of regressors from model (3). Due to space limitations, only marginal effects for GDP (2nd row of Fig. 3), unemployment (3rd row) and renewables (last row) are shown here. Overall, we can see that model (3) is robust against minor changes in spatial definitions.

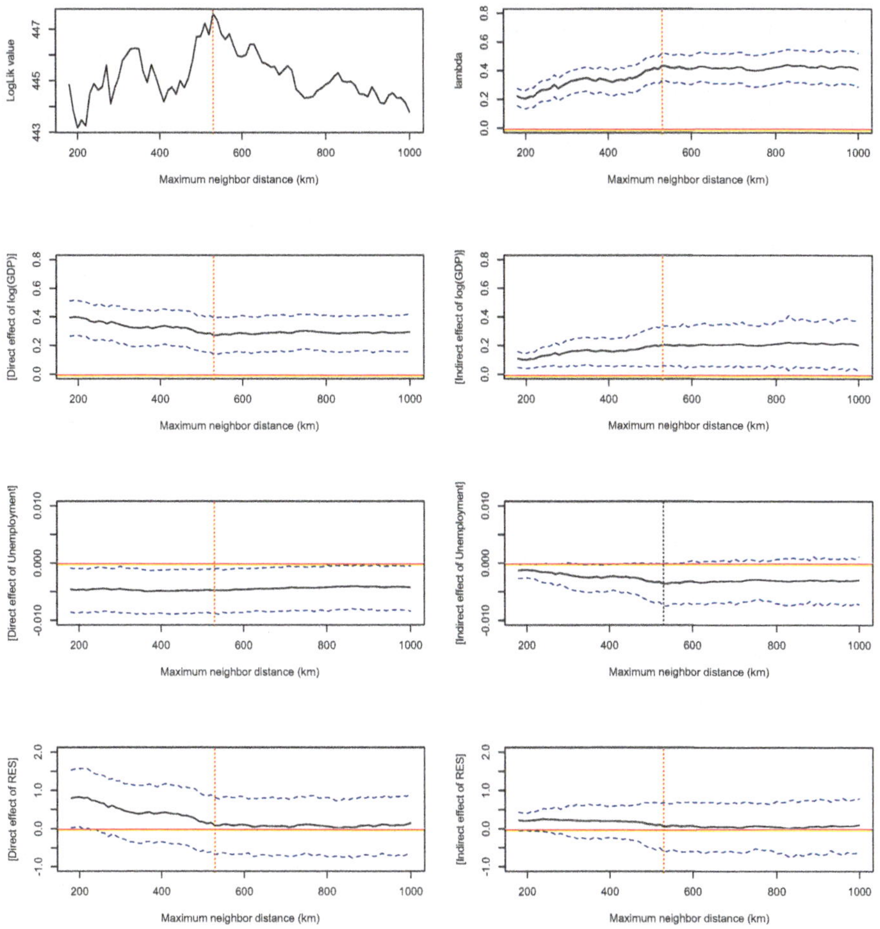

Fig. 3. Model robustness with respect to changing maximum neighbor distance thresholds (GFCF total)

4 Conclusions

As we model the dynamics of fixed gross capital formation, spatial panel models provide a unique possibility of quantifying ceteris paribus effects of diverse economic factors – those may be discerned from both individual effects (individual heterogeneities) and spatial effects (spatial autocorrelation, spillover effects).

The empirical analysis presented in this contribution is based on balanced panel of NUTS2-level data (110 regions in Austria, Belgium, Czech Republic, Germany, Hungary, Luxembourg, the Netherlands, Poland, Slovakia and Slovenia), with annual observations covering the period 2012–2017. A complex evaluation of model stability with respect to altering the assumed spatial structures (inherently unknown) is also provided.

Overall, it may be concluded that the increased share of renewable energy consumption does not interfere with investment activities (measured by gross fixed capital formation). Also, this contribution points at possible paths of future research, which could be targeted at evaluating the impact of renewables on different types of investment activities – e.g. disaggregated along different industrial sectors.

Acknowledgement. Supported by the grant No. IGA F4/19/2019, Faculty of Informatics and Statistics, University of Economics, Prague. Geo-data source: GISCO - Eurostat (European Commission), Administrative boundaries: © EuroGeographics.

References

1. Abdoli, G., Farahani, Y., Dastan, S.: Electricity consumption and economic growth in OPEC countries: a cointegrated panel analysis. OPEC Energy Rev. **39**, 1–16 (2015)
2. Baltagi, B.H.: Econometric Analysis of Panel Data. John Wiley, New York (2005)
3. Elhorst, J.P.: Spatial Econometrics: From Cross-Sectional Data to Spatial Panels. Springer-Briefs in Regional Science. Springer, Heidelberg (2014). https://doi.org/10.1007/978-3-642-40340-8
4. Formánek, T.: Semiparametric spatio-temporal analysis of regional GDP growth with respect to renewable energy consumption levels. Appl. Stoch. Models Bus. Ind. **36**(1), 145–158 (2020). https://doi.org/10.1002/asmb.2445
5. Formánek, T.: Spatially augmented analysis of macroeconomic convergence with application to the Czech Republic and its neighbors. In: Silhavy, R., Silhavy, P., Prokopova, Z. (eds.) Applied Computational Intelligence and Mathematical Methods, CoMeSySo 2017. Advances in Intelligent Systems and Computing, vol. 662, pp. 1–12. Springer, Cham (2018). https://doi.org/10.1007/978-3-319-67621-0_1
6. Lee, L., Yu, J.: Some recent developments in spatial panel data models. Reg. Sci. Urban Econ. **40**(5), 255–271 (2010). https://doi.org/10.1016/j.regsciurbeco.2009.09.002
7. Lequiller, F., Blades, D.: Understanding National Accounts, 2nd edn. OECD Publishing (2014). https://doi.org/10.1787/9789264214637-en
8. LeSage, J.P., Pace, R.K.: The biggest myth in spatial econometrics. Econometrics **2**(4), 217–249 (2014). https://doi.org/10.3390/econometrics2040217
9. Michaelides, E.E.: Alternative Energy Sources. Springer, Heidelberg (2012). https://doi.org/10.1007/978-3-642-20951-2
10. Millo, G., Piras, G.: splm: Spatial panel data models in R. J. Stat. Softw. **47**(1), 1–38 (2012). https://doi.org/10.18637/jss.v047.i01
11. Ou, B., Zhao, X., Wang, M.: Power of Moran's I test for spatial dependence in panel data models with time varying spatial weights matrices. J. Syst. Sci. Inf. **3**(5), 463–471 (2015). https://doi.org/10.1515/JSSI-2015-0463
12. Piras, G., Arbia, G.: Convergence in per-capita GDP across EU-NUTS2 regions using panel data models extended to spatial autocorrelation effects. Statistica **67**(2), 157–172 (2007). https://doi.org/10.6092/issn.1973-2201/3513

A Method to Prove the Existence of a Similarity

Mahyuddin K. M. Nasution(✉)

Information Technology Department, Fakultas Ilmu Komputer dan Teknologi Informasi (Fasilkom-TI), Universitas Sumatera Utara, USU, Padang Bulan, 20155 Medan, Sumatera Utara, Indonesia
mahyuddin@usu.ac.id

Abstract. Each object will build its own vector space through naming and weighting its characteristics. Based on the vector space, it is possible to measure the similarity between objects where certain parts of an object no longer need to decipher its behavior by simply referring to the measured part of another object. In this case, the similarity measurement tool can streamline the process or reduce the complexity of processing data by referring to other object characteristics. However, there are many similarities in the used measurement. Each measurement has a meaning, and this requires proof of each measurement in common so that the meaning indicates the measurement function either in theory or computationally in simulation.

Keywords: Object · Dissimilarity · Distance · Measurement · Difference · Simulation

1 Introduction

The similarity is a measurement to determine the proximity of one object to another object [1]. Similarity involves the process of characterizing each object by doing an ontology-taxonomy or describing in detail the features of the object [2, 3]. The process results what we call it as naming and weighting to characteristics of objects in such there is a way to provide a method enabled for measuring their similarity [4,5]. The method is determined by placing a comparison of what is the same and what is different about two objects. In principle, it involves the concept of overlapping between the characteristics of the two objects [6]. Characterizing and weighting of objects based on the vector space as a target of the optimization of the measurement process is to reduce the complexity of its measurement. With the reason that each space has its own measurement complexity while the method can be adjusted again for it [7].

In general, measuring the similarity of two objects is to avoid the duplication of other measurements many times to similar objects. The characteristics of objects however can be measured by various similarities so that gives different meanings [8]. Therefore, the existence of measurement of similarity determines

R. Silhavy et al. (Eds.): CoMeSySo 2020, AISC 1295, pp. 243–252, 2020.
https://doi.org/10.1007/978-3-030-63319-6_21

the meaning of the measurement results. Thus, the purpose of this paper is to outline the evidence about the existence of a similarity.

2 Problem Definition

Each object or entity that is in the study space (universe of discourse) $\mathcal{U}$ has characteristics that are features and weights. When an object's behavior is known, surely the behavior of other's objects can be efficiently determined based on measuring the similarity or closeness of those objects [9,10].

Suppose $X \subset \mathcal{U}$ as a set, there is a mapping of s from X to the set of numbers, for example the set of real numbers R, whereby $R \subset \mathcal{U}$ [8,11].

Definition 1. *Let X is a set. The function $s : X \times X \to R$ is called a* similarity *for X if for all $x, y \in X$ apply*

A11 $s(x, y) \geq 0$ *(non-negativity);*
A12 $s(x, y) = s(y, x)$ *(symmetry);*
A13 $s(x, x) \geq s(x, y)$.

The term *similarity* is also called as *proximity*. The measurement result of the function s is in the range $[0, 1]$: The minimum value is 0 and the maximum value is 1. Thus, for every $x, y \in X$, $x \neq y$, applies $0 \leq s(x, y) \leq 1$ and $s(x, x) = 1$ [12].

Proposition 1. *Suppose $s(x, y)$ is a similarity for $x, y \in X$. If $s(x, y) \in [0, 1]$, then there is an inverse of $s(x, y)$ called the* distance *$d(x, y)$ such that $d(x, y) = 1 - s(x, y) \in [1, 0]$.*

Proof. Assuming that $s(x, y) \in [0, 1]$. By involving the lower limit of the range $[0, 1]$, i.e. $s(x, y) = 0$, which means that $0 - s(x, y)$ can be minus if $s(x, y) > 0$, but $s(x, y) = 0$ and based on Definition 1 axiom **A11**, it also means that $0 - s(x, y) = 0$. By involving the upper limit of the range $[0, 1]$, i.e. $s(x, y) = 1$, which means that $1 - s(x, y)$ can be minus if $s(x, y) > 1$, but $s(x, y) = 1$ and based on Definition 1 axiom **A11** is obtained $1 - s(x, y) = 0$, or $1 - s(x, y)$ can be positive if $s(x, y) < 1$, or $1 - s(x, y) = 1$ if $s(x, y) = 0$. Based on that, generally, it can be stated that

$$\begin{aligned} [0, 1] &= [0, 1] - s(x, y) \\ &= 1 - s(x, y), \end{aligned} \tag{1}$$

or suppose $d(x, y) = [0, 1]$, the Eq. (1) becomes

$$d(x, y) = 1 - s(x, y) \tag{2}$$

or the invers of a similarity. So $d(x, y) \in [1, 0]$.

Lemma 1. *If $s(x, y) \in [0, 1]$, then a pair of vector $x, y \in X$ on points $s(x, y)_i$, $i = 1, \ldots, n$, form a straight line.*

Proof. Suppose the vertical axis in a coordinate system is a range $[0,1]$, and horizontal axis in a coordinate system is a range i. The points of $s(x,y)$ are expressed at the same distance from the two axes. The shared values of x and y, written as $x \cap y$, $x, y \in X$, apply $x \cap y \leq x$ and $x \cap y \leq y$. In other words, $x \cap y \leq x + y$. If $s(x,y) = 0$, then $x \cap y = 0$ and one of x and y is not zero. Based on the assumption, that the value of $s(x,y)$ or point is between two equal parts of the graphic, that is $\frac{1}{2}s(x,y)$ applies to a range value of $[0,1]$ and for all values of $\frac{1}{2}s(x,y)_i$, $i = 1, 2, \ldots, n$, $i = 1, 2, \ldots, n$, a collection of points from $s(x,y)$ is on a straight line from 0 to 1 diagonally.

Lemma 2. *If* $s(x,y) \in [0,1]$, *then the index of points* $s(x,y)_i$, $i = 1, \ldots, n$ *form a square together with* $[0,1]$.

Proof. A rectilinear unit is four sides of the same size. The opposite sides of $[0,1]$ have formed a unit between 0 and 1. If the index i is divided by the largest index value of the number of pairs $x, y \in X$, then i can be written back as $i = 0/n, 1/n, \ldots, n/n$, or $i = 0, 1/n, \ldots, 1$, a range that is equal to or in $[0,1]$. Thus, the vertical axis and the horizontal axis are as low as 0 and as high as 1 respectively, and the points forming a straight line are in $[0,1]^2$, a rectilinear.

Proposition 1 reveals that each similarity has the opposite which is referred to as *dissimilarity* which is stated as follows [8].

Definition 2. *Suppose* X *is a set. The function* $d : X \times X \rightarrow R$ *is called* dissimilarity *on* X *if for all* $x, y \in X$ *apply:*

A21 $d(x,y) \geq 0$ *(non-negativity);*
A22 $d(x,y) = d(y,x)$ *(symmetry);*
A23 $d(x,x) = 0$.

In topology, x and y are vectors that have magnitudes and directions, and X is vector space, and d is a distance function for $x, y, z \in X$ applies $d(x,y) = d(x,z) - d(y,z)$ and $x \neq y$. So, (X,d) represents *space distance* for vector space X with a distance function d. The relationship between similarity and dissimilarity as in Eq. (2) reveals that they are complementary, i.e. $s(x,y)^c = 1 - s(x,y)$. By referring back to Proposition 1, the points $s(x,y)$ on the straight line L have dual, so if $s(x,y) \in [0,1]$ and $d(x,y) \in [1,0]$, where the straight line is $L : \{s(x,y)_i | i = 1, \ldots, n\} \rightarrow [0,1]$ while the dual is straight line $L^c : \{d(x,y)_i | i = 1, \ldots, n\} \rightarrow [1,0]$. Two lines that are formed will reveal the nature of similarity as the identity or meaning of the similarity itself.

3 An Approach

For example, for every $x, y \in X$, there is the shared features that are stated as $x \cap y \in X$. The features indicate the existence of a relationship between x and y. So, $x \cap y$ is also a vector in the vector space X, which has weight and direction. The difference between x and y is the distance that expresses $d(x,y)$ as applicable in Definition 2, which indicates the existence of x and y respectively in the vector space X, namely that $x \leq y$ or $x \cap y > 0$.

In theory, a measurement different from other measurements in the same space indicates the existence of a measurement. Measurement formula like *Jaccard coefficient* J_c [13], *mutual information* M_i [14], *dice coefficient* D_c [15], *overlap coefficient* O_c [16], etc. are measurements that conceptually involve the same vectors [17], namely $x, y, x \cap y \in X$, each of which is formulated as follow.

Definition 3. *Let there are vectors $x, y \in X$ and $x \cap y \in X$.* Jaccard coefficient *J_c as a measurement of $s(x, y)$ if it meets*

$$J_c = \frac{x \cap y}{x \cup y} \tag{3}$$

where $x \cup y = x + y - (x \cap y)$.

If $x \cap y = 0$ means $J_c = 0$. For $J_c = 1$. Equation (3) becomes $x \cap y = x + y - (x \cap y)$ or

$$x \cap y = \frac{x + y}{2} \tag{4}$$

If $x = y$, then Eq. (3) becomes $x \cap y = x$ or $x \cap y = y$.

Definition 4. *Let $x, y \in X$ and $x \cap y \in X$. A measurement of $s(x, y)$ is declared as* mutual information *if applicable*

$$M_i = \log \frac{m(x \cap y)}{xy} \tag{5}$$

where m is a multiple constant.

For example, $M_c = 0$, $10^0 = m(x \cap y)/xy$, or $1 = m(x \cap y)/xy$, $xy = m(x \cap y)$ generates $\log 1 = 0$. Instead, $M_c = 1$, $10^1 = m(x \cap y)/xy$ or $10xy = m(x \cap y)$, and the distribution results to Eq. (5): $\log 10 = 1$.

Definition 5. *Let $x, y \in X$ and $x \cap y \in X$. A measurement of $s(x, y)$ is called* dice coefficient *if it satisfies*

$$D_c = \frac{2(x \cap y)}{x + y} \tag{6}$$

As behavior of J_c, $D_c = 0$ when $x \cap y = 0$. For $D_c = 1$, a equation will be Eq. (4), Therefore, D_c is a different treatment from J_c.

Definition 6. *Let $x, y \in X$ and $x \cap y \in X$. A measurement of $s(x, y)$ is* overlap coefficient, *i.e.*

$$O_c = \frac{x \cap y}{\min(x, y)} \tag{7}$$

As behavior of J_c, $O_c = 0$ when $x \cap y = 0$. For $O_c = 1$, Eq. (7) becomes $x \cap y = \min(x, y)$. In other words, if $x < y$, then $\min(x, y) = x$ and $x \cap y$ approaches x or $x \cap y = x$, the results is $O_c = 1$.

Definition 7. *Let $x, y \in X$ and $x \cap y \in X$. A measurement of $s(x, y)$ is* cosine similarity *if*

$$C_c = \frac{x \cap y}{\sqrt{xy}} \tag{8}$$

Corresponds to J_c, $C_c = 0$ when $x \cap y = 0$. For $C_c = 1$, Eq. (8) becomes $\sqrt{xy} = x \cap y$.

Formulations in Eqs. (3), (5), (6), (7), and (8) shows all the measurements one by one has differences, although basically the difference is no visible. For example, for a maximum value of $s(x, y) = 1$,

$$x \cap y = \frac{x + y}{2} = \frac{10xy}{m} = x = \sqrt{xy}, x < y. \tag{9}$$

Of course, the difference requires proof in theory and computation. Computational verification involves simulating calculations that are determined based on sequential values to construct the intended straight line, and can certainly illustrate the differences stated as follows:

$$\delta = |s_1(x, y) - s_2(x, y)| \tag{10}$$

where $\delta \geq 0$. δ is the distance between one similarity to another in the same vector space. That difference, in this case, has a proven distance through theory, but also has evidence in calculations. Therefore, this distance requires a watch line that is also straight and is in $[0, 1]^2$.

Technically, the steps to measure similarity and a distance between two or more similarities are as follows:

```
similarity function(x, y, x ∩ y, c)
    if c = 1 then
        s = x ∩ y/(x + y − x ∩ y)
    if c = 2 then
        s = 2(x ∩ y)/(x + y)
    ...
return s
```

$\delta = |\text{similarity function}(x, y, x \cap y, 1) - \text{similarity function}(x, y, x \cap y, 2)|$

Lemma 3. *If $d(x, y) \in [0, 1]$, then a pair of vectors $x, y \in X$ to the index points $d(x, y)_i$, $i = 1, \ldots, n$ form a straight line that intersects a straight line from $s(x, y)_i$.*

Proof. Every point formed from $d(x,y)_i$, $i = 1, \ldots, n$ is from $1 - s(x,y)_i$ for all $x, y \in X$, i.e. for $s(x,y) = 0$ obtained $d(x,y) = 1$, and so on until $s(x,y) = 1$ gets $d(x,y) = 0$. So the two ends of the two lines do not meet in one point. $[0,1]^2$ or $[0,1] \times [1,0]$ forms a square of the same size, a straight line L is formed from the lower left corner of 0 to the upper right corner of 1 and bisect equal parts of $[0,1]^2$, but another straight line of L^c is formed from the upper left corner of 1 to the lower right corner of 0 and bisecting the same portion of $[0,1]^2$. Thus L and L^c intersect at the midpoint of each line.

The relationship between Lemma 1, Lemma 2, and Lemma 3 can be presented in Fig. 1.

4 Proof

In theory and computing, for every $x, y \in X$ applies

$$x \geq x \cap y \tag{11}$$

and

$$y \geq x \cap y. \tag{12}$$

Conceptually, the similarities in Eqs. (3), (5), (6), (7), and (8) is locked in the $x \cap y$ numerator, while the denominator is composed of various variations of the vectors $x, y, x \cap y \in X$ [18]. Proof of the existence of each similarity is equivalent to the meaning of each of the measurements as follows.

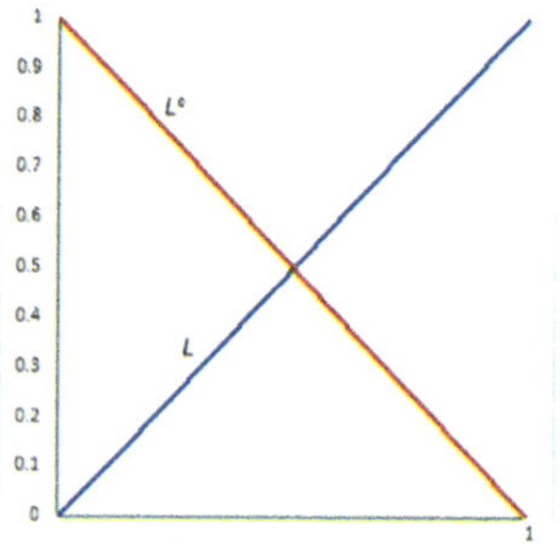

Fig. 1. L and L^c lines

4.1 Difference

The difference between each numerator in each of Eqs. (3), (5), (6), (7), and (8) is the presence of a statement that states that one of the numerators is bigger or smaller than the other numerator.

Case 1. The difference between J_c and D_c is corrected by the two denominators, that is $x+y-(x \cap y)$ and $x+y$ whereby $x+y-(x \cap y) < x+y$. $x+y-(x \cap y) = x+y$ if $x \cap y = 0$ for $x > 0$ or $y > 0$, which means that $J_c = D_c = 0$. Whereas if $x = y = (x \cap y)$, $(x \cap y)/(x+y-(x \cap y)) = x/(x+x-x) = x/x$ or $(2(x \cap y)/(x+y)) = 2x/(x+x) = 2x/2x$, means that $J_c = D_c = 1$. So the lines formed by J_c and D_c meet at a point with a value 0 and a point with a value 1, but it will be different when $x \cap y > 0$, which is based on consideration of $x \cap y < 2(x \cap y)$ and $x+y-(x \cap y) < x+y$, which means that $J_c < D_c$. All this, in general, is written with $J_c \leq D_c$.

Case 2. The difference between J_c and O_c focuses on the denominator: $x+y-(x \cap y)$ and $\min(x,y)$. If $x < y$, then $\min(x,y) = x$, or $x \cap y \leq x$, and consequently $x+y-(x \cap y) = x+y-x = y$. J_c and O_c are worth 0 if $x \cap y = 0$, whereas if $x = y$ then J_c and O_c are worth 1. Thus $J_c \leq O_c$.

Case 3. The difference between D_c and O_c has to do with the numerator and denominator. There is a difference between D_c and O_c in the numerator as a result of the multiplier constant in the Eq. (6), but for $x \cap y = 0$, also $D_c = O_c = 0$. The denominator between D_c and O_c reveals that $x+y > \min(x,y)$ or $\frac{x+y}{2} > \min(x,y)$, $x < y$. So, it is $D_c \leq O_c$.

Cases 1, 2, and 3 show that the difference between one similarity with another is expressed by the difference between δ. For example, $J_c \leq D_c \leq O_c$, by which the difference of δ can be measured based on the dual-line L^c. Other similarities will generally show the same behavior as being.

Proposition 2. *Suppose $s_i(x,y)$, $i = 1, \ldots, n$ are similarities based on the concept of $x \cap y$, then the difference between $s_i(x,y)$ lies in $\delta > x \cap y$.*

Proof. Based on Case 1, Case 2, and Case 3 it is found that all similarity measurements will be the same on the results of 0 and results of 1. The Eq. (9) expresses the difference in each measurement if $x \cap y < \delta = \frac{x+y}{2}$, $x \cap y < \delta = x+y$, $x \cap y < \delta = \frac{10xy}{m}$, $x \cap y < \delta = x$, or $x \cap y < \delta = \sqrt{xy}$, so $sim_i(x,y) \in [0,1]$. Likewise, measuring other similarities with the concept of $x \cap y$, there will be a difference of δ.

In theory, Proposition 2 shows the existence of each similarity formula.

4.2 Simulation

Computationally, the relationship between x and y satisfies $x \leq y$ or $x \geq y$, which means that $x < y$, $x > y$ or $x = y$. Thus, indirectly the value of $x \cap y$ maximum is x. Suppose the maximum value is 10, or $x = y = 10$, the value of $x < y$ will be in the range of $[0,10]$, and likewise the value of $x \cap y \in [0,10]$ or $n = 10$, as assumed in Table 1.

The simulation line constructed by the measurement J_c, using Eq. (3), Fig. 1 and Fig. 2, where $L = J_c$ and $L^c = d = 1 - J_c$. This reveals that there is

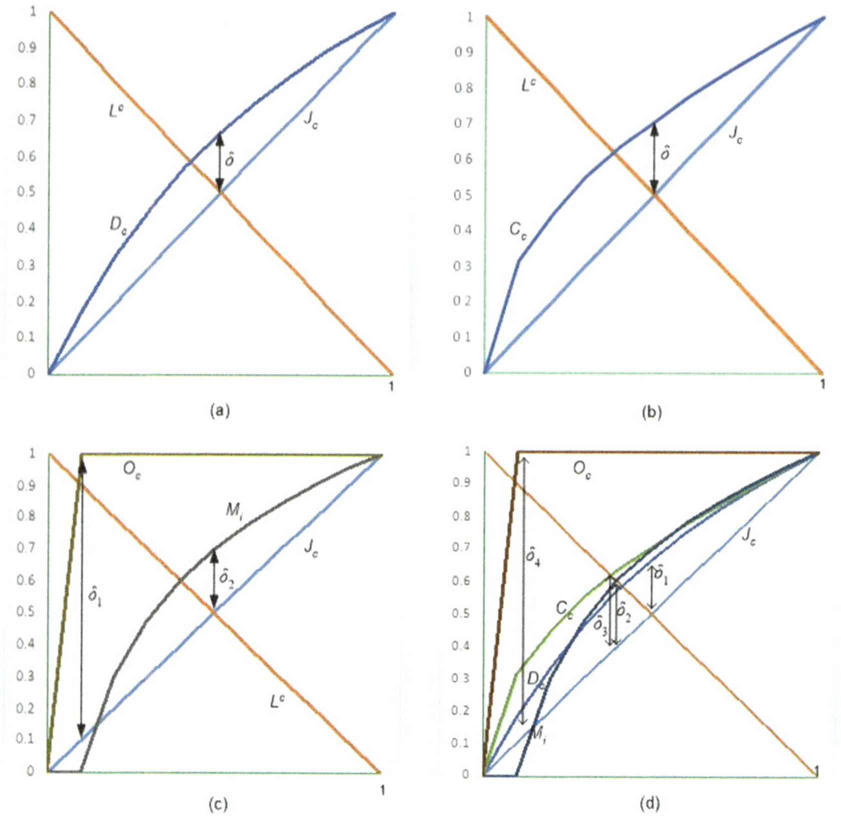

Fig. 2. L, L^c, J_c, D_c, C_c, O_c, and M_i lines

Table 1. The values of x, y, and $x \cap y$.

Variables	Values										
x	0	1	2	3	4	5	6	7	8	9	10
y	10	10	10	10	10	10	10	10	10	10	10
$x \cap y$	0	1	2	3	4	5	6	7	8	9	10

a symmetrical strength of the relationship between two vectors in vector space. The L^c distance line, which bisects the straight line from J_c into two different forces: low and high. For example from Table 1, $x = 5$, $y = 10$, $x \cap y = 5$, $J_c = 0.5 \in [0.1]$, while $d = 1 - 0.5 = 0.5 \in [0.1]$. In this case, the point J_c coincides with the point d where the L line and the L^c line intersect at the same point.

Constructing the simulation line is to measure D_c by involving Eq. (6) and the curve of data in Table 1 concretely parallels to the L line, as in Fig. 2(a).

The purpose of this measurement is to maintain the heterogeneous nature of sensitive information. For example from Table 1, $x = 5$, $y = 10$, $x \cap y = 5$. In computing D_c is 0.69897 while $d = 1 - D_c = 0.30103$. In other words, two points are opposite each other in the form of a mirror with an angle of 45°. Whereas the simulation line that is constructed measures C_c by involving Eq. (8) and the data in Table 1 produces a concave curve towards the line L, as in Fig. 2(b). This measurement serves to determine the position of the two vectors in the vector space. Suppose from Table 1, $x = 5$, $y = 10$, $x \cap y = 5$, the calculation using C_c yields 0.70711 and $d = 1 - C_c = 0.29289$. Thus, the two-points position concerning the L line is further than the two points of the D_c measurement. The simulation line constructed by measuring M_i by involving Eq. (5) and the data in Table 1 form different curved lines that start from the coordinate point 1 (not 0) so that it intersects the line L at the beginning but ends at the same end. The m factor as an additional constant that changes determine the position of the line following the gridline of L. Thus, the value of the distance d to this measurement is also inversely proportional to L. Based on that, the measurement of similarity to two vectors is aimed at determining *mutual dependence* between two vectors in vector space. Of course, as an amplifier of mutual independence between the two vectors, it is needs to equip with the size of the overlap of the two vectors. In this case, O_c is worth 1 if $x \cap y = x$ and $x < y$, for example $x = 5$, $y = 10$, $x \cap y = 5$, then $O_c = 1$. The value of O_c will be between 0 and 1 if $x \cap y < x < y$ or $x \cap y < y < x$. See Fig. 2(c). The simulation shows that computationally $O_c \geq C_c \geq D_c \geq J_c = L$, whereas M_i has a different position in measuring similarity. By dividing the two measurements by L^c into the low value and the high value, the low value of M_i is partially below L, the other part is above L, while the high value of M_i is generally greater from D_c and C_c. The existence of each similarity is clearly stated by their respective distances with $L = J_c$, i.e. $\delta_4 \geq \delta_3 \geq \delta_2 \geq \delta_1$. See Fig. 2(d).

5 Conclusion

In theory, each similarity measurement results are in a certain position in the space of $[0, 1]^2$. In theory, each similarity based on the formula differs from one another, although one formula can be derived from another formula. However, the existence of similarity can also be expressed through computing in a simulation, in addition to the differences between them. Therefore, methods involving differences in theory and simulations reveal the existence of a similarity.

References

1. Santini, S., Jain, R.: Similarity measures. IEEE Trans. Pattern Anal. Mach. Intell. **21**(9), 871–883 (1999). https://doi.org/10.1109/34.790428
2. Saruladha, K., Aghila, G., Raj, S.: A survey of semantic similarity method for ontology based information retrieval. In: 2010 Second International Conference on Machine Learning and Computing, pp. 297-301 (2010). https://doi.org/10.1109/ICMLC.2010.63

3. Nasution, M.K.M.: Ontology. J. Phys. Conf. Ser. **1116**, 2 (2018). https://doi.org/10.1088/1742-6596/1116/2/022030
4. Gabbay, D.M., Malod, G.: Naming worlds in modal and temporal logic. J. Logic Lang. Inf. **11**, 29–65 (2002)
5. Dharmendra, S., Modha, W., Spangler, S.: Feature weighting in k-means clustering. Mach. Learn. **52**, 217–237 (2003)
6. Werdiningsih, I., Hendradi, R., Purbandini, Nuqoba, B., Ana E: Identification of risk factors for early childhood diseases using association rules algorithm with feature reduction. Cybern. Inf. Technol. **19**(3), 154–167 (2019). https://doi.org/10.2478/cait-2019-0031
7. Rusydi, M.I., Huda, S., Rusydi, F., Sucipto, M.H., Sasaki, M.: Pattern recognition of overhead forehand and backhand in badminton based on the sign of local Euler angle. Indones.J. Electr. Eng. Comput. Sci. **2**(3), 625–635 (2015). https://doi.org/10.11591/ijeecs.v2.i3.pp625-635
8. Deza, E., Deza, M.-M.: Dictionary of Distances. Elsevier, Boston (2006). https://doi.org/10.1016/B978-0-444-52087-6.X5000-8
9. Nasution, M.K.M.: Kolmogorov complexity: clustering objects and similarity. Bull. Math. **3**(1), 1–16 (2011)
10. Elekes, A., Englhardt, A., Schäler, M., Böhm, K.: Toward meaningful notions of similarity in NLP embedding models. Int. J. Digit. Libr. (2018). https://doi.org/10.1007/s00799-018-0237-y
11. Nasution, M.K.M., Sitompul, O.S., Nasution, S., Ambarita, H.: New similarity. Conf. Ser. Mater. Sci. Eng. **180**(1), 012297 (2017). https://doi.org/10.1088/1757-899X/180/1/012297
12. Mathisen, B.M., Aamodt, A., Bach, K., Langseth, H.: Learning similarity measures from data. Prog. Artif. Intell. **9**, 1–15 (2019). https://doi.org/10.1007/s13748-019-00201-2
13. Cheng, J., Zhang, L.: Jaccard coefficient-based bi-clustering and fusion recommender system for solving data sparsity. In: LNCS (including subseries LNAI and LNBI), vol. 11440, LNAI, pp. 369–380 (2019). https://doi.org/10.1007/978-3-030-16145-3_29
14. Takagi, K.: Principles of mutual information maximization and energy minimization affect the activation patterns of large scale networks in the brain. Front. Comput. Neurosci. **13**, 86 (2020). https://doi.org/10.3389/fncom.2019.00086
15. Stephanie, C., Sarno, R.: Detecting business process anomaly using graph similarity based on dice coefficient, vertex ranking and spearman method. In: Proceedings - 2018 International Seminar on Application for Technology of Information and Communication: Creative Technology for Human Life, iSemantic 2018, pp. 171–176 (2018). https://doi.org/10.1109/ISEMANTIC.2018.8549830
16. Zhou, Y., Chen, T., Zhao, Q., Jiang, T.: Testing the equality of two double-parameter exponential distributions via overlap coefficient. Commun. Statist. Theory Methods **49**(5), 1248–1260 (2020). https://doi.org/10.1080/03610926.2018.1563169
17. Hussain, M.J., Wasti, S.H., Huang, G., Wei, L., Jiang, Y., Tang, Y.: An approach for measuring semantic similarity between Wikipedia concepts using multiple inheritances. Inf. Process. Manage. **57**(3), 102188 (2020). https://doi.org/10.1016/j.ipm.2019.102188
18. Nasution, M.K.M., Noah, S.A.: Comparison of the social network weight measurements. In: IOP Conference Series: Materials Science and Engineering, vol. 725 (2020)

Parametric Methods and Algorithms of Volcano Image Processing

Sergey Korolev[1(✉)], Igor Urmanov[1], Aleksandr Kamaev[1], and Olga Girina[2]

[1] Computing Center of Far East Branch of the Russian Academy of Sciences, Khabarovsk, Russia
serejk@febras.net

[2] Institute of Volcanology and Seismology of Far East Branch of the Russian Academy of Sciences, Petropavlovsk-Kamchatsky, Russia

Abstract. A key problem of any video volcano surveillance network is an inconsistent quality and information value of the images obtained. To timely analyze the incoming data, they should be pre-filtered. Additionally, due to the continuous network operation and low shooting intervals, an operative visual analysis of the shots stream is quite difficult and requires the application of various computer algorithms. The article considers the parametric algorithms of image analysis developed by the authors for processing the shots of the volcanoes of Kamchatka. They allow automatically filtering the image flow generated by the surveillance network, highlighting those significant shots that will be further analyzed by volcanologists. A retrospective processing of the full image archive with the methods suggested helps to get a data set, labeled with different classes, for future neural network training.

Keywords: Image · Algorithm · Information system · Volcano

1 Introduction

The research and monitoring of hazardous natural sites, such as volcanoes, is a complex interdisciplinary science and technology problem. An important part in this research is given to computer systems and technology providing for the collection, systematizing [1] and a specialized information processing [2, 3], numerical modeling of various processes [4], etc. The main data source for the research is instrumental surveillance networks, which are actively used to monitor volcanic activity, too.

The unique nature of volcanoes conditioned by their geographic location, the equipment used, and scientific problems being resolved requires an individual approach towards the development of computer systems intended inter alia for the image processing. They should provide image filtering to eliminate non-informative or spoiled data as well as the search for the signs of volcano activity. At the initial stage it is important to get the sets of high-quality images which can be used for the operative control of the volcano state and for further research of certain historical events. To implement these functions, one need a complex approach and the application of various methods and algorithms of computer vision.

R. Silhavy et al. (Eds.): CoMeSySo 2020, AISC 1295, pp. 253–263, 2020.
https://doi.org/10.1007/978-3-030-63319-6_22

The quality of the obtained shots depends on various factors. For example, the presence of fog, cloudiness or precipitation, light-striking because of sunlight as well as video equipment technical issues and data communications breaks. As a result, a corrupted image is generated, where volcano could be not clearly visible, and the evaluation of its state could be difficult or impossible. Usually, the volcano video observation systems are built with specialized hardware (for example, thermal cameras [5]). Certain research dwell upon the development of the algorithms of shot analysis made by cameras operated in the visible spectrum [6]. However, the solutions developed on their basis are usually limited by the requirement of clear object visibility. The obvious method of detecting anomalies on such shots is the search for the areas with the brightness is higher than some certain threshold [7]; still, such areas can correspond to some extrinsic illuminated objects. The neural nets that have been widely applied for image processing [8] require a large labeled training dataset and an individual approach towards the selection of optimal architectures and solutions. The manual generation of a training data is an extremely time-consuming task.

The authors are developing the algorithms and computer systems on their basis for processing volcano images generated by Kamchatka volcanoes video observation network [9]. This paper presents a partial result of this work. Using them, it is possible to conduct a basic classification for the images made during the daylight (hereinafter referred to as day shots) and reveal potentially hazardous thermal anomalies in the shots made at night with the cameras equipped with infrared-cut filters (hereinafter referred to as night shots).

2 The Volcano Day Shot Analysis Algorithm

To analyze the day shots, the estimate of the natural object contour visibility is suggested; these contours are represented as open polygons with the branch points in the vertexes [10]. First, the Canny edge detector [11] is used to calculate a discrete map of edges. Then, the depth-first search and breadth first search are used to extract branch points and contour ends (Fig. 1a) which are further connected by curves (Fig. 1b). The example of the parametric contours built is given in Fig. 2.

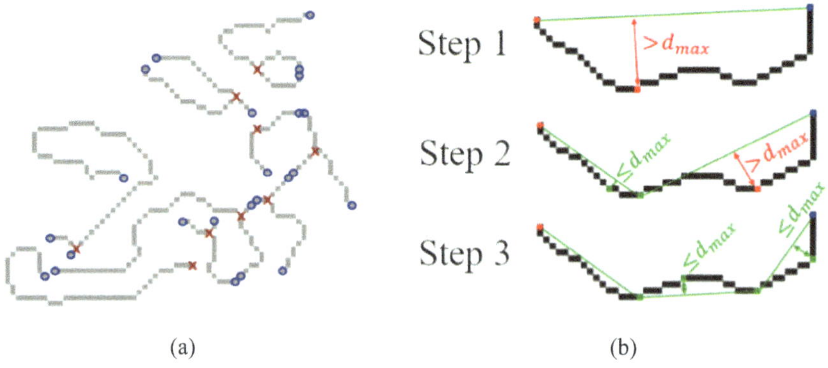

Fig. 1. Building parametric contours: a) intersection points (red crosses) and end points (blue circles); b) recursive edges build process.

Fig. 2. Contours built for Klyuchevskoy volcano example image.

For each camera, the set of reference shots must be compiled from shots obtained in good weather and illumination conditions at various seasons. The number of reference shots in a set should be at least 10, what is significantly smaller than a necessary training set size for a neural net. The reference shot set is used to extract the reference volcano contours.

For any new given image, the object parametric contours are extracted. Then they reduced to subset, that is common for at least γ contours from reference subset (Fig. 3). To obtain the most precise estimates, a parameter γ is selected separately for each camera with the account of the number of the obtained reference contours of the considered volcano.

(a) (b)

Fig. 3. Reference contours examples for different γ.

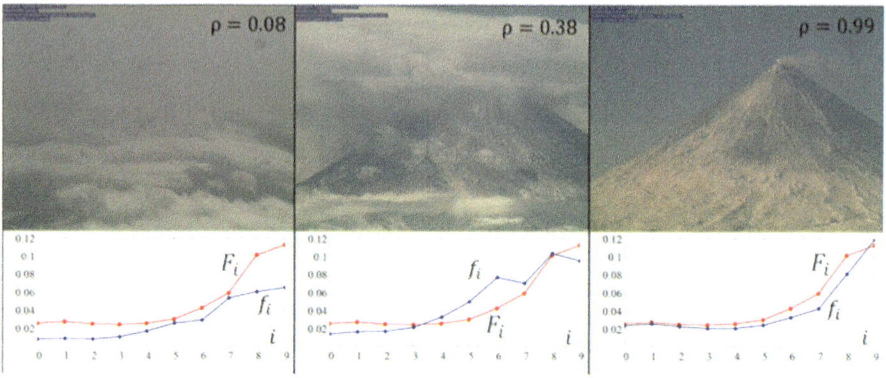

Fig. 4. Examples of frequency characteristics comparison for Klyuchevskoy volcano images taken in different weather conditions; F – octave frequency contribution vector for reference images, f – for the analyzed image.

The contours in the shots obtained the same camera can have some displacement relative to each other because of the external factors, for example, camera tremor due to the wind. To overcome that, a discrete contour comparison method is suggested [12], based on the distance map [13].

The total estimate σ of volcano contours visibility is defined by the equation:

$$\sigma = \max\left(\min\left(1, \frac{\sigma'_{ext}}{\min_{i=1,2,\ldots,m}\overline{\sigma}_{ext,i}}\right), \min\left(1, \frac{\sigma'_{int}}{\min_{i=1,2,\ldots,m}\overline{\sigma}_{int,i}}\right)\right), \tag{1}$$

where σ'_{ext} and σ'_{int} – estimates obtained by the comparison of test shot contours with external and internal reference contours respectively, while $\overline{\sigma}_{ext,i}$ and $\overline{\sigma}_{int,i}$ – estimates obtained at the comparison of contours of the i-th reference image with external and internal reference contours respectively, m – number of reference shots.

Despite some cloudiness, certain shots show clear volcano contours thus giving a possibility to detect some volcano activity signs. The estimate σ for such shots can be understated due to a poor visibility of contours considered as the reference ones. In such cases the estimate σ is corrected by the image frequency characteristics estimate $\rho \in [0, 1]$, calculated with the octave frequency contribution vector for the image brightness component [10]. As is it was done for contours, the frequency characteristics are calculated for the reference images and further compared with the corresponding characteristics of the images under analysis. Figure 4 shows the examples of the calculated frequency characteristics for Klyuchevskoy volcano images made in different weather conditions and the results of their comparison with the reference parameters.

The correction is made for the shots estimate σ which is in the Δ - vicinity of a given threshold σ, and result volcano visibility estimate is defined as:

$$\alpha = \sigma f(\sigma) + \rho(1 - f(\sigma)), \tag{2}$$

where

$$f(\sigma) = \min\left(1, \frac{1}{\Delta^2}(\sigma - \tau)^2\right). \quad (3)$$

The developed algorithm is tested at image dataset for volcanoes Sheveluch, Klyuchevskoy and Kizimen (3,000 shots for each volcano). For Sheveluch and Klyuchevskoy volcanoes the final estimate was calculated wrong (too low or too high) for 1% of images, for Kizimen – for 2% [10]. The experiments showed, that the error in volcano visibility estimate may be caused, at first, by the peculiarities of camera settings, such as resolution (the higher the resolution is, the larger number of contours can be detected for the shot) as well as amount of space which the volcano takes in the image (camera zoom).

3 Thermal Anomaly Detection Algorithm

The methods and approaches based on the contour detection are not applicable for thermal anomalies detection on night shots (taken by IR-cut equipped cameras) because of noise presence. Besides, the geometric form of such anomalies varies greatly from shot to shot and cannot be compared with some previously selected reference. Moreover, there could be bright spots in the image that are not related to volcano (moonlight, industrial objects light, etc.). Due to this, a special thermal anomaly detection algorithm for night images is developed.

The anomaly is interpreted as a part of a shot which brightness is higher than the brightness of the surrounding area and fades from the center to edges (Fig. 5). Such areas can correspond to possible signs of the volcano activity (for example, lava outflows from the craters).

The algorithm uses of a multi-scale DoG (Difference of Gaussian) detector [14]. It first finds the maximum points in DoG layers to locate anomaly centers. After that, it calculates the anomaly areas around each highlighted center. To do this, a breadth-first search of neighboring pixels is conducted given that the brightness value is not less than 0.1 of the central value. Because of the noise in the shots some highlighted areas can be identified as several different anomalies what produced incorrect results. That is why if the absolute brightness values for the centers of adjacent areas differ not more than 0.1, such areas are combined and considered as one.

For each anomaly found an attribute vector is calculated: the value of the DoG function in the center, anomaly elongation, the ration between a perimeter and a smallest possible perimeter (edge complexity), asymmetry of edge values, the difference between center brightness value and border brightness average value, the central brightness value and the number of a broad-scale layer where the given anomaly was found. At the final stage, the previously obtained data set is divided into classes: “thermal anomaly”, “false anomaly”, using the SVM-classifier [15] and a radial basis function [16], A average image analysis time is about 5 s. The paper [17] contains detailed algorithm description.

Fig. 5. Thermal anomaly example for Sheveluch volcano night image.

The tests of the suggested method were performed on a labeled training set of Sheveluch volcano night images produced by the Axis P1343 camera, with total amount of 5068 images. The 2% of images were classified incorrectly.

4 Image Processing Tools

To conduct the automatic analysis of the volcano images using the tools suggested, a special set of software tools was developed. Its operation diagram is shown in Fig. 6. At the first stage, the image is classified to be day or night shot, by comparing the pixel values in three channels (R, G, B): for night shots (grayscale) they will be the same, and for day shot – will not. After defining the shot type, the corresponding analysis algorithm is applied (Sect. 2 and 3), and the obtained results are presented. Program output has JSON format. For day shots, both contours visibility estimation and frequency characteristics are populated, as well as result estimate. This metadata is used for flexible search through the image archive and allows to reduce whole image flow to most informative dataset (Fig. 7).

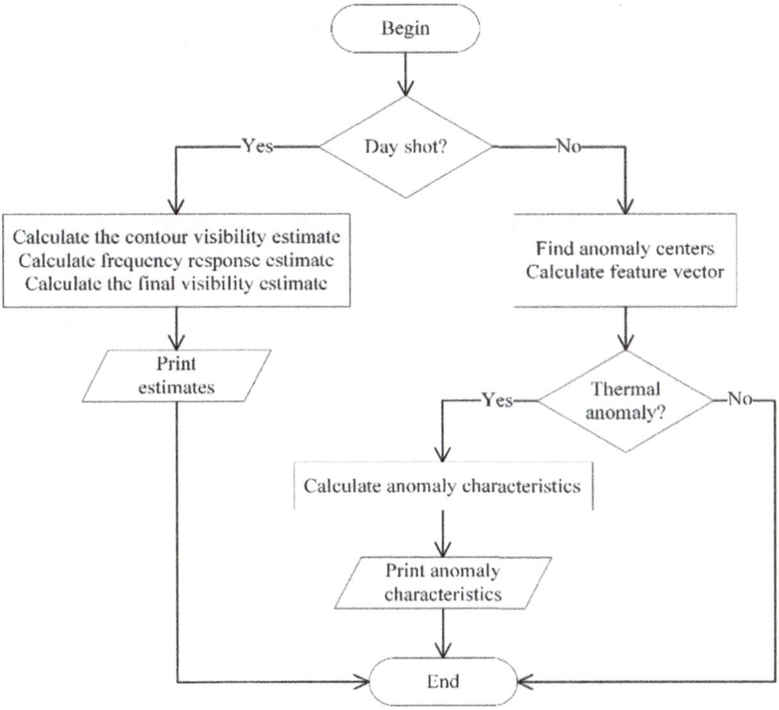

Fig. 6. Operation of the algorithm of volcano shot analysis.

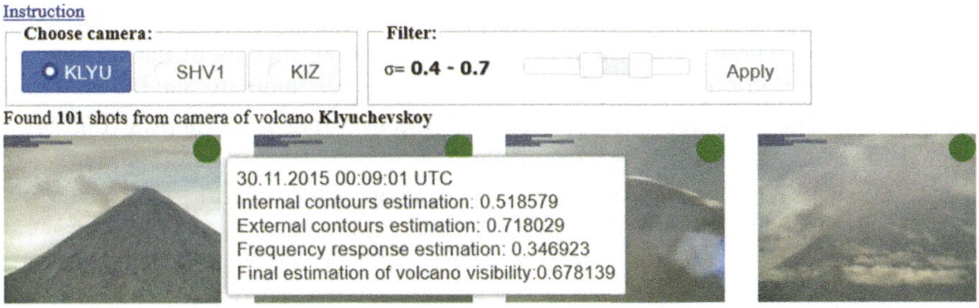

Fig. 7. Expert user interface to browse images classified by volcano visibility.

Data output for day shots (Fig. 8):

{“result”:0.72963, “contours”:0.72963, “frequency”:1}

Fig. 8. Klyuchevskoy volcano 21.02.2016 00:19 UTC.

The following values are calculated for thermal anomalies detected on night shots: size in pixels, average and standard deviations of the area brightness and center brightness value. These values help to search appropriate images in the archive and then to track changing in time the intensity of the possible anomaly and thus to define volcano state. The program output for night shot processing is as follows (Fig. 9):

{"night":[{"data":{"size":209, "mean":0.626231,"sd":0.174559,"maximum_value":1}}]}

Fig. 9. A night shot of Sheveluch volcano 25.02.2014 18:49 UTC with the detected thermal anomaly.

5 Conclusion

The developed parametric methods and algorithms for volcano shots analysis allow to pre-filter large image flow by elimination of non-informative images and help to detect a possible volcano activity. However, their use in the automated system requires manual parameter setting and unique control for each video camera in the observation network. Therefore, the next work proposes the transition to more adaptive methods of image analysis, such as convolution neural nets, which are widely used today.

Using the possibilities of the developed tools, the accumulated historical archive of the images of Kamchatka volcanoes [9] are being processed (17 million shots). This will make it possible to produce a labelled dataset with various classes of shots and various volcano state captured. Thus, the obtained training dataset makes up the basis for the next work stage including:

- approbation of algorithms for unsupervised clustering of photo images based upon the analysis of the original images with deep neural nets – autoencoders;
- training of a convolution neural net for automatic classification of new volcano images.

Acknowledgements. The reported study was funded by RFBR, project number 20-37-70008. Computations were performed with the methods and techniques developed under the RFBR, project number 18-29-03196.

References

1. Girina, O.A., Gordeev, E.I.: KVERT project: reduction of volcanic hazards for aviation from explosive eruptions of Kamchatka and Northern Kuriles volcanoes. Vestnik FEB RAS **132** (2), 100–109 (2007)
2. Gordeev, E.I., Girina, O.A., Loupian E.A., Sorokin, A.A., Kramareva, L.S., Efremov, V.Yu., Kashnitskii, A.V., Uvarov, I.A., Burtsev, M.A., Romanova, I.M., Melnikov, D.V., Manevich, A.G., Korolev, S.P., Verkhoturov, A.L.: The VolSatView information system for monitoring the volcanic activity in Kamchatka and on the Kuril Islands. J. Volcanol. Seismolog. **10**(6), 382–394 (2016). https://doi.org/10.1134/S074204631606004X
3. Sorokin, A.A., Korolev, S.P., Malkovsky, S.I.: The signal automated information system: research and operational monitoring of dangerous natural phenomena in the Russian Far East. Sovremennye Problemy Distantsionnogo Zondirovaniya Zemli iz Kosmosa **16**(3), 238–248 (2019). https://doi.org/10.21046/2070-7401-2019-16-3-238-248
4. Malkovsky, S.I., Sorokin, A.A., Girina, O.A.: Development of an information system for numerical modelling of the propagation of volcanic ash from Kamchatka and Kuril volcanoes. Comput. Technol. **24**(6), 79–89 (2019). https://doi.org/10.25743/ICT.2019.24.6.010
5. Ando, B., Pecora, E.: An advanced video-based system for monitoring active volcanoes. Comput. Geosci. **32**(1), 85–91 (2006). https://doi.org/10.1016/j.cageo.2005.05.004
6. Viteri, F., Barrera, K., Cruz, C., Mendoza, D.: Using computer vision techniques to generate embedded systems for monitoring volcanoes in Ecuador with trajectory determination. J. Eng. Appl. Sci. **13**(3), 3164–3168 (2018). https://doi.org/10.3923/jeasci.2018.3164.3168
7. Rabal, H.J., Braga, J.R.A.: Dynamic Laser Speckle and Applications. CRC Press, Boca Raton (2009)
8. Kramareva, L.S., Andreev, A.I., Bloshchinskiy, V.D., Kuchma, M.O., Davidenko, A.N., Pustatintsev, I.N., Shamilova, Y.A., Kholodov, E.I., Korolev, S.P.: The use of neural networks in hydrometeorology problems. Comput. Technol. **24**(6), 50–59 (2019). https://doi.org/10.25743/ICT.2019.24.6.007
9. Sorokin, A., Korolev, S., Romanova, I., Girina, O., Urmanov, I.: The Kamchatka volcano video monitoring system. In: Proceedings of 2016 6th International Workshop on Computer Science and Engineering (WCSE 2016), pp. 734–737. The Science and Engineering Institute, LA, CA, USA (2016)
10. Kamaev, A.N., Urmanov, I.P., Sorokin, A.A., Karmanov, D.A., Korolev, S.P.: Images analysis for automatic volcano visibility estimation. Comput. Opt. **42**(1), 128–140 (2018). https://doi.org/10.18287/2412-6179-2018-42-1-128-140. (in Russian)
11. Canny, J.: A computational approach to edge detection. IEEE Trans. Pattern Anal. Mach. Intell. **8**(6), 679–698 (1986)
12. Urmanov, I., Kamaev, A., Sorokin, A.: Computer methods of image processing of volcanoes. In: Proceedings of the IV International Research Conference "Information Technologies in Science, Management, Social Sphere and Medicine" (ITSMSSM 2017), vol. 72, pp. 371–374. Atlantis Press, Paris (2017). https://doi.org/10.2991/itsmssm-17.2017.77
13. Borgefors, G.: Distance transformations in digital images. CVGIP **34**(3), 344–371 (1986). https://doi.org/10.1016/S0734-189X(86)80047-0

14. Lowe, D.G.: Distinctive image features from scale-invariant keypoints. Int. J. Comput. Vision **60**(2), 91–110 (2004). https://doi.org/10.1023/B:VISI.0000029664.99615.94
15. Shmilovici, A.: Support vector machines. In: Data Mining and Knowledge Discovery Handbook. Springer, Boston (2009). https://doi.org/10.1007/0-387-25465-X_12
16. Schölkopf, B., Tsuda, K., Vert, J.P.: Kernel Methods in Computational Biology. MIT Press, Cambridge (2004)
17. Kamaev, A.N., Korolev, S.P., Sorokin, A.A., Urmanov, I.P.: Detection of thermal anomalies in the images of volcanoes taken at night. J. Comput. Syst. Sci. Int. **59**(1), 95–104 (2019). https://doi.org/10.1134/S106423071906008X

Interval Valued Markov Integrated Rhotrix Optimization Using Genetic Algorithm for Predictive Modeling in Weather Forecasting

G. Kavitha(✉), Desai Manish, and S. Krithika

Hindustan Institute of Technology & Science,
Kelambakkam 603 103, Tamil Nadu, India
kavithateam@gmail.com

Abstract. This paper emphasizes on the development of a new predictive modeling based on interval valued Markov integrated Rhotrix optimization using genetic algorithm. Interval valued partitioning was used for development of predictive modeling of future behavior based on the most important genetic algorithm optimizer. The proposed predictive model is trained and tested on the datasets taken from multiple websites for short and decadal long weather prediction. The initialization of the weather parameters is done in first phase using interval valued Markov integrated Rhotrix predictive model. Optimized estimation of weather parameters is performed in second phase using Genetic Algorithm improving the error in predictive modeling. The experimental result shows that the increase in the greenhouse gases per year is 1.915 ppm, the average increase in temperature for each year is 0.055 °C and the average increase in temperature for each year due to the impact of greenhouse gases is −0.137 °C. Further it was also analyzed and estimated that for 2020–2029, the average concentration of greenhouse gases will be 414.1015 ppm, the average increase in temperature will be 13.696 °C, the average increase in temperature due to greenhouse gases will be 13.6413 °C. The actual global temperature was computed by adding 12.7 °C (20th century global average temperature) to the temperature anomaly.

Keywords: Predictive modeling · Rhotrix · Markov model · Genetic algorithm · Weather forecasting

1 Introduction

Predictive modeling is a learning algorithm based on data to create a robust accurate model to make predictions. Uncertainty and Randomness lead to high fluctuations making it difficult for decision makers to make decisions. Optimization takes its role in this crucial phase of uncertainty when one or more of the input parameters are subject to randomness. In today's world, both structured and unstructured publicly available data from open data sources are too huge and complex to analyze and extract information for building a model, generating the result, optimizing the parameters and validating the results.

R. Silhavy et al. (Eds.): CoMeSySo 2020, AISC 1295, pp. 264–277, 2020.
https://doi.org/10.1007/978-3-030-63319-6_23

A Markov chain is an essential part of stochastic process satisfying the Markov ageless property which means that knowing the current state of the process, future prediction can be done in a best possible way with no background of its past states. The probabilities of state transitioning events into other states or into the same state can be computed using Markov chains and summarized in a matrix. The matrix of the Markov chain is taken as input to return the steady state vector containing the long-term probabilities in each state. The model proposed aims at partitioning the states, determining the transition probabilities and long run probability vector.

A rhotrix is an array of numbers in rhomboidal form. A special form of matrix called coupled matrix is integrated with Markov to solve the problem involving $n \times n$ state transitions and $(n - 1) \times (n - 1)$ state transitions simultaneously. Conversion of rhotrix to coupled matrix is made by 45° anti-clockwise rotation of a rhotrix. Using coupled rhotrix, an integrated Markov Rhotrix model is developed for the state matrix.

Genetic Algorithm is an optimization technique for optimizing the unknown parameters. The Initial set of parameter population selected is used to compute the fitness of the function. On repeated selection process using crossover and mutation operators of genetic algorithm the fitness is computed till optimal population convergence is reached. The proposed model involves Genetic Algorithm for Markov integrated Rhotrix optimization.

The focus of this study is to develop an integrated model in a novel way for weather forecasting application. The advantage of the model is that it involves both classical as well learning algorithm model for predictive modeling in accurate weather forecast using a unique modeling structure mathematically. The significance of the study is involving Genetic Algorithm for optimizing the parameters in the fitness error function.

There are six sections in this paper. A brief literature review is provided in Sect. 2. Section 3 deals with interval valued Markov integrated Rhotrix predictive model and Markov integrated Rhotrix optimization using Genetic Algorithm predictive model. Section 4 gives the experimental result and discussion of the developed model in weather forecasting. Section 5 concludes briefly the overview of the model.

2 Review of Literature

Markov chain is a powerful mathematical tool highly significant in stochastic process. Markov chain study began in the year 1906 by a Russian Mathematician A. A. Markov [1]. Sheldon M Ross et al. [2] introduced continuous-time Markov chain and related them to the discrete-time Markov chain. Many useful Markov Chain techniques for numerical computations were presented and analyzed.

Ajibade, A.O. [3] introduced the concept of rhotrices for mathematical consideration and has a unique approach with wide range of applications in various domains. Doug M. Smith et al. [4] have proposed a modeling system focusing on both internal and external force variations to forecast surface temperature both globally and in specific regions. J. David Neelin et al. [5] have proposed an approach to guide and aid the parameter choices, inter-comparison of sensitivity properties among climate models. The study also pointed out the underlying nature of the system having an enormous impact on the most suitable strategies that are available for the evaluation of sensitivity and optimization. Robert Fildes et al. [6] combined standard time series methods with the structure of atmospheric-ocean general circulation models for higher forecasting accuracy in decadal prediction focusing on carbon dioxide emissions alone. David G. McMillan et al. [7] have investigated the evidences from a short and very long dataset highlighting the danger and relationship between temperature and carbon dioxide emissions. Siva Venkadesh et al. [8] developed more accurate ANN models for predicting air temperature for each prediction horizon by applying a genetic algorithm to each environmental variable. Eckehard Specht et al. [9] have developed a mathematical model to understand the effect of carbon dioxide on the mechanism of global warming. The model predicted a temperature increase of more than 0.4 K for the future. Hossein Hassani et al. [10] modeled the nonlinearity into the relationship between global temperature and carbon dioxide giving a conclusive evidence that carbon dioxide can predict global temperature. Peter C. Young [11] provided a model for forecasting global temperature anomaly by analyzing the dynamic relationship between global averaged measures of Total Radiative Forcing and surface temperature measured by global temperature anomaly. Gabriele C Hegerl et al. [12] addressed the causes of observed climate variations from 1750 to present focusing on long-term changes in ocean and atmosphere.

The existing literature survey discloses diversified techniques, their merits and demerits in modeling as well as accuracy. The highlight of the proposed study is in handling interval valued partitioning for Markov integrated Rhotrix optimization by Genetic Algorithm for predictive modeling in weather forecasting. The main objective of the proposed research is to focus on experimental results of Global warming and its effects on temperature affecting the climate for short and decadal long weather forecasts. Simple and classical forecasting techniques are least accurate compared to complex forecasting techniques in predicting the average global temperature. Building such complex and accurate models are a great challenge and has important consequences in environmental planning.

3 Materials and Methods

In this study, interval valued Markov integrated Rhotrix predictive modelling using Genetic Algorithm for weather forecasting is presented. Genetic Algorithm is used for Global search. Interval valued Markov chain integrated Rhotrix was used in the initial phase for local search and the best parameters obtained were used for predictive modeling in weather forecasting. The general framework of the proposed model is shown in Fig. 1.

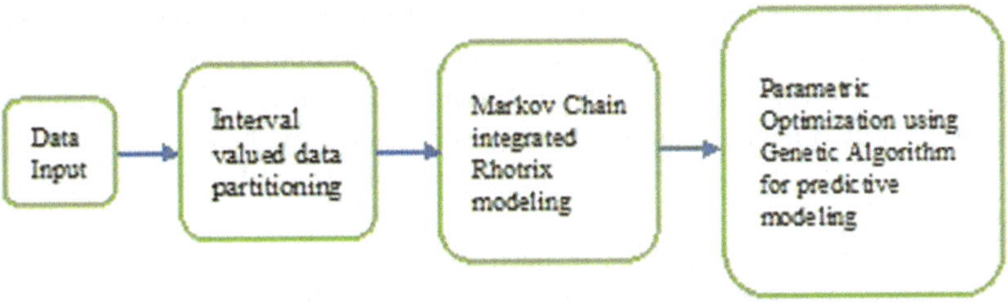

Fig. 1. General framework of the interval valued Markov integrated Rhotrix predictive modelling using Genetic Algorithm model.

The proposed models were experimented with historical temperature and greenhouse gases data obtained from various websites.

3.1 Interval Valued Markov Integrated Rhotrix Predictive Model

Consider a Markov chain as a stochastic process $\{SP_n, n = 0, 1, 2, \ldots\}$, if $Likelihood\ of\ \{SP_{n+1} = j/SP_n = i\} = Likelihood_{i,j}$ where (i,j) indicates the movement between the location pairs over a given time interval from t to $t+1$. The Continuous numerical dataset undergoes a process of partitioning the continuous variables to discretized intervals of equal lengths forming an interval valued partitioned vector. For the given dataset vector $\{d_1, d_2, \ldots, d_n\}$, determine the minimum, maximum and the range choosing the number of equal subintervals to be partitioned. Let us define and denote it as $min\{d_1, d_2, \ldots, d_n\} = a$, $max\{d_1, d_2, \ldots, d_n\} = b$, range $\{d_1, d_2, \ldots, d_n\} = \frac{max\{d_1, d_2, \ldots, d_n\} - min\{d_1, d_2, \ldots, d_n\}}{number\ of\ subintervals} = \frac{b-a}{si}$, where $'si'$ denotes the number of subintervals. The interval valued partitioned vector formed will have finite set of points according to the number of subintervals chosen. For instance, if the number of subintervals is taken as $'si'$, then the number of points in the interval valued partitioning vector will be $'si+1'$ and the union of all the subintervals must be equal to the original interval valued partitioned vector set and the intersection of these subintervals will be a null set. The interval valued partitioned vector set states are denoted by $IVPV\{ivpv_1, ivpv_2, \ldots, ivpv_{si}\}$ such that $ivpv_r$ takes a value in the interval (a_r, a_{r+1}) for $1 \leq r \leq si$, hence forming an interval valued partition of the dataset. The midpoints of $IVPV\{ivpv_1, ivpv_2, \ldots, ivpv_{si}\}$ is determined for symmetry of the balanced intervals. According to the number of movements of the location pairs (i,j) over a given time interval from t to t + 1, the Markov chain transition probability matrix $'MCTPM'$ is determined. The steady state probability matrix $'Pi'$ for the corresponding $'MCTPM'$ is calculated. This is the likelihood probability of long run. The midpoints of $IVPV\{ivpv_1, ivpv_2, \ldots, ivpv_{si}\}$ in interval valued partition vector set state having the highest probability is taken for prediction and analysis. Consider feature 1 in the dataset, calculate Markov chain transition probability of $(n \times n)$ dimension and steady state probability matrix in the long run of $(1 \times n)$ dimension for the feature 1 selected for prediction. Here, for feature 1, there are $(n \times n)$ unknown parameters to be optimized, considered as case 1. Now consider feature 2 in the dataset, calculate Markov chain transition probability of $\{(n-1) \times (n-1)\}$ dimension and steady state probability in the long run of

$[1 \times (n-1)]$ dimension for the feature 2 selected for prediction. Here, for feature 2, there are $\{(n-1) \times (n-1)\}$ unknown parameters to be optimized, considered as case 2. Now the 2 matrices formed are with 2 different dimensions for the features considered and the two matrices are probabilistically related by their states.

A Rhotrix (3) is defined as a special matrix called coupled matrix *'cpldmtx'* of dimension $\{[n+(n-1)] \times [n+(n-1)]\}$, given two matrices of $(n \times n)$ dimension and $\{(n-1) \times (n-1)\}$ dimension respectively.

For instance, A Rhotrix R of dimension n can be written as

$$R_n = \left\langle \begin{array}{ccccccccc} & & & & a_{11} & & & & \\ & & & a_{21} & c_{11} & a_{12} & & & \\ \cdots & \cdots & \cdots & \cdots & \cdots & \cdots & \cdots & \cdots & \cdots \\ \cdots & \cdots & \cdots & \cdots & \cdots & \cdots & \cdots & \cdots & \cdots \\ a_{t1} & \cdots & \cdots & \cdots & \cdots & \cdots & \cdots & \cdots & a_{1t} \\ \cdots & \cdots & \cdots & \cdots & \cdots & \cdots & \cdots & \cdots & \cdots \\ \cdots & \cdots & \cdots & \cdots & \cdots & \cdots & \cdots & \cdots & \cdots \\ & & & a_{tt-1} & c_{t-1t-1} & a_{t-1t} & & & \\ & & & & a_{tt} & & & & \end{array} \right\rangle$$

The element $a_{ij}(i,j = 1,2,\ldots,t)$ and $c_{kl}(k,l = 1,2,\ldots,t-1)$ are called major and minor entries of R respectively. Here for forming the Rhotrix, the two matrices are calculated by Markov Chain transition probability and then integrated with Rhotrix for predictive modelling. Calculate Markov chain integrated Rhotrix transition probability of $\{[n+(n-1)] \times [n+(n-1)]$ dimension and steady state probability in the long run of $\{1 \times [n+(n-1)]\}$ dimension for combined impact of feature 2 on feature 1 selected for prediction. Here, for combined impact of feature 2 on feature 1, there are $\{[n+(n-1)] \times [n+(n-1)]\}$ unknown parameters to be optimized, considered as case 3. All the three cases with unknown parameters to be optimized enter the Genetic Algorithm optimization procedure in the next phase for predictive modeling.

3.2 Markov Integrated Rhotrix Optimization Using Genetic Algorithm Predictive Model

Genetic Algorithm optimized models are built in less time and highly significant for its global search. Combining the Genetic Algorithm with Interval valued Markov integrated Rhotrix, its implementation and analysis had given a key asset for predictive modeling in weather forecast. Following are the steps for the optimized GA algorithm

Step 1: Load the input dataset and partition it into equal sub intervals.
Step 2: Calculate Markov chain transition probability of $(n \times n)$ dimension and steady state probability in the long run of $(1 \times n)$ dimension for the feature 1

selected for prediction. Here, for feature 1, there are $(n \times n)$ unknown parameters to be optimized, considered as case 1.

Step 3: Calculate Markov chain transition probability of $\{(n-1) \times (n-1)\}$ dimension and steady state probability in the long run of $[1 \times (n-1)]$ dimension for the feature 2 selected for prediction. Here, for feature 2, there are $\{(n-1) \times (n-1)\}$ unknown parameters to be optimized, considered as case 2.

Step 4: Calculate Markov chain integrated Rhotrix transition probability of $\{[n+(n-1)] \times [n+(n-1)]$ dimension and steady state probability in the long run of $\{1 \times [n+(n-1)]\}$ dimension for combined impact of feature 2 on feature 1 selected for prediction. Here, for combined impact of feature 2 on feature 1, there are $\{[n+(n-1)] \times [n+(n-1)]$ unknown parameters to be optimized, considered as case 3.

Step 5: Apply GA algorithm for parameter optimization of case 1, case 2, case 3 as mentioned in Steps 2–4. The Genetic operators used are Selection, Crossover and Mutation operators.

Step 6: Perform training and testing on the entire dataset using the above Genetic Algorithm operators and the error measures are computed.

Step 7: Compare the result obtained with the actual output using GA fitness function.

Step 8: Optimization terminates when stopping criteria, the average change in the fitness value less than options is reached.

Step 9: Display the predictive modeling output.

4 Experimental Result and Discussion of Interval Valued Markov Integrated Rhotrix Optimization Using Genetic Algorithm Predictive Model in Weather Forecasting

The proposed approach has been simulated in MATLAB 2020(a). The dataset for the experimental study is taken from the websites as given below.

https://www.ncdc.noaa.gov/sotc/global/202003

https://www.esrl.noaa.gov/gmd/dv/data/index.php?perpage=200&pageID=1&category=Greenhouse%2BGases&parameter_name=Carbon%2BDioxide

The experimental study reveals the fact that an increase in greenhouse gases leads to an increase in the average global temperature. Understanding the relation between the greenhouse gases and average global temperature is the key step in weather forecasting. The average global temperature has increased drastically in the past 100 years. This sudden increase was referred to as global warming. There are many reasons for the change in global temperature. Some of the main reasons are due to the increase in the concentration of greenhouse gases, deforestation, the fluctuation in suns energy, volcanic eruptions. The earth has faced these temperature fluctuations in the past. But, since the industrialization, there is a drastic increase in temperature. According to Robert Fildes et al. [6], the global average temperature has increased at a rate of 0.07 per decade from 1880 to 1980 but from 1981 it geared up to 0.18 per decade, which resulted in an overall rise of 2 °C.

Today, the main reason for global warming is due to the presence of greenhouse gases in the atmosphere. As the sun's radiation reaches the earth only 70–75% of the radiation passes through the atmosphere and reaches the surface. This radiation is absorbed by the Earth's surface and gets warmed. During nights these radiations are emitted back to the atmosphere in the form of infrared radiation (700 nm to 1000 nm) from the Earth and radiate back to the surface. This heat is trapped in the atmosphere leading to global warming. According to Hassani et al. [10], the carbon dioxide content in the atmosphere before the industrial revolution was 200 ppm and in the year 2018 it was 407.8 ppm. From this data it is evident that an increase in the global temperature is influenced by the increase in the Greenhouse gas content on a great scale.

The proposed methodology has been trained and tested on datasets and optimized using an interval valued Markov integrated Rhotrix optimization using Genetic Algorithm, for short and long decadal weather forecasts. Figures 2, 3, 4 and Figs. 5, 6, 7 show the experimental results of the three cases studied namely temperature prediction, greenhouse gases prediction and temperature affected by greenhouse gases prediction for short-term yearly weather forecast and long-term decadal weather forecast respectively. The variables are optimized using Genetic Algorithm and the error measures for accuracy validation has been noted and tabulated after Genetic Algorithm optimization. Figure 8 gives the results of Actual values and Predicted case 1, case 2, case 3 values after genetic algorithm optimization for short-term yearly weather forecast from 1959 to 2020.

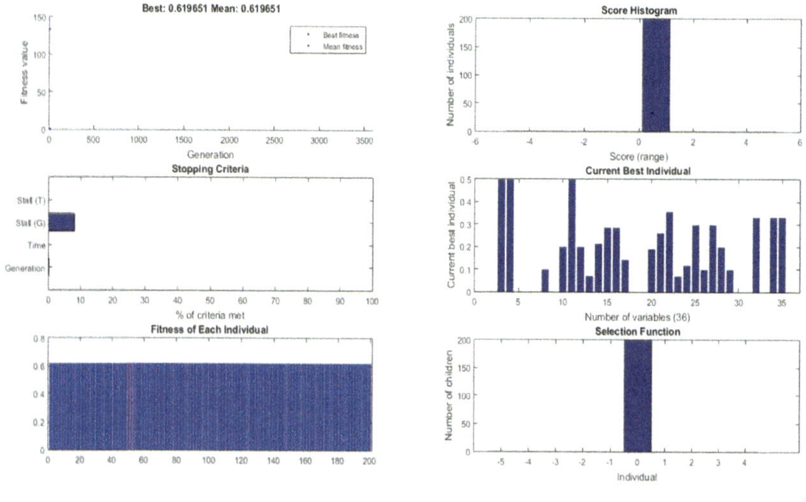

Fig. 2. Results of Markov Chain transition probability matrix of (6 × 6) dimension optimization using Genetic Algorithm for temperature prediction of short-term yearly weather forecast

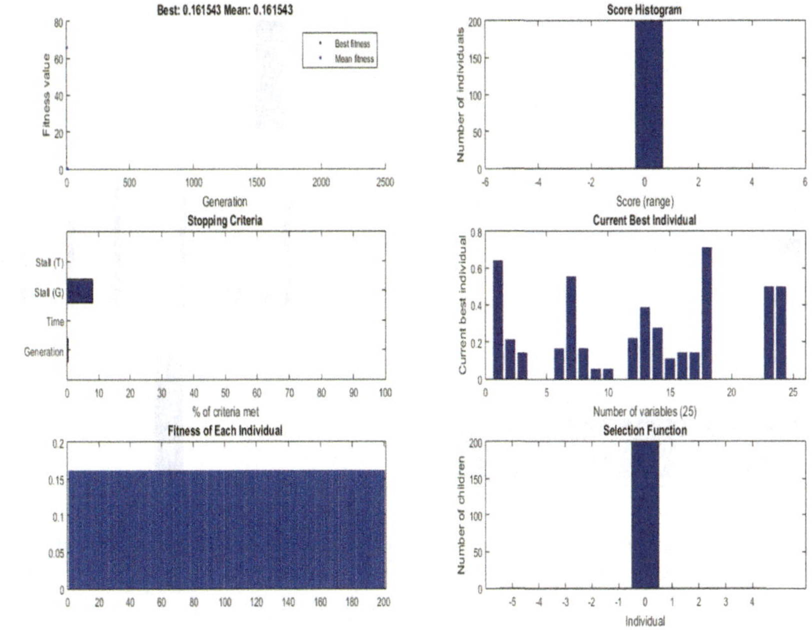

Fig. 3. Results of Markov Chain transition probability matrix of (5×5) dimension optimization using Genetic Algorithm for greenhouse gases prediction of short-term yearly weather forecast

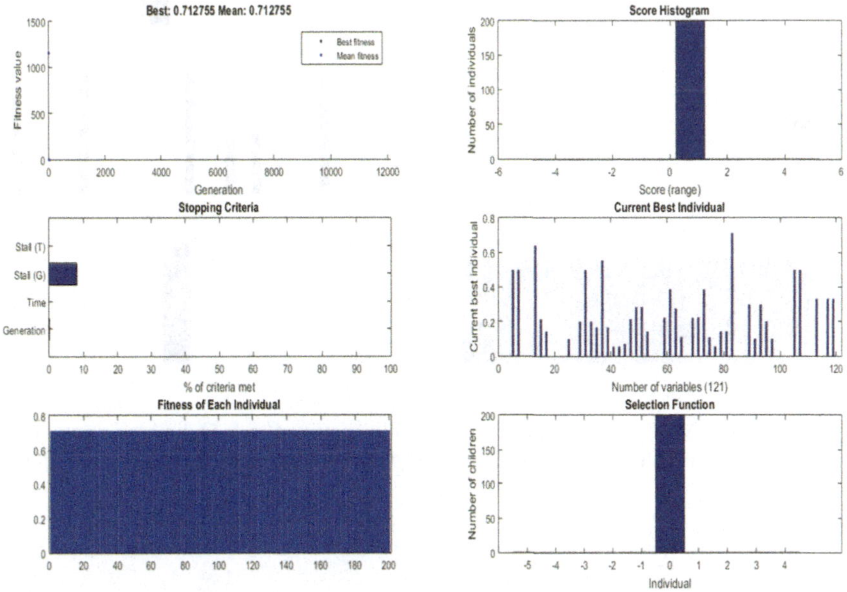

Fig. 4. Results of Markov Chain integrated Rhotrix transition probability matrix of (11×11) dimension optimization using Genetic Algorithm for temperature affected by greenhouse gases prediction of short-term yearly weather forecast

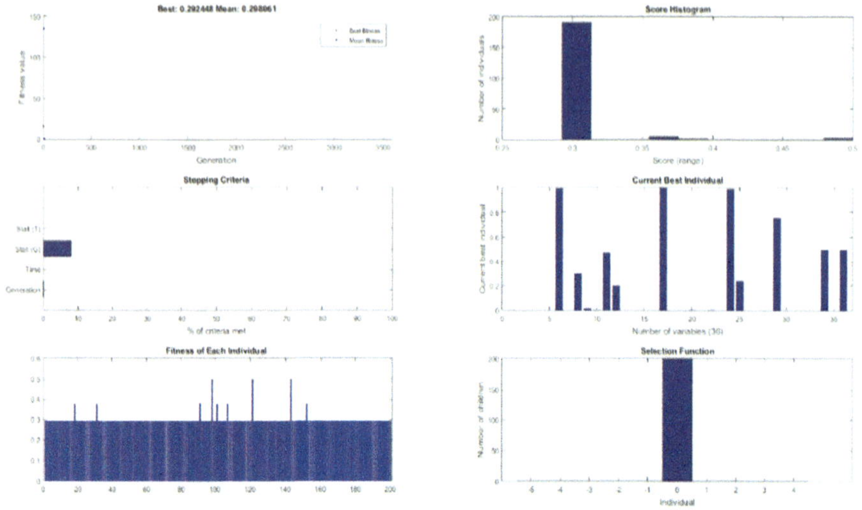

Fig. 5. Results of Markov Chain transition probability matrix of (6 × 6) dimension optimization using Genetic Algorithm for temperature prediction of long-term decadal weather forecast

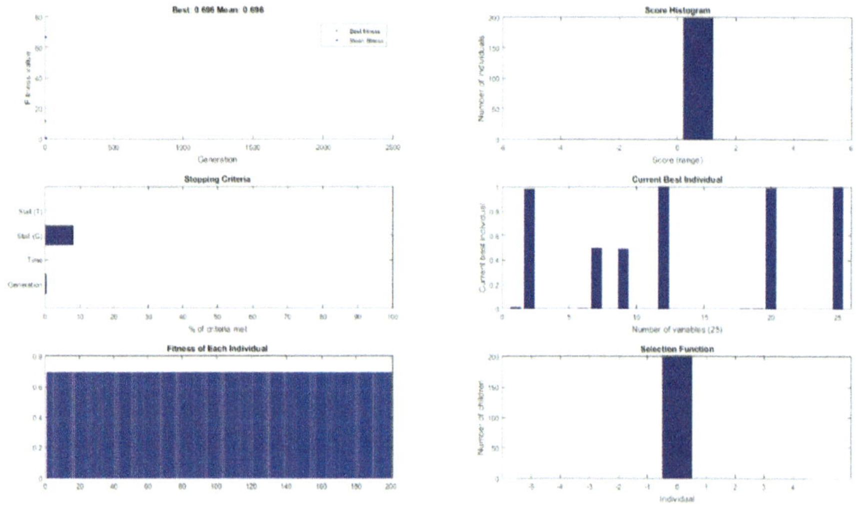

Fig. 6. Results of Markov Chain transition probability matrix of (5 × 5) dimension optimization using Genetic Algorithm for greenhouse gases prediction of long-term decadal weather forecast

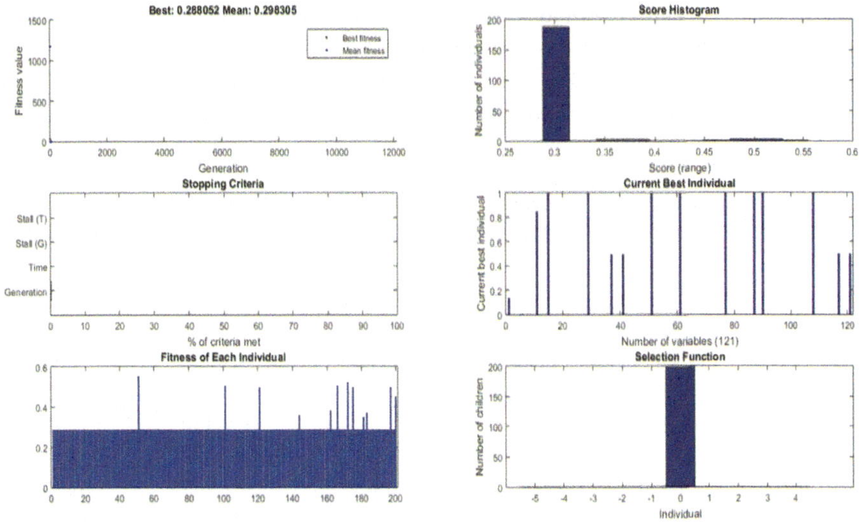

Fig. 7. Results of Markov Chain integrated Rhotrix transition probability matrix of (11 × 11) dimension optimization using Genetic Algorithm for temperature affected by greenhouse gases prediction of long-term decadal weather forecast

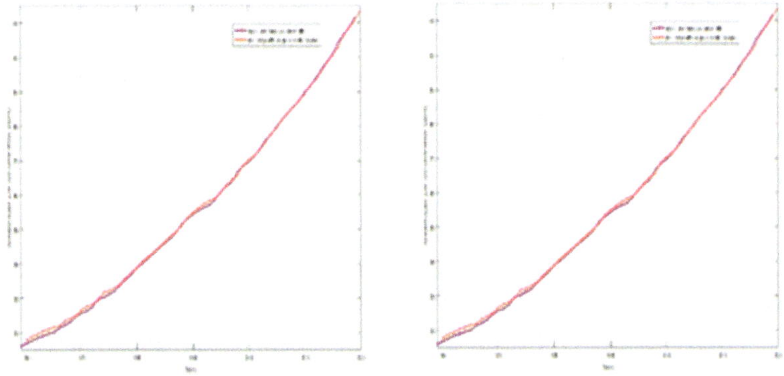

Fig. 8. Results of actual values and predicted case 1, case 2, case 3 values after genetic algorithm optimization for short-term yearly weather forecast from 1959 to 2020

Table 1 and Table 2 gives the result of global temperature, the greenhouse gases concentration, the temperature changes due to greenhouse gases for long-term decadal forecast and short-term yearly forecast from 1959 to 2020 respectively. The output obtained gives better results after optimizing using Genetic Algorithm for all the input dataset considered. Then the temperature anomaly was added with the average global temperature to find out the global temperature and for the predicted year 2020, it was found to be 1.005 °C for case 1 and 0.81273 °C for case 3 prediction.

Table 1. Results of long decadal forecast of global temperature, greenhouse gases concentration and temperature impacted by greenhouse gases from 1960 to 2029.

Years	Actual average temperature in °C for each 10 years	Actual average greenhouse gases in ppm for each 10 years	Case 1 prediction for temperature in °C	Case 2 prediction for green house gases in ppm	Case 3 prediction for temperature in °C impacted by greenhouse gases
1960–1969	12.724	320.285			
1970–1979	12.767	330.853	12.921	334.1725	12.86627
1980–1989	12.978	345.543	12.964	344.7405	12.90927
1990–1999	13.11	360.462	13.175	359.4305	13.12027
2000–2009	13.298	378.581	13.307	374.3495	13.25277
2010–2019	13.499	400.214	13.495	392.4685	13.44027
2020–2029			**13.696**	**414.1015**	**13.64127**

The forecasting accuracy was validated with the Error Measures such as MSE - the mean square error, RMSE - the root mean square error, MAPE - the mean absolute percentage error during Genetic Algorithm parameter optimization, thereby evaluating the efficacy of the proposed model. Table 3 gives the calculated error measure values and forecasting accuracy details obtained during the experimental study for both short-term yearly weather forecast and long-term decadal weather forecast. The traditional Markov chain integrated with Rhotrix was combinedly deployed with machine driven Genetic Algorithm that result in a robust and significant predictive modeling for weather forecasting.

Table 2. Results of short-term yearly forecast of global temperature, the greenhouse gases concentration, the temperature changes due to greenhouse gases from 1959 to 2020.

Year	Actual values TMP	Actual values GHG	Predicted case 1	Predicted case 2	Predicted case 3	Year	Actual values TMP	Actual values GHG	Predicted case 1	Predicted case 2	Predicted case 3
1959	12.8	316.0				1990	13.2	354.4	13.0	355.0	12.9
1960	12.8	316.9	12.8	317.9	12.6	1991	13.1	355.6	13.2	356.3	13.0
1961	12.8	317.6	12.8	318.8	12.6	1992	12.9	356.5	13.1	357.5	13.0
1962	12.8	318.5	12.9	319.6	12.7	1993	13.0	357.1	13.0	358.4	12.8
1963	12.8	319.0	12.9	320.4	12.7	1994	13.0	358.8	13.0	359.0	12.8
1964	12.6	319.6	12.9	320.9	12.7	1995	13.2	360.8	13.1	360.7	12.9
1965	12.6	320.0	12.6	321.5	12.4	1996	13.0	362.6	13.2	362.7	13.0
1966	12.7	321.4	12.7	322.0	12.5	1997	13.2	363.7	13.1	364.5	12.9
1967	12.7	322.2	12.7	323.3	12.6	1998	13.4	366.7	13.3	365.6	13.1
1968	12.7	323.0	12.8	324.1	12.6	1999	13.1	368.4	13.4	368.6	13.2
1969	12.8	324.6	12.7	325.0	12.5	2000	13.1	369.6	13.2	370.3	13.0

(*continued*)

Table 2. (*continued*)

Year	Actual values		Predicted case			Year	Actual values		Predicted case		
	TMP	GHG	1	2	3		TMP	GHG	1	2	3
1970	12.8	325.7	12.9	326.5	12.7	2001	13.3	371.1	13.2	371.5	13.0
1971	12.6	326.3	12.8	327.6	12.6	2002	13.3	373.3	13.3	373.1	13.1
1972	12.7	327.5	12.7	328.2	12.5	2003	13.3	375.8	13.4	375.2	13.2
1973	12.9	329.7	12.8	329.4	12.6	2004	13.3	377.5	13.4	377.7	13.2
1974	12.6	330.2	13.0	331.6	12.8	2005	13.4	379.8	13.3	379.4	13.2
1975	12.7	331.1	12.7	332.1	12.5	2006	13.3	381.9	13.4	381.7	13.2
1976	12.6	332.0	12.8	333.0	12.6	2007	13.3	383.8	13.4	383.8	13.2
1977	12.9	333.8	12.7	334.0	12.5	2008	13.3	385.6	13.4	385.7	13.2
1978	12.8	335.4	13.0	335.7	12.8	2009	13.4	387.4	13.3	387.5	13.1
1979	12.9	336.8	12.9	337.3	12.7	2010	13.4	389.9	13.4	389.3	13.2
1980	13.0	338.8	13.0	338.8	12.8	2011	13.3	391.7	13.5	391.8	13.3
1981	13.0	340.1	13.0	340.7	12.8	2012	13.3	393.9	13.3	393.6	13.1
1982	12.9	341.5	13.1	342.0	12.9	2013	13.4	396.5	13.4	395.8	13.2
1983	13.1	343.1	12.9	343.4	12.8	2014	13.4	398.7	13.4	398.4	13.2
1984	12.9	344.7	13.1	345.0	12.9	2015	13.6	400.8	13.5	400.6	13.3
1985	12.9	346.1	12.9	346.6	12.7	2016	13.7	404.2	13.7	402.7	13.5
1986	12.9	347.4	12.9	348.0	12.7	2017	13.6	406.6	13.8	406.2	13.6
1987	13.1	349.2	13.0	349.3	12.8	2018	13.5	408.5	13.7	408.5	13.5
1988	13.1	351.6	13.1	351.1	12.9	2019	13.7	411.4	13.6	410.4	13.4
1989	13.0	353.1	13.1	353.5	13.0	2020			13.7	413.3	13.5

TMP Temperature in °C, GHG Greenhouse gas concentration in ppm
Predicted Case: 1-TMP in °C, 2-GHG in ppm, 3–TMP in °C due to GHG

Table 3. The error measures and forecast accuracy of the proposed model after GA optimization

Forecast term	Prediction case	Error measure			Function count	Number of variables optimized
		MSE	RMSE	MAPE		
Short-term yearly weather forecast	Temperature prediction	0.0126	0.1121	0.6197	52200	36
	Greenhouse gases prediction	0.5344	0.7310	0.1615	52200	25
	Temperature affected by greenhouse gases prediction	0.0161	0.1268	0.7128	52200	121
Decadal long-term weather forecast	Temperature prediction	0.0048	0.0694	0.2924	52200	36
	Greenhouse gases prediction	14.282	3.7792	0.6960	52200	25
	Temperature affected by greenhouse gases prediction	0.0067	0.0816	0.4528	408600	121

5 Conclusion

The proposed interval valued Markov integrated Rhotrix optimization using Genetic Algorithm for predictive modeling in weather forecasting technique is better for all kind of datasets from short to long decadal forecasts compared to other classical techniques. By optimizing the parameters of the Markov integrated Rhotrix using genetic algorithm operators like Selection, Crossover and Mutation, a better forecasting accuracy is obtained. The outputs were simulated by splitting the entire dataset into training and testing datasets. Optimization of different combinations of training and testing datasets have been tried to view the accuracy of weather forecasting in predictive modeling. The experimental result showed good efficiency in training and testing, thereby improving the average prediction error in less time. The error metrics calculated indicated that the increase in the greenhouse gases per year is 1.915 ppm, the average increase in temperature for each year is 0.055 °C and the average increase in temperature for each year due to the impact of greenhouse gases is −0.137 °C. Further it was also analyzed and estimated that for 2020–2029, the average concentration of greenhouse gases will be 414.1015 ppm, the average increase in temperature will be 13.696 °C, the average increase in temperature due to greenhouse gases will be 13.6413 °C and due to this the average global temperature anomaly will be increased by 0.81273 °C relative to the average global temperature, thereby affecting the climatic change. Future direction is to use different optimization techniques for parameter optimization, study multi factor impact on global temperature and implement Markov integrated even dimensional Rhotrix model.

References

1. Markov, A.A.: Rasprostranenie zakona bol'shih chisel na velichiny, zavis yaschie drug ot druga. Izvestiya Fiziko-matematicheskogo obschestva pri Ka zanskom universitete, 2-ya seriya, tom, vol. 15, pp. 135–156 (1906)
2. Ross, S.M., et al.: Stochastic Processes. Wiley, New York (1996)
3. Ajibade, A.O.: The concept of Rhotrix in mathematical enrichment. Int. J. Math. Educ. Sci. Technol. **34**(2), 175–179 (2003)
4. Smith, D.M., et al.: Improved surface temperature prediction for the coming decade from global climate model. Science **317**, 796–799 (2007)
5. Neelin, J.D., et al.: Considerations for parameter optimization and sensitivity in climate models. Proc. Natl. Acad. Sci. **107**(50), 21349–21354 (2010)
6. Fildes, R., et al.: Validation and forecasting accuracy in models of climate change. Int. J. Forecast. **27**, 968–995 (2011)
7. McMillan, D.G., et al.: The relationship between temperature and CO_2 emissions: evidence from a short and very long dataset. Appl. Econ. **45**, 3683–3690 (2013)
8. Venkadesh, S., et al.: A genetic algorithm to refine input data selection for air temperature prediction using artificial neural networks. Appl. Soft Comput. **13**(5), 2253–2260 (2013)
9. Specht, E., et al.: Simplified mathematical model for calculating global warming through anthropogenic CO_2. Int. J. Therm. Sci. **102**, 1–8 (2016)

10. Hassani, H., et al.: Predicting global temperature anomaly: a definitive investigation using an ensemble of twelve competing forecasting models. Physica A (2018). https://doi.org/10.1016/j.physa.2018.05.147
11. Peter, C.Y.: Data-based mechanistic modelling and forecasting globally averaged surface temperature. Int. J. Forecast. **34**, 314–335 (2018)
12. Hegerl, G.C., et al.: Causes of climate change over the historical record. Environ. Res. Lett. **14**(12), 1–25 (2019)

Association of Cardiovascular Events and Blood Pressure and Serum Lipoprotein Indicators Based on Functional Data Analysis as a Personalized Approach to the Diagnosis

N. G. Plekhova[1,2(✉)], V. A. Nevzorova[1,2], T. A. Brodskay[1], K. I. Shakhgeldyan[3,4], B. I. Geltser[3,4], L. G. Priseko[1], I. N. Chernenko[1], and K. L. Grunberg[1]

[1] Pacific State Medical University, 2 Ostryakova Ave, Vladivostok 690002, Russia
pl_nat@hotmail.com.com

[2] Institute of Chemistry, Far Eastern Branch of the Russian Academy of Sciences, 159 Pr. 100th Anniversary of Vladivostok, Vladivostok 690022, Russia

[3] Vladivostok State University of Economics and Service, 41 Gogolya St., Vladivistok 690014, Russia

[4] Far East Federal University, Sukhanova St. 8, 690091 Vladivostok, Russia

Abstract. The development of trends and practice-oriented approaches to personalized programs for the diagnosis and correction depending on the clinical and phenotypic variants of the person is relevant. A software application was created for data mining from respondent profiles in a semi-automatic mode; libraries with data preprocessing were analyzed. The anthropometric measurements and serum lipoprotein spectrum of 2131 volunteers (average age 45.75 ± 11.7 years) were studied. To estimate the association of blood pressure and cardiovascular events markers was carried out by means of multivariate analysis of data by the methods of selection and classification significant signs. The machine learning was used to predict cardiovascular events. Depends on gender there was found the significant difference in atherogenic index of plasma (AIP) ($F < 0.05$). In young women (20–30 y.o.), the lipoproteins did not correlate with the presence of hypertension, whereas for older women the statistically significant markers were higher, such as cholesterol (CH, $F = 0.03$), low-density lipoproteins (LDL, $F = 0.03$) and AIP ($F = 0.02$). In men for identifying the risk of hypertension developing lipoproteins should be considered depending on age. Accuracy of the risk recognition for the cardiovascular disease (CVD) model was more than 89% with an average confidence of the model in each forecasted case of 90%. The markers for diagnosing the risk of CVD, the following indicators can be used according to their degree of significance: AIP, CH and LDL. Thus, the data obtained indicate the importance of risk factor phenotyping using anthropometric markers and biochemical profile for determining their significance in the top 17 predictors of CVD. The machine learning provides CVD prediction according to standard risk assessments.

Keywords: Machine learning · Cardiovascular diseases · Arterial hypertension

R. Silhavy et al. (Eds.): CoMeSySo 2020, AISC 1295, pp. 278–293, 2020.
https://doi.org/10.1007/978-3-030-63319-6_24

1 Introduction

Cardiovascular disease (CVD) associated with atherosclerosis is the main cause of adult mortality in both economically developed and developing countries. In the development and progression of CVD, the accompanying criteria, called risk factors (RF), play a leading role. Today, more than 200 RF of the CVD are known, and their number annually increases [1, 2]. RFs are divided into two subgroups: non-modifiable, impossible to influence, and modifiable, amenable to both multimodal behavioral interventions and medical therapy. Moreover, it is necessary to determine the total cardiovascular risk that means – the probability of developing a cardiovascular event connected with atherosclerosis over a specific period. This is the key to selecting preventive strategies and specific interventions for patients. The prevention and management of CVD increasingly demand effective diagnostic testing. Consensus defines a diagnostic as a method and an associated device that performs a physical measurement from a patient or associated biological sample and produces a quantitative or descriptive output, known as a biomarker. The definition of a biomarker, in turn, encompasses "a characteristic that is objectively measured and evaluated as an indicator of normal biological processes, pathogenic processes, or pharmacologic responses to a therapeutic intervention" [3]. Diagnostics, because of their strategic position at the intersection between patients and their clinically actionable data, directly affect the patient experience and the quality of care that individuals receive. The methods of the biochemical and cellular biofluid analysis advanced, the portfolio of available tests expanded and central laboratories emerged to standardize sample acquisition and measurement [4]. Today, technology is expanding the number of diagnostic tests that can reach beyond the walls of centralized laboratories and back to the point of care for use across a broad range of clinical settings.

Established risk factors for CVD – such as AH, high levels of low-density lipoprotein cholesterol (LDL-C), low levels of high-density lipoprotein cholesterol (HDL-C), smoking, male gender, and old age do not entirely account for CVD risk [5]. Since treating modifiable risk factors is known to reduce the risk of CVD [6, 7], improving CVD risk stratification would enable better allocation of prevention resources [8]. And one approach to improving risk prediction is to consider the risk of CVD associated with the size distribution of a patient's lipoprotein particles. The Atherogenic Index of Plasma (AIP) is easily calculated from a standard lipid profile. It is a logarithmically transformed ratio of molar concentrations of triglycerides (TG) to high-density lipoprotein cholesterol (HDL). The strong correlation of AIP with lipoproteins may explain its high predictive value [9]. However, the determination of TC, HDL, and low-density lipoprotein cholesterol (LDL) concentrations is not sufficient for appropriate medical therapy. LDL and HDL should be sub-fractionated to measure concentrations of large, anti-atherogenic HDL [10]; less atherogenic HDL [11]; and small, atherogenic particles of HDL [12], as well as less atherogenic LDL [9] and atherogenic LDL lipoproteins component – ApoA component – ApoB [13].

Again, the basis of a specialized medical information system consists of the instrument-computer complexes. Such use of a computer in combination with measuring technology in clinical and laboratory diagnostics allows creating new effective means for providing automated information, collection on a patient's condition and processing of this data in real-time. Data used for medical diagnosis have several features, such as the qualitative nature of information, the presence of data gaps, a large number of variables with a relatively small number of observations [14, 15]. Moreover, the significant complexity of the object of observation (disease) often does not allow building even a verbal description of the doctor's diagnosis procedure. The creation of medical device-computer complexes allows us to approach from the new positions to the understanding of instrumental diagnostic methods in the cumulative consideration of all parameters to establish an accurate diagnosis [16]. Traditionally, when modeling the course of the disease, probabilistic prediction of the value of binary variables is used [15], which is made based on regression analysis or using automated systems built based on neural network analysis. The optimal approach is the use of a scoring scale, such as, for example, the SCORE cardiovascular risk assessment system, the Framingham scale or the mathematical model of PROCAM [1, 17, 18]. The disadvantage of this approach is that the scales known today have been developed regarding a certain population and nosological forms of the disease. The important direction in this area is developing a more universal scale, with a specific application for analyzing any feature, interested to the researcher, described by the binary variable.

Objective: To assess the prospects of using artificial intelligence technologies in predicting the outcomes and risks of cardiovascular diseases in patients with hypertension.

2 Materials and Methods

2.1 Study Design

A cross-sectional study of the impact of factors on the development of CVD is a prospective population study made on a representative sample of the population of Vladivostok in the multicenter observational Russian program Epidemiology of Cardiovascular Diseases (ESSE-RF) in Primorsky Krai to identify the prevalence and frequency of chronic noncommunicable Diseases and their attendant risk factors. The data were published earlier [19]. The procedures were approved by the Ethics Committee of the Pacific State Medical University (agreement no. 46/23.11.2014). Written informed consent was obtained from all the subjects. Beginning with 2014, three basic surveys (2012–2019) with follow up 3 observation phases were conducted at intervals of 2 years. Since people from 24 to 65 years old were included in the ESSE-RF study, volunteers aged 20–23 (n = 245) were additionally examined whose survey protocol was closely match to the ESSE-RF. The 901 volunteers (502 women and 399 men) 692 healthy individuals of them were included in the first representative sample and patients in the second, with a diagnosed arterial hypertension (AH, 209 people). The age of the volunteers was from aged 20 to 44 years, which according to the WHO classification corresponded to young age.

In research, the method of questioning and clinical research, including anthropometric, instrumental (investigation of arterial pressure, pulse, ECG recording, were used. In this group anthropometric parameters such as height, body weight, body mass index (BMI), and waist circumference were monitored. To determine the body weight standard stand-on scales were used. The height measurement was performed using altimeter, the measured person was always without shoes. BMI was calculated as the ratio of body weight in kg to the squared height in meters. The waist circumference was measured at half the distance between the bottom edge of the lower rib and iliac crest of the hip bone at a horizontal level. Waist circumference values were defined according to the cardio-metabolic risk. There is a moderate risk in waist circumference >94 cm, possibly 80 cm (risk level 1), and a high risk (risk level 2) in waist circumference higher than 102 cm in men and 88 cm in women.

For the biochemical examination, following an overnight fast, blood samples were drawn into tubes and centrifuged the same day to separate serum, which was stored frozen (−80 °C) for subsequent analysis. A venous blood sample was withdrawn on an empty stomach and parameters were determined in certified laboratories using standard laboratory methods. From biochemical parameters the following values were monitored: glycaemia, uric acid, total cholesterol, LDL-cholesterol, and components of low (LDL) and high-density lipoproteins (HDL), and triglycerides (TG) was carried out in a colorimetric method using an automatic biochemical analyzer Mindray BS-200 (Shenzhen Mindray Bio-Medical Electronics, Chine) and reagents from Alpha Diagnostics (San Antonio, TX, US). The atherogenic index of plasma (AIP) and the atherogenic coefficient (AC) were computed as log (TG/HDL) and non-HDL/HDL, respectively.

2.2 Machine Learning

For the machine learning the Neural network data processing was carried out using the NeuralNetworkTool software package, which is part of the Matlab R2010b software (Mathworks, USA). The software product data was selected due to its modernity, accuracy of results and user support policy. Due to their capability to solve complex problems by manipulation of high volume data the designation, training and usage of NeuralNetworkTool requires computer environment. The network was trained according to the Bayesian regularization algorithm, since it gave the smallest error equal to 0.01. Bayesian regularization minimizes the linear combination of quadratic errors and weights. Modification is carried out in such a way that as a result a network with high generalizing properties is obtained.

2.3 Statistical Analysis

All statistical analyses were performed using SPSS Statistics 22 (IBM, Armonk, NY, United States). Testing for normality was performed by the Kolmogorov Smirnov test. Differences between group means were calculated using a two-sample t-test, assuming or not assuming equal variances (based on Levene's Test for Equality of Variances). The strength of the linear relationship between the two variables was expressed by the Pearson correlation coefficient; a p-value of <0.05 was assumed to be statistically significant. Means and standard deviations are reported in terms of the original distributions.

Step 1. $\chi 2$ test was used to analyze the associations between data (the sex of the subjects (0, 1), the body mass index (BMI), the presence of a smoking factor (yes, no), arterial hypertension (AH, yes, no) and interval variables were included; age, systolic pressure, serum cholesterol, TG, HDL, LDL, ApoA, ApoB, C reactive protein (CRP), thyroid-stimulating hormone (TSH), leptin, adiponectin, and insulin.
Step 2. Analyze the relation of smokers and AH, gender status, age status and another. Generate receiver-operating characteristic (ROC) curve, calculate the Youden index, and found cutoff points for AH, gender status, age status.
Step 3. Binary logistic regression with backward: conditional method was performed. Dependent variables were AH, gender status, age status, and systolic pressure. Covariates were the related clinicopathological data from step1 and the AH cut-off points from step 2. Borrmann type and grading were exceptional because some cases had no available Borrmann type or grading data.
Step 4. If Borrmann type or grading factor was one of the related factors, binary logistic regression was then performed again using AH, gender status, age status or systolic pressure from step 2 as dependent variables and the related clinicopathological data from step 1 and the AH.
Step 5. If Borrmann type or grading factor was not one of the related factors obtained through binary logistic regression from step 3, the results in step 2 were the final result. However, if Borrmann type or grading factor was the related factor, the results in step 3 were the final result.

3 Result

3.1 Patient Characteristics

The average age in the group of patients with AH was 34.5 ± 2.8 years (from 25 to 44 years), the same indicator in the group of healthy individuals was 32 ± 1.7 years (from 20 to 44 years). In the group of patients with AH, there was a significant increase of indicators as blood pressure (BP), weight, and triglyceride levels compared with the control group (Table 1). Significant differences in indicators of total cholesterol, HDL, and LDL between the subjects with normal pressure and those with the presence of AH was not detected ($P \leq 0.007$ for overall trend for each variable).

3.2 Analyses of Biochemical Parameters

For a more accurate study of the distribution of the lipid spectrum, we separated the indicators of healthy subjects and patients with AH, respectively, of age and sex (Tables 2 and 3). The indicators of concentration of total cholesterol increase with age and reaches a level of 5.3 ± 0.7 mmol/L for the older age group (over 40 years) of healthy women, and patients with AH – 5.5 ± 0.5 mmol/L (Table 3). For men, a similar trend was noted (Table 2). In all age groups and regardless of gender, the average index of HDL was higher than 1.0 mmol/L for men and 1.2 mmol/L for women, which corresponds to the normal values accepted in the population (>1.0 for men and >1.2 for women). Significant differences of HDL surveyed with normal

Table 1. Comparisons of risk factors, and biochemical parameters between groups

Indicators	Group I (healthy volunteers) (n = 692)	Group II (patients with AH) (n = 209)
Age (years)	34,5 ± 2,8	32 ± 1,7
Weight (kg)	69 ± 5,8	87* ± 7,5
AP (ASP and ADP), mm Hg	118 ± 14,5 79 ± 6,9	151 ± 11,5* 91 ± 7,2*
Heart rate, beats per minute	72 ± 5,4	80 ± 7,5
Smokers, n (%)	55 ± 4,2	92 ± 6,5
Smoking Person Index, pack/years	4,75	5,25
Total cholesterol, mmol/L	4,95 ± 0,6	5,0 ± 0,4
ApoA, gr/L	1,79 ± 0,3	1,81 ± 0,6
ApoB, gr/L	0,92 ± 0,03	0,83 ± 0,05
LDL, mmol/L	3,13 ± 0,47	3,35 ± 0,27
HDL, mmol/L	1,31 ± 0,6	1,28 ± 0,4
TG, mmol/L	1,21 ± 0,09	1,87* ± 0,06
AIP	2,26 ± 0,06	3,0 ± 0,2
AC	0,46 ± 0,02	0,55 ± 0,01

Note: values are represented as n (%) or mean ± SD/median (IQR).
Abbreviations: BMI – body mass index, AP – blood arterial pressure, ASP – arterial systolic pressure, ADP – arterial diastolic pressure (measured in sitting position using a mercury sphygmomanometer), ApoA – apolipoprotein A, ApoA – apolipoprotein B, LDL and HDL – low and high density lipoproteins, TG – triglycerides, AIP – atherogenic index of plasma, AC– atherogenic coefficient, conventional units.
* $P < 0.05$, compared with Group I.

pressure and individuals with the presence of AH was not detected ($P \leq 0.005$ for overall trend for each variable). Concerning LDL, there is a slight increase correlated with age. In healthy volunteers, the maximum value of the concentration LDL was found in persons of the age group 40 and more years: for men – 3.2 mmol/L, for women – 3.3 mmol/L. In persons with hypertension, this indicator did not exceed 3.8 mmol/ L in men and 3.4 mmol/L in women.

The atherogenic index of plasma (AIP) and the atherogenic coefficient (AC) is not recognized by all researchers and in existing clinical guidelines, it is not included in asses of the risk of cardiovascular accidents [20]. At the same time, there are opinions about the possibility of its use to determine the risk of developing CVD. The AIP should not exceed 2.5 in healthy men between 20 and 30 years old, and 2.2 in healthy women of the same age, in people of both sexes 31–40 years old 3.0, and in people over 40 years old without clinical manifestations of atherosclerosis 3.5. We found that in men with increasing age AIP decreased. So, if in the age group from 20 to 30 years it was 2.4 ± 0.7, then in the rest AIP had average values of 1.8 ± 0.4 and 1.1 ± 0.6, respectively ($p < 0.05$, Table 2). While, AIP rates in men with AH in all age groups significantly exceeded those for healthy individuals ($p < 0.05$, Table 2). In women,

Table 2. Biochemical parameters of different age men's groups

Indicators	Group I (healthy volunteers, n = 298), years old			Group II (patients with AH, n = 92), years old		
	25–30 (n = 96)	31–40 (n = 112)	40 - more (n = 90)	25–30 (n = 20)	31–40 (n = 48)	40 - more (n = 24)
Total cholesterol, mmol/L	4,7 ± 0,5*	5,0 ± 0,7	5,1 ± 0,4	4,5 ± 0,5	5,4 ± 0,5	5,6 ± 0,6
LDL, mmol/L	1,3 ± 0,03	1,3 ± 0,05	1,3 ± 0,03	1,2 ± 0,04	1,2 ± 0,02	1,2 ± 0,1
HDL, mmol/L	3,1 ± 0,4	3,3 ± 0,5	3,2 ± 0,4	3,1 ± 0,35	3,7 ± 0,4	3,8 ± 0,4
AIP	2,4 ± 0,06	1,8 ± 0,06*	1,1 ± 0,6*	2,7 ± 0,05	3,3 ± 0,07	3,6 ± 03
AC	0,4 ± 0,03*	0,5 ± 0,15	0,5 ± 0,07	0,6 ± 0,07	0,6 ± 0,06	0,7 ± 0,5

Note: values are represented as n (%) or mean ± SD/median (IQR).
Abbreviations: LDL and HDL – low and high density lipoproteins, AIP – atherogenic index of plasma, AC– atherogenic coefficient.
* P < 0.05, compared with Group II.

Table 3. Biochemical parameters of different age women's groups

Indicators	Group I (healthy volunteers, n = 394), years old			Group II (patients with AH, n = 117), years old		
	25–30 (n = 127)	31–40 (n = 164)	40 - more (n = 103)	25–30 (n = 21)	31–40 (n = 64)	40 - more (n = 32)
Total cholesterol, mmol/L	4,6 ± 0,5*	5,0 ± 0,	5,3 ± 0,3	4,3 ± 0,5	5,1 ± 0,4	5,5 ± 0,7
LDL, mmol/L	1,3 ± 0,04	1,4 ± 0,03	1,4 ± 0,03	1,3 ± 0,05	1,4 ± 0,03	1,4 ± 0,06
HDL, mmol/L	2,7 ± 0,5	3,2 ± 0,4	3,3 ± 0,25	2,8 ± 0,4	3,3 ± 0,35	3,4 ± 0,5
AIP	2,6 ± 0,05	2,8 ± 0,04	2,9 ± 0,05	2,7 ± 0,05	2,8 ± 0,05	2,9 ± 0,02
AC	0,4 ± 0,02	0,5 ± 0,05	0,5 ± 0,07	0,4 ± 0,07	0,5 ± 0,05	0,5 ± 0,09

Note: values are represented as n (%) or mean ± SD/median (IQR).
Abbreviations: LDL and HDL – low and high density lipoproteins, AIP – atherogenic index of plasma, AC– atherogenic coefficient.
* P < 0.05, compared with Group II.

both with normal blood pressure values and in the group with AH, the AIP values increased with age (F < 0.05, Table 3). Thus, with the indicator for healthy women, the rate of 2.2 in our study, AIP for this age group was 2.6 ± 0.05, and in patients with AH 2.7 ± 0.05. From our point of view, this fact testifies to the variability of this indicator for different regions. Probably, the "normal" value of the cholesterol coefficient for this group of women (20–30 years old) requires adjustment. As for the AIP values for older women, this indicator did not differ and did not exceed the established normal values, both in the group of healthy women and in the group of women with AH (Table 3). AC is a sensitive marker for detecting the risk of CVD and normally does not exceed 1.0

[21]. We found that this coefficient did not exceed a critical value for both healthy individuals and those with AH. The average AC values for all age groups of men and women ranged from 0.4 to 0.6 (Table 2, 3). Thus, our study showed that the values of lipid profile total cholesterol, HDL, LDL, AC in men and women do not have statistically significant differences in age and sex, whereas a significant difference between the indicators of men and women is determined only concerning the AIP ($P < 0.05$).

The direct relationship between indicators lipid profile and AH at women in the age group of 20–30 years no found ($P < 0.05$). For indicators, total cholesterol, HDL, LDL, AIP, and AC the Fisher's correlation coefficients were 0.39, 0.20, 0.32, 0.28 and 0.81, respectively. In women of 31–40 years old, a correlation was found between the indicators of LDL and AIP and the presence of AH ($F = 0.04$, $P < 0.05$). As for women in the age group of 40 years or more, significant values of the Fisher criterion were identified for three, except for HDL and AC, the values of total cholesterol, LDL and AIP and were 0.03, 0.03 and 0.02, respectively ($P < 0.05$). In men of the age group of 20–30 years, the AIP was determined as a significant marker of the presence of AH. The levels of total cholesterol, HDL and AC showed the value of the Fisher coefficient more critical. In men over 40 years old, a positive correlation was found between cholesterol ($F = 0.04$), LDL ($F = 0.04$), AC ($F = 0.05$) and AH.

3.3 Numerical Results

Optimal Scaling (CATREG) was chosen as the regression model, this model operates with categorical variables, all included interval and order predictors were categorized, taking into account categorization. The binary variable was the presence of AH, 12 out of 18 potential predictors were included in the regression model. The "importance" coefficients (importance) calculated by the regression analysis are presented in Table 4, their values are proportional to the degree of the predictor's contribution. The values of the dependent variable were calculated for each of the predictors, included in the regression model by multiplying the absolute value of the corresponding importance factor by 100 and rounding to the integers.

The two groups of healthy individuals and patients with AH were compared by the values of each of the 21 potential predictors. For nominal variables, analysis of contingency tables was used, for ordinal and interval tests the Kruskal-Wallis test was used. The result of the analysis is shown in Table 5. With a significance level of 0.05, they were reliably associated only with the dependent variable. Predictors that had a statistical relationship with the dependent variable with a significance level of $p = 0.15$ or more were then included in the regression model.

A threshold total score was determined, after which the dependent variable assumes a value with an empirical probability of developing unwanted development of CVD. To calculate the threshold score, a regression analysis was initially carried out, in which the total score of each patient served as a predictor, and the dependent variable remained the same. By definition, the dependent variable used binary logistic regression. The equation

Table 4. Statistical relationship of the dependent variable with potential predictors

n	Predictor	Group I (n = 692)	Group II (n = 209)	P
1	Gender (male 0, female 1)	n = 298 (0); n = 394 (1)	n = 92 (0); n = 117 (1)	0,148
2	Age, Med (HKв, BKв)	34,5 (20; 42)	32,0 (26; 44)	0,582
3	Smoking (no 0, yes 1)	n = 312 (0) n = 380 (1)	n = 17 (0); n = 192 (1)	0,763
4	Systolic pressure, mm Hg, Med (HKв, BKв)	118 (90; 140)	151,8 (120; 230)	0,2833
5	BMI, kg/m^2, Med (HKв, BKв)	22 (33; 39)	28 (32; 37)	0,099
6	Glucose, mmol/l Med (HKв, BKв)	5,3 (5,0; 6,0)	6,0 (4,9; 6,8)	0,072
7	Total cholesterol, mmol/L	4,9 ± 0,47	5,26 ± 0,15	0,645
8	HDL, mmol/L	1,365 ± 0,09	1,33 ± 0,085	0,047
9	LDL, mmol/L	3,27 ± 0,12	3,32 ± 0,28	0,051
10	TG, mmol/L	1,35 ± 0,38	1,4 ± 0,3	0,049
11	ApoA-I gr/L	1,785 ± 0,13	1,68 ± 0,15	0,067
12	ApoB gr/L	0,79 ± 0,01	0,82 ± 0,04	0,174
13	TH mmol/L	1,14 ± 0,1	1,39 ± 0,08	0,182
14	Leptin, ng/ml (HKв, BKв)	9,75 (6,7; 15,6)	14,7 (13,8; 16,7)	0,002
15	Adiponectin, μg/ml	8,38 ± 2,21	10,13 ± 3,26	0,020
16	CRP, gr/L, Med (HKв, BKв)	1,74 ± 0,6	1,64 ± 0,6	0,031
17	Insulin	2,97 ± 0,4	7,7 ± 0,7	0,079
18	TSH, mEd/L	1,8 ± 0,08	1,43 ± 0,2	0,161

Note: values are represented as n (%) or mean ± SD/median (IQR).
Abbreviations: BMI – body mass index, LDL and HDL – low and high density lipoproteins, TG – triglycerides, ApoA – apolipoprotein A, ApoB – apolipoprotein B, TH – thyroid hormone, CRP – C reactive protein, TSH – thyroid-stimulating hormone.

$$p = 1/1 + e3.698 * \ e\ 0.045$$

was obtained, where p is the theoretical probability of the presence of AH (dependent variable), and x is the value of the total score for a particular patient. We calculated with the help of this equation, the theoretical values of the probability of conjunction with AH. The scattering diagram, which reflects this dependence (see Fig. 1A).

When calculating the average probability values in the group of patients with the value of the dependent variable “no” and in the group of patients with the value of the dependent variable “yes” it turned out, that the diagnosis of AH was noted if the theoretical probability of its development was in the range from 0.124 to 0.151. The graph (see Fig. 1B) shows that the lower limit of this range (0.124) corresponds to an interval of 10 to 40 points. Moreover, the actual frequency of AH is noted at patients with a total score of 10 or less. It turned out to be equal to 4.93% (about 5%).

Table 5. The fragment of the resulting regression analysis table with the optimal scaling and the total score of the CVD risk

n	Predictor	Standardized coefficients		F	Correlations			Importance	Points
		Beta	Std. error		Zero order	Partial	Part		
1	Gender	9,823E−02	0,047	4,364	0,082	0,106	0,097	0,048	+5
2	BMI	2,298E−02	0,064	0,130	0,132	0,018	0,017	0,018	+2
3	Glucose	8,015E−02	0,049	2,709	0,129	0,084	0,077	0,062	+3
4	HDL	0,0239	0,051	2,209	0,300	0,233	0,219	0,0427	+4
5	LDL	0,111	0,047	0,544	0,144	0,118	0,109	0,015	+2
6	TG	3,514E−02	0,050	0,497	0,084	0,036	0,03	0,018	+2
7	ApoA	3,679E−02	0,063	0,346	0,102	0,030	0,027	0,022	+2
8	TH	0,141	0,051	0,683	0,099	0,014	0,013	0,017	+2
9	Leptin	6,955E−02	0,052	0176	0,014	0,007	0,006	0,006	+1
10	ADP	0,14	0,013	0,005	0,588	0,016	0,012	0,013	+1
11	CRP	0,11	0,0013	0,043	0,088	0,02	0,009	0,011	+1
12	Insulin	0,139	0,048	0,712	0,095	0,009	0,008	0,019	+2

Abbreviations: BMI – body mass index, LDL and HDL – low and high density lipoproteins, TG – triglycerides, ApoA – apolipoprotein A, TH – thyroid hormone, ADP – adiponectin, CRP – C reactive protein.

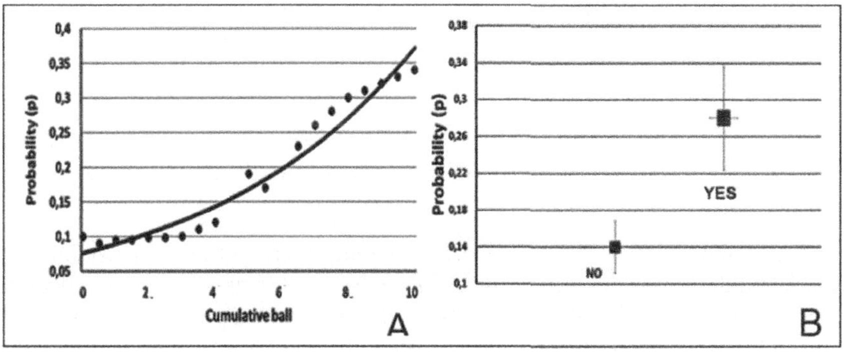

Fig. 1. The theoretical probability presence of AH. A. – the scattering diagram the dependence of the theoretical probability of the AH on the value total score (sensitivity 0.688, specificity 0.673 range from 0.124 to 0.151); B. – range of theoretical probability risk AH in healthy individuals with the absence (NO) and presence of this diagnosis (mean value ± error of mean).

The machine learning was carried out using NeuralNetworkTool software package, which is part of the Matlab R2010b software (Mathworks, USA). The network was trained according to the Bayesian regularization algorithm, since it gave the smallest error equal to 0.01. With the input-output method, a hidden layer and an output layer are created, which is a multilayer neural network. Next, the number of training set, test set and test set is selected. The training set contains a set of data that have pre-classified target and predictor variables. To determine how well the model works with data outside the training set, a test data set or test set is used. The test set contains data obtained from the preliminary profile, but they are not used when the data from the test set passes through the model to the end, when the compared data is compared with the model results. Using the randomization function X_train, X_test, y_train, y_test = train_test_split (X, Y, test_size = 0.40, random_state = 42), 2 samples were formed from the total data array: training (488 people) and test (245 people), which included data from patients with established the diagnosis of hypertension. Of all the subjects with hypertension (n = 733), the number of smokers was 144 people, 170 smoked and quit smoking, 41 non-smokers. As input, 17 of the most important variables were used, which constituted the input forecast layer model (Table 4, Fig. 2).

Hidden layers were determined empirically: the first layer includes 26 neurons (positions where multiplication weighting matrix and matrix input data of previous neurons), output layer consisted of 1 neuron, which corresponded with AH.

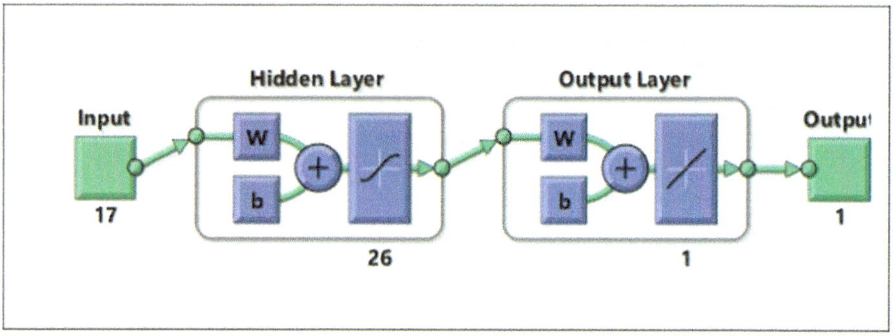

Fig. 2. Neural network model.

Training and optimization of the machine learning were carried out according to the Bayesian regularization algorithm, since it gave the smallest error equal to 0.01. Neural network testing was carried out on indicators of 288 people not included in the training sample, 144 of which were with the presence of hypertension, the rest without the presence of hypertension. The neural network predicts the presence of hypertension in young people at a given time with an accuracy of 76.06%, which is a satisfactory result. Since office blood pressure measurement can't determine availability disguised hypertension common in this cohort and in general population the probability of a correct assessment by a doctor is approximately 70% [20]. The change in the accuracy of the machine learning in the process of training and testing is presented in Fig. 3. The sample size for the machine learning was 66.6% of all subjects with hypertension. Training and optimization was carried out in 1000 eras, the volume of data submitted at a time amounted to 32 units. As a result of testing using the Bayesian regularization algorithm, the prognostic accuracy reached 97.9%, and the loss value was in the range 10^{-7}–10^{-8} (Fig. 3). During testing, the accuracy of the network decreased to 95.5%.

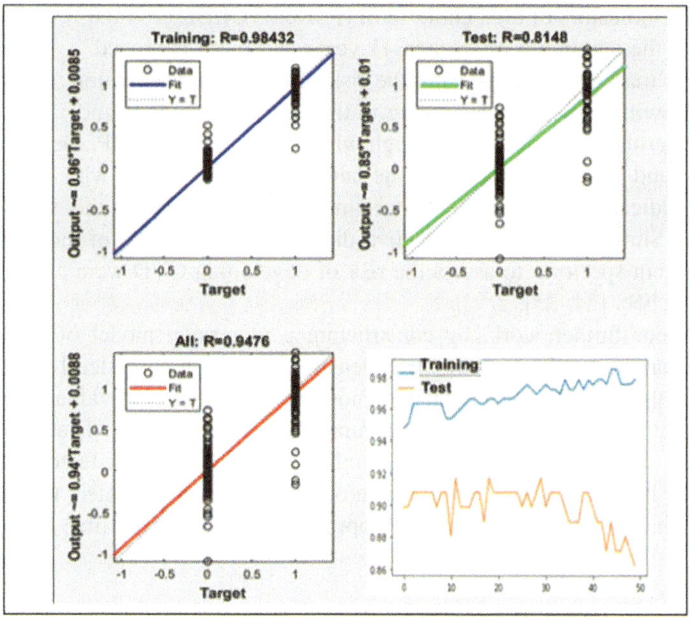

Fig. 3. Testing of neural network in the learning process.

4 Discussion

In 2012, the Ministry of Health of the Russian Federation initiated a multicenter observational study "Epidemiology of cardiovascular diseases in various regions of the Russian Federation (ESSE-RF)" to study the "traditional" and "new" risk factors CVD for implement preventive programs [22]. About 20 000 participants were included in the research: representative samples from the unorganized male and female population aged 20–64 years from 13 regions of the Russian Federation, including Primorsky Krai, in which Pacific State Medical University took part [23]. The results of this study will provide objective information on the prevalence of major CVDs in the population and predict the health of Russians. In the present study, we evaluated the significance of the atherogenic index's blood spectrum and also revealed the degree correlation between their and AH. Depending on the gender and age of the examined persons, the most significant markers for diagnosing of AH were determined. As a result, it was found that in young men of the age group of 20–30 years, the AIP can be considered as a risk factor combined with AH. A statistically significant index of correlation is determined between LDL, AIP and AH for the age group of 31–40 years and for the age group of 41 and more years total cholesterol (F = 0.04), LDL (F = 0.04) and AIP (F = 0,05). In young women, no indicator of the lipoprotein blood spectrum correlated with the existence of AH. Probably, additional extended studies of this group of persons are needed. Whereas, in the age group of 31–40 years, as well as for men, LDL, AIP can serve as marker conjunction with the AH. Also, a statistically significant correlation between the indicators of total cholesterol (F = 0.03), LDL (F = 0.03), AIP (F = 0.02) and AH for the age group of women 41 years and more is proved.

Thus, as markers for diagnosing the risk of CVD, in our opinion, it is possible to use the following indicators, according to their degree of significance: AIP, LDL, and total cholesterol. Moreover, even though when calculating the AIP, the values of total cholesterol and HDL were used and the latter does not correlate with the presence of AH, this indicator, is of fundamental importance concerning the serum level of cholesterol. Similar data on the selective diagnostic significance of individual indicators lipoprotein spectrum to assess the risk of developing CVD were also obtained by other researchers [24, 25].

During our further work, by constructing a regression model of risk, in which additional parameters, such as the high-density lipoprotein cholesterol level, the body mass index, the adiponectin content and another we can make individual characteristics of the specific patients to improve the accuracy of predictions [26]. These functions are multivariate algorithms that combine the information in CVD risk factors such as sex, age, systolic blood pressure, total cholesterol, high-density lipoprotein cholesterol, and smoking behavior to estimate risk developing CVD over a fixed time [27, 28].

5 Conclusion

Taking into attention about of the content certain substances norms in the body are the average values characteristic for the majority of healthy people, their correction is necessary for each case. So, patients suffering from diabetes, obesity and other diseases that usually accompany the change in lipid metabolism, are recommended to maintain the level of total cholesterol at the lowest level for the prevention of CVD, while for healthy people, these values may be slightly increased. Besides, when evaluating research data, not only the figures obtained for different indicators are important, but also their ratio among themselves. The conditional norm of the total cholesterol content is 2.97–8.79 mmol/L (for middle-aged people - up to 5.2 mmol/L) stays in a rather wide range. So, for individuals younger than 40 years, this indicator should be considered in conjunction with other factors, namely age, sex, smoking status and the values of systolic pressure. Such an approach from the position of multifactorial analysis allows diagnosing the state of lipid metabolism in healthy individuals more accurately, considering an objective assessment of the developing CVD risk for earlier treatment of lipid-regulating therapy.

The value of the coefficient for assessing the overall risk of developing CVD specific for the Russian population, allows us to estimate also the relative risk (RR) since it establishes a monotonous numerical scale, low values of which indicate low relative risk values, and high scales indicate a high relative risk. The advantage of the study is the consideration of a set of anthropometric data, the results of laboratory tests and other important predictors of CVD development. Thus, the machine learning in combination with extended phenotyping increases the accuracy of predicting cardiovascular events in the population of subjects with the presence of such developmental RF as hypertension. The developed approaches allow us to approach a more accurate understanding of the markers of subclinical diseases without a priori assumptions about the causality of their occurrence.

Acknowledgments. The study was supported by a grant from the Russian Foundation for Basic Research 19-29-01077 and is part of the Ministry Health the Russian Federation state task «Clinical and phenotypic variants and molecular genetic features of vascular aging in people of different ethnic groups».

Declaration of financial and other relationships. All authors participated in the development of the concept, the design of the study and the writing of the manuscript. The final version of the manuscript was approved by all authors.

References

1. Boersma, E., Pieper, K.S., Steyerberg, E.W., Wilcox, R.G., Chang, W., Lee, K.L., Akkerhuis, K.M., Harrington, R.A., Deckers, J.W., Armstrong, P.W. et al.: Predictors of outcome in patients with acute coronary syndromes without persistent St-segment elevation. Results from an international trial of 9461 patients. Circulation 101(22), 2557–2567 (2000)
2. Pollack Jr., C.V., Sites, F.D., Shofer, F.S., Sease, K.L., Hollander, J.E.: Application of the TIMI risk score for unstable angina and non-St elevation acute coronary syndrome to an unselected emergency department chest pain population. Academic. Emergency Med. **13**(1), 13–18 (2006)
3. Biomarkers and surrogate endpoints: preferred definitions and conceptual framework. Biomarkers definitions working group. Downing GJ, ed. Clin. Pharmacol. Ther. 69, 89–95 (2001)
4. Rosenfeld, L.: Clinical chemistry since 1800: growth and development. Clin. Chem. **48**, 186–197 (2002)
5. Wilson, P.W.F., D'Agostino, R.B., Levy, D., Belanger, A.M., Silbershatz, H., Kannel, W.B.: Prediction of coronary heart disease using risk factor categories. Circulation **97**(18), 1837 (1998)
6. Thompson, P.D., Buchner, D., Pina, I.L., Balady, G.J., Williams, M.A., Marcus, B.H., et al.: Exercise and physical activity in the prevention and treatment of cardiovascular disease. Circulation **107**, 3109–3116 (2003)
7. Stone, N.J., Robinson, J.G., Lichtenstein, A.H., Bairey Merz, C.N., Blum, C.B., Eckel, R.H., et al.: American college of cardiology/American heart. Circulation **25**(Suppl 2), S1–45 (2014)
8. Superko, H.R., King, S. Lipid management to reduce cardiovascular risk: a new strategy is required. 3rd. Circulation **117**(4), 560–568 (2008)
9. Dobiásová, M.: AIP-atherogenic index of plasma as a significant predictor of cardiovascular risk: From research to practice. Vnitr. Lek. **52**, 64–71 (2006)
10. Pearson-Stuttard, J., Bandosz, P., Rehm, C.D., Afshin, A., Peñalvo, J.L., Whitsel, L., Danaei, G., Micha, R., Gaziano, T., Lloyd-Williams, F., et al.: Comparing the effectiveness of mass media campaigns with price reductions targeting fruit and vegetable intake on US cardiovascular disease mortality and race disparities. Am. J. Clin. Nutr. **106**, 199–206 (2017)

11. Perk, J., De Backer, G., Gohlke, H., Graham, I., Reiner, Z., Verschuren, M., Albus, C., Benlian, P., Boysen, G., Cifkova, R., et al.: European guidelines on cardiovascular disease prevention in clinical practice (version 2012). The fifth joint task force of the European society of cardiology and other societies on cardiovascular disease prevention in clinical practice (constituted by representatives of nine societies and by invited experts). Eur. Heart J. **33**, 1635–1701 (2012)
12. Roth, G.A., Johnson, C., Abajobir, A., Abd-Allah, F., Abera, S.F., Abyu, G., Ahmed, M., Aksut, B., Alam, T., Alam, K., et al.: Global, regional, and national burden of cardiovascular diseases for 10 causes, 1990 to 2015. J. Am. Coll. Cardiol. **70**, 1–25 (2017)
13. Yusuf, S., Hawken, S., Ounpuu, S., Dans, T., Avezum, A., Lanas, F., McQueen, M., Budaj, A., Pais, P., Varigos, J., et al.: Effect of potentially modifiable risk factors associated with myocardial infarction in 52 countries (the INTERHEART Study): Case-control study. Lancet **64**, 937–952 (2004)
14. Perez, L., Dragicevic, S.: An agent-based approach for modeling dynamics of contagious disease spread. Int. J. Health Geogr. **8**(1), 50–54 (2009)
15. Hernández, A.I., Le Rolle, V., Defontaine, A., Carrault, G.A.: Multiformalism and multiresolution modelling environment: application to the cardiovascular system and its regulation. Philos. Transact. Math. Phys. Eng. Sci. **367**(1908), 4923–4940 (2009)
16. Antman, E.M., Cohen, M., Bernink, P.J., McCabe, C.H., Horacek, T., Papuchis, G., Mautner, B., Corbalan, R., Radley, D., Braunwald, E.: The TIMI risk score for unstable angina/non-St elevation Mi: a method for prognostication and therapeutic decision making. J. Am. Med. Assoc. **284**(7), 835–842 (2000)
17. Eagle, K.A., Lim, M.J., Dabbous, O.H., Pieper, K.S., Goldberg, R.J., Van de Werf, F., Goodman, S.G., Granger, C.B., Steg, P.G., Gore, J.M.: A validated prediction model for all forms of acute coronary syndrome. Estimating the risk of 6-month postdischarge death in an international registry. J. Amer. Medical Assoc. **291**(22), 2727–2733 (2004)
18. Conroy, R.M., Pyorala, K., Fitzgerald, A.P.: Estimation of ten-year risk cardiovascular disease in Europe: the SCORE project. Eur. Heart J. **24**, 987–1003 (2003)
19. Sakovskaia, A., Nevzorova, V., Brodskaya, T., Chkalovec, I.: Condition aortic stiffness and content of adipokines in the serum of patients with essential hypertension in young and middle-aged. J. Hypertension **33**(N e-suppl.1), 182–187 (2015)
20. Ni, W., Zhou, Z., Liu, T., Wang, H., Deng, J., Liu, X., Xing, G.: Gender-and lesion number-dependent difference in "atherogenic index of plasma" in Chinese people with coronary heart disease. Sci Rep. 16,7(1), 13207 (2017)
21. Gunay, S., Sariaydin, M., Acay, A.: New predictor of atherosclerosis in subjects with COPD: atherogenic indices. Respir Care. **61**(11), 1481–1487 (2016)
22. Scientific and Organizing Committee of the ESSE-RF project. Epidemiology of cardiovascular diseases in various regions of Russia (ESSE - RF). Justification and design of the research. Preventive medicine. 6, pp. 25–34 (2013)
23. Nevzorova, V.A., Shumatov, V.B., Nastradin, O.V.: The state of the function of the vascular endothelium in people with risk factors and patients with coronary heart disease. Pacific Med. J. **2**, 37–44 (2012)
24. Odden, M.C., Tager, I.B., Gansevoort, R.T., Bakker, S.J.L., Fried, L.F., Newman, A.B., Katz, R., Satterfield, S., Harris, T.B., Sarnak, M.J., Siscovick, D., Shlipak, M.G.: Hypertension and low HDL cholesterol were associated with reduced kidney function across the age spectrum: a collaborative study. Ann. Epidemiol. **23**(3), 106–111 (2013)
25. Al-Naamani, N., Palevsky, H.I., Lederer, D.J., Horn, E.M., Mathai, S.C., Roberts, K.E., Tracy, R.P., Hassoun, P.M., Girgis, R.E., Shimbo, D., Post, W.S., Kawut, S.M.: Prognostic significance of biomarkers in pulmonary arterial hypertension. Ann. Am. Thorac. Soc. **13**(1), 25–30 (2016)

26. Fowkes, F.G., Murray, G.D., Butcher, I., Heald, C.L., Lee, R.J., Chambless, L.E.: Ankle brachial index combined with Framingham risk score to predict cardiovascular events and mortality: a meta-analysis. JAMA **300**, 197–208 (2008)
27. D'Agostino, R.B., Pencina, M.J., Massaro, J.M., Coady, S.: Cardiovascular disease risk assessment: insights from Framingham. Global Heart **8**(1), 11–23 (2013)
28. Steyerberg, E.W., Vickers, A.J., Cook, N.R., Gerds, T., Gonen, M., Obuchowski, N., Pencina, M.J., Kattan, M.W.: Assessing the performance of prediction models: a framework for some traditional and novel measures. Epidemiology **21**(1), 128–138 (2010)

Identification of Similarities in Approaches to Paired Comparisons in Visual Arts Education

Milan Cieslar[1], Tomas Koudela[1], Gabriela Pienias[1], and Tomas Barot[2(✉)]

[1] Department of Visual Arts Education, Faculty of Education, University of Ostrava, Fr. Sramka 3, 709 00 Ostrava, Czech Republic
{Milan.Cieslar,Tomas.Koudela,Gabriela.Pienias}@osu.cz
[2] Department of Mathematics with Didactics, Faculty of Education, University of Ostrava, Fr. Sramka 3, 709 00 Ostrava, Czech Republic
Tomas.Barot@osu.cz

Abstract. In some applied sciences, the type of the qualitative research has been frequently appeared. However, the quantitative research can be also the useful methodology; even though, the statistical significance can increase the guarantee of the research. In this paper, the quantitative type of the methodology is presented on the applied practically realized research of the student's results in the area of the visual arts education. On the data, the statistical approaches for purposes of the paired comparisons are applied. Firstly, the generally used statistical techniques of the paired tests are realized with an obtainment of *p* values, the test criterions and the effect sizes. As the second, the correlation coefficients and the descriptive characteristics are determined. Both approaches are provided with regards to the sample size. In the achieved results, the similarities of these different types of approaches are identified by the cluster analysis with regards to the changing the sample size.

Keywords: Paired comparisons · Paired tests · Testing hypotheses · Correlation analysis · Cluster analysis · Applied research · Visual arts education

1 Introduction

In the visual arts education [1], the qualitative type of research has been widely appeared. The utilization of the quantitative research [2, 3] can be frequently considered as rare approach. In this paper, the possibilities of using the research data in the quantitative sense [2–5] are presented; even thought, the analysis of the behaviour of the statistical parameters is demonstrated on the practically based research.

In the field of the arts, the research can have the similar research questions and structure of methodology as the quantitative research in the social sciences (e.g. [6–8]), technical sciences (e.g. [9]) or in the medical sciences (e.g. [10–12]).

As the one of the possible research topics in the field in the visual arts education [1], the progress of the knowledge and experiences evaluated by experts can be appropriate. The time collected data can have the significant role for these purposes.

R. Silhavy et al. (Eds.): CoMeSySo 2020, AISC 1295, pp. 294–303, 2020.
https://doi.org/10.1007/978-3-030-63319-6_25

As an example of this type of the applied quantitative research, the progress of the artist's results can be compared, e.g. the improving the principles from the drawing the lines through the understanding the application of colors to the practical proof of the skills in the nature; those achievement can be advantageously statistically analyzed.

The generally used descriptive comparisons using the correlation analysis [3] has been widely appeared in the practice in case of comparisons of the numbered (cardinal) variables in the social sciences. However, the utilization of methods of the mathematical induction [2], with principles of testing the hypotheses, brings the guarantee of the statistical significance [3]; in addition, if the results are complemented by the effect sizes [4].

Especially, the paired based comparisons of the achieved evaluations (e.g. of the artist's study results) can be suitable data set for the Paired T-test [2] or for the Wilcoxon paired test [2]. Both tests are able to provide the results with regards to the guarantee of the statistical significance. A difference between these tests is the condition of using them according to the normality property of data [2, 3]. The paired test can be also suitable applicable for the situations occurred e.g. in the numerical methods (e.g. [9]) or also in the signal processing (e.g. [13]).

The utilization of the methods of the mathematical induction [2, 3] can bring the advantages also in the research in the visual arts. Using the descriptive statistical characteristics in comparison with conclusions of the statistical methods of the mathematical induction has not been so widely discussed in the applied research. Therefore, the applied research can be also a suitable background to verification the behaviour of the particular statistical parameters. The importance of using the methods of mathematical induction have the irreplaceable role beside the descriptive methods. In this paper, a behaviour of the parameters of these both approaches (correlation coefficients across the test criterions, p value and effect sizes) are discussed with regards to the changing sample size using the cluster analysis [5].

2 Principles of Paired Comparisons Using Testing Hypotheses

With regards to the principle of the paired comparison [2], the statistically significant paired differences [2] can be proven using the method of the mathematical induction [2, 3]. Especially, the ordered pairs of values are considered in both states occurred before some situation and after changes of the state.

In the frame of the quantitative research, two appropriate statistical methods are determined for solving this type of problems. Considering the statistical significance level α, the software computing of the paired tests can be realized with resulting the p value [2].

In case of $p > \alpha$, the zero hypothesis (1) for the paired comparison is failed to reject on the significance level α. In the opposite case, the zero hypothesis (1) is rejected in favor of the alternative hypothesis (2) on the significance level α [2].

$$H_0 : \theta_1 = \theta_2 \tag{1}$$

$$H_1 : \theta_1 \neq \theta_2 \tag{2}$$

According to the normality of data [2], the parameter θ can be represented as the mean value of the i–th (i =1, 2) data file. In case of unfulfilling the normality of data, the parameter θ expresses the medians of each data sequence. The normality of data has an influence on the selection of the concrete method aimed for the purposes of the testing the hypotheses (1)–(2). In the positive case, the Paired T-test [2] should be selected. In the opposite case, the Wilcoxon paired test [2] should be applied. After the own testing the hypotheses, the zero hypothesis (1) is failed to reject in case of $p > \alpha$ or the zero hypothesis is rejected in favor of the alternative hypothesis (2) in case of $p < \alpha$. Both conclusions provide information about the statistically significant paired differences between the paired set of data on the significance level α. The significance level can be declared as 0.05 (as in social sciences, e.g. [6, 7]), 0.01 or 0.001 (as in technical sciences, e.g. [9] or as in medical sciences, e.g. [2, 3, 10–12].

The strength of the rejecting the zero hypothesis should be complemented by information about the effect size. For the Paired T-test, the effect size can be expressed by the Cohen d (3). In computation of the Cohen d, the arithmetical averages $(\overline{x_1}, \overline{x_2})$ of two data series and the standard deviation σ of whole data are considered [4].

$$d = \frac{\bar{x}_1, \bar{x}_2}{\sigma} \tag{3}$$

In case of utilization of the Wilcoxon paired test, the effect size r (4) should be determined. The z score and a number of whole data N are appeared in this equation [4].

$$r = \frac{z}{\sqrt{N}} \tag{4}$$

For purposes of the further analyses, the monitored parameters of the paired tests will be p values, the test criterions (t for the Paired T-test and W for the Wilcoxon paired test) and the effect sizes. These parameters can be obtained in the statistical software in the realization of the particular method [2–4].

3 Analysis of Similarities in Results of Paired Comparisons

As the proposal of analysis of similarities, the comparison of results between different types of the statistical tests is considered in this paper. In favor of this comparison, the cluster analysis [5] is applied on data measured due to changing the sample size. In this paper the size of the observed interval is proposed as an interval between 75% samples to 100% considered obtained samples.

At first, the testing hypothesis is realized in the following structure of the zero and the alternative hypotheses for purposes of paired comparison of r paired sets of data, as can be seen in Table 1.

Table 1. Structure of considered hypotheses in realized quantitative research

kH	k-th Zero Hypothesis	k-th Alternative Hypothesis
$1H$	$\theta_1 = \theta_2$	$\theta_1 \neq \theta_2$
$2H$	$\theta_2 = \theta_3$	$\theta_2 \neq \theta_3$
$\vdots$	$\vdots$	$\vdots$
$(r-1)\ H$	$\theta_{r-1} = \theta_n$	$\theta_{r-1} \neq \theta_n$

As the carrier parameter θ in testing the hypotheses, the mean value is considered in case of fulfilling the normality of data using the Shapiro-Wilk test. In the opposite case, the medians are represented using this parameter. In both cases, pairs of values are bound together for each respondent [2].

In the zero hypothesis, the non-existence of the statistically significant paired differences is assumed on the significance level α. In the alternative hypothesis, the negation of the zero hypothesis is defined.

In addition; p values, the test criterions of the realized statistical methods t (in case of the Paired T-test), W (in case of the Wilcoxon paired test) and the effect sizes (3) or (4) will be monitored for changing the sample size n (from 75% to 100% of considered obtained samples size) [2, 3].

As the second, the descriptive approach will be implemented in the proposed analysis. Concretely, the Pearson correlation coefficient (5) is computed for the correlation analysis of the existence of dependences of the numbered results of the first data sequence (with arithmetical average $\bar{x}_1$ and standard deviation σ_1) on the second data sequence (with arithmetical average $\overline{x_2}$ and standard deviation σ_2). Each considered ordered pair of values can be geometrically seen as a point in the cartesian system with value $[x_{i1}, x_{i2}]; \mathrm{i} \in 1, n$.

$$\rho = \frac{\frac{1}{n}\sum_{i=1}^{n}(x_1 - \bar{x}_1)(x_2 - \bar{x}_2)}{\sigma_1\sigma_2} \tag{5}$$

The behaviour of parameters of methods of testing the hypotheses and the statistical description, which are the different statistical approaches, will be concretely complemented by difference of the arithmetical averages $\bar{x}_2, \bar{x}_1$ (in case of fulfilling the normality of data) or by the difference of medians $\theta_2 - \theta_1$ (in case of unfulfilling the normality of data).

Analysis of progresses of the behaviour of all described parameters will be measured with regards to the changing the sample size. After these computations, the cluster analysis [5] will be applied. The cluster analysis could identify the similarities between proposed groups of data according the achieved parameters, which can be near or far respecting the Euclidian distances. Obtained results of the cluster analyses can be discussed with regards to the concrete statistical approaches to paired comparisons for the methods of induction and also for the descriptive option.

4 Results

The practical realized quantitative research was provided on the study results of 110 students. As the study field of students, the Preparation of the future primary school teachers was selected across several years focusing the visual arts education. The long-time based progress of the study results was statistical processed by the methods of the mathematical induction and the correlation analysis. Both approaches were further analyzed by the proposed approach to their comparison.

As the ordered set of results, the 5 carrier courses in the visual arts education were considered with the following structure: 1st Course (Studies of Lines), 2nd Course (Studies of Colors), 3rd Course (Studies of 3rd Dimension), 4th Course (Architecture and History of Visual Arts) and 5th Course (Practical Course of Painting in Nature). Each course was realized in both forms of study (in the full-time and in the part-time form) at the Faculty of Education at University of Ostrava. Since 2015, each three years' flow of courses was recorded.

Data files were considered for the changing sample size from 85 to 110 (as 75% of $n = 110$ respondents). Selection of respondents for the eliminated sample file was implemented randomly. After testing the normality of data using the Shapiro-Wilk test, the testing the 4 hypotheses (Table 2) was provided on the significance level $\alpha = 0.05$ with the conclusions written also in Table 2. The statistical computations were realized in IBM SPSS Statistics 26 and PAST Statistics.

Table 2. Conclusions about testing hypotheses in realized quantitative research

kH	Comparison of Paired Differences between Courses on α	Conclusion about *k*-th Hypothesis
1*H* (Not-Normal Data)	1st Course and 2nd Course (Wilcoxon Paired Test)	$p<\alpha \Rightarrow \theta_1 \neq \theta_2$ (Medians)
2*H* (Not-Normal Data)	2nd Course and 3rd Course (Wilcoxon Paired Test)	$p > \alpha \Rightarrow \theta_2 = \theta_3$ (Medians)
3*H*(Not-Normal Data)	3rd Course and 4th Course (Wilcoxon Paired Test)	$p > \alpha \Rightarrow \theta_3 = \theta_4$ (Medians)
4*H*(Not-Normal Data)	4th Course and 5th Course (Wilcoxon Paired Test)	$p<\alpha \Rightarrow \theta_4 \neq \theta_5$ (Medians)

In the first part with the approach of the analysis of the mathematical induction, parameters of the Wilcoxon paired test for 1*H*–4*H*: *p* values, the test criterions *W* and the effect sizes r with regards to changing the sample size, can be seen in Tables 3 and 4. In $1H$ and $4H$ was proved, that there are the statistically significant paired differences between the measured results on $\alpha = 0.05$. In case of $2H$ and $3H$, the statistically significant paired differences between the measured results were not appeared on $\alpha = 0.05$.

Table 3. Behaviour of obtained parameters of approach of mathematical induction

Sample Size	W (1H)	W (2H)	W (3H)	W (4H)	r (1H)	r (2H)	r (3H)	r (4H)
85	1828.0	1581.0	1842.0	2563.5	0.411	0.079	0.188	0.487
86	1896.0	1581.0	1842.0	2642.0	0.422	0.079	0.188	0.496
87	1965.0	1581.0	1888.5	2721.0	0.433	0.079	0.188	0.505
88	2014.0	1626.0	1935.0	2795.0	0.431	0.081	0.188	0.510
89	2060.0	1676.0	1954.5	2795.0	0.428	0.085	0.175	0.510
90	2093.0	1743.0	1995.5	2865.0	0.418	0.096	0.172	0.514
91	2126.0	1811.0	1995.5	2943.5	0.407	0.108	0.172	0.521
92	2194.5	1823.5	2042.0	3020.5	0.415	0.093	0.172	0.527
93	2223.0	1850.5	2042.0	3082.0	0.403	0.085	0.150	0.525
94	2286.5	1877.5	2088.5	3161.0	0.407	0.078	0.150	0.531
95	2337.5	1926.5	2135.0	3241.0	0.405	0.080	0.149	0.537
96	2394.0	1969.0	2165.5	3310.0	0.406	0.078	0.142	0.538
97	2465.0	1985.5	2196.0	3380.0	0.413	0.066	0.134	0.539
98	2539.0	1998.0	2226.5	3451.0	0.421	0.052	0.126	0.540
99	2612.0	2014.5	2226.5	3537.5	0.428	0.041	0.126	0.546
100	2672.5	2057.0	2226.5	3625.0	0.429	0.039	0.126	0.553
101	2709.5	2131.0	2273.0	3711.0	0.419	0.050	0.125	0.559
102	2764.5	2183.0	2292.5	3711.0	0.417	0.052	0.113	0.559
103	2836.0	2211.0	2339.0	3798.0	0.422	0.045	0.112	0.564
104	2871.5	2289.0	2380.0	3881.0	0.412	0.056	0.109	0.568
105	2952.5	2301.5	2392.5	3951.0	0.420	0.042	0.094	0.567
106	2996.5	2330.5	2405.0	4021.0	0.413	0.035	0.080	0.566
107	3079.5	2343.0	2446.0	4107.0	0.421	0.022	0.077	0.570
108	3133.5	2402.0	2458.5	4178.0	0.417	0.026	0.063	0.568
109	3133.5	2402.0	2499.5	4266.0	0.417	0.026	0.060	0.572
110	3191.5	2457.0	2499.5	4362.5	0.415	0.028	0.060	0.579

In the second approach of the descriptive statistics, the difference of medians $d\theta = \theta_2 - \theta_1$ and the Pearson correlation coefficients (5) were computed for the samples of data (the same interval 85–110 respondents). The medians were determined due to reason of unfulfilling the normality of data. The obtained results are written in Table 5. In brackets, a number of the corresponding data set used for testing the hypothesis is used, e.g. $d\theta$ (1) for data set used for $1H$.

Table 4. Behaviour of obtained p values of approach of mathematical induction

Sample Size	p (1*H*)	p (2*H*)	p (3*H*)	p (4*H*)
85	1.6×10^{-5}	0.41	0.05	3.3×10^{-7}
86	9.5×10^{-6}	0.41	0.05	1.9×10^{-7}
87	5.7×10^{-6}	0.41	0.05	1.2×10^{-7}
88	6.1×10^{-6}	0.40	0.05	8.6×10^{-8}
89	7.1×10^{-6}	0.37	0.07	8.6×10^{-8}
90	1.2×10^{-5}	0.31	0.07	6.9×10^{-8}
91	1.9×10^{-5}	0.26	0.07	4.7×10^{-8}
92	1.3×10^{-5}	0.33	0.07	3.3×10^{-8}
93	2.4×10^{-5}	0.37	0.11	3.6×10^{-8}
94	1.9×10^{-5}	0.42	0.12	2.5×10^{-8}
95	2.1×10^{-5}	0.40	0.12	1.8×10^{-8}
96	2.1×10^{-5}	0.41	0.14	1.7×10^{-8}
97	1.5×10^{-5}	0.49	0.16	1.6×10^{-8}
98	1.0×10^{-5}	0.58	0.18	1.5×10^{-8}
99	7.2×10^{-6}	0.67	0.18	9.9×10^{-9}
100	6.9×10^{-6}	0.68	0.18	6.6×10^{-9}
101	1.1×10^{-5}	0.60	0.19	4.6×10^{-9}
102	1.2×10^{-5}	0.58	0.23	4.6×10^{-9}
103	9.6×10^{-6}	0.64	0.24	3.3×10^{-9}
104	1.6×10^{-5}	0.56	0.25	2.6×10^{-9}
105	1.1×10^{-5}	0.66	0.32	2.7×10^{-9}
106	1.5×10^{-5}	0.71	0.40	2.9×10^{-9}
107	1.0×10^{-5}	0.81	0.42	2.4×10^{-9}
108	1.2×10^{-5}	0.79	0.51	2.6×10^{-9}
109	1.2×10^{-5}	0.79	0.53	1.9×10^{-9}
110	1.3×10^{-5}	0.77	0.53	1.3×10^{-9}

By the setting the Euclidian distance method, the cluster analysis was provided for the observed parameters of the mathematical induction (Tables 3 and 4) and for the observed parameters of the descriptive approach (Table 5) for each hypothesis. The graphical results of the cluster analysis can be seen in Figs. 1, 2, 3 and 4. Scale of the figures is focused on the most detailed view; therefore, the longer Euclidian distances are not displayed (not important for the analysis). Greek symbols in the figures are replaced by following notations: dθ as dM (difference of medians) and ρ as rho.

As can be seen in the obtained results in the frame of the cluster analyses, the similarities between the Pearson coefficients and p values were identified in both cases of rejecting the zero hypotheses 1*H* and 4*H*. In both cases of failed to rejecting the zero hypotheses 2*H* and 3*H*, the similarities were identified between the Pearson correlation coefficients and the effect sizes. Relations were not identified for the test criterions *W*.

Table 5. Behaviour of obtained parameters of approach of statistical description

Sample Size	$d\theta$ (1)	$d\theta$ (2)	$d\theta$ (3)	$d\theta$ (4)	ρ (1)	ρ (2)	ρ (3)	ρ (4)
85	4.42	0.42	−0.54	2.54	0.07	0.06	0.20	0.13
86	4.73	0.42	−0.53	2.62	0.08	0.06	0.21	0.12
87	5.03	0.41	−0.52	2.68	0.08	0.06	0.21	0.11
88	4.97	0.42	−0.50	2.68	0.08	0.06	0.21	0.11
89	4.89	0.44	−0.45	2.65	0.08	0.06	0.20	0.11
90	4.77	0.50	−0.42	2.64	0.07	0.07	0.20	0.11
91	4.65	0.56	−0.42	2.66	0.06	0.07	0.20	0.11
92	4.70	0.46	−0.40	2.66	0.07	0.07	0.20	0.11
93	4.55	0.41	−0.28	2.62	0.06	0.08	0.18	0.11
94	4.54	0.36	−0.27	2.63	0.06	0.08	0.18	0.11
95	4.48	0.37	−0.25	2.63	0.06	0.08	0.18	0.11
96	4.45	0.35	−0.22	2.61	0.06	0.08	0.18	0.11
97	4.47	0.28	−0.19	2.60	0.06	0.07	0.18	0.11
98	4.52	0.18	−0.15	2.58	0.07	0.07	0.17	0.11
99	4.55	0.11	−0.15	2.60	0.07	0.06	0.18	0.11
100	4.51	0.10	−0.15	2.61	0.07	0.06	0.18	0.11
101	4.41	0.16	−0.14	2.61	0.06	0.07	0.18	0.11
102	4.35	0.17	−0.10	2.59	0.06	0.07	0.17	0.11
103	4.35	0.13	−0.09	2.59	0.06	0.07	0.17	0.11
104	4.24	0.19	−0.07	2.59	0.06	0.07	0.17	0.11
105	4.29	0.10	−0.02	2.55	0.06	0.07	0.17	0.11
106	4.20	0.07	0.03	2.52	0.07	0.07	0.16	0.11
107	4.24	−0.02	0.05	2.51	0.07	0.06	0.16	0.11
108	4.19	0.00	0.09	2.48	0.07	0.06	0.16	0.10
109	4.15	0.00	0.11	2.48	0.07	0.06	0.16	0.10
110	4.10	0.01	0.11	2.49	0.07	0.07	0.16	0.10

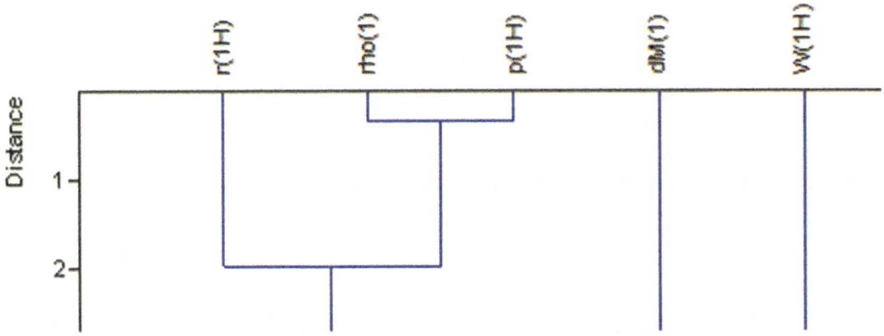

Fig. 1. Cluster analysis for parameters of comparisons between 1st and 2nd course

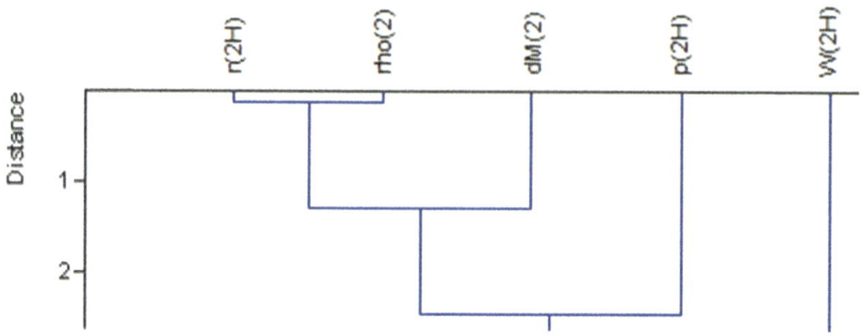

Fig. 2. Cluster analysis for parameters of comparisons between 2nd and 3rd course

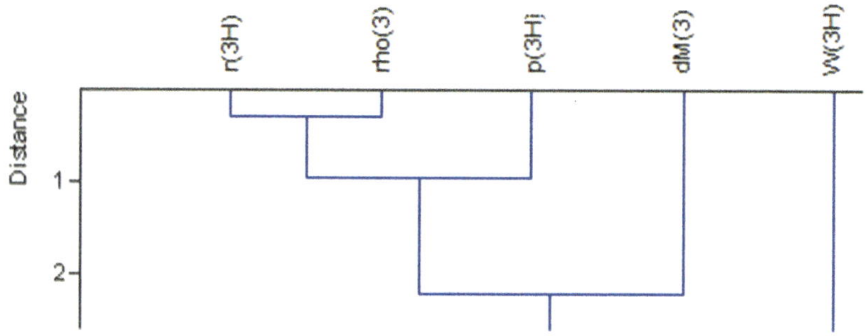

Fig. 3. Cluster analysis for parameters of comparisons between 3rd and 4th course

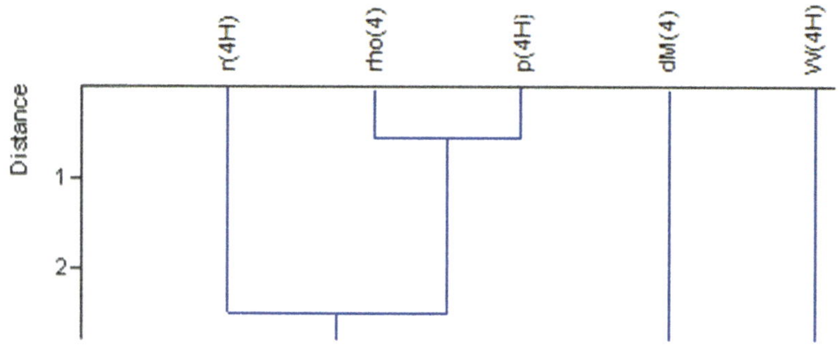

Fig. 4. Cluster analysis for parameters of comparisons between 4th and 5th course

5 Conclusion

In the practical realized quantitative research, the potential similarities across the behaviour of the different types of the statistical parameters were analysed with regards to the changing sample size. Two different approaches of the mathematical induction and of the descriptive statistics were compared in the table form and using the cluster analysis. In case of the mathematical induction, p value, the test criterion and effect size were monitored in the frame of the testing the hypotheses. For the descriptive statistics, the difference of medians and the Pearson correlation coefficient were computed. Similar situations were identified in the same cases based on type of conclusions about the testing the hypotheses. Using the cluster analysis, the comparisons were provided for the hypotheses of the realized practical research in the field of the visual arts education. In this paper, the motivation for the realization of the quantitative research and its analysis in the field of the visual arts is appeared.

References

1. Knight, L., Lasczik Cutcher, A. (eds.) Arts-Research-Education. Springer (2018)
2. Kitchenham, B., Madeyski, L., Budgen, D., et al.: Robust Statistical Methods for Empirical Software Engineering. Empirical Software Engineering, pp. 1–52. Springer (2016)
3. Gauthier, T.D., Hawley, M.E.: Statistical Methods. In: Introduction to Environmental Forensics: Third Edition, 99–148. Elsevier (2015). https://doi.org/10.1016/b978–0-12-404696-2.00005-9
4. Tomczak, M., Tomczak, E.: The need to report effect size estimates revisited. An overview of some recommended measures of effect size. TRENDS Sport Sci. **1**(21), 19–25. TRENDS in Sport Sciences (2014). ISSN 2299-9590
5. Bijnen, E.J.: Cluster analysis: Survey and Evaluation of Techniques. Springer (1973)
6. Lackova, L.: Protective factors of university students. New Educ. Rev. **38**(4), 273–286 (2014). ISSN 1732–6729
7. Sikorova, Z., Barot, T., Vaclavik, M., et al.: Czech university students‘ use of study resourcesin relation to the approaches to learning. New Educ. Rev. **56**(2), 114–123 (2019). ISSN 1732-6729
8. Korenova, L.: GeoGebra in teaching of primary school mathematics. Int. J. Technol. Math. Educ. **24**(3), 155–160 (2017)
9. Barot, T., Krpec, R., Kubalcik, M.: Applied Quadratic Programming with Principles of Statistical Paired Tests. Computational Statistics and Mathematical Modeling Methods in Intelligent Systems, Advances in Intelligent Systems and Computing, vol. 1047, pp. 278–287. Springer (2019). ISBN 978-3-030-31361-6. https://doi.org/10.1007/978-3-030-31362-3_27
10. Svoboda, Z., Honzikova, L., Janura, M., et.al.: Kinematic gait analysis in children with valgus deformity of the hindfoot. Acta Bioeng. Biomech. **16**(3), 89–93. Institute of Machine Design and Operation (2014). ISSN 1509-409X
11. Jandacka, D., Silvernail, J.F., Uchytil, J., et al.: Do athletes alter their running mechanics after an Achilles tendon rupture? **10**(1). J. Foot Ankle Res. (2017). https://doi.org/10.1186/s13047-017-0235-0
12. Bendickova, K., Tidu, F., De Zuani, M., et al.: Calcineurin inhibitors reduce NFAT-dependent expression of antifungal pentraxin-3 by human monocytes. J. Leukocyte Biol. **107**(3), 497–508. Wiley-Blackwell (2020). https://doi.org/10.1002/jlb.4vma0318-138r
13. Barot, T., Burgsteiner, H., Kolleritsch, W.: Comparison of Discrete Autocorrelation Functions with Regards to Statistical Significance. Advances in Intelligent Systems and Computing. Springer. ISSN 2194-5357 (in Print)

Impact of Mobility on Performance of Distributed Max/Min-Consensus Algorithms

Martin Kenyeres[1(✉)] and Jozef Kenyeres[2]

[1] Institute of Informatics, Slovak Academy of Sciences, Dubravska cesta 9, 845 07 Bratislava, Slovakia
martin.kenyeres@savba.sk
[2] Sipwise GmbH, Europaring F15, 2345 Brunn am Gebirge, Austria
jkenyeres@sipwise.com
http://www.ui.sav.sk/w/en/dep/mcdp/

Abstract. Distributed consensus-based algorithms are an important part of many multi-agent systems as they ensure agreement among the entities of these systems without any centralization. In this paper, we address the max-consensus algorithm and the min-consensus algorithm over mobile systems represented as stationary Markovian evolving graphs. We compare their performance over static networks to the performance over mobile systems with a varied number of mobile entities in order to identify the impact of the mobility on the algorithm rate and the estimation precision of the mentioned algorithms.

Keywords: Distributed computing · Consensus algorithms · Data aggregation · Max-consensus algorithm · Min-consensus algorithm · Mobility

1 Introduction

Over the past decades, significant progress in multiple areas including communicating, computing, sensing, and actuation has allowed the application of large-scale multi-agent systems in various fields [1]. An important term in multi-agent coordination is consensus, which means that all the entities in a multi-agent system achieve an agreement on certain quantities by communicating with each other via a networked system [2,3]. Thus, the consensus algorithms enable the autonomously operating entities in multi-agent systems to coordinate themself and to achieve the consensus without the need to transmit enormous data amount to a fusion center [4]. These algorithms are bio-inspired approaches, appearing in various forms in nature, e.g., swarm behavior (birds, ants, fish, bees, etc.), physical and chemical processes, etc. [5,6]. The principle of these algorithms lies in iterative exchanges of information among the neighbors in

R. Silhavy et al. (Eds.): CoMeSySo 2020, AISC 1295, pp. 304–313, 2020.
https://doi.org/10.1007/978-3-030-63319-6_26

multi-agent systems and updating the inner states by applying local rules [7,8]. The algorithms do not require the presence of any fusion center for collecting data, resulting in numerous benefits such as enhanced robustness, easy scalability, absence of routing, etc. [9]. As shown in [10], the consensus algorithms have found applicability in various areas over the last years, e.g., cooperative control of multivehicle networks, flocking problems, distributed control of robotic systems, control of sensor systems, etc.

Recently, mobile multi-agents have gained significant attention since the mobility results in numerous advantages, e.g., entities can move with a tracked target, redeployment is simplified, energy consumption is significantly optimized, the capacity of the transmission channels is increased, etc. [11]. Therefore, mobile multi-agent systems are applied in numerous areas, e.g., medical treatment, navigation systems, natural monitoring, etc. [11].

This paper is concerned with the max-consensus algorithm and the min-consensus algorithm, which are proposed for extrema finding over multi-agent systems (more specifically, for finding the maximum and the minimum from the initial inner states). We analyze the impact of the mobility on the performance of these algorithms, i.e., we compare their rate and the estimation precision over static networks with the performance over networks with a varied number of moving entities. Our goal is to identify whether the mobility of the entities results in a performance enhancement and how the number of mobile entities affects the algorithm rate and the estimation precision.

In Sect. 2, we provide theoretical insight into the topic, i.e., we focus on how to model mobile multi-agent systems and the analyzed algorithms. The next section is concerned with the applied research methodology and the used metrics for performance evaluation. Section 4 consists of the experimentally obtained results in Matlab2018b and a consecutive discussion about observed phenomena. The last section briefly summarizes our contribution.

2 Theoretical Background

2.1 Model of Mobile Multi-agent Systems

We model a mobile-multi agent system as the initial graph G_0 and a stationary Markovian evolving graph (SMEG) both determined by the vertex set $\mathbf{V}$ and the edge set $\mathbf{E}$ by the definition [9]. Two arbitrary vertices ($v_i \in \mathbf{V}$ and $v_j \in \mathbf{V}$) are adjacent to one another in the case of being linked by an edge (labeled as $e_{ij} \in \mathbf{E}$) [9]. Subsequently, the set of all the neighbors of an arbitrary vertex v_i can be determined as $\mathcal{N}_i = \{v_j : e_{ij} \in \mathbf{E}\}$ [9]. In general, evolving graphs do not capture how graph topologies evolve but are a product of random processes, however, SMEGs can capture the underlying process if this is a Markov chain [12]. In this paper, a mobile multi-agent system are represented by the initial graph G_0 with the stationary distribution of the corresponding Markov chain and a sequence of random graphs G_1, G_2, ... with the Markov property, i.e., the current, the past, and the future graphs are independent of each other [12].

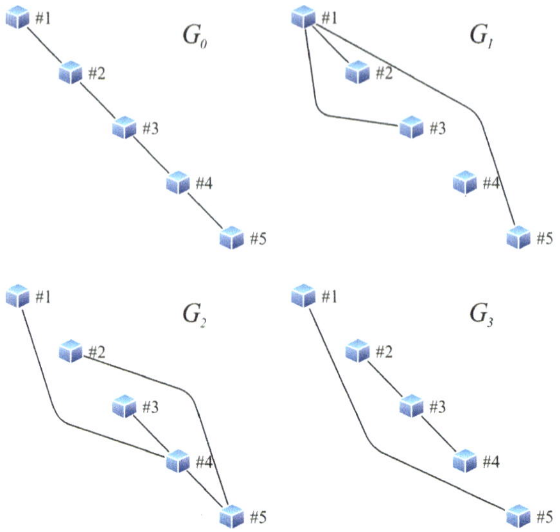

Fig. 1. Example of Stationary Markovian evolving graph (including initial graph) with order $n = 5$

Thus, SMEG is a Markov chain with state-space $\mathbf{G}^1$, and so, $\mathbf{M} = \{G_k : k \in \mathrm{N}\}$ [13]. In our analyses, the existence of an edge at the kth iteration is conditioned only by the probability p^2 and is not affected by its existence/nonexistence at the previous and the next iterations [12]. See Fig. 1 for an example of SMEG (including the initial graph G_0) with five vertices.

2.2 Max/Min-Consensus Algorithm

In this paper, we address the max-consensus and the min-consensus algorithm, which are synchronous distributed algorithms for finding the maximum/the minimum from the inner states of all the entities in a multi-agent system [9]. Each entity has its own inner state $(x_i(k))^3$, which is a scalar quantity initiated by for example local measurement [9]. The principle of these algorithms lies in iterative exchanges of the current inner states among the adjacent neighbors and updating the inner state for the next iteration in a distributed fashion by applying the following update rules ((1) is the update rule of the max-consensus algorithm, and (2) of the min-consensus algorithm) [14]:

$$x_i(k+1) = \max(x_i(k), \max_{j \in \mathcal{N}_i}(x_j(k)). \tag{1}$$

$$x_i(k+1) = \min(x_i(k), \min_{j \in \mathcal{N}_i}(x_j(k)). \tag{2}$$

[1] $\mathbf{G}$ is a set of graphs with the same order n.
[2] Determines the probability of the existence of each edge at an iteration.
[3] $x_i(k = 0)$ represents the initial inner states of v_i.

Thus, each entity communicates only with its neighbors in order to collect their current inner state. From the received states from neighbors and the current inner state, the entities find the maximum/the minimum and set this value as their inner state for the next iteration [14]. The inner states are iteratively updated until consensus on the maximum/the minimum from all the initial states is achieved, meaning the inner state of each entity in a multi-agent system is equal to the maximum/the minimum. As stated in [9], the algorithm rate over static systems can be analytically bounded by the graph diameter D, i.e.:

$$D = \max_{i,j}(d(v_i, v_j)) \geqslant \max_{i}(d(v^{ext}, v_i)). \tag{3}$$

Here, d poses the graph distance between two vertices, i.e. the shortest path between these two vertices, and v^{ext} is the vertex whose initial inner state is the wanted extreme (the right expression in (3) poses the rate over static graphs) [9]. Thus, the number of the iterations necessary for all the entities in a multi-agent system to achieve the consensus on the maximum/the minimum over unchanging graphs can be bounded by the graph diameter D, which is the longest shortest path in a graph [9]. In Fig. 2, we show an example of the max-consensus algorithm over a graph with the line topology formed by four vertices. The principle of the min-consensus algorithm is the same with the difference that the minimum is found instead of the maximum.

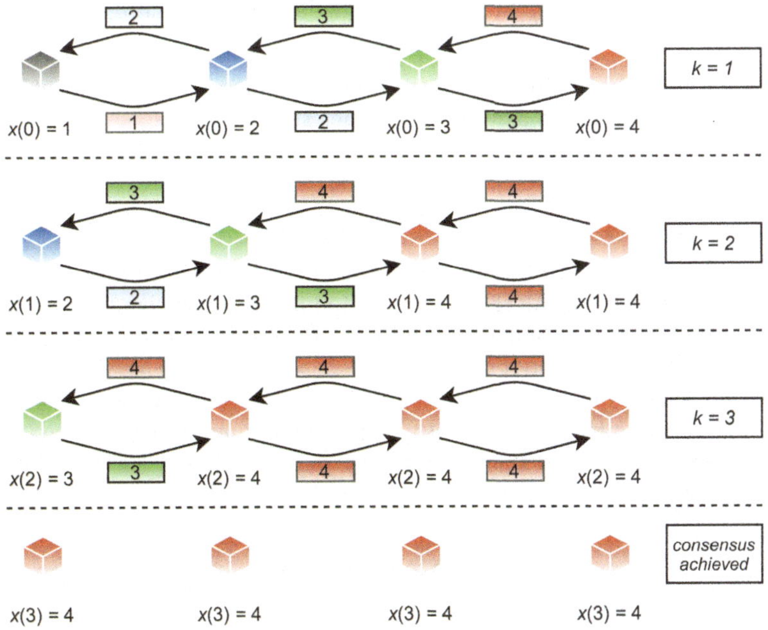

Fig. 2. Execution of max-consensus algorithm over graph of line topology with $n = 4$ - consensus is achieved after three iterations

3 Applied Research Methodology

As mentioned above, we model the mobility of multi-agent systems by applying SMEGs formed by 200 vertices (i.e., $n = 200$) and with a varied probability p in order to generate graphs of various connectivity. For the sake of high research credibility, 1000 unique SMEGs are generated for each p, which takes these following values in our analyses:

- ☐ $p = 2\%$
- ☐ $p = 3\%$
- ☐ $p = 4\%$
- ☐ $p = 5\%$
- ☐ $p = 20\%$ (chosen in order to analyze the algorithm also over densely connected graphs)

Since our goal is to analyze the impact of the mobility on the performance on the examined algorithms, we vary the number of mobile entities (mobile nodes are selected randomly) and carry out these five scenarios:

- ☐ *Scenario 1* (S1) – all the entities are static
- ☐ *Scenario 2* (S2) – 10% of the entities are mobile, i.e., 20 entities can move
- ☐ *Scenario 3* (S3) – 40% of the entities are mobile, i.e., 80 entities can move
- ☐ *Scenario 4* (S4) – 70% of the entities are mobile, i.e., 140 entities can move
- ☐ *Scenario 5* (S5) – all the entities are mobile, i.e., 200 entities can move

In our experiments, the entities in a multi-agent system are allocated random independent and identically distributed random values of the standard Gaussian distribution, which are set as the initial inner state, i.e.:

$$x_i(0) \sim N(0,1),\ for\,\forall v_i \in \mathrm{V}. \tag{4}$$

Furthermore, we apply two metrics for performance evaluation:

- ☐ *Algorithm rate* – expressed as the number of the iteration for the consensus achievement among all the entities in a multi-agent system. The algorithms are stopped at the first iteration when the inner states of all the entities are equal to the wanted extreme.
- ☐ *Mean Square Error over iterations* – is a reasonable metric applied also in other areas, labeled as MSE(k), applied for quantifying the deviation of the inner states from the estimated extreme at the {1st, 2nd, 3rd, 4th, 5th} iteration, and defined as follows [15,16]:

$$\mathrm{MSE}(k) = \frac{1}{n} \cdot \sum_{i=1}^{n} \left(x_i(k) - \mathbf{1}^{\mathrm{T}} \times \frac{\mathbf{x}(0)}{n} \right)^2. \tag{5}$$

Here, $\mathbf{1}$ is an all-ones vector, $\mathbf{1}^{\mathrm{T}}$ is its transposed variant, and $\mathbf{x}(0)$ is a vector gathering all the initial inner states. We show and furthermore analyze the average of both metrics computed from their values obtained over 1000 SMEGs (the average for various p/scenarios is depicted and analyzed separately).

4 Experimental Part

In this section, we provide and discuss the numerically obtained results in Matlab2018b.

In the first experiment, we analyze the rates of the algorithms expressed as the number of the iterations for the consensus achievement (Fig. 3, Fig. 4). From the results shown in Fig. 3 (the max-consensus algorithm), it can be seen that an increase in p results in a decrease in the number of the iterations necessary for the consensus achievement in each analyzed scenario; therefore, the algorithm rate is increased as connectivity increases. Also, we can see that the lowest rate is achieved when the algorithm is executed over static graphs, and an increase in the number of the mobile entities ensures a greater rate, i.e., the highest rate is achieved in Scenario 5, when all the entities are mobile. The only exception is Scenario 2 (i.e., only 10% of the entities are mobile during the algorithm execution), where the algorithm achieves a lower rate for p = 2% and p = 3%

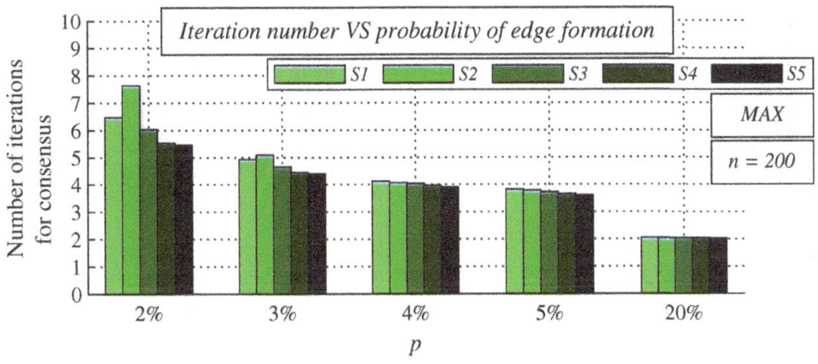

Fig. 3. Algorithm rate expressed as number of iterations for consensus achievement in analyzed scenarios and for varied p – max-consensus algorithm

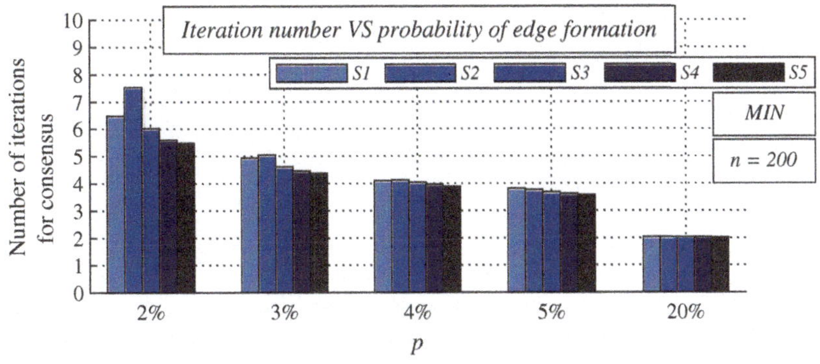

Fig. 4. Algorithm rate expressed as number of iterations for consensus achievement in analyzed scenarios and for varied p – min-consensus algorithm

(i.e., in the graphs with low connectivity) than in the static graphs. With an increase in connectivity, the difference in the rates among the analyzed scenarios decreases and is almost negligible for $p = 20\%$. In Fig. 4, the same experiment is repeated but now with the min-consensus algorithm. It can be observed from the results that the course of the depicted functions and their numerical values are very similar to those shown in Fig. 3. The major difference is only that the rate in Scenario 2 is lower than in the static graphs also for $p = 4\%$.

The next experiment is concerned with the estimation precision quantified by MSE. In Fig. 5, the max-consensus algorithm is examined in the same scenarios as in the previous analysis. From the results, we can see that an increase in the iteration number results in a decrease in MSE for each p and in each scenario. Also, an increase in graph connectivity and an increase in the number of the mobile entities ensure a decrease in MSE even though the number of mobile

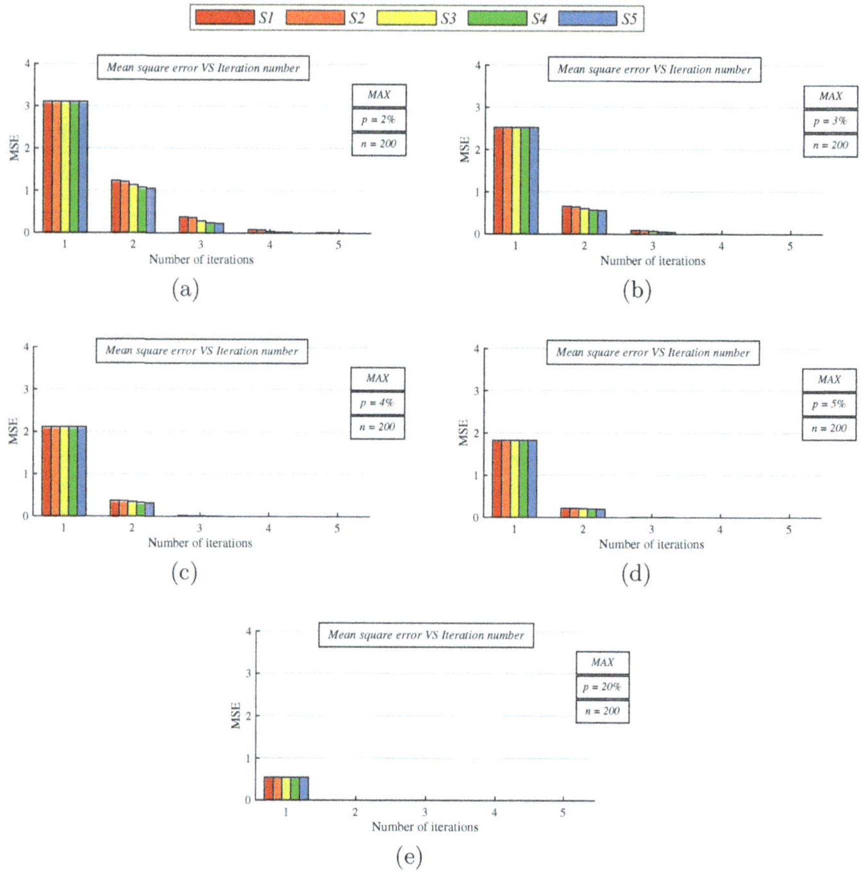

Fig. 5. Estimation precision quantified by mean square error over iterations in analyzed scenarios and for varied p – max-consensus algorithm

entities has only a marginal impact on MSE. The only exception can be seen in the case of Scenario 2, which is slightly outperformed by Scenario 1 at later iterations. Furthermore, the difference in MSE between the analyzed scenarios decreases as the iteration number and p is increased. As seen in each scenario and for each p, MSE approaches zero (as the iteration number increases) and finally takes this value, and thus, the algorithm is able to determine the exact value of the estimated aggregate function in contrast to numerous consensus-based algorithms that can determine only an estimate of an aggregate function [11,17,18]. In Fig. 6, MSE of the min-consensus algorithm is analyzed. Like in the previous analysis, an increase in connectivity, in the iteration number, and in the number of the mobile entities results in lower MSE. Again, MSE in Scenario 2 is a bit greater than in Scenario 1 at later iterations. In general, MSE is a bit smaller than MSE of the max-consensus algorithm at each iteration, for each value of p, and in each scenario.

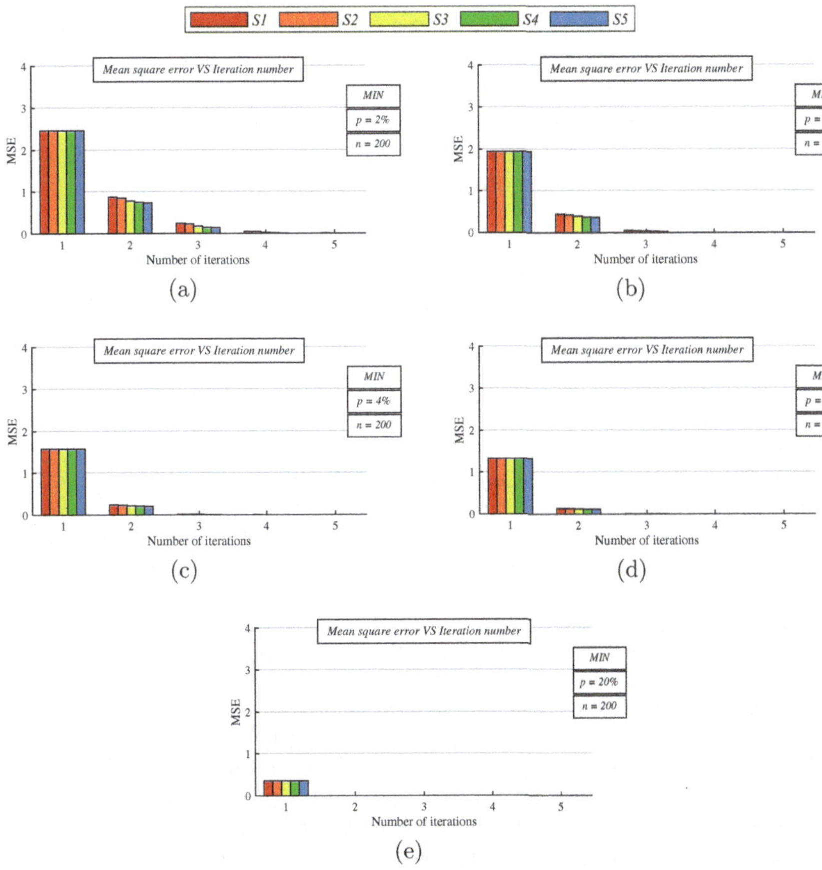

Fig. 6. Estimation precision quantified by mean square error over iterations in analyzed scenarios and for varied p – min-consensus algorithm

5 Conclusion

In this paper, we address the max-consensus algorithm and the min-consensus algorithm for extrema finding over mobile systems represented as the initial graphs of the stationary distribution of the corresponding Markov chain and SMEGs with various connectivity. We compare the algorithms over static graphs to their performance over mobile systems with a varied number of mobile entities by applying two metrics, namely MSE(k) and the algorithm rate expressed as the number of the iterations for the consensus achievement. From the results, it can be seen that an increase in connectivity and in the number of the mobile entities ensure that the algorithm rate is higher and MSE is lower. The only exception is Scenario 2 (i.e., only 10% of the entities are mobile), when the algorithm rate is lower when connectivity of the graphs is low and MSE is higher at later iterations than in static graphs. Also, we can see that an increase in the number of iterations results in a decrease in MSE regardless of the probability p, the analyzed scenario, and the found extreme. In contrast to many consensus-based algorithms, the analyzed algorithms are able to determine the exact value of the estimated aggregate function. Moreover, it is observed that there is no significant difference between the performance of the max-consensus algorithm and the performance of the min-consensus algorithm over both static and mobile systems. Thus, our contribution identifies that the mobility, in general, ensures an increase in the performance of the analyzed algorithms.

Acknowledgment. This work was supported by the VEGA agency under the contract No. 2/0155/19 and by COST: Improving Applicability of Nature-Inspired Optimisation by Joining Theory and Practice (ImAppNIO) CA 15140. Since 2019, Martin Kenyeres has been a holder of the Stefan Schwarz Supporting Fund.

References

1. Li, Z., Duan, Z.: Cooperative Control of Multi-Agent Systems: A Consensus Region Approach. CRC Press, Boca Raton (2017). https://doi.org/10.1201/b17571
2. Wang, Z., Xu, J., Song, X., Zhang, H.: Consensus conditions for multi-agent systems under delayed information. IEEE Trans. Circuits Syst. II Exp. Briefs **65**, 1773–1777 (2018). https://doi.org/10.1109/TCSII.2017.2788564
3. Zhu, W., Jiang, Z.P., Feng, G.: Event-based consensus of multi-agent systems with general linear models. Automatica **50**, 552–558 (2014). https://doi.org/10.1016/j.automatica.2013.11.023
4. Sardellitti, S., Giona, M., Barbarossa, S.: Fast distributed average consensus algorithms based on advection-diffusion processes. IEEE Trans. Signal Process. **58**, 826–842 (2010). https://doi.org/10.1109/TSP.2009.2032030
5. Tang, H., Yu, F.R., Huang, M., Li, Z.: Distributed consensus-based security mechanisms in cognitive radio mobile ad hoc networks. IET Commun. **6**, 974–983 (2012). https://doi.org/10.1049/iet-com.2010.0553
6. Barbarossa, S., Scutari, G.: Bio-inspired sensor network design. IEEE Signal Process. Mag. **24**, 26–35 (2007). https://doi.org/10.1109/MSP.2007.361599

7. Merezeanu, D., Nicolae, M.: Consensus control of discrete-time multi-agent systems. U. Politeh. Buch. Ser. A **79**, 167–174 (2017)
8. Stamatescu, G., Stamatescu, I., Popescu, D.: Consensus-based data aggregation for wireless sensor network. Control Eng. App. Inf. **19**, 43–50 (2017)
9. Kenyeres, M., Kenyeres, J.: Synchronous distributed consensus algorithms for extrema finding with imperfect communication. In: IEEE 18th World Symposium on Applied Machine Intelligence and Informatics (SAMI 2020), pp. 157–164. IEEE Press, New York (2020)
10. Amelina, N., Granichin, O., Granichina, O., Ivanskiy, Y., Jiang, Y.: Adjustment of consensus protocol step-size in a network system with different task priorities via SPSA-like algorithm under the cost constraints. IFAC-PapersOnLine **51**, 100–105 (2018). https://doi.org/10.1016/j.ifacol.2018.12.018
11. Kenyeres, M., Kenyeres, J.: Performance analysis of generalized metropolis-hastings algorithm over mobile wireless sensor networks. In: 30th International Conference on Cybernetics and Informatics (K and I 2020), pp. 1–6. IEEE Press, New York (2020)
12. Avin, C., Koucky, M., Lotker, Z.: How to explore a fast-changing world (cover time of a simple random walk on evolving graphs). In: 35th International Colloquium on Automata, Languages and Programming (ICALP 2008), pp. 121–132. Springer, Heidelberg (2008). https://doi.org/10.1007/978-3-540-70575-8_11
13. Clementi, A., Monti, A., Pasquale, F., Silvestri, R.: Information spreading in stationary Markovian evolving graphs. IEEE Trans. Parallel Distrib. Syst. **22**, 1425–1432 (2011). https://doi.org/10.1109/TPDS.2011.33
14. Muniraju, G., Tepedelenlioglu, C., Spanias, A.: Analysis and design of robust max consensus for wireless sensor networks. IEEE Trans. Signal Inf. Process. Netw. **5**, 779–791 (2019). https://doi.org/10.1109/TSIPN.2019.2945639
15. Pereira, S.S., Pages-Zamora, A.: Mean square convergence of consensus algorithms in random WSNs. IEEE Trans. Signal Process. **58**, 2866–2874 (2010). https://doi.org/10.1109/TSP.2010.2043140
16. Skorpil, V., Stastny, J.: Back-propagation and k-means algorithms comparison. In: 8th International Conference on Signal Processing (ICSP 2006), pp. 1871–1874. IEEE Press, New York (2006). https://doi.org/10.1109/ICOSP.2006.345838
17. Kenyeres, M., Kenyeres, J.: Average consensus over mobile wireless sensor networks: weight matrix guaranteeing convergence without reconfiguration of edge weights. Sensors **20**, 3677 (2020). https://doi.org/10.3390/s20133677
18. Schwarz, V., Matz, G.: On the performance of average consensus in mobile wireless sensor networks. In: IEEE 14th Workshop on Signal Processing Advances in Wireless Communications (SPAWC 2013), pp. 175–179. IEEE Press, New York (2013)

An Integrated Approach to Assessing the Risk of Malignant Neoplasms for Adults

Natalia V. Efimova(✉)

East-Siberian Institute of Medical and Ecological Research, Angarsk, Russia
medecolab@inbox.ru

Abstract. The incidence and mortality from oncological pathology of the population of Siberia is characterized by a more pronounced increase than in other regions of Russia. The purpose of the study is to give an epidemiological assessment of the realized carcinogenic risk for workers exposed to carcinogens in everyday life and at work. Estimation of the incidence and mortality is given taking into account the place of permanent residence and the main place of work based on the carcinogenic register of the Oncologycal Clinic. An analysis of the epidemiological (realized) risk of oncological pathology was carried out according to relative risk indicators (RR). The influence of the studied factors on the risk of developing of malignant neoplasms was determined using the Bayes method, evaluating the information content of each of the factors. The carcinogenic hazardous workers are characterized by high risks of morbidity of malignant neoplasms: according to the sum of localizations - RR = 6.6, malignant respiratory organs RR = 10.1, digestion RR = 7.8, genital organs RR = 5.7, urinary tract RR = 5, 4, lymphoid and hematopoietic tissues RR = 5.1 and melanoma RR = 4.2 times. Tumors of these localizations have a large production condition (76.2–90.1%). The data obtained determine the further direction of the activities of interested bodies and institutions within their competencies in monitoring carcinogenic risk for the population and reducing the level of exposure of substances with a blastomogenic effect to risk groups.

Keywords: Complex assessment · Bayes method · Malignant neoplasms · Risk · Carcinogenic factors

1 Introduction

Malignant neoplasms (MN) is one of the leading places among all causes of death and the burden of disease [1]. Among the leading causes of MN include not only genetic factors [2, 3], exposure to chemical carcinogens by inhalation and alimentary routes [4], but also working conditions [5–8]. At the same time, there are very few comprehensive works that allow not only to evaluate the level of health losses associated with malignant neoplasms, but also to identify blastomogenic factors. Modern machine data processing methods are a powerful and flexible tool for analyzing large information arrays and research results in the era of precision medicine [9, 10]. An important problem remains the justification of approaches to the examination of the connection between MN and the impact of environmental factors [6].

R. Silhavy et al. (Eds.): CoMeSySo 2020, AISC 1295, pp. 314–321, 2020.
https://doi.org/10.1007/978-3-030-63319-6_27

The long-term dynamics of oncological pathology among the population of Siberia is characterized by a more pronounced increase in mortality from MN, total oncological morbidity, hematopoiesis and the genitourinary system; endocrine system than in other regions of Russia [11, 12]. This determined the interest in studying the problem of oncological pathology in the cities of Siberia.

Our studies have shown that in the industrial and administrative center, a multi-medium carcinogenic risk for the population is unacceptable for the population. The share contribution to the total multicarcinogenic carcinogenic risk of substances delivered by inhalation was 67.1%, and alimentary - 32.9%. Priority carcinogens entering the environment of the city: in the air - formaldehyde, benz (a) pyrene; in drinking water - arsenic, cadmium and lead; in food products - cadmium, arsenic and lead [8]. Most often, malignant neoplasms of the lungs (33.3 cases per 100 thousand of the population) and stomach (28.1 per 100 thousand of the population) are recorded. The employees of the main and auxiliary professions of carcinogenic hazardous enterprises have carcinogenic risk indicators that are ten times higher than for the population of the city, even with less than 5–10 years of experience. ICR for workers is associated with exposure to chromium VI, nickel, formaldehyde, benzene [13]. Despite the high level of ICR, production control does not fully monitor carcinogens.

The purpose of the study was to provide an epidemiological assessment and identify carcinogenic risk factors for an adult urban population based on an integrated approach.

2 Materials and Methods

Estimation of the incidence of MN among the population is given taking into account the place of permanent residence and the main place of work based on the carcinogenic register of the Oncologycal Clinic. The code for malignant neoplasms is used according to the International Classification of Diseases 10. A total of 8206 units of individual information on patients and 7412 on deaths from MN from 2006 to 2016 were analyzed. The data on the place of residence, work, habitual household intoxications (tobacco smoking, alcohol consumption) were taken into account. In order to identify the peculiarities of the prevalence of malignant neoplasms among the population of the industrial and administrative center of Siberia, which is exposed to multiple environmental effects in industrial and domestic conditions, an in-depth analysis of the health losses of wagon repair plant workers for 2006–2016 was carried out.

A comparison of intensive indicators (morbidity and mortality) was carried out according to the Fisher criterion, extensive (structure) - according to the χ^2 criterion, the level $p < 0.05$ was considered significant. An analysis of the epidemiological (realized) risk of MN was carried out according to relative risk indicators with the calculation of a 95% confidence interval. To prove the connection with the industrial exposure to carcinogens, the relative risk was calculated in comparison with the population groups of the region, whose residents are not predominantly employed in carcinogenic hazardous enterprises. A value with a lower limit of > 1 was taken as a statistically significant risk. The indicator of the "etiological share" was calculated, to identify the relationship of the frequency of oncological pathology with the influence of the production factor. The share of the factor > 50% was considered a high level of significance, with RR $\geq$ 2 [14].

The role of factors in the formation of MN was studied using the Bayes method. We analyzed individual information about patients according to the materials of the oncological clinic (n = 8206). The study included data from 367 people with a confirmed diagnosis and 399 people with no MN. The main risk factors were evaluated according to the dichotomous option (yes/no). Factors included: individual carcinogenic risk, smoking, alcohol abuse, a history of etiologically significant infectious diseases (viral hepatitis B and C, human papillomavirus, tuberculosis, HIV infection), type 2 diabetes mellitus. The factor of hereditary predisposition to MN was taken into account at the established diagnosis in parents, children, siblings. Individual carcinogenic risk reflected the total pollution of atmospheric air (ICRa) and air of the working area (ICRw), which is presented in detail in our works [4, 13]. For a comparative assessment of the share contribution of individual factors to the general level of their significance, the χ^2 criterion was used, the differences were considered statistically significant at $p < 0.05$.

3 Results

The average mortality rate for men is 1.5 times higher than for women, but the differences are not statistically significant (p = 0,308) (Table 1).

Table 1. The average mortality and incidence of malignant neoplasms of adults (18–60 years).

Average indicator in 2006–2016	Gender groups	per 100 000 (CI)	p =
Mortality	Male	0.014(0.009–0.020)	0,308
	Female	0.009(0.006–0.014)	
Incidence	Male	136.0(115.6–156.8)	0,000
	Female	268.7(240.1–297.3)	

Another important indicator of health loss is the newly diagnosed incidence of MN. The incidence of men is less than that of women. In the structure of morbidity, diseases of the digestive system (22.9%), genital organs (18.8%), respiratory system (11.6%), urinary tract (8.5%), skin melanoma (5.4%) predominate. cancer of lymphoid and hematopoietic tissue - 3.6%. It should be noted that the structure does not significantly differ from that in the subpopulation of residents of the background region (χ^2 = 6.9 with a critical value of χ^2 = 7.815, p = 0.068).

Epidemiological risks for employees of the enterprise are 2–6.3 times higher than for the population unexposed in production, which we consider to be the "background" level of mortality from MN. For workers exposed to a "double" carcinogenic exposure, the risk of developing malignant tumors in general (RR = 4.6) and including localizations such as the respiratory system (RR = 4.8), digestion (RR = 4.6), and genital (RR) is statistically significant = 3.9) and urinary (RR = 6.3) organs (Table 2).

Table 2. Relative risk of mortality and incidence of malignant neoplasms among employees of a carcinogenic hazardous enterprise.

Indicator	Total	Tumor localization					
		C16–C26	C30–C39	C56–59* C60–C63**	C64–68	C69–72	C81–96
Mortality							
RR CI	4,63, 3-6,4	4,6 2,8-7,7	4,8 2,3-10,0	3,9 2,9-5,0	6,3 4,2-9,1	5,5 0,6-60,6	2,0 0,9-4,3
EF, %	78,3	78,3	79,2	74,4	84,1	81,8	50,0
Incidence							
RR CI	6,6 6,1-7,2	7,8 6,5-9,3	10,1 7,7-13,4	5,7 4,7-6,9	5,4 4,1-7,1	2,0 0,8-5,3	5,1 3,3-7,8
EF, %	84,8	87,2	90,1	82,5	81,5	50,0	80,4

Note * - MN of female genital organs; ** MN of the male genital organs; C16–C26- Digestive organs; C30–C39- Respiratory organs; C56–59, C60–C63 - Genital organs; C64–68 - Urinary tract; C69–72 - eye, brain and other organs of the central nervous system; C81–96 – Lymphoid and hematopoietic tissues.

The etiological share of additional cases of MN was in the range of 50–84.1%, which indicates a high association of the indicator with the production of carcinogens. Given the gender differences in mortality rates from MN, we consider the relative risks by sex. For men, the risks of death from tumors of the kidneys and urinary tract (RR = 4 CI (2.5–6.4), and genital organs are statistically confirmed (RR = 3.1 CI (2.3–5.4), respiratory (RR = 2.9 CI (1.3–6.6) and digestive organs (RR = 2.6 CI (1.4–5). that the risks for women exposed to carcinogens under production conditions are significantly higher than for men, so in general, the risk of mortality in women is 13.3 CI (11.2–15.2), and in men 2.7 CI (1.8–4.3). The highest mortality risks are typical for such localization of MN as the urinary tract, especially for women (RR = 12.7 CI (6.2–26.1)), whose rates are 3.2 times higher than in men.

It has been established that in terms of total MH, the risk indicators for women are higher than for men by 4.9 times, for the individual groups under consideration, the risks for women are higher than for men. In addition, it was found that for women workers, the risk of MN of lymphoid and hematopoietic tissues is statistically significant (RR = 6.8 CI (2.6–17.9)), and for men it can only be regarded as a trend (RR = 1.4 CI (0.6–6.5)), and requires further research. It should be noted that the following is despite the fact that for melanoma of the skin and MN of the brain and other organs of the central nervous system RR = 5.5, but the lower risk limit is below unity, which indicates the absence of statistically significant differences with the background level. The incidence of MN among women working in carcinogenic hazardous enterprises is higher than among men both in the sum of all localizations ($p = 0.000$), and in the main localizations: digestive organs ($p = 0{,}000$), urination ($p = 0{,}039$), lymphoid and hematopoietic tissues ($p = 0{,}000$), skin melanoma ($p = 0{,}000$), genital organs ($p = 0{,}000$).

To identify the role of production factors in the occurrence of MN, the epidemiological risks were calculated for employees of the enterprise as a whole, as well as for men and women.

Higher risks of MN morbidity are characteristic for employees of the enterprise under consideration than for the population: in all localizations - 6.6 times, MN of respiratory organs - 10.1 times, digestion - 7.8 times, genital organs - 5.7 times, urinary tract - 5.4 times, lymphoid and hematopoietic tissues - 5.1 times and melanoma - 4.2 times. Tumors of these localizations have a large production condition (from 76.2% for melanoma, up to 90.1% for respiratory organs MN). Only the risk of the incidence of MN of the eyes, brain and other organs of the central nervous system is not confirmed statistically for employees of the enterprise relative to residents of the background region (RR = 2 CI (0.8-5.3), EF = 50%).

For men, employees of the wagon repair enterprise in all locations, the risk of MN is 4.1 CI (3.6-4.7). The rank row has the following order: MN of respiratory organs RR = 6.7, lymphoid and hematopoietic tissues - RR = 4.5, digestion - RR = 3.9 times, kidney and urinary tract - RR = 3.7, genital organs - RR = 2.7 times. For women working in the enterprise in all localizations, the risk of MN is higher than for men by 1.6 times and amounts to 6.6 CI (6.4–7.2). Localization morbidity risks also vary in level and significance. So, the first ranks have: genital MN - RR = 18.6, respiratory organs RR = 16.9, lymphoid and hematopoietic tissues - RR = 12.5. The relative risk of the incidence of MN of the digestive system is 7.8 times higher than in the population, and 5.4 times higher in the kidneys and urinary tract, and 4.2 times higher in melanoma and skin cancer. We also note that for women exposed to the complex effects of carcinogens in domestic and industrial conditions, the relative risk of breast MN also increases, both in terms of the newly detected incidence rate (RR = 19.5 CI (16.1–23.7)), so and mortality (RR = 15.2 CI (10.2–21.2)).

Using the Bayesian method, the most significant risk factors were identified: industrial carcinogens > tobacco smoking > burdened heredity > gender > alcohol abuse > living in areas with high ICRa (Table 3). The share of the studied risk factors in their total information content was: industrial carcinogens - 37.2%, smoking - 19.8%, the presence of MN in relatives - 16.6%, gender - 8.0%, alcohol abuse - 6.8%, carcinogens contained in atmospheric air - 4.5%. The history of oncogenic diseases was estimated as follows: viral hepatitis B - 3.5%, diabetes mellitus - 2.7%; tuberculosis - 0.77% of the total information content.

4 Discussion

The experience of an integrated approach to the assessment of population health loss associated with MN has shown an interesting result. At the first stage, we estimated the level of realized risk of MN for the adult working population. The carcinogenic hazardous workers are characterized by high risks of morbidity of malignant neoplasms: according to the sum of localizations - RR = 6.6, malignant respiratory organs RR = 10.1, digestion RR = 7.8, genital organs RR = 5.7, urinary tract RR = 5, 4, lymphoid and hematopoietic tissues RR = 5.1 and melanoma RR = 4.2 times. Many researchers

Table 3. Informational characteristics of risk factors for malignant neoplasms of urban adults (18–60 years).

Risk factors	Presence of factor Yes/no	Frequency ratio (group MN/comparison group)	Informative factor	
			Depending on the presence of the factor	The total
ICR w	yes	3,40	32,49	36,10
	no	0,87	3,61	
Smoking	yes	2,15	16,69	19,23
	no	0,89	2,55	
Burdened by heredity	yes	2,49	14,82	16,14
	no	0,92	1,33	
ICRa	yes	1,09	1,24	4,36
	no	0,78	3,39	
Alcohol abuse	yes	1,82	6,08	6,60
	no	0,95	0,52	
Female group	yes	1,18	3,29	7,78

have noted the lack of monitoring programs for substances with a blastomogenic effect in the atmosphere and at work for [5, 8].

At the second stage, using the Bayesian method, we estimated the significance of environmental factors for the formation of MN. Our results on the frequency of tumor localization in different gender groups do not coincide with some data [15]. Lezhnin V. L. et al. [15] showed that the risk of developing lung cancer was higher in men living in an industrial center and was directly dependent on the age, duration and intensity of smoking, as well as on the degree of alcohol abuse. It is known that the structure of localizations and the prevalence of MN have gender differences, the severity of which is heterogeneous, according to various authors [16–18]. We have identified gender differences in both the incidence rate and the mortality rate from MN. They can be associated with a different level of exposure of carcinogenic factors in the production environment to representatives of certain professions, differences in employment in carcinogenic production processes of men and women, as well as a greater sensitivity of the female body to the negative influence of blastomogenic factors [2, 6, 18].

Currently, it is universally recognized that primary prevention of malignant tumors plays a priority in the fight against cancer, the measures of which are aimed at eliminating adverse environmental factors of a person and increasing the nonspecific resistance of the body. When comparing our results with the ranking of the most significant factors for people with MN in the United States [19], some differences were revealed, primarily related to the different set of studied risk factors for MN. However, the fractional contribution of factors has practically no differences for smoking, alcohol consumption, and a history of some infectious diseases. The results presented in this study have great uncertainties, primarily due to the lack of industrial control over the content of many carcinogenic substances in the air of the working area were noted in their works by researchers [5, 8]. The uncertainty of the estimates is associated with a genetic predisposition realized against the background of the dominant role in the etiology of malignant tumors of environmental factors, the working environment, and human lifestyle [6, 20]. Observational studies according to the cancer registry provide important information, but they have some limitations [9, 21]. The heterogeneity of the data, the variety of causes and conditions of the occurrence and development of MN, the analytical complexity in interpreting the results, makes it easy to make mistakes and, according to Carmona-Bajonas A. [21], requires improving the statistical methodology.

5 Conclusion

Complex method provide a robust and flexible analytic approach to the challenge of health datasets. These multifactorial health and environmental condition datasets pose specific analytic challenges because of missing data, large size, and complexity, changing populations, and nonlinear relationships between exposures and biological effects. Using the epidemiological methods of evidence-based medicine, it was possible to assess the level of realized risk of MN for the adult working population and to prove a connection with the working environment. And Bayesian methods can generate a risk forecast at the individual level, which is relevant and relevant at the local level. The magnitude of the information content determines the factors of increased risk of MN development: work at carcinogenic enterprises; smoking; burdened heredity; gender; alcohol abuse, living under the influence of chemical carcinogens.

The data obtained determine the further direction of the activities of interested bodies and institutions within their competencies in monitoring carcinogenic risk for the population and reducing the level of exposure of substances with a blastomogenic effect to risk groups.

Acknowledgment. The work was carried out as part of the state order of VSIMEI. The author is grateful to Professor S.N. Vasiliev (Trapeznikov Institute of Control Sciences of RAS) for a consultation in mathematical and statistical data processing.

References

1. Naghavi, M., Abajobir, A.A., Abbafati, C., Abbas, K.M.: Global, regional, and national age-sex specific mortality for 264 causes of death, 1980–2016: a systematic analysis for the Global Burden of disease study 2016. Lancet **390**, 1151–1210 (2017)
2. Ito, K., Miki, Y., Suzuki, T., McNamara, K.M., Sasano, H.: In situ androgen and estrogen biosynthesis in endometrial cancer: focus on androgen actions and intratumoral production. Endocr. Relat. Cancer **23**(7), 323–335 (2016)
3. Backofen, R., Engelhardt, J., Erxleben, A., Fallmann, J.: RNA-bioinformatics: tools, services and databases for the analysis of RNA-based regulation. J. Biotechnol. **261**, 76–84 (2017)
4. Efimova, N.V., Khankharev, S.S., Motorov, V.R., Madeeva, E.V.: Assessment of the carcinogenic risk for the population of Ulan-Ude. Hyg. Sanit. **98**(1), 90–93 (2019)
5. Gurvich, V.B., Kuzmin, S.V., Kuzmina, E.A., Adrianovsky, V.I., Kochneva, N.I.: A systematic approach to the assessment and management of the carcinogenic hazard of business entities using the example of the Sverdlovsk region. Bull. Ural Med. Acad. Sci. **2** (53), 40–43 (2015)

6. Serebryakov, P.V.: Features of the examination of occupational carcinogenic risk. Hyg. Sanit. **2**, 69–72 (2015)
7. Partovi, E., Fathi, M., Assari, M.J., Esmaeili, R., Pourmohamadi, A., Rahimpour, R.: Risk assessment of occupational exposure to BTEX in the National Oil Distribution Company in Iran. Chronic Dis. J. **4**, 48–55 (2018)
8. Efimova, N.V., Rukavishnikov, V.S., Pankov, V.A., Perezhogin, A.N., Shayakhmetov, S.F., Meshchakova, N.M., Lisetskaya, L.G.: Assessment of carcinogenic risks to workers of the main enterprises of the Irkutsk region. Hyg. Sanit. **95**(12), 1163–1167 (2016)
9. Ten Haaf, K., Jeon, J., Tammemägi, M.C., Han, S.S., Kong, C.Y., Plevritis, S.K., Feuer, E. J., de Koning, H.J., Steyerberg, E.W., Meza, R.: Risk prediction models for selection of lung cancer screening candidates: a retrospective validation study. PLoS Med. **14**(4), e1002277 (2017). https://doi.org/10.1371/journal.pmed.1002277
10. Arora, P., Boyne, D., Slater, J.J., Gupta, A., Brenner, D.R., Druzdzel, M.J.: Bayesian networks for risk prediction using real-world data: a tool for precision medicine. Value Health **22**(4), 439–445 (2019). https://doi.org/10.1016/j.jval.2019.01.006
11. Pisareva, L.F., Odintsova, I.N., Ananina, O.A., Boyarkina, A.P.: Cancer incidence among population of Siberia and Russian Far East. Sib. Oncol. J. **1**, 68–75 (2015)
12. Efimova, N.V., Motorov, V.R., Mylnikova, I.V., Blokhin, A.A.: Malignant neoplasms in the Republic of Buryatia: a retrospective analysis. Hyg. Sanit. **97**(10), 881–886 (2018)
13. Efimova, N.V., Sudeikina, N.A., Motorov, V.R., Kurenkova, G.V., Lemeshevskaya, E.P.: Comparative assessment of the dynamics of individual carcinogenic risk for workers of the main professions of wagon repair production. Occup. Med. Ind. Ecol. **59**(5), 260–265 (2019)
14. Izmerov, N.F., Kasparov, A.A.: Occupational Medicine. Introduction to the Specialty. Medicine (2002)
15. Lezhnin, V.L., Kazantsev, V.S., Polzik, E.V.: Assessment of the multifactorial effect of technogenic pollution on the development of lung cancer in the population. Hyg. Sanit. **93** (3), 26–30 (2014)
16. Hohenadel, K., Raj, P., Demers, P.A., Zahm, S.H., Blair, A.: The inclusion of women in studies of occupational cancer: A review of the epidemiologic literature from 1991–2009. Am. J. Ind. Med. **58**, 276–281 (2015)
17. Scarselli, A., Corfiati, M., Di Marzio, D., Marinaccio, A., Iavicoli, S.: Gender differences in occupational exposure to carcinogens among Italian workers. BMC Public Health **18**(1), 413 (2018)
18. Global battle against cancer won't be won with treatment alone Effective prevention measures urgently needed to prevent cancer crisis. WHO. Lyon/London (2014). https://www.iarc.fr/wp-content/uploads/2018/07/pr224_E.pdf. Accessed 05 Oct 2020
19. Islami, F., Goding Sauer, A., Miller, K.D., Siegel, R.L., Fedewa, S.A., Jacobs, E.J., et al.: Proportion and number of cancer cases and deaths attributable to potentially modifiable risk factors in the United States. CA Cancer J. Clin. **68**(1), 31–54 (2018). https://doi.org/10.3322/caac.21440
20. Havet, N., Penot, A., Morelle, M., Perrier, L., Charbotel, B., Fervers, B.: Varied exposure to carcinogenic, mutagenic, and reprotoxic (CMR) chemicals in occupational settings in France. Int. Arch. Occup. Environ. Health **90**, 227–241 (2017)
21. Carmona-Bayonas, A., Jimenez-Fonseca, P., Fernández-Somoano, A., Álvarez-Manceñido, F., Castañón, E., Custodio, A., et al.: Top ten errors of statistical analysis in observational studies for cancer. Res. Clin. Transl. Oncol. **20**(8), 954–965 (2018). https://doi.org/10.1007/s12094-017-1817-9

State-of-Health Estimation of Lithium-Ion Batteries with Attention-Based Deep Learning

Shengmin Cui, Jisoo Shin, Hyehyun Woo, Seokjoon Hong, and Inwhee Joe(✉)

Department of Computer Science, Hanyang University, Seoul, South Korea
{shengmincui,sjrevers,whh1015,daniel379,iwjoe}@hanyang.ac.kr

Abstract. Lithium-ion batteries are most commonly used in electric vehicles (EVs). The battery management system (BMS) assists in utilizing the energy stored in the battery more effectively through various functions. State of health (SOH) estimation is an essential function in a BMS. The accurate estimation of SOH can be used to calculate the remaining lifetime and ensure the reliability of batteries. In this paper, we propose a data-driven deep learning method that combines Gate Recurrent Unit (GRU) and attention mechanism for SOH estimation of lithium-ion batteries. Real-life datasets of batteries from NASA are used for evaluating our proposed model. The experimental results show that the proposed deep learning model has higher accuracy than conventional data-driven models.

Keywords: Lithium-ion battery · State of health · Gated recurrent unit · Attention

1 Introduction

Lithium-ion batteries are very important in many places because lithium-ion batteries have higher energy density, lightweight, and very long charge-discharge cycle [1]. However, owing to complex physical and chemical changes during use, the performance of lithium ion batteries may deteriorate or malfunction, resulting in an accident or economic loss [2]. For this reason, a lot of research has been conducted on topics related to the battery management system (BMS). In addition, problems may arise because the expected life of lithium-ion batteries depends on the environment and the strength of use. Moreover, even in the case of a battery of the same model, the battery life may be different in actual use. Therefore, it is necessary to study the state of health (SOH) of the battery, which affects the remaining useful life of the battery. SOH is determined by measurement parameters of the lithium ion battery, such as terminal voltage, current, and battery capacity.

There are two types of approaches for SOH monitoring. They are model-based methods and data-driven methods. Model-based methods are principally

R. Silhavy et al. (Eds.): CoMeSySo 2020, AISC 1295, pp. 322–331, 2020.
https://doi.org/10.1007/978-3-030-63319-6_28

to analyze the physical and chemical theorems of the battery and to establish a mathematical and physical model to characterize the deterioration process of the lithium-ion battery.

Data-driven methods making short-term predictions on the same battery based on data. Recently, many methods of predicting the performance of a lithium-ion battery using various battery data using an artificial intelligence model have been applied.

In this paper, we propose an attention-based Gated Recurrent Unit (GRU) [3] method for monitoring the SOH of lithium-ion batteries. This method is a model that applies attention mechanism based on GRU. SOH is predicted by applying the proposed algorithm. The dataset uses NASA lithium-ion battery dataset. As a result of the experiment, it can be seen that the method using the proposed algorithm has a reduced and root mean squared error (RMSE) error rate compared to the general Recurrent Neural Network (RNN) [4] and variant methods.

The rest of this paper is organized as follows: We review related work about SOH estimation in Sect. 2. Then, in Sect. 3 we describe our proposed attention-based GRU model. Next, we present and analyze experimental results in Sect. 4. Finally, we conclude in Sect. 5.

2 Related Work

2.1 Model-Based Methods

Ashwin et al. [5] developed an electrochemical battery model. This model is developed for aging under cyclic loading conditions. This model can constitute a process for reducing the capacity of lithium-ion batteries. Pola et al. [6] estimated state of charge (SOC) and estimated the discharge time of the lithium-ion battery with particle filtering based prognosis framework that utilizes statistical characterization of using profiles.

The model-based method yielded good results in battery condition prognosis. However, there are some problems still exist. The proposed model is only applicable to specific battery materials and running environments. In addition, the model parameters depend on the physical characteristics of the battery.

The method of [6] is founded on particle filters (PF) and Kalman filters (KF). However, this method is vulnerable to noise and environmental interference, thence it is difficult to track the dynamic characteristics of the load. However, this kind of method is vulnerable to noise and environmental interference.

For this reason, considering the complexity of the physical and chemical changes in lithium-ion batteries, noise, and environmental diversity, it is very difficult to establish an accurate model for SOH estimation in lithium-ion batteries.

2.2 Data-Driven Methods

In data-driven methods, SOH monitoring and RUL prediction of lithium-ion batteries are generally materialized by analyzing previous data such as current, voltage, capacitance and impedance. Data-driven methods are faster, more convenient, and less complex when compared to model-based methods.

Weng et al. [7] developed a model parameterization and adaptation framework using a simple structure of support vector regression (SVR) representation along with the determined support vector (SV) to enable model parameters to be estimated in real time. This model is based on experimental battery aging data. Nuhic et al. [8] used SVM to estimate the SOH and RUL of lithium ion batteries and included diagnosis and prognosis of system status. However, SVM has some limitations, including the need for kernel functions to meet Mercer conditions, prone to local optimization, and difficulty in determining loss functions and penalty factors.

Recently, many researchers are trying to build models using neural networks (NNs). Ren et al. [9] proposed deep neural network model combined an auto-encoder for predict RUL of lithium-ion batteries. This model using a long short-term memory (LSTM) [10] and recurrent neural network (RNN) because of the superiority of time series prediction. Liu et al. [11] proposed an adaptive recurrent neural network for predict the system dynamic state for estimate lithium-ion battery RUL. Zhang et al. [12] proposed LSTM model for build RUL model of lithium-ion batteries. Although the above method worked well, few of them can capture temporal dependencies appropriately and did not pay attention to the difference of contribution of each time step to the final prediction.

3 Methodology

In this section, we propose our model for SOH estimation. We first introduce the basic GRU cell structure, and then explain our attention-based model in detail.

3.1 Gated Recurrent Unit

GRU is one of the variants of RNN that is a kind of neural network for learning the pattern of time series data. A GRU can be used to process a sequential values $x_1, x_2, \ldots, x_t$. The traditional RNN has been applied in many fields, but the problems of gradient vanishing and gradient explosion have affected the performance of the traditional RNN. LSTM solved this problem by adding several gating systems in the cell. Although LSTM has achieved good performance in many fields, GRU can achieve better performance on small data sets through simpler gating systems. Unlike the LSTM cell, which has input gate, forget gate, and output gate, the GRU cell only has update gate and reset gate. The update gate and reset gate both can decide to ignore parts of the previous information. However, the update gate not only has the function of determining how much past information should be forgotten, but also has the function of determining how much new information can be added to the state. The structure of GRU cell is shown in Fig. 1 and update equations are as follows:

$$u_t = \sigma(W_u \cdot [h_{t-1}, x_t]) \tag{1}$$

$$r_t = \sigma(W_r \cdot [h_{t-1}, x_t]) \tag{2}$$

$$\widetilde{h}_t = \tanh(W \cdot [r_t * h_{t-1}, x_t]) \tag{3}$$

$$h_t = (1 - u_t) * h_{t-1} + u_t * \widetilde{h}_t, \tag{4}$$

where u_t and r_t are update gate and reset gate, respectively. W_u, W_r, and W are parameters to be trained. σ and tanh are sigmoid function and tanh function, respectively. h_t and x_t are hidden state and input of time step t, respectively.

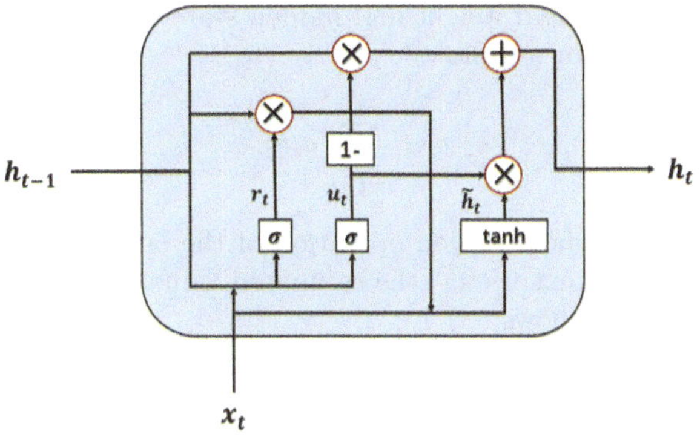

Fig. 1. Structure of an GRU cell.

3.2 Proposed Attention Model

Attention mechanism can be said to be one of the most powerful technologies in the field of deep learning nowadays. It is based on the characteristics of human attention and focuses on a certain part in processing long sequence data. It turns out that attention mechanisms are often used to deal with natural language processing problems, however, many recent work on time series has added attention mechanisms. In general, when dealing with the problem of SOH estimation, RNN or a variant of RNN will be used to solve it. However, as the time step length increases, the model needs to deal with too much hidden state information. For the prediction of the current time step, we believe that the most important information is the hidden state of the current time step. Therefore, in order to make an effective estimate, we use the attention mechanism to learn which previous hidden states are closely related to the current hidden state.

Figure 2 shows the overview structure of our proposed model. Our work is mainly inspired by the attention mechanism by Luong [13]. First, we apply a scoring function to calculate the relationship between the hidden state of each time step and the current hidden state. The scoring function is as follows:

$$score(h_t, h_i) = h_t^T W_s h_i, \tag{5}$$

where W_s is trainable parameters matrix, h_t is the current hidden state h_i is the previous hidden state.

After calculating the score, we assign weights to each hidden state to indicate its correlation with the current hidden state. The weight of hidden state h_i is calculated as follows:

$$a_i = \frac{\exp(score(h_t, h_i))}{\sum_{j=t-k}^{t} \exp(score(h_t, h_j))}, \tag{6}$$

where k is the time step length of our prediction model.

Next, use the calculated weight and hidden state at each time point to calculate the context vector as follows:

$$c = \sum_{i=t-k}^{t} a_i h_i. \tag{7}$$

Finally, after the concatenation operation of the current hidden state value and the calculated context vector, the estimated value y_t is calculated by using the linear function as follows:

$$h_t^{'} = concat(c, h_t) \tag{8}$$

$$y_t = W_o h_t^{'} + b_o \tag{9}$$

4 Experiments

In this section, we first introduce the data set, then compare the results of the conventional GRU model and the our proposed model with attention mechanism, next compare our model to general methods on each battery dataset, and finally we use B0005 and B0006 as training set and test with B0007 and compare our model to conventional methods.

4.1 Dataset

Lithium-ion battery dataset from the NASA Ames Prognostics Center of Excellence (PCoE) is used for validating our proposed model. We select B0005, B0006, and B0007 battery data sets and each battery was charged and discharged at room temperature. The detailed information of batteries are enumerated in the Table 1.

In this paper, we use root mean squared error (RMSE) as the evaluation criterion for our proposed model. The calculation of RMSE is as follows:

$$RMSE = \sqrt{\frac{1}{N}\sum_{t=1}^{N}(\hat{y}_t - y_t)^2} \tag{10}$$

where $\hat{y}_t$ and y_t are estimated and actual SOH values at time step t respectively.

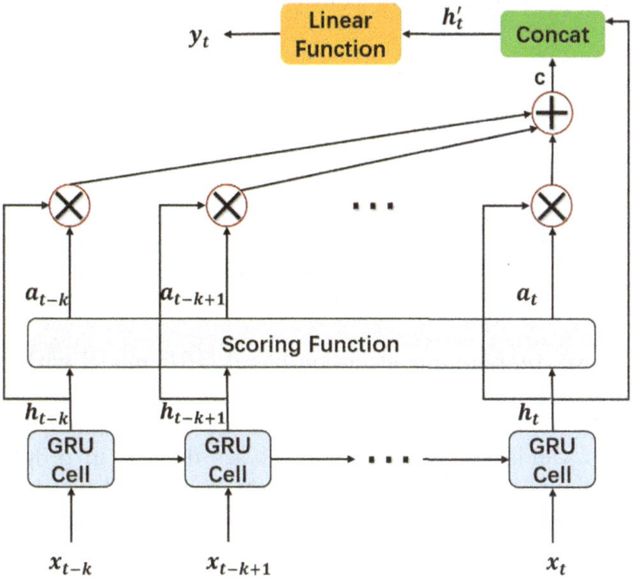

Fig. 2. Structure of proposed attention-based GRU model.

Table 1. Detailed experimental information of battery dataset

Battery	B0005	B0006	B0007
Charge cutoff voltage (V)	4.2	4.2	4.2
Discharge cutoff voltage (V)	2.7	2.5	2.2
Charging current (A)	1.5	1.5	1.5
Discharge current (A)	2	2	2
Temperature (°C)	24	24	24

4.2 Comparison of General GRU and Attention-Based GRU

In order to verify the effectiveness of the attention mechanism, we compare our attention-based GRU model with the general GRU model. For each dataset, the data is divided into a training set, a validation set, and a training set at a ratio of 4: 1: 5. We train the model with training set and store the best performing model on the validation set to make predictions on the test set. For fair comparison, our parameter settings for two models are as follows: hidden units size are among 5, 10, 15, 20; time step length are among 5, 10, 15, 20; batchsize is 32. RMSprop are used as optimizer to train the model. The comparison results of RMSE on test set are tabulated in Table 2.

As shown in Table 2, we can find that our model performs better than the general GRU in the three data sets. This experimental results show that by adding attention mechanism, the accuracy rate can be improved.

Table 2. Comparison of general GRU and attention-based GRU.

Dataset	GRU	Attention+GRU
B0005	0.0202	0.0163
B0006	0.0173	0.0158
B0007	0.0196	0.0154

4.3 Comparison of Attention-Based GRU and Conventional Methods

In this section, we compare our attention-based GRU model with conventional model. These models are general RNN, LSTM, and independent RNN (IndRNN) [14]. In this experiment, we still use three data sets for comparison. Similarly, the training set, validation set, and test set are divided in proportion to the previous experiment. The hyperparameter settings of each model are also the same as the previous experiment. The RMSE of these models are tabulated in Table 3 and shown in Fig. 3.

Table 3. Comparison of RMSE with different models on each battery dataset.

Dataset	RNN	LSTM	IndRNN	Attention+GRU
B0005	0.0276	0.0271	0.0355	0.0163
B0006	0.0198	0.0288	0.0296	0.0158
B0007	0.0330	0.0282	0.0262	0.0154

As shown in Table 3 and Fig. 3, our model outperforms all other methods in the four data sets. It can be seen from the test results of the four data sets that the GRU results are better than other general methods. Then by adding the attention mechanism on the GRU the accuracy are improved.

4.4 Training on B0005/B0006 and Test on B0007

The above experiments are to train and test each data set separately. In order to verify the generality of our method, we use batteries B0005 and B0006 as the training set, and divide 20% of the data into validation set, and finally use B0007 battery is tested and compared with the general method. The results are tabulated in Table 4 and Fig. 4.

As shown in Table 4 and Fig. 4, our proposed model performs better than conventional models that conclude RNN, LSTM, IndRNN, and GRU. This result shows that our model can better learn the dynamic relationship between current, voltage, and time and SOH.

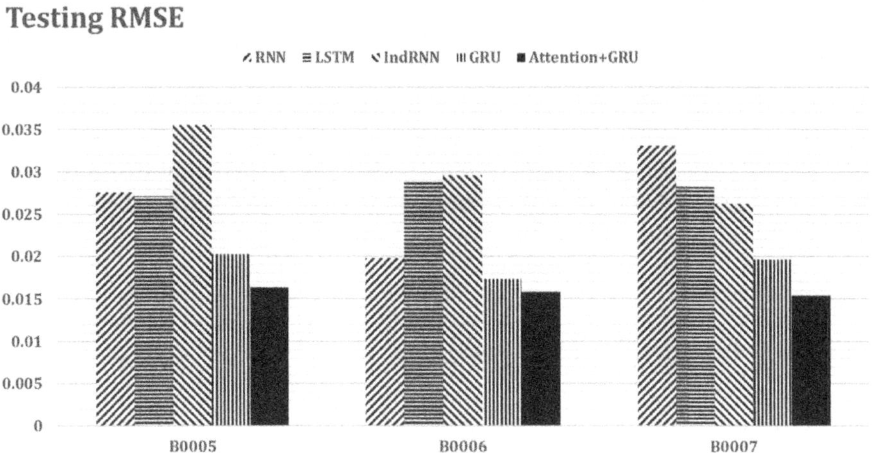

Fig. 3. RMSEs of different models on testing data.

Table 4. Comparison of RMSE with different models on B0007 dataset

Dataset	RNN	LSTM	IndRNN	GRU	Attention+GRU
B0007	0.0250	0.0165	0.0196	0.0167	0.0161

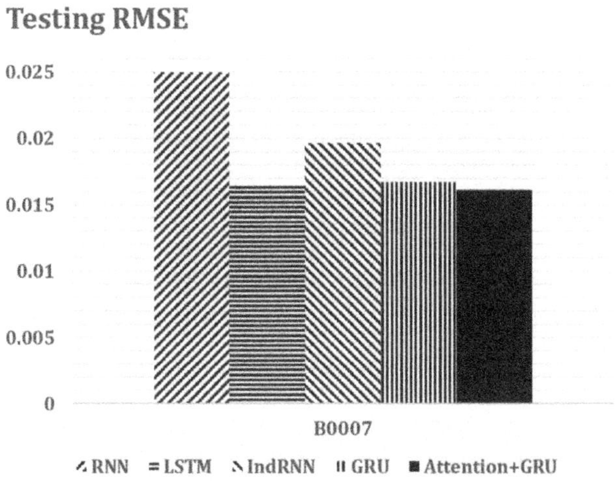

Fig. 4. RMSEs of different models on B0007 dataset

5 Conclusions

In this paper, we have proposed an Attention-based GRU model for SOH estimation of Lithium-ion battery. Four battery data sets of NASA such as B0005, B0006, and B0007 at room temperature are trained and validated. We have also used B0005 and B0006 batteries as the training set, and B0007 battery as the test set to verify the generality of our model. The experimental results show that our proposed model is more accurate in terms of RMSE compared to RNN, LSTM, IndRNN, and GRU networks.

Acknowledgment. This research was supported by Basic Science Research Program through the National Research Foundation of Korea (NRF) funded by the Ministry of Education (NRF-2019R1I1A1A01058964).

References

1. Hu, X., Zou, C., Zhang, C., Li, Y.: Technological developments in batteries: a survey of principal roles, types, and management needs. J. IEEE Power Energy Mag. **15**(5), 20–31 (2017)
2. Goebel, K., Saha, B., Saxena, A., Celaya, J.R., Christophersen, J.P.: Prognostics in battery health management. J. IEEE Instrum. Meas. Mag. **11**(4), 33–40 (2008)
3. Cho, K., Van Merriënboer, B., Gulcehre, C., Bahdanau, D., Bougares, F., Schwenk, H., Bengio, Y.: Learning phrase representations using RNN encoder-decoder for statistical machine translation. arXiv preprint arXiv:1406.1078 (2014)
4. Rumelhart, D.E., Hinton, G.E., Williams, R.J.: Learning representations by back-propagating errors. Nature **323**(6088), 533–536 (1986)
5. Ashwin, T.R., Chung, Y.M., Wang, J.: Capacity fade modelling of lithium-ion battery under cyclic loading conditions. J. Power Sources **328**, 586–598 (2016)
6. Pola, D.A., Navarrete, H.F., Orchard, M.E., Rabié, R.S., Cerda, M.A., Olivares, B.E., Silva, J.F., Espinoza, P.A., Pérez, A.: Particle-filtering based discharge time prognosis for lithium-ion batteries with a statistical characterization of use profiles. J. IEEE Trans. Rel. **64**(2), 710–720 (2015)
7. Weng, C., Cui, Y., Sun, J., Peng, H.: On-board state of health monitoring of lithium-ion batteries using incremental capacity analysis with support vector regression. J. Power Sources **235**, 36–44 (2013)
8. Nuhic, A., Terzimehic, T., Soczka-Guth, T., Buchholz, M., Dietmayer, K.: Health diagnosis and remaining useful life prognostics of lithium-ion batteries using data-driven methods. J. Power Sources **239**, 680–688 (2013)
9. Ren, L., Zhao, L., Hong, S., Zhao, S., Wang, H., Zhang, L.: Remaining useful life prediction for lithium-ion battery: a deep learning approach. J. IEEE Access **6**, 50587–50598 (2018)
10. Hochreiter, S., Schmidhuber, J.: Long short-term memory. Neural Comput. **9**(8), 1735–1780 (1986)
11. Liu, J., Saxena, A., Goebel, K., Saha, B., Wang, W.: An adaptive recurrent neural network for remaining useful life prediction of lithium-ion batteries. In: Annual Conference of the Prognostics and Health Management Society, PHM (2010)
12. Zhang, Y., Xiong, R., He, H., Pecht, M.G.: Long short-term memory recurrent neural network for remaining useful life prediction of lithium-ion batteries. J. IEEE Trans. Veh. Technol. **67**(7), 5695–5705 (2018)

13. Luong, M.T., Pham, H., Manning, C.D.: Effective approaches to attention-based neural machine translation. arXiv preprint arXiv:1508.04025 (2015)
14. Li, S., Li, W., Cook, C., Zhu, C., Gao, Y.: Independently recurrent neural network (IndRNN): building a longer and deeper RNN. In: Proceedings of the IEEE Conference on Computer Vision and Pattern Recognition, pp. 5457–5466 (2018)

Model for Choosing Rational Investment Strategies, with the Partner's Resource Data Being Uncertain

V. Lakhno[1], V. Malyukov[1], D. Kasatkin[1](✉), G. Vlasova[2], P. Kravchuk[2], and S. Kosenko[3]

[1] Department of Computer Systems and Networks, National University of Life and Environmental Sciences of Ukraine, Kiev, Ukraine
{valss21,dm_kasat}@ukr.net
[2] Private Higher Educational Institution "European University", Kiev, Ukraine
[3] State Scientific Research Forensic Center of the Ministry of Internal Affairs of Ukraine, Kiev, Ukraine

Abstract. A model is considered for searching rational strategies for players' investments in information technologies (IT), with their being insufficiently informed. The financial resources of one of the investors are supposed to be a member of a fuzzy set. The model is designed to be used for creating a mathematical module of the decision-making support system when determining rational options for investment. The apparatus of the theory of multistage quality games with several terminal surfaces has enabled adopting a new approach to solving the problem in question. An important feature of the model that distinguishes it from other models is the assumption that Player 1 has no information about the strategies and financial resources of Player 2, which makes the problem under consideration more realistic. The operability of the model has been confirmed experimentally in the MathCad environment.

Keywords: Investment · Multistage game theory · Financial strategies · Fussy set · Decision-making support

1 Introduction

In real life, as we know, situations where decision makers are fully aware of the characteristics affecting the decision-making process practically do not exist, which has a significant impact on determining the best decision in this situation. For example, in [1, 2] it is stated that many companies and organizations are not fully aware of the necessity to find a solution of a permanent task of investment, for example, in information technologies (IT). This results in the fact that, under the growing need in the development of promising technologies and systems [3], situations occur with high risks due to improper reactions to uncertain information when investing in the IT industry.

In finding rational strategies for investments in the IT sphere, a number of problems arise [3, 4], namely: there is no methodological support for determination of investment

R. Silhavy et al. (Eds.): CoMeSySo 2020, AISC 1295, pp. 332–341, 2020.
https://doi.org/10.1007/978-3-030-63319-6_29

related risks; suggested models and approaches are difficult to implement; there is no methodology for assessing the investor's control actions, in particular, when the information is not complete or the resources for investment are not limited, etc.

This list of reasons, which is far from being exhaustive, allows us to believe that the problem of further research on new approaches and models for finding strategies for investments in information technologies and systems remains relevant.

2 Literature Survey and Analysis

As shown in [3, 4], decision-making on IT funding is a permanent task. However, some papers, in particular [5–7], have a shortcoming consisting in the absence of tangible recommendations on the development of IT investment strategies. In particular, there is no research on models that would take into account the strategies of active financial counter-measures on the second investor's part. A new trend is represented by studies devoted to various expert [8, 9] and decision-making support systems [10, 11] for choosing IT investment strategies. The shortcoming of the studies presented in [7–9, 11] is the lack of unambiguous modeling results, and thus, the investor cannot obtain a reliable forecast. The greater part of the models described do not enable finding effective recommendations and strategies for funding complex informatization objects and the segment of high-risk but promising IT trends, for example, those related to cybersecurity, Smart City and others. The models suggested in [5, 7, 8, 11] also do not allow assessing the risk of losing the financial resources by the main investor. In [4, 12–14] models are suggested that employ game theory for assessing the investment effectiveness. However, a lot of factors have been neglected by the authors. For example, changes in the financing components of the second investor have not been taken into account in the event of mutual investment. The weakness of different authors' works can be removed by using the techniques of the theory of differential and multistage quality games with several terminal surfaces [4, 12]. Besides, it seems preferable to commit bulky computations and inference generation on the choice of a rational strategy for the investor to a computer program, for example, by using a decision-making support system.

Thus, our analysis of the studies performed shows that the problem of further development of models for decision-making support systems in investment tasks, e.g. in the IT sphere, remains relevant. In particular, situations should be considered with uncertain information about the investor's financial position in the process of finding sets of preference and rational strategies for investment funding.

3 The Research Aim

The objective of the paper is designing a model for systems supporting decision-making on choosing rational strategies for mutual investments in IT, with information about the financial resources of the parties being uncertain.

4 Methods and Models

It should be noted that in this paper we traditionally employ the game theory apparatus for solving the problem in question. We, therefore, continue the course of research presented in [4, 12], where an interaction is presented between two players who are investors in IT: Player 1 – first investor – (INV_1); Player 2 – second investor – (INV_2).

Both players use their financial resources to accomplish their goals.

Statement of the Problem. There are two investing players. The time interval is specified as $\{0, \ldots, T\}$, (T – natural number). The financial resources of the players are: $z_1(0)$ (first investor) и $z_2^{\xi}(0)$(second investor).

There is an interaction between the players. The interaction is considered within the bilinear multistage game with simultaneous moves, the information being uncertain. The difference between the game under consideration and a game with full information is that Player 1 is not fully aware of the initial state of Player 2. But it is known that his state belongs to a fuzzy set $\{X, m(.)\}$, where X is a subset of, R_+ ,$m(.)$ is the membership function of the state $z_2^{\xi}(0)$ in the set $X, m(z_2^{\xi}(0)) \in [0, 1]$ for $z_2^{\xi}(0) \in X$. Besides, the states $z_1(\tau)$ are known at any time $t(t = 0, 1, \ldots, T)$, with $\tau \leq t$. Here, the following conditions are satisfied: $z_1(\tau) > 0$ with $z_1(\tau) > 0$, the certainty being $\geq p_0 (0 \leq p_0 \leq 1)$ and $z_1(\tau) < 0$ with $z_1(\tau) < 0$ the certainty being $< p_0$. Also, the degrees of implementation of Player 1's strategies $u(\tau)(\tau \leq 1)$ for interaction with Player 2 are known.

A solution will be found from the position of Player 1. In so doing, it is assumed that there are no assumptions regarding the awareness of Player 2 about the project, which means that Player 2 may possess any information. Both investors make their steps simultaneously.

The dynamics of their interaction can be described as:

$$\begin{aligned} z_1(t+1) &= g_1 \cdot z_1^+(t) - u(t) \cdot g_1 \cdot z_1^+(t) - f_2 \cdot v(t) \cdot g_2 \cdot z_2^{+,\xi}(t); \\ z_2^{\xi}(t+1) &= g_2 \cdot z_2^{+,\xi}(t) - v(t) \cdot g_2 \cdot z_2^{+,\xi}(t) - f_1 \cdot u(t) \cdot g_1 \cdot z_1^+(t); \end{aligned} \tag{1}$$

Here, $t \in \{0, \ldots, T\}$,f_1, f_2- investment elasticity factors of Players 1 and 2 (or investors INV_1 and INV_2) accordingly; g_1, g_2- the players' financial resources growth rates; $u(t), v(t)-$ the players' strategies implementations at time t.

$$z^+ = \left\{ \begin{array}{c} z, z \geq 0 \\ 0, z \prec 0 \end{array} \right\}$$

The end of the interaction is specified by the conditions:

$$((z_2^{\xi}(t) < 0), (z_1(\tau) > 0)), \text{ with certainty } \geq p_0 \tag{2}$$

$$((z_2^{\xi}(t) < 0), (z_1(\tau) < 0)), \text{ with certainty } < p_0 \tag{3}$$

The game takes place as follows:

At a certain time t Player 1 multiplies by factor g_1 (change rate, growth rate) and chooses a value $u(t)(u(t) \in [0, 1]$ that defines the part of his financial resource $g_1 \cdot z_1(t)$ allocated to investments (e.g. in IT) at the time t. Player 2 acts in a similar way. At the time t Player 2 multiplies $z_2^{\xi}(0)$ by factor g_2 (change rate, growth rate) and chooses a value $v(t)(v(t) \in [0, 1])$ that defines the part of his financial resource $g_2 \cdot z_2^{\xi}(t)$ allocated to investments at the time t. Additional financial resources are allocated by the players because of the interaction (elasticity) between the investments. Then the states of the players at the time $t+1$ are determined from (1).

If condition (2) is satisfied, we will say that the investment has brought INV_1 the desired result, with certainty being $p \geq p_0$, and the procedure is over.

If condition (3) is satisfied, we will say that the investment has brought INV_2 the desired result, with certainty being $p > 1 - p_0$, and the procedure is over.

If neither condition (2) nor (3) is satisfied, the investment procedure continues.

We define the function as follows: $F(.) : X \to R_+, F(x) = \{\sup m(y), with\, y \leq x\}$.

We introduce the following designations: Φ – a set of such functions, $T^* = \{0, \ldots, T\}$, – a set of natural numbers, including zero, not exceeding T.

Definition. A pure strategy $u(.,.,.)$ of Player 1 refers to a function $u(.,.,.) : T^* \times R_+ \times \Phi \to [0, 1]$, such that $u(t, z_1, F) \in [0, 1], (F \in \Phi)$.

In other words, the strategy of Player 1 is a rule that, on the basis on the information available, allows him to determine the amount of financial resource that he allocates to investments (e.g. in IT).

Player 2 chooses his strategy based on any information.

Player 1 seeks to find a set of his initial states that satisfy the condition below.

Condition: If the game begins from the initial states, Player 1 can, by choosing his strategy $u_*(.)$, ensure that at a certain time t condition (2) be satisfied. In so doing, the strategy chosen by Player 1 prevents Player 2 from satisfying condition (3) at the previous times.

We refer to a set of such states as Player 1's *preferability set* – Q_1. The strategies of Player 1 that satisfy the conditions above are referred to as his optimal (rational) strategies.,1. Player 1 has an objective to construct preferability sets and to find strategies by using which he will have condition (2) satisfied. According to the decision-making theory classification, this game model can be viewed as a decision-making problem under uncertain information. We will note that such a model is a bilinear multistage quality game with several terminal surfaces with simultaneous moves. The construction of Player 1's preferability sets and rational strategies depends on many parameters. For describing the preferability sets of Player 1, we will introduce the following variable: $\phi(0) = \inf\{\phi'\}, \quad F(\phi') \geq p_0$.

The problem stated is solved using the apparatus of the theory of multistage quality games [4]. It allows finding a solution with any relationships of game parameters.

The paper provides a solution, i.e. preferability sets Q_1 and rational strategies $u_*(.,.)$ for all relationships of game parameters. We will note that Q_1 is a union of sets Q_1^i. This is a set of such initial states $(z_1(0), \phi(0))$ that, if the game starts from them, there exists a strategy of Player 1 which, for any implementation of Player 2's strategy, leads the state of the system $(z_1(0), \phi(0))$, at a time $t = i$, to such that condition (2) is satisfied. Herewith, Player 2 has no strategy that could have condition (3) satisfied at any of the preceding point in time.

Case A: $g_1 \leq g_2$.

$Q_1^i = \{(z_1(0), c(0)) : k(i-1) \cdot g_2 \cdot c(0) \leq f_1 \cdot g_1 \cdot z_1(0) < k(i-2) \cdot g_2 \cdot c(0)\}$, $i = 1, \ldots u_* = \{u_*(0, (z_1, c)), \ldots, u_*(t-1, (z_1, c))\} = \{[1 - (f_2 \cdot g_2 \cdot c)/(g_1 \cdot z_1)]\}$ with $(z_1, c) \in R_+^2, f_1 \cdot z_1 > f_2 \cdot g_2 \cdot c$ and is not defined otherwise;

$$t = 0, 1, \ldots, i-1.$$

Here

$$k(i) = 1 + f_1 \cdot f_2 - (f_1 \cdot g_1 \cdot f_2)/(g_2 \cdot k(i-1));$$
$$k_{-1} = 0, k_0 = 1 + f_1 \cdot f_2; Q_1 = \bigcup_{j=1}^{\infty} Q_1^j.$$

The beam

$$(f_1 \cdot g_1/g_2) \cdot z_1(0) = \{1 + f_1 \cdot f_2 + ((1 + f_1 \cdot f_2)^2 - 4 \cdot f_1 \cdot f_2 \cdot g_1/g_2)^{0.5}\}/2 \cdot c(0)$$

Is a barrier [4]. A barrier is a case where Player 1 cannot attain his goal from the states $(z_1(0), c(0))$:

$(f_1 \cdot g_1/g_2) \cdot z_1(0) \leq \{1 + f_1 \cdot f_2 + ((1 + f_1 \cdot f_2)^2 - 4 \cdot f_1 \cdot f_2 \cdot g_1/g_2)^{0.5}\}/2 \cdot c(0)$ at any time.

Case B: $g_1 > g_2, f_1 \cdot f_2 \geq 1$.

In this case, the preferability set of Player 1 – Q_1 is a union of a finite number of sets Q_1^i, namely $(N+2)$ sets, where

$$N : k(i) > f_1 \cdot f_2 \cdot g_1/g_2, i = 0, \ldots, N-1; k(N) \leq f_1 \cdot f_2 \cdot g_1/g_2,$$

$$Q_1^i = \{(z_1(0), c(0)) : k(i-1) \cdot g_2 \cdot c(0) \leq f_1 \cdot g_1 \cdot z_1(0) \prec k(i-2) \cdot g_2 \cdot c(0)\},$$
$$i = 1, \ldots, N+1;$$

$$Q_1^{N+2} = \{(z_1(0), c(0)) : f_1 \cdot f_2 \cdot g_2 \cdot c(0) \leq f_1 \cdot g_1 \cdot z_1(0) < k(N) \cdot g_2 \cdot c(0)\}.$$

The rational (optimal) strategy $u_* = (u_*(0, (z_1, c), \ldots, u_*(N+1, (z_1, c)))$ is determined as:

$u_*(0, (z_1, c)) = \{0, (z_1, c) \in R_+^2, g_1 \cdot z_1 > f_2 \cdot g_2 \cdot c$ and is not defined otherwise$\}$,

$u_*(0, (z_1, c)) = \{[1 - (f_2 \cdot g_2 \cdot c)/(g_1 \cdot z_1)], (z_1, c) \in R_+^2, g_1 \cdot z_1 > f_2 \cdot g_2 \cdot c$ and is not defined otherwise; $t = 1, \ldots, N+1\}$.

Case C: $g_1 > g_2, f_1 \cdot f_2 < 1$.

In this case, the preferability set of Player 1 – Q_1 is also a union of a finite number of sets Q_1^i. But here it contains $N + i_* + 2$ sets, where

$N : k(i) > (g_1/g_2), i = 0, \ldots, N-1; k(N) \le (g_1/g_2); i_*-$ the least non-negative integer defined by the inequality

$$k(N) \cdot (g_2/g_1)^{i_*+1} < f_1 \cdot f_2.$$

Then

$$Q_1^i = \{(z_1(0), c(0)) : k(i-1) \cdot g_2 \cdot c(0) \le f_1 \cdot g_1 \cdot z_1(0) < k(i-2) \cdot g_2 \cdot c(0)\}, i = 1, \ldots, N+1;$$

If $i_* = 0$, then

$$Q_1^i = \{(z_1(0), c(0)) : k(i-1) \cdot g_2 \cdot c(0) \le f_1 \cdot g_1 \cdot z_1(0) < k(i-2) \cdot g_2 \cdot c(0)\}, \quad i = 1, \ldots,$$

$$N+1; \; Q_1^{N+2} = \{(z_1(0), c(0)) : f_1 \cdot f_2 \cdot g_2 \cdot c(0) \le f_1 \cdot g_1 \cdot z_1(0) < k(N) \cdot g_2 \cdot c(0)\}.$$

The rational strategy here can be presented in the same way as in Case B.

If $i_* > 0$, then

$$Q_1^{N+1+j} = \{(z_1(0), c(0)) : k(N) \cdot (g_2/g_1)^j \cdot g_2 \cdot c(0) \le f_1 \cdot g_1 \cdot z_1(0) < k(N) \cdot (g_2/g_1)^{j-1} \cdot g_2 \cdot c(0)\},$$

$i = 0, \ldots, i_*$;

$$Q_1^{N+1+i_*} = \{(z_1(0), c(0)) : f_1 \cdot f_2 \cdot g_2 \cdot c(0) \le f_1 \cdot g_1 \cdot z_1(0) < k(N) \cdot (g_2/g_1)^{i_*} \cdot g_2 \cdot c(0)\},$$

The rational strategy $u_* = (u_*(0, (z_1, c), \ldots, u_*(N+1, (z_1, c)))$ in this case is determined as:

$u_*(0, (z_1, c)) = \{0, with (z_1, c) \in R_+^2, g_1 \cdot h_1 > f_2 \cdot g_2 \cdot c$ and is not defined otherwise, $i = 0, \ldots, i_*\}, u_*(0, (z_1, c)) = \{[1 - (f_2 \cdot g_2 \cdot c)/(g_1 \cdot z_1)], with (z_1, c) \in R_+^2, g_1 \cdot z_1 > f_2 \cdot g_2 \cdot c, i \ge i_* + 1$ and is not defined otherwise; $t = 1, \ldots, N+1\}$.

The solution of this problem enables representation of a positive orthant in a surface $(z_1(0), c(0))$ in terms of three sets (cones with the vertex in the point (0,0)). But the set (cone) adjacent to the $0Z_1$ axis is a set preferable for Player 1. A second set (cone) is a set preferable for Player 2. And a third set (cone) is neutral with regard to both players. Actually, this set characterizes the equilibrium of investing players. In other words, for states belonging to this set, the players have strategies allowing them to continue their investing activities indefinitely.

In other words, the conditions $(z_1(0), c(0))$ will be satisfied for any time t. We will note that the beams that are the boundaries of the cones are specified by coefficients representing a combination of parameters defining the investment dynamics. Therefore, if the initial values $(z_1(0), c(0))$ of the financial resources of the investing parties are specified, these parameters can, for example, be varied. In particular, it is possible to demand that the parameters specifying the financial resource change rate should be such that the point $(z_1(0), c(0))$ is within the equilibrium region. Or, on the equilibrium beam if the cone dividing two preferability sets is a beam. If some of the parameters defining the financial resource change rate are fixed, it is possible to demand that the values $(z_1(0), c(0))$ and some of the parameters that are not fixed should be such that the point $(z_1(0), c(0))$ gets into the equilibrium region. This may, in turn, influence both the investment process itself and recommendations for the investors on the choice of strategies. If nothing can be changed, the presented above solution of the game will indicate a possible result of the investment procedure within the assumptions under which the problem is considered.

5 Computational Experiment

The experiment was performed in Mathcad. The model was also implemented in a decision-making support system software module [4]. Three experiment tests have been realized (Fig. 1, 2 and 3). The experiment was aimed at defining the sets of the rational strategies of the players (first and second investors) and, therefore, at defining the risks for the players to lose their financial resources. Cases are considered where the players' strategies bring them to the appropriate terminal surfaces, i.e. defined by conditions (2) and (3). In the course of the experiment, sets of the objects' initial states are determined that allow the objects to bring the system to this or that terminal surface.

The Z_1 – axis on the surface means the financial resources of the first investor. The C-axis – financial resources of the second investor. The region below the beam – "preferability" region of the first investor. The region above the beam – "preferability" region of the second investor. The equilibrium beam is represented by a solid line with round markers. The points are received from the experiment. The players' trajectories are represented by dotted lines with triangular markers. The trajectories lie in the players' preferability region.

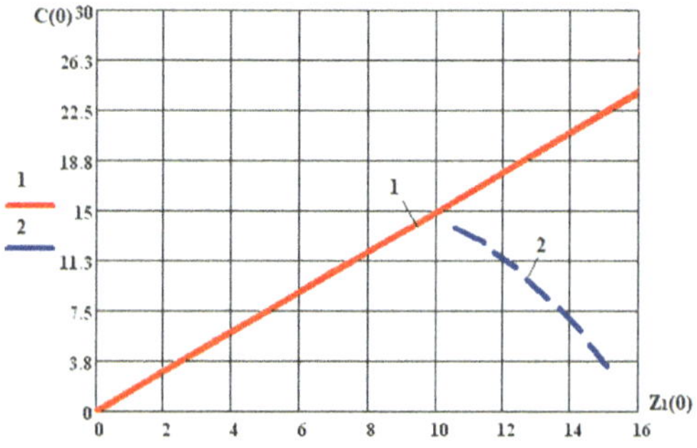

Fig. 1. Experiment 1 Results. Trajectory of Player 1

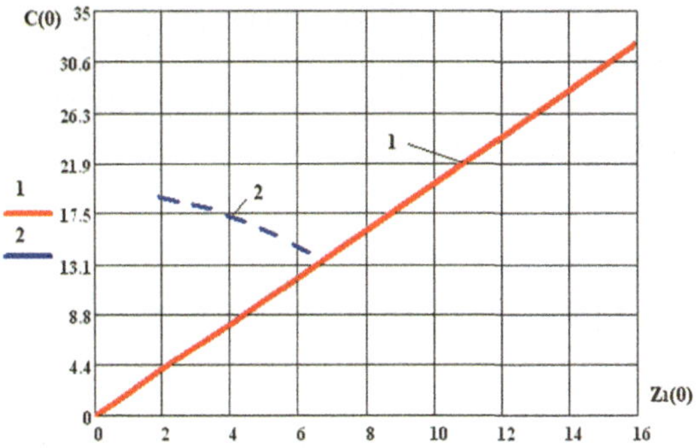

Fig. 2. Experiment 2 Results. Trajectory of Player 2

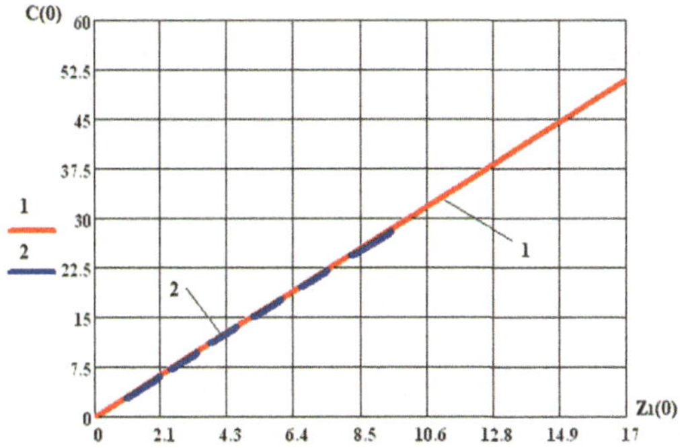

Fig. 3. Experiment 3 Results. ("Stability" of the System)

The results obtained show the effectiveness of the suggested approach. During the testing of the model, the correctness of the obtained results has been proved.

Figure 1 shows the situation where Player №1 has a better ratio of the initial financial resources, i.e. they are in the preferability set of Player 1. In this case, Player 1 will, using his optimal strategy, attain his goal, namely, he will bring the state of the system to "his" terminal surface. We take a positive orthant on the surface. Further, beams proceeding from the point (0,0) are considered in this orthant. These beams are specified by the expression: $c = (1.5 - 1/n) \cdot z_1$. They specify the preferability sets of Player 1 in n steps. For example, the set Q_1^n is:

$\{(z_1(0), c(0)) : (z_1(0), c(0)) \in R_+^2, (1.5 - 1/(n-1) \cdot z_1(0) \le c(0) < (1/5 - 1/n) \cdot z_1(0)\}$. For example, for $n = 1$ we have

$$Q_1^1 = \{(z_1(0), c(0)) : (z_1(0), c(0)) \in R_+^2, 0 \le c(0) < (0.5) \cdot z_1(0)\}.$$

The beam $c(0) = (1.5) \cdot z_1(0)$ is an equilibrium beam.

Figure 2 shows the situation where Player 2, taking advantage of Player 1's non-optimal behaviour at the initial time, succeeds in bringing the state of the system to "his" terminal surface. In Fig. 2 a positive orthant on the surface is shown. In this orthant, beams proceeding from the point (0,0) are considered. These beams are specified by the expression: $c = (2 + 1/n) \cdot z_1$. They specify the preferability sets of Player 2 in n steps. For example, the set Q_1^n is

$$\{(z_1(0), c(0)) : (z_1(0), c(0)) \in R_+^2, (2 + 1/(n-1)) \cdot z_1(0) \leq c(0) < (2 + 1/n) \cdot z_1(0)\}.$$

For $n = 1$ we have:

$$Q_1^1 = \{(z_1(0), c(0)) : (z_1(0), c(0)) \in R_+^2, 0 \leq c(0) < 3 \cdot z_1(0)\}.$$

The beam $c(0) = 2 \cdot z_1(0)$ is an equilibrium beam.

Figure 3 represents the case where the initial state of the system is in the equilibrium beam, and both players "move" along this beam using their rational strategies. This "satisfies" them both.

We will note that the suggested model describes the process of predicting the results of investing in IT. The discovered shortcoming of the model is the fact that predicted data obtained when choosing investment strategies do not always coincide with actual ones.

6 Discussion of the Experiment Results

In the course of the computational experiments and based on the practical approbation data [12], it has been established that the suggested model, within the frames of a bilinear differential quality game for a system supporting decision-making in managing the procedure of mutual investment in IT, allows adequate description of dependent movements by means of bilinear functions. It provides investors with an effective tool. As compared to the existing models, the suggested solution provides the investor with better effectiveness and predictability by an average of 11–12% [3, 6–10, 13, 14].

7 Conclusions

The paper deals with the problem of further development of models for systems supporting decision-making in investing activities, investing in IT taken as an example. The model considered allows finding rational investment strategies when one investor does not have complete information about the financial resources and strategies of the other. Incompleteness of information is considered in terms of fuzzy set theory, which makes this kind of model more realistic as clear information does not practically exist in real life. The problem under consideration has been solved using the apparatus multistage quality games with several terminal surfaces with simultaneous moves.

A computational experiment has been performed. The greatest deviations from actual data have not exceeded 12%, which proves the serviceability of the model.

References

1. Holden, K., El-Bannany, M.: Investment in information technology systems and other determinants of bank profitability in the UK. Appl. Financ. Econ. **14**(5), 361–365 (2004)
2. Nguyen, T.H., Newby, M., Macaulay, M.J.: Information technology adoption in small business: Confirmation of a proposed framework. J. Small Bus. Manage. **53**(1), 207–227 (2015)
3. Fu, J.R., Chen, J.H.: Career commitment of information technology professionals: the investment model perspective. Inf. Manage. **52**(5), 537–549 (2015)
4. Akhmetov, B.B., Lakhno, V.A., et al.: The choice of protection strategies during the bilinear quality game on cyber security financing. Bull. Nat. Acad. Sci. Repub. Kaz. **3**, 6–14 (2018)
5. Lee, I., Shin, Y.J.: Fintech: ecosystem, business models, investment decisions, and challenges. Bus. Horiz. **61**(1), 35–46 (2018)
6. Nazareth, D.L., Choi, J.: A system dynamics model for information security management. Inf. Manage. **52**(1), 123–134 (2015)
7. Altuntas, S., Dereli, T.: A novel approach based on DEMATEL method and patent citation analysis for prioritizing a portfolio of investment projects. Expert Syst. Appl. **42**(3), 1003–1012 (2015)
8. Dincer, H., Hacioglu, U., Tatoglu, E., Delen, D.: A fuzzy-hybrid analytic model to assess investors' perceptions for industry selection. Decis. Support Syst. **86**, 24–34 (2016)
9. Chourmouziadis, K., Chatzoglou, P.D.: An intelligent short term stock trading fuzzy system for assisting investors in portfolio management. Expert Syst. Appl. **43**, 298–311 (2016)
10. Dincer, H., Hacioglu, U., Tatoglu, E., Delen, D.: A fuzzy-hybrid analytic model to assess investors' perceptions for industry selection. Decis. Support Syst. **86**, 24–34 (2016)
11. Chiang, W.C., Enke, D., Wu, T., Wang, R.: An adaptive stock index trading decision support system. Expert Syst. Appl. **59**, 195–207 (2016)
12. Lakhno, V., Malyukov, V., et al.: Funding model for port information system cyber security facilities with incomplete hacker information available. J. Theor. Appl. Inf. Technol. **96**(13), 4215–4225 (2018)
13. Wang, Q., Zhu, J. Optimal information security investment analyses with the consideration of the benefits of investment and using evolutionary game theory. In: 2016 2nd International Conference on Information Management (ICIM), pp. 105–109. IEEE (2016)
14. Sachs, J.D.: Investing in social capital. In: World Happiness Report, pp. 152–166 (2015)

Modified Method of Ant Colonies Application in Search for Rational Assignment of Employees to Tasks

Vladimir A. Sudakov and Yurii P. Titov(✉)

Plekhanov Russian University of Economics, Stremyanny Lane, 36, Moscow 117997, Russian Federation
sudakov@ws-dss.com, kalengul@mail.ru

Abstract. The approach to determining the rational appointment of employees for the tasks of an innovative project is considered. It is proposed to apply a modification of the ant colony method to solve the assignment problem for cases when the task completion time for an employee is determined by a fuzzy set, taking into account the loss of time for employee interaction. To modify the ant colony method, recommendations on the choice of parameters are proposed: the number of agents in the group, the evaporation coefficient, the parameters of the elite and rank algorithm. The problem of "looping" the ant colony algorithm due to the allocation of one path by all agents is considered. To solve the problem, it is proposed to reset the solution graph and return to its initial state. Improving the algorithm when resetting the decision graph is proposed to introduce weights from the best solutions.

Keywords: Ant colony optimization · Fuzzy set · Assignment task · Loop problem · Employee interaction

1 Introduction

The modern approach associated with the management of work at the enterprise usually operates at times when certain stages of the work and its tasks are performed. To determine this time, an estimate of the time to complete such work is used, the manager has the experience of performing or managing it. Operating with the duration of the entire work is carried out using the Critical Path Method (CPM) [1–3], PERT [4–6] and Gantt charts [7, 8]. In the framework of such work, the head operates with units and staffed teams. Changes in the execution time of a task or stage associated with the different staffing of these teams are usually determined by risks, the calculation of which is a separate task [9].

In modern approaches, fuzzy sets have been applied to the CPM method to set the time to complete a task [10–16]. Inaccuracy in determining the time to complete a task or stage is designed to take into account possible risks and delays. In innovative projects, it is not always possible to evaluate the time taken by a manager to complete a task, even in the form of a fuzzy set. This is usually due to a lack of experience in the development of such projects. In such cases, the employee or team leader is

R. Silhavy et al. (Eds.): CoMeSySo 2020, AISC 1295, pp. 342–348, 2020.
https://doi.org/10.1007/978-3-030-63319-6_30

interviewed as an expert to evaluate the project's lead time. But such an approach implies a limitation that the employee performs the task alone, otherwise it is impossible to evaluate the time with sufficient accuracy.

A mathematical apparatus is described which, on the basis of fuzzy sets of task execution time by each of the workers, makes it possible to calculate the total task execution time [17]. There are various algorithms for accounting for the need for interaction between employees assigned to one task. Such solutions allow in innovative projects to conduct an employee survey on the time to perform various tasks, and, after collecting the necessary data, to solve the problem of assigning employees to tasks taking into account delays in interaction. According to the results of the calculations, you can get the terms fuzzy functions of the time it takes to complete individual tasks and the stages of work and work in general with a specific purpose of employees. At the same time, when interviewing employees, one may be interested in his ability to perform not only the tasks that he does best, but also other types of tasks that the employee is able to perform. As a result, you can get a flexible system for planning the assignment of workers to tasks.

To solve the assignment problem, various heuristic or meta-heuristic algorithms are often used. In this paper, we consider the possibility of applying a modification of the meta-heuristic algorithm of ant colonies [18–22].

2 Application of the Ant Colony Method to Solve the Assignment Problem

Unlike the original ant colony optimization method proposed by Italian researcher Marco Dorigo in 1992, the proposed algorithm looks for a path in a decision tree, not a Hamiltonian path in a graph [23, 24]. For the algorithm to work, a decision graph is created (Fig. 1) in which each employee corresponds to a set of nodes that determine the assignment of an employee to a task.

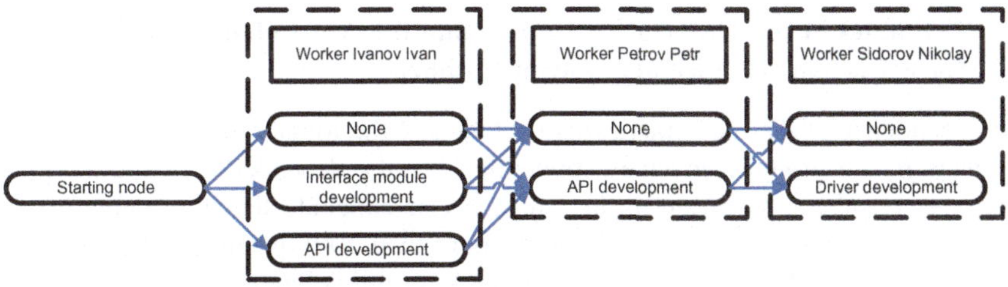

Fig. 1. Graph of solutions for the task of assigning workers to tasks.

To find a way in this column, it is necessary to modify the method of ant colonies [25, 26]. In addition to entering weights (pheromones) on the nodes of the graph, and not on the arcs, the objective function also underwent modifications. To find the best option for assigning all employees to tasks, it is necessary to determine the value of the criterion that determines how much this appointment is better or worse than others. As such a criterion, one of the many methods of defuzzification of the fuzzy function "task execution" can be used. The method of ant colonies in this case allows for directed enumeration of various options for appointing employees to work. In this case, it is necessary to take into account the direction of search, in order to maximize or minimize the criterion. If you use the task execution time as a criterion, then this is definitely a minimized criterion. Various parameters of the modification of the ant colony method affect the speed of finding a solution and its accuracy. These parameters include: the number of agents before the state of the graph changes, the coefficient of "evaporation" of the balance, the parameter of the introduced pheromone.

To collect and analyze statistics, it is necessary to determine the moment the module stops working that implements modifications of the ant colony method. In fact, there are 2 criteria for stopping the modification of the ant colony method: On reaching a certain number of iterations; By finding at least one solution satisfying the constraints.

To conduct tests, the task of appointing 35 employees for 15 tasks was considered. Each employee could perform a different number of tasks, but it is necessary to assign it to only one task. In total, the test considered more than 180 functions of accessories "task execution by a specific employee." The interaction of employees assigned to one task was taken into account according to the principle of mentoring, when the interaction of all employees with only the most experienced employee was taken into account. The generalized criterion was calculated as the sum of the execution times for all tasks. If there are no workers assigned to the task, then its execution time is equated to conditional infinity. The best value of the criterion (Krit) for this problem is slightly less than 352. The graphs show the estimate of the mathematical expectation and the confidence interval of this estimate for a confidence probability of 0.99. The score is calculated from 500 implementations of the ant colony method.

3 Estimates of the Criteria the Ant Colony Method

3.1 Estimates of the Criteria for Completing the Ant Colony Method for a Certain Number of Implementations

The dependence of the mathematical expectation of the found criterion on the number of agents in the group is inverse-exponential, but the dependence of the number of solutions considered is linear. From this we can conclude that the increase in the accuracy of the solution found with large (more than 100) quantities of agents in the group is much less than the time spent searching for this solution. It is recommended to select the number of agents in the interval (Number of layers; Number of layers * 2), where the Number of layers determines the number of nodes groups in the decision graph, i.e. Number of employees. Since the layers for our task determine the purpose of the employee, this value depends on the number of employees and varies in the interval (35; 35 * 2). But in the case when the accuracy of the solution is more critical for the time of its search, this indicator should be increased, since it most strongly affects the accuracy of the solution found. It is recommended to choose the value of the evaporation coefficient in the range (0.8; 0.95) depending on the speed of calculating the criterion for new solutions.

The parameter of the ant colony method, which is responsible for the amount of weight introduced by the agents, does not greatly affect the effectiveness of the methods. This is due to the relative nature of the scales when agents search for a route. But at the initial stage, the values of the introduced changes in the weights of agents can compete with the initial value of the weights at the nodes. The initial value is not equal to 0, since otherwise the initial probabilistic choice of nodes would have been impossible.

With an increase in the number of iterations of the ant colony algorithm, not only will the number of solutions increase and the estimate of the mathematical expectation of the solution found will improve, but the confidence interval of this estimate will also increase. This means that increasing the number of iterations does not always lead to better solutions. When setting a limit of more than 300 iterations, neither the number of considered solutions, nor the estimate of the mathematical expectation of the criterion practically change.

3.2 Estimation of Criteria for Completion of the Ant Colony Method to Achieve Limitations

The criterion for stopping the algorithm associated with the search for a solution that satisfies the constraints is called upon to correct this defect. When using the ant colony method, situations may arise when several agents from a group, after following a good route, carry so much weight that further deviation of other agents from this route becomes minimal. As a result, most agents move along close routes, but there is no guarantee that this route will satisfy restrictions. This situation can be called "looping." To cut off the looping situation, we establish a rather large restriction on the number of iterations of the algorithm. For our problem, we set the looping parameter of the algorithm for 1000 iterations (Fig. 2).

Consider the effect of the parameters of the ant colony algorithm: the number of agents and the evaporation coefficient for making recommendations. The histogram in the graphs of the number of iterations shows the percentage of successful runs, i.e. runs for finding solutions in which the number of iterations did not reach the "loop" criterion. For histograms, a separate, right axis on the graphs is provided. Diagrams with rhombuses mark similar characteristics when using a certain number of iterations as a stop criterion.

Recommendations for choosing parameters do not depend on the criterion for stopping the ant colony algorithm. So the number of agents in one group should be chosen as the greatest, based on the time constraints (number of solutions), and the evaporation parameter should be chosen within (0.8; 0.95). These recommendations are also due to the low loop coefficient of the algorithm. The problem of looping the ant colony algorithm is not solved by increasing the restriction on iteration. As a result, for our task, in any case, about 25% of iterations will not be able to find a solution that satisfies the established required constraints. The only parameter that can be adequately changed to reduce the percentage of looping the algorithm is the number of agents in

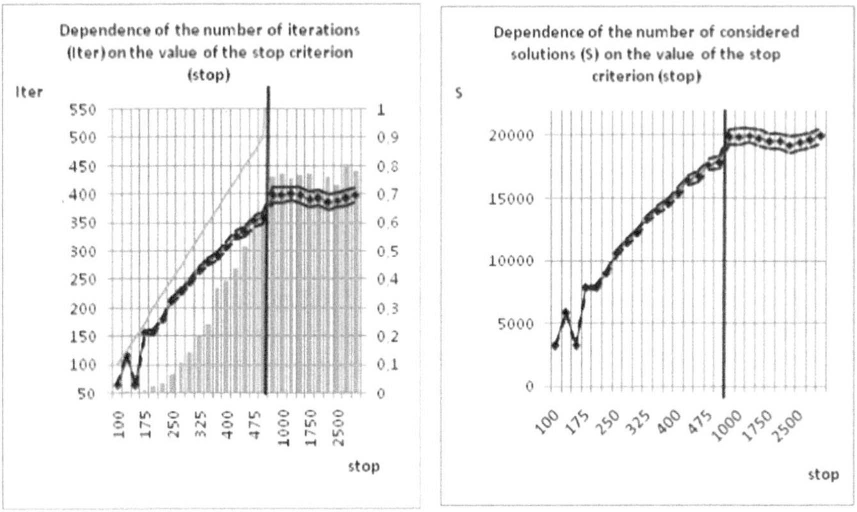

Fig. 2. Results of the modified ant colony method with an alternative stop criterion.

the group, but increasing this parameter in any case increases the number of iterations considered and, therefore, the running time of the algorithm.

When the stop criterion is improved, in our case, the total execution time of all the work is reduced, the number of algorithm loops will increase until the loop coefficient becomes equal to 1. For our task with a constraint of 360, no iteration found a solution that satisfies the constraint. The main problem is that this does not mean the impossibility of the existence of a solution that satisfies these constraints, but only that with the current parameters of the algorithm it is almost impossible to unambiguously achieve the desired solution.

To solve the looping problem, it is proposed to reset the state of the decision graph. As a result of resetting the graph, it returns to its initial state. As a result, the problem of looping the algorithm is almost completely solved. In this case, one can consider various algorithms for determining the moment of the graph reset. To improve the initial state of the decision graph, weights from the best found solutions can be entered during the operation of the algorithm.

4 Discussions

The article proposes the solution of appointing workers to the tasks of a high-tech project. A feature of the input data is multidimensional assignment due to the possibility of assigning each individual employee one of many tasks. Setting the time to complete a task by a specific employee is carried out using fuzzy sets. The time required for interaction is also taken into account if several employees are assigned to one task. For such tasks, genetic algorithms, random graphs, and other organizational and heuristic methods are usually used.

An approach based on a modification of the heuristic method of ant colonies is considered. To solve the assignment problem, a decision graph is constructed. Based on the test results, the following recommendations were determined on the established parameters of the ant colony method: Number of agents - from the interval (Number of employees; Number of employees * 2), evaporation coefficient - in the range (0.8; 0.95). It is recommended to apply the ranking algorithm for ant colonies with a rank value 3–4 times less than the number of employees. In addition to "the stop" criterion of the algorithm for a certain number of iterations, the possibility of applying "the stop" criterion when searching for a solution satisfying the constraints is considered.

There is a problem of "looping" the algorithm in cases where many agents move along the same route. Based on the reset of the decision graph, that is, its transfer to the initial state, an approach is proposed that allows overcoming the problem of "looping". The calculated average value of the criterion and the number of new solutions found were used to determine the moment of "looping". The best indicators were when detecting the "looping" of the algorithm by the average value of the criterion at the iteration. Setting up the initial weights was implemented depending on the time when the employee completed the task, and the possibility of taking additional weights to the schedule for the rational routes found at the moment.

The results obtained make it possible to use the ant colony method to solve the problem of appointing employees to work. Due to the simplicity of the path selection mechanism, it is possible to introduce fairly complex objective functions and limitations. The disadvantage of this method is its heuristic: finding the optimal solution is guaranteed only with an infinite time of its operation. As a result, finding a rational solution may require a lot of time for the algorithm to work, which greatly depends on the given parameters.

This study was financed by a grant from the Plekhanov Russian University of Economics.

References

1. Kiran, D.R.: Production Planning and Control: A Comprehensive Approach. Butterworth-Heinemann, Oxford (2019). ISBN 978-0-12-818364-9
2. George, E.: Project Management in Product Development: Leadership Skills and Management Techniques to Deliver Great Products. Butterworth-Heinemann, Oxford (2016). ISBN 978-0-12-802322-8
3. Kim, K., De La Garza, J.M.: Critical path method with multiple calendars. J. Constr. Eng. Manag. **131**(3), 522–532 (2005). ISSN: 1943-7862
4. Naili, M., Naili, M., Tari, A.A.K.: Uncertainty in the pert's critical path. Int. J. Sci. Technol. **4**(1), 1–9 (2018). ISSN: 2454-5880
5. Pritchard, C.L.: The Project Management Drill Book: A Self-Study Guide. ESI International, New York (2018). ISBN 978-0-20-374176-4
6. Bondarenko, A.N., Shavrin, A.V.: PERT method in project management. Proj. Prog. Manag. **1**, 68–78 (2016)
7. Nahmias, S.: Encyclopedia of Operations Research and Management Science. Springer, Boston (2013). ISBN 978-1-4419-1154-4

8. Wilson, J.M.: Gantt charts: a centenary appreciation. Eur. J. Oper. Res. **149**(2), 430–437 (2003). ISSN: 1872-6860
9. Gephart Jr, R.P., Miller, C.C., Helgesson, K.S.: The Routledge Companion to Risk, Crisis and Emergency Management. Routledge, New York (2018). ISBN: 9781138208865
10. Chen, S.P., Hsueh, Y.J.: A simple approach to fuzzy critical path analysis in project networks. Appl. Math. Model. **32**, 1289–1297 (2008). ISSN: 0307-904X
11. Liu, S.T.: Fuzzy activity times in critical path and project crashing problems. Cybern. Syst. Int. J. **34**, 161–172 (2003). ISSN: 1087-6553
12. Popescu Costin-Ciprian: On critical path with fuzzy weights. Econ. Comput. Econ. Cybern. Stud. Res. **52**(4), 49–60 (2018). ISSN 1842–3264
13. Shih, P.C.: Analysis of critical paths in a project network with fuzzy activity times. Eur. J. Oper. Res. **183**(1), 442–459 (2007). ISSN: 0377-2217
14. Pawe, Z.: On computing the latest starting times and Xoats of activities in a network with imprecise durations. Fuzzy Sets Syst. **150**(1), 53–76 (2005). ISSN: 0165-0114
15. Nasution, S.H.: Fuzzy critical path method. IEEE Trans. Syst. Man Cybern. **24**, 48–57 (1994). ISSN: 1558-2442
16. Chen, T.C., Sue, F.H.: Applying fuzzy method for measuring criticality in project network. Inform. Sci. **177**(12), 2448–2458 (2007). ISSN: 0020-0255
17. Sudakov, V.A.: Titov, Y.P.: The solution to the problem of determining the time to complete the work by a group of employees using fuzzy sets. Open Educ. **23**(5), 74–82 (2019). ISSN: 2079-5939
18. Colorni, A., Dorigo, M., Maniezzo, V.: Distributed optimization by ant colonies. In: Proceedings of the First European Conference on Artificial Life, ECAL'91. Elsevier, Paris, France, pp. 34–142 (1992)
19. Dorigo, M., Birattari, M., Stutzle, T.: Ant colony optimization. IEEE Comput. Intell. Mag. **1** (4), 28–39 (2006). ISSN: 1556-6048
20. Ying, P., Wenbo, W., Song, Z.: Basic ant colony optimization. In: IEEE 2012 International Conference on Computer Science and Electronics Engineering, Mar 2012. ISBN: 978-0-7695-4647-6
21. Al-Salami, N.M.A.: Ant colony optimization algorithm. J. UbiCC **4**(3), 823–826 (2009). ISSN: 1992-8424
22. Wang, J., Fan, X., Wan, S.: A graph-based ant colony optimization approach for process planning. Swarm Intelligence and Its Applications (2014). ISSN: 1537-744X
23. Wang, J.F., Chu, K.Y., Wang, Q.Y.: An approach of process planning using ant colony optimization algorithm. Adv. Mater. Res. **978**, 209–212 (2014). ISSN: 1662-8985
24. Titov, Y.P.: Experience in modeling supply planning using modifications of the ant colony method in high availability systems. High Availab. Syst. **14**(1), 27–42 (2018). ISSN: 2072-9472
25. Titov, Y.P.: Modifications of the ant colony method for aviation routing problems. Autom. Remote Control. T. **76**(3), 458–471 (2015). ISSN: 1608-3032
26. Titov, Y.P., Davydkina, E.A.: Expanding the capabilities of the ant colony method through the use of fuzzy sets. Trends Dev. Sci. Educ. **54**(2), 16–19 (2019). SPLN: 001-000001-0641-LJ

Cognitive Maps of Knowledge Diagnosis as an Element of a Digital Educational Footprint and a Copyright Object

Uglev Viktor(✉), Zakharin Kirill, and Baryshev Ruslan

Siberian Federal University, Krasnoyarsk, Russia
Uglev-v@yandex.ru

Abstract. The paper discusses the problem of using the digital educational footprint (DEF), formed during the interaction of a student with intelligent automated educational systems (IAES), as a copyright object. The DEF factors and data obtained during the educational process are highlighted. It is proposed to accumulate and visualize them in the form of a cognitive map of knowledge diagnosis (CMKD) by performing sequential statistical, metric, semantic and logical concentration of knowledge. The use of CMKD in the IAES decision-making mechanisms allows not only to increase the degree of individualization of the educative impact on the student, but also to model the process of reflective governance (according to Lefebvre). The possibility of displaying various aspects of CMKD and putting the maps together into the individual and group atlases is indicated.

The DEF alienation as a copyright object entails the need to single out the aggregate of the descriptive components. A table of correspondence of the DEF components, their display formats and specifications is given. To aggregate these components, it is also proposed to use the CMDK assembly mechanism. The necessity of registering the aggregate copyright object and each of its parts through the deposit mechanism is indicated. For this purpose, it is proposed to use the digital platform for knowledge sharing and copyright management (IPUniversity) as a platform solution.

#COMESYSO1120.

Keywords: E-learning · Intelligent automated educational systems · Cognitive maps of knowledge diagnosis · Digital footprint · Copyright object · IPUniversity platform

1 Introduction

The modern economic system assumes a high level of monetization of intangible assets and their inclusion in the amount of turnover using communication technologies. These assets include copyright objects obtained both as a result of the goal-oriented human activity and indirectly, for example, when operating the automated systems. It should be noted that not only the technology for registering the copyright objects shall be provided, but also the means for their description, verification and turnover shall be developed based on the applicable regulatory/legal framework.

R. Silhavy et al. (Eds.): CoMeSySo 2020, AISC 1295, pp. 349–357, 2020.
https://doi.org/10.1007/978-3-030-63319-6_31

One of the insufficiently developed areas of managing the copyright objects is the automated education, or rather, the student's activity logs recording student's work with didactic material. Such data is accumulated in automated educational systems that have developed monitoring and control functions (as a rule, operating on the basis of artificial intelligence algorithms and virtualization technologies). An aggregate of such materials constitutes a digital educational footprint (DEF), which shall be recorded, aggregated and stored for the purpose of its subsequent use. In the digital economy, the consumers of the DEF data are educational institution staff, employers, interview panels, government agencies, etc., that evaluate the student's reputation in order to make their decisions [1]. It is obvious that this data is alienable and constitutes the personal data. Therefore, one has to know how to record, summarize and protect such data.

Let us consider the genesis of the digital educational footprint by intelligent automated educational systems (IAES) and methods for its registering in the form of the copyright object based on the technology of cognitive maps of knowledge diagnosis.

2 Methods

2.1 Digital Educational Footprint

A student's digital footprint is a set of active and passive "traces". The active traces include personal data used when enrolling for electronic courses, grade record books, correspondence between students, correspondence with a real or virtual teacher/tutor, etc. The passive DEF components contain the pattern of student's browsing through the didactic material (preferences, sequence of moves, their frequency and time indicators), actions in a virtual environment (virtual and augmented reality devices, digital laboratories and game spaces), and the effectiveness of the response to stimuli (both explicit and implicit). All of them can be included in the student's profile and form the basis for his/her model [2] in the IAES memory to be used by IAES to make decisions. The more detailed the student model is, the more qualitatively the operation of the IAES algorithms can be organized and the more effective the education can be [3].

Having received the initial data on the DEF, it is necessary to process and save it. And here it turns out that the data from the footprint mean little if they are not supplemented by the information about the electronic course and are not summarized, since they constitute an extensive array of initial lowly connected data. We will show the main stages of the DEF formation as an independent (alienated) copyright object (Fig. 1). Combining the student model and the model of an individualized electronic course within the framework of one object, one can notice that it is the electronic course that is the invariant throughout the entire educational process in the IAES. The consequence is an increase in the dimensionality of the space taken by the IAES factors that need to be recorded and analyzed.

The challenge of processing high-dimensional data can be solved both by reducing the dimension, and by changing its display format. In fact, this is a combination of data concentration (summarization) mechanisms and its visualization (mapping). The

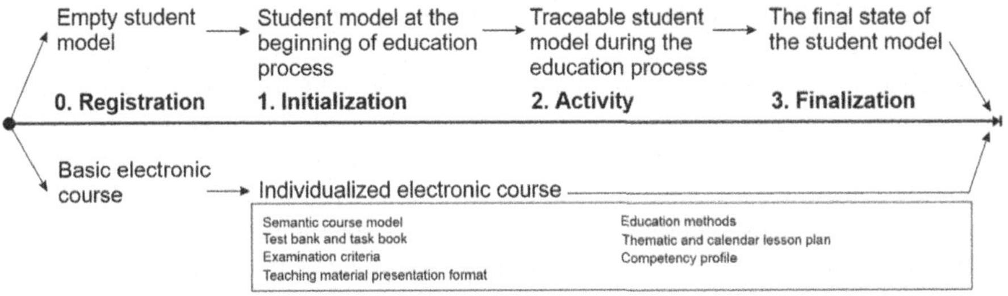

Fig. 1. Main stages of the digital educational footprint formation as a copyright object

combination of these two approaches is widely used in science both for the presentation of information and knowledge [4–6], and for mapping of cognitive processes [7–11].

2.2 Cognitive Maps of Knowledge Diagnosis

An increase in the computer hardware power and development of artificial intelligence algorithms lead to a situation where the digital educational footprint begins to be used by the developed automated educational systems as a basis for individualizing the control of the educational process and even reflection [12–14], thus increasing the concentration of information during several stages, for example, due to the formation of a Cognitive Map of Knowledge Diagnosis (CMKD) shown in Fig. 2 below.

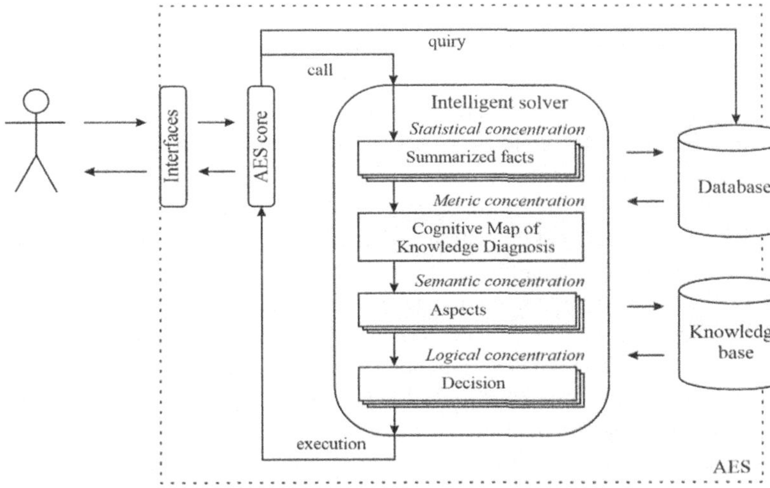

Fig. 2. Stages of summarizing information about the DEF for decision-making in the IAES as per [10]

CMKD is a map that reflects the result of the data summarization stage (metric concentration) in the IAES operation logic, that automatically prepares (translates) data about the educational process into knowledge in order to simplify a comprehensive expert analysis of the educational case study and develop an effective system response to the student actions [10]. When build up a map, invariable indicators from the DEF and variable indicators (depending on the current moment) are registered. Moreover, the map can be visualized with respect to various aspects of the analysis (competency, target, regulatory, etc.) and the models present in the IAES (student, teacher, subject tutor and methodologist). Further, depending on the tested hypotheses, the maps allow quick extraction of data for making management decisions by the intelligent core of the educational system (solver or planner).

Getting back to the need of registering and alienating the CMKD, it shall be noted that the amount of data underlying the map shall be presented in its entirety (see Fig. 1). This leads to the necessity of describing both the DEF components separately and building up an assembly from it (a composite copyright object). In this regard, maps can be arranged in an atlas, reflecting both the factor space of the student model and the dynamics of its change (within the framework of one or several academic disciplines). Thus, the CMKD combines both the basic knowledge about the student's digital footprint during the educational process and upon its completion, as well as additional data that is important for the IAES "here and now" (individualized course model). With skillful summarization of data, maps can be used both for the analysis of group (including project) work, and for "accompanying" a student during the educational process comprising more than one discipline.

3 Results

In order to illustrate the process of compiling the CMKD and to describe the general sequence of its conversion into the copyright object, let us consider an example. Let there be a basic structure of an electronic course with a variety of topics divided into separate didactic units (di), characterized not only by the initial sequence of presentation, but also by inter-element cause-and-effect links of the education material. Figure 3 (left) shows the basic configuration of the electronic course "Simulation Modeling" (target aspect) taught within a master's program "Computer Science and Computer Engineering" (Department of Applied Physics and Space Technologies at Siberian Federal University). Such a course structure is taken as the basis for constructing the metric space of the map, coding not only communications (small world model), but also importance (color), function (form), and variability of implementation.

Having data from the invariable (questionnaire when registering for the course) and variable (admission test) parts of the DEF, the basic course structure enters the IAES solver, which, based on the reflective governance model and dialogue with the user, synthesizes the course configuration from the perspective of the student model (Fig. 3, central). Through the mechanism of compromise search (subject tutor model), the course structure shown in Fig. 3 (right) is built up at the output of the individualization algorithm: in the given example the number of topics of the material studied was reduced by 24%, but its complexity (the depth of study and examination) slightly increased, because the student indicated the importance of individual topics of the course for his professional growth. Based on the updated structure of the didactic

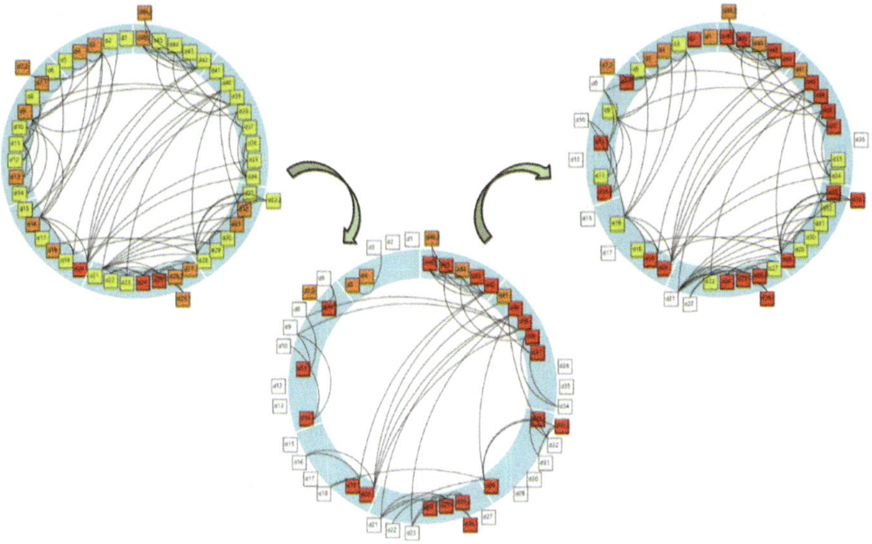

Fig. 3. Reconfiguration of the electronic course content in the process of its individualization (initial stage of the educational process) as per [15]

material of the electronic course, which reflects a compromise between the goals of the student, teacher and methodologist models, the examination criterion is specified (in terms of composition and depth), as well as the data for the student's DEF are updated, which allows the IAES to put more appropriate emphasis while managing the educational process and synthesizing the dialogues. In fact, the map can be synthesized at any time during the student's interaction with the educational system and it constitutes the basis for making individualized decisions. Without going into the details of these processes (see the references to the article for more details), we will discuss the possibility of alienating and using the CMKD as an element of the DEF.

4 Discussions

Modern IAES and LMS use their own course storage formats. They are not suitable for describing the alienable copyright objects associated with the DEF. Let us single out the data on the electronic course and the user models in the form of an aggregate of connected structures, the characteristics of which are given in Table 1. The analysis shows that it is possible, but rather problematic, to build up a homogeneous copyright object to obtain an alienated DEF, even if all the information is encoded into a single data format (for example, into CSV or database tables). Moreover, the computing capability for storing and processing the aggregate of such DEFs will become large, due to unreasonably excessive storage.

How, then, to "assemble" the CMKD and the DEF into a single entity? To solve this issue, one can use a composite (fragmented) copyright object [16]: individual

Table 1. DEF components that allow describing the CMKD in the form of an aggregate copyright object

Component	Composition	Display	Specification
The basic model of the electronic course	The list of didactic units, their characteristics for the basic, individual and resulting compositions	Object model or table	Electronic course model (independent type of the copyright object)
Student profile	Questions, answers, parameters	Table	Initial data arrays
Test results	Tasks, answers, parameters	Table	Initial data arrays
The pattern of browsing through the course material	"Coordinates", time, actions	Table	Initial data arrays
Log of dialogs with the system	Questions, answers, response	Table	Initial data arrays
IAES hypotheses and decisions verification report	Event, old status, new status, basis, response type	Table	Initial data arrays
Atlas of cognitive maps of knowledge diagnosis	Data table for visualization and processing of the map/atlas	Table	Knowledge bases

components are independent entities, and a separate copyright object containing only the data that is important for the assembly combines them. Storage of each component listed in Table 1 in the form of an independent copyright object has the following advantages:

- flexibility to describe each component;
- compactness and atomicity of data storage;
- high degree of depersonalization of personal data;
- the ability to set different access rights to individual DEF components;
- variability of the export of the DEF data to external sources;
- acceleration of computational processes during group monitoring and analysis of individual educational patterns;
- the ability to make links between the components (preservation of semantic integrity without duplication of data);
- the ability to refer to each component individually (as well as to count their scientometric parameters);
- the ability to integrate additional footprint components (expand) by interacting with various educational systems (openness to a composition).

To put into practice such an approach to the copyright object decomposition, it is necessary to have:

- the standardized data description template for each deposited component;

- the information system for description, storage and processing;
- the data exchange modules that are compatible or integrated with modern IAES and LMS for loading the aggregate of information about the educational process and building up the DEF (including CMDK).

The "Digital platform for knowledge sharing and copyright management" [17] can be used as the basic information system for processing the DEF: it allows to deposit the digital copyright objects of various types, form links between previously deposited objects and provide access to them for the copyright holders. Table 1 (the right edge column) contains the copyright object types (except for the first one), which are already available for depositing on the platform. Based on them, it is possible to make a composite copyright object corresponding to the DEF. Data from various IAES and LMS can be exported both by internal means of each system (for example, scripts), and using external enquiries. Thus, the DEF formation is potentially possible on the basis of any educational system whose functions support detailed logging of data on the educational process.

The legislative component is one of the remaining unresolved issues concerning the use of the DEF. At the moment, there is no legal regulation of the following issues:

- What, from a legal point of view, can be considered a digital footprint and its components?
- At what level of depersonalization does the footprint data cease to have the status of personal data?
- What legal status will individual CMDK or educational profiles have?
- What footprint data is it legitimate to transfer between educational systems when a student transits between them?
- What footprint components does the institution and higher authorities have the right to use while carrying out their functions?
- How should a student consent to the use of his/her DEF?
- Can a graphic representation of CMDK be freely distributed (for example, as illustrative material), if it is only depersonalized?

All this only slows down the process of recognition of the DEF as an object of copyright and prevents the legalization of its turnover. We are also actively working in this direction.

5 Conclusion

The mechanisms of operation of modern intelligent automated educational systems encompass an increasing volume of factors for effective decision-making. The use of a digital educational footprint as a single entity shall lead to the standardization of individual mechanisms of LMS operation. As a result, the formation of a cognitive map of knowledge diagnosis allows solving a number of issues both within a single educational system and during student transition between educational platforms [18].

Legalization of cognitive maps and the entire digital educational footprint (including student activity logs recording student's work with virtualization devices) is a complex task that requires not only algorithmic solutions, but also a deep methodical study. From this point of view, such projects as IPUniversity.ru can become those points of growth that will form a new culture of using digital data while implementing the distance and automated education.

Acknowledgements. This research is supported by the Ministry of Science and of Higher Education the Russian Federation (research theme code №. FSRZ-2020-0011).

References

1. Bates, C.: Take charge of your online reputation. https://er.educause.edu/articles/2018/10/take-charge-of-your-online-reputation. Accessed 25 May 2020
2. Karpenko, A., Dobryakov, A.: model for automated training systems. Overview Sci. Educ. **7**, 1–63 (2011)
3. Uglev, V.: New generation of intelligent automated educational systems: main attributes and principles of organization. In: Perspective Methods and Instruments of the Intellectual Systems: Proceedings of the All-Russian Science School for Yang Scientists, NSTU, Novosibirsk, pp. 37–40 (2015)
4. Blymenau, D.: Problems of the Scientific Data Compression. Nauka, Moscow (1982)
5. Zenkin, A.: Cognitive Computer Graphic. Nauka, Moscow (1991)
6. Tufte, E.: The Visual Display of Quantitative Information, 2nd edn. Graphics Press, Cheshire (2001)
7. Tolman, E.: Cognitive maps in rats and men. Psychol. Rev. **55**, 189–208 (1948). https://doi.org/10.1037/h0061626
8. Brusilovsky, P., Rus, V.: Social navigation for self-improving intelligent educational systems. In: Design Recommendations for Intelligent Tutoring Systems, US Army Research Laboratory, Orlando, vol. 7, pp. 131–145 (2019)
9. Uglev, V., Kovaleva, T.: An application of cognitive visual representation as a tool to support individual education. In: Science and Education 3 (2014), http://technomag.bmstu.ru/en/doc/700661.html. Accessed 25 May 2020
10. Uglev, V.: Implementation of decision-making methods in intelligent automated educational system focused on complete individualization in learning. AASRI Procedia **6**, 66–72 (2014). https://doi.org/10.1016/j.aasri.2014.05.010
11. Uglev, V., Suchinin, D.: Automated education: tendency for scientific approaches convergence. In: ICASSR2014: International Proceedings, pp. 20–23. Atlantis Press, Paris (2014). https://doi.org/10.2991/icassr-14.2014.6
12. Uglev, V.A.: On the specifics of individualization of learning in automated educational systems. Philos. Educ. **2**, 68–74 (2010)
13. Lefebvre, V.: Lectures' About the Theory of Reflexive Games. Cogito-Centre, Moscow (2009)
14. Uglev, V.: The base model of reflexive process in the intellectual automation educational systems. Math. Struct. Model. **1**, 111–121 (2018)
15. Uglev, V.D.: The results of an experiment on individualizing the route for studying discipline. In: Proceedings II International Eurasian Pedagogical Conference, Science and Education, Penza, pp. 26–29 (2018)

16. Uglev, V., Feodorov, Y.: Evaluation of the indicator of the copyright contribution, copyright "cleanliness" and citation of fragmented resources. In: XXVI All-Russian Seminar Neuroinformatics, its Applications and Data Analysis, ICM SO RAS, Krasnoyarsk, pp. 133–137 (2018)
17. IPUniversity Homepage. https://ipuniversity.ru. Accessed 25 May 2020
18. Uglev, V., Cholodilov, S., Cholodilova, V.: Map as a basis for decision-making in the automated learning process. In: Applied Methods of Statistical Analysis. Nonparametric Methods in Cybernetics and System Analysis: Proceedings of the International Workshop, NSTU, Novosibirsk, pp. 325–334 (2017)

Distributions of the Collision Times Between Two Atoms That Have Overcome the Potential Barrier on the Surface

Sergey Zheltov[1] and Leonid Pletnev[2(✉)]

[1] Tver State University, Tver, Russia
zheltov_s@mail.ru
[2] Tver State Technical University, Tver, Russia
pletnevl@yandex.ru

Abstract. The evaporation of atoms from the surface of the condensed phase is of interest both from the practical and theoretical point of view. The existence of a potential barrier on the surface of the condensed phase is a necessary condition for solving the problem of the movement of atoms over the surface of the condensed phase. The size of the potential barrier, the surface temperature of the condensed phase, and the weight and size of the atoms affect the density of atom collision distributions over time and the average collision times.

This study establishes the densities for the distributions of collisions between two atoms over time and the average values of these distributions for different values of these parameters. It determines the regularities connecting the distributions of two colliding atoms by velocity as a function of these parameters. The Monte Carlo method was used in computer experiments. The algorithm of the method was adapted to parallel computing on GPUs with the CUDA architecture.

Keywords: Collision of atoms · Monte Carlo · Potential barrier

1 Introduction

The process of substance evaporation from the surface of a condensed phase is closely related to the process of heat and mass transfer of a substance in systems. One of the current problems is the process of heat and mass transfer in microsystems [1–5]. Atoms or molecules have to overcome the forces of intermolecular interaction with surface molecules. It is related to overcoming a potential barrier U on the surface of the condensed phase. Evaporation of atoms from the surface of the condensed phase is the simplest case of evaporation, since in this case the internal degrees of freedom that complicate the description of the evaporation process are not considered. The analytical microscopic approach is related to the solution of the Boltzmann equation or model equations taking into account boundary conditions [6–10]. More complex boundary conditions were obtained for the evaporation of Lennard-Jones liquids and the evaporation of binary mixtures in [11–14]. Of considerable interest is the work [15] in which the method of molecular dynamics is used to obtain boundary conditions during evaporation.

R. Silhavy et al. (Eds.): CoMeSySo 2020, AISC 1295, pp. 358–367, 2020.
https://doi.org/10.1007/978-3-030-63319-6_32

Studies [16, 17] conducted detailed research on overcoming the potential barrier U by atoms on the surface of the condensed phase and applying the results to the analysis of the heat and mass transfer process in open cylindrical systems by the Monte Carlo method [18]. When the atoms overcame the potential barrier, a constant value equal to U was subtracted from the component of the atom's kinetic energy, perpendicular to the surface. The functions of atom velocity distributions, the average values of the velocity components and the energy of the atoms that have overcome the potential barrier were determined using the Monte Carlo method and theoretically confirmed. It was found that with the increase of the dimensionless parameter $r = U/kT$, where k is the Boltzmann constant and T is the surface temperature of the condensed phase, the value of the average velocity component, perpendicular to the surface, increases and tends to an asymptotic value. At $r = 0$, the distribution of outgoing atoms is equally probable, and at $r \to \infty$ tends to the cosine distribution law.

2 Modeling

Simulating the evaporation of atoms from the surface of a condensed phase is possible only from a limited microscopic area of the surface. Even in this case, there are problems with modeling the evaporation process, because a large number of atoms evaporate and it is necessary to take into account their mutual collisions over the surface. Some of the atoms may return to the condensed phase after the collisions. In this regard, the problem of the collision time of two escaped atoms above the surface, the density of distribution and the average collision time of these atoms in time is relevant.

In this paper, studies are carried out to determine the densities of collision distributions and the average collision times of two atoms in time, as a function of the parameter r, the temperature of the system T, and the length of the evaporation area a. Model experiments were performed for atoms with radii $R = 1.5 \times 10^{-10}$ m and masses $m = 40$ a. e. m. Densities for the distributions of velocity components v_z of colliding atoms are obtained as a function of the parameters under consideration.

The Monte Carlo simulation algorithm was as follows. The positions of two atoms on the surface of the condensed phase ($z = 0$) were played out using a uniformly distributed random variable sensor. The minimum distance between the centers of the atoms was chosen to be greater than their diameter. Then the components of the velocities of these atoms and the ability to overcome the potential barrier were played out, and the possibility of collision for these atoms was determined. A diagram of the evaporation area and possible positions of atoms on the surface is shown in Fig. 1.

In each computer experiment, for the specified parameters, departures of pairs of atoms were played out until the number of atom pair collisions reached 300,000,000. This task has data parallelism and can be effectively solved on NVIDIA CUDA GPUs. The new parallel algorithm developed taking into account the features of the computing platform made it possible to increase efficiency and reduce actual time costs [19]. To generate random variables distributed according to normal and uniform laws, use was made of the CURAND library and generators with corresponding parameters. The results of numerical simulation were obtained on the NVIDIA Tesla K80 GPU of the heterogeneous "HybriLIT" platform [20].

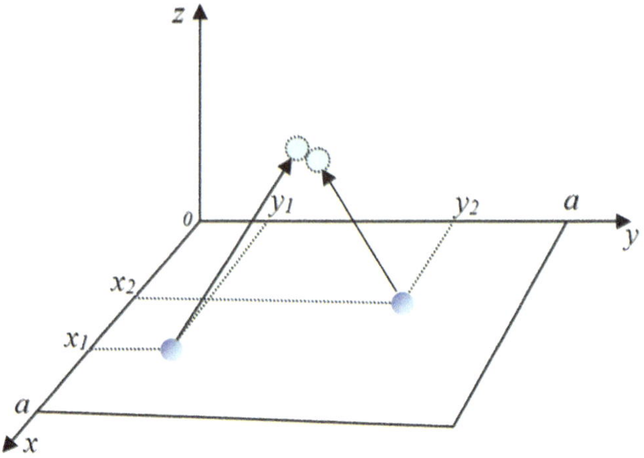

Fig. 1. Scheme of departure of two atoms from the surface of the condensed phase.

3 Analysis of the Results Obtained

Figure 2 and 3 present the results of calculations for the density distributions of atom collision times as a function of the parameter *r* and surface temperatures of the condensed phase $T = 50$ K and $T = 100$ K respectively. It was found that the minimum and maximum times between the collisions of two atoms differ by several tens of millions of times. All time density distributions have the same form – in extreme cases they tend to zero, and have a maximum that depends on the parameters of the condensed phase.

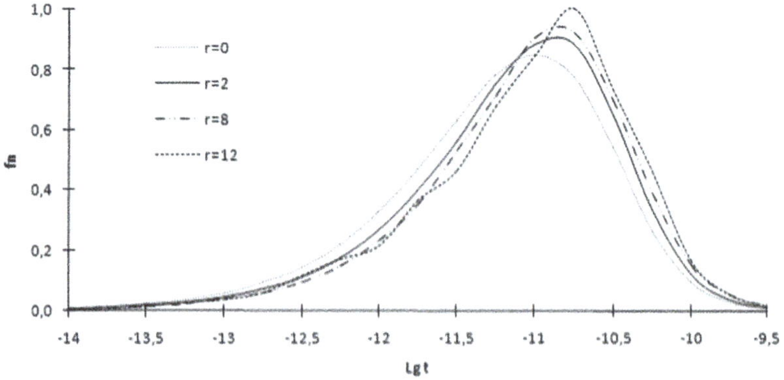

Fig. 2. Densities of distributions of atom collision times as a function of the parameter r. T = 50 K.

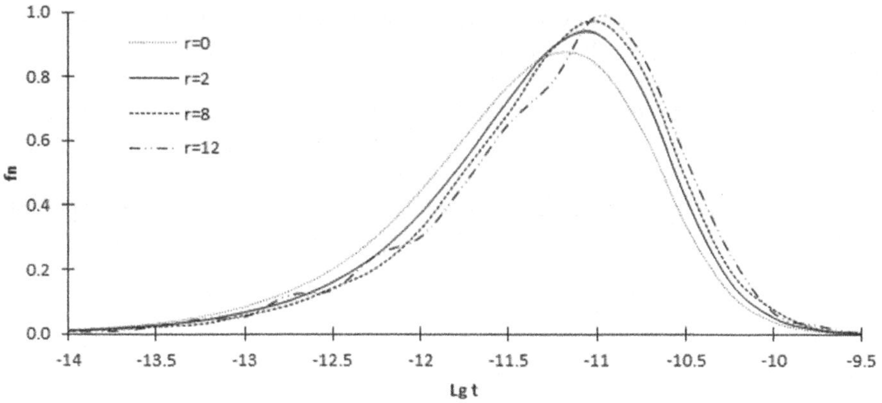

Fig. 3. Densities of distributions of atom collision times as a function of the parameter r. T = 100 K.

As the r parameter increases, both in the first and second cases, the maximum values increase and shift to the right. This is because as the parameter r increases, the average velocity components perpendicular to the surface of the condensed phase increase, while the other components remain constant.

This leads to the fact that the contribution of small collision times decreases, while the contribution of large ones increases. It is important to note that the calculation results obtained for the value $r = 12$ differ from the limit for $r \to \infty$ by less than a few hundredths of a percent [17].

Figure 4 shows the results of calculations for the average collision times between two atoms as a function of the parameter r for two values of surface temperatures. Two factors affect the appearance of these distributions. On the one hand, as the parameter r increases, the average velocity of a component perpendicular to the surface increases. On the other hand, as the surface temperature increases, the velocity components of the departing atoms increase, which leads to a decrease in the travel time before the collision. The positions of the maxima shift slightly with increasing parameter r.

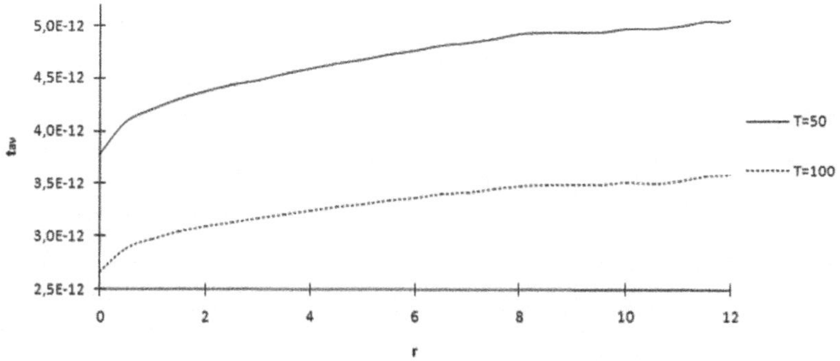

Fig. 4. Distribution of average atom collision times as a function of the parameter r.

Figure 5 and 6 represent the densities of distributions of atom collision times as a function of the temperature T for the parameter values $r = 2$ and $r = 8$. All curves have the same type of distributions, with the same maxima, but the positions of the maxima depend on the surface temperature of the condensed phase. It was found that for both parameter values, with increasing temperature the maxima of distributions shift to the left, toward smaller time intervals for atom collisions. This can be explained by the fact that, as the temperature increases, the average values of the velocity components, perpendicular to the surface, of the ejected atoms increase. This reduces the travel time of the atoms before the collision.

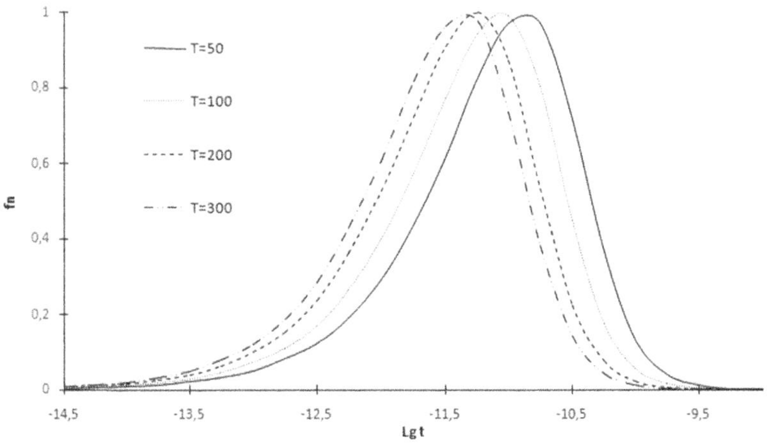

Fig. 5. Densities of distributions of atom collision times as a function of the temperature T. $r = 2$.

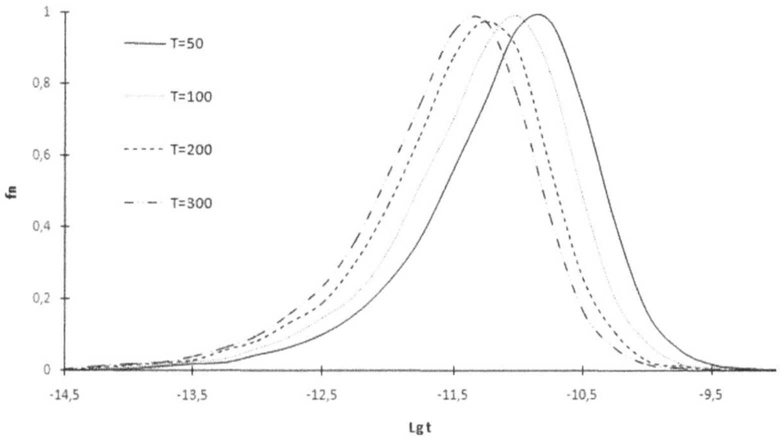

Fig. 6. Densities of distributions of atom collision times as a function of the temperature T. $r = 8$.

The data relating to the regularities of the average atom collision times as a function of the temperature, shown in Fig. 7, are of considerable interest. As the temperature increases, the average values of collision times decrease, which is associated with the increase in the average velocity components, perpendicular to the surface, for escaped pairs of atoms. It is the case that, for a certain value of the parameter *r*, the average values of collision times equal to the square root of the inverse ratio of temperatures.

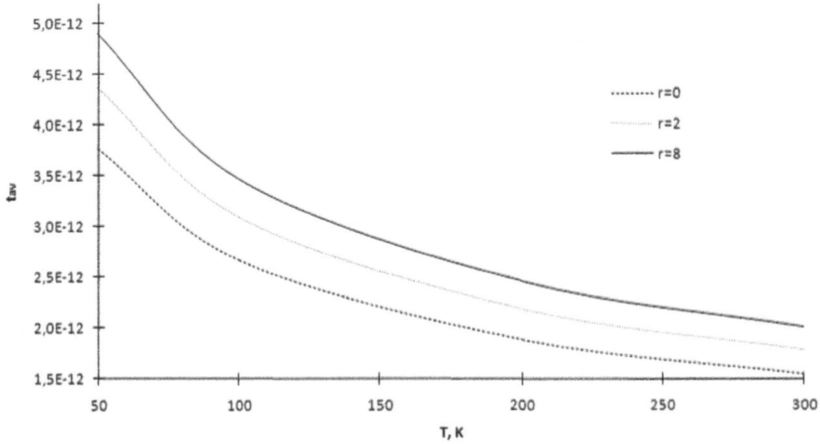

Fig. 7. Distribution of average atom collision times as a function of the temperature *T*.

The computer experiments conducted established the regularities associated with determining the densities of distributions of atom collision times as a function of the evaporation area *a*. The calculation results are shown in Fig. 8 and 9. The size of the condensed phase evaporation area changed more than 60 times compared to previous calculations. Similar distribution patterns are obtained for both calculations with parameter $r = 2$ and calculations with parameter $r = 8$. As the size of the evaporation area *a* increases, the distribution maxima shift to the right, toward larger collision times by an order of magnitude. Distribution maxima are reduced in size by almost half.

These regularities in distributions can be explained by the fact that the probability of collisions decreases as the area of evaporation increases, since the distances between the escaped atoms and the collision times between the atoms increase.

The distribution of the average atom collision times as a function of the evaporation area *a* is shown in Fig. 10. As the evaporation area increases, the average collision times of two atoms increase almost linearly. Regularities in the arrangement of distribution curves relative to each other for different values of the parameter *r* are explained by the fact that, as the parameter *r* increases, the average velocity components, perpendicular to the surface, increase, but the probability of collisions at short distances from the surface decreases.

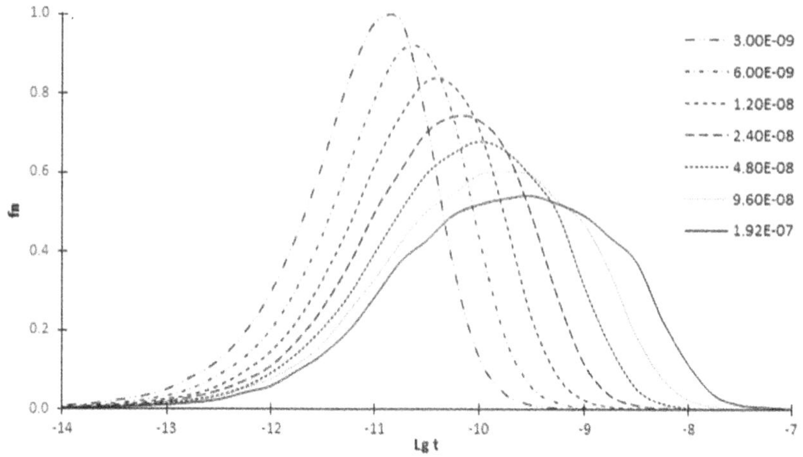

Fig. 8. Densities of distributions of atom collision times as a function of the evaporation area *a*. $T = 50$ K. $r = 2$.

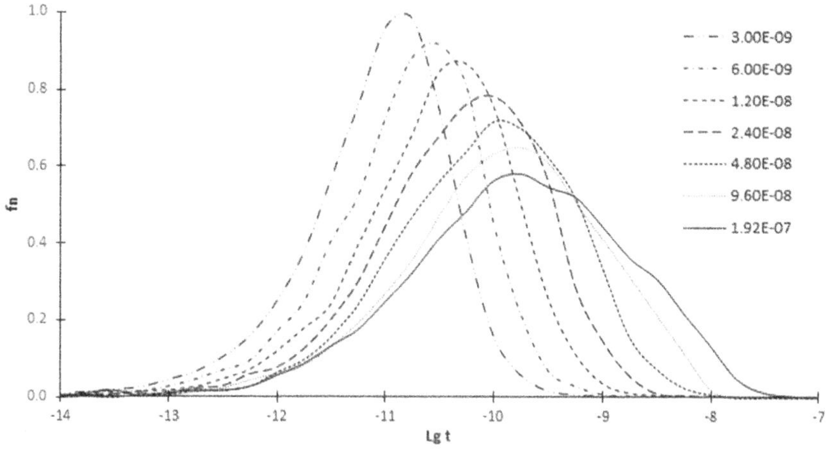

Fig. 9. Densities of distributions of atom collision times as a function of the evaporation area *a*. $T = 50$ K. $r = 8$.

The time between atom collisions depends on the velocity components of these atoms. The regularities of distributions of the average v_z components as a function of the parameter r for two temperature values are shown in Fig. 11.

The distributions represent almost parallel lines that do not depend on the parameter r. The ratios of the average speeds of the components are equal to the square root of the inverse ratio of temperatures. A more detailed consideration of the distributions of the mean v_z component values allowed us to establish more complex dependencies on the parameter r, shown in Fig. 12. The distributions have corresponding maxima. The difference between the maximum and minimum values does not exceed 0.5 m and is about 0.6%.

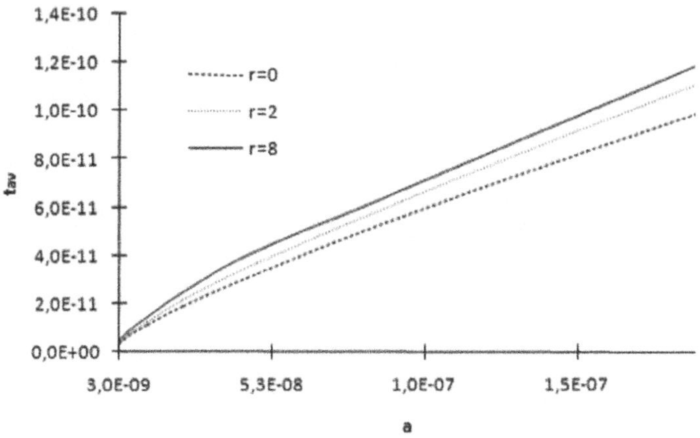

Fig. 10. Distribution of average atom collision times as a function of the evaporation area *a*. $T = 50$ K.

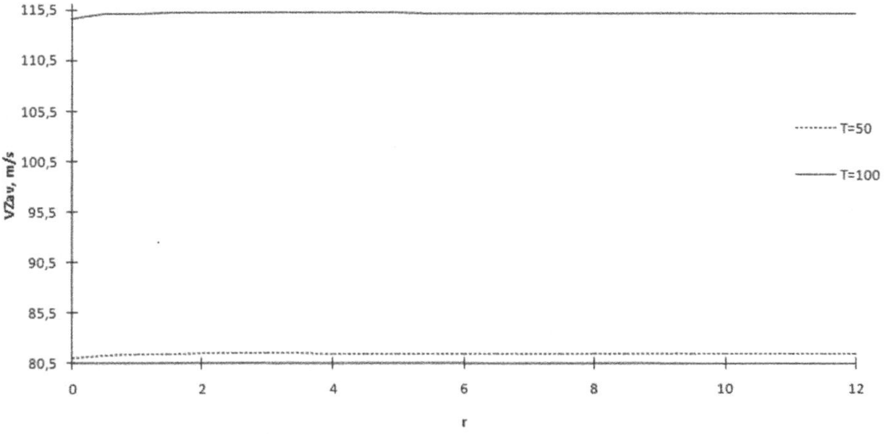

Fig. 11. Distribution of average atom velocities as a function of the parameter *r*. $T = 50$ K.

In calculations using the Monte Carlo method or molecular dynamics, the question arises as to the accuracy of the results obtained. As the number of atoms played out increases, the accuracy increases. Figure 13 shows the results of calculations for 20 computer experiments to determine the average values of the velocity components v_z, perpendicular to the surface of the condensed phase. All data are in the band 0.016 m/s, i.e. the accuracy of the results obtained is about 0.01%. These data confirm the correctness of the results obtained for individual computer experiments and the possibility of similar calculations with a given number of atoms.

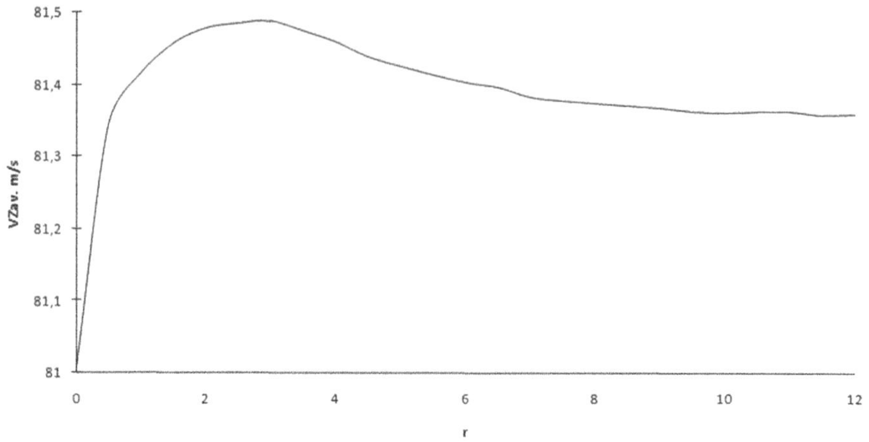

Fig. 12. Distribution of average atom velocities as a function of the parameter r. $T = 50$ K.

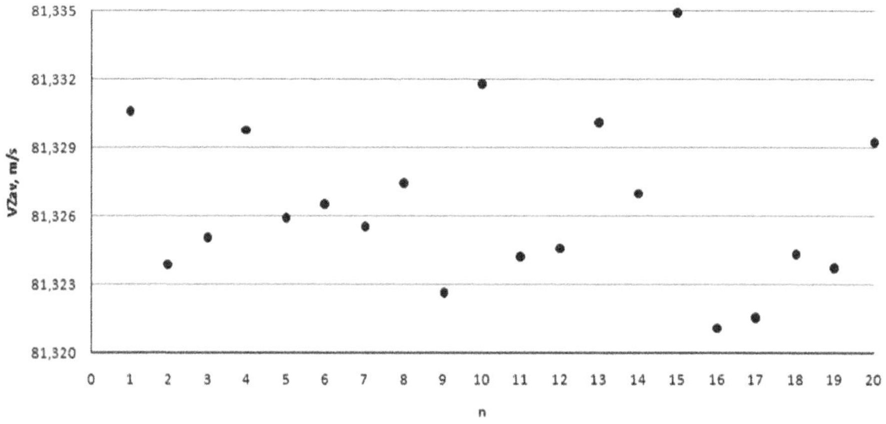

Fig. 13. Distributions of average velocity values for 20 computer experiments. $r = 0$, $T = 50$ K.

4 Conclusions

This paper determines the regularities of collisions between pairs of atoms that have overcome the potential barrier on the surface of the condensed phase and establishes dependences for the density distributions of atom pair collisions and the average collision times as a function of the parameter r, the surface temperature T and the size of the evaporation area a. The new regularities obtained for distributions of collisions between two atoms in time suggest that collisions of these atoms in space above the surface of the condensed phase will also have a complex form, different from the exponential distribution law. The proposed approach to obtaining collision times for two atoms can be extended to obtain collision distributions of three, four, etc. atoms.

References

1. Lee, J., Laoui, T., Karnik, R.: Nanofluidic transport governed by the liquid/vapour interface. Nat. Nanotechnol. **9**, 317–324 (2014)
2. Li, Y., Alibakhshi, M.A., Zhao, Y., Duan, C.: Exploring ultimate water capillary evaporation in nanoscale conduits. Nano Lett. **17**, 4813–4819 (2017)
3. Wilke, K.L., Barabadi, B., Lu, Z., Zhang, T., Wang, E.N.: Parametric study of thin film evaporation from nanoporous membranes. Appl. Phys. Lett. **111**, 171603 (2017)
4. Xiao, R., Maroo, S.C., Wang, E.N.: Negative pressures in nanoporous membranes for thin film evaporation. Appl. Phys. Lett. **102**, 123103 (2013)
5. Lu, Z., Narayanan, S., Wang, E.N.: Modeling of evaporation from nanopores with nonequilibrium and nonlocal effects. Langmuir **31**, 9817–9824 (2015)
6. Frezzotti, A.: Boundary conditions at the vapor-liquid interface. Phys. Fluids **23**, 030609 (2011)
7. Tcheremissine, F.G.: Solution to the Boltzmann kinetic equation for high-speed flows. Comput. Math. Phys. **46**, 315–329 (2006)
8. Ishiyama, T., Fujikawa, S., Kurz, T., Lauterborn, W.: Nonequilibrium kinetic boundary condition at the vapor-liquid interface of argon. Phys. Rev. E **88**, 042406 (2013)
9. Kon, M., Kobayashi, K., Watanabe, M.: Liquid temperature dependence of kinetic boundary condition at vapor–liquid interface. Int. J. Heat Mass Transfer **99**, 317–326 (2016)
10. Persad, A.H., Ward, C.A.: Expressions for the evaporation and condensation coefficients in the Hertz-Knudsen relation. Chem. Rev. **116**, 7727–7767 (2016)
11. Cheng, S., Lechman, J.B., Plimpton, S.J., Grest, G.S.: Evaporation of Lennard-Jones fluids. J. Chem. Phys. **134**, 224704 (2011)
12. Kon, M., Kobayashi, K., Watanabe, M.: Method of determining kinetic boundary conditions in net evaporation/condensation. Phys. Fluids **26**, 072003 (2014)
13. Kon, M., Kobayashi, K., Watanabe, M.: Kinetic boundary condition in vapor-liquid two-phase system during unsteady net evaporation/condensation. Eur. J. Mech. B Fluids **64**, 81–92 (2017)
14. Kobayashi, K., Sasaki, K., Kon, M., Fujii, H., Watanabe, M.: Kinetic boundary conditions for vapor–gas binary mixture. Microfluid Nanofluid **21**, 53–59 (2017)
15. Kobayashi, K., Hori, K., Kon, M., Sasaki, K., Watanabe, M.: Molecular dynamics study on evaporation and reflection of monatomic molecules to construct kinetic boundary condition in vapor–liquid equilibrium. Heat Mass Transf. **52**, 1851–1859 (2016)
16. Pletnev, L.V., Gamayunov, N.I., Zamyatin, V.M.: Computer simulation of evaporation process into the vacuum. In: Uvarova, L.A., Arinstein, A.E., Latyshev, A.V. (eds.) Mathematical Models of Non-Linear Excitations, Transfer, Dynamics, and Control in Condensed Systems and Other Media, Boston, Dordrecht, London, Moscow , pp. 153–156. Kluveer Academic/Plenum Publishers, New York (1999)
17. Pletnev, L.V.: Monte Carlo simulation of evaporation process into the vacuum. J. Monte Carlo Methods Appl. **6**(3), 191–203 (2000)
18. Pletnev, L.V., Gvozdev, M.A., Samartsau, K.S.: Computer modeling of particles transport stationary process in open cylindrical nanosystems by Monte Carlo method. J. Monte Carlo Methods Appl. **6**(2), 191–203 (2009)
19. Zheltov, S.A.: Features of adaptation of computational algorithms to the CUDA architecture. In: Proceedings of the Mathematical Methods of Controlm, Tver State University, Tver, pp. 33–36 (2011)
20. The heterogeneous “HybriLIT” platform. http://hlit.jinr.ru/. Accessed 2 Mar 2020

Graphical Method of Intellectual Simulation Models' Analysis on the Basis of Technical Systems' Testing Results

Olga Isaeva(✉), Ludmila Nozhenkova, Nikita Kulyasov, and Sergey Isaev

Institute of Computational Modelling SB RAS, Krasnoyarsk, Russia
isaeva@icm.krasn.ru

Abstract. In this article we present graphical tools for models' analysis using expert knowledge for solving functional tasks. We have studied special models of technical systems' work logics simulation used in spacecraft onboard equipment production. We also took a close look at the modern approaches to intellectual models' assessment and verification. We suggest our own method of model assessment on the basis of technical systems' testing. The unicity of our method is that it allows not only to compare the quantitative characteristics of the equipment, but also to study logical sequences of the model's actions. In the method, model's structural elements (blocks) describing technical devices, switching interfaces and junctions defining the ways of information interaction – have been formalized. We provide an example of the onboard equipment's command-and-software control model's architecture. We suggest an algorithm for creating the control points and criteria of model analysis. We present graphical tools allowing to build functional dependencies of model's elements visually and in an interactive mode. We provide tools for test monitoring. The results of tests are saved in a data storage and are used for automated comparison with simulation tests in the chosen control points. We suggest the criteria for interpretation of the results of data comparison. Our method substitutes a complicated manual process of knowledge base studying with the tools of automated digital support. This method is highly potential for scaling for analysis of big models.

Keywords: Simulation modeling · Spacecraft onboard equipment · Knowledge base · Testing · Model analysis

1 Introduction

In the modern digital world, complex systems study methods on the basis of computer modeling are becoming more popular. They provide efficient support to decision making in different spheres of human activity, replacing laborious and expensive experiments with digital models of real objects' operation [1]. Simulation allows to create abstracts of physical reality which not only reduces the time of studying and extends the intervals of observations, but also solves the problems related to the necessity of studying the processes that are difficult to repeat multiple times in real conditions.

R. Silhavy et al. (Eds.): CoMeSySo 2020, AISC 1295, pp. 368–376, 2020.
https://doi.org/10.1007/978-3-030-63319-6_33

Choice of the method of modeling and the necessary detailing of models depend much on the stage of a product's life cycle and on the purposes of model creation. Speaking about high-tech production such as spacecraft onboard equipment, besides math models, it is necessary to build simulation models describing systems' functions on the basis of different and often hard-to-be-formalized factors influencing the object's behavior in whole. In this case, models must provide expert knowledge in a compact and comfortable way. The models based on use of expert knowledge for solving technological and functional tasks got the name of intellectual simulation systems (intellectual simulation models) [2]. Application of expert knowledge allows to simulate dynamic behavior of analyzed objects. It is used at building of the concept of a spacecraft's mission, during onboard systems design for their efficiency checking and also during operational control and for diagnostics of equipment failure [3].

For support of the onboard equipment design, the Institute of Computational Modeling has developed a software complex "Software-and-mathematical model of the command-and-measurement system of a spacecraft" [4]. It is designed for graphical configuration of the onboard systems, creation of knowledge bases with a logical inference production model, simulation of the onboard equipment operation and analysis of the equipment on the basis of simulation modeling precedents [5]. The complex has been implemented at a company-manufacturer of satellite systems.

Application of methods and technologies of the design support deals with use of large formal data volumes, diversity and complexity of model structures and knowledge bases. All this plus the human factor sets significant requirements for provision of the specified level of trust [6] to a model while performing the reality. Literature provides different approaches to assessment of the quality of simulation models. Article [7] provides the method of models validation based on application of empirical data and knowledge from the related areas. Knowledge plays the role of a benchmark during verification of a model and varies from a real system's data to a quality experts' experience. Article [8] provides a simulation modeling environment with the architecture where the methods of model verification and test generation are embodied. Article [9] suggests an intellectual method of model verification based on the analysis of similarity between the time series of modeling from a computerized model and the time series observed from the real system. The task is formulated as a problem of classification including many similarities to a complex measure allowing to forecast the level of confidence on the basis of training samples. The authors of this study developed the methods of structural and graphical analysis that formally automate the search of errors by heuristic criteria and build dependencies of model's elements, reveal blank spots in knowledge bases, missing and excess data and structures for which the rules had not been defined, providing control over full functional presentation [10]. Although that production knowledge model and automation of formal search of models' errors are simple and clear, the problem of analysis of a model's function in full has not been solved. In our article, we suggest original graphical tools allowing to compare the results of a model's work with the data received during autonomous testing of the onboard equipment. Such control will provide building of quality simulation models which is an important scientific component of an efficient support of complex technical systems design.

2 Intellectual Simulation Model

In order to create a graphical method of analysis of intellectual simulation models, let's formalize structural elements, determine the control points and set the criteria of models' assessment on the basis of the spacecraft onboard equipment test results.

Main structural elements of models are intellectual agents – the objects that, on the basis of input data and external influence, use the embedded knowledge and get the results changing the condition of a model [11]. Figure 1 shows the structure of the model's element: X – the input parameters (influences); K – the control commands; Y – the output parameters (observations), F – methods of model's functioning, $R: A \rightarrow Z$ – knowledge base's rules, A – conditions for rule completion, Z – actions caused during rule completion, V – program library (implementation of mathematical methods). Entry/exit points for commands and parameters' transmission form the number of switching interfaces I.

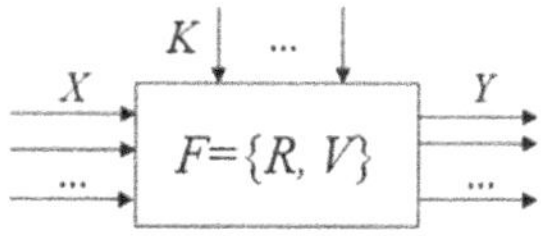

Fig. 1. Structure of model element B_i.

B – the set of model's elements, $B_i \in B$, $B_i = \langle N_i, T_i, I_i, X_i, Y_i \rangle$, $i = [1, \ldots, |B|]$, $|B|$ – the set of model's elements, N_i – the name of the i-th element, T_i – the type of device, $I_i \subseteq I$ – the subset of switching interfaces, $X_i \subseteq X$, $Y_i \subseteq Y$ – the subset of parameters. R^i – rules of B_i functioning. For interaction of the model's elements, a set of commutation links C is introduced: $C_{ij} \subseteq C$, $C_{ij} = \langle I_i, I_j, \tau_{ij} \rangle$, I_i – interface of B_i, I_j – interface of B_j, τ_{ij} – time of signal passing between interfaces. An example of the structure of a model with commutation links is shown in Fig. 2.

In figure: B_1 – command-and-measurement system's model; B_2 and B_3 – models of the ground and the onboard control complexes, B_4 – remote indication model, B_5 and B_6 – models of receivers and transmitters. Commutation links $\{C_{ij}\}$ describe the ways and directions of interaction and allow us to combine simulators into a complex system for modeling of spacecraft onboard equipment operation. This structure is designed for modeling of the onboard equipment with command control. Commands are impacts on the spacecraft's onboard systems, they allow us to change settings of the receiving and transmitting path, switch between the working sets of equipment, turn on the modes of range measuring, telemetry information transmission, authentication, etc.

The model's work logics is the following: functioning begins from creation and transmission of the commands from the ground control complex to the onboard systems; via data reception-transmission devices, the commands enter the command-and-measurement system; on board of a spacecraft, the commands are processed, analyzed and executed; in accordance with the results of the actions the commands are acknowledged, and the receipts as well as the information about the onboard systems' status are gathered by the onboard remote indication equipment and transmitted to

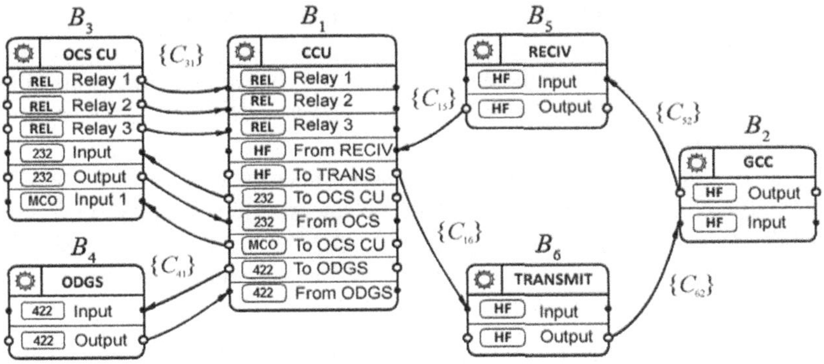

Fig. 2. An example of the onboard equipment operation simulation model.

Earth in telemetry data packages; the command execution control is performed by the ground control complex in accordance with the telemetry received. Onboard systems' configuration, their work parameters, order of interaction, structures of the command and telemetry data packages are specified in technical documentation.

3 Graphical Method of Model Analysis

Methods of simulation modeling are set in the knowledge base on the basis of condition-action rules. During modeling, a logical inference is completed [11], rules applicable to the current state of the model are chosen, and the actions described in these rules are completed. Model analysis and assessment of its compliance with the real equipment demands comparison of the results of simulation tests with the data received during onboard systems' testing.

For making the analysis, it is necessary to choose the rules describing command interaction of the onboard systems and the command execution results. The algorithm is as follows:

1. Set the commands incoming to simulators through commutation interfaces. For this the function *Projection* $Pr(A, K)$ of set A to the set of commands K is used: $Pr(A, K) = \{k \in K \mid k \text{ is included in } A\}$. Let's define $K' = Pr(A, K)$.
2. Highlight the rules where the methods of command completion are set. For this the function *Selection* $Sel(R, \theta)$ is used. It forms the subset of elements from R satisfying the condition θ. In our case $R' = Sel(R, k \in K')$.
3. For each rule of R', build chains of rules $Rch^i \subseteq R^i$, completed in the process of logical inference for elements of the model B_i.
4. For each chain $Rch^i \subseteq R^i$, find chains $Rch^j \subseteq R^j$ that are dependent, i.e. completion of Rch^i for an element of model B_i leads to completion of rule chain Rch^j for an element of model B_j. Let's define it as $Dep(Rch^i, Rch^j)$. The dependency is not symmetrical: $Dep(Rch^i, Rch^j) \neq Dep(Rch^j, Rch^i)$.

5. Combine the dependent rule chains for the elements of model $B_1,\ldots, B_p$, i.e. build the set $FMod(B_1,\ldots, B_p) = Rch^1 \cup Rch^2 \cup \ldots \cup Rch^p$, where $\forall\, i\, \exists\, j\; Dep(Rch^i, Rch^j)$, $i, j, p \in [1, |B|]$. Combine the dependent rule chains for the elements of model $B_1,\ldots, B_p$, i.e. build the set $FMod(B_1,\ldots, B_p) = Rch^1 \cup Rch^2 \cup \ldots \cup Rch^p$, where $\forall\, i\, \exists\, j\; Dep(Rch^i, Rch^j)$, $i, j, p \in [1, |B|]$.
6. For all $FMod(B_i, B_j) \neq \varnothing$, find commutation links by which data is transmitted during different interactions of the model's elements. They form the set $C' \subseteq C$. C' includes $C^{ij} = \langle I^{\,i}, I^{\,j}, \tau\rangle$, $I^{\,i} \in \Pi p(Z^i, I)$ and $I^{\,j} \in \Pi p(A^j, I)$, where Z^i – consequences of rules from Rch^i for element B_i, A^j - antecedents of rules from Rch^j for element B_j and $Dep(Rch^i, Rch^j)$.

An example of graphical presentation is provided in Fig. 3. An example of the rules creating the shown in Fig. 3 dependencies is provided in Fig. 4.

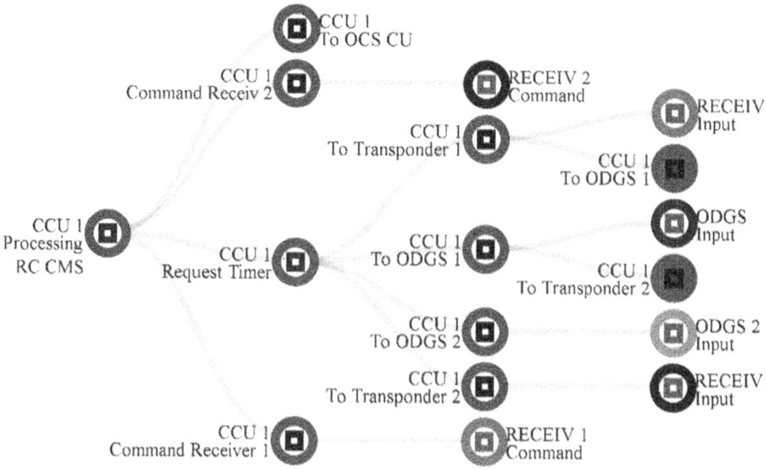

Fig. 3. A fragment of dependencies for the model's elements.

The chosen rules describe actions of the model during different ways of command-and-software control of the onboard equipment. Commutation links determine the control points for analysis of the simulation model's work. It is necessary to compare the data received through them with the results of spacecraft onboard systems' testing. We created graphical tools allowing to build dependent rule chains visually and in an interactive mode.

Graphical interface allows to choose the related elements for which the chains are built and to demonstrate the obtained dependencies. The color of the unit's outer circle determines the model's element, the color of the inner square – the type of the logical element (transmitting/receiving interface/timer). In addition, for each unit of the diagram there is an advanced list of rules.

Rule 19
Conditions:
the timer worked Processing RC CMS AND the value of the field Identifier of the command-pulse output in the variable Accepted RC CCU is 1C
Actions:
on the pulse interface On the control unit CU issue a pulse with a duration of 200 ms. and an amplitude of 1 V.
Rule 20
Conditions:
the timer worked Processing RC CMS AND the value of the field Identifier of the command-pulse output in the variable Accepted RC CCU is 1D
Actions:
on the pulse interface On the control unit CU issue a pulse with a duration of 200 ms. and an amplitude of 5 V.

Fig. 4. Example of the rules in dependent chains.

Physical devices of the onboard equipment are digitally externalized in the model's elements. During tests, the transmitted data packages and the received telemetry are continuously monitored. Figure 5 shows how the information is displayed during tests.

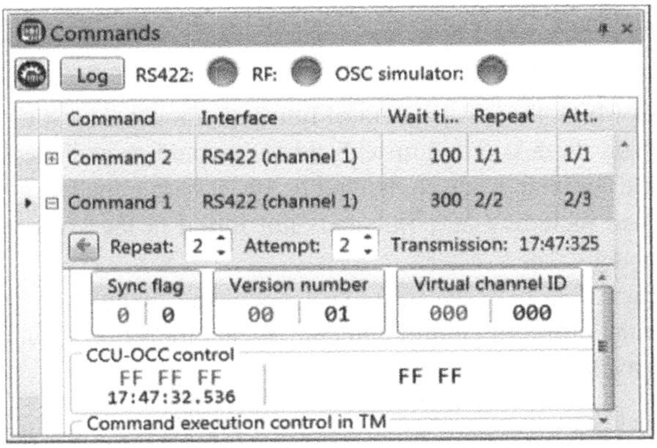

Fig. 5. Test data monitoring.

Software visualizes the list of commands with color and graphical identifiers of control of their execution. Let's define all the commands that have been tested as set K''. The picture shows the parameters of command sending, time of data transmission and telemetry reaction, as well as the control values. The results of tests are saved in a data storage and are used for simulation model analysis.

Set K' is compared with the set of the tested commands K''. Further, for all switching interfaces from C', simulation test data is compared with the onboard equipment test data. An example of the results of this comparison is shown in Fig. 6.

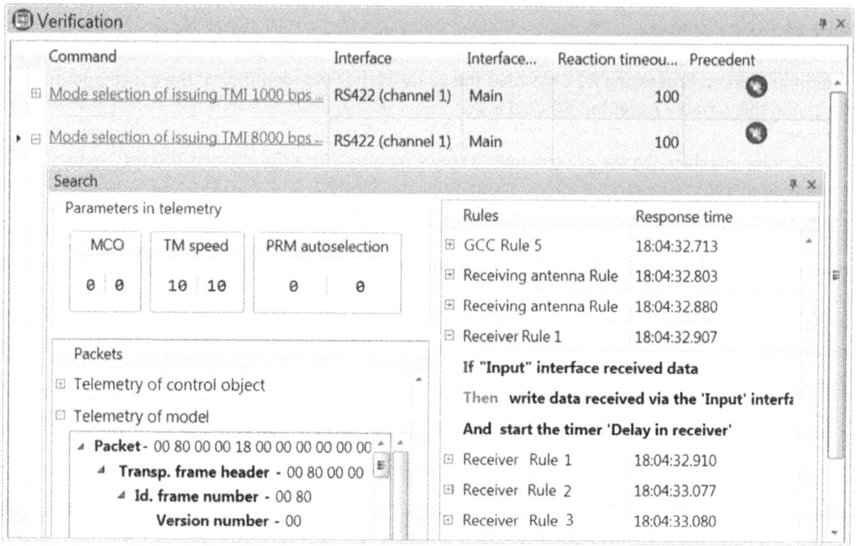

Fig. 6. Comparison of the test and simulation data.

For interpretation of the results of the model's analysis, the following criteria is used:

1. If $\exists\, k \in K'$, for which there is no data of finished equipment testing, i.e. $k \notin K''$, it is necessary to add the test program with this command or exclude k from K'.
2. If $\exists\, k \in K''$ – the command was completed in the test program, but $k \notin K'$, then a list of errors is formed including all commands not found in the knowledge base of the simulation model.
3. If in the chosen rule chain the results of command completion in the simulation model and in the object of control coincide, then a transition to analysis of the next rule chain is made.

If in the model the values corresponding with the object of control's telemetry are not found, a list of errors is created including the chosen rule chains.

If the knowledge base does not include the rules of command completion, or the results of modeling and the object of control's test data do not coincide, the simulation model needs to be revised by creating new rules of the knowledge base on the basis of test procedures.

4 Discussion

The described method is used for analysis of intellectual simulation models of spacecraft onboard systems' operation. Simulation models are built using expert onboard equipment designers' knowledge and experience. As a rule, analysis of logical functions of a simulation model demands high qualification from the subject area's specialists and comprehensive expertise in specific features of the simulated object. A wide range of commands and telemetry values demonstrating completion of the command-and-software control of a spacecraft make it impossible to test a model without digital support. The new method allows to simplify the process of simulation models' analysis.

Interactive graphical tools provide comparison of the simulation test results with the data obtained during testing of devices. They also allow to manually check compliance of the structure and elements of models with technical descriptions of the design documentation. Presently, the simulation model's knowledge base contains several hundred rules. Thanks to automatic subdividing of the model into logical chains providing different work modes of the simulated object, the method is highly potential for scaling for the support of big models. The suggested automation provides clear criteria of assessment of the simulation experiments' authenticity.

The method was included in the software implemented at a company-manufacturer of satellite systems. Application of intellectual methods for simulation modeling and onboard equipment operation analysis provides scientific basis for efficient engineering in space industry.

Acknowledgment. The reported study was funded by RFBR and Government of Krasnoyarsk Territory according to the research project № 18-47-242007 «The technology of intellectual support of the spacecraft onboard systems' design on the basis of the heterogeneous simulation models».

References

1. Koo, C.: Development of simulation infrastructure compatible with ESA SMP for validation of flight software and verification of mission operation. In: Proceedings of Simulation and EGSE for Space Programmes, pp. 1–8. ESA/ESTEC, Amsterdam (2012)
2. Luger, G.: Artificial Intelligence: Structures and Strategies for Complex Problem Solving, 6th edn. Pearson Education, Boston (2009)
3. Zanon, O.: The SimTG simulation modeling framework a domain specific language for space simulation. In: Proceedings of the 2011 Symposium on Theory of Modeling & Simulation, pp. 16–23. TMS-DEVS, Boston (2011)
4. Nozhenkova, L.: Tools of computer modeling of the space systems' onboard equipment functioning. SPIIRAS Proc. **1**(56), 144–168 (2018). https://doi.org/10.15622/sp.56.7
5. Nozhenkova, L.: Creation of the base of a simulation model's precedents for analysis of the spacecraft onboard equipment testing results. In: Advances in Intelligent Systems Research, vol. 151, pp. 78–81 (2018). https://doi.org/10.2991/cmsa-18.2018.18
6. Chung, C.: Simulation Modeling Handbook. CRC Press, London (2004)
7. Min, F.: Knowledge-based method for the validation of complex simulation models. Simul. Model. Pract. Theory **18**(5), 500–515 (2010)
8. Garavel, H.: OPEN/CÆSAR: an open software architecture for verification, simulation, and testing. In Proceedings of the First International Conference on Tools and Algorithms for the Construction and Analysis of System, pp. 2–18. INRIA, France (1998)

9. Zhou, Y.: An intelligent model validation method based on ECOC SVM. In Proceedings of the 10th International Conference on Computer Modeling and Simulation, pp. 67–71. Association for Computing Machinery, New York (2018). https://doi.org/10.1145/3177457.3177487
10. Kulyasov, N.: Method of creation and verification of the spacecraft onboard equipment operation model. In: IOP Conference Series: Materials Science and Engineering, vol. 537, pp. 1–6 (2019). https://doi.org/10.1088/1757-899X/537/2/022042
11. Russel, S., Norvig, P.: Artificial Intelligence: A Modern Approach, 3rd edn. Prentice Hall, New Jersey (2010)

On One Problem for Equation of Oscillator Motion with Viscoelastic Damping

Temirkhan Aleroev[1] and Alexey Bormotov[2(✉)]

[1] National Research Moscow State University of Civil Engineering (NRU MGSU), Yaroslavskoye Shosse, 26, 129337 Moscow, Russia
[2] Penza State Technological University (PenzSTU), pr. Baidukova/ul. Gagarina, 1a/11, 440039 Penza, Russia
aleks21618@yandex.ru

Abstract. In this paper we study point-to-point Dirichlet problem for equation of oscillator motion with viscoelastic damping for damping order $1 < \alpha < 2$. It was shown, that operator, generated by this problem, is dissipative operator of Keldysh type with oscillation properties. The authors of the work proved that the linear operator accompanying a mechanical system in which there is energy dissipation is dissipative. The article it follows that the linear operator is oscillatory and since the linear operator describes the motion of the oscillator, this operator has a complex of oscillatory properties. In addition, it is shown that in the case when the $0 < 0 < 1$ linear operator accompanying point-to-point Dirichlet problem has the above properties, and the authors have determined the eigenfunctions of problem point-to-point Dirichlet and it is proved that the system of functions of point-to-point Dirichlet problem forms a basis in L^2 (0,1).

Keywords: Oscillator · Damping · Dirichlet problem

1 Introduction

Many problems of mathematical physics [1, 2], associated with perturbations of normal operators with discrete spectrum, lead to consideration in Hilbert space H of a compact operator

$$A = (I + S)H,$$

called, for a compact S, as a week perturbation H or as operator of Keldysh type.

Let's consider operator B, generated by differential expression

$$l(u) = u'' - \varepsilon D_{0x}^{\alpha} u \tag{1}$$

and boundary conditions

$$u(0) = 0, \quad u(1) = 0. \tag{2}$$

R. Silhavy et al. (Eds.): CoMeSySo 2020, AISC 1295, pp. 377–384, 2020.
https://doi.org/10.1007/978-3-030-63319-6_34

Here D_{0x}^{α} – is operator of fractional (in Sturm-Liouville sence) differentiation of order $1 < \alpha < 2$

$$D_{0x}^{\alpha}u = \frac{1}{\Gamma(2-\alpha)}\frac{d^2}{dx^2}\int_0^x \frac{u(t)dt}{(x-t)^{\alpha-1}}, \qquad (3)$$

where $n = [\alpha] + 1$, denotes the integer part of number α.

Note, that problem (1)–(2) is the main focus of many authors [3–5]. This is due primarily to the fact that problem (1)–(2) simulates various physical processes, in particular, the oscillator motion under the action of elastic forces, characteristic for viscoelastic medium [6]. We note, in particular, that the solution of the first boundary value problem for the equation of string vibration in the medium with fractal geometry, i.e. for problem

$$\frac{\partial^2 u}{\partial t^2} = \frac{\partial^2 u}{\partial x^2} - \frac{\partial^\alpha u}{\partial x^\alpha}, \quad (0<\alpha<1), \qquad (4)$$

$$u(0,t) = u(1,t) = 0, \qquad (5)$$

$$u(x,0) = \varphi(x)\,;\; u_t'(x,0) = \psi(x) \qquad (6)$$

also will be reduced (by Fourier's method) to problem (1)–(2).

2 Methods

This paper, mainly, devoted to studying of follow problems:

1) is the operator B of Keldysh type?
2) is the operator B dissipative (since any linear operator B, associated with mechanical system, and which have dissipation of energy, will to satisfy to condition [13] $\mathrm{Re}(Bf, f) \leq 0$)?
3) is the operator B oscillational (since operator B describes the motion of oscillator, then this will have a whole complex of oscillational properties)?

In case, for $0 < \alpha < 1$, in [4] was proved, that operator B, associated with problem (1)–(2), is a) Keldysh type; б) dissipative; в) oscillational.

Also in [4] was written eigenfunctions of problem (1)–(2), and was proved that problem (1)–(2) doesn't generate associated functions, and the system of eigenfunctions of problem (1)–(2) forms a basis in $L^2(0,1)$.

In present paper problem (1)–(2) studied, mainly, for $0 < \alpha < 1$. This case is poorly studied and in this connection we note, that the any mechanical system, described by problem (1)–(2), is very sensitive to fractional order changes. For example, if we consider a fractional damped van der Pol equation [8].

$$x''(t) + \mu(x^2 - 1)D_{0t}^{\alpha}x(t) + x(t) = \sin(at), \tag{7}$$

then [8] periodic, quasi-periodic and chaotic motions existed, when the order of fractional damping is less than 1. When the order of fractional damping is $1 < \alpha < 2$, then there are chaotic motions only.

3 Results

Let's show, that operator B, associated with problem (1)–(2) is operator of Keldysh type. As in [11, 12], let denote:

$$Mu = \begin{cases} u'' \\ u(0) = 0, \quad u(1) = 0 \end{cases}, \tag{8}$$

$$G(x,t) = \begin{cases} t(x-1), & t \neq x \\ x(t-1), & t > x \end{cases}. \tag{9}$$

Lemma 1. Problem (1)–(2) is equal to equation.

$$u + \varepsilon M^{-1}D_{0x}^{\alpha}u - 2M^{-1}u = 0. \tag{10}$$

Here

$$M^{-1}D_{0x}^{\varepsilon}u = \frac{1}{\Gamma(2-\alpha)}\left[\int_0^x (x-t)^{1-\alpha}u(t)dt - \int_0^1 x(1-t)^{1-\alpha}u(t)dt\right]. \tag{11}$$

Proof. First suppose that $0 < \alpha < 1$, then

$$\begin{aligned} M^{-1}D_{0x}^{\varepsilon}u &= \tfrac{1}{\Gamma(1-\alpha)}\int_0^1 G(x,t)D_{0t}^{\alpha}udt \\ &= \tfrac{1}{\Gamma(1-\alpha)}\int_0^x t(x-1)D_0^{\alpha}udt + \tfrac{1}{\Gamma(1-\alpha)}\int_x^1 x(1-t)D_{0t}^{\alpha}udt. \end{aligned} \tag{12}$$

Using the Dirichlet transposition formulas we can see that

$$\begin{aligned} &\tfrac{1}{\Gamma(1-\alpha)}\int_0^x t(x-1)D_0^{\alpha}udt = \tfrac{1}{\Gamma(1-\alpha)}x(x-1)\int_0^x \tfrac{u(\zeta)}{(x-\zeta)^{\alpha}}d\zeta \\ &-\tfrac{1}{(2-\alpha)}(x-1)\int_0^x (x-\zeta)^{1-\alpha}u(\zeta)d\zeta, \end{aligned}$$

$$\frac{1}{\Gamma(1-\alpha)}\int_x^1 x(1-t)D_{0t}^{\alpha}u\,dt = \frac{-1}{\Gamma(1-\alpha)}x(x-1)\int_0^x \frac{u(\zeta)}{(x-\zeta)}d\zeta$$
$$+\frac{x}{\Gamma(2-\alpha)}\int_0^x\left[(1-\zeta)^{1-\alpha}-(x-\zeta)^{1-\alpha}\right]u(\zeta)d\zeta \quad ,$$

which proves Lemma 1 for $0 < \alpha < 1$. For we have $0 < \alpha < 1$

$$M^{-1}D_{0x}^{\alpha}u = \frac{1}{\Gamma(2-\alpha)}\int_0^1 G(x,t)\left\{\frac{d^2}{dt^2}\int_0^t \frac{u(\zeta)}{(t-\zeta)^{\alpha-1}}d\zeta\right.$$
$$= \frac{1}{\Gamma(2-\alpha)}\int_0^x t(x-1)\left\{\int_0^t \frac{u(\zeta)}{(t-\zeta)^{\alpha-1}}d\zeta\right\}''dt + \frac{1}{\Gamma(2-\alpha)}\int_x^1 x(1-t)\left\{\int_0^t \frac{u(\zeta)}{(t-\zeta)^{\alpha-1}}d\zeta\right\}''dt$$

$$= \frac{1}{\Gamma(2-\alpha)}\left[t(x-1)d\left\{\int_0^t \frac{u(\zeta)d\zeta}{(t-\zeta)^{\alpha-1}}\right\}' + \int_x^1 x(1-t)d\left\{\int_0^t \frac{u(\zeta)}{(t-\zeta)^{\alpha-1}}d\zeta\right\}'\right]$$

$$= \frac{1}{\Gamma(2-\alpha)}\left[t(x-1)\left\{\int_0^t \frac{u(\zeta)d\zeta}{(t-\zeta)^{\alpha-1}}\right\}'\Big|_0^x - (x-1)\int_0^x\left\{\int_0^x \frac{u(\zeta)d\zeta}{(t-\zeta)^{\alpha-1}}\right\}'dt\right.$$

$$+x(1-t)\left\{\int_0^t \frac{u(\zeta)d\zeta}{(t-\zeta)^{\alpha-1}}\right\}'\Big|_x^1 + x\int_x^1\left\{\int_o^t \frac{u(\zeta)d\zeta}{(t-\zeta)^{\alpha-1}}\right\}'dt$$
$$= \frac{1}{\Gamma(2-\alpha)}\left[-(x-1)\int_0^t \frac{u(\zeta)d\zeta}{(t-\zeta)^{\alpha-1}}\Big|_0^x + x\int_0^t \frac{u(\zeta)d\zeta}{(t-\zeta)^{\alpha-1}}\Big|_x^1\right]$$

$$= \frac{1}{\Gamma(2-\alpha)}\left[-x(x-1)\left(\int_0^x \frac{u(\zeta)d\zeta}{(x-\zeta)^{\alpha-1}} - 0\right) + x\left(\int_0^1 \frac{u(\zeta)d\zeta}{(1-\zeta)^{\alpha-1}}d\zeta\right)\right]$$

$$+x\left[\int_0^1 \frac{u(\zeta)}{(1-\zeta)^{1-\alpha}}d\zeta - \int_0^x \frac{u(\zeta)}{(t-\zeta)^{\alpha-1}}d\zeta\right] = \frac{1}{\Gamma(2-\alpha)}\left[\int_0^x (x-t)^{1-\alpha}u(t)dt - \int_0^1 x(1-t)^{1-\alpha}u(t)dt\right].$$

that proves Lemma 1 completely. From the fact that problem (1)–(2) is equal to Eq. (10), follows, that operator B is operator of Keldysh type.

Thus, consideration of problem (1)–(2) is reduces to consideration of linear operator of a sheaf $L(\lambda) = I + \varepsilon T - \lambda N$, where $Tu = M^{-1}D_{0x}u$; $Nu = M^{-1}u$.

In [12], for $0 < \alpha < 1$ using Keldysh theorem, was solved problem on the distribution of the characteristic numbers of operator B, and there was established the completeness of system of eigen- and associated functions for this problem. As for $0 \le \alpha \le \frac{3}{2}$ operator $T \in G_{\infty}$, (operator T has a trace) [15], then from Lemma 1 and Keldysh theorem follows.

Corollary 1. The system of eigen- and associated functions of operator B, for $0 \leq \alpha \leq \frac{3}{2}$, is complete in $L^2(0,1)$.

Let's denote as $n(r, B)$ the number of characteristic values of operator B, lying in circle $|\lambda| \leq r$. The problem of allocation of characteristic values of operator A is posed as investigation of asymptotic properties for $n(r, B)$ when $r \to \infty$. In [12] this problem was solved for order of operator D_{0x}^{α} less than 1. As in [12] the studying of function $n(r, B)$ reduces to study of the spectrum for linear operator sheaf $L(\lambda) = I + T - \lambda N$, where

$$Tu = \int_0^1 G(x,t) D_{0x}^{\alpha} u du, \quad Nu = \int_0^1 G(x,t) u(1) dt \quad (1 < \alpha) < \frac{3}{2}.$$

Clearly, that $T \in G_{\infty}$, and N is positive operator. Thus, if we may to set for distribution function $n(r, B)$ of characteristic values of operator N the nondecreasing function $\varphi(r)$ $(0 \leq r \leq \infty)$ with properties [12]:

1) $\varphi(r) \to \infty$ *for* $r \to \infty$;
2) $\lim_{r \to \infty} [\ln \varphi(r)] < \infty$;
3) $\lim_{r \to \infty} \frac{n(r,N)}{\varphi(r)} = 1$;

then by Keldysh theorem [1] we have

$$\lim_{r \to \infty} \frac{n(r, B)}{n(r, N)} = 1. \tag{13}$$

Obviously, that in our case we shall to take as $\varphi(r)$ the function $\sqrt{r}$.

In applied problems the most interesting is usually determining the first eigenvalues. Equation (13) allows to use the technique given in [11] of estimation for first eigenvalues for $1 < \alpha < 2$. Actually, form (13) follows, that in this case, i.e. for $1 < \alpha < 2$ we may to determine.

$$D_B(\lambda) = \prod_{j=1}^{\infty} (1 - \lambda \lambda_j(B)).$$

Complete implementation of this method is associated with large computing and therefore estimates for the first eigenvalues of the problem (1)–(2) for $1 < \alpha < 2$ will be developed in another paper.

Theorem 1. If $\varepsilon > 0$, then operator B is dissipative.

Proof. Actually

$$(Bf, f) = (f'', f) - \varepsilon (D_{0x}^{\alpha} f, f)$$

We know, that operator M is sectorial. The sectoriality of operator D_{0x}^{α} was proved in papers of one of the co-authors [11]. That's prove the Theorem 1.

As we found out later, the solution of problem 3, which was mentioned at pp. 2, involves considerable analytical difficulties. In this paper, we restrict ourselves to the proof of the following theorem.

Theorem 2. If $0 \le \varepsilon \le \frac{1}{3}$, then first eigenvalue of problem (1)–(2) is real and prime, and main tone has no nodes.

Proof. It's known, that [5, pp. 143] problem (1)–(2) is equal to equation

$$u(x) = \lambda \int_0^1 G_2(x,\tau)u(\tau)d\tau. \tag{14}$$

where

$$G_2(x,\tau) = G_1(x,\tau) - \frac{\varepsilon}{\mathrm{E}_{\frac{1}{2}}(\varepsilon,2)} \int_\tau^1 \mathrm{E}_\rho[\varepsilon(\eta-\tau)]^\beta d\eta \cdot \int_0^1 G(x,t) D_{0t}^{\alpha-1} \mathrm{E}_\beta\left[\varepsilon t^\beta\right] dt, \tag{15}$$

$$G_1(x,\tau) = \begin{cases} (1-x)\int_\tau^x \mathrm{E}_\beta[\varepsilon(t-\tau)]dt \\ -x - \int_x^1 \mathrm{E}_\beta\left[\varepsilon(t-\tau)^\beta\right]dt, & x \ge \tau \\ -x\int_\tau^1 \mathrm{E}_\beta\left[\varepsilon(t-\tau)^\beta\right]dt, & x \le \tau \end{cases} \tag{16}$$

Here

$\mathrm{E}_\rho(z,\mu) = \sum_{\infty}^{\kappa=0} \frac{z^\kappa}{\Gamma(\mu+\kappa\rho^{-1})}$ – is function of Mittag-Leffler type, and $\mathrm{E}_\rho[z] = \sum_{\infty}^{\kappa=0} \frac{z^\kappa}{\Gamma(1+\kappa\rho^{-1})}$ – is Mittag-Leffler function.

Let's show, that Green's function of problem (1)–(2) is function of fixed sign. Clearly, that for $x \le \tau\, G_2(x,\tau)$, is negative. It is enough to show that

$$x + \int_x^1 \mathrm{E}_\rho\left[\varepsilon(t-\tau)^\beta\right]d\tau \ge (1-x)\int_\tau^x \mathrm{E}_\beta\left[\varepsilon(t-\tau)^\beta\right]dt. \tag{17}$$

Since $0 < \varepsilon < \frac{1}{3}$ then $1 < \mathrm{E}_\beta[\varepsilon, z] < \frac{2}{3}$ (here taken into account the fact that $z < 1$).

Thus

$$x+1\int_{x} \mathrm{E}_{\beta}\left[\varepsilon(t-\tau)^{\beta}\right]d\tau > 1, \tag{18}$$

and expression

$$(1-x)\int_{\tau}^{x} \mathrm{E}_{\beta}\left[\varepsilon(t-\tau)^{\beta}\right]d\tau < \frac{3}{4}, \tag{19}$$

That's proves the statement (17). From this should be sign-definition for $G_2(x,t)$. That's proved Theorem 2.

4 Discussions

From the proof of Theorem 2 follows, that estimation $0<\varepsilon<\frac{1}{3}$ which has been used for proving fixed-sign kernel $G_2(x,t)$ is very rough and, certainly, it can be clarified. But at this stage, such problem not intended.

5 Conclusion

The authors of the work proved that the linear operator B accompanying a mechanical system in which there is energy dissipation is dissipative. According to Theorems 1 and 2, it follows that the operator B is oscillatory and since the operator B describes the motion of the oscillator, this operator B has a complex of oscillatory properties.

In addition, it is shown that in the case when the $0 < \alpha < 1$ operator B accompanying problem (1–2) has the above properties, and the authors have determined the eigenfunctions of problem (1–2) and it is proved that the system of functions of problem (1–2) forms a basis in L^2 (0, 1).

Remark
The authors declares that there is no conflict of interest regarding the publication of this paper.

References

1. Keldysh, M.V.: On eigenvalues and eigenfunctions of some classes of non-self-adjoined equations. Theses USSR Acad. Sci. **77**(1), 11–14 (1951)
2. Keldysh, M.V.: On completeness of eigenfunctions for some classes of non-self-adjoined linear operators. Uspekhi Mat. Nauk **26**(4), 15–41 (1971)

3. Ingman, D., Suzdalnitsky, J.: Iteration method for equation of viscoelastic motion with fractional differential operator of damping. Comp. Methods Appl. Mech. Engrg. **190**, 5027–5036 (2001)
4. Aleroev, T.S., Aleroeva, H.T.: On the Eigenfunctions and Eigenvalues of a Class of non-selfadjoint operators. Lobachevskii J. Math. **37**(3), 227–230 (2016)
5. Nakhushev, A.M.: Fractional Calculus and its Applications. Fizmatlit, Moscow (2003)
6. Aleroev, T.S., Kirane, M., Tang, Y.-F.: Boundary-value problems for differential equations of fractional order. J. Math. Sci. **10**(2), 158–175 (2013)
7. Aleroev, M., Aleroev, T., Kirane, M., Tang, Y.-F.: On one class of persymmetric matrices generated by boundary value problems for differential equations of fractional order. Appl. Math. Comput. **268**, 151–163 (2015)
8. Chen, J.-H., Chen, W.-C.: Chaotic dynamics of the fractionally damped van der Pole equation. Chaos Solutons Fractals **35**, 188–198 (2008)
9. Aleroev, T.S., Aleroeva, H.T.: On the basis of Eigenfunctions of boundary value problems for second order differential equations with fractional derivative. NRNU MEPHI Bull. **4**(6), 646–648 (2015)
10. Larionov, E.A., Zveriaev, E.M., Aleroev, T.S.: On Theory of Normal Operators Weak Perturbations. Keldysh Institute Preprints 14, Moscow, Russia (2014)
11. Aleroev, T.S.: Boundary value problems for differential equations with fractional derivatives. Doctorate thesis (doctor of sciences thesis) physics and mathematics sciences, Lomonosov MSU, Moscow, Russia (2000)
12. Aleroev, T.S.: On eigenfunctions and eigenvalues for one non-self-adjoined operator. Diff. Equ. **11**(25), 1996–1997 (1989)
13. Gantmacher, F.R., Krein, M.G.: Oscillation Matrices and Kernels and Small Vibrations of Mechanical Systems. AMS Chelsea Publishing, New York (2003)
14. Aleroev, T.S., Aleroeva, H.T.: The letter to Edition. Lobachevskii J. Math. **37**(6), 227–230 (2016)
15. Gohberg, I.M., Krein, M.G.: Introduction to the Theory of Linear Nonselfadjoint Operators in Hilbert Space. AMS Chelsea Publishing, New York (1969)

Addressing a Problem of Regional Socio-Economic System Control with Growth in the Social and Engineering Fields Using an Index Method for Building a Transitional Period

Karolina V. Ketova(✉) and E. A. Saburova

Izhevsk State Technical University,
Studncheskaya Street 7, Izhevsk 426069, Russia
ketova_k@mail.ru, saburovaea@inbox.ru

Abstract. In the paper, a problem of socio-economic system management, as exemplified by one of the regions of the Russian Federation, is solved. An algorithm of an optimal control using an index method for building a transitional period and with an ability to take into account growth rates in the engineering and social fields is created. Terms of reference are given on the basis of a regional economic system macro-level model, with production capital and human capital viewed as development factors. A human capital modeling underlying hypothesis is an assumption that it is built on three main aggregative premises: healthcare, education and culture. Including human capital factor in the macro model and an ability to consider progress growth rates are two distinctive features of a given management problem statement. The problem solving algorithm comprises two steps: building the quasi magistral itself and figuring an optimized trajectory which would propel and economic system to the quasi magistral. A gateway period is a transitional period where an optimal control is built using an index method. A transitional period can be diminished by speeding up progress growth rates in the social and engineering fields. In this paper, a two-stage way to build an optimized investment distribution is used upon a model which contains more than to factors for the first time. In this case, a quasi-stationary state is an optimized balanced growth half line. As a statistical base for the analysis, a demographic data and an Udmurt Republic human capital and production capital investment volume data are used. For an unknown parameters' identification, a period of years 1998–2018 is drawn upon. Optimized investment rates which enable an economic system to reach balanced growth trajectory by the year 2025 are calculated. The proposed method can be used for an optimal control of socio-economic systems, as well as for comparing and estimating their growth rates in the social and engineering fields.

Keywords: Socio-economic system · Mathematical modeling · Region · Investments · Optimal control · Production capital · Technological progress growth rates · Human capital · Socio-educational progress growth rates

R. Silhavy et al. (Eds.): CoMeSySo 2020, AISC 1295, pp. 385–396, 2020.
https://doi.org/10.1007/978-3-030-63319-6_35

1 Introduction

Socio-economic system state indicators stable growth is set out while planning its development strategy, which defines financing volume of production and social activities. A development strategy should be planned with the application of economical and mathematical modeling methods, which enables us to obtain results which are both economically feasible and mathematically verified.

In this paper, the problem of socio-economic system management using an index method for building a transitional period is solved. As an initial regional model, we used the model presented in [1]. The distinctive feature of a given problem statement is an introduction of a human capital factor as a key factor which contributes to the creation of the final product.

A human capital (HC) includes intelligence, health, knowledge, productive labor and life quality [2–5]. A concept of human capital has a long history of being discussed. This issue was raised as early as in 17–18 centuries by such renowned economists as W. Petty [6] and A. Smith [7]. Further on, in the 19th century, A. Marshall [8] and I. Fisher [9] devoted their works to that topic. Concepts of the modern human capital theory were generally introduced in the middle of the previous century. At the time, a human capital was studied by mathematicians and economists G. Becker [10], T. Schultz [11], W. Bowen [12], B. Weisbrod, J. Mincer, M. Fisher [13], Y. Ben-Porat [14]. An estimation of a human capital is also presented in the works of Russian researchers, such as R. I. Kapelushnikov [15], A. I. Dobrynin [16], A. V. Koritzky [17]. At the present time, there is a constant search for a new knowledge in the human capital field [18–21].

It should be noted that human capital estimation is an exceptionally challenging task. Building general calculation methods for a human capital is obstructed by a larger fraction of subjective estimations in the modeling of this parameter. A human capital can be analyzed from several perspectives. For example, it can be researched as a human and society life quality matter [22], as an ability of a person to perform innovative activities [23, 24], as a volume of his earnings [25], as his production value [26] or, finally, as an investment cost of this valuable factor of production – the human capital factor [27]. Further, the human capital can be researched on the basis of one particular parameter or by the more complicated approach which considers different aspects of the human life and the society.

Nevertheless, regardless of the method chosen by the researcher, one thing is certain – from the perspective of a behavior of the complicated system of economic relations and inevitable environmental complexity growth, the human capital factor is crucial for achieving social and economic progress growth rates.

In the paper, the human capital parameter is obtained by means of the model derived from the transfer equation which considers budgetary and private investments into the human capital development (education, healthcare and culture) [1].

The problem statement includes an account of engineering and social progress as well. An existing production capital determines scientific and technological progress (STP) growth rates, while realized human capital determines social and educational progress (SEP).

Since human capital is a trait carried by demographic elements, human capital parameter estimation should be performed in accordance with the demographic structure. In this context, a distribution of demographic elements by age stages becomes important. A problem of modeling and forecasting demographic dynamics is closely presented in [28] paper. Forecasting functions for different age stages are to be introduced in the model explicitly.

Thereby, as in many other applied problems, in this setting, the actual forecasting information is clearly tied to the calendar time. Models of such kind are non-homogenous. A homogenous model has a stationary point of an optimal strategy, while non-homogenous one contains a quasi-stationary trajectory (a quasi magistral) [1].

The optimal control problem solving algorithm comprises two main stages: a stage of building a quasi magistral and a stage of building an optimized trajectory of an economic system which propels the system towards the quasi magistral. Let us call a gateway period a transitional period. A quasi magistral achievement time can be reduced by an accelerated accumulation of STP and SEP growth rates. The position of a quasi magistral itself for the certain economic system can also be shifted towards the high economic indicators value range by means of changing STP and SEP growth rates.

The reasoning of an optimal control in non-homogenous models is based on the idea that homogenous model solving is taken as a null approximation. This is approach is correct if the prognostic curves which are entered into the model change smoothly (which is true in this case due to inertia of demographic processes). An exposed structure of homogenous model solution is transitioned to non-homogenous setting, subsequently forming a quasi-stationary trajectory. A solution structure in homogenous setting is derived from the Bellman's principle of optimality [29]. Correctness of the transition is supported by the Pontryagin's maximum principle [30, 31]. For a model with an arbitrary number of factors, it is quite difficult to fomulate an equation of general form. The approach is implemented to the full extent for a two-factor model. This paper is the first to use this method to formulate an optimal equation for a model with more than two factors. To formulate an optimal equation in the transitional period, an investment distribution index method was used [1].

The solution for the problem is exemplified by the statistical data of a regional economic system of the Udmurt Republic (UR).

2 Materials and Research Methods

The Fundamental Principles are as Follows:

1. At a macro level, when modeling an economic dynamics of a region, we are going to consider generalized parameters: gross regional product (GRP) Y, production capital (fixed capital assets FCA) K, investments into FCA I, human capital H, investments into HC J, tax allocation to the federal budget N_F, grants, transfers and subventions T, and, finally, a consumption C. The produced product equals $E = Y + T - N_F$.

2. The coefficient ω reflects an interaction between the region and the external economic environment: $E = \omega F(K,H)$, $\omega = 1 + \upsilon r_F \lfloor \nu(r_R/r_F) - 1 \rfloor$; $D = (1+\nu) r_R \upsilon Y$, $T = \nu r_R \upsilon Y$, $N_F = r_F \upsilon Y$; where r_F and r_R – shares of federal and regional budget taxes in the total taxes amount, the ν coefficient defines a share of regional taxes which regains to the regional budget in the form of financial assets (transfers, grants and subventions), D – consolidated revenue of the region, υ – share which defines a volume of taxes in a sales volume Y. These parameters are presented schematically in Fig. 1.

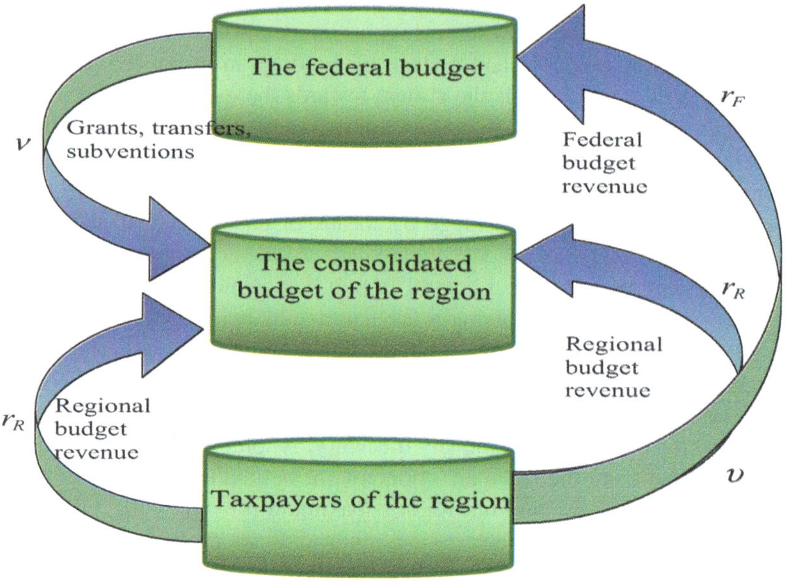

Fig. 1. Budgetary interaction between the region and external environment.

3. Out of a total population of the region $P(t)$ let's save out an economically active population group $L(t)$, which contributes to the GRP production. Consumption in an economic system is distributed throughout the population $P(t)$. Working-age population share in total population equals $\lambda = L/P \in (0,1)$, т.к. $0 < L(t) < P(t)$. Curves $P(t)$ and $L(t)$ are obtained from solving the demographic dynamics problem and are introduced in the model exogenously. Relation calculation λ is of the following form:

$$\lambda(t) = \frac{L(t)}{P(t)} = \left(\int_0^{\tau_m} \varepsilon_{м}(t,\tau)\rho_{м}(t,\tau)d\tau + \int_0^{\tau_m} \varepsilon_{ж}(t,\tau)\rho_{ж}(t,\tau)d\tau \right) \Big/ \int_0^{\tau_m} \rho(t,\tau)d\tau ,$$

where $\rho(t,\tau)$ is the function of a population of the age τ distribution in the year of t (density), $\varepsilon_{м}(t,\tau)$ and $\varepsilon_{ж}(t,\tau)$ are the shares of men and women of the age of τ who are engaged in the GRP production in the year of t, τ_m is the survival time, $\rho_{м(ж)}(t,\tau)$ is the male (female) population distribution density by age stages. Let us also introduce $\bar{\varepsilon}$, which is an averaged by all ages share of the population engaged in the GDP production.

4. A management problem is treated in a continuous time with a finite planning horizon $[t_0,\ t_T]$; δ is a discounting coefficient.
5. Two regional economic system development scenarios are taken into consideration: inertial and innovational. STP growth rates are defined by the β coefficient. SEP growth rates are defined by the κ coefficient.
   During the inertial scenario implementation, there is an enlarged GRP reproduction within an economic system. There is room for the production capital $K_1(t)$ with the rate of $\beta = \beta_1 = 0$ and the human capital $H_1(t)$ with the rate of $\kappa = \kappa_1 = 0$.
   An innovational scenario of economic system development is being implemented from the moment of t_0, when the production capital $K_2(t)$ is being formed, with the rate of $\beta_2 > 0$, and the human capital $H_2(t)$ with the rate of $\kappa_2 > \kappa_1$.
   Let's define I_1 and I_2 as the volume of investments in a production capital K_1 and K_2 respectively. Let's also define an investment volume in H_1 and H_2 as J_1 and J_2.
6. FCA production capital dynamics are described by the differential equation with the capital withdrawal coefficient η: $\dot{K}_i = e^{\beta_i(t-t_0)} s_{ki} E - \eta_i K_i, \quad i = 1,2;$ $K = K_1 + K_2;\ K_{10} = K(t_0),\ K_{20} = 0,\ K_{iT} = K_i(t_T)$.
7. HC dynamics are described by the differential equation with the capital withdrawal coefficient χ:$\dot{H}_i = e^{\kappa_i(t-t_0)} \bar{\varepsilon} s_{hi} E - \chi_i H_i, \quad i = 1,\ 2, \quad H = H_1 + H_2; \quad H_{i0} = H_i(t_0)$, $i = 1,\ 2,\ H_{iT} = H_i(t_T)$.
8. Every year $\forall t \in [t_0,\ t_T]$, the produced products are distributed $E = Y + T - N_F = I_1 + I_2 + J_1 + J_2 + C$ onto 5 parts: investments $I_1,\ I_2, J_1,\ J_2$ in the production factors $K_1,\ K_2,\ H_1,\ H_2$ and the consumption C. Economic system management functions according to the vector $s = (s_0, s_{k1}, s_{k2}, s_{h1}, s_{h2})$, where $s_0 = C/E$ is a consumption rate, $s_{ki} = I_i/E$ is a rate of investment in K_i; $s_{hi} = J_i/E$ is a rate of investment in H_i; wherein $s_0 + s_{k1} + s_{k2} + s_{h1} + s_{h2} = 1,\ s_0 = const.$
9. A production volume is defined by a production $Y = F(K,H)$. It's an upward-convex function which increases steadily upon each parameter. Moreover, a production volume is defined by the linear and homogeneous function $F(K,H) = LF(K/L,\ H/L) = LF(k,h)$, where $k = K/L$ and $h = H/L$ are unit (per one worker) values of production and human capital respectively.
10. A criterion for a management task optimality is a unit (per one person) discounted maximally accumulated consumption for the entire planning horizon $[t_0,\ t_T]$: $Cr = \int_{t_0}^{t_T} s_0 \lambda \omega F(k,h)\, e^{-\delta(t-t_0)} dt \to \max_{s\in\Omega}$. An admissible control set is of the form:

$$\Omega = \left\{ s = (s_l) = (s_{k1}, s_{k2}, s_{h1}, s_{h2}) \ :\ s_l \in [0,1],\ \sum_l s_l = 1 - s_0 \right\}.$$

The stated problem is a problem of an optimal economic system management with production and social fields' progress growth rates taken into account.

3 Problem Solving Algorithm

Let's discuss two parts of an optimal system trajectory motion. The first section represents transition period movement till the moment of reaching quasi-stationary section (a transfer time thereon can change due to the change of FCA and SEP rates), while the second section represents motion along the quasi magistral.

The model under consideration falls under the class of economic dynamics models with linear and homogenous production function (Gale models) [31, 32]. In this case, quasi-stationary state is the half line which is defined by a relation between factors of the model. In this regard, let's introduce the following nomenclature: $w = (k_1 + k_2)/(h_1 + h_2)$; $w_{k1} = k_1/(h_1 + h_2)$; $w_{h1} = h_1/(h_1 + h_2)$; $w_{k2} = k_2/(h_1 + h_2)$; $w_{h2} = h_2/(h_1 + h_2)$. Therefore, $f(w) = Aw^{\alpha}$.

The Following Equation Represents Production and Human Capital Unit Values:

$$\dot{k}_i = s_{ki}\omega e^{\beta_i(t-t_0)}F(k,h) - \gamma_{ki}k_i,\ \dot{h}_i = s_{hi}\bar{\varepsilon}\omega e^{\kappa_i(t-t_0)}F(k,h) - \gamma_{hi}h_i, \tag{1}$$

where $\gamma_{ki} = \eta_i + \dot{L}/L$, $\gamma_{hi} = \chi_i + \dot{L}/L$.

Let us denote the phase variables' vector as $x = (k_1, k_2, h_1, h_2)$, the dual variables' vector as $\psi = (\psi_{k1}, \psi_{k2}, \psi_{h1}, \psi_{h2})$, the control variables' vector $s = (s_0, s_{k1}, s_{k2}, s_{h1}, s_{h2})$. A Hamiltonian of the matter $\mathrm{H}(\psi, s, x, t)$ is of the form:

$$\begin{aligned} \mathrm{H}(\psi, s, x, t) &= s_0\lambda\omega F(k,h)\, e^{-\delta(t-t_0)} + \psi_{k1}[s_{k1}\omega F(k,h) - \gamma_{k1}k_1] \\ &+ \psi_{k2}\left[s_{k2}\omega e^{\beta(t-t_0)}F(k,h) - \gamma_{k2}k_2\right] + \psi_{h1}[s_{h1}\bar{\varepsilon}\omega F(k,h) - \gamma_{h1}h_1] \\ &+ \psi_{h2}\left[s_{h2}\bar{\varepsilon}\omega e^{\kappa(t-t_0)}F(k,h) - \gamma_{h2}h_2\right], \end{aligned} \tag{2}$$

According to the Pontryagin's maximum principle, with a new dual variable notation taken into account, we can obtain:

$$s^o = \arg\max_{s\in\Omega} \mathrm{H}(q, s, x, t) = \arg\max_{s\in\Omega} \left[\begin{array}{l} s_0\lambda\omega F(k,h) + q_{k1}s_{k1}\omega F(k,h) \\ + q_{k2}s_{k2}\omega e^{\beta(t-t_0)}F(k,h) + q_{h1}s_{h1}\bar{\varepsilon}\omega F(k,h) \\ + q_{h2}s_{h2}\bar{\varepsilon}\omega e^{\kappa(t-t_0)}F(k,h)\ . \end{array}\right] \tag{3}$$

Let us also introduce the following notation: $Q = (\lambda, Q_{k1}, Q_{k2}, Q_{h1}, Q_{h2}) = (\lambda, q_{k1}, q_{k2}e^{\beta(t-t_0)}, q_{h1}\bar{\varepsilon}, q_{h2}\bar{\varepsilon}e^{\kappa(t-t_0)})$, $Q_m = \max(Q_{k1}, Q_{k2}, Q_{h1}, Q_{h2})$.

The determination of an optimal control comes down to solving a linear problem of a mathematical programming for any point in time $t \in [t_0, t_T]$. Therefore, in accordance with the notation, the following can be written:

$$\begin{aligned} &\lambda s_0 + q_{k1}s_{k1} + q_{k2}e^{\beta(t-t_0)}s_{k2} + q_{h1}\bar{\varepsilon}s_{h1} + q_{h2}e^{\kappa(t-t_0)}\bar{\varepsilon}s_{h2} = Qs \to \max_{s(t)\in\Omega} \\ &\max_{s(t)\in\Omega} = \lambda s_0 + Q_m(1 - s_0); \end{aligned} \tag{4}$$

$$s_0 + s_{k1} + s_{k2} + s_{h1} + s_{h2} = 1 \tag{5}$$

Let us denote $s = (s_l)$, where $l = (l_i) = (k_1, k_2, h_1, h_2)$, and $l_m = \operatorname{Ind}\max_i Q_{l_i}$. The equation structure can be written as:

$$s = (s_{l_i}) = \begin{cases} \sum_{l_i} \tilde{s}_{l_i} = 1 - s_0, & l_i = l_m; \\ 0, & l_i \neq l_m. \end{cases} \tag{6}$$

An adjoint system of equations is of the form:

$$\dot{k}_i = s_{ki}\omega e^{\beta_i(t-t_0)}F(k,h) - \gamma_{ki}k_i,\ k_{i0} = k_i(t_0),\ k_{iT} = k_i(t_T) \Rightarrow k_i^*,\ i = 1,\ 2; \tag{7, a}$$

$$\dot{h}_i = s_{hi}\bar{\varepsilon}\omega e^{\kappa_i(t-t_0)}F(k,h) - \gamma_{hi}h_i,\ h_{i0} = h_i(t_0),\ h_{iT} = h_i(t_T) \Rightarrow h_i^*,\ i = 1,\ 2; \tag{7, b}$$

$$\dot{q}_{ki} = (\delta + \gamma_{ki})q_{ki} - [\lambda s_0 + Q_m(1 - s_0)]\omega F_k'(k,h),\ i = 1,\ 2; \tag{8, a}$$

$$\dot{q}_{hi} = (\delta + \gamma_{hi})q_{hi} - [\lambda s_0 + Q_m(1 - s_0)]\omega F_h'(k,h),\ i = 1,\ 2; \tag{8, b}$$

By means of combining Eqs. (7, a, 7, b) and using a linear homogeny of the production function, we can obtain:

$$\begin{aligned} \dot{w} = &\left[s_{k1} + s_{k2}e^{\beta(t-t_0)} - \left(s_{h1} + s_{h2}e^{\kappa(t-t_0)}\right)\bar{\varepsilon}w\right]\omega f(w) - (\gamma_{k2} - \gamma_{h2})w \\ &-[(\gamma_{k1} - \gamma_{k2})w_{k1} - (\gamma_{h1} - \gamma_{h2})ww_{h1}], \end{aligned} \tag{9, a}$$

$$\begin{aligned} \dot{w}_{k1} = &\left[s_{k1} - \left(s_{h1} + s_{h2}e^{\kappa(t-t_0)}\right)\bar{\varepsilon}w_{k1}\right]\omega f(w) - \gamma_{k1}w_{k1} \\ &+ [\gamma_{h2}w_{k1} + (\gamma_{h1} - \gamma_{h2})w_{k1}w_{h1}], \end{aligned} \tag{9, b}$$

$$\begin{aligned} = \dot{w}_{h1} = &\left[s_{h1} - \left(s_{h1} + s_{h2}e^{\kappa(t-t_0)}\right)w_{h1}\right]\bar{\varepsilon}\omega f(w) - \gamma_{h1}w_{h1} \\ &+ [\gamma_{h2}w_{h1} + (\gamma_{h1} - \gamma_{h2})w_{h1}w_{h1}], \end{aligned} \tag{9, c}$$

$$w_{k1} + w_{k2} = w,\ \ w_{h1} + w_{h2} = 1 \tag{9, d}$$

In a similar fashion, Eqs. (8, a, 8, b) can be rearranged as follows:

$$\begin{aligned} \dot{Q}_m &= (\delta + \gamma_{k1})Q_m - [\lambda s_0 + Q_m(1 - s_0)]\omega F_k'(k,h) \\ \dot{Q}_m &= (\delta + \beta + \gamma_{k2})Q_m - [\lambda s_0 + Q_m(1 - s_0)]\omega e^{\beta(t-t_0)}F_k'(k,h) \end{aligned} \tag{10, a}$$

$$\begin{aligned}\dot{Q}_m &= (\delta + \gamma_{h1})Q_m - [\lambda s_0 + Q_m(1 - s_0)]\omega\bar{\varepsilon}F'_h(k,h) \\ \dot{Q}_m &= (\delta + \kappa + \gamma_{h2})Q_m - [\lambda s_0 + Q_m(1 - s_0)]\omega\bar{\varepsilon}e^{\kappa(t-t_0)}F'_h(k,h)\end{aligned} \quad (10,\text{b})$$

A quasi magistral $(k_i^*(t),\, h_i^*(t))$ which represents quasi-stationary sector of an optimal trajectory, is defined by the conditions $Q_l = \underset{l}{idem} = Q_m$, $\dot{Q}_l = \underset{l}{idem} = \dot{Q}_m$.

From (10, a, 10, b), considering the fact that $F'_k(k,h) = f'(w)$ and $F'_h(k,h) = f(w) - wf'(w)$, we can obtain the following:

$$Q_m = \frac{\lambda s_0 \omega\{f'(w) - \bar{\varepsilon}[f(w) - wf'(w)]\}}{(\gamma_{k1} - \gamma_{h1}) - (1 - s_0)\omega\{f'(w) - \bar{\varepsilon}[f(w) - wf'(w)]\}} = Q_m(w_t, \lambda_t) \quad (11)$$

By means of the system (10, a, 10, b) and considering the fact that $\dot{Q}_m = \frac{\partial Q_m}{\partial w}\dot{w} + \frac{\partial Q_m}{\partial \lambda}\dot{\lambda}$, we can obtain the quasi magistral:

$$\dot{Q}_m = (\delta + \beta + \gamma_{k2})Q_m - [\lambda s_0 + Q_m(1 - s_0)]\omega e^{\beta(t-t_0)}f'(w); \quad (12,\text{a})$$

$$\dot{Q}_m = (\delta + \gamma_{h1})Q_m - [\lambda s_0 + Q_m(1 - s_0)]\omega\bar{\varepsilon}[f(w) - wf'(w)]; \quad (12,\text{b})$$

$$\dot{Q}_m = (\delta + \kappa + \gamma_{h2})Q_m - [\lambda s_0 + Q_m(1 - s_0)]\omega\bar{\varepsilon}e^{\kappa(t-t_0)}[f(w) - wf'(w)], \quad (12,\text{c})$$

where function equation $w^* = w^*(t)$ is of the form (9, a).

A quasi magistral is calculated under the stipulation that values of the parameters w^*, w^*_{k1} and w^*_{h1} at starting point of time $t = t_0$ are known, $w^*_{k1} + w^*_{k2} = w^*$, $w^*_{h1} + w^*_{h2} = 1$, $s^*_{k1} + s^*_{k2} + s^*_{h1} + s^*_{h2} = 1 - s_0$. To perform the calculations necessary for building the quasi magistral, a stepwise approximation method is used.

To build the transitional period before reaching the quasi magistral, time dependent Eqs. (7, a, 7, b) and (8, a, 8, b) are used, which are solved in a reverse time using the shooting method. Throughout the calculations, the modified Euler method with a correction [33] was used. On account of initial values of variables $w(t_0)$, $w_{k1}(t_0)$ and $w_{h1}(t_0)$, during the solution, a certain moment in time t^* is picked out, at which economic system motion trajectory reaches the quasi magistral. Simultaneously during the solution, values of $\{Q_{k1}, Q_{k2}, Q_{h1}, Q_{h2}, Q_m\}_t$ and $\{s_{k1}, s_{k2}, s_{h1}, s_{h2}\}_t$ variables restitute. In the final phase, using an index method [1], a problem of optimal distribution of investments is solved in a forward fashion.

4 Research Results

A problem of managing a regional socio-economic system by means of transitional period building index method while taking into account social, educational and technological progress growth rates as an indicator of effectiveness of investments in corresponding fields is solved on the basis of programming and computing suite« Solving the problem of an optimal regional economic system management» [34, 35].

The suite includes database on demographic, social and economic indicators of the Udmurt Repiblic. The database is derived from statistical data presented at the Goskomstat [36] website, in the Official statistics section ("National accounts", "Population", "Entrepreneurship" subsections) and at the Federal Treasury [37] web-site, in the Budget performance section (subsections "Russian Federation territorial entities' consolidated budgets" and "Community-based state non-budgetary funds of Russian Federation territorial entities"). The programming and computing suite implements a mathematical model of human capital dynamics and forecasts demographic indices of the region (see Fig. 2).

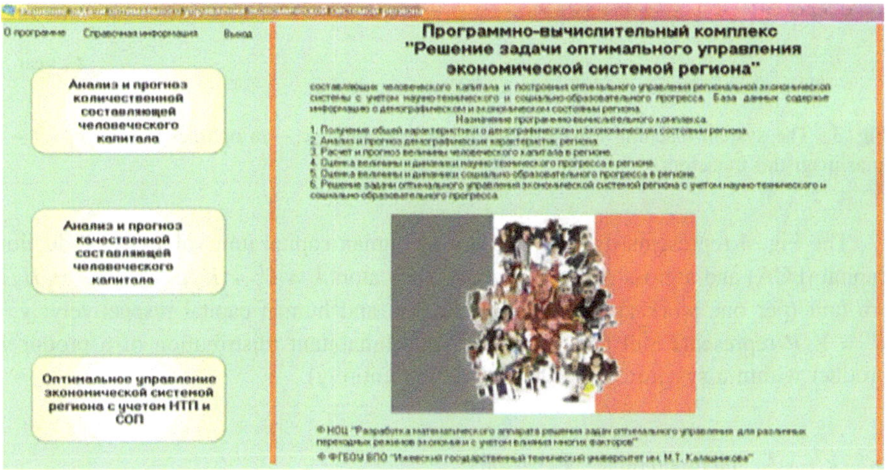

Fig. 2. The menu of the programming and computing suite.

The parameters which appear in optimal management problem statement and ought to be defined, are calculated for the period of years 1998–2018 on the basis of statistical data of the Udmurt Republic [35, 36] in accordance with an unknown parameters identification algorithm [1]. The following values were obtained: $s_0 = 0,65$; $\chi = 0,07$; $\eta = 0,14$; $\omega = 0,75$; $\bar{\varepsilon} = 0,\ 76$; $\nu = 0,15$; $\upsilon = 0,4$; $Y = F(K,H) = 0,43K^{0,4}H^{0,6}$. The discounting coefficient is $\delta = 0,05$.

Figure 3 and 4 represent several results of socio-economic system management problem solution. Calculations were made in comparable data normalized to the year 2018. The Fig. 3 represents the period of a quasi magistral attainment, where reaching a balanced growth trajectory will take place in the year 2025. Changing SEP and STP rates enables us to change the time necessary for the system to attain a quasimagistral trajectory. Determining current SEP and STP rates is an individual social and economic task.

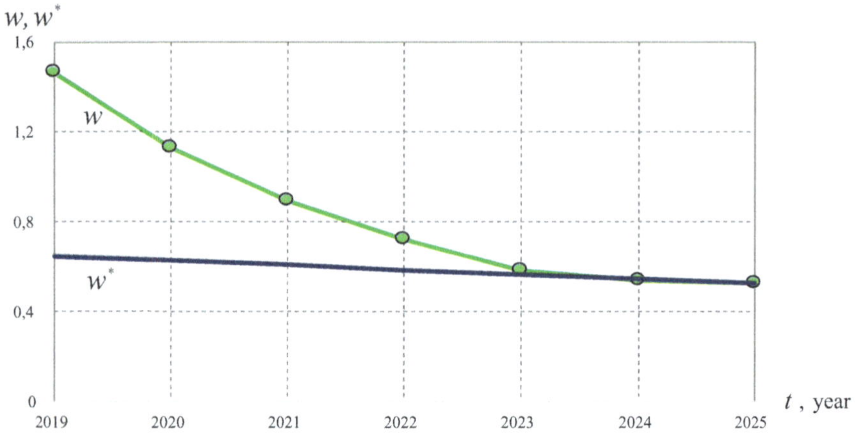

Fig. 3. The system reaching a balanced growth trajectory: w – an optimal trajectory, $w*$ – a quasimagistral trajectory.

The Fig. 4 represents dynamic pattern of human capital unit values, a production capital (FCA) and a gross regional product. Indicators $k = k^L = K/L$ и $h = h^L = H/L$ are unit (per one worker) values of production and human capital respectively; $y = y^P = Y/P$ represents unit GRP value per one inhabitant (distribution of a produced product within a system covers the population entirely).

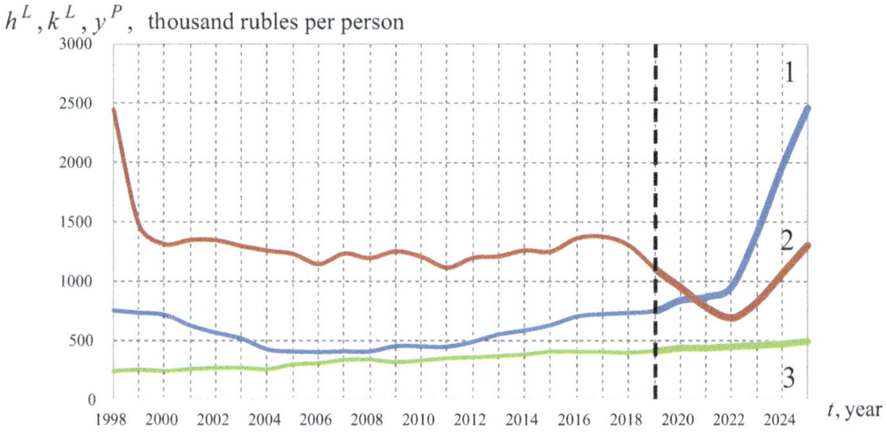

Fig. 4. The macro button chooses the correct format automatically. Dynamics of unit values: 1 – the human capital, 2 – the production capital, 3 – the gross regional product.

According to the calculations, during an optimal control scenario implementation, the production capital diminishes at the first stage due to necessity of an extraction of stale funds, which have poor performance and defray high maintenance cost. This policy provides an opportunity to increase the human capital by 1.7 times by the year 2025. Starting from the year 2023, production capital accumulates. An optimal distribution of investments between production and social fields enables us to increase unit gross regional product by 1.2 times by the year 2025.

5 Conclusion

The algorithm introduced in the paper enables us to solve the problem of an optimal socio-economic system management and allows us to account growth rates of the technical and socio-educational fields.

The algorithm is implemented via the programming and computing suite which enables us to solve the problem of modeling human capital and forecasting demographic indices as well. Handling the problem of an optimal control is possible through combining analytical approach with numerical analysis.

Solving the management problem as exemplified by a socio-economic system of the Udmurt Republic enabled us to obtain optimal values of macroeconomical indices of the region. It was demonstrated that the system is able to reach the balanced economic growth trajectory while implementing an optimal management scenario by the year 2025, which enables us to increase GRP by that time by 20%. It is also shown that at this point, the prime objective is to develop human capital since it's the factor of regional development which enables us to achieve the fastest growth of economic indices possible.

References

1. Ketova, K.V.: Mathematical Models of Economic Dynamics. ISTU Publishing House, Izhevsk (2013)
2. Haken, G., Plath, P., Ebeling, V., Romanovskii, U.: General self-organization principles in nature and society. On the history of synergetics. Institute of computer research, Moscow-Izhevsk, Russia (2018)
3. Goldin, C.: Human Capital. Springer-Verlag, Heidelberg (2016)
4. Ketova, K.V., Romanovskii, U.M., Rusiak, I.G.: Mathematical modeling of human capital dynamics. Model. Comput. Res. **11**(2), 329–342 (2019)
5. Ustinova, O.E.: Human capital and its impact on the innovative development of regions in Russia. Econ. Relat. **9**(2), 987–1008 (2019)
6. Petty, W.: Works on Economics and Statistics. Socekgiz, Moscow (1940)
7. Smith, A.: An inquiry into the nature and causes of the wealth of nations. Library an anthology of economics classics. 1 (edn.), Ekonov, Moscow, Russia (1993)
8. Marshall, A.: Principles of Economics. Progress, Moscow (1993)
9. Fisher, I.: Senses of Capital. Econ. J. **7**, 201–202 (1897)
10. Becker, G.S.: Human Capital: A Theoretical and Empirical Analysis with Special Reference to Education. Columbia University Press, N.Y. (1975)
11. Schultz, T.W.: Economic Value of Education. Columbia University Press, New York (1963)
12. Bowen, H.R.: Investment in Learning. Jossey-Bass publisher, San Francisco (1978)
13. Forrester, J.: World Dynamics. Nauka, Moscow (1978)

14. Ben-Porath, Y.: The Production of Human Capital and the Life Cycle of Earning. J. Polit. Econ. **75**, 352–365 (1987)
15. Kapelushnikov, R.I.: A note on national human capital. Higher school of economics State university, Moscow, Russia (2008)
16. Dobrynin, A.I., Diatlov, S.A., Tsyrenova, E.D.: Human Capital in Transitional Economics: Development, Estimation, Utilization Efficiency. Nauka, Saint-Petersburg (1999)
17. Koritsky, A.V.: An Introduction to Human Capital Theory. Siberian University of Consumer Cooperation Publishing House, Novosibirsk (2000)
18. Kildyiarova, G.R.: The Impact of Human Capital on Innovational Processes and Gross Domestic Product. Creative Econ. **9**(12), 1647–1656 (2015)
19. Gontmakher, E.: Russian human capital: current status and trends. World Econ. Int. Relat. **61** (3), 15–24 (2017)
20. Fiodorova, O.I., Zuev, E.G.: The human capital accumulation and implementation in the new environment: limits and possibilities. Creative Econ. **12**(10), 1649–1660 (2018)
21. Marginson, S.: Limitations of human capital theory. Stud. High. Educ. **44**(3), 1–15 (2017)
22. Aivazyan, S.A.: An analysis of synthetic categories of Russian territorial entities' population life quality: their estimation, dynamics, main tendencies. Qual. Life Russ. Reg. **11**, 5–40 (2002)
23. Vladimirova, D.C.: Challenges of information economics: a human capital development. Labor Econ. **6**(3), 1029–1042 (2019)
24. Borsh, L.M., Zharova, A.R.: A methodology for developing human capital from the perspective of information economics. Creative Econ. **13**(11), 2141–2158 (2019)
25. German, M.V., Pomuleva, N.S.: The human capital as a key factor of innovational development. Tomsk State Univ. Herald **17**, 149–153 (2012)
26. Troitskaia, A.A.: A competitive human capital of a worker: development and implementation issues. Labor Econ. **6**(2), 647–658 (2019)
27. Gabdullin, N.M.: A human capital: modern approach and estimation methods. Issues Innov. Econ. **8**(4), 785–798 (2018)
28. Ketova, K.V.: The development of regional economic system strategy research and optimization methods. Doctor of science in physics and mathematics doctoral dissertation. Federal State-Funded Educational Institution of Higher Professional Education Izhevsk State Technical University, Izhevsk, Russia (2008)
29. Intriligator, M.: Mathematical Optimization and Economic Theory. Airis Press, Moscow (2002)
30. Belenkii, V.Z.: Economic Dynamics Optimal Models. A Conceptual Framework. One-dimensional Models. Nauka, Moscow (2007)
31. Pontryagin, L.S.: The Mathematical Theory of Optimal Processes. In: Pontryagin, L.S.V., Boltyanskiy, G. (eds.) et al.: Nauka, Moscow (1961)
32. Gale, D.: Pure exchange equilibrium of dynamic economic models. J. Econ. Theory **6**, 12–36 (1973)
33. Kalitkin, N.N.: Numerical Analysis. Nauka, Moscow (2011)
34. Ketova, K.V., Rusiak, I.G., Derendyaeva, E.A.: Regional economic system optimal control problem solution. Programming and computing unit. A registration approval certificate for ECM №. 2013615414. Application №. 2013612795. Registered in the ECM programs register on 6 June 2013
35. Ketova, K.V., Rusyak, I.G., Derendyaeva, E.A.: Solution of the problem of optimum control regional economic system in the conditions of the scientific and technical and social and educational progress. Math. Model. **10**(25), 65–79 (2013)
36. Russian state statistics service official website. http://www.gks.ru. Accessed 11 May 2020
37. Federal treasury official website. http://www.roskazna.ru. Accessed 11 May 2020

Creating of Feature Dictionary Using Contour Analysis, Moments and Fourier Descriptors for Automated Microscopy

S. V. Chentsov[1], Inga G. Shelomentseva[1,3](✉), and N. V. Yakasova[1,2]

[1] Siberian Federal University, 79, Svobodny Ave., Krasnoyrsk 660041, Russia
inga.shell@yandex.ru
[2] Khakassia State University Named After N. F. Katanov, 90, Lenin Ave., Abakan, Republic of Khakassia 655017, Russia
[3] Krasnoyarsk State Medical University Named After Professor V.F. Voino-Yasenetsky, 1, Partizan Zheleznyak Ave., Krasnoyrsk 660022, Russia

Abstract. The diagnosis of tuberculosis using automated systems in a quick and inexpensive method is relevant now, especially because tuberculosis is considered one of the most important public health problems worldwide. One of the components of this process is the task of creating a feature dictionary for the subsequent classification of mycobacterium tuberculosis on digital images of sputum stained by the Ziehl-Neelsen method. This paper considers the main principles of central moments, normalized central moments and hu moments, and discrete Fourier transform in relation to the contour analysis of studied objects. To compare the effectiveness of the selected methods, the neuro-fuzzy model of the Takagi Sugeno Kang algorithm on system AnFis is used. The feature vector was formed based on color descriptors (based on RGB model), shape descriptors (eccentricity and compactness), and contour descriptors. As comparison criteria, we used indicators of the regression coefficient, standard error, accuracy, specificity and sensitivity.

Keywords: Descriptor · Fourier · Contour analysis · Moment · Ziehl-Neelsen

1 Introduction

Pulmonary tuberculosis continues to be one of the major threats to global public health. One of the ways to diagnose this disease is to detect mycobacterium tuberculosis in the sputum preparations of patients using the Ziehl-Neelsen method. This method involves such a coloring of the biopreparation, in which acid-resistant bacteria (which include mycobacterium tuberculosis) is colored in various shades of red, and the non-acid-resistant medium of the biopreparation is painted in blue. Current research is aimed at automating this process and involves the differentiation of acid-resistant bacteria from non-acid-resistant bacteria using the theory of contour analysis.

The basic task of the theory of pattern recognition is the classification problem, which contains questions about the method to identify ROI (regions of interest) and the formation of a feature dictionary. Analysis of the works of authors such as S. A. Raza, R. Khutlang, and Roa Osama Awad Altayeb has shown that one of the ways to describe

R. Silhavy et al. (Eds.): CoMeSySo 2020, AISC 1295, pp. 397–403, 2020.
https://doi.org/10.1007/978-3-030-63319-6_36

the ROI of mycobacteria is contour analysis and their description using the theory of moments and Fourier transforms [1–3]. The purpose of this research is to study the basic theory of moments and Fourier transforms as descriptors of objects of study in automated sputum microscopy, and the efficiency of the use of these descriptors by means of neuro-fuzzy networks.

2 Materials of research

The study materials are images of sputum analyzes of patients of a TB dispensary obtained by microscopy using the Ziehl-Nielsen method with a ToupCam digital camera with a resolution of 0.3 MP (Fig. 1).

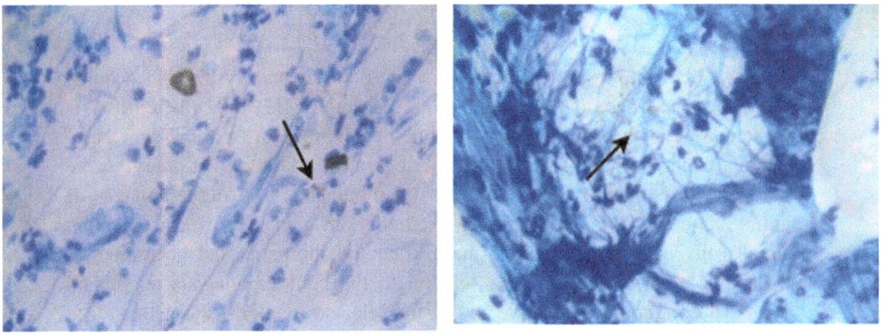

Fig. 1. Examples of images of sputum analyses stained by the Ziehl-Neelsen technique

The images of the study go through the stages of filtering and segmentation. The preprocessing process is necessary in order to exclude background areas from the image and select ROI. In the current research we use linear filtering using convolution as a filtering method, and LOG filter as the segmentation method. The result of image pre-processing and the object of study are presented in Fig. 2.

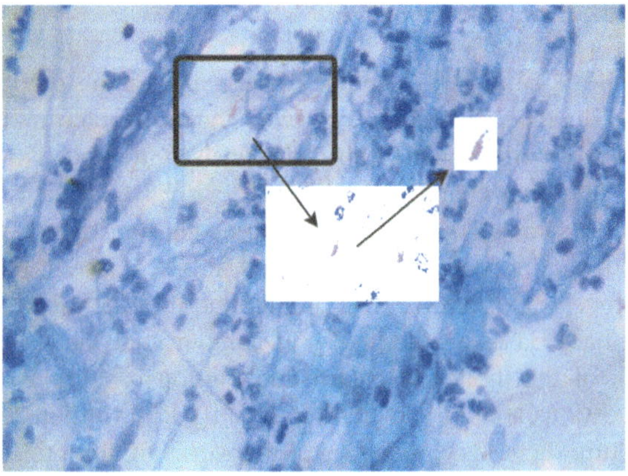

Fig. 2. Stages of image preprocessing and object of research

3 Methods of research

In the theory of image recognition, the moment is a weighted average value that characterizes a certain area of the image [4]. Descriptors are the structural and/or mathematical description of the entire image or its individual parts. The output of the descriptor is the feature vector of the input image or its singular point [5].

There are a number of requirements for descriptors - the generated mathematical description must be invariant to morphological changes and scaling. Moments and descriptors are useful for describing objects after filtering and segmentation operations, at the level of local ROI attributes.

3.1 Moments

From the point of view of contour analysis, the moment is a characteristic of the contour, calculated by integrating (summing) all the pixels of the contour. The basic moments of the contour are such characteristics as the centroid of the region, intensity, size, perimeter and orientation.

For contour analysis, the formula (1) is used to calculate moments [6].

$$m_{p,q} = \sum\nolimits_{i=1}^{n} I(x,y)x^{p}y^{q}, \tag{1}$$

where *I (x, y)* is the intensity of the pixel with coordinates *x* and *y*, *p* and *q* is the power of the parameter of summing, *n* is the number of contour pixels.

To ensure invariance to the contour position, use the central moments, which are calculated using the formula (2).

$$\mu_{p,q} = \sum\nolimits_{i=0}^{n} I(x,y)(x - x_c)^{p}(y - y_c)^{q}, \tag{2}$$

where x_c, y_c is the center of mass.

To ensure scale invariance, central moments are normalized (formula 3) [4].

$$\eta_{p,q} = \frac{\mu_{p,q}}{m_{00}^{\frac{p+q}{2}+1}} \tag{3}$$

In order to ensure invariance not only to the scale, but also to operations such as rotation and reflection, normalized moments are reduced to hu-moments (formulas 4–11) [4].

$$h_1 = \eta_{20} + \eta_{02} \tag{4}$$

$$h_2 = (\eta_{20} - \eta_{02})^2 + 4\eta_{11}^2 \tag{5}$$

$$h_3 = (\eta_{30} - 3\eta_{12})^2 + (3\eta_{21} - \eta_{03})^2 \tag{6}$$

$$h_4 = (\eta_{30} + \eta_{12})^2 + (\eta_{21} + \eta_{03})^2 \tag{7}$$

$$\begin{aligned} h_5 = {} & (\eta_{30} - 3\eta_{12})(\eta_{30} + \eta_{12})\left[(\eta_{30} + \eta_{12})^2 - 3(\eta_{21} + \eta_{03})^2\right] \\ & + (3\eta_{21} - \eta_{03})(\eta_{21} + \eta_{03})\left[3(\eta_{30} + \eta_{12})^2 - (\eta_{21} + \eta_{03})^2\right] \end{aligned} \tag{8}$$

$$h_6 = (\eta_{20} - \eta_{02})\left[(\eta_{30} + \eta_{12})^2 - (\eta_{21} + \eta_{03})^2\right] + 4\eta_{11}(\eta_{30} + \eta_{12})(\eta_{21} + \eta_{03}) \tag{9}$$

$$\begin{aligned} h_7 = {} & (3\eta_{21} - \eta_{03})(\eta_{21} + \eta_{03})\left[3(\eta_{30} + \eta_{12})^2 - (\eta_{21} + \eta_{03})^2\right] \\ & - (\eta_{03} - 3\eta_{12})(\eta_{21} + \eta_{03})\left[3(\eta_{30} + \eta_{12})^2 - (\eta_{21} + \eta_{03})^2\right] \end{aligned} \tag{10}$$

where $\eta_{p,q}$ are the corresponding central moments.

3.2 Fourier Transforms and Descriptors

The Fourier transform refers to the replacement of a function of a real variable by its representation, built on the basis of the amplitudes of harmonic vibrations when decomposing the function into components. There are several varieties of Fourier transforms - integral, discrete, etc. A discrete transformation is defined as replacing a finite quantity of real numbers with a limited quantity of Fourier coefficients [7].

Any Fourier transform is based on the concept of the amplitude spectrum (formula 11) and its power (formula 12), $i = 0, \pm 1, \pm 2, \ldots$ [8, 9].

$$p_i = \sqrt{P_i} \tag{11}$$

$$P_i = |c_i|^2 \tag{12}$$

where c_i are the required Fourier coefficients.

Discrete Fourier Transform calculates its coefficients according to formulas (13)–(15).

$$c_i = a_0 + \sum_{i=1}^{N/2} a_i \cos\left(\frac{2\pi ij}{N}\right) + \sum_{i=1}^{N/2} b_i \sin\left(\frac{2\pi ij}{N}\right) \tag{13}$$

$$a_0 = \frac{1}{N}\sum_{j=1}^{N-1} x_j,\ a_{N/2} = \frac{1}{N}\sum_{i=0}^{N-1} x_i(-1)^i,\ a_k = \frac{2}{N}\sum_{i=0}^{N-1} x_i \cos\left(\frac{2\pi ik}{N}\right) \tag{14}$$

$$b_k = \frac{2}{N}\sum_{i=0}^{N-1} x_i \sin\left(\frac{2\pi ik}{N}\right),\ i \le k < N/2 \tag{15}$$

To ensure the invariance of the Fourier descriptors, the resulting amplitude spectrum is normalized to zero harmonic, i.e.

$$P_{fourier} = \left[\frac{p_1}{p_0}, \frac{p_2}{p_0}, \ldots, \frac{p_{\frac{N}{2}}}{p_0}\right] \quad (16)$$

The considered Fourier descriptor is invariant to rotation and scaling.

4 Computational Experiment

The computational experiment took place in 4 stages - preparatory, the stage of digitizing objects, computational and resulting. The computational experiment took place in 4 stages – the preparatory stage, the stage of digitizing objects, the computational stage, and the resulting stage.

Stage 1. At this stage, filtering and segmentation operations used to obtain regions of interest. The result was about 5084 ROI (region of interest) with Mycobacterium tuberculosis and 93197 ROI without Mycobacterium tuberculosis. Further, augmentation technology (data reproduction) applied to the ROI with mycobacteria to obtain a balanced data set as a result.
Stage 2. At this stage, researchers selected the ROI, restored contours of the objects, and obtained arrays of points that describe the contours.
Stage 3. At this stage, researchers calculated the moments, converted the coordinates of the contour points into complex numbers, and performed a discrete Fourier transform. Tables 1, 2 and 3 shows examples of determining invariant moments and Fourier descriptors for contour analysis of the studied image. Tables 1, 2 and 3 shows examples of determining invariant moments and Fourier descriptors for contour analysis of the image under study.

Table 1. Normalized central moments

η_{20}	η_{11}	η_{02}	η_{30}	η_{21}	η_{12}	η_{03}
0.5267	−0.673	1.2858	−0.232	0.1880	−0.1117	-0.17296

Table 2. Hu – moments

h_1	h_2	h_3	h_4	h_5	h_6	h_7
1.8124	2.3897	0.5537	0.1188	−0.0002	−0.0758	−0.0304

Table 3. Fourier descriptors

P_1	P_2	P_3	P_4	P_5	P_6	P_7
0.3033	0.4757	0.2851	0.0611	0.01717	0.1382	0.0353

Stage 4. The effectiveness of the selected features was tested on the basis of neuro-fuzzy networks (Table 4).

A model based on the Takagi-Sugeno-Kang algorithm implemented in the ANFis system used as the classification model. The ANFis system is currently one of the most popular neuro-fuzzy systems, which is due to its flexibility and versatility, and the convenient graphical interface provided by the Matlab system [10, 11].

In addition to contour descriptors, the feature vector included color descriptors (the value of the red, blue, and green colors of the RGB model) and shape descriptors (eccentricity and compactness, which is the ratio of the width and length of the rectangle bounding the ROI).

The comparison criteria were accuracy, sensitivity, specificity, mean square error MSE, and regression coefficient R. The results of the computational experiment are presented in Table 4.

Table 4. The results of computational experiment

Feature dictionary	Accuracy, %	Sensitivity, %	Specificity, %	MSE	R
Normalized central moments	75,6	78,1	73,5	0,24	0,38
Hu – Moments	82	75,2	93,9	0,13	0,69
Fourier descriptors	89	82	93,03	0,08	0,85

5 Conclusions

The feature vector based on Fourier descriptors shows the highest accuracy value (89%) and the highest sensitivity (82%). The feature vector based on hu-moments shows the highest specificity value −93.9%. From a mathematical point of view, moments are the projection of a contour on a polynomial basis, and Fourier descriptors are the projection of the same contour on the basis of harmonic functions. When comparing feature dictionaries using neuro-fuzzy network models, the projection on the basis of harmonic functions gives the best values of classification quality criteria.

References

1. Raza, S.A., Arif, M., Marjan, M.Q., Butt, F.: Anisotropic tubular filtering for automatic detection of acid-fast bacilli in digitized microscopic images of ziehl-neelsen stained sputum smear samples. SPIE Med. Imaging Digital Pathol. 1–8 (2015)
2. Khutlang, R., Krishnan, S., Dendere, R., Whitelaw, A., Veropoulos, K., Learmonth, G., Douglas, T.S.: Classification of mycobacterium tuberculosis in images of ZN-stained sputum smears. IEEE Trans. Inf. Technol. Biomed. **14**(4), 949–957 (2010)
3. Altayeb, R.O.A.: Automatic Method for Tuberculosis Bacilli Identification in Sputum Smear Microscopic Images Using Image Processing Techniques, Sudan (2016)

4. Hu, M.K.: Visual pattern recognition by moment invariants. IEEE Trans. Inf. Theory **8**, 179–187 (1962)
5. Gonzales, R.C., Woods, R.E.: Digital Image Processing. Pearson Education, London (2002)
6. Flusser, J., Suk, T., Zitova, B.: What are Moments? Moments and Moment Invariants in Pattern Recognition. Wiley, Hoboken (2009)
7. A professional information and analytical resource dedicated to machine learning, pattern recognition, and data mining. machinelearning.ru. Accessed 15 2020 15
8. Ahmed, N., Rao, K.R.: Orthogonal Transforms for Digital Signal. Springer, Heidelberg (1975)
9. Makarov, M.A.: Contour analysis in solving problems of description and classification of objects. Mod. Probl. Sci. Educ. **3**, 44–50 (2014)
10. Terano, T., Asai, K., Sugeno, M.: Applied Fuzzy Systems. Academic Press, Cambridge (1989)
11. Omisore, M.O., Samuel, O.W., Atajeromavwo, E.J.: A genetic-neuro-fuzzy inferential model for diagnosis of tuberculosis. Appl. Comput. Inform. **13**(1), 27–37 (2017)

Applying an Integral Algorithm for the Evoked P300 Potential Recognition to the Brain-Computer Interface

S. N. Agapov[1], V. A. Bulanov[1], A. V. Zakharov[2(✉)], and V. F. Pyatin[2]

[1] IT Universe LLC, Samara, Russia
sergeyagapov@it-universe.ru
[2] Samara State Medical University of the Ministry of Health of Russia, Federal State Budgetary Educational Institution of Higher Education, Samara, Russia
zakharov1977@mail.ru, pyatin_vf@list.ru

Abstract. *Objective of the Research:* Developing an integral recognition algorithm for the response potential (ERP), evoked by the target visual stimulus, and verification of the suggested algorithm operability on the wireless 5-channel Emotiv Insight electroencephalography headset with dry sensor technology.

Materials and Methods: The object of the research was EEG records of five volunteers. The research was carried out with the wireless 5-channel Emotiv Insight electroencephalography headpiece with dry sensor technology, the eSpeller software developed by the research authors, MathWork® MATLAB software environment version R2015a.

Findings. The developed integral algorithm for recognition of the electrical response of the brain cortex to a target visual stimulus manifests the recognition accuracy ranging from 71.5% to 90.6%, with the average value of 80.1 ± 7.2%.

Conclusion: The presented algorithm demonstrates high-level recognition accuracy of the potential evoked by the target visual stimulus, but does not require any large computation capacities, sophisticated classification methods or machine learning. The test of the suggested algorithm has proven the applicability of the Emotiv Insight electroencephalography headpiece with the dry sensor technology to a neurocomputer interface.

Keywords: EEG · Evoked potentials · P300 · Integral algorithms · Neurocomputer interface

1 Introduction

A the present moment, the neurocomputer interfaces (NCI) use the evoked reaction of synchronization/desynchronization (ERS/ERD) of the sensomotor rhythms in motor imagery [1, 2], slow cortical potentials (SCP) [3], steady state visually evoked potentials (SSVEP) [4], P300 component of ERP, event-related potential, [5] as the control signals.

The event-related potentials (ERP, or ERP response) are generated in the brain cortex to a stimulus presentation and consist of several components. P300 component

R. Silhavy et al. (Eds.): CoMeSySo 2020, AISC 1295, pp. 404–412, 2020.
https://doi.org/10.1007/978-3-030-63319-6_37

of the ERP wave manifests itself when the tested person is doing various comprehension-related tasks. One of the simplest experiments for detection of the brain reaction to the target visual stimulus uses the oddball paradigm and consists in presenting two types of stimuli, a target and a non-target one, to the tested person. The proportion of presenting the target and non-target stimuli is approximately 20% to 80% respectively. The P300 component manifests itself in EEG in the most prominent way to the presentation of the target stimulus.

There are various techniques used for ERP response detection, from averaging the EEG epochs to machine learning methods [6]. The modern significant system detection algorithms used in NCI are quite sophisticated and consist of several stages of signal processing [7].

Not so long ago, simple, convenient and affordable electroencephalographers including the wireless 5-channel Emotiv Insight headset with dry sensor technology (USA, https://emotiv.com/insight) (see Fig. 1) were introduced.

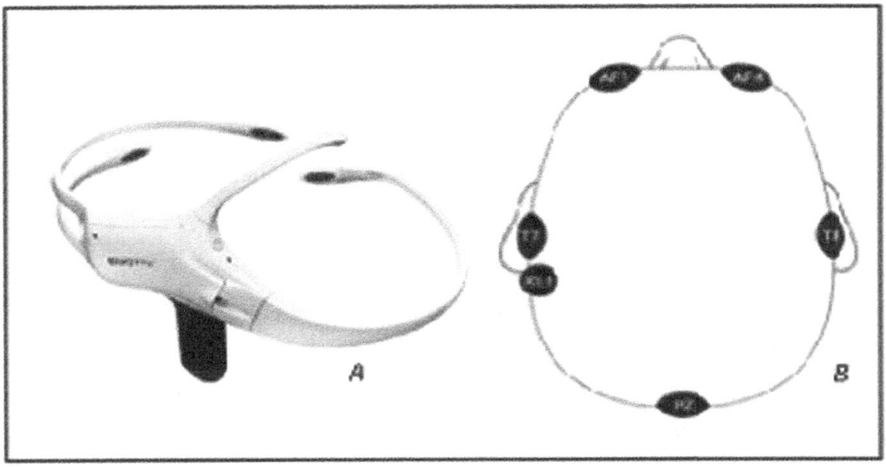

Fig. 1. Wireless Emotiv Insight electroencephalography headset with the dry sensor technology: 1) appearance; b) arrangement of the EEG electrodes on the head.

Due to the simplicity of this EEG device structure, it can be embedded in the NCI technology. However, there occurs the need for the development of algorithms and software solutions with respect to the specificity of such devices, particularly, the high noise rate in the EEG signal, low sensitivity and low signal sampling interval.

2 Objective

Development of the integral algorithm for recognition of the event-related potential (ERP response) evoked by the target visual stimulus and verification of the suggested algorithm on the wireless 5-channel Emotiv Insight electroencephalography headset with the dry sensor technology.

3 Materials and Methods

The research was carried out from May 23 to June 20, 2016, in compliance with the ethical principles of the Declaration of Helsinki enacted by the World Medical Association (WMA) (2013 year's edition). The research subjects were five men aged from 29 to 44 year old (35.8 ± 7.2) who had provided their informed consent to the test. In this paper, they are identified as *subjA, subjB, subjK, subjP, subjS*. EEG registration was carried out in daylight in an office room, using the wireless Emotiv Insight EEG headset with the dry sensor technology. The electrodes were arranged on the head of the tested persons according to the international 10–20 system (see Fig. 1). The data were recorded with the sampling interval of 128 sec-1. The tested volunteers were seated in front of the computer monitor standing 50–70 cm away from their eyes.

Visual Stimulation. For displaying the stimuli on the computer screen, the eSpeller software developed by the research authors was used.

(platform Java 1.8 and above, OS Windows 7 and above). The software operation was based on the single character paradigm [8]. On the computer screen, the tested volunteer was demonstrated a square with a side length of 250 mm divided into nine cells. As the visual stimuli, the gray and red colored circles demonstrated in every cell against the black background were used (see Fig. 2).

The cells were highlighted in a random order: the circle diameter would increase by 1.5 times (from 33 mm to 50 mm) and the circle color was changed from dark to bright (RGB colors 179.179.179 to 255.0.0). Each of the nine cells was highlighted one time per cycle. The highlight duration was 120 ms, the intervals between the highlights being 180 ms; the interstimulus interval (ISI) was 300 ms. The cells were highlighted in a random order.

Within one session, the number of cell highlighting cycles was 20. The intersession interval was 10 s. During one session, the tested volunteer was requested to subsequently concentrate his or her attention on the cells from 1 to 9.

Simultaneously with the stimulus visualization, the EEG signal was recorded into a file, registering the cell identification marks and the stimulus presentation timing marks. In each experiment, nine files of EEG signals were recorded by the number of the cells the tested volunteer concentrated his or her attention on. For each tested volunteer, three experimental sessions were carried out. This way, 135 EEG records were made.

Data processing sequence. The EEG record files were converted from the CSV format to the matrix form. In the process, the cell identifying and timing marks were noted.

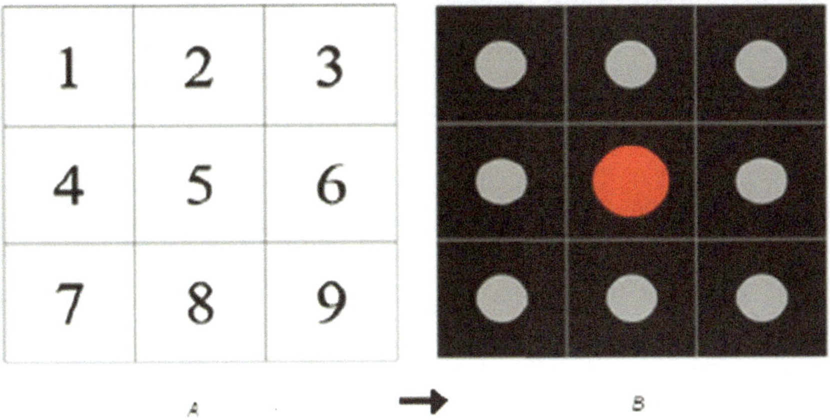

Fig. 2. ESpeller software for visualization of the stimuli on the computer screen: a) sequence of concentration of the tested volunteer on the cells during the experimental session; b) the visualization window: in the presented example, cells 1–4 and 6–9 are not highlighted, cell 5 is shown in the highlighted state.

For the EEG signal analysis, out of the five leads of the Emotiv Insight EEG headset, the PZ lead was selected for providing the most prominent ERP response, as proven by the published research data [5]. We used a fourth-order Butterworth low pass filter under 30 Hz.

The EEG signal was divided into epochs of 1 s from the moment of the visual stimulus presentation. In each EEG record, 180 epochs were identified (by the number of visual stimuli presentations within one session: 9 cells □ 20 cycles), out of which 20 epochs corresponded to the target stimulus presentation (target epochs) and 160 epochs corresponded to non-target stimulus presentation (non-target epochs).

To eliminate the inclined linear trend, the amplitude values in every acquired epoch are aligned.

Summing and averaging the amplitude values in the target and non-target epochs separately enhance the significant signal and compensate for the random noise. *Computing the temporal window size within the epoch.* The P300 component of the ERP wave is quite prominently localized in time. To generalize the temporal limits of the ERP waves, the epochs were averaged in all the session to acquire the ERP wave temporal localization limit values. The averaging was done for 2700 target and 21600 non-target epochs (see Fig. 3). The temporal limits of the analysis window constituted 322 and 591 ms from the start of the visual stimulus presentation. The acquired limit values were applied to further analysis.

Classification of epochs. For differentiation of the target and non-target epochs, the EEG signal amplitude values were summarized in the analysis window after every cell highlighting cycle had been completed. The maximum value of the amplitude range within an epoch of the cycle was interpreted as a target epoch.

For the EEG analysis, the MathWork® MATLAB version R2015a software was used (www.mathworks.com).

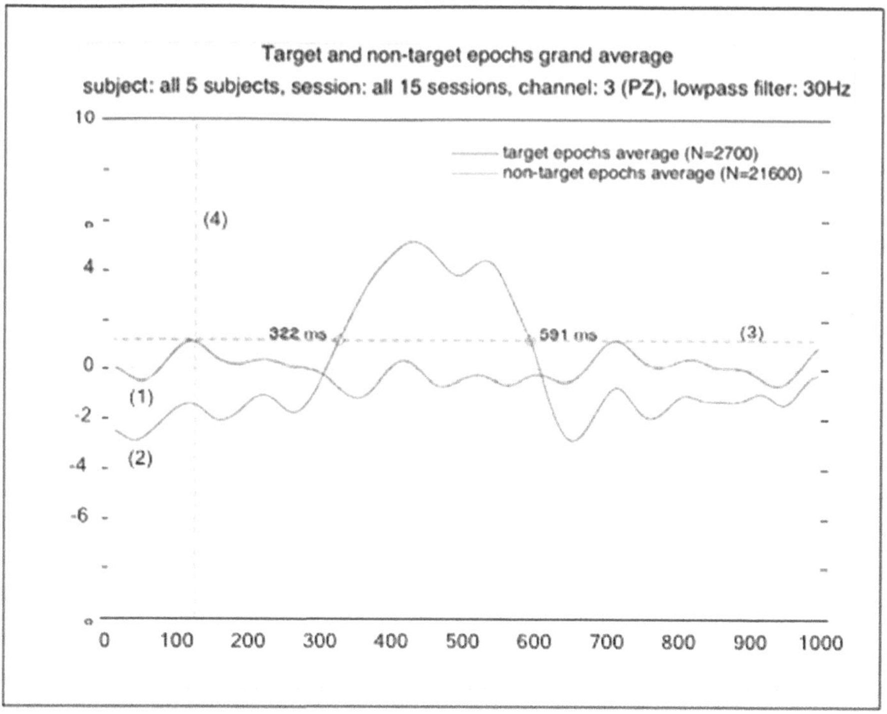

Fig. 3. Average diagram of the target and non-target EEG epochs: (1) non-target epoch diagram; (s) target epoch diagram; (3) maximum positive value of the non-target epochs; (4) time of termination of the visual stimulus demonstration is 120 ms. At the intersection of the maximum positive value of non-target epochs (3) and target epoch diagram (2), the time values taken as the temporal 'limits of the EEG signal analysis window are indicated.

4 Results and Discussion

The differentiation analysis (confidence of recognition) of the target and non-target stimuli in the integral algorithm of recognizing the P300 component in ERP response revealed a significant range of confidence level even for one tested subject throughout different sessions (Table 1).

For example, for *subjB*, the range of confidence varied from 68.5% to 94.0% in three sessions (range spread 25.5%). On the other hand, for *subjS*, the confidence range varied from 87.7% to 93.9% (range spread 6.2%). Such an uneven distribution of the confidence level value could be caused by an unstable fit of the contacts on the subject‘s head during the session. It may also be related to a different level of attention of the tested persons and initially unstable connection between the willful effort of the person and its manifestation in EEG records.

Table 1. Confidence of visual stimuli recognition, %.

Number of epochs used for averaging the EEG signal	Sessions grouped by the tested subjects															Average accuracy value by epoch
	subjA			subjB			subjK			subjP			subjS			
1	39.2	60.2	60.7	64.8	76.2	42.9	79.3	72.7	41.9	65.8	78.5	76.0	66.0	70.1	76.0	64.7
2	45.4	55.2	62.5	65.1	83.4	64.7	85.1	71.2	46.8	63.3	74.7	81.2	74.4	72.9	83.8	68.6
3	46.0	61.0	67.0	48.3	93.3	69.5	83.2	74.9	49.1	71.9	71.7	78.8	73.5	83.8	82.3	70.3
4	53.7	62.5	70.0	50.0	94.3	71.4	79.2	85.7	48.3	77.6	64.8	91.8	80.8	87.4	75.9	72.9
5	54.1	63.0	65.4	56.7	90.9	81.5	75.2	79.8	47.3	74.4	66.3	92.0	80.8	92.7	78.3	73.2
6	58.0	60.8	65.4	58.5	90.0	80.3	81.7	83.6	49.5	78.1	69.4	91.8	91.1	91.1	78.2	75.2
7	58.3	67.9	60.3	62.6	94.3	75.8	79.1	79.9	53.6	79.4	69.7	90.7	96.4	95.3	83.6	76.5
8	65.5	67.9	62.2	65.8	95.3	73.3	85.6	84.0	56.5	77.2	72.3	91.6	97.5	97.4	85.6	78.5
9	67.9	71.8	62.8	62.4	93.7	76.0	81.9	86.8	59.0	75.7	69.0	90.7	98.2	99.2	87.1	78.8
10	65.9	78.0	71.4	65.8	92.0	80.3	81.3	86.1	62.0	80.5	76.2	90.9	99.0	97.1	90.0	81.1
11	74.2	74.9	72.5	66.0	93.7	85.2	77.9	92.8	65.8	81.7	76.9	90.6	96.9	98.5	88.2	82.4
12	72.1	77.4	75.5	70.9	95.4	83.8	80.2	90.6	67.7	83.2	78.9	91.2	97.4	99.4	84.9	83.2
13	72.1	79.2	83.4	73.0	96.2	82.4	79.6	92.1	67.3	82.2	85.5	91.5	96.0	99.7	87.9	84.5
14	74.7	77.9	84.2	75.7	99.0	86.1	76.1	92.0	67.7	85.8	85.2	91.7	94.9	98.6	92.2	85.5
15	77.4	86.7	84.5	74.3	99.0	89.6	76.3	94.8	68.6	84.4	86.2	90.7	95.9	98.3	95.0	86.8
16	78.2	86.4	84.1	79.0	98.7	89.8	76.1	95.0	71.6	78.6	89.7	88.5	93.6	98.6	94.6	86.8
17	83.8	89.6	85.4	82.0	98.9	87.7	78.3	95.5	75.2	79.5	86.1	90.5	95.1	99.2	95.9	88.2
18	82.2	86.0	85.9	83.0	98.0	86.1	79.1	96.1	73.3	78.9	87.3	90.8	94.7	99.3	96.4	87.8
19	82.5	85.4	87.6	83.9	98.7	87.5	75.9	97.6	72.6	78.1	88.5	93.4	91.4	99.7	98.7	88.1
20	82.4	86.9	89.8	81.6	98.6	86.4	76.9	96.2	75.4	78.8	91.1	97.5	91.9	99.7	99.4	88.8
Average accuracy value by session	66.7	73.9	74.0	68.5	94.0	79.0	79.4	87.4	61.0	77.8	78.4	89.6	90.3	93.9	87.7	80.1
Average accuracy value by tested subject	71.5±4.2			80.5±12.8			75.9±13.5			81.9±6.7			90.6±3.1			80.1±7.2

Recognition confidence values of 80% and above.
Recognition confidence values of 90% and above.
Recognition confidence values of 95% and above.

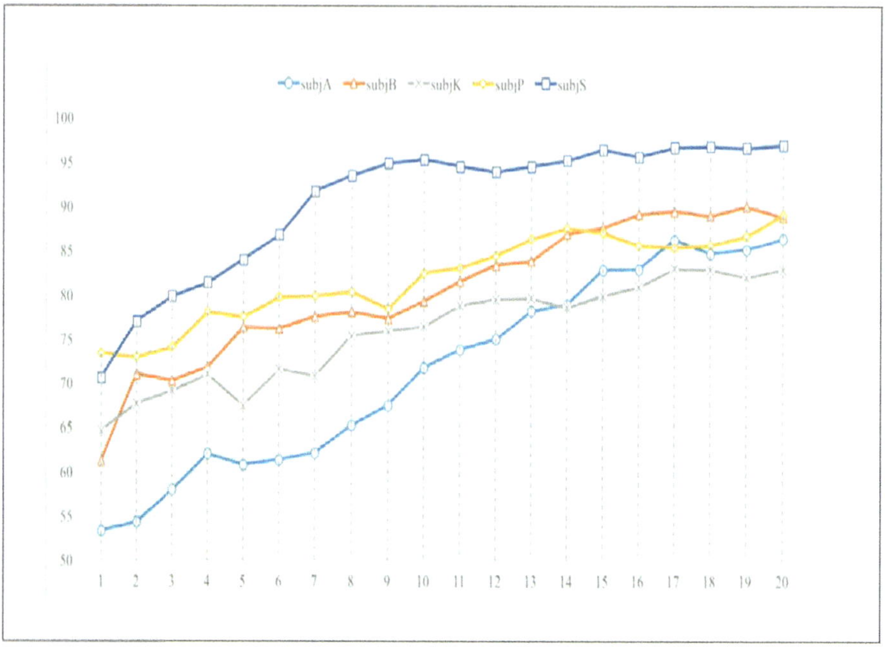

Fig. 4. Diagram of target visual stimulus recognition confidence depending on the number of stimulus presentations.

The collected data analysis demonstrated a significant range of the differentiation confidence levels target and non-target visual stimuli for different tested subjects. For example, for the tested volunteer subjA, the average confidence level constituted 71.5%, while for subjS this level reached 90.6%. This range of differences can be explained by the individual features of the tested persons. The increase in the target and non-target visual stimuli recognition confidence was observed to depend on the number of the stimulus presentations (see Fig. 4). After 14–17 presentations of the target stimulus, all the tested volunteers have reached their own "confidence limit" after which no recognition confidence growth was found.

The recognition confidence rate analysis showed that the number of target stimulus presentations ranging from 2 to 10 causes a significant growth in the recognition confidence: on the average, from 2% to 4% with every next presentation of the stimulus (see Fig. 5).

Further increase of the target stimulus number from 11 to 17 causes a minor growth in recognition confidence (not exceeding 1.5% with every new presentation of the stimulus) and even less (under 1%) when the target stimulus is presented from 18 to 20 times. Interestingly, after the 18th presentation of the target stimulus, the recognition confidence changes in a negative way. It can be explained both by a random deviation or growing fatigue of the tested volunteers.

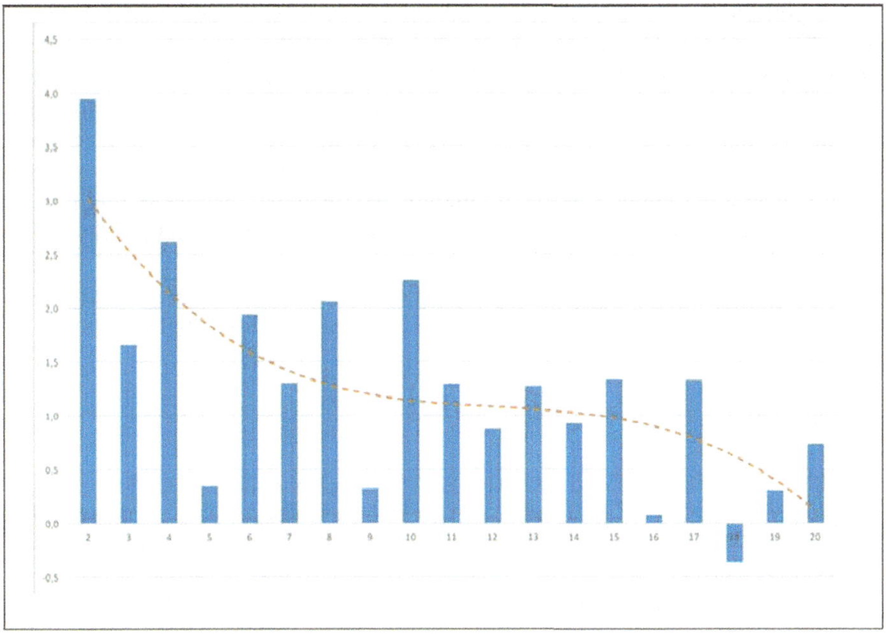

Fig. 5. Diagram of recognition confidence level growth rate depending on the number of presentations of the target visual stimuli: the columns show the confidence level increment to the previous confidence level value; the dashed line indicates the third order polynomial trend.

5 Conclusion

The developed integral algorithm of recognizing the ERP response to the target visual stimulus shows good results and can be used in practice. The target and non-target visual stimuli differentiation confidence level varied from 71.5% to 90.6% for different tested subjects average value 80.1 ± 7.2%.

It should be remarked that every tested volunteer manifested a 100 per cent confidence of recognition of the given cell at any number of target stimulus presentation from 1 to 20; the number of such sessions counted 13 of 135, or 9.6%. The presented integral algorithm manifested a 100 per cent confidence of recognition of the target visual stimulus after a single presentation of the stimulus as well; the number of such sessions constituted 31 of 135, or 22.9%.

The test of the suggested algorithm has proven the applicability of the Emotiv Insight electroencephalography headpiece with the dry sensor technology to a neurocomputer interface.

The developed integral recognition algorithm requires no additional hardware except for the EEG device and a computer with the necessary software installed. Moreover, no significant computation capacities, sophisticated classification methods or machine learning are required. At the same time, the algorithm requires several cell highlighting cycles, which takes some time to enter one control command in the NCI.

The collected results require additional experimental studies and further improvement of the integral algorithm for recognition of the ERP response to a target visual stimulus. It is possible to study the degeneralization of the suggested algorithm and the effect of applying adaptive decision-making methods to the recognition of the target stimulus. In this case, the integral algorithm can be adjusted to every tested person to make the target stimulus recognition decision in a flexible manner. This approach allows reducing the number of target stimulus presentations for recognition (by 2–5 times), at the same time retaining the high confidence level (90% and above).

Acknowledgements. The study was implemented with the financial support of the Ministry of Science and Higher Education of the Russian Federation (grant RFMEFI60418X0208).

References

1. Pyatin, V.F., Kolsanov, A.V., Sergeeva, M.S., Zakharov, A.V., Antipov, O.I., Korovina, E.S., Tyurin, N.L., Glazkova, E.N.: Information possibilities of using mu and beta rhythms of the EEG dominant hemisphere in the construction of brain-computer interface. Fundam. Res. **2** (5), 975–978 (2015)
2. Edlinger, G., Allison, B.Z., Guger, C.: How many people can use a BCI system? In: Clinical Systems Neuroscience, pp. 33–66 (2015)
3. Birbaumer, N., Hinterberger, T., Kuebler, A., Neumann, N.: The thought-translation device (TTD): neurobehavioral mechanisms and clinical outcome. IEEE Trans. Neural Syst. Rehabil. **11**(2), 120–123 (2003)
4. Norcia, A.M., Appelbaum, L.G., Ales, J.M., Cottereau, B.R., Rossion, B.: The steady-state visual evoked potential in vision research. A review. J. Vis. **15**(6), 4 (2015)
5. Luck, S.J.: An Introduction to the Event-Related Potential Technique, vol. 78, no. 3 (2005)
6. Krusienski, D.J., Sellers, E.W., Cabestaing, F., Bayoudh, S., McFarland, D.J., Vaughan, T. M., Wolpaw, J.R.: A comparison of classification techniques for the P300 Speller. J. Neural Eng. **3**(4), 299–305 (2006)
7. Sun, S., Zhou, J.: A review of adaptive feature extraction and classification methods for EEG-based brain-computer interfaces. In: Proceedings of the International Joint Conference on Neural Networks (IJCNN), pp. 1746–1753 (2014)
8. Fazel-rezai, R., Gavett, S., Ahmad, W., Rabbi, A., Schneider, E.A.: Comparison among several P300 brain-computer interface speller paradigms. Clin. EEG Neurosci. **42**(4), 209–213 (2011)

Fractal Analysis of EEG Signals for Identification of Sleep-Wake Transition

A. V. Zakharov(✉) and S. S. Chaplygin

Samara State Medical University of the Ministry of Health of Russia, Federal State Budgetary Educational Institution of Higher Education, Samara, Russia
zakharov1977@mail.ru

Abstract. *Objective* is the combination of frequency filtering and nonlinear analysis methods for generating hypnograms through the analysis of electroencephalographic (EEG) signals in the somnological studies.

Methods. The frequency filtration methods are used for the primary preparation of the EEG signals for further nonlinear analysis. Among the nonlinear analysis methods, such fractal deterministic chaos methods as the Hurst standardized range method, approximate entropy method, and correlation integral computation with the Grassberger-Procaccia algorithm. To apply the two latter methods, we used the pseudo-phase space reconstruction method based on Taken's theorem. Relying upon the nonlinear analysis results, and as a result of the somnology examination of patients, the hypnograms of sleep stage transitions were made up. To verify the collected results, they were compared with the hypnograms generated with the classical method based on the Rechchaffen and Kales parameters. Moreover, the problems related to various disturbing factors are considered, the ways of mitigation of their influence on the final results are suggested.

Conclusion. With the properly selected method parameters, the accurate standardization of the input data and correct averaging of results, the given methods can be used to produce a hypnogram showing complete coincidence of the identified sleep phases for almost a half of the epochs registered by EEG. Interestingly, these results can be achieved with one EEG registration channel only.

Keywords: Fractals · Deterministic chaos · Hypnogram · Frequency analysis · Electroencephalography

1 Introduction

Timely diagnostics of sleep disorders may reveal and prevent the development of many serious diseases [1, 2]. Since many pathological processes may emerge or, vice versa, diminish during sleep, in last years the medicine of sleep studying the pathogenesis, clinic and treatment of the pathologies that emerge in the period of sleep has been rapidly evolving [3–5]. The commonly known sleep stage classification system was developed by Rechchaffen and Kales in 1968 [1, 2]. According to this method, the expert manually analyses electrophysiological parameter records of approximately eight hours length. For every thirty-second fragment of the record, the determination

R. Silhavy et al. (Eds.): CoMeSySo 2020, AISC 1295, pp. 413–420, 2020.
https://doi.org/10.1007/978-3-030-63319-6_38

characteristics used to classify the segment as belonging to a certain sleep stage are subsequently computed. The Rechchaffen and Kales parameters-based hypnogram generation method is still very popular despite a number of significant limitations, such as high labor intensity and subjectivity of the interpretation. For this reason, at the present moment there is an urge for objective automated methods for sleep phase recognition which, combined with an electrophysiological signal registration device, would make a sleep disorder diagnostic system. Due to the fractal nature of the EEG signals, using fractal measures in this case appears the most natural [6]. Consequently, this method would produce more accurate results with less input data. In this paper, the sleep phase division problem was solved by fractal measures applied to the EEG channel analysis. In particular, this study requires the record of one EEG only (without EOG or EMG) and only one channel to formulate the results.

2 Nonlinear Analysis Methods

The fractal methods described below have been successfully used by the authors in their previous works, applied to a wide range of problems [7–13]. Particularly, the Hurst exponent computation-based method [11–13], the Grassberger-Procaccia method [13], the Takens' theorem-based method [11, 13], the approximate entropy computation method [11–13] and other nonlinear analysis methods [14] have been used in various ways [15]. The mathematical grounds of the mentioned methods are briefly presented below.

The Hurst Standardized Range Method Applied to the EEG Signal Time Sampling Computation. At the first stage of the Hurst exponent computation, the mean value of signal $\langle U \rangle_N$ for N time cycles:

$$\langle U \rangle_N = \frac{1}{N} \sum_{n=1} U(n)$$

Then the accrued deviation $U(n)$ from its mean value $\langle U \rangle_N$ is determined with the sum:

$$X(n, N) = \sum_{p=1}^{n} \{U(p) - \langle U \rangle_N\}$$

The deviation range is defined as follows:

$$R\langle N \rangle = \max_{1 \le n \le N} X(n, N) - \min_{1 \le n \le N} X(n, N)$$

Standard deviation can be calculated with the square root formula from the dispersion:

$$S(N) = \sqrt{\left(\frac{1}{N}\sum_{n=1}^{N}\{U(n) - \langle U\rangle_N\}^2\right)}$$

As demonstrated in the Hurst's studies, for the majority of temporal series, the observed standardized range can be described with the empirical relation [16]:

$$R/S = (aN)^{H}$$

where H is the Hurst exponent, and α is the arbitrary constant. It should be mentioned that the range is referred to as standardized as it has to be divided by the square root from the dispersion.

The Phase Space Reconstruction Method and Takens' Theorem. Takens' theorem [17] can be used to calculate the correlation integral (to be described below) and the fractal dimensionality based on the temporal sequence measurements made for one component only. According to Takens, it is necessary

$$X_i = X(t_i) = \{x(t_i), x(t_i - \tau), \ldots, x(t_i - (m-1)\tau)\}$$

The Grassberger-Procaccia Method for Correlation Integral Computation. With the delay method described above, let us use the studied series to formulate the attractors in the m-dimensional pseudo-phase spaces for m =1, 2, 3, ... Then, for every attractor in the m space, let us calculate the correlation integral with the formula [18]:

$$Ce(\varepsilon, N) = lim_{N\to\infty}\frac{1}{N(N-1)}\sum\nolimits_i\sum\nolimits_j \theta(\varepsilon - |x_i - x_j|), i \neq j,$$

where N is the number of attractor points, $\angle x_i - x_j\angle$ is the absolute distance between the i -th and j-th attractor points in the m-dimensional space, ε is the resolution cell size, θ is the Heaviside function. Basically, $Ce(\varepsilon, N)$ is the dependency of the number of attractor points in the m-dimensional space with the distance between them being $<\varepsilon$, on the size of the resolution cell, related to the total number of paired points, i.e. $\sim N^{-2}$ $(N(N-1)$ in the formula denominator due to the condition of $Ce(\varepsilon, N)$ $i \neq j$). The generated dependencies are plotted in double logarithmic scale on the plane (theoretically, it can be done with any logarithm base, but 10 is used for better visualization). Then, the linear segments of the plotted curves are indicated, and the least square method is applied to find the lines approximating them. For all the generated curves, $Ce(\varepsilon, N)$ the first derivative of the approximating line D_C is calculated and plotted as a function of m.

Approximate Entropy. Being a measure of deterministic chaos, approximate entropy is intended to collect information on the complexity of processes occurring in the system based on short time series

$$X = [x(1), x(2), \ldots, x(N)]$$

where N is the length of the studied series, consisting of approximately 75–5000 readings [19]. The approximate entropy value depends on the dimensionality of the pseudo-phase space m based on the Takens method, "filtration factor" r and the studied series length N, calculated with the equation:

$$ApEn(m, r, N) = \Phi^m(r) - \Phi^{m+1}(r)$$

Here, $\Phi^m(r)$ and $\Phi^{m+1}(r)$ are derived from the equations:

$$\Phi^m(r) = \frac{1}{N-m}\sum_{i=1}^{N-m} ln(C_i^m(r)),\ \Phi^{m+1}(r) = \frac{1}{N-m}\sum_{i=1}^{N-m} ln(C_i^{m+1}(r)),$$

just as for the correlation integral, $C_i^m(r)$ and $C_i^{m+1}(r)$ are determined with the sums:

$$C_i^m(r) = \frac{1}{N-m+1}\sum_{j=1}^{N-m+1} \theta(r - |x(i) - x(j)|),$$

$$C_i^{m+1}(r) = \frac{1}{N-m}\sum_{j=1}^{N-m} \theta(r - |x(i) - x(j)|).$$

As a result, the approximate entropy calculation process can be limited to the acquisition of the *ApEn* value with the general equation [20]:

$$ApEn(m, r, N) = \frac{1}{N-m}\left[\sum_{i=1}^{N-m} ln\left(\frac{C_i^m(r)}{C_i^{m+1}(r)}\right)\right],$$

followed by the computation of $C_i^m(r)$ values for every $C_i^{m+1}(r)$ i.

3 Frequency Filtration Methods as Essential Artifact Combating Methods Used in Nonlinear Analysis

Studies of the EEG signals frequency parameters are successfully used both in modern medicine and in clinical studies [21]. The subject matter here is selecting certain fractal methods for solving particular tasks, selection of channels for analysis, and selection of frequency diapasons for filtration and particular method application specificities. Below, let us elaborate on the Zenkov artifact classification system [22] and the suggested solutions on elimination of artifacts in using the fractal method.

As remarked in [22], *the EMG potentials* are the pointed-shaped high-frequency activity of irregular frequency lying within the diapason of 15–100 Hz. For EEG signals, we suggest using a digital filter based on the fast Fourier transformation (FFT) method to crop the frequencies exceeding 40 Hz, as the peak of EMG activity dominates in the 70 Hz area.

The ECG potentials occur, as a rule, in the reference electrodes and are identified through a distinctive electrocardiogram shape [22]. This is explained by the fact that the pair of electrodes transmitting the signal are not equally spaced from their generation source, the heart. As such disturbance factors are classified as quasi-periodic, they make less impact on the fractal measures than spectral analysis.

The EOG potentials are associated with the eyeball moves and, therefore, the orientation of the electrical axis of the eye determined with the corneoretinal potential [22]. We found a partial solution of the problem through filtering the frequencies below 0.5 Hz to get rid of the eyeball move-related high-range EEG jumps. However, to minimize the influence made by these disturbance factors on automated recognition, the leads remote from the forehead area should be selected.

The swallowing-induced electrical potentials are represented by the high-range slow two and polyphased waves with the period of 0.5–2 s that normally occur in unipolar leads on all the channels [22]. According to the presented method, it is suggested to stick to one 30 s epoch throughout the entire period of sleep. As far as the produced hypnogram is concerned, the influence of such artifacts is expected to be quite short, 1–2 epochs long, and if during the automatic recognition the transition from the current sleep phase occurs, it cannot be registered according to Rechchaffen and Kales.

Network interference is the consequence of electromagnetic setting on the patient-electrodes-conductors system. The way of eliminating such interference is the same for any situation: they are eliminated by reducing the impedance between the electrode and the patient's skin to the required level during standard skin preparation routine and narrow-band filtering to the frequency area of the municipal illumination network, i.e. the so-called notch filtering. Our practice shows that in the majority of cases excluding the 47–53 Hz zone by means of FFT digital filtration is enough to get rid of such interference.

Skin condition change-related electrical potentials are mostly determined by the difference of potentials between the superficial and deep layers of the skin, perspiratory activity, skin circulation fluctuations and, therefore, skin resistance variations. As the frequency of their occurrence is 1–5 s, it is possible to partially get rid of them by filtering out the frequencies under 0.5 Hz. Further shift of the bottom filter threshold value towards the delta diapason will cause a significant distortion in the sleep station interpretation, as such disturbances appear to be the most problematic when they occur.

Summing up, there is a number of key recommendations for applying deterministic chaos fractal measures to the sleep stage identification in polysomnography:

- for various reasons, it is better to use the leads closest to the sagittal line in the bregma and central point area;
- for EEG signals, the digital filtration should be carried out with the FFT and the interpreted signal should be in the frequency diapason from 0.5 to 40 Hz.

- during the statistical processing of the fractal computation output, the obviously artifact-related deviations should be filtered out;
- to mitigate the short-term transitions from the sleep stage which may be related to artifacts of various origins, the considered epoch length shall not exceed 20–30 s.

4 Example of Combining the Frequency Filtration and Nonlinear Analysis Methods in Hypnogram Recording

The calculations were carried out on a computer using a specially developed program based on Borland C++ Builder medium. The program calculation results were verified by means of comparing the results with those acquired by the authors in other researches and with the results collected by other authors [23–25]. Figure 1 presents the hypnogram recorded by the specialists and the output of the EEG computations by fractal methods for the given epochs.

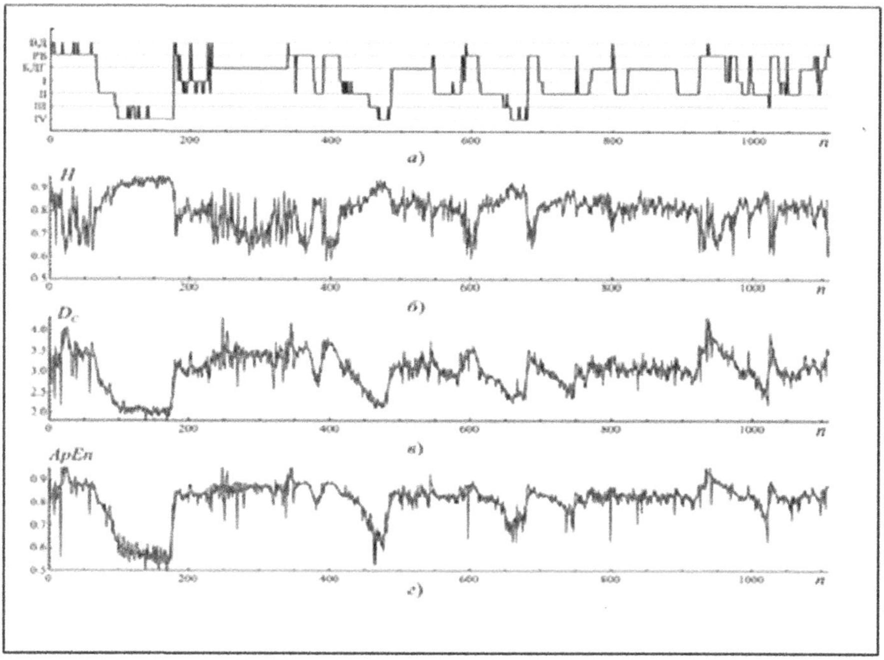

Fig. 1. Hypnogram comparison with different fractal measure values for the respective epochs: a – hypnogram recorded by the specialists; b – Hurst exponent value, c – correlation dimensionality value, d – approximate entropy value. For the fractal measures, the thin lines present the unprocessed values; the bold lines present the values after the averaging method has been applied.

The hypnograms produced with all the three methods presented in this paper are shown in Fig. 2 as dashed lines. The hypnogram recorded by the specialists is presented in the same figure as a solid line.

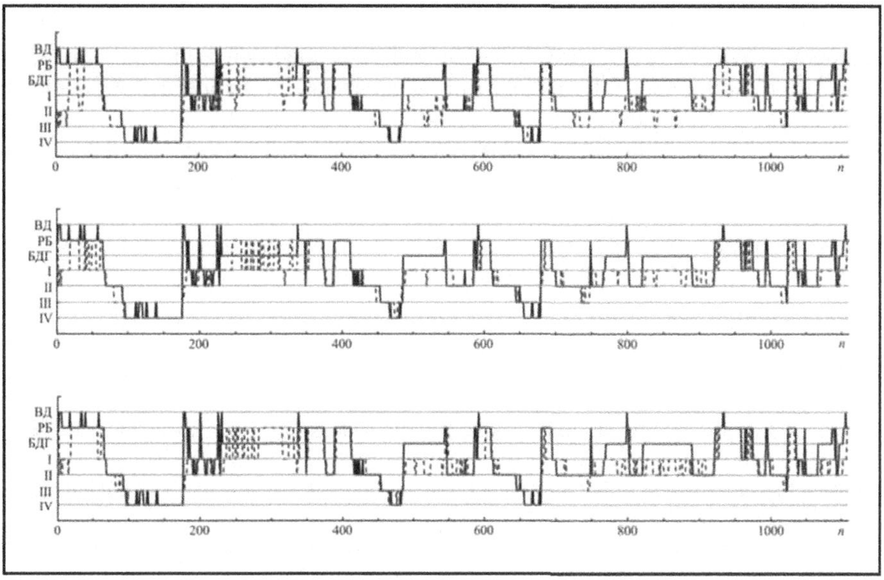

Fig. 2. Comparison of the hypnogram recorded by the specialists (solid line) with the hypnograms recorded with the following fractal deterministic chaos methods (dashed line): a – Hurst standardized range method; b – Grassberger-Procaccia method; c – approximate entropy method.

On the horizontal axis, the numbers of epochs since the EEG recordings are plotted. On the vertical axis, the following phases are plotted [26]: MT – motion time (unrecognizable phase associated with the presence of motion artifacts of approximately 25% of the epoch duration long), RW – relaxed wake, REM – rapid eye movement phase (paradoxical sleep phase), I, II, III, IV – 1st, 2nd, 3rd and 4th stages of sleep respectively.

5 Conclusion

As a result of the combined application of the frequency filtration methods for the preliminary processing of the EEG signals and the nonlinear dynamics methods as a way of assessing the depth of asynchronization of the brain activity processes during sleep, some satisfactory hypnograms were generated. As we can see from the collected results, the combination of all the three fractal nonlinear analysis methods made up the complete picture of sleep, as the deep delta sleep and the transition to the RW phase were clearly differentiated. The superficial sleep recognition was only complicated by the absence of the paradoxical sleep phase recognition. The qualitative characteristics of the coincidence between the hypnograms made up by means of different methods used in the program and the hypnograms manually recorded by the specialists were the following: the Hurst standardized range method results – 52.2%, Grassberger-Procaccia correlation integral – 47.8%, approximate entropy – 48.5%. The percentage indicates the numbers of epochs that show the complete coincidence of the sleep phases with the hypnogram recorded by the specialists. This indicates good differentiation of the sleep phases in the little input information. Both approaches, the frequency and nonlinear ones, should be considered not as competitive but as mutually supplementary; the combination of the methods opens wider perspectives than separate and independent use.

Acknowledgements. The study was implemented with the financial support of the Ministry of Science and Higher Education of the Russian Federation (grant RFMEFI60418X0208).

References

1. Levin, Ya.I.: Somnology: sleep, its structure and functions; insomnia. Russ. Med. J. **15**(15), 11–30 (2007)
2. Gusev, E.I., Konovalov, A.N., Skvortsova, V.I., Gekht, A.B.: Neurology. National guidelines, GEOTAR-Media, Moscow, Russia (2009)
3. Zakharov, A.V., Antipov, O.I., Khivintseva, E.V.: Polysomnographic characteristics of sleep in patients with fibromyalgia. Russ. J. Pain **1**(42), 58–59 (2014)
4. Zakharov, A.V., Afrosina, E.Yu., Khivintseva, E.V., Antipov, O.I.: The quality of nocturnal sleep in patients with multiple sclerosis. S.S. Korsakov Neurol. Psychiatry J. **116**(2-2), 41–43 (2016)
5. Zakharov, A.V., Vlasov, Ya.V., Poverennova, I.E., Khivintseva, E.V., Antipov, O.I.: Posture disorders in patients with multiple sclerosis. S.S. Korsakov Neurol. Psychiatry J. **114**(2-2), 55–58 (2014)
6. Khaken, G.: Brain activity principles: Synergetic approach to brain activity, behavior and activity cognition, PER SE, Moscow, Russia (2001)
7. Antipov, O.I., Neganov, V.A.: The study of dynamic chaos in impulse surge voltage inverting type with the influence of the active losses using fractal calculation measures. Nonlinear World Moscow Russia **6**(7), 364–377 (2008)

Investigation of Robust Stability for Fractional-Order LTI Systems with Multilinear Structure of Ellipsoidal Parametric Uncertainty

Radek Matušů[1(✉)] and Bilal Şenol[2]

[1] Centre for Security, Information and Advanced Technologies (CEBIA–Tech), Faculty of Applied Informatics, Tomas Bata University in Zlín, nám. T. G. Masaryka 5555, 760 01 Zlín, Czech Republic
rmatusu@utb.cz

[2] Department of Computer Engineering, Faculty of Engineering, Inonu University, 44280 Malatya, Turkey
bilal.senol@inonu.edu.tr

Abstract. The contribution focuses on the investigation of robust stability for fractional-order linear time-invariant (LTI) systems with the multilinear structure of ellipsoidal parametric uncertainty, i.e., the analyzed family of fractional-order polynomials has the multilinear uncertainty structure and an ellipsoid-shaped uncertainty bounding set. The robust stability test is based on the numerical calculation and subsequent plot of the value sets, and the application of the zero exclusion condition. Unlike the previously published works, this contribution shows that, contrary to the case of a two-dimensional ellipse of parameters, the internal points of a three-dimensional ellipsoid of parameters cannot create the boundary of the value set in the complex plane even under more complicated uncertainty structures, such as the multilinear one.

Keywords: Robust stability · Fractional-order systems · Ellipsoidal parametric uncertainty · Multilinear uncertainty · Value set

1 Introduction

The precise knowledge of all physical parameters is only an idealized scenario in the majority of practical control applications. Moreover, the simplifications and linearizations made during the modeling contribute also to the typical mismatch between the mathematical model and the true behavior of the controlled plant. That is why the use of uncertain models and robust control techniques has been extremely popular during the last decades.

Since stability is the critical requirement of control loops, we are interested in robust stability from the viewpoint of models with uncertainty. It means that the stability of the control system must be secured not only for one nominal plant but also for the whole group of plants, which is usually called the family of plants. A common way of defining such family consists in the use of parametric uncertainty [1, 2], i.e., the

R. Silhavy et al. (Eds.): CoMeSySo 2020, AISC 1295, pp. 421–429, 2020.
https://doi.org/10.1007/978-3-030-63319-6_39

structure is "known" and fixed, but the parameters may "slowly" vary within prescribed bounds. The commonly used engineering approach is to consider these bounds as real intervals; in other words, the bound has a shape of a box. However, there are some other alternative shapes available, such as a diamond, a sphere, or, more generally, an ellipsoid [2–10].

The popularity of fractional-order systems and their application in control engineering has been growing rapidly [11–13] for the sake of their improved performance compared with the conventional integer-order systems. It is understandable that great research attention is also paid to the robust stability of fractional-order systems under various uncertainty conditions [14, 15].

The contribution deals with the investigation of robust stability for fractional-order LTI systems with the multilinear structure of ellipsoidal parametric uncertainty. In other words, the studied family of fractional-order (characteristic) polynomials has the multilinear uncertainty structure and an ellipsoid-shaped uncertainty bounding set. The robust stability analysis takes advantage of the numerical calculation and subsequent plot of the value sets and the application of the zero exclusion condition [2].

More specifically, this contribution is intended to be a follow-up to the previously published works [8–10]. In [8], the families of integer-order polynomials with multilinear uncertainty structure were studied, but the example dealt only with a two-dimensional uncertainty bounding set (ellipse). Then, in [9], the fractional-order version of the polynomial family with multilinear uncertainty structure was studied, among other structures, but only for the elliptical uncertainty bounding set as well. The case of a three-dimensional uncertainty bounding set (ellipsoid) was outlined in [9] in combination with the polynomic uncertainty structure. In [10], the three-dimensional uncertainty bounding set was combined with the independent uncertainty structure. Unlike the previous works, this contribution demonstrates that, contrary to the case of a two-dimensional ellipse of parameters [8, 9], the internal points from three-dimensional ellipsoid of parameters cannot create the boundary of the value set in the complex plane even under more complicated uncertainty structures, such as the multilinear uncertainty structure.

Since this contribution is example-oriented and the level of presented theory and literature survey is restricted to a minimum, the interested readers are referred to the authors' previous works, especially [9].

2 Mathematical Description of Analyzed Family

Suppose that a family of systems has a corresponding family of fractional-order characteristic polynomials with three uncertain parameters and that this family of fractional-order polynomials is defined by the multilinear structure of uncertainty, according to:

$$\begin{aligned} p(s,q) &= s^{3.1} + (2q_1q_2q_3 + 3q_1q_2 - q_2q_3)s^{2.2} + \cdots \\ &+ (5q_1q_2q_3 + 0.1q_1 + 0.2q_2 + 0.3q_3)s^{0.9} + (q_1q_2 + q_2q_3 + q_1q_3) \end{aligned} \tag{1}$$

and by the uncertainty bounding set Q in a shape of a (three-dimensional) ellipsoid, described by:

$$\begin{aligned} & \|q - q^0\|_{2,W} \leq r = 1 \\ & q^0 = [q_1^0, q_2^0, q_3^0] = [2, 1, 0.5] \\ & W = \begin{bmatrix} w_1^2 & 0 & 0 \\ 0 & w_2^2 & 0 \\ 0 & 0 & w_3^2 \end{bmatrix} = \begin{bmatrix} 2 & 0 & 0 \\ 0 & 5 & 0 \\ 0 & 0 & 10 \end{bmatrix} \end{aligned} \quad (2)$$

Alternatively, the same uncertainty bounding set can be written in the form:

$$\begin{aligned} & 2(q_1 - q_1^0)^2 + 5(q_2 - q_2^0)^2 + 10(q_3 - q_3^0)^2 \leq r = 1 \\ & q^0 = [q_1^0, q_2^0, q_3^0] = [2, 1, 0.5] \end{aligned} \quad (3)$$

The graphical representation of this uncertainty bounding set, given either by (2) or (3), is depicted in Fig. 1. The axes are plotted in an equal aspect ratio. Unsurprisingly, the uncertainty bounding set really has the ellipsoidal shape, and thus, all admissible triplets of parameters q_1, q_2, and q_3 have to be located inside or on the surface of this ellipsoid.

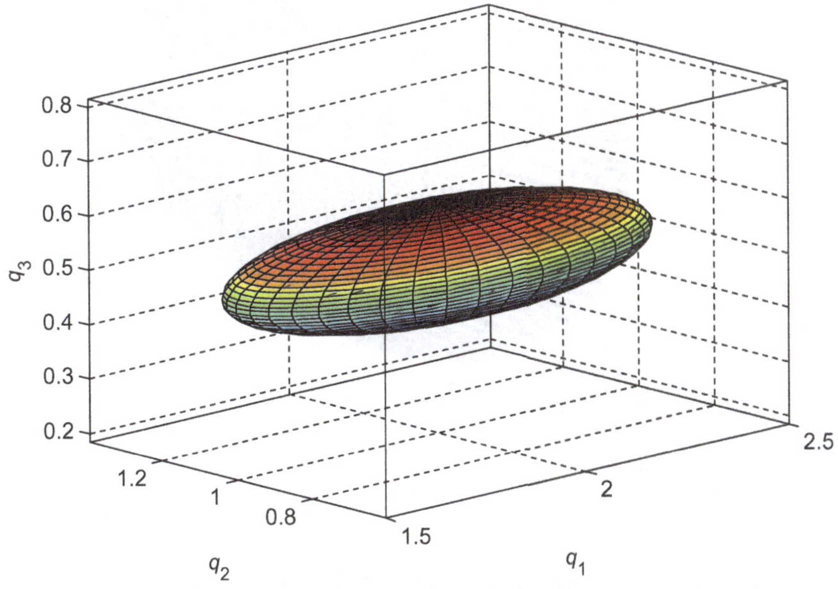

Fig. 1. Ellipsoidal uncertainty bounding set (2), (3).

3 Value Set – Construction and Properties

The applied graphical test of robust stability is based on plotting the value sets and verifying the condition for zero exclusion. Therefore, a discussion on the value set of the family (1), (2) follows.

The value set $p(j\omega, Q) = \{p(j\omega, q) : q \in Q\}$ [2] for a chosen fixed frequency $\omega = 2$ and various "gridding" is shown in Figs. 2, 3 and 4. As can be seen in [8, 9], in the case of multilinear or more complicated uncertainty structure and a two-dimensional uncertainty bounding set, the boundaries of the value set in the complex plane may be mapped not only from the boundaries in two-dimensional parameter space (ellipse) but possibly also from some internal points. This holds true both for the integer-order [8] and for the fractional-order [9] polynomials. However, for the case of a three-dimensional uncertainty bounding set, it is enough to grid just the surface area of the ellipsoid, because no internal point from the ellipsoid of parameters (Fig. 1) may be mapped to the boundary of the value set. In other words, no internal points can comprise the value set boundary. Contrary to the statement in [10], this holds true not only for simple uncertainty structures but even for the complicated ones (such as multilinear uncertainty structure in this paper).

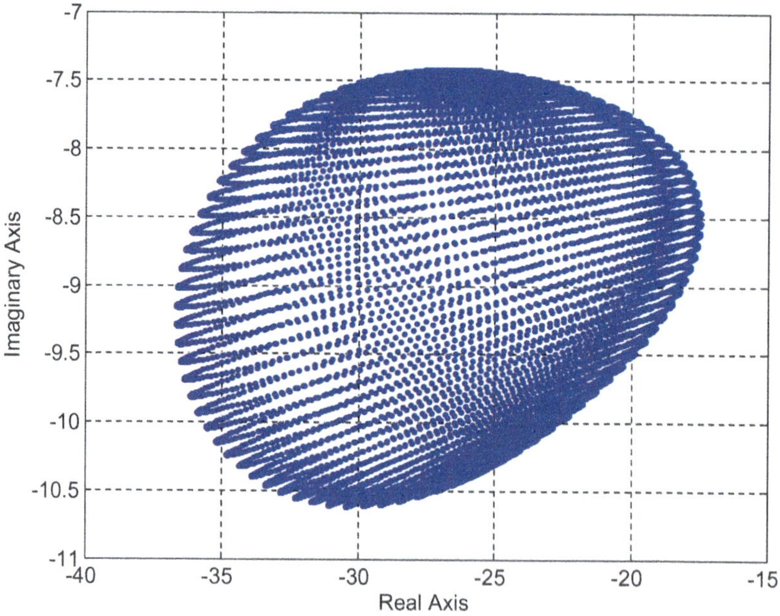

Fig. 2. Value set of family (1), (2) at $\omega = 2$ for 90 samples of both θ and φ

Thus, the surface area of the ellipsoid from Fig. 1 will be gridded, and subsequently, the related value set points will be directly computed and plotted. The surface area can be parametrized, e.g., by using the set of parametric equations:

$$\begin{aligned} q_1 &= \left(\frac{r}{w_1^2}\right) \cos\theta \cos\varphi + q_1^0 \\ q_2 &= \left(\frac{r}{w_2^2}\right) \cos\theta \sin\varphi + q_2^0 \\ q_3 &= \left(\frac{r}{w_3^2}\right) \sin\theta + q_3^0 \end{aligned} \tag{4}$$

where $\theta \in [0, 2\pi]$ and $\varphi \in [0, \pi]$.

First, both parameters θ and φ were sampled by 90 equidistant samples within the whole range of aforesaid intervals. The value set for $\omega = 2$ and this, relatively sparse, gridding can be seen in Fig. 2.

Next, the number of equidistant samples of θ and φ was tripled, i.e., 270 samples within given intervals are used. The resulting value set (again for $\omega = 2$) is shown in Fig. 3. Obviously, it is much denser compared to the value set from Fig. 2. However, the cost for it is a higher computational effort.

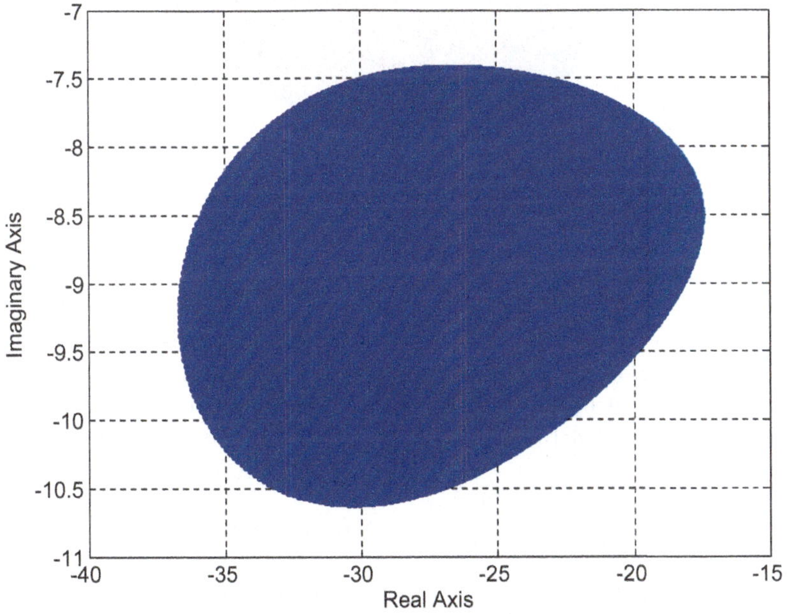

Fig. 3. Value set of family (1), (2) at $\omega = 2$ for 270 samples of both θ and φ

Note that some points from the surface area of the three-dimensional ellipsoid of parameters (Fig. 1) are mapped to the boundary of the value set in the complex plane while the others are mapped to the interior of the value set (Figs. 2 and 3). Nevertheless, as discussed above, no internal point from the three-dimensional ellipsoid of parameters can be mapped to the boundary of the value set, even for the family of fractional-order polynomials with multilinear uncertainty structure. For illustration, Fig. 4 presents the same value set as in Fig. 3, but it also contains 10,000 extra points (red x-marks) that are mapped from 10,000 randomly chosen points from the interior of the ellipsoid of parameters.

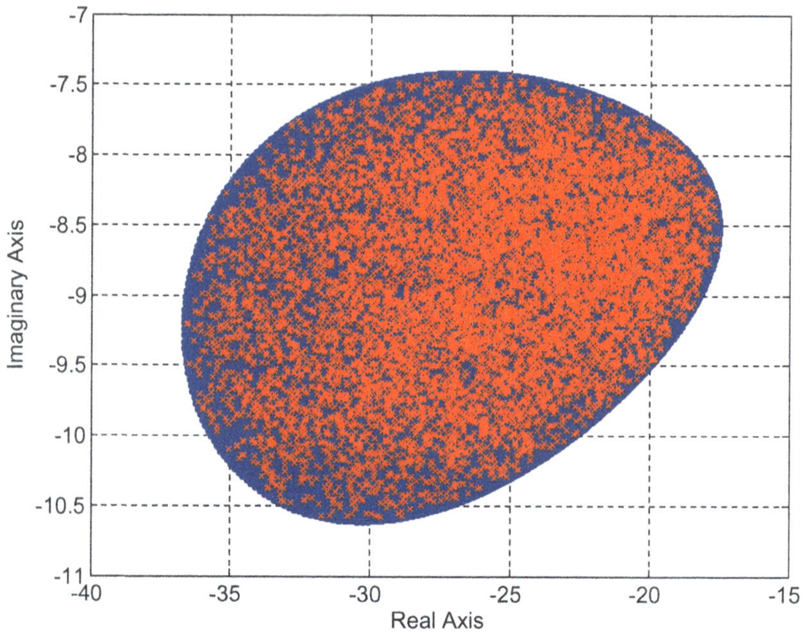

Fig. 4. Value set of family (1), (2) at $\omega = 2$ with 10,000 extra random points

As can be seen, Fig. 4 is not much convincing, because some red points lie very close to the value set boundary, even though not in the boundary itself. Thus, the procedure is repeated, but 10,000 random points are taken from the inner space of the ellipsoid according to:

$$2(q_1 - 2)^2 + 5(q_2 - 1)^2 + 10(q_3 - 0.5)^2 \leq 0.9 \tag{5}$$

The value set with 10,000 extra points (red x-marks) mapped from 10,000 randomly selected points in the parameter space that comply with (5) is depicted in Fig. 5. It demonstrates that, contrary to the case of a two-dimensional ellipse of parameters [8, 9], the internal points from a three-dimensional ellipsoid of parameters cannot create the boundary of the value set even under more complicated uncertainty structures, such as the multilinear one.

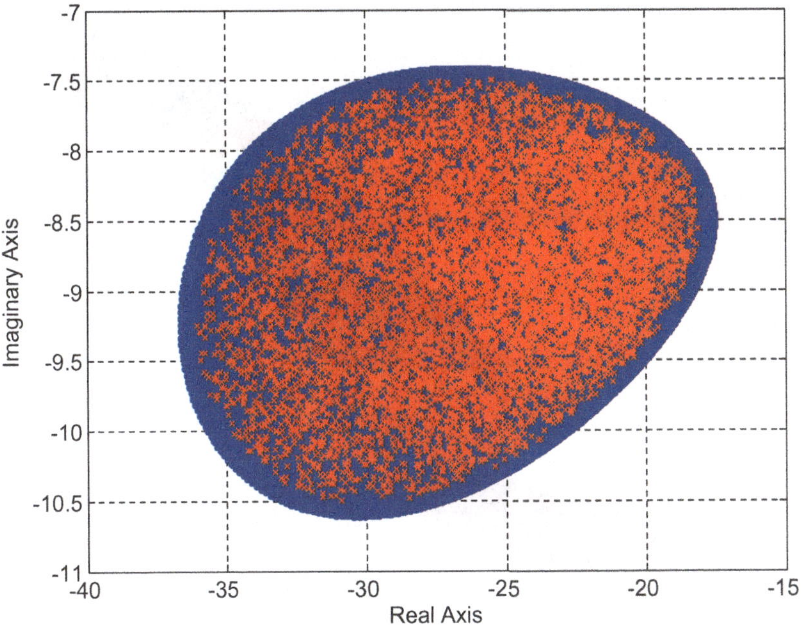

Fig. 5. Value set of family (1), (2) at $\omega = 2$ with 10,000 extra random points that fulfill (5)

4 Robust Stability Test

In the final phase, the value sets will be utilized for the simple graphical test of robust stability. The procedure of obtaining the value set at one fixed frequency, e.g., $\omega = 2$, and for 270 samples of θ and φ (Fig. 3) will be repeated for a selected range of nonnegative frequencies. The Fig. 6 shows the value sets for the range of frequencies from 0 to 2.5 with the step size 0.1 The value sets would enter the fourth quadrant for even higher frequencies, but its current range is selected in order to have a good view of the most important area, which is the neighborhood of the zero (complex plane origin).

In accordance with the zero exclusion condition [2, 9], the family of fractional-order polynomials with the multilinear structure of ellipsoidal parametric uncertainty (1), (2) is robustly stable because the zero point is excluded from the value sets and all necessary preconditions (invariant degree of the polynomial family, pathwise connected Q, continuous coefficient functions, and at least one stable member of the family) are fulfilled.

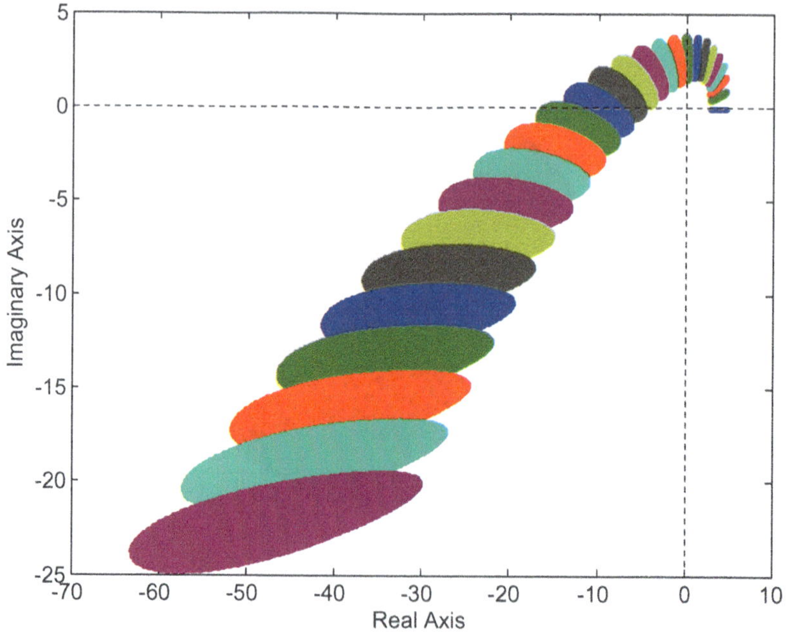

Fig. 6. Value sets of family (1), (2) for $\omega = 0 : 0.1 : 2.5$

5 Conclusion

This example-oriented contribution was focused on graphical testing of robust stability for fractional-order LTI systems with the multilinear structure of ellipsoidal parametric uncertainty via the numerical calculation of the value sets, and the use of the zero exclusion principle. It is intended as a follow-up to some previously published works. Unlike them, this contribution showed that the internal points of a three-dimensional ellipsoid of parameters could not produce the boundary of the value set even under more complicated uncertainty structures, such as the multilinear one, which is not true for the case of a two-dimensional ellipse of parameters.

Acknowledgments. This work was supported by the Ministry of Education, Youth and Sports of the Czech Republic within the National Sustainability Programme project No. LO1303 (MSMT-7778/2014).

References

1. Bhattacharyya, S.P.: Robust control under parametric uncertainty: an overview and recent results. Ann. Rev. Control **44**, 45–77 (2017)
2. Barmish, B.R.: New Tools for Robustness of Linear Systems. Macmillan, New York (1994)
3. Sadeghzadeh, A., Momeni, H., Karimi, A.: Fixed-order H_∞ controller design for systems with ellipsoidal parametric uncertainty. Int. J. Control **84**(1), 57–65 (2011)

4. Sadeghzadeh, A.: Identification and robust control for systems with ellipsoidal parametric uncertainty by convex optimization. Asian J. Control **14**(5), 1251–1261 (2012)
5. Sadeghzadeh, A., Momeni, H.: Robust output feedback control for discrete-time systems with ellipsoidal uncertainty. IMA J. Math. Control Inf. **33**(4), 911–932 (2016)
6. Matušů, R.: Spherical families of polynomials: a graphical approach to robust stability analysis. Int. J. Circuits Syst. Signal Process. **10**, 326–332 (2016)
7. Matušů, R.: Value sets of ellipsoidal polynomial families with affine linear uncertainty structure. In: Cybernetics and Automation Control Theory Methods in Intelligent Algorithms: Proceedings of 8th Computer Science On-line Conference 2019, vol. 3 – Advances in Intelligent Systems and Computing, vol. 986, pp. 255–263. Springer, Cham (2019)
8. Matušů, R., Şenol, B.: Application of value set concept to ellipsoidal polynomial families with multilinear uncertainty structure. In: Computational Statistics and Mathematical Modeling Methods in Intelligent Systems: Proceedings of 3rd Computational Methods in Systems and Software 2019, vol. 2 – Advances in Intelligent Systems and Computing, vol. 1047, pp. 81–89. Springer, Cham (2019)
9. Matušů, R., Şenol, B., Pekař, L.: Value-set-based approach to robust stability analysis for ellipsoidal families of fractional-order polynomials with complicated uncertainty structure. Appl. Sci. **9**(24) (2019). Article No. 5451
10. Matušů, R., Şenol, B., Vašek, V.: Robust stability of fractional-order LTI systems with independent structure of ellipsoidal parametric uncertainty. In: Proceedings of the 30th DAAAM International Symposium, Vienna, Austria, pp. 161–166 (2019)
11. Machado, J.A.T., Kiryakova, V., Mainardi, F.: Recent history of fractional calculus. Commun. Nonlinear Sci. Numer. Simul. **16**(3), 1140–1153 (2011)
12. Chen, Y., Petráš, I., Xue, D.: Fractional order control – a tutorial. In: Proceedings of the 2009 American Control Conference, St. Louis, MO, USA (2009)
13. Xue, D.: Fractional-Order Control Systems: Fundamentals and Numerical Implementations. De Gruyter, Berlin (2017)
14. Tan, N., Özgüven, Ö.F., Özyetkin, M.M.: Robust stability analysis of fractional order interval polynomials. ISA Trans. **48**(2), 166–172 (2009)
15. Jiao, Z., Zhong, Y.: Robust stability for fractional-order systems with structured and unstructured uncertainties. Comput. Math. Appl. **64**(10), 3258–3266 (2012)

Monitoring the Characteristics of Human Emotional Reactions Based on the Analysis of Attractors Reconstructed According to EEG Patterns

Konstantin V. Sidorov(✉) and Natalya I. Bodrina

Tver State Technical University, Afanasy Nikitin Quay 22, Tver 170026, Russia
bmisidorov@mail.ru

Abstract. The paper considers a dynamic model of emotional reactions that allows monitoring the change in the direction of testee's emotional development during an audiovisual stimulation, which activates the auditory and visual analyzer of a person. This problem was solved by forming a new system of features that characterizes the morphology of attractors restored from the electroencephalogram (EEG) time series. There is a description of a bioengineering system for specific experiments and the composition of the generated multimodal database with the patterns of recorded signals. The paper discusses the issues of selecting parameter values (time delay, embedding dimension, time series length) that are necessary for the correct attractor reconstruction. Changes in human emotional reactions are assessed by fuzzy estimates of three characteristics of two-dimensional projections of an attractor reconstructed from EEG patterns. The results of monitoring emotions are presented in as a matrix of increment estimates of attractor characteristics determined by an additional index scale, which allows generating a numerical estimate for each fuzzy set element using the term index and a membership function. The created algorithm allows processing a sequence of EEG patterns recorded with a time shift and forming verbal conclusions about the direction of the emotional reaction development. Further research might involve the creation of a tool for the tasks of monitoring and correcting emotional reactions and cognitive activity of a person.

Keywords: Emotional response · Emotion · Emotion recognition · Bioengineering system · Biomedical signal · EEG · Attractor · Fuzzy set

1 Introduction

The mathematical formalization of the process of formation and development of human emotional reactions remains an important topic of modern research. It is of interest to many specialists in robotics and decision support systems, as well as to developers of medical diagnostic systems [1]. An objective assessment of testee's emotional reactions can help to diagnose somatic disorders and to identify the phase of destabilization in patients suffering from depression when all drugs are abruptly cancelled [2].

R. Silhavy et al. (Eds.): CoMeSySo 2020, AISC 1295, pp. 430–443, 2020.
https://doi.org/10.1007/978-3-030-63319-6_40

The results of interdisciplinary studies conducted in recent decades have made it possible to determine the fact that human emotional reactions correlate with certain changes in biomedical signals recorded during the observation of testees. Nowadays, the assessment and monitoring human emotions involves physiological signal patterns [3], including speech signals [4, 5] and video images with samples of facial expressions and gestures [6]. All these solutions are based on the analysis of signals of functional systems controlled by the brain and, therefore, they are executive. Obviously, to describe the dynamic process of the formation and development of emotional reactions, it is necessary to analyze the signals of the brain's electrical activity, i.e. electroencephalograms (EEG).

The most common EEG processing methods are spectral and coherent analysis. Wavelet analysis [7] and nonlinear dynamics methods [8, 9] are used less frequently.

In the past 10 years, there have been a number of papers on the analysis of EEGs converting them into attractors [10–13]. These studies say that the properties of attractors are correlated with the properties of dynamical systems, which generate them. This interconnection has become the basis for the classification algorithms of biomedical signals (EEG, EMG, ECG, etc.) that single out individual pathologies, as well as the speech recognition algorithms according with the sign of an emotional reaction demonstrated by a person [14–17]. It should be noted that more reliable and accurate means of recognizing a sign of emotions are integrated systems that make decisions by analyzing information from several parallel channels (for example, auditory and visual).

Despite the appeared applied solutions, many issues in the field remain unresolved. In particular, it is necessary to increase the sensitivity of monitoring systems, since existing solutions allow creating reliable interpretations of emotions only when the level of an emotional reaction is high (pronounced) and when biomedical signal patterns are fairly long.

2 Methods

2.1 Bioengineering System and Multimodal Database

We propose a new method for monitoring the characteristics of human emotional reactions based on the analysis of attractors reconstructed using EEG patterns. The experiments involved using a multichannel bioengineering system EEG/Speech+, which was described in detail in [18, 19]. This system allows recording testees' EEG and speech signals during audiovisual stimulation that activates human auditory and visual analyzers.

At this stage of the research, the experiments involved 10 testees (5 women and 5 men), whose age ranged from 19 to 26 years. The testees were the students and postgraduates of the Tver State Technical University. Each testee gave voluntary consent to participate in the experiments. During audiovisual stimulation (emotionally differentiated audio and video stimuli) and when recording responses (EEG patterns and speech signal samples), the subjects were sitting in a comfortable chair.

EEG were recorded according to the international 10–20 system [20] (19 leads, sampling frequency 250 Hz: O2-A2, O1-A1, P4-A2, P3-A1, C4-A2, C3-A1, F4-A2, F3-A1, Fp2-A2, Fp1-A1, T6-A2, T5-A1, T4-A2, T3-A1, F8-A2, F7-A1, Pz-A1, Cz-A2, Fz-A1). Voice signals were recorded in .wav format (sampling frequency 22 050 Hz, resolution 16 bits). According to the experimental results, we have created a multimodal database with 1200 artifact-free signal samples (600 EEG patterns and 600 control phrases recorded in parallel), lasting 6 s (1 500 samples for EEG, 132 300 samples for speech signals). All patterns are divided into 6 classes (neutral state, joy, surprise, anger, sadness, fear) according with the testees' emotions experienced during the perception of audio and video stimuli.

This work reflects the research results of only EEG patterns, which were analysed by reconstructing attractors and building signal models in the fuzzy feature space.

2.2 Attractor Reconstruction

In order to construct attractors based on time series obtained by digitizing physiological signals, we use the provisions arising from F. Tackens theorem [21], which allow reconstructing an attractor based on a time series (x_1, x_2, ..., x_N) using the coordinate delay method:

$$y_n = (x_n, x_{n+\tau}, \ldots, x_{n+(m-1)\tau}), n = 0, \ldots, s-1, s = N-(m-1)\tau, \qquad (1)$$

where y_n is the reconstructed attractor vector; N is the total number of elements (points) of the time series; τ is the time delay between the time series elements (time lag); m is the embedding dimension (lag space dimension).

It is worth highlighting that the uncertainty of various authors when using the terms of nonlinear dynamics has caused situations where one term can be used in opposite senses. Therefore, in this work, the terms of nonlinear dynamics mean the following: 1) the space for restoring the attractor from the time series using the coordinate delay method is the embedding space; 2) the set of points that generates the original attractor is the restored attractor; 3) the restored attractor projection is the attractor reconstruction. Figure 1 shows a three-dimensional reconstruction of an attractor using an EEG pattern lasting 6 s (1 500 counts).

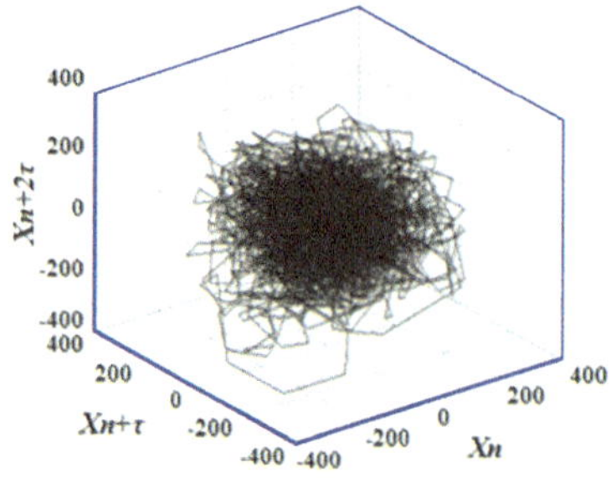

Fig. 1. Attractor reconstruction (1) by an EEG pattern (F8-A2 lead) when N = 1500, τ = 20 and m = 3 (x-axis – Xn, μV; y-axis – $Xn + \tau$, μV; z-axis – $Xn + 2\tau$, μV).

When creating programs for the automatic processing of biomedical signals using nonlinear dynamics, the problems of correct determining of N, τ and m (1) values arise. To estimate them, there are several well-known algorithms that can be modified when creating profession-oriented applications.

2.2.1 Time Delay Selection

The time delay τ (1) is chosen to minimize the dependences of neighboring reconstructed vectors $\{y_i\}$ on each other. The reasons for these dependences might be random errors or poor measurement accuracy. Typically, the choice of τ is based on the autocorrelation function (ACF) [22] or the mutual information function [23]. Let us consider the problems related to their calculating and possible ways to overcome them.

The value of τ is determined based on the idea that if the points forming the time series are mutually independent, then the reconstructed vectors $\{y_i\}$ contain the maximum amount of information about the time series. Moreover, the value of τ is selected so that the correlation between the time series elements x_n and x_{n-1} is minimal. This is achieved due to the autocorrelation function $B(\tau)$. Then the delay τ corresponds to the value of the time to reach the first zero of the function $B(\tau)$ (see Fig. 2, a) [24].

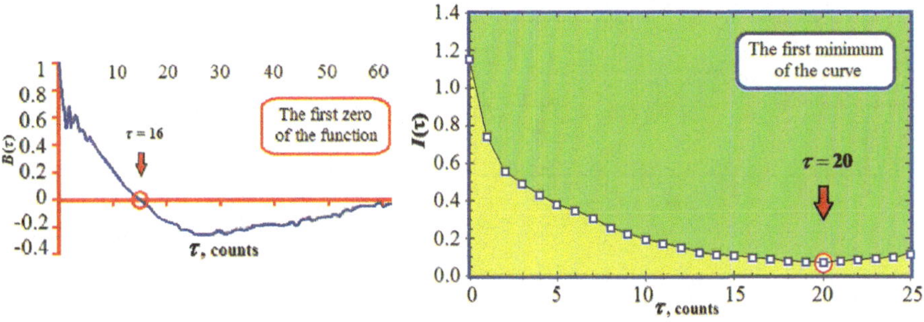

(a) (b)

Fig. 2. Choosing τ time delay value (1) for the EEG pattern: a is the autocorrelation function $B(\tau)$; b is the mutual information function $I(\tau)$.

An alternative way to determine τ is using a mutual information function that reflects a linear or non-linear relationship between two variables. Mutual information means the following: let a segment equal to the time series amplitude range be conventionally divided into the i-th and j-th intervals, then the mutual information function $I(\tau)$ will look as following [23]:

$$I(\tau) = -\sum_{i,j} p_{ij}(\tau) \cdot \ln(p_{ij}(\tau)/p_i p_j), \quad (2)$$

where p_i is the probability of a time series element to be in the i-th interval; p_j is the probability of a time series element to be in the j-th interval; $p_{ij}(\tau)$ is the joint probability of one time series element to be in the i-th interval, and another element taken with τ delay to be in the j-th interval.

The delay value τ for this approach is selected on the x-axis (see Fig. 2, b) when reaching the first minimum on the curve $I(\tau)$ (2).

Each of the above methods for calculating τ delay can produce different values of the required value (the spread can be either minimal or colossally large).

2.2.2 Selecting an Embedding Dimension Value

The methods based on the concept of False Nearest Neighbors (FNN) have shown high efficiency in calculating m value (1). The concept consists in the fact that as the m value decreases, the number of *FNN* in the neighbor of any embedding space point increases (see Fig. 3, a). This implies the simplest method of defining *FNN* as a function of dimension equal to zero in the limit [24]. At the attractor reconstruction, we randomly select two neighboring points (x_i and x_j), find the distance between the following two points (x_{i+1} and x_{j+1}), after that we determine the indicator K:

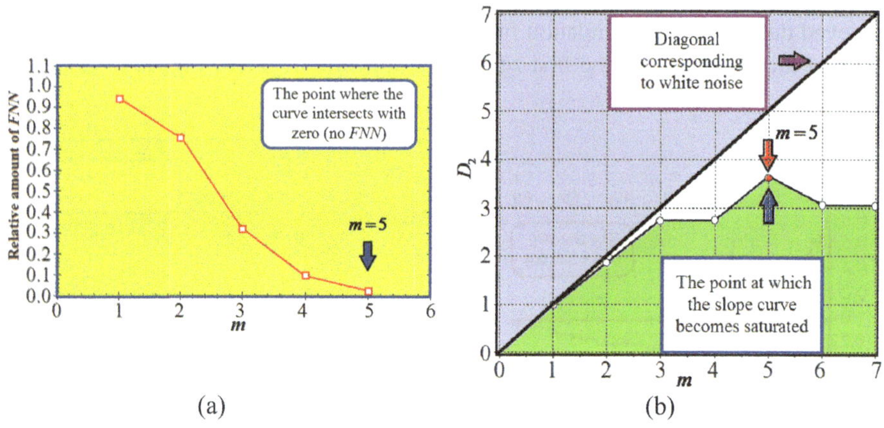

Fig. 3. The choice of the value of the embedding dimension m (1) on the example of the EEG pattern: a – the false nearest neighbor method (*FNN*); b – the saturation method.

$$K = \left|x_{i+1} - x_{j+1}\right| / \left|x_i - x_j\right|. \quad (3)$$

If the obtained K value (3) exceeds a certain threshold, which has an empirically selected value, then the points x_i and x_j are false nearest neighbors.

The embedding dimension m can also be determined from the saturation position using the algorithm for calculating the values of the correlation integral $C(\varepsilon)$ and the correlation dimension D_2 [25]. The correlation integral $C(\varepsilon)$, which displays the relative number of attractor point pairs (x_i, x_j) located at a distance not bigger than ε, is defined as following:

$$C(\varepsilon) = \lim_{M\to\infty} \frac{1}{M(M-1)} \sum_{i,j=1}^{M} \theta(\varepsilon - r(x_i, x_j)), D_2 = \lim_{\varepsilon\to 0} \frac{\log C(\varepsilon)}{\log \varepsilon}, \quad (4)$$

where i, j = 1, …, M; M is the number of considered states x_i (a number of x_i points on an attractor); r is the distance between attractor points; ε is the attractor reconstruction partition cell size; $\Theta(x)$ is the Heaviside step function ($\theta(x) = \left\{ \begin{matrix} 0, x \leq 0 \\ 1, x > 0 \end{matrix} \right\}$).

The embedding dimension m must be not less than the double attractor correlation dimension D_2 (4), i.e. the condition $m \geq 2D_2 + 1$ must be fulfilled [21, 26].

After determining $C(\varepsilon)$ and D_2, we build the dependence of the correlation dimension D_2 on the embedding dimension m and determine the point for the slope curve to be saturated (see Fig. 3, b). If the time series under study is chaotic, then the graph of D_2 against m will increase at first, and then will become saturated. If the process under study is stochastic, then saturation will not occur.

2.2.3 Choosing the Minimum Time Series Length

There are various criteria for estimating the minimum length of the time series N (1) (e.g., $N > 10^{2+0.4D2}$, $N > 10^{D2(D2+1)}$, $N > 10^{D2/2}$, $N > 2\tau^{m/2}$) [27].

When recording signals with a low sampling frequency (for example, an EEG sampling frequency is 250 Hz), we introduce additional modifications. One of them is the Theiler window (w) [28] based on the assumption that points located close to each other in the time series are correlated. If the time series is finite, this leads to systematic errors in calculating the correlation integral $C(\varepsilon)$ (4). To reduce them, it is necessary not to take into account points located at a distance less than w steps in the time series when calculating $C(\varepsilon)$. The recommended ($w \gg \tau$) and minimum ($w > \tau\ (2/N)^{2/m}$) values of this window are given in [28]. The studies in [24, 27] have shown that the recommended value of the Theiler window (w) may be insufficient or too high. In such case, there is no guarantee that w compensates for the short length of the time series (N). In general, each researcher seeks to select acceptable values of N, τ, and m (1) that will make it possible to solve the assigned tasks successfully.

3 Results

3.1 The Relationship Between the Attractor and Emotional Reaction Characteristics

The papers [15, 17] propose a model for interpreting the three main characteristics of emotional reactions specified by the rules for their assessing. The rules are based on experimentally established correlation between the indicated emotion characteristics and the morphology of attractors [29, 30], reconstructed from EEG patterns and speech signal samples. The same patterns in attractors reconstructed using different biomedical signals recorded in parallel in different channels experimentally confirm the legitimacy of using the selected set of quantitative attributes to interpret the sign (valency) and the emotional reaction level both separately and generalized in speech samples and EEG patterns.

The emotion interpretation model, which includes three components, can be represented as following:

$$Emotion = \langle Z, U, D \rangle, \tag{5}$$

where Z is the emotion sign (positive/negative); U is the emotion level (low, medium, high); D is the emotional response dynamics (emotion intensifies/emotion weakens).

To evaluate the literals from (5), we use the biomedical signal pattern model, which is based on attractor reconstruction by the speech sample or EEG pattern:

$$Mod\_sig = \left\langle \bigcup_h (f(R_i),\ Mp,\ k_0) \right\rangle, R_i = \sqrt{x(t_i)^2 + x(t_i + \tau)^2 + \ldots + x(t_i + m \cdot \tau)^2}, \tag{6}$$

where R_i is the length of the vector restored from the origin of coordinates to the i-th point of the two-dimensional attractor projection; $x(t_i)$ is the biomedical signal value at t_i time; $f(R_i)$ are linear transformations; h is the number of attractor projections ($h = 3$); Mp is the density matrix of the attractor trajectories; k_0 is the number of zero cells in the matrix Mp.

A sequential scan of all two-dimensional attractor projections is carried out within the algorithm framework (three projections are analyzed: $Xn - Xn + \tau$ ($i = 1$), $Xn - Xn + 2\tau$ ($i = 2$) and $Xn + \tau - Xn + 2\tau$ ($i = 3$)). Each projection has four selected quadrants and the determined vector R_i (6). We selected the maximum vector of the j-th quadrant of the i-th projection of the attractor ($R^{i,j}_{\max}$), and then the averaged vector of the i-th attractor projection ($\bar{R}^{i}_{\max}$), which is the base for rough estimates of the two-dimensional attractor projection area.

The experiments have revealed the relationship between the density (p) of the trajectories near the attractor projection center with the sign of testee's emotional reaction. To determine p, the two-dimensional attractor projection (see Fig. 4, a) is covered with a regular grid with a constant step d. For each cell of this grid, we can find the ratio of the number of trajectory points on the selected projection section to the cell area.

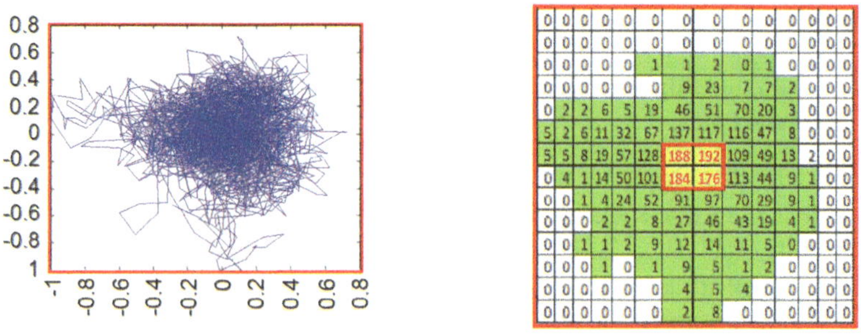

0	0	0	0	0	0	0	0	0	0	0	0	0	0
0	0	0	0	0	0	0	0	0	0	0	0	0	0
0	0	0	0	0	1	1	2	0	1	0	0	0	0
0	0	0	0	0	0	9	23	7	7	2	0	0	0
0	2	2	6	5	19	46	51	70	20	3	0	0	0
5	2	6	11	32	67	137	117	116	47	8	0	0	0
5	5	8	19	57	128	188	192	109	49	13	2	0	0
0	4	1	14	50	101	184	176	113	44	9	1	0	0
0	0	1	4	24	52	91	97	70	29	9	1	0	0
0	0	0	2	2	8	27	46	43	19	4	1	0	0
0	0	1	1	2	9	12	14	11	5	0	0	0	0
0	0	0	1	0	1	9	5	1	2	0	0	0	0
0	0	0	0	0	0	4	5	4	0	0	0	0	0
0	0	0	0	0	0	2	8	0	0	0	0	0	0

(a) (b)

Fig. 4. Attractor characteristics: a is the two-dimensional attractor projection for the EEG pattern (F8-A2 lead); b is the density matrix for the attractor projection shown in Fig. 4, a.

The set of estimates ($p_{i,j}$) for all grid cells can be represented by the matrix (see Fig. 4, b):

$$Mp = \lfloor p_{i,j} \rfloor_{n \times m}, \tag{7}$$

where $p_{i,j}$ is the density of the cell at the intersection of the i-th row and the j-th column in the Mp matrix; n is the number of matrix rows (n = 14); m is the number of matrix columns (m = 14).

The analysis of the distribution of density estimates over the rows and columns of the density matrix Mp (7) has shown that the four central cells are the most sensitive (see Fig. 5). Their total score is used further as a quantitative characteristic $p_{i,j}$.

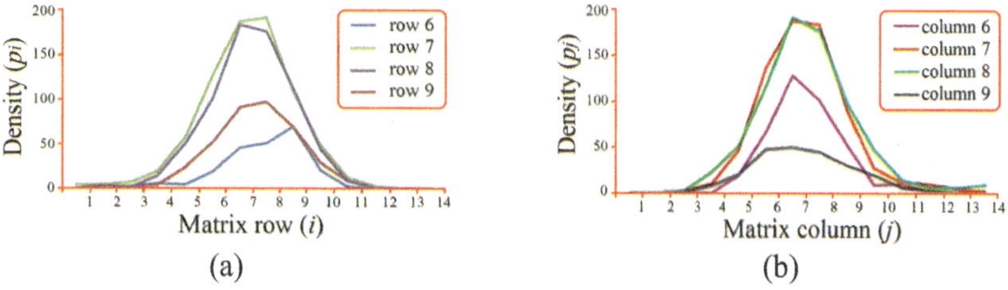

Fig. 5. The density distribution over the rows and columns of the matrix Mp (7): a is the density distribution in separate rows; b is the density distribution in separate columns.

The nature of the trajectories on the two-dimensional attractor projection depends on the time series sections selected for its reconstruction, therefore, the $x(t_i)$ analysis (6) uses a sliding window, which saves the signs of chaos in the analyzed signal. The attractor is reconstructed for each window, and its characteristics are determined. The integral estimates for the entire biomedical signal are the average values of the characteristics for all calculation windows. The experiments have revealed a relationship between an increase in the density of attractor points in the center of the matrix Mp (7), which covers the figure projection, and a negative reaction (anger, sadness, fear) to an external stimulus. An inverse relationship was observed for positive emotions (joy, surprise). These results have made it possible to use central density monitoring $p_{i,j}$ to assess the emotion development during sequential recording of biomedical signals in parallel with marked emotional stimuli.

A series of experiments was carried out to build the dynamic component of the emotion model (5). The experiments involved sequential presenting of video stimuli to a testee, which caused positive or negative emotional assessments. EEG was recorded at 19 leads; the right hemisphere lead F8-A2 was selected for further processing. This decision is coherent with the results of physiological studies [16, 31].

Figure 6 shows how a picture of changing $p_{i,j}$ estimates (7), combined with a diagram of changes in external stimuli presented to testees.

The graph (see Fig. 6) illustrates the change in testee's emotional state during the perception of audio and video stimuli. The subsets of $p_{i,j}$ estimates associated with a

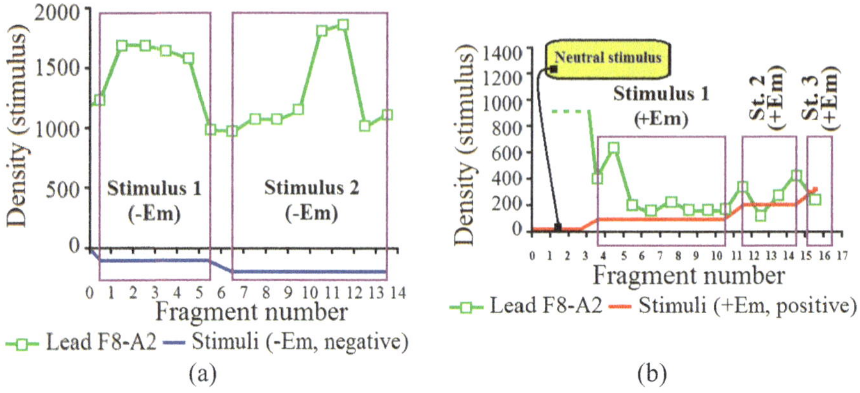

Fig. 6. The change in the point density in the center of EEG attractors (by F8-A2 lead) when a person emotionally reacts to a sequence of stimuli: a is negative stimuli (sadness); b is positive stimuli (joy).

single stimulus are outlined. The stimuli were evaluated on a conditional scale: each stimulus is rated at 10 points (taking into account the induced emotion sign). For each subsequent stimulus of one sign, the estimate increases by 10 points. A neutral stimulus is rated at 0 points. When perceiving the first negative stimulus (see Fig. 6, a), the average value of $p_{i,j}$ exceeded the level corresponding to the neutral stimulus.

The transition to the second negative stimulus did not cause a strong reaction; in fact, $p_{i,j}$ remained at a neutral level. The perception of a sequence of 3 positive stimuli (see Fig. 6, b) is accompanied by a sharp decrease in density. Similar conclusions are for most testees, however, to create an interpreter of emotional reactions, it is reasonable to switch to fuzzy variables [32], which will reduce the influence of subjectivity in the testees' perception of external stimuli.

3.2 Fuzzy Estimates of the Direction of Emotional Reaction Development

To assess the direction of emotional reaction development in the interval (t_k–t_j, $k > j$), we use the phazification of $p(t)$ values (6, 7) at selected time points and the subsequent transition from fuzzy estimates ${}_jY_i \in T_j$ to index values $W(p(t_j))$. The central density $p(t)$ phazification $p(t)$ was performed using triangular membership functions [32].

The idea of using an artificial index scale was proposed in [33] for problems of recognizing fuzzy trends in medical diagnostics. To transit to the index scale, we use the following formula:

$$W_j(y_k) = ind_{Ti} + \mu_{Ti+1}(y_k), \tag{8}$$

where $W_j(y_k)$ is a numerical representation of a linguistic density estimate $(T_i/\mu_{Ti}(y_k))$; ind_{Ti} is the index of a term with a fuzzy set intersecting the left extension of the fuzzy set corresponding to the term T_{i+1}; $\mu_{Ti+1}(y_k)$ is the correspondence degree of the density estimate (y_k) to the fuzzy set T_{i+1}.

If we analyzed a time series representing the results of monitoring emotions and including several sections for which these calculations were performed, then all results are combined as a square matrix *M_otr*. Each element of this matrix represents an estimate on the index scale of density increment for a certain time interval $(t_{j+1}-t_j)$:

$$M\_otr = [m_{i,j}], m_{i,j} = 0, \quad j = j+1, \quad m_{i,j} = W(p_j) - W(p_{j+1}). \tag{9}$$

The length of the time interval $(t_{j+1} - t_j)$ is constant. All descriptions on the index scale of density increments, which are observed during the transition from one signal registration point to the next one (in fact, these are density increments related to emotional reactions to external stimuli), are located on the matrix diagonal (to the right of the main diagonal). The remaining rows are trends options that can be built from each diagonal point, sequentially skipping one element of this row.

Figure 7, a shows the results of calculating density increments (8, 9). Similar calculations make it possible to build a matrix of density increments (see Fig. 7, b); the estimates of graph sections are highlighted in yellow. The matrix is symmetrical regarding the main diagonal. The analysis of the full matrix or individual sections allows estimating the direction of emotional reaction development.

Fragment number	*pt,j*	*W*	*del_W*
0	1205	3.001	
1	1244	3.1	-0.099
2	1698	4.23	-1.13
3	1696	4.225	0.005
4	1653	4.18	0.045
5	1582	3.22	0.96
6	997	2.52	0.7
7	973	2.46	0.06
8	1093	2.78	-0.32
9	1071	2.7	0.08
10	1159	2.35	0.35
11	1809	4.47	-2.12
12	1867	4.55	-0.08
13	1023	2.65	1.9
14	1120	2.8	-0.15

(a)

	0	1	2	3	4	5	6	7	8	9	10	11	12	13	14
0	0	**-0.099**	1.229	-1.224	-1.179	-0.219	0.481	0.541	0.221	0.301	0.651	-1.469	-1.549	0.351	0.201
1	-0.099	0	**-1.13**	-1.125	-1.08	-0.12	0.64	2.4	0.32	0.4	0.75	-1.37	-1.45	0.45	0.3
2	-1.229	-1.13	0	**0.005**	0.05	1.005	1.71	1.77	1.45	1.53	1.88	-0.24	-0.32	1.58	1.43
3	-1.224	-1.125	0.005	0	**0.045**	1.71	1.705	1.765	1.445	1.525	1.875	-0.245	-0.325	1.575	1.425
4	-1.179	-1.08	0.05	0.045	0	**0.96**	1.66	1.72	1.4	1.48	1.83	-0.29	-0.37	1.53	1.38
5	-0.219	-0.12	1.005	1.71	0.96	0	**0.7**	0.76	0.44	0.52	0.87	-1.25	-1.33	0.57	0.42
6	0.481	0.64	1.71	1.705	1.66	0.7	0	**0.06**	-0.26	-0.18	0.17	-1.95	-2.03	-0.13	-0.28
7	0.541	2.4	1.77	1.765	1.72	0.76	0.06	0	**-0.32**	-0.24	0.11	-2.01	-2.09	-0.19	-0.34
8	0.221	0.32	1.45	1.445	1.4	0.44	-0.26	-0.32	0	**0.08**	0.43	-1.69	-1.77	0.13	-0.02
9	0.301	0.4	1.53	1.525	1.48	0.52	-0.18	-0.24	0.08	0	**0.35**	-1.77	-1.85	0.05	-0.1
10	0.651	0.75	1.88	1.875	1.83	0.87	0.17	0.11	0.43	0.35	0	**-2.12**	-2.2	-0.3	-0.45
11	-1.469	-1.37	-0.24	-0.245	-0.29	-1.25	-1.95	-2.01	-1.69	-1.77	-2.12	0	**-0.08**	1.82	1.67
12	-1.549	-1.45	-0.32	-0.325	-0.37	-1.33	-2.03	-2.09	-1.77	-1.85	-2.2	-0.08	0	**1.9**	1.75
13	0.351	0.45	1.58	1.575	1.53	0.57	-0.13	-0.19	0.13	0.05	-0.3	1.82	1.9	0	**-0.15**
14	0.201	0.3	1.43	1.425	1.38	0.42	-0.28	-0.34	-0.02	-0.1	-0.45	1.67	1.75	-0.15	0

(b)

Fig. 7. Estimates of the central density of attractors by the base, linguistic, and index scales: a – attractor density estimates for the graph sections in Fig. 6, a; b – increment matrix *M_otr* (9) of the central density of attractors for separate trend sections in Fig. 6, a.

We can distinguish the section $(t_0 - t_1 - t_2 - t_3 - t_4 - t_5)$ in the experiment under consideration. At the beginning of this section (t_1), we start demonstrating the first video from the archive of negative stimuli to a testee. The sign of the ratio of finite differences $signal(\Delta W/\Delta t)$ changes, i.e. there is a change in the direction of emotion development. Since $m_{0,1} < 0$, $m_{1,2} < 0$, according to the rules [34], the graph section corresponds to an increase of negative emotions. To evaluate the degree of correspondence of the section $(t_0 - t_5)$ to the fuzzy set "Negative emotions, -Em", we use the formulas from the papers [33–35]:

$$W\_TO = Sm/r, \quad Sm = \left(\sum\sum m\_ij\right)/2, r = n(n-1)/2. \tag{10}$$

After substituting the values found by the matrix *M_otr* (9), we get the following in the Eq. (10): $Sm = (\sum\sum m\_ij)/2 = -7.26/2 = -3.63$; $r = n(n - 1)/2 = 5*(5 - 1)/2 = 10$; $W\_TO = -0.363$.

Based on the rules, the conclusion is the following: The graph section $p(t_0 - t_5)$ is an element of the fuzzy set "Increasing negative emotions, -Em" with the correspondence function 0.363.

4 Discussions

Nowadays, the so-called fractal dimensions are used for quantitative estimates and comparing attractors. Despite the great variety of formulas for their estimation, the analysis of attractors constructed from experimental data is typically based on the correlation dimension D_2 (4) or the Lyapunov dimension (*DL*). However, the results of the experiments with speech signal samples and EEG patterns described in a number of works [2, 10, 11, 16, 27] have shown that fractal dimensions *DL* and D_2 in most cases allow differentiating only a neutral state of a person from emotional arousal, but show too weak reactions to a change in the emotional reaction sign (valency) and do not allow evaluating the dynamics. The solution proposed in this work makes compensates these shortcomings.

The mathematical apparatus, which is based on fuzzy estimates of the two-dimensional attractor projection characteristics (8–10), provides more accurate levels of generalization and refinement of EEG pattern descriptions in the given classes of emotions (neutral state, joy, anger, sadness, fear, surprise). Additionally, there is a fuzzy assessment of the direction of emotion development, which allows identifying and analyzing weak and pronounced emotional reactions according to EEG patterns.

5 Conclusion

The research has shown the effectiveness of a new method for monitoring the characteristics of human emotional reactions based on analyzing attractors reconstructed by EEG patterns. The mathematical support and software proposed in this paper allow analyzing the sequence of EEG patterns ($S(t_1)$, $S(t_2)$, …, $S(t_n)$) recorded with a time shift ($(t_2 - t_1) > 0$) and generating verbal conclusions about the valency, level and direction of development of testees' emotional reactions.

Further research might involve the creation of a tool for the tasks of monitoring and correcting human emotional reactions and cognitive activity.

Acknowledgements. The work has been done within the framework of the grant of the President of the Russian Federation for state support of young Russian PhD scientists (MK-1398.2020.9).

Ethical Approval. All procedures performed in studies involving human participants were in accordance with the ethical standards of the institutional and/or national research committee and with the 1964 Helsinki declaration and its later amendments or comparable ethical standards.

References

1. Filatova, N.N., Dmitriev, G.A., Grigorieva, O.M.: Methods and algorithms for classifying graphic objects in problems of medical diagnostics. Tver State Technical University Publ., Tver (2011). (in Russian, Metody i Algoritmy Klassifikatcii Graficheskikh Obektov v Zadachakh Meditcinskoi Diagnostiki)
2. Lebedeva, N.N., Karimova, E.D.: The acoustic characteristics of the speech signal as an indicator of the human functional state. Adv. Physiol. Sci. **45**(1), 57–95 (2014). (in Russian, Uspekhi Fiziologicheskikh Nauk)
3. Shu, L., Xie, J., Yang, M., Li, Z., Li, Z., Liao, D., Xu, X., Yang, X.: A review of emotion recognition using physiological signals. Sensors **18**(7), 2074 (2018). https://doi.org/10.3390/s18072074
4. Abhang, P.A., Gawali, B.W.: Correlation of EEG images and speech signals for emotion analysis. Br. J. Appl. Sci. Technol. **10**(5), 1–13 (2015). https://doi.org/10.9734/BJAST/2015/19000
5. Farhoudi, Z., Setayeshi, S., Rabiee, A.: Using learning automata in brain emotional learning for speech emotion recognition. Int. J. Speech Technol. **20**(3), 553–562 (2017). https://doi.org/10.1007/s10772-017-9426-0
6. Samadiani, N., Huang, G., Cai, B., Luo, W., Chi, C.H., Xiang, Y., He, J.: A review on automatic facial expression recognition systems assisted by multimodal sensor data. Sensors **19**(8), 1863 (2019). https://doi.org/10.3390/s19081863
7. Rangayyan, R.M.: Biomedical Signal Analysis, 2nd edn. Wiley-IEEE Press, New York (2015). https://doi.org/10.1002/9781119068129
8. Malineckij, G.G., Potapov, A.B.: Modern Problems of Nonlinear Dynamics. Editorial URSS Publ., Moscow (2000). (in Russian, Sovremennye Problemy Nelinejnoj Dinamiki)
9. Anishhenko, V.S.: Introduction to Nonlinear Dynamics Handbook. LKI Publ., Moscow (2008). (in Russian, Znakomstvo s Nelinejnoj Dinamikoj: Uchebnoe Posobie)
10. Rabinovich, M.I., Muezzinoglu, M.K.: Nonlinear dynamics of the brain: emotion and cognition. Adv. Phys. Sci. **180**(4), 371–387 (2010). https://doi.org/10.3367/UFNr.0180.201004b.0371. (in Russian, Uspekhi Fizicheskikh Nauk)
11. Starchenko, I.B., Perervenko, J.S., Borisova, O.S., Momot, T.V.: Nonlinear dynamics methods for biomedical applications. Izvestiya SFedU. Eng. Sci. **110**(9), 42–51 (2010). (in Russian, Izvestija JuFU. Tehnicheskie Nauki)
12. Kalauzi, A., Vuckovic, A., Bojic, T.: Topographic distribution of EEG alpha attractor correlation dimension values in wake and drowsy states in humans. Int. J. Psychophysiol. **95**(3), 278–291 (2015). https://doi.org/10.1016/j.ijpsycho.2014.11.008
13. Hutt, A., Graben, P.B.: Sequences by metastable attractors: interweaving dynamical systems and experimental data. Front. Appl. Math. Stat. **3**(11), 1–14 (2017). https://doi.org/10.3389/fams.2017.00011
14. Lokannavar, S., Lahane, P., Gangurde, A., Chidre, P.: Emotion recognition using EEG signals. IJARCCE **4**(5), 54–56 (2015). https://doi.org/10.17148/IJARCCE.2015.4512
15. Filatova, N.N., Sidorov, K.V., Iliasov, L.V.: Automated system for analyzing and interpreting nonverbal information. Int. J. Appl. Eng. Res. **10**(24), 45741–45749 (2015)

16. Lan, Z., Sourina, O., Wang, L., Liu, Y.: Real-time EEG-based emotion monitoring using stable features. Vis. Comput. **32**(3), 347–358 (2015). https://doi.org/10.1007/s00371-015-1183-y
17. Filatova, N.N., Sidorov, K.V.: Computer models of emotions: construction and methods of research. Tver State Technical University Publ., Tver (2017). (in Russian, Kompyuternye modeli emotsy: postroenie i metody issledovaniya)
18. Filatova, N.N., Bodrina, N.I., Sidorov, K.V., Shemaev, P.D.: Organization of information support for a bioengineering system of emotional response research. In: Proceedings of the XX International Conference on Data Analytics and Management in Data Intensive Domains "DAMDID/RCDL 2018", Moscow, 9–12 October 2018, pp. 90–97. CEUR-WS (2018)
19. Sidorov, K.V., Filatova, N.N., Bodrina, N.I., Semaev, P.D.: Analysis of biomedical signals as a way to assess cognitive activity during emotional stimulation. Proc. Southwest State Univ. Ser. Control Comput. Eng. Inf. Sci. Med. Instrum. Eng. **9**(1), 74–85 (2019). (in Russian, Izvestiya Yugo-Zapadnogo gosudarstvennogo universiteta. Serija: Upravlenie, vychislitel'naja tekhnika, informatika. Meditsinskoe priborostroenie)
20. Jasper, H.H.: Report of the committee on methods of clinical examination in electroencephalography. Electroencephalogr. Clin. Neurophysiol. **10**(2), 370–375 (1958). https://doi.org/10.1016/0013-4694(58)90053-1
21. Takens, F.: Detecting strange attractors in turbulence. In: Rand, D., Young, L.S. (eds.) Dynamical Systems and Turbulence, Warwick 1980. Lecture Notes in Mathematics, vol. 898, pp. 366–381. Springer, Heidelberg (1981). https://doi.org/10.1007/BFb0091924
22. Fraser, A.M.: Reconstructing attractors from scalar time series: a comparison of singular system and redundancy criteria. Phys. D **34**(3), 391–404 (1989). https://doi.org/10.1016/0167-2789(89)90263-7
23. Fraiser, A.M., Swinney, H.L.: Independent coordinates for strange attractors from mutual information. Phys. Rev. A **33**(2), 1134–1140 (1986). https://doi.org/10.1103/PhysRevA.33.1134
24. Gorshkov, V.A., Kasatkin, S.A.: Identification of time series of aviation events by methods and algorithms for nonlinear dynamics. Blank Design Publ., Moscow (2008). (in Russian, Identifikacija Vremennyh Ryadov Aviacionnyh Sobytij Metodami i Algoritmami Nelinejnoj Dinamiki)
25. Grassberger, P., Procaccia, I.: Measuring the strangeness of strange attractors. Phys. D **9**(1–2), 189–208 (1983). https://doi.org/10.1016/0167-2789(83)90298-1
26. Eckmann, J.P., Ruelle, D.: Ergodic theory of chaos and strange attractors. Rev. Mod. Phys. **57**(3), 617–656 (1985). https://doi.org/10.1103/RevModPhys.57.617
27. Mekler, A.A.: The software package for the analysis of electroencephalograms using dynamic chaos theory. Ph.D. thesis. N.P. Bechtereva Institute of the Human Brain of the Russian Academy of Sciences, Saint-Petersburg (2006). (in Russian, Programmnyj Kompleks dlya Analiza Elektroencefalogramm Metodami Teorii Dinamicheskogo Haosa)
28. Theiler, J.: Spurious dimension from correlation algorithms applied to limited time series data. Phys. Rev. A **34**(3), 2427–2432 (1986). https://doi.org/10.1103/PhysRevA.34.2427
29. Filatova, N.N., Sidorov, K.V., Shemaev, P.D., Iliasov, L.V.: Monitoring attractor characteristics as a method of objective estimation of testee's emotional state. JEAS **12** (11), 9164–9175 (2017). https://doi.org/10.36478/jeasci.2017.9164.9175
30. Sidorov, K.V., Filatova, N.N., Shemaev, P.D.: An interpreter of a human emotional state based on a neural-like hierarchical structure. In: Abraham, A., et al. (eds.) Proceedings of the Third International Scientific Conference "Intelligent Information Technologies for Industry" (IITI 2018). Advances in Intelligent Systems and Computing, vol. 874, pp. 483–492. Springer, Switzerland (2019). https://doi.org/10.1007/978-3-030-01818-4_48

31. Lapshina, T.N.: Psychophysiological diagnosis of human emotions based on EEG indices. PhD thesis. M.V. Lomonosov Moscow State University, Moscow (2007). (in Russian, Psihofiziologicheskaya Diagnostika Emocij Cheloveka po Pokazatelyam EEG)
32. Filatova, N.N., Sidorov, K.V., Shemaev, P.D.: Prediction properties of attractors based on their fuzzy trend. In: Abraham, A., et al. (eds.) Proceedings of the Second International Scientific Conference "Intelligent Information Technologies for Industry" (IITI 2017). Advances in Intelligent Systems and Computing, vol. 679, pp. 244–253. Springer, Switzerland (2018). https://doi.org/10.1007/978-3-319-68321-8_25
33. Filatova, N.N., Milovidov, A.A.: The program complex for linguistic interpretation of graphic dependencies with fuzzy coordinates of central points. Softw. Syst. **3**, 110–114 (2011). (in Russian, Programmnye Produkty i Sistemy)
34. Filatova, N.N., Sidorov, K.V., Terekhin, S.A., Vinogradov, G.P.: The system for the study of the dynamics of human emotional response using fuzzy trends. In: Abraham, A., et al. (eds.) Proceedings of the First International Scientific Conference "Intelligent Information Technologies for Industry" (IITI 2016). Advances in Intelligent Systems and Computing, vol. 451, pp. 175–184. Springer, Switzerland (2016). https://doi.org/10.1007/978-3-319-33816-3_18
35. Filatova, N.N., Milovidov, A.A.: Analysis and interpretation of audiograms in the space of fuzzy attributes. Fuzzy Syst. Soft Comput. **6**(1), 55–66 (2011). (in Russian, Nechetkie Sistemy i Myagkie Vychisleniya)

A Software Package for Monitoring Human Emotional Reactions and Cognitive Activity by Analyzing Biomedical Signals

Konstantin V. Sidorov(✉) and Natalya I. Bodrina

Tver State Technical University, Afanasy Nikitin Quay 22, Tver 170026, Russia
bmisidorov@mail.ru

Abstract. Human speech samples and in bioelectric signal patterns (EEG, EMG) contain non-verbal information. The paper discusses software for its interpretation. It provides quantitative and qualitative estimates of human emotional reaction characteristics and determines the level of human cognitive activity. There is a description of the structure of a software package, which allows calculating power spectra and reconstructing attractors for signal samples. The features for the attractor analysis are: the length of maximum vectors describing the contours of attractor projections, the density of the trajectories of attractor projections, the degree of matrix sparsification density of attractor projection trajectories. The models of speech patterns, EEG and EMG signals are based on these features and power spectra. The models of biomedical signals have a transition to fuzzy features. The module of an emotion interpreter and cognitive activity evaluates three characteristics: an emotion sign, its level, a tendency of development (emotion dynamics) and a cognitive activity level. The relationship between these characteristics and the signal model components is described as a set of fuzzy rules. The paper also describes testing of the software system. The purpose of the experiments was to study the separation power of the characteristics of the biomedical signal pattern model and the emotion interpreting rules created on their basis. The software package was tested on a sample of Russian and German speech samples, as well as on EEG signal samples. For Russian speech, the accuracy of emotion sign interpretation was 96%, for German speech (from the Emo-DB database), it was 92%.

Keywords: Emotion interpreter · Speech analysis · EEG signal analysis · Power spectrum · Fuzzy set · Nonlinear dynamics · Attractor

1 Introduction

The need to create new tools for quick processing of poorly structured information remains relevant in robotics and in the development of automated control systems. Recently, interdisciplinary studies related to the formalization of models of the mechanism for the rapid analysis of information by a person have been highly active.

One of the approaches to solving this problem is to imitate the biological mechanism of emotional reactions, which is naturally inherent in humans. We can consider

R. Silhavy et al. (Eds.): CoMeSySo 2020, AISC 1295, pp. 444–459, 2020.
https://doi.org/10.1007/978-3-030-63319-6_41

emotions as non-verbal assessments of information processed by the brain. Emotional reactions usually precede reactions associated with verbal assessments [1].

One of the first tasks to solve in this direction is the creation of models and algorithms for recognition of individual characteristics of emotional reactions (sign, strength, emotion orientation). For this purpose, there are software systems for recognizing emotions by integrating the results of the analysis of testee's speech, gestures, facial expressions, as well as bioelectric signals (EEG, ECG, EMG, etc.) [2–7]. Such approach is necessary due to the integration of a large amount of information on the one hand, and a rather limited set of gestures and facial expressions of a civilized person on the other hand. Nowadays, the problems of recognizing the sign of emotions have been successfully solved, especially when they are clear. However, these solutions are linked with the features of a specific functional system of a person (speech, motor apparatus, etc.), which does not allow creating a generalized model of the emotion mechanism.

One of the ways to solve a common problem is to create multi-channel automated systems for studying emotions, which enable recording human reactions to external emotionally significant stimuli through several channels, including the mandatory monitoring of brain electrical activity. The EEG/Speech+ system [7, 8] allows solving the following problems:

- creating a database with bioelectric signal samples recorded with objective confirmation of the sign and level of testee's emotional reaction;
- forming and testing the models of the interpreter of emotion characteristics based on the analysis of speech samples, electroencephalogram (EEG) or electromyogram (EMG) signal samples;
- studying the influence of gender, age, and also national characteristics of languages on the emotion model parameters.

2 Methods

2.1 EEG/Speech+ System Software

The system is based on the principle of openness, which allows combining software tools supplied with the equipment, well-known problem-oriented programs, as well as original proprietary solutions. The general software structure for the bioengineering system EEG/Speech+ is shown in Fig. 1. The system software includes separate software packages (SP). SP-I, SP-II and SP-III are for recording and pre-processing of bioelectric signals.

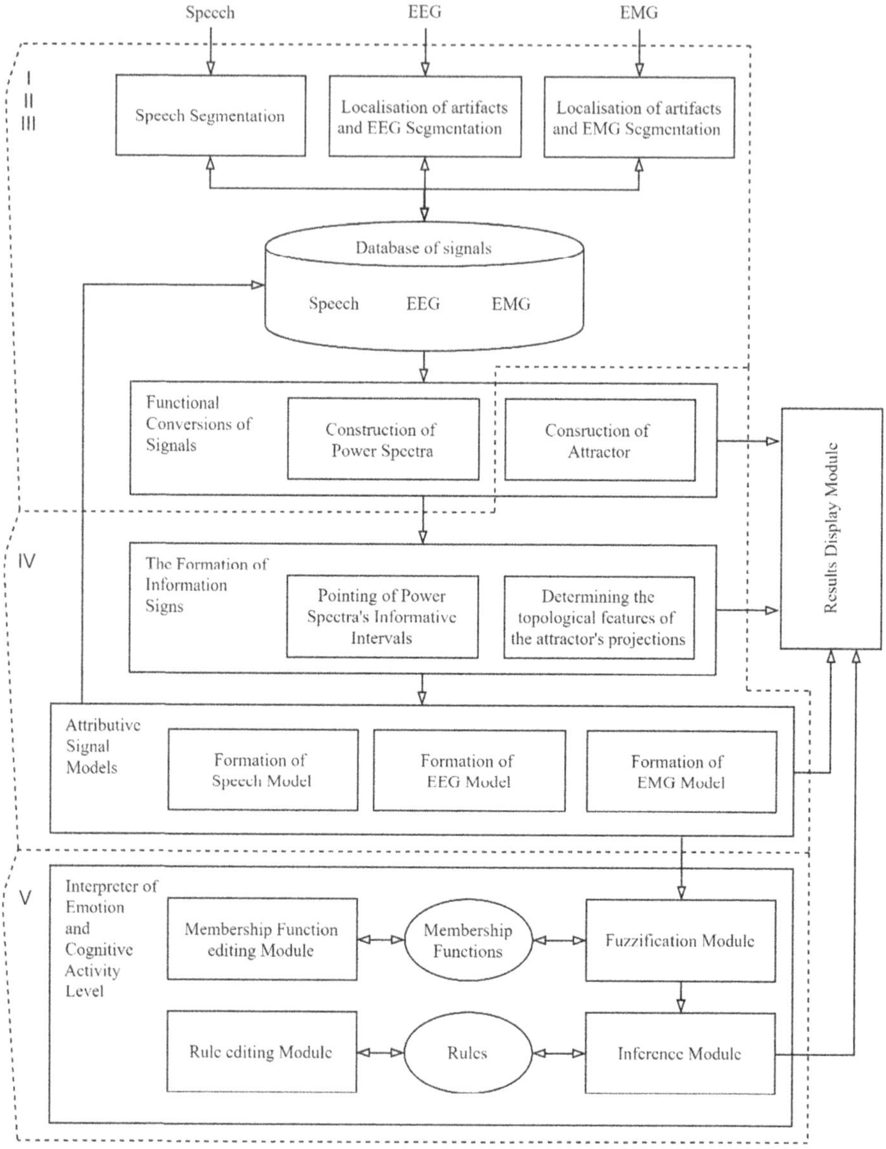

Fig. 1. Basic software modules of EEG/Speech+ system

SP-I, which works in conjunction with a computer encephalograph-analyzer (Encephalan-131-03), solves the problems of EEG recording by 19-lead in the ASCII format, visualization, filtering and segmentation (in the interactive mode) of signals, as well as calculation of power spectra and a complex of 15 associated signs.

SP-II includes the CoolEdit sound editor, which records testee's speech samples in ASCII format, as well as signal visualization and segmentation (in the interactive mode).

SP-III works with a computer myograph (Neuro-MVP-4) to record EMG via two channels in ASCII format, provides visualization, segmentation, calculation of power spectra and related features.

As a result, SP-I, SP-II and SP-III do an automatic signal segmentation in order to extract fragments *x(t)*:

- without artifacts (EEG and EMG signals);
- corresponding to a separate vowel phoneme (speech samples);
- with the same duration (*L*).

SP-IV and SP-V implement algorithms for analyzing and interpreting emotional reaction characteristics and the cognitive activity level. These software systems can work in combination with others and autonomously. Currently, they process and analyze three types of signals (speech signal, EEG and EMG). Input data is loaded from ASCII files. The software modules are implemented in MATLAB and in C# 3.0 for the.NET Framework 3.5 and higher.

SP-IV modules allow defining a set of discrete features to form attributive models of biomedical signals. A set of discrete features has two types of characteristics: features based on the spectral power density (SPD) and morphological features of two-dimensional attractor projections.

SP-V modules allow monitoring emotions and a person's cognitive activity level using linguistic interpretation and inference.

This paper discusses the methods of biomedical signal analysis used in the EEG/Speech+ system, as well as the issues of interpreting emotions from speech signals and EEG.

2.2 The Methods of Analyzing Biomedical Signal Fragments

The selected signal fragments can be analyzed by spectral methods [9, 10] and/or nonlinear dynamics methods [11].

Spectral Analysis Methods. For each *x(t)*, we calculate the power spectrum *S(f)*; its further analysis is carried out using a neural hierarchical structure (NHS classifier) [9] or using cluster analysis methods. For this purpose, a description of each spectrum is forms in as a vector $S(f) = <sx_1, sx_2, \ldots, sx_u>$. The dimension of such vector depends on the method of setting its components. The NHS classifier allows creating rules and recognizing objects with descriptions including from 10 to 1 000 features. If we consider the relative value of the area under the power spectrum curve in a limited frequency range (Δf) as sx_j and take $\Delta f = f_x/F_w$ (where f_x is the sampling frequency, F_w is the FFT window width), then speech sample description will require 300 features [9]. It is possible to set Δf and reduce the description dimension. Therefore, if we use frequency interval values associated with certain EEG rhythms (alpha, beta, etc.) to describe the power spectra of EEG signals, the description dimension will be reduced to 5–9 features.

The use of NHS for graph analysis and classification (for example, power spectra) enables automatical selection of frequency intervals, within which the primary feature values can be considered as elements of one α level fuzzy set (with α = 0.5). Therefore, we can automatically determine the maximum permissible values of Δf (the module for pointing power spectra's informative intervals in Fig. 1), while reducing the number of features in the vector: $S(f) = \langle sx_1, sx_2, \ldots, sx_u \rangle$ [10].

Nonlinear Dynamics Analysis Methods. We use the multidimensional phase portrait reconstruction, which should reflect the phase space of a dynamic biosystem (the attractor constructing module in Fig. 1). According to F. Tackens' theorem [12], initial signal samples shifted by τ delay time can form the reconstructed space:

$$y_n = x_n, x_{n+\tau}, \ldots, x_{n+(m-1)\tau}, n = 0, \ldots, s-1; s = N - (m-1)\tau,$$

where N is the total number of the time series elements (points); τ is the time delay between the time series elements (time lag); m is the embedding dimension (lag space dimension).

The parameters m and τ have a significant effect on the attractor shape and projections. The time delay is determined by the autocorrelation function (ACF): τ is equal to the time of the first zero crossing by ACF. The algorithm for searching for false nearest neighbors determines the size of m embedding dimension. When studying speech, EEG and EMG signals, $m \geq 2$. Attractor projections are based on the attractor reconstruction (Fig. 2).

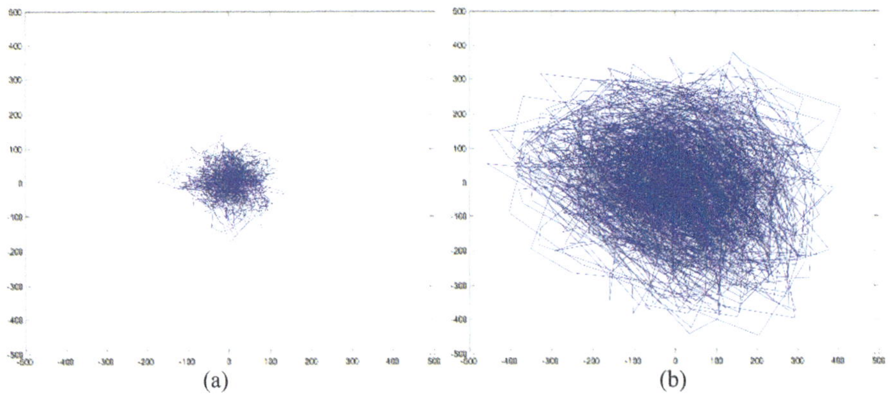

Fig. 2. Attractors's reconstruction when $m = 2$ with the lead F4-A2 for negative (a) and positive (b) emotions

Usually the geometry of attractors from experimental data is estimated by the correlation dimension [13]. This feature differentiates a person's neutral state from emotional arousal, but shows weak reaction to a change in an emotion sign.

The module for determining the attractor topological features, which is part of SP-IV, provides quantitative estimates of new features [14]:

- the lengths of maximal R vectors describing the contours of attractor projections and characterizing biomedical signals by amplitude composition;
- the densities of γ attractor projection trajectories;
- the sparsity degrees of the density matrix of attractor projection trajectories.

The algorithm for their estimation is based on a sequential scanning of all attractor projections, each of which is connected to its own coordinate system. For each projection, we select four quadrants and determine R vector [11, 15], for example, for $m = 3$:

$$R_i^{k=1} = \sqrt{x(t_i)^2 + x(t_i + \tau)^2},$$

$$R_i^{k=2} = \sqrt{x(t_i)^2 + x(t_i + 2\tau)^2},$$

$$R_i^{k=3} = \sqrt{x(t_i + \tau)^2 + x(t_i + 2\tau)^2},$$

where R_i is a vector for the i-th attractor point; $x(t_i)$ is a bioelectric signal value in the t_i time point.

The subsets R^j are distinguished from the R set for each phase plane quadrant:

$$R^1 = \left\{R_i \in R|(\forall i)x_i, x_{i+\tau} \geq 0 \rightarrow R(x(t_i), x(t_i + \tau))\bigcup R^1\right\}$$

$$R^2 = \left\{R_i \in R|(\forall i)x_i \leq 0, x_{i+\tau} \geq 0 \rightarrow R(x(t_i), x(t_i + \tau))\bigcup R^2\right\} \quad \text{etc.}$$

$$R^3 = \left\{R_i \in R|(\forall i)x_i \leq 0, x_{i+\tau} \leq 0 \rightarrow R(x(t_i), x(t_i + \tau))\bigcup R^3\right\} \quad \text{etc.}$$

$$R^4 = \left\{R_i \in R|(\forall i)x_i \geq 0, x_{i+\tau} \geq 0 \rightarrow R(x(t_i), x(t_i + \tau))\bigcup R^4\right\} \quad \text{etc.}$$

A vector $R_{max}^{k,j}$ (the maximum vector of the j-th quadrant of the k-th attractor projection), and then the averaged vector of the k-th attractor projection ($\bar{R}_{max}^k$) is formed. It has four components and provides rough estimates of the attractor projection area.

In order to determine the density γ, the attractor projection is covered by a regular grid with a constant step d. For each grid cell, we can find the ratio of the number of points in the cell (P) to the cell area (S); the points at the cell borders are distributed evenly between neighboring elements:

$$\gamma_{i,j} = P_{i,j} * S_{i,j}^{-1} \text{ при } P_{i,j} = h_{i,j} + r_{i,j}/k, S_{i,j} = \delta^2,$$

where $h_{i,j}$ is the number of points inside the cell with coordinates (i, j); $r_{i,j}$ is the number of points at the cell boundaries; k is the number of neighboring cells having common points with the cell (i, j).

The set of estimates $\{\gamma_{i,j}\}$ found for all grid cells is represented by the following matrix:

$$M_\gamma = \lfloor m_{ij} \rfloor, \forall(i,j) m_{ij} = \gamma_{ij}$$

Each element characterizes the attractor density in the (i, j)-th projection section (Fig. 3). The number of zero cells k0 of the Mγ matrix is used to estimate the density matrix sparseness, as well as to assess the uniformity of the distribution of attractor trajectories along the projection plane.

Using the above features and settings Δf, Fw, the user can create various model versions of speech patterns and biomedical signals (attributive signal models in Fig. 1).

γ	1	2	3	4	5	6	7	8	9	10	11	12	13	14
1	0	0	0	0	0	0	0	0	0	0	0	0	0	0
2	0	0	0	0	0	1	0	0	1	0	0	0	0	0
3	0	0	0	0	1	1	0	0	1	0	0	0	0	0
4	0	0	0	1	1	6	6	8	7	3	0	0	0	0
5	0	0	1	2	11	26	58	39	25	9	3	0	0	0
6	0	1	1	5	19	86	131	142	95	42	9	0	2	0
7	0	0	1	8	38	119	256	213	129	46	5	2	0	0
8	0	0	1	5	39	130	219	226	156	45	6	1	0	0
9	0	0	0	6	21	68	115	144	78	19	4	0	0	0
10	0	0	0	4	7	22	28	34	29	7	4	0	0	0
11	0	0	0	0	1	4	14	6	4	0	0	0	0	0
12	0	0	0	0	0	0	2	1	1	0	0	0	0	0
13	0	0	0	0	0	0	1	1	0	0	0	0	0	0
14	0	0	0	0	0	0	0	0	0	0	0	0	0	0

(a)

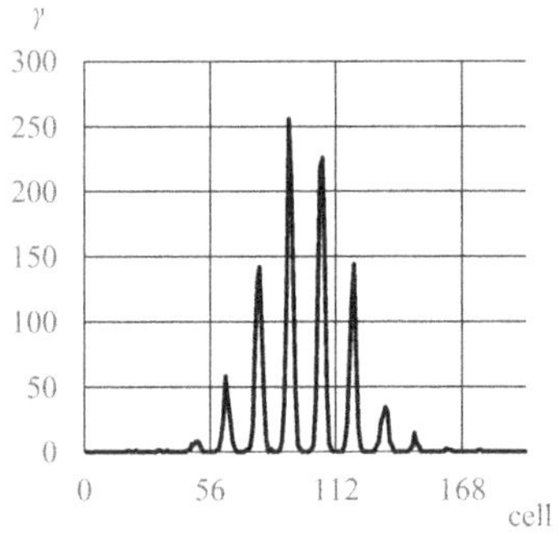

(b)

Fig. 3. A density matrix of attractors' trajectories: (a) – a matrix sample of a testee's positive state, EEG sample; (b) – a density variation graph depending on matrix element number

2.3 The Interpreter of Emotions and a Cognitive Activity Level as a Part of EEG/Speech+

The interpreter uses the transition to fuzzy features in models of biomedical signals (fuzzification module in Fig. 1). For this purpose, using the following unified term-set (Feature value Small (T1_), … Medium (T2_), … Large (T3_)), fuzzy sets are created by linking membership functions to the boundaries of universal feature value sets on the corresponding signal scale (speeches in a specific language, EEG or EMG). The level cognitive activity is estimated only by EEG patterns. The fuzzification of any feature uses a triangular conorm:

$$Max(\mu_A, \mu_B) \Rightarrow A \bigcup B$$

The emotion interpretation by signal patterns is based on estimates of three characteristics: the sign of the emotion (Z), its level (U), and the tendency of development (emotion dynamics, *D*). The relationship between *Z, U, D* and the components of signal models is described as a set of rules.

The inference module forms the final solution to the problem of recognizing a class of emotions based on the results of applying the rules for all found attractor projections. If one class is determined from all attractor projections, for example (*Lf*), its name is used to determine the corresponding emotion characteristic ($Z = Lf$). The correspondence degree of this conclusion is considered equal to the minimum membership function of fuzzy sets constructed for these projections. If the conclusions match only p projections ($p < (m - 1)$), then the next signal sample is selected for analysis.

3 Results and Discussion

The tests of the bioengineering system EEG/Speech+ included using a specific multimodal base of emotions that contained speech signals and EEG patterns [8]. There were experiments for studying the separation power of the features of the biomedical signal pattern model and the rules for interpreting emotions created on their basis.

There are generated experimental samples of biomedical signals (Table 1). The EEG/Speech+ system has automatically generated a description for each object in the form of attribute models.

Table 1. Experimental samples of biomedical signals

Signal samples	Cluster 1 – «negative»	Cluster 2 – «neutral»	Cluster 3 – «positive»
Speech, training samples	124	116	200
Speech, test samples	156	96	148
EEG, training samples	160	160	160
EEG, test samples	160	160	160

3.1 The Effect of Emotional State on the Spectral Power Density of Biomedical Signals

The studies of spectral power density (SPD) of signals based on a neural hierarchical classifier [9, 10] have shown the possibility of distinguishing intervals of the most informative features, which provide a good level of generalization and detailing object descriptions in clusters. For EEG patterns, these were the features from the right hemisphere leads. Each of the eight leads (lead Fp2-A2 is not analyzed) is described by 41 features. The most informative SPD features for a speech signal are in the range of 1100–5100 Hz. Thus, each speech signal object was described by 400 features ($x_{100} - x_{500}$) (Fig. 4); the EEG object representation was considered by a vector consisting of 328 features ($x_1 - x_{328}$) (Fig. 5).

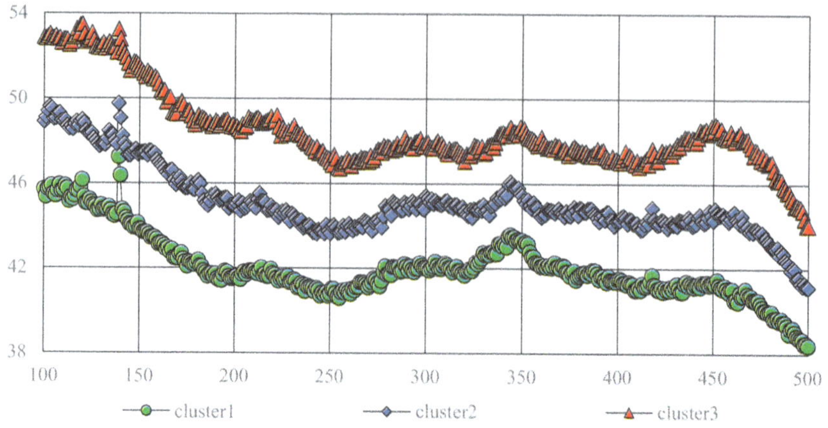

Fig. 4. Cluster centers by SPD features in a speech signal ($x_{100} - x_{500}$)

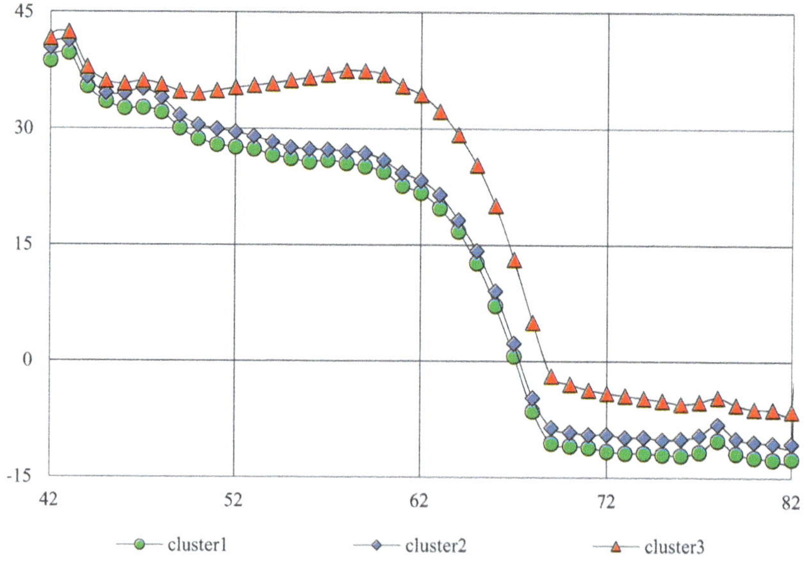

Fig. 5. Cluster centers by SPD features by 8 EEG leads ($x_{42} - x_{82}$)

The obtained research results have shown acceptable accuracy in the classification of speech samples and EEG patterns in accordance with an emotion sign (Cluster 1, Cluster 2, Cluster 3) (Table 2). The combination of the classification results shows that incorrectly classified speech samples and EEG patterns belong to the same people. Two far classes (Cluster 1 and Cluster 3) are accurately separated.

Table 2. Classifications of EEG and speech signals by SPD features

	EEG, training samples	EEG, test samples	Speech, training samples	Speech, test samples
Classification of Cluster 1, %	100	71	100	87
Classification of Cluster 2, %	100	73	100	83
Classification of Cluster 3, %	100	86	100	91
Classification success, %	100	76	100	88

Conclusion. When using the SPD apparatus, we can observe multiple intersection of biomedical signal clusters. The results of the classification of speech signal and EEG objects have the same convergence.

3.2 The Studies of the Separation Force of the Maximum Vectors ($R_1 - R_{15}$) of Attractor Projections on a Speech Signal

We studied Russian speech samples (Fig. 6). Each speech signal object is described by 15 features (($R^{k,j}_{max}$, $\bar{R}^{k}_{max}$) or ($R_1 - R_{15}$)).

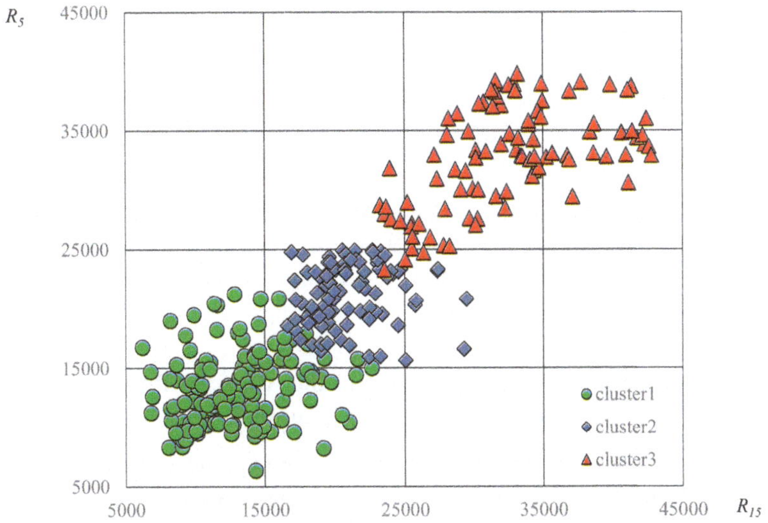

Fig. 6. The maximum vectors of attractor projection (R_{15}, R_5) in Russian speech samples

It can be noted that there is no linear separation of clusters with positive and neutral emotions, and there are significant intersections of neutral and negative emotion clusters. However, the centers of these clusters (Fig. 7) differ significantly in all the features from the feature vector (R_1 – R_{15}).

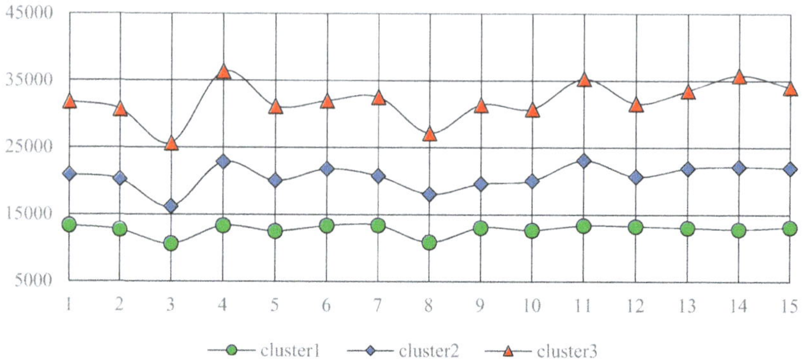

Fig. 7. The cluster centers by features (R_1 – R_{15}) in a speech signal

Table 3 shows the results of the speech sample classification based on attractor features.

Table 3. Speech signal classification by attractor features

	Training samples	Test samples
Classification of Cluster 1, %	100	92
Classification of Cluster 2, %	100	100
Classification of Cluster 3, %	100	100
Classification success, %	100	96

Conclusion. The comparison of the obtained results with the results given in Sect. 3.1 shows that the system of features (R_1 – R_{15}) is superior in terms of separation properties to the features based on SPD (x_{100} – x_{500}), and successfully works with the samples of Russian speech.

3.3 The Studies of the Separation Force of the Attractor Projection Maximum Vectors (R_1 – R_{120}) on EEG Samples

The analysis of the distribution pattern of attributive EEG models presented in the form (R_1 – R_{120}) illustrates a good separation of the cluster center recorded at positive emotions (Cluster 3) from the neutral state (Cluster 2) and negative emotions (Cluster 1) (Fig. 8). There is a significant intersection of neutral and negative emotion clusters in almost all components and in all 8 leads of the right hemisphere. As in Sect. 3.2, the clusters centers significantly differ in all features (R_1 – R_{120}).

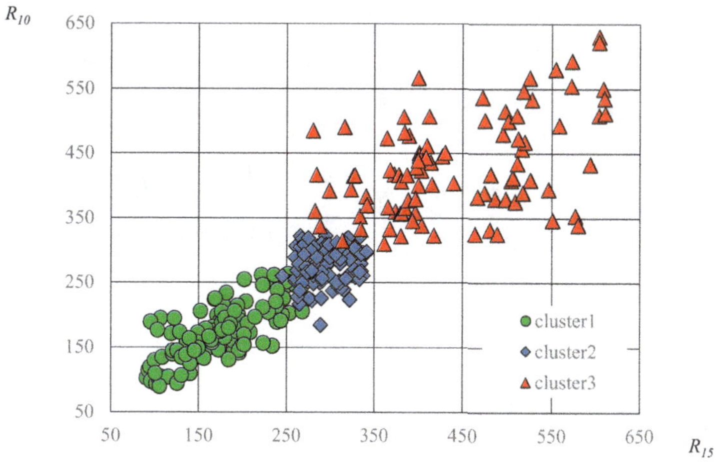

Fig. 8. The maximum vectors of attractor projection (R_{15}, R_{10}) in EEG samples

There is an obvious relationship between the values of the feature vector (R_1 – R_{120}) and the lead number, i.e. the localization point of the cerebral cortex signal. The estimates of the mean square deviation (MSD) of the difference in the signal amplitudes recorded by the *j*-th and *k*-th leads allowed identifying biomedical signals that are the most sensitive to emotional reactions (Fig. 9 and Fig. 10).

The MSD dependence character for different pairs of leads (j, k = 1.8) made it possible to identify right hemisphere leads as the most sensitive (F4-A2, F8-A2).

The experimental results of the in Sects. 3.2 and 3.3 have shown good convergence at a qualitative level: the objects located at the intersection of speech signal clusters correspond to EEG objects of the same testee with a similar localization in the feature space.

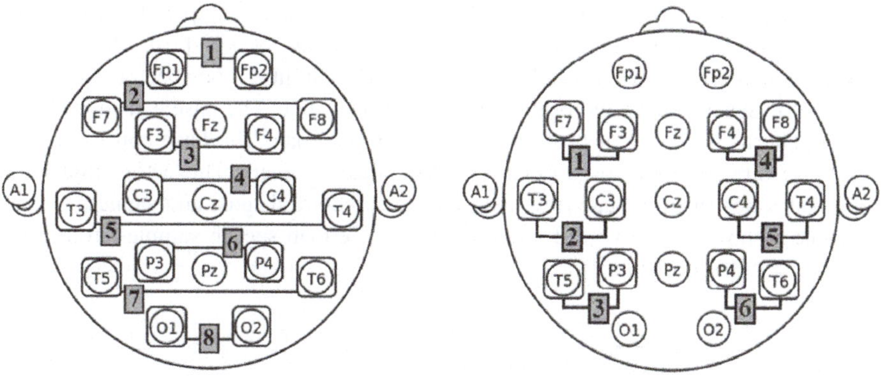

Fig. 9. Lead pairs selection options

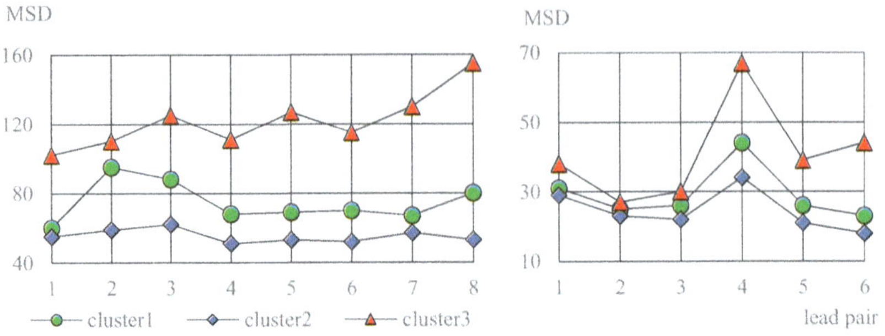

Fig. 10. MSD of the amplitude difference in the *j*-th and *k*-th leads

The created classifier has shown high success of interpretation on EEG patterns (Table 4).

Table 4. EEG classification by attractor features

	Training samples	Test samples
Classification of Cluster 1, %	100	75
Classification of Cluster 2, %	100	75
Classification of Cluster 3, %	100	100
Classification success, %	100	83

Conclusion. The system of features ($R_1 - R_{120}$) is successful for an EEG signal, and also shows higher separation properties compared with SPD-based features ($x_1 - x_{328}$).

3.4 Interpretation of Emotions by Emo-DB Speech Samples

For an independent estimation of the EEG/Speech+, we used samples from the international Emo-DB database (Berlin Database of Emotional Speech) [16]. This database is very popular with many researchers due to its accessibility; it has been used to form and develop systems for monitoring human emotional state. The base is a collection of speech patterns (phrases) of professional actors in German. The base consists of 535 phrases of speech from 10 broadcasters (5 women, 5 men). The actors imitated 7 emotional states (neutral state, joy, fear, anger, boredom, disgust and sadness).

Table 5 shows the results of interpretation of German speech samples from Emo-DB using EEG/Speech+.

Table 5. The results of interpretation of speech samples from Emo-DB

Classes	Emotions	Interpretation error, %
3	neutral, negative (anger + disgust + fear + sadness + boredom), positive (joy)	8
7	anger, disgust, fear, sadness, boredom, joy, neutral	17

Conclusion. The emotion interpretation results based on the analysis of speech samples from Emo-DB suggest that the EEG/Speech+ system is as good as the known analogues by an interpretation mistake.

4 Conclusion

The convergence of the emotion interpretation results on many EEG patterns and on many speech signal samples from a multimodal base of emotions, as well as on speech recordings from the Emo-DB base, confirms the reliability and effectiveness of the proposed interpreter of emotions and cognitive activity for the EEG/S bioengineering system.

The software package was tested on a sample of Russian and German speech samples, as well as on EEG signal samples. For Russian speech, the accuracy of interpretation of emotion characteristics was 96%, for German (from Emo-DB) it was 92%. The experiment has proved the coincidence of the results of emotion interpretation on speech and EEG samples recorded from the same testee. The implemented solutions make it possible to record intermediate and final research results, are sensitive to small changes in testee's emotional state, do not require long-term testing, and help reduce errors in diagnosing testee's emotional state.

Acknowledgements. The work has been done within the framework of the grant of the President of the Russian Federation for state support of young Russian PhD scientists (MK-1398.2020.9)

Ethical Approval. All procedures performed in studies involving human participants were in accordance with the ethical standards of the institutional and/or national research committee and with the 1964 Helsinki declaration and its later amendments or comparable ethical standards.

References

1. Baars, B.J., Gage, N.M.: Cognition, Brain, and Consciousness: Introduction to Cognitive Neuroscience. 2nd edn. Elsevier, AP (2010). https://doi.org/10.1016/c2/009-0-01556-6
2. Chao, H., Dong, L., Liu, Y., Lu, B.: Emotion recognition from multiband EEG signals using CapsNet. Sensors (Basel) **19**(9), 2212 (2019). https://doi.org/10.3390/s19092212

3. Umamaheswari, J., Akila, A.: An enhanced human speech emotion recognition using hybrid of PRNN and KNN. In: Proceedings of International Conference on Machine Learning, Big Data, Cloud and Parallel Computing (COMITCon), Faridabad, India, pp. 177–183 (2019). https://doi.org/10.1109/comitcon.2019.8862221
4. Vithanawasam, T.M.W., Madhusanka, B.G.D.A.: Face and upper-body emotion recognition using service robot's eyes in a domestic environment. In: Proceedings of International Research Conference on Smart Computing and Systems Engineering (SCSE), Colombo, Sri Lanka, pp. 44–50 (2019). https://doi.org/10.23919/scse.2019.8842658
5. Wang, K., Ho, Y., Huang, Y., Fang, W.: Design of intelligent EEG system for human emotion recognition with convolutional neural network. In: Proceedings of IEEE International Conference on Artificial Intelligence Circuits and Systems (AICAS), pp. 142–145. IEEE, Hsinchu (2019). https://doi.org/10.1109/aicas.2019.8771581
6. Bodyanskiy, Y., Kulishova, N., Malysheva, D.: The multidimensional extended neo-fuzzy system and its fast learning for emotions online recognition. In: Proceedings of IEEE Second International Conference on Data Stream Mining & Processing (DSMP), pp. 473–477. IEEE, Lviv (2018). https://doi.org/10.1109/dsmp.2018.8478564
7. Filatova, N.N., Sidorov, K.V., Shemaev, P.D., Rebrun, I.A.: Emotion and cognitive activity monitoring system. In: Proceedings of the 3rd Russian-Pacific Conference on Computer Technology and Applications (RPC 2018), pp. 1–4. IEEE, Vladivostok (2018). https://doi.org/10.1109/rpc.2018.8482220
8. Filatova, N.N., Bodrina, N.I., Sidorov, K.V., Shemaev, P.D.: Organization of information support for a bioengineering system of emotional response research. In: Proceedings of the XX International Conference on Data Analytics and Management in Data Intensive Domains (DAMDID/RCDL 2018), pp. 90–97. CEUR-WS (2018). http://ceur-ws.org/Vol-2277/paper18.pdf
9. Sidorov, K.V., Filatova, N.N., Shemaev, P.D.: An interpreter of a human emotional state based on a neural-like hierarchical structure. In: Abraham, A., et al. (eds.) Proceedings of the Third International Scientific Conference "Intelligent Information Technologies for Industry" (IITI 2018). IITI 2018. Advances in Intelligent Systems and Computing, vol. 874, pp. 483–492. Springer, Cham (2019). https://doi.org/10.1007/978-3-030-01818-4_48
10. Sidorov, K.V., Filatova, N.N., Shemaev, P.D.: A neural-like hierarchical structure in the problem of automatic generation of hypotheses of rules for classifying the objects specified by sets of fuzzy features. In: Abraham, A., et al. (eds.) Proceedings of the Third International Scientific Conference "Intelligent Information Technologies for Industry" (IITI 2018). IITI 2018. Advances in Intelligent Systems and Computing, vol. 874, pp. 511–523. Springer, Cham (2019). https://doi.org/10.1007/978-3-030-01818-4_51
11. Filatova, N.N., Sidorov, K.V., Shemaev, P.D.: Prediction properties of attractors based on their fuzzy trend. In: Abraham, A., et al. (eds.) Proceedings of the Second International Scientific Conference "Intelligent Information Technologies for Industry" (IITI 2017). IITI 2017. Advances in Intelligent Systems and Computing, vol. 679, pp. 244–253. Springer, Heidelberg (2018). https://doi.org/10.1007/978-3-319-68321-8_25
12. Takens, F.: Detecting strange attractors in turbulence. In: Rand, D.A., Young, L.-S. (eds.) Dynamical Systems and Turbulence, Warwick 1980, vol. 898, pp. 366–381. Springer, Heidelberg (1981). https://doi.org/10.1007/bfb0091903
13. Anishhenko, V.S.: Lectures on Nonlinear Dynamics. Regulyarnaya i khaoticheskaya dinamika Publ., Moscow–Izhevsk (2011). (in Russ., Lektsii po nelineynoy dinamike)
14. Filatova, N.N., Sidorov, K.V.: Computer models of emotions: construction and methods of research. Tver State Technical University, Tver (2017). (in Russian, Kompyuternye modeli emotsy: postroenie i metody issledovaniya)

15. Filatova, N.N., Sidorov, K.V., Bodrina, N.I., Voronkov, M.E., Klyuev, P.S.: Monitoring the level of trainee's cognitive activity by EEG signal fragmentary analysis. Proc. Southwest State Univ. Ser. Control Comput. Eng. Inf. Sci. Med. Instrum. Eng. **9**(4), 8–23 (2019). (in Russian, Izvestiya Yugo-Zapadnogo gosudarstvennogo universiteta. Serija: Upravlenie, vychislitel'naja tekhnika, informatika. Meditsinskoe priborostroenie)
16. Burkhardt, F., Paeschke, A., Rolfes, M., Sendlmeier, W., Weiss, B.: A database of German emotional speech. In: Proceedings International Conference Interspeech, Lissabon (2005). http://pascal.kgw.tu-berlin.de/emodb/index-1280.html

Ergodicity and Consistency of Statistical Estimates of Poisson Flow Intensity and Stochastic Properties of Medianta

Gurami Tsitsiashvili(✉)

Institute for Applied Mathematics, Far Eastern Branch of Russian Academy Sciences, Vladivostok, Russian Federation
guram@iam.dvo.ru

Abstract. This paper raises the question of how to transform the observation procedure to replace a long series of observations by several short and, conversely, several short series by one long. This problem is solved to assess the intensity of the stationary Poisson flow and to estimate the average time of operation of technical systems for failure, using the concept of ergodicity of the observation process. We find out the conditions, under which random noise strongly affects the average of the fraction sequence, and why the medianta is a fairly stable characteristic.

Keywords: Ergodicity · Poisson flow · Exponential distribution · Mean failure time · Medianta

1 Introduction

When statistical data is processed, problems arise in replacing a long series of observations for several shorter and several short series of observations for one longer. The task of replacing a long series of observations with short ones is largely technical in nature, since long-term observations are difficult for technical reasons. Such problems appear, for example, in optoelectronics [1], in the statistical estimation of the failure time of technical systems [2], etc.

The problem of replacing several short series of observations with one longer series is closely related to the problem of small samples. This task occurs, when processing observations of the behaviour of ecological, sociological, biological, geological, and technical systems, as well as observations of flows in queuing systems [3–6]. Difficulties arise, when obtaining statistically reliable estimates: unbiased, consistent, effective, and asymptotically normal. The transition to a longer series of observations allows in some cases to overcome these difficulties.

This paper asks, whether it is possible to replace a long series of observations with several shorter ones and, conversely, several short series with one longer one. It is important to know how such substitutions will affect the quality of statistical estimates obtained. It turns out that this question rests on the concept

R. Silhavy et al. (Eds.): CoMeSySo 2020, AISC 1295, pp. 460–466, 2020.
https://doi.org/10.1007/978-3-030-63319-6_42

of ergodicity of the observation process: "the average of the ensemble coincides with the average of the trajectory". A large number of both theoretical and applied papers are devoted to the ergodic theory of dynamical systems, see, for example, [7–9].

However, to use the concept of ergodicity to solve problems about small samples and long series of observations, it is necessary to choose a convenient probabilistic model of observations. In this paper, a model of a stationary Poisson flow and a model of a sequence of times of failure of technical systems with an exponential distribution, having a large average value are selected for this purpose. Statistical estimates of the intensity of the Poisson flow and the parameter of the exponential distribution are constructed. Convenient methods of conducting observations are chosen based on the ergodicity property.

The estimation of the Poisson flow intensity is closely related to the concept of medianta introduced by A. Y. Khinchin [10, 11] in the theory of chain fractions. The medianta of fractions $\frac{a}{b}$, $\frac{c}{d}$ refers to the fraction $\frac{a+c}{b+d}$. The medianta is most widely used in applied mathematics, in particular in mathematical statistics [12–14], and currently in demography and biology [15, 16]. Maurice Cline [17], in fact, rediscovered the medianta by suggesting the "football arithmetic" of adding fractions. This addition was used by M. Kline to determine the average performance of a football player-a striker for two games. It also considers cases of determining the average efficiency of trade and the average speed of the car.

In [15], the opinion based on processing a large amount of empirical information is expressed that the average value of a sequence of fractions is determined mainly by random noise. On the contrary, medianta, understood as the ratio of the average numerator to the average denominator, is very stable statistics, its even called "invariant". The paper provides an explanation for the similar results of data processing, using the classic probabilistic models. We find out the conditions, under which random noise strongly affects the average of the fraction sequence, and why the medianta is a fairly stable characteristic.

2 Estimation of Poisson Flow Intensity

Consider a stationary Poisson flow Λ with intensity θ. Let $A = [a, b)$, $a < b$, $B = [b, c)$, $b < c$, – disjoint half-intervals on the real axis, $C = A \cup B$. It is known that random variables $n(A)$, $n(B)$, defining the number of points of a stationary Poisson flow on disjoint half-intervals A, B, are independent [19, p.12, the property 2^0 of the lack of after-effect]. Then the random variables $n(A)$, $n(B)$, $n(C)$ have a Poisson distribution with parameters

$$\theta\alpha,\ \theta\beta,\ \theta\gamma,\ \gamma = \alpha + \beta,\ \alpha = b - a,\ \beta = c - b.$$

From the Central Limit Theorem for renewal processes (see, for example, [18, Chapter 7, theorem 2]) it follows a uniform by t, $-\infty < t < \infty$, convergence

$$P\left(\frac{n(C)/\gamma - \theta}{\sqrt{\theta/\gamma}} < t\right) = P\left(\frac{n(C) - \theta\gamma}{\sqrt{\theta\gamma}} < t\right) \to \Phi(t),\ \gamma \to \infty, \quad (1)$$

where $\Phi(t)$ is the standard normal distribution (with zero expectation and unit variance).

As an estimate of the θ parameter, we take $\widehat{\theta} = \dfrac{n(C)}{\gamma}$. The mean and variance of this estimate satisfy the equalities

$$M\widehat{\theta} = \theta,\ D\widehat{\theta} = \frac{\theta}{\gamma}. \tag{2}$$

The first equality in (2) means unbiased, the relations (1), (2) entail consistency and asymptotic normality of the estimate $\widehat{\theta}$ for $\gamma \to \infty$.

Let the half-intervals $C_1, \ldots, C_m$ not intersect in pairs, have the length $\gamma_1, \ldots, \gamma_m$, and let the equality $\gamma = \sum_{k=1}^{m} \gamma_k$ holds . Put $n(C_1), \ldots, n(C_m)$ – the number of points of the Poisson flow Λ on these half -intervals and construct a statistical estimate of $\widehat{\theta}'$ of the parameter θ :

$$\widehat{\theta}' = \frac{\sum_{k=1}^{m} n(C_k)}{\sum_{k=1}^{m} \gamma_k}. \tag{3}$$

Since $n(C)$ matches the distribution of $\sum\limits_{k=1}^{m} n(C_k)$, then the statistics $\widehat{\theta}'$, defined by equality (3), matches the distribution with the statistics $\widehat{\theta}$.

Thus, the parameter θ can be estimated as observing a number of points flow Λ on the disjoint union of half intervals $C_1, \ldots, C_m$, and the observation points of the flow Λ on half interval C of length $\gamma \gg 1$. And the number m may be small. From the formula (3) it follows that the estimate $\widehat{\theta}'$ of the parameter θ may be considered as the medianta [10, p. 23], [11, p. 631], introduced for a pair of fractions, but allowing generalization for a finite number of fractions.

These properties of the estimate $\widehat{\theta}'$ of the intensity θ of the Poisson flow Λ are close to the concept of ergodicity, which allows us to find a solution to the problem of small samples, consisting of the number of points of the Poisson flow on disjoint half-intervals with a large total length.

Note that the Lebesgue measurable sets C_k, $k = 1, \ldots, m$ may belong not only to a straight line, but also to a plane or three dimensional space. In this case, the intensity of the Poisson flow of points in these sets obeys the same relations as in the one dimensional case. At the same time, such a probabilistic model may be applied in problems of earth sciences, where Boolean models of random sets are used, defined via a Poisson flow of points [6].

3 Estimation of a Large Parameter of the Exponential Distribution

Assume that there are n technical systems, whose failure times $y_1, \ldots, y_n$ are independent and equally distributed random variables (i.i.d.r.v's) with an exponential distribution $P(y_1 > t) = \exp(-t/T)$, $t \geq 0$. Let the parameter $T = My_1$

be large: $T \gg 1$. Our task is to build a method for organizing tests and evaluating the parameter T, in which an increase in the number of tested systems by m times leads to a decrease in the average test time by $m-1$ times.

Denote $\widehat{T}_n = \dfrac{y_1 + \ldots + y_n}{n}$, when constructing the estimate $\widehat{T}_n$ total test time U_n with probability one satisfies the inequality $U_n \geq \max(y_1, \ldots, y_n)$. Therefore, the following relations are satisfied

$$M\widehat{T}_n = T,\ D\widehat{T}_n = \frac{T^2}{n},\ MU_n \geq M\max(y_1, \ldots, y_n) \geq My_1 = T \gg 1. \tag{4}$$

To reduce the total test time, we will construct the following procedure. Let nm systems start working at time 0. Tests end, when the first n systems fail. Assume that the moments of failure of these systems satisfy the relations $0 \leq T_1 \leq \ldots \leq T_n$. Then the independent random variables $z_1 = T_1,\ z_2 = T_2 - T_1, \ldots, z_n = T_n - T_{n-1}$ have exponential distributions with parameters $nmT,\ (nm-1)T, \ldots, (nm-n+1)T$. Hence, the independent random variables $nmz_1,\ (nm-1)z_2, \ldots, (nm-n+1)z_n$ have the exponential distribution with the parameter T. Using observations $z_1, \ldots, z_n$, we construct estimate $\overline{T}_n$ of the parameter T :

$$\overline{T}_n = \frac{nmz_1 + (nm-1)z_2 + \ldots + (nm-n+1)z_n}{n}.$$

Thus, the random variable $\overline{T}_n$ coincides in distribution with the random variable $\widehat{T}_n$ and, therefore, the following relations are fulfilled

$$M\overline{T}_n = T,\ D\overline{T}_n = \frac{T^2}{n}. \tag{5}$$

In turn, the total test time $V_n = z_1 + \ldots + z_n$ consists of z_k terms such that $Mz_k = \dfrac{T}{mn-k+1},\ k = 1, \ldots, n$, satisfies the following relations

$$\frac{T}{m} = \frac{T}{mn}n < MV_n = \sum_{k=1}^{n} \frac{T}{mn-k+1} < \frac{T}{mn-n}n = \frac{T}{m-1}.$$

This means that for the value Mv_n, the inequalities are valid

$$\frac{T}{m} < MV_n < \frac{T}{m-1}. \tag{6}$$

Comparing the formula (4) with the formulas (5), (6) shows that increasing the number of tested systems by m times allows us to reduce the average test time from MU_n to MV_n by about $m-1$ times by constructing a statistical estimate of $\overline{T}_n$. This method of organizing tests and processing their results may be useful, when evaluating the parameter $T \gg 1$, that characterizes a large average time to failure. This technique is close to the use of ergodicity in estimating the intensity of a Poisson flow.

4 Comparing the Medianta and Average of a Sequence of Fractions

Let there be a sequence of independent and identically distributed random vectors (z_k, c_k, d_k), $k = 1, \ldots, n$, with independent components, such that for some a, $0 < a < 1$, the inequalities $-a < c_k, d_k < a$, $0 < z_k$ are almost certainly true. We assume that there are finite mathematical expectations and variances

$$Mz_1 = Z,\ Dz_1 = \sigma_z^2,\ M\left(\frac{1}{1+d_1}\right) = D, \tag{7}$$

$$Mc_1 = 0,\ Md_1 = 0,\ Dc_1 = \sigma_c^2,\ Dd_1 = \sigma_d^2. \tag{8}$$

Let's set the sequences of

$$x_k = \theta z_k(1+c_k),\ y_k = z_k(1+d_k),\ k = 1, \ldots, n, \tag{9}$$

and build a sequence of random fractions

$$\frac{x_k}{y_k} = \frac{\theta(1+c_k)}{1+d_k},\ k = 1, \ldots, n. \tag{10}$$

The formulas (7) – (10) define a probabilistic model for a sequence of random fractions that depends on the parameter θ.

Let's now consider two estimates of the parameter θ.

$$\widehat{\theta}_1 = \frac{\widehat{x}_n}{\widehat{y}_n},\ \widehat{\theta}_2 = \frac{1}{n}\sum_{k=1}^{n}\frac{x_k}{y_k},\ \frac{1}{n}\sum_{k=1}^{n} x_k = \widehat{x}_n,\ \frac{1}{n}\sum_{k=1}^{n} y_k = \widehat{y}_n.$$

The estimate $\widehat{\theta}_1$ is the medianta, and the estimate $\widehat{\theta}_2$ is the average of a sequence of random fractions $\frac{x_k}{y_k}$, $k = 1, \ldots, n$. We will conduct a comparative analysis of these estimates.

Let's first examine the estimate $\widehat{\theta}_1$. Due to the strengthened law of large numbers (see, for example, [18]), the limit relations are valid $\widehat{x}_n \Rightarrow \theta Z$, $\widehat{y}_n \Rightarrow Z$, $n \to \infty$. Here the symbol $\Rightarrow$ indicates convergence with probability one. Thus, the limiting ratio is valid $\widehat{\theta}_1 \Rightarrow \frac{\theta Z}{Z} = \theta$ and so $\frac{\widehat{\theta}_1}{\theta} \Rightarrow 1$, $n \to \infty$. More accurate analysis of medianta behaviour may be constructed, using methods of [20].

Let's now go to the evaluation analysis of $\widehat{\theta}_2$. Due to the strengthened law of large numbers and independence of random variables c_k, d_k, with probability one, the limit relation is valid for $n \to \infty$:

$$\frac{\widehat{\theta}_2}{\theta} \Rightarrow M\frac{x_1}{y_1} = M\frac{\theta z_1(1+c_1)}{\theta z_1(1+d_1)} = M(1+c_1)M\left(\frac{1}{1+d_1}\right) = M\left(\frac{1}{1+d_1}\right) = D(a).$$

Assuming, for example, that the random variable d_1 has a uniform distribution on the segment $[-a, a]$, $0 < a < 1$, and calculate $D(a)$:

$$D(a) = \int_{-a}^{a} \frac{1}{1+u} \cdot \frac{du}{2a} = \frac{1}{2a} \ln \frac{1+a}{1-a}.$$

For $a \to 0$, the value $D(a) \to 1$, and for $a \to 1$: $D(a) \to \infty$. Thus, due to d_k, $k = 1, \ldots, n$, deviation $\frac{\widehat{\theta}_n^{(2)}}{\theta}$ from unit may increase sharply.

5 Conclusion

This article discusses three different problems of applied statistics, the solution of which is based on the properties of ergodicity, and on the construction of the medianta. These properties of statistical estimates are useful both in constructing the estimates themselves and in analyzing their compatibility with the estimated parameters. All the problems considered arise from various applications in optoelectronics, environmental, sociological, and biocenotic observations, and observations of flows in queuing systems, that model various technical systems.

References

1. Kulchin, Yu.N.: Photonics of self- organizing biomineral nanostructures. UPhN **181**(8), 891–896 (2011). (in Russian)
2. Monsik, V.B., Skrynnikov, A.A.: Evaluating the parameter of an exponential distribution for a truncated sample. Sci. Bull. MSTUCA. Ser. Appl. Math. Comput. Scie. **105**, 134–139 (2006). (in Russian)
3. Ilyasov, F.N.: Inverse sampling problem and motivation in the Forex market. Sociol. Res. **3**, 112–116 (2011). (in Russian)
4. Ilyasov, F.N.: Representativeness of survey results in marketing research. Soc. Res. **2**, 49–59 (2016). (in Russian)
5. Semkin, B.I.: On the invariance of average ratios of values (on the example of some morphological features of laminaria japonica Aresh layers) from the sublitoral of the Northern Primorye. Izvestiya TINRO **171**, 313–320 (2012). (in Russian)
6. Stoyan, D.: On some qualitative properties of the Boolean model of stochastic geometry. ZAMM **59**, 447–454 (1979)
7. Neumann, J.: Physical applications of the ergodic hypothesis. Proc. Natl. Acad. Sci. U.S.A. **18**, 263–266 (1932)
8. Khinchin, A.Ya.: Mathematical foundations of statistical mechanics. OGIZ, State Technical-Theoretical Publishing House, Moscow (1943). (in Russian)
9. Kornfeld, Ya., Sinai, I.P., et al.: Ergodic theory. Nauka, Moscow (1980). (in Russian)
10. Khinchin, A.Ya.: Chain fractions. Fizmatlit, Moscow (1961). (in Russian)
11. Editor-in-chief Vinogradov, I.M.: Mathematical encyclopedia, 3. Soviet encyclopedia, Moscow (1982). (in Russian)
12. Selton, G.A.: Automatic processing, storage and search of information. Sov. Radio, Moscow (1973). (in Russian)

13. Crane, M., Lemoine, O.: Introduction to the regenerative method for analysis of models. Science, Moscow (1982). (in Russian)
14. Schwartz, G.: Selective method. Guidance on the application of statistical methods of assessment. Statistics, Moscow (1978). (in Russian)
15. Semkin, B.I., Gorshkov, M.V., et al.: On changes in water content in annual shoots of coniferous woody plants in the temperate climate zone. Siberian Ecol. J. **15**, 537–544 (2008). (In Russian)
16. Semkin, B.I.: The elementary theory of similarity and its application in biology and geography. Multi-site measures of similarity and dissimilarity. Pattern Recogn. Image Anal. **25**(1), 1–9 (2015)
17. Klein, M.: Mathematics. Loss of certainty. World, Moscow (1984)
18. Borovkov, A.A.: Course in probability theory. Nauka, Moscow (1972). (in Russian)
19. Khinchin, A.Ya.: Works on the theory of mathematical queuing theory. Fizmatlit, Moscow (1963). (in Russian)
20. Kendall, M., Stewart, A.: Theory of distributions. Science, Moscow (1966). (in Russian)

Signature Detection and Identification Algorithm with CNN, Numpy and OpenCV

Zhanna S. Afanasyeva(✉) and Alexander D. Afanasyev

Irkutsk National Research Technical University, Irkutsk, Russia
a_zhanna2016@mail.ru, afanasiev-ad2009@yandex.ru

Abstract. The purpose of the article is to present a simple signature detection algorithm and its subsequent signature identification using a deep learning model for processing images based on a convolutional neural network. To solve the task of the image recognition, a binary classification has been performed to predict text or signature and signature classifications to determine the author of this signature. The proposed algorithm is interesting in the preliminary processing of scanned documents with signatures in order to extract the area with the signature and transfer it to the trained model. The research results are presented for documents of the same type, in which the signature is located in the same place. To select a specific element in the document we are using the tensor-slicing operations on Numpy arrays. To extract areas with text and signature, OpenCV tools are used. The results on the ready-made neural network model studies on a small dataset are presented in this article. Good results have been achieved in recognizing the famous writers' signatures. The proposed algorithm demonstrates the possibility of using the classical convolution network model for solving specific practical problems. The studies can be recommended to students in the study of neural networks to understand the basics of deep learning and apply a ready-made model as a template for solving practical problems in the field of computer vision.

Keywords: Image classification · Signature recognition · Deep learning

1 Introduction

Signature detection and identification are an actual practical task both for document work and for security services. The preliminary task that must be solved before identifying the signature is to extract the signature from a document and prepare it for recognition. The task of signature detection in a document is not trivial. One of the common ways to extract a signature on a document is to use the OpenCV computer vision library [1]. Signature recognizing task on a document is similar to the image segmentation task, for example, a face, any object, a barcode, etc. The segmentation algorithms are based on processing pixels belonging to the corresponding region, for example, marking pixels to obtain a mask for a segmented image (WaterShed algorithm), clustering pixels according to certain criteria (MeanShift algorithm), etc. [2].

The choice of the algorithm is determined based on a specific task. Moreover, the signature detection result on scanned documents of different types is not stable. For

R. Silhavy et al. (Eds.): CoMeSySo 2020, AISC 1295, pp. 467–479, 2020.
https://doi.org/10.1007/978-3-030-63319-6_43

example, in the case of documents with tables, some horizontal and vertical lines remain. For documents where the signature is located in different places, there is a problem with drawing the boundaries of the contours and regulating their number. In this article, we consider signature detection for documents of the same type - for the case when the signature is located at the same place in the document. This allows you to extract a document object which contains the signature to simplify the subsequent processing of this extracted object. The scanned document was previously processed using the Numpy library to extract the needed part of the document for subsequent processing it by OpenCV tools. When converting a document with a signature into an image, its processing and recognition can be reduced to the classical task of the computer vision – image recognition. This problem can be solved using a multilayer perceptron, a convolutional neural network [3, 4], autoencoders [5], etc.

Image processing using convolutional neural networks as a type of deep learning model is considered to be the best approach. Since 2015, rapid progress has been observed in the field of artificial intelligence. Including the availability of the Keras framework, a breakthrough in the development of deep learning technologies was made. As noted by Keras developer François Chollet, the creation of deep learning models has become similar to the LEGO constructor, where it is important to efficiently assemble compatible layers into a pipeline for data processing [3]. The Keras library is quite simple to use and a large number of ready-made models have already been created on its basis that allow the use of neural networks for solving real-world problems.

In this work, we have conducted research how to use "Keras CNN Dog or Cat Classification" convolutional neural network model [6] as a template for solving signature and text classification task, authors of signature classification and its subsequent recognition on an extracted part of a scanned document.

We have developed the signature detection algorithm using Numpy and OpenCV and subsequent signature identification using deep learning models for image processing based on a convolutional neural network.

2 Methodology

2.1 Signature Detection on a Scanned Document and Subsequent Identification Algorithm

To solve the recognizing signature detection problem on a scanned document and subsequent identification algorithm was developed as follows:

1. Prepare data for two types of image classification:
    - for text and signature classification: training and test samples with text and signature;
    - for authors' signatures classification: training and test samples with signatures of the respective authors.
2. Apply a model «Keras CNN Dog or Cat Classification» for training on data to classify text and signature.

3. Test the trained model on test data (text and signature).
4. Apply a model «Keras CNN Dog or Cat Classification» for training on data to classify authors' signatures.
5. Do the tensor-slicing operations on the Numpy array to get a part of a scanned document with a signature.
6. Extract areas of text and signature on the resulting part of the scanned document in the form of rectangles using OpenCV.
7. Apply the trained model for text and signature classification to the extracted text and signature parts to save parts with signatures in a special directory for subsequent signature author recognition.
8. Apply a trained model for authors' signature classifying to identify a signature.

2.2 Data Preparation

The preparation data task for training and testing a neural network model is as important as building an architecture model.

A dataset for text and signature classification has been created. It is made up of 265 unique text and signatures images, respectively. An open dataset «CDAR2013 – Handwriting Stroke Recovery from Offline Data» has been used to form the training and test datasets with signatures. This dataset contains 200 writers' signatures images [7].

The text dataset is an image containing 3–8 words. This dataset has been created by extracting a part of text from an arbitrary document using a slice between two given indices along each tensor axis. Previously, the text document was converted to an array Numpy. Figures 1 and 2 show random images from datasets.

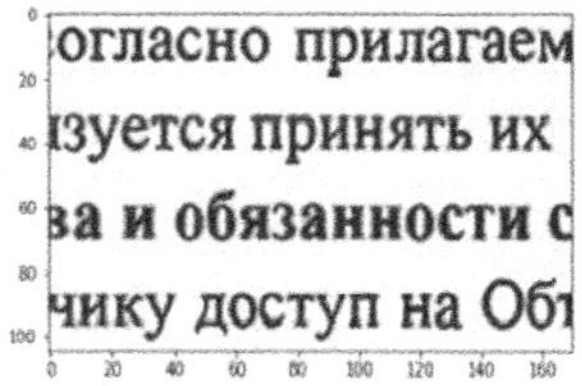

Fig. 1. A random text image.

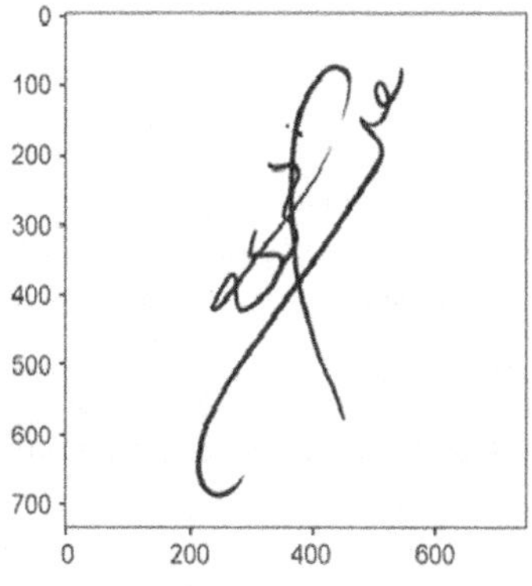

Fig. 2. A random signature image.

The task of authors' signatures classification has been solved for three authors. A dataset containing 252 signature images of three authors has been created. Initial images have also been taken from a dataset «CDAR2013 – Handwriting Stroke Recovery from Offline Data». This dataset consists of 6 unique signatures of each author. The training dataset has been formed from 4 signatures of each of the three authors through its augmentation. The test dataset is made up of 2 remaining signature samples of each author – a total of 6 signatures.

It should be noted that when training the model, the 'accuracy' metric has been used, so special attention is drawn to the balance of datasets (Fig. 3). We split the data into a training set and validation set automatically using the train_test_split () function of the sklearn.model_selection module. A validation set has been created by setting apart 20% from the original training data. The balance of the sets has been controlled by the random_state parameter:

```
train_df, validate_df = train_test_split(df, test_size = 0.20, random_
state = 47)
```

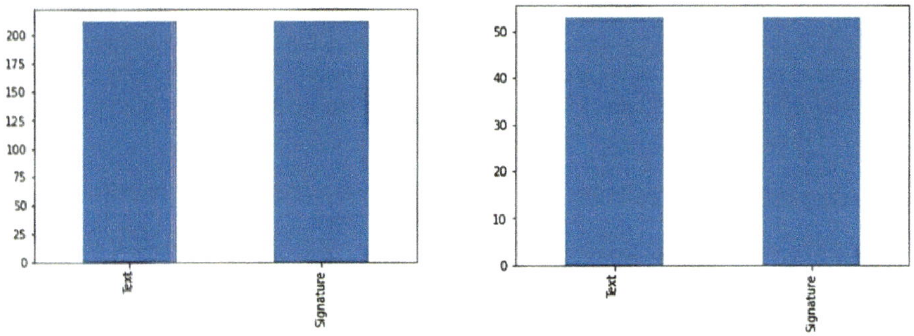

(a) (b)

Fig. 3. The distribution of the dataset on the training (a) and validation (b) data for the text and signature classification.

2.3 Model Training

We use a technique to artificially expand the size of a training dataset by generating existing training samples during the model training with ImageDataGenerator module, the so-called data augmentation [3]. It should be noted that for a small dataset it is necessary to carefully select parameters for random transformation image to be closer to the reality.

The built-in keras.callbacks module (EarlyStopping and ReduceLROnPlateau) has been used during the model training to get rid of overfitting. We have added a callback ModelCheckpoint [8].

Training stops if the val_loss on validation set or the target metric val_acc stops improving. As a result, the callbacks to the keras.callbacks module have been used in the training model such as follows:

- EarlyStopping (early stop) to interrupt training when the validation loss val_loss improvement stops for 3 epochs;
- ReduceLROnPlateau - dynamic adjustment of the optimizer's training step—reducing the learning rate when val_acc improvement stops for 2 epochs;
- ModelCheckpoint - saving the current model weights during training which allows saving the best model.

Overfitting has already been observed during the first epochs for signature-classification model (Fig. 4). As a result, the value of validation accuracy has reached 100% faster than training accuracy. Due to the callback ModelCheckpoint, the best result has been saved. However, this result is far from reality. It has been confirmed on test data - most of the images were not recognized correctly. For example, the spread in the value of 'accuracy' ranged from 0.74 to 0.89, for 'val_accuracy' - from 0.48 to 1.0.

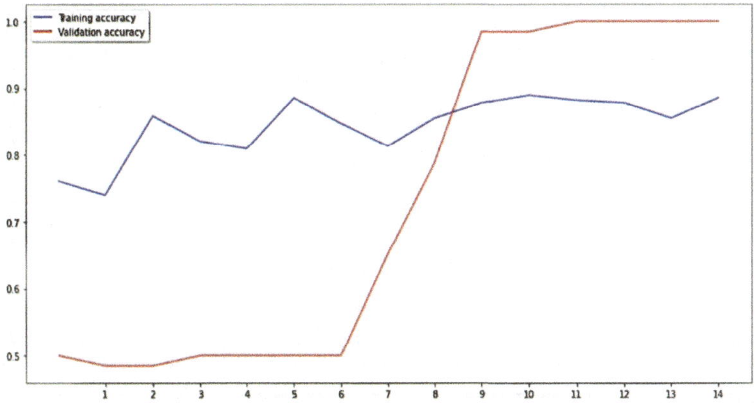

Fig. 4. The signature-classification model Overfitting.

In addition, the callback EarlyStopping stops training if for more than 3 epochs the accuracy on the test sample does not improve. As a result, the training stops without reaching acceptable accuracy. Accuracy does not improve even without the callback EarlyStopping - it does not rise above 40%.

The main reason of overfitting, in our view, is the small dataset. However, overfitting has been overcome by simplifying the signature-classification model and optimal the number of epochs. The architecture of the signature-classification model is shown in Fig. 5. The BatchNormalization() and Dropout() layers were deleted from the original model.

Model training for text and signature classification have been performed for 15 epochs (Fig. 6). We can see, the training accuracy and the validation accuracy are such as follows:

- 'accuracy': [0.9413203001022339, … 1.0];
- 'val_accuracy': [0.4952380955219269, … 0.9904761910438538]

The test accuracy is 97.9%.

Model training for signature classification has been performed for 5 epochs (Fig. 7). In this case, the training accuracy and the validation accuracy are such as follows:

- 'accuracy': [0.5771144032478333, … 1.0];
- 'val_accuracy': [0.9607843160629272, … 1.0]

All test data have been recognized correctly (Fig. 8).

```
Model: "sequential_1"
_________________________________________________________________
Layer (type)                 Output Shape              Param #
=================================================================
conv2d_3 (Conv2D)            (None, 126, 126, 32)      896
_________________________________________________________________
max_pooling2d_3 (MaxPooling2 (None, 63, 63, 32)        0
_________________________________________________________________
conv2d_4 (Conv2D)            (None, 61, 61, 64)        18496
_________________________________________________________________
max_pooling2d_4 (MaxPooling2 (None, 30, 30, 64)        0
_________________________________________________________________
conv2d_5 (Conv2D)            (None, 28, 28, 128)       73856
_________________________________________________________________
max_pooling2d_5 (MaxPooling2 (None, 14, 14, 128)       0
_________________________________________________________________
flatten_1 (Flatten)          (None, 25088)             0
_________________________________________________________________
dense_2 (Dense)              (None, 512)               12845568
_________________________________________________________________
dense_3 (Dense)              (None, 3)                 1539
=================================================================
Total params: 12,940,355
Trainable params: 12,940,355
Non-trainable params: 0
```

Fig. 5. The architecture of signature-classification model.

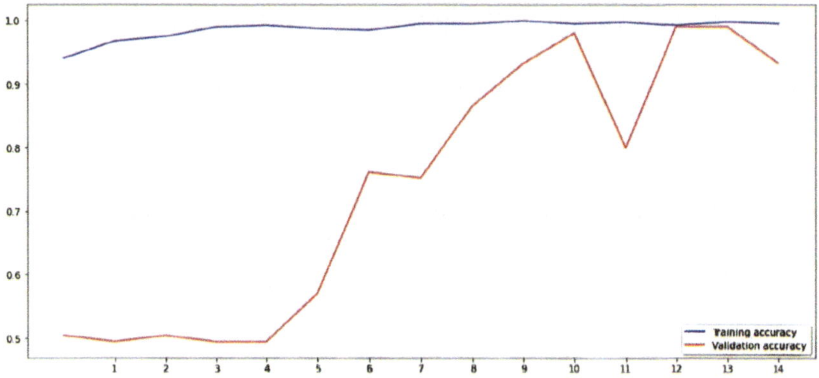

Fig. 6. Training and validation accuracy on text and signature.

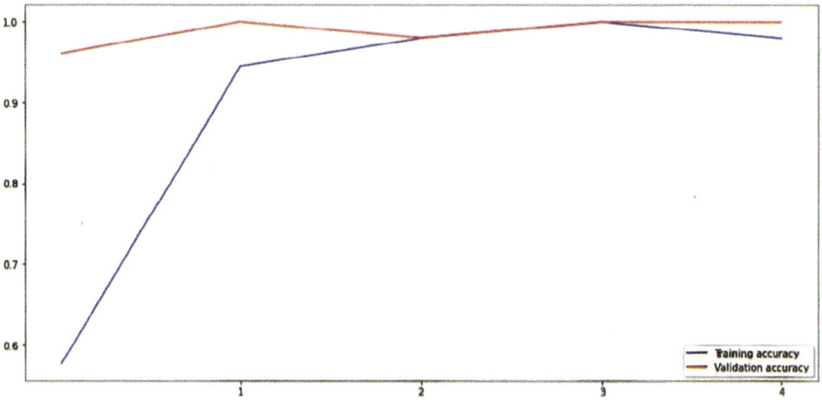

Fig. 7. Training and validation accuracy on authors' signature.

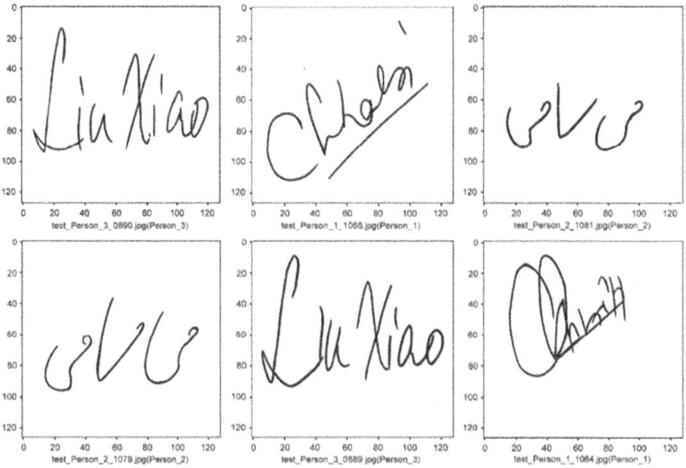

Fig. 8. Identification of authors' signatures on test data.

3 Image Classification and Signature Identification Results

Samples with text and signatures extracted from the scanned document have been identified by the models correctly both in the case of text and signature classification as well as authors' signature classification.

It should be noted that text and signature classification model have been trained on parts of text in Russian. At the same time the trained model successfully recognizes not only images with text in Russian but also samples with text in English (Fig. 9). This confirms that the model has achieved a sufficient degree of generalization regarding the style of signature and text. The authors of the signature on the extracted parts of the scanned document have also been identified correctly.

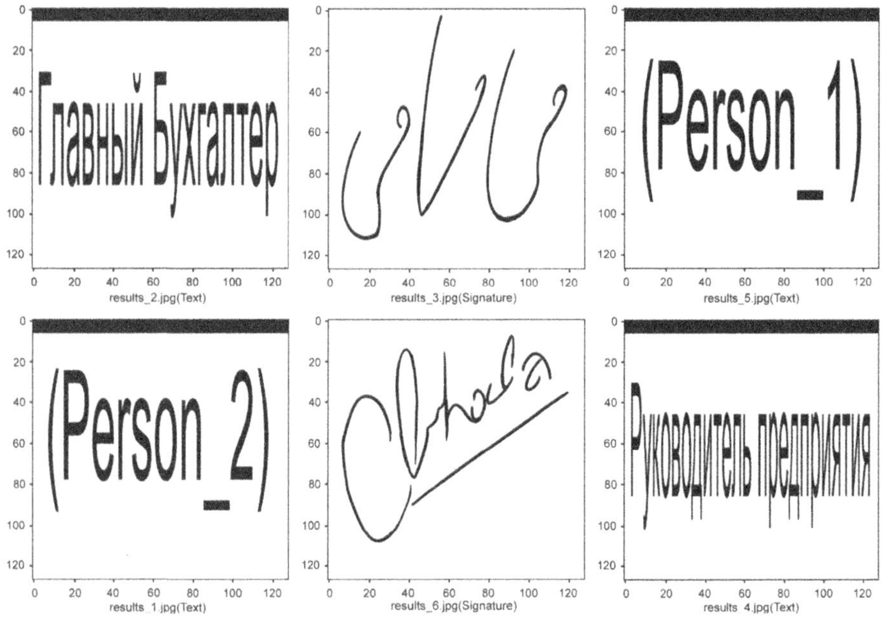

Fig. 9. Results of recognition of text and signature parts.

4 Discussion

Signature recognition and identification have been performed using the «Keras CNN Dog or Cat Classification» model based on CNN layers, tensor manipulation using the Numpy library and OpenCV tools.

4.1 Application of the «Keras CNN Dog or Cat Classification»

The "Keras CNN Dog or Cat Classification" model is published on the «Dogs vs. Cats» competition Kaggle for predicting the image of a dog (Fig. 10). This model is as a template for text and signature recognition in our research.

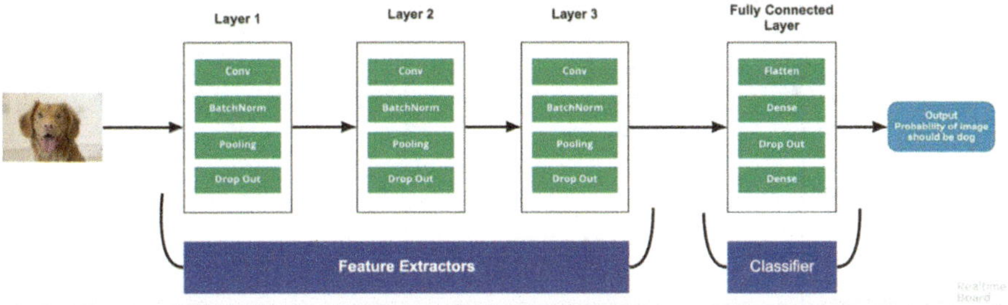

Fig. 10. CNN model for the cats and dogs classification [6].

The presented model is a linear stack of layers and is created using the Sequential class.

The neural network model consists of a convolutional base and a fully connected classifier. Feature extraction is done using the convolutional basis of the model. The convolutional neural network is a stack of alternated Conv2D (with relu activation) and MaxPooling2D layers. The BatchNormalization layer follows after each Conv2D convolutional layer. The Dropout layer is used to regularize the neural network model. The number of filters for each convolutional layer doubles, respectively: 1st layer - 32, 2nd layer - 64, 3rd layer - 128.

Such a neural network architecture is one of the most popular for solving computer vision problems. A similar model is considered as a practical example for the cats and dogs classification by F. Chollet [3]. In addition, the well-known VGG16 model has a similar convolution base [9].

This neural network architecture has been effective on small datasets. However, the model quickly reaches overfitting. Research has shown that the results of the image classification greatly depend on the dataset size, the quality of the source images, the options of image random transformations by augmentation, and the model capacity.

4.2 Manipulating Tensors in Numpy

For documents of the same type, the task of extracting a part of a document with a signature is greatly simplified. As a rule, signatures are located either at the beginning or at the end of a document, then using slices so you can extract a part of the document with the intended signature location. Then the extracted part is processed by OpenCV tools. In the case of the standard signature location, it is possible to choose the optimal size of the area for its subsequent processing, which has been demonstrated in this article. Selecting a specific part of the scanned document has been performed by manipulating tensors in Numpy. The scanned image is converted to a Numpy array. Next, the tensor-slicing operations have performed on a Numpy array to extract the specific document part (Fig. 11 and 12).

```
plt.imshow(Image.fromarray(img_array).convert('RGB'))
plt.show()
```

Fig. 11. The initial image of the scanned document.

```
slice_part = img_array[560:2500, 15:] # slice
plt.imshow(slice_part, plt.cm.binary)
plt.show()
```

Fig. 12. A slice of a part of a scanned document.

4.3 OpenCV Image Processing Tools

To extract areas with text and signature, in our case, the signature detection problem on the scanned document has been solved. Extracting text and signature areas have included two stages: preliminary image filtering and selecting rectangular contours of the signature and text [10].

Image Filtering Step
A Gaussian kernel has been used as a filter. Its configuration has been customized by the cv2.GaussianBlur() function. The text and signature areas have been defined using gradient operations. Then, unwanted noise has been reduced to focus on the text and signature areas. The main steps for pre-filtering a scanned document are listed in Table 1. It should be noted that the setting of the image rectangle to select signature and text objects is well regulated by the getStructuringElement() function parameters.

Table 1. Image Filtering step.

Action	Function
Gaussian Filtering	blur = cv2.GaussianBlur(image, (11,11),0)
Morphological transformations tuning with Erosion and Dilation operators	kernel = cv2. getStructuringElement(cv2. MORPH_CROSS,(8,8))
Morphological Gradient	grad = cv2.morphologyEx(blur, cv2. MORPH_GRADIENT, kernel)
Binarization using Threshold	_, thresh = cv2.threshold(grad, 0, 255, cv2.THRESH_BINARY + cv2. THRESH_OTSU)
Additional use of Dilation to join broken parts of an object after filters	dilated = cv2.dilate(thresh, kernel, iterations = 2)
Cross-shaped Kernel tuning	kernel = cv2. getStructuringElement(cv2. MORPH_CROSS,(7,7))
Morphological operation CLOSE	connected = cv2.morphologyEx (thresh,cv2.MORPH_CLOSE,kernel)

The Selecting Rectangular Contours Step
Signature and text contours have been searched by the cv2.findContours() function. The found contours areas have been analyzed. The cv2.boundingRect () function has been used to draw rectangular contours of the image. The extracted signature and text parts have were recorded in a specially created directory. The image processing results by OpenCV tools are shown in Figs. 13, 14, 15 and 16.

Fig. 13. Initial image.

Fig. 14. Binarization using threshold.

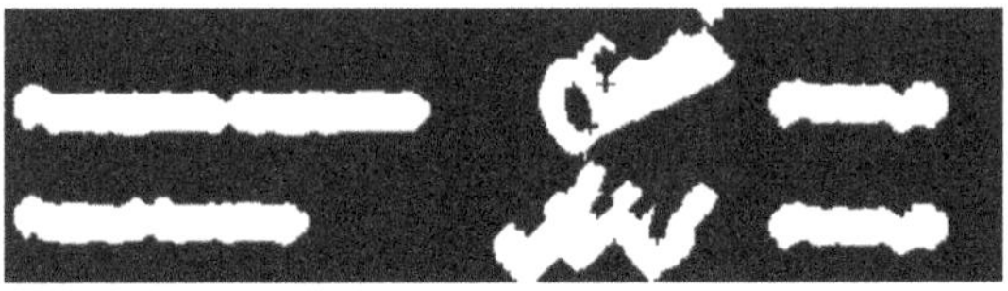

Fig. 15. CLOSE filtering result.

Fig. 16. Extracted text and signature parts from a scanned document.

5 Conclusion

The proposed algorithm is a simple solution to detect and identify signatures in the same type documents and counting the number of signatures on a document page.

Manipulating tensors of scanned image array in the Numpy and using the Open CV computer vision library tools allow to get a stable result for extracting a signature from a standard document. The building deep-learning model requires a thorough analysis to solve the problem of the image recognition on a small dataset.

The studies on the use of a ready-made model for the classification problem include all the basic steps for applying deep learning to solve real-world problems, such as follows: preparing the data, including augmentation; image preprocessing for training model; building optimal network; evaluating a model on validation and test data; the processing of fighting overfitting. These studies can be recommended to students in the study of neural networks to understand the basics of deep learning and apply a ready-made model as a template to solve real-world problems of computer vision.

Acknowledgments. We are grateful to Oleg Konorev for support in the algorithm development.

References

1. The computer vision library OpenCV. https://opencv.org. Accessed 25 Dec 2019
2. Song, Y., Yan, H.: Image segmentation techniques overview. In: 2017 Asia Modelling Symposium (AMS), Kota Kinabalu, pp. 103–107 (2017). https://doi.org/10.1109/ams.2017.24
3. Chollet, F.: Deep Learning with Python. Manning, Shelter Island (2017)
4. Rai, R.D., Lather, J.S.: Handwritten signature verification using TensorFlow. In: 2018 3rd IEEE International Conference on Recent Trends in Electronics, Information & Communication Technology (RTEICT), Bangalore, India, pp. 2012–2015 (2018). https://doi.org/10.1109/rteict42901.2018.9012273
5. Sasirekha, K., Ravikumar, R., Thangavel, K.: Online signature denoising using deep autoencoder. Int. J. Comput. Intell. Inform. **7**(1) (2017). https://pdfs.semanticscholar.org/2d28/de72efb60fa73ae4bb7becf85cb89a201c2a.pdf. Accessed 02 Nov 2019
6. Keras CNN Dog or Cat Classification. https://www.kaggle.com/uysimty/keras-cnn-dog-or-cat-classification/notebook#Import-Library. Accessed 02 Nov 2019
7. CDAR2013, Handwriting Stroke Recovery from Offline Data. https://www.kaggle.com/c/icdar2013-stroke-recovery-from-offline-data/data. Accessed 02 Nov 2019
8. Keras documentation. https://keras.io. Accessed 02 Nov 2019
9. Keras Applications. VGG16. https://keras.io/api/applications/vgg/#vgg16-function. Accessed 02 Nov 2019
10. Build software better, together. https://github.com/topics/signature-recognition. Accessed 02 Nov 2019

Software for Structure Selection of an Artificial Neural Network to Control the Induction Soldering Process

Anton Milov[1(✉)], Vadim Tynchenko[1,2], Vladimir Bukhtoyarov[1,2], Valeriya Tynchenko[1,2], and Vladislav Kukartsev[1,2]

[1] Reshetnev Siberian State University of Science and Technology, Krasnoyarsk, Russia
antnraven@ieee.org
[2] Siberian Federal University, Krasnoyarsk, Russia

Abstract. This article presents algorithmic and software for determining the optimal structure of an artificial neural network, designed to solve the problem of controlling the technological process of induction soldering of spacecraft's thin-walled aluminum waveguide paths. The algorithm is designed for experimental determination of the optimal structure of an artificial neural network designed to solve the task of controlling the induction soldering process. The result of the algorithm is a trained neural network model with the best structure. The software product itself is an application for Windows operating system. To implement a software system the Python programming language was chosen. The Pandas, Matplotlib, Keras libraries, and PyQt framework were also used as tools for building a graphical user interface. The software application represents an automated control system module for induction soldering automated system. Experimental verification of the module showed that it allows training the neural network models with the best structure to ensure high quality control of the induction soldering process. With this, the process of finding the best neural network structure is greatly facilitated.

Keywords: Induction soldering · Artificial neural networks · Neural network structure · Automation

1 Introduction

One of the most common and effective methods of creating permanent joints within the framework of the technological process for the production of spacecraft's waveguide paths is the method based on induction heating - induction soldering [1]. The greatest negative impact in the framework of this technological process is exerted by regulatory and non-normative errors of measuring instruments [2]. In the framework of the above works, control was carried out only along one circuit - according to the temperature of the product. One way to improve the control quality of this technological process is the introduction of another control loop - positioning the workpiece in the inductor window [3]. However, the introduction of an additional control loop is not enough to solve the

R. Silhavy et al. (Eds.): CoMeSySo 2020, AISC 1295, pp. 480–490, 2020.
https://doi.org/10.1007/978-3-030-63319-6_44

above problems associated with the influence of non-standard errors on the quality of technological process control. On the other hand, intelligent methods such as artificial neural networks (ANN), fuzzy controller, neuro-fuzzy controller are well suited to reduce the impact of non-standard errors. As previous studies [4] show, these intelligent methods are quite effective for detecting and correcting non-standard errors of measuring instruments. Also, the use of intelligent methods can reduce the complexity of the technological process initial setup, which also has a significant impact on the quality of the process control [5].

Previous studies [5, 6] have clearly shown that artificial neural networks are the most effective intelligent method for controlling the technological process of induction soldering. In this case, the greatest difficulty is the choice of the artificial neural network optimal structure for solving the stated problem [7–12]. Moreover, the choice of the artificial neural network structure together with the choice of the most suitable training method make the most impact on the quality of the neural network model.

As such, a universal algorithm for determining the optimal structure of an artificial neural network for solving any problem does not exist [13–18]. However, the practical application of artificial neural networks for various tasks, including the tasks of process control, involves determining the optimal structure that will allow us to solve the problem with the highest accuracy.

As part of the problem solution of the induction soldering control, the ANN optimal structure was determined manually before through experimental studies. This paper presents the development of algorithmic and software for automating the determination of the optimal structure of an artificial neural network to solve a specific problem - controlling the technological process of induction soldering of thin-walled aluminum waveguide paths.

The software product being developed is a module of an existing automated control system (ASC) for the induction soldering of waveguide paths ACS "Soldering". The algorithm implements experimental studies of the effectiveness of artificial neural networks with different structures. The result of the module is a trained neural network model for solving the induction soldering control problem.

Using this software module will significantly facilitate the task of finding the optimal structure of an artificial neural network to solve the problem of controlling the technological process of aluminum waveguide paths induction soldering.

2 Structure and Algorithms of the Developed Software

In general, the software product is an application for Windows operating systems. The performance has been reliably confirmed on versions of Windows 7, Windows 8, and Windows 10. Since the program module uses libraries and a framework available on various platforms, it is possible to build the module under linux-based and unix-based operating systems. Python is chosen as the programming language [19, 20], since many libraries have been implemented for the Python language that facilitate working with machine learn-ing tools such as Pandas [21] and Keras [22]. For visualization, the Matplotlib library is used [23]. To implement the graphical user interface, the PyQt framework is used [24]. As a development methodology, object-oriented programming was used [25–27].

The purpose of the software product is to automate the selection of the optimal structure of an artificial neural network to solve the problem of controlling the technological process of induction soldering of spacecraft's thin-walled aluminum waveguide paths. The module allows conducting experimental studies of the artificial neural networks' effectiveness with various structures, both in manual and automated mode. The module is supposed to be used as part of a general information system for controlling the technological process of induction soldering.

Figure 1 shows the structure of the software module.

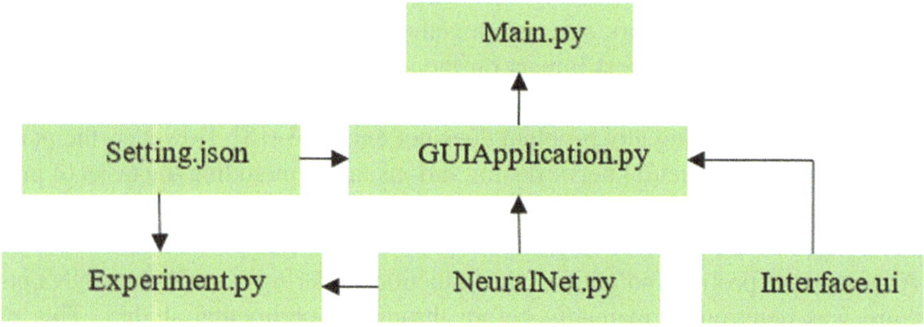

Fig. 1. Block diagram of a software product.

Structurally, the module consists of the following elements:

- Main module - a class responsible for the logic of the software product.
- Class GUI Application - a class that implements the processing of the graphical user interface elements.
- The NeuralNet class - a class responsible for building and training neural network models.
- Class Experiment - a class that implements the logic of determining the optimal structure of an artificial neural network.
- Interface class - a class containing graphical interface elements.
- Setting.json - configuration file for experimental research.

Figure 2 shows a block diagram of the software product functioning.

At the initial stage, all application objects are initialized. Next, the user is given a choice of the application operating mode. The choice of the optimal structure is possible both in manual and automated mode.

When choosing the manual mode for determining the optimal structure, the user must manually set the parameters of the neural network model in the elements of the graphical interface each time. The result of the model training is saved in a file that is downloaded to disk. Then, based on the obtained results, the user himself must choose the most suitable neural network model.

If the user has chosen an automated mode for determining the optimal structure of an artificial neural network, then the settings are taken from the configuration file. The

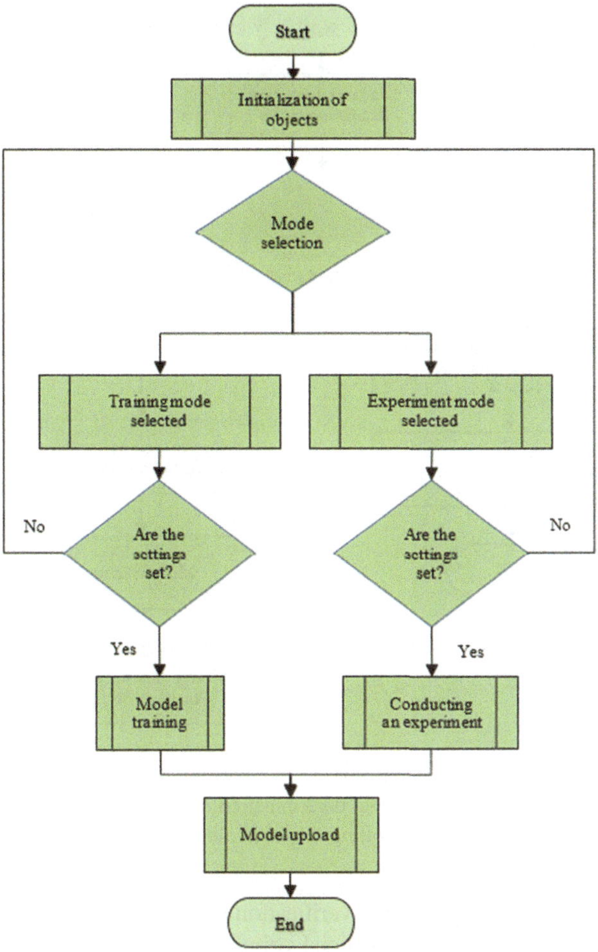

Fig. 2. The block diagram of the software module functioning.

configuration file is specified in the JSON format, in which settings for a series of experiments are set, such as: the number of training epochs, the number of hidden layers, the number of neurons on each layer, activation and regularization functions, etc.

Figure 3 shows a flowchart of the experimental determination process of the artificial neural network optimal structure to solve the problem of controlling the technological process of induction soldering.

As can be seen from the flowchart, the process of conducting experiments is to obtain experimental data from a configuration file, followed by training models from the input experimental data. The results of the experiments are recorded in a file for their further analysis. Upon completion of the experiments, a model with the best recognition quality is selected for uploading.

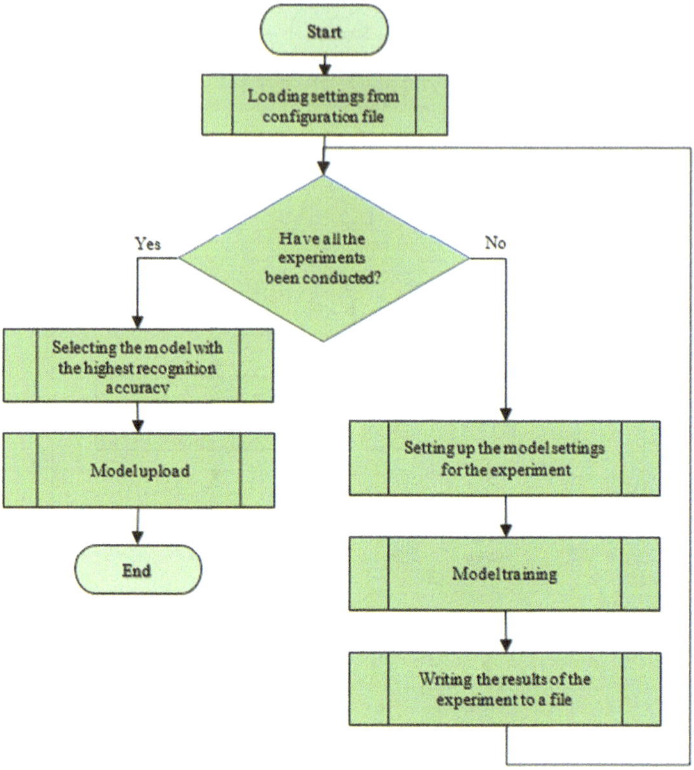

Fig. 3. The flowchart of the experimental determination of the artificial neural network optimal structure to solve the problem of controlling the technological process of induction soldering.

The statement of the experimental verification problem is a training and assessment of the recognition accuracy of an artificial neural network. The inputs of an artificial neural network are:

- E_t - temperature difference between the soldered elements.
- E_m - difference in the heating rate of the soldered elements.
- A_{pr} - previous algorithm for controlling the heating of the product.
- K_1 and K_2 - coefficients of the previous algorithm.

The outputs of this artificial neural network are:

- Dir - direction of movement of the workpiece.
- A_t - heating control algorithm of the product.
- $K_{t1} \ldots K_{t2}$ - coefficients of the heating control algorithm.
- A_m - displacement control algorithm.
- $K_{m1} \ldots K_{m2}$ - coefficients of the displacement control algorithm.

Settings for hidden layers are set in the configuration file as experiment settings. The structure of an artificial neural network is shown in Fig. 4.

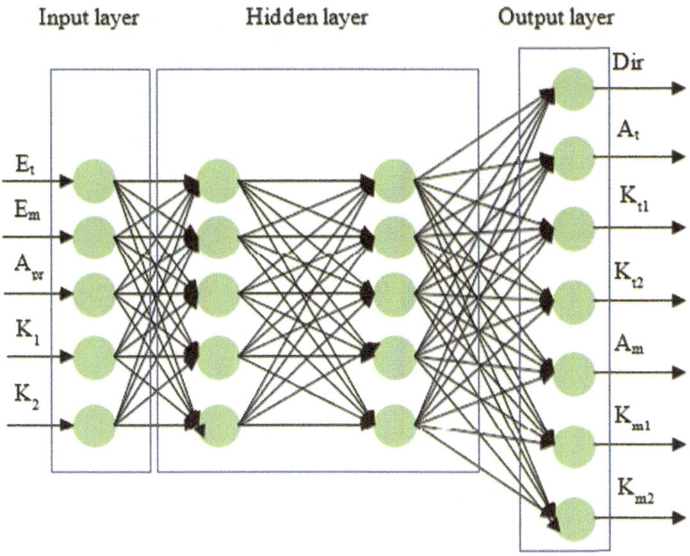

Fig. 4. Structure of an artificial neural network.

The neurons of the input and output layers are the same for all experiments. The hidden layer neurons are configured according to the configuration file.

3 Description of the Proposed Software System

The software system provides a graphical interface for manual and automated selection of the optimal structure of an artificial neural network to solve the problem of controlling the technological process of induction soldering of spacecraft's thin-walled aluminum waveguide paths.

Figure 5 shows the screen form of the main application window. From this figure it is seen that the main window of the software module can be divided into 5 zones:

- Zone 1 - settings for working in manual mode.
- Zone 2 - settings for working in automated mode.
- Zone 3 - paths to the files of training and test samples, which are used in any mode.
- Zone 4 - main active buttons, responsible for starting the manual mode, automated mode and uploading the model.
- Zone 5 - logging the learning process.

If you select a manual mode of operation after training an artificial neural network, the graphs of the training error and recognition accuracy for the current model will be displayed, examples of which can be seen in Fig. 6 and Fig. 7, respectively.

If the automated mode is selected, the graph output will be omitted, while only the model with the optimal structure of the artificial neural network will be saved. The model is uploaded in the model.h5 file.

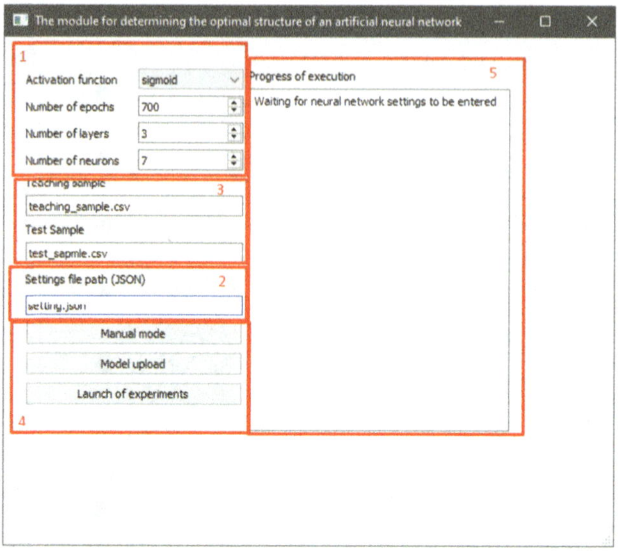

Fig. 5. The main window of the software module.

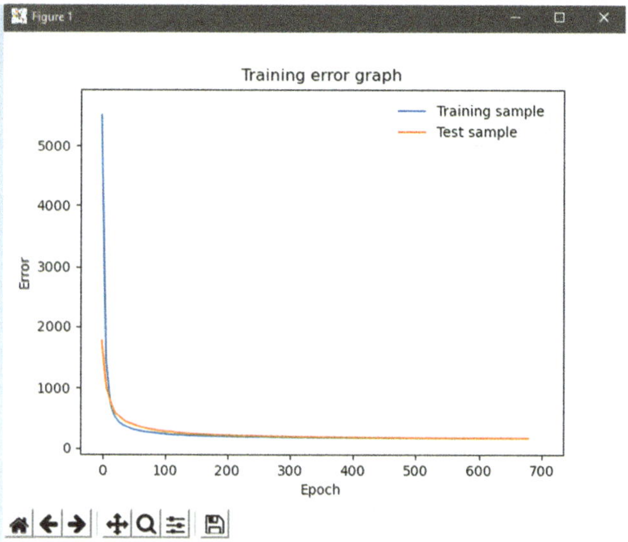

Fig. 6. A window with a training error graph of the neural network model.

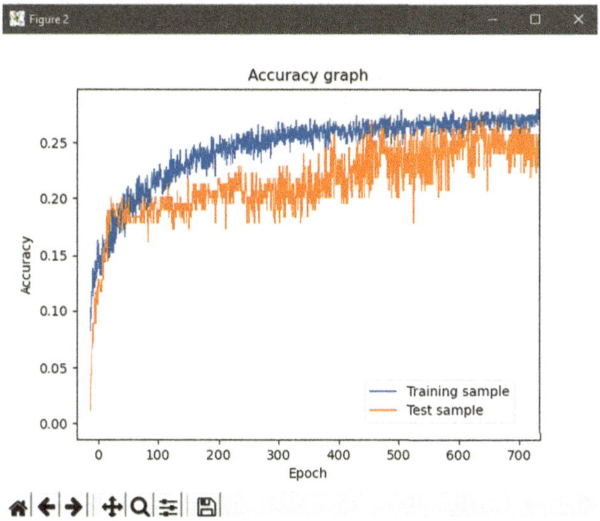

Fig. 7. A window with a recognition accuracy graph of the neural network model.

4 Experimental Study of an Automated System Efficiency

The effectiveness of the software module was verified on the basis of data from real technological processes of induction soldering of thin-walled aluminum waveguide paths.

Software verification was carried out in 10 series of experiments with 30 experiments each. The results of experimental studies are presented in Table 1 and in Fig. 8.

Table 1. Results of the experimental study.

Series of experiments	Assembly size	Layers × neurons	Accuracy
Series #1 (manual)	22 × 11 mm	3 × 3	81%
Series #2 (automated)	22 × 11 mm	5 × 5	96%
Series #3(automated)	22 × 11 mm	7 × 7	93%
Series #4 (manual)	19 × 9,5 mm	3 × 3	82%
Series #5 (automated)	19 × 9,5 mm	5 × 5	94%
Series #6 (automated)	19 × 9,5 mm	7 × 7	92%
Series #7 (manual)	35 × 15 mm	3 × 3	80%
Series #8 (automated)	35 × 15 mm	5 × 5	94%
Series #9 (automated)	35 × 15 mm	7 × 7	92%

The results of an experimental study show that the developed software module for determining the optimal structure of an artificial neural network allows us to find the best structure of a neural network model in both manual and automated mode.

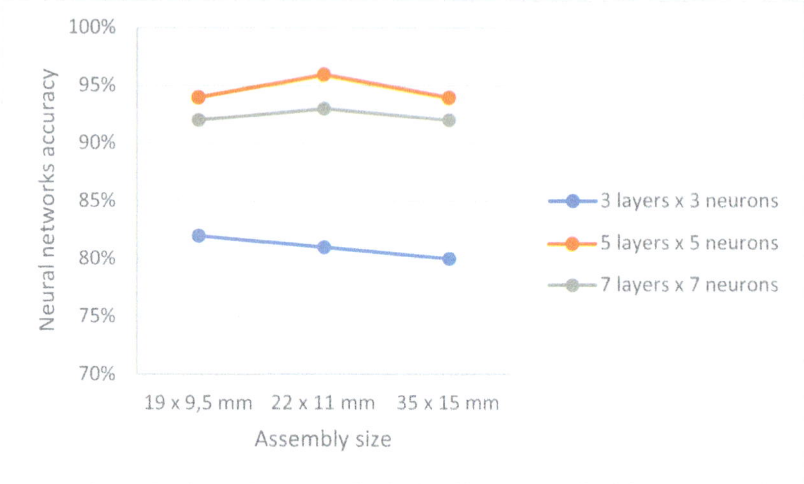

Fig. 8. Experimental study of the neural networks' accuracy.

In the framework of the current problem statement, the model with 5 hidden layers and 5 neurons per layer turned out to be the best structure of an artificial neural network, since models with a large number of layers reach the state of retraining on the available set of training data, and smaller structures cannot learn enough. A larger amount of training data will allow us to build better models with more difficult structures.

5 Conclusion

This paper presents the algorithmic and software for determining the optimal structure of an artificial neural network to solve the problem of controlling the technological process of induction soldering of spacecraft's thin-walled aluminum waveguide paths.

The software product is an application for the Windows operating system, implemented in the Python programming language from libraries for working with machine learning tools such as Pandas and Keras.

The result of the software product is a trained neural network model with the best structure, ready for implementation in the main induction soldering process control system.

An experimental verification of the module efficiency showed that it allows training neural network models with the best structure to ensure high quality control of the induction soldering process.

Acknowledgments. This research was financially supported by Reshetnev Siberian State University of Science and Technology project "Mathematical and software of the induction solder-ing of spacecraft's waveguide paths".

References

1. Murygin, A.V., Tynchenko, V.S., Laptenok, V.D., Emilova, O.A., Bocharov, A.N.: Complex of automated equipment and technologies for waveguides soldering using induction heating. In: IOP Conference Series: Materials Science and Engineering, vol. 173, no. 1, p. 012023 (2017)
2. Milov, A.V., Tynchenko, V.S., Murygin, A.V.: Experimental verification of flux effect on process of aluminium waveguide paths induction soldering. In: International Russian Automation Conference, pp. 282–291. Springer, Cham (2019)
3. Tynchenko, V.S., Murygin, A.V., Petrenko, V.E., Seregin, Y.N., Emilova, O.A.: A control algorithm for waveguide path induction soldering with product positioning. In: IOP Conference Series: Materials Science and Engineering, vol. 255, no. 1, p. 012018 (2017)
4. Milov, A.V., Tynchenko, V.S., Kukartsev, V.V., Tynchenko, V.V., Bukhtoyarov, V.V.: Use of artificial neural networks to correct non-standard errors of measuring instruments when creating integral joints. J. Phys: Conf. Ser. **1118**(1), 012037 (2018)
5. Tynchenko, V.S., Milov, A.V., Tynchenko, V.V., Bukhtoyarov, V.V., Kukartsev, V.V.: Intellectualizing the process of waveguide tracks induction soldering for spacecrafts. Int. Rev. Aerosp. Eng. **12**(6), 280–289 (2019)
6. Milov, A., Tynchenko, V., Petrenko, V.: Intellectual control of induction soldering process using neuro-fuzzy controller. In: 2019 International Russian Automation Conference (RusAutoCon), pp. 1–6. IEEE Press, New York (2019)
7. Khosravi, A., Nahavandi, S., Creighton, D.: A prediction interval-based approach to determine optimal structures of neural network metamodels. Expert Syst. Appl. **37**(3), 2377–2387 (2010)
8. Xu, S., Chen, L.: A novel approach for determining the optimal number of hidden layer neurons for FNN's and its application in data mining. In: 5th International Conference on Information Technology and Applications, pp. 683–686. University of Tasmania, Cairns (2008)
9. Sexton, R.S., Dorsey, R.E., Sikander, N.A.: Simultaneous optimization of neural network function and architecture algorithm. Decis. Support Syst. **36**(3), 283–296 (2004)
10. Murata, N., Yoshizawa, S., Amari, S.: Network information criterion-determining the number of hidden units for an artificial neural network model. IEEE Trans. Neural Netw. **5** (6), 865–872 (1994)
11. Kavzoglu, T.: Determining optimum structure for artificial neural networks. In: 25th Annual Technical Conference and Exhibition of the Remote Sensing Society, pp. 675–682. Remote Sensing Society Nottingham, Cardiff (1999)
12. Henrique, H.M., Lima, E.L., Seborg, D.E.: Model structure determination in neural network models. Chem. Eng. Sci. **55**(22), 5457–5469 (2000)
13. Goodfellow, I., Bengio, Y., Courville, A.: Deep Learning. MIT press, Cambridge (2016)
14. Buduma, N., Locascio, N.: Fundamentals of Deep Learning: Designing Next-Generation Machine Intelligence Algorithms. O'Reilly Media, Sebastopol (2017)
15. Livingstone, D.J. (ed.): Artificial Neural Networks: Methods and Applications. Humana Press, Totowa (2009)
16. Suzuki, K. (ed.): Artificial Neural Networks: Architectures and Applications. IntechOpen, London (2013)
17. Levine, D.S.: Introduction to Neural and Cognitive Modeling. Routledge, Abingdon (2018)
18. Rutkovskaya, D., Pilinsky, M., Rutkovsky, L.: Neural Networks, Genetic Algorithms and Fuzzy Systems. Telekom, Moscow (2006)
19. Lutz, M.: Programming Python. O'Reilly Media, Sebastopol (2001)

20. Van Rossum, G.: Python programming language. USENIX Ann. Tech. Conf. **41**, 36 (2007)
21. McKinney, W.: Pandas: a foundational Python library for data analysis and statistics. Python High Perf. Sci. Comput. **14**(9), 1–9 (2011)
22. Ketkar, N.: Introduction to Keras. Deep learning with Python, pp. 97–111. Apress, Berkeley (2017)
23. Tosi, S.: Matplotlib for Python Developers. Packt Publishing Ltd., Birmingham (2009)
24. Summerfield, M.: Rapid GUI Programming with Python and Qt: The Definitive Guide to PyQt Programming. Pearson Education, London (2007)
25. Lutz, M.: Learning Python: Powerful Object-Oriented Programming. O'Reilly Media, Sebastopol (2013)
26. Gunter, C.A. (ed.): Theoretical Aspects of Object-Oriented Programming: Types, Semantics, and Language Design. MIT press, Cambridge (1994)
27. Bruce, K.B.: A paradigmatic object-oriented programming language: design, static typing and semantics. J. Funct. Program. **4**(2), 127–206 (1994)

Intelligent Real-Time Management of Agrotechnologies

I. M. Mikhailenko(✉) and V. N. Timoshin

Federal State Budgetary Scientific Institution "Agrophysical Research Institute", Grazhdansky 14, 195220 St. Petersburg, Russia
ilya.mihailenko@yandex.ru

Abstract. The programmed management of agricultural technologies is aimed to obtain a given crop yield at the end of the growing season by forming a sequence of technological operations (optimal control programs) that are implemented when the most characteristic phenophases of the cultivated crop change. Optimal control programs are formed for the average long-term values of climatic factors and parameters of the used mathematical models obtained in the previous growing season. Actual deviations of the climatic factors values from the mean long-term values and perturbations of the parameters of mathematical models lead to the need to correct optimal control programs in real time. To make decisions on the correction of optimal programs, the sensitivity functions of the criterion of optimality of program control to perturbed parameters are used. Moreover, due to the complexity of the algorithm for generating optimal programs, such sensitivity functions are preliminarily constructed by repeatedly solving the program control problem for various variants of conditions and parameters. In addition to correcting optimal control programs in time, an important task of real-time control is the spatial correction of the parameters of technological operations, the need for which is associated with the presence of spatial heterogeneity of the parameters of the state of crops and soil environment. To implement the correction of this type, a linear model of the corrector is developed, the parameters of which are estimated by the results of multiple solutions to the program control problem for various options for the parameters of the state of crops and soil environment. Real-time management intellectualization consists in the transition to expert systems and the use of cloud information technology. Moreover, the knowledge base of the expert system includes segments of variants of optimal control programs and models of spatial correctors. This knowledge base is transferred to local expert systems, which also include blocks for local assessment of the parameters of the state of crops and soil environment, as well as a block for the formation of local control based on the obtained estimates and the knowledge base.

Keywords: Optimal control programs · Correction in space and time · Expert control systems · Knowledge bases · Mathematical models · Parameter estimation

R. Silhavy et al. (Eds.): CoMeSySo 2020, AISC 1295, pp. 491–504, 2020.
https://doi.org/10.1007/978-3-030-63319-6_45

1 Introduction

An important area of technological breakthrough in the agro-industrial sector is the transition to "smart agriculture". The Smart Agriculture Conference (SAC 2020), recently held in Beijing, presented a report on the topic of "smart farming" published by CLEAR consulting company. This report clarifies for the first-time definitions related to digital agriculture, precision farming and intelligent agricultural production. Intelligent agricultural production is a completely new mode of production, it is an agro ecosystem based on digital agriculture and precision farming. They create new production scenarios to deal with many negative factors. Intelligent technologies help optimize the costs of various resources: seeds, fertilizers and pesticides along with human labor.

In the long run, "smart agriculture" will allow better organization of processes that affect the end indicators. Thanks to the introduction of an intellectual component into the processes of technology design and control of robotic machines, data obtained from sensors and measuring systems are combined with the knowledge of specialists. Russian farmers are just beginning to introduce these technologies, now only about 5-10% of manufacturers use certain elements of intelligent systems. At present, Russia is witnessing a period of active development of almost the entire spectrum of precision farming elements, where global satellite positioning, Earth remote sensing (ERS) and GIS are most in demand, as well as systems for monitoring and control of equipment and quality of work performed. The combination of all these components into a single intelligent agrotechnology management system is possible only based on a developed theoretical base. The scientific and methodological foundations of the intellectualization of agricultural technology management have already been considered by us in [2, 3]. Here, attention was paid to the tasks of the strategic and program management levels. At a strategic level, crop cultivation technologies are coordinated among themselves as part of crop rotation. At the level of program management, technological operations are formed that are implemented during one growing season, considering the change in the phenophases of the crop.

Such optimal control programs are formed for long-term average climatic conditions. If real climatic conditions always corresponded to long-term average values, and the parameters of the mathematical models used, as well as the time of the phenophases onset, remained unchanged, then the optimal control programs obtained before the start of the growing season would be unchanged in real time. But in fact, climatic conditions do not always correspond to long-term average values, due to internal unpredictable changes, the parameters of all the mathematical models used are indignant, and in addition, the state of the soil environment, and with it the state of crops, is always heterogeneous in the field area. Real-time control is designed to compensate for all the above mentioned disturbances and uncertainties. More strictly, its tasks are:

- clarification of the optimal control programs upon detection of a significant deviation of the climatic conditions from the calculated values by which the current version of the optimal control program was synthesized;
- clarification of optimal control programs when detecting the fact of a significant deviation of the parameters of the used mathematical models;

- the implementation in time of technological operations upon the actual onset of phenological phases;
- correction of the parameters of technological operations, in connection with the spatial heterogeneity of the state parameters of crops and soil environment.

The aim of this work is to present a new approach to real-time management of agricultural technologies, which includes justification of the structure of the control system, models and algorithms aimed at solving the above mentioned problems.

2 Materials and Methods

The formation of optimal agricultural technology management programs is carried out in two stages [1, 3]. At the first stage, the optimal parameters of the soil state are found, combined into the vector $V(t) = [v_N(t), v_K(t), v_P(t), v_{Mg}(t), v_5(t)]$, v_N is the nitrogen content in soil, $kg \cdot ha^{-1}$, v_K is the potassium content in the soil, $kg \cdot ha^{-1}$; v_P is the phosphorus content in the soil, $kg \cdot ha^{-1}$; v_{Mg} is the magnesium content in the soil, $kg \cdot ha^{-1}$; v_5 - moisture content in the soil, mm. At the second stage, the optimal sequence of technological operations is found, which provides the best approximation to the optimal program for changing the state parameters of the soil medium obtained in the first stage. This sequence is the dose of mineral fertilizers and irrigation, combined in the vectors D (*Ti*), $D = [d_N, d_K, d_P, d_{Mg}, d_W]$, applied at the time of occurrence of the given phenophases *Ti*.

It is the optimality criterion for the first stage of the problem that meets the requirements of the ultimate control goal and has the following form

$$J(T) = [X(T) - X^*]^T G[X(T) - X^*] \tag{1}$$

where: $X^{*T} = [x_1^* \quad x_2^* \quad x_3^*]$ vector of the set values of the parameters of the sowing biomass (in the considered example of spring wheat), the components of which are: total biomass x_1, wet weight x_2, grain weight x_3, (crop); X(T) is a vector of optimized (predicted) parameters of sowing biomass with the same components, $G_u = \begin{bmatrix} g_{11} & 0 & 0 \\ 0 & g_{22} & 0 \\ 0 & 0 & g_{33} \end{bmatrix}_u$ - weight matrix of mass and qualitative components of the criterion by which it is possible to change the final ratio of biomass components; T is the time point of the end of the growing season.

With deviations of climatic conditions characterized by the vector $F = [f_1, f_2, f_3]$, where f_1 is the average daily air temperature, °C; f_2 is the average daily radiation level, $W \cdot (m_2 \cdot h)^{-1}$; f_3 is the average daily rainfall, mm; from the average long-term values, or from the last option used in the synthesis of the control program, there is a loss of optimality of the control program. It seems that the control is realized now of the occurrence of the given phenophases, then perturbations of climatic conditions should be analyzed at interphase intervals (T_i, T_{i+1})

$$\Delta F(t) = \tilde{F}(t) - F(t), \quad t \in (T_i, T_{i+1}) \tag{2}$$

where: $\tilde{F}(t)$ - is the vector of the last variant of the values of climatic conditions used in the synthesis of control programs, $F(t)$ - is the vector of the current predicted values of climatic factors, i = 1, 2,… I are the indices of phenophases.

Predicted loss of optimality due to mismatch of climatic conditions to calculated values can be determined as follows

$$\Delta J(t_i) = \frac{\partial J(T)}{\partial F(t_i)} \Delta F(t_i) = 2G[X(T) - X_i^*] \frac{\partial X(T)}{\partial F(t_i)} \Delta F(t_i) \tag{3}$$

Due to the two-stage procedure for optimizing control programs and the presence of several phenophases, it is not possible to analytically determine the partial derivatives of the vector of biomass parameters at the end of the growing season from climatic conditions of each phenophases $\frac{\partial X(T)}{\partial F(t_i)}$. Therefore, the value of the predicted loss of optimality should be determined by empirical expression

$$\Delta J(t_i) = K^T \Delta F(t_i), \tag{4}$$

where: K is the vector of parameters estimated by the results of multiple solutions to the program control problem for various values of the vector of climatic conditions, when the numerical values of variations of the criterion (1) are obtained. Based on these variations and variations of the vector of climatic conditions in the solutions, the parameters of the linear model are estimated (4).

Given the threshold value of the loss of optimality due to the mismatch of the climatic parameters with the calculated values $(-\delta)$, one can make decisions on the refinement of control programs on individual phenophases,

$$\text{specify if } -\Delta J(T|\Delta F(t_i)) \leq -\delta, \tag{5}$$

$$\text{do not specify if } -\Delta J(T|\Delta F(t_i)) \rhd -\delta.$$

Perturbations of the parameters of mathematical models lead to errors in predicting the parameters of sowing biomass and loss of optimality of control programs. For this case, you can apply the approach described above

$$\Delta J(t_i) = \frac{\partial J(T)}{\partial X(t_i)} \Delta X(t_i) = 2G[X(T) - X_i^*] \frac{\partial X(T)}{\partial X(t_i)} \Delta X(t_i). \tag{6}$$

However, for the above reasons, it is impossible to determine the partial derivatives of the vector of biomass parameters at the end of the growing season from the perturbations of the state parameter vectors in individual phenophases.

Therefore, as in the first case, it is advisable to resort to the empirical determination of optimality losses under perturbations of the parameters of the mathematical model of the parameters of the state of crops

$$\Delta J(t_i) = P^T \Delta X(t_i), \quad (7)$$

where P is the vector of parameters estimated by the results of multiple solutions to the program control problem for various values of the parameters of the mathematical model of the parameters of the state of biomass at different phenophases, when the numerical values of variations of criterion (1) are obtained.

In this case, the decision to re-solve the program control problem is made according to rule (5).

In the formation of optimal control programs, the time of occurrence of phenophases, as well as the duration of interphase periods, is taken equal to the long-term average values for specific crops and their varieties. Different timelines (e.g. Zadox scale) are most often used here. Due to the dependence of the time and speed of the occurrence of phenophases on climatic conditions, the actual moments of the occurrence of phenophases, as well as the duration of the interphase periods, may not coincide with the calculated value, which leads to additional losses in the optimality of control programs. This situation can be corrected if moments of the actual occurrence of the phenophases are detected, which can be realized according to Earth remote sensing data. According to its content, throughout the growing season, such a detection task always has two alternatives - "the next phenophase has come", or "the current phenophase continues". The change in phenophases is always accompanied by a significant change in the spectral composition of the Earth remote sensing data. Since the change of phenophases on individual sowing plants does not occur simultaneously, the general transition to the next phenophase takes a certain time, until the time when the number of plants that have replaced the phenophase exceeds 70%. Such a threshold corresponds to a certain combination of reflection parameters on all used spectral channels. To identify such a combination can be used linear decision rules (LDR) of the following form

$$\begin{aligned} i(t) &= i+1(t), \ \textit{if} \ \ L^T Z(t) \triangleright \delta_{i+1} \\ i(t) &= i(t), \ \textit{if} \ \ L^T Z(t) \leq \delta_{i+1} \end{aligned} \quad (8)$$

where: L is the vector of the parameters of the decision maker, estimated from experimental data; Z is the vector of reflection parameters for the used channels of the remote sensing system, is the phenophase index.

Spatial correction of optimal control programs is carried out considering the real state of crops and soil environment in individual elementary areas and can be presented in the following form

$$U_i(t,x,y) = V_i^*(t) + B_i^T(X_i^*(t) - \widehat{X}_i(t,x,y)) - \widehat{V}_i(t,x,y), \quad (9)$$

where: V*(t) is the average over the field area vector of technological impacts (operations), representing a real-time adjusted control program; X*(t) is the average field average optimal development program for the vector of sowing condition parameters corresponding to the optimal control program; $\widehat{X}(t,x,y)$ - assessment of the vector of parameters of the state of sowing at the time of the technological operations; $\widehat{V}(t,x,y)$ - assessment of the vectors of parameters of the state of the soil at the time of the technological operations; Bi – matrix of corrective regulators of technological operations; (x, y) – spatial coordinates of the elementary site on which the technological impacts are realized.

As can be seen from expression (9), local control on a separate elementary site (with an area of 2.0–3.0 m^2) consists of three components. The first, the same throughout the area - the optimal program; the second component is the correction of the program for assessing the real state of the sowing parameters; the third component is the correction of the program for the actual content of nutrients and soil moisture reserve.

To implement such a 3-component control, the parameters of the matrices of Bi correcting regulators are necessary. To construct the matrices of these parameters, it is necessary to repeatedly solve the program control problem with different values of the vectors of the state parameters of the crops in the phenophases on which the program control is carried out. The variations of these vectors and the variations of the parameters of technological operations are used to evaluate the parameters of the Bi matrices.

In addition, for the implementation of local control, it is necessary to construct estimates of the state parameters of crops in elementary areas $\widehat{X}_i(t,x,y)$ to compare them with the optimal program for the development of crops $X_i^*(t)$. The algorithm for constructing such estimates according to remote sensing data has the following form [4–6]

$$\begin{aligned}\dot{\widehat{X}}_i(t,x,y) &= A_i\widehat{X}_i(t,x,y) + B_i\widehat{V}_i(t,x,y) + C_iF(t)\\ &+ R_i(t,x,y)P_i\frac{\partial W^T(\widehat{X}_i)}{\partial \widehat{X}_i}K_{zi}^{-1}(Z_i(t,x,y) - W(\widehat{X}_i(t,x,y))),\\ \dot{R}_i(t,x,y) &= R_i(t,x,y)A_i^T + A_iR_i(t,x,y)\\ &-R_i(t,x,y)P_i\frac{\partial W^T(\widehat{X}_i)}{\partial \widehat{X}_i}K_{zi}^{-1}\frac{\partial W(\widehat{X}_i)}{\partial \widehat{X}_i}P_i^TR_i(t,x,y),\end{aligned} \tag{10}$$

where: A_i, B_i, C_i, P_i are the matrices of the parameters of mathematical models of the parameters of the state of crops and the remote sensing model on the i-th phenophase; W (X_i) - remote sensing model operator; K_{zi}– covariance matrix of the ERS model errors; R_i is the covariance matrix of estimation errors.

With a large field area under crops (more than 500 hectares), the number of elementary plots can exceed 2 million units. This causes great difficulties in the direct implementation of algorithm (10), where each elementary segment with the spatial coordinate (x, y) forms its own modeling cycle. This circumstance forces us to resort to approximation simplifications of the form

$$\dot{\hat{X}}_i(t,x,y) = \hat{X}_i(t) + S_i(Z_i(t) - Z_i(t,x,y)) \tag{11}$$

where: is the average over the area field estimate of the parameters of the state of sowing at the i-th phenophase, $\hat{X}_i(t)$ is the average over the area of the field is the value of the vector of reflection parameters for all used channels of the remote sensing system, and S_i is the matrix of parameters for the corrector of estimates of the parameters of the state of sowing, formed by repeatedly solving the problem (10) for different values of the vector of reflection parameters $Z_i(t,x,y)$, the subsequent obtaining of variations of estimates of the parameters of the state of sowing $\Delta X_i(t,x,y)$ and variations of the vector of reflection parameters, from which the parameters $\Delta Z_i(t,x,y)$ of the matrix S_i are estimated.

Estimation of the vector of soil state parameters $\hat{V}(t,x,y)$ is carried out according to a more complex two-stage algorithm scheme [5–9]. Here, at the first stage, the parameters of the chemical state of the sowing biomass are estimated, and at the second stage, the parameters of the state of the soil environment themselves. Therefore, in this case, it is advisable to use approximations of the form (11)

$$\hat{V}_i(t,x,y) = \hat{V}_i(t) + H_i(Z_i(t) - Z_i(t,x,y)) \tag{12}$$

where: $\hat{V}_i(t)$ is the average over the area of the field estimate of the vector of parameters of the soil state at the i- th phenophases $Z_i(t)$ is the average over the area of the field is the value of the vector of reflection parameters for all used channels of the remote sensing system, Hi is the matrix of parameters for the corrector of estimates of the parameters of the state of the soil, formed by repeatedly solving the two-stage estimation problem for different values of the vector of reflection parameters $Z_i(t,x,y)$, obtaining variations of estimates of the vector of soil state parameters $\Delta\hat{V}_i(t,x,y)$ and variations of the vector of reflection parameters $\Delta Z_i(t,x,y)$, from which the parameters of the matrix H_i are estimated.

The block diagram of the control algorithm, reflecting all the above procedures, is presented in Fig. 1.

For the functioning of the algorithm, it is necessary to carry out preliminary studies of the control object, according to the results of which the parameters of mathematical models are estimated, as well as the parameters of spatial correctors (9), (11), (12). Before the start of the next growing season, based on the parameters of mathematical models and the average long-term values of meteorological factors, an optimal control program is synthesized, the form of which is shown in Fig. 1. With the beginning of the growing season, periodically (after 3 days), real weather parameters are monitored at the weather station, and on the sowing test sites (the number of which is 10–15 pcs., and the area of each site is 20–30 m^2), the real sowing biomass parameters and soil parameters are measured. These measurements are carried out both by instrumentation and by sampling. Simultaneously with these measurements, with the same frequency

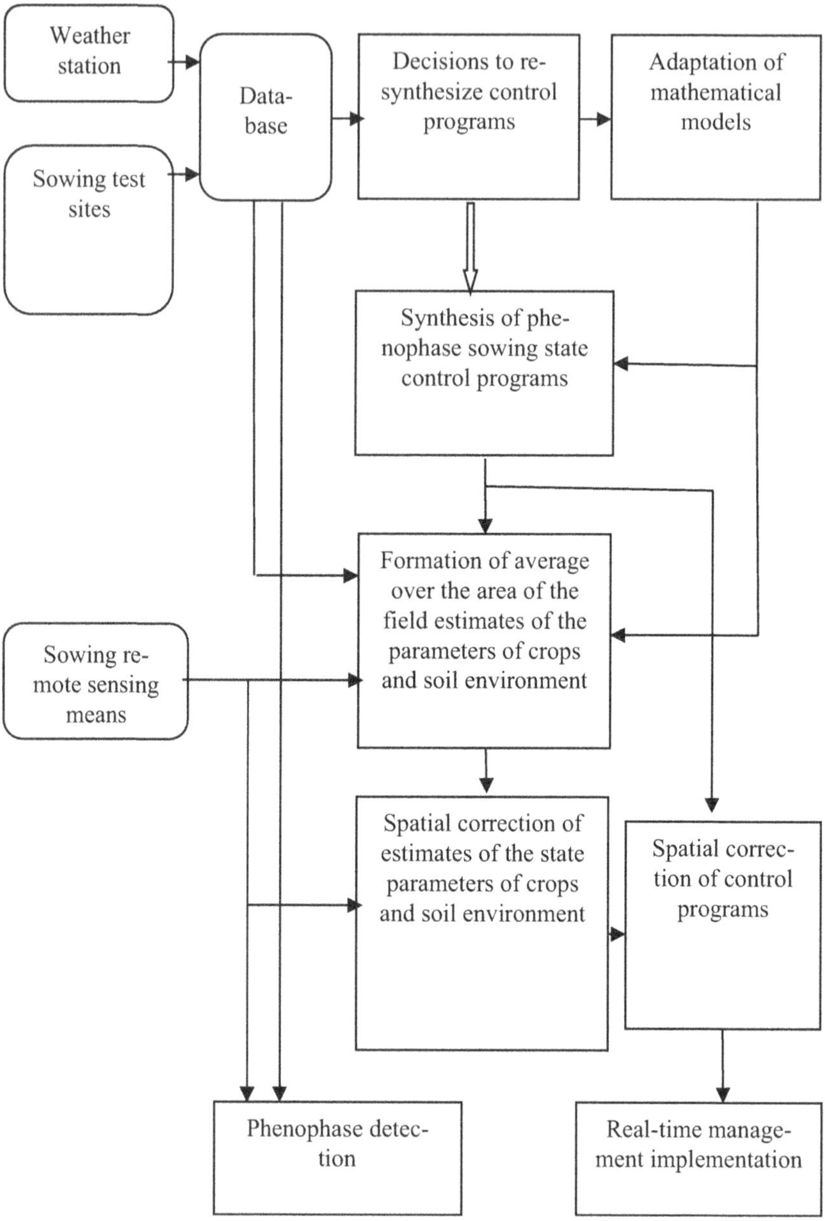

Fig. 1. The block diagram of the control algorithm of agricultural technology in real time.

using remote sensing, aerial photography of the sowing surface is carried out together with the test sites.

The obtained information about real meteorological factors, parameters of the state of sowing and soil, as well as reflection parameters in remote sensing data are compared in the database (DB) with predicted or simulated values. If, because of the comparison in the database, the deviations fit into the allowable tolerances, then the optimal control program is reproduced in accordance with the law (9), and these controls are worked out by robotic agricultural machines, as indicated in [1]. If the deviations of the real values of the meteorological factors from the predicted values exceed the specified tolerance, then the synthesis of the optimal control program for the updated forecast of meteorological factors is carried out. In those cases when the real parameters of the state of the sowing and soil environment, as well as the reflection parameters in the remote sensing data, differ from the modeled values by values that exceed the specified tolerance, the parameters of the corresponding mathematical models are first clarified, and then the optimal control program is re-synthesized.

When working out the control according to the law (9), on the basis of mathematical models of the dynamics of the parameters of sowing, taking into account the phenophases, dynamics of the parameters of the soil and models of optical measurements of remote sensing, the parameters of the sowing and soil environment are evaluated, first on average over the area of the field, and then their spatial correction is corrected (11), (12). Based on spatial estimates of the parameters of sowing and soil environment, final control is formed in accordance with law (9), which requires the presence of robotic machines with the possibility of local exposure to elementary areas. In the absence of such machines, only control is carried out on average over the area of the field or over homogeneous zones determined by the parameters of the soil environment. Direct technological effects are produced upon the occurrence of the corresponding sowing phenophases, the detection of which is carried out according to remote sensing data at the moments of the flight of an unmanned aerial vehicle over the sowing field, the state of which is controlled by the system under consideration.

3 Results

Intellectualization of management consists in the transition to expert systems in which decisions are made based on knowledge bases (KB) [2, 3]. At the same time, knowledge bases are formed in the cloud by repeatedly solving the control problem for various variants of conditions and are transmitted to local systems, where users independently find the most effective solutions. For the strategic and programmatic levels of management, the formation and transfer of knowledge bases does not present serious difficulties, since they solve one problem and use the same algorithms to solve them. As already shown above, the real-time control level consists of several intertwined tasks, which greatly complicates the process of creating knowledge bases, which can have a more complex segment structure.

To form the structure of the knowledge base, it is first necessary to separate the management functions and tasks presented in Fig. 1 into those that can be implemented by local automated agro-technology management systems (AATMS) and those that should be transferred to cloud data centers serving all local AATMS-s of the region. Such a separation is shown in the block diagram of Fig. 2.

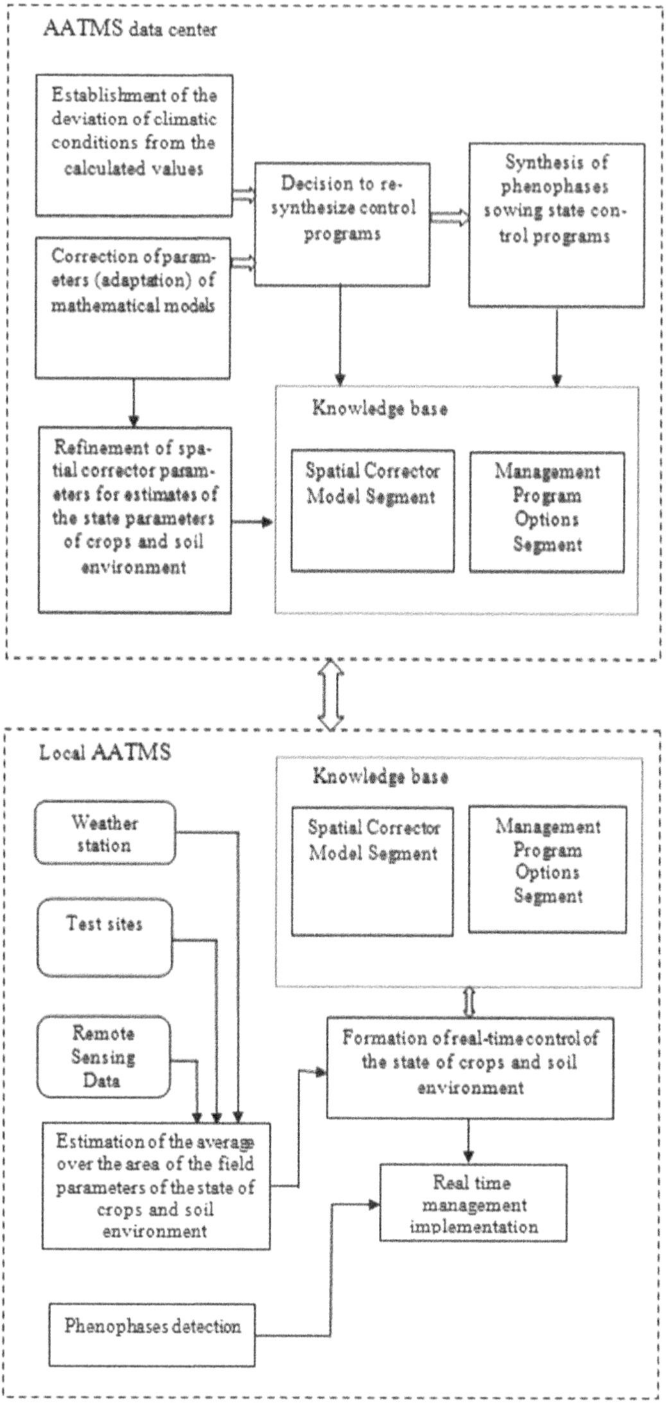

Fig. 2. Block diagram of an expert system for managing agricultural technology in real time.

Here, tasks and functions are carried out to the cloud data processing center, the implementation of which requires the presence of a complex mathematical software apparatus and highly qualified personnel. Such tasks include making decisions on the formation and the formation of the knowledge base segment according to the options for optimal control programs, including the adaptation functions of the mathematical models used. In addition, functions to refine the parameters of spatial corrector models of estimates of crop parameters and soil environment are taken to the level of the processing center. The formed knowledge base is transferred to the level of the local automated control system, where local management of agricultural technologies is formed on its basis. For its formation, in addition to the optimal version of the control program on an average over the field area, functions are used to evaluate the parameters of the state of sowing and soil environment, both on average over the field area and in all local areas. Based on these estimates, real-time control is formed considering the spatial heterogeneity of the state of crops and soil environment. The implementation of this control occurs at the time of phenophases detection, when it comes in the form of tasks for automated technological machines that operate according to the law (9) and form local doses of fertilizer application and irrigation rates [1].

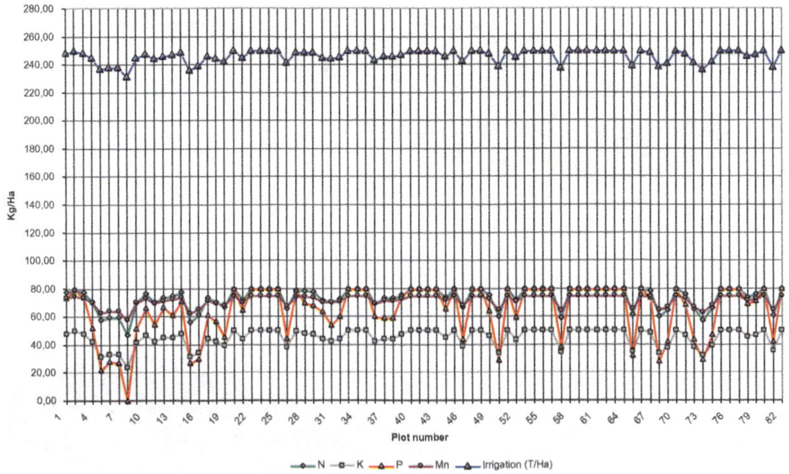

Fig. 3. Graphs for setting the parameters of the correcting controller for all 5 control channels.

Figure 3 shows the graphs for setting the parameters of the correcting controller (9) for all 5 control channels.

Here, the geometric icons denote the experimental values of the corrective controls, and the continuous lines indicate the simulated values. As you can see from the graphs, the tuning accuracy is quite high, which allows you to generate reliable estimates of local controls for all 5 channels. They are presented in Fig. 4, 5, 6, 7 and 8 for the flowering phase of sowing spring wheat. For the rest of the planting phenophases, the spatial distributions of the controls are formed similarly.

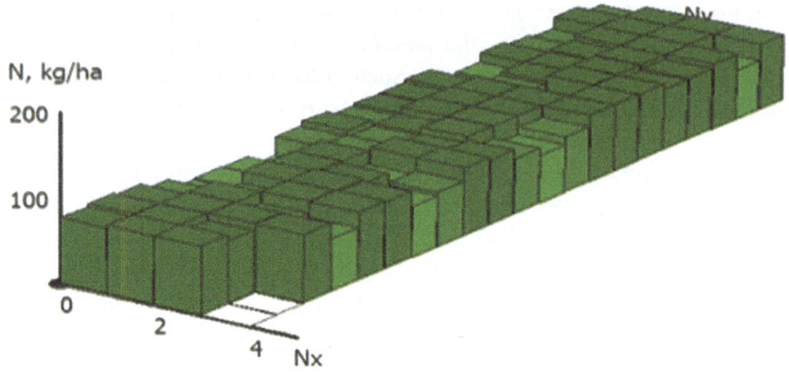

Fig. 4. Distribution of nitrogen doses over the field area.

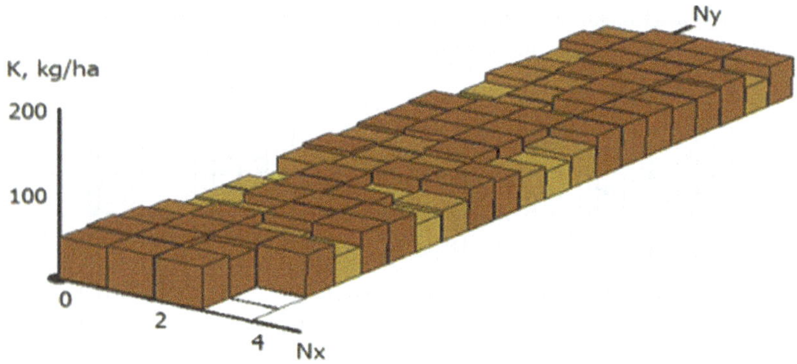

Fig. 5. Distribution of potassium doses over the field area.

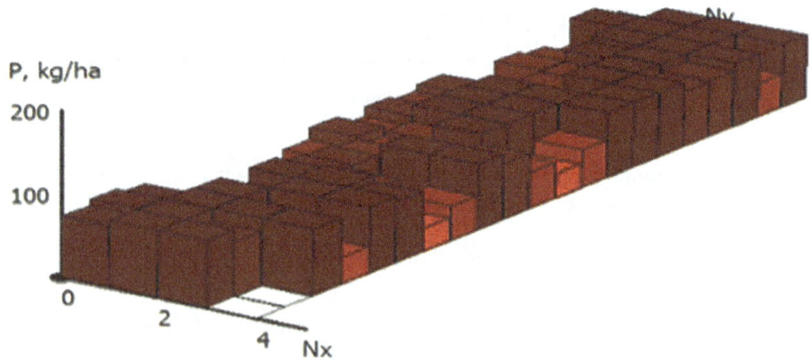

Fig. 6. Distribution of phosphorus doses over the field area.

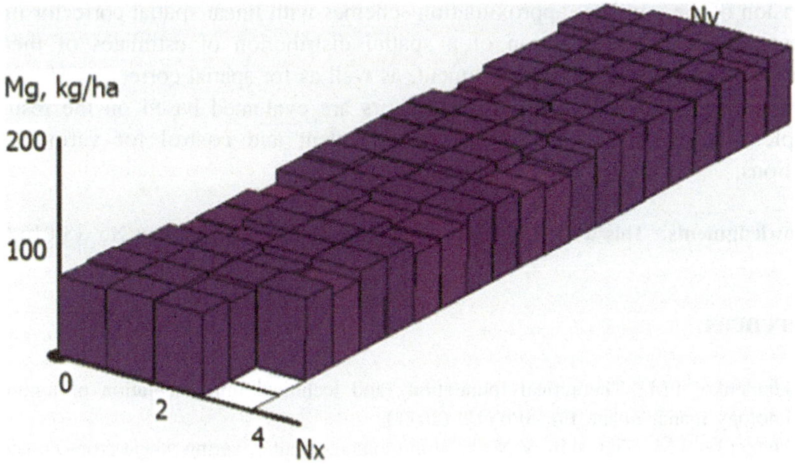

Fig. 7. Distribution of magnesium doses over the field area.

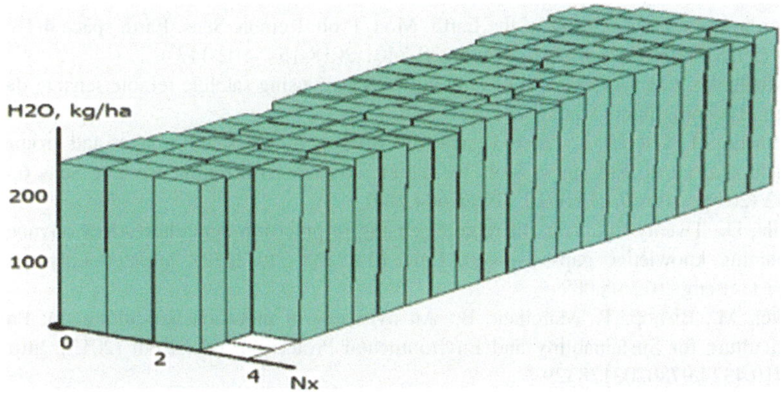

Fig. 8. Distribution of irrigation norms over the field area.

4 Conclusions

A new approach to the creation of intelligent control systems in precision farming is proposed. It is based on the use of cloud information technology and expert management systems, in which the knowledge base is formed in the cloud and transferred to local management systems. In this case, the knowledge generator is an algorithm for optimal control of the state of the crop. The control algorithm includes procedures for correcting optimal control programs in time and space. Correction in time is associated with real deviations from the calculated values of climatic factors and the parameters of the mathematical models used. The correction in space is aimed at eliminating the spatial heterogeneity of the parameters of the state of sowing and soil environment, which are estimated from the data of remote sensing of the Earth. Due to the large dimension of the problem, approximation schemes with linear spatial corrector models are proposed for the formation of a spatial distribution of estimates of the state parameters of crops and soil environment, as well as for spatial correction of controls. In this case, the parameters of the correctors are evaluated based on the results of multiple solutions to the problems of assessment and control for various initial conditions.

Acknowledgments. This work was partially supported by the RFBR grant No. 18-016-0008.

References

1. Mikhailenko, I.M.: Theoretical foundations and technical implementation of agricultural technology management. Ed. SPbSTU (2017)
2. Mikhailenko, I.M., Timoshin, V.N.: Program management of spring wheat crops considering phenophases. Eur. J. **8**(65)part 4, 12–18 (2019). https://doi.org/10.31618/esu.2413-9335.2019.4.65.274
3. Kazakov, I.E.: Methods for Optimizing Stochastic Systems. Nauka, Moscow (1987)
4. Mikhailenko, I.M., Timoshin, V.N.: Estimation of the chemical state of the soil environment according to remote sensing of the Earth. Mod. Prob. Remote Sens. Earth space **4**(18), 125–134 (2018). https://doi.org/10.21046/2070-7401-2018-15-7-102-113
5. Mikhailenko, I.M.: Assessment of crop and soil state using satellite remote sensing data. Int. J. Inf. Technol. Oper. Manag. **5**(1), 41–52 (2013)
6. Quemada, M., Gabriel, J., Zarco-Tejada, P.: Airborne hyperspectral images and ground-level optical sensors as assessment tools for maize nitrogen fertilization. Remote Sens **6**, 2940–2962 (2014). https://doi.org/10.3390/rs6042940
7. Mulla, D.: Twenty-five years of remote sensing in precision agriculture: Key advances and remaining knowledge gaps. Biosyst. Eng. **114**, 358–71 (2012). https://doi.org/10.1016/j.biosistemseng.2012.08.009
8. Oliver, M., Bishop, T., Marchant, B.: An overview of precision agriculture. In: Precision Agriculture for Sustainability and Environmental Protection. Eds. Rout (2013). https://doi.org/10.4324/9780203128329
9. Zarco-Tejada, P.J., Guillén-Climent, M.L., Hernández-Clemente, R., Catalina, A., González, M.R., Martín, P.: Estimating leaf carotenoid content in vineyards using high resolution hyperspectral imagery acquired from an unmanned aerial vehicle (UAV). Agric. Forest Meteorol. **171–172**, 281–294 (2013). https://doi.org/10.1016/j.agrformet.2012.12.013

Applying Peltier Thermoelectric Coolers to Compensate Temperature Errors of the MEMS Used in UAVs

A. R. Bestugin, I. A. Kirshina(✉), P. A. Okin, and O. M. Filonov

Saint Petersburg State University of Aerospace Instrumentation (SUAI), Saint-Petersburg, Russia
ikirshina@mail.ru

Abstract. This article presents the results of theoretical studies related to the application of Peltier film microelements to compensate the temperature drift of fundamental frequency typical for the MEMS resonators with a frequency-modulated output. It is shown that the proposed method can ensure a correction within the range limited by the amount of Joule heat released during the sensor operation.

Keywords: Silicon resonator · MEMS · Peltier microelement · Internal friction · Microtechnology

1 Introduction

An ever-growing interest in the application of unmanned aerial vehicles and an increased range of tasks resolved with their use require their permanent technical improvement to be adapted for new application conditions.

This, of course, requires improvements, first of all, in flight control systems and flight navigation equipment.

The compliance with the requirements for weight and size characteristics included in this set of devices is achieved by using commercial MEMS designed to measure physical quantities ensuring UAV control, including that in autonomous mode [1, 2].

The analysis of problems arising during the synthesis of a flight navigation complex shows that in case of decreased UAV dimensions it is reasonable to focus on MEMS that are common in design and have one single physical principle of operation and a structure of overall operational error.

A part of the UAV flight navigation complex designed to measure motion parameters dependent on stationary pressure at the point where the aircraft is located (vertical and air speeds, relative and absolute altitudes) can be implemented based on micromechanical absolute pressure sensors. As of today, micromechanical absolute and different pressure sensors with secondary resonant measuring transducers with a frequency-modulated output mostly meet the requirements for measurement accuracy, temporal stability and mass-size characteristics [3, 4].

This type of sensors available from Yokogawa (EJA and EJX series) has now the smallest relative errors among the existing MEM pressure meters and is widely used in

R. Silhavy et al. (Eds.): CoMeSySo 2020, AISC 1295, pp. 505–511, 2020.
https://doi.org/10.1007/978-3-030-63319-6_46

energy transport and storage systems. But these sensors have one disadvantage, which is the complex technological process of synthesizing a polycrystalline capsule that vacuums the microresonator and provides a precision level of static pressure measurement (from scale ranges) [5].

2 Methods

Our team developed and patented several designs of H-type sensors and micromechanical pressure sensors with an uncapsulated microresonator. A structure diagram is given in Fig. 1.

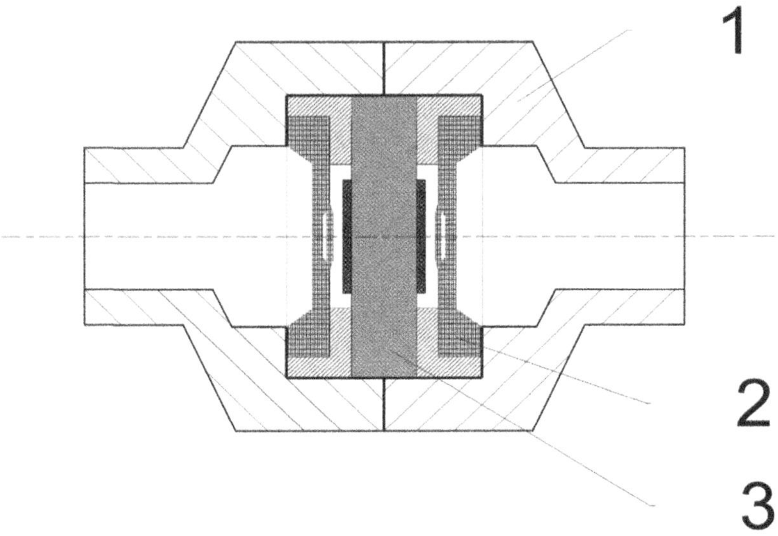

Fig. 1. A structure diagram: 1—body, 2—membrane with resonator, 3—permanent magnetic exciting circuit.

The patented structures [6, 7] have a number of fundamental technological advantages as compared to Yokogawa's sensors and can be considered basic in the production of any pressure meters used in UAV motion control systems.

Since the sensing element (membrane) and the secondary measuring transducer (microresonator) are a monolithic single-crystal structure, the effect of external factors on the sensor operation and the operational component of an overall error is minimized.

A special feature of the microresonator as a bridge beam that distinguishes it from other types of measuring transducers is its inevitable heating during operation (natural oscillations with frequencies from 100 kHz to 1 MHz depending on dimensions) due to the dissipation of kinetic energy in the beam through various internal friction mechanisms.

When heated up to 450–550 K, the temperature expansion of the microresonator beam leads to axial pressing forces that result in the drift of fundamental frequency of the system oscillations and, therefore, in an error of the measured pressure.

For the purpose of theoretical studies of the dissipation effect on sensor errors, we used the one-dimensional Fourier thermal conductivity equation for steady-state temperature distribution:

$$\frac{\partial^2 T}{\partial x^2} - \frac{\varepsilon\sigma_0 S_Б}{\lambda S}\left(T^4 - T_0\right) + \frac{\sigma^2 \varpi \alpha^2}{E}\frac{\varpi\tau}{1+\omega^2\tau^2}T = 0 \tag{1}$$

By putting the following expressions in: $\frac{\varepsilon\sigma_0 S_Б}{\lambda S} = D_1$; $\frac{\sigma^2 \varpi \alpha^2}{E}\frac{\varpi\tau}{1+\omega^2\tau^2} = D_2$; $\frac{\varepsilon\sigma_0 S_Б}{\lambda S}T_0 = D_3$

where:

α—silicon linear extension coefficient;
$\omega = \frac{\pi^2}{l^2}\sqrt{\frac{EI}{\rho wt}}$—fundamental frequency of resonator natural oscillations;
$I = \frac{wt^3}{12}$—moment of inertia of resonator beam cross section;
E—silicon modulus of elasticity in the direction <100>;
l, w, t—beam length, width and thickness;
$\tau = \frac{t^2 C_V}{\pi^2 \lambda}$—time of temperature relaxation in the beam in the plane of bending oscillations;
C_V—volumetric specific silicon heat capacity;
λ—silicon heat conductivity coefficient;
ε—silicon emissivity;
$S = w \cdot t$—cross-sectional area;
$S_Б = 2l(w+t)$—area of the side surface;
$\sigma = 0.2[\sigma]_{Si}$—Peierls-Nabarro stress of silicon;
ρ—silicon density;
$T = T(x)$—temperature distribution at different points of the beam;
T_0—temperature at points of beam fastening.

Let us rearrange the Eq. (1) in the form as follows:

$$\frac{\partial^2 T}{\partial x^2} - D_1 T^4 + D_2 T + D_3 = 0 \tag{2}$$

The first integral of this equation is:

$$\left(\frac{\partial T}{\partial x}\right)^2 - \frac{2}{5}D_1 T^5 + D_2 T^2 + 2D_3 T + C = 0 \tag{3}$$

boundary conditions: $T = T_0$ at $\begin{cases} x = 0 \\ x = 2l \end{cases}$ and $\frac{\partial T}{\partial x} = 0$, at $\begin{cases} x = 0 \\ x = 2l \end{cases}$

For this beam, the temperature distribution will be symmetrical in relation to its middle $x = l$, i.e. it is possible to set $T_{x=l} = T_{\max} = T_l$.

We find the constant integration value C from the boundary condition $\frac{\partial T}{\partial x} = 0$, at $x = 0$:

$$C = -\frac{2}{5} D_1 T_l^5 + D_2 T_l^2 + D_3 T_l$$

then:

$$\frac{\partial T}{\partial x} = \sqrt{\frac{2}{5} D_1 \left(T^5 - T_l^5\right) + D_2 \left(T^2 - T_l^2\right) + 2D_3 (T - T_l)} \tag{4}$$

from here after integration in accordance with T:

$$x = \int\limits_{T_0}^{T} \frac{1}{\sqrt{\frac{2}{5} D_1 \left(T^5 - T_l^5\right) + D_2 \left(T^2 - T_l^2\right) + 2D_3 (T - T_l)}} \tag{5}$$

From integral (5) after rather time-consuming calculations for the beam of $2l = 500\mathcal{MKM}$ in length, we obtained temperature distributions in the resonator (Fig. 2) under boundary conditions (for $x = 0$ and $x = 2l$) for $T_0 = 300\,\text{K}, T_0 = 273\,\text{K}, T_0 = 250\,\text{K}$.

A simple familiarization with the diagrams presented in Fig. 2 shows that a decrease of T_0 at the points of the resonator-to-membrane fastening leads to a decrease in pressing forces and thus contributes to compensating the drift of the resonator fundamental frequency.

3 Results and Discussion

An established dependence of temperature distribution in the microresonator beam must be taken into account when designing with instruments for the compensation of operational errors of MEMS of the considered type related to the drift of the resonator fundamental frequency during the beam temperature expansion.

As the analysis shows, from the existing set of technical facilities the only acceptable solution for cooling small-sized devices as applicable to MEMS is the one based on Peltier film thermoelements built into the sensing element. The diagram of H-resonator with Peltier cooling film elements is given in Fig. 3.

As a means of cooling the Peltier elements were first considered by A.F. Ioffe in the 1930s, but their technical implementation was constrained by technological factors related to the synthesis of planar semiconductor elements of complex chemical composition.

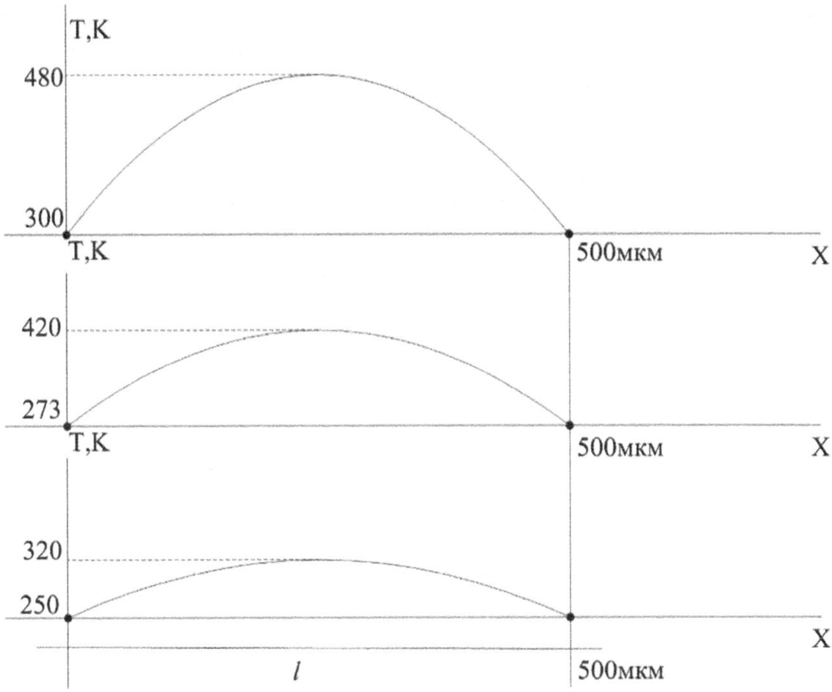

Fig. 2. Familiarization with the diagrams.

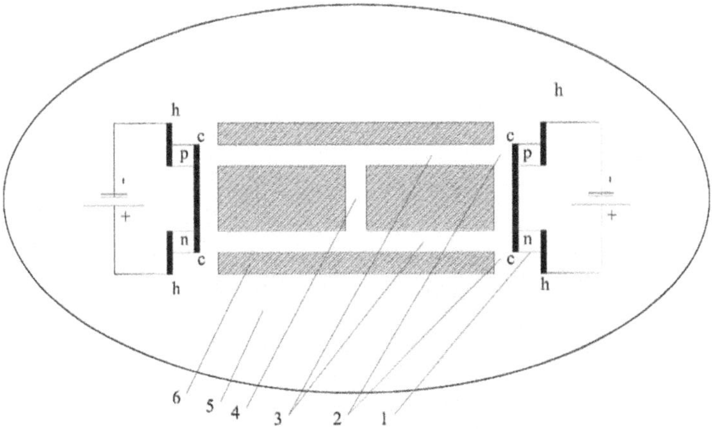

Fig. 3. Diagram of H-resonator with Peltier cooling film elements: H-resonator on the membrane with cooling elements, where 1 – Peltier element, 2 – anchors, 3 – resonator beam, 4 – dielectric brace, 5 – membrane, 6 – working cavity.

Today, Peltier 3-D elements are used to cool different devices, but there is no evidence of their application in micromechanical systems.

According to Fig. 3, the Peltier film planar element represents closed circuits of two p-type film conductors $Bi_2Te_3Sb_{0.3}$—bismuth biteluride doped with antimony and n-type $Bi_2Te_3Se_{0.3}$—bismuth biteluride doped with selenium that includes a power source. These semiconductors are chosen because of the maximum difference in the Peltier coefficients π_1 and π_2 for these materials, which is the main factor determining the theoretical efficiency of elements in general [8].

The polarity of connecting the source to the circuit determines the nature of thermal processes at semiconductor contact points (heating or cooling) and allows Peltier elements to be used as a heating and cooling device depending on the direction of a field in the circuit.

Cooling efficiency is characterized by efficiency coefficient Z [9] that is represented by the expression:

$$Z = \frac{(\alpha_1 + \alpha_2)^2}{\left(\sqrt{\rho_1\lambda_1} + \sqrt{\rho_2\lambda_2}\right)^2} \tag{6}$$

where: α_1 and α_2—thermal EMF coefficients of the semiconductors (p and n),

ρ_1 and ρ_2—their specific resistances, λ_1 and λ_2—thermal conductivity coefficients.

A maximum temperature decrease will be

$$\Delta T = (T_h - T_c) = \frac{1}{2} Z \cdot T_c^2 \tag{7}$$

where T_h—hot contact temperature, and T_c—cold contact temperature. Coefficient Z, α_1 and α_2—temperature functions and this makes it difficult to directly apply Peltier elements in MEMS without the use of additional specialized integrated circuits (ASICs).

Taking into account the Joule heat released in the circuit, the thermal power q removed from the resonator will be

$$q = (\pi_1 - \pi_2)i - \frac{i^2R}{2} \tag{8}$$

where i—current strength in the circuit, R—resistance of the element circuit.

According to [10], a temperature difference during cooling will be

$$(T_h - T_c) = \frac{(\pi_1 - \pi_2)i - \frac{i^2R}{2}}{\lambda} \tag{9}$$

From here it follows that the required temperature difference can be provided by varying the current in the circuit, which will naturally be a limitation of the method since Joule heat increases as grows and decreases the heat withdrawn from the resonator.

4 Conclusions

1. The principal possibility of decreasing a temperature component of the resonator drift due to a temperature decrease at the points of resonator fastening to the sensor membrane (sensing element) is confirmed.
2. The technical possibility of resolving this task by planar Peltier elements synthesized on the membrane is justified and theoretically confirmed.
3. It is shown that the cooling process can be controlled by changing the current in the circuit with a limitation due to Joule heat released when the current passes through the circuit during operation.

Acknowledgements. This work was financially supported by the Russian Foundation for Basic Research (Grant No. 18-07-01005-a).

References

1. Ryu, J.H., Gankhuyag, G., Chong, K.T.: Navigation system heading and position accuracy improvement through GPS and INS data fusion. J. Sens. **2016**, 1–6 (2016)
2. Jekeli, C.: Inertial Navigation Systems with Geodetic Applications. Walter de Gruyter, Berlin (2001)
3. Martin, H., Groves, P., Newman, M.: The limits of in run calibration of MEMS inertial sensors and sensor arrays. Navig. J. Inst. Navig. **63**(2), 127–143 (2016)
4. Quinchia, A.G., Falco, G., Falletti, E., Dovis, F., Ferrer, C.: A comparison between different error modeling of MEMS applied to GPS/INS integrated systems. Sensors **13**(8), 9549–9588 (2013)
5. Shen, K., Xia, Y., Wang, M., Neusypin, K.A., Proletarsky, A.V.: Quantifying observability and analysis in integrated navigation. Navig. J. Inst. Navig. **65**(2), 169–181 (2018)
6. Filonov, O.M., Bardin, A.V., Metkin, N.P., Okin, P.A.: The method of measuring different pressures by a sensor with a frequency-modulated output and the sensor for implementing this method. Patent for invention No. 2548583, Published on April 20, 2015, Bulletin No. 11 (2015)
7. Bestugin, A.R., Filonov, O.M., Kirshina, I.A., Okin, P.A.: Analysis of the dependence of the kinetic energy dissipation process in microresonators on ambient temperature. Achievements Mod. Radioelectronics **12**, 125–128 (2017)
8. Goldman, B.M., Dashevsky, Z.M., et al.: Film Thermoelements. Science, Moscow (1985)
9. Thielmann, I.: Thermoelastic Cooling Technology. https://www.binder-world.com/en/tips-tricks/peltier-technology. Accessed 06 June 2020
10. Kandasamy, S., Kalantar-zadeh, K., Rosengarten, G., Wlodarski, W.: Modelling of a thin film thermoelectric micro-peltier module. In: IEEE Region 10 Conference: Analog and Digital Techniques in Electrical Engineering, TENCON, Chiang Mai, Thailand, vol. D, pp. D310–D313. IEEE (2004)

Context-Aware Smart-Contracts for Service Bundles

Tatiana Levashova(✉) and Michael Pashkin

St. Petersburg Institute for Informatics and Automation of the Russian Academy of Sciences, 39, 14th Line, St. Petersburg 199178, Russia
tatiana.levashova@iias.spb.su, michaelpashkin@mail.ru

Abstract. A conceptual framework that supports configuration of service bundles and automatic provision of services from these bundles according to the consumer context is proposed. The framework adheres the idea of ontology-based product configuration where service bundles play the role of products. Service bundle is defined as a group of related customized services that are provided selectively in context-aware way on terms favorable to the consumer. Smart contracts regulate service provisioning/consumption. These contracts are represented in terms of the product/service ontology using SWRL-rules. The use case from the Internet services domain demonstrates the main ideas behind the framework.

Keywords: Service bundle · Smart contract · Product configuration · Ontology · Automated service provision · Semantic rules

1 Introduction

The practice of selling product packages is widespread in the market. When a convergence of products and services happened, the concept of combining products and services in so-called product-service systems as a bundled offer was developed [1, 2]. In the product-service systems with the highest level of service content (service-centric systems), products have been replaced by services [3] and product packages have been replaced by service bundles. Service bundles simplify service selection and configuration. Usually, a price for a package is lower than if the customer bought all the services from this package separately.

Smart contracts are a step towards automatic regulation of services provisioning/consumption. In 1994, Nick Szabo proposed the idea of smart contract. He had defined it as "a set of promises, specified in digital form, including protocols within which the parties perform on these promises." [4]. In scope of the current level of information systems, smart contracts are viewed as decentralized applications that are available to all sides of the contract through the cloud of in decentralized way. Due to the use of Turing-complete language for contract description, it is possible to implement rather complex algorithms. At the same time, it is mandatory to have conditions under which the contract must be executed as well as the list of actions assigned to the submitted conditions. All conditions of a smart contract must be described in a strong mathematical way and provide clear execution logic. The smart contract monitors the

R. Silhavy et al. (Eds.): CoMeSySo 2020, AISC 1295, pp. 512–521, 2020.
https://doi.org/10.1007/978-3-030-63319-6_47

specified conditions of achievement or violation of the points and makes independent decisions based on the programmed conditions. The main principle of a smart contract is the complete automation and reliability of the performance of contractual relations between the parties [5].

The aim of this paper is presentation of a conceptual framework that supports configuration of service bundles and automatic provision of services from these bundles according to the consumer context. The framework adheres the idea of ontology-based product configuration where service bundles play the role of products. Service bundle is defined as a group of related customized services that are provided selectively in context-aware way on terms favorable to the consumer. Smart contracts regulate service provisioning/consumption. Logic representation of smart contracts is created when configuring service bundles and compatible with the product configuration ontology to enable automatic service provision regulated by these contracts.

The proposed framework is beneficial in several ways. Firstly, in a new situation for the consumer, he/she does not need to search and request a required service if such a service is one of the preferable ones, i.e. it is included in the service bundle. Second, there is no need to enter into a service agreement since the logical representation for the smart contract formalizes all the commitments to regulate provisioning/consumption of services from the bundle. Finally, the service bundle is configured in a cost beneficial way for the consumer.

The novelty of the proposed framework is in combining ontology-based product configuration to create service bundles with semantic rules formalizing smart contracts to regulate the automated service provisioning/consumption.

The remainder of this paper is structured as follows. Section 2 provides an overview of related research. The conceptual framework supporting context-aware smart-contracts for service bundles is presented in Sect. 3. Section 4 addresses a simple use case illustrating the main ides behind the proposed framework. Some concluding remarks are summarized in the Conclusion.

2 Related Research

Two research directions can be considered as related to the present research. They are ontology-based product/service configuration and smart contracts on services provision/consumption.

Research on ontology-based product configuration deal with construction of product ontologies [6, 7], creation of ontology-based models of product-service systems [8–11], and development of ontology-based representations for product configuration knowledge [12–14]. The ontology models developed as a result of these approaches are extensive knowledge sources for creating a service configuration ontology.

With reference to the aim of this paper, a methodology of ontology-based product configuration [15] later adapted to service bundle configuration [16] is the most noticeable. This methodology represents general terms and relations common to the configuration domain using OWL (Web Ontology Language) [17]. As well, the methodology offers configuration constraints represented in SWRL (Semantic Web Rule Language) [18]. Actual configuration processes are carried out using JESS, a rule engine for the Java platform, by mapping OWL-based configuration facts and SWRL-based configuration constraints into JESS facts and JESS rules, respectively. The authors of the methodology define service bundle as a set of one or more service elements (single services) that can be provisioned together as a whole compatible with the rules.

Concerning compatibility of formal representation of smart contracts and ontologies, a number of approaches [19, 20] that deal with semantic smart contracts can be mentioned. The motivation of these approaches is that smart contracts are coded commitments. Utilization of such contracts for a wide range of usages is challenging. A unified ontology for smart contracts is believed to reduce the semantic ambiguity of the contracts. The ontologies proposed in such approaches represent the smart contracts independent on their functional behaviour. Kim and Laskowski [21] offer some ideas how to provide functionality for smart contracts using domain ontology. They map ontology axioms represented in first-order logic into smart contracts coded in Solidity language (an object-oriented programming language) to track provenance of supply chains.

Semantic rules in conjunction with ontologies brings ones closer to creation of context aware smart-contracts. In the frameworks coupling these two components, the ontologies represent domain-specific knowledge and the rules formalize domain constraints for the smart-contracts. These rules are specified with respect to ontology concepts. In real situations, contextual information instantiate the ontologies and provides instance-specific constraints to the rules as a result of which the rules change according to the context. Approaches as auto-generation of smart contracts from domain ontologies [22], conditional formalization of smart contracts [23], ontology aided smart contract execution for unexpected situations [24] belong to this category. In these researches, the ontologies are specified by means of OWL, the rules are represented in SWRL.

In the research to auto-generation of smart contracts from domain ontologies and semantic rules [22], rules specify domain constraints. The smart contracts are generated based on domain-specific templates, which capture all possible manifestations of the ontology. These templates are translated into an abstract syntax for the source code of a computer program. The actual constraints on the properties are inserted into a source code file, thus generating a smart contract, which is the combination of the ontology-derived template and the dynamic values extracted from specific settings.

Rules represent constraints for the transactions in the methodology for ontology aided smart contract execution [24]. Actual information about attributes of assets and participants involved in a transaction is used to verify whether this transaction will be completed or not. In unexpected situations – the situations that require going beyond pre-defined rules of the smart contracts – such information is introduced to the rules that is considered as the rules change.

In the research on conditional formalization of smart contracts [23], rules formalize conditions for the smart contracts and the ontology provides values of data properties for these contracts. The data from the ontology are inserted in the conditional part of a semantic rule and the smart contract checks whether the transaction is suitable for the rule. If it is, then the transaction is completed and the data resulting from this transaction are saved in the ontology.

3 Conceptual Framework

The conceptual framework follows the idea of solving the problem of creating service bundles as an ontology-based product configuration problem. Ontology-based product configuration enables efficient and effective development of product configuration systems by reusing configuration knowledge [1]. The configuration problem aims to find a configuration solution from a given set of predefined components to satisfy the individual needs of customers without violating any constraint imposed on the components [25]. Applying to the research purpose the components are services and a configuration solution is a service bundle.

In the conceptual framework (Fig. 1), the product configuration ontology represents knowledge of service bundles configuration. This knowledge among other thing comprises representations for services, services consumers, services providers, and various configuration constraints. *Service* is a product of mostly intangible nature that a provider offers to its consumers. *Consumer* is an entity (organization, agent, human, etc.) that receives services. *Provider* is an entity (usually, a company) that supplies services. Configuration constraints are encoded into the ontology, configuration rules are formalized in SWRL. The methodology of service bundle configuration [16] defines most of such constraints and rules, they are proposed to be adapted in the conceptual framework.

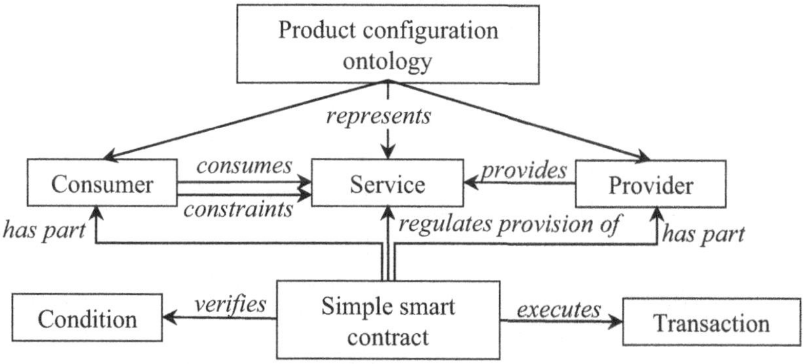

Fig. 1. Conceptual framework: interrelations between the product configuration ontology and smart contracts.

Smart contracts regulate agreements between the parties involved in service delivery and consumption. These contracts are activated (or process *transactions*) when certain *conditions* are met [4]. When a consumer agrees to consume a service, a *simple smart contract* is created. This contract regulates agreements on provision of this particular service. SWRL is used to express rules represented by the contracts. The product configuration ontology provides the vocabulary for the SWRL-rules.

Profiles of the consumers represent information about the services that these consumers usually get. When a consumer requests some service, the *configuring system* reads the information about the consumer's interests in his/her profile and creates an initial configuration of a *service bundle*. The service bundle combines services related to the requested service by the configuration constraints (if any) and other related services corresponding to the consumer interests. At that, services of different providers can be included into the bundle. The initial configuration is submitted to the consumer; the consumer could either approve this configuration or propose some changes.

After the consumer has agreed on the configuration, a *smart contract* for the service bundle is created (Fig. 2). Contradictory to procedure-based contract representation, where the order of the instructions in the procedure is important, the order of rules in rule-based contracts does not matter [26]. This enables to create a smart contract through a mere enumeration of the rules expressing the simple smart contracts for the services included into the bundle. Then, contextual rules are introduced in the smart contract. These rules are designed to enable services that meet the consumer needs in a specific situation (*context*) and disable services that are out of the current consumer interests. If the services are not free, the price for the service consumption is calculated accordingly. The smart contact regulates all the problems associated with the provision of services by different providers.

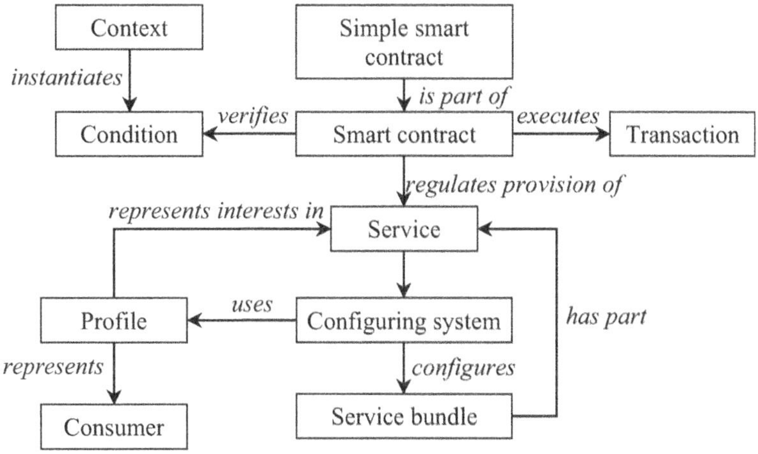

Fig. 2. Conceptual framework: context aware service provision.

Context represents the actual settings around the consumer. These settings instantiate the conditions of the smart contract. The conditions determine services that the consumer needs in the current situation. Consequently, the consumer is automatically provided with the necessary services.

4 Use Case

Often, consumers use different Internet service providers and Internet plans at different locations-times. They make contracts with multiple providers and switch between these providers depending on their contexts. If a service provider available at the current consumer location can be defined automatically, automatic costs customization for the services provided by different providers is not supported. The consumers have to contact their providers (through the personal account or in an office) to the providers suspend/resume their services at a certain time instant. After that, the providers independently of each other act in accordance with the appropriate contract on customer service and calculate the service cost accordingly. An ontology-based service bundle coupled with smart contracts enable to customize services from different providers automatically.

The proposed use case deal with the context-aware smart contracts for a bundle of Internet services provided by different providers. The case is very simple since it is intended to illustrate the ideas behind the conceptual framework only. Real situations would require models that are more complex. The problem of ontology-based bundle configuration is not included in the use case since the methodology on service bundle configuration [16] describes this problem in detail.

One requests a mobile operator a service to provide mobile Internet. The profile of the requester represents information that the person uses the wired Internet from another provider. The product configuration ontology combines the information from the person profile and the information on the requested service. In this ontology, the class *Service* is represented by two subclasses: *Mobile Internet* and *Wired Internet.* Mobile Internet is defined as the Internet as accessed by means of any mobile device; Wired Internet is a network that provides access to the Internet by being hard wired from the consumer's premises to the provider, within the premises the consumers can access this network via Wi-Fi. *Mobile operator* and *Digital lines* are providers of the services above (Fig. 3). The same SWRL-rule can be used to represent simple smart contracts between both providers and consumers of their services:

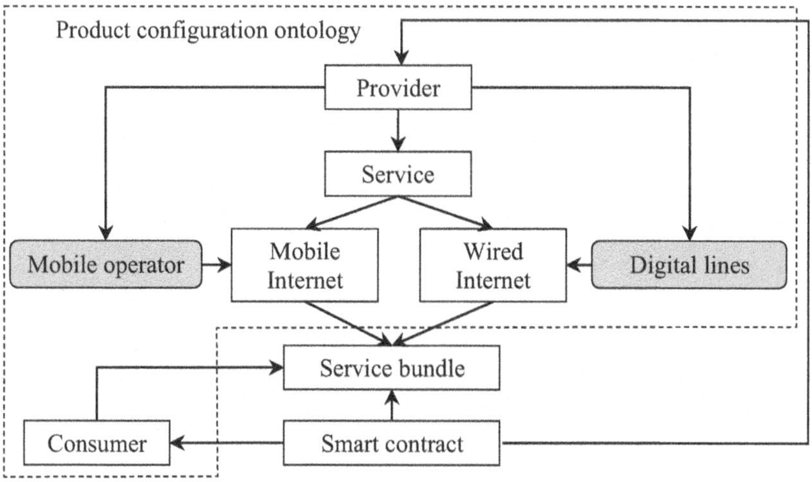

Fig. 3. Conceptual framework: use case.

```
Provider(?p) ^ Service(?s) ^ Consumer(?c) ^
swrlb:lessThan(CurrentDay, SettlementDay) ^
PaymentStatus(?s, true) ->
Provider(?p) ^ is_enabled(?s, true) -
```

Provider enables the service if the service has been paid and the current date is not the payment day.

The provider "Digital Lines" has the option to disable Internet service for some time with corresponding recalculation of the amount to be paid but the consumers have to activate this option by themselves. The plan that the person chooses for the mobile Internet assumes a monthly payment without the possibility of suspension for a number of days within a month.

The mobile operator suggests to the person a service bundle of two related services – the wired Internet and the mobile Internet (Fig. 3). The suggestion is about when the consumer uses the mobile Internet he/she does not use the wired Internet or Wi-Fi; the cost of the wired Internet service is calculated only for the days when it is actually consumed. The smart contract for this bundle automatically disables the wired Internet if the person does not authorize to the Wi-Fi zone of this Internet for a certain time and enables it when the person is in this zone. The person pays only for the days when he/she logged in to the network. The set of SWRL-rules representing the smart contract for this offering is as follows:

```
R1: Provider(Mobile_operator) ^ Consumer(Alex) ^
Service(Mobile_Internet) ^ swrlb:lessThan(CurrentDay,
SettlementDay) ^ PaymentStatus(Mobile_Internet, true) ->
Provider(Mobile_operator) ^ is_enabled(Mobile_Internet,
true)
R2: Provider(Digital_lines) ^ Consumer(Alex) ^
Service(Wired_Internet) ^ swrlb:lessThan(CurrentDay,
SettlementDay) ^ PaymentStatus(Wired_Internet, true) ->
Provider(Mobile_operator) ^ is_enabled(Wired_Internet,
true) -
```

The providers "Mobile Operator" and "Digital Lines" enable the service on Internet provision to consumer Alex if this service has been paid and the current date is not the payment day. Rules R1 and R2 are the rules that represent simple smart contracts on service provision.

```
R3: Provider(Digital_lines) ^ Service(Wired_Internet) ^
Consumer(Alex) ^ LogIn(Alex, Logged_in) ^
swrlb:subtract(?x, CurrentDay, Logged_in) ^
swrlb:greaterThan(?x, 3) -> is_enabled(Wired_Internet,
false) ^ swrlb:equal(st_dis, ?x) -
```

The provider “Digital Lines” disables the service on provision of wired Internet for consumer Alex and saves the last access date if Alex has not been online for more than 3 days.

```
R4: Provider(Digital_lines) ^ Service(Wired_Internet) ^
Consumer(Alex) ^ LogIn(Alex, Logged_in) ^
swrlb:subtract(?x, CurrentDay, Logged_in) ^
swrlb:equal(?x, 0) ^ is_enabled(Wired_Internet, false) ^
swrlb:subtract(?dis_d, CurrentDay, st_dis) ^
swrlb:multiply(?y, ?dis_d, day_sum) ->
is_enabled(Wired_Internet, true) ^
swrlb:subtract(AmountToPay, ?y) -
```

The provider “Digital Lines” enables the service on provision of wired Internet for consumer Alex and deducts the amount for the days when Alex did not access the Internet from the amount to be paid according to the plan.

5 Conclusion

The conceptual framework proposed in this paper enables to configure customized bundles of related services and automatically regulate provision and consumption of these services in context-aware way. Ontology-based representation of services allows ones to use methods of ontology-based product configuration to configure service bundles. Representations of the services compatible with representation of smart contracts makes it possible to regulate automatic service provision with reference to the consumer contexts. The use case from the Internet services domain demonstrates the main ideas behind the framework.

Acknowledgements. The research is funded mainly by the grant from the Russian Foundation for Basic Research (RFBR) (project no. 19-07-00630); the part of services customization based on the consumer profile is due to the grant from RFBR (project no. 20-07-00490); the use case is motivated partly by the research issue of State Research no. 0073–2019–0005.

References

1. Morelli, N.: Designing product/service systems: a methodological exploration. Des. Issues **18**, 3–17 (2002). https://doi.org/10.1162/074793602320223253
2. Baines, T.S., Lightfoot, H.W., Evans, S., Neely, A., Greenough, R., Peppard, J., Roy, R., Shehab, E., Braganza, A., Tiwari, A., Alcock, J.R., Angus, J.P., Bastl, M., Cousens, A., Irving, P., Johnson, M., Kingston, J., Lockett, H., Martinez, V., Michele, P., Tranfield, D., Walton, I.M., Wilson, H.: State-of-the-art in product-service systems. Proc. Inst. Mech. Eng. Part B J. Eng. Manuf. **221**, 1543–1552 (2007). https://doi.org/10.1243/09544054JEM858
3. Yang, L., Xing, K., Lee, S.-H.: A new conceptual life cycle model for result-oriented product-service system development. In: Proceedings of 2010 IEEE International Conference on Service Operations and Logistics, and Informatics, pp. 23–28. IEEE (2010). https://doi.org/10.1109/SOLI.2010.5551621
4. Szabo, N.: Smart contracts: building blocks for digital markets. EXTROPY J. Transhumanist Thought **16** (1996). https://www.fon.hum.uva.nl/rob/Courses/InformationInSpeech/CDROM/Literature/LOTwinterschool2006/szabo.best.vwh.net/smart_contracts_2. Accessed 25 Nov 2020
5. Smirnov, A., Sheremetov, L., Teslya, N.: Fuzzy cooperative games usage in smart contracts for dynamic robot coalition formation: approach and use case description. In: Proceedings of the 21st International Conference on Enterprise Information Systems, pp. 361–370. SCITEPRESS - Science and Technology Publications (2019). https://doi.org/10.5220/0007763003610370
6. Lee, H.: Product configuration strategy based on product family similarity. Int. J. Ind. Manuf. Eng. **7**, 1709–1713 (2013)
7. Akmal, S., Batres, R., Shih, L.H.: An ontology-based approach for product-service system design. In: The Philosopher's Stone for Sustainability, pp. 67–72. Springer, Heidelberg (2013). https://doi.org/10.1007/978-3-642-32847-3_11
8. Annamalai, G., Hussain, R., Cakkol, M., Roy, R., Evans, S., Tiwari, A.: An ontology for product-service systems. In: Functional Thinking for Value Creation, pp. 231–236. Springer, Heidelberg (2011). https://doi.org/10.1007/978-3-642-19689-8_41
9. Klingner, S., Becker, M.: Formal modelling of components and dependencies for configuring product-service-systems. Enterp. Model. Inf. Syst. Archit. **7**, 44–66 (2012). https://doi.org/10.1007/BF03345922
10. Hajimohammadi, A., Cavalcante, J., Gzara, L.: Ontology for the PSS lifecycle management. Procedia CIRP. **64**, 151–156 (2017). https://doi.org/10.1016/j.procir.2017.03.016
11. Medini, K., Boucher, X.: Specifying a modelling language for PSS Engineering – a development method and an operational tool. Comput. Ind. **108**, 89–103 (2019). https://doi.org/10.1016/j.compind.2019.02.014
12. Soininen, T., Tiihonen, J., Männistö, T., Sulonen, R.: Towards a general ontology of configuration. Artif. Intell. Eng. Des. Anal. Manuf. **12**, 357–372 (1998). https://doi.org/10.1017/S0890060498124083
13. Felfernig, A., Friedrich, G., Jannach, D., Stumptner, M., Zanker, M.: Configuration knowledge representations for Semantic Web applications. Artif. Intell. Eng. Des. Anal. Manuf. **17**, 31–50 (2003). https://doi.org/10.1017/S0890060403171041
14. Xuanyuan, S., Li, Y., Patil, L., Jiang, Z.: Configuration semantics representation: a rule-based ontology for product configuration. In: 2016 SAI Computing Conference (SAI), pp. 734–741. IEEE (2016). https://doi.org/10.1109/SAI.2016.7556062

15. Yang, D., Dong, M., Miao, R.: Development of a product configuration system with an ontology-based approach. Comput. Des. **40**, 863–878 (2008). https://doi.org/10.1016/j.cad.2008.05.004
16. Dong, M., Yang, D., Su, L.: Ontology-based service product configuration system modeling and development. Expert Syst. Appl. **38**, 11770–11786 (2011). https://doi.org/10.1016/j.eswa.2011.03.064
17. McGuinness, D.L., van Harmelen, F.: OWL Web Ontology Language Overview. https://www.w3.org/TR/owl-features/. Accessed 25 Mar 2020
18. Horrocks, I., Patel-Schneider, P.F., Boley, H., Tabet, S., Grosof, B., Dean, M.: SWRL: A Semantic Web Rule Language Combining OWL and RuleML. https://www.w3.org/Submission/SWRL/. Accessed 20 Feb 2020
19. de Kruijff, J., Weigand, H.: Ontologies for commitment-based smart contracts (2017). https://doi.org/10.1007/978-3-319-69459-7_26
20. Baqa, H., Truong, N.B., Crespi, N., Lee, G.M., le Gall, F.: Semantic smart contracts for blockchain-based services in the Internet of Things. In: 2019 IEEE 18th International Symposium on Network Computing and Applications (NCA), pp. 1–5. IEEE (2019). https://doi.org/10.1109/NCA.2019.8935016
21. Kim, H.M., Laskowski, M.: Toward an ontology-driven blockchain design for supply-chain provenance. Intell. Syst. Account. Financ. Manag. **25**, 18–27 (2018). https://doi.org/10.1002/isaf.1424
22. Choudhury, O., Rudolph, N., Sylla, I., Fairoza, N., Das, A.: Auto-generation of smart contracts from domain-specific ontologies and semantic rules. In: 2018 IEEE International Conference on Internet of Things (iThings) and IEEE Green Computing and Communications (GreenCom) and IEEE Cyber, Physical and Social Computing (CPSCom) and IEEE Smart Data (SmartData), pp. 963–970. IEEE (2018). https://doi.org/10.1109/Cybermatics_2018.2018.00183
23. Hyun Na, S., Hyun An, J., Yang, J.S., Park, Y.B.: Conditional formalization of smart contract using semantic web rule language. J. Eng. Appl. Sci. **13**, 8716–8721 (2018). https://doi.org/10.36478/jeasci.2018.8716.8721
24. Mohsin, F., Zhao, X., Hong, Z., de Mel, G., Xia, L., Seneviratne, O.: Ontology Aided Smart Contract Execution for Unexpected Situations. Troy, Hursley (2019)
25. Mittal, S., Frayman, F.: Towards a generic model of configuration tasks. In: Proceedings of the Eleventh International Joint Conference on Artificial Intelligence, pp. 1395–1401 (1989)
26. Idelberger, F., Governatori, G., Riveret, R., Sartor, G.: Evaluation of logic-based smart contracts for blockchain systems (2016). https://doi.org/10.1007/978-3-319-42019-6_11

Method of Synthesis of the Information Model of the Judicial Decision Support System

L. E. Mistrov and A. V. Mishin(✉)

Central Branch FGBOUVO «Russian State University of the Justice», 20 Years of October 95, 394006 Voronezh, Russia
odpvo@mail.ru

Abstract. The method of synthesis of the information model of the system of support of judicial decisions aimed at increasing the efficiency and confidence in the work of judges is proposed. The method is based on the methods of decision-making theories in fuzzy conditions, fuzzy sets, and sequential improvement of decision options.

1 Introduction

In modern conditions of development of the information society, the development and implementation of court decision support systems (DSSS) is an urgent direction for improving the quality of justice. Currently, we have accumulated experience in developing such systems-advisers in various fields of activity [1–5], including in courts [6, 7].

In this regard, we propose a method for synthesizing the information model of the DSSS, which provides a logical-heuristic representation of the judicial process in the form of formalized procedures for resolving conflicts (offenses) in the socio-economic sphere on the basis of formal criteria for evaluating the validity of judicial decisions.

2 Materials and Methods

The judge's decision on the case in question represents the procedure for choosing a solution to resolve the conflict based on the criteria of legality and justice in the form of

$$K = K_1 \cdot K_2,$$

where K_1 – is the criterion for the judge to make a legal decision to resolve the conflict in accordance with the articles of current legislation;

K_2 – the criterion of "justice" the decision of the judge, characterizing the magnitude of the acceptable risk when making.

Representation of the judge's decision in the form of this dependence allows taking into account objective (legality) and fair (subjectivity) circumstances when making a decision, which are guided by the judges when making decisions.

The most common form of formalization of these criteria is their representation in the form of a display φ [8, 9], which corresponds to each element of the set of

R. Silhavy et al. (Eds.): CoMeSySo 2020, AISC 1295, pp. 522–530, 2020.
https://doi.org/10.1007/978-3-030-63319-6_48

acceptable options for the judge's decisions V_D [10] a subset of rational V^{opt} options: $\varphi: V_D \rightarrow V^{opt}$, $V^{opt} \in V_D$. At the same time, the display φ must correspond to the functional purpose of the judge when considering cases, be relatively easy to determine, have a clear physical meaning, and be resistant to minor changes in the conflict conditions.

In [8, 9, 11, 12] it is shown that the criteria for making decisions of judges on conflict resolution depend to a certain extent on the nature of the uncertain factors taken into account. This makes it possible to distinguish between the synthesis of judges decisions in terms of certainty, risk, uncertainty and vagueness (in the sense of Zadeh [13]). Synthesis of solutions for these first three conditions is carried out with a well-defined set of solutions V_D and unambiguity of the target function display φ.

The synthesis of decisions in fuzzy conditions is associated with the judge's fuzzy ideas about the tasks set, the set of acceptable options for resolving the case V_D and the strategies of behavior of the subjects of the conflict Y, as well as with the ambiguity of the display φ. Depending on the conditions of decision synthesis, the judge's decision-making criterion is formally presented on the basis of the maximum utility principle in a mathematical form of the form

$$Max < V_D, R > = Arg \max_{v \in V_D} U(v). \tag{1}$$

With a certain meaningful interpretation of the utility of the judge's decision, the maximum utility principle allows you to reduce the synthesis problem to a mathematical programming problem.

In a typical case, the task of decision-making by the judge under certainty is determined by the set of feasible options V_D and strategies of the actors of the conflict; from the set selected (in the General case, there may be several), with maximum usefulness. Formally, this problem is a mathematical programming problem and is represented as

$$V^{opt} = V^{opt}(Y) = Arg \max_{v \in V_D} U(v, Y). \tag{2}$$

It follows from Eq. (2) that in the synthesis of a legal solution in terms of certainty, the principle of preference is used, which coincides in form with the principle of maximum utility (1).

The synthesis of the judge's decisions under risk conditions takes place when the uncertain factors of the conflict subjects behavior strategy are random variables with known distribution laws. In this case, the risk is understood as the degree of discrepancy between the properties and characteristics of the judge's decision and the behavior strategies of the subjects. The essence of the problem in these conditions is as follows: a set of acceptable solutions V_D and strategies for interaction of the conflict subjects are defined Y, the laws of distribution of random factors $\tilde{Y} \subset Y$ of these conditions are known; a variant (or several variants) is selected from the set of V_D that minimizes the risk of making a judge's decision in a real situation. The minimum risk is formally equivalent in a certain probabilistic sense to the maximum utility of the solution. Therefore, depending on one or another meaningful interpretation of the synthesis of the probabilistic meaning of the utility of the solution, the maximum utility principle can take a different form.

Analysis of the above criteria (principles of preference) for the decision of the judge on conflict resolution in relation to the factors taken into account allows using the principle of maximum utility, implemented in the form of a mathematical programming problem, to find a legitimate solution. In these conditions, the task of making a legal and fair decision by the judge to resolve the conflict consists in choosing the preferred solutions $V^{opt} \in V_D$ from a set of acceptable strategies V_D for $y \in Y$ the behavior of subjects that meets the Hurwitz criterion and is presented in the form of

$$Arg \max_{v \in V_D} [K_2 \min_{y \in Y} K_1(v, y) + (1 - K_2) \max_{y \in Y} K_1(v, y)], \tag{3}$$

where $K_1(v, y)$ – is the target function (criterion) of the judge's decision-making.

In general, the judicial process for reviewing cases is iterative, aimed at consistently parrying the information uncertainty by the judge at the stages of the trial in order to make a legitimate and methodically justified decision, clarified from the position of fairness. The decision-making process of the judge is carried out at discrete points in time $t = \overline{1, T}$, in which the step of discretization of the decision-making process can vary from stage to stage. Part of the decisions, mainly at the initial stages of the trial, are aimed at clarifying the initial data about the conflicting system "plaintiff – defendant" ("prosecutor – defendant"), while the other part only coordinates the work of representatives of the parties. In this case, the judge can coordinate the target functions of representatives of the conflicting parties (in the task, it is assumed that they are defined). When making decisions, the main goal of the judge is to find the decision vector $x_i^o(t)$ at the i-th ($i = \overline{1, N}$) stage of the trial, which provides the maximum of the vector of its target function $K_1(v, y, x_1^o, \ldots, x_i^o, \ldots, x_N^o)$ with the coordinating goal $x_{N+1}^o(t)$ defined by the judge. The found decision of the judge $\{x_i^o(t)\}$ must belong to a subset of acceptable decisions – meet the requirements of the relevant procedural codes, be legal and fair.

The use of a single criterion $K_1(v, y, \overline{x}_N^o)$ with subsequent clarification of its stages of trial complicates the task of the judge to resolve the conflict, but allows you to take into account the target functions of representatives of the parties (lawyers, prosecutors). In addition, the objective function of judge $K_1(v, y, \overline{x}_N^o)$ does not explicitly depend on the decisions proposed by the representatives of the parties, which makes it difficult for them to work in analyzing and justifying the behavior strategies of the parties to the conflict. Therefore, it is assumed that each j-th representative at the i-th stage of the trial has goals $K_1(v, y, \overline{x}_{ij})$ defined on the sets of their admissible decisions, and the integral target function $K_1(v, y, \overline{x}_1, \ldots, \overline{x}_N)$ is a vector function and depends on both the judge's criterion $K_1(v, y, \overline{x}_N^o)$ and the target functions of the representatives of the parties $K_1(v, y, \overline{x}_{ij})$.

If we present $\overline{x}_N$ in the form of $\overline{x}_N = \{\overline{x}_{ij}\}$, $i = \overline{1, N}, j = \overline{1, M}$, where $M = m_i$ - is the number of participants in the trial at the i-th stage (in a particular case, it is assumed that $M = 3$: $j = 1$ - is the judge, $j = 2$ and $j = 3$ – are representatives of the subjects of the conflict), then it is assumed that when finding solutions $\overline{x}_{(i-1)j}$ ($j = \overline{1, m_{i-1}}$), all decisions $\bar{x}_{ij}^{o}$ ($j=\overline{1,m_i}$) have already been taken. Therefore, it is possible to use methods of group optimization, for example [7], and invariant immersion [9] to optimize the judge's decisions. This allows us to consider a family of problems with $N=1,2,3,\ldots;\ x_{N+1}^{o}=[0,\ \infty]$ instead of a problem with a specific solution (plan) x_{N+1}^{o} at the $(N+1)$-th stage of the trial and a fixed number of stages N. In this case, the maximum of the $K_1(v,y,\bar{x}_1,\ldots,\bar{x}_N)$ function depends on the values of N and $\bar{x}_i$ – and this dependence must be set. Based on this, problem (3) is decomposed by using the invariant immersion method into a set of problems related to the decisions of the judge at the stages of the trial. Thus as characteristics of the decisions made j-e participants in the trial used membership function variants of their decisions to the subset of acceptable ($C\subset X$), effective (appropriate punishment for the offense) and coordinated (meet the prosecutor (plaintiff) and the defendant (respondent)) decision $\mu_R(\bar{x}_N)$. This function is not limited to determining only the quantitative (crisp) values of the function $K_1(v,y,\ \bar{x}_N)$, as it includes not only the characteristic of effective solutions, but also the characteristics of its validity (the degree of membership of a solution $\bar{x}_N$ to the subset of acceptable solutions C) and coordination with the purposes and limitations of the representatives of the parties (the degree of belonging of the solution $\bar{x}_N$ to the subset of the coordinated decisions of the representatives of the parties K) based on the methods of vector optimization. Membership functions are defined on the set of solutions $X=\{x_{ij}\}$.

Effective solutions of fuzzy sets objective $G\subset X$ as a fuzzy subset with membership function $\mu_G(x_{ij})$, the admissibility decisions – fuzzy subset of $C\subset X$ with membership function $\mu_C(x_{ij})$ and coordination (similar to [9]) – fuzzy set $K\subset X$ with membership function $\mu_K(x_{ij})$. The resulting effect of fuzzy goals G, constraint C and coordination K on the choice of solution x_{ij} can be represented by the intersection of $G\cap C\cap K$. Then the function of the judge's decision belonging to the set of admissible solutions for the intersection is represented by the relation

$$M_{G\cap C\cap K}(x_{ij})=\min\{\mu_G(x_{ij}),\mu_C(x_{ij}),\mu_K(x_{ij})\},\ x_{ij}\in X. \tag{4}$$

Based on this, a subset of $R=G\cap C\cap K$ is represented by a vector (fuzzy) solution of the j-th representatives of the parties at the i-th stage, and

$$M_R(x_{ij})=M_{G\cap C\cap K}(x_{ij}). \tag{5}$$

When resolving a material dispute between the parties of conflict or the presence of a variety of balance equations describing the conflict, the decision of judge r-th level decision ($j=1$) to $(i+1)$-th stage of the trial may be due to the decision of the representatives of the parties at the i-th stage equation $x_{(i+1)r}=K_1(v,y,x_{i1},\ldots,x_{iM})$.

If the set of solutions X consists of only one parameter, the equation is represented as

$$x_{(i+1)r}=\sum_{j=1}^{M}x_{ij}. \tag{6}$$

However, solutions are more complex and are characterized by a vector of parameters $\bar{x}_{ij}$, such as a property dispute, expenditure of funds and etc.

The decision of the judge coordinated with the representatives of the parties at the N-th stage in accordance with [10] is represented by the expression

$$R(x_N) = [\prod_{k=1}^{N-1} \prod_{l=1}^{M} R(x_{kl})] \cap C(x_N) \cap G(x_N), \tag{7}$$

from which it is easy to get a simple recurrent procedure for making decisions by the judge at the N-th stage of conflict resolution. Due to the invariance of the entered description for the stages of conflict resolution, it follows from expression (7) that

$$R(x_{ij}) = [\prod_{k=1}^{N-1} \prod_{l=1}^{M} R(x_{kl})] \cap C(x_{ij}) \cap G(x_{ij}), \tag{8}$$

in particular,

$$R(x_{(i+1)r}) = [\prod_{k=1}^{N} \prod_{l=1}^{M} R(x_{kl})] \cap C(x_{(i+1)r}) \cap G(x_{(i+1)r}), \tag{9}$$

Due to the fact that the decision on the i-th trial depends only on the installed data on the characteristics of the environment and strategies of behaviour of the actors of the conflict at this stage; the decisions taken by the judge on the $(i-1)$-th stage and the decisions of the representatives of the parties, the decision $R(x_{ij})$ takes into account all strategies of the representatives of the parties in accordance with the expression (8). Then the expression (9) can be represented through fuzzy solutions of the previous stage as

$$R(x_{(i+1)r}) = [R(x_{11}) \cdot \ldots \cdot R(x_{1M})] \cap C(x_{(i+1)r}) \cap G(x_{(i+1)r}). \tag{10}$$

Moreover, in order for the judge to make a decision at the $(i-1)$ -th stage, he must first convert the decisions of the representatives of the parties $R(x_{i1}), \ldots, R(x_{iM})$ into his own decision $R(x_{i1}, \ldots, x_{iM})$. Then the expression (10) will take the form

$$R(x_{(i+1)r}) = R(x_{11}, \ldots, x_{1M}) \cap C(x_{(i+1)r}) \cap G(x_{(i+1)r}). \tag{11}$$

The presence of such a recurrent procedure is due to the peculiarity of decision-making by the judge and the presence of legally enshrined rights to obtain missing information and make decisions. However, in contrast to the traditional representation of decision parameters by numbers, the decisions of the representatives of the parties in the process of their coordination are fuzzy and are set by membership functions $\mu(x)$. Therefore, for the membership functions of the considered subset of effective and admissible solutions, Eq. (11) takes the form

$$\mu_R(x_{(i+1)r}) = \mu_R(x_{i1}, \ldots, x_{iM}) \otimes \mu_G(x_{(i+1)r}) \otimes \mu_C(x_{(i+1)r}), \tag{12}$$

where

$$\mu_R(x_{i1}, \ldots, x_{iM}) = \mu_R(x_{i1}) \otimes \ldots \otimes \mu_R(x_{iM}). \tag{13}$$

A binary operation $\otimes$ can be represented by an algebraic product or addition. The choice of a particular type of operation is determined by the type of operation on sets, the physical meaning of membership functions, and the axiomatics of fuzzy sets.

To make a decision at the (i + 1) -th stage, taking into account Eq. (6), it is necessary to design the coordinating decision of the judge (13) on the axis $x_{(i+1)r}$, i.e. to present it as

$$\mu_K(x_{(i+1)r}) = \mu_R(\sum_{j=1}^{M} x_{ij}). \tag{14}$$

Equation (14) shows that the judge's decision $x_{(i+1)r}$ is based on taking into account the decisions of the representatives of the parties $x_{i1}, \ldots, x_{iM}$ and will be most effective when each representative at the i-th stage makes optimal decisions (maximizes its membership function), and the judge in determining his decision seeks to maximize the projection of these decisions on the axis $x_{(i+1)r}$, that is

$$\mu_K(\sum_{j=1}^{M} x_{ij}) = \max_{\{x_{ij}\}}[\mu_R(x_{i1}) \otimes \ldots \otimes \mu_R(x_{iM})]. \tag{15}$$

The judge's decisions have the property that whatever the state and decision at the previous stage of the trial, subsequent decisions must be optimal in relation to this decision. This makes it possible to use the Bellman optimality principle [14] when implementing the most effective method of coordinating decisions taken at individual stages of the trial, to minimize the exchange of information between stages and to ensure Autonomous work with information from representatives of the parties.

In this regard, the recurrent decision-making procedure by the judge at the (i + 1) -th stage includes two stages.

Stage 1. The optimization problem is solved on the Cartesian product of sets $x_{i1} \times \ldots \times x_{iM}$ with the target function (15) at the (i + 1) -th stage, taking into account the restriction (6) and restrictions of the form:

$$\overline{M}(x_{i1}, \ldots, x_{iM}) = 0, \ \overline{W}(x_{i1}, \ldots, x_{iM}) \geq 0, \tag{16}$$

describing the relationship of strategies for the behavior of representatives of the parties at the i-th stage.

The result of solving the problem is the membership function (14).

Stage 2. On the basis of Eqs. (12) and (13), the fuzzy decision $\mu_K(x_{(i+1)r})$ of the judge at the (i + 1) -th stage is found. In order for the judge to make a clear decision at the N-th stage, it is necessary to choose a decision x_N^o for which the degree of belonging to a subset of effective and acceptable solutions $\mu_R(x_N)$ is the maximum, i.e. $\mu_R(x_N) = \max_{\{x_N\}} \mu_R(x_N)$.

After making a clear decision at the N-th stage (based on the solution of the optimization problem (15) and (16)) for $\sum_{j=1}^{M} x_{(N-1)j} = x_N^o$ there are clear solutions at the $(N-1)$-th stage: $x^o_{(N-1)j}, \ldots, x^o_{(N-1)M}$. This procedure is repeated until all decisions are clearly and quantitatively made by the judge $\{x_{ij}\}$ $(i = \overline{1,N}, \quad j = \overline{1,M})$.

Note that if condition (16) is not met, the final decision of the judge may be appealed by representatives of the parties on appeal or cassation.

3 Results

Thus, the decision of the representatives of the parties at the i-th stage are described by membership functions $\mu_R(x_{ij})$, and the decision of the judge to coordinate decisions in the trial – function $\mu_K(x_{(i+1)r})$. Based on this, the tasks solved by the representatives of the parties in the process are coordinated among themselves at time t only when

$$\forall j\, \exists x_{ij}\, \exists x_{(i+1)r} : \mu_R(x_{ij}) \neq 0 \wedge \mu_K(x_{(i+1)r} \neq 0. \tag{17}$$

Therefore, the coordination of the tasks of the representatives of the parties requires that the problem in question has a solution for at least one $x_{(i+1)r}$, and for this solution of the set of particular tasks R, solved by the representatives of the parties, also has a solution.

In general, the membership function $\mu(x_{ij})$ can be conditional, that is, it depends on some parameter y_{ij} – $\mu(x_{ij}|y_{ij})$. For example, if x_{ij} you take into account the time parameters of the conflict at the i-th stage then y_{ij} – the amount of information received about the parameters of the conflict in question.

4 Discussion

So, there is a fundamental possibility of automating the process of making court decisions. There remains another, equally important question: what should be the degree of automation – full or partial? To answer this question, it is necessary to take into account the factors of social protection of judges, the effectiveness and efficiency of their decisions. Automation, even partial, will, on the one hand, reduce social tension in the work of judges due to the possibility of formalized consideration of their decisions. On the other hand, it will allow you to quickly respond to clarifying changes in evidence in the court process and speed up the process of making an effective decision. And, third, it will increase the social justice of the decisions made, since each of them is based on the interests and destinies of specific citizens, and their perception of the justice of the state on behalf of which judges make decisions.

These factors make it necessary to partially automate judicial decision – making, which ensures the development of recommendations on the conflict to be resolved, leaving it to the judge to make a final objective and fair decision.

The use of the method will reduce the emotional and psychological stress of judges work, increase the effectiveness of their decisions and improve the attitude of citizens to the judicial authorities.

5 Conclusion

The proposed method meets the basic requirements of system analysis [8, 15, 16], since it provides for modeling: the integrity of the representation of the judge's functioning in conflict resolution in the process of analyzing the entire conflicting system (Prosecutor, plaintiff, defendant, defendant and lawyers) from the position of a single goal; comprehensive consideration and consideration of the main elements and processes of the conflicting system and the relationships between them; a sufficient degree of simplification in modeling, which allows us to adequately reflect the judicial process and take into account the determining factors in it – this is due to the fact that when developing solutions, we do not study the judicial process itself, but its model. On its basis, it is possible to build numerical procedures that form the basis of the information model of the DSSS by a judge for the legal and fair resolution of conflicts.

References

1. Smirnov, A.V., Levashova, T.V., Ponomarev, A.V.: Decision support based on human-machine collective intelligence: state-of-the-art and conceptual model. Inf. Control Syst. **2** (105), 60–70 (2020)
2. Pakhomova, K.I., Korobko, A.V.: Application of formal concept analysis for intellectual decision support. Artif. Intell. Decis. Making **4**, 37–46 (2019)
3. Smirnov, A.V., Levashova, T.V.: Models of decision support in socio-cyber-physical systems. Inf. Control Syst. **3**(100), 55–70 (2019)
4. Dai, Y.N., Viken, G., Joo, E., Bente, G.: Risk assessment in e-commerce: how sellers' photos, reputation scores, and the stake of a transaction influence buyers' purchase behavior and information processing. Comput. Hum. Behav. **84**, 342–351 (2018)
5. Rybina, G., Blokhin, Yu.: Methods and software of intelligent planning for integrated expert systems construction. Artif. Intell. Decis. Making **1**, 12–28 (2018)
6. Chinese judges will be replaced by computer programs. http://www.ultrachip.ru/news/1925-1829.html
7. The Chinese judges in sentencing will help the computer. http://mc.by/publication/369/
8. Mistrov, L.E., Serbulov, Y.S.: Methodological Bases of Synthesis of Information-Providing Functional Organizational and Technical Systems. Scientific Book, Voronezh (2007)
9. Mistrov, L.E.: Method for synthesizing strategies for resolving socio-economic conflicts. Dyn. Complex Syst.-XXI Century **4**, 41–65 (2016)
10. Novikova, N.M., Pospelova, I.I., Zenyukov, A.I.: Method of convolution in multicriteria problems with uncertainty. J. Comput. Syst. Sci. Int. **56**(5), 774–795 (2017)
11. Noghin, V.D.: Multicriteria choice based on fuzzy information. Artif. Intell. Decis. Making **2**, 50–61 (2019)

12. Sedakov, A.A., Zhen, M.: Opinion dynamics game in a social network with two influence nodes. Vestnik Saint Petersburg Univ. Appl. Math. Comput. Sci. Control Process. **15**(1), 118–125 (2019). https://doi.org/10.21638/11702/spbu10.2019.109
13. Luhandjula, M.K.: Fuzzy optimization: milestones and perspectives. Fuzzy Sets Syst. **74**, 4–11 (2015)
14. Azanov, V.M., Kan, Y.S.: Bilateral estimation of the Bellman function in the problems of optimal stochastic control of discrete systems by the probabilistic performance criterion. Autom. Remote Control **79**(2), 203–215 (2018)
15. Alasheev, M.A., Bresler, I.B., lyaskovsky, V.L.: Methods and models for selecting system-technical solutions for creating (developing) distributed information and control systems of organizational type. Proc. Russ. Acad. Sci. Theory Control Syst. **2**, 111–127 (2020)
16. Baranov, V.V., Tsvirkun, A.D.: Development control: structural analysis, problems, stability. Autom. Remote Control **79**(10), 1780–1796 (2018)

Predictive Model of Energy Consumption of a Home

Michal Mrazek[1(✉)], Daniel Honc[1], Eleonora Riva Sanseverino[2], and Gaetano Zizzo[2]

[1] Faculty of Electrical Engineering and Informatics, University of Pardubice, Pardubice, Czech Republic
st36200@student.upce.cz

[2] Engineering Department, University of Palermo, Palermo, Italy
eleonora.rivasanseverino@community.unipa.it

Abstract. The usage of the renewable energy sources and electric vehicles is rising which might result in a problem with the energy grid stabilization. Some works state that buildings contribute to the overall energy consumption by 20–40% which makes them a great subject to study. Finding solutions for better energy management can have a significant impact on the energy sector. This work focuses on proposing a simplified model of a home considering main appliances, heating and cooling, a photovoltaic system, an electric vehicle and weather and solar irradiance forecasts. The model is created in MATLAB as a simulation script with defined parameters. It uses a 5 day weather forecast to predict energy demands of a home as well as expected energy generation by the photovoltaic panels. This model can be used to better understanding the energy demand of either a single building or a community of buildings put together based on the information about the outside environment. Another possibility is to use the model for predictive optimization in order to find better ways of energy usage in a building.

Keywords: Home simulation · Energy demand · Weather forecast

1 Introduction

In some works it is pointed out that buildings account for 20–40% of the overall energy consumption in the developed countries out of which 50% corresponds to HVAC energy consumption [1, 2]. As the penetration of renewable energy sources rises the demands on the power grid are growing as the stabilization of the grid becomes more difficult by incorporating these resources [3, 4]. Electric vehicles will also become a bigger part of the grid which will lead to overall higher energy demands as well as greater peaks of this demand during the days [5]. That might then lead to even greater instability in the grid.

There are several studies that propose solutions for better energy usage in homes or complexes of bigger buildings. Some consider better predictive models of energy consumption of the consumers (which can be either residential or industrial) with regards to the Demand Response methods [6]. These methods are mainly used from the

R. Silhavy et al. (Eds.): CoMeSySo 2020, AISC 1295, pp. 531–540, 2020.
https://doi.org/10.1007/978-3-030-63319-6_49

perspective of aggregators to shave the peaks of energy demand. The energy demands of industrial customers seem to be more predictable than that of the residential customers [7]. Other studies focus on the predictive models from the perspective of the consumers. These methods focus on finding solutions of better energy management in a home for example. Better strategy consists of scheduling of appliances, calculation of desired temperatures as heating and cooling which play a big role in the overall energy consumption, or scheduling of battery usage. This is achieved by using optimization methods such as variations of Model Predictive Control method that takes into account a model of the building and its appliances, weather and solar irradiance forecasts as well as pricing and several constraints for which the optimal solution should be found [1, 3, 4].

All of these methods are based on a precise model that can be used for the simulations and predictions. This work focuses on creating a simplified model of a home with all the main components to understand their effect on the overall power consumption. The aim is to obtain a model that could be used for an optimization. Such an optimization as well as the model should focus on understanding what is sensible to optimize in order to achieve the highest impact on the better energy management in a home.

This model could also be used to simulate how a block of houses or flats with different properties would behave together. Such results could be used to better understand a Renewable Energy Community [8].

2 Materials and Methods

The model of a home contains several building blocks. Each one represents a part that affects the energy consumption. Firstly, the model contains appliances that have a fixed power consumption when turned on. These appliances are built into the model through specifying a timeframe when these should be turned on in any given day in a week. Examples of such appliances are a fridge, kitchen appliances, a washing machine, TV and PC or lighting. Secondly, the model contains heating and cooling which have a great effect on the overall power consumption. These are connected with a weather forecast, mainly the temperature and solar irradiation forecasts which have the biggest effect on the temperature indoors. Thirdly, the model contains a photovoltaic system that takes into account the solar irradiation forecast and expected levels of cloudiness. And lastly, a model of a battery of an electric vehicle is built into the model.

2.1 Simple Appliances

Such appliances in the model are a fridge, kitchen appliances, a washing machine, TV and PC and lighting. All of these are considered to have a constant power consumption so they are defined by their timing during the week.

2.2 Weather Forecast

Weather forecast uses the OpenWeatherMap API that has different types of forecasts available via a personal key that is obtained after a free registration [9]. The 5 day forecast with 3 h timestep was used in this work. The forecast contains information about outside temperature, pressure, humidity, cloudiness, wind speed and direction, rain and snow.

2.3 Sunshine Irradiation Forecast

For the irradiation forecast a system from the European Union Science Hub called PVGIS (Photovoltaic Geographical Information System) was used [10]. It contains historical data of irradiation in different areas around the world measured for several years. The monthly irradiation data were chosen for this work as they account for the typical day of a given month for a given place. In this case Palermo was chosen as the city of the interest. Such irradiation profile was then recalculated based on the expected cloudiness in the upcoming days by the formula

$$I_{\text{clouded}} = 0.1 \cdot I_{\text{forecasted}} + 0.9 \cdot (1 - C/100) \cdot I_{\text{forecasted}} \tag{1}$$

where

I_{clouded} is the irradiation with cloudiness taken into account, W/m^2,
$I_{\text{forecasted}}$ is the typical irradiation on a clear day, W/m^2,
C is the level of cloudiness, %.

The attenuation of the irradiation by clouds is considered to be 90%. That means that even when the expected cloudiness level is 100% there is still some part of the irradiation that gets to the surface of the Earth.

2.4 Heating and Cooling Subsystem

Both are connected with the weather forecast. The model takes into account the desired temperature indoors and the outside environment, or the forecast of all of these. Heating and cooling both have a defined maximum power consumption and efficiency of transforming electrical energy into heating or cooling power. Energy balance is described by the equation

$$P_{\text{heating}} + P_{\text{Sun}} - P_{\text{cooling}} = K \cdot (T_{\text{in}} - T_{\text{out}}) \tag{2}$$

where P_{heating} is the heating power, W,
P_{Sun} is the heating power of the Sun, W,
P_{cooling} is the cooling power, W,
K is the loss coefficient, $W/°C$,
T_{out} is the outside temperature, $°C$,
T_{in} is the inside temperature, $°C$.

Inside temperature is considered to be equal to the desired temperature as long as the heating or cooling power is less than their defined maximums. The amount of power that needs to be added or removed in order to keep the indoor temperature on desired value can be described by the formula

$$P_{\text{heating}} - P_{\text{cooling}} = K \cdot (T_{\text{desired}} - T_{\text{out}}) - P_{\text{Sun}}. \tag{3}$$

Based on the result the next steps can be described by the following algorithm, where heating and cooling power are calculated taking into account the maximal power and actual inside temperature

```
if P_heating - P_cooling > 0
    if P_heating > P_heating_max
        P_heating_power_consumption = P_heating_pc_max;
        P_heating = P_heating_max;
    else
        P_heating = K * (T_desired - T_out) - P_Sun;
        P_heating_power_consumption = P_heating/k_heating_efficiency;
    end
    T_inside = (P_heating + P_Sun)/K + T_out;
else
    if P_cooling > P_cooling_max
        P_cooling_power_consumption = P_cooling_pc_max;
        P_cooling = P_cooling_max;
    Else
        P_cooling = K * (T_desired - T_out) + P_Sun;
        P_cooling_power_consumption = P_cooling/k_cooling_efficiency;
    end
    T_inside = (P_Sun - P_cooling)/K + T_out;
end
```

Heating power of sunshine is defined by the area affected by the Sun during the day and the solar irradiation forecast by the formula

$$P_{\text{Sun}} = I_{\text{clouded}} \cdot A_{\text{exposed}} \cdot k_{\text{solar\_heat}} \tag{4}$$

where P_{Sun} is the heating power from the Sun that is transferred indoors, W,
A_{exposed} is the area that is exposed to the Sun, m^2,
$k_{\text{solar\_heat}}$ is the coefficient of irradiation transformation to heating power.

2.5 Photovoltaic Subsystem

The photovoltaic is built into the model in a similar way as the heating power of the Sun. It is described by an area of the photovoltaic panels, the irradiation forecast with cloudiness levels and the efficiency of turning the solar power into electrical power. Electrical power obtained from the photovoltaic panels is described by the formula

$$P_{pv} = I_{clouded} \cdot A_{pv} \cdot k_{pv\_efficiency} \tag{5}$$

where P_{pv} is the electrical power generated by the photovoltaics, W,
A_{pv} is the area of the photovoltaic panels, m^2,
$k_{pv\_efficiency}$ is the efficiency factor of photovoltaic panels.

2.6 Electric Vehicle and Battery Subsystem

Battery subsystem in this model is considered only as an appliance in the form of an electric vehicle. State of charge (SOC) of the battery in charging mode is given by the equation

$$SOC(k) = SOC(k-1) + P_{charging} \cdot \varepsilon_{ch} / C_{max} \cdot 100 \tag{6}$$

where $SOC(k)$ is the state of charge in the current step, %,
$SOC(k-1)$ is the state of charge in the previous step, %,
$P_{charging}$ is the charging power, kW,
ε_{ch} is the charging efficiency,
C_{max} is the maximum possible battery capacity, kWh.
The discharging mode is described by the equation

$$SOC(k) = SOC(k-1) - P_{discharged} / \varepsilon_d / C_{max} \cdot 100 \tag{7}$$

where $P_{discharged}$ is the discharged power on the output of the battery, kW,
ε_d is the discharging efficiency.

Charging and discharging have their own timing in the simulation to model when the battery is charged and when it is discharged. The energy from the battery in the model is only used for driving the electric vehicle and not as a supporting energy system for the home, which could be the next feature implemented in the model.

Both modes also have a maximum power that can be charged or discharged in any given moment. It is considered that when either mode is active the maximum amount of power is charged/discharged unless the battery is close to its limits of the State of charge, either maximum or minimum, in which case the charge or discharged power is calculated. This can be shown by the algorithm

```
if charging
    SOC(k) = SOC(k-1) + Pcharging*epscharging/Cmax*100;
    if SOC(k) > 100
        SOC(k) = 100;
        Pcharging(k) = (SOC(k) - SOC(k-1))/epscharging/100*Cmax;
    else
        Pcharging(k)=Pcharging_max;
    end
end

if discharging
    SOC(k) = SOC(k-1) - Pdischarged/epsdischarging/Cmax*100;
    if SOC(k) < 0
        SOC(k) = 0;
        Pdischarged(k) = (SOC(k-1) -
SOC(k))/100*epsdischarging*Cmax;
    else
        Pdischarged(k)=Pdischarging_max;
    end
end
```

3 Results and Discussion

The simulation model was built in MATLAB to see how it would behave with an actual weather forecast. On the following figures are the results of all the subsystems implemented in the model.

3.1 Weather Forecast

First part of the figure is the expected outside temperature for the upcoming 5 days. The second part shows the solar irradiation forecast that takes into account the expected levels of cloudiness. The third part shows the overall power consumption as a stacked area plot where all the appliances are taken into account (Fig. 1).

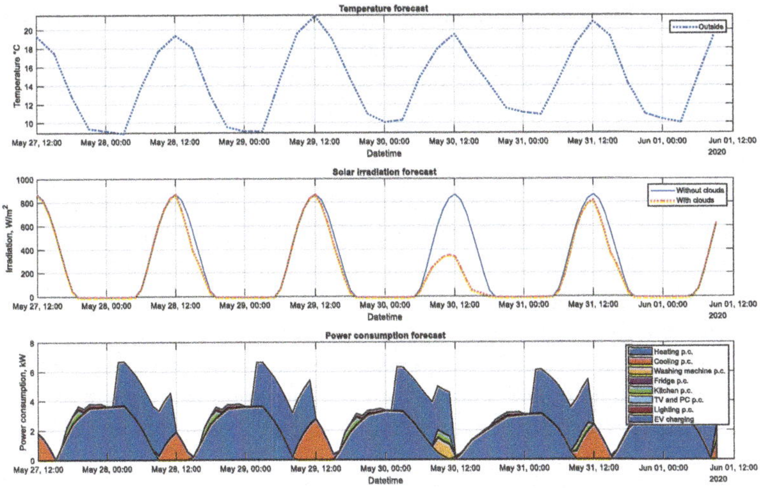

Fig. 1. Weather forecast and power consumption forecast

3.2 Heating and Cooling

Heating and cooling are affected by the outside temperature and the heating power of the Sun during the day. All of these are accounted for in the model in order to achieve the desired temperature indoors and the resulting power that will be consumed (Fig. 2).

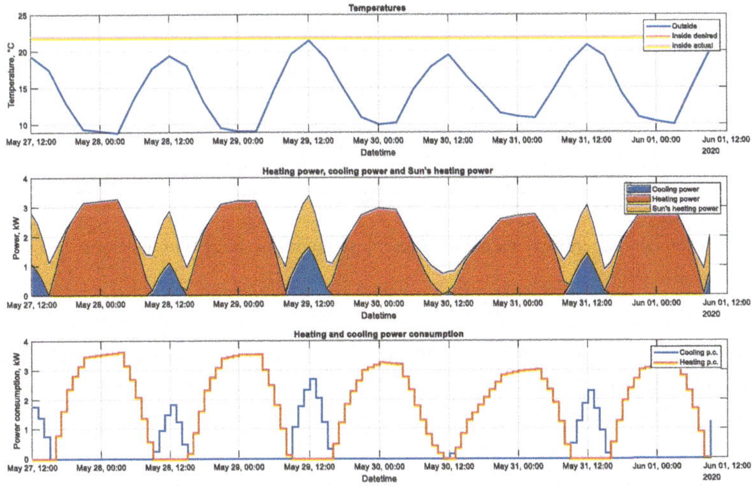

Fig. 2. Forecast of heating and cooling power consumptions with regards to the desired and outside temperatures

These results were obtained for considering the building's loss coefficient of 0.25 W/°C, heating maximal power consumption of 4 kW with heating efficiency 90%, maximal power consumption of cooling of 4 kW and the efficiency of cooling 60%.

3.3 Photovoltaics

First part of the figure shows the overall predicted power consumption and the predicted power generation from the photovoltaic panels. By combining both the second part is obtained where the overall power consumption of the home is lowered by the power generated by the photovoltaic panels. The surplus of generated energy results in the possibility to sell it back to the grid. The third part of the figure shows the amount of money that is forecasted to be paid in each step as well as a cumulated sum of the money that will be paid on the forecast horizon. The fourth part represents the amount of money that is forecasted to be received if the energy surplus is sold back to the grid (Fig. 3).

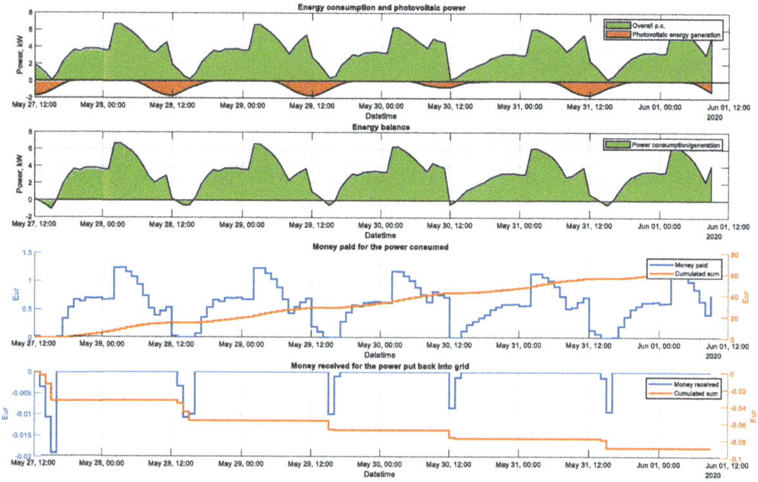

Fig. 3. Forecast of the energy balance taking into account photovoltaics and the forecast of money paid and received

These results were obtained for the area of 10 m^2 of photovoltaic panels and the efficiency of 20%. The pricing was considered to be 0.185 € for a kWh consumed from the grid and 0.018 € for a kWh sold back to the grid.

3.4 Electric Vehicle Battery

First part of the figure shows the charging cycles that were fixed in the week. Second part shows the discharging which reflects the car being driven. Also these timings were fixed for the testing purposes. The third part shows the State of charge of the battery on the forecast horizon (Fig. 4).

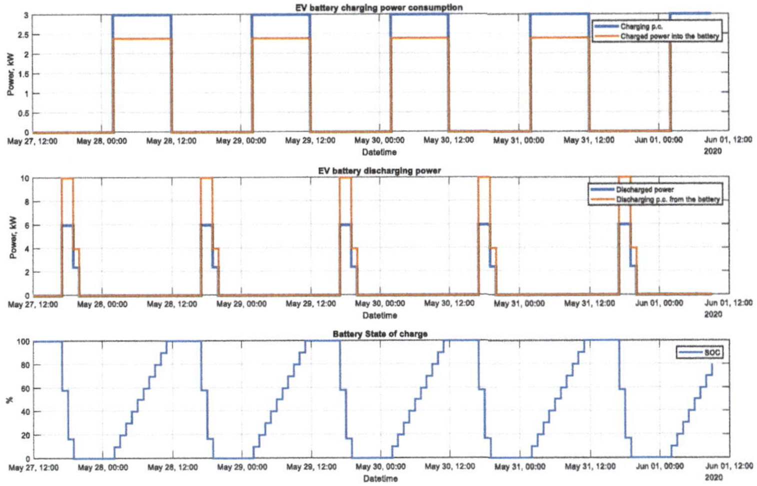

Fig. 4. Forecast of the electric vehicle battery charging, discharging and its state of charge

These results were obtained for the battery defined with a maximum capacity of 24 kWh, maximal charging power of 3 kW, charging efficiency 80%, maximal discharging power 6 kW and discharging efficiency of 60%.

4 Conclusion

A simplified model of a home was created to obtain a better idea about its predicted energy demands. The model consists of simple appliances such as a washing machine, fridge, kitchen appliances, TV and PC and lighting which are described by their power consumption and timing during the week. The more complicated parts of the model describe the heating and cooling which are described by the energy balance with regards to the inside desired and outside temperatures. Weather forecast is used mainly to obtain the expected outside temperature and cloudiness which is used to calculate the expected solar irradiance. Solar irradiance is also used to predict the power generation from the photovoltaics which is used for energy balance calculation of the whole building. The amount of money that is expected to be paid on the forecast horizon is calculated from this energy balance and the prices for buying the energy from the grid and selling it back to the grid. Lastly there is a subsystem of the battery of an electric vehicle.

Even though the model is simplified it provides useful information when all of the subsystems are put together. Especially when the weather forecast is put in place.

There are several next steps to take this model further. Firstly, the amount of heating energy from the Sun should depend on the situation of the sunlit area and the time of day. Then there is no accumulation considered in the heating and cooling which might play quite an important role. The next aim would be to use optimization on the model to optimize the usage of appliances throughout the days to reflect the energy generated from the photovoltaic panels or use the battery in the electric vehicle to support the energy management or to tune the heating and cooling process.

Acknowledgment. This research was supported by SGS grant at Faculty of Electrical Engineering and Informatics, University of Pardubice.

References

1. Nagpal, H., Staino, A., Basu, B.: Application of predictive control in scheduling of domestic appliances. Appl. Sci. **10**, 1627 (2020). https://doi.org/10.3390/app10051627
2. Pérez-Lombard, L., Ortiz, J., Pout, C.: A review on buildings energy consumption information. Energy Build. **40**, 394–398 (2008). https://doi.org/10.1016/j.enbuild.2007.03.007
3. Cecilia, A., Carroquino, J., Roda, V., Costa-Castelló, R., Barreras, F.: Optimal energy management in a standalone microgrid, with photovoltaic generation, short-term storage, and hydrogen production. Energies **13**, 1454 (2020). https://doi.org/10.3390/en13061454
4. Simmons, C.R., Arment, J.R., Powell, K.M., Hedengren, J.D.: Proactive energy optimization in residential buildings with weather and market forecasts. Processes **7**, 929 (2019). https://doi.org/10.3390/pr7120929
5. Ghotge, R., Snow, Y., Farahani, S., Lukszo, Z., van Wijk, A.: Optimized scheduling of EV charging in solar parking lots for local peak reduction under EV demand uncertainty. Energies **13**, 1275 (2020). https://doi.org/10.3390/en13051275
6. Mohajeryami, S., Doostan, M., Asadinejad, A.: An investigation of the relationship between accuracy of customer baseline calculation and efficiency of peak time rebate program (2016). https://doi.org/10.1109/peci.2016.7459237
7. Mohajeryami, S., Doostan, M., Asadinejad, A., Schwarz, P.: Error analysis of customer baseline load (CBL) calculation methods for residential customers. IEEE Trans. Ind. Appl. (2017). https://doi.org/10.1109/tia.2016.2613985
8. Roberts, J., Frieden, D., d'Herbemont, S.: Energy community definitions. https://www.compile-project.eu/wp-content/uploads/Explanatory-note-on-energy-community-definitions.pdf. Accessed 29 May 2020
9. OpenWeatherMap 5 day forecast API. https://openweathermap.org/forecast5. Accessed 29 May 2020
10. Photovoltaic Geographical Information System. https://re.jrc.ec.europa.eu/pvg_tools/en/tools.html. Accessed 29 May 2020

Features of Territorial Distribution of Population in Russia

Vsevolod V. Andreev[1,2(✉)]

[1] Russian Institute for Advanced Study, Moscow Pedagogical State University, Moscow 119991, Russia
andreev_vsevolod@mail.ru
[2] Chuvash State University, Cheboksary 428015, Russia

Abstract. Knowledge of territorial distribution of population and trends of its change are very important for design of socio-economic development plan for separate regions and for whole countries. In this paper features of territorial distribution of population on settlements in Russia are studied. For study of population territorial distribution the data of all-Russian population census of 2010 are used. In total, an information about 21276 settlements are included on these data. Test with using of Pearson's criterion of hypothesis about normal distribution law of random variable, which is a decimal logarithm of settlements population size, showed, that such hypothesis must be rejected at significance level $\alpha = 0.005$. Thus, Gibrat law is not fulfilled for Russian settlements. It is established, that one entrenched system of population territorial distribution by settlements in Russia is absent currently. The largest cities Moscow and Saint Petersburg are developing separately from the rest Russian territories. Excluding Moscow and Saint Petersburg there are up to 26 different clusters of settlements. Within each of such clusters a people distribution by separate settlements is obeyed to Pareto. And Pareto distribution parameters differ from each other in the different clusters. For the largest Russian cities Zipf law does not fulfilled.

Keywords: Gibrat law · Pareto distribution · Zipf law · Russia · Clusters of settlements · City size · City rank

1 Introduction

Knowledge of territorial distribution of population and trends of its change are very important for design of socio-economic development plan for separate regions and for whole countries. Previously, territorial distribution of population in Russia [1], as well as in regions of the Volga Federal District of Russia [2], were investigated. As a starting point data of the all-Russian population census of 2010 were used [3]. It was found that in all cases the distribution of population by cities and settlements of Russia is not lognormal, i.e. does not obey Gibrat law [4]. Also, in accordance with results of the above studies for the largest cities of Russia Zipf law is not fair [5]. At the same time, similar studies performed for the USA [6], showed that the lognormal distribution law of residents by settlements is kept well. Although there was a discussion about

R. Silhavy et al. (Eds.): CoMeSySo 2020, AISC 1295, pp. 541–553, 2020.
https://doi.org/10.1007/978-3-030-63319-6_50

accuracy of the lognormal distribution law for the USA cities and other settlements [7, 8]. It should be noted, that establishment of a precise law of random variable distribution is a very difficult problem. In this connection in many cases empirical distribution law of the random variable is not established for the entire sample, but for its separate parts.

Berry and Okulicz-Kozaryn showed [9], that current conflict in the literature, devoted to study of city sizes distribution, is a consequence of poorly-selected units of observation. In case of properly defined urban regions the USA urban growth obeys Gibrat's Law.

In paper [10] implemented systematic review of literature on Zipf's law application for study of city size distribution. The authors noted, that available empirical data show that Zipf's law for some territories is not correct even for cities in upper tail of distribution. However, conflict with empirical data often arises by reason of choice errors of sample elements, methodological drawbacks and data narrowness. It should be noted, that debate here is still far from being final. In addition, the paper identifies four important areas for study of Zipf's law and of city size distribution:

- distribution of city size in the lower tail;
- distribution of city size in subnational regions;
- alternative forms of Zipf's law;
- relationship between Zipf's law and coherence property of urban systems [11].

In paper [12] confirmed that the small USA cities on coordinate plane Log (Rank) - Log (Settlement size) are described with high accuracy by inverse and inverse generalized Pareto distributions.

Chinese and Indian cities size distribution throughout 1950 to 2010 was studied using the lognormal, Pareto, and generalized Pareto distributions [13]. It is established, that in different time periods cities sizes were described by different distribution laws. In early time periods in both countries the cities sizes distribution was characterized by lognormal distribution. In 2010 Chinese cities by population were distributed according to Pareto. Indian cities in 2000 and 2010 were distributed by size according to Zipf's law.

In paper [14] provided new comparative evidence on city size distribution in OECD (the Organization for Economic Co-operation and Development) countries and hypothesis of power law of city size distribution was not rejected in most of the OECD countries.

The application of Gibrat's law, and also the power distribution laws, Zipf's law is a special case of them, is very diverse in studies of population territorial distribution in different countries [15–21].

In this paper features of territorial distribution of population on settlements in Russia are studied.

2 Materials and Methods

In this paper for study of population territorial distribution the data of all-Russian population census of 2010 [3] are used. In total there are 21276 settlements in Russia. According to these data, a histogram of population distribution by settlements in Russia was obtained (see Fig. 1). On the histogram along a horizontal axis the decimal logarithms $\lg N$ of population size N in various settlements are indicated. Whole interval of values $\lg N$ from minimal $\lg N_{\min}$ to maximal $\lg N_{\max}$ value is divided into $m = 100$ subintervals with uniform step h (see Eq. (1)).

$$h = (\lg N_{\max} - \lg N_{\min})/m. \tag{1}$$

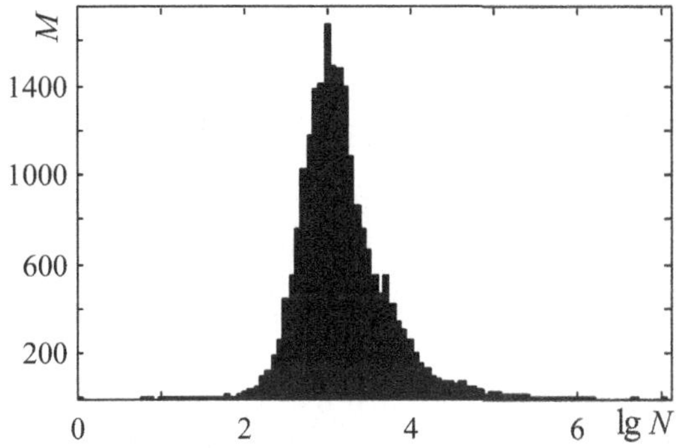

Fig. 1. The histogram of the Russian Federation population distribution on settlements.

Then for each subinterval

$$[\lg N_{\min} + (i-1)h,\ \lg N_{\min} + ih),\ i = 1, \ldots, m$$

amount M of settlements, the values $\lg N$ for which belong to indicated subinterval, is determined. At the same time, a last subinterval for $i = m$ is closed, i.e.

$$[\lg N_{\min} + (m-1)h,\ \ \lg N_{\min} + mh].$$

For the histogram (see Fig. 1) the hypothesis about a normal distribution law of random values $\lg N$ is tested using Pearson's criterion (χ^2- criterion) on a significance level $\alpha = 0.005$. In this case the random values N are distributed lognormally. The probability density function for random values N is given by Eq. (2).

$$\rho(N) = \frac{1}{\sqrt{2\pi}\sigma N}\exp\left(-\frac{(\ln N - \mu)^2}{2\sigma^2}\right). \tag{2}$$

Here $\sigma > 0$ and $\mu > 0$ are the distribution parameters.

A test of population territorial distribution of Russia for compliance with Pareto law (see Eq. (3)) is also carried out.

$$\rho(N) = \begin{cases} \frac{kN_m^k}{N^{k+1}}, & N \geq N_m \\ 0, & N < N_m \end{cases}. \tag{3}$$

Here $N_m > 0$ and $k > 0$ are the Pareto distribution parameters. Thereby, the cumulative distribution function of random variable N has a form of Eq. (4).

$$F(N) = 1 - \left(\frac{N_m}{N}\right)^k. \tag{4}$$

Let the parameter M_m is the number of settlements with a population not less than N_m, and a value R is the settlement rank. Then, the settlement rank R, taking into account of Eq. (4), is determined by Eq. (5).

$$R = M_m(1 - F(N)) = M_m\left(\frac{N_m}{N}\right)^k. \tag{5}$$

After taking the logarithm of Eq. (5) a following linear regression Eq. (6) is obtained.

$$\lg N = A - a \lg R. \tag{6}$$

Here $a = 1/k$ and $A = a \lg M_m + \lg N_m$. If $a = 1/k = 1$, then Zipf's law is obtained. Taking into account Eq. (6) Zipf's law is represented by Eq. (7).

$$\frac{\text{Population number of city A}}{\text{Population number of city B}} = \frac{\text{Rank of city B}}{\text{Rank of city A}}. \tag{7}$$

In this paper a linear regression analysis of the Log (Rank) – Log (Settlement size) distribution is also performed.

3 Results and Discussion

Test of hypothesis about normal distribution law of random variable $\lg N$ with using of Pearson's criterion showed, that such hypothesis must be rejected at significance level $\alpha = 0.005$. Thus, Gibrat's law is not fulfilled for Russian settlements.

In the case of lognormal law of population territorial distribution (Gibrat's law), the number of settlements M_m with a population equal to or greater than N_m, is determined by Eq. (8) [2, 6].

$$\lg R = \lg\left[M_m \middle/ \left\{1 - \Phi\left(\frac{\lg N_m - \mu_N}{\sigma_N}\right)\right\}\right] + \lg\left\{1 - \Phi\left(\frac{\lg N - \mu_N}{\sigma_N}\right)\right\}. \tag{8}$$

Here μ_N is the expected value and σ_N is the standard deviation of normally distributed random variable $\lg N$; $\Phi(x)$ is the cumulative distribution function of standard normal random variable.

Solid curve 2 in Fig. 2 is obtained by Eq. (8) for following parameters: $N_m = 1$, $M_m = 21276$, $\mu_N = 3$, $\sigma_N = 1.045$. In Fig. 1 it can be found, that such value of parameter σ_N does not correspond to the histogram. If we take $\sigma_N = 0.5$, which corresponds more to the histogram in Fig. 1, a curve, shown in Fig. 3, is obtained. In this case the rank values belong to an interval $-11.5 < \lg R < 4.5$. This result does not correspond to an empirical dependence of $\lg N$ from $\lg R$, represented in Fig. 2 (curve 1, "asterisks"). Empirical data of territorial distribution of Russia population (see Fig. 2) are not possible to approximate by Eq. (8) with a necessary accuracy.

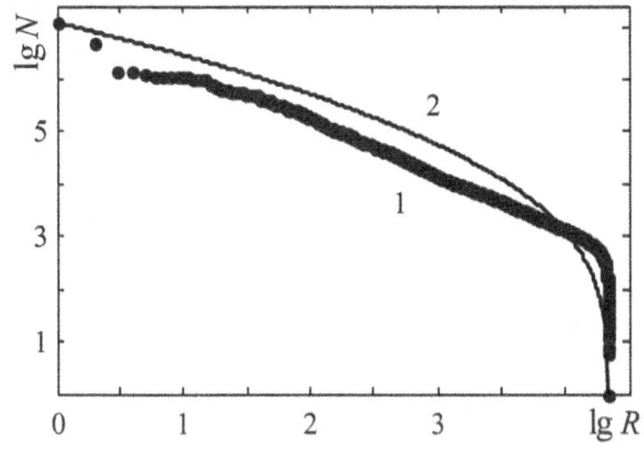

Fig. 2. Curve 1 is the diagram Log (Rank) – Log (Settlement size) for Russia. A solid curve 2 was obtained by Eq. (7).

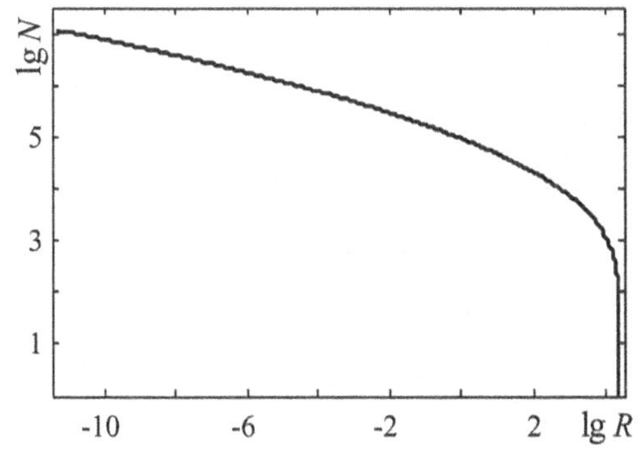

Fig. 3. A curve obtained by Eq. (7) for following parameters: $N_m = 1$, $M_m = 21276$, $\mu_N = 3$, $\sigma_N = 0.5$.

However, in Fig. 2 the empirical dependence of the $\lg N$ from the $\lg R$ behaves almost linearly on separate parts of curve 1. Consequently, this dependence can be approximated by Eq. (6) on linear parts of curve 1.

In Fig. 2 it can be seen, that two extremely left-handed "asterisks", corresponding to Moscow and Saint Petersburg, are located separately. Thus, these two cities constitute a separate subsystem among all settlements of Russia. So, according to the All-Russia Population Census 2010 data [3] in Moscow lived 11503501 people, and in Saint Petersburg – 4879566. At the same time, in Novosibirsk, which was the third city by rank, lived 1473754 inhabitants. This value is 7.81 and 3.31 times less than in Moscow and Saint Petersburg respectively. Inhabitants in Moscow was 2.36 times higher than in Saint Petersburg. If Zipf's law is valid (Eq. 7), the result would be equal to 2.

Linear regression equations are represented in Table 1 for different ranges of ranks, in which the curve 1 in Fig. 2 is most linear. The information is presented graphically in Fig. 4. In general, excluding individual cases, the "asterisks" are well approximated by the corresponding linear regression equations. So, two extreme right-hand "asterisks", denoting cities Saratov and Krasnodar with ranks R = 16 and R = 17 respectively, are located separately with deviation from the straight line (see Fig. 4(1)). But, if theirs to move on Fig. 4(2), they would be also located separately with deviation from the corresponding straight line. For settlements with ranks R from 3 to 21085 it is possible to say about good approximation of "asterisks" by straight lines, given by equations of a view (6) with the appropriate parameters A and a for each range of ranks (column 3 and rows from 1 to 23 of Table 1; Figs. 4(1)–(23)). This range of ranks corresponds to 21083 Russian settlements population from 1473754 ($R = 3$) to 153 ($R = 21085$) inhabitants. In range of ranks from $R = 21086$ to $R = 21276$, approximation of "asterisks" by straight lines, given by equations of the form of Eq. (6) (column 3 and rows from 24 to 26 of Table 1), is worse (Fig. 4(24)–(26)). The specified range of ranks corresponds to 191 Russian settlements population from 153 ($R = 21086$) to 1 ($R = 21276$) inhabitants.

Table 1. Pareto distribution parameters for different ranges of ranks for Russian settlements.

№	Ranks	Linear regression equations	Settlements	Pareto distribution parameter $k = 1/a$
1	$3 \le R \le 17$	$\lg N = 6.3139 - 0.2923 \lg R$	Novosibirsk ($R = 3$), Ekaterinburg, Nizhny Novgorod, Samara, Omsk, Kazan, Chelyabinsk, Rostov-on-Don, Ufa, Volgograd, Perm, Voronezh, Krasnoyarsk, Saratov, Krasnodar ($R = 17$)	$k = 3.4211$
2	$18 \le R \le 38$	$\lg N = 6.4632 - 0.4913 \lg R$	Tolyatti ($R = 18$; 719632 inhabitants), …, Kirov ($R = 38$; 498381 inhabitants)	$k = 2.0354$
3	$39 \le R \le 85$	$\lg N = 7.033 - 0.8645 \lg R$	Cheboksary ($R = 39$; 464208 inhabitants), …, Balashiha ($R = 85$; 225381 inhabitants)	$k = 1.1567$
4	$86 \le R \le 153$	$\lg N = 7.7653 - 1.2472 \lg R$	Blagoveshchensk ($R = 86$; 219847 inhabitants), …, Novoshakhtinsk ($R = 153$; 111075 inhabitants)	$k = 0.8018$
5	$154 \le R \le 197$	$\lg N = 6.7632 - 0.785 \lg R$	Novokuybyshevsk ($R = 154$; 110729 inhabitants), …, Gelendzhik ($R = 197$; 91126 inhabitants)	$k = 1.2739$

(*continued*)

Table 1. (*continued*)

№	Ranks	Linear regression equations	Settlements	Pareto distribution parameter $k = 1/a$
6	$198 \leq R \leq 248$	$\lg N = 8.1264 - 1.3824 \lg R$	Dolgoprudny (R = 198; 90956 inhabitants), …, Nadymsky municipal region (R = 248; 66530 inhabitants)	$k = 0.7234$
7	$249 \leq R \leq 313$	$\lg N = 6.6788 - 0.7739 \lg R$	Volsk (R = 249; 66508 inhabitants), …, Fryazino (R = 313; 55381 inhabitants)	$k = 1.2922$
8	$314 \leq R \leq 670$	$\lg N = 7.4666 - 1.0903 \lg R$	Uzlovaya (R = 314; 55282 inhabitants), …, Urban settlement Bezenchuk (R = 670; 23454 inhabitants)	$k = 0.9172$
9	$671 \leq R \leq 1115$	$\lg N = 7.7219 - 1.1887 \lg R$	Ershov (R = 671; 23414 inhabitants), …, Uyar (R = 1115; 12665 inhabitants)	$k = 0.8412$
10	$1116 \leq R \leq 2982$	$\lg N = 6.9679 - 0.9402 \lg R$	Korablino (R = 1116; 12658 inhabitants), …, Shimorskoye (R = 2982; 5091 inhabitants)	$k = 1.0636$
11	$2983 \leq R \leq 8100$	$\lg N = 7.6296 - 1.1291 \lg R$	Rural settlement Ishchersky (R = 2983; 5088 inhabitants), …, Rural settlement Chernovka (R = 8100; 1675 inhabitants)	$k = 0.8857$
12	$8101 \leq R \leq 9900$	$\lg N = 7.1747 - 1.0108 \lg R$	Rural settlement Sagunovskoe (R = 8101; 1674 inhabitants), …, Rural settlement Urussinskoe (R = 9900; 1363 inhabitants)	$k = 0.9893$
13	$9901 \leq R \leq 12885$	$\lg N = 7.9255 - 1.1986 \lg R$	Rural settlement Obvinskoe (R = 9901; 1363 inhabitants), …, Rural settlement Chapaevsky (R = 12885; 1002 inhabitants)	$k = 0.8343$
14	$12886 \leq R \leq 13855$	$\lg N = 8.8229 - 1.4164 \lg R$	Rural settlement Krasnomylsky sel'sovet (R = 12886; 1002 inhabitants), …, Rural settlement Vorobyevskoye (R = 13855; 903 inhabitants)	$k = 0.706$
15	$13856 \leq R \leq 16170$	$\lg N = 9.9447 - 1.6873 \lg R$	Rural settlement Krasnosakmarsky sel'sovet (R = 13856; 903 inhabitants), …, Rural settlement Zhidkinskoe (R = 16170; 694 inhabitants)	$k = 0.5927$
16	$16171 \leq R \leq 17420$	$\lg N = 12.068 - 2.1922 \lg R$	Rural settlement Turginskyoe (R = 16171; 694 inhabitants), …, Rural settlement Kenaiskoe (R = 17420; 588 inhabitants)	$k = 0.4562$
17	$17421 \leq R \leq 17710$	$\lg N = 14.313 - 2.722 \lg R$	Rural settlement Beryozovskoe (R = 17421; 587 inhabitants), …, Rural settlement Mochegaevsky sel'sovet (R = 17710; 562 inhabitants)	$k = 0.3674$
18	$17711 \leq R \leq 18620$	$\lg N = 14.911 - 2.8626 \lg R$	Rural settlement Ermakovskoe (R = 17711; 562 inhabitants), …, Rural settlement Oreshensky sel'sovet (R = 18620; 486 inhabitants)	$k = 0.3493$
19	$18621 \leq R \leq 19370$	$\lg N = 19.445 - 3.9246 \lg R$	Rural settlement Ivanovsky sel'sovet (R = 18621; 486 inhabitants), …, Rural settlement Sursovaiskoe (R = 19370; 415 inhabitants)	$k = 0.2548$
20	$19371 \leq R \leq 19933$	$\lg N = 26.841 - 5.6498 \lg R$	Rural settlement Maresevsky sel'sovet (R = 19371; 415 inhabitants), …, Rural settlement Arangasskiy nasleg (R = 19933; 353 inhabitants)	$k = 0.177$
21	$19934 \leq R \leq 20506$	$\lg N = 35.31 - 7.6194 \lg R$	Rural settlement Troitsky nasleg (R = 19934; 353 inhabitants), …, Rural settlement Krasnokholmskoe (R = 20506; 283 inhabitants)	$k = 0.1312$

(*continued*)

Table 1. (*continued*)

№	Ranks	Linear regression equations	Settlements	Pareto distribution parameter $k = 1/a$
22	$20507 \leq R \leq 20910$	$\lg N = 74.8 - 16.78 \lg R$	Rural settlement Mikshinskoye (R = 20507; 283 inhabitants), …, Rural settlement Kurlychenskoe (R = 20910; 203 inhabitants)	$k = 0.0596$
23	$20911 \leq R \leq 21085$	$\lg N = 152.3 - 34.71 \lg R$	Rural settlement Belenky sel'sovet (R = 20911; 203 inhabitants), …, Rural settlement Kos'yu (R = 21085; 153 inhabitants)	$k = 0.0288$
24	$21086 \leq R \leq 21204$	$\lg N = 388.5 - 89.34 \lg R$	Rural Settlement Sergeevsky sel'sovet (R = 21086; 153 inhabitants), …, Rural settlement Chaibuha (R = 21204; 92 inhabitants)	$k = 0.0112$
25	$21205 \leq R \leq 21253$	$\lg N = 1546.2 - 356.9 \lg R$	Rural settlement Khoyskoe (R = 21205; 90 inhabitants), …, Inter-settlements territories – rural population (R = 21253; 38 inhabitants)	$k = 0.0028$
26	$21254 \leq R \leq 21276$	$\lg N = 8607.3 - 1988.6 \lg R$	Inter-settlements territories – rural population (R = 21254; 33 inhabitants), …, Inter-settlement territory – rural population (R = 21276; 1 inhabitant)	$k = 0.0005$

Thus, the entire range of ranks of Russian settlements (see Fig. 2) can be divided into 26 parts, on each of them dependence lg N from lg R is almost linear. Consequently, on each of such ranges inhabitants distribution by settlements is described according to Pareto. Pareto distribution parameters on each ranges of ranks differ (Table 1). However, it possible to see several pairs of ranges, on which Pareto distribution parameters values are close. There are rows 5 and 7 with $k = 1.2739$ and $k = 1.2922$ respectively, rows 9 and 13 with $k = 0.8412$ and $k = 0.8343$ respectively in the Table 1. For these cases the ranges of ranks are not adjacent. Range of ranks R from 8101 to 9900, on which Zipf's parameter a is very close to 1 and equal to 1.0108, is also available (row 12 in Table 1). In this case Zipf's law is fulfilled well. The specified above range of ranks corresponds to inhabitants number of 1800 Russian settlements from 1674 (R = 8101) to 1363 (R = 9900).

If Zipf's parameter a is greater than 1, then large cities are overpopulated in comparison with case, when Zipf's law is valid. In the opposite case of $a < 1$ population is dispersed in small settlements disproportionately more. This follows from properties of Pareto distribution [2]. For the largest Russian cities with ranks from 3 to 17 Zipf's parameter is equal to $a = 0.2923$ (see the first row in Table 1). In further this parameter increases from 0.4913 for cities with ranks in range $18 \leq R \leq 38$ (see the second row in Table 1) to 0.8645 for cities with ranks in range $39 \leq R \leq 85$ (see the third row in Table 1). Thus, for Zipf's law satisfaction the inhabitants number must be higher in the largest Russian cities. Moreover, the higher city rank, the more times it is necessary to increase the city population. It should be noted, that in an equilibrium case Zipf's law takes place for the largest cities of country. Population distribution on all settlements in country is lognormal for equilibrium case [4, 6, 8]. From this point of view, the population distribution by Russian settlements is far from equilibrium (see Fig. 2).

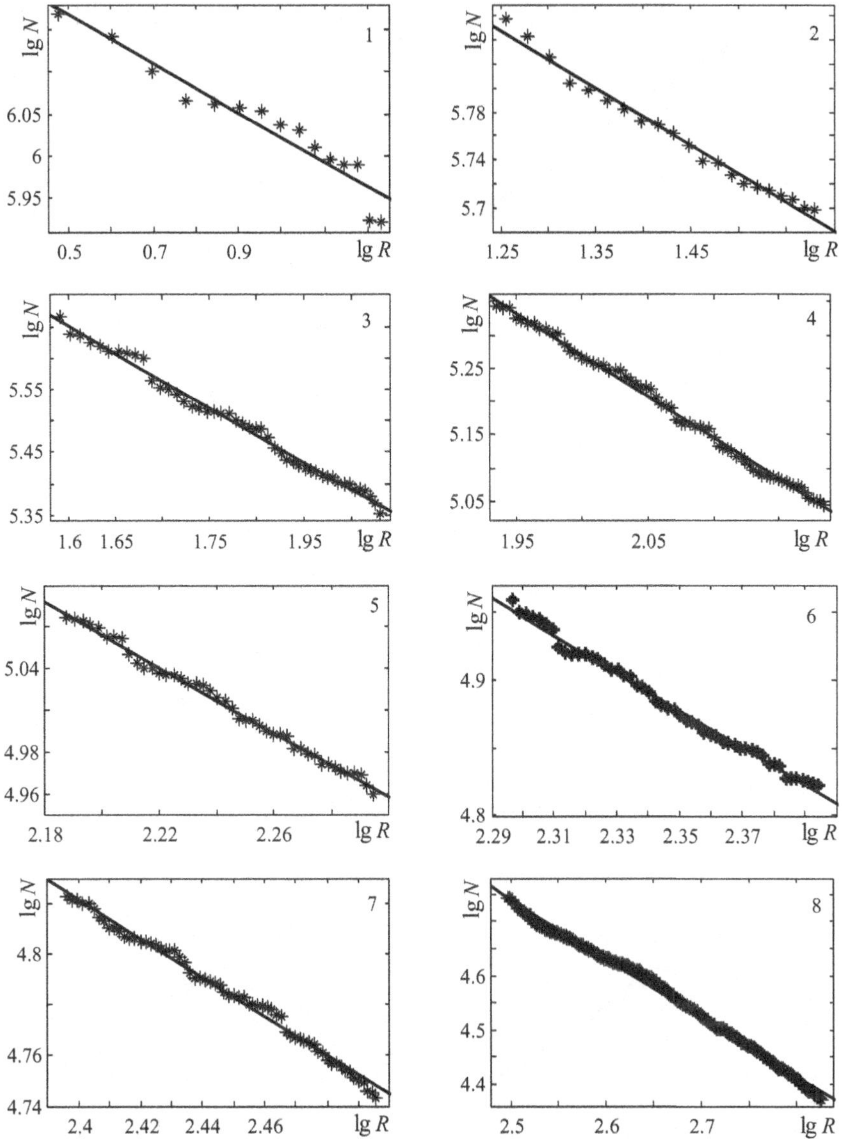

Fig. 4. Linear regressions for different ranges of ranks logarithms for Russian settlements. Numbering in an upper right corner in Figures corresponds to the ordinal numbers in Table 1. "Asterisks" correspond to settlements. Solid lines are obtained by linear regression equations in corresponding column and rows of Table 1.

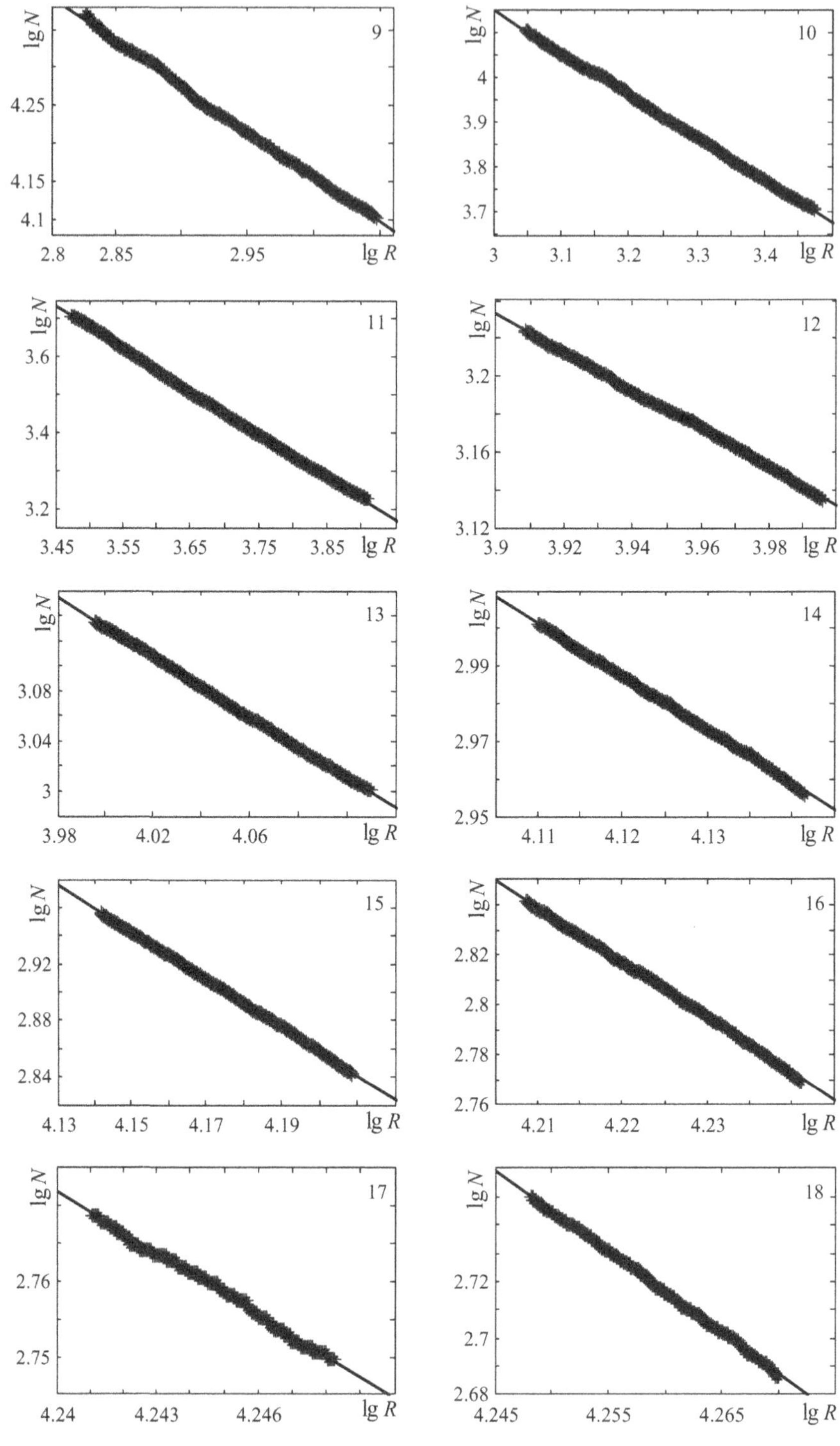

Fig. 4. (*continued*)

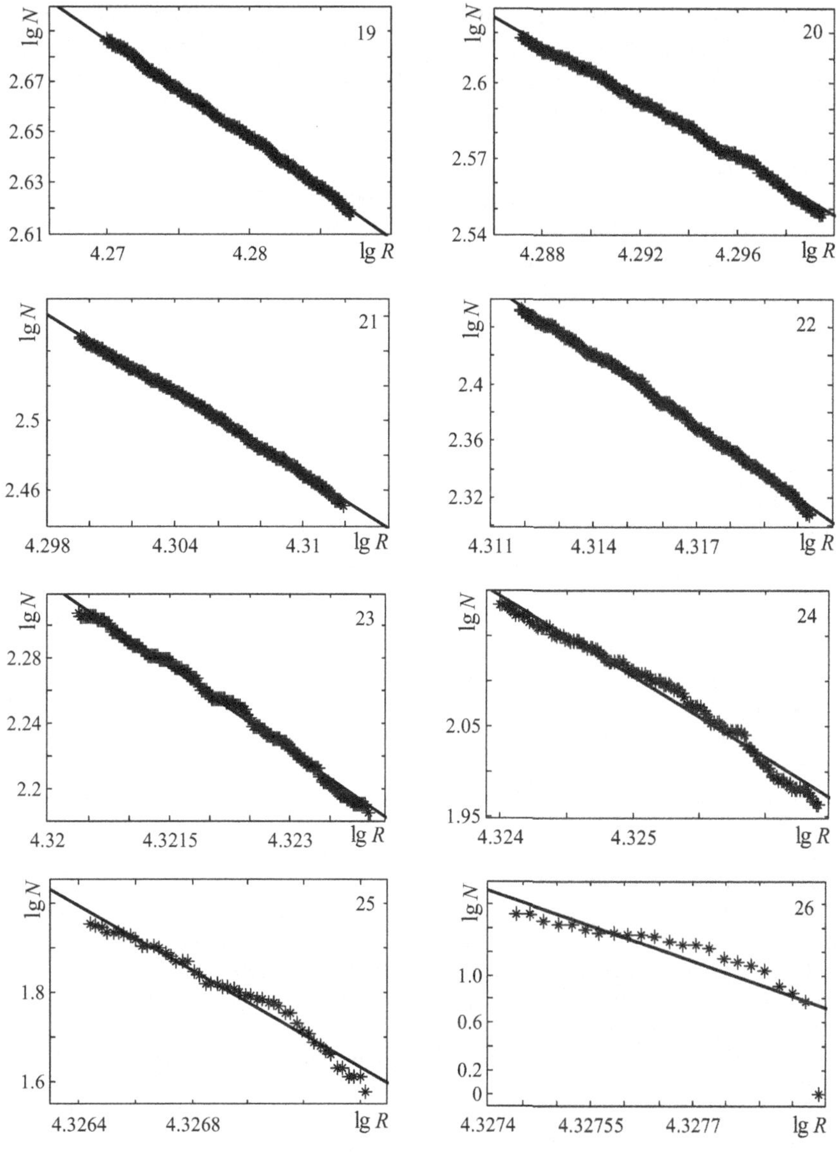

Fig. 4. *(continued)*

4 Conclusions

Thus, there is not one entrenched system of population territorial distribution by settlements in Russia. The largest cities Moscow and Saint Petersburg are developing separately from the rest Russian territories. The main population migration flows from the remaining Russian settlements are directed to these two cities. Excluding Moscow and Saint Petersburg there are up to 26 different clusters of settlements. Within each of such clusters a people distribution by separate settlements is obeyed to Pareto. And Pareto distribution parameters differ from each other in the different clusters (see Table 1). For the largest Russian cities Zipf's law does not fulfilled.

Acknowledgments. The reported study was funded by Russian Foundation for Basic Research (RFBR) according to the research project № 19-010-00631.

I express my deep gratitude to the International Program of scientific grants of Russian Institute for Advanced Study (RIAS) of Moscow Pedagogical State University for supporting in 2018 study on project: "Development and progress of mathematical methods for dynamics analysis of socio-economic processes and systems".

References

1. Andreev, V.V.: Territorial distribution of the population in the Russian Federation. Econ. Reg. **13**(3), 803–811 (2017)
2. Andreev, V.V., Lukiyanova, V.Yu., Kadyshev, E.N.: Analysis of people territorial distribution in regions of the Volga Federal District on the base of Zipf and Gibrat laws. Appl. Econ. **48**(4), 97–121 (2017)
3. Portal of the 2010 All-Russian Census. http://www.gks.ru/free_doc/new_site/perepis2010/croc/perepis_itogi1612.htm. Accessed 11 June 2020
4. Gibrat, R.: Les Inégalités Économiqués; Applications: Aux Inégalités des Richesses, a la Concentration des Entreprises, Aux Populations des Villes, Aux Statistiques des Familles, etc., d'une Loi Nouvelles, La Loi de l'Effet Proportionnel. Librairie du Recueil Sirey, Paris (1931)
5. Zipf, G.K.: Human Behavior and the Principle of Least Effort. Addison-Wesley Press, Cambridge (1949)
6. Eeckhout, J.: Gibrat's law for (all) cities. Am. Econ. Rev. **94**(5), 1429–1451 (2004)
7. Levy, M.: Gibrat's law for (all) cities: comment. Am. Econ. Rev. **99**(4), 1672–1675 (2009)
8. Eeckhout, J.: Gibrat's law for (all) cities: reply. Am. Econ. Rev. **99**(4), 1676–1683 (2009)
9. Berry, B.J.L., Okulicz-Kozaryn, A.: The city size distribution debate: resolution for US urban regions and megalopolitan areas. Cities **29**, S17–S23 (2012)
10. Arshad, S., Hu, S., Ashraf, B.N.: Zipf's law and city size distribution: a survey of the literature and future research agenda. Physica A Stat. Mech. Appl. **492**, 75–92 (2018)
11. Cristelli, M., Batty, M., Pietronero, L.: There is more than a power law in Zipf. Sci. Rep. **2**, 812 (2012)
12. Devadoss, S., Luckstead, J.: Size distribution of U.S. lower tail cities. Physica A Stat. Mech. Appl. **444**, 158–162 (2016)
13. Luckstead, J., Devadoss, S.: A comparison of city size distributions for China and India from 1950 to 2010. Econ. Lett. **124**(2), 290–295 (2014)
14. Veneri, P.: City size distribution across the OECD: does the definition of cities matter? Comput. Environ. Urban Syst. **59**, 86–94 (2016)
15. Batty, M.: Rank clocks. Nature **444**, 592–596 (2006)
16. Jażdżewska, I.: Spatial and dynamic aspects of the rank-size rule method. Case of an urban settlement in Poland. Comput. Environ. Urban Syst. **62**, 199–209 (2017)
17. Jiang, B., Yin, J., Liu, Q.: Zipf's law for all the natural cities around the world. Int. J. Geogr. Inf. Sci. **29**(3), 498–522 (2015)

18. Justo, W.R.: Zipf's law and the Gibrat's law: what do the facts have to say about the Brazilian cities? J. Financ. Econ. **2**(5), 136–144 (2014)
19. Li, H., Wei, Y.D., Ning, Y.: Spatial and temporal evolution of urban systems in China during rapid urbanization. Sustainability **8**, 651–667 (2016)
20. Pérez-Campuzano, E., Guzmán-Vargas, L., Angulo-Brown, F.: Distributions of city sizes in Mexico during the 20th century. Chaos Solitons Fractals **73**, 64–70 (2015)
21. Rozenfeld, H.D., Rybski, D., Gabaix, X., Makse, H.A.: The area and population of cities: new insights from a different perspective on cities. Am. Econ. Rev. **101**(5), 2205–2225 (2011)

Study of Development of the Largest Now Russian Cities Since the End of XIX Century to the Present Time

Vsevolod V. Andreev[1,2(✉)]

[1] Russian Institute for Advanced Study, Moscow Pedagogical State University, Moscow 119991, Russia
andreev_vsevolod@mail.ru
[2] Chuvash State University, Cheboksary 428015, Russia

Abstract. In this study the dynamics of development since 1897 of the 15 largest currently Russian cities with a population of more than a million is studied. Understanding the territorial distribution laws for the population and trends in its change is a key element in the implementation of socio-economic policies. Data analysis is carried out in terms of "City Rank - City Population". Also, Zipf law is applied for data analysis. Following main results are obtained. Moscow and St. Petersburg constitute a separate subsystem among Russian cities. This status has been maintained since at least 1897. Another considered cities have not generated during the transformations since 1897 a clearly defined structure that is an equal part of the subsystem of the above two cities. At last time two new systemically important centers in Russia, represented by the cities Novosibirsk and Yekaterinburg, arise. The proposed methods for analysis the territorial distribution of the population may be useful for structures and authorities involved in the elaboration of programs for socio-economic development of both individual regions and the country as a whole.

Keywords: Zipf law · Pareto distribution · Temporal trajectories · City rank · Clustering · Agglomeration · Transformation of territorial distribution of population · Gibrat law

1 Problem Statement

Earlier, using the Gibrat [1] and Zipf [2] laws, the territorial distribution features for the population in the Chuvash Republic [3], the Russian Federation [4] and the population in the constituent entities of the Volga Federal District [5] were investigated. In these studies data from the 2010 All-Russian Population Census [6] were used. Among the studies of the population size in Russian cities carried out using the Zipf law, in papers [7, 8] important results were obtained. These studies show that the Gibrat and Zipf laws are not satisfied in the cases considered.

In 1931, Robert Gibrat established that in an equilibrium condition the number of workers in firms and city dwellers follows the lognormal distribution [1]. In accordance with Gibrat law, the growth rate of firms' employees or towns' people does not depend on their size. According to Zipf law [2], when ranking cities in descending order of

R. Silhavy et al. (Eds.): CoMeSySo 2020, AISC 1295, pp. 554–566, 2020.
https://doi.org/10.1007/978-3-030-63319-6_51

residents living in them, the ratio of the population in any two of them is inversely proportional to the ratio of their ranks. However, Zipf law corresponds to the Pareto distribution [5], and not the lognormal. This contradiction was analysed in detail in papers [9–11]. In a study [9] it was shown that for the largest USA cities, the right "tail" of the Pareto distribution and the lognormal distribution coincide quite well. Therefore, for such cities Zipf law is approximately satisfied. Although it was shown [11] that the largest USA cities are distributed on the inhabitant number according to the Pareto. In response, paper of Eeckhout [10] appeared with the justification that the results and conclusions obtained in paper [9] are correct. The reconciliation of the Gibrat and Zipf laws was also discussed in detail by [5].

Understanding the territorial distribution laws for the population and trends in its change is a key element in the implementation of socio-economic policies. The economic activity of the regions depends not only on the mineral resource availability, land fertility, a favourable geographical location but also on many other factors [12, 13]. In particular, to understand the reasons that contribute to the emergence and development of agglomerations, it is important to study the dynamics of changes in the territorial distribution of the population over long time interval [14], as well as the driving forces of population and labor mobility [15–17].

In my study the dynamics of development since 1897 of the 15 largest currently Russian cities with a population of more than a million (see Table 1) is studied. The period under consideration includes three stages of the evolution of these cities: in the Russian Empire, then as part of the Russian Soviet Federative Socialistic Republic (RSFSR) and, finally, in the Russian Federation. In this regard, of the most important tasks of the study is to analyse the transformations that took place in the development of the cities under consideration during the change in the State structure.

Despite the fact that the so-called "the center dictate" in the USSR was strong enough, it cannot be denied that each the Union Republic, having many attributes of statehood, was a fairly independent administrative-territorial unit. Therefore, there is substantial reason to believe that the territorial distribution of the population that took shape in the territory of the present Russian Federation after 1897 as a result of the extremely complex conglomerate of various incentives and actions emanating from millions of people and a huge number of enterprises and organizations had many similarities at three stages of development. Therefore, although not all of the 15 largest Russian cities were such at the end of the century before last, their evolution over more than a century should reflect the general tendencies in the formation of the current structure of cities in the dynamic system "Russian Empire – RSFSR – Russian Federation".

2 Theory and Research Methodology

An extensive review of publications in scientific periodicals where the validity of Zipf's law application for analysis of the territorial distribution of population, labor resources and the development of firms was studied, was generalized in papers [4, 5].

In paper [14] it was indicated as the main drawback of most studies carried out using the Gibrat and Zipf laws that they are often carried out on the basis of data

Table 1. The population of the 15 largest currently Russian cities from 1897 to 2017.

№	Cities	Years											
		1897	1926	1939	1959	1965	1970	1979	1982	1989	2002	2010	2017
1	Moscow	1039000/2	2029000/1	4137000/1	5085000/1	6423000/1	7077000/1	8011000/1	8302000/1	8967000/1	10126424/1	11503501/1	12380664/1
2	St. Petersburg	1265000/1	1690000/2	3191000/2	3321000/2	3641000/2	4027000/2	4588000/2	4722000/2	5024000/2	4661219/2	4879566/2	5281579/2
3	Novosibirsk	–	120000/15-16	405000/7	886000/four	1029000/four	1161000/four	1312000/four	1357000/four	1437000/four	1425508/3	1473754/3	1602915/3
4	Yekaterinburg	43000/34	136000/13	426000/6	779000/6	919000/6	1025000/6	1211000/6	1252000/5	1365000/5	1293537/5	1383179/4	1455514/4
5	Nizhny Novgorod	90000/8-9	185000/5	644000/3	942000/3	1085000/3	1170000/3	1344000/3	1373000/3	1438000/3	1311252/4	1259738/5	1261666/5
6	Samara	90000/8-9	176000/8	390000/9	806000/5	948000/5	1045000/5	1216000/5	1243000/6	1254000/6	1157880/6	1164814/6	1169719/9
7	Omsk	37000/39	162000/10	289000/13	581000/12-13	721000/10	821000/10	1014000/8	1061000/8	1148000/7	1134016/7	1154116/7	1178391/8
8	Kazan	130000/4	179000/6	398000/8	647000/8	762000/9	869000/8	993000/10	1023000/10-11	1094000/9	1105289/8	1143535/8	1231878/6
9	Chelyabinsk	20000/68-72	59000/48-49	273000/16	689000/7	805000/7	875000/7	1030000/7	1066000/7	1142000/8	1077174/9	1130132/9	1198858/7
10	Rostov-on-Don	119000/5	308000/3	503000/4	600,000/10	720000/11	789000/12	934000/12	966000/12	1019000/12	1068267/10	1089261/10	1125299/10
11	Ufa	49000/26-27	99000/21-23	250000/20-21	547000/14	665000/14	780000/13	979000/11	1023000/10-11	1082000/11	1042437/11	1071640/11	1115560/11
12	Volgograd	55000/17	148000/12	445000/5	592000/11	700000/12	815000/11	929000/13	956000/13	999000/13	1011417/12	1021215/12	1015586/15
13	Perm	45000/32	85000/28	255000/18	629000/9	764000/8	850000/9	999000/9	1028000/9	1091000/10	1001653/13	991170/13	1048005/13
14	Voronezh	81000/10	120000/15-16	327000/11	448000/15	576000/15	660000/15	783000/16	820000/16	887000/16	848752/16	975373/14	1039801/14
15	Krasnoyarsk	27000/51-53	72000/38-39	190000/26	412000/16	541000/16	648000/16	796000/15	833000/15	913000/14	909341/14	974591/15	1082933/12

relating to a fixed point in time. But, as a rule, the size of the initial data, the processing and analysis of which gives the main results of the study, actually varies over time. Moreover, over time, variability and turbulent nature in the micro-dynamics of the distributions may appear due to the fact that individual objects in the source data array can quickly change their characteristics (for example, rank). Moreover, the cumulative distributions related to certain points in time can be quite stable at the macrolevel.

Manifestations of Zipf law in growing systems with multiplicative interaction were studied in paper [18]. It follows from the results that with moderate feedback that spontaneously adjusts the parameters, the Pareto-Zipf distribution is universal.

The development and decline of cities are inextricably linked with spatial factors, among which the proximity to natural resources is important. By Dentinho [19] results were obtained for various countries showing the decisive role of natural resources in the spatial concentration of cities and their development. The research methodology was to evaluate the elasticity of Zipf curve for a group of countries.

Based on a double logarithmic regression model and on data for urban population of China from 1990 to 2010, trend of the Zipf index was studied in paper [20]. It is concluded that the population distribution among the largest cities in China is relatively balanced, and it is possible to say that Zipf law is practically implemented. At the same time, within the separate provinces of China Zipf law as a whole is not implemented.

Due to the fact that the Zipf distribution is very widely used in practice for analysis of variety processes, both natural and artificial, attempts are made to its various generalizations [21–23], including those based on the application of physical principles [21]. In study [23] the Pareto-Positive Stable (PPS) distribution [22] was used as a new model for analysis the size of cities for six census periods from 1951 to 2001 in the Kerala State (India). The classical Pareto and Zipf distributions are included as special cases in the PPS distribution. It is concluded that the PPS distribution is a more flexible model for the entire data range and provides higher accuracy in describing the size of cities in comparison with the Gibrat and Zipf laws.

The size distribution of Chinese and Indian cities from 1950 to 2010 was studied using lognormal, Pareto, and generalized Pareto distributions in paper [24]. It has been established that, at different time periods, the sizes of cities are obeyed to different distribution laws. In the early periods of time in both countries the size distribution of cities is characterized by a lognormal distribution. In 2010 population in Chinese cities was distributed according to Pareto. In 2000 and 2010 Indian cities were distributed by sizes according to Zipf law.

In paper [25] it was found that small USA cities on the coordinate plane log (Rank) – log (Population number) are very accurately described by the inverse and inverse generalized Pareto distributions. But for the largest USA cities Zipf law is respected.

Thus, the methodology, theoretical and practical aspects of Zipf law applying in studies devoted to the evolutionary dynamics of cities are very diverse.

3 Initial Data

The initial data from 1897 to 2017 on the population size in the 15 largest currently Russian cities are presented in Table 1. The data from 1897 to 1965 were taken from a collection "The Size, Composition and Movement of the Population in the USSR. Statistical materials", prepared in 1965 by the Central Statistical Office under the Council of Ministers of the USSR [26] (data are given starting on page 154). Data for 1970, 1979 and 1982 are presented in a handbook [27]. Information on the results of 1989 census in the USSR was taken from the statistical compilation [28]. 2002 and 2010 All-Russian Census Data are available on the Internet [6, 29]. Estimates of the population of Russian cities for last years are available on the website of the Federal State Statistics Service [30] and in a more concise form - on the site about countries, cities, population statistics, etc. [31]. In Table 1 in the cells containing data on inhabitants number, the cities ranks also indicated for the corresponding years. If, for example, the values 8–9 are indicated, then this means that in the corresponding year the city on population number shared 8 and 9 positions with another city. In such cases in this paper the rank is determined as the arithmetic mean.

4 Results and Analysis

In Fig. 1 the decimal logarithms of the population number are represented for the 13 largest cities (with the exception of Moscow and St. Petersburg) for the time interval from 1897 to 2017. Since 1989, in the all cities, except for Moscow (1), Kazan (8), Rostov-on-Don (10) and Volgograd (12), fluctuations in the direction of decreasing and subsequent restoration of the population number growth trend are observed on some time intervals (see Table 1 and Fig. 1). In Volgograd (12) tendency towards a decrease in the residents number arose since 2010 and at the beginning of 2017 it still took place.

Processes of population decline and subsequent restoration of the growth trend in the cities under consideration could also take place during the World War II and post-World War II years. However, due to the lack of reliable data between 1939 and 1959, they are not shown in Fig. 1.

From Fig. 1 analysis it follows that after the World War II by 1959 from the temporal trajectories of development of the largest at present Russian cities 3 clusters were formed. The first group includes trajectories 14 (Voronezh) and 15 (Krasnoyarsk). The second group formed by trajectories 7 (Omsk), 8 (Kazan), 9 (Chelyabinsk), 10 (Rostov-on-Don), 11 (Ufa), 12 (Volgograd) and 13 (Perm). The third group formed from trajectories 3 (Novosibirsk), 4 (Yekaterinburg), 5 (Nizhny Novgorod) and 6 (Samara). Within each of the groups, the temporal trajectories behave rather randomly. They mix intensively and not reveal any features of structuring and emergence of stable configurations, corresponding to the population distribution according to Zipf law or, at least, according to some other regular law. After 1989, the collapse of the clusters began. Since 2002 certain features of formation again of new 2 clusters of the trajectories arose. One from them formed from trajectories 5 (N. Novgorod), 6 (Samara), 7 (Omsk), 8 (Kazan), 9 (Chelyabinsk), 10 (Rostov-on-Don), 11 (Ufa), 12 (Volgograd), 13 (Perm), 14 (Voronezh) and 15 (Krasnoyarsk). The second formed from trajectories 3

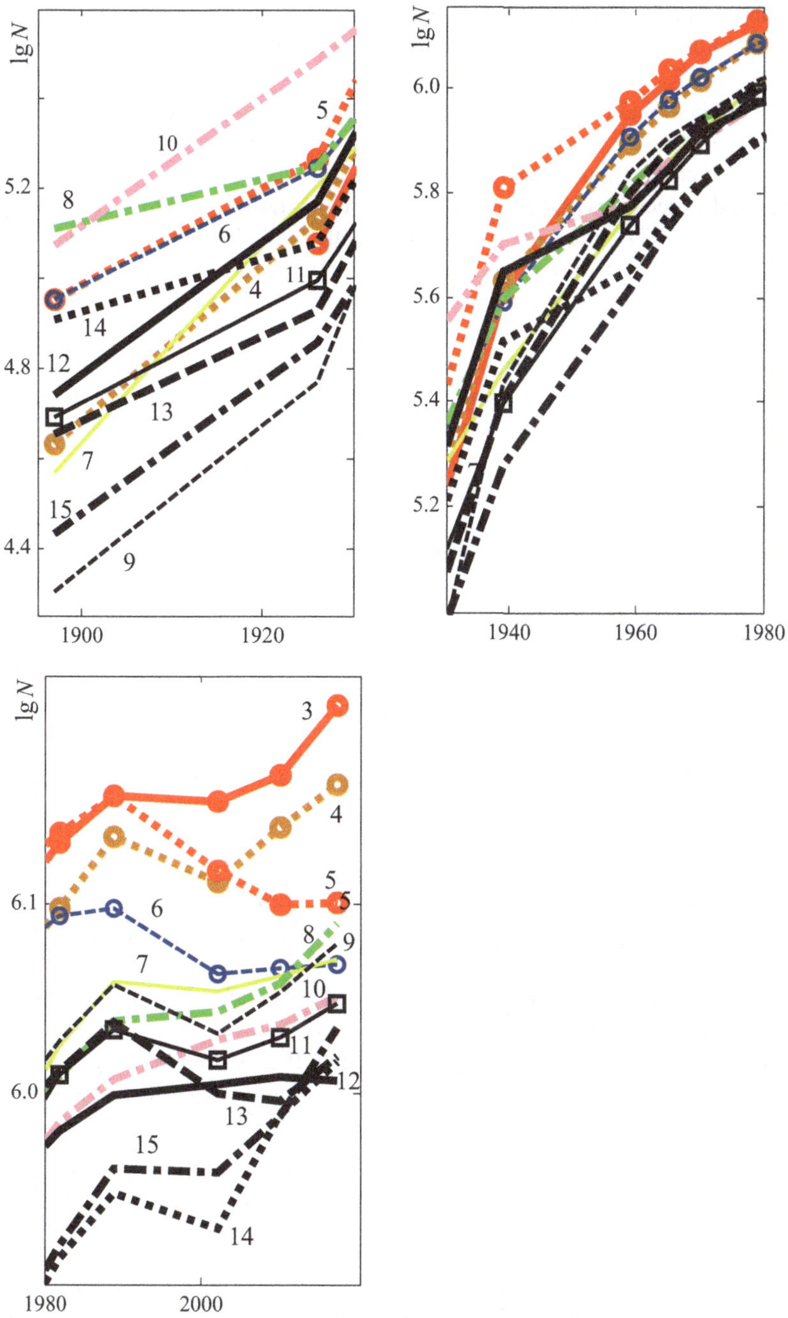

Fig. 1. Dependencies on time (years) of decimal logarithms from the number of inhabitants in cities. The curves numeration corresponds to the serial numbers of cities in Table 1. The curves corresponding to Moscow and St. Petersburg are not presented here.

(Novosibirsk) and 4 (Yekaterinburg). If in the second cluster two trajectories 3 and 4 develop without mutual intersections, then in the first group their intensive mixing occurs.

It should be noted that there are the cities Moscow and St. Petersburg which are not included in any of the clusters and not marked in Fig. 1. These two cities hold a special position, as it can be seen in Fig. 2 (curves 1 and 2, respectively). From an analysis of the temporal trajectories corresponding to the cities ranks, it follows that until 1959 they were mixing very intensively and were chaotic (with the exception of Moscow and St. Petersburg). This indicates on the transformation processes in the economic and social field of the country in the first half of the last century. But the observed chaotic trajectories of the city ranks indicate on the absence of a clear strategy for urban agglomerations development. This, perhaps, was matter-of-course within of the huge country and the existing government policy of uniform development of all territories. By 1959, 2 groups of the cities formed, excluding Moscow and St. Petersburg. The first includes the following cities Novosibirsk (3), Yekaterinburg (4), Nizhny Novgorod (5) and Samara (6), which rank trajectories are fairly regular with rare intersections until around 2010. The second group consists the cities with trajectories of ranks 7-15 that mixed intensively until the end of the 1980s. The second group trajectories then are structured into 2 groups of rank trajectories. Cities with serial numbers from 7 to 9 (see Table 1) formed a separate group, and since 2010 the city with serial number 6 got into the same cluster. Moreover, within these two forming groups, the rank trajectories continue to mix quite intensively. Consequently, on the whole, the urban agglomeration formation process is far from its completion in the country.

In Fig. 2 an interesting rank trajectory 16 corresponding to the city Saratov is present. If at the beginning of the last century it was the third most populated city within the borders of the current Russian Federation, now it is only the 17th.

In accordance with Zipf law, we have

$$\frac{\text{The population number of the city A}}{\text{The population number of the city B}} = \frac{\text{The rank of the city B}}{\text{The rank of the city A}}. \tag{1}$$

Substituting the data from Table 1 into Eq. (1), we can verify that as a whole the relation (1) is not satisfied with desirable accuracy.

Zipf law is a special case of Pareto distribution. If a random variable X follows the Pareto distribution, then its probability density function is determined by the formula

$$\rho_p(x) = \begin{cases} \frac{kx_m^k}{x^{k+1}}, & x \geq x_m \\ 0, & x < x_m \end{cases}. \tag{2}$$

Here $k > 0$, $x_m > 0$ and the distribution function for random value X has a form

$$F(x) = 1 - \left(\frac{x_m}{x}\right)^k. \tag{3}$$

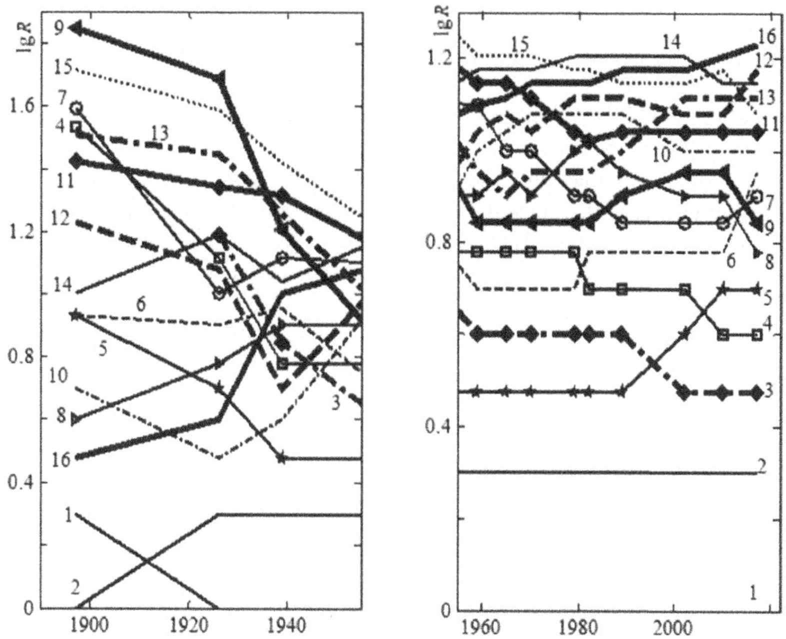

Fig. 2. Dependencies on time (years) of the decimal logarithms of city ranks. The curves numeration corresponds to the serial numbers of cities in Table 1. Here, curve 16 corresponds to the city Saratov which is not presented in Table 1.

Let us denote the population number of a settlement by N and the truncation point of the Pareto distribution ($x_m = N_m$) by N_m. Let M_m is the number of settlements with population number is not less than N_m. Then, taking into account Eq. (3), the rank R of the settlement is calculated as follows

$$R = M_m(1 - F(N)) = M_m\left(\frac{N_m}{N}\right)^k. \tag{4}$$

Let us calculate the logarithm of Eq. (4). As a result, the following Eq. (4) is obtained.

$$\lg N = A - a \lg R. \tag{5}$$

Here $a = 1/k$ and $A = a \lg M_m + \lg N_m$. Therefore, in the case of Zipf law validity following equality $a = 1$ we have. Then Eqs. (2) and (3) take the forms determining by Eqs. (6) and (7), respectively.

$$\rho_P(N) = \begin{cases} \frac{N_m}{N^2}, & N \geq N_m \\ 0, & N < N_m \end{cases}. \tag{6}$$

$$F(N) = 1 - \frac{N_m}{N}. \tag{7}$$

As a rule, a relatively small part of settlements are either small or large in terms of the number of their inhabitants. Such settlements belong to the left and right "tails" of the distribution. The remaining "moderately populated" settlements form a "hump" of the distribution probability density function. However, the functions determined by Eqs. (2) and (6) are monotonically decrease for $N \geq N_m$. Consequently, they can only describe the right descending section of the probability density function for the distribution of inhabitants by population centers (i.e., by cities and settlements with a large number of inhabitants). Therefore, Zipf law describes of the population distribution by large cities.

In Fig. 3 the regression equations of the form (5) are represented, respectively having the forms defining by Eqs. (8).

$$\begin{aligned} \lg N &= 5.9318 - 0.9084 \lg R;\ \lg N = 6.8347 - 0.8047 \lg R; \\ \lg N &= 5.2201 - 0.4265 \lg R;\ \lg N = 6.2021 - 0.1708 \lg R. \end{aligned} \tag{8}$$

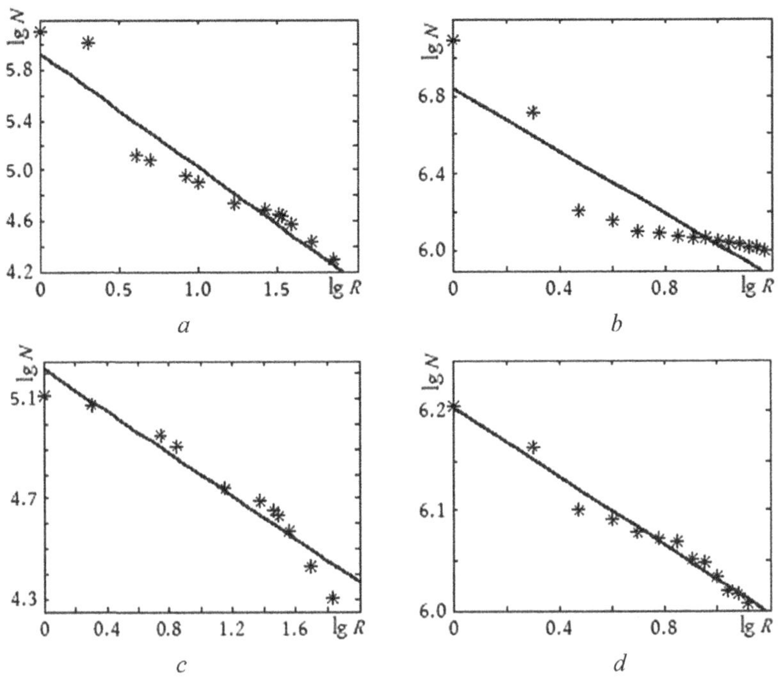

Fig. 3. Dependencies $\lg N$ from $\lg R$ for Russian cities presented in Table 1: taking into account Moscow and St. Petersburg (*a* for 1897 year and *b* for 2017 year); excluding Moscow and St. Petersburg (*c* for 1897 year and *d* for 2017 year).

In the upper left corner of Figs. 3*a* and *b* two “asterisks” correspond to the cities Moscow and St. Petersburg. On the Fig. 3*a* corresponding to 1897, St. Petersburg is located to the left of Moscow, and on the Fig. 3*b* corresponding to 2017, Moscow is located to the left of St. Petersburg. The remaining “asterisks” correspond to other cities from Table 1. It should be noted that the cities are located from left to right in increase order of their ranks indicated in columns of Table 1 corresponding to the indicated years. In Figs. 3*a* and *b* it can be noted that the “asterisks” corresponding to the cities Moscow and St. Petersburg form a separate independent subsystem. The behaviour trend of the remaining “asterisks” is completely different, and they form a separate subsystem. Moreover, the same behaviour takes place for the data from Table 1 for 1926, 1939, 1959, 1965, 1970, 1979, 1982, 1989, 2002 and 2010, i.e. Figures 3*a* and *b* show the general trend of the last 120 years. In Figs. 3*a* and *b* it is even approximately impossible to draw a satisfactory straight line through “asterisks”. Therefore, taking into account Eq. (5), population distribution on the 15 cities included in Table 1 never was obey to the Pareto and Zipf distributions since 1897. However, in case of exclusion from consideration of Moscow and St. Petersburg, the “asterisks” corresponding to 13 other cities are described well by regression equation defining by Eq. (5). A similar trend occurs throughout temporal interval from 1897 until now. The graphs of such regression equations describing by the last two Eqs. (8) and corresponding to 1897 and 2017 years are presented in Figs. 3*c* and *d*. Nature of changes of the parameters A and a of a regression equation (see Eq. 5) since 1897 are presented in Fig. 4. It can be seen that the value a does not exceed 0.45. Consequently, the residents distribution on the 13 cities under consideration follows with accuracy, depending on the accuracy of values A and a determination, to Pareto distribution with the parameter $k = 1/a$ on the entire temporal interval from 1897 until now. In this case, the parameter a is significantly less than 1 everywhere in the studied time interval. Thus, at the last 120 years, a disproportionately large dispersal of the population on small and medium-sized settlements and cities is observed [5].

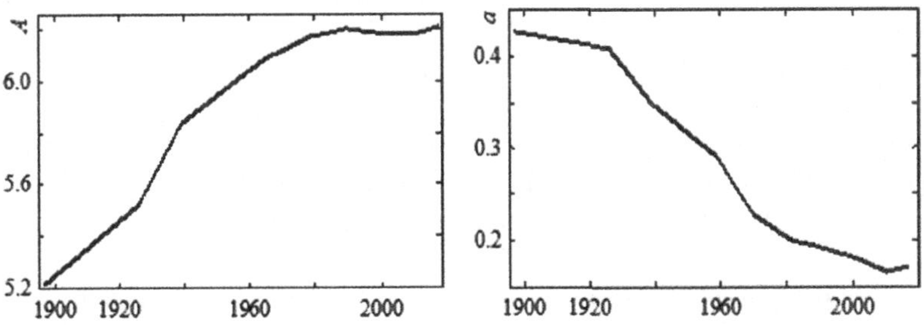

a *b*

Fig. 4. Dependencies from time (years) of coefficients A (on the left) and a (on the right) for regression Eq. (5).

Table 2. Verification of Zipf's law (see Eq. (1)) for Moscow and St. Petersburg from 1897 to 2017

Years	1897	1926	1939	1959	1965	1970	1979	1982	1989	2002	2010	2017
The ratio of the population in Moscow to the same indicator in St. Petersburg	0.821	1.201	1.296	1.531	1.764	1.757	1.746	1.758	1.785	2.172	2.357	2.344

5 Conclusions

Moscow and St. Petersburg constitute a separate subsystem among Russian cities. This status is maintained at least since 1897. For these two cities the Zipf law was best observed at the end of 1990s (see Table 2). Until 1939, these cities differed no more than 1.3 times in the number of inhabitants. At the same time, at the end of 19th and the beginning of 20th centuries, residents number in St. Petersburg was approximately 1.2 times higher. After 1918, when Moscow became a capital of the country, to 1926 the opposite situation emerged: inhabitants number in Moscow became 1.2 times higher than in St. Petersburg. In the post-World War II years, Moscow developed at a faster rate and by 1959 1.5 times more residents lived in the city. The trend of faster increase in inhabitants number in Moscow is observed to the present. At the same time, Moscow and St. Petersburg significantly surpass all other Russian cities in terms of population number. From the Figs. 1, 2 and 3 analysis it follows that Moscow and St. Petersburg represent the actually formed structural units in the process of formation in Russia of large cities network that obey to Zipf law in terms of residents number.

Although the other 13 considered cities are the largest at present time in terms of inhabitants number after Moscow and St. Petersburg, they didn't generate during the transformations since 1897 a clearly defined structure, that would be an equivalent part of subsystem consisting of the abovementioned two cities, and for which the Zipf law would be implemented.

By 2017, Novosibirsk and Yekaterinburg showed a tendency to be structured into systemically important cities. After 2002, a fairly rapid increase of residents number is observed in them - by 1.12 times up by 2017. The ranks of Novosibirsk and Yekaterinburg are equal to 3 (since 2002) and 4 (since 2010) respectively.

During the post-World War II years in transformation processes of the 13 largest cities (excluding Moscow and St. Petersburg) three clusters from temporal trajectories of population number appeared (see Fig. 1). Formation of structures from trajectories occurs only in a cluster that includes the largest cities — Novosibirsk, Yekaterinburg, Nizhny Novgorod, and Samara. In other clusters, the temporal trajectories behave rather randomly. Clusters began to collapse at the end of 1980s, but since the beginning of 2000s the signs of formation of new clusters appeared. In case of analysis of the temporal trajectories of the city ranks (see Fig. 2), two clusters formed at the end of 1950s can be detected. One of the clusters, containing temporal trajectories of city ranks with serial numbers from 7 to 15, divided into two separate parts at the end of 1980s. It should be noted that the clusters formation from temporal trajectories of population number begins earlier than from temporal trajectories of city ranks. The lag time is about 20 years. This is a result of the fact that change in the cities ranks is more slow process in comparison with changes of urbans residents number.

Population distribution by cities included in Table 1 with serial numbers from 3 to 15 is described well on the time range from 1897 to the present by Pareto distribution with a parameter a lying in the range from 0.15 to 0.45. It should also be noted, that here Moscow and St. Petersburg were excluded from consideration. The parameter a value steadily decreases until 2010. This indicates that from 1897 to 2010, none of the cities with serial numbers from 3 to 15 claimed purposely for status of the third largest Russian city. A slight increase of the value a, observed at the last years (see Fig. 4), may indicate on tendency to formation of two more systemically important centers in Russia, represented by cities Novosibirsk and Yekaterinburg.

The reason for the Zipf law failure for the largest cities is, apparently, in the spatial scales of Russia. Population migration, including labor migration, occurs mainly in two directions:

- from small and medium-sized by inhabitants number settlements of the Federation subject to the largest cities within the same subject:
- from the Federation subjects to Moscow and, to a lesser extent, to St. Petersburg.

Thus, there is almost no population migration from some Federation subjects to others.

The proposed methods for analysis of population territorial distribution may be interesting for government institutions, involved in the development of programs for social and economic development of both individual regions and the country as a whole.

Acknowledgments. The reported study was funded by Russian Foundation for Basic Research (RFBR) according to the research project № 19-010-00631.

I express my deep gratitude to the International Program of scientific grants of Russian Institute for Advanced Study (RIAS) of Moscow Pedagogical State University for supporting in 2018 study on project: “Development and progress of mathematical methods for dynamics analysis of socio-economic processes and systems”.

References

1. Gibrat, R.: Les Inégalités Économiqués; Applications: Aux Inégalités des Richesses, a la Concentration des Entreprises, Aux Populations des Villes, Aux Statistiques des Familles, etc., d’une Loi Nouvelles, La Loi de l’Effet Proportionnel. Librairie du Recueil Sirey, Paris (1931)
2. Zipf, G.K.: Human Behavior and the Principle of Least Effort. Addison-Wesley Press, Cambridge (1949)
3. Andreev, V.V., Lukiyanova, V.Y.: Test of Gibrat’s law for populated localities of Chuvash Republic. Vestnik Chuvashskogo Universiteta [Bull. Chuvash Univ.] **1**, 154–161 (2015)
4. Andreev, V.V.: Territorial distribution of the population in the Russian Federation. Econ. Reg. **13**(3), 803–811 (2017)
5. Andreev, V.V., Lukiyanova, V.Y., Kadyshev, E.N.: Analysis of people territorial distribution in regions of the Volga Federal District on the base of Zipf and Gibrat laws. Appl. Econ. **48**(4), 97–121 (2017)

6. Portal of the 2010 All-Russian Census. http://www.gks.ru/free_doc/new_site/perepis2010/croc/perepis_itogi1612.htm. Accessed 11 June 2020
7. Rastvortseva, S., Manaeva, I.: Zipf's law appearance in the Russian cities. Reg. Sci. Inq. VIII **1**, 51–59 (2016)
8. Manaeva, I., Rastvortseva, S.: Zipf's law as assessment tool of urban inequality. Reg. Sci. Inq. VIII **3**, 19–30 (2016)
9. Eeckhout, J.: Gibrat's law for (all) cities. Am. Econ. Rev. **94**(5), 1429–1451 (2004)
10. Eeckhout, J.: Gibrat's law for (all) cities: reply. Am. Econ. Rev. **99**(4), 1676–1683 (2009)
11. Levy, M.: Gibrat's law for (all) cities: comment. Am. Econ. Rev. **99**(4), 1672–1675 (2009)
12. Rastvortseva, S.: Agglomeration economics in regions: the case in the Russian industry. Reg. Sci. Inq. IX **2**, 45–54 (2017)
13. Rastvortseva, S.N.: Economic activity in Russian regions. Econ. Soc. Changes Facts Trends Forecast **11**(1), 84–99 (2018)
14. Batty, M.: Rank clocks. Nature **444**, 592–596 (2006)
15. Khavinson, M.Y., Kulakov, M.P., Mishchuk, S.N.: Prediction of foreign labor migration dynamics at the regional level. Stud. Russ. Econ. Dev. **24**(2), 170–178 (2013)
16. Mishchuk, S.N.: Domestic and international migration in the Russian Far East in the mid-19th-early 21st century. Reg. Res. Russia **3**(4), 348–355 (2013)
17. Mishchuk, S.N.: Russian-Chinese agricultural cooperation in the Russian Far East. Reg. Res. Russia **6**(1), 59–69 (2016)
18. Ohtsuki, T., Tanimoto, S., Sekiyama, M., Fujihara, A., Yamamoto, H.: Pareto-Zipf law in growing systems with multiplicative interactions. Phys. Lett. A **382**(22), 1456–1459 (2018)
19. Dentinho, T.: Urban concentration and spatial allocation of rents from natural resources. A Zipf's curve approach. Region **4**(3), 77–86 (2017)
20. Ziqin, W.: Zipf law analysis of urban scale in China. Asian J. Soc. Sci. Stud. **1**(1), 53–58 (2016)
21. Ausloos, M., Cerqueti, R.: A universal rank-size law. PLoS ONE **11**(11), e0166011 (2016)
22. Sarabia, J.M., Prieto, F.: The Pareto-positive stable distribution: a new descriptive model for city size data. Physica A Stat. Mech. Appl. **388**(19), 4179–4191 (2009)
23. Vallabados, C.A., Arumugam, S.A.: An evaluation of pareto, lognormal and PPS distributions: the size distribution of cities in Kerala, India. J. Mod. Appl. Stat. Methods **15**(1), 825–835 (2016)
24. Luckstead, J., Devadoss, S.: A comparison of city size distributions for China and India from 1950 to 2010. Econ. Lett. **124**(2), 290–295 (2014)
25. Devadoss, S., Luckstead, J.: Size distribution of U.S. lower tail cities. Physica A Stat. Mech. Appl. **444**, 158–162 (2016)
26. Project "Historical Materials". http://istmat.info/files/uploads/31715/naselenie-sssr-1965.pdf. Accessed 11 June 2020
27. Volkov, A.G., Dmitrieva, R.M., Zayonchkovskaya, Zh.A., Isupov, A.A., Kostakov, V.G., Labutova, T.S.: Naselenie SSSR: Spravochnik [Population of the USSR: Handbook]. In: Volodarskiy, L.M. (ed.). Politizdat, Moscow (1983)
28. Population of the USSR: According to the 1989 All-Union Population Census/Goskomstat of the USSR. Finance and Statistics, Moscow (1990)
29. All-Russian Census. http://www.perepis2002.ru/index.html?id=11. Accessed 11 June 2020
30. Federal State Statistics Service. http://www.gks.ru. Accessed 11 June 2020
31. Site about countries, cities, population statistics. http://www.statdata.ru/goroda-millionniki-rossii-po-naseleniu. Accessed 11 June 2020

Numerical Optimization of the Galactomannan Sulfation Process with a Sulfamic Acid-Urea Complex

Aleksandr S. Kazachenko[1,2(✉)], Natalya Yu. Vasilyeva[1,2], and Yuriy N. Malyar[1,2]

[1] Institute of Chemistry and Chemical Technology SB RAS, Federal Research Center "Krasnoyarsk Science Center SB RAS", Akademgorodok, 50/24, Krasnoyarsk 660036, Russia
leo_lion_leo@mail.ru

[2] Siberian Federal University, Svobodny Av., 79, Krasnoyarsk 660041, Russia

Abstract. The process of sulfation of galactomannan of herbaceous plants with complex of sulfamic acid and urea, was optimized. The influence of varying factors was established - the ratio of sulfamic acid: urea (SAA: U), the ratio of galactomannan: sulfating complex (GM: SC) and the duration of the process on the degree of substitution in the obtained galactomannan sulfate. The data of analysis of variance for the obtained mathematical models indicate good prognostic properties. The optimal conditions for galactomannan sulfation, which provide water-soluble sulfated galactomannan with a high (0.89) degree of substitution, have been established. The optimal conditions for the process of sulfation of galactomannan of herbaceous plants amount of urea (mol) per 1 mol of SAA – 2.26 mol, amount of SC (mol) per 1 mol of GM – 3.9 mol and process time – 118 min. Sulfated galactomannan of herbaceous plants can be used as a substitute for animal-derived anticoagulant heparin.

Keywords: Galactomannan · Sulfation · Optimization · Sulfamic acid · Urea

1 Introduction

Recently, interest in polysaccharide materials has increased due to their wide distribution and a range of unique properties [1, 2]. Polysaccharides are polymer derivatives of complex carbohydrates containing chains of monosaccharide subunits with intermediate bonds. They are found in plants [3], microbes and animal sources [4]. Plant biomass is a widely available, sustainable, and virtually inexhaustible source for the production of various types of polysaccharides [5]. In addition, polysaccharides are biologically active and non-toxic substances, which led to their widespread biomedical use [6].

The role of polysaccharides in human nutrition is diverse. They are harmless to the human body and are absorbed by 69–95%, depending on the structure. Polysaccharides serve as a source of energy, affect lipid metabolism, play the role of enterosorbents,

R. Silhavy et al. (Eds.): CoMeSySo 2020, AISC 1295, pp. 567–574, 2020.
https://doi.org/10.1007/978-3-030-63319-6_52

lower cholesterol, sorb microflora and salts of heavy metals [7–9]. Polysaccharides of higher plants, including galactomannans, have immunomodulating properties [10].

Galactomannans (composed of galactose and mannose) are non-toxic compounds widely used as food additives, stabilizers, flocculants, thickeners and gelling agents [11–14]. Pharmacological studies have shown that galactomannans and their derivatives exhibit hepatoprotective, analgesic [15, 16] and anticoagulant properties [17]. It has been established that the use of partially hydrolyzed galactomannan positively affects lipid and carbohydrate metabolism and normalizes the intestinal microflora [16].

One of the important and interesting directions for the modification of galactomannans is the production of their sulfated derivatives. Due to the introduction of a sulfate group into the biopolymer macromolecule, water solubility and biodegradability increase [18, 19]. In addition, sulfated derivatives of galactomannans can be used in pharmaceuticals as potential antioxidants, antiviral drugs, and a new class of anticoagulants [20–22].

The aim of this work was to numerically optimization of the process of galactomannan sulfation with the sulfamic acid-urea complex. This complex is environmentally safer in comparison with traditional methods (SO_3, H_2SO_4, and $ClSO_3H$ compounds), and eliminates the use of toxic solvents, which is an important step towards environmental safety of this chemical process.

2 Experimental

The source of a raw material was galactomannan from Cyamopsis tetraganoloba (Sigma). The mannose/galactose (M/G) ratio was 1.0:1.8.

2.1 Features of the Galactomannan Sulfation

2.1.1 Sulfation of Galactomannan

Sulfation of galactomannan (GM) was performed using the sulfamic acid–urea complex (SAA:U) according to the modified procedure from [23, 24]. To do this, the sulfating complex (SC) and starch were triturated to obtain a homogeneous mass. The sulfating complex was obtained by preliminary mixing 7.2 g of sulfamic acid (75 mmol) and 4.5 g (75 mmol), 9.0 g (150 mmol) and 13.5 g (225 mmol) of urea (to obtain sulfamic acid/urea: 1:1, 1:2, 1:3, respectively). The ratio of galactomannan / sulfating complex was varied between 1:2, 1:3 and 1:4 (mol/mol).The resulting reaction mixture was thermostated with constant stirring and at temperatures 80 °C, with a process duration of 30, 60, 120 min. In this temperature range, the reaction mass melted.

2.1.2 Dissolution and Neutralization

At the end of the thermostating process, the melt was cooled to room temperature. The obtained solid product was dissolved in 50 ml of water and the unreacted sulfamic acid was neutralized with a 10-% aqueous sodium hydroxide solution to pH 7–8. The resulting solution was evaporated in vacuum to a volume of 10–15 ml using a water jet pump.

2.1.3 Dialysis of Sulfated Galactomannan

The product was dialyzed against water in a MF-503-46 MFPI plastic bag (US) with a pore size of 3.5 kDa for 10 h to remove the excess reactants. Water was changed every hour. In the dialysis process, the high molecular weight compounds were purified from the low molecular weight impurities, including inorganic compounds. The dissolved high molecular weight fraction remained in the dialysis bag. The low molecular weight products passed through the membrane of the dialysis bag to the environment (distilled water).

The degree of substitution (DS) was calculated as [25]:

$$DS = \frac{1.62 * S\%}{32 - 1.02 * S\%}, \tag{1}$$

where S% is the sulfur content determined by the elemental analysis.

2.2 Calculation

The numerical optimization of the galactomannan sulfation process with sulfamic acid was carried out using the Statgraphics Centurion XVI software, according to the procedure described in [26, 27].

3 Results and Discussion

The mechanism of sulfation of polysaccharides by sulfamic acid is not well understood. However, it is assumed [28–30] that upon alcohols sulfation reaction proceeds by the first order for sulfamic acid and zero order for alcohol.

Two factors are included in the study as independent variables (in parentheses the levels of their variation): X_1 - Amount of urea (mol) per 1 mol of SAA (1, 2, 3 mol); X_2 - Amount of SC (mol) per 1 mol of GM (2, 3, 4 mol) and X_3 - Time (30, 60, 120 min). The result of the sulfation process was characterized by the output parameter: Y_1—degree of substitution (DS). Each experiment was carried out in two parallels. The designations of the variables are given in Table 1.

Table 1. Designations of independent variables (factors) and output parameters (experimental results)

Factors and parameters	Notation in equations	Range of variation
Amount of urea (mol) per 1 mol of SAA, mol	X_1	1–3
Amount of SC (mol) per 1 mol of GM, mol	X_2	2–4
Time, min	X_3	30–120
Degree of substitution	Y_1	

In the study of galactomannan sulfation by sulfamic acid the process time and reagent ratio were varied. Data on the sulfur content in galactomannan sulfate obtained under these experimental conditions is shown in Table 2.

Table 2. Sulfur content of galactomannan sulfate obtained by sulfamic acid-urea complex

№	Amount of urea (mol) per 1 mol of SAA	Amount of SC (mol) per 1 mol of GM	Time, min	DS
1	1	2	30	0.08
2	1	2	60	0.13
3	1	2	120	0.21
4	1	3	30	0.19
5	1	3	60	0.26
6	1	3	120	0.45
7	1	4	30	0.27
8	1	4	60	0.35
9	1	4	120	0.57
10	2	2	30	0.23
11	2	2	60	0.33
12	2	2	120	0.56
13	2	3	30	0.35
14	2	3	60	0.47
15	2	3	120	0.81
16	2	4	30	0.43
17	2	4	60	0.58
18	2	4	120	0.89
19	3	2	30	0.15
20	3	2	60	0.25
21	3	2	120	0.40
22	3	3	30	0.29
23	3	3	60	0.41
24	3	3	120	0.63
25	3	4	30	0.36
26	3	4	60	0.52
27	3	4	120	0.71

The dependence of the output parameter (Y_1) on variable process factors was approximated by second-order regression equations. The results of analysis of variance are given in Table 3 and Fig. 1.

Analysis of variance showed that, within the limits of the accepted experimental conditions, the factor—process time of sulfation of galactomannan—makes the greatest contribution to the total dispersion of the output parameter. This is indicated by high values of the dispersion F-ratios for the main effects, also called influence efficiencies.

Table 3. The result of analysis of variance

Sources variance	Statistical characteristics	
	F-ratio	P-value
X_1	47.99	0.0000
X_2	178.37	0.0000
X_3	260.04	0.0000
X_1^2	98.49	0.0000
X_1X_2	0.02	0.8925
X_1X_3	2.56	0.1277
X_2^2	5.12	0.0370
X_2X_3	7.75	0.0127
X_3^3	0.00	0.9501
Df	26	
R^2	97.3	
R^2_{adj}	95.8	

The information contained in columns P of the table is interpreted similarly (Table 3). The effect of the dispersion source on the output parameter is considered statistically significant if the significance level is less than a given critical value.

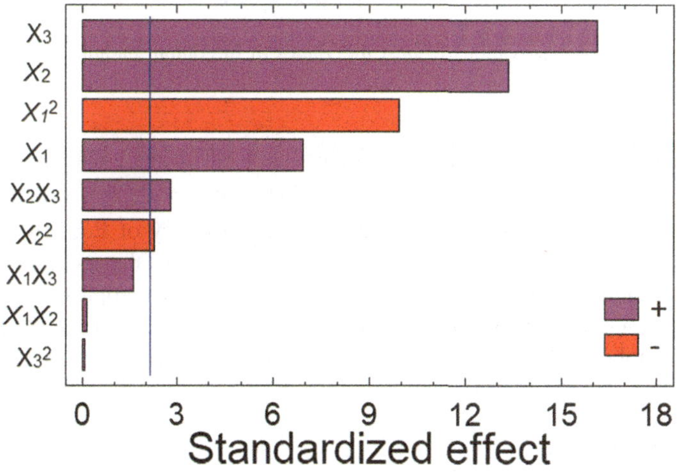

Fig. 1. Pareto analysis data of galactomannan sulfation process.

The dependence of the degree of substitution in galactomannan sulfates (Y_1) on variable process factors is approximated by the regression equation:

$$Y_1 = -1.036 + 0.7147X_1 + 0.308X_2 + 0.0004X_3 - 0.1706X_1^2 + 0.0017X_1X_2 \\ + 0.000425X_1X_3 - 0.039X_2^2 + 0.00074X_2X_3 + 6.1 * 10^{-7}X_3^2 \quad (2)$$

The prognostic properties of Eq. (2) are demonstrated in Fig. 2, which compares the values of the output parameter Y_1 obtained in the experiment with the values calculated by Eq. (2). A straight line corresponds to the calculated values of Y_1, points - to the results of observations. The proximity of the "experimental points" to the line confirms the good prognostic properties of Eq. (2).

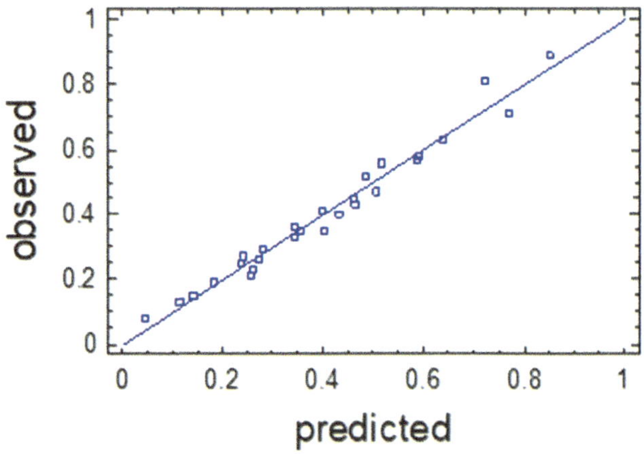

Fig. 2. The results of observations against the values of the output parameter Y_1 predicted by the mathematical model (2).

The quality of approximation is also characterized by the coefficient of determination R^2_{adj}. In the considered problem, the value of R^2_{adj} = 95.8%, which indicates a good quality of approximation. This indicates the adequacy of Eq. (2) to the observation results and allows using it as a mathematical model of the process under study.

Optimal conditions for sulfation of galactomannan with a complex of sulfamic acid with urea are amount of urea (mol) per 1 mol of SAA – 2.26 mol, amount of SC (mol) per 1 mol of GM – 3.9 mol and process time – 118 min.

4 Conclusions

For the first time, numerical optimization of the process of sulfation of galactomannan of herbaceous plants by the sulfamic acid-urea complex in the absence of a solvent was carried out.

This method eliminates the use of both toxic solvents and aggressive agents.

For obtaining galactomannan sulfate with maximum sulfur content the optimal conditions are amount of urea (mol) per 1 mol of SAA – 2.26 mol, amount of SC (mol) per 1 mol of GM – 3.9 mol and process time – 118 min.

Acknowledgments. The reported study was funded by RFBR, project number 20-33-70256.

References

1. Gopinath, V., Saravanan, S., Al-Maleki, A.R., Ramesh, M., Vadivelu, J.: A review of natural polysaccharides for drug delivery applications: special focus on cellulose, starch and glycogen. Biomed. Pharmacother. **107**, 96–108 (2018)
2. Xiong, Y., Qu, X., Liu C., Bentley, W.E., Shi, X.-W., Payne, G.F.: Polysaccharide-based smart materials. In: Chemoresponsive Materials: Stimulation by Chemical and Biological Signals, pp. 208–241. The Royal Society of Chemistry (2015)
3. Liu, Y., Sun, Y., Huang, G.: Preparation and antioxidant activities of important traditional plant polysaccharides. Int. J. Biol. Macromol. **111**, 780–786 (2018)
4. Yu, Y., Shen, M., Song, Q., Xie, J.: Biological activities and pharmaceutical applications of polysaccharide from natural resources: a review. Carbohyd. Polym. **183**, 91–101 (2018)
5. Parikka, M.: Global biomass fuel resources. Biomass Bioenergy **27**(6), 613–620 (2004)
6. Liu, J., Willför, S., Xu, C.: A review of bioactive plant polysaccharides: biological activities, functionalization, and biomedical applications. Bioact. Carbohydr. Diet. Fibre **5**(1), 31–61 (2015)
7. Gracher, A.H.P., Santana, A.G., Cipriani, T.R., Lacomini, M.: A procoagulant chemically sulfated mannan. Carbohyd. Polym. **136**, 177–186 (2015)
8. Reis, R.L., Neves, N.M., Mano, J.F., Gomes, M.E., Marques, A.P., Azevedo, H.S.: Natural-Based Polymers for Biomedical Applications. Elsevier, Lisboa (2008). 832 p.
9. Goun, E.A., Petrichenko, V.M., Solodnikov, S.U., Suhinina, T.V.: Anti-cancer and anti-thrombin activity of Russian plants. J. Ethnopharmacol. **81**, 337–342 (2002)
10. Silveira, J.L.M., Bresolin, T.M.B.: Pharmaceutical use of galactomannans. Quim. Nova **34** (2), 292–299 (2011)
11. Filatova, A.V., Azimova, L.B., Turaev, A.S.: Study of the process of gelation of galactomannan from the seeds of Styphnolobium japonicum (Fabaceae). Chem. Plant Raw Mater. **1**, 33–39 (2020)
12. Mercier, T., Guldentops, E., Lagrou, K., Maertens, J.: Galactomannan, a surrogate marker for outcome in invasive aspergillosis: finally coming of age. Front. Microbiol. **9**, 661 (2018)
13. Perera, N., Yang, F.L., Chang, C.M., Lu, Y.T., Zhan, S.H., Tsai, Y.T., Hsieh, J.F., Li, L.H., Hua, K.F., Wu, S.H.: Galactomannan from antrodia cinnamomea enhances the phagocytic activity of macrophages. Org. Lett. **19**(13), 3486–3489 (2017)
14. Tolstenkov, A.C., Drozd, H.H., Lapikova, E.U., Makarov, V.A., Mestechkina, N.M., Bannikova, G.E., Ilyina, A.B., Varlamov, V.P.: Effect of galactomannan from seeds of Cyamopsis Tetragonoloba (L.) taub on anticoagulant activity of rat plasma with intravenous administration. Clin. Hematol. Hemorheol. Cardiovasc. Surg. **7**, 242–243 (2007)
15. Mestechkina, N.M., Dovletmuradov, K., Scherbukhin, V.D.: Galactomannan common licorice seeds (Glyzyrrhizaglabra). Appl. Biochem. Microbiol. **27**(3), 435–441 (1991)
16. Krishtanova, N.A., Safonova, M.J., Bolotova, V.Ts.: Prospects for the use of plant polysaccharides as therapeutic and therapeutic agents. Proc. Voronezh St Univ. Ser. Chem. Biol. Pharm. **1**, 212–221 (2005). (in Russian)

17. Mestechkina, N.M., Anulov, O.V., Scherbukhin, V.D.: Study of galactomannan seed Amorphafruticosa L. Appl. Biochem. Microbiol. **34**(5), 549–552 (1998)
18. Caputo, H.E., Strau, J.E., Grinstaff, M.W.: Design, synthesis, and biomedical applications of synthetic sulphated polysaccharides. Chem. Soc. Rev. **48**, 2338–2365 (2019)
19. Vasilyeva, N.Y., Levdanskiy, A.V., Kazachenko, A.S., Djakovich, L., Pinel, K., Kuznetsov, B.N.: Sulfation of mechanoactivated arabinogalactan with sulfuric anhydride–pyridine complex in pyridine medium. J. Sib. Fed. Univ. Chem. **6**(2), 158–169 (2013)
20. Mestechkina, N.M., Egorov, A.V., Shcherbukhin, V.D.: Synthesis of galactomannan sulfates. Appl. Biochem. Microbiol. **42**, 326 (2006)
21. Wang, X., Wang, J., Zhang, J., Zhao, B., Yao, J., Wang, Y.: Structure-antioxidant relationships of sulfated galactomannan from guar gum. Int. J. Biol. Macromol. **46**(1), 59–66 (2009)
22. Zhang, Z., Wang, H., Chen, T., Zhang, H., Liang, J., Kong, W., Yao, J., Zhang, J., Wang, J.: Synthesis and structure characterization of sulfated galactomannan from fenugreek gum. Int. J. Biol. Macromol. **125**, 1184–1191 (2019)
23. Sirviö, J.A., Ukkola, J., Liimatainen, H.: Direct sulfation of cellulose fibers using a reactive deep eutectic solvent to produce highly charged cellulose nanofibers. Cellulose **26**(4), 2303–2316 (2019)
24. Akman, F., Kazachenko, A.S., Vasilyeva, N.Y., Malyar, Y.N.: Synthesis and characterization of starch sulfates obtained by the sulfamic acid-urea complex. J. Mol. Struct. **1208**, 127899 (2020)
25. Wang, J., Zhao, B., Wang, X., Yao, J., Zhang, J.: Structure and antioxidant activities of sulfated guar gum: homogeneous reaction using DMAP/DCC catalyst. Int. J. Biol. Macromol. **50**(5), 1201–1206 (2012)
26. Kazachenko, A.S., Vasilyeva, N.Y., Sudakova, I.G., Levdansky, V.A., Lutoshkin, M.A., Kuznetsov, B.N.: Numerical optimization of the process of abies ethanol lignin sulfation with sulfamic acid in 1,4-dioxane medium in the presence of urea. J. Sib. Fed. Univ. Chem. **13**(2), 232–246 (2020)
27. Pen, R.Z.: Planning an experiment at Statgraphics Centurion. SibSTU, Krasnoyarsk (2014). 293
28. Vasilyeva, N.Y., Levdansky, A.V., Kuznetsov, B.N., Skvortsova, G.P., Kazachenko, A.S., Djakovitch, L., Pinel, C.: Sulfation of arabinogalactan by sulfamic acid in dioxane. Russ. J. Bioorg. Chem. **41**(7), 725–731 (2015)
29. Al-Horani, R.A., Desai, U.R.: Chemical sulfation of small molecules—advances and challenges. Tetrahedron **66**(16), 2907–2918 (2010)
30. Spillane, W., Malaubier, J.B.: Sulfamic acid and its N- and O-substituted derivatives. Chem. Rev. **114**(4), 2507–2586 (2014)

Handwritten Digit Recognition by Deep Learning for Automatic Entering of Academic Transcripts

Houssem Eddine Nouri(✉)

LARIA, Ecole Nationale des Sciences de l'Informatique, Université de la Manouba, Manouba, Tunisia
houssemeddine.nouri@gmail.com,
houssemeddine.nouri@ensi-uma.tn

Abstract. Recently, deep learning has presented a new field of investigation for researchers in machine learning and has made it more artificial intelligence. Many domains of application are concerned by deep learning, such as health, education, robotics, surveillance, sports, etc., and more precisely Convolutional Neural Network (CNN). In fact, the CNN has become a standard to be used broadly in text classification, speech recognition, face detection, object detection, document analysis, and manuscript digit recognition. In this paper, we presented a case study consisting to recognize handwritten digits from academic transcripts. A new six-layered convolutional neural network is developed, trained using the known MNIST dataset, allowing to attain a training accuracy of 98,45% and a validation accuracy of 98,01%. Its efficiency is evaluated based on two experimental scenarios.

Keywords: Handwritten digits recognition · Deep learning · Convolutional Neural Network · Academic transcripts

1 Introduction

Handwritten digit recognition, in university context, is a process that converts scanned academic transcripts into an electronically understandable Excel format. Thus, enabling computers to recognize manuscript digits presented in images. The latest advancements in technology have highlighted the need for robust methods of automatic handwritten digit recognition. This article uses deep learning concepts, and more precisely Convolutional Neural Network (CNN), for handwritten digit recognition in academic transcripts.

Actually, the number of fields in which the convolutional neural network can be applied is increasing. The CNN [8, 10] is widely used for analysis of visual imagery as well as image processing applications such as text classification, speech recognition, face detection, object detection, document analysis, and manuscript digit recognition, and it is about this last application that we are interested in this paper. The importance of these fields including handwritten digits recognition using convolutional neural networks has attracted human level attention. The first CNN framework for handwritten

R. Silhavy et al. (Eds.): CoMeSySo 2020, AISC 1295, pp. 575–584, 2020.
https://doi.org/10.1007/978-3-030-63319-6_53

digits recognition has been designed by LeCun et al. [11], with seven layers of convolutional neural networks, and where the gradient descent and back propagation techniques are used to train the model [12]. The goal of this article is to observe the influence of using CNN on handwritten digit recognition. We have developed a new type of Convolutional Neural Network algorithm based on Modified National Institute of Standards and Technology (MNIST) dataset. The most known Google framework named tensorflow, written in python, is used to implement the CNN model.

2 Related Works

The convolutional neural network has presented an important impact in many different areas such as image processing. Handwritten digit recognition has become an issue of interest among researchers. A large number of papers and articles are being published these days about this topic, where it have been shown that deep learning CNN performed better when they are trained, using Keras with Theano and Tensorflow, in comparison with the most widely used machine learning algorithms like SVM, KNN & RFC. New applications of CNN are developing day by day with many kinds of research. Even though the implementation changes, the issue still remains the same and open for anyone to solve and specially to minimize the error rates.

An approximation based KNN classification algorithm is developed by Jain and Zongker [7] for handprinted digit recognition. They performed matching on two character's deforming edges and dissimilarities attaining an accuracy of 99.25%. Zhang et al. [21] utilized the ASSOM technique to provide numerical stability that can predict precise digits. For each digit, individual ASSOM was constructed and compared with ten several construction-related errors to minimize the misclassification. A new agent-oriented approach is proposed by Alhajj and Elnagar [1] to separating handwritten connected digits. This approach appointed agents to each character where their only purpose is to identify hills and valleys, which are nothing but blacks and whites. Overlapping digits are identified based on the cut-points and using the intersection of agent paths allowing to obtain an accuracy of 97%. Wu and Zhang [18] proposed 3-NN model, based on MNIST for Handwritten digit classification, attaining an error rate of 1.19%. An NCFM technique is introduced by Yin et al. [19] to avoid drawbacks of traditional convolutional layer's allowing better performance using MNIST datasets with an accuracy of 99.81%. Tavanaei and Maida [16] developed a new bio-inspired spiking convolutional neural network. The new model is evaluated on the MNIST digit dataset using clean and noisy images, reaching an accuracy of 98% and loss range 0.1% to 8.5%. Ghosh and Maghari [5] elaborated comparisons between CNN, DNN, DBN approaches to determine which neural network is useful in the field of computer vision, specifically in image classification like OCR. Their analysis claims that DNN's (Deep Belief Network) accuracy beats other neural networks with 98.08% accuracy but falls behind CNN in terms of execution time. Aly and Mohamed [2] proposed a technique for handwritten numerical strings of arbitrary length recognition using SVM and PCA, addressing the major challenge in word detection, which is overlapping characters. Their method uses hybrid PCA called PCANet for segmentation and SVM for segmentation classification together called PCA- SVMNet. Yin et al. [20] presented a study showing that the CNN is the best technique, out of all neural network implementations, to use for OCR based image classification systems.

3 CNN Model for Handwritten Digits Recognition in Academic Transcripts

3.1 Problem Description

In university context, the manuscript digit recognition from academic transcripts (MDR-AT) is a real case of application of the handwritten digit recognition problem (MDR). The MDR-ET consists to create an automatic process that converts scanned academic transcripts into an electronically understandable Excel format. Thus, enabling computers to recognize manuscript digits presented in images.

To solve this problem, we chose to use the deep learning, which has continued to be presented as a new field of investigation for researchers. According to the literature, many domains of application are concerned by deep learning, such as health, education, robotics, surveillance, sports, etc., and more precisely Convolutional Neural Network (CNN). In fact, the CNN has become a standard to be used broadly in text classification, speech recognition, face detection, object detection, document analysis, and manuscript digit recognition which is the problem to be treated in this paper.

We developed a web application integrating a CNN model allowing an auto-fill of manuscript digits, handwrote by teachers, from academic transcripts to be automatically registered in an understandable Excel format file, see Fig. 1.

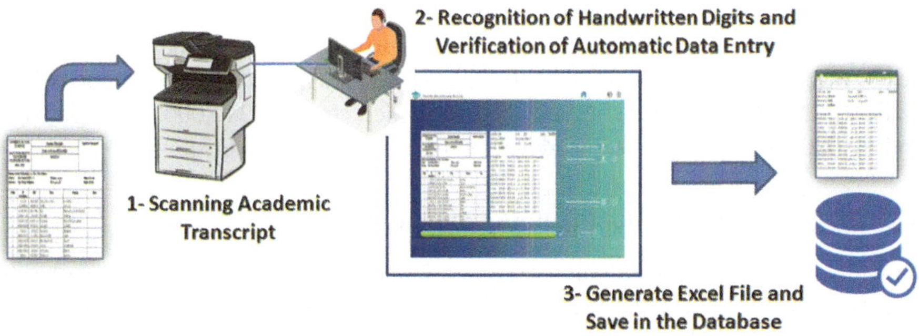

Fig. 1. Solution presentation

3.2 Dataset Selection

To train the proposed CNN model, we used the MNIST dataset (Modified National Institute of Standards and Technology dataset) see Fig. 2, which is a large database of handwritten digits that is commonly used for training various image processing systems [4, 17]. The database is also widely used for training and testing in the field of machine learning [14, 15]. It was created by “re-mixing” the samples from NIST’s original datasets [6]. The creators felt that since NIST’s training dataset was taken from

American Census Bureau employees, while the testing dataset was taken from American high school students, it was not well-suited for machine learning experiments [13]. Furthermore, the black and white images from NIST were normalized to fit into a 28 × 28 pixel bounding box and anti-aliased, which introduced grayscale levels [13].

Fig. 2. Sample images from MNIST dataset

The MNIST database contains 60,000 training images and 10,000 testing images [9]. Half of the training set and half of the test set were taken from NIST's training dataset, while the other half of the training set and the other half of the test set were taken from NIST's testing dataset [22]. The original creators of the database keep a list of some of the methods tested on it [13]. In their original paper, they use a support-vector machine to get an error rate of 0.8% [12]. An extended dataset similar to MNIST called EMNIST has been published in 2017, which contains 240,000 training images, and 40,000 testing images of handwritten digits and characters [3].

3.3 Preprocessing for the Input and Output Data

This is one of the most important parts of training a model, where we have to make sure that our model looks at the dataset the way it wants to, so we must to reshape our dataset inputs (X_train and X_test) to the shape that our model expects when we train the model. The first number is the number of images X_train → 60000 and X_test → 10000. Then, we have the shape of each image 28 * 28 and the last value of 1 signified that the image is grayscale.

Furthermore, by attaining the final layer of the CNN, we will obtain exactly 10 neurons, each of them corresponding to the respective digit from 0 to 9. So, when we will enter an image into the model, a prediction vector will be returned containing probabilities about the identified digit according to each neuron. That is why, we need to hot encode the generated variables. For example, if the image is of the number 5, it will have a value of 1 in column 6 and 0 in the remaining columns, like [0, 0, 0, 0, 0, 1, 0, 0, 0].

3.4 A Six-Layered CNN for Handwritten Digits Recognition

In fact, to recognize the handwritten digits, a six-layered convolutional neural network is created, composed by one input layer, four hidden layers and one output layer, see Fig. 3.

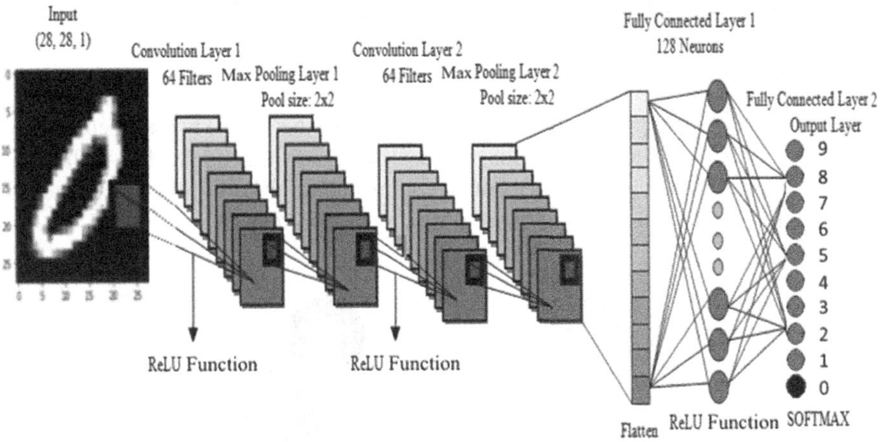

Fig. 3. A six-layered CNN model for handwritten digits recognition

The first 2D convolutional layer is the input layer of the CNN, taking an input shape of 28 * 28 pixel images signifying that it contained 784 neurons as input data, allowing to assign a value of 1 specifying the use of grayscale. A selection of 64 filters is integrated based on a kernel of 3 * 3. The ReLU activation function is used to keep only the positive values and to replace the negative values by 0 in order to guarantee the non-linearity. In addition, the hidden layers are processed in four steps: firstly, a Maxpooling layer with a shape of 2 * 2 is applied in order to browse all pixels of each image allowing to keep the greatest values on its dimension to generate a new minimized version. Then, a second 2D convolutional hidden layer is added with a selection of 64 filters based on a kernel of 3 * 3 and the ReLU activation function to change the negative values of the image to 0. Thirdly, the next hidden layer is second Maxpooling layer with a shape of 2 * 2 to minimize again its dimension keeping only the greatest values on all pixels. Finally, a flatten effect is applied to transform neurons into a single vector. This vector is linked to a first fully connected layer of 64 + 64 = 128 neurons activated by the ReLU function. The output layer is a second fully connected layer linked to exactly 10 neurons, where each one of them took a value successively from 0 to 9, based on the Softmax activation function allowing to change the output values to a series of probabilities.

The compilation process is one of the most important steps to create the CNN model. So, we have to define three main parameters: the first parameter is the optimizer which is responsible to control the learning rate. The "adam" optimizer is selected allowing to execute the compilation processes, then it is inspired both from Stochastic gradient and RMSprop optimizers. The second parameter is the loss function, where the "categorical crossentropy" is used, knowing as the most common choice for multi-classification models. The "metrics" is the last parameter, where we chose the "accuracy" value in order to see the accuracy score on the validation set while training the model.

4 Experimental Results and Discussions

4.1 Experimental Setup

The CNN model is implemented in python 3.x using the most known deep learning framework of Google tensorflow 2.x on Google Colab notebook. Then, the main program is combined with some OpenCV commands in order to add specific rectangular colored forms and to get better real time detection results, see Fig. 4.

Fig. 4. The used implementation frameworks

The execution process of the model is detailed in Fig. 6, where we fixed three epochs to train the CNN model with 60000 steps for training and 10000 steps for validation during each epoch, presenting a training accuracy of 98,45% and a validation accuracy of 98,01%, see Fig. 5.

4.2 Experimental Tests

First Scenario. To test our developed model, we made a prediction test on the first four images from the testing set of the MNIST dataset, see Fig. 7.
Then, the prediction results showed that our model succeeded to recognize all the selected images with a set of probabilities as follows in Fig. 8:

In addition, to show the real label value of each obtained vector, we converted each one of them in a 0–1 vector allowing to replace all the probabilities by 0, expect the highest one which it is replaced by 1, see Fig. 9.

Second Scenario. For the second scenario, we tested a real case of digits recognition from a scanned image containing two-tabled foal numbers 05,00 and 18,50, see Fig. 10, and where we can see that our model gave good results by identifying all presented digits in the image, see Fig. 11.

```
Train on 60000 samples, validate on 10000 samples
Epoch 1/3
60000/60000 [==============================] - 87s 1ms/step - loss: 0.2636 - accuracy: 0.9534 - val_loss: 0.0604 - val_accuracy: 0.9813
Epoch 2/3
60000/60000 [==============================] - 87s 1ms/step - loss: 0.0617 - accuracy: 0.9820 - val_loss: 0.0666 - val_accuracy: 0.9791
Epoch 3/3
60000/60000 [==============================] - 87s 1ms/step - loss: 0.0504 - accuracy: 0.9845 - val_loss: 0.0665 - val_accuracy: 0.9801
```

Fig. 5. Execution history

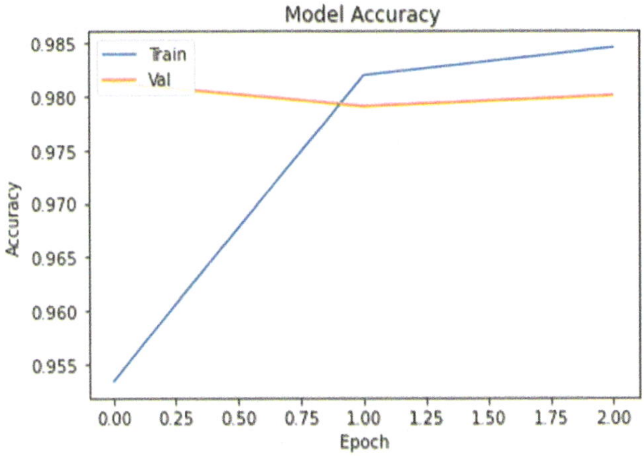

Fig. 6. Model accuracy diagram

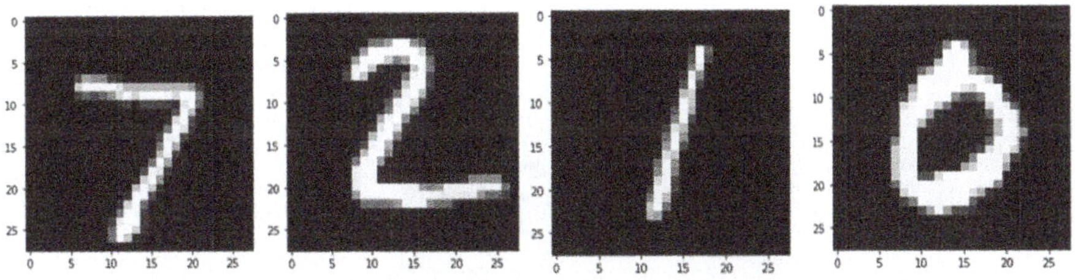

Fig. 7. The first four images from the testing set of the MNIST dataset

```
array([[9.06114128e-10, 1.54364184e-08, 1.48705499e-06, 1.14800787e-05,
        2.21797798e-08, 9.78754073e-08, 3.90101189e-12, 9.99986291e-01,
        4.12618618e-07, 2.37716037e-07],
       [2.53696602e-08, 5.58285584e-10, 9.99999881e-01, 5.70497927e-10,
        4.82447285e-12, 4.66532407e-15, 7.83199852e-08, 7.73096198e-10,
        1.97607930e-10, 1.18318062e-13],
       [1.16322490e-05, 9.99926925e-01, 6.70282077e-07, 2.72636291e-09,
        1.87862582e-07, 3.70722745e-07, 5.55171173e-05, 3.26033557e-07,
        4.22702715e-06, 1.99567065e-08],
       [9.99995828e-01, 9.36795246e-08, 1.32053062e-08, 1.24489976e-11,
        2.70391237e-10, 7.88297372e-10, 3.96886253e-06, 1.47958943e-07,
        3.58920467e-08, 9.86130910e-09]], dtype=float32)
```

Fig. 8. The predictions results of the selected images

```
[0. 0. 0. 0. 0. 0. 0. 1. 0. 0.]

[0. 0. 1. 0. 0. 0. 0. 0. 0. 0.]

[0. 1. 0. 0. 0. 0. 0. 0. 0. 0.]

[1. 0. 0. 0. 0. 0. 0. 0. 0. 0.]

[7 2 1 0]
```

Fig. 9. The conversion structure of each obtained vector of values

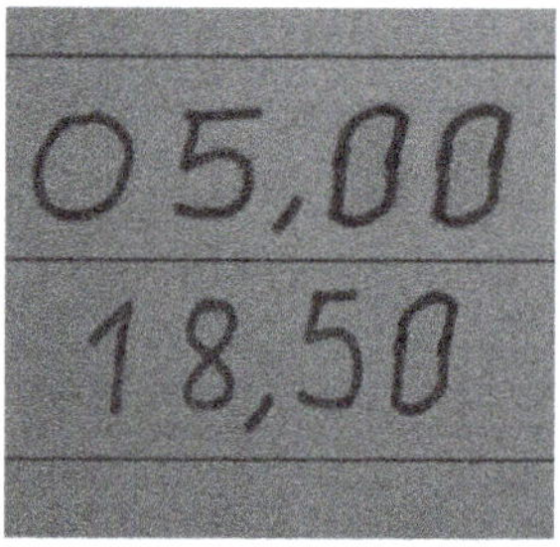

Fig. 10. A two-tabled float number image

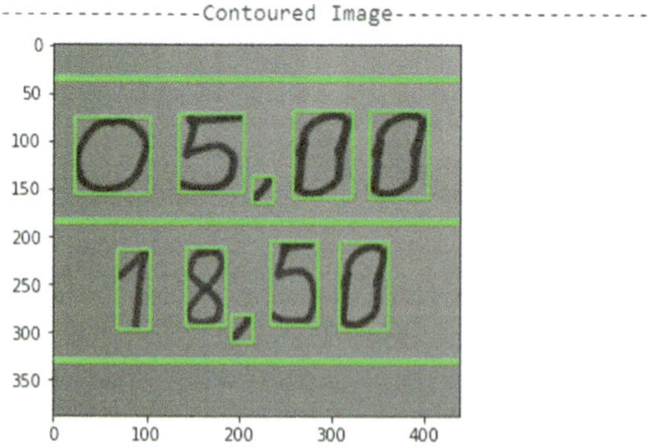

Fig. 11. The detection result of the considered image

Then, it predicted the correct value of each detected manuscript digit, see Fig. 12, which showed its efficiency to recognize handwritten digits from a real scanned copy of paper.

Fig. 12. The prediction results of the detected forms

5 Conclusion

In this paper, we presented a real case application of handwritten digit recognition in academic transcripts. A new six-layered convolutional neural network is developed, trained based on the known MNIST dataset, allowing to attain a training accuracy of 98,45% and a validation accuracy of 98,01%. The experimental tests are made on two scenarios, a first scenario to test the capability of the model to recognize images from the testing set of the MNIST dataset, and the second scenario to evaluate it on a real case of a scanned copy of paper containing two tabled float numbers. The obtained results showed that our model succeeded to recognize the images extracted from the training set in the first scenario, and then, to predict the correct values of the identified handwritten digits in the second scenario, which validated its efficiency to solve an important part of the main treated problem. In the perspectives, we will make more experimental tests on full real examples and will compare the generated results with others machine learning or deep learning models to evaluate its performance relatively to the best approaches of the literature.

References

1. Alhajj, R., Elnagar, A.: Multiagents to separating handwritten connected digits. IEEE Trans. Syst. Man Cybern.-Part A Syst. Hum. **35**(5), 593–602 (2005)
2. Aly, S., Mohamed, A.: Unknown-length handwritten numeral string recognition using cascade of PCA-SVMNet classifiers. IEEE Access **7**(1), 52024–52034 (2019)
3. Cohen, G., Afshar, S., Tapson, J., Van Schaik, A.: EMNIST: an extension of MNIST to handwritten letters. Comput. Vis. Pattern Recogn. **1**(1), 1–10 (2017)

4. Gangaputra, S.: Handwritten digit database (2013). http://cis.jhu.edu/~sachin/digit/digit.html
5. Ghosh, M.M.A., Maghari, A.Y.: A comparative study on handwriting digit recognition using neural networks. In: The International Conference on Promising Electronic Technologies, pp. 77–81 (2017)
6. Grother, P.J.: NIST Special Database 19 - Handprinted Forms and Characters Database. National Institute of Standards and Technology (2016). https://www.nist.gov/system/files/documents/srd/nistsd19.pdf
7. Jain, A.K., Zongker, D.: Representation and recognition of handwritten digits using deformable templates. IEEE Trans. Pattern Anal. Mach. Intell. **19**(12), 1386–1390 (1997)
8. Krizhevsky, A., Sutskever, I., Hinton, G.E.: ImageNet classification with deep convolutional neural networks. In: The International Conference on Neural Information Processing Systems, vol. 1, pp. 1097–1105 (2012)
9. Kussul, E., Baidyk, T.: Improved method of handwritten digit recognition tested on MNIST database. Image Vis. Comput. **22**(12), 971–981 (2004)
10. LeCun, Y., Boser, B., Denker, J.S., Henderson, D., Howard, R.E., Hubbard, W.: Backpropagation applied to handwritten zip code recognition. Neural Comput. **1**(1), 541–551 (1989)
11. LeCun, Y., Boser, B., Denker, J.S., Henderson, D., Howard, R.E., Hubbard, W.: Handwritten digit recognition with a back-propagation network. In: The International Neural Information Processing Systems Conference, vol. 2, pp. 396–404 (1990)
12. LeCun, Y., Bottou, L., Bengio, Y., Haffner, P.: Gradient-based learning applied to document recognition. Proc. IEEE **86**, 2278–2324 (1998)
13. LeCun, Y., Cortez, C., Burges, C.C.J.: The MNIST Handwritten Digit Database (2020). http://yann.lecun.com/exdb/mnist/
14. Platt, J.C.: Using analytic QP and sparseness to speed training of support vector machines. In: The International Conference on Advances in Neural Information Processing Systems, vol. 2, pp. 557–563 (1999)
15. Qiao, Y.: THE MNIST DATABASE of handwritten digits (2007). https://www.gavo.t.u-tokyo.ac.jp/~qiao/database.html
16. Tavanaei, A., Maida, A.S.: Multi-layer unsupervised learning in a spiking convolutional neural network. In: The International Joint Conference on Neural Networks, pp. 2023–2030 (2017)
17. Vision Systems Design: Support vector machines speed pattern recognition - Vision Systems Design (2004). https://www.vision-systems.com/home/article/16737424/support-vector-machines-speed-pattern-recognition
18. Wu, M., Zhang, Z.: Handwritten digit classification using the MNIST data set. Course Project CSE802: Pattern Classification & Analysis (2010)
19. Yin, Y., Wu, J., Zheng, H.: Ncfm: accurate handwritten digits recognition using convolutional neural networks. In: The International Joint Conference on Neural Networks, pp. 525–531 (2016)
20. Yin, Y., Zhang, W., Hong, S., Yang, J., Xiong, J., Gui, G.: Deep learning-aided OCR techniques for Chinese uppercase characters in the application of Internet of Things. IEEE Access. **7**(1), 47043–47049 (2019)
21. Zhang, B., Fu, M., Yan, H., Jabri, M.A.: Handwritten digit recognition by adaptive-subspace self-organizing map (ASSOM). IEEE Trans. Neural Netw. **10**(4), 939–945 (1999)
22. Zhang, B., Srihari, S.N.: Fast k-nearest neighbor classification using cluster-based trees. IEEE Trans. Pattern Anal. Mach. Intell. **26**(4), 525–528 (2004)

Orchestration of Clusters of IoT Devices with Erlang

Jorge Coelho[1,2](✉) and Luís Nogueira[1]

[1] School of Engineering (ISEP), Polytechnic Institute of Porto (IPP), Porto, Portugal
jmn@isep.ipp.pt

[2] LIACC Research Centre, University of Porto, Porto, Portugal

Abstract. The ubiquity of Internet of Things (IoT) devices and the need for novel approaches to maximize their use is the motivation for the work we describe in this paper. Here, we propose an Erlang-based orchestration layer for enabling the use of several IoT devices as a transparent distributed system. The orchestration layer is responsible for task distribution, execution monitoring and failure management. It coordinates a cluster of IoT devices that can dynamically grow or shrink as devices enter or leave the cluster. We believe that this approach is of great importance to maximize the usage of spare capacity of IoT devices allowing other devices to rely on the cluster for extending their processing power.

1 Introduction

Internet of Things (IoT) devices [11] place computing devices almost everywhere, introducing new challenges [12] and making it possible to offload at least part of the computation from the cloud to the edge [13].

The rational use of IoT devices makes the orchestration approach a useful one [7,14]. With this approach, one can use the computational power of several devices easily and without having to deal with each device's idiosyncrasies. Erlang [2,4] is a functional programming language that fits particularly well in this domain. It is simple, network-oriented and provides support for fault-tolerance [9].

Orchestration in distributed systems is a common approach to create an abstraction layer between the several devices of the system. With the orchestration layer, devices that constitute the distributed system are "hidden" and their details and behavior is managed by the orchestrator (also referred as coordinator), providing a simplified interface to those devices and centered in the use of their resources without the need to know other operational details. This is particularly relevant in Service Oriented Architectures (SOA) [5] with obvious applications when using clusters of IoT devices [3].

Here, we propose an Erlang-based orchestration layer for coordination of IoT devices. The idea is to provide a simple to use framework such that the

R. Silhavy et al. (Eds.): CoMeSySo 2020, AISC 1295, pp. 585–594, 2020.
https://doi.org/10.1007/978-3-030-63319-6_54

programmer can explore a cluster of IoT devices without being bothered with the details.

This paper is organized as follows. In the next section, we present the IoT devices and the details behind the ones used in our cluster. Then, we present the orchestration layer and how the coordination works. In the following sections, we present some implementations details and the evaluation of our implementation. Finally, we conclude and present future work.

2 IoT Devices

IoT devices (from now on, also referred as nodes) are all the devices with computing capabilities and able to communicate over a network. For our purpose, we will focus on one type of IoT devices, namely Single Board Computer (SBC) and, particularly, Raspberry Pis (RPi). This type of IoT devices is very versatile, with multiple applications [8]. Nevertheless, we believe that the methods presented here can be applied to a broader set of devices, if they are open and extendable enough to allow it.

Raspberry Pi devices are similar to a full-fledged computer in the sense they have similar components, such as a CPU, GPU, memory banks and several interfaces. The common operating system in RPis is Linux. This operating system enables two important features we need in order to achieve our goal: (i) a resource allocation policy such that the device's original functionality is not compromised; and (ii) to determine the device's relative performance to wisely distribute work among the available nodes. Both of these topics are discussed next.

2.1 Dynamic Resource Allocation

Resource allocation is one of the most complex problems in large multi-processor and distributed systems, and in general it is considered NP-hard. Computation platforms now integrate hundreds to thousands of processing cores, running complex and dynamic applications that make it difficult to foresee the amount of load they can impose to those platforms.

Elementary combinatorics provides us with evidence of the problem of scale. For a simple formulation of the problem of allocating jobs to processors (one-to-one allocation), one can see that the number of allocations grows with the factorial of the number of jobs and processors.

To cope with dynamism, a dynamic approach to resource management is the most obvious choice, aiming to dynamically learn and react to changes to the load characteristics and to the underlying computing platform. A static allocation decided before deployment, based on the (nearly) complete knowledge about the load and the platform, is no longer viable. In traditional embedded systems, the workload is usually allocated in terms of its worst-case behaviour, but static allocations that take such characterisation into account tend to produce under-utilized platforms. It is then evident that optimal resource allocation algorithms cannot cope with this type of problem, and that lightweight heuristic solutions are needed. A comprehensive survey of the kinds of resource allocation heuristics that can cover different levels of dynamicity, while coping with the scale and complexity of high-density many-core platforms, is available in [6].

Linux has a strong momentum in the embedded software industry and has, in the past years, become the prevalent choice as operating system for new platforms. A paradigm based on resource reservation can endow applications with timing and throughput guarantees, independently of the good or malicious behavior of other applications, and can be employed across all system resources, including processor cycles, communication bandwidth, disk bandwidth, and storage.

2.2 Benchmark

Due to the wide range of applications, devices have very different hardware configurations, resulting in different performance levels. Even in our scenario, where we only use RPis SBCs, and enumerating just basic technical specifications, they can vary from a Raspberry Pi Zero with only 1 GHz, single core cpu and 512 MB of RAM to a Raspberry Pi 4B with a 1.5 GHz, quad-core CPU and 8 GB of RAM. Clearly, some devices will outperform others and it is important to know each one's performance level in order to assign them with tasks of proportional complexity.

Since devices can enter and leave the cluster at any time, and be available for an unknown and maybe short periods of time, it is important to use a relatively fast method to compute some performance score that helps making decisions on the amount of work they are able to process.

In this work, we use *sysbench* [10] that allows us to get several performance metrics from the hardware in a quite fast way. *sysbench* is bundled with predefined tests for file I/O, CPU and memory, among others. It is possible to make tests with different numbers of threads, exploring performance in multicore architectures, while being able to impose a time limits on their execution. Tests end up with minimum, maximum and average time of execution in milliseconds and these values can be combined to get a global performance score.

3 Orchestrator

The orchestrator plays a central role in the system. It is responsible for the coordination of the different participants, dealing with the details of each device and providing the programmer with an API that abstracts the use of the distributed system. Thus, its main features are:

- Communication between all the participants;
- Adding nodes to the cluster and removing nodes from the cluster;
- Task distribution;
- Fault-tolerance.

The coordinator relies on the host that needs to offload work to other nodes. We now describe in more detail the ideas behind each of its features and the relevant details will be clarified in the implementation section.

3.1 Communication

Communication is done by message passing. All the different participants behave like actors [1]. All the messaging relies on the built-in features of the Erlang language that provide high-level approaches to message passing and code distribution, facilitating the whole process.

3.2 Adding and Removing Nodes

IoT devices can enter and leave the cluster at anytime. When they enter, they are available to accept tasks to execute. This process presented next. The node starts by sending a registration message to the coordinator and, after acknowledgment, sends its score that results from computing a benchmarking analysis in the node. This will allow the coordinator to rank that specific node within the cluster.

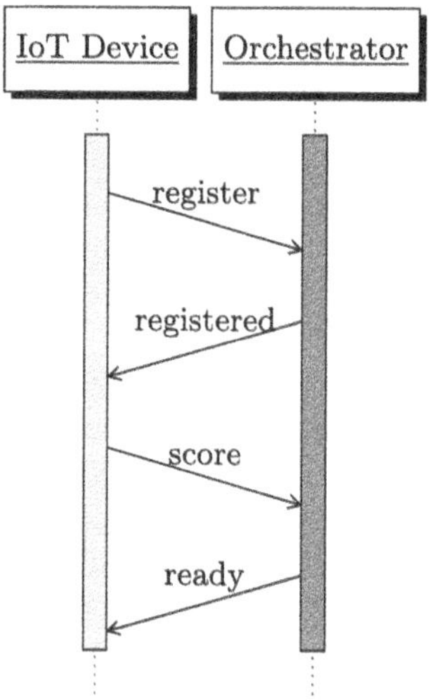

Nodes can leave the cluster in two different scenarios: (i) when the coordinator is shutdown; and (ii) when they stop, for example, due to power failure. In the first case, a message is sent from the coordinator to the node, terminating the collaboration process. In the second one, the coordinator detects the node's failure and removes it from the list of available nodes. Again, these features rely strongly in the built-in features of Erlang.

3.3 Task Distribution

The API that the coordinator provides accepts code and data, and returns the result of applying the code to the data. Its main goal is to distribute the data by

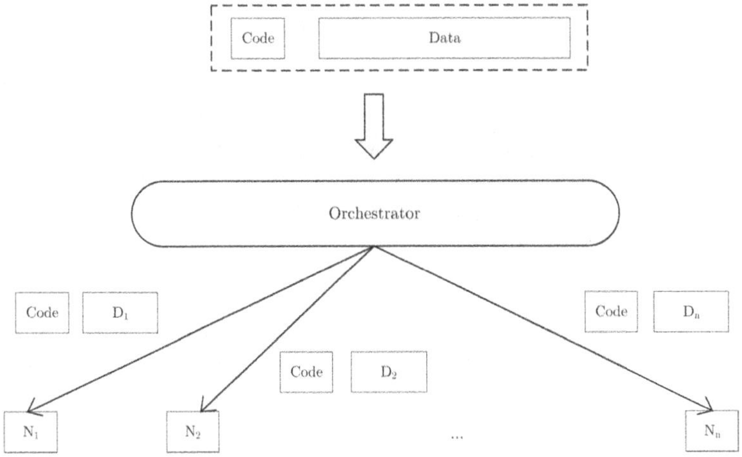

Fig. 1. Orchestration process

the nodes. It starts by splitting the data by the different nodes using their rank (given by the score obtained by benchmarking) to create partitions of data that each one will process. This is illustrated in Fig. 1, where we have an input to the orchestrator consisting of code and data, it distributes the code by the different nodes $\{N_1, N_2, \ldots, N_n\}$ and data $\mathcal{D} = \{D_1, D_2, \ldots, D_n\}$, such that they can execute it locally. For each N_i, the size of D_i varies accordingly with N_i's rank.

3.4 Fault-Tolerance

It is also the role of the orchestrator to provide fault-tolerance. Here, fault-tolerance consists in guaranteeing that all the data is processed. Since data is split among different nodes, failure in one or more than one node during execution results in loosing the corresponding partial result. To guarantee the completion of the designated task, the orchestrator maintains a permanent link with each node in the cluster and detects any failure. In case of failure, the task given to that node is rescheduled for execution in the highest ranking available node.

4 Implementation

We now present some relevant aspects of the implementation that will help to clarify how we put to practice the features described in the previous section.

4.1 IoT Node Implementation

The IoT node must have installed the Erlang VM to allow code transfer, execution and communication. Whenever an IoT device is available for collaboration, it searches for a registered orchestrator (in Erlang registered processes are those that have a name associated with them) and uses the Erlang built-in instruction $net\_adm : ping('orchestrator\_name')$ to register with the orchestrator. On success, the ping will add the IoT device to the list of knows neighbors in the orchestrator process. Then, the device will compute its performance score using *sysbench*. It will compute the average value of the CPU and Memory tests and the following formula will be applied.

$$\mathcal{R} = \alpha * CPU_{avg} + \beta * MEM_{avg}$$

The α and β are percentage values that can be tuned in order to weigh the memory and cpu values accordingly with some specific application's criteria.

Example 1. Given values $\alpha = 40\%$ and $\beta = 60\%$, and given a device D_1 with CPU score of $C_1 = 12.3$ and memory score of $M_1 = 10.2$, and given a device D_2 with CPU score of $C_1 = 9.2$ and memory score of $M_1 = 14.0$, the score of D_1 is $R_1 = 11.04$ and the score of D_2 is $R_2 = 12.08$. In this scenario, D_2 will be ranked first and D_1 will be ranked second. With different values for α and β the position of each node in the ranking can change.

After computing its performance score, the device sends a message to the orchestrator with that value, allowing the orchestrator to rank it. From now on, it is ready to be used as part of the cluster.

4.2 Orchestrator Implementation

The orchestrator lies on the node that needs to offload data. This node, as all the others, runs an Erlang VM and the orchestrator is activated whenever it needs to offload work to others. After activated, the orchestrator is able to build a cluster of IoT devices and knowing each one's performance score. Therefore, it can split tasks accordingly. In more detail, after knowing the list of available nodes $\mathcal{A}$ (nodes that successfully registered and sent their performance score), the first step is to create a ranked list. Given a list of nodes $\mathcal{A} = \{N_1, \dots, N_n\}$, they will be ordered in a list from the one with highest rank to the one with lowest rank. Their relative computational power will be used to compute the partition they must handle. Next, we present a simplified example to illustrate this. For the sake of clarity, it is intentionally simpler than a real world example.

Example 2. Given a cluster with $\mathcal{A} = \{N_1, N_2, N_3\}$ such that their score is $score(N_1) = 60$, $score(N_2) = 25$ and $score(N_3) = 15$, this means that N_1 will get 60% of the work, N_2 will get 25% and N_3 will get 15%. If the data consists in 1000 000 numbers to process then it will be partitioned with 600 000 numbers being processed by N_1, 250 000 by N_2 and 150 000 by N_3.

The orchestrator then proceeds with the transfer of code $\mathcal{C}$ and data $\mathcal{D}$ to nodes in the cluster. Erlang provides a easy way to do this. Given a module

Mod on the device where the orchestrator is running, the Erlang instruction $c : nl(Mod)$ will transfer it to the nodes in $\mathcal{A}$. After this step, all the nodes have the same version of the code and data and know the partition they will work on (which is also sent by the orchestrator).

Please note that this approach does nos escalate well when the problem being decomposed does not have an even distribution of work and it is unable to split data that has dependencies. Nevertheless, these type of problems can also be handled by the orchestrator, but data will not be partitioned. Instead, all the application will be transferred to the node with the highest score.

After all the work has been distributed, the orchestrator waits for results and handles failure. Waiting for results means it will wait for an answer from each node with the result of its particular execution. In case there is some failure, for example one node disconnects, it detects this because it relies on the underlying Erlang mechanism that generates a message whenever one element of $\mathcal{A}$ stops working. Thus, it is easy to know if one node stopped working and which part of the data it was processing. In this case, the orchestrator reschedules this exact execution to another node. The criteria implemented is to wait until the end of the current execution and rescheduled the uncompleted one to the highest ranked node.

4.3 API

The programmer does not need to know the specific details about the cluster or how the orchestrator handles them. It must only know the public functions provided by the orchestrator. The most common use is described next:

$$orchestrator : process(Module, Function, SizeData)$$

The parameters of this function are described next:

Module: The Erlang module where the code and the data are;
Function: The function to apply to the data;
SizeData: Size of data, used to compute the partitions.

Example 3. Given a module *primes* where a function $sum\_primes(B, E, List)$ returns the sum of the prime numbers in the interval from B to E in $List$ and given a list of 1000 000 numbers, summing all the primes in the list can be done using the cluster of IoT devices available, by submitting the following instruction to the orchestrator:

```
orchestrator:process(primes,sum_primes,1000000)
```

By calling this function, all the features previously described are used to create a collaborative effort of handling the problem.

5 Evaluation

We carried several tests using the cluster presented in Fig. 3. The device on the left is the one offloading work and executing the orchestrator. The devices on the right provide their computational power. All the devices are connected wireless

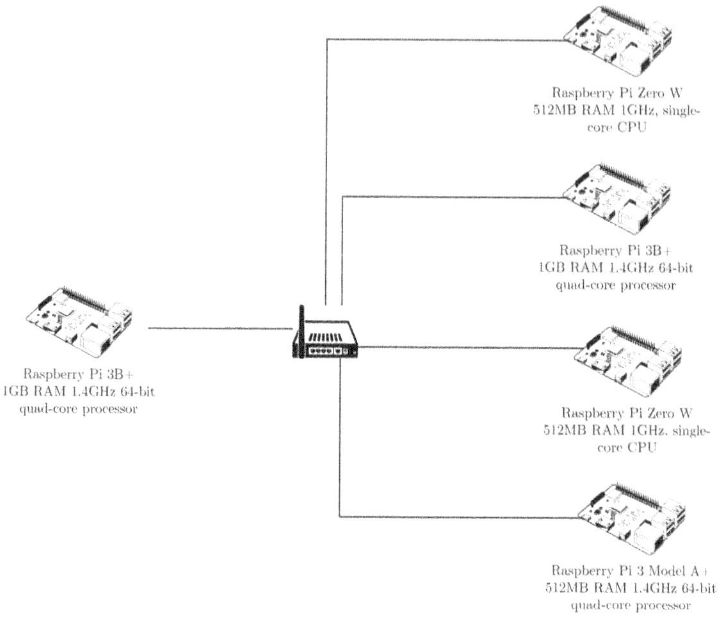

Fig. 2. Test setup

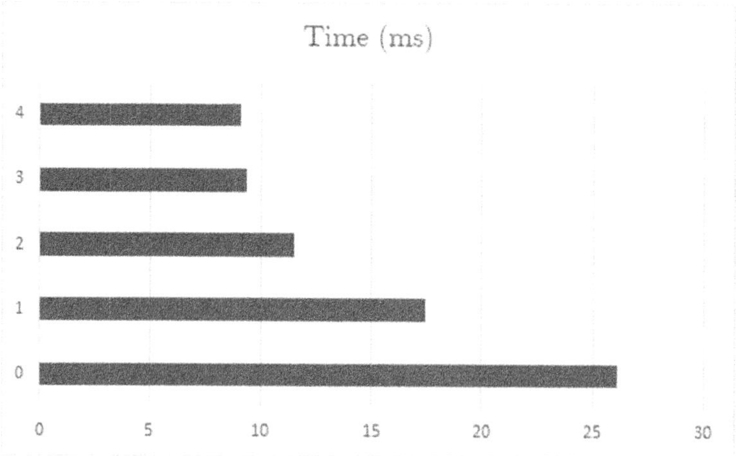

Fig. 3. Test results

to the same WiFi router (Fig. 2). To test our framework, we used a battery of tests based on data decomposable problems. Here, we present one of them, consisting in counting prime numbers in the interval [1, 50000]. The results are summarized in Fig. 3, where one can see that there is a considerable gain on using such approach when compared to the single device execution. The gain is more perceptible when the amount of data that the nodes process presents a good tradeoff between that amount and the communication overhead. In Table 1, one can see the percentage of gain in terms of execution time when adding, 1, 2, 3 and 4 devices to the cluster. The advantage is evident, although the gain when adding 4 devices when compared to three devices is marginal. One thing we notice is that problems have generally an ideal number of devices in the cluster to get the better tradeoff between the size of the problem and the cluster setup and communication overhead.

Table 1. Percentage of gain by adding devices

Devices added	Gain when compared to single device
1	33.2%
2	55.8%
3	64.1%
4	65.2%

6 Conclusion and Future Work

In this paper, we presented an Erlang-based approach to the orchestration of a cluster of IoT devices. The goal is to maximize their use by enabling the offload of code from overloaded devices to neighbors, dynamically forming a cluster and collaborating to solve problems. Our implementation is directed to a specific type of IoT devices, particularly Raspberry Pis, due to their properties that ease the whole process. We strongly believe that with the proliferation of IoT devices optimizing their usage will be an important topic of research, allowing to create a distributed computing platform that sums the power of several devices and enables the execution of tasks that otherwise would not be possible to solve with the available resources in a non-collaborative approach. Our next goal is to further develop the proposed approach, dynamically determining the ideal number of nodes for the problem being solved, as well as to study how this approach can be generalized to a broader type of devices, aiming to create a real world, consumer electronics-based distributed computer for use at home and office.

Acknowledgments. This work was partially supported by LIACC (FCT/UID/CEC/0027/2020) through Programa de Financiamento Plurianual of FCT (Portuguese Foundation for Science and Technology).

References

1. Agha, G.: Actors: A Model of Concurrent Computation in Distributed Systems. MIT Press, Cambridge (1986)
2. Armstrong, J.: A history of erlang. In: Proceedings of the Third ACM SIGPLAN Conference on History of Programming Languages, HOPL III, pp. 6–1–6–26. ACM, New York (2007). https://doi.org/10.1145/1238844.1238850
3. Atzori, L., Iera, A., Morabito, G.: The Internet of Things: a survey. Comput. Netw. **54**, 2787–2805 (2010)
4. Cesarini, F., Vinoski, S.: Designing for Scalability with Erlang/OTP: Implement Robust, Fault-Tolerant Systems, 1st edn. O'Reilly Media Inc., Sebastopol (2016)
5. Erl, T.: Service-Oriented Architecture: Concepts, Technology, and Design. Prentice Hall PTR, Upper Saddle River (2005)
6. Indrusiak, L., Dziurzanski, P., Singh, A.: Dynamic Resource Allocation in Embedded, High-Performance and Cloud Computing. River Publishers, Gistrup (2016)
7. Jawabreh, R., Darawshe, N., Othman Othman, M.M., Awad, A.: FogNetwork orchestration for heterogeneous networks. In: Proceedings of the 3rd International Conference on Future Networks and Distributed Systems, ICFNDS 19. Association for Computing Machinery, New York (2019) https://doi.org/10.1145/3341325.3342006
8. Johnston, S.J., Basford, P.J., Perkins, C.S., Herry, H., Tso, F.P., Pezaros, D., Mullins, R.D., Yoneki, E., Cox, S.J., Singer, J.: Commodity single board computer clusters and their applications. Future Gener. Comput. Syst. **89**, 201–212 (2018). http://www.sciencedirect.com/science/article/pii/S0167739X18301833
9. Kopestenski, I., Van Roy, P.: Erlang as an enabling technology for resilient general-purpose applications on edge iot networks. In: Proceedings of the 18th ACM SIGPLAN International Workshop on Erlang, Erlang 2019, pp. 1–12. ACM, New York (2019). https://doi.org/10.1145/3331542.3342567
10. Kopytov, A.: sysbench (2020). https://github.com/akopytov/sysbench. Accessed 22 June 2020
11. Li, S., Xu, L.D., Zhao, S.: The Internet of Things: a survey. Inf. Syst. Front. **17**(2), 243–259 (2015). https://doi.org/10.1007/s10796-014-9492-7
12. Shi, W., Cao, J., Zhang, Q., Li, Y., Xu, L.: Edge computing: vision and challenges. IEEE Internet Things J. **3**(5), 637–646 (2016)
13. Shi, W., Dustdar, S.: The promise of edge computing. Computer **49**(5), 78–81 (2016)
14. Velasquez, K., Abreu, D.P., Assis, M.R.M., Senna, C., Aranha, D.F., Bittencourt, L.F., Laranjeiro, N., Curado, M., Vieira, M., Monteiro, E., Madeira, E.: Fog orchestration for the internet of everything: state-of-the-art and research challenges. J. Internet Serv. Appl. **9**(1), 14 (2018). https://doi.org/10.1186/s13174-018-0086-3

Computational Complexity of Optimal Blocking of a Subset of Vertices in a Directed Graph

Gurami Tsitsiashvili[1(✉)], Marina Osipova[1,2], Alex Losev[1], and Yuri Kharchenko[1]

[1] Institute for Applied Mathematics, Far Eastern Branch of Russian Academy Sciences, Vladivostok, Russian Federation
guram@iam.dvo.ru

[2] Far Eastern Federal University, Vladivostok, Russian Federation

Abstract. This paper provides a complete solution to the problem of minimizing the number of edges in the directed graph, the removal of which blocks all paths that pass through a set of undesirable nodes. The minimization problem is reduced to the problem of minimum cut and maximum flow in a specially constructed weighted digraph with a source and a drain. The initial problem appears in analysis of protein networks with some tainted proteins. It may be used in transport systems also.

Keywords: Computational complexity · Weighted directed graph · Minimal cut · Maximal flow

1 Introduction

We consider a certain protein network with several affected (corrupted) proteins that deform the signals passing through them. Biotechnologist V.P. Bulgakov set a task to localize damaged proteins, blocking the pathways of signals passing through them. To do this, it is proposed to cut the pathways entering or leaving the set of affected proteins, minimizing the number of cut paths. This minimization deforms the structure of the protein network in the smallest way.

It should also be noted that isolation measures, that restrict the usual mode of functioning of various communication networks, such as transport, economic, educational, etc., have recently invaded our lives. Therefore, the considered biotechnological problem acquires a more general meaning, which requires the construction of cost-effective algorithms for its solution. A special feature of this problem is that if it is set incorrectly, it can easily becomes an NP problem.

The mathematical equivalent of this task is to consider a directed graph (digraph) G with finite sets of vertices U and edges V. Here, the vertices of the G digraph are the proteins, and the edges are the channels for passing signals between them.

R. Silhavy et al. (Eds.): CoMeSySo 2020, AISC 1295, pp. 595–600, 2020.
https://doi.org/10.1007/978-3-030-63319-6_55

Let's denote U_1 the set of corrupted vertices, G_1 – a subgraph of the digraph G with the set of vertices U_1 and the set of edges V_1 between them. Let U_1', $U_1' \subseteq U_1$, – the set of input vertices, i.e. vertices, in which edges income from the set $U \setminus U_1$, and U_1'', $U_1'' \subseteq U_1$, – the set of output vertices, i.e., vertices from which edges outcome into the set $U \setminus U_1$. We call a path starting with the edge, coming from the set $U \setminus U_1$ in the set U_1', passing through the set U_1 and ending with the edge, coming from the set U_1'' in the set $U \setminus U_1$, by passing through the set of vertices U_1. Let V_1' (let V_1'') the set of edges from the vertex set $U \setminus U_1$ into the set of vertices U_1 (the set of edges of U_1 into $U \setminus U_1$) (Fig. 1).

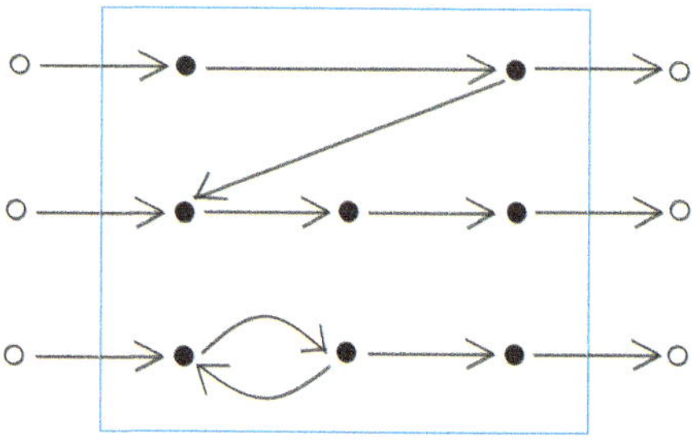

Fig. 1. Digraph G with subgraph G_1 inside a rectangle marked with blue lines.

Our task is to select in the set of edges $V_1' \cup V_1''$ subsets with the minimum number of edges, removing which results in breaking all paths, that pass through the set of $U_1 vertices$. In this paper, we managed to find such a transformation of the digraph G, which allows us to reduce this problem to the problem of the minimum section and maximum flow in the bipolar [1,2] with power complexity. This transformation consists of the following steps: 1) clusterization of the digraph G_1 with respect to cyclic equivalence, 2) converting a clustered digraph to a bipartite digraph, 3) converting a bipartite digraph to a bipolar by inserting a source and a drain into it, 4) an introducing weights of the bipolar edges, so that the edges of a bipartite digraph become unloaded. As a result of such transformations, the original problem turns into the problem of maximizing the flow in a specially constructed two-pole and determining the minimum cut by the maximum flow. Note that this problem may also occur in transport networks.

As a result of such transformations, the original problem turns into the problem of maximizing the flow in a specially constructed two-pole (two-poles) and determining the minimum section by the maximum flow. Decomposing a bipolar into components that are not connected by edges reduces the computational complexity of determining the minimum cut.

2 Recurrent Clustering Algorithm

Renumber the vertices $U_1 = \{1, ..., n\}$ of the digraph G_1 and denote its adjacency matrix $||d_{ij}|/_{i,j=1}^n$, while putting $d_{ii} = 1,\ 1 \le i \le n$. Let's say that two vertices of a digraph are equivalent if there is a loop containing these vertices. We divide the set U_1 into equivalence classes (components of connectivity) and denote K the set of these classes. On the set K, we define the binary relation: $[k_1] \succeq [k_2]$, $[k_1]$, $[k_2] \in K$, if there is a path from a vertex of cluster $[k_1]$ to a vertex of cluster $[k_2]$. Obviously, that this binary relation is reflexive, transitive, and antisymmetric, and is therefore a partial-order relation.

At step 1, there is a single vertex 1, forming a cluster [1], and a set of clusters $K = \{[1]\}$. Enter the matrix $a = ||a([p], [q])||_{[p],[q]\in K}$ that characterizes the partial order relation "$\succeq$": $a([p], [q]) = 1$, if $[p] \succeq [q]$, otherwise $a([p], [q]) = 0$. At step 1, this matrix is defined by the equality $a([1], [1]) = 1$.

Let the set of clusters K form a partition of the set of vertices $\{1, ..., t-1\}$ into disjoint subsets at step $t-1$, and the matrix a is given. Each cluster $[k] \in K$ is indexed by the maximum number of the vertex k, that is included in it. At the step t we define the sets

$$B_1 = \{k \in \{1, ..., t-1\} :\ t \succeq [k]\},\ B_2 := \{k \in \{1, ..., t-1\} :\ [k] \succeq t\}$$

$$A := B_1 \cap B_1,\ A_1 = B_1 \setminus A,\ A_2 := B_2 \setminus A,\ B := K \setminus (B_1 \cup B_2)\,.$$

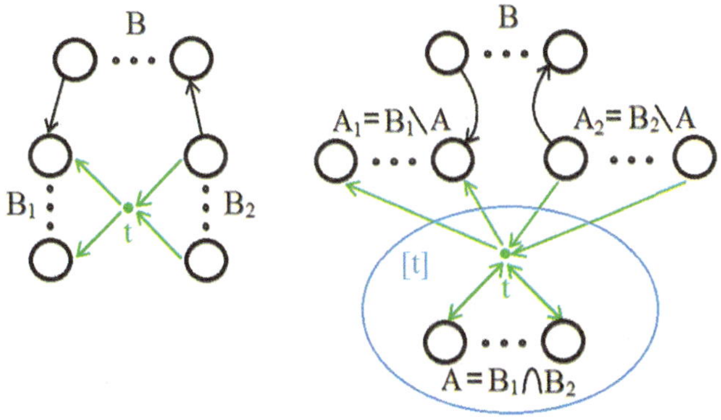

Fig. 2. Algorithm for switching from step $t-1$ to step t for clusters.

The new vertex t and the vertices of the set A form a new cluster, and the set of clusters K is also transformed (Fig. 2):

$$[t] := \{t\} \cup A,\ K := (K \setminus A) \cup \{[t]\}.$$

The matrix a is recalculated on the updated set of clusters K according to the rule:

$$a([t], [i]) := 1,\ [i] \in A_1 \cup \{[t]\},\ a([i], [j]) := 1,\ [i] \in A_2,\ [j] \in A_1 \cup \{[t]\},$$

$$a([i],[j]) := 0,\ [i] \in A_1,\ [j] \in A_2 \cup \{[t]\} \cup B,\ a([i],[j]) := 0,\ [j] \in A_2,\ [i] \in B \cup \{[t]\},$$
$$a([t],[i]) = a([i],[t]) := 0,\ [i] \in B.$$

All other values of the matrix a are defined in the previous step $t-1$. So the matrix a at the step t has the following cellular structure [3] (Table 1):

Table 1. Matrix a of partial order at the step t.

Matrix a	Clusters of the set A_1	Cluster $[t]$	Clusters of the set A_2	Clusters of the set B
Clusters of the set A_1	Meanings at the step $t-1$	0		
Cluster $[t]$			0	
Clusters of the set A_2	1		Meanings at the step $t-1$	
Clusters of the set B	Meanings at the step $t-1$	0		Meanings at the step $t-1$

As a result of this clustering, the digraph G_1 is transformed into a digraph G_2, with a set of vertices U_2, coinciding with the set of clusters K, built on the n-th step of the algorithm. Between two clusters $u,\ u' \in U_2$ an edge is drawn in the digraph G_2 if there is an edge between them in the digraph G (Fig. 3).

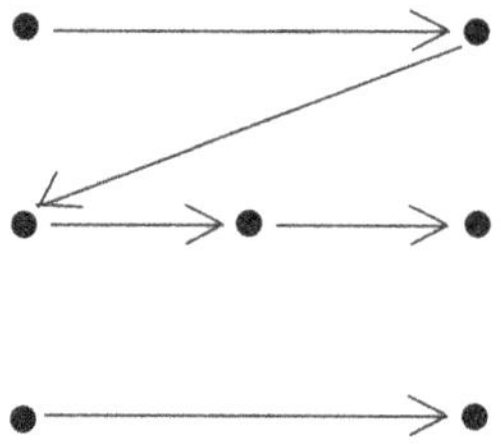

Fig. 3. Clustered digraph G_2.

3 Minimizing the Number of Deleting Edges

In the digraph G_2, we select the set U_2' of input clusters and the set U_2'' of output clusters. The cluster $u \in U_2$ is called input (output) if the digraph G has an input vertex (has is an output vertex) that belongs to this cluster. Let's construct a bipartite digraph G_3 with a set of vertices $U_2' \cup U_2''$. If $u' \in U_2',\ u'' \in U_2''$ and $u' \succeq u''$, then the bipartite digraph G_3 holds the edge $u' \to u''$. It is obvious that in this case, in the digraph G there is a path γ, passing through the set of vertices U_1 (Fig. 4).

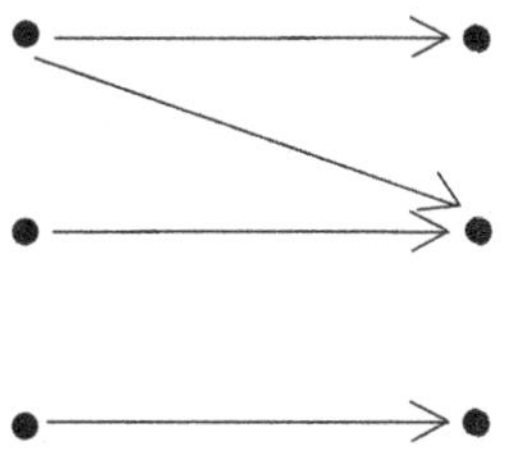

Fig. 4. Bipartite digraph G_3.

Now we decompose the digraph G_3 by removing the orientation of its edges and turning it into an undirected graph, then selecting the connectivity components and restoring the original orientation of the edges. This decomposition splits the bipartite digraph G_3 into the minimum number of disjoint subgraphs. As a result, the original problem of selecting the minimum number of edges, blocking in the digraph G all paths passing through the set of vertices U_1, is divided into subtasks for each of the sub-graphs in the digraph G_3 [4].

For simplicity of notation, assume that the digraph G_3 consists of a single a sub-graph, resulting from decomposition. Let's add it to the bipolar G_4, entering the source s, drain l and drawing edges $s \to u' : u' \in U_2', u'' \to l : u'' \in U_2''$. We will make the resulting digraph weighted by setting the throughput n (weight) of its edges as follows. Let $n(s \to u')$ be equal to the number of edges of the digraph G, included from the set of vertices $U \setminus U_1$ in the cluster $u' \in U_2'$, and $n(u'' \to l)$ is equal to the number of edges of the digraph G, leaving the cluster $u'' \in U_2''$ into the set of vertices $U \setminus U_1$. For edges of the form $u' \to u''$, $u' \in U_2'$, $u'' \in U_2''$ the throughput is determined by the equality $n(u' \to u'') = n(s \to u') + 1$ [5] (Fig. 5).

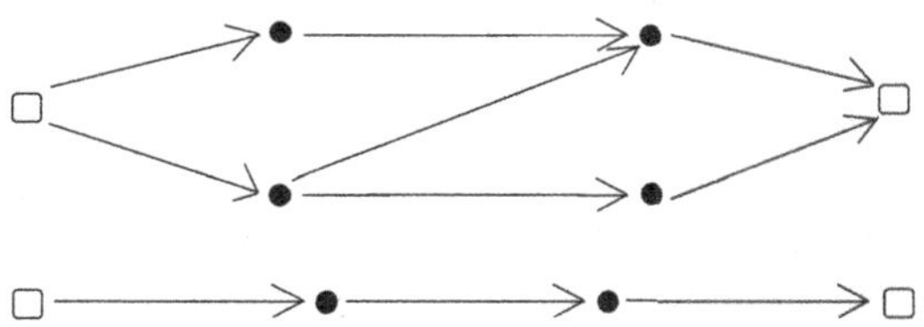

Fig. 5. Bipolar G_4 with a source on the left and a drain on the right.

It is easy to show that, when deleting edges that are part of the $u' \in u_2'$ cluster, it should either delete all edges or not delete any edges. Similarly, when deleting edges, that exit the $u'' \in u_2''$ cluster, it should either delete all edges or not delete any edges. If the digraph G is only a part of the edges, included in the cluster $u' \in U_2'$ from the vertex set $U \setminus U_1$, or a part of edges, leaving the cluster $u'' \in U_2''$ in the vertex set $U \setminus U_1$, then it can't be a solution to the original problem. In this case, the number of deleted edges is not minimized, since it is possible to further reduce the number of deleted edges by excluding the edges, that are included in the cluster $u' \in U_2'$ and/or the edges. that come out of the cluster $u'' \in U_2''$.

Now our task is to build a minimum cut in the weighted digraph G_4. To find the minimum cut in the weighted digraph G_4, we use the Ford – Fulkerson theorem on the equality of the minimum cut and the maximum flow f [1,2]. It follows from this theorem that the edges, included in the minimal cut may only be saturated, i.e. edges of the form $u' \to u''$, $u' \in U_2'$, $u'' \in U_2''$ cannot be included in the minimum cut. This means that the problem of finding the minimum cut in a weighted bipolar G_4 gives a solution to the original problem, since its solution may only include edges of the form $s \to u'$, $u' \in U_2'$ and/or edges of the form $u'' \to l$, $u'' \in U_2''$.

All known algorithms for finding the maximum flow are based on the search for a complementary path, where you can increase the flow that has already been found. The most commonly used of them are the Edmonds-Karp algorithm [6] and Dinits algorithm [7], which differ in the way of finding a complementary path. The shortest complementary path is selected in the Edmonds-Karp algorithm. In the Dinitz algorithm, a "bundl" of the graph is made to select a complementary path. If N is the number of vertices, and M is the number of edges in the bipolar, then the computational complexity for the Edmonds–Karp algorithm is $O(NM^2)$, a for the Dinitz algorithm is $O(N^2M)$.

Let's assume that the known methods found the maximum flow in the two-pole G_4. Then the following algorithm is used to find the minimum cut [1,2]. The set U_4 of vertices of the bipolar G_4 is divided into subsets X, $U_4 \setminus X$, and the minimal cut is constructed from the edges, connecting these subsets. The set X is constructed recursively: $X = \{s\}$ is assumed at the first step, next, if $x \in X$ and $f(x \to y) < n(x \to y)$, then $X := X \cup \{y\}$. if $x \in X$ and $f(y \to x) > 0$, then $X := X \cup \{y\}$. Thus, all edges $x \to y$, that are included into the minimal cut are saturated, i.e. $f(x \to y) = n(x \to y)$.

The authors thank V.P. Bulgakov for large help in setting the problem.

References

1. Ford Jr., L.R., Fulkerson, D.R.: Maximal flow through a network. Can. J. Math. **8**, 399–404 (1956)
2. Ford Jr., L.R., Fulkerson, D.R.: A simple algorithm for finding maximal network flows and an application to the Hitchcock problem. Can. J. Math. **9**, 210–218 (1957)
3. Tsitsiashvili, G.Sh.: Sequential algorithms of graph nodes factorization. Reliab. Theory Appl. **8**(4), 30–33 (2013)
4. Tsitsiashvili, G.Sh.: Improved algorithm of blocking the selected edges in the digraph. Ann. Biostatistics Biom. Appl. **3**(3) (2019). https://doi.org/10.33552/ABBA.2019.03.000561
5. Tsitsiashvili, G.Sh.: Optimal blocking of undesired nodes in digraph. Ann. Biostatistics Biom. Appl. **4**(1) (2020). https://doi.org/10.33552/ABBA.2020.04.000578
6. Edmonds, J., Karp, R.M.: Theoretical improvements in algorithmic efficiency for network flow problems. J. ACM **19**(2), 248–264 (1972)
7. Dinic, E.A.: Algorithm for solution of the problem of maximum flow in networks with power estimation. Rep. USSR Acad. Sci. **194**(4), 754–757 (1970). (in Russian)

Computational Analysis and Classification of Road Surface Using Naïve Bayes Classifiers

Aditya R. Moesya, P. H. Gunawan(✉), and Indwiarti

School of Computing, Telkom University,
Jl. Telekomunikasi no 1, Terusan Buah Batu, Bandung 40275, Indonesia
dealtron@studenttelkomuniversity.ac.id,
{phgunawan,indwiarti}@telkomuniversity.ac.id

Abstract. Road becomes one of the main paths for land transportation, therefore the quality of the road plays a big role in order to achieve the convenience and safety for the community. This research works on a solution to identify the quality of road into two categories (smooth and rough). The datasets are made possible by Phyphox mobile application producing up to 300 datasets. The extracted datasets originally have three axis such as x-axis, y-axis, and z-axis graphs. After conducting individual and multiple axis with combination of scenarios, it is found that the combination of y-axis and z-axis scenario is the best fit for current experiment. The methods that are being proposed include the classification models under Naïve Bayes classifiers. The models include Gaussian Naïve Bayes, Multinomial Naïve Bayes with Min-max scalar, Complement Naïve Bayes with Min-max scalar, Multinomial Naïve Bayes with absolute number, and Complement Naïve Bayes with absolute number. The accuracy of all models are being calculated and compared in this paper. Nevertheless, the Complement Naïve Bayes with Min-max scalar shows the best accuracy value up to 91.4% and 91.6% for model validation and performance measurement respectively.

Keywords: Road surface · Naïve Bayes · Classification · Accelerometer

1 Introduction

Roads have been connecting people from one place to another place for some daily activities. Currently, land transportation is the most chosen mode of travel if it is compared with air and sea transportation mode. This popularity develops since all social and economics activities are mainly done on the land. Therefore, high and a good quality of roads is very important in economy, infrastructure development, education, tourism and other aspects in society. Moreover, roads can connect every actors which played in society for instance producers to consumers.

R. Silhavy et al. (Eds.): CoMeSySo 2020, AISC 1295, pp. 601–611, 2020.
https://doi.org/10.1007/978-3-030-63319-6_56

Since roads have been used very frequently, there is a high chance of deteriorating the quality and durability of the road itself. Indeed, regular roods need maintenance after some periods of time. Therefore, it is important to detect the condition of the road whether the damage can be tolerated or not. Here, there are two types of road condition can be extremely severed, lightly damaged or just minor bumps found along the way.

Fig. 1. Example of damaged road in Sukapura.

Fig. 2. Example of smooth road in Soekarno-Hatta road.

For instance, Fig. 1 and Fig. 2 illustrate damage and smooth quality of road respectively. Figure 1 shows the sample of a damaged road in Sukapura Street area, Bandung Indonesia. The general characteristics of a damaged road are the presence of potholes, some bumps, uneven surfaces and etc. Meanwhile, Fig. 2 shows the sample of a smooth road in Soekarno-Hatta Street area, Bandung Indonesia.

There are some researches regarding road surfaces detection and prediction. For example, in the paper of Aljaafreh et al. [1], the speed bump can be detected by fuzzy inference system (FIS) based on speed and acceleration that have been detected by their accelerometer. Their work was utilized with the embedded sensor of accelerometer in the Smartphone. Meanwhile, Sattar et al. [11] utilizes smartphone sensors to give a review on road surface monitoring. The study compares and reviews about the present strategies for using smartphones for road surface anomaly observation. The comparison for the current strategies implements divided to some primary criteria and have performance results.

In this paper, some Naïve Bayes methods for classification problem will be used to classify the condition of the road surfaces. Indeed, the reason of using Naïve Bayes methods is because of the method can handle small number of datasets. Phyphox application will be used to obtain dataset which is installed in survey tool. Therefore, the prediction of the road surface condition is also influenced by the data taken from Phyphox. The raw datasets extracted from Phyphox consisted of vibration waves in three dimensions axis.

2 Literature Review

2.1 Related Works

There has been several researches regarding road bumps detection and road surface anomalies monitoring. Pavement surface characteristics are generally assessed by means of skid resistance tests, texture depth and roughness measurements as explained by Gianluca Cossale et al. [5].

Kaiyue Zang et al. [13] had successfully combined two tools, GPS and Accelerometer sensors to analyse the level smoothness of road on bicycle-mounted smartphone. The paper uses the International Roughness Index (IRI) for indexing the level of smoothness of road. Some measurements have been used for input parameters in the algorithm, for instance, vertical acceleration, time game and etc. Finally the algorithm produces a collection of bumpy areas that are composed of the timestamps for producing result.

Meanwhile, Bello-Salau et al. [2] proposed the use of a new measure termed "z-difference square' to enhance accelerometer data analysis. The "z" is attributed to the z-axis of the accelerometer output. The concept behind the "z-difference square" measure is to ensure that the derived statistic is positive definite, which leads to the use of a single threshold system. The work introduces a filtering effect which naturally reduces the noise content in the data.

Celaya-Padilla et al. [4] detects road abnormalities using genetic algorithm approach. The model is able to discover a linear logistic model able to detect road abnormalities precisely with cross-validation method. Moreover, using blind evaluation, the method had an accuracy of 0.9714.

2.2 Naïve Bayes Classifier

A Naïve Bayes classifier is a classification technique based on Bayes theorem. Generally, Naïve Bayes classifiers can perform well on a wide range of classification problems including text classification, multi-class prediction, recommendation system, etc. Compared with more advanced and complex system, Naïve

Bayes classifiers often perform better, because of it's ability of handling much variables either discrete or continuous and also a lot of datasets [3,7]. Bayes theorem gives an insight on how to calculate prior probability of an event, $P(l \mid o)$, from $P(l)$, $P(o)$, and $P(o \mid l)$ [9] as follow,

$$P(l \mid o) = \frac{P(o \mid l)\, P(l)}{P(o)} \tag{1}$$

where $P(l \mid o)$, the posterior probability of class given predictor (attribute) can be calculated using, $P(o \mid l)$ as the probability of predictor given class, $P(l)$ as the prior probability of class, and $P(o)$ as the prior probability of predictor.

There are few types of Naïve Bayes models such as Gaussian, Multinomial, Complement, and Bernoulli. In this paper, a Gaussian Naïve Bayes algorithm for continuous-valued features is proposed. Gaussian Naïve Bayes assumes that the continuous values corresponding to each feature are distributed according to Normal distribution. Basically, the prior probability of predictor of the given class is assumed to be Gaussian, also known as the conditional probability is often modelled with the normal distribution as follows:

$$P(X_j|C = c_i) = \frac{1}{\sigma_{ji}\sqrt{2\pi}} exp\left(-\frac{(X_j - \mu_{ji})^2}{2\sigma_{ji}^2}\right) \tag{2}$$

where μ_{ji} is mean (average) of feature values X_j and σ_{ji} is the standard deviation of feature values X_j.

The multinomial Naïve Bayes implements multinomial distribution in which it is practical to model feature vectors where each value represents. The multinomially distributed data is parameterized by vectors such that $\theta_x = (\theta_{x1}, ..., \theta_{xn})$ for each class x, where n is the number of features and θ_{xi} is the conditional probabilities $P(o_i \mid x)$ of feature i appearing in a sample belonging to class x.

Then the conditional probabilities are computed with a relative frequency count which corresponds to applying a smoothed maximum likelihood approach such that:

$$\hat{\theta} = \frac{N_{xi} + \alpha}{N_x + \alpha n} \tag{3}$$

where N_x is total amount of all features for class x. Where N_{xi} as the amount of feature i appears in a sample of class x in the training dataset, α as the alpha parameter (Laplace smoothing factor) $\alpha = 1$ to prevent the model from resulting in null probabilities when the frequency is zero.

Complement Naïve Bayes is best used to deal with skewed training data in which the data having more training examples for one class than another [10]. Skewed training data essentially causes the decision boundary weights to be biased and complement Naïve Bayes helps to mitigate the bias in the weight estimates. Furthermore, the method for calculating the weights is in the following:

$$\hat{\theta}_{\tilde{x}i} = \frac{N_{\tilde{x}i} + \alpha_i}{N_{\tilde{x}} + \alpha} \tag{4}$$

where $\hat{\theta}_{\tilde{x}i}$ as the occurrences of a feature in the complement to the class, $N_{\tilde{x}}$ as the total number of features in the opposite class, $N_{\tilde{x}i}$ as the number of repetitions of a feature in the opposite class, and α as the smoothing Laplace parameter.

After the calculation of the occurrences, the weight for each feature is calculated with the following equation,

$$\hat{m}_{xi} = \frac{log\,\hat{\theta}_{xi}}{\sum_z |log\,\hat{\theta}_{xi}|}, \tag{5}$$

where $\hat{m}_{xi}$ as the weight for a feature i in the complement to the class x. Essentially, the final decision can be calculated with the following equation:

$$\hat{x} = arg\ min_x \sum_i t_i m_{xi}. \tag{6}$$

It is concluded that the classification result is taken from the class with the minimum value of the sum of weights for each feature in the class.

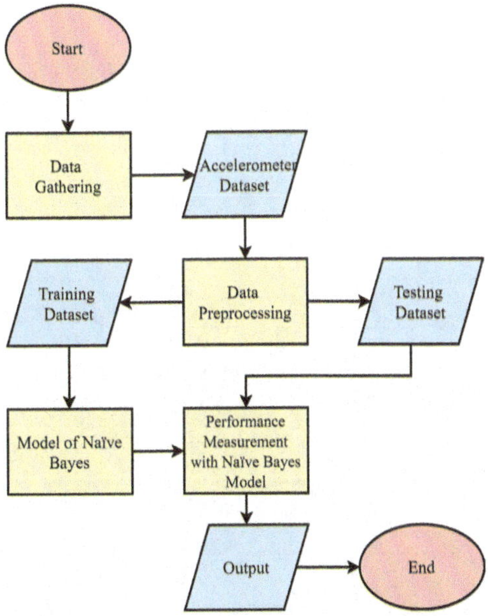

Fig. 3. Design process for road classification

3 Experiments Process

3.1 System Design

The system design for this study can be seen in Fig. 3. Based on Fig. 3, the first step that is taken for this research is data gathering. The data collected consists of accelerometer data that will then be used as the raw accelerometer dataset.

3.2 Data Gathering

The raw datasets are gathered by using Phyphox mobile application and taken by riding motorcycle. The intended velocity of the motorcycle is in the range of 25 km/h to 27 km/h. The intended duration of each raw dataset is 60 s. The mobile device is strapped horizontally on the motorcycle riding position with the screen facing upward. With such data gathering method, the acceleration

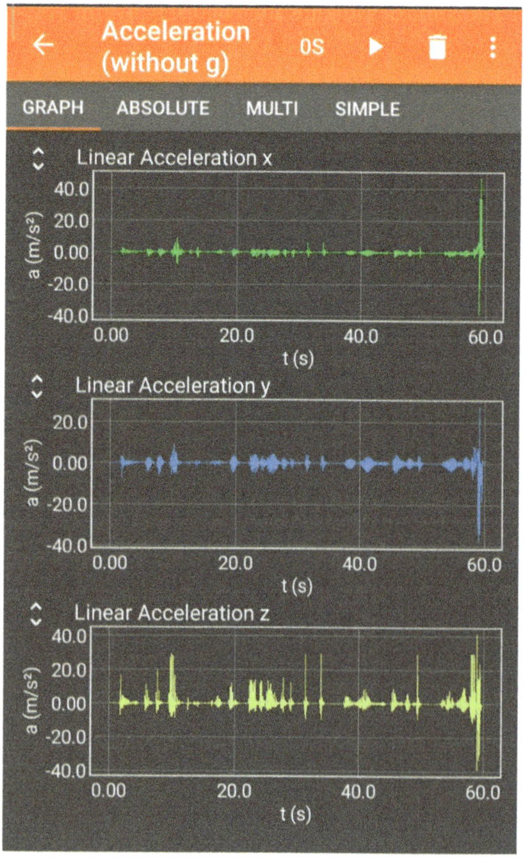

Fig. 4. The interface of Phyphox Accelerometer.

data from each axis can be taken. The x-axis acceleration data can be captured by turning left as negative value and turning right as positive value. The y-axis acceleration data can also be obtained by the forward acceleration as positive value and backward acceleration as negative value. Lastly, the z-axis acceleration data can be acquired if there is an acceleration from sudden shock upward as positive value and downward as negative value.

Phyphox facilitates a wide range of integrated sensors that can be accessed from mobile devices [12]. In this case, the sensor used for this project is Accelerometer, a micro-electro-mechanical system technology instrument for measuring acceleration [6]. The interface of accelerometer when taking raw dataset is shown in Fig. 4.

3.3 Preprocessing

The raw accelerometer datasets from each road originally was separated. In every separable dataset contain the time as the duration in seconds, x as the acceleration towards x-axis in m/s^2, y as the acceleration towards y-axis in m/s^2, and z as the acceleration towards z-axis in m/s^2.

This process will convert the raw data attributes with statistical descriptive into new attributes including max, min, lower quartile, upper quartile, mean (negative), mean (positive), standard deviation, and variance. The mean are calculated from the each axis attribute of raw accelerometer dataset. Then every positive mean and negative mean is collected into a same file using the following formula,

$$\mu_{neg} = \frac{\sum_{i=1}^{N_{neg}} x_i}{N_{neg}}, x < 0, \tag{7}$$

$$\mu_{pos} = \frac{\sum_{i=1}^{N_{pos}} x_i}{N_{pos}}, x \geq 0. \tag{8}$$

Based from formula above (7–8), μ_{neg} as mean negative, can be calculated by using N_{neg} as the number of negative data (negative population) and μ_{pos} as Mean positive, can be calculated by N_{pos} as the number of positive data (positive population) with x_i as the individual values in the data x_1, x_2, x_3, to data x_N.

Standard deviation and variance are also used as features, with the formula as follows:

$$\sigma^2 = \frac{1}{N}\sum_{i=1}^{N}(x_i - \mu)^2 \tag{9}$$

where σ as standard deviation and σ^2 as variance can be calculated by N as the number of data (population), x_i as the individual values in the data $x_1, x_2, x_3, \cdots, x_N$, and μ as the mean of the data. If the standard deviation or variance is low, it means that most of the data points are close to the average (mean). Whereas, if the standard deviation or variance is high, then the data points are spread out over a larger set of values.

For Multinomial and Complement Naïve Bayes models, the feature scaling method is also used such as Min-max scalar and absolute number techniques since the values of the datasets vary widely. The Min-max scaling method scales the data to a fixed range from 0 to 1, 0 to 5, 0 to 100, 0 to 25, and 0 to 50. The range that has the optimum result is within 0 to 50. Min-max scaling can be calculated as follows:

$$Z_i = \frac{X_i - X_{min}}{X_{max} - X_{min}} \tag{10}$$

Since the best Min-max scalar ranges from 0 to 50, the result of Z_i needs to be multiplied by 50 and the result will be the new value of the dataset. Whereas, the absolute number technique is used to eliminate the negative values of the data. The multinomial Naïve Bayes assumes that the features of the dataset contain multinomial distribution, therefore it cannot contain negative values. After the preprocessing process, the next step is to implement the classification process.

3.4 Classification

In this classification process, Gaussian Naïve Bayes, Multinomial Naïve Bayes with Min-Max scalar, Multinomial Naïve Bayes with absolute number, Complement Naïve Bayes with Min-Max scalar, and Complement Naïve Bayes with absolute number, are chosen to compare their accuracy each other to find the best Naïve Bayes classifier to be used in this case. The result of the proposed model should be binary (smooth as 0 and rough as 1).

In operations on each Naïve Bayes classification, the dataset need to be split into training set, used to build and train the model, and test set, used to evaluate the efficiency of the model, with the proportion of 80% and 20% respectively. Then calculate the prior probability of the class attribute in training set and calculate the likelihood probability with each attribute for each class. And then putting the calculated values in Bayes formula and calculating the posterior probability and evaluate the accuracy of the model by using a Confusion matrix.

3.5 K-Fold Cross Validation

In order to evaluate our classification models, we implement k-fold cross validation. The technique divides the training datasets into folds then each fold is used as a testing set. The scenario which yields the best performance in this research uses 10-fold cross validation ($k = 10$). The average accuracy of 10-fold cross validation is labeled as validation accuracy. Finally, this project will be able to analyse the performance and the accuracy of the Road Surface classification.

3.6 Accuracy

Confusion matrix can be used to evaluate the performances of classification model as shown in Table 1. Here in confusion matrix, there are four values can be observed. Value of TP and TN are the number of positive and negative instances that are correctly classified respectively. Meanwhile value of FP and FN are the number of positive and negative instances that are miss classified respectively [8].

The evaluation of the optimal performance of a classification model can be defined based on confusion matrix as shown in Table 1. The row table indicates the predicted class, whereas the column indicates the actual class. The number of positive and negative instances that are correctly classified are consecutively denoted as TP and TN. Whereas, the number of miss classified negative and positive instances are consecutively denoted as FP and FN [8].

Table 1. Confusion matrix

		Predicted	
		Actual positive class	Actual negative class
Actual	Predicted positive class	True positive (TP)	False negative (FN)
	Predicted negative class	False positive (FP)	True negative (TN)

The threshold metric that is used in this classification is accuracy (*Acc*) as obtained in the following [8]:

$$Acc = \frac{TP + TN}{TP + FP + TN + FN} \tag{11}$$

Here, we use validation accuracy as the one being implemented with 10-fold cross validation for training dataset and predictive accuracy as the one being implemented with training and testing dataset.

4 Results

In order to achieve the highest accuracy, the experiment is executed several testing scenarios of individual axis and the combination of the axis. Since the accuracy is slightly different in each iteration, therefore the scenario is done 10 times for each individual combination of axis. The final accuracy for each scenarios is based on the average accuracy from 10 iterations. The combination of axis scenario that has the most influence is determined by the highest average accuracy.

In this paper, we also compare some types of Naïve Bayes models such as Gaussian Naïve Bayes, Multinomial Naïve Bayes with Min-max scalar, Complement Naïve Bayes with Min-max scalar, Multinomial Naïve Bayes with absolute number, and finally Complement Naïve Bayes with absolute number. The comparison of results can be seen in Table 2.

Table 2. The accuracy comparison of individual and combination axis scenario

	x-axis	y-axis	z-axis	x-axis and y-axis	x-axis and z-axis	y-axis and z-axis	x-axis, y-axis, and z-axis
Model validation							
GaussianNB	0.871	0.896	0.91	0.873	0.883	0.903	0.894
MultinomialNB (Min-max Scalar)	0.87	0.895	0.912	0.895	0.896	0.913	0.909
ComplementNB (Min-max Scalar)	0.869	0.894	0.91	0.896	0.896	0.914	0.91
MultinomialNB (Absolute number)	0.816	0.85	0.846	0.887	0.847	0.851	0.86
ComplementNB (Absolute number)	0.873	0.866	0.842	0.886	0.845	0.853	0.852
Performance measurement							
GaussianNB	0.86	0.891	0.906	0.891	0.892	0.903	0.892
MultinomialNB (Min-max Scalar)	0.854	0.845	0.898	0.892	0.905	0.916	0.899
ComplementNB (Min-max Scalar)	0.854	0.841	0.895	0.892	0.905	0.916	0.897
MultinomialNB (Absolute number)	0.823	0.842	0.849	0.896	0.873	0.846	0.869
ComplementNB (Absolute number)	0.881	0.852	0.844	0.884	0.868	0.847	0.87

Based on the comparison in Table 2, the combination of y-axis and z-axis scenario has the highest model validation up to 91.4% using Complement Naïve Bayes with Min-max scalar classifier and performance measurement up to 91.6%. Therefore, Complement Naïve Bayes with Min-max scalar model is preferred to classify the road surface datasets since it has the highest accuracy compared with the other models in the scenario.

During the data gathering, y-axis moves forward alongside the direction of the bike, while z-axis itself is positioned perpendicular to the screen of the phone, therefore the vibration caused by the uneven surface of the road is likely to be shown in z-axis graph. In this case, x-axis does not play significant role in the scenario since the vehicle moves mostly in straight line. Taken from the accuracy result in Table 2, combination of y-axis and z-axis scenario is used in this research.

5 Conclusion

Based on the conducted research, it is concluded that Complement Naïve Bayes with Min-max scalar model has the highest accuracy value than the other Naïve Bayes models used in the scenario. The model validation and performance measurement reach up to 91.4% and 91.6% consecutively. The target feature consists of 2 labels that indicate rough road surface and smooth road surface. The datasets are retrieved using accelerometer android sensor which then generates x, y, and z axis graphs. After various combination testing, it is concluded that the combination of y-axis and z-axis dataset has the most optimal performance. Moreover, this paper can be improved by developing a device that can apply our algorithm to read and classify the road surface data in real time. Finally, this research is expected to assist further technology development related to road surface condition monitoring.

References

1. Aljaafreh, A., Alawasa, D.K., Aljaafreh, S., Alabadleh, A.: Fuzzy inference system for speed bumps detection using smart phone accelerometer sensor. J. Telecommun. Electron. Comput. Eng. **9**, 133–136 (2017)
2. Bello Salau, H., Aibinu, A., Onwuka, L., Dukiya, J., Bima, M., Onumanyi, A., Folorunso, T.: A new measure for analysing accelerometer data towards developing efficient road defect profiling systems. J. Sci. Res. Rep. **7**, 108–116 (2015)
3. Bouckaert, R.R.: Naive bayes classifiers that perform well with continuous variables. In: Proceedings of the 17th Australian Joint Conference on Advances in Artificial Intelligence. AI 2004, pp. 1089–1094. Springer, Heidelberg (2004). https://doi.org/10.1007/978-3-540-30549-1_106
4. Celaya Padilla, J., Galván Tejada, C., López-Monteagudo, F., Alonso-González, O., Moreno-Baez, A., Martínez-Torteya, A., Galván Tejada, J., Arceo, J., Luna-Garcia, H., Gamboa-Rosales, H.: Speed bump detection using accelerometric features: a genetic algorithm approach. Sensors **18**, 443 (2018)
5. Cossale, G., Elliott, R., Widyatmoko, I.: The importance of road surface texture in active road safety design and assessment. In: International Conference Road Safety and Simulation, pp. 22–25 (2013)
6. Dadafshar, M.: Accelerometer and gyroscopes sensors: operation, sensing, and applications. Maxim Integrated [online] (2014)
7. Domingos, P., Domingos, P., Pazzani, M.: On the optimality of the simple Bayesian classifier under zero-one loss. Mach. Learn. 29(2–3), 103–130 (1997). https://doi.org/10.1023/A:1007413511361
8. Hossin, M., Sulaiman, M.N.: A review on evaluation metrics for data classification evaluations. Int. J. Data Min. Knowl. Manage. Process **5**, 01–11 (2015)
9. Kaviani, P., Dhotre, S.: Short survey on Naive Bayes algorithm. Int. J. Adv. Res. Comput. Sci. Manage. **04**, 11 (2017)
10. Rennie, J.D., Shih, L., Teevan, J., Karger, D.R.: Tackling the poor assumptions of Naive Bayes text classifiers. In: Proceedings of the 20th International Conference on Machine Learning (ICML-03). pp. 616–623 (2003)
11. Sattar, S., Li, S., Chapman, M.: Road surface monitoring using smartphone sensors: a review. Sensors **18**, 3845 (2018)
12. Staacks, S., Hütz, S., Heinke, H., Stampfer, C.: Advanced tools for smartphone-based experiments: phyphox. Phys. Educ. 53(4), 045009 (2018). https://doi.org/10.1088/1361-6552/aac05e
13. Zang, K., Shen, J., Huang, H., Wan, M., Shi, J.: Assessing and mapping of road surface roughness based on GPS and accelerometer sensors on bicycle-mounted smartphones. Sensors **18**, 914 (2018)

A K-means Bat Algorithm Applied to the Knapsack Problem

Leonardo Pavez[1,2](✉), Francisco Altimiras[3], and Gabriel Villavicencio[4]

[1] Instituto de Ciencias Naturales, Universidad de las Américas, Av. Manuel Montt 948, Providencia, 7500972 Santiago, Chile
lpavez@udla.cl
[2] Departamento de Ciencias Químicas y Biológicas, Universidad Bernardo O'Higgins, General Gana 1702, 8370854 Santiago, Chile
[3] Facultad de Ingeniería y Negocios, Universidad de las Américas, Santiago, Chile
fjaltimiras@gmail.com
[4] Escuela de ingeniería en construcción, Pontificia Universidad Católica de Valparaíso, Valparaíso, Chile
gabriel.villavicencio@pucv.cl

Abstract. Combinatorial problems are frequent in the industry and in the engineering field. Many of these problems are NP-hard and can hardly be addressed with complete techniques. Therefore, the design of binary algorithms based on continuous metaheuristic swarm intelligence is an area of interest in operational research. In this article we use a general binarization mechanism based on the k-means technique. This technique is applied to the bat algorithm with the objective of solving the problem of the multidimensional backpack (MKP). The experiments are designed to demonstrate the utility of the k-means technique in binarization. In addition, we verified the efficiency of our algorithm through reference instances, showing that the k-means binary bat (BKBA) algorithm obtains adequate results when evaluated against another next-generation algorithm.

Keywords: Combinatorial optimization · KnapSack · Metaheuristics · k-means

1 Introduction

The combinatorial problems have great relevance at industrial level, we find them in various areas such as Civil Engineering [1,2], sustainability [3], Operational Research [4–6], resource allocation [7–9], scheduling problems [10–12], routing problems [13–16] among others. On the other hand, in recent years, algorithms inspired by nature phenomena to solve optimization problems have been generated. As examples of these algorithms we have Cuckoo Search, Black Hole, and Bat Algorithm,among others. Many of these algorithms work naturally in continuous spaces and therefore must be adapted to solve combinatorial problems.

R. Silhavy et al. (Eds.): CoMeSySo 2020, AISC 1295, pp. 612–621, 2020.
https://doi.org/10.1007/978-3-030-63319-6_57

In the process of adaptation, the mechanisms of exploration and exploitation of the algorithm can be altered, having consequences in the efficiency of the algorithm.

Several binarization techniques have been developed to address this situation. In a literature search, the main binarization methods used correspond to transfer functions, angle modulation and quantum approach, for more detail refer to [17]. In this article we present a new binarization method that uses the k-means technique to group the solutions and then perform the binarization process. To verify the efficiency of our method, we used the bat algorithm.

To check our k-means binary bat algorithm (KBBA), we use the well-known knapsack problem. Experiments were developed using a random operator to validate the contribution of the k-means technique in the binarization process of the bat algorithm. In addition, local search operator is used to strengthen the results. Moreover, the binary artificial algae (BAAA) algorithm was used to compare our results. BAAA was developed in [18] and uses transfer functions to perform the binarization process. The results show that the k-means technique obtains results superior to those obtained by the random operator and that our KBBA algorithm shows competitive results against the BAAA algorithm.

2 KnapSack Problem

The multidimensional knapsack problem (MKP), models resource allocation situations. The goal is to find the subset of objects that produce the greatest benefit by satisfying a set of constraints. MKP is one of the most studied NP-hard class combinatorial problems, however it remains as a challenge due to the difficulties encountered in solving medium and large size instances. A search for the last few years in the literature, we find that MKP has been solved for example in [19] using a quantum binarization technique, in [20] applied the k-means clustering technique to perform binarization, in [21] used a differential algorithm with transfer functions, and in [22] a modification of the PSO equations was used. MKP can be set as:

$$\text{maximize} \sum_{j=1}^{n} p_j x_j \tag{1}$$

$$\text{subjected to} \sum_{j=1}^{n} c_{ij} x_j \leq b_i \ , \ i \in \{1, ..., m\}. \tag{2}$$

With $x_j \in \{0, 1\}$, $j \in \{1, ..., n\}$. p_j corresponds to the profit of element j. c_{ij} represents a cost associated with dimension i and element j. The constraints in each dimension i are represented by b_i. The solution can be modelled using a binary representation, in this representation a 0 means the element is not included in the knapsack.

3 Binary Bat Algorithm

The application of statistics and data mining, can be found in many areas such as transports, smart cities, agriculture and computational intelligence [20,23–26]. In this section, we explore the application of the k-means technique to the binarization of continuous metaheuristics.

The k-means binary bat algorithm(KBBA) uses 4 operators to achieve its goal. An initialization operator, A local search operator, the binary operator using the k-means technique, and a repair operator for the case of solutions that do not meet any of constraints. The general scheme of the algorithm is shown in Fig. 1. In the first step, KBBA initializes the solutions, this is detailed in Sect. 3.1. Once we have all the solutions initialized, we check the compliance of the stopping criterion of the algorithm associated to the maximum number of iterations. The bat algorithm is then run in conjunction with the k-means operator, this is described in Sect. 3.2. Subsequently once the binarization was performed, the results are compared with the best solution obtained so far. In the case of finding a superior solution, it replaces the previous solution and is submitted to a local search operator. In the following subsection we will explain in detail the initialization method, the binary operator and the repair operator.

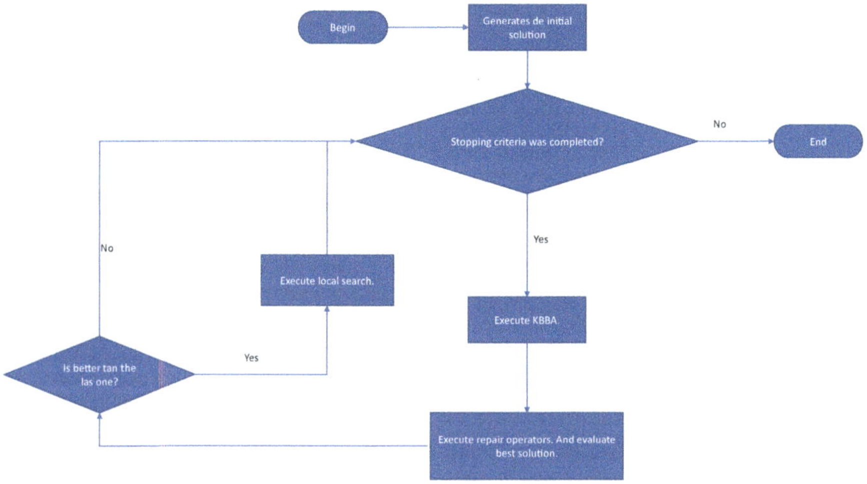

Fig. 1. Flowchart of the binary bat algorithm.

3.1 Initialization and Element Weighting

As KBBA is a swarm algorithm, to begin the exploration and exploitation of the search space, the list of solutions must be initialized. For the generation of each solution, an element is randomly chosen first. The next step is to check if other elements can be incorporated, for this we must evaluate the constraints of our problem. To select the new element, we generate a list of possible elements that comply with the constraints. Each element is calculated a weight and the best weight element is selected. This procedure is repeated until no additional element can be incorporated.

To calculate the weight of each element, several methods have been developed. In [27] a pseudo-utility in the surrogate duality approach was proposed. The way to calculate it is shown in Eq. 3. In this equation the variable w_j corresponds to the surrogate multiplier whose value is between 0 and 1. This multiplier can be interpreted as a shadow prices of the j-th constraint.

$$\delta_i = \frac{p_i}{\sum_{j=1}^{m} w_j c_{ij}} \tag{3}$$

A more intuitive measure focused on the average resource occupancy was proposed by [28]. It is shown in Eq. 4.

$$\delta_i = \frac{\sum_{j=1}^{m} \frac{c_{ij}}{m b_j}}{p_i} \tag{4}$$

In this article we used a variation of this last measure focused on the average occupation and proposed in [20]. This variation considers the elements that exist in knapsacks to calculate the average occupancy. In each iteration depending on the selected items in the solution the measure is calculated again. The expression of this new measure is shown in Eq. 5.

$$\delta_i = \frac{\sum_{j=1}^{m} \frac{c_{ij}}{m(b_j - \sum_{i \in S} c_{ij})}}{p_i} \tag{5}$$

3.2 k-means Transition Ranking Operator

Consider our metaheuristic is a continuous and population based. Due to its iterative nature, it needs to update the position of particles at each iteration. When the metaheuristic is continuous, this update is performed in $\mathbb{R}^n$ space. In Eq. 6, the update position is presented in a general manner. The x_{t+1} variable represents the x position of the particle at time $t{+}1$. This position is obtained from the position x at time t plus a Δ function calculated at time $t{+}1$. The function Δ is proper to each metaheuristic and produces values in $\mathbb{R}^n$. For example in Cuckoo Search $\Delta(x) = \alpha \oplus Levy(\lambda)(x)$, in Black Hole $\Delta(x) = \text{rand} \times (x_{bh}(t) - x(t))$ and in the Firefly, Bat and PSO algorithms Δ can be written in simplified form as $\Delta(x) = v(x)$.

$$x_{t+1} = x_t + \Delta_{t+1}(x(t)) \tag{6}$$

In the k-means transition ranking operator, a model for transitions in a discrete space is proposed. This model is based on considering the movements generated by the metaheuristic in each dimension for all particles. $\Delta^i(x)$ corresponds to the magnitude of the displacement $\Delta(x)$ in the i-th position. Subsequently these displacement are grouped using the magnitude of the displacement $\Delta^i(x)$. This grouping is done using the k-means technique where k represents the number of clusters used. Finally, a generic P_{tr} function given by the Eq. 7 is proposed to assign a transition probability.

$$P_{tr} : \mathbb{Z}/k\mathbb{Z} \rightarrow [0, 1] \tag{7}$$

A transition probability through the function P_{tr} is assigned to each group. Naturally, this P_{tr} function is modelled as a cumulative probability function. For the case of this study, we particularly use the linear function given in Eq. 8. In this equation, N(x^i) indicates the location of the group to which $\Delta^i(x)$ belongs. The α coefficient, corresponds to the transition probability and β to the transition separation coefficient. Both parameters are estimated in each metaheuristic. For our particular case, N(x^i) = 0 corresponds to elements belonging to the group that has the lowest Δ^i values. N(x^i) = 7 corresponds to the group of elements that have the greatest Δ^i values.

$$P_{tr}(x^i) = P_{tr}(N(x^i)) = \alpha + \beta N(x^i)\alpha \tag{8}$$

The k-means transition ranking operator starts calculating Δ^i for each of the particles. This step is specific in each metaheuristic. Subsequently the particles are grouped using k-means clustering technique and the magnitude of Δ^i as distance. With the group assigned to each particle we obtain the probability of transition using Eq. 8. Afterwards the transition of each particle is performed. In the case of bat algorithm the rule 9 is used to perform the transition, where $\hat{x}^i$ is the complement of x^i. Finally, each solution is repaired using the repair operator.

$$x^i(t+1) := \begin{cases} \hat{x}^i(t), & \text{if } rand < P_{tg}(x^i) \\ x^i(t), & \text{otherwise} \end{cases} \tag{9}$$

3.3 Repair Operator

Local search and binary k-means technique can generate infeasible solutions. There are different mechanisms to address these infeasible solutions. In this article, the repair of the solutions was considered. To perform the repair, the measure described in the Eq. 5 was used. If the solution requires repair, the element with the maximum measure is chosen and it is removed from the solution, this process is iterated until a valid solution is obtained. The solution is then improved by exploring the possibility of incorporating new elements. This stage of improvement is iterated until there are no elements that can be incorporated without violating the constraints.

4 Results

4.1 Insight of KBBA Algorithm

This section aims to find out the contribution of the binary k-means operator to the performance of the algorithm. To carry out the comparison problems cb.5.250 from the OR-library were selected. Violin charts and the Wilcoxon non-parametric signed-rank test were used to perform the statistical analyzes. In the charts, the X axis identifies the studied instances and the Y axis the % -Gap described in the Eq. 10. The Wilcoxon test is run to determine if the results obtained by KBBA have significant difference with respect to other algorithms. The parameter settings and browser ranges are shown in Table 1.

$$\% - Gap = 100\frac{BestKnown - SolutionValue}{BestKnown} \tag{10}$$

Table 1. Setting of parameters for Binary bat Algorithm.

Parameters	Description	Value	Range
N	Number of solutions	30	[20, 25, 30]
G	Number of clusters	5	[4,5,6]
Iteration Number	Maximum iterations	1000	[1000]
cmax	Maximum for c coefficient	1	1
cmin	Minimum for c coefficient	0.00001	0.00001

Evaluation of k-means Binary Operator. A random operator was designed to evaluate the contribution of the k-means binary operator. This random operator has the peculiarity of performing the transitions with a fixed probability of 0.5 without considering in which cluster the variable is located. Two configurations were studied: The first one where the local search operator is included and the second where the local operator is not considered. KBBA corresponds to our standard algorithm. *random.ls* is the random variant that includes the local search operator. *BKGOA.wls* corresponds to the version with k-means binary operator without local search operator. Finally *random.wls* describes the random algorithm without local search operator.

When we compared the Best Values between KBBA and *random.ls* which are shown in Table 2. KBBA outperforms to *random.ls*. However the Best Values between both algorithms are very close. In the Average comparison, KBBA outperforms *random.ls* almost in all problems. The comparison of distributions is shown in Fig. 2. We see the dispersion of the *random.ls* distributions are bigger than the dispersions of KBBA. In particular this can be appreciated in the problems 1, 2, 3, 5, 6 and 7. Therefore, the k-means binary operator together

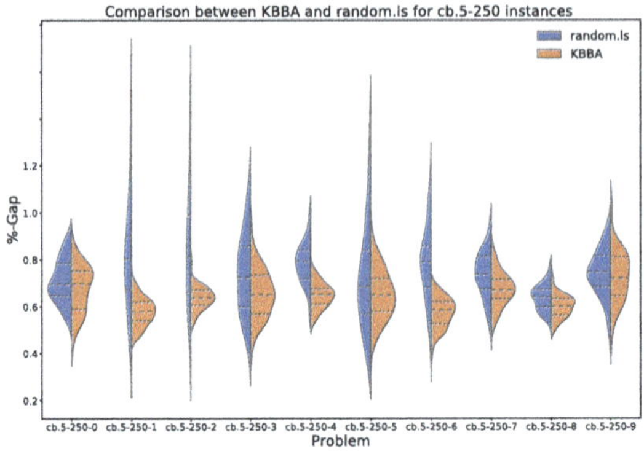

Fig. 2. Evaluation of k-means binary operator with Local Search operator

Table 2. Evaluation of k-means binary operator

Set	Best Known	best *random.ls*	best KBBA	best *random.wls*	best *wls*	avg *random.ls*	avg KBBA	avg *random.wls*	avg *wls*
cb.5.250-0	59312	59211	59225	59158	59175	59132.1	59151.4	59071.8	59132.1
cb.5.250-1	61472	61435	61435	61409	61409	61324.6	61389.1	61288.3	61370.7
cb.5.250-2	62130	62036	62074	61969	61969	61894.4	61988.5	61801.6	61924.6
cb.5.250-3	59463	59367	59446	59365	59349	59257.8	59332.2	59136.1	59269.3
cb.5.250-4	58951	58914	58951	58883	58930	58725.6	58814.4	58693.6	58758.5
cb.5.250-5	60077	60015	60056	59990	60015	59904.6	59942.6	59837.8	59945.9
cb.5.250-6	60414	60355	60355	60348	60349	60208.2	60345.1	60230.6	60311.1
cb.5.250-7	61472	61436	61436	61407	61407	61290.8	61336.5	61233.9	61342.3
cb.5.250-8	61885	61829	61885	61790	61782	61737.1	61782.2	61644.9	61738.1
cb.5.250-9	58959	58832	58866	58822	58787	58769.1	58786.6	58653.7	58770.4
Average	60413.5	60343	60372.9	60314.1	60317.2	60224.4	60285.15	60157.2	60256.3
p-value							1.37 e−04		2.62 e−05

with local search operators, contribute to the precision of the results. Finally, the KBBA distributions are closer to zero than *random.ls* distributions, indicating that KBBA has consistently better results than *random.ls*. When we evaluate the behaviour of the algorithms through the Wilcoxon test, this indicates that there is a significant difference between the two algorithms.

4.2 KBBA Comparisons

In this section we evaluate the performance of our KBBA with the algorithm BAAA developed in [18]. BAAA uses transfer functions as a general mechanism of binarization. In particular BAAA used the tanh $= \frac{e^{\tau|x|}-1}{e^{\tau|x|}+1}$ function to perform the transference. The parameter τ of the tanh function was set to a value 1.5. Additionally, a elite local search procedure was used by BAAA to improve solutions. As maximum number of iterations BAAA used 35000. The computer configuration used to run the BAAA algorithm was: PC Intel Core(TM) 2 dual

CPU Q9300@2.5 GHz, 4 GB RAM and 64-bit Windows 7 operating system. In our KBBA algorithm, the configurations are the same used in the previous experiments. These are described in the Table 1.

The results are shown in Table 3. The comparison was performed for the set cb.5.500 of the OR-library. The results for KBBA were obtained from 30 executions for each problem. In black, the best results are marked for both indicators the Best Value and the Average. For the best value indicator BAAA was higher in 4, and KBBA in 26. In the Average indicator BAAA was higher in 2 instances, and KBBA in 28. We should also note that the standard deviation in most problems was quite low, indicating that KBBA has good accuracy.

Table 3. OR-Library benchmarks MKP cb.5.500

Instance	Best known	BAAA Best	Avg	KBBA Best	Avg	Time(s)	Std
0	120148	120066	120013.7	120096	119478.1	452	44.8
1	117879	117702	117560.5	117656	116972.3	501	48.3
2	121131	120951	120782.9	120923	119622.1	527	47.1
3	120804	120572	120340.6	120683	120290.2	506	48.6
4	122319	122231	122101.8	122212	121203.7	518	45.2
5	122024	121957	121741.8	121982	120881.1	531	61.3
6	119127	119070	118913.4	118956	117674.8	514	45.8
7	120568	120472	120331.2	120463	119055.5	502	52.3
8	121586	121052	120683.6	121295	119909.2	541	51.3
9	120717	120499	120296.3	120467	120093.6	508	57.1
10	218428	218185	217984.7	218296	216654.4	481	52.9
11	221202	220852	220527.5	220951	219199.4	494	54.8
12	217542	217258	217056.7	217356	216875.6	471	49.5
13	223560	223510	223450.9	223516	221487.4	495	48.3
14	218966	218811	218634.3	218848	216934.6	546	45.1
15	220530	220429	220375.9	220410	218589.7	514	48.5
16	219989	219785	219619.3	219858	218228.3	501	68.7
17	218215	218032	217813.2	218010	215853.7	479	57.3
18	216976	216940	216862	216866	215700.6	430	39.4
19	219719	219602	219435.1	219631	219338.4	494	36.9
20	295828	295652	295505	295688	294366.3	408	38.6
21	308086	307783	307577.5	307924	306974.1	446	38.4
22	299796	299727	299664.1	299796	298497.3	401	47.8
23	306480	306469	306385	306480	302992.1	407	39.1
24	300342	300240	300136.7	300245	298713.2	414	38.8
25	302571	302492	302376	302481	301441.2	395	30.8
26	301339	301272	301158	301284	299230.8	401	29.6
27	306454	306290	306138.4	306331	303179.1	388	30.4
28	302828	302769	302690.1	302771	301320.4	366	31.9
29	299910	299757	299702.3	299828	298504.1	378	29.5
Average	214168.8	214014.2	213862.0	214040.8	213981.2	462.1	54.1

5 Conclusions

In this article, the k-means technique is used to perform the binarization of the bat algorithm. It should be noted that the k-means technique can be applied in the binarization of any continuous swarm-intelligence algorithm. We used the knapsack problem to evaluate our binary algorithm. The contribution of the k-means binary operator was studied, observing that this operator contributes to the precision and quality of the solutions obtained. Additionally, we develop the comparison with the BAAA algorithm.

As a future line of research, it is interesting to compare the performance of the k-means operator with the transfer functions used in BAAA, all applied in the binarization of the bat algorithm. Also, it is interesting to use the k-means operator to binarize other algorithms in addition to solving other NP-hard problems. From the theoretical point of view, it would also be interesting to understand how the exploration and exploitation of space are altered when we introduce the k-means operator. Another interesting research line is to explore adaptive techniques to automate the selection of parameters.

References

1. García, J., Martí, J.V., Yepes, V.: The buttressed walls problem: an application of a hybrid clustering particle swarm optimization algorithm. Mathematics **8**(6), 862 (2020)
2. García, J., Yepes, V., Martí, J.V.: A hybrid k-means cuckoo search algorithm applied to the counterfort retaining walls problem. Mathematics **8**(4), 555 (2020)
3. Yepes, V., Martí, J.V., García, J.: Black hole algorithm for sustainable design of counterfort retaining walls. Sustainability **12**(7), 2767 (2020)
4. Crawford, B., Soto, R., Monfroy, E., Astorga, G., García, J., Cortes, E.: A meta-optimization approach for covering problems in facility location. In: Workshop on Engineering Applications, pp. 565–578. Springer (2017)
5. García, J., Crawford, B., Soto, R., Astorga, G.: A percentile transition ranking algorithm applied to binarization of continuous swarm intelligence metaheuristics. In: International Conference on Soft Computing and Data Mining, pp. 3–13. Springer (2018)
6. García, J., Moraga, P., Valenzuela, M., Pinto, H.: A db-scan hybrid algorithm: an application to the multidimensional knapsack problem. Mathematics **8**(4), 507 (2020)
7. Garcia, J., Crawford, B., Soto, R., Astorga, G.: A percentile transition ranking algorithm applied to knapsack problem. In: Proceedings of the Computational Methods in Systems and Software, pp. 126–138. Springer (2017)
8. García, J., Altimiras, F., Peña, A., Astorga, G., Peredo, O.: A binary cuckoo search big data algorithm applied to large-scale crew scheduling problems. Complexity **2018**, 15 (2018)
9. García, J., Lalla-Ruiz, E., Voß, S., Droguett, E.L.: Enhancing a machine learning binarization framework by perturbation operators: analysis on the multidimensional knapsack problem. Int. J. Mach. Learn. Cybern. **11**, 1–20 (2020)

10. García, J., Crawford, B., Soto, R., García, P.: A multi dynamic binary black hole algorithm applied to set covering problem. In: International Conference on Harmony Search Algorithm, pp. 42–51. Springer (2017)
11. García, J., Crawford, B., Soto, R., Astorga, G.: A clustering algorithm applied to the binarization of swarm intelligence continuous metaheuristics. Swarm Evol. Comput. **44**, 646–664 (2019)
12. Crawford, B., Soto, R., Astorga, G., García, J.: Constructive metaheuristics for the set covering problem. In: International Conference on Bioinspired Methods and Their Applications, pp. 88–99. Springer (2018)
13. García, J., Peña, A.: Robust optimization: concepts and applications. In: Nature-Inspired Methods for Stochastic, Robust and Dynamic Optimization, p. 7 (2018)
14. Astorga, G., Crawford, B., Soto, R., Monfroy, E., García, J., Cortes, E.: A meta-optimization approach to solve the set covering problem. Ingeniería **23**(3), 274–288 (2018)
15. García, J., Moraga, P., Valenzuela, M., Crawford, B., Soto, R., Pinto, H., Peña, A., Altimiras, F., Astorga, G.: A db-scan binarization algorithm applied to matrix covering problems. Comput. Intell. Neurosci. **2019**, 16 (2019)
16. Garcia, J., Măntoiu, M.: Localization results for zero order pseudodifferential operators. J. Pseudo Diff. Oper. Appl. **5**(2), 255–276 (2014)
17. Crawford, B., Soto, R., Astorga, G., García, J., Castro, C., Paredes, F.: Putting continuous metaheuristics to work in binary search spaces. Complexity **2017**, 1–19 (2017)
18. Zhang, X., Wu, C., Li, J., Wang, X., Yang, Z., Lee, J.-M., Jung, K.-H.: Binary artificial algae algorithm for multidimensional knapsack problems. Appl. Soft Comput. **43**, 583–595 (2016)
19. Haddar, B., Khemakhem, M., Hanafi, S., Wilbaut, C.: A hybrid quantum particle swarm optimization for the multidimensional knapsack problem. Eng. Appl. Artif. Intell. **55**, 1–13 (2016)
20. García, J., Crawford, B., Soto, R., Carlos, C., Paredes, F.: A k-means binarization framework applied to multidimensional knapsack problem. Appl. Intell. **48**, 1–24 (2017)
21. Liu, J., Wu, C., Cao, J., Wang, X., Teo, K.L.: A binary differential search algorithm for the 0–1 multidimensional knapsack problem. Appl. Math. Modell. **40**(23–24), 9788–9805 (2016)
22. Bansal, J.C., Deep, K.: A modified binary particle swarm optimization for knapsack problems. Appl. Math. Comput. **218**(22), 11042–11061 (2012)
23. García, J., Pope, C., Altimiras, F.: A distributed k-means segmentation algorithm applied to lobesia botrana recognition. Complexity **2017**, 14 (2017)
24. Graells-Garrido, E., García, J.: Visual exploration of urban dynamics using mobile data. In: International Conference on Ubiquitous Computing and Ambient Intelligence, pp. 480–491. Springer (2015)
25. Peredo, O.F., García, J.A., Stuven, R., Ortiz, J.M.: Urban dynamic estimation using mobile phone logs and locally varying anisotropy. In: Geostatistics Valencia 2016, pp. 949–964. Springer (2017)
26. Graells-Garrido, E., Peredo, O., García, J.: Sensing urban patterns with antenna mappings: the case of Santiago, Chile. Sensors **16**(7), 1098 (2016)
27. Pirkul, H.: A heuristic solution procedure for the multiconstraint zero? One knapsack problem. Naval Res. Logistics **34**(2), 161–172 (1987)
28. Kong, X., Gao, L., Ouyang, H., Li, S.: Solving large-scale multidimensional knapsack problems with a new binary harmony search algorithm. Comput. Oper. Res. **63**, 7–22 (2015)

A K-means Bat Optimisation Algorithm Applied to the Set Covering Problem

Leonardo Pavez[1,2](✉), Francisco Altimiras[3], and Gabriel Villavicencio[4]

[1] Instituto de Ciencias Naturales, Universidad de las Américas, Av. Manuel Montt 948, Providencia, 7500972 Santiago, Chile
lpavez@udla.cl

[2] Departamento de Ciencias Químicas y Biológicas, Universidad Bernardo O'Higgins, General Gana 1702, 8370854 Santiago, Chile

[3] Facultad de Ingeniería y Negocios, Universidad de las Américas, Santiago, Chile
fjaltimiras@gmail.com

[4] Escuela de ingeniería en construcción, Pontificia Universidad Católica de Valparaíso, Valparaíso, Chile
gabriel.villavicencio@pucv.cl

Abstract. Optimization at the industry level is a fundamental area since it allows reducing costs and being more sustainable. Many of these problems are combinatorial and NP-hard. On the other hand, swarm intelligence metaheuristics have been able to successfully address these types of problems, however, many of these techniques work in searching continuous spaces. In this article, we explore a general binarization mechanism of continuous metaheuristics based on the k-means technique. In particular, we applied the k-means technique to the bat algorithm with the aim of addressing the set covering problem (SCP). Experiments were designed to evaluate the contribution of the k-means technique in the binarization process. In addition, we verify the effectiveness of our algorithm through reference instances. The results indicate that the k-means binary bat algorithm (KBBA) gets adequate results when evaluated with a combinatorial problem like SCP.

1 Introduction

The combinatorial problems have great relevance at industrial level, we find them in various areas such as Civil Engineering [1,2], sustainability [3], Operational Research [4,5], resource allocation [6–8], scheduling problems [9–11], routing problems [12–15] among others.

Several binarization techniques have been developed to address this situation. In a literature search, the main binarization methods used correspond to transfer functions, angle modulation and quantum approach. In this article we present a binarization method that uses the k-means technique to group the solutions and then perform the binarization process. To verify the efficiency of our method, we used the KBBA.

R. Silhavy et al. (Eds.): CoMeSySo 2020, AISC 1295, pp. 622–632, 2020.
https://doi.org/10.1007/978-3-030-63319-6_58

On the other hand, in recent years, algorithms inspired by natural phenomena have been able to solve NP-hard combinatorial problems of significant size. As an example of algorithms inspired by nature, we have Cuckoo Search [16], Black Hole [17], Bat Algorithm [18], among others. Many of these algorithms work naturally in continuous spaces and, therefore, must be adapted to solve combinatorial problems. In the process of adaptation, the mechanisms of exploration and exploitation of the algorithm can be altered, having consequences in the efficiency of the algorithm. Several binarization techniques have been developed to address this situation. In a bibliographic search, the main binarization methods used are: transfer functions, angle modulation, and quantum approach. In this article, we develop a binarization method that uses the k-means technique to cluster the solutions and then perform the binarization process. To verify the effectiveness of our method, we use KBBA.

To check our k-means binary bat algorithm (KBBA), we use the well-known set covering problem. Experiments were developed using a random operator to validate the contribution of the k-means technique in the binarization process of the bat algorithm. In addition, to verify our results, the JPSO algorithm developed by [19] and KBBA developed in [9] were chosen. The results show that KBBA algorithm obtain competitive results.

2 Set Covering Problem

The set covering problem corresponds to a classical problem in combinatorics, and complexity theory. The problem aims to find subsets of a set. Given a set and its elements, we want to find subsets that completely cover the set at a minimum cost. The SCP is one of the oldest and most studied optimization problems. It is well-known to be NP-hard [20].

SCP is an active problem because medium and large instances often become intractable and cannot be solved any more using exact algorithms. Additionally, SCP due to its large number of instances, is used to verify the behavior of proposed new algorithms. In recent years, SCP has been approached by various continuous Swarm intelligence metaheuristics and using different methods to binarize metaheuristics. In [21], they used the teaching-learning-based optimisation with a specific binarization scheme to solve medium and large size instances. An SCP in fault diagnosis application was developed by [22]. In this article, the Gravitational Search algorithm was used and transfer functions were applied to perform the binarization. A rail scheduling problem application using SCP was developed in [9]. In [19] a Jumping PSO was used to solve SCP and in [6] CS metaheuristics were binarized using a k-means algorithm.

SCP has many practical applications in engineer, e.g., vehicle routing, facility location, railway, and airline crew scheduling [23–26] problems.

The SCP can be formally defined as follows. Let $A = (a_{ij})$, be a $n \times m$ zero-one matrix, where a column j cover a row i if $a_{ij} = 1$, besides a column j is associated with a non-negative real cost c_j. Let $I = \{1, ..., n\}$ and $J = \{1, ...m\}$, be the row and column set of A, respectively. The SCP consists in searching a minimum cost subset $S \subset J$ for which every row $i \in I$ is covered by at least one column $j \in J$, i.e.:

$$\text{Minimize } f(x) = \sum_{j=1}^{m} c_j x_j \tag{1}$$

$$\text{Subject to } \sum_{j=1}^{m} a_{ij} x_j \geq 1, \forall i \in I, \text{ and } x_j \in \{0,1\}, \ \forall j \in J \tag{2}$$

where $x_j = 1$ if $j \in S$, $x_j = 0$ otherwise.

3 K-means Binary Bat Algorithm

The application of statistics and machine learning techniques, are used in many areas such as transports, smart cities, agriculture and computational intelligence, examples of applications are found in [27–30,30,31]. In this article, we explore the k-means technique applied to binarization of continuous metaheuristics. As an algorithm to binarize, bat is used and as problem to solve, SCP. The Proposed KBGOA has two modules. The first module corresponds to the initialization of the feasible solutions, and is detailed in the Sect. 3.1. Once the initialization of the particles is performed, it is consulted if the maximum if iterations is satisfied. Subsequently if the criterion is not satisfied, The KBGOA is then run in conjunction with the k-means technique, this is detailed in Sect. 3.2). Once the transitions of the different solutions are made, we compare the resulting solutions with the best solution previously obtained. In the event that a superior solution is found, this replaces the previous one. The general algorithm scheme is detailed in Fig. 1.

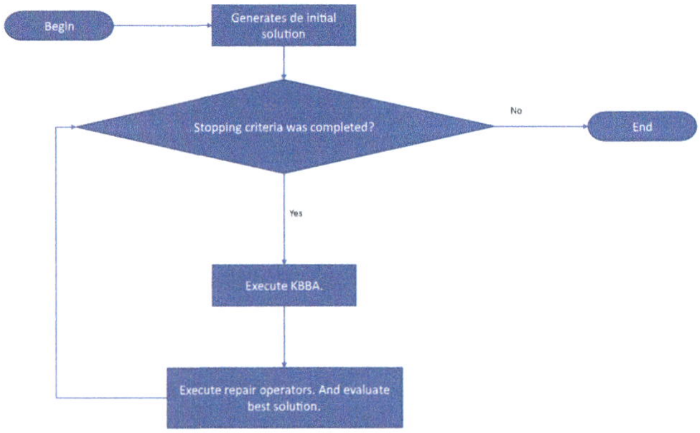

Fig. 1. Flowchart of the KBBA algorithm.

3.1 Initialization

For initialization of a new solution, a column is randomly chosen. It is then queried whether the current solution covers all rows. In the case of the solution does not meet the coverage condition, the heuristic operator is called (Sect. 3.3). This operator aims to select a new column. This heuristic operation is iterated until all rows are covered. Once the coverage condition is met, the solution is optimized. The optimization consists of eliminating columns where all their rows are covered by more than one column. The detail of the procedure is shown in Algorithm 1.

Algorithm 1. Initialization Operator

```
Function Initialization()
Input
Output Initialized solution S_out
S ← SRandomColumn()
while All row are not covered do
    S.append(Heuristic(S))
end while
S ← dRepeatedItem(S)
S_out ← S
return S_out
```

3.2 k-means Transition Ranking Operator

Consider our metaheuristic is a continuous and population based. Due to its iterative nature, it needs to update the position of particles at each iteration. When the metaheuristic is continuous, this update is performed in $\mathbb{R}^n$ space. In Eq. 3, the update position is presented in a general manner. The x_{t+1} variable represents the x position of the particle at time t+1. This position is obtained from the position x at time t plus a Δ function calculated at time t+1. The function Δ is proper to each metaheuristic and produces values in $\mathbb{R}^n$. For example in Cuckoo Search $\Delta(x) = \alpha \oplus Levy(\lambda)(x)$, in Black Hole $\Delta(x) = \text{rand} \times (x_{bh}(t) - x(t))$ and in the Firefly, Bat and PSO algorithms Δ can be written in simplified form as $\Delta(x) = v(x)$.

$$x_{t+1} = x_t + \Delta_{t+1}(x(t)) \tag{3}$$

In the k-means transition ranking operator, a model for transitions in a discrete space is proposed. This model is based on considering the movements generated by the metaheuristic in each dimension for all particles. $\Delta^i(x)$ corresponds to the magnitude of the displacement $\Delta(x)$ in the i-th position. Subsequently these displacement are grouped using the magnitude of the displacement $\Delta^i(x)$.

This grouping is done using the k-means technique where k represents the number of clusters used. Finally, a generic P_{tr} function given by the Eq. 4 is proposed to assign a transition probability.

$$P_{tr} : \mathbb{Z}/k\mathbb{Z} \rightarrow [0, 1] \tag{4}$$

A transition probability through the function P_{tr} is assigned to each group. Naturally, this P_{tr} function is modelled as a cumulative probability function. For the case of this study, we particularly use the linear function given in Eq. 5. In this equation, $\mathrm{N}(x^i)$ indicates the location of the group to which $\Delta^i(x)$ belongs. The α coefficient, corresponds to the transition probability and β to the transition separation coefficient. Both parameters are estimated in each metaheuristic. For our particular case, $\mathrm{N}(x^i) = 0$ corresponds to elements belonging to the group that has the lowest Δ^i values. $\mathrm{N}(x^i) = 7$ corresponds to the group of elements that have the greatest Δ^i values.

$$P_{tr}(x^i) = P_{tr}(N(x^i)) = \alpha + \beta N(x^i)\alpha \tag{5}$$

The k-means transition ranking operator starts calculating Δ^i for each of the particles. This step is specific in each metaheuristic. Subsequently the particles are grouped using k-means clustering technique and the magnitude of Δ^i as distance. With the group assigned to each particle we obtain the probability of transition using Eq. 5. Afterwards the transition of each particle is performed. In the case of bat algorithm the rule 6 is used to perform the transition, where $\hat{x}^i$ is the complement of x^i. Finally, each solution is repaired using the repair operator.

$$x^i(t+1) := \begin{cases} \hat{x}^i(t), & \text{if } rand < P_{tg}(x^i) \\ x^i(t), & \text{otherwise} \end{cases} \tag{6}$$

3.3 Heuristic Operator

The goal of the Heuristic operator is to select a new column for cases where a solution needs to be built or repaired. As input variables, the heuristic operator considers the incomplete solution S_{in} which must be completed. The operator obtains the columns that belong to S_in, then obtains the rows R that are not covered by the solution to S_in. With the set of rows not covered and using Eq. 7 we obtain in line 4 the best 5 rows to be covered. With this list of rows ($lRows$) on line 5 we obtain the list of the best columns according to the heuristic indicated in Eq. 8. Finally randomly in line 6 we obtain the column to incorporate.

$$WeightRow(i) = \frac{1}{L_i}. \tag{7}$$

where L_i is the sum of all ones in row i

$$WeightColumn(j) = \frac{c_j}{|R \cap M_j|}. \tag{8}$$

where M_j is the set of rows covered by Col j

Algorithm 2. Heuristic operator

1: **Function** Heuristic(S_{in})
2: **Input** Input solution S_{in}
3: **Output** The new column *colOut*
4: $lRows \leftarrow$ getBestRows(S_{in}, N=10)
5: $listcolumsnOut \leftarrow$ getBestCols($LRows$, M=5)
6: $colOut \leftarrow$ getCol($lcolsOut$)
7: **return** *colOut*

4 Results

In this section we detail the behavior of KBBA when it is applied to SCP. The contribution of the binary k-means operator was studied when solving the different SCP instances. Additionally, KBBA was compared with other algorithms that have recently resolved SCP. To solve the different SCP instances, a PC with Windows 10, core i5 processor and 16 GB in RAM was used. The program was coded in Python 3.5. For the statistical analysis, the non-parametric Wilcoxon signed-rank test was used. The analysis of violin charts is performed by comparing the dispersion, median and the interquartile range of the distributions.

4.1 Insight of KBBA Algorithm

In this module we developed experiments that allowed us to study the contribution of the binary k-means operator with respect to the quality of the solutions obtained and the iterations used when solving SCP instances. To perform this evaluation, we used the instances of Balas and Carrera. In the comparison, a random operator was designed which replaces our binary k-means operator. This random operator instead of executing transitions assigning transition probabilities per group, uses a fixed value of 0.5. To compare the distributions of the results of the different experiments we use violin Chart. The horizontal axis X corresponds to the problems, while Y axis uses the measure % - Gap defined in Eq. 9

$$\% - Gap = 100\frac{BestKnown - SolutionValue}{BestKnown} \quad (9)$$

Furthermore, a non-parametric test, Wilcoxon signed-rank test is carried out to determine if the results of KBBA with respect to the random algorithm have significant difference or not. The parameter settings and browser ranges are shown in Table 1.

Table 1. Setting of parameters for k-means Binary Bat Search Algorithm.

Parameters	Description	Value	Range
N	Number of solutions	25	[20, 25, 30]
G	Number of clusters	5	[4,5,6]
Iteration number	Maximum iterations	500	[400,500,600]
cmax	Maximum for c coefficient	1	1
cmin	Minimum for c coefficient	0.00001	0.00001
f	f parameter for s function	0.5	0.5
l	l parameter for s function	1.5	1.5

KBBA corresponds to our standard algorithm. *random.0.5* is the random variant. The results are shown in Table 2. When we compared the Best Values between KBBA and *rnd.0.5* which are shown in Table 2. In the Average comparison, KBBA outperforms *rnd.0.5* in all problems. When we evaluate the behaviour of the algorithms through the Wilcoxon test, this test indicates that there is a significant difference between the two algorithms.

Table 2. Balas and Carrera instances.

Instance	Row	Col	Density	Best known	KBBA (avg)	Std	(Secs)	*rnd.0.5*	Std
AA03	106	8661	4.05%	33155	33354.2	121.4	206	33693.5	241.1
AA04	106	8002	4.05%	34573	34988.7	62.1	197	34931	231.2
AA05	105	7435	4.05%	31623	31723.6	65.8	205	31889.2	103.2
AA06	105	6951	4.11%	37464	37674.4	54.2	205	37651.3	97.1
AA11	271	4413	2.53%	35384	35802.8	131.1	180	35901.2	197.1
AA12	272	4208	2.52%	30809	31427.4	152.5	208	31631.3	210.2
AA13	265	4025	2.60%	33211	34032.6	136.4	189	34201.4	231.4
AA14	266	3868	2.50%	33219	33766.3	124.7	146	33967.3	121.1
AA15	267	3701	2.58%	34409	35010.6	113.8	140	34846.8	161.3
AA16	265	3558	2.63%	32752	33205.1	137.4	187	33601.3	183.9
AA17	264	3425	2.61%	31612	32002.5	86.1	165	32386.2	149.3
AA18	271	3314	2.55%	36782	37174.8	81.2	131	37417.6	148.1
AA19	263	3202	2.63%	32317	33202.2	79.8	144	33109.1	136.1
AA20	269	3095	2.58%	34912	35847.5	90.1	140	35963.1	97.8
BUS1	454	2241	1.88%	27947	28405.3	87.4	192	28756.1	231.6
BUS2	681	9524	0.51%	67760	68828.7	82.4	174	68678.9	101.2
Average				35495.56	36028.2	102.1	174.2	36164.1	165.1
p-value								1.3e7−05	

4.2 KBBA Compared with JPSO and MDBBH

In this section, we detail the comparisons made with the objective of evaluating the performance of our KBBA algorithm. For the evaluation, the larger problems of the OR-library were chosen. To develop the comparison, two algorithms were selected. The first one corresponds to a discretization Particle Swarm Optimization (PSO) technique called Jumping PSO (JPSO), [19]. The second one, is a binarization of the Black Hole technique called Multi Dynamic Binary Black Hole (MDBBH) Algorithm, [9]. JPSO uses a discrete PSO based on the frog jump. JPSO works without the concept of velocity, replacing this one by a component of random jump which allows to perform the movement in the discrete search space. On the other hand, MDBBH uses a binarization mechanism specific to the Black Hole algorithm. This is based on the concept of closeness to the Black Hole (BH). BH corresponds to the solution that has obtained the best value. When a solution is close to BH, the transitions are less likely to be performed. In the case of being away from the BH, the probability of transition is greater.

For instances E and F which are of medium size, the results of the MDBBH and KBBA algorithms are similar when comparing their Best Value. Only in instance E.2 KBBA was superior to MDBBH. When analyzing the average,

Table 3. OR-Library benchmarks.

Instance	row	col	Density	Best Known	JPSO Avg	MDBBH Best	Avg	KBBA Best	Avg	Time(s)
E.1	500	5000	10%	29	29	29	29	29	29.7	9.8
E.2	500	5000	10%	30	30	31	31.6	30	30.8	14.6
E.3	500	5000	10%	27	27	27	27.4	27	27.1	16.2
E.4	500	5000	10%	28	28	28	29.1	28	28.2	18.4
E.5	500	5000	10%	28	28	28	28	28	28.5	17.4
F.1	500	5000	20%	14	14	14	14.3	14	14.5	16.9
F.2	500	5000	20%	15	15	15	15.5	15	15.5	18.2
F.3	500	5000	20%	14	14	14	14.6	14	15.8	18.8
F.4	500	5000	20%	14	14	14	14.7	14	14.2	22.4
F.5	500	5000	20%	13	13	14	14.3	14	14.8	23.1
G.1	1000	10000	2%	176	176	177	178.9	176	178.3	185.9
G.2	1000	10000	2%	154	155	157	160.2	155	156.2	176.5
G.3	1000	10000	2%	166	167.2	168	170.8	167	170.3	174.4
G.4	1000	10000	2%	168	168.2	169	170.7	169	172.3	178.5
G.5	1000	10000	2%	168	168	168	169.6	168	170.4	158.2
H.1	1000	10000	5%	63	64	64	64.9	64	65.4	157.1
H.2	1000	10000	5%	63	63	64	64	64	64.5	154.4
H.3	1000	10000	5%	59	59.2	59	60	60	60.7	145.1
H.4	1000	10000	5%	58	58.3	59	60.4	59	59.9	168.2
H.5	1000	10000	5%	55	55	55	56.4	55	56.4	141.2
Average				67.1	67.30	67.7	68.71	67.5	69.14	88.9

JPSO obtained better results, however, the values of KBBA were close, being MDBBH the one that obtained the worse performance. For the case of problems G and H, KBBA was superior in problems G.1, G.2 and G.3 and MDBBH in problem H.3 when comparing their Best Value. In the case of averages, JPSO was superior in all cases. However, KBBA obtained results quite close, leaving MDBBH behind. In this point we must emphasize that the k-means technique used in KBBA allows binarizing any continuous swarm intelligence algorithm unlike JPSO that is specific for PSO (Table 3).

5 Conclusions

In this work, the KBBA algorithm based on the k-means technique was proposed to perform binarization of GOA metaheuristics. We must emphasize that this k-means technique can be applied in the binarization of any continuous metaheuristics of the swarm intelligence type. To evaluate the performance of Our k-means technique the set covering problem was used together with the bat algorithm. Experiments were designed to evaluate the contribution of the binary operator k-means of the algorithm. It was found that the operator contributes significantly to improve the accuracy and quality of the solutions, all these visualized through violin graphics. Finally, compared to the best metaheuristic algorithm that has solved SCP, our algorithm had a lower yield of 1.26%, which is not a big difference considering that JPSO uses a particular adaptation mechanism for PSO and the k-means technique It can be easily adapted to any technique. As future works, we believe that the autonomous search for efficient configuration is a little-explored area and it could help to improve the performance of the binarizations both in the quality of the solutions and in the times of convergence. Additionally, it is interesting to see how this k-means technique works with other metaheuristics and other NP-hard problems. Finally, we believe that using machine learning we can generate indicators or obtain correlations that allow us to understand how the k-means algorithm translates the exploration and exploitation properties from the continuous space to the binary space.

References

1. García, J., Martí, J.V., Yepes, V.: The buttressed walls problem: an application of a hybrid clustering particle swarm optimization algorithm. Mathematics **8**(6), 862 (2020)
2. García, J., Yepes, V., Martí, J.V.: A hybrid k-means cuckoo search algorithm applied to the counterfort retaining walls problem. Mathematics **8**(4), 555 (2020)
3. Yepes, V., Martí, J.V., García, J.: Black hole algorithm for sustainable design of counterfort retaining walls. Sustainability **12**(7), 2767 (2020)
4. García, J., Moraga, P., Valenzuela, M., Pinto, H.: A db-scan hybrid algorithm: an application to the multidimensional knapsack problem. Mathematics **8**(4), 507 (2020)

5. García, J., Crawford, B., Soto, R., Astorga, G.: A percentile transition ranking algorithm applied to binarization of continuous swarm intelligence metaheuristics. In: International Conference on Soft Computing and Data Mining, pp. 3–13. Springer (2018)
6. Garcia, J., Crawford, B., Soto, R., Astorga, G.: A percentile transition ranking algorithm applied to knapsack problem. In: Proceedings of the Computational Methods in Systems and Software, pp. 126–138. Springer (2017)
7. García, J., Altimiras, F., Peña, A., Astorga, G., Peredo, O.: Abinary cuckoo search big data algorithm applied to large-scale crewscheduling problems. Complexity **2018**, 15 (2018)
8. García, J., Lalla-Ruiz, E., Voß, S., Droguett, E.L.: Enhancing a machine learning binarization framework by perturbation operators: analysis on the multidimensional knapsack problem. Int. J. Mach. Learn. Cybern. **11**(19), 1–20 (2020)
9. García, J., Crawford, B., Soto, R., García, P.: A multi dynamic binary black hole algorithm applied to set covering problem. In: International Conference on Harmony Search Algorithm, pp. 42–51. Springer (2017)
10. García, J., Crawford, B., Soto, R., Astorga, G.: A clustering algorithm applied to the binarization of swarm intelligence continuous metaheuristics. Swarm Evol. Comput. **44**, 646–664 (2019)
11. Crawford, B., Soto, R., Astorga, G., García, J.: Constructive metaheuristics for the set covering problem. In: International Conference on Bioinspired Methods and Their Applications, pp. 88–99. Springer (2018)
12. García, J., Peña, A.: Robust optimization: concepts and applications. In: Nature-Inspired Methods for Stochastic, Robust and Dynamic Optimization, p. 7 (2018)
13. Astorga, G., Crawford, B., Soto, R., Monfroy, E., García, J., Cortes, E.: A meta-optimization approach to solve the set covering problem. Ingeniería 23(3), 274–288 (2018)
14. García, J., Moraga, P., Valenzuela, M., Crawford, B., Soto, R., Pinto, H., Peña, A., Altimiras, F., Astorga, G.: A db-scan binarization algorithm applied to matrix covering problems. Comput. Intell. Neurosci. **2019**, 16 (2019)
15. Garcia, J., Măntoiu, M.: Localization results for zero order pseudodifferential operators. J. Pseudo Diff. Oper. Appl. **5**(2), 255–276 (2014)
16. Yang, X.-S., Deb, S.: Cuckoo search via lévy flights. In: World Congress on Nature & Biologically Inspired Computing, 2009. NaBIC 2009, pp. 210–214. IEEE (2009)
17. Hatamlou, A.: Black hole: a new heuristic optimization approach for data clustering. Inf. Sci. **222**, 175–184 (2013)
18. Yang, X.-S.: A new metaheuristic bat-inspired algorithm. In: Nature Inspired Cooperative Strategies for Optimization (NICSO 2010), pp. 65–74 (2010)
19. Balaji, S., Revathi, N.: A new approach for solving set covering problem using jumping particle swarm optimization method. Nat. Comput. **15**(3), 503–517 (2016)
20. Gary, M.R., Johnson, D.S.: Computers and intractability. In: A Guide to the Theory of NP-Completeness (1979)
21. Lu, Y., Vasko, F.J.: An or practitioner's solution approach for the set covering problem. Int. J. Appl. Metaheuristic Comput. (IJAMC) **6**(4), 1–13 (2015)
22. Li, Y., Cai, Z.: Gravity-based heuristic for set covering problems and its application in fault diagnosis. J. Syst. Eng. Electron. **23**(3), 391–398 (2012)
23. Crawford, B., Soto, R., Monfroy, E., Astorga, G., García, J., Cortes, E.: A meta-optimization approach for covering problems in facility location. In: Workshop on Engineering Applications, pp. 565–578. Springer (2017)
24. Kasirzadeh, A., Saddoune, M., Soumis, F.: Airline crew scheduling: models, algorithms, and data sets. EURO J. Transp. Logistics **6**(2), 111–137 (2017)

25. Horváth, M., Kis, T.: Computing strong lower and upper bounds for the integrated multiple-depot vehicle and crew scheduling problem with branch-and-price. Cent. Eur. J. Oper. Res. **27**(1), 1–29 (2017)
26. Stojković, M.: The operational flight and multi-crew scheduling problem. Yugoslav J. Oper. Res. **15**(1), 25–48 (2016)
27. García, J., Crawford, B., Soto, R., Carlos, C., Paredes, F.: A k-means binarization framework applied to multidimensional knapsack problem. Appl. Intell. **48**(2), 1–24 (2017)
28. García, J., Pope, C., Altimiras, F.: A distributed k-means segmentation algorithm applied to lobesia botrana recognition. Complexity **2017**, 14 (2017)
29. Graells-Garrido, E., García, J.: Visual exploration of urban dynamics using mobile data. In: International Conference on Ubiquitous Computing and Ambient Intelligence, pp. 480–491. Springer (2015)
30. Graells-Garrido, E., Peredo, O., García, J.: Sensing urban patterns with antenna mappings: the case of Santiago, Chile. Sensors **16**(7), 1098 (2016)
31. Peredo, O.F., García, J.A., Stuven, R., Ortiz, J.M.: Urban dynamic estimation using mobile phone logs and locally varying anisotropy. In: Geostatistics Valencia 2016, pp. 949–964. Springer (2017)

The Error Analysis for Enterprise Software Application Using Analytic Hierarchy Process and Supervised Learning: A Hybrid Approach on Root Cause Analysis

Hoo Meng Wong(✉) and Sagaya Sabestinal Amalathas

Taylor's University, Subang Jaya, Selangor, Malaysia
hoomeng@hotmail.com

Abstract. Many businesses today still very much depend on software application service to process the daily transactions. Under such heavy dependency, it is frustrated whenever the software application service is unavailable. For the errors which can cause the software application service malfunction, it can be possible to arise mostly in either within the software application layer, or other factors which are falling outside the software application layer. In this complex situation, a lot of time consume to identify the root cause is unavoidable. The objective is not seeing from the angle of only solving the problem. More importantly, it aims to reduce the total time spent on root cause analysis, and to decide the preferred resolution for the root cause. Indeed this is crucial to propose an approach toward to develop a logic model, that makes a precise decision on identifying the root cause, as well as on identifying the preferred resolution to the root cause. The proposed logic model will consist of the algorithm incorporated with both Analytic Hierarchy Process (AHP) and Supervised Learning.

Keywords: Analytic hierarchy process · Application log file analysis · Application log file debugging · Machine learning · Supervised learning

1 Introduction

The enterprise software application has become an important service that is used by today's business operations. More importantly this service is required to be available 24 h a day, and 7 days a week. This makes the expectation of service availability is high and the toleration of service downtime is lower and lower. With such high expectation of the enterprise software application service, the root cause analysis activity must not be prolonged, and a high level of accuracy is required. However, dealing with an enterprise software application that is involving multiple tiers, the root cause analysis may not be straight forward. This is because the multiple tiers are Front-end tier, Web Server tier, Application Server tier, and Database Server tier. The software application running in each tier has its log file. Especially in the Application Server tier, there can be more than one middleware (software application that is having integration to another software application) running. Which means more log files are

R. Silhavy et al. (Eds.): CoMeSySo 2020, AISC 1295, pp. 633–643, 2020.
https://doi.org/10.1007/978-3-030-63319-6_59

being produced. According to Management Logic (2012), more evidences are required when conducting root cause analysis. Evidences are crucial to identify the root cause. This is very time consuming to collect the evidences especially the evidences are the involved log files across multiple tiers. These logs have to be read through event by event to identify and filter out the alert and warning. Then finally it is still depending on the person who is experience enough to make a judgment on the ultimate error which causes the root cause at the end of analysis activity.

2 Motive of the Study

The are several reasons have caused the motivation of this proposed research. Mainly there are four of them. First of all, the enterprise software application itself has more than one log file spanned across multiple tiers. However, with all the enterprise software application error log files, it may not be adequate for the root cause analysis. This is because what if the actual cause is falling outside the enterprise software application boundary. On the next reason, which is the time consuming to read through all the enterprise software application log files manually during root cause analysis activity. Thirdly, due to the root cause analysis activity is conducted by human, it is very subjective to the person who makes judgment and point out the actual error that causes the failure to enterprise software application. Lastly, whenever the operation cost to run the enterprise software application support operation is increasing, it is always facing high expectation from the company's management to have slim down-time on the enterprise software application.

3 Statement of the Problem

To conduct the root cause analysis activity on enterprise software application error, it is time consuming to identify the actual error as it involves multiple tiers and different layers.

There are several events and behavior leading to the above statement of problem should be stated out as follows. At the same time, please note that the different layers here are referring to Enterprise Software Application, Operating System, Networking, Virtual Hardware/Physical Hardware, and connectivity to the Database. The enterprise software application behaves in an unexpected manner due to error occurred within or even beyond enterprise software application layer. This can be possibly that enterprise software application defect escapes from software application testing, and it is successfully released to production environment. On the other hand, whenever the root cause analysis activity is conducted manually, human would overlook the important clues and these clues are crucial to identify the actual error. With that, if it is in the situation that the human is lack of analysis experience. This would even waste much more time and cause the entire root analysis activity prolong. Furthermore, due to the root cause analysis activity is conducted by human, it is still very subjective to the human who can really make a good judgment on the actual error. Under such circumstance, the business users are facing difficulty to continue their daily tasks as the enterprise software application is unavailable or malfunction. The human who conducts the root analysis activity will be under pressure on how rapid to rectify the actual root cause.

4 Statement of the Objectives

The first objective is to mitigate prolonging time duration on conducting the root cause analysis activity. The second objective is to improve accuracy on identifying the root cause whenever error is occurred.

Both objectives are initiated based on the statement of the problem. The contribution of fulfilling both objectives will benefit the business operations which are highly depending on the service provided by the enterprise software application. Therefore, to mitigate the enterprise software application's downtime that creates risk to the business operation, it will improve the reliability of the enterprise software application used in the business operations.

To further explain on how to mitigate prolonging time duration on conducting root cause analysis, a simple mathematics approach is used (Table 1).

Table 1. The description of the variables.

Variable	Description
SLA	Assume that Service Level Agreement (SLA) would only allow 8 h in total for downtime
W	The total time taken is given to resume the software application service
X	The root cause analysis activity duration with the initial value assigned, 2 h
Y	The software coding effort in the resolution, testing the resolution along with the sanity and regression test. Please note that Y has to be a constant value and it is assigned to be 6 h
p	p = Prescriptive Analytical Logic Model (PAL), also known as the proposed logic model

Let $X = 2$

Let $Y = 6$

If $W = X + Y$, then $W = 8$ h.

Please note that X may not be always 2 (hours) as it is depending on how rapid the root cause is analyzed and identified manually. With that, it is required to consider how accurate to identify the actual error under the manual activities. In the worse case, X can be more than 2 h due to the knowledge and experience of the personnel who conducts the analysis activity manually. This will cause the SLA to breach.

$X > 2 \leftarrow$ SLA is breached.

Hence, if X can be shorten or kept below 2 h, this will impact the hours in total for W.

Which means, p as the Prescriptive Analytical Logic Model (PAL) to reduce the time spent on root cause analysis activity and keep X to be 2 or lesser.

$X <= 2$ if $X-p$, then

$W = (X-p) + Y$, $W < SLA \leftarrow$ With that, W can be kept below 8 h.

5 Literature Review

By referring to Valdman (2001), it is stated that some features are presented in more detail while others are just intentions or suggestions. On the other hand, Vaidya and Kumar (2004) had reviewed Analytic Hierarchy Process (AHP) with 150 application papers in total, and there is no knowledge in applying AHP on identifying valid software application error. This means a research gap, and it would allow to contribute the area of proposed research to incorporate the log data searching technique (introduced in the past research) into this research of prescriptive analytical logic. Apart of adopting techniques of previous reaches, this research introduces AHP to carry out decision making actions such as shortlisting the valid software application error among multiple software application errors are found, and shortlisting the final resolution to the software application error identified. With the AHP is applied, it can help the proposed logic model to be more accurate to evaluate and identify the valid software application error, and of course using AHP again to be more accurate to evaluate and identify the preferred resolution to the software application error.

Neves and Camanho (2015), proposes a case study to assist in IT project prioritization by adopting AHP. This case study has two aims

- To explore the expectations of a company's senior management and to identify the challenges during the implementation of the selected method that is decided by the company.
- To present a case study for the adoption of AHP along with multi-criteria method (or so called multi-criteria decision analysis of operation research) that explicitly evaluates multiple conflicting criteria in decision making by a large Brazilian multinational company in the oil and gas industry.

Both aims will derive a well defined structured process of decision making analysis to assist on IT project prioritization which would contribute to those companies that are in the Oil and Gas industry. In the above case study, projects were classified into three categories, which are Innovation (technology projects), Growth Productivity (business projects), and Continuity (projects that are for improvements and maintenance on systems). The case study evaluated both the domestic and foreign markets in the oil and gas industry, and based on that defined the criteria for the prioritization of portfolio projects under the consideration of the company's investments. Then, based on the given categories, the criteria is prioritized under the configuration of the multi-criteria tree in a hierarchy was applied as a support tool. After the configuration of the multi-criteria tree was built, the weights were assigned to the criteria and all the sub sub-criteria in the tree. After all, AHP plays an important role to estimate the relative magnitudes of factors through pair-wise comparisons. With that, it allowed the company to obtain a micro view of decision-making as a full picture that came into effect for the departments within the company, to address the strategic directions on their IT projects. With the given case study, it is learned that AHP can actually be applied on analyzing the software application errors, as well as those errors related to the incident of software application failure. Whenever the root cause of software application error is identified, by adopting AHP again to evaluate the possible resolutions, the preferred resolution will help to resolve the software application error under a structured technique.

With the comparison of the proposed logic model to Machine Learning, it has certain similarities based on Klass (2018). However, this proposed logic model is not completely under the Machine Learning approach. One of the similarities is that the proposed logic model has to be trained with training data which consist of inputs paired with the correct outputs. This means the algorithm of proposed logic model will search for the defined patterns in all required log files that correlate with the desired outputs. It contains a little similarity to Machine Learning's Supervised Learning which is applied to predict the correct label for newly present input data, such as $O = f(i)$. O is the predicted output which is determined by a mapping label, and i is the input value that is assigned to a class. Wilson (2019) explained that the Machine Learning's Supervised Learning predicts the output based on the predefined features as label applied on the newly present input data under Classification subcategory. On the other hand, Regression subcategory is a predictive statistical process which attempts to find out the relationship between dependent and independent variables. However, the proposed logic model cannot predict the resolution if the new found error has no straight forward resolution defined in the configuration file.

The reason behind is that when the proposed logic model scan through all the required log files, it is based on the keywords as the label defined in the configuration file. It will require to put the newly present input data into the category of classification (such as *Required-for-Instruction*) in its configuration file, if the data is not matched with the existing defined label found in the configuration file. This is because in a situation that if the error is from the software application layer, and the resolution is required the software development cycle to do the fix. There is no immediate resolution can be applied during the software application execution. Furthermore, of course there are situations that are having immediate resolution such as restarting the database connectivity or software application process. These resolutions will be stored into the configuration file of proposed logic model. Indeed, there are multiple layers such as CPU, Memory and Storage (i.e. Hard Disk), OS, Network Connectivity, and Database can possibly contribute to the actual root cause. The first level of decision making is to identify which layer is having the cause of error, followed by the second level of decision making is to identify the real root cause within the layer. Therefore, by understanding this circumstance, the AHP is at the suitable position to conduct the weight evaluation and decision making for the valid error of the root cause.

As mentioned earlier, Supervised Learning has another subcategory that is Linear Regression. According to Brownlee (2019a), it is required to identify well on the relationship between input variable and output variable, and both variables must be linear. A good example provided by Milan (2019) is that in a tropical climate location, if the humidity increases, then the air is holding much more water, and eventually it leads to raining possibility is high. However, in the scenario of software application running in a multiple-tier environment, there is difficult to have a linear relationship. This is because (by example) if end-users accessing to the software application via the Internet and receive a failure page in response, this doesn't always mean that the root cause is because of the software application is failing itself. It can be possibly that:-

- The server resources on the Web server is running low,
- The messaging queue (MQ) is pilling up,
- The backend agents are not responsive to the requests sent from the Web container,
- The network connectivity to the database is intermittent, and etc.

Linear regression is expecting a straight forward outcome which means if it is not in one case then it will be another case. There is no third case in the outcome under such prediction. Therefore, the proposed logic model can adopt this approach under the Regression subcategory for straight forward expectation of the outcome if the linear relationship is clear. Otherwise, AHP will be applied to close the gap of decision making. This also means, there is possibility to design and to develop an algorithm of proposed logic model to be a hybrid of Linear Regression under Supervised Learning along side with AHP for the decision making.

6 Proposed Logic Model

Analytic Hierarchy Process (AHP) was developed by Thomas L. Saaty stated in R. W. Saaty (1987). The PAL is proposed to consist of algorithm that is having a hybrid approach of both AHP and Supervised Learning to fulfill the statement of the objectives. This proposed Prescriptive Analytical Logic Model does not perform any predictive analytic, but it provides speedy root cause analysis activity by the pre-configured relationship among the related software applications within the enterprise level. Then it performs the analysis based on the exact time of software application error with cross-check the log events of all the related software applications at the time of software application error occurred. Therefore, the proposed algorithm is proposing the following features which highly differentiate from the market products:-

i. Using the "time" of software application error event when it occurred as the key to extract out log events from all the related logs. The logs can be from software applications, Operating System, Network, and even Database to form a dataset for root cause analysis. This would leave no foot print and no interference to all the existing software applications and database operation.
ii. Extract only the exact amount of input information is important. It is unnecessary to extract excess information and to waste time on root cause analysis. Therefore, the log extraction will be 5 min before the time of software application error event when it occurred, and 5 min after the time of software application error event when it occurred. Then, scan the error only from the extracted amount of log rows. This will allow the root cause analysis to understand any symptom or pattern before and after the software application error event occurs. The rule of 5-minutes-before-and-after log extraction is not only applied on software application log itself. It is applied across all the error log events extracted from the related

various software applications and database. This would allow to understand any external factor at other layers could possibly impact to the software application layer.

iii. Standard error events are easily classified based on keyword, except for those unknown error events. As far as the error detection concern, this area can be applied Machine Learning's supervised learning to conduct the pattern searching activities on those collected log events. Then for these unknown error events, the proposed logic model will produce a report consists of the compilation of all the related log events from related software applications and database. The log events are before and after 5 min of the point of software application error was detected. Although this report is still required human interference to make judgment and decision to put the final action into the configurations of complex analysis module. With the improvement of the configurations of complex analysis module, the proposed model's intelligent level is increasing until the judgment and decision making can be solely handled by the purposed model itself. Crucial factor is the specific error log events extracted at the right timing for analysis and the compilation of the error symptom and pattern. Unlike the market products which only allow to set trigger for specific action based on the total number of same error events or specific error event occurred.

iv. The proposed logic model's standard analysis module does the specific action based on the specific error event occurred. Under the proposed design, it will not allow the same resolution action to be carried out if the same resolution action had been carried in the past. This is because when the same error event re-occurs, but the knowledge-based database of the proposed model identifies the same resolution action had been carried in the past, the proposed model will trigger complex analysis module to carry out further analysis action.

v. Situation can be happened such as multiple software application errors are detected at the same timing. The extracted software application error events can possibility be either false alarm or warning, low impact error or high impact error. The proposed logic model's analysis activity is based on the configurations of standard and complex analysis modules together with the studies of past analysis. The analysis module (based on which module is selected) will then apply the resolution retrieved from the configuration files. However for the decision-making activity to identify the chosen software application error, this activity will leave it to AHP. This means for each error event is identified by the configuration of the analysis module (in this situation, it is the complex analysis module), and which is the crucial software application error to be fixed first, this is all on the AHP to find out the highest weight assigned to the software application error. AHP is introduced in the proposed model and it is applied to evaluate and to identify the software application error which is carrying the highest weight. Note that the standard analysis module is having straight forward resolution steps, and the complex analysis module will perform deep dive analysis activity by analyzing the log files.

vi. Another situation can be happened such as more than one possible resolution is applicable to resolve the identified root cause. Decision making is required.

Hence, AHP is in the great position to evaluate and to identify the preferred resolution against the identified root cause.

vii. All the current analysis activities and resolution steps will be stored into a knowledge-based database for the future reference. As well as it is for the input information of the future analysis.

The PAL is designed to have two configuration files, one is designed to record down the common errors and room to cater for new found error. Then, another one is designed to record down the resolution steps for the common errors as well as room to cater for human to provide new resolution steps based on the new found error. These preferred resolution actions under the resolution are manual, semi-automatic, and fully automatic. The PAL must obey the configurations in order to carry out the resolution under the preferred actions. Especially for the "Complex Analysis", it is required to store the analysis result and the resolution action into a knowledge based database which is associated to the PAL for future reference. Regardless the analysis activity is under simple analysis or complex analysis, the knowledge based is catered for upcoming analysis activity and fine tuning the logic. It is mainly for improving the decision making on selecting the final resolution to be applied to the new found error in the near future.

PAL is required to be trained. This is because the PAL cannot be intelligent by itself without the initial training. Human interaction will still involve at the beginning stage to handle the fine tuning of the configurations of PAL. Since training is crucial, it is required real data as the input of the training. With all the important features stated above, the proposed algorithm inside the PAL is derived and shown as follows (Table 2).

Table 2. Proposed algorithm inside the Prescriptive Analytical Logic Model.

AHP	Supervised learning	Algorithm
		1. Collect or Integrate various log files obtained from different software application databases with/without a predefined database schema
		2. Identify whether the newly reported software application error is first time occurrence or re-occurrence by cross-checking the database which is associated to the prescriptive analytical logic
	*	3. Identify possible log data and select the necessary log data for analysis under the defined software application error classification
*		4. Allocate weight to each possible software application error based on *Analytic hierarchy process* (AHP)
*		5. Shortlist the software application error under the highest weight
		6. Analyze the selected log data for shortlisted software application errors and define possible resolution option
*		7. Allocate weight to each possible resolution option based on AHP

(*continued*)

Table 2. (*continued*)

AHP	Supervised learning	Algorithm
*		8. Shortlist the preferred resolution option under the highest weight
		9. Deploy the preferred resolution option to fix the software application error under the predefined condition if the resolution does not involve SDLC. Otherwise, produce the Analysis Report
		10. Store the analysis result and resolution action into a database which is associated to the prescriptive analytical logic for future reference and knowledge base activities

7 Conclusion

The enterprise software applications come along with the event logging feature for record trial and analysis purposes. This contributes to the proposed methodology on identifying the root cause analysis if there is any possibility no interruption or footprint during the execution of these enterprise software applications. Although there are third party tools available in the market which are utilized for analyzing the enterprise software application log files, these tools are mainly conducting analysis based on the collected log files from the enterprise software applications. Such an analysis approach is reasonable if the hardware resources of the system are under healthy stage, while the enterprise software application is still running in a good condition. However, it is unavoidable that whenever the hardware memory is running low, or the central processing unit (CPU) resource is saturated, or even hard disk space is running low. All these causes would lead to the software application crashes or unstable stage. Indeed, in the event that such circumstance happens, the enterprise software applications will never able to log the exact error event into their log files. Hence, if the root cause analysis activity is solely depending on software application error log files, it is insufficient to identify the actual error accurately. In conclusion, this is the good opportunity and potential to introduce PAL as the new approach for the root cause analysis of the enterprise software application.

References

da Silva Neves, A.J., Camanho, R.: The use of AHP for IT project priorization – a case study for oil & gas company. Procedia Comput. Sci. **55**, 1097–1105 (2015). ScienceDirect. ELSEVIER. https://www.sciencedirect.com/science/article/pii/S1877050915015513. Accessed 20 Apr 2020

Valdman, J.: Log file analysis (2001). https://www.kiv.zcu.cz/site/documents/verejne/vyzkum/publikace/technicke-zpravy/2001/tr-2001-04.pdf. Accessed 18 Jan 2017

Murínová, J.: Application Log Analysis (2015). http://is.muni.cz/th/374567/fi_m/thesis_murinova.pdf. Accessed 18 Jan 2017

Brownlee, J.: Linear regression for machine learning. Machine Learning Mastery Pty. Ltd. (2019). https://machinelearningmastery.com/linear-regression-for-machine-learning/. Accessed 24 Feb 2020

Management Logic: Root-Cause Analysis (2012). http://www.management-logic.com/toolbox/sales/Root-Cause%20Analysis/Index.html. Accessed 28 Nov 2015

Omkarprasad, S.V., Kumar, S.: Analytic hierarchy process: an overview of applications. Department of Mechanical Engineering, Army Institute of Technology, Pune 411 015, India. National Institute of Industrial Engineering (NITIE), Vihar Lake, Mumbai 400 087, India (2004). http://ac.els-cdn.com/S0377221704003054/1-s2.0-S0377221704003054-main.pdf?_tid=39850adc-5d78-11e7-b5bf-00000aab0f26&acdnat=1498815840_e0c9a10c99c46ad30db8da4ef17e817b. Accessed 12 Sept 2018

Saaty, T.L.: Decision making with the analytic hierarchy process. Int. J. Services Sciences 1(1). http://citeseerx.ist.psu.edu/viewdoc/download?doi=10.1.1.409.3124&rep=rep1&type=pdf. Accessed 12 Sept 2018

Saaty, R.W.: The analytic hierarchy process - what it is and how it is used. Pergamon Journals Ltd. Great Britain (1987). http://ac.els-cdn.com/0270025587904738/1-s2.0-0270025587904738-main.pdf?_tid=55956250-6b90-11e7-b75e-00000aab0f02&acdnat=1500365511_5a54a3ecf035ad3e45ab72b909b7632e. Accessed 12 Sept 2018

Wilson, A.: A brief introduction to supervised learning. Towards Data Science (2019). https://towardsdatascience.com/a-brief-introduction-to-supervised-learning-54a3e3932590. Accessed 21 Feb 2020

Hoo Meng, W., Amalathas, S.S.: A new approach towards developing a prescriptive analytical logic model for software application error analysis. In: Silhavy, R., Silhavy, P., Prokopova, Z. (eds.) Intelligent Systems Applications in Software Engineering. CoMeSySo 2019 2019. Advances in Intelligent Systems and Computing, vol. 1046. Springer, Cham (2019). https://doi.org/10.1007/978-3-030-30329-7_24. Print ISBN:978-3-030-30328-0

Wong, H.M., Amalathas, S.S.: An Approach towards developing an Algorithm for Software Application Error Analysis. David Publishing Company (2019). https://doi.org/10.17265/2328-2185/2019.04.006

Wong, H.M., Amalathas, S., Zitkova, T.: A prescriptive logic model for software application root cause analysis. Eur. J. Electr. Eng. Comput. Sci. 3, 5 (2019). https://doi.org/10.24018/ejece.2019.3.5.133

Wong, H.M., Amalathas, S.S., Zitkova, T.: A Business Operation Stability by Improving the Speed of Restoration on Software Application Service. David Publishing Company (2020). https://doi.org/10.17265/2328-2185/2020.01.008

Wong, H.M., Amalathas, S.S., Zitkova, T.: An approach towards designing a prescriptive analytical logic model for software application root cause analysis. Int. J. Adv. Res. Publ. (IJARP) (2019). http://www.ijarp.org/published-research-papers/nov2019/An-Approach-Towards-Designing-A-Prescriptive-Analytical-Logic-Model-For-Software-Application-Root-Cause-Analysis.pdf

Wong, H.M., Amalathas, S.: A new analysis approach incorporated with analytic hierarchy process for the software application in a multiple tiers environment. Eur. J. Electr. Eng. Comput. Scie. **3**, 6 (2019). https://doi.org/10.24018/ejece.2019.3.6.160

Wong, H.M., Amalathas, S.: The root cause analysis algorithm design incorporated with analytic hierarchy process for software application error. Eur. J. Electr. Eng. Comput. Sci. 4, 1 (2020). https://doi.org/10.24018/ejece.2020.4.1.166

Peng, W.W., Dolores, R.: Wallace. Software Error Analysis. NIST Special Publication (1993). http://www.geocities.ws/itopsmat/SoftwareErrorAnalysis.pdf. Accessed 12 Dec 2015

Cmielowski, L.: Adoption of machine learning to software failure prediction. IBM Watson Data (2017). https://medium.com/ibm-data-science-experience/adoption-of-machine-learning-to-software-failure-prediction-e8d85ed0338f. Accessed 2 Apr 2019
Klass, L.: Machine Learning - Definition and application examples. MM International (2018). https://www.spotlightmetal.com/machine-learning–definition-and-application-examples-a-746226/?cmp=go-aw-art-trf-SLM_DSA-20180820&gclid=CjOKCQjwxYLoBRCxARIsAEf16-vBblTDXGp3-EuK0AsxQu90ckOqGbTyXEvjQBWSA3D3MHB51TxDhqUaAlPDEALw_wcB. Accessed 21 Feb 2020
Brownlee, J.: Linear Regression for Machine Learning. Machine Learning Mastery Pty. Ltd. (2019). https://machinelearningmastery.com/linear-regression-for-machine-learning/. Accessed 24 Feb 2020
Guru99: Unsupervised Machine Learning: What is, Algorithms, Example (2020). www.guru99.com. https://www.guru99.com/unsupervised-machine-learning.html. Accessed 24 Feb 2020
Roman, V.: Unsupervised machine learning: clustering analysis. towards data science (2019). https://towardsdatascience.com/unsupervised-machine-learning-clustering-analysis-d40f2b34ae7e. Accessed 9 Mar 2020

Self-tuning Dichotomy and Bonuses for Renovation

Vladimir Tsyganov(✉)

Institute of Control Sciences, Profsoyuznaya 65, 117997 Moscow, Russia
bbc@ipu.ru

Abstract. The management technology related to model-based manufacturing system engineering and processing of random data is considered. This technology is based on dichotomy models and their applications. If the parameters of the random process are known, the primary dichotomy model is used to process the data. If these parameters are unknown, the self-tuning procedure is used. In secondary dichotomy, two such procedures are combined into a complex self-tuning procedure using system engineering. Such complex procedure is used to manage an agent in an environment of incomplete awareness. However, the visionary agent can manipulate his indicators in such a way as to maximize own target function, which depends on the results of the secondary dichotomy. This may result in poor management. A formal method for solving this problem is proposed, based on an ontology-based management model that includes both self-tuning of secondary dichotomy and corresponding stimulation. The conditions are determined under which the ontology-based model increases the quality of management due to the greater interest of the agent. Based on this model, data processing and analysis of the performance of renovation of railway locomotives are carried out.

Keywords: Mathematical modeling · Systems engineering · Ontology · Management · Dichotomy · Performance

1 Introduction

Improving the manufacturing performance according to Industry 4.0 concept is based on advances in intelligent systems [1]. In turn, artificial intelligence involves adaptation, learning and self-organization. Mathematical modeling of adaptive, learning and self-organizing systems occupy an important place in the theory of cybernetics and information [2].

The other important aspect of improving the performance of the industry in the face of change is the disclosure of internal reserves and resources through the activation of the human factor. To do this, we need to consider enterprises and companies as multi-agent systems, and take into account the conflicting interests of people in the process of interaction [3].

Intelligent management of multi-agent systems relies on data science [3]. Models and computational methods of mathematical statistics are gaining ground in the face of change. For sustainable development, adaptive models and statistical processing

R. Silhavy et al. (Eds.): CoMeSySo 2020, AISC 1295, pp. 644–656, 2020.
https://doi.org/10.1007/978-3-030-63319-6_60

mechanisms are needed [4]. On their basis, regulation and intelligent control of multi-agent system development can be implemented [5]. Such models and methods are used in machine learning procedures in the management of both large-scale industrial systems [6] and their enterprises [7]. Similar adaptive models and methods are also used in econometrics, for example, for adaptive identification of innovative production function of corporation [8]. In advanced human-machine models and methods of training and management of multi-agent systems, expert hints are used - tutor [9] or coach [10].

To combine a variety of mathematical models, system engineering is used [11]. System and data engineering models and their application are being developed. For example, designing adaptive information models for manufacturing management are considered in [12].

This paper discusses the system engineering of self-tuning dichotomy models, data processing, and staff reward, with their application for renovation railway locomotives.

2 Methods of Awareness Dichotomy

Dichotomy (Greek. Διχοτομία: δῖχῆ, "in two" + τομή, "division") - bifurcation, sequential division into two parts, more connected inside than among themselves. The method of such logical division is that the divisible concept is completely divided into two mutually exclusive concepts. For example, the scope of the concept of "assessment" can be divided into two complementary grades: satisfactory or unsatisfactory. The concepts of satisfactory or unsatisfactory are complementary to each other.

Dichotomous division, as a way of forming subsections of a concept or term, facilitates systems engineering. Dichotomous division is attractive because of its simplicity. Indeed, in a dichotomy, we always deal with only two grades that exhaust the scope of a divisible concept. Thus, dichotomous division is always proportionate; members of the division complement each other, since each object of the divisible set falls into only one of the grades *a* or not *a*; the division is carried out on one basis - the presence or absence of some attribute. Denoting the divisible concept by the letter *a* and highlighting some kind of, say, *b* in its volume, we can divide the volume *a* into two parts - *b* and not *b*.

2.1 Primary Dichotomy

Many tasks of the management body (briefly - management) are reduced to dichotomy - assigning one of two grades to the controlled object. Depending on its result, control is carried out. Examples include the definition of "strengths" and "weaknesses" with a view to its development and investment, the introduction of innovative technical means, the determination of objective and subjective reasons for non-fulfillment of plans, the adoption of personnel decisions, bonuses, etc.

Primary dichotomy is obtained by first dividing the whole into two parts (for example, satisfactory or unsatisfactory). For dichotomy, a crucial rule is needed.

Let t denote the time period, $t = 0.1, \ldots$ Let z_t be a random variable characterizing the state of the object in the period t, $z_t \in \Delta \subset R^1$. The primary dichotomy involves assigning z_t to one of the two areas that make up the set Δ. Incorrect dichotomy leads to losses.

Suppose first that there is enough statistical data to determine the probability density $q(z)$ of a random variable z. We denote $\{\Delta_1, \Delta_2\}$ some partition of the set Δ into 2 regions, $\bigcup_{k=1}^{2} \Delta_k = \Delta$. The challenge is to define a partition that minimizes the average risk associated with a dichotomy. We introduce for each so far unknown region Δ_k, $k = \overline{1, 2}$, the dichotomy loss function:

- $L_1(c, z)$- losses in case of graduation of an object as grade 1, while $z \in \Delta_2$;
- $L_2(c, z)$- losses in case of graduation of an object as grade 2, while $z \in \Delta_1$,

where c is an unknown parameter. Then the affiliation z of one or another region is determined by the sign of the crucial rule

$$\mu_{12}(c, z) = L_1(c, z) - L_2(c, z) : z \in \Delta_1 \quad \text{for } \mu_{12}(c, \xi) < 0, \text{ and } z \in \Delta_2 \\ \text{for } \mu_{12}(c, z) \geq 0 \tag{1}$$

The task is to determine the parameter c with the crucial rule (1) that minimizes the average risk of dichotomy:

$$J(c) = \sum_{k=1}^{2} \int_{\Delta_k} L_k(c, z) q(z) dz \underset{c}{\mapsto} \min \tag{2}$$

Now, knowing $q(z)$, we can find the parameter c as a solution to problem (2).

2.2 Self-tuning Primary Dichotomy

However, statistics are often insufficient. Suppose that $q(z)$ is unknown, so that the direct determination of the parameter c that solves the optimization problem (2) is impossible. Then we can adjust the parameter c of the crucial rule for observations z_t to minimize risk $J(c)$.

We write the condition for the minimum average risk of dichotomy (2) in the form:

$$M_z \left\{ \sum_{k=1}^{2} \widehat{B}_k(c, z) \partial L_k(c, z) / \partial c = 0 \right\}, \ \widehat{B}_k(c, z) = \begin{cases} 1 & \text{if } z \in \Delta_k, \\ 0 & \text{else} \end{cases}, \tag{3}$$

where M_z is the mathematical expectation operator. For simplicity of calculations, we will consider linear loss functions below:

$$L_1(c, z) = z - vc, \ 0 < v < 1, \ L_2(c, z) = d(c - z), \ d > 0. \tag{4}$$

Substituting these expressions in (1), we obtain the crucial rule in the form:

$$z \in \Delta_1 \text{ if } z < (d+v)c/(d+1), \text{ and } z \in \Delta_2 \text{ if } z \geq (d+v)c/(d+1). \quad (5)$$

Now, using the stochastic approximation method for solving (2), and taking into account (5), it is easy to show that the procedure for self-tuning the parameter estimate of the crucial rule of the primary dichotomy has the form:

$$c_{t+1} = S(c_t, z_t) = \begin{cases} c_t + \gamma_t v & \text{if} \quad z_t < (d+v)c_t/(d+1), \\ c_t - \gamma_t d & \text{if} \quad z_t \geq (d+v)c_t/(d+1), \end{cases} c_0 = c^0,\ t = 0, 1, \ldots \quad (6)$$

where γ_t is the gain in the period t, $\gamma_t > 0$, $\sum_{t=0}^{\infty} \gamma_t < \infty$ [2]. Using the parameter of the crucial rule (6), a dichotomy is performed: when

$$Z_i \geq (d+v)c_i/(d+1) \quad (7)$$

the object belongs to grade 2, otherwise - to grade 1.

2.3 Self-tuning Secondary Dichotomy

Dichotomous division has a drawback: when dividing the volume of a concept into two concepts, each time its part to which the "not" particle belongs is undetermined. If, after the primary dichotomy, we divide the assessments into satisfactory or unsatisfactory, then the first group is very unclear. In addition, if at the beginning of a dichotomous division it is usually quite easy to establish the presence of a contradictory concept, then as we move away from the first pair of concepts, it becomes more difficult to find it. This prompts the study of the secondary dichotomy obtained by successively dividing the whole into two parts, and then each of them again into two parts.

Namely, we consider the secondary dichotomy, in which:

- the first division of the whole into two parts uses a normative approach (for example, division in relation to the implementation of the norm or plan);
- when dividing each of the obtained parts into two parts, the primary dichotomy method with self-tuning is used, described in Sect. 2.2.

As a result of the secondary dichotomy, 4 grades of indicator z_t of the object's performance are determined. This makes the results of the secondary dichotomy more visual and convenient for management.

We will assume that the first division of the whole into two parts is based on the sign of some indicator z_t, calculated as the deviation of the observed actual state of the object from its norm. Therefore, the value of z_t can take both positive and negative values. On its basis, a secondary dichotomy is carried out.

An example is dividing by the indicator z_t the work of an object, determined depending on its implementation of the plan. If the plan is completed, then the work of an object is considered satisfactory, otherwise - no. On this basis, a secondary dichotomy is carried out. At the same time, an object whose work is found to be satisfactory can receive a good or excellent grade. And an object whose work is considered unsatisfactory may receive a mediocre or poor grade.

In self-tuning of the secondary dichotomy, management uses 2 crucial rules with 2 customizable parameters estimates based on the procedure (6). We will assume that grade 4 corresponds to an excellent grade, grade 3 - good, grade 2 - mediocre and grade 1 - bad, and the higher the indicator z_t, the higher the grade. Let's give a formal description of Procedure of a Self-Tuning Secondary Dichotomy.

2.4 Procedure of a Self-tuning Secondary Dichotomy

1. If indicator is negative ($z_t < 0$), then the object can be assigned to grade 1 or 2. The corresponding loss functions are determined by formulas (4), and the parameter of the crucial rule is determined by formula (6). If (7) holds, then the object belongs to grade 2, otherwise - to grade 1.

2. If indicator is non-negative ($z_t \geq 0$), then the object can be assigned to grade 3 or 4. Moreover, the loss functions have a form similar to (4):

$$L_3(C, z) = z - VC,\ 0 < V < 1,\ L_4(C, z) = D(C - z),\ D > 0, \tag{8}$$

where $L_i(C, z)$ is the loss function for an erroneous assignment to an object of grade i, $i = \overline{3, 4}$, C is an unknown parameter. To minimize losses, its estimate of C_t is tuned according to a formula similar to (6):

$$C_{t+1} = Q(C_t, z_t) = \begin{cases} C_t + \gamma_t V & \text{if} \quad z_t < (D+V)C_t/(D+1), \\ C_t - \gamma_t D & \text{if} \quad z_t \geq (D+V)C_t/(D+1), \end{cases} C_0 = C^0, \tag{9}$$
$$t = 0, 1, \ldots$$

Further, by estimating the parameter of the crucial rule C_t, like (7), a dichotomy is carried out: if

$$z_t \geq (D+V)C_t/(D+1), \tag{10}$$

then the object belongs to grade 4, otherwise - to grade 3.

The described procedure of self-tuning of the secondary dichotomy, which includes calculations according to formulas (6)–(10), will be formally designated as $W(c_t, C_t, z_t)$. This procedure allows the intellectual analysis of data about the object.

3 Observational-Based Dichotomy and Agent Activity

The elements of manufacturing are people, as well as the management processes by which they carry out their activities. Therefore, when researching and developing manufacturing system, it is necessary to take into account the socio-psychological aspects of undesirable activity of its subsystem. Such an active subsystem will be called agent.

3.1 Self-tuning Observational-Based Secondary Dichotomy

Consider a two-tier manufacturing system with management at the top and agent at the bottom. Assume that z_t is the maximum random value (capability) of the agent actual performance y_t in period t: $y_t \leq z_t$. The capability z_t is unknown to management. But the capability z_t becomes known to the agent before choosing the actual performance y_t in period t. Based on the condition that actual performance cannot exceed capability, an agent can manipulate y_t to its advantage.

Management, on the other hand, observes only performance y_t, which does not necessarily coincide with the capability of the agent (because $y_t \leq z_t$). Therefore, management is forced to form the agent grade under uncertainty, caused not only by stochastic capability, but also by undesired agent activity.

For this, management uses the self-tuning procedure of secondary dichotomy $W(c_t, C_t, z_t)$, performed according to formulas (6)–(10), substituting the observed performance y_t (instead of the unknown capability z_t) in them. The resulting estimates of the parameters of the crucial rules will differ from c_t and C_t. Below we give a formal description of this Secondary Dichotomy Procedure.

3.2 Secondary Dichotomy Procedure

1. If performance is negative ($y_t < 0$), then the agent can be assigned to grade 1 or 2. Moreover, the loss functions have the form similar to (4): $L_1(a, y) = z - va$, $0 < v < 1$, $L_2(a, y) = d(a - y)$, $d > 0$. There $L_i(a, y)$ is the loss function for the erroneous assignment of an agent of grade i when observing y, $i = \overline{1, 2}$, and a is an unknown parameter. To minimize losses, its estimate a_t is tuned basing on the observation of y_t, according to a formula whose structure is similar to structure (6):

$$a_{t+1} = S(a_t, y_t) = \begin{cases} a_t + \gamma_t v & \text{if} \quad y_t < (d+v)a_t/(d+1), \\ a_t - \gamma_t d & \text{if} \quad y_t \geq (d+v)a_t/(d+1), \end{cases} \quad a_0 = c^0,\ t = 0, 1, \ldots \tag{11}$$

Further, by analogy with (7), a dichotomy is performed using the parameter of the crucial rule (11): if

$$y_t \geq (d+v)a_t/(d+1), \tag{12}$$

then agent belongs to grade 2, otherwise it belongs to grade 1. Meaningfully, the value $(d+v)a_t/(d+1)$ is the lower limit of performance y_t corresponding to satisfactory operation of the agent. Therefore, the value $(d+v)a_t/(d+1)$ will be called the assessment of the satisfaction threshold.

2. If performance $y_t \geq 0$ is non-negative, then the agent can be assigned to grade 3 or 4. Moreover, the loss functions have a form similar to (8): $L_3(A, y) = y - AV$, $0 < V < 1$, $L_4(A, y) = D(A - y)$, $D > 0$. There $L_i(A, y)$ is the loss function for the

erroneous assignment of an agent to grade i, $i = \overline{3,4}$, A is an unknown parameter. To minimize losses, its estimate A_t is tuned according to a formula similar to (9):

$$A_{t+1} = Q(A_t, y_t) = \begin{cases} A_t + \gamma_t V & \text{if} \quad y_t < (D+V)A_t/(D+1), \\ A_t - \gamma_t D & \text{if} \quad y_t \geq (D+V)A_t/(D+1), \end{cases} A_0 = C^0, \quad (13)$$
$$t = 0, 1, \ldots$$

With the aid of estimate A_t, like (10), a dichotomy is carried out: if

$$y_t \geq (D+V)A_t/(D+1), \quad (14)$$

then agent belongs to grade 4, otherwise it belongs to grade 3. Meaningfully, the value $(D+V)A_t/(D+1)$ is the lower limit of performance y_t, which corresponds to the excellent work of the agent. Therefore, the value $(D+V)A_t/(D+1)$ will be called the assessment of the threshold of excellence.

3. Taking into account conditions (12) and (14), the result of the secondary dichotomy in period t is presented in the form:

$$w_t(a_t, A_t, y_t) = \begin{cases} 4 & \text{if} \quad y_t \geq (D+V)A_t/(D+1) \\ 3 & \text{if} \quad 0 \leq y_t < (D+V)A_t/(D+1) \\ 2 & \text{if} \quad (d+v)a_t/(d+1) \leq y_t < 0 \\ 1 & \text{if} \quad y_t < (d+v)a_t/(d+1) \end{cases}, \quad (15)$$

where $w_t(a_t, A_t, y_t)$ is the agent grade, as a function of its performance y_t and estimates a_t and A_t.

4. The Secondary Dichotomy Procedure based on the agent performance y_t, and including calculations by formulas (11), (13), (15), is formally denoted as $W(a_t, A_t, y_t)$.

3.3 Activity and Ontology-Based Management Model

Management is interested in constantly using the capabilities of the agent, i.e. $y_t = z_t$, $t = 0, 1, \ldots$ In the general case $y_t \neq z_t$. Meaningfully, the reason is that management is not able to take into account random interferences and other factors that become known to the agent during the production process. This allows the agent to underestimate performance y_t, and then $y_t < z_t$. To avoid this, management must ensure that $y_t = z_t, t = 0, 1, \ldots$. To do this, management may appoint a bonus for agent. For example bonus is an amount of money added to wages on a seasonal basis, especially as a reward for good performance in period t. Such bonus can be linked to the results of the Secondary Dichotomy Procedure. Since it is accepted that grade 4 corresponds to an excellent grade, 3 - good, 2 - mediocre and 1 - bad (Sect. 2.3), then the bonus should be the greater, the higher the agent grade.

Suppose that to ensure $y_t = z_t$ management establishes the order of management of the manufacturing system in period t formally described by the following ontology.

Before the beginning of the period t management knows the estimates a_t and A_t. At the beginning of the period t agent recognizes the value z_t, $z_t \in \Delta$, and selects the preferred output y_t^*, $y_t^* \leq z_t$. This output y_t^* becomes known to management. After that, knowing the estimates a_t and A_t, and using the Secondary Dichotomy Procedure (see Sect. 3.2), management determines the grade w_t (15) and the corresponding agent bonus $b_t = B(w_t)$, where $B(w_t)$ is the monotonically increasing function w_t. In addition, management forms estimates a_{t+1} and A_{t+1} for the next period $t + 1$ according to (11) and (13), respectively.

Substantively, the corresponding ontology management involves the formation of parameters of crucial rules using self-tuning procedures, a secondary dichotomy based on their comparison with the observed performance, and the agent reward. Formally, the Secondary Dichotomy Procedure $W(a_t, A_t, y_t)$ and bonus assignment procedure $b_t = B(w_t)$ can be combined into an ontology-based management model $M = (W, B)$.

3.4 Agent Behavior

Agent can manipulate y_t in order to achieve their interests. Formally, knowing the ontology-based management model, the agent selects its output in such a way as to increase his own target function G_t, depending on bonuses $b_\tau = B[w\ (a_\tau, A_\tau, y_\tau)]$, $\tau = \overline{t, t+T}$:

$$G_t = \sum_{\tau=t}^{t+\theta} \rho^{\tau-t} B[w(a_\tau, A_\tau, y_\tau)], \tag{16}$$

where ρ is the discount coefficient, $0 < \rho < 1$, θ is the number of periods taken into account by the agent. We assume that agent is aware that $y_\tau \leq z_\tau$, $z_\tau \in \Delta$, in each period τ, $\tau = \overline{t+1, t+\theta}$. To make a choice under such uncertainty, agent focuses on the guaranteed value of the target function (16), which depends on the current choice of agent performance y_t: $v_t(y_t) = \min\limits_{z_\tau \in \Delta,\ \tau=\overline{t+1, t+\theta}} \min\limits_{y_\tau \leq z_\tau,\ \tau=\overline{t+1, t+\theta}} G_t$. Then the set of possible agent choices is $Y_t(M, z_t) = \{y_t^* | y_t^* \leq z_t,\ v_t(y_t^*) \geq v_t(y_t),\ y_t \leq z_t\}$. Suppose that the set $Y_t(M, z_t)$ includes z_t. In this case, we say that the agent's benevolence hypothesis with respect to management is valid: if $z_t \in Y_t(M, z_t)$ then the agent chooses performance $y_t^* = z_t$. $t = 0, 1,\ldots$

4 Optimal Ontology-Based Management Model

Let us turn to the practice of manufacturing management in a large-scale corporation. Firstly, evaluation and reward procedures are usually designed so that bonuses grow as performance increases, compared with their current ratings, norms, plans. Usually bonuses are assigned when they are exceeded [13]. Therefore, the higher the current norm or plan, the more difficult it is to get a bonus at the present time.

Secondly, the planning procedure in a large-scale corporation is often organized in such a way that the plan in each subsequent period is growing by a certain percentage of the result achieved today (the so-called "planning from the achieved level"). Then the future plan (norm) will be the higher, the higher the performance achieved today. Therefore, the agent may not be interested in exceeding the plan, as this will increase the future plan, and it will be more difficult to get a bonus [13].

Thus, the problem arises of the lack of interest of the visionary agent in using his capabilities. In this case $y_t < z_t$. Consider how this problem can be solved by selecting the appropriate ontology-based management model $M = (W, B)$.

4.1 Sufficient Optimality Conditions

Consider the following ontology of the management and agent moves in period t. The first move in period t is made by management, setting the ontology-based management model $M = (W, B)$. The second move is made by the agent, choosing y_t^*. Then these moves are repeated in the period t + 1, t = 0, 1,...

Theorem. To unleash the capability: $y_t^* = z_t$, $t = 0, 1, \ldots$, it is enough to specify an ontology-based management model $M = (W, B)$ with the Secondary Dichotomy Procedure $W(a_t, A_t, z_t)$ determined according to (11), (13), (15).

Proof. The agent target function $G_t = \sum_{\tau=t}^{t+\theta} \rho^{\tau-t} B[w(a_\tau, A_\tau, y_\tau)]$ determined according to (16) depends on both current and future bonuses $b_\tau = B[w\,(a_\tau, A_\tau, y_\tau)]$, $\tau = \overline{t, t+T}$.

According to (11), (13), (15), with growth y_t, the agent grade $w\,(a_t, A_t, y_t)$ grows (does not reduces). Consequently, the current bonus $b_t = B[w\,(a_t, A_t, y_t)]$ is growing (does not reduces).

Further, management uses self-tuning procedures (11) and (13), in which the estimates a_τ and A_τ, accordingly, reduce (do not increase) with increasing y_t at $\tau = \overline{t+1, t+T}$. Therefore, according to (15), future agent bonuses $b_\tau = B[w\,(a_\tau, A_\tau, y_\tau)]$ increase (do not reduce) with increasing y_t at $\tau = \overline{t+1, t+T}$.

Further, $G_t = \sum_{\tau=t}^{t+\theta} \rho^{\tau-t} B[w(a_\tau, A_\tau, y_\tau)]$ is monotonically growing in $B[w\,(a_\tau, A_\tau, y_\tau)]$, $\tau = \overline{t, t+T}$. And since $b_\tau = B[w(a_\tau, A_\tau, y_\tau)]$ is monotonically growing (does not reduces) in y_t, $\tau = \overline{t, t+T}$, then the agent target function (16) also is growing (does not reduces) with growth y_t. Since $y_t \le z_t$, then the maximum of the agent target function $G_t = \sum_{\tau=t}^{t+\theta} \rho^{\tau-t} B[w(a_\tau, A_\tau, y_\tau)]$ is achieved at $y_t = z_t$. Consequently, by virtue of the benevolence hypothesis, $y_t^* = z_t$, QED.

4.2 Discussion

The theorem admits a simple interpretation. Management observes the value y_t^* characterizing the agent's performance in the period t, $y_t^* \le z_t$, where z_t is the unknown random maximal performance. Management tunes the parameters of the crucial rule

(11) and (13). In addition, management forms the agent grade (15), and assigns agent bonus in such a way that the higher the grade, the higher the bonus.

For example, according to (15), when $y_t^* \geq (D+V)A_t/(D+1)$ agent refers to the most successful: $w_t(a_t, A_t, y_t) = 4$. So agent bonus is maximal. From the other side, if $y_t^* < (d+v)a_t/(d+1)$ then agent refers to the outsider $(w_t(a_t, A_t, y_t) = 1)$ and is punished. Any of such decisions are associated with certain risks for management. In the first case, the losses increase with a reduce in performance y_t (undeserved bonus of agent). In the second case, these losses increase with increasing performance y_t and unfair punishment of the agent. The value $(d+v)a_t/(d+1)$ corresponds to an assessment of the lower limit of the agent's satisfactory performance.

Note that, according to (11) and (13), the higher is the performance y_t^* the lower are the estimates a_{t+1} and A_{t+1} in the next period. But, according to (15), a reduce in these estimates leads to a reduce in the threshold assessments of satisfaction and of excellence. Consequently, the agent becomes easier to receive greater grade $w_{t+1}(a_{t+1}, A_{t+1}, y_{t+1})$ and greater bonus $b_{t+1} = B[w\ (a_{t+1}, A_{t+1}, y_{t+1})]$ in the period t + 1, even with a smaller value of the random capability z_t.

In other words, with increased performance y_t^*, an agent receives not only a higher bonus. With y_t^* growth, future bonuses can be greater. This additionally interests the agent in using its capabilities, i.e. in choice $y_t^* = z_t$.

5 Model-Based Performance Analyses for Renovation of Locomotives

Ontology-based management model $M = (W, B)$ satisfying the conditions of the theorem allows revealing the agent capability. Let us consider an example of the application of this model for performance analyses and management of renovation locomotives of Russian Railways Corporation. Locomotive energy consumption is up to 80% of the total energy consumption of this large-scale transport monopoly [14]. The head of its regional department is in charge of railway locomotives renovation facility. He leads the regional management that oversees the work of a subordinate renovation facility which acting as an agent.

The regional management monthly calculates renovation facility performance y_t as the difference between the actual and planned number of renovated locomotives, divided by that planned number. There t is the month number during the year, $t = \overline{1, 12}$. Based on y_t, the regional management performs a secondary dichotomy and assigns monthly bonus to the staff of the renovation facility.

In order to regularly perform these functions, the head of regional department forms an ontology-based management model $M = (W, B)$. So parameter estimates of crucial rules, assessments of the threshold of excellence and satisfaction, and results of secondary dichotomy are calculated according to formulas (11), (13), (15). The results of these calculations are illustrated in Figs. 1 and 2.

In Fig. 1, the graph of performance y_t is shown with a black line. The graph of the assessment of the threshold of excellence $(D+V)A_t/(D+1)$ is shown of the blue (upper) smooth curve. To determine it, the estimate of the parameter of the crucial rule

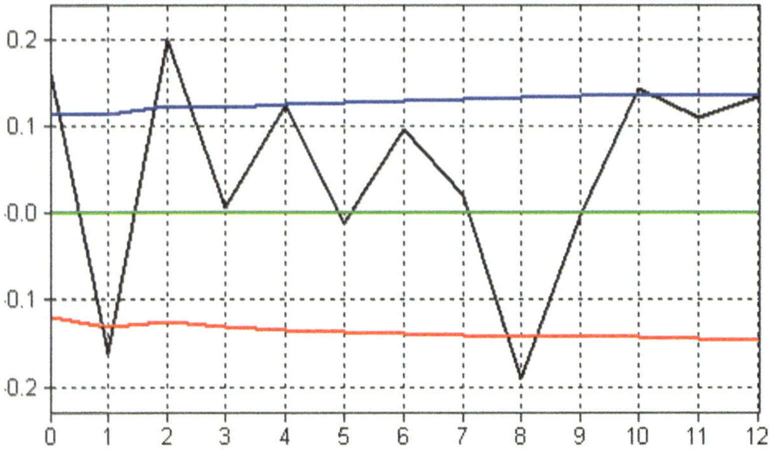

Fig. 1. Monthly performances y_t (black line), assessments of the thresholds of excellence $(D + V)A_t/(D + 1)$ (blue line) and satisfaction $(d + v)a_t/(d + 1)$ (red line) during the year.

A_t was calculated according to formula (13). Further, the graph of satisfaction threshold assessments $(d + v)a_t/(d + 1)$ is shown in Fig. 1 of the red (lower) smooth curve. In this case, the estimate a_t was calculated by formula (11).

Knowing the assessments of the thresholds of excellence and satisfaction, management defines the renovation facility grade by the formula (15). The pink line in Fig. 2 shows a graph of the change in this grade. This graph helps to test the renovation facility performance.

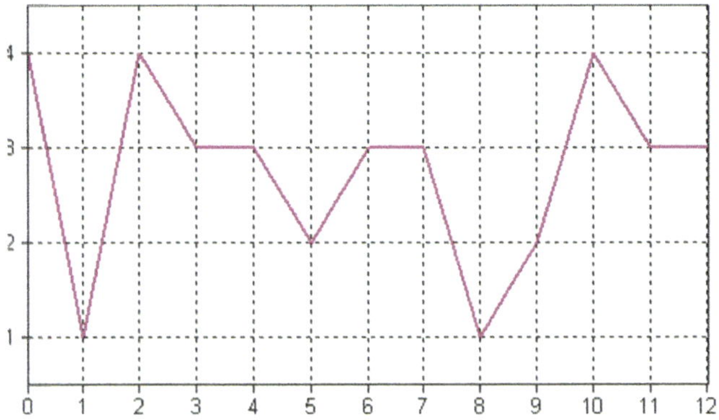

Fig. 2. Renovation facility grades during the year.

Depending on the grade assigned to the renovation facility, its administration and personnel are stimulated. In the event that the renovation facility does not fulfill the plan (i.e. performance $y_t < 0$), its administration loses the bonus. Moreover, if the renovation facility belongs to grade 1 (i.e., performance is below the assessment of the satisfaction threshold), then its personnel also lose the bonus. In addition, an audit is conducted to determine whether it is necessary to punish the administration of the renovation facility or to modernize its manufacturing processes, etc. If the renovation facility belongs to grade 2 (i.e., performance is above the assessment of the satisfaction threshold), then only its personnel can receive a bonus.

If the renovation facility fulfills the plan, then both the administration and the personnel receive the bonuses. Moreover, the higher is the grade, the greater is the monthly bonuses. Formally, the bonus assignment procedure is such that the bonus grows with the grade. In addition, according to (11) and (13), the higher the performance y_t, the lower the assessments of the thresholds of satisfaction and excellence for the next period. Thus, with an increase in performance, the renovation facility receives not only a higher current bonus, but also "reward thresholds" for such bonuses will reduce in the future. As a result, the ontology-based management model $M = (W, B)$ ensures that both the administration and the personnel are interested in using the resources and hidden reserves of renovation railway locomotives.

6 Conclusion

A connection is established between the theory, design, development and evaluation of dichotomy ideas in multi-agent manufacturing systems using tools, methods and applications of mathematical statistics and econometrics.

Further dichotomy will allow combining a larger number of mathematical models. This will increase the degree of intellectualization of manufacturing systems.

System engineering of multi-agent manufacturing systems associated with the imposition of mathematical models allows the development of advanced methods in the theory of cybernetics and automation.

References

1. Blanchet, M., Rinn, T., Thaden, G., Thieulloy, G.: Industrie 4.0 - the new industrial revolution. How Europe will succeed. Roland Berger Strategy, München (2014)
2. Schuppen, J.: System identification with information theoretic criteria. In: Bittanti, S., Picci, G. (eds.) Identification, Adaptation, Learning 1996, pp. 289–338. Springer, London (1996)
3. Burkov, V., Gubko, M., Kondratiev, V., Korgin, N., Novikov, D.: Mechanism synthesis and management. In: Mathematical Methods for Smart Organizations. NOVA Publishers, New York (2013)
4. Borodin, D., Gurlev, I., Tsyganov, V.: Adaptive mechanisms for sustainable development. Syst. Sci. **30**, 89–95 (2004)
5. Tsyganov, V.: Regulation of decentralized active system development and intelligent control mechanisms. IFAC Proc. Vol. (IFAC-PapersOnline) **9**, 94–98 (2010)
6. Tsyganov, V.: Learning mechanisms in digital control of large-scale industrial systems. In: Proceedings of the Global Smart Industry Conference, Chelyabinsk, pp. 1–6. IEEE (2018). https://doi.org/10.0509/glosic.2018.8570093

7. Tsyganov, V.: Decision making and learning in wagon-renovation. In: 12th International Conference on Management of Large-Scale System Development, Moscow, pp. 1–6. IEEE (2019). https://doi.org/10.1109/mlsd.2019.8911042
8. Tsyganov, V.: Adaptive identification of innovative production function of corporation. FME Trans. **47**(4), 691–699 (2019)
9. Tsyganov, V.: Tutoring mechanisms of business management. In: 21st International Conference on Business Informatics, Moscow, pp. 60–67. IEEE (2019)
10. Tsyganov, V.: Coaching in energy efficiency control of manufacturing. IOP Conf. Ser. Mater. Sci. Eng. **862**, 042010. https://doi.org/10.1088/1757-899x/862/4/042010
11. Kossiakoff, A., Sweet, W., Seymour, S., Biemer, S.: Systems Engineering. Principles and Practice. John Wiley, New York (2011)
12. Tsyganov, V.: Designing adaptive information models for production management. Procedia CIRP **84**, 1088–1093 (2019)
13. Shishkin, G., Tsyganov, V.: Mechanism of adaptation of microelectronics manufacturing to market. In: Proceedings of the 6th International Conference on Experience of Designing and Application of CAD Systems in Microelectronics, pp. 119–120. LSU, Lvov (2001)
14. Tsyganov, V.: Optimization of transport monopoly control. IFAC-PapersOnLine **51**(32), 698–703 (2018)

On the Question of Modifying Membership Functions

Nickolay Barchev and Vladimir Sudakov(✉)

Plekhanov Russian University of Economics, Stremyanny Lane, 36, Moscow 117997, Russian Federation
sudakov@ws-dss.com

Abstract. Membership functions are a tool for reflecting expert preferences when solving problems in an unclear information environment. Procedures of accessory functions modification allow to receive new functions on the basis of available ones without additional attraction of experts reducing the general labor input of carried out examinations. Existing and new ways of performance of such modifications are investigated in this work. The modifier representing a pair of linear transformations is defined on the basis of available accessory functions. Approbation of the given technique has shown absence of unsatisfactory results and reception of results acceptable to experts in most cases of its application.

Keywords: Membership function · Fuzzy set · Lexical variable · Expert estimates · Linear transformations

1 Introduction

One of the difficulties in the practical application of the apparatus of the fuzzy sets theory is that the membership functions are subjective in nature, and traditional approaches to their construction involve preliminary identification of expert opinions. The corresponding methods have been known for quite a long time [1–3].

If in real problems such approaches are extended to all functions of an accessory which describe fuzzy initial data, the volume of necessary examinations can be unfeasible. Meanwhile, many fuzzy sets that formalize close linguistic concepts have the same area of definition of their membership functions and communicate with each other using some standard in terms of content transitions such as "very/not very", "more/less", "stronger/weaker" and the like.

Formalization of this or that transition means reconstruction (modification) of initial accessory functions and is performed by corresponding operators, which are usually called accessory function modifiers [4] or logical-linguistic modifiers [5].

The use of modifiers to obtain new functions on the basis of previously found allows to significantly reduce the amount of necessary expertise. It is also of practical interest to process with the help of modifiers notions which are not initially connected with those on which basis these modifiers were obtained.

An appropriate heuristic justification may be that transitions used in expert assessments usually have a sufficiently clear universal meaning, they are stable when

R. Silhavy et al. (Eds.): CoMeSySo 2020, AISC 1295, pp. 657–662, 2020.
https://doi.org/10.1007/978-3-030-63319-6_61

used in expert judgments and are weakly dependent on the subject area where they are used. This allows to assume the invariance of the modifiers under discussion, both in relation to different experts and in relation to different subject areas under certain conditions.

2 Literature Review

The interest in constructing the functions of belonging to linguistic terms using logical-linguistic modifiers is not new and is systematically reflected in publications. For example, it is often recommended to formalize the transition of the "very" type by constructing the values of the initial belonging function into a square (concentration operation [5, 7, 8]) or into higher integer degrees (amplification operation [9]). Transition of the type "not very" in this idea is formalized by extracting the square root (stretching operation [5–7]) or the root of a higher degree (weakening operation [10]) from the corresponding values of the original function. The use of such mechanisms may require imposing additional restrictions on the types of statements to be modified and the scales on which these statements are considered [11].

Comparison of the results of the application of these transformations with actual expert assessments often demonstrates significant inconsistencies. The reason of observed results when using concentration/stretch (gain/loosen) operations can be compression or stretching in the process of modifying the graphs of the original accessory functions, without changing the position of their maximums on the abscissa axis. In content, this corresponds to a change in the degree of uncertainty in the description of a particular concept (in essence, the degree of confidence of the expert) [12]. In practice, transitions of the "very/not very" type do not always reflect such changes.

In addition to the above mentioned, the practice of formalizing expert judgments using the functions of belonging having a triangular form is widespread. This is explained by the fact that the Expert Advisor with minimum effort can specify three values $\{x_i; i = 1, 3\}$ of the area of the belonging function definition, one of which necessarily belongs to the chosen fuzzy concept from the point of view of the Expert Advisor (linguistic term, value of the belonging function $\mu(x_2) = 1$), while the other two absolutely do not belong to it ($\forall x \leq x_1 : \mu(x) = 0; \forall x \geq x_3 : \mu(x) = 0$) [1]. After finding the three points, they are connected by segments in the plane, which easily defines simple functions that have a piecewise linear character. Practice shows that a fairly large number of concepts can be quite satisfactorily described with such functions of belonging.

The use of a specific type of accessory function allows you to unify and simplify their processing. Thus, it is logical to assume the desire of the researcher to get the same kind of functions as the original ones when modifying them. It in addition gives the chance of uniform semantic interpretation of formed results. However, application of nonlinear transformations obviously leads to occurrence of nonlinear functions on their output (except for separate special cases). When using triangular form as initial functions, it also means losing the properties of piecewise linearity, which is undesirable and sometimes unacceptable.

One of the alternative ways to modify the accessory functions is the heuristic method proposed in [2], designated as a parametric approach to their construction. Its idea is a preliminary description of a transition of the "very/not very" type using fractional linear transformations on a set of automorphic functions. The method assumes a pair of accessory functions, one of which corresponds to the initial concept (linguistic term) and the other to a modified one (e.g. "cold water" and "very cold water" [4]). The transformation (modifier) obtained on the basis of the existing function pair is further applied to perform similar in semantic modifications (e.g. "very cold water" and "very, very cold water" [4]), while the transformation coefficients are calculated using rather cumbersome formulas.

Unfortunately, there is no substantial justification for the procedures described in [2], and attempts to apply them to functions different from those demonstrated in according examples have shown a very serious difference between the results of modifications and functions obtained by the expert method.

3 Method

The membership functions of fuzzy concepts (linguistic terms) associated with the help of transitions of the "very/not very" type differ from each other in the general case with their displacement in one direction or another along the abscissa axis, and also the size of the intervals on which their graphs are based [1, 8]. Thus, formal operations defined by modifiers are reduced to transferring and compressing/stretching the graphs of original functions.

Let's assume that the membership functions are triangular. Let's suppose also that in areas of definition of these functions there are points where the function has both zero values and values equal to one, as the membership functions used in practice are exactly such.

To simplify the further presentation, let us call the initial function a preimage and the result of its modification - an image. The existing pair of membership functions corresponding to the linguistic terms A and B, one of which is a modification of the other, will be called a prototype. Description of the transition, following the idea [1], we will search using a pre-known prototype, which serves as a sample of the expert's judgments on the semantic transition of a specific type from the original concept to the modified one.

Let the functions of the prototype $\mu_A(x)$ and $\mu_B(x)$ (preimage and image, respectively) be defined (obtained by the Expert Advisor) in their area of definition X using two triples of different numbers:

$$\{x_{A_i} \in X; i = 1, 3\} : \mu_A(x_{A_1}) = \mu_A(x_{A_3}) = 0, \mu_A(x_{A_2}) = 1 \tag{1}$$

$$\{x_{B_i} \in X; i = 1, 3\} : \mu_B(x_{B_1}) = \mu_B(x_{B_3}) = 0, \mu_B(x_{B_2}) = 1 \tag{2}$$

Then the transformation sought providing a prototype-defined function rebuild should be a procedure, which on the basis of the triple of numbers defining the preimage lets to find the triple of numbers defining the image.

Since the result of a modification of some function of the $\mu_C(y)$ membership corresponding to the linguistic term C and given (for example, also obtained by an expert way) in its Y-definition region is searched also among the triangular functions, the construction on the basis of the existing function a new one ($\mu_D(y)$ membership function corresponding to the modified term D) is reduced to the calculation with the help of the found transformation (modifier) of the three numbers $\{y_{D_i} \in Y; i = 1,3\}$, based on the triple $\{y_{C_i} \in Y; i = 1,3\}$ that defines the function $\mu_C(y)$.

The behavior of the membership function reflects the dynamics of changes in expert judgments with respect to the degree to which the values from the area of the function definition belong to some concept as we approach the areas of "clarity", where the function is equal to one or zero, and also as we retreat from these areas. Therefore, it is expedient to demand to keep in the course of modification the ratio of dynamics of change of the initial and transformed function on corresponding sites determined by the available prototype.

Based on the stated and interpreting modification mainly as a transition from one dynamics of change of expert judgments (speed of change of the membership function) to another, it is possible to draw a conclusion that the transformation should be carried out separately for the areas of piecewise linearity of functions. In this case, the prototype will determine some set of transformations, which will be the desired modifier. These transformations will have to be applied when constructing the modified terms also in the aggregate to the original functions of the same structure (each transformation will be applied to the corresponding fragment of the function).

It is obvious that for piecewise linear functions the rate of their change is piecewise constant at the corresponding areas of linearity, with the area of values of any considered functions of the triangular species membership at these areas coinciding with the segment [0; 1]. Consequently, for two such membership functions the modification preserving the ratio of their change rates is determined only by the change in the lengths of the segments of the abscissa axis on which the fragments of the graphs of functions corresponding to the segments of their linearity in pairs rest. The coefficients of changes in the lengths of the sections can be easily found using a prototype.

The algorithm for solving the problem consists of the following steps:

1. On the basis of the existing pair of functions (preimage of prototype and image of prototype) $\mu_A(x)$ and $\mu_B(x)$ defined by the three numbers (1) and (2), must be found a modifier which is a pair of linear transformations:

$$f_j(x) = a_j x + b_j; j = 1,2 \tag{3}$$

each of which provides conversion of one of the side fragments of the $\mu_A(x)$ preimage into the corresponding fragment of the $\mu_B(x)$ image. The a_j coefficients are calculated as length ratios of pairs of corresponding segments of the abscissa axis:

$$a_1 = \frac{x_{B_2} - x_{B_1}}{x_{A_2} - x_{A_1}}; a_2 = \frac{x_{B_3} - x_{B_2}}{x_{A_3} - x_{A_2}} \tag{4}$$

and b_j values are determined after substitution in (3) of known and obtained values.

2. Subsequent modification of some $\mu_C(y)$ function (preimage) defined by three numbers:

$$\{y_{C_i} \in Y; i = 1, 3\} : \mu_C(y_{C_1}) = \mu_C(y_{C_3}) = 0, \mu_C(y_{C_2}) = 1 \tag{5}$$

is performed by applying the found transformations to the corresponding parts of the area of definition of this function. At that, the end points of the triplet defining the modified function $\mu_D(y)$ (image) are unambiguously located:

$$y_{D_1} = f_1(y_{C_1}) = a_1 y_{C_1} + b_1; y_{D_3} = f_2(y_{C_3}) = a_2 y_{C_3} + b_2 \tag{6}$$

for the middle point y_{C_2}, the found transformations generally allow to obtain different values:

$$y'_{D_2} = f_1(y_{C_2}) = a_1 y_{C_2} + b_1; y''_{D_2} = f_2(y_{C_2}) = a_2 y_{C_2} + b_2 \tag{7}$$

3. Determining the position of the y_{D_2} midpoint on the found segment $[y_{D_1}; y_{D_3}]$ is carried out by selecting the y_{D_2} value from the two obtained values (7), which can be implemented in various heuristic ways.
The situation is trivial when one of the found values (7) does not fall on the interval $[y_{D_1}; y_{D_3}]$. Experiments show that such situation can be observed at formal application of the found transformations for modification of functions-preimages which form essentially differs from form of the preimage of prototype. This usually means that the choice of modification variant for work with functions of the certain form is not too good. In this case the value (7) located inside $[y_{D_1}; y_{D_3}]$ is chosen as the value y_{D_2}.
When both values (7) are located within $[y_{D_1}; y_{D_3}]$, one option may be to select a value obtained by transforming f_j, which performs a more substantial transformation of the corresponding function fragments. This preference reflects the subjective assumption of the researcher that the chosen transformation better reflects the expert difference between the original and modified terms, and therefore closer to reality. In practice this option consistently demonstrates acceptable good results.
If the researcher subjectively seeks to smooth the difference between the original and modified concept as the value of y_{D_2}, on the contrary, can be chosen one was obtained by transforming f_j that performs a smaller transformation. The neutral position can find expression in the definition of y_{D_2} value as the arithmetic mean of y'_{D_2} and y''_{D_2}.
The found triple of numbers:

$$\{y_{D_i} \in Y; i = 1, 3\} : \mu_D(y_{D_1}) = \mu_D(y_{D_3}) = 0, \mu_D(x_{D_2}) = 1 \tag{8}$$

defines a modified $\mu_D(y)$ membership function.

4 Conclusion

Verification of the practical applicability of the outlined methodology shows that in the majority of its application cases in the absence of clearly unsatisfactory results, expertly acceptable results are obtained. Research of possibility of the considered approach application to modification of membership functions of S-shaped and trapezoidal kind, and also generalization of its ideas on a case of piecewise smooth membership functions of any form is perspective.

Acknowledgements. This study was financed by a grant from the Plekhanov Russian University of Economics.

References

1. Borisov, A., Krumberg, O.: Theory of possibility for decision-making. Fuzzy Sets Syst. **9**(1), 13–23 (1983)
2. Alexeyev, A., Borisov, A., Glushkov, V., Krumberg, O., Merkuryeva, G., Popov, V., Slyadz, N.: A linguistic approach to decision-making problems. Fuzzy Sets Syst. **22**(1–2), 25–41 (1987)
3. Kozielecki, J.: Psychologiczna teoria decyzji. PWN, Warszawa (1975)
4. Dutov, A., Nesterov, V., Sudakov, V., Sypalo, K.: Fuzzy preference domains and their use for selecting an electronic flight bag for flight crews. J. Comput. Syst. Sci. Int. **57**(2), 230–238 (2018)
5. Atanassov, K., Vassilev, P.: Intuitionistic fuzzy sets and other fuzzy sets extensions representable by them. J. Intell. Fuzzy Syst. **38**(1), 525–530 (2020)
6. Bělohlávek, R., Dauben, J., Klir, G.: Fuzzy Logic and Mathematics: A Historical Perspective. Oxford University Press, Oxford (2017)
7. Bentkowska, U.: Fuzzy sets and their extensions. In: Interval-Valued Methods in Classifications and Decisions. Studies in Fuzziness and Soft Computing, vol. 378. Springer, Cham (2020)
8. Bustince, H., Barrenechea, E., Pagola, M., Fernandez, J., Xu, Z., Bedregal, B., Montero, J., Hagras, H., Herrera, F., De Baets, B.: A historical account of types of fuzzy sets and their relationships. IEEE Trans. Fuzzy Syst. **24**(1), 179–194 (2016)
9. Bentkowska, U., Bustince, H., Jurio, A., Pagola, M., Pękala, B.: Decision making with aninterval-valued fuzzy preference relation and admissible orders. Appl. Soft Comput. **35**, 792–801 (2015)
10. Taneja, S.: A new approach for data classification using fuzzy logic. In: 6th International Conference – Cloud System and Big Data Engineering, Noida, India (2016)
11. Pękala, B.: Uncertainty Data in Interval-Valued Fuzzy Set Theory. Properties, Algorithms and Applications. Studies in fuzziness and soft computing. Springer, Cham (2019)
12. Kreinovich, V., Kosheleva, O., Shahbazova, S.: Why triangular and trapezoid membership functions: a simple explanation. In: Recent Developments in Fuzzy Logic and Fuzzy Sets. Studies in Fuzziness and Soft Computing, vol. 391. Springer, Cham (2020)

A Percentil Gravitational Search Algorithm an Aplication to the Set Covering Problem

Leonardo Pavez[1,2](✉), Francisco Altimiras[3], and Gabriel Villavicencio[4]

[1] Instituto de Ciencias Naturales, Universidad de las Américas, Av. Manuel Montt 948, Providencia, 7500972 Santiago, Chile
lpavez@udla.cl
[2] Departamento de Ciencias Químicas y Biológicas, Universidad Bernardo O'Higgins, General Gana 1702, 8370854 Santiago, Chile
[3] Facultad de Ingeniería y Negocios, Universidad de las Américas, Santiago, Chile
fjaltimiras@gmail.com
[4] Escuela de ingeniería en construcción, Pontificia Universidad Católica de Valparaíso, Valpara íso, Chile
gabriel.villavicencio@pucv.cl

Abstract. The study and understanding of algorithms that solve combinatorial problems based on swarm intelligence continuous metaheuristics, is an area of interest at the level of basic and applied science. This is due to the fact that many of the problems addressed at industrial level are of a combinatorial type and a subset no less than these are of the NP-hard type. In this article, a mechanism of binarization of continuous metaheuristics that uses the concept of the percentile is proposed. This percentile concept is applied to the An Lion optimization algorithm, solving the set covering problem (SCP). Experiments were designed to demonstrate the importance of the percentile concept in the binarization process. Subsequently, the efficiency of the algorithm is verified through reference instances. The results indicate that the binary percentile gravitational search algorithm (BPGSA) obtains adequate results when evaluated with a combinatorial problem such as the SCP.

1 Introduction

In recent years, there is a research line of algorithms inspired by natural phenomena focused on solving optimization problems. As examples of these algorithms we have Cuckoo Search [1], Black Hole [2], Bat Algorithm [3] and Sine Cosine Algorithm [4] among others. Due to the type of phenomena in which these algorithms are inspired, a large part of these algorithms operate naturally in continuous spaces. On the other hand, the combinatorial problems have a great relevance at the scientific and industrial level. We find them in different areas, such as Civil Engineering [5,6], sustainability [7], Operational Research [8–11], Big Data [12], resource allocation [13–15], programming problems [16,17],

R. Silhavy et al. (Eds.): CoMeSySo 2020, AISC 1295, pp. 663–673, 2020.
https://doi.org/10.1007/978-3-030-63319-6_62

routing problems [11,18–20] among others. If we want to apply continuous algorithms to combinatorial problems, these algorithms must adapt. In the process of adaptation, the mechanisms of exploration and exploitation of the algorithm can be altered, having consequences in the efficiency of the algorithm.

Several binarization techniques have been developed to address this situation. In a literature search, the main binarization methods used correspond to transfer functions, angle modulation and quantum approach. In this article we present a binarization method that uses the percentile concept to group the solutions and then perform the binarization process. To verify the efficiency of our method, we used the BPGSA. This algorithm was proposed in [3].

To check our binary percentile gravitational search algorithm (BPGSA), we use the well-known set covering problem. Experiments were developed using a random operator to validate the contribution of the percentile technique in the binarization process of the sine-cosine algorithm. In addition, to verify our results, the JPSO algorithm developed by [21] and MDBBH developed in [16] were chosen. The results show that BPGSA algorithm obtain competitive results.

2 Set Covering Problem

The set covering problem corresponds to a classical problem in combinatorics, and complexity theory. The problem aims to find subsets of a set. Given a set and its elements, we want to find subsets that completely cover the set at a minimum cost. The SCP is one of the oldest and most studied optimization problems. It is well-known to be NP-hard [22].

SCP is an active problem because medium and large instances often become intractable and cannot be solved any more using exact algorithms. Additionally, SCP due to its large number of instances, is used to verify the behavior of proposed new algorithms. In recent years, SCP has been approached by various continuous Swarm intelligence metaheuristics and using different methods to binarize metaheuristics. In [23], they used the teaching-learning-based optimisation with a specific binarization scheme to solve medium and large size instances. An SCP in fault diagnosis application was developed by [24]. In this article, the Gravitational Search algorithm was used and transfer functions were applied to perform the binarization. A rail scheduling problem application using SCP was developed in [16]. In [21] a Jumping PSO was used to solve SCP and in [13] CS metaheuristics were binarized using a percentile algorithm.

SCP has many practical applications in engineer, e.g., vehicle routing, facility location, railway, and airline crew scheduling [25–28] problems.

The SCP can be formally defined as follows. Let $A = (a_{ij})$, be a $n \times m$ zero-one matrix, where a column j cover a row i if $a_{ij} = 1$, besides a column j is associated with a non-negative real cost c_j. Let $I = \{1, ..., n\}$ and $J = \{1, ...m\}$, be the row and column set of A, respectively. The SCP consists in searching a minimum cost subset $S \subset J$ for which every row $i \in I$ is covered by at least one column $j \in J$, i.e,:

$$\text{Minimize } f(x) = \sum_{j=1}^{m} c_j x_j \tag{1}$$

$$\text{Subject to } \sum_{j=1}^{m} a_{ij} x_j \geq 1, \forall i \in I, \text{ and } x_j \in \{0,1\}, \ \forall j \in J \tag{2}$$

where $x_j = 1$ if $j \in S$, $x_j = 0$ otherwise.

3 Binary Percentile Gravitational Search Algorithm

The application of statistics and machine learning techniques, are used in many areas such as transports, smart cities, agriculture and computational intelligence, examples of applications are found in [29–32, 32, 33]. In this article, we explore the percentile concept applied to binarization of continuous metaheuristics. As an algorithm to binarize, gravitational search algorithm is used and as problem to solve, SCP. The Proposed BPGSA has two modules. The first module corresponds to the initialization of the feasible solutions, and is detailed in the Sect. 3.1. Once the initialization of the particles is performed, it is consulted if the maximum if iterations is satisfied. Subsequently if the criterion is not satisfied, The BPGSA is then run in conjunction with the percentile operator, this is detailed in Sect. 3.2). Once the transitions of the different solutions are made, we compare the resulting solutions with the best solution previously obtained. In the event that a superior solution is found, this replaces the previous one. The general algorithm scheme is detailed in Fig. 1.

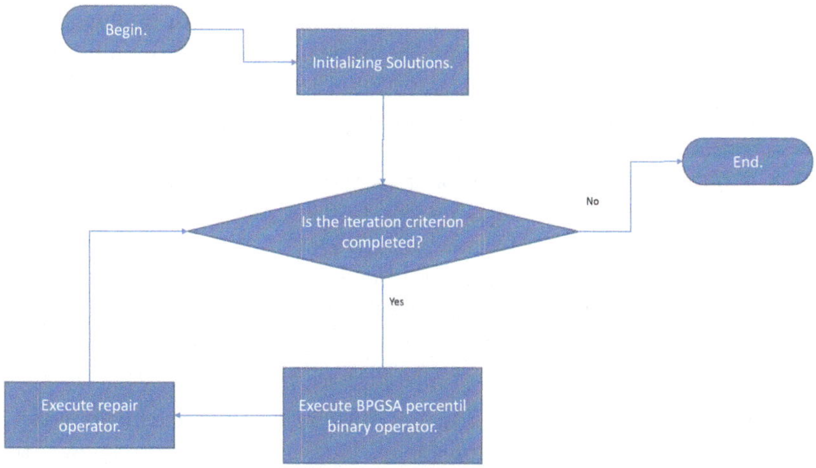

Fig. 1. Flowchart of the BPGSA algorithm.

3.1 Initialization

For initialization of a new solution, a column is randomly chosen. It is then queried whether the current solution covers all rows. In the case of the solution does not meet the coverage condition, the heuristic operator is called (Sect. 3.3). This operator aims to select a new column. This heuristic operation is iterated until all rows are covered. Once the coverage condition is met, the solution is optimized. The optimization consists of eliminating columns where all their rows are covered by more than one column. The detail of the procedure is shown in Algorithm 1.

Algorithm 1. Initialization Operator

Function Initialization()
Input
Output Initialized solution S_{out}
$S \leftarrow$ SRandomColumn()
while All row are not covered **do**
  S.append(Heuristic(S))
end while
$S \leftarrow$ dRepeatedItem(S)
$S_{out} \leftarrow S$
return S_{out}

3.2 Percentile Binary Operator

Since our gravitational search algorithm is continuous swarm intelligence metaheuristic, it works in an iterative way by updating the position and velocity of the particles in each iteration. As gravitational search is continuous, the update is done in $\mathbb{R}^n$. In Eq. 3, the position update is written in a general way. The $x(t+1)$ variable represents the x position of the particle at time $t+1$. This position is obtained from the position x at time t plus a Δ function calculated at time $t+1$. The function Δ is proper to each metaheuristic and produces values in $\mathbb{R}^n$. For example in Cuckoo Search $\Delta(x) = \alpha \oplus Levy(\lambda)(x)$, in Black Hole $\Delta(x) = \text{rand} \times (x_{bh}(t) - x(t))$ and in the Firefly, gravitational search and PSO algorithms Δ can be written in simplified form as $\Delta(x) = v(x)$.

$$x(t+1) = x(t) + \Delta_{t+1}(x(t)) \tag{3}$$

In the percentile binary operator, we considering the movements generated by the algorithm in each dimension for all particles. $\Delta^i(x)$ corresponds to the magnitude of the displacement $\Delta(x)$ in the i-th position for the particle x. Subsequently these displacement are grouped using $\Delta^i(x)$, the magnitude of the displacement. This grouping is done using the percentile list. In our case the percentile list used the values $\{20, 40, 60, 80, 100\}$.

The percentile operator has as entry the parameters percentile list (pList) and the list of values (vList). Given an iteration, the list of values corresponds to the magnitude $\Delta^i(x)$ of the displacements of the particles in each dimension. As a first step the operator uses the vList and obtains the values of the percentiles given in the pList. Later, each value in the vList is assigned the group of the smallest percentile to which the value belongs. Finally, the list of the percentile to which each value belongs is returned (pGroupValue). The algorithm is shown in Algorithm 2.

A transition probability through the function P_{tr} is assigned to each element of the vList. This assignment is done using the percentile group assigned to each value. For the case of this study, we particularly use the Step function given in Eq. 4.

$$P_{tr}(x^i) = \begin{cases} 0.1, & \text{if } x^i \in \text{group } \{0,1\} \\ 0.5, & \text{if } x^i \in \text{group } \{2,3,4\} \end{cases} \tag{4}$$

Afterwards the transition of each particle is performed. In the case of gravitational search the Eq. 5 is used to perform the transition, where $\hat{x}^i$ is the complement of x^i. Finally, each solution is repaired using the heuristic operator.

$$x^i(t+1) := \begin{cases} \hat{x}^i(t), \text{ if } rand < P_{tg}(x^i) \\ x^i(t), \qquad \text{otherwise} \end{cases} \tag{5}$$

Algorithm 2. percentile binary operator

1: **Function** percentileBinary(vList, pList)
2: **Input** vList, pList
3: **Output** pGroupValue
4: percentileValue = getPValue(vList, pList)
5: **for each** value in vList **do**
6:     pGroupValue = getPGroupValue(pValue,vList)
7: **end for**
8: **return** pGroupValue

3.3 Heuristic Operator

The goal of the Heuristic operator is to select a new column for cases where a solution needs to be built or repaired. As input variables, the heuristic operator considers the incomplete solution S_{in} which must be completed. The operator obtains the columns that belong to S_in, then obtains the rows R that are not covered by the solution to S_in. With the set of rows not covered and using Eq. 6 we obtain in line 4 the best 5 rows to be covered. With this list of rows ($lRows$) on line 5 we obtain the list of the best columns according to the heuristic indicated in Eq. 7. Finally randomly in line 6 we obtain the column to incorporate.

$$WeightRow(i) = \frac{1}{L_i}. \tag{6}$$

where L_i is the sum of all ones in row i

$$WeightColumn(j) = \frac{c_j}{|R \cap M_j|}. \quad (7)$$

where M_j is the set of rows covered by Col j

Algorithm 3. Heuristic operator

```
1: Function Heuristic(S_in)
2: Input Input solution S_in
3: Output The new column colOut
4: lRows ← getBestRows(S_in, N=10)
5: listcolumsnOut ← getBestCols(LRows, M=5)
6: colOut ← getCol(lcolsOut)
7: return colOut
```

4 Results

In this section we detail the behavior of BPGSA when it is applied to SCP. The contribution of the binary percentile operator was studied when solving the different SCP instances. Additionally, BPGSA was compared with other algorithms that have recently resolved SCP. To solve the different SCP instances, a PC with Windows 10, core i5 processor and 16 GB in RAM was used. The program was coded in Python 3.5. For the statistical analysis, the non-parametric Wilcoxon signed-rank test was used in addition to violin charts. The analysis of violin charts is performed by comparing the dispersion, median and the interquartile range of the distributions.

4.1 Insight of BPGSA Algorithm

In this module we developed experiments that allowed us to study the contribution of the binary percentile operator with respect to the quality of the solutions obtained and the iterations used when solving SCP instances. To perform this evaluation, we used the instances of Balas and Carrera. In the comparison, a random operator was designed which replaces our binary percentile operator. This random operator instead of executing transitions assigning transition probabilities per group, uses a fixed value of 0.5. To compare the distributions of the results of the different experiments we use violin Chart. The horizontal axis X corresponds to the problems, while Y axis uses the measure % - Gap defined in Eq. 8

$$\% - Gap = 100\frac{BestKnown - SolutionValue}{BestKnown} \quad (8)$$

Furthermore, a non-parametric test, Wilcoxon signed-rank test is carried out to determine if the results of BPGSA with respect to the random algorithm have significant difference or not. The parameter settings and browser ranges are shown in Table 1.

Table 1. Setting of parameters for Binary percentile gravitational search Algorithm.

Parameters	Description	Value	Range
N	Number of solutions	20	[20, 25, 30]
G	Number of percentiles	5	[4,5,6]
Iteration Number	Maximum iterations	800	[700,800,900]

BPGSA corresponds to our standard algorithm. *random.0.5* is the random variant. The results are shown in Table 2. When we compared the Best Values between BPGSA and *rnd.0.5* which are shown in Table 2. In the Average comparison, BPGSA outperforms *rnd.0.5* in all problems. When we evaluate the behaviour of the algorithms through the Wilcoxon test, this test indicates that there is a significant difference between the two algorithms.

Table 2. Balas and Carrera instances.

Instance	row	col	Density	Best known	BPGSA (avg)	std	(Secs)	*rnd.0.5*	std
AA03	106	8661	4.05%	33155	33576.1	128.4	209	33693.5	246.2
AA04	106	8002	4.05%	34573	34869.2	67.8	200	34931	234.8
AA05	105	7435	4.05%	31623	31887.3	69.1	203	31889.2	111.3
AA06	105	6951	4.11%	37464	37659.5	61.4	201	37651.3	101.4
AA11	271	4413	2.53%	35384	35769.1	138.4	190	35901.2	198.4
AA12	272	4208	2.52%	30809	31457.4	156.7	210	31631.3	215.8
AA13	265	4025	2.60%	33211	33992.1	137.1	190	34201.4	233.8
AA14	266	3868	2.50%	33219	33867.2	124.5	145	33967.3	126.8
AA15	267	3701	2.58%	34409	34826.1	122.1	142	34846.8	166.5
AA16	265	3558	2.63%	32752	33329.7	137.6	188	33601.3	183.5
AA17	264	3425	2.61%	31612	32001.8	101.6	164	32386.2	153.6
AA18	271	3314	2.55%	36782	37297.1	96.4	134	37417.6	149.8
AA19	263	3202	2.63%	32317	33096.4	87.2	147	33109.1	138.5
AA20	269	3095	2.58%	34912	35843.1	99.1	142	35963.1	107.2
BUS1	454	2241	1.88%	27947	28481.9	104.2	190	28756.1	235.9
BUS2	681	9524	0.51%	67760	68549.4	94.4	175	68678.9	105.6
Average				35495.56	36031.5	107.9	176.8	36164.08	169.3
p-value								1.72e−04	

4.2 BPGSA Compared with JPSO and MDBBH

In this section, we detail the comparisons made with the objective of evaluating the performance of our BPGSA algorithm. For the evaluation, the larger problems of the OR-library were chosen. To develop the comparison, two algorithms were selected. The first one corresponds to a discretization Particle Swarm Optimization (PSO) technique called Jumping PSO (JPSO), [21]. The second one, is a binarization of the Black Hole technique called Multi Dynamic Binary Black Hole (MDBBH) Algorithm, [16]. JPSO uses a discrete PSO based on the frog jump. JPSO works without the concept of velocity, replacing this one by a component of random jump which allows to perform the movement in the discrete search space. On the other hand, MDBBH uses a binarization mechanism specific to the Black Hole algorithm. This is based on the concept of closeness to the Black Hole (BH). BH corresponds to the solution that has obtained the best value. When a solution is close to BH, the transitions are less likely to be performed. In the case of being away from the BH, the probability of transition is greater.

For instances E and F which are of medium size, the results of the MDBBH and BPGSA algorithms are similar when comparing their Best Value. Only in instance E.2 BPGSA was superior to MDBBH. When analyzing the average, JPSO obtained better results, however, the values of BPGSA were close, being MDBBH the one that obtained the worse performance. For the case of problems G and H, BPGSA was superior in problems G.1, G.2 and G.3 and MDBBH in

Table 3. OR-Library benchmarks.

Instance	row	col	Density	Best Known	JPSO Avg	MDBBH Best	Avg	BPGSA Best	Avg	Time(s)
E.1	500	5000	10%	29	29	29	29	29	31.39	20.3
E.2	500	5000	10%	30	30	31	31.6	30	34.44	21.7
E.3	500	5000	10%	27	27	27	27.4	27	29.02	29.3
E.4	500	5000	10%	28	28	28	29.1	28	31.19	21.1
E.5	500	5000	10%	28	28	28	28	28	30.88	24.4
F.1	500	5000	20%	14	14	14	14.1	14	15.16	28.3
F.2	500	5000	20%	15	15	15	15.3	15	18.40	28.3
F.3	500	5000	20%	14	14	14	14.8	14	16.23	27.2
F.4	500	5000	20%	14	14	14	14.9	14	16.02	27.9
F.5	500	5000	20%	13	13	14	14.1	14	17.21	32.7
G.1	1000	10000	2%	176	176	177	178.5	177	179.71	210.5
G.2	1000	10000	2%	154	155	157	160.6	156	160.91	222.9
G.3	1000	10000	2%	166	167.2	168	170.4	168	171.83	241.9
G.4	1000	10000	2%	168	168.2	169	170.9	170	173.26	238.7
G.5	1000	10000	2%	168	168	168	169.8	168	172.63	268.7
H.1	1000	10000	5%	63	64	64	64.9	65	67.74	206.8
H.2	1000	10000	5%	63	63	64	64	65	66.80	195.1
H.3	1000	10000	5%	59	59.2	59	60	60	61.55	197
H.4	1000	10000	5%	58	58.3	59	60.4	59	62.91	190.2
H.5	1000	10000	5%	55	55	55	56.4	56	58.31	179.5
Average				67.1	67.295	67.7	68.71	67.85	70.78	120.625

problem H.3 when comparing their Best Value. In the case of averages, JPSO was superior in all cases. However, BPGSA obtained results quite close, leaving MDBBH behind. In this point we must emphasize that the percentile technique used in BPGSA allows binarizing any continuous swarm intelligence algorithm unlike JPSO that is specific for PSO.

5 Conclusions

In this work, the BPGSA algorithm based on the percentile concept was proposed to perform binarization of gravitational search metaheuristics. We must emphasize that this percentile concept can be applied in the binarization of any continuous metaheuristics of the swarm intelligence type. To evaluate the performance of Our percentile concept the set covering problem was used together with the gravitational search algorithm. Experiments were designed to evaluate the contribution of the binary operator percentile of the algorithm. It was found that the operator contributes significantly to improve the accuracy and quality of the solutions, all these visualized through violin graphics. Finally, compared to the best metaheuristic algorithm that has solved SCP, our algorithm had a lower yield of 1.93%, which is not a big difference considering that JPSO uses a particular adaptation mechanism for PSO and the percentile concept It can be easily adapted to any technique. As future works, we believe that the autonomous search for efficient configuration is a little-explored area and it could help to improve the performance of the binarizations both in the quality of the solutions and in the times of convergence. Additionally, it is interesting to see how this percentile technique works with other metaheuristics and other NP-hard problems. Finally, we believe that using machine learning we can generate indicators or obtain correlations that allow us to understand how the percentile algorithm translates the exploration and exploitation properties from the continuous space to the binary space.

References

1. Yang, X.-S., Deb, S.: Cuckoo search via lévy flights. In: World Congress on Nature & Biologically Inspired Computing, 2009. NaBIC 2009, pp. 210–214. IEEE (2009)
2. Hatamlou, A.: Black hole: a new heuristic optimization approach for data clustering. Inf. Sci. **222**, 175–184 (2013)
3. Yang, X.-S.: A new metaheuristic bat-inspired algorithm. In: Nature Inspired Cooperative Strategies for Optimization (NICSO 2010), pp. 65–74 (2010)
4. Mirjalili, S.: SCA: a sine cosine algorithm for solving optimization problems. Knowl. Based Syst. **96**, 120–133 (2016)
5. García, J., Martí, J.V., Yepes, V.: The buttressed walls problem: an application of a hybrid clustering particle swarm optimization algorithm. Mathematics **8**(6), 862 (2020)
6. García, J., Yepes, V., Martí, J.V.: A hybrid k-means cuckoo search algorithm applied to the counterfort retaining walls problem. Mathematics **8**(4), 555 (2020)

7. Yepes, V., Martí, J.V., García, J.: Black hole algorithm for sustainable design of counterfort retaining walls. Sustainability **12**(7), 2767 (2020)
8. García, J., Moraga, P., Valenzuela, M., Pinto, H.: A db-scan hybrid algorithm: an application to the multidimensional knapsack problem. Mathematics **8**(4), 507 (2020)
9. Crawford, B., Soto, R., Astorga, G., García, J.: Constructive metaheuristics for the set covering problem. In: International Conference on Bioinspired Methods and Their Applications, pp. 88–99. Springer (2018)
10. García, J., Peña, A.: Robust optimization: Concepts and applications. In: Nature-Inspired Methods for Stochastic. Robust and Dynamic Optimization. IntechOpen (2018)
11. García, J., Moraga, P., Valenzuela, M., Crawford, B., Soto, R., Pinto, H., Peña, A., Altimiras, F., Astorga, G.: A db-scan binarization algorithmapplied to matrix covering problems. Comput. Intell. Neurosci. **2019**, 16 (2019)
12. García, J., Altimiras, F., Peña, A., Astorga, G., Peredo, O.: Abinary cuckoo search big data algorithm applied to large-scale crewscheduling problems. Complexity **2018**, 15 (2018)
13. García, J., Crawford, B., Soto, R., Astorga, G.: A percentile transition ranking algorithm applied to knapsack problem. In: Proceedings of the Computational Methods in Systems and Software, pp. 126–138. Springer (2017)
14. Astorga, G., Crawford, B., Soto, R., Monfroy, E., García, J., Cortes, E.: A meta-optimization approach to solve the set covering problem. Ingeniería **23**(3), 274–288 (2018)
15. García, J., Lalla-Ruiz, E., Voß, S., Droguett, E.L.: Enhancing a machine learning binarization framework by perturbation operators: analysis on the multidimensional knapsack problem. Int. J. Mach. Learn. Cybern. **11**, 1–20 (2020)
16. García, J., Crawford, B., Soto, R., García, P.: A multi dynamic binary black hole algorithm applied to set covering problem. In: International Conference on Harmony Search Algorithm, pp. 42–51. Springer (2017)
17. García, J., Crawford, B., Soto, R., Astorga, G.: A percentile transition ranking algorithm applied to binarization of continuous swarm intelligence metaheuristics. In: International Conference on Soft Computing and Data Mining, pp. 3–13. Springer (2018)
18. Crawford, B., Soto, R., Astorga, G., García, J., Castro, C., Paredes, F.: Putting continuous metaheuristics to work in binary search spaces. Complexity **2017**, 19 (2017)
19. García, J., Crawford, B., Soto, R., Astorga, G.: A clustering algorithm applied to the binarization of swarm intelligence continuous metaheuristics. Swarm Evol. Comput. **44**, 646–664 (2019)
20. Garcia, J., Măntoiu, M.: Localization results for zero order pseudodifferential operators. J. Pseudo Diff. Oper. Appl. **5**(2), 255–276 (2014)
21. Balaji, S., Revathi, N.: A new approach for solving set covering problem using jumping particle swarm optimization method. Nat. Comput. **15**(3), 503–517 (2016)
22. Gary, M.R., Johnson, D.S.: Computers and intractability. In: A Guide to the Theory of NP-Completeness (1979)
23. Lu, Y., Vasko, F.J.: An or practitioner's solution approach for the set covering problem. Int. J. Appl. Metaheuristic Comput. (IJAMC) **6**(4), 1–13 (2015)
24. Li, Y., Cai, Z.: Gravity-based heuristic for set covering problems and its application in fault diagnosis. J. Syst. Eng. Electron. **23**(3), 391–398 (2012)

25. Crawford, B., Soto, R., Monfroy, E., Astorga, G., García, J., Cortes, E.: A meta-optimization approach for covering problems in facility location. In: Workshop on Engineering Applications, pp. 565–578. Springer (2017)
26. Kasirzadeh, A., Saddoune, M., Soumis, F.: Airline crew scheduling: models, algorithms, and data sets. EURO J. Transp. Logistics **6**(2), 111–137 (2017)
27. Horváth, M., Kis, T.: Computing strong lower and upper bounds for the integrated multiple-depot vehicle and crew scheduling problem with branch-and-price. CEJOR **27**(1), 1–29 (2017)
28. Stojković, M.: The operational flight and multi-crew scheduling problem. Yugoslav J. Oper. Res. **15**(1), 25–48 (2016)
29. García, J., Crawford, B., Soto, R., Carlos, C., Paredes, F.: A k-means binarization framework applied to multidimensional knapsack problem. Appl. Intell. **48**(2), 1–24 (2017)
30. García, J., Pope, C., Altimiras, F.: A distributed k-means segmentation algorithm applied to lobesia botrana recognition. Complexity **2017**, 14 (2017)
31. Graells-Garrido, E., García, J.: Visual exploration of urban dynamics using mobile data. In: International Conference on Ubiquitous Computing and Ambient Intelligence, pp. 480–491. Springer (2015)
32. Graells-Garrido, E., Peredo, O., García, J.: Sensing urban patterns with antenna mappings: the case of Santiago, Chile. Sensors **16**(7), 1098 (2016)
33. Peredo, O.F., García, J.A., Stuven, R., Ortiz, J.M.: Urban dynamic estimation using mobile phone logs and locally varying anisotropy. In: Geostatistics Valencia 2016, pp. 949–964. Springer (2017)

Principal Component Analysis Enhanced Multi-layer Perceptron for Multi-economic Indicators in Southern African Economic Prediction

Moses Olaifa(✉) and Tranos Zuva

Department of ICT, Faculty of Applied and Computer Science, Vaal University of Technology, Andries Potgieter Blvd, Vanderbijlpark 1900, South Africa
newmosesolaifa@gmail.com, tranosz@vut.ac.za

Abstract. Economic growth may be determined based on different economic parameters. The main purpose of this research work is to predict economic growth based on several economic indicators. In this study, an enhanced Multi-layer Perceptron (MLP) model is built and the prediction performances are evaluated and compared with existing models. Using the Principal Component Analysis technique, major contributing indicators where identified and passed as input to tne MLP input layer. Results show a better performance by the PCA enhanced MLP with consistent accuracy as time-step increases. Furthermore, drop in GDP over the first 3 time-step and improvement in the last 3 time-step can be observed.

Keywords: Economic indicators · Multilayer perceptron · Economic prediction · Economic growth

1 Introduction

Prediction of events into the future has become an approach adopted by different groups and organizations to get an idea of the possible occurrences. When such predicted information is available, nations are able to evaluate their policies and make necessary preparations to either avert unfavorable or facilitate desired events. In other cases, concerned cadres in the industries or governments are able to make informed decisions in the respective fields. Like most other areas, predicting economic growth proves useful in informing better and timely economic policies, as delayed implementation of the right policies may impact the economy negatively.

In order to establish the right premises for economy related policy creation, the UN recommended in her 1992 Rio de Janeiro Conference that nations should identify indicators capable of fostering sustainable economic developments [1]. Over the years, countries measure their economy growth and development using

R. Silhavy et al. (Eds.): CoMeSySo 2020, AISC 1295, pp. 674–684, 2020.
https://doi.org/10.1007/978-3-030-63319-6_63

different methods including gross national product (GNP), gross domestic product (GDP), production and export rate, consumption and import rate. Understanding economic growth is associated with the challenge task of determining how different economic indicators might influence economic values [2,3], such as the Gross Domestic Product (GDP). To address this challenge, various methods and tools are required to make economic and development predictions. A great deal of research has been done on various economic indicators using different metrics and tools for predicting possible direction of business and economic growth.

In Africa, South Africa is considered to have maintained a buoyant economy which is evaluated based on the real GDP [4]. The real GDP gives the sum of goods and services produced over a time period without considering the impact of inflation. Even though a number of studies have applied different tools for economic forecast, few studies have considered suitable models for Africa economy and important indicators that may influence accurate prediction of the economic growth for well informed policy making in African nations. Most of the previous studies have largely focused on developed countries with more reliable macroeconomic aggregates. Since economic growth can be a complex macroeconomic phenomenon [5], there is a great need for suitable economic growth prediction models to help enhance the efficiency of economic systems in developing countries. This will answer questions revolving around future economic trends in African nations.

Moreover, most related studies on Africa economic growth are based on the effect of specific economic parameters rather than a holistic economic forecast using a spread of influential economic indicators. The impact of foreign capital inflow on economic growth is presented in [6]. Results from the study shows a causal relationship between foreign capital flow and economic growth. A causality analysis using ELG hypothesis in [7] reveals a unidirectional causality from Foreign Direct Investment to non-oil exports and economic growth. In [8] this study examines the causal and long-run relationship between the performance of the stock market and Nigeria economic growth. Using Exponential Generalized Autoregressive Conditional Heteroskedasticity and Vector Auto-regression (VAR) technique respectively, [9,10] determined the impact of interest rate and exchange rate on Nigeria economic growth over the period of 1970 to 2010. Rangan et al. [11] adopted the use of Factor Models (FMs) to predict South Africa economic growth rate based on per capital growth rate, inflation and nominal short term interest rate. A major challenge with these studies is that several other indicators that contributes to the economic growth are not considered. This study adopts machine learning techniques for economic growth forecast in sub-Saharan Africa using South Africa as a case study. Multiple economic indicators from 1981 to 2019 are subjected machine learning models. This is a preliminary study of an ongoing project on Africa economic growth. The remainder of this article is organized into; Further background study; Model background description; data description; performance evaluation and conclusion.

2 Further Background Study

Over the past few decades, different approaches have been proposed such as time series techniques [12–14], According to the work in [13], a method to achieve stationary time series that reduces the difference between elimination of short-run fluctuations and timeliness for real time GDP growth estimation is presented. The method adapted in their work was based on large-scale generalized dynamic factor model (GDFM) using information from large panel of time series. Using time-series and scenario planning, an analytical approach to economic development for the 13th Five-Year plan in Schenzhen is presented in [14]. Their work adopts the Box-Jenkins method – ARIMA, to identify models, determine model's order and fit.

The application of the widely adopted linear regression in time series prediction has generated numerous research outputs in a wide range of fields [15–19]. However, a major downside to the use of linear regression is the assumption that the economic indices data set under consideration are linear.

The models in this study are evaluated for accuracy in long-run predictions and more influential economic indicators. This will build general knowledge of the capacity of the different models using several economic indicators as time step increases.

3 Model Background Description

3.1 Principal Component Analysis

Principal Component Analysis [20] is one of the effective dimensionality reduction techniques used in identifying relevant data variables by using orthogonal transformation of the feature set. The dimension of the data attributes is reduced.

In this study, PCA is used to merge the economic indicators into few distinctive attributes. It enables the selection of the most sensitive and effective indicators of economic development. Through this, attributes contributing the most to changes in economic growth are projected for a more accurate prediction.

Given the data set $X[m * n]$, where m is the dimension of attributes and n represents the number of instances, then the covariance of $X[m * n]$ is given as $CV[m * m]$. The next step involves the computation of the eigenvalues for the covariance as λ_i, where $i = 1, 2, 3, ..., m$. This generates an ordered corresponding eigenvectors $\mathcal{U}$ for λ_i. Through this, a new data space $DS[m * n] = \mathcal{U} * X[m * n]$. The new data space $DS[m * n]$ encompasses the *Principal Components* (PCs). The first set of *PCs* should represent the greatest spread of the data with the lower part of the order considered as noise components. The major contributors should be in the first set of PCs and these will constitute the new economic indicators $X[m * n]$.

In this study, the purpose of applying *PCA* is to generate influential indicators by producing one or two *PCs* that contains major economic growth attributes that can be used to produce more accurate predictions. Using the PCs indicates that the prediction is not based only on the real GP values but influential indicators were also factored into the prediction.

3.2 Multi-layer Perceptron

Neural networks [21] can predict the direction of growth of an economy by learning the relationships between the indicators. The ability of the MLP to predict is derived from the multiple layered structure of the network of perceptrons. A perceptron is composed of the input, activation function and the output. Each of the input parameter is weighted and the sum of the weighted input is combined with the bias that is passed to the activation function. The activation function maps the weighted input to the output. Figure 1 depicts a simple perceptron model.

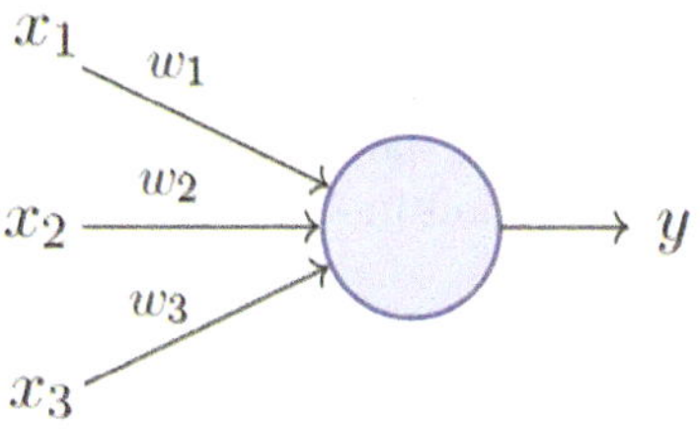

Fig. 1. Simple perceptron model (Minsky-Papert 1969)

Input data accepted into the perceptron are real numbers which in turn produces numerical output values. The sum of input values with each associated weight is combined with the bias as $\sum_{i=1}^{n} x_i \cdot w_i + b$. This is passed to an activation function that computes the input-output mapping. For a simple perceptron, the activation function will be a threshold function which generates an output given as y in Eq. 1:

$$y = f(\sum_{i=1}^{n} x_i \cdot w_i + b) \tag{1}$$

where y is the predicted output; $f(.)$ is the activation function which is the threshold function in this instance and $\sum_{i=1}^{n} x_i \cdot w_i + b$ is the input propagated to the activation function. In more complex setup with multiple layers of perceptrons, other activation functions such as the sigmoid function are used. An output y generated through a three-layered network can be computed using Eq. 2

$$y = f(b_0 + \sum_{k=1}^{m} \phi(\psi_k + \sum_{i=1}^{n} x_i \cdot w_{ik}) z_k) \tag{2}$$

where b_0 and z_k denote the output bias and weight associated with input propagated from hidden layer to output layer respectively, $\phi()$ and $f(.)$ are hidden layer transfer function and nonlinear transfer function of inputs x_i respectively, y is the predicted output. Any typical training process involves forward pass, error calculation and backpropagation.

4 Data Description

The data set used to evaluate the performance of the prediction models contains economic indicators used to determine South Africa's real GDP growth and real per Capital GDP growth rate. Sourced from the Humanitarian Data Exchange for African Development Bank Group, the data provides update on African Economic Outlook update. The data contains two dependent variables as mentioned above and thirteen independent variables including, the general government final consumption expenditure, household final consumption expenditure, gross capital formation private sector, gross capital formation public sector, exports of goods and services, imports of goods and services among others. Yearly interval of record was presented and spanned over a period of 39 years from 1981 to 2019 with 40 records.

For proper linear scaling of the data set. Input variables are normalized to [0, 1] value range using the equation in

$$x_{new} = \frac{x - x_{min}}{x_{max} - x_{min}} \tag{3}$$

where x_{new} is the new value between 0 and 1, x_{min} and x_{max} are minimum and maximum values respectively.

5 Performance Evaluation

For the purpose of prediction performance comparison, three models are constructed namely: Linear Regression, Multi-layer Perceptron and PCA-MLP. These three models are compared to determine the prediction performances on short and long term predictions. LR and MLP models are used to make predictions on the initial data, while the PCA-MLP is used on the PCA analysed data set.

Initial weight of 0.3 is used for the network with a learning rate of 0.1 and momentum of 0.1. The network contains 11 hidden layers. Four different variations of test data including 30%, 20%, 40% and 50% are examined to determine the level of accuracy of the different models. The experiments are conducted for the prediction of six (6) time steps beyond the end of the data. Predictions are compared to expected outputs to compute the Mean Absolute Error (MAE) and Root Mean Square Error (RMSE).

According to the performances of the models on the training data set, the three models show low prediction errors. Table 1 shows the performances of the three models on the training data. For time step 1, LR, MLP and PCA-MLR gives 1.2489, 0.1466 and 0.3874 MAEs (1.5754, 0.1617 and 0.4233 RMSE) respectively. Though the PCA-MLP presents the best performance, the other two models followed closely behind. At time step 5, there is only a slight difference in LR performance with lower MAE and higher MAEs for MLP and PCA-MLP respectively. That is, LR decreased in MAE from 1.2489 to 1.2047, while MLP and PCA-MLP slightly increased from 0.1466 to 0.4707 and 0.3874 and 1.2603

Table 1. Summary of prediction error on training data

	LR		MLP		PCA-MLP	
	MAE	RMSE	MAE	RMSE	MAE	RMSE
Step-1	1.2489	1.5754	0.1466	0.1617	0.3874	0.4233
Step-2	1.2511	1.6697	0.1776	0.2229	0.4103	0.5137
Step-3	1.1949	1.6641	0.2089	0.258	0.4388	0.6575
Step-4	1.1753	1.7044	0.2509	0.3359	0.8014	1.0577
Step-5	1.1812	1.6784	0.3177	0.4968	1.0902	1.57
Step-6	1.2047	1.6878	0.4707	0.6959	1.2603	1.9594

respectively. This is depicted in Figs. 2 and 3. On the training data set, the LR seem to improve in prediction accuracy over longer time steps but the MLP and PCA-MLP produced slightly less accurate prediction in longer time steps.

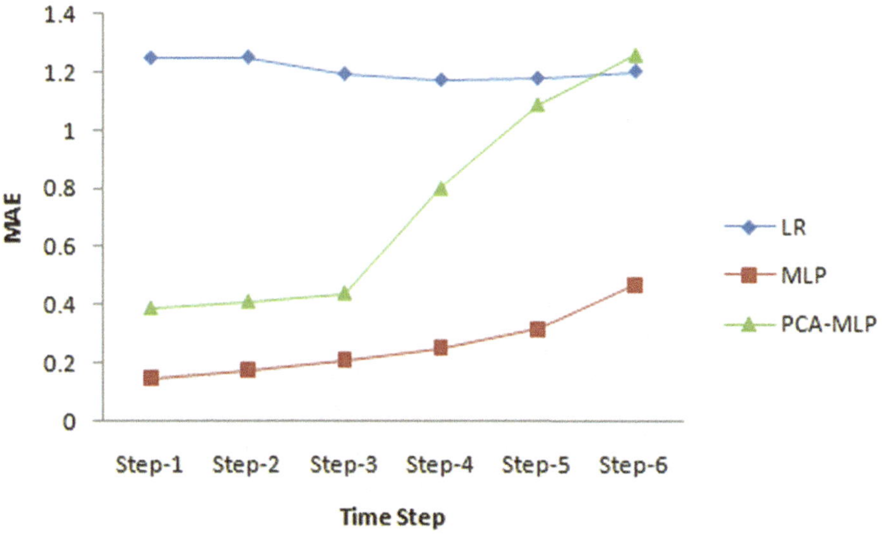

Fig. 2. Mean absolute error on training

Different percentages of 30%, 20%, 40% and 50% are used for testing the models. Results produced from the 30% is used as the base performance result for the 6 Step prediction periods. For the immediate time step, PCA-MLR produced the least error, followed by LR and MLP respectively. However, as the models step into higher time steps, prediction error produced by LR increase exponentially; MLP produces a steady to slightly lower prediction errors and

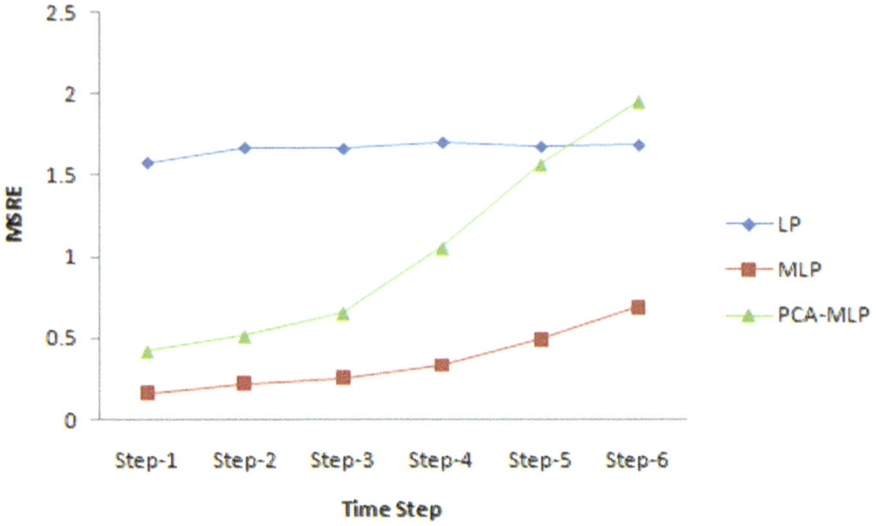

Fig. 3. Mean root square error on training

PCA-MLP a more steady performance. In terms of values, PCA-MCA produces the least error over longer time period, followed by MLP and LR respectively.

Table 2. Prediction performance on 30% test data

	LR		MLP		PCA-MLP	
	MAE	RMSE	MAE	RMSE	MAE	RSME
Step-1	3.0834	3.3304	5.1396	5.8544	1.2846	1.6231
Step-2	6.6013	6.9425	7.7992	8.4938	1.602	1.8872
Step-3	11.5551	12.5445	8.5771	9.2056	1.6386	2.1669
Step-4	18.9566	20.9855	5.317	5.4955	2.1801	2.7868
Step-5	30.0063	33.7979	5.2098	5.367	1.9006	2.3195
Step-6	44.1608	49.9381	4.9372	4.9939	0.6774	0.8077

Results from PCA-MLP shows improved prediction over longer time steps as opposed to LR that shows higher prediction errors with increase in time step. The tables below shows the results for 20%, 40% and 50% of test data. In general, the tables show that PCA-MLR out performed the MLR and LR with its lower MAE and RMSE.

The 6-time step chart for the PCA-MLR prediction over a period of 6 years is presented in Table 5. It can seen from the chart that the GDP dropped from time step 1 to 3 (year 2020 to 2022) before rising from time step 4–6 (year 2023 to 2025)

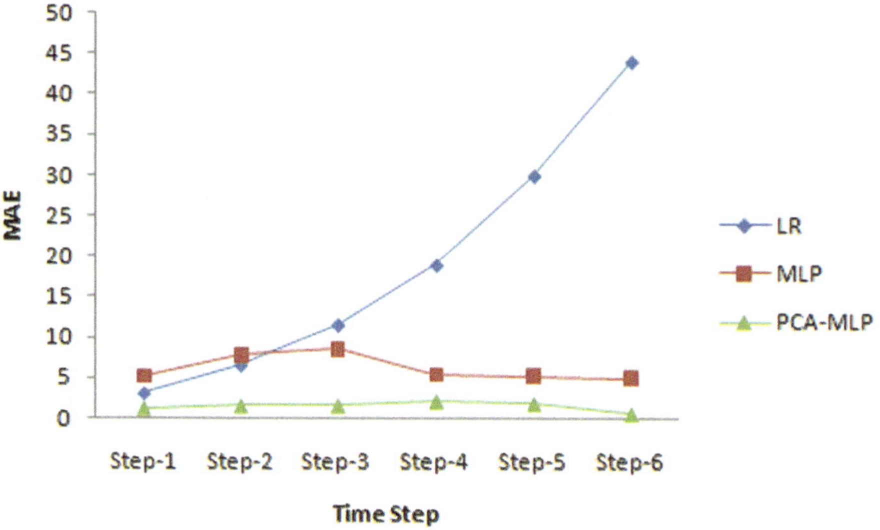

Fig. 4. Mean absolute error on 30% of test data

Table 3. Prediction performance on 20% test data

	LR		MLP		PCA-MLP	
	MAE	RMSE	MAE	RMSE	MAE	RMSE
Step-1	3.4962	4.2409	2.4055	2.8154	1.1903	1.4984
Step-2	5.8288	6.5039	3.1841	3.3974	0.7286	1.0216
Step-3	6.5806	6.9283	4.4061	4.4549	1.0403	1.3804
Step-4	5.2494	5.9835	4.1346	4.2914	1.8609	2.0638
Step-5	5.4325	6.1394	3.7406	3.9217	1.5741	1.6067
Step-6	6.0161	8.8645	6.1969	7.3427	1.1586	1.7115

Table 4. Prediction performance on 40% test data

	LR		MLP		PCA-MLP	
	MAE	RMSE	MAE	RMSE	MAE	RMSE
Step-1	4.1661	5.2362	8.261	9.8608	2.966	3.2513
Step-2	13.5195	16.1474	12.6811	14.8728	2.9306	3.1459
Step-3	40.8586	46.5983	12.229	13.8978	3.1294	3.3933
Step-4	126.9199	145.0148	12.1772	13.5215	1.6974	1.9047
Step-5	389.1566	463.0056	12.3616	12.7179	1.0281	1.1463
Step-6	1108.2889	1364.2483	14.6742	14.9222	1.4397	1.6016

Table 5. Prediction performance on 50% of test data

	LR		MLP		PCA-MLP	
	MAE	RMSE	MAE	RMSE	MAE	RMSE
Step-1	10.4333	15.0044	3.7369	4.5741	1.3477	1.5166
Step-2	10.9607	15.4139	6.1855	8.1523	2.2014	2.6358
Step-3	11.3952	15.8371	6.7437	8.7484	1.4436	2.0721
Step-4	9.3855	14.4293	5.5497	6.9967	1.9628	2.3672
Step-5	9.9132	14.8977	3.7303	4.1676	1.5939	1.8465
Step-6	10.3405	15.3848	3.52	3.8924	2.4984	2.7436

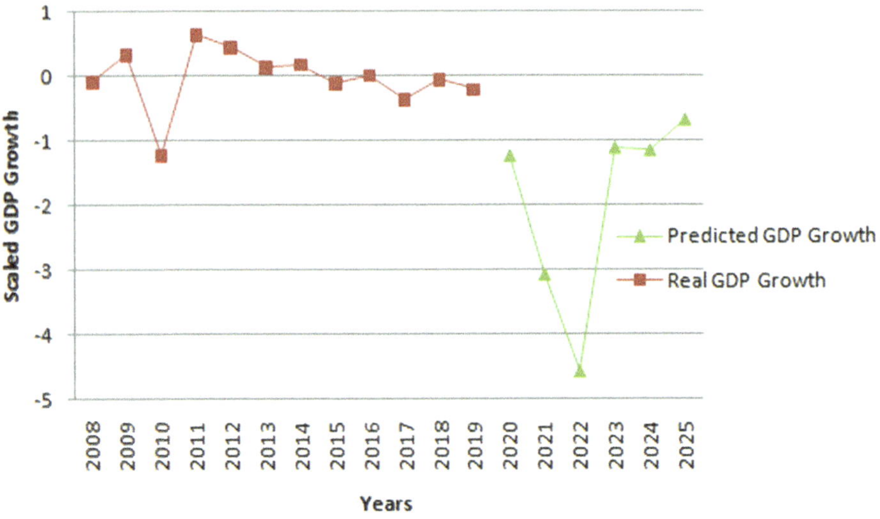

Fig. 5. PCA-MLR economic projection chart over 6 time-step period (2020–2025)

6 Conclusion

This study evaluated the performances of two machine learning techniques namely; Multi-layer perceptron (MLR) and Regression Analysis on economic parameters to forecast 6 time-step trend in South Africa economic growth. The above application of the techniques on South African multiple economic development indicators shows that the PCA enhanced Multi-layer perceptron produces a better performance than the ordinary MLP and regression analysis. The PCA based MLR established in this study reflects a better accuracy when considering many economic indicators as opposed to studies that focused on few indicators. Results of the prediction accuracy are shown in Tables 1, 2, 3, 4 and 5 and a chart of the prediction in Fig. 5. Further improvements can be realized by enhancing the MLR and similar algorithms for more accurate long-term predictions achievable in many economic indicators of African countries for better decision making.

References

1. Schlör, H., Fischer, W., Hake, J.F.: Methods of measuring sustainable development of the German energy sector. Appl. Energy **101**, 172–181 (2013). https://doi.org/10.1016/j.apenergy.2012.05.010
2. Sözen, A., Arcaklioglu, E.: Prediction of net energy consumption based on economic indicators (GNP and GDP) in Turkey. Energy Policy **35**(10), 4981–4992 (2007)
3. Dalevska, N., Khobta, V., Kwilinski, A., Kravchenko, S.: A model for estimating social and economic iindicators of sustainable development. Entrepreneurship Sustain. Iss. **6**(4), 1839–1860 (2019)
4. Department of Statistics, South Africa. https://www.statssa.gov.za/economic_growth/. Accessed 30 Apr 2020
5. Skare, M., Rabar, D.: Measuring economic growth using data envelopment analysis. Amfiteatru Econ. J. **18**(42), 386–406 (2013). https://doi.org/10.1016/j.apenergy.2012.05.010
6. Nkoro, E., Furo, A.O.: Foreign capital inflows and economic growth in Nigeria: an empirical approach. Acad. J. Interdiscip. Stud. **1**(2), 55–71 (2012)
7. Olayiwola, K., Okodua, H.: Foreign direct investment, non-oil exports, and economic growth in Nigeria: a causality analysis. Asian Econ. Finanacial Rev. **3**(11), 1479–1496 (2013)
8. Anigbogu, U.E., Nduka, E.K.: Stock market performance and economic growth: evidence from Nigeria employing vector error correction model framework. Econ. Financ. Lett. **1**(9), 90–103 (2014)
9. Babatunde, O.A.: Stock market volatility and economic growth in Nigeria (1980–2010). Int. Rev. Manage. Bus. Res. **2**(1), 201 (2013)
10. Obansa, S.J., Okoroafor, O.D., Aluko, O.O., Eze, M.: Percieved relationship between exchange rate, interest rate and economic growth in Nigeria: 1970–2010. Am. J. Human. Soc. Sci. **1**(3), 116–124 (2014)
11. Gupta, R., Kabundi, A.: A large factor model for forecasting macroeconomic variables in South Africa. Int. J. Forecast. **27**(4), 1076–1088 (2011)
12. Hong-xia, C.: Forecast study on Shanxi 12th five-year economic growth. J. Stat. Inf. Forum. 10 (2010)
13. Altissimo, F., Cristadoro, R., Forni, M., Lippi, M., Veronese, G.: New Eurocoin: tracking economic growth in real time. Rev. Econ. Stat. **92**(4), 1024–1034 (2010)
14. Wang, T.: Forecast of economic growth by time series and scenario planning method-a case study of Shenzhen. Mod. Econ. **7**(02), 212 (2016)
15. Song, K.B., Baek, Y.S., Hong, D.H., Jang, G.: Short-term load forecasting for the holidays using fuzzy linear regression method. IEEE Trans. Power Syst. **20**(1), 96–101 (2005)
16. Koegst, T., Tränckner, J., Franz, T., Krebs, P.: Multi-regression analysis in forecasting water demand based on population age structure. In: 11th International Conference on Urban Drainage, Ediburgh, Scotland, UK (2008)
17. Yasar, A., Bilgili, M., Simsek, E.: Water demand forecasting based on stepwise multiple nonlinear regression analysis. Arab. J. Sci. Eng. **37**(8), 2333–2341 (2012)

18. Kaytez, F., Taplamacioglu, M.C., Cam, E., Hardalac, F.: Forecasting electricity consumption: a comparison of regression analysis, neural networks and least squares support vector machines. Int. J. Electr. Power Energy Syst. **67**, 431–438 (2015)
19. Ghinea, C., Drăgoi, E.N., Comăniţă, E.D., Gavrilescu, M., Câmpean, T., Curteanu, S., Gavrilescu, M.: Forecasting municipal solid waste generation using prognostic tools and regression analysis. J. Environ. Manage. **182**, 80–93 (2016)
20. Wold, S., Esbensen, K., Geladi, P.: Principal component analysis. Chemometr. Intell. Lab. Syst. **2**(1–3), 37–52 (1987)
21. Haykin, S.: Neural Networks: A Comprehensive Foundation. Prentice Hall PTR (1994)

An Approach for Journal Summarization Using Clustering Based Micro-Summary Generation

Hammed A. Mojeed(✉), Ummu Sanoh, Shakirat A. Salihu, Abdullateef O. Balogun, Amos O. Bajeh, Abimbola G. Akintola, Modinat A. Mabayoje, and Fatimah E. Usman-Hamzah

Department of Computer Science, University of Ilorin, Ilorin, Nigeria
{mojeed.ha, balogun.aol, bajehamos, akintola.ag, mabayoje.ma, usman-hamza.fa}@unilorin.edu.ng, sanohumu67@gmail.com, salihusal980@gmail.com

Abstract. Researchers are continuously faced with the task of reviewing numerous journal articles from different domains to understand the current trends in their study. This justifies the need for a summarization system that could aid a quick review of articles. The major challenge in summarization lies in distinguishing the more informative parts of a document from the less informative ones. Previous works focused majorly on impact-based summarization approach that depends on external data such as citation sentences, conference talks transcripts and references. This work presents a new full content-based approach for summarization of journal articles based on micro-summary generation to produce sufficiently informative summaries and to aid article review process. The approach focused on the segmentation of the journal articles into sections based on its peculiar structure and the application of an information retrieval sentence similarity-based clustering approach with k-means algorithm to produce micro summaries for each section. The generated micro-summaries are combined systematically to generate the final summary. Results from the evaluation of the proposed approach on seven selected journals proved promising with a good precision value close to 0.8, recall of 1.00 and F-measure of 0.9. The proposed approach also outperformed a similar clustering-based technique without microsummary in accuracy and speed. The proposed approach is expected to aid the article review process of researchers.

Keywords: Text summarization · Information retrieval · Sentence similarity · Clustering

1 Introduction

Scientific or journal articles are structured, peer-reviewed and published write-ups that report verifiable research results to give new insight about a particular subject. Teufel and Moens [1] highlights distinct characteristics of journal articles that make them different from other types of general texts including time-independence (journal articles do not follow any time-dependent events like news articles and documentaries rather

R. Silhavy et al. (Eds.): CoMeSySo 2020, AISC 1295, pp. 685–699, 2020.
https://doi.org/10.1007/978-3-030-63319-6_64

they report the studies carried out by researchers), diverse writing style (which varies significantly based on the journal requirements) and bias (journal articles are biased by author's scientific and objective point of view). Another important characteristic presented in Lev et al. [2] are length and complexity which usually require substantial comprehending effort from experts. The increasing rate of journal articles' publication in different repositories (such as Google Scholar, Scopus, Web of Science, etc.) makes it difficult for researchers to keep up with relevant research. Researchers are constantly and increasingly faced with the daunting task of having to quickly digest large amounts of journal articles in these repositories while conducting their studies. Therefore, there is a dire need for summarizing scientific articles in digital repositories to facilitate quick access to necessary information that could aid upcoming research.

Automatic Summarization is the process of reducing a text document with a computer program in order to create a summary that preserves key information content overall meaning of the original document. Automatic summarization simplifies the tedious task of manually summarizing text documents to quickly retrieve meaningful and useful information. According to Kaynar, Isik, and Gormez [3], the main idea of summarization is to find a representative subset of text, which contains the information of the entire set. Text summarization is a challenging subproblem in the field of Natural Language Processing (NLP) because of the difficulty in interpreting every point in the text. It requires analyzing various aspects of the text such as lexical relations, similarity analysis, semantic analysis, entity recognition, and so on [4]. These kinds of analysis can only be achieved with a great deal of word knowledge. In the field of automatic text summarization, it is customary to differentiate between extraction based summaries where the summary is composed selected fragments from the source text and abstraction based summaries where the source text is transcribed into some formal representations and from this it is regenerated in a shorter and more concise form [5]. Abstraction involves paraphrasing the sections of the source document. In general, abstraction can condense a text more strongly than extraction. However, the programs that can do this are harder to develop as they require the use of natural language generation technology, which itself is still a growing field [6]. The Extraction based summarization works by selecting a subset of existing words, phrases, or sentences in the original text to form the summary. In this summarization task, the automatic system extracts salient sentences from the entire collection without modifying the texts themselves.

The major challenge in extractive summarization lies in distinguishing the more informative parts of a document from the less ones. Though there have been instances of research describing the automatic creation of abstracts, most works presented in the literature focused on direct extraction of sentences to address the problem of text summarization [7]. To achieve this, various methods have been employed in the literature to summarize different types of text. keyword extraction-based summarization technique was used in [8] to summarize essay articles. Kernel-based statistical approach applied to online document in html format has also been reported [9]. Clustering technique (k-Means) based on similarity measures has also been applied in the summarization of simulated essay texts [10]. Semantic analysis methods such as LSA and Graph based methods such as TextRank and Support Sets were used in [11] to summarize news article, film scripts and documentary subtitles. Fuzzy approach based on pre-summarization has also been reported in [12]. These summarization methods applied on unstructured texts (i.e. essays, online news articles, film scripts and documentary subtitles) cannot be directly applied on structured texts such as journal articles as summarizing them requires selecting representative information from different sections of the journal.

To address this problem, various approaches have been designed in the literature. Yasunaga et al. [13] presented "ScisummNet" an impact-based summarization method that integrates only article abstracts and citation sentences collected from other works which cite the target article to generate hybrid summaries for scientific journals. The same citation-based approach has been previously used in Qazvinian et al. [14], Abu-Jbara and Radev [15] and Mei and Zhai [16]. These studies do not consider the full text of the articles but only abstracts and the citation sentences. Citations, however, represent the personal view of the citing author on the article with focus only on information that is useful to his subject of discourse. Some salient information in the original article may be neglected by the citing author. Moreover, findings have revealed that different authors discuss different contributions when citing an article [14].

Some studies employ other external sources for summarizing journal articles. Lev et al. [2] presented "TalkSumm" an annotation method for journal summarization based on conference talks transcription. Their approach employed unsupervised alignment algorithms to map the transcripts of video contents of conference talks to the corresponding papers' text, and generate extractive summaries. The approach could not be generalized as a substantive number of scientific articles do not pass through conference presentation and moreover, presentation transcripts are not readily available. Erera et al. [17] presented "IBM Science Summarizer" a summarization system for computer science journal articles based on the user query. The system extracts summary from a selected section of the article using Cross-Entropy multi-criteria optimization algorithm taking user query and summary length as additional input. This approach is, however, limited to summarizing a selected section of the article based on user's query as an external source. Xu et al. [18] proposed an approach to summarize related papers into a survey utilizing attention mechanism to capture relations among document structure based on the full article text and references. The approach is based on a hierarchical attention model built with deep neural network LSTM to learn document structure from all the papers for summarization. The approach however requires huge amount of dataset to train the model for accurate performance which is a very challenging task.

Recently, Parveen [19] proposed a full content-based approach towards summarizing journal articles employing bipartite graph representation with PageRank algorithm to extract important, non-redundant and coherent sentences. The method used integer liner programming to select optimal sentences for the summary. The graph-based representation model, however, is not informative enough to comprehend the text and the pagerank algorithm requires journal articles to be in XHTML format. These actually limit the approach's ability in capturing all salient information in the articles and its usability.

There is still a need for a full content-based and lightweight summarization approach for scientific journal articles with accepted level of informativeness. Therefore, this work focuses on designing a model for journal summarization based on micro-summaries generation using clustering-based approach to capture salient information and contributions of the journal articles. Our main objective is to produce an adequately informative summary from the full text of journal articles.

2 Methods

This section presents the proposed approach for Journal summarization based on micro-summary generation. It involves the steps followed in the development of the summarization model. The steps begin with the preprocessing of the article, then splitting it into various sections- each of which is viewed as a single segment on its own. The idea is to take a single document; that is the journal article, split it into sections based on its structure to derive multiple documents that the summarization process will be performed on. After the splitting, information retrieval technique is used to carry out similarity measures within each section and assign numerical weight to each sentence. After that, clustering based summarization is applied to cluster sentences approach using the simple k-Means algorithm. Then candidate sentences are selected to obtain the micro-summaries. The micro-summaries are, thereafter, systematically combined to generate the actual summary taking into consideration, informativeness, redundancy, and coherence. Figure 1 presents the flow diagram for the proposed approach.

In text mining, input texts are usually preprocessed to remove non-useful terms and to prepare the text in a format usable by the algorithms. In this study, the input/text document is preprocessed by using tokenization, stop word removal and word net stemming. Tokenization breaks the input text into recognizable terms/words, referred to as tokens. Removal of stop word is implemented based on stop word list which includes list of terms (such as punctuations, articles, etc.) to be excluded during information retrieval process. Word stemming involves reducing a word to its root form to enable similar words to be represented using the same term. It particularly handles prefixes and suffixes. It is implemented in this work using word net database. The pre-processing contributes to the accuracy of the similarity measures and consequently reduced computational complexity of the summarization approach.

2.1 Text Splitting

After preprocessing the text is split into subsection based on the general structure of the journals. All journals are categorized into sections using section number or label. This feature is leveraged on to partition the whole journal text into various sections that can be handled efficiently. The text is scanned from the beginning to the end to search for a partitioning pattern. Specifically, one or two new line characters (as the case may be) followed by a number or a capital letter, followed a space or tab and a string designate a section. This pattern is extracted using the regular expression:

^\n\n?([1–9]|[A-Z]) \\.? \s|\t.* \n$

The process continues until the whole text is partitioned.

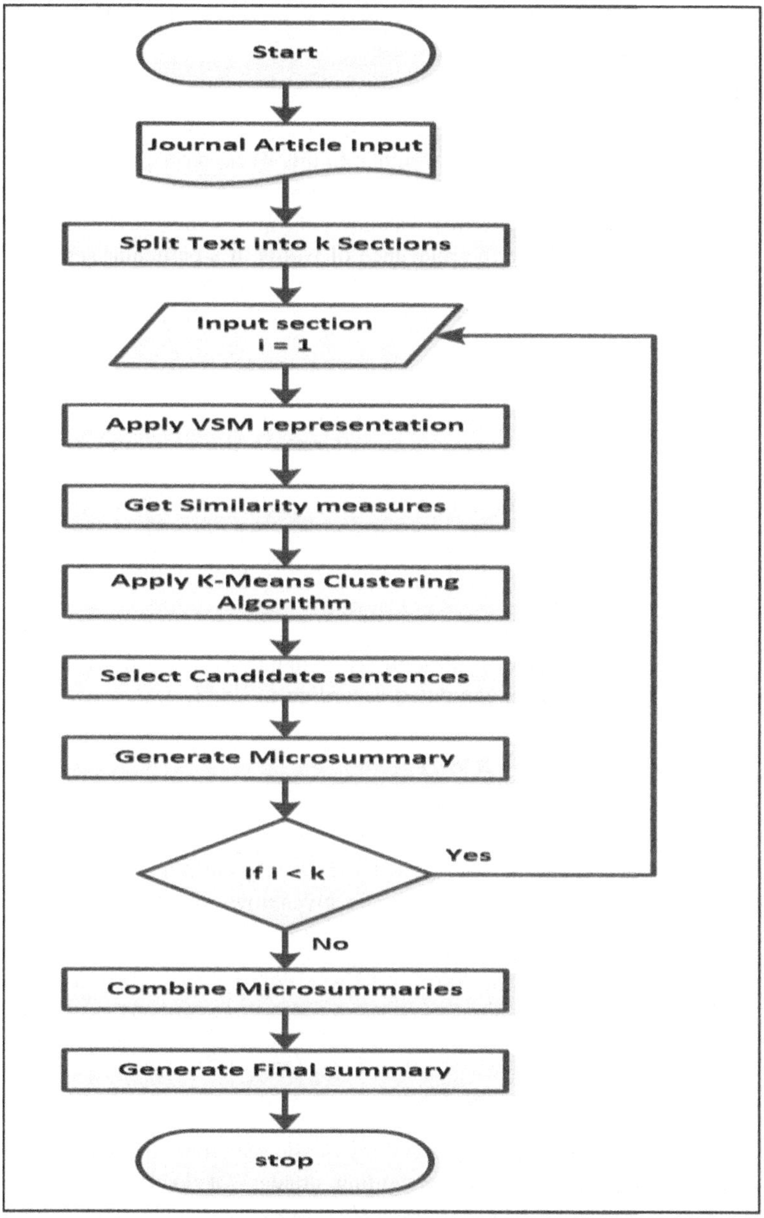

Fig. 1. Flow chart of the proposed journal summarization approach

2.2 Information Retrieval-Based Sentence Similarity Measure

Information Retrieval (IR), is concerned with retrieving of relevant information from databases. It involves techniques that can be employed to generate similarities betweendocument in a corpus. In this work information retrieval is used to calculate the similarities between sentences in a subsection of an article. The IR method Vector Space Model with Cosine similarity are used.

Vector Space Model (VSM)
Vector space model is an algebraic model for representing text documents as vectors of index terms with their corresponding weights. In this work, term weight that constitute the VSM representation is computed for each sentence in a section using TF-IDF. TF-IDF contains two components: term frequency (TF) and inverse document frequency (IDF). It is adapted to measure the relevance of words in a particular sentence within the article.

TF is the frequency of occurrence of a term in a sentence. It is computed as given in Eq. 1.

$$TF(t,s) = \frac{F_{ts}}{\#terms} \tag{1}$$

Where F_{ts} denotes the number of occurrence s of a term t in a sentence s, and $\#terms$ represents the number of terms in the section. The basic idea of TF is that the more frequently a term occurs within an article, the higher the probability that the term describes the article's content.

IDF reflects the measure of how significant a term is in the entire section by computing the inverse frequency of sentence containing the term within the entire section to be summarized. It is computed according to Eq. 2.

$$IDF(t) = log\left(\#s/_{s_t}\right) \tag{2}$$

Where $\#s$ denotes the total number of sentences in the section and s_t denotes the number of sentences containing term t. The basic assumption is that a term appearing in most sentences within the section should be given low weight as they may likely represent common terms that can be omitted from consideration in the summary.

The overall weight of each term is computed as the product of its TF and IDF as given in Eq. 3.

$$w(t) = \frac{F_{ts}}{\#terms} \times log\left(\#s/_{s_t}\right) \tag{3}$$

While TF-IDF was originally developed for classic search in information retrieval, it is one of the most widely used term-weighting strategy for document modelling. For instance, TF-IDF is used by 83% of surveyed text-based research-paper recommender systems [20].

For example, if a 30-word sentence S in a section X of a journal article contains the term "software" 5 times, the TF for the word 'software' is

TF(software, S) = 5/30 = 0.16

Let's assume there are 15 sentences that contain the term "software" out of 45 sentences in X, then IDF is calculated as:

IDF (software) = log (45/15) = 0.47

Lastly, the final weight the term software is given as:

w(software) = 0.16 * 1.47 = 0.075

The term weight of each term in the sentence is calculated and this forms the vector representation of the sentence.

Cosine Similarity

Cosine similarity is a widely implemented measure applied on document representations based on vector of weights as used in this work. The similarity between two sentences can be derived by calculating cosine value between their term weight vectors. One advantage of cosine similarity is its low-complexity. The cosine similarity of two non-zero vector of term weights can be derived by using the Euclidean dot product of two vectors A and B given in Eq. 4.

$$A.B = |A||B|\cos\theta \tag{4}$$

Given two vectors of term weights, A and B, the cosine similarity, cos(θ), is given in Eq. 2.

$$similarity(\mathrm{A.B}) = \sum\nolimits_{i=1}^{n} A_i B_i / \sqrt{\sum\nolimits_{i=1}^{n} A_i^2} \sqrt{\sum\nolimits_{i=1}^{n} B_i^2} \tag{5}$$

Where A_i and B_i are components of vector A and B respectively.

After this process every sentence in a subsection is given score according to their relevance and importance in the section. The cosine similarity score is used by the clustering phase to group similar sentences within each section into clusters representing distinct information to be included in the summary.

2.3 Clustering and Micro-summary Generation

Clustering is an unsupervised approach that group data objects based only on information that is found in the data which describes the objects and their relationships such as similarity score. In this context, the higher the similarity within a cluster and the wider the difference between clusters, the optimal and more distinct the clusters generated. The clustering algorithm used in this work is the widely known k-Means algorithm. The clusters generated by k-Means clustering algorithm are fully dependent on the selection of the initial centroids [21]. K number of data elements are selected as initial centers; then distances of all data elements are calculated based on a distance measure[22, 23]. Then, it continuously finds means value for each cluster and define new centroid to allocate data points until no change occurs in the clusters [22]. In this work, k-Means is used to group the sentences in VSM representation into clusters so that related sentences are grouped in the same cluster based on their cosine similarity scores. To adapt the k-Means algorithm to our summarization goal (adequately informative summary), the two important factors which determines the efficiency and accuracy of the clustering (the number of clusters k and the initial centroid sentences) need to be heuristically selected.

For the number of clusters to generate by k-Means, the value of k is computed to be dependent on the number of sentences and the total number of terms in the section as shown in Eq. 6.

$$k = \text{round}\left(\frac{\#\text{terms}}{\#\text{s}}\right) \quad (6)$$

Where $\#terms\, and\, \#s$ represents the number of terms and the total number of sentences in the section respectively. The value is rounded up to the nearest integer value.

The formula caters for good compression and estimation of terms-sentence relationship to decide the number of clusters.

For the initial k centroids, sentences are selected based on the magnitude of their term weights vectors. The vectors are ranked according to their magnitude values and the first k vectors are selected as the centroid. The idea behind this approach is that highly relevant sentences are likely to contain the main points which other related sentences support and corroborate. The specific procedure of the adapted k-Means algorithm for generating micro-summaries is presented in Algorithm 1.

The next step is to select candidate sentences for the micro-summary generation. The goal of the selection procedure is to choose a set of sentences that contain important information. The task is to identify which sentences in the source section are worth taking into the micro-summary. In this work, we used centrality-based algorithm where sentences are ranked based on their similarity with other sentences in the same cluster as processed by the k-Means algorithm. The intuition behind this approach is that the core information will be repeated or referred to in many sentences. Thus the sentences that are similar to many other sentences in the cluster are likely to contain this information and therefore should be included in the extract. In this manner, the most similar sentence to the final centroid of each cluster are selected as candidate sentences for the microsummary. The clusters with higher centroid weight from the clustering output should contain sentences that are more relevant in the section than the one with lower centroid value. Consequently, more sentences are extracted from the cluster with higher centroid value than the ones with lower centroid values.

A micro-summary is obtained after candidate sentences have been selected. Each sentence from a cluster selected to form a micro-summary has higher scores in comparison to other sentences in the same cluster. This process is iterated for all sections or partitions generated from the source journal article and all sections have their own micro-summaries.

Algorithm 1: k-Means Clustering Algorithm for Microsummary Generation

Input: A set of terms weight vectors V representing sentences in the section
Output: A set of k Clusters
1: Initialize k as: round $(\frac{\#terms}{\#s})$
2: Choose C_k centroids from V:
2.1: **for** each V_i in V:
2.2: compute the magnitude $|V_i| = \sqrt{\sum_1^n x^2}$ for $x \in V_i$, n $dim. of$ V_i
2.3: **end for**
2.4: sort all V_i in V in descending order of their magnitude
2.5: select the first k V_i in V as C_k (initial clusters)
3: **repeat:**
3.1: Assign other V_i in V to closest cluster in C_k using Equation 5
3.2: Compute new cluster centroids by calculating mean magnitude $|V_i|$
4: **until:**
4.1: No change in cluster center; OR
4.2: No object changes its clusters.
5: **end**

2.4 Final Summary Generation

The micro-summaries generated are combined together to form a reduced text representation of the original article. However, the important informationin the journal article are known to be scattered all over it and contain numerous redundant information [19]. For example, the introduction, methodology and conclusion section contain the motivation of the study and its approach. To deal with this redundancy, the clustering process is applied again on the combined microsummary to remove redundant sentences and generates the final summary. Another important characteristic of any good summary is coherence. To ensure coherence of the final summary, sentences a placed according to the order in which they appear in the original journal text. This way the summary flows normally like the original article from section to section.

2.5 Performance Measures Used

Since summarization is an information retrieval process, standard information retrieval performance measures precision, recall and F-measure are used to evaluate the performance of the proposed approach. These measures are based on the co-selection summary evaluation methods [4] that rely on the co-occurrence of terms in system generated summary in relation to a reference summary.

Precision: Precision related to reproducibility and repeatability, is the degree to which repeated measurements under unchanged conditions show the same results. It is also called Positive Predictive Value. As shown in Eq. 3, it is computed in this work as the fraction of relevant sentences among the retrieved sentences in the summary.

$$precision = \frac{\sum_{s \in sys} \sum_{gramN \in s} Count_{match}(gramN)}{\sum_{s \in sys} \sum_{gramN \in s} Count(gramN)} \tag{7}$$

Where sys denoted system summary and $Count(gramN)$ is the number of N-gram in system summary. $Count_{match}(gramN)$ is the number of N-grams co-occurring in system summary and reference summary.

Recall: Recall measures the sensitivity of the summarization process which represents the probability that relevant sentences are retrieved by the summary. As presented in Eq. 4, it is the fraction of relevant sentences that is retrieved among the total relevant sentences in the summary.

$$recall = \frac{\sum_{s \in sys} \sum_{gramN \in s} Count_{match}(gramN)}{\sum_{s \in ref} \sum_{gramN \in s} Count(gramN)} \tag{8}$$

Where *ref* represents the reference summary, s for sentence, $Count(gramN)$ for is the number of N-gram in reference summary and $Count_{match}(gramN)$ for the number of N-grams co-occurring in system summary and reference summary.

F-Measure: The F measure (F1 score or F score) is a measure of a test's accuracy and is defined as the weighted harmonic mean of the precision and recall of the test. It's the computed as the square of the geometric mean divided by the arithmetic mean as shown in Eq. 5.

$$F - measure = 2 \times \frac{(precision \times recall)}{(precision + recall)} \tag{9}$$

3 Results

The proposed approach was implemented in Python using the libraries NLTK, Scikit-Learn, NumPy, SciPy and Sympy. Experiments on selected journals are performed on an intel core i5 processors of 2.26 GHz speed each and a RAM of 4 GB size PC running windows 10. Figure 3 and 4 show the screenshots of the original journal in text format and the resulting summary generated from the proposed approach.

To evaluate the performance of the proposed summarization method, the summary obtained is compared with manual extracts which are provided by three different human experts on seven different journals selected from different fields based on majority vote. Table 1 shows the comparison between the number of sentences obtained by human expert using majority count and the number of sentences obtained by the automatic summarizer.

Table 1. Number of sentences obtained by human and proposed summarizer

Text	#Sentences in original text	#Sentences human expert	#Sentences automatic Summarizer
JN1	120	38	35
JN2	97	33	24
JN3	105	19	22
JN4	126	37	28
JN5	165	38	21
JN6	80	8	14
JN7	105	32	25

As a test of efficacy, the precision, recall and F-measure are calculated based on the seven considered journal articles. Table 2 presents these results. Relevant sentences used in computing the measures are as selected from the extracts produced by two human experts and the article's abstract based on Eq. 10.

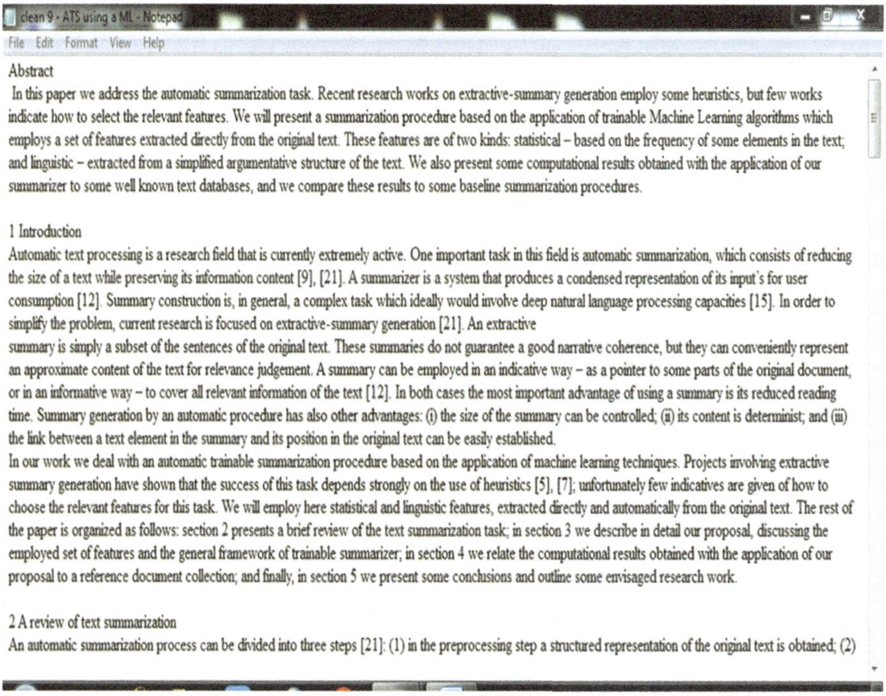
clean 9 - ATS using a ML - Notepad

File Edit Format View Help

Abstract

In this paper we address the automatic summarization task. Recent research works on extractive-summary generation employ some heuristics, but few works indicate how to select the relevant features. We will present a summarization procedure based on the application of trainable Machine Learning algorithms which employs a set of features extracted directly from the original text. These features are of two kinds: statistical – based on the frequency of some elements in the text; and linguistic – extracted from a simplified argumentative structure of the text. We also present some computational results obtained with the application of our summarizer to some well known text databases, and we compare these results to some baseline summarization procedures.

1 Introduction

Automatic text processing is a research field that is currently extremely active. One important task in this field is automatic summarization, which consists of reducing the size of a text while preserving its information content [9], [21]. A summarizer is a system that produces a condensed representation of its input's for user consumption [12]. Summary construction is, in general, a complex task which ideally would involve deep natural language processing capacities [15]. In order to simplify the problem, current research is focused on extractive-summary generation [21]. An extractive

summary is simply a subset of the sentences of the original text. These summaries do not guarantee a good narrative coherence, but they can conveniently represent an approximate content of the text for relevance judgement. A summary can be employed in an indicative way – as a pointer to some parts of the original document, or in an informative way – to cover all relevant information of the text [12]. In both cases the most important advantage of using a summary is its reduced reading time. Summary generation by an automatic procedure has also other advantages: (i) the size of the summary can be controlled; (ii) its content is determinist; and (iii) the link between a text element in the summary and its position in the original text can be easily established.

In our work we deal with an automatic trainable summarization procedure based on the application of machine learning techniques. Projects involving extractive summary generation have shown that the success of this task depends strongly on the use of heuristics [5], [7]; unfortunately few indicatives are given of how to choose the relevant features for this task. We will employ here statistical and linguistic features, extracted directly and automatically from the original text. The rest of the paper is organized as follows: section 2 presents a brief review of the text summarization task; in section 3 we describe in detail our proposal, discussing the employed set of features and the general framework of trainable summarizer; in section 4 we relate the computational results obtained with the application of our proposal to a reference document collection; and finally, in section 5 we present some conclusions and outline some envisaged research work.

2 A review of text summarization

An automatic summarization process can be divided into three steps [21]: (1) in the preprocessing step a structured representation of the original text is obtained; (2)

Fig. 2. Input journal text

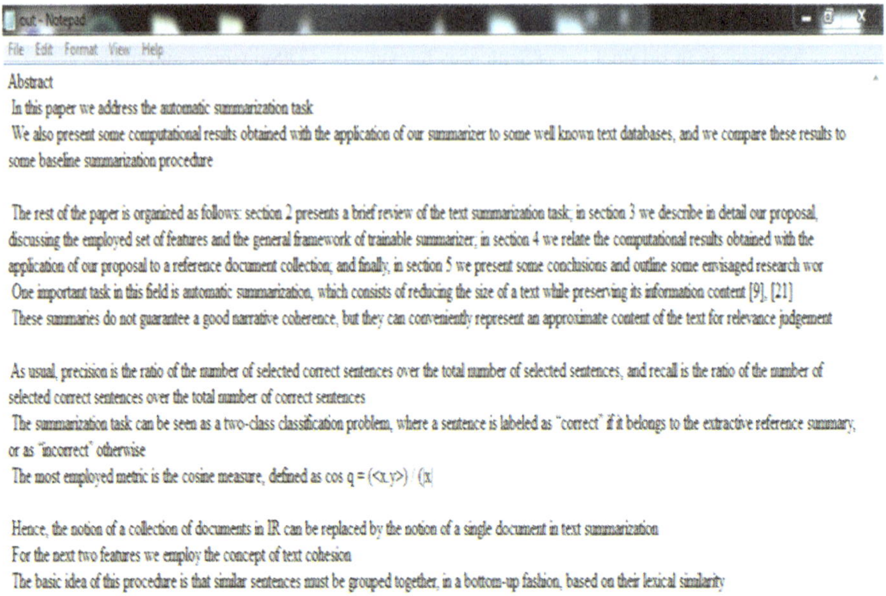

Fig. 3. Summary output

$$Relevant\ sentences = summary1 \cap summary2 \cap abstract \quad (10)$$

In other words, sentences common to the summaries extracted by the two experts and the article's abstract are selected as relevant sentences. Theses sentences form the reference summary with which the system generated summary is evaluated.

Table 2. Precision, Recall and F-Measure results

S/N	Precision	Recall	F-Measure
JN1	0.825	1.000	0.904
JN2	0.580	0.711	0.638
JN3	0.591	0.867	0.702
JN4	0.714	0.869	0.783
JN5	0.857	1.000	0.922
JN6	0.643	0.900	0.750
JN7	0.680	0.894	0.772

For sanity check, we compared our proposed approach with the same summarization model without micro-summary. Table 3 shows the average precision, recall and F-measure comparison of the two approaches.

We also carried out execution time analysis to measure the computational efficiency of our approach compared to clustering approach without micro-summary. The result we got is presented in Fig. 4.

Table 3. Comparizon result of the proposed approach and a clustering model without Microsummary

S/N	Precision	Recall	F-Measure
Clustering with microsummary	0.71	0.89	0.78
Clustering without microsuammary	0.61	0.72	0.64

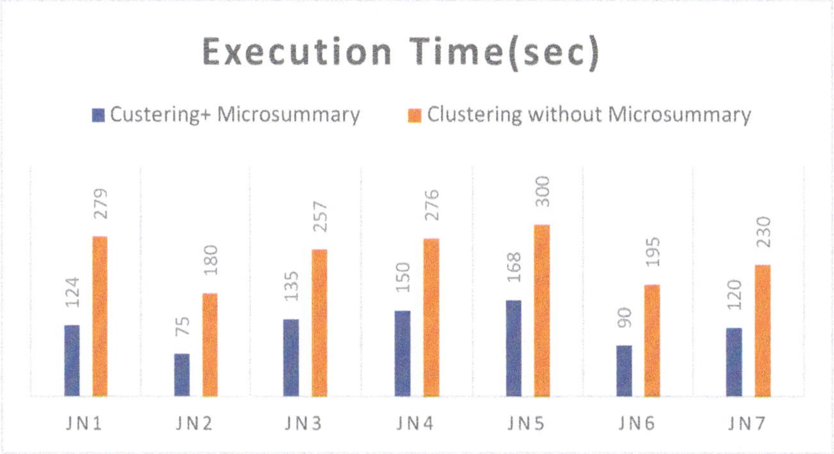

Fig. 4. Execution time of proposed approach and clustering approach without microsummary

4 Discussion

It can be seen from the results that the summarizer was able to produce good number of sentences in the summary generated with high compression ratio as expected for effective summary. It was able to generate lower number of relevant sentences in 5 out of 7 considered journals giving 71.4% performance rate.

From the performance evaluation results, it is obvious that the precision of the proposed approach is effective with the value reaching 0.825 a showing high informativeness. This shows that the proposed approach was able to produce relevant sentences from the journal. The precision performance in journals 2 and 3 is relatively low with 0.58 and 0.59 precision values. This may be due to the fact that the journals are relatively shorter in size. The performance in terms of recall is high in all instances reaching up to 1.00 value. It can be observed that the recall compensates for the shortfall in the precision performance. This is clearly manifested in the f-measure values. The lowest F-measure value of 0.638 is recorded for journal 2, this is due to its short length as manifested in the number of sentences it comprised of. It is also important to note that the proposed approach will improve the relevance of journal summaries if applied on long journals with good precision, recall and F-measure values as shown in the results.

The effect of using micro-summary generation approach is evaluated and from Table 3, it is clear that influence of the micro-summary is significant in the final summary as it recorded higher average precision, recall and F-measure across all the considered journal articles when compared with a clustering approach without micro-summary generation. This significant influence is due to the fact that, micro-summaries allow the final summary to capture all relevant information or contribution present in the original journal article and therefore higher informativeness.

As shown in Fig. 4, the efficiency of the proposed approach is underscored with lower execution time compared to clustering approach without micro-summary. In all journal articles, the execution time is considerably lower with the proposed algorithm buying up to about 44.44% execution time.

5 Conclusion

This work presents a new approach to extractive journal summarization based on micro-summary generation. It proposed splitting of the journal text into sections and applying IR sentence similarity-based clustering technique to generate micro-summaries for each section. The generated summaries are then combined using the same approach to produce the final summary taking informativeness, non-redundancy and coherence into consideration. The approach was evaluated on carefully selected seven journal articles s from different fields in computing using standard IR measures; precision, recall and F-measure. It was observed from the results that the proposed algorithm ranks some sentences that are relevant higher than human extracts. With performance measures result, the proposed approach recorded good precision values rising close to 0.8, recall rising up to 1.00 and F-measure of 0.9. This shows the ability and effectiveness of the proposed approach in handling large sized text which are difficult to handle by the present summarization approaches.

In the future, we plan to compare the performance of the proposed algorithm empirically with other state of the art approaches.

References

1. Teufel, S., Moens, M.: Summarizing scientific articles: experiments with relevance and rhetorical status. Comput. Linguist. **28**(4), 409–445 (2002)
2. Lev, G., Shmueli-Scheuer, M., Herzig, J., Jerbi,A., Konopnicki, D.: TalkSumm: a dataset and scalable annotation method for scientific paper summarization based on conference talks. arXiv preprint arXiv:1906.01351 (2019)
3. Kaynar, O., Isik, Y.E., Gormez, Y.: Graph based automatic document summarization with different similarity methods. In: Proceedings of IEEE Conference on Signal Processing and Communications Applications (2017). https://doi.org/10.1109/siu.2017.7960613

4. Verma, P., Verma, A.: A review on text summarization techniques. J. Sci. Res. **64**(1), 251–257 (2020)
5. Hassel, M.: Exploitation of named entities in automatic text summarization for Swedish. In: Proceedings of NODALIDA 2003 - 14th Nordic Conference on Computational Linguistics, Reykjavik, Iceland (2004)
6. Zadbuke, A., Pimenta, S., Padwal, D., Wangikar, V.: Automatic summarization of news articles using TextRank. In: International Conference on Electronics & Computing Technologies. vol. 6, p. 3 (2016)
7. Zhang, J., Zhou, Y., Zong, C.: Abstractive cross-language summarization via translation model enhanced predicate argument structure fusing. IEEE/ACM Trans. Audio Speech Lang. Process. **24**(10), 1842–1853 (2016)
8. Jayashree, R., Murthy, S.K., Sunny, K.: Keyword extraction based summarization of categorized Kannada text documents. Int. J. Soft Comput. **2**(4), 81 (2011)
9. Gupta, V., Chauhan, P., Garg, S., Borude, A., Krishnan, S.: An statistical tool for multi-document summarization. Int. J. Sci. Res. Publ. **2**(5), 1–5 (2012)
10. Agrawal, A., Gupta, U.: Extraction based approach for text summarization using k-means clustering. Int. J. Sci. Res. Publ. **4**(11), 782–787 (2014)
11. Aparício, M., Figueiredo, P., Raposo, F., de Matos, D.M., Ribeiro, R., Marujo, L.: Summarization of films and documentaries based on subtitles and scripts. Pattern Recogn. Lett. **73**, 7–12 (2016)
12. Yadav, S., Shah, D.: News summarization using text mining. Int. Res. J. Eng. Technol. (IRJET) **05**(11), 202–206 (2018)
13. Yasunaga, M., Kasai, J., Zhang, R., Fabbri, A.R., Li, I., Friedman, D., Radev, D.R.: ScisummNet: a large annotated dataset and content-impact models for scientific paper summarization with citation networks. In: AAAI 2019. arXiv preprint arXiv:1909.01716 (2019)
14. Qazvinian, V., Radev, D.R., Mohammad, S.M., Dorr, B., Zajic, D., Whidby, M., Moon, T.: Generating extractive summaries of scientific paradigms. J. Artif. Intell. Res. **46**, 165–201 (2013)
15. Abu-Jbara, A., Radev, D.: Coherent citation-based summarization of scientific papers. In: Proceedings of the 49th Annual Meeting of the Association for Computational Linguistics: Human Language Technologies, vol. 1, pp. 500–509 (2011)
16. Mei, Q., Zhai, C.: Generating impact-based summaries for scientific literature. In: Proceedings of ACL-08: HLT, pp. 816–824 (2008)
17. Erera, S., Shmueli-Scheuer, M., Feigenblat, G., Nakash, O.P., Boni, O., Roitman, H., Lev, G.: A summarization system for scientific documents. arXiv preprint arXiv:1908.11152 (2019)
18. Xu, H., Wang, Z., Weng, X.: Scientific literature summarization using document structure and hierarchical attention model. IEEE Access **7**, 185290–185300 (2019)
19. Parveen, D.: A graph-based approach for the summarization of scientific articles. Doctoral dissertation, Faculty of Modern Languages, Heidelberg University (2020)
20. Beel, J., Gipp, B., Langer, S., Breitinger, C.: Paper recommender systems: a literature survey. Int. J. Digit. Libr. **17**(4), 305–338 (2016)
21. Jiawei, H., Kamber, M., Pei, J.: Data Mining: Concepts and Techniques. Morgan Kaufmann Publishers, San Francisco California (2012)
22. Madhulatha, T.S.: An overview on clustering methods. arXiv preprint arXiv:1205.1117 (2012)
23. Marapareddy, R., Saripalli, S.W.: Runway detection using k-means clustering method using UAVSAR data. In: Proceedings of the International Conference on Artificial Intelligence and Applications (AIAPP), Geneva, Switzerland, pp. 25–36 (2017)

EEG Recognition Based on Parallel Stacked Denoise Autoencoder and Convolutional Neural Network

Tao Xie[1], Desong Kong[2(✉)], Qing Liu[3], Zhenfu Yan[2], and Xianlun Tang[1]

[1] College of Automation, Chongqing University of Posts and Telecommunications, Chongqing 400065, China
[2] School of Computer Science, Chongqing University of Posts and Telecommunications, Chongqing 400065, China
874148831@qq.com
[3] College of Mobile Telecommunications, Chongqing University of Posts and Telecommunications, Chongqing 401520, China

Abstract. As an emerging interactive technology, brain-computer interface (BCI) has been widely used in various fields. The study of electroencephalogram (EEG) can not only improve people's cognition of the brain, but also establish a new way for the brain to connect with the outside world. On the basis of the high dimensional, non-linear and spatiotemporal features of EGG signals, a novel method called parallel stacking encoded convolutional network is proposed for feature extraction and recognition of motor imagery EEG signals. The proposed method combines the unsupervised learning of stacked denoise autoencoder (SDA) with the supervised learning of convolutional neural network (CNN). First, the EEG data are applied to complete the parallel unsupervised training of multiple SDAs. The pre-training process can individually extract EEG signal features of each channel and avoid the interference between the various channels of EEG signals. Then the supervised convolutional training of the network based on gradient descent algorithm effectively fuses the features of multi-channel EEG signals. The experimental results indicate that the proposed algorithm has better accuracy than mainstream algorithms of EEG recognition.

Keywords: Convolutional neural network · EEG recognition · BCI motor imagery

1 Introduction

Brain-computer interface [1, 2] is a direct communication pathway between an enhanced and an external device. It has been proven to have potential value in the medical, the military and entertainment. In a BCI system, the feature extraction of EEG data is the key technology. Methods which are commonly used in EEG signal feature extraction include the autoregressive model [3, 4], common spatial pattern (CSP) [5] adaptive common spatial pattern (ACSP) [6]. Traditional EEG signal feature extraction

R. Silhavy et al. (Eds.): CoMeSySo 2020, AISC 1295, pp. 700–713, 2020.
https://doi.org/10.1007/978-3-030-63319-6_65

methods with characteristics such as artificial feature extraction will lead to the loss of some hidden useful information during data processing. The extracted features cannot be well analyzed for EEG signal processing with traditional feature extraction methods.

In recent years, deep learning has achieved great success in image recognition [7], natural language processing [8] and speech recognition [9]. Deep learning also has been used to analyze EEG signals because it has advantages with regard to processing non-linear and high-dimensional data. Jirayucharoensak used a deep learning network with principal components to obtain effective results in human emotional states recognition [10]. An used a deep belief network (DBN) to improve the performance of the AdaBoost algorithm by combining multiple weak classifiers [11]. Zhicheng Jiao structured a deep convolutional neural network (DCNN) for left and right hand motor imagery EEG feature extraction and recognition [12]. Jin Zhang trained an efficient CNN to classify EEG signal transformed by Short-time fourier transform (STFT) [13]. Lngkvist summed up a series of deep learning methods and applied them to feature learning of time series signals, which solved the problems caused by the characteristics of time-series data [14].

This paper proposes a new model to achieve accurate recognition of motor imaging EEG signals. The model is used for efficient feature extraction of raw EEG data, and it has powerful classification capability. Wavelet analysis is used to extract beta data (14–36 Hz) which can improve the recognition accuracy to a certain extent from EEG data. In order to make full use of spatial information to strengthen EEG feature extraction, the parallel SDAs are performed on beta data to individually extract EEG features of each channel. The CNN is used to perform feature classification. At the same time, 1-D convolution which is friendly to temporal features are used for CNN. In this paper, the performance of the proposed method is analyzed based on the EEG sets collected by Emotiv and BCI IV sets, and is promising for EEG recognition in a BCI system. Comparisons with some popular feature extraction methods show that the proposed method can achieve better performance, and good generalization ability.

2 Stacked Denoise Autoencoder

SDA is an improved version of autoencoder (AE) by adding noise to the input. Compared with traditional AE, DAE weight has more robust. Such improvement can be used to clean noisy input [15].

If training samples are x, noise output y is expressed as

$$y = x + \varepsilon, \varepsilon \sim (0, \delta^2 I) \tag{1}$$

Here, ε is the noise.

Hidden layer output h is mapped by the encoding function. The encoding operation is

$$h = f(x) = f_s(wy + b) \tag{2}$$

Here, f_s is activate function, w is weight, b is bias.

Reconstructed output Y is mapped by the decoding function. The decoding operation is

$$Y = G(h) = f_s(\hat{w} h + \hat{b}) \tag{3}$$

Here, $\hat{w}$ is forward mapping transposition, f_s is activate function, $\hat{b}$ is bias.

The loss function of the DAE is defined as the reconstruction error. The network parameters $\{w, b\}$ can be obtained based on the gradient error method, which is used to minimize the reconstruction error. The reconstruction error is

$$\begin{gathered} J_{DAE}(w, b) = \sum_{x \in s} L(x, g(f(y))) = \\ -\sum_{i=1}^{n} \left[x_i(g(f(y_i))) + (1 - x_i) \ln(1 - g(f(y_i)))\right] \end{gathered} \tag{4}$$

Relative to DAE network, deep SDA network with multiple layers can extract more data features. The SDA network is constructed based on DAE so that the upper encoding layer output is used as the lower encoding layer input.

3 Convolutional Neural Network

CNN optimizes the network architecture by using local receptive field and weight sharing. The bias and weights learned in the feature extraction layers are trained by gradient descent optimization and back propagation [12]. As shown in Fig. 1, a traditional CNN has different types of layers: the input layer, the convolutional layer, the pooling layer and the output layer.

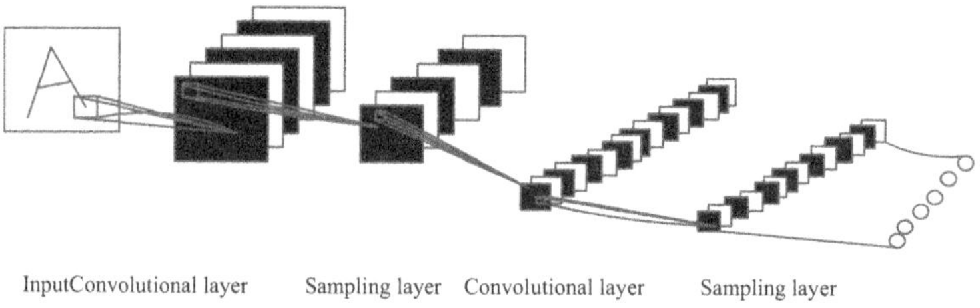

Fig. 1. Structure of CNN.

The convolutional layers consist of convolution processing and non-linear operation. Convolution processing extracts some primary signal features by using local connection and weight sharing. The convolutional formula is

$$y_m = f(\sum_{J=0}^{J-1}\sum_{I=0}^{I-1} x_{m+I,n+J} w_{IJ} + b) \tag{5}$$

Here, $0 \leq m < M, 0 \leq n < N$, w is the convolution kernel of $J * I$ and b is bias.

The pooling layers propagate the value within the local region to the next layer and reduce the variability in the hidden activations. The pooling operation formula is

$$y_m = f(\sum_{J=0}^{J-1}\sum_{I=0}^{I-1} x_{m+I,n+J} w_{IJ} + b) \tag{6}$$

Here, S_1 and S_2 are sample sizes.

4 EEG Recognition Based on Parallel SDA and CNN

EEG signals triggered by the imagination of brain or environmental stimulation include temporal information, spatial feature, and inherent noise invariance. Traditional EEG classification algorithms cannot achieve good classification results because the spatiotemporal features of EEG data are not considered and cannot handle the high-dimensional characteristic of EEG data well. Reasonable use of EEG data with rich hidden spatiotemporal features is helpful in improving classification accuracy. In this paper, a network called parallel stacking encoded convolutional network (PSECN), which integrates the parallel SDA to select feature data and uses the supervised learning of CNN to learn nonlinear feature representations, shown in Fig. 2. Because there are correlations between different channels of EEG, if feature extraction is performed directly on the channel data, it is easy to confuse features of EEG data. Multiple SDAs are performed on each EEG channel data to individually extract EEG signal features of each channel and avoid the interference between the various channels of EEG signals. After the EEG date processed by wavelet processing and the SDAs are trained with enough epochs until the training process converges, the feature data are extracted for CNN. Once the CNN classifier fulfils its training, it will carry out the classification task. Meanwhile, unlike the image data, the EEG data have the features of time series in initial signals. 1-D convolution being friendly to temporal features is used in CNN.

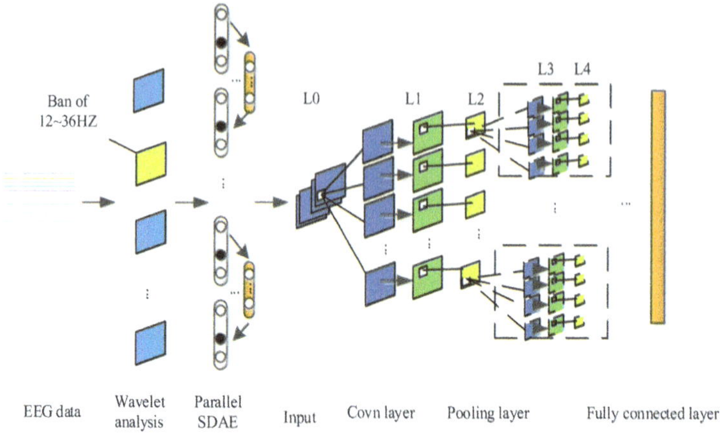

Fig. 2. The topology of PSECN. It consists of three parts: (1) Beta bands which are fed into multiple SDAs are extracted for each EEG channel by wavelet analysis. (2) Feature data matrix is fed into a CNN. (3) The Softmax is defined as a classifier of the fully connected layer.

Convolution operation which is essentially a process of weighted averaging can enhance feature of EEG signals and reduce noise of EEG data. Its main mechanism, weight sharing, can reduce a large number of parameters. In the past EEG researches, multi-D convolution is usually used for convolution mapping to achieve data balance, but there is a problem of destroying the temporal feature of EEG. Taking the 1*5 and 5*5 convolution kernel as an example, the local receptive field of 5*5 convolution kernel spans multiple EEG channels, which is likely to confuse features of EEG data. However, for the 1*1 convolution kernel, the local receptive field is on single channel of EEG data, and the direction of movement is the temporal direction of EEG data. Therefore, using 1-D convolution in CNN has better feature extraction effect for EEG data. In this paper, the 1*3 convolution kernel is applied to the supervised training of CNN.

The new input matrix y is comprised of input and hidden layer features obtained by SDAs. The calculation formula of y is

$$y = (x, y) = [x, s(wx + b)] \tag{7}$$

Sigmoid is usually used as an activation function in convolution operations, but it is easy to fall into local optimum. The new activation function is performed on PSECN to map the EEG features. Unlike the Sigmoid function, the Swish function [16] is a smooth and non-monotonic function. The Swish formula is

$$f(x) = x * sigmoid(x) \tag{8}$$

PSECN training steps are as follows:

Step 1: Construct SDAs with two hidden layers, determine the number of nodes $[m, n, o]$ and noise figure α is set.
Step 2: SDAs network are trained in accordance with Sect. 2.
Step 3: SDAs is trained with enough epochs until the loss function converges. The parameters (w, b) are updated by gradient descent method. The flow chart is as follows:

Parameters $\{\Delta w, \Delta b\}$ are updated by the following formula:

$$\Delta w = \Delta w + \nabla L_w(x, g(f(y)))$$

$$\Delta b = \Delta b + \nabla L_b(x, g(f(y)))$$

Parameters (w, b) are updated by the following formula:

$$w = w - \alpha(\frac{1}{m}\Delta w)$$

$$b = b - \alpha(\frac{1}{m}\Delta b)$$

Step 4: Channel features extracted from parallel SDAs network is mixed input and transformed to the input of convolutional network.
Step 5: Construct CNN with four hidden layers and network weights are initialized.
Step 6: CNN with the feature layer is trained with enough epochs until the training process converges, the feature vector layer is extracted for Softmax classifier.

5 Experiments on Motor Imagery EEG Classification

5.1 Experimental Data

Two experimental data sets are used as confirmation data sets in this paper. Data set 1, collected by Emotiv, comprises of the left and right hand motor imagery EEG signals, as shown in Fig. 3. The Emotiv EEG acquisition instrument contains 16 electrodes, as shown in Fig. 4. These subjects are 3 graduate students, aged from 22 to 24 years old. They are right-handed and have no history of brain diseases. The process of data collection is shown in Fig. 5. At the beginning of the experiment (t = 0 s), the subject keeps calm and remains relaxed. At t = 2 s, the subject performs a motion imagery task according to the prompt on the computer screen. The 240 experimental samples collected contain 120 trials of left hand motor imagery and 120 trials of right hand motor imagery. The sampling frequency of the signal is 128 Hz. Figure 6 shows the motor imagery EEG signals of F3 channel and imagery EEG map.

Data set 2 is BCI Competition 2008–Graz data set B. It consist of 9 subjects and each subject contains 720 trials of left and right hand motor imagery EEG data, in which the sampling frequency of channels is 250 Hz.

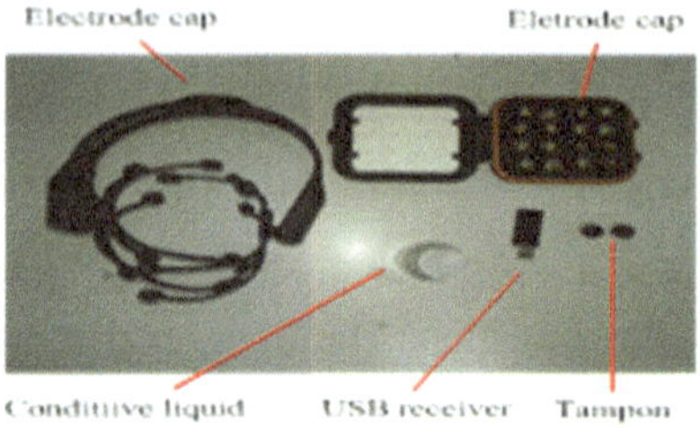

Fig. 3. Emotiv EEG signal acquisition instrument.

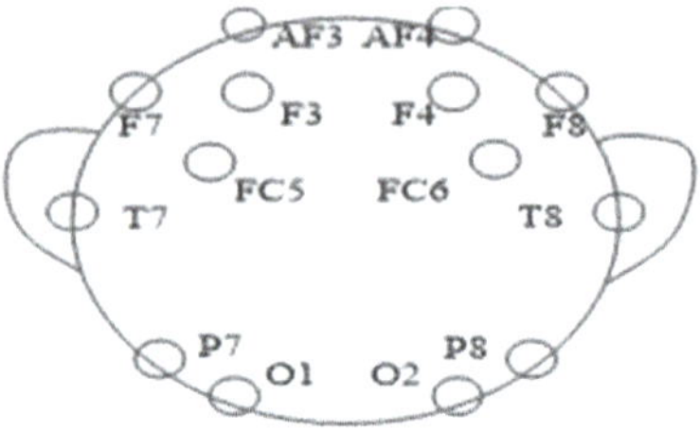

Fig. 4. The placement position of Emotiv electrodes

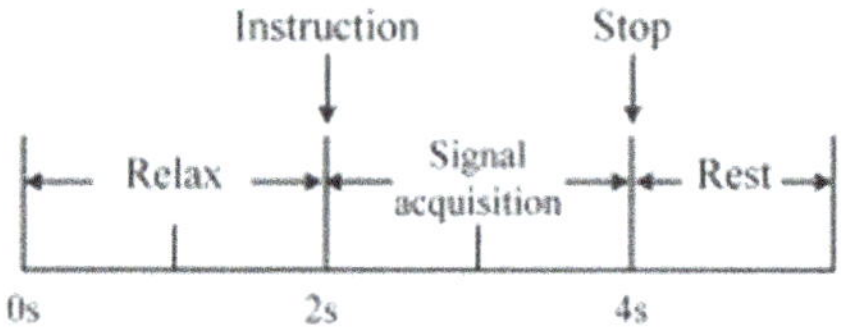

Fig. 5. Experimental procedure for one trial

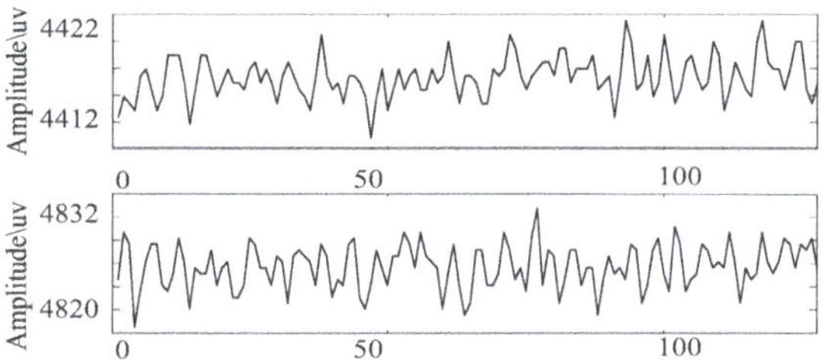

A) EEG signal in F3 channel

B) imagery EEG map

Fig. 6. EEG signals of Emotiv

5.2 Wavelet Preprocessing

According to multi-resolution analysis, the number of signal decomposition layers is related to the sampling frequency and useful components of the signal. The motor imagery EEG task related to the synchronization/desynchronization phenomenon is mainly demonstrated in 14–36 Hz. Therefore, data set 1 (128 Hz) are decomposed into three layers and the data set 2 (250 Hz) are decomposed into four layers by using fast wavelet decomposition. The wavelet coefficients and corresponding sub bands of data set 1 are shown in Table 1. The wavelet coefficients and corresponding sub bands of data set 2 are shown in Table 2.

Table 1 shows that the wavelet coefficient D2 (16–32 Hz) is in the range of the beta wave. Table 2 shows that the wavelet coefficient D3 (15.625–31.5 Hz) is in the range of the beta wave. Therefore, D2 and D3 are reconstructed by using the wavelet reconstruction. The experimental results of wavelet analysis of FC5channel are given in Fig. 7 and Fig. 8.

Table 1. The wavelet coefficients and corresponding sub bands of data set 2

Wavelet coefficients	Band range/Hz	Number of layers
D1	62.5–125	1
D2	31.5–62.5	2
D3	**15.625–31.5**	**3**
D4	7.8125–15.625	4
A4	0–7.8125	4

Table 2. The wavelet coefficients and corresponding sub bands of data set 1

Wavelet coefficients	Band range/Hz	Number of layers
D1	32–64	1
D2	**16–32**	2
D3	8–16	**3**
A3	0–8	3

5.3 Results and Analysis

For data set 1, EEG signal channels (F3, F4, FC5, FC6, T7, and T8) in the motor sensation zone are selected and five experiments are implemented with each experiment containing 200 training samples and 40 test samples. The number of hidden layer nodes each SDA is 128 and the noise figure is linear. Figure 9 shows the accuracy rate of single channel EEG signal and multi-channel EEG signals on data set 1. It can be seen from Fig. 9 that the recognition rate of each subject rises significantly with the increase of channels. The single EEG channel signal obtains lower recognition rate as hidden information contained in other EEG channel signals are lost. Thereby it is not feasible to distinguish between the left and right hand motor imagery EEG with the single

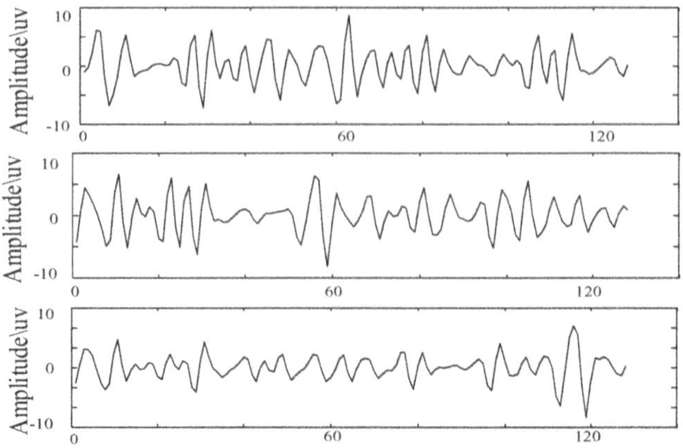

Fig. 7. Beta waves in data set1

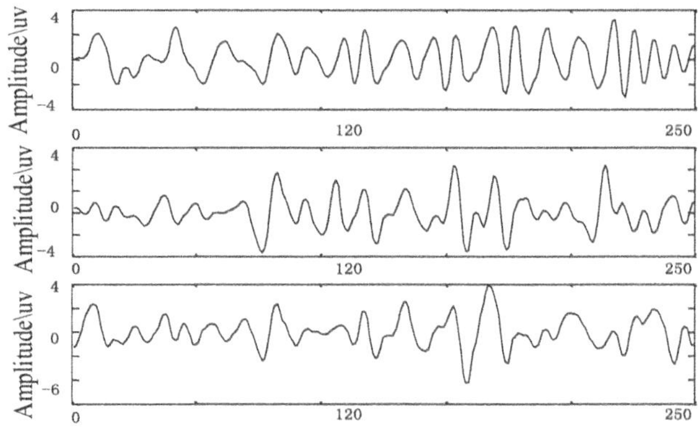

Fig. 8. Beta waves in data set 2

channel feature. Meanwhile, the experiment proves that PSECN has a good effectiveness on multi-channel EEG signals feature extraction.

To evaluate the performance of PSECN on data set 1, a comparison experiment has been conducted, including extreme energy difference (EED), CSP, ACSP, hidden-layer visible deep stacking network (HVDSN), extreme energy ratio (EER), AE and DBN. For the features extracted by CSP, ACSP, EER and EED, the support vector machine (SVM) is as the classifier. For stack hidden visible deep network, the number of hidden-layer nodes is set at 50.

The recognition accuracy of different algorithms is given in Table 3, and the highest recognition accuracy for each subject is highlighted in black bold face. As Table 3 shows, compared with the other methods, PSECN can achieve superior performance in motor imagery EEG recognition. EED, EER and CSP are utilized for different criteria for feature extraction and ACSP reduces the influence of time variation on non-stationary signals by adding a variable factor. Because these four algorithms are

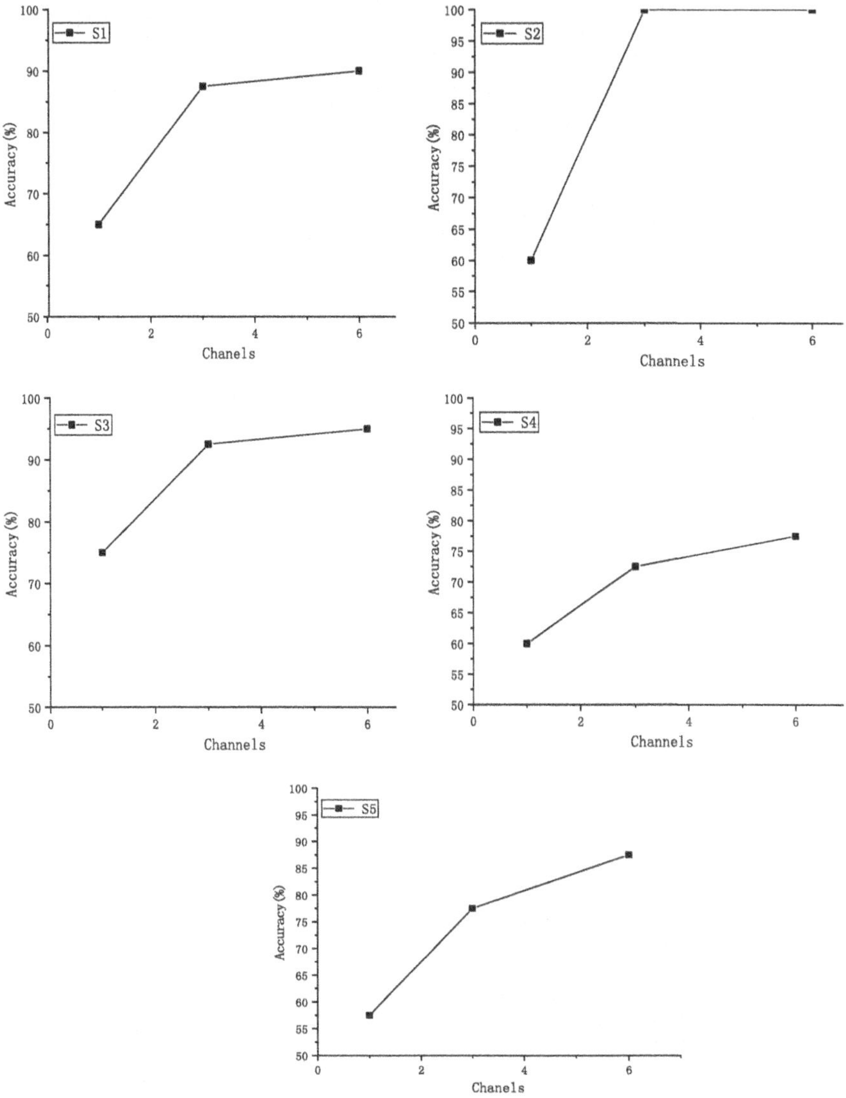

Fig. 9. Recognition rate of single channel and multi-channel EEG signals

artificial feature extraction methods and extract spatial features, it can't extract EEG features well. Supervised training for each module of HVDSN, resulting in waste of unlabeled samples. DBN have good feature extraction capabilities, but destroy the independent characteristics of the channel. The proposed method is utilized for independent pre-training for multi-channel EEG signal and used 1-D convolution in CNN, which makes full use of hidden spatiotemporal features and avoid the problem that manual feature selection is easy to lose important features.

Table 3. The recognition accuracy (%) of different algorithms for five subjects

Subject	EER	EED	CSP	ACSP	DBN	HVDSN	AE	PSECN
S1	86.67	87.50	89.17	87.50	77.50	86.67	78.33	**90.00**
S2	96.67	98.33	96.67	91.67	84.17	92.50	86.67	**100.00**
S3	86.67	75.50	89.17	93.33	89.16	85.00	90.00	**95.00**
S4	70.83	72.5	58.33	60.83	87.50	91.67	**91.66**	77.50
S5	75.00	77.50	70.83	84.17	70.83	77.50	75.83	**87.50**
S1	86.67	87.50	89.17	87.50	77.50	86.67	78.33	**90.00**

To evaluate the performance of PSECN on data set 2, some classical EEG feature extraction algorithms like Coyle, CSP, EED and ACSP are compared. As Table 4 shows, the present method achieves better classification results in BCI competition IV for most subjects and it can obtain relatively high recognition accuracy and stability, which indicate the proposed method has shown robust EEG feature extraction ability.

Table 4. The recognition accuracy (%) of different algorithms on BCI competition IV

Subject	EED	CSP	ACSP	Coyle	PSECN
S1	56.56	66.56	67.50	60.00	**80.56**
S2	52.14	57.86	55.36	56.00	**60.91**
S3	58.13	61.25	**62.19**	56.00	56.94
S4	89.69	94.06	94.69	98.00	**100.00**
S5	64.69	80.63	76.88	79.00	**81.08**
S6	71.88	75.00	75.94	75.00	**81.94**
S7	56.25	72.50	71.25	69.00	**75.00**
S8	80.00	89.38	89.38	**93.00**	84.21
S9	70.63	**85.63**	81.25	73.00	83.33

5.4 Online Experiment

In order to verify the superiority of the proposed algorithm on the BCI system, online experiment is conducted on intelligent wheelchair BCI system. It includes EEG signal acquisition, signal processing, communication module, control System and intelligent wheelchair, as shown in Fig. 10. In online experiment, EEG and Electromyography (EMG) which are collected by Emotiv simultaneously are performed on intelligent wheelchair BCI system. The stop state of the wheelchair is controlled by a high-frequency wave (EMG) which is generated by clenching teeth in the F8 channel. The forward direction of the wheelchair is controlled by a medium-frequency wave (EMG) which is generated by blinking in the F7 channel. The rotation direction of wheelchair is determined by the classification results of the proposed algorithm(EEG). The accuracy of online recognition (straight, left turn and right turn) is shown in Table 5.

Table 5. The recognition accuracy (%) of online experiment

Direction of rotation	Accuracy (%)
Left	70
Right	75
Straight	90

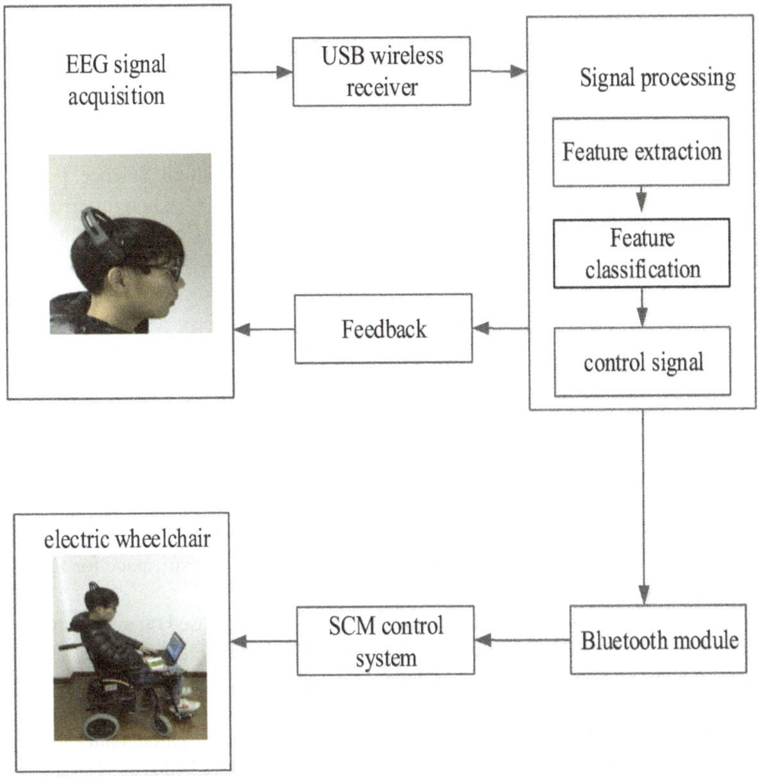

Fig. 10. BCI system architecture deployment

It can be seen from Table 5 that the online recognition rate of EMG signals is higher than that of EEG signals. This is because EMG signals have more obvious features than EEG signals and are easier to recognize. Comparing the recognition accuracy rates in Table 3, Table 4 and Table 5, it can be found that the overall offline recognition rate of EEG data is higher than the online recognition rate, which is due to various factors such as environmental impact, subject caused by fatigue, etc. In general, the intelligent wheelchair BCI system designed in this paper can better complete the forward, left and right turn functions, which verifies the effectiveness of the algorithm in this paper.

6 Conclusion

Traditional EEG feature extraction algorithms with artificial feature selection are easy to lose hidden information and do not consider the spatiotemporal features of EEG data. The proposed method makes full use of the spatiotemporal features of EEG data for pre-training and the supervised network is used for feature classification. The test results show that: the proposed method which automatically extracts abstract features avoids that the manual feature selection losing hidden features, Furthermore, the multiple SDAs process can individually extract EEG signal features of each channel and avoid the interference between the various channels of EEG signals. In an actual BCI system application, it is necessary to identify multi-classification of EEG signals to obtain more external functions for controlling external devices.

Acknowledgement. This work is supported by the National Natural Science Foundation of China under Project 61673079, the Natural Science Foundation of Chongqing under Project cstc2018jcyjAX0160 and the Science and Technology Project of Chongqing Education Committee under Project KJQN201902404.

References

1. Ang, K.K., Chua, K.S.G., Phua, K.S., et al.: A randomized controlled trial of EEG-based motor imagery brain-computer interface robotic rehabilitation for stroke. Clin. EEG Neurosci. **46**(4), 310–320 (2015)
2. Vansteensel, M.J., Pels, E.G.M., Bleichner, M.G., et al.: Fully implanted brain–computer interface in a locked-in patient with ALS. N. Engl. J. Med. **375**(21), 2060–2066 (2016)
3. Li, P., Wang, X., Li, F., et al.: Autoregressive model in the Lp norm space for EEG analysis. J. Neurosci. Methods **240**, 170–178 (2015)
4. Zhang, Y., Liu, B., Ji, X., et al.: Classification of EEG signals based on autoregressive model and wavelet packet decomposition. Neural Process. Lett. **45**(2), 365–378 (2017)
5. Wu, W., Chen, Z., Gao, X., et al.: Probabilistic common spatial patterns for multichannel EEG analysis. IEEE Trans. Pattern Anal. Mach. Intell. **37**(3), 639–653 (2015)
6. Owen, F., Kenneth, P.C., Joseph, M.: The analytic common spatial patterns method for EEG-based BCI data. J. Neural Eng. **9**(4), 045009.1–045009.13 (2012)
7. Wei, Y., Zhao, Y., Lu, C., et al.: Cross-modal retrieval with CNN visual features: a new baseline. IEEE Trans. Cybern. **47**(2), 449–460 (2017)
8. Young, T., Hazarika, D., Poria, S., et al.: Recent trends in deep learning based natural language processing. IEEE Comput. Intell. Mag. **13**(3), 55–75 (2018)
9. Gan, M., Wang, C.: Construction of hierarchical diagnosis network based on deep learning and its application in the fault pattern recognition of rolling element bearings. Mech. Syst. Signal Process. **72**, 92–104 (2016)
10. Jirayucharoensak, S., Pan-Ngum, S., Israsena, P.: EEG-based emotion recognition using deep learning network with principal component based covariate shift adaptation. Sci. World J. **2014**, 1–10 (2014)
11. An, X., Kuang, D., Guo, X., et al.: A deep learning method for classification of EEG data based on motor imagery. In: 2014 International Conference on Intelligent Computing, pp. 203–210. Springer, Cham (2014)

12. Jiao, Z., Gao, X., Wang, Y., et al.: Deep convolutional neural networks for mental load classification based on EEG data. Pattern Recogn. **76**, 582–595 (2018)
13. Jin, Z., Chuangang, Y., Xiaoliang, G.: Deep convolutional neural network for decoding motor imagery based brain computer interface. In: IEEE International Conference on Signal Processing, Communications and Computing, pp. 1–5. Piscataway, USA (2017)
14. Längkvist, M., Karlsson, L., Loutfi, A.: A review of unsupervised feature learning and deep learning for time-series modeling. Pattern Recogn. Lett. **42**, 11–24 (2014)
15. Chen, M., Weinberger, K.Q., Xu, Z., et al.: Marginalizing stacked linear denoising autoencoders. J. Mach. Learn. Res. **16**(1), 3849–3875 (2015)
16. Li, D., Yang, H., Yuan, C.A., et al.: Leaf classification utilizing densely connected convolutional networks with a self-gated activation function. In: International Conference on Intelligent Computing, pp. 370–375. Springer, Cham (2018)

Personalized Automated Itineraries Generator for Tourism

Ki Ageng Satria Pamungkas(✉) and Dade Nurjanah

School of Computing, Telkom University,
Telekomunikasi St. No. 1, Bandung, Indonesia
iampamungkas@students.telkomuniversity.ac.id,
dadenurjanah@telkomuniversity.ac.id

Abstract. In this present era, people want everything to be quick and easy. In tourism, for example, people want an easy way to get information about attractions, events or the best city itinerary. With the huge data on the internet, a personalized recommender system is needed to filter the information and make it useful. In recent years, not only the recommendation for attractions are developed, but there is also the urgency of itinerary recommendation. In this paper, we would like to propose a system that can generate an itinerary that matches user preferences. The itinerary generation process will be dealt with using Ant Colony Optimization which is capable to solve the Traveling Salesman Problem in a good manner. On the other hand, to match user preferences, Personalized Technique will be introduced to create a recommendation that will be used to create a good itinerary that considers user preferences, optimal route, and preferred time to visit attractions.

Keywords: Tourism · Itinerary · Recommender system · Ant Colony Optimization · Personalized Technique

1 Introduction

In the last few years, people all over the world are experiencing the trend of digital traveling because of the fourth industrial revolution. This trend also resulted in an explosion of volumes of information on the internet about the best tourist attractions and travel itineraries. However, the traveling itineraries are mostly made of only based on the most famous attractions and sometimes the tourist itself is not comfortable with the type of attractions or the pace of the traveling itineraries on the internet. Hence, there is a need for personalized travel itineraries tools to help tourists plan their traveling activities.

The first step in the making of personalized traveling itineraries tools is to filter out tourism attractions based on what the user wants. This step involved taking user preferences and then find a list of tourist attractions that matched it. User preferences that commonly used for tourism are attraction types such as Historical Site [7], Cultural, Wildlife, and Amusement Park. One of the techniques that are commonly used in this task is Personalization techniques [5]. It is a technique to cluster an information object based on user preferences. This technique is commonly used for the

R. Silhavy et al. (Eds.): CoMeSySo 2020, AISC 1295, pp. 714–727, 2020.
https://doi.org/10.1007/978-3-030-63319-6_66

recommender system. In the case of tourism, User preferences usually are attraction type, the location area of attraction [7], popularity level and lastly price of the tickets. These user preferences are used to filter all possible attractions by calculating the similarity value of the attractions and user preferences.

After having a list of user-preferred attractions, itineraries also need a good route in the manner of the distance between each location and the preferred time for each attraction. The step of creating an optimal route, including multiple destinations with restrictions is a superclass of TSP (Traveling Salesman Problem) which is known as an NP-hard problem [9]. The most known algorithm to semi-solves the problem is Ant Colony Optimization [7] and Genetic Algorithm [9]. Ant Colony Optimization(ACO) is an optimization algorithm to create a path/route and solve TSP by using the behavior of ant when they are traveling [4]. It spread out a so-called pheromone through its path and identify which path has the most pheromone which will be the optimum path to be taken. In the recent study of ACO, is now has been improved and implemented to a humanoid to solve a complex combinatorial problem [11].

While it is possible to create an optimal route in the manner of distance using ACO, we must also consider whether each attraction is placed at the right time or not. For example, a good time to visit farmland is in the morning to afternoon, such that we must allocate this attraction at the morning slot of time in the itinerary. This thing can be done by modifying Ant Colony Optimization to also user preferred time as one of the variables upon taking a decision where to go next. The preferred time will act as an increase of chance to be visited by ant collaborating with the amount of pheromone are there.

With all the components to create an itinerary are set, the goal here is to create a good itinerary. A good itinerary can be described as an itinerary that considers what attractions users are preferred to, then placed in such a route that has optimal traveling distance yet also considers the timeline in which attractions should be visited.

2 Literature Review

2.1 Personalization Technique

Personalization Technique is a technique to analyze data of the user to know what they want and what they like and use it to show another similar object from what they like [5]. There two most commonly used algorithms for personalization techniques are Collaborative Filtering and Content-Based [3].

Collaborative filtering is a technique that uses the data of a user and searches the other user that has similar interest with it by comparing the rating given by the user [2]. Then it will recommend objects that rated good by this similar user. This technique is commonly used for movies, books, and e-commerce goods. This technique excels for huge data and experienced users. But if new users come and they have not rated any object, this technique will be kind of difficult.

While collaborative filtering match users with users, content-based match the user with the object itself [2]. How it works is by defining a value of category that users like into an object then calculate the similarity of its value. This technique commonly used for new users, online advertising by consuming data from its past activities. Although it is good for new users, if this technique meets a huge number of objects, its performance will decrease as it needs to calculate the similarity of each data.

SigTur [1] is a recommendation system that uses Collaborative Filtering, Content-Based, and Demographic Clustering to get a sing overall rating as the recommendation. The idea is the combination of all algorithms will make a perfect balance rating based on user preferences, user ratings, and user demographic location. As it balances, the algorithm may overcome cons of another algorithm.

SPETA [6] propose another unique system where it uses collaborative filtering and knowledge-based sequentially. First, a collaborative filter will be used to deal with all data and get user preferences based on the filter. Then user preferences will be passed to knowledge-based to create a recommendation.

REJA [8] on the other hand propose a system that uses these algorithms for a different state of user. Demographic for new users as it does not have data to use other algorithms. Content-Based the used as the user search for more detail attractions. And lastly, the Collaborative filtering used as the user has already give rating to any attractions. This kind of implementation kind of optimize each of the algorithms pros and overcoming its cons on the process.

2.2 Ant Colony Optimization

Ant Colony Optimization is an artificial intelligence algorithm that inspired by ant behavior of ants organizing their colony [1]. When they are traveling around, and usually trace their path by marking its path with so-called pheromone. The use of pheromone is to give mark that it is the right way to go to the food [1].

The implementation of Ant Colony Optimization itself is it acts as post-processing for a recommendation system on a domain that needs a step-by-step route upon doing it. In tourism, for example, it [2] is used as one of the agents that consume the result of a recommendation to create an itinerary. The use of ACO in tourism is because it plans the best possible route among all routes possible. It is one of the most reliable ways to solves the TSP problem in the right manner.

So technically Ant Colony Optimization will compute matrices for pheromone and distance, then on each iteration, it will update pheromone value on the matrices. Every time an ant comes to a certain state, it will then calculate a probability where the ant will go next and lay a pheromone. This idea of exploration or exploitation is used with a rule such that the itinerary grows.

$$f(x) = \begin{cases} \arg\max\limits_{u \in j_k(r)} \left([\tau(r,u)] \cdot [\eta(r,u)]^{\beta} \right), \; if\, q \leq q_0 \\ S, Otherwise \end{cases} \tag{1}$$

where $\eta(r,u) = \frac{1}{\delta(r,u)}$, $\delta(r,u)$ is the distance between cities r and u, Jk(r) is the set of cities that remained to be visited by ant k positioned on city r, β is a parameter which determines the relative importance of pheromone versus distance and is over zero. The probability calculation use the epsilon greedy algorithm where firstly it declares a certain value called gamma (q_0), then on each iteration it randoms new float (q). If

$q \leq q_0$, then exploration must be done. Else exploitation must be done. Exploration means that the ant will pick a random route while exploitation means ant will go to the best nearby node. $\tau(r,u)$ is the amount of pheromone in the edge from r to u, S is a random route that is selected using:

$$P_k(r,s) = \begin{cases} \frac{[\tau(r,s)]\cdot[\eta(r,s)]^\beta}{\sum_{u \in j_k(r)}[\tau(r,u)]\cdot[\eta(r,u)]^\beta}, & if\ s \in j_k(r) \\ 0, & if\ otherwise \end{cases} \quad (2)$$

$P_k(r,s)$ is the probability that ant k in city r will go city s. The pheromone that will be laid out have a function of:

$$\tau(r,s) \leftarrow (1-\varphi)\cdot\tau(r,s) + \varphi\cdot\Delta\tau_k(r,s) \quad (3)$$

φ is the decaying rate of pheromone and $\Delta\tau_k(r,s) = \tau_0$, τ_0 is the initial value of pheromones if $\Delta\tau_k(r,s)! = \tau_0$:

$$\Delta\tau_k(r,s) = \begin{cases} \frac{1}{L_{gb}}, & if\ (r,s) \in Ktours \\ 0, & if\ otherwise \end{cases} \quad (4)$$

L_{gb} in the equation is the length in which the global-best tour occurs. It means that each iteration, the pheromone will decay in value based on the time spent. This behavior mimics how in real ant pheromone will also decay time by time. The purpose of this behavior is such that the itinerary grows by time to find the optimal route.

Below is the summary of every parameters that ant colony optimization use (Table 1):

Table 1. Ant Colony Optimization parameters

Parameters	Descriptions
Epsilon greedy (q_0)	The rate in which decide whether ant will do an exploration or exploitation
Decaying rate (φ)	The rate in which pheromone value will be decreased in each iteration
Distance relevancy (β)	Distance impotency level. Used to make distance value more impactful upon next node probability calculation
Number of Ants	How many ants will be deployed on each iteration
Number of Iterations	How many iterations it will took to finished creating a route

3 Proposed System

3.1 Attraction Data

The attractions data are gathered through two steps. First, a list of Bandung Attractions is gathered from the internet. The data that are taken consist of name, address, coordinate, open hour, close hour, price, recommended duration and description. Then, the list is passed through a questionnaire to Telkom university student that has identity from Bandung to tags and give the suitable time to visit the attractions. In the database, there is 21 attractions located around Bandung. The attractions name was gathered from Disbudpar of West Java web bit.ly/2FxYYXd. Then another attribute where gathered using google. The example of data gathered is such (Table 2):

Table 2. Attraction data examples

Attribute	Value	Description
Name	Floating Market	Attractions Name
Latitude	−6.818960	Latitude Coordinate used to calculate distance
Longitude	107.618518	Longitude Coordinate used to calculate distance
Opening Hour	09.00	Time in which the attraction open
Closing Hour	19.00	Time in which the attraction closed
Recommended Duration	150 min	Duration should be spent in the attraction
Recommended Hour	09.00–12.00	Time in which the attraction is optimally visited
Tags	Culture, Shopping, and Popular	Attractions characteristics to match user preference

For the tags, there are several other kinds of tag namely Culture, Shopping, Hidden Paradise, Popular, Wildlife, Museum, Mountain, Beach, History, and Amusement.

3.2 User Preferences

Table 3. Categories and tags

Categories	Tags
Popularity	Popular, Hidden Paradis
Attraction Type	Culture, Shopping, Wildlife, Amusement, History
Physical Appearance	Waterfall, Mountain, Museum

Based on the tags, there are several groupings called categories such as popularity that covers popular and hidden paradise; attractions type that covers culture, shopping, wildlife, amusement, and history; attractions physical appearances such as mountain, waterfall, and museum. These categories will be shown to users like grouping in which users select their preferences. Any combination is possible except for the popularity categories; user can only select one kind of popularity. User then need to select their combination of tags to set the user preferences that will be used to filter and customize their itinerary.

3.3 Personalization Technique

Based on the preferences that previously selected by the user, we then filter out all the attraction data using that preference such that only attractions that have at least one matched tag left. The filtering is being done to prune the data such that the similarity distance is not quite far from the user preferences (Fig. 1).

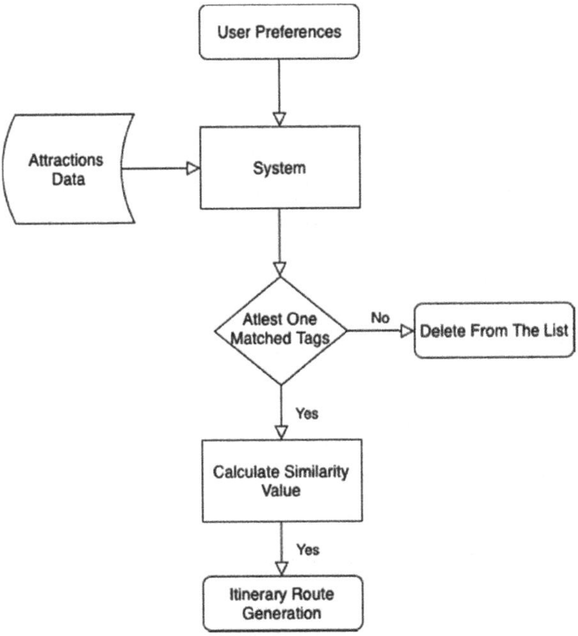

Fig. 1. Personalization technique flowchart

Then each attraction data assigned by a map in which use tags as the key and Boolean as the value. If an attraction has the tags, the map value of that tags will be true, otherwise it will be false. The list then ranked by calculating the similarity between each attraction to the user preferences. This method is called Content-Based filtering. It is simply calculated the distance between user preferences and attractions tags. As we already have maps that represent attractions tags, we then need to calculate the distance using Euclidean distance.

$$Similarity = \sqrt{\sum\nolimits_{i=1}^{m} (UserPreferences_i - AttractionsTags_i)^2} \quad (5)$$

Using Euclidean distance, attractions that have their tags matched will get smaller value as it is considered much more like user preferences. The more least tags matched, the greater its value. Keep in mind that attractions may have one matched tag, but they may have another tag with them.

For example, we want to calculate the floating market (Table 3) similarity distance with user preference that chooses shopping, mountain, and history. Then compare the tags and the user preferences. Then we count how many tags and user preferences are different. We found that both are different in mountain, history, popular and culture such that we have four different tags. To get the similarity distance, we only need to square root the number of different tags. So, the similarity distance of floating market to that user preference is 2.

3.4 Itinerary Route Generation

After we filter the list and get the similarity distance, the list then passed to the route generation engine. As we use ant colony optimization here to find the route, two matrices are needed. The first one will hold the distance between attractions. Then the second one will hold pheromone value. Both matrices will have the size of N×N with N is the number of attractions on the list. Each element of matrices will represent the edge from one attractions to another.

The pheromone matrix is much easier to build as we only need to initialize the value of each element to be zero. To build the distance matrix, we will use the Haversine formula to calculate the circular distance of two coordinates. We use this formula because the coordinates that are saved are the earth coordinates. Haversine formula can be used to calculate the distance between two coordinates in the sphere shape. The distance formula is as follow:

$$2r \arcsin\left(\sqrt{sin^2\left(\frac{\varphi_2 - \varphi_1}{2}\right) + \cos(\varphi_1)\cos(\varphi_2)sin^2\left(\frac{\lambda_2 - \lambda_1}{2}\right)}\right) \tag{6}$$

Where φ_1, φ_2 are the latitude of both points in radiant and λ_1, λ_2 are the longitude of both points in radiant. This formula uses trigonometric to project booth coordinate into sphere shapes and then calculate the circular distance between both. We cannot simply reuse eucledian distance again because the earth is a sphere.

Because one of our goals is to create an optimal route, distances must take a significant role in the algorithm. Variable β is the control value that determines the relative importance of distance when calculating node probability and deciding between exploitation and exploration. We set β to be 0.8 such that distance has an impact (Fig. 2).

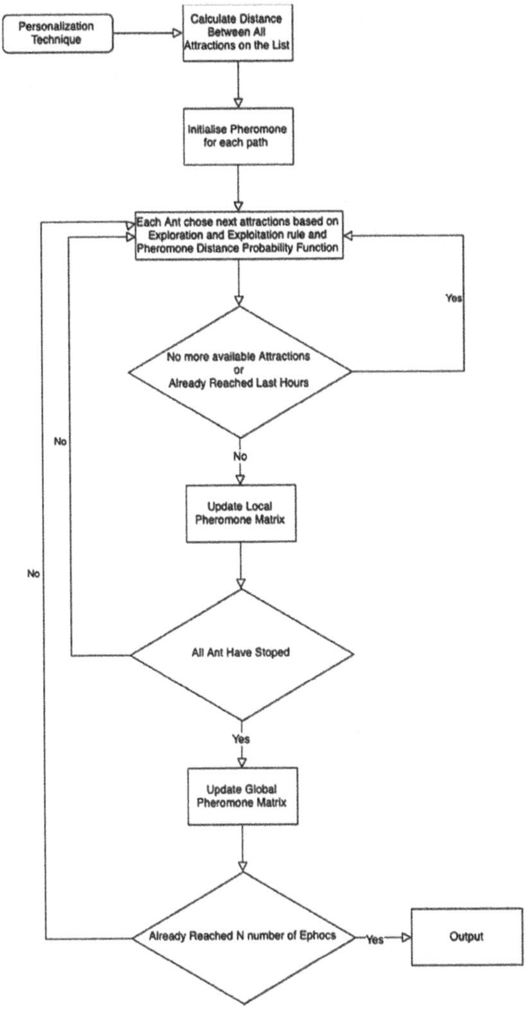

Fig. 2. Ant Colony Optimization flowchart

Another goal that we need to achieve is to make sure that attraction should be visited on its recommended duration. What we are going to experiment here is to modify the pheromone value for attractions that if the ants visit them, it is inside the recommended duration range when they are arrived after calculating the traveling time. The modification rule is as follow:

$$\tau(r,s) = \begin{cases} \tau_{max}, \; if \; ArrivalTime \cap RecommendedHours \\ \qquad [\tau(r,s)], \; if \; otherwise \end{cases} \tag{7}$$

τ_{max} is the maximum value of pheromone of the edge that connected to r. This experiment will make that if an attraction is on its recommended duration upon arrival, only distance that will be used to calculate the probability for each next destination (Fig. 3).

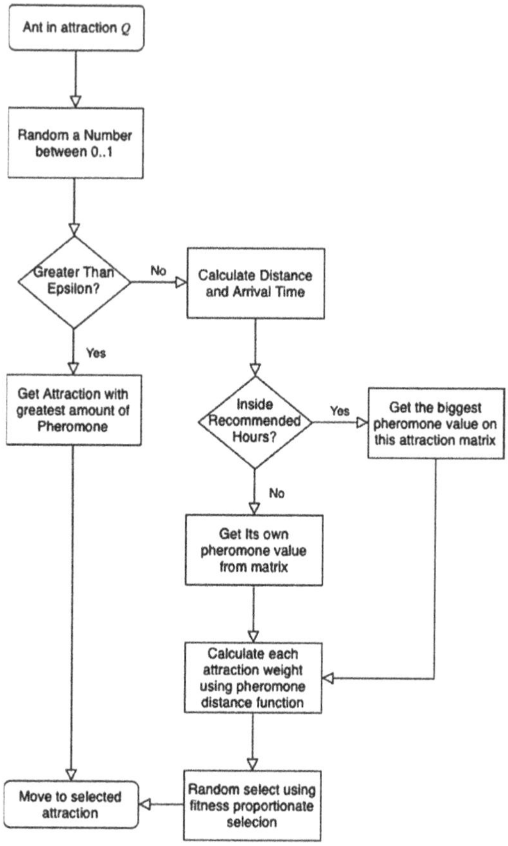

Fig. 3. Decision making flowchart

To accommodate personalization technique inside the route creation, we are experimenting by modifying the amount of pheromone laid if an edge is visited as follows:

$$\Delta\tau_k(r,s) = \begin{cases} \frac{\frac{1}{2}MaxSimilarity - Similarity_s}{L_{gb}}, \; if\ (r,s) \in Ktours \\ 0, \; if\ otherwise \end{cases} \tag{8}$$

With $Similarity_s$ is the similarity value of the destination and MaxSimilarity is the similarity value if an attraction only has one matched tag with user preference. By doing this experiment, we hope that user preference also makes an impact while creating the route for the itinerary.

We also consider opening and closing of the attractions while deciding the next destination. We calculate the time of arrival for all possible next destination and then if it has not open or it has already closed, we exclude them from the possible next destination list.

Based on all of this rule, ant will be deployed to circulate until there is no more possible next destination. This can be achieved by setting a hard limit on the hours that the itinerary should stop. We are experimenting by setting up that the last attraction must be visited before 7:00 PM upon arrival.

4 Experiment and Evaluation

We are experimenting with two different user preferences as follows (Table 4):

Table 4. Test case preferences

Name	Preferences
Preference 1	Hidden Paradise, Mountain, Waterfall, Wildlife
Preference 2	Popular, History, Museum, Shopping
Preference 3	History, Waterfall

The first and second user preferences are selected to evaluate how the personalization technique handles multiple tags and categories combination with emphasize on the popularity as each preference specified on each one of them. Note that we here do not try to test the tags because the tags can easily be swapped with other tags. We want to test whether the system can handle multiple tags preference. Then the third user preferences is selected to evaluate how the system would handle such an input that may have far geographical distance as history and waterfall attraction are usually not close in distance.

For the Ant Colony Optimization options, we set the parameters to be (Table 5):

Table 5. Experiment parameters

Parameters	Value
Epsilon greedy (q_0)	0.5
Decaying rate (φ)	0.2
Distance relevancy (β)	0.8
Number of Ants	50
Number of Iterations	2000

The epsilon greedy are set to be as it is such that the rate of exploitation and exploration are the same [7]. As we also modified the ant colony to accommodate the personalization technique, we set the distance relevancy to be only 0.8 such that it does not dominate our modification completely. As we know from past studies [4, 7, 10] the number of ant will impact the quality of the route as more ant means more route can be explored. With the number of iterations set to be two thousand, there will be a hundred thousand route explored by the colony.

4.1 Experiment Result

Preference 1 result is as follow (Table 6):

Table 6. Experiment one result

Attractions	Similarity distance	Recommended time	Database preferred time
Curug Batu Templek	0	08.04–0.04	09.00–16.00
The Lodge Woodland Maribaya	1	10.54–13.54	09.00–16.00
Taman Lembah Dewata	1	14.50–16.50	07.00–12.00
Gunung Batu	1	18.18–20.18	17.00–20.00

This itinerary traveling distance is around 32 km starting from Telkom University at 6 AM and ending in Gunung Batu at 8 PM to watch Bandung scenery in the night. From the table, we can see that the itinerary consisted of items in which has the lowest possible similarity distance from this user preference with the matching score of 81,25%. The name of attractions is also match with the user preference because it is dominated by mountain, wildlife and waterfall type attractions. The itinerary also has a quite good timeline comparison with each attraction recommended hours to visit with score of 75%. Nearly every attraction is visited inside its recommended hours except the third attraction, Taman Lembah Dewata. This might be happening because it has low similarity distance combined with the actual distance traveled from previous attraction compared to other attractions with matched recommended hours. The traveling time compared to attraction time are at 5 h on the road and 9 h on the location. Although it seems that the traveling time is taking so long, it is the first trip from Telkom to Curug Batu Templek. This happened probably because the closest possible attraction that already open at 6 PM is only Curug Batu Templek.

Preference 2 result is as follow (Table 7):

Table 7. Experiment two result

Attractions	Similarity distance	Recommended time	Database preferred time
Dago Dream Park	1.4	08.20–12.20	09.00–16.00
Taman Hutan Raya Ir. H. Djuanda	1.4	12.33–14.33	10.00–19.00
Gedung Sate	1	15.27–18.27	08.00–16.00
Cihampelas Walk	1.4	18.43–21.13	15.00–19.00

The second itinerary has a traveling distance of around 22 km starting from Telkom University at 6 AM and ending with a shopping place of Cihampelas Walk at 9 PM. In this itinerary, no attraction matched user preference perfectly. The similarity distance varies from one to two matched tags for each attraction. From the name of the attractions, we can also see that the type of variety is much more spread but still matched with the user preference with the score of 56,25%. Every single preference is accommodated with at least one attraction. In this itinerary, the timeline is perfectly matched with the recommended hours for each attraction with score of 98,4%. Traveling time is still high because of the first trip from Telkom University to the first attractions because of the same reason as the previous itinerary.

Preference 3 result is as follow (Table 8):

Table 8. Experiment three result

Attractions	Similarity distance	Recommended time	Database preferred time
Curug Batu Templek	1	08.04–10.04	09.00–16.00
Gedung Sate	1	11.23–14.23	08.00–16.00
Museum Konferensi Asia Afrika	1	14.44–17.44	08.00–16.00

Our third preference are made to explore an extreme value of traveling distance. The route has a traveling distance of 23 km starting from Telkom University. As we set the user preferences to be so extreme, no attraction can perfectly have matched the user preferences with the score of 50%. The route also accommodates each tag at least one attractions. Although there are limited number of attraction that matched, the route still preserve its quality with two attraction visited on their recommended hours.

5 Conclusion

Based on the experiment result, the proposed system, personalized itinerary generator successfully generates itineraries. They use personalization techniques to matched user preferences in selecting attractions. Both itineraries show a low number of similarity distances for each attraction inside them. Therefore, our combination is performing well upon creating the attraction list. The experiment to involve similarity distance upon route creation also increase the performance of the recommendation.

The route experimentation using modified ant colony optimization is also performing well judging by the amount of traveling time from one attraction to another. The number of attraction visited during recommended hours shows that the modification experiment working well and making the itinerary good itinerary since it makes ant colony consider when is the best time to visit an attraction. We also have tried an extreme user preference that has a high traveling distance between any of its attraction on the recommendation list. The system still creates the most optimum route while still preserving the quality of the itinerary. Hence, the personalized itinerary generator can suggest a good personalized itinerary.

References

1. Borràs, J., de la Flor, J., Pérez, Y., Moreno, A., Valls, A., Isern, D., Orellana, A., Russo, A., Anton-Clavé, S.: SigTur/E-Destination: a system for the management of complex tourist regions. In: Law, R., Fuchs, M., Ricci, F. (eds.) Information and Communication Technologies in Tourism 2011, pp. 39–50. Springer Vienna, Vienna (2011)
2. Borràs, J., Moreno, A., Valls, A.: Intelligent tourism recommender systems: a survey. Expert Syst. Appl. **41**, 7370–7389 (2014). https://doi.org/10.1016/j.eswa.2014.06.007
3. Chiang, H.-S., Huang, T.-C.: User-adapted travel planning system for personalized schedule recommendation. Inf. Fusion **21**, 3–17 (2015). https://doi.org/10.1016/j.inffus.2013.05.011
4. Dorigo, M., Di Caro, G.: Ant colony optimization: a new meta-heuristic. In: Proceedings of the 1999 Congress on Evolutionary Computation-CEC99 (Cat. No. 99TH8406), Washington, DC, USA, pp. 1470–1477. IEEE (1999)
5. Gao, M., Liu, K., Wu, Z.: Personalisation in web computing and informatics: theories, techniques, applications, and future research. Inf. Syst. Front. **12**, 607–629 (2010). https://doi.org/10.1007/s10796-009-9199-3
6. García-Crespo, A., Chamizo, J., Rivera, I., Mencke, M., Colomo-Palacios, R., Gómez-Berbís, J.M.: SPETA: social pervasive e-Tourism advisor. Telematics Inform. **26**, 306–315 (2009). https://doi.org/10.1016/j.tele.2008.11.008
7. Lee, C.-S., Chang, Y.-C., Wang, M.-H.: Ontological recommendation multi-agent for Tainan City travel. Expert Syst. Appl. **36**, 6740–6753 (2009). https://doi.org/10.1016/j.eswa.2008.08.016
8. Martinez, L., Rodriguez, R.M., Espinilla, M.: REJA: a georeferenced hybrid recommender system for restaurants. In: 2009 IEEE/WIC/ACM International Joint Conference on Web Intelligence and Intelligent Agent Technology, Milan, Italy, pp. 187–190. IEEE (2009)
9. Maruyama, A., Shibata, N., Murata, Y., Yasumoto, K., Ito, M.: P-tour: a personal navigation system for tourism 12

10. Pedemonte, M., Nesmachnow, S., Cancela, H.: A survey on parallel ant colony optimization. Appl. Soft Comput. **11**, 5181–5197 (2011). https://doi.org/10.1016/j.asoc.2011.05.042
11. Yang, J., Zhuang, Y.: An improved ant colony optimization algorithm for solving a complex combinatorial optimization problem. Appl. Soft Comput. **10**, 653–660 (2010). https://doi.org/10.1016/j.asoc.2009.08.040

Virtual Training Environments in a Printing Company

Pavel Pokorný[1(✉)] and Michal Birošík[2]

[1] Department of Computer and Communication Systems, Faculty of Applied Informatics, Tomas Bata University in Zlín, Nad Stráněmi 4511, 760 05 Zlín, Czech Republic
pokorny@utb.cz

[2] Department of Informatics and Artificial Intelligence, Faculty of Applied Informatics, Tomas Bata University in Zlín, Nad Stráněmi 4511, 760 05 Zlín, Czech Republic
michal.birosik@gmail.com

Abstract. Although Virtual Reality technology has existed for decades, it has currently been experiencing a boom. This is - in particular, ensured by significant improvements in the availability and quality of Virtual Reality (VR) hardware that this technology requires. Virtual Reality is quickly expanding in many fields of human interests - scientific visualisation, the military, entertainment, healthcare, construction, etc. It is also increasingly being used in education and training. With the help of VR, industrial companies can train their employees to work more effectively, to improve their performance or to perform safe training programmes in high-risk scenarios. This paper sets out to describe proposed and implemented solutions for the use of VR for the new employee training process in a printing company. This solution includes the choice of a suitable VR headset and corresponding software tools. After that, the necessary graphical assets were prepared - (e.g. 3D models and their textures); these were then imported into the software in which the VR application was created. Its functionality was then tested in our laboratory and will be used in the relevant industrial environment.

Keywords: Virtual Reality · Modelling · Visualisation · Application

1 Introduction

The term Virtual Reality (VR) represents the computer-generated simulation of a 3D environment, which seems very real to the person experiencing it, using special electronic equipment. Its main objective is to achieve a strong sense of being present in the virtual environment. Because the user needs to view stereoscopic 3D scenes, today's VR technology involves wearing a head-mounted display (headset) [1].

VR is based on five key elements – a virtual world, immersion, interactivity, as well as participants and creators [2]. A Virtual World represents an imaginary space that describes a group of objects in a given space and relationships among them. Immersion means the possibility of perceiving something besides the non-physical (normal) world.

R. Silhavy et al. (Eds.): CoMeSySo 2020, AISC 1295, pp. 728–738, 2020.
https://doi.org/10.1007/978-3-030-63319-6_67

Interactivity is the virtual world's response to users' action. These responses are generated in order for a user to feel fully immersed. The participant is a very important element because everything that happens in the VR world, happens in their mind. Hence, every Virtual Reality experience is different for each person- These experiences are given to creators - (one person or team), who design and create the virtual reality application.

Although consumer VR is only a few years old and was primarily used in the computer gaming field; over time, we are already beginning to see a staggering range of applications in many other areas of human activities. Developers are trying to build everything in the VR field. Understandably, with mixed success so far [3]. Currently, there are several domains in which VR finds applications; these include engineering, the military, robotics, architectural design, entertainment and leisure, tourism, as well as medicine, education and training, and many more.

The use of VR in education and training represents high potential. The highest motivation for using head-mounted displays is that a student/person can be exposed to various educational situations that allow them to repeatedly practice new experiences and skills in an environment that enables their correction and non-dangerous failure. Moreover, research shows that learners are generally very positive about using head-mounted displays - although there are still substantial barriers to their use, especially in regards to cyber-sickness symptoms, the technical limitations of peripheral devices and the lack of appropriate software [4, 5].

This paper deals with the proposed and implemented solution of the use of VR in a printing company for the new employees training process. Employees are one of the key factors in the development of a company - and therefore, it is important to improve their skills and their further development in order to ensure the operation of an efficient production system. New and existing employees expect to be trained, retrained and educated in the technologies and manufacturing processes used so as to understand the system, while gaining practical experience. The company's management team expect quick and effective training - which will not be unduly expensive and will not reduce the production process' profitability, efficiency and profitability. The use of Virtual Reality in this situation thus seems appropriate and advantageous.

2 Description of the Printing Technology

The core of the production process is a GALLUS RCS330 printing machine. In addition to the operation of the printing machine itself, the complete production process also requires the operation of other components and devices. In order to attain successful printing, the process and procedure also require the implementation of the necessary decisions and steps from print preparations to the setting of the printing machine. The following lines briefly describe the work of the operator - which was realised using Virtual Reality.

The production process begins with preparation. This consists in decision-making activities; which components should be used including the appropriate size of the printing cylinders. The production operator goes to the first device - the stand, where the printing cylinders are stored. The cylinders differ in circumference. The production operator chooses a suitable cylinder according to its size. The requirement is that the cylinder should print a surface with a single turn. The production operator takes the suitable printing cylinder and transfers it to another device - the first machine.

This machine glues the printing polymers onto the printing cylinder (Fig. 1). An essential task in the printing process is the transfer of the required colour onto the printing material. The size of the printing polymer is chosen by the production operator based on the size of the order and the size of the printing cylinders. The polymer gluing device works on the principle of inserting a printing cylinder into the space between two arms. Subsequent pressing of a button activates the hydraulic piston, which closes the inserted cylinder between the arms.

Fig. 1. The machine that glues the printing polymers on the printing cylinder

The polymer gluing process proceeds progressively from the initial application, alignment and gluing of the upper starting side. The operator checks the correctness of the initial gluing. If the polymer has not adhered correctly, the glued part is peeled off and the procedure is repeated from the beginning. If the polymer is glued correctly, the operator starts the rotation of the arms, which initiate the inserted printing cylinder's rotation. This rotation ensures that the rest of the printing polymer is glued. When this process is finished, the operator removes the printing cylinder with the glued printing polymer, releases it from the arms and goes to the printing machine with this cylinder.

This machine is used for the actual printing of patterns onto the printing material. The printable material is stored in its input hopper, which - after the start of the printing process moves through the individual machine parts to the printing head. The head assures the printing of parts of the final print pattern in one specific colour. The operator puts the polymer printing cylinders onto the rotating arm and secures them with screws.

The operator must also make the selection of the correct ink cartridge. Therefore, they must go to the shelf where filled and prepared ink tanks are stored, take a selected tank and return with it to the printing machine. They put the ink tank into the specified place of the printing unit of the printing machine.

Before the printing process starts, the printing material must be also put in the input part of the machine. Its type and size is chosen based on the requirements in the order.

Finally, the operator visually checks that everything is prepared correctly and goes to the printing machine control unit. Here, they activate the printing and inking (raster) component, which ensures the contact of the individual cylinders. They then check the machine once again, verify the completeness of all the previous steps of the process and starts the printing process.

Fig. 2. An Oculus Rift head-mounted display with its controllers

3 The Suggested Solution

This chapter will presented the whole suggestion of the solution itself. This includes the selection of the individual hardware and software solutions needed to create and run the planned VR applications. The main focus is on the choice of the suitable VR hardware, since the performance and quality of VR is essential for the final VR application and its practical use. Choosing the right software can speed up the preparation of partial 3D models, their textures and the final application.

3.1 Hardware

In order to create the VR application, it was decided to use a desktop form of Virtual Reality. The advantages of this solution include the fact that it is the most powerful form of VR, including price and efficiency in the use of hardware devices. There are a number of manufacturing companies that are occupied with desktop forms of Virtual Reality technology.

Based on the requirements, Oculus Rift (Fig. 2) was finally selected because it offers a combination of quality and powerful technologies that promise VR at a high level. It uses the OLED display technology, supports 2160 × 1200 resolution, 90 Hz refresh rate and 110 degrees of (diagonal) field-of-view. It also includes an integrated audio VR that provides 3D audio effects, and positional tracking is accomplished with the help of infrared sensors that allow one to use it while standing, sitting or walking around the room. This device is compatible with the Microsoft Windows platform, and the possible development of other applications for this device [6].

3.2 Software Tools

In order to create the planned VR application, three software tools were used in their latest versions – Blender, Substance Painter and Unity. Blender was used to perform the preparation of all necessary 3D models with their corresponding UV maps. Then, Substance Painter was used to create all textures. All models and their textures were then imported into the Unity engine. Unity was used as the developer environment, in which the final application was programmed.

Blender is a free and open-source 3D creation suite. It supports the entirety of the 3D pipeline – modelling, rigging, animation, simulation, rendering, compositing and motion tracking – and even allows video editing and game creation [7]. It provides all the requisite tools for creating models and scenes that may be made into still images or animated movies. In order to attain effective work, Blender supports a wide variety of graphic tools. The created models and scenes can then be exported into many graphic formats - including images (jpeg, jpeg2000, png, tga, tiff), video (avi, mpeg, quicktime) and 3D formats (3ds, dae, dxf, fbx, obj, stl, svg, vrml, x3d), and many others [8].

Textures were created in the Substance Painter suite [9]. This tool encapsulates all necessary commands for the texturing of 3D assets. It supports smart materials and smart masks that are adapted to a user's models, performs real-time painting in 3D space using smart brushes, projection tools and even particles, fast renderer, extracting and baking any material from any 3D object with a GPU accelerated baker, imports assets from many commonly-used graphic formats and then exports them to any game engine [10].

Unity [11] is a cross-platform real-time engine that provides an excellent entry-point into game and 3D application development, balancing features and functionality with price point. The free Unity version allows people to experiment, learn, develop, and sell games before committing any of their hard-earned money. Thanks to its huge user community, Unity removes the elitist separation between programmers, game designers and artists that is typical of high-priced game engines. It makes it possible for anyone to take the first step to bringing their ideas to life [12]. The Unity engine offers primary scripting in the C# programming language, but many scripts are prepared in templates and are easier to use without a deeper knowledge of this programming language [13].

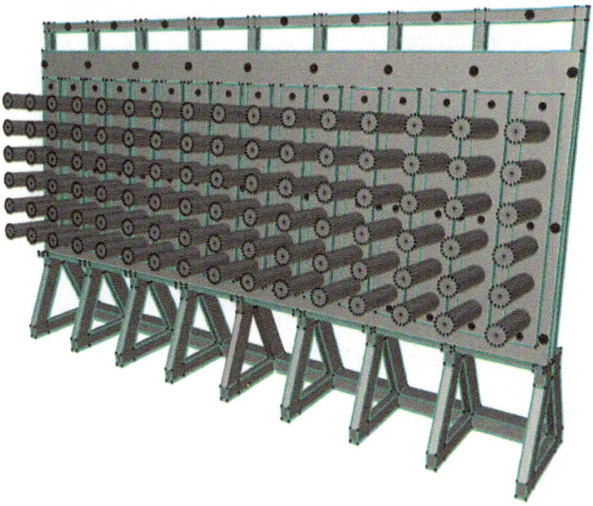

Fig. 3. The 3D model of the printing cylinder stand

4 3D Models and Their Textures

All models created in Blender use polygonal representation – a collection of vertices, edges and faces that define the geometric shapes of objects. Preference was given to the so-called box modelling process for model creation. This is a technique, where primitive shapes are used to begin with - (Planes, Circles, Cubes, Tubes, Spheres, etc.) and these objects are then shaped to the final model using standard modelling tools. Blender offers many editing, transformation and modelling tools and these then often generate new vertices, edges and faces so they allow one to make more detailed and better shaped models. Examples of these tools are Extrude, Bevel, Loop Cut, Knife, Subdivide or Insert Face [14]. Figure 3 shows the model of a printing cylinders stand which was modelled in this way.

Should one need a rounded shape, Blender performs this in the Sculpt Mode. This mode is used for shape-forming, where brushes manipulate an object's whole geometry in the brush region of influence and shapes it based on the dependency of brushes type and its set parameters.

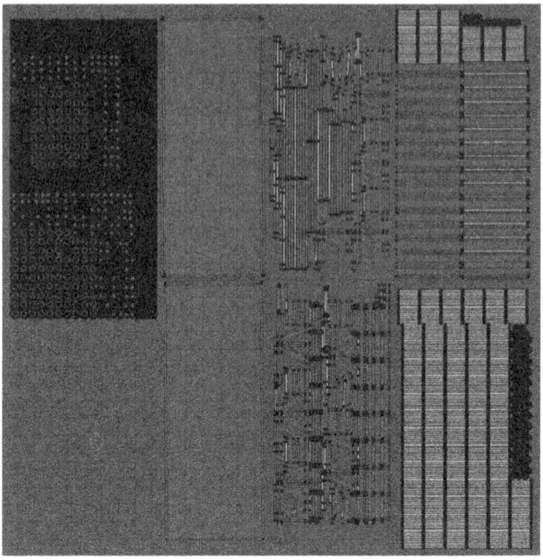

Fig. 4. UV map for the stand from Fig. 2

When the shapes of all mesh models have been created, the UV mapping phase is started. The main goal in this phase is to make UV maps for all of them and their entire faces. In order to obtain UV maps, Blender supports the Unwrap command. Before this command runs, it is necessary to mark the edges of the modelled objects, where the faces must be ripped in order to obtain the correct 2D image. If the UV maps do not correspond to our ideas - (all UV maps should cover image areas as much as possible), they are improved manually. An example of such a UV map is shown in Fig. 4.

When the modelling and UV map have been completed, all created models are then exported via the .fbx graphic format and the UV maps then exported in the .tiff graphic format with a 1024 × 1024 pixel resolution.

All textures were drawn in the Substance Painter application. The technique Relief Mapping was used, because it performs accurate and effective render surface detail of 3D objects. Depending on what texture affects - (colours, transparency, bumps, glossy, etc.), they were then separated into multiple layers. An example of the draw colour layer for the UV map in Fig. 4 is shown in Fig. 5. The final appearance of the surface was specified by a combination of multiple layers of materials or textures.

When all of the textures for the 3D models had been created, they were exported to the Unity engine. Their resolution was increased to 4096 × 4096 pixels in order to attain the very high quality of 3D objects later in the Virtual Reality application and the output texture format was set to .tiff.

Fig. 5. UV texture for the stand from Fig. 2 (Substance Painter Output)

5 An Application

When all graphic assets had been prepared, the next step was to install drivers and the necessary software for Oculus Rift. The requisite Oculus Rift Software installation package was downloaded from the manufacturer's website [15], and then installed. Since we also wanted to use some sounds in our application, the requisite sounds in . mp3 format were downloaded from the website [16]. We planned to use sound effects for the user-interface (select commands from menus, button press, button hover), as well as sounds reproducing walking, running, grabbing, dropping the object, and the sounds of the printing machine while the printing process runs. The appropriate sounds for these actions were downloaded.

Then a new 3D project was created in the Unity engine. The Oculus Integration asset package was added during this creation process in order to have Oculus device support in this project. The next step was to add all prepared assets - (i.e. 3D models, textures, sounds) to this project. After that, the 3D models - with their textures, were placed in the scene. They were transformed - (moved, rotated or scaled) if necessary. The whole VR environment was made in this way. Figure 6 shows the model of the stand with printing cylinders in this phase.

This process was then augmented by the addition of colliders - (simple objects like boxes), which were provided by Unity itself. These were added to all added 3D models and their individual components. Then, colliders that serve as sensors were added in order to perform the running of programmed functions in the event of a collision with another collider in the environment. It was also necessary to add objects based on the

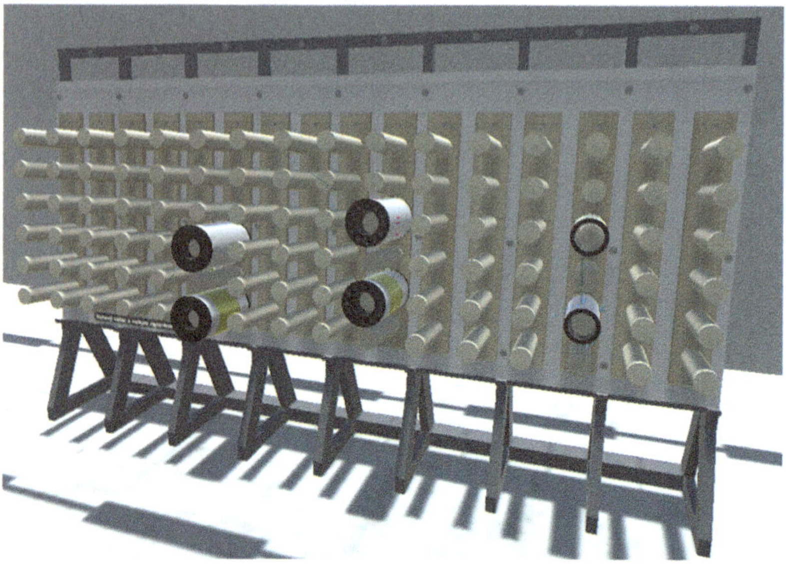

Fig. 6. The stand with mapped textures in the unity engine

user's choice to allow one to highlight the requisite space or object in the scene. The Unity ParticleSystem object was used for this purpose.

A control panel was the last physical object created in this phase. This implies the need for buttons for operating the device for the printing polymer. The buttons were then animated - (using the Animator component in Unity), in order to simulate their pushing. Then, sounds were added to the appropriate objects and their properties set - (e.g. volume, situations in which they should be played, if they are to be played in cycles, etc.).

Before the application was first tested, it was necessary to insert the most important VR objects into the scene hierarchy. Mainly, this includes the OVRPlayerController object - which ensures the whole functionality of all VR devices - (i.e. capturing and tracking the position of the headset, capturing controls, controlling character movement and representing the main camera). CustomHandLeft and CustomHandRight were other necessary VR objects; these represent virtual hands - (communication tools between the user and virtual environment). These hands provide base functionality via scripts and were set as necessary in this application.

After the first testing stage and the elimination of minor mistakes, the creation of the user interface - (main menu, settings, control of devices, etc.) began. Its base is built on the Canvas object that can contain sub-objects like Quads, Panels or Buttons. These individual items were then set up for interactions with the user, whose interventions can run the created scripts. Text objects to characterise corresponding commands have been added as needed. All texts displayed in the developed VR application were written in the Czech language, since this application was developed for a Czech manufacturing company.

Since our VR application was focused on the new employee training and education processes, the scene was further expanded by auxiliary objects labelled with simple text descriptions that simply describe the individual machine parts and their purposes. Figure 7 shows the printing machine with these text descriptions in the Unity engine.

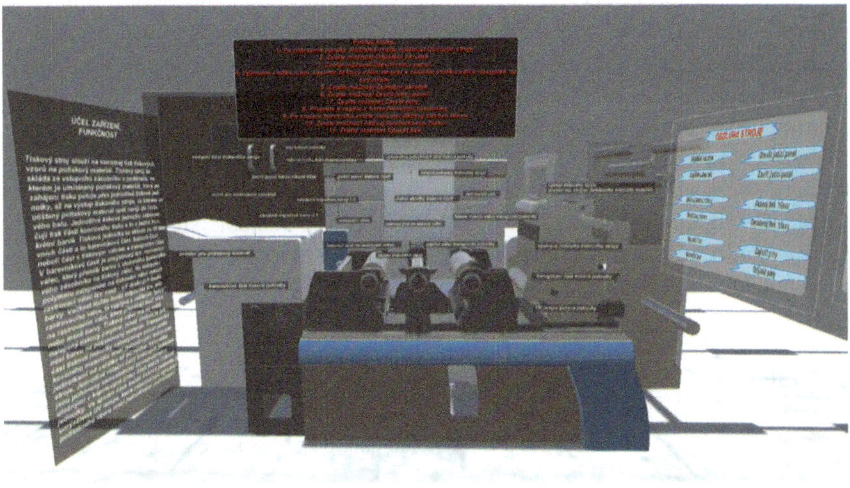

Fig. 7. The printing machine in the unity engine with descriptions and instructions (in Czech)

With these activities completed, the 3D scene was finished and was ready to create the application logic. This included the creation of individual devices scripts and functionalities, by user interactions. Some scripts were already integrated (the basic motion control, headset interaction with the scene, sound playback). The missing scripts were written directly in Unity; and programme source codes were then created in the C# programming language in the Microsoft Visual Studio 2017 environment.

After creating all parts of the application - including application logic and testing its functionality, the standalone VR application was built. The target platform was set into the Windows x86_64 architecture. The application was tested under laboratory conditions and some bugs that appeared during its development were fixed.

6 Conclusion

This paper describes the use of VR for the new employee training process in a printing company environment. Current research shows that its use in the education and training processes offers huge potential. The benefits can include lower costs, the acceleration of training, better quality - (it is possible to learn in a natural "trial and error" way), its attractiveness, simplification - (the reduction of the cognitive complexity of training), and streamlining - (the possibility of creating simulations and training according to their own ideas and needs and further possible expansions or modifications of the same). The shortcomings include the possibility of cyber-sickness symptoms and technical limits of the hardware devices.

The main outcome of this work is a fully functional VR application that simulates the printing production process. The most difficult - and longest part of the work, was the most accurate creation of all 3D models, the preparation of their UV maps, and the texturing process during the creation of the VR application. This application is fully functional and capable of being implemented into the real production process. It also provides feedback that will allow the created application to be optimised/improved - and should also prove the success of the use of VR in this area.

In the future, it would be possible to expand the application with a detailed layout and analysis of individual machines and their parts, with the addition of animations and information about the most common faults. This could also essentially complement the application with another way to use VR – as a service manual. The application could also be extended in order to simulate other production processes and their procedures in this manufacturing company - or other manufacturing companies.

References

1. Linowes, J.: Unity Virtual Reality Projects, 2nd edn. Packt Publishing, Birmingham (2018). ISBN 978-1-78847-880-9
2. Sherman, W.R., Craig, A.B.: Understanding Virtual Reality: Interface, Application and Design. Morgan Kaufmann, Cambridge (2019). ISBN 978-0-12-818399-1
3. Parisi, T.: Learning Virtual Reality: Developing Immersive Experiences and Applications for Desktop, Web, and Mobile, 1st edn. O'Reilly Media, Sebastopol (2015). ISBN 978-1-49192-283-5
4. Jensen, L., Konradsen, F.: A review of the use of virtual reality head-mounted displays in education and training. Educ. Inf. Technol. **23**(4), 1515–1529 (2018)
5. Wang, P., Wu, P., Wang, J., Chi, H.L., Wang, X.: A critical review of the use of virtual reality in construction engineering education and training. Int. J. Environ. Res. Public Health **15**(6), 1204 (2018)
6. Oculus Rift: VR Headset for Computers prepared for VR. https://oculus.com/rift-s/. Accessed 18 Jan 2020
7. Blender.org – Home of the Blender project. https://www.blender.org/. Accessed 18 Jan 2020
8. Blain, J.M.: The Complete Guide to Blender Graphics, 3rd edn. CRC Press, Boca Raton (2016). ISBN 978-1-4987-4647-2
9. Substance Painter. https://www.substance3d.com/products/substance-painter/. Accessed 18 Jan 2020
10. Toudou: Substance Painter for Blender user. 1st edn. Amazon.com Services LLC (2017)
11. Unity Real-Time Development Platform. https://unity.com/. Accessed 18 Jan 2020
12. Blackman, S.: Unity for Absolute Beginners, 1st edn. Apress, Berkeley (2014). ISBN 978-1-4302-6779-9
13. Blackman, S., Tuliper, A.: Learn Unity for Windows 10 Game Development. APress, Berkeley (2016). ISBN 978-1-4302-6758-4
14. Guevarra, E.T.M.: Modeling and Animation Using Blender: Blender 2.80: The Rise of Eevee, 1st edn. Apress, Berkeley (2020). ISBN 978-1-4842-5339-7
15. Oculus Setup: Rift S, Rift, Quest and Go. https://oculus.com/setup. Accessed 18 Jan 2020
16. Freesound – Freesound. https://freesound.org/. Accessed 18 Jan 2020

Sentiment Analysis of Amazon Product Reviews

Steven Brownfield(✉) and Junxiu Zhou

Northern Kentucky University, Highland Heights, KY 41099, USA
{brownfiels2, zhouj2}@nku.edu

Abstract. Sentiment analysis allows computer programs to identify subjective information from a document with the help of text analysis, computational linguistics, and machine learning methods. Sentiment analysis models are trained to detect polarity within a text (e.g. positive or negative opinion), whether it's a whole document, paragraph, or a single sentence. In this research, both traditional machine learning methods, such as Naïve Bayes analysis and Support Vector Machines, as well as deep learning models such as Multilayer Perceptron (MLP) and Recurrent Neural Network (RNN) are built, tested and compared in terms of their effectiveness to detect the correlation between Amazon product reviews and the star rating given by customers. By comparing these models, we gain a comprehensive understanding of the role they can play in sentiment analysis. This understanding can be used to better understand people's feelings when making business decision in applications like e-commerce.

Keywords: Sentiment analysis · Machine learning · Neural networks

1 Introduction

Sentiment is a view of or attitude toward something. It is an emotion, an opinion that you have before or after an event in your life. It can be your opinion about the service you receive from a business or it can be your personal rating of a movie that you just watched. Sentiment analysis is the discovery and study of those personal opinions by interested parties. Some of the researchers interested in your options are businesses that just sold you their product or it could be a business trying to develop new products and they want to know your thoughts and feelings about similar products on the market. For many reasons, there are a lot people studying ways to discover and analysis your opinions.

The objective of the work here is to classify positive and negative reviews given by Amazon customers about products. Our dataset consists of Amazon.com product reviews that are labeled with star ratings ranging from one to five. Features were extracted in three different ways. We built traditional models such as Multinomial Naive Bayes, SVM, etc. and also neural network models including MLP and RNN. Accuracy was compared between all the models using the different feature types and also with different neural network optimizations. As you will soon see, the dataset is imbalanced, having many more samples in the positive class. In an attempt to increase

R. Silhavy et al. (Eds.): CoMeSySo 2020, AISC 1295, pp. 739–750, 2020.
https://doi.org/10.1007/978-3-030-63319-6_68

accuracy, we explored strategies of over-sampling with several different algorithms. These comparisons allow us to see the efficiencies of these models at classifying the product opinions of customers.

2 Details of the Approach

2.1 Data Preprocessing

For our dataset, five subsets from Amazon.com 2018 reviews [1] were merged. Each set contains at least five reviews per product and five reviews overall per reviewer. These subsets, from a selection of product categories, were merged into a single dataset. The product categories include Appliances, Amazon Fashion, Magazine Subscriptions, All Beauty, and Software. Each data sample includes several fields but only the 'reviewtext' column, which is the customer submitted text review, and the 'overall' column, which is the star-rating by the customer, are used. The star-rating field contains values in the range of 1 to 5. These values represent the customer's over all opinion about a product and can be defined as 1: negative, 2: slightly negative, 3: neutral, 4: slightly positive, and 5: positive. The combined number of samples is 25902 and after removing nulls we ended up with 25879 samples to analyze.

Before performing sentiment analysis, a number of linguistics preprocessing steps were applied on the dataset. A typical NLP approach was taken to normalizing the reviews to prepare them for processing by machine learning algorithms. We began by stripping all of the html tags from the documents, and then removed text between brackets, replaced contractions with words, removed non-ASCII text, removed punctuations, replaced numbers with words, removed stopwords, and finally lemmatized the verbs in all of the samples.

2.2 Features

A traditional strategy for extracting features for developing sentiment analysis models is to use a unigram model, also known as a bag of words (BOW) model. These unigram models transform documents into frequency word-vectors where each element in a vector is assigned a number that represents how many times a word appears in the document. This feature representation is very inefficient because the length of the vectors is determined by the number of unique words in all of your samples (Fig. 1).

	Unique words from all samples				
	apple	like	run	bike	...
sample 1	0	0	0	0	...
sample 2	2	0	3	0	...
sample 3	0	1	0	0	...

Fig. 1. Unigram model, also called Bag of Words (BOW) model

Our samples have 30475 unique words and the majority of the samples are less than 350 words long. In the unigram model, the majority of feature vectors have 30125 zeros. To help lower the number of features and thus remedy this inefficiency we set a minimum occurrence threshold during feature extraction to be a minimum of 20 samples. This means that the feature vectors are only composed of words that occur in a minimum of 20 reviews. We ended up with 4246 features. This threshold was chosen because it produced a similar number of features used in another study, by Wanliang Tan et al. [2], who took a similar approach and extracted 4223 features.

Another popular approach to feature representation is the use of pretrained or learned word vectors using algorithms such as GloVe [3]. When using GloVe vectors to represent a sequence such as our text reviews, each word is represented by a vector, see Fig. 2. A third approach to feature representation is the use of mean vectors of the GloVe vectors. To create these new vectors, the mean of the individual GloVe vectors is calculated for each word. In this way, each review is represented by the mean vector of the multidimensional GloVe vectors of all individual words making up the review as suggested by Wanliang Tan et al. [2]. In addition to the 50d-GloVe vectors that they used in their study, we also trained and compared models with GloVe vectors of 100, 200, and 300-dimensional GloVe vectors pretrained on Wikipedia.

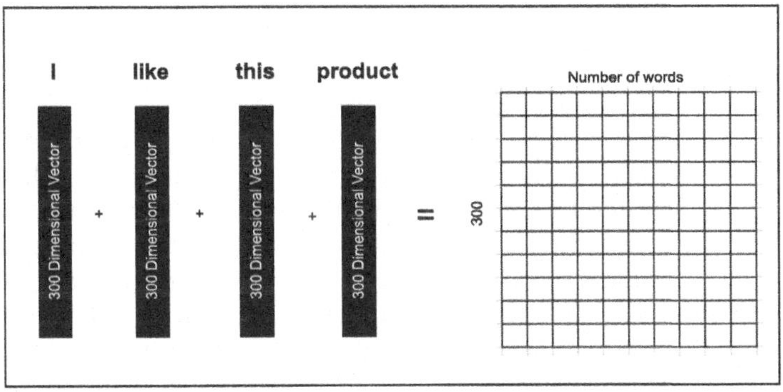

Fig. 2. GloVe vector feature matrix

Before utilizing the means of the GloVe vectors, we needed to consider the length of the reviews. The reviews need to be the same length so padding/trimming will be necessary. When determining the best length, it is desirable to include as much of the review text when trimming as we could while keeping the padding to minimum. Figure 3 is a distribution plot of the number of words per sample. It was determined that 98% of the 25879 reviews have 350 words or less so the max sequence length has been set to 350 when testing with GloVe.

2.3 Building Models

The traditional models in this research such as Naive Bayes and SVM are implemented via Scikit-Learn [4] and the neural networks, both MLP and RNN are implemented via

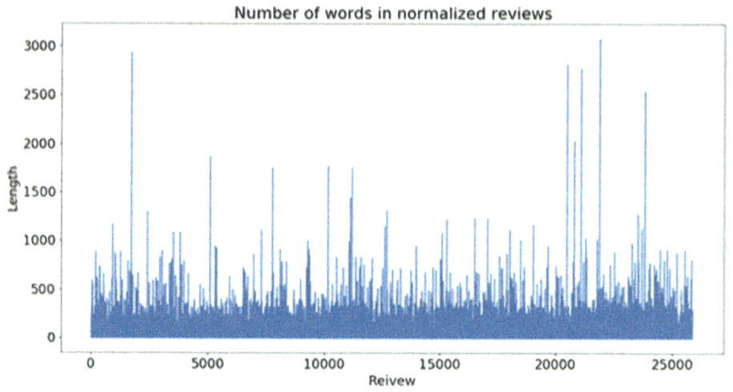

Fig. 3. Number of words per sample

Keras with a Tensorflow backend [5]. The Naïve Bayes model was trained with the BOW features but because the GloVe feature vectors include negative values the Gaussian model was tested with the GloVe vectors.

The SVM model with a Linear kernel is a linear separator that, given samples of different classes, tries to separate them with a straight line. The SVM optimization problem is trying to maximize the margin between the decision boundary and the support vectors. Along with the SVM with a Linear kernel, other kernels tested included RBF, Polynomial, and Sigmoid kernels all of which transform the data points to higher dimensions to classify. All of the SVM models were trained on both the BOW features and the mean vectors derived from the GloVe vectors.

The first MLP neural network models were built with an input layer of 4246 nodes, three hidden layers with 5 nodes each and an output layer of 5 nodes. Each hidden layer used 'relu' activation function and the output layer applied the 'softmax' activation. Models were fit using a batch size of 32 in 20 epochs to match that of the Wanliang Tan et al. research. At first, the Stochastic Gradient Decent (SGD) optimizer was used with a learning rate of .001. The MLP models were trained on the BOW features, all dimensions of mean vectors, and also using the GloVe vectors themselves.

When training the MLP models with the GloVe vectors an embedded layer set to not trainable, that had the GloVe vectors input as a weight matrix was used. In the dataset, 98% of the texts have less than 350 words so it was necessary to pad samples with less than 350 words and trim the samples with more than 350 words. The output of the embedded layer went into three dense layers with 5 nodes each, activation equal to 'relu', and then finally to an output layer of 5 nodes with softmax activation. The accuracies, using the BOW features, in Table 1 include the training and testing results before applying any strategy to improve accuracy.

We can see in Table 1 that the MLP and the SVMs had the greatest accuracies in the training and testing sets. Taking a closer look at the MLP you can see that the testing accuracy is nearly 9% lower than training accuracy. This may be caused by the imbalanced data or poor model which means that the model doesn't work well on this task; this is why we use multiple models. With the neural networks in this research we try different optimization strategies and also over-sampling options to address the imbalanced data.

Table 1. Off-the-shelf classifiers using BOW features.

Model	Training accuracy	Testing accuracy
Multinomial Naïve Bayes	63.95%	60.57%
Gaussian Naïve Bayes	21.73%	18.69%
SVM Linear Kernel	90.04%	72.21%
SVM RBF Kernel	88.16%	73.15%
SVM Polynomial Kernel	82.64%	66.85%
SVM Sigmoid Kernel	61.29%	59.41%
MLP Neural Network	81.92%	73.14%

2.4 Optimizations

To start we tried different optimizers with a learning rate of .001. The training and testing accuracies with different optimizers is listed in Table 2. We could see that there was a significant improvement in the 'training' accuracy but the improvements in the 'testing' accuracies where modest. The Adam optimizer had the highest training and testing accuracy but the gap between them was 24%. On the other hand, the Adadelta optimizer, highlighted in the table, actually classified at a lower training and testing accuracy, the difference was only by 1%. It would be optimal to see high accuracy on both the training and the testing data but obtaining highest accuracy possible on the testing set is a good indicator for model quality.

Table 2. Different optimizers with MLP model on BOW features.

Optimizer	Train acc.	Change	Test acc.	Change	Gap
SGD	81.92%	–	73.14%	–	8.4%
Adam	98.78%	17.35%	74.86%	1.83%	23.92%
Adagrad	85.29%	3.86%	73.95%	.92%	11.34%
Adamax	98.71%	17.28%	73.26%	.23%	25.45%
Adadelta	65.08%	−16.35%	64.43%	−8.6%	0.65%
RMSprop	98.72%	17.29%	74.81%	1.7%	23.91%

Next, dropout layers and early stopping was added to the MLP training. By turning off some of the nodes in the network and stopping before the validation accuracy deviated negatively from the training accuracy, we can possibly improve accuracy. Dropout layers were added with of .2 and also of .5 to the models and run the same train/test as before with and without early stopping with patience of 3. Table 3 has the accuracies of these models. As you can see there was no improvement in the accuracies over the previous models without dropout or early stopping. As can be expected the large gaps between training and testing sets was closed.

You will soon see that the dataset is skewed and how we applied over-sampling strategies in an attempt to improve testing accuracy. Before applying over-sampling,

Table 3. MLP with dropouts and early stopping on BOW features.

Optimizer	Train acc.	Test acc.	Dropout	Early stop
SGD	61.55%	61.34%	0.2	TRUE
Adam	**72.37%**	**68.03%**	**0.5**	**TRUE**
Adagrad	63.16%	62.94%	0.2	FALSE
Adamax	70.14%	67.72%	0.2	TRUE
Adadelta	61.52%	61.30%	0.2	TRUE
RMSprop	68.70%	64.12%	0.2	FALSE

we tried representing the reviews with the means of the GloVe vectors and the GloVe vectors themselves. The Wanliang Tan et al. study also tested these features alongside of the BOW features [2]. They used 50-d GloVe vectors which were pretrained on Wikipedia. In addition to the 50-d, here we also employed 100, 200, and 300-dimensional GloVe vectors.

As mentioned in Sect. 2.2, 98% the reviews are 350 words or less so they were adjusted to be 350. After padding and trimming the reviews, vectors were then created for each review by taking the mean of the GloVe vector for the respective word. Words not found in the pretrained GloVe vectors became zero. These 'mean' vectors are then used to train/test the models.

Pretrained vectors provided by the GloVe project that were trained on Wikipedia 2014 + Gigaword were used [6]. In Table 4 we can see that there is an increase in accuracy when using the mean vectors with Gaussian Naïve Bayes (GNB) but a drop in the accuracy can be seen in Table 5 and Table 6 when testing the SVM and MLP models respectively, with the same features. Table 7 shows that when, embedding the GloVe vectors themselves into the MLP models as features, there is no significant improvement in testing accuracy over the MLP with the mean vectors.

Table 4. Gaussian Naïve Bayes with means of the GloVe vectors.

Feature type	Training accuracy	Testing accuracy
Means of 50d-GloVe	**46.89%**	**46.29%**
Means of 100d-GloVe	45.68%	45.67%
Means of 200d-GloVe	45.41%	45.37%
Means of 300d-GloVe	45.07%	44.78%

2.5 Imbalanced Data

Next, oversampling strategies were applied in an attempt to balance the sample set and improve accuracies. In Fig. 4 you can see that the dataset is left-skewed imbalanced because the 5-star rating counts are more than twice the count of 4-star ratings and many multiples more than 1, 2, or 3-star ratings. The models above could be learning the majority class and doing poorly at recognizing the minorities.

Table 5. Highest SVM accuracies with means of the GloVe vectors.

SVM Kernel	Train acc.	Test acc.	Dim.
Sigmoid	54.81%	54.37%	100
Linear	68.92%	67.99%	300
RBF	71.15%	69.61%	300
Polynomial	**72.94%**	**70.83%**	**300**

Table 6. Highest MLP accuracies with means of the GloVe vectors.

Optimizer	Train acc.	Test acc.	Dropout	Stop	Dim
RMSprop	61.99%	62.27%	0.2	'True'	50
Adam	66.25%	65.40%	0.2	'False'	100
Adam	68.18%	66.50%	0.2	'True'	200
Adam	**68.66%**	**66.95%**	**0.5**	**'False'**	**300**

Table 7. Highest MLP accuracies with embedded GloVe vectors.

Optimizer	Train acc.	Test acc.	Dropout	Stop	Dim
RMSprop	74.84%	66.21%	0.2	'True'	50
Adamax	**81.33%**	**66.55%**	**0.5**	**'False'**	**100**
Adamax	85.24%	66.45%	0.2	'False'	200
RMSprop	78.50%	64.46%	0.5	'False'	200

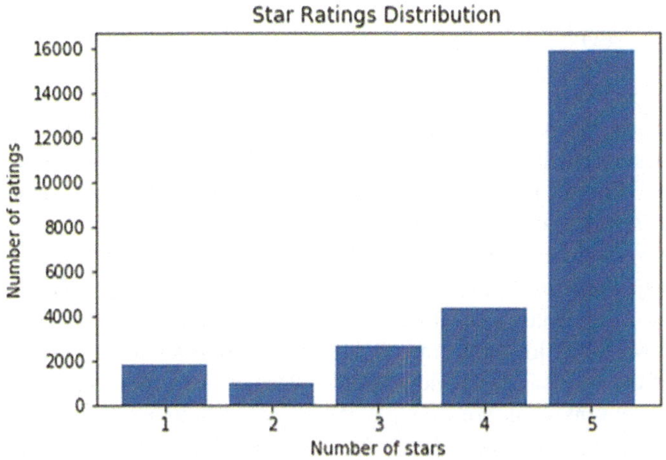

Fig. 4. Distribution of star ratings.

Many different strategies have been researched to address imbalanced datasets. One way to approach the problem of an imbalanced dataset is to generate new samples in the classes which are under-represented. The simplest strategy is to generate new samples by randomly sampling, with replacement, from the current available samples [7]. After doing the random oversampling the training samples went from 15527 to 47714. The accuracies after randomly oversampling can be seen in Table 8. Please note that oversampling should only be done after the split into training and testing to avoid having some of the samples appear in both the training and testing sets.

Table 8. Highest accuracies of MLP random oversampling.

Optimizer	Train acc.	Test acc.	Dropout	Early stop	Features
SGD	62.29%	24.55%	0.5	FALSE	GloVe
Adam	46.62%	32.02%	0.5	FALSE	GloVe
Adagrad	28.23%	31.05%	0.2	TRUE	GloVe
Adamax	65.67%	25.64%	0.5	FALSE	GloVe
Adadelta	32.35%	33.67%	0.2	FALSE	GloVe
RMSprop	59.46%	23.95%	0.2	TRUE	GloVe
SGD	81.52%	64.94%	0.5	FALSE	BOW
Adam	90.78%	70.28%	0.5	FALSE	BOW
Adagrad	59.06%	64.70%	0.5	FALSE	BOW
Adamax	**89.30%**	**71.56%**	**0.5**	**TRUE**	**BOW**
Adadelta	28.03%	42.16%	0.2	FALSE	BOW
RMSprop	85.04%	61.60%	0.5	TRUE	BOW

Different than the naïve random approach of over-sampling, which duplicates some of the original samples of the minority class, the SMOTE and ADASYN algorithms [7] generate new samples in the data by interpolation. However, the approaches used by these two algorithms to interpolate/generate new synthetic samples differ. In fact, ADASYN focuses on generating samples next to the original samples which are wrongly classified using a k-Nearest Neighbors classifier while the basic implementation of SMOTE will not make any distinction between easy and hard samples to be classified using the nearest neighbors rule. Therefore, the decision function found during training will be different among these two algorithms [7] (Tables 9 and 10).

Sometimes there are limitations with an MLP when learning from text sequences because all of the features are treated independent of each other. The connections between the words are lost in the input and expected to be learned by the MLP network. The order of the words in our review samples encodes a lot of information that may be useful to predict sentiment. For example, the phases, "eat to live" and "live to eat" have the same words but have completely different meanings. The BOW model and the GloVe features do nothing to encode word ordering into our features. This is where the Recurrent Neural Network (RNN) is helpful.

An RNN is a popular way to analysis sequences of data. RNNs lend themselves nicely to the analysis of dynamic data such as video/audio streams and of course, text

Table 9. Highest accuracies of MLP using SMOTE oversampling.

Optimizer	Train acc.	Test acc.	Dropout	Early stop	Features
SGD	25.63%	44.42%	0.2	TRUE	GloVe
Adam	33.29%	45.74%	0.2	TRUE	GloVe
Adagrad	23.92%	49.39%	0.2	TRUE	GloVe
Adamax	45.89%	48.92%	0.5	FALSE	GloVe
Adadelta	20.97%	49.68%	0.5	TRUE	GloVe
RMSprop	32.86%	44.72%	0.5	TRUE	GloVe
SGD	66.89%	61.60%	0.5	FALSE	BOW
Adam	**74.25%**	**65.59%**	**0.2**	**FALSE**	**BOW**
Adagrad	51.68%	63.06%	0.2	TRUE	BOW
Adamax	73.88%	65.40%	0.5	FALSE	BOW
Adadelta	35.94%	51.52%	0.2	TRUE	BOW
RMSprop	63.46%	64.12%	0.2	TRUE	BOW

Table 10. Highest accuracies of MLP using ADASYN oversampling.

Optimizer	Train acc.	Test acc.	Dropout	Early stop	Features
SGD	39.13%	36.94%	0.2	FALSE	GloVe
Adam	33.02%	50.18%	0.2	TRUE	GloVe
Adagrad	40.34%	42.30%	0.5	FALSE	GloVe
Adamax	36.37%	51.36%	0.5	TRUE	GloVe
Adadelta	27.25%	30.74%	0.2	FALSE	GloVe
RMSprop	34.01%	49.84%	0.5	TRUE	GloVe
SGD	58.42%	63.78%	0.5	FALSE	BOW
Adam	68.62%	63.09%	0.2	TRUE	BOW
Adagrad	49.64%	60.41%	0.5	FALSE	BOW
Adamax	71.20%	63.64%	0.2	FALSE	BOW
Adadelta	23.91%	54.26%	0.5	TRUE	BOW
RMSprop	**64.10%**	**65.59%**	**0.2**	**TRUE**	**BOW**

analysis. In order to reduce the vanishing (and exploding) gradient problem that vanilla RNNs have, an LSTM (Long Short-Term Memory) model was implemented [8].

LSTM models benefit from memory of the sequence ordering. The recurrent connections in the network add information about the sequence of words which helps with classifying. In order to use the GloVe Vectors directly the LSTM was designed with an input embedding layer that had an input dimension equal to each of the GloVe vector dimensions. The vectors were put into the layer as non-trainable weights since they are the feature space. Next, a fully connected dense layer with 5 nodes was added with activation function of 'relu'. Lastly, a dense layer with 5 nodes and 'softmax' activation was added to predict the probably of our sentiment labels from 1 to 5. The LSTM was trained 20 epochs, Adam optimizer, and a batch size of 32. It was also trained with the means of the GloVe vectors. In or to train on the mean vectors from all four of the pretrained GloVe vectors, the LSTM was built with an input layer with 15 output units, a batch size of 32, and input units equal to each of the dimensions of the mean vectors. Tables 11, 12, and 13 have the results of training and testing both of these LSTM models.

Table 11. LSTM accuracy using BOW features.

Feature type	Training accuracy	Testing accuracy
Unigram (BOW)	**98.25%**	**74.87%**

3 Results

The entire dataset of 25879 reviews was divided into a training set of size 20703 (80%), and a testing set of 5176 (20%). In the training set, 16562 samples (80% of train data) was trained on and 4141 (20% of train data) was used to validate.

Out of the box, the SVM with RBF kernel and the MLP performed the best. Through regularization with early stopping and dropout layers there was an improvement in the MLP models with BOW features. The MLP models trained with Adam optimizer outperformed the others with and without early stopping and dropout layers.

The Gaussian Naïve Bayes models performed very poor on the BOW features with a testing accuracy of only 18.69%. By training the model with the mean of GloVe vectors features, the accuracy improved to 46.29%; this is an increase of 28%. That is not a high accuracy but the improvement is significant. Also note that the Gaussian accuracy improvement was gained while training with the means of the 50d-GloVe vectors. The performance of the MLP on the mean of GloVe vectors was highest with the 300d-GloVe vectors at 66.95% on test samples. The RANDOM, SMOTE, and ADASYN over-sampling algorithms did not help the accuracy of the MLP and was lower when training on the GloVe vectors themselves. Although the testing accuracies of the models using the oversampling strategies were not on top, it is worth noting that the testing accuracy of the MLP model using the SMOTE oversampling algorithm, with dropout at .5, and early stopping with patience of 3, had a testing accuracy of 71.56%, which was less than 3% lower than the top accuracy.

The LSTM trained on the means of the GloVe vectors had an accuracy of 72.37 on the 300d vectors; this is slightly better than the Wanliang Tan et al. study that obtained 70.2% on the 50d-GloVe vectors. The LSTM trained with the BOW features had an accuracy of 74.87 on the testing set which was the highest accuracy achieved from all the models which was the top scoring model/feature type of the Wanliang Tan et al. study [2].

Table 12. LSTM accuracy with Embedded GloVe vector features.

Feature type	Training accuracy	Testing accuracy
Means of 50d-GloVe	61.75%	61.55%
Means of 100d-GloVe	61.83%	61.63%
Means of 200d-GloVe	**62.46%**	**61.75%**
Means of 300d-GloVe	62.05%	61.68%

Table 13. LSTM accuracy with means of the GloVe vectors.

Feature type	Training accuracy	Testing accuracy
Means of 50d-GloVe	70.97%	67.75%
Means of 100d-GloVe	73.42%	69.25%
Means of 200d-GloVe	76.42%	70.29%
Means of 300d-GloVe	**80.81%**	**72.37%**

4 Discussion and Conclusions

The following models, features and combinations thereof were trained, tested and compared to gain an understanding of the role they can play in sentiment analysis.

Models

- Multinomial and Gaussian Naïve Bayes
- SVM with Linear, RBF, Sigmoid, and Polynomial kernels
- MLP and RNN neural networks

Features

- Bag of Words (BOW)
- Pretrained GloVe Vectors
- Mean Vectors of the Pretrained GloVe Vectors

Some of the accuracies in this study were low. One possible reason for lower accuracy could be because of the data imbalance problem. In this study, a minute attempt was made to balance the data. The random over-sampling strategy showed that simpler was better compared with SMOTE and ADASYN but the accuracy was still not acceptable. Future work may formally address and include other over-sampling and under-sampling strategies in order to make a noteworthy attempt to solve imbalance in the product reviews.

In an attempt to remedy the inefficiency of the BOW features, a minimum occurrence threshold was set during feature extraction to be a minimum of 20 samples. This means that the feature vectors were only composed of words that occurred in a minimum of 20 reviews. As we know from Shannon, and the study of information theory, rare events, such as a low frequency of a word occurrence in a document, carries more information than commonly occurring words [9]. By only keeping words that occur in

20 or more reviews, the least occurring words are removed and at the same time, possibly valuable information is removed. Future research may run more tests with different values to determine how a lower or higher threshold affects accuracy.

The simple Bag of Words (BOW) features performed well in this study; the GloVe features did not significantly help the accuracy. The highest accuracy was found with the LSTM on the BOW features. The LSTM was expected to perform well because the recurrent model adds context back into the problem, something often lacking in machine learning [10]. We can see that more complex features do not necessarily equal improved accuracy and successful sentiment analysis is largely determined by selecting a model that builds and utilizes context.

References

1. Ni, J.: Justifying recommendations using distantly-labeled reviews and fine-grained aspects. In: Proceedings of the 2019 Conference on Empirical Methods in Natural Language Processing and the 9th International Joint Conference on Natural Language Processing (EMNLP-IJCNLP), pp. 188–197. Association for Computational Linguistics (2019)
2. Wanliang, T., Wang, X., Xu, X.: Sentiment Analysis for Amazon Reviews (2018)
3. Pennington, J., Socher, R., Manning, C.D.: GloVe: global vectors for word representation. In: Proceedings of the 2014 Conference on Empirical Methods in Natural Language Processing (EMNLP), pp. 1532–1543 (2014)
4. Pedregosa, F., Varoquaux, G., Gramfort, A., Michel, V., Thirion, B., Grisel, O., Blondel, M., Prettenhofer, P., Weiss, R., Dubourg, V., Vanderplas, J., Passos, D., Brucher, M., Perrot, M., Duchesnay, E.: Scikit-learn: machine learning in python. J. Mach. Learn. Res. **12**, 2825–2830 (2011)
5. Aurélien, G.: Hands-on Machine Learning with Scikit-Learn, Keras, and TensorFlow: Concepts, Tools, and Techniques to Build Intelligent Systems. O'Reilly Media, Inc., Sebastopol (2019)
6. GloVe Project. https://nlp.stanford.edu/projects/GloVe/. Accessed 01 Apr 2020
7. Amin, A., Anwar, S., Adnan, A., Nawaz, M., Howard, N., Qadir, J., Hawalah, A., Hussain, A.: Comparing oversampling techniques to handle the class imbalance problem: a customer churn prediction case study. IEEE Access **4**, 7940–7957 (2016)
8. Goodfellow, I., Bengio, Y., Courville, A.: Deep Learning. The MIT Press, Cambridge (2016)
9. Stone, J.V.: Information Theory: A Tutorial Introduction. Sebtel Press (2016)
10. Fox, R., Brownfield, S.: Applying context to hand-written character recognition. In: Silhavy, R. (ed.) Advances in Intelligent Systems and Computer, vol. 985, pp. 40–50. Springer, Cham (2019)

Attendance System with Face Recognition

Maulana Dimas Iffandi, Rangga Nata Adiningrat, Jeremia Rizki Pandapota, Jihad Fahri Ramadhan, Bayu Kanigoro(✉), and Edy Irwansyah

School of Computer Science, Bina Nusantara University, Jakarta, Indonesia
{bkanigoro,eirwansyah}@binus.edu

Abstract. For every organization, today attendance is the most important thing to record the presence of someone. The presence of someone in an organization is a sign that the person is carrying out their obligations to come to the agency or organization. Usually, attendance is done manually. It can be signed or called one by one. In this digital age, there must be a change from this absence to be able to accelerate and provide time efficiency. We can use face recognition to record attendance from everyone present in an organization. In this face recognition, many algorithms are performed to dissect and capture images of someone's face, such as Machine Learning and Deep Learning. With this algorithm, the system can recognize a person's face and record attendance from that person so that attendance activities are more efficient and faster.

Keywords: Attendance system · Face recognition · Attendance management · Machine Learning · Face capturing

1 Introduction

Face Recognition is a technology capable of identifying or verifying a person's face with digital masking from image or video. So many methods in face recognition systems. Face recognition uses biometrics to map facial features from an image or video. Face recognition can help verify the personal identity from the face of a person. With these features of face recognition, we think face recognition can help people to verify attendance. In today's digital age, face recognition is very helpful in this era. Especially for work areas that require attendance verification. Maybe some parts are now relying on technology to verify attendance. But some still use traditional methods that take a long time. Therefore face recognition is very helpful in terms of verifying attendance to speed up the process of recording and verifying the person.

Face recognition is one of the most intensively studied technologies in computer vision, with new approaches and encouraging results reported every year. Face recognition approaches are generally classified as feature-based and holistic approaches. In holistic based approaches, recognition is done based on global features from faces, whereas in feature-based approaches, faces are recognized using local features from faces [6].

R. Silhavy et al. (Eds.): CoMeSySo 2020, AISC 1295, pp. 751–757, 2020.
https://doi.org/10.1007/978-3-030-63319-6_69

A face recognition system is a perfect way to solve these problems. Using face detection, student's faces will be taken in realtime while the student is learning in class without them noticing and the learning process will be smooth. Students also can be attentive without being disturbed to sign for their attendance and lost some of the information that is given by the lecturer. As for the lecturer, there is no hassle to keep all the student's attendance registers for reports later because the attendance is automatically generated by the system. The data generated for the report will be accurate and there will be no more misleading data since the chances for students to forge the documents are eliminated [7].

2 Theoretical Background

According to Literature, Student Attendance System by Face Detection, maintaining attendance is very important and compulsory in all the institutes for checking the performance of students. Every institute has its method in this regard. Some are taking attendance manually using the old paper old file-based approach and some have adopted methods of automatic attendance using some biometric techniques [3].

An automated attendance system based on face recognition is a biometric system where typically, it registers the attendance of each student present in a class by detecting and identifying all of their faces, and then this recorded information is ideally transmitted to a server device which may compute the attendance of each student and store and update the corresponding data in a database. Automated attendance systems are more reliable, rigid, and efficient than the traditional attendance systems and other biometric attendance systems, leading to better productivity and output of both the teachers and students, as well as better consumption of time [13].

An Automatic Attendance System Using Image Processing, maintaining attendance is very important and compulsory in all the institutes for checking the performance of students. Every institute has its method in this regard. Some are taking attendance manually using the old paper or file-based approach and some have adopted methods of automatic attendance using some biometric techniques. There are many automatic methods available for this purpose i.e. biometric attendance. All these methods also waste time because students have to make a queue to touch their thumb on the scanning device [2].

Face detection and recognition section detect face from the image capture by the camera, and the image of the face is crop and store. The element recognizes the images of student's faces, which have been registered manually with their names and ID code in the record. Face recognition data and face identification data are verification into the record. Automatic face recognition (AFR) technology has seen a remarkable improvement in presentation over the past years, and such systems are now widely used for safety and marketable applications. An automatic system for human face recognition in a real-time environment for a university to mark the attendance of its employees [14].

3 Methodology

This paper uses the method of a systematic PRISMA review. By using this PRISMA method we can get focus reports from various researches. This method helps in making paper to review attendance technology using facial recognition. Figure 1 is a flow diagram report from the amount of research, journals, and articles used in this paper.

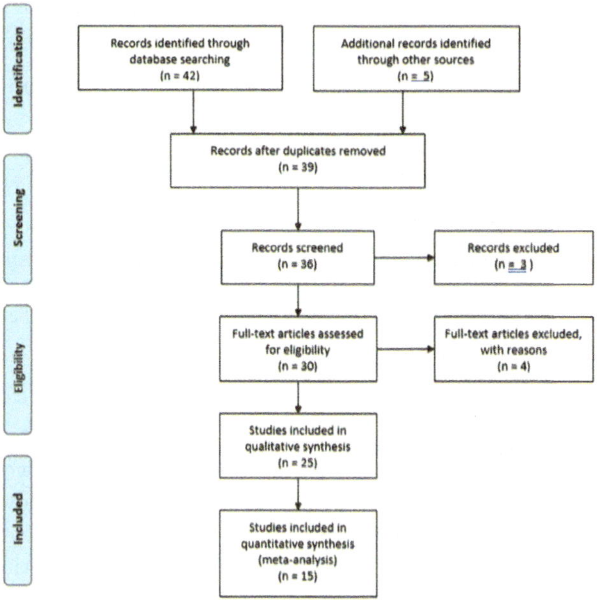

Fig. 1. PRISMA checklist methodology.

The criteria of the paper that we collected are those that meet the attendance system keywords using facial recognition. The journals, research, and articles that we collect are those of the last five years, namely 2015. From all the papers we collect, we prioritize papers that use English. After each person collects papers then we begin to do this PRISMA method to help us review papers with the theme of attendance using facial recognition.

In this PRISMA method, there are four steps that we do. The first is the identification stage, at this stage, we are looking for papers that fit our theme using the internet, and through platforms such as Science Direct, Springer Link, etc. The second stage is the screening stage, at this stage, we separate which papers fit the criteria and non-criteria. The third stage is eligibility, at this stage we begin to focus on papers that are in line with the theme that we will review. The last stage is included, the last stage is to focus this paper review using only paper that meets the final criteria.

The data that we take after reading all the records from the paper then collected and felt that the data is relevant to the theme. Table 1 shows that it has been extracted and analyzed from each relevant record.

Table 1. Data extraction.

Extracted data	Description
ID	Unique ID based on the record that we take
Reference	Authors, year of publication, title of the paper
Context	The context of why this paper was written
Methodologies	The methods, models, algorithms, and datasets that we use in the process of reviewing this paper
Factor	The topic that is related to paper review

Based on the information in the table that has been explained, the main context of the Table 1 is the data that has been extracted by the theme of the attendance system using facial recognition.

4 Result

We used two FaceNet models for training purposes. Initially, we trained both the models with a non augmented dataset and achieved the following results.

The facial recognition process begins with a camera, installed on any compatible device in communication with said camera. This application is then able to use computer vision and a deep neural network to find a prospective face within its stream. There are two primary effective ways to do so: The first is the TensorFlow object detection model and the second is Caffe face tracking. Both these methods have functioned well, and are a part of the OpenCV library. Once a face has been captured, the cropped image will be relayed with an HTTP form-data request to the back end. This facial image is then saved by the API, both on the local file system and in the detection log, appended with a personID.

Based on some papers that we have reviewed, we can conclude that the more datasets that are given to the computer, the higher the success rate will be in identifying face objects.

Based on the search results that have been done by the author, we find out that under a controlled environment, the project shows promising results. Some of the controlled environment includes:

1. **Background** – For the face recognition to work at it best, the background needs to be static. It is harder to recognize a face in a dynamic background.
2. **Lighting** – Lighting also imposes an important aspect of recognizing a face. An environment with adequate light shows more accurate results rather than a dim environment.

3. **Facial Changes** – Some changes to the facial area may cause inconsistency in recognizing the student's faces. For example, students with veil and spectacle are harder to recognize if they did not wear their veil or spectacle [13].

Table 2. Identification table to answer RQ2.

Algorithm	Definition
One Shot Learning [8]	One Shot Learning is algorithm object categorization problem, this algorithm found mostly in computer vision, especially in this system using face recognition
CNN [10]	Convolutional Neural Network or usually called CNN is Machine Learning method for manipulate two dimensional object
Viola Jones Algorithm [5]	Viola Jones Algorithm is algorithm that use for object recognition framework that allows the detection of image features in real time
Scikit-image [15]	Scikit-image is a collection of algorithms for image processing
FaceNet [15]	Directly learns a mapping from face images to a compact Euclidean space where distances directly correspond to a measure of face similarity
Float Boost Algorithm [11]	FloatBoost backtracks after the newest weak classifier is added and delete unfavorable weak classifiers from the ensemble, following the idea of Floating Search
HAAR Classifier [4]	A machine learning object detection algorithm used to identify objects in an image or video and based on the concept of features proposed by Paul Viola and Michael Jones in their paper "Rapid Object Detection using a Boosted Cascade of Simple Features" in 2001
Kernel Method [9]	Using the kernel tricks that are often used in SVMs, we extend the conventional methods to kernel space where we can extract nonlinear features among three or more pixels
Principal Component Analysis [1]	PCA is a statistical approach used for reducing the number of variables in face recognition. In PCA, every image in the training set is represented as a linear combination of weighted eigenvectors called Eigenfaces
Linear Discriminant Analysis [12]	Their classification accuracy is affected by the "small sample size" (SSS) problem which is often encountered in FR tasks

Based on several papers that we have read, the system can recognize human faces by capturing many examples of faces whose data will be trained as an attempt to recognize faces according to their names. This method is called Deep Learning. So the collection of faces will be studied by the system to experiment to match the name according to his face. This constitutes for RQ1: How the system can identify the face?

We attach the identification as shown in Table 2 be able to answer RQ2. We obtained ten algorithms that were identified based on the results of the main studies that had been extracted. We classify the main research so that it will provide information about the algorithm for face recognition. Each algorithm has its use for face recognition. Based on the paper that has been studied, face recognition has a variety of ways to detect a person's face, so the use of algorithms is needed as needed. Each algorithm also has advantages and disadvantages. Until now the level of accuracy of the algorithm that is best used for face detection is the deep-learning algorithm. This constitutes for RQ2: What algorithm is used for face detection?

5 Conclusion

This systematic literature review observes and identifies the Attendance System with Face Recognition. This systematic literature review based on the Deep Learning method, starting with many initial studies collected from online databases. Through the article selection process obtained 37 articles are eligible then analyzed. This study aims at developing and understanding of the Attendance System with Face Recognition. Knowledge of the framework and critical success factors in the implementation of Deep Learning is critical to the successful implementation. FaceNet returns a 128-dimensional vector embedding for each face. The face uses the concept of triplet loss. For RQ to answer the algorithm for Face detection their's ten algorithms are used. We obtained ten algorithms that were identified based on the results of the main studies that had been extracted. We classify the main research so that it will provide information about the algorithm for face recognition. Each algorithm has its use for face recognition. And until now the level of accuracy of the algorithm that is best used for face detection is a Deep Learning algorithm.

References

1. Agrawal, S., Khatri, P.: Facial expression detection techniques: based on Viola and Jones algorithm and principal component analysis. In: 2015 Fifth International Conference on Advanced Computing & Communication Technologies, pp. 108–112. IEEE (2015)
2. Ahmedi, A., Nandyal, S.: An automatic attendance system using image processing. Int. J. Eng. Sci. (IJES) **4**(11), 1–8 (2015)
3. Bodhe, V.M., Bhakre, S.M., Ikhar, S.D.: Student attendance system by face detection. Int. J. Innov. Res. Comput. Commun. Eng. **5**(3), 3958 (2017)

4. Chintalapati, S., Raghunadh, M.: Automated attendance management system based on face recognition algorithms. In: 2013 IEEE International Conference on Computational Intelligence and Computing Research, pp. 1–5. IEEE (2013)
5. Deshpande, N.T., Ravishankar, S.: Face detection and recognition using Viola-Jones algorithm and fusion of PCA and ANN. Adv. Comput. Sci. Technol. **10**(5), 1173–1189 (2017)
6. D'Silva, K., Shanbhag, S., Chaudhari, A., Patil, M.P.: Spot me-a smart attendance system based on face recognition. Int. Res. J. Eng. Technol. (IRJET) **6**(3), 4239 (2019)
7. Elias, S.J., Hatim, S.M., Hassan, N.A., Abd Latif, L.M., Ahmad, R.B., Darus, M.Y., Shahuddin, A.Z.: Face recognition attendance system using local binary pattern (LBP). Bull. Electr. Eng. Inf. **8**(1), 239–245 (2019)
8. Fei-Fei, L., Fergus, R., Perona, P.: One-shot learning of object categories. IEEE Trans. Pattern Anal. Mach. Intell. **28**(4), 594–611 (2006)
9. Jee, H., Lee, K., Pan, S.: Eye and face detection using SVM. In: Proceedings of the 2004 Intelligent Sensors, Sensor Networks and Information Processing Conference, 2004, pp. 577–580. IEEE (2004)
10. Lawrence, S., Giles, C.L., Tsoi, A.C., Back, A.D.: Face recognition: a convolutional neural-network approach. IEEE Trans. Neural Networks **8**(1), 98–113 (1997)
11. Li, S.Z., Zhang, Z.: FloatBoost learning and statistical face detection. IEEE Trans. Pattern Anal. Mach. Intell. **26**(9), 1112–1123 (2004)
12. Lu, J., Plataniotis, K.N., Venetsanopoulos, A.N.: Regularization studies of linear discriminant analysis in small sample size scenarios with application to face recognition. Pattern Recogn. Lett. **26**(2), 181–191 (2005)
13. Shrivastava, K., Manda, S., Chavan, P., Patil, T., Sawant-Patil, S.: Conceptual model for proficient automated attendance system based on face recognition and gender classification using Haar-Cascade, LBPH algorithm along with LDA model. Int. J. Appl. Eng. Res. **13**(10), 8075–8080 (2018)
14. Shriwastav, S., Jain, D.C.: A review on face recognition attendance system. Int. J. Comput. Appl. **143**(8), 19–22 (2016)
15. Stelea, G.A., Gavrilă, C., Zamfir, S., Curpen, R.: Face recognition for education in the cloud. eLearning Softw. Educ. **2**, 181–188 (2017)

Predicting the Result of a Cricket Match by Applying Data Mining Techniques

Fahim Ahmed Shakil(✉), Abu Hasnat Abdullah, Sifat Momen, and Nabeel Mohammed

Department of Electrical and Computer Engineering, North South University, Plot-15, Block-B, Bashundhara, Dhaka, Bangladesh
{fahim.shakil,hasnat.abdullah,sifat.momen,nabeel.mohammed}@northsouth.edu

Abstract. Many different factors influence the outcome of a cricket match. This paper presents the prospect of applying machine learning algorithms in predicting the outcome of an international cricket match. This study has two goals - identifying influential features that affects the outcome of a cricket match and predicting the outcome of a cricket match using machine learning algorithms. To achieve the objectives, feature selection algorithm Recursive Feature Elimination and machine learning algorithms - ZeroR, Decision Tree, Random Forest and XGBoost have been applied. To evaluate the performance of the models, dataset containing international ODI and T20 cricket matches from 2004–2018 has been used. An accuracy of 85.48% has been achieved by applying XGBoost algorithm on the test dataset.

Keywords: Classification · Data mining · Features selection · Cricket · XGBoost

1 Introduction

Cricket is one of the most popular sports in the world with most of the followers coming from the United Kingdom and Commonwealth nations. The recently held ODI cricket world cup in England and Wales in 2019 had reached an astounding 1.6 billion average viewers in live coverage [6]. There are three different types of formats according to which international cricket is played - Test, ODI and T20. Even though Test is the oldest format of this sport, the shorter versions of the game - T20 and ODI, are more beloved among the fans [8]. International One Day Cricket (ODI) has been played since 1971 and T20 since 2004. Even though few rules have been changed or introduced over time, the general rules of the game remain the same. Cricket is a team game where two teams compete with each other having eleven members each. The result of a cricket match may depend on various factors including team members, venue, pitch condition and so on. Team selection is one of the most important factors that influences the outcome of a cricket match and selecting the best players may not always result in the best combination. The best team combination may depend on many other

R. Silhavy et al. (Eds.): CoMeSySo 2020, AISC 1295, pp. 758–770, 2020.
https://doi.org/10.1007/978-3-030-63319-6_70

aspects such as the opponent's strength and weakness, the venue of the game, the weather condition, etc. It is difficult to determine the best combination for a team as the outcome of a cricket match is influenced by plenty of factors. To solve this problem machine learning algorithms can be applied to determine the outcome of a cricket match which then can be used to determine the best combination of players for team selection. Cricket boards can also use the machine learning algorithms to determine the most suitable venues and pitch conditions to ensure victory for their teams.

In this paper, we have used machine learning algorithms to predict the winner of international T20 and ODI cricket matches. We have used several machine learning techniques to get higher accuracy. There has been very little research conducted on predicting international cricket match results. Among them, most of the work has been done on ODI matches [2] and some used statistical calculation to predict the outcome [19]. But we have worked on T20 as well as ODI matches using machine learning algorithms to determine the most accurate results. We have collected and selected more interesting and effective attributes like average rating of players in a team, team rank, home team, etc. which helps our model to minimize the errors and produce more accurate results.

One of the important outcomes of this research work is that it facilitates, through the development of predictive models, the managers and coaches of different teams to plan and select team members that will favor winning a particular match.

The paper is outlined as follows: In Sect. 1, the problems, objectives and research methodologies that have been followed are presented. Section 2 reviews related articles on predicting the outcome of a cricket match using machine learning models. Section 3 discusses the methodology of the research and data and features used. Section 4 analyses the experiments and results. In Sect. 5 conclusion and future work are discussed.

2 Literature Review

Tejinder, Vishal and other colleagues worked on the prediction of score and result of a cricket match where they obtained 70% to 91% accuracy as the match progress. They have used two separate models for first and second innings respectively Linear Regression and Naïve Bayes classifier though they interestingly concluded that the game like cricket is highly unpredictable [19].

Multiple linear regression can be a fantastic method to predict probabilities of the competing teams in ODI matches showed by Bailey and Clarke where they gathered information of 2200 ODI matches played prior to January 2005 [2].

Muthuswamy and Lam creatively used a neural network approach using Backpropagation Network (BPN) and Radial Basis Function Network (RBFN) to predict Indian cricket team bowlers performance for ODI matches [14].

For Indian premier league (IPL), Singh and Kaur have predicted the outcomes up to 71% accuracy using supervised KNN with k = 4 where non-relational database HBase was used for scalability purpose [18].

Iyer and Sharda predicted the athlete's performance using neural networks. They have tried to obtain the batsman's and bowler's performances. Here Neural networks were able to identify 87% of actual performance before the world cup [9]. For One Day International matches Passi and Pandey forecasted the success of players by analyzing their characteristics and stats using supervised machine learning techniques. For this, they have predicted the performance of batsmen and bowlers separately as how many runs a batsman would score and how many wickets a bowler would take in a given match [15].

Another model was created by Sankaranarayanan and other colleagues using a combination of linear regression and nearest neighbor clustering algorithms. They showed the success of their algorithms in predicting the number of runs scored and the determinants of match results. They got the highest 70% accuracy according to their paper [17].

Haseeb et al. worked on predicting the top ten list of rising cricketers based on a weighted average, performance patterns and growing star ratings by applying BN, NB, CART and SVM [1].

The most interesting paper regarding the issue was written by Mustafa and other colleagues where they have collected crowd opinions from social networks and surprisingly got 75% correct prediction with the power of Support Vector Machine (SVM) over other classifiers (Naïve Bayes and Linear Regression) [13]. Munir et al. tried to predict T20 cricket match result during on going match and at the end of the game they attained 85.5% accuracy with Multiple Linear Regression algorithm [12].

3 Methodology

The goal of this paper is to build a model which can be used to predict the outcome of a cricket match. In this section, the data collection process, data pre-processing, selected features and applied machine learning models are described elaborately.

Figure 1 describes the complete workflow of the task. The raw dataset of international cricket matches has been collected from Cricsheet [16] where information about each single match is presented in separate YAML files. Then a csv file has been created containing information about all matches together. Team ranking and average of player rating have been scrapped from Reliance ICC Ranking [7]. After applying various data pre-processing methods, machine learning algorithms have been applied to generate classification models.

3.1 Data Collecting and Pre-processing

The initial raw dataset has been collected from Cricsheet [16]. The raw dataset contains information about international ODI and T20 cricket matches from 2004 to 2018. In the raw dataset, each single match is represented in a single YAML file which contains detailed information about a cricket match such as teams participating in the match, venue, city, date of the match, type of the match,

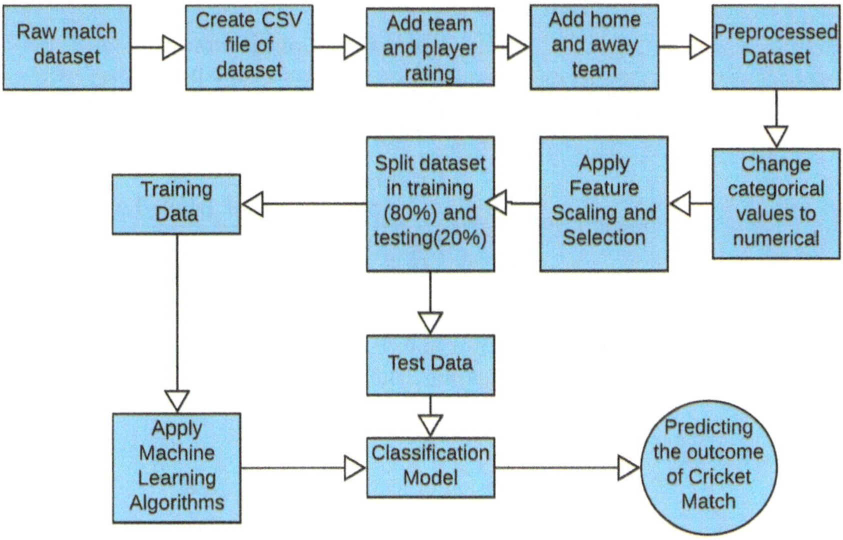

Fig. 1. Proposed framework for predicting outcome of a cricket match

toss winner, umpires, ball by ball information for both batting and bowling, tournament type, margin of winning and outcome of the match. To construct a complete dataset that contains useful information about all cricket matches in a single csv file, firstly a R package - Yorkr [4], has been used to convert YAML files to Rdata files and then convert the Rdata files to csv files. After creating single csv files representing each single match, Python Pandas packages has been used to create single csv file containing information of all the matches which contains - teams participating in the match, venue, date, home and away team, batting first and batting second team, match type and winner of the match. Using the "date" attribute, a new attribute has been generated called "season". Then to add team rating and player average rating, web scrapper has been used in Wikipedia International Cricket Season pages [20] and Reliance ICC Player Ranking website [7] respectively.

Team ranking and rating point has been collected from Wikipedia [20] according to the defined season and match type from the previously created csv file. The team ranking and rating points in the Wikipedia site has been accumulated from archived ICC website [5]. To collect average rating point of batsmen for a team in a single match, firstly the name of all players that batted for the team in that match has been collected from the previously created single match csv file. Secondly, the rating points of each of these players has been scraped from Reliance ICC Ranking [7] for the specified date from the "Date" attribute of the csv file. Finally, the total rating points of all the players that batted has been divided by the number of players that batted to determine the average batsmen rating for the team. Similar approach has been used to figure out average rating for bowlers.

$$\text{Avg. batsmen rating} = \frac{\text{Total rating of players of the team that batted}}{\text{Number of players that batted in the innings}} \tag{1}$$

$$\text{Avg. bowlers rating} = \frac{\text{Total rating of players of the team that bowled}}{\text{Number of players that bowled in the innings}} \tag{2}$$

Equations (1) and (2) have been used to calculate batting and bowling average of each team in each innings of a match.

After completing the pre-processing, the dataset have features described in Table 1.

Table 1. Description of features in the dataset

Attribute name	Description
TeamA	Name of one participating team
TeamB	Name of other participating team
Venue	Stadium name where the match was held
Season	Cricket season in which the match was played
TeamA Rank	Ranking of TeamA in given season
TeamA Rating	Rating of TeamA in given season
TeamB Rank	Ranking of TeamB in given season
TeamB Rating	Rating of TeamB in given season
Home Team	Name of home team from TeamA and TeamB or neutral
Away Team	Name of away team from TeamA and TeamB or neutral
TeamA Batting Average Rating	Average rating of batsmen of TeamA that batted
TeamA Bowling Average Rating	Average rating of bowlers of TeamA that bowled
TeamB Batting Average Rating	Average rating of batsmen of TeamB that batted
TeamB Bowling Average Rating	Average rating of bowlers of TeamB that bowled
Match Type	Either ODI or T20
Winner	Name of the winning Team

3.2 Feature Selection

Recursive Feature Elimination or *RFE* of Sklearn has been used to select the best features for predicting the outcome of a match. *Recursive Feature Elimination* works by eliminating less relevant features and then building model on the remaining features. By using model accuracy, it determines which features and combination of features contribute the most in determining the class attribute. By applying *RFE* on the dataset with *Decision Tree Classifier*, 7 features have been selected as the best features for maximizing the accuracy of prediction. The selected features are - participating teams, home team, batting and bowling averages of both teams.

3.3 Feature Scaling and Splitting Dataset

As categorical values do not work properly with many machine learning algorithms, firstly the attributes in the dataset have been converted to integers using Sklearn package - *Label Encoder*. Then another Sklearn package - *Standard Scalar* has been used to scale or standardize the values of the attributes according to the Eq. (3).

$$z = (x - u)/s \tag{3}$$

The pre-processed dataset contains 1861 instances. The dataset has been split into train set and test set where train set contains 80% of the total dataset and test set contains the remaining 20% data. To ensure the same representation of all classes in the train and test dataset, stratified cross validation has been applied.

3.4 Applied Machine Learning Algorithms

Supervised machine learning algorithms have been applied to the pre-processed dataset to determine the outcome of a international cricket match. The performance of the machine leaning algorithms have been evaluated by first applying the four learning algorithms on the dataset with all features and then by applying the learning algorithms on the dataset using only the features selected by *Recursive Feature Elimination* algorithm.

ZeroR is the simplest classification method which relies on the target and ignores all predictors [10]. ZeroR classifier predicts the target which appears the most i.e. it simply predicts the majority class. Even though the ZeroR classifier does not have any actual capacity for prediction, it is very effective to establish a baseline performance. ZeroR classifier can be used as a benchmark for other classifiers.

Decision Tree is one of the most popular and used algorithm for classification problems as it is very robust against noise in dataset and generally performs well for predicting categorical outcome. In brief, decision tree algorithm chooses the best feature as root which partitions the dataset best. To choose a feature that partitions the dataset best, the concept of entropy is used. Entropy means randomness and the goal of decision tree algorithm is to eventually partition dataset such that each partition contains only one class i.e. to reduce randomness (entropy). By using *Iterative Dichotomiser 3* (ID3), decision tree algorithm calculates gain of each feature and selects the feature with highest gain as the feature with highest information gain reduces entropy the most.

$$E = \sum_{i=1}^{c} -p_i \log_2 p_i \tag{4}$$

$$Gain(S, A) = Entropy(S) - \sum \frac{\|S_v\|}{\|S\|} Entropy(S_v) \tag{5}$$

Equations (4) and (5) are used for calculating entropy and information gain respectively.

Random Forest or random decision forests are an ensemble learning algorithm. It splits dataset to multiple subset and train learning algorithms on the subsets to create multiple models. It then applies these multiple models on test set and through voting determines the class of unknown instance. In case of decision tree, classifier learns on the whole dataset whereas in case of random forests classifier learns over several different subsets. Diversity increases as the models are built using different dataset and variance decreases as all the models are ensemble together for predicting the target. As a result, the resultant ensemble model performs much better.

XGBoost stands for *eXtreme Gradient Boosting*. This algorithm is a scalable end-to-end tree boosting system and is an implementation of gradient boosting machines presented by Tianqi Chen in paper *XGBoost: A Scalable Tree Boosting System* [3]. XGBoost algorithm handles missing data automatically and support parallelization of tree construction which increases the execution speed. Boosting is an ensemble technique where new models are sequentially added to fix the errors made by previous models and it is continued until no more improvement can be achieved. In gradient boosting, the new model firstly predict the errors of existing models and then added together to make prediction over dataset. As the XGBoost algorithm uses gradient descent algorithm to minimize the loss while adding new models, it is also called gradient boosting algorithm [11].

4 Result and Analysis

The experiments have been conducted in Python environment using packages such as - Sklearn, Pandas and XGBoost. To assess the performance of the models, evaluation metrics - accuracy, precision, recall, f1-score, AUC (area under the curve); have been used.

The dataset contains total 1861 instances which have been split into train set (80%) and test set (20%). The machine learning algorithms have been applied to the dataset initially containing all the features and then containing only the selected 7 features by RFE algorithm. Table 2 shows the obtained train and test accuracy. In Fig. 2, comparison of accuracy among the learning classifier algorithms for the selected best 7 features has been shown.

Table 2. Accuracy on training and test dataset

Algorithms	Train accuracy for all features	Test accuracy for all features
ZeroR	11.90 %	11.29 %
Decision Tree	77.2 %	65.32 %
Random Forests	92.20 %	81.18 %
XGBoost	93.39 %	85.48 %

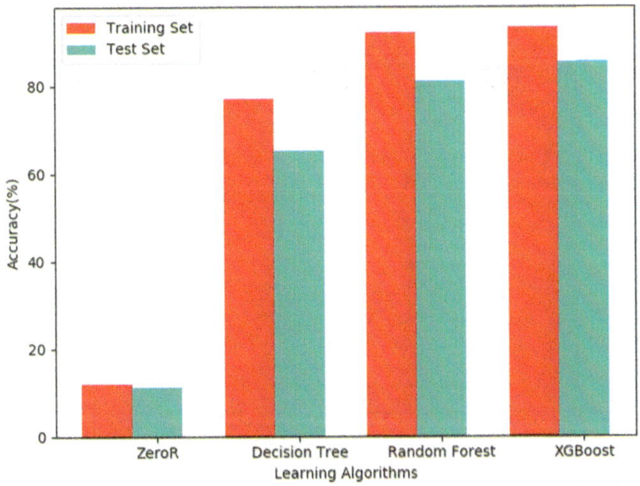

Fig. 2. Comparison between model performances for 7 selected attributes on train and test dataset

The models performances have also been evaluated by other metrics - precision, recall, f1-score, auc. Precision is a measure of correctly positive prediction rate. Precision is calculated by determining the rate of true positive (TP) from the predicted total positive that is the summation of true positive (TP) and false positive (FP).

$$Precision = \frac{TP}{TP + FP} \tag{6}$$

Recall is a measure of correctly-predicted positive class among all the positive class instances present in the dataset.

$$Recall = \frac{TP}{TP + FN} \tag{7}$$

F1 Score is the weighted average of Precision and Recall. It is calculated by using the formula (8)

$$F1 - Score = \frac{2 * Recall * Precision}{Recall + Precision} \tag{8}$$

AUC or area under curve is the area under ROC curve and the bigger the value of auc the better the performance of the model. By using these metrics, performances of the models have evaluated. In Fig. 3, the precision-recall curve or PR-curve for the applied algorithms have been illustrated. As this is a multi-class classifier problem, to generate precision-recall curve for all the target attributes *OneVsAll* method has been used.

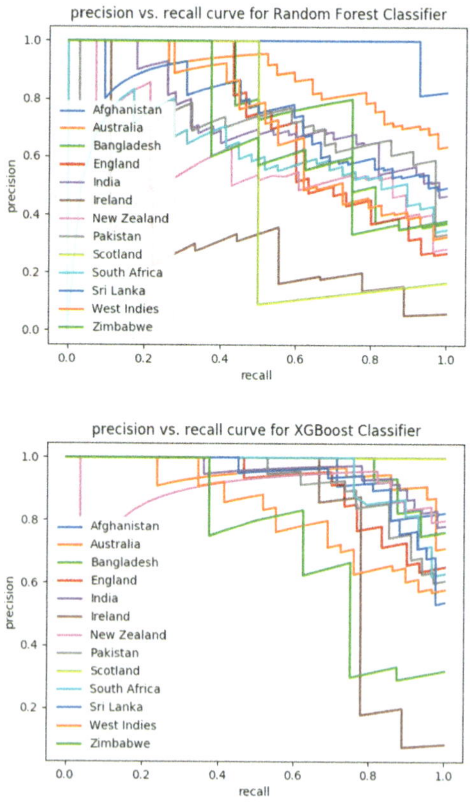

Fig. 3. PR curve for Random Forest classifier and XGBoost classifier

In Fig. 4, the receiver operating characteristic curve for Random Forest and XGBoost algorithm has been presented. By analyzing Fig. 4, it appears that area under the ROC curve for XGBoost model is larger than the area under for Random Forest model.

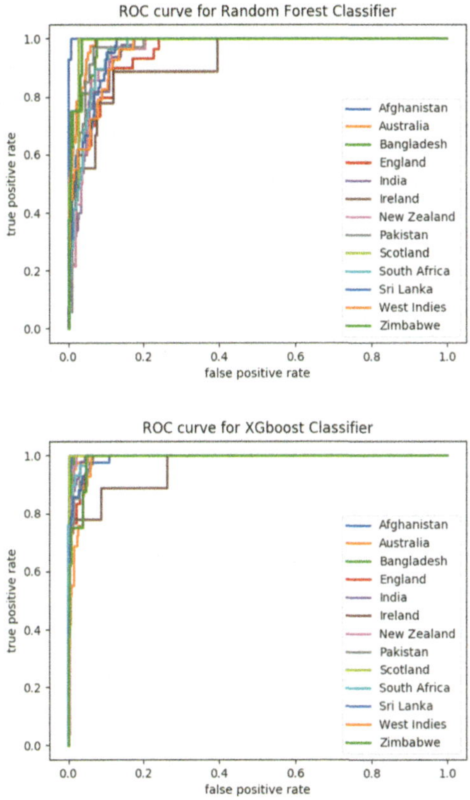

Fig. 4. ROC curve for Random Forest classifier and XGBoost classifier

As XGBoost learning algorithm has performed the best for predicting the target class, in Table 3 value of precision, recall, f1-score, AUC for XGBoost classifier obtained by using *OneVsAll* approach is presented.

The precision-recall curves for XGBoost classifier for several target attributes has been presented in Fig. 5 which illustrates that XGBoost classifier performs well in determining class attributes that is to predict which team is going to win a cricket match.

Figure 6 displays the receiver operatic characteristic curves for XGBoost classifier for different class variables.

From the presented data, it can be analyzed that the accuracy of the models can be improved by reducing the number of classes. From Table 3, we can see

Table 3. Detailed performance evaluation on test dataset for XGBoost learning algorithm

Class	Precision	Recall	F1-Score	Area under ROC
Afghanistan	0.76	0.93	0.84	0.9587
Bangladesh	1.00	0.81	0.90	0.9063
England	0.85	0.93	0.89	0.9593
Australia	0.91	0.95	0.93	0.9701
India	0.89	0.94	0.91	0.9606
Ireland	1.00	0.78	0.88	0.8888
Netherlands	1.00	0.83	0.91	0.9166
Pakistan	0.78	0.82	0.80	0.8999
Scotland	1.00	1.00	1.00	1.0000
Sri Lanka	0.87	0.93	0.90	0.9551
Zimbabwe	0.67	0.75	0.71	0.8708
New Zealand	0.93	0.89	0.91	0.9435
South Africa	0.90	0.90	0.90	0.9439
West Indies	0.76	0.76	0.76	0.8691
Canada	1.00	0.50	0.67	0.7500
Kenya	0.57	1.00	0.73	0.9959
Hong Kong	0.33	1.00	0.50	0.9973
UAE	0.75	0.60	0.67	0.7986

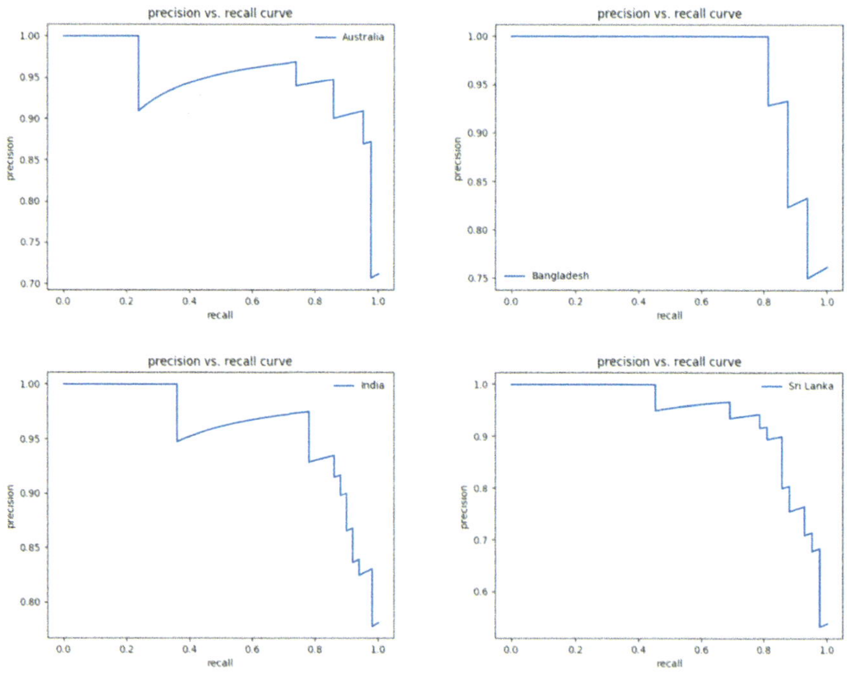

Fig. 5. PR curve for XGBoost classifier for Australia, Bangladesh, India and Sri Lanka

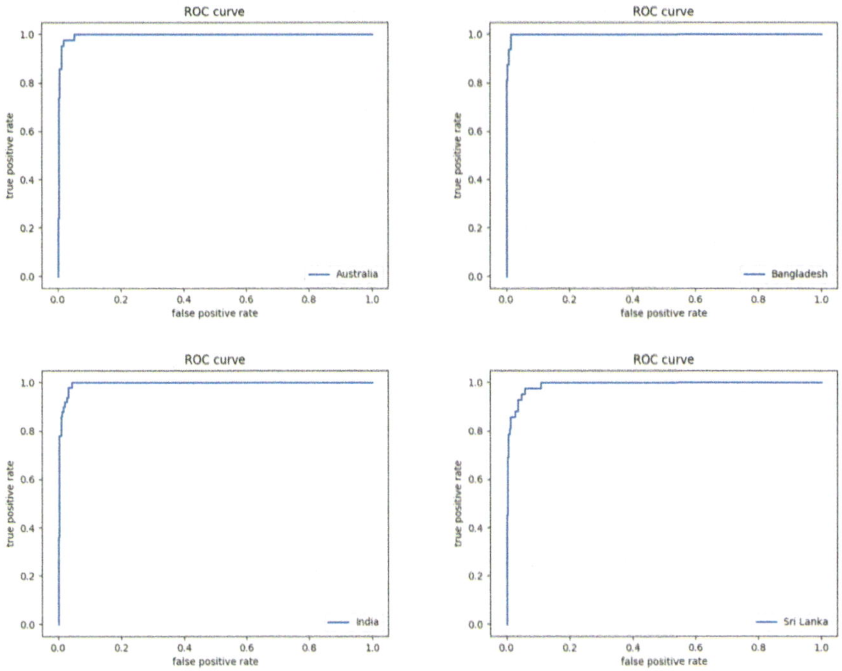

Fig. 6. ROC curve for XGBoost classifier for Australia, Bangladesh, India and Sri Lanka

that classes like - Kenya, Hong Kong, Zimbabwe; have very low precision. There are few classes similar to the mentioned ones which are very under represented and also ratings for players of these teams are not available which results in inaccurate predictions and thus decreases the average prediction of the model.

5 Conclusion and Future Work

Identifying the influential features set for predicting the outcome of a cricket match is still an ongoing study. In this paper we have presented a machine learning approach to predict the outcome of a cricket match. For selecting the influential features, *Recursive Feature Elimination* algorithm has been used. Machine learning algorithms - ZeroR, Decision Tree, Random Forest and XGBoost; have been applied over dataset to predict the result of a match and XGBoost learning algorithm has performed the best with a test accuracy of 85.48%.

Based on what we have completed in this research study, improvements can still be made. This study contains international cricket matches played between 2004–2018; by increasing the time-frame the length of the dataset can be increased which may improve the test accuracy further. The accuracy of prediction may also be increased by implementing neural network model and using different feature selection model.

References

1. Ahmad, H., Daud, A., Wang, L., Hong, H., Dawood, H., Yang, Y.: Prediction of rising stars in the game of cricket. IEEE Access **5**, 4104–4124 (2017)
2. Bailey, M., Clarke, S.R.: Predicting the match outcome in one day international cricket matches, while the game is in progress. J. Sports Sci. Med. **5**(4), 480 (2006)
3. Chen, T., Guestrin, C.: XGBoost: a scalable tree boosting system. In: Proceedings of the 22nd ACM SIGKDD International Conference on Knowledge Discovery and Data Mining, pp. 785–794 (2016)
4. Ganesh, T.: Yorkr package. https://www.rdocumentation.org/packages/yorkr. Accessed 22 Feb 2020
5. ICC archived site (2004). http://web.archive.org/web/20070301143620/http://www.icc-cricket.com/icc/test/archive/2004.html. Accessed 10 Feb 2020
6. 2019 men's cricket world cup most watched ever. https://www.icc-cricket.com/media-releases/1346930. Accessed 18 Feb 2020
7. International cricket council reliance ranking. http://www.relianceiccrankings.com. Accessed 22 Feb 2020
8. India constitutes 90 percent of one billion cricket fans: ICC research (2018). https://economictimes.indiatimes.com/news/sports/india-constitutes-90-percent-of-one-billion-cricket-fans-icc-research/articleshow/64760726.cms?from=mdr. Accessed 18 Feb 2020
9. Iyer, S.R., Sharda, R.: Prediction of athletes performance using neural networks: an application in cricket team selection. Expert Syst. Appl. **36**(3), 5510–5522 (2009)
10. Kaushik, S., Choudhury, A., Dasgupta, N., Natarajan, S., Pickett, L.A., Dutt, V.: Evaluating frequent-set mining approaches in machine-learning problems with several attributes: a case study in healthcare. In: International Conference on Machine Learning and Data Mining in Pattern Recognition, pp. 244–258. Springer (2018)
11. Kumar, A.: Xtreme gradient boosting algorithm (XGBoost) (2018). http://www.ashukumar27.io/xgboost/. Accessed 26 Feb 2020
12. Munir, F., Hasan, M., Ahmed, S., et al.: Predicting a T20 cricket match result while the match is in progress. Ph.D. thesis, Brac University (2015)
13. Mustafa, R.U., Nawaz, M.S., Lali, M.I.U., Zia, T., Mehmood, W.: Predicting the cricket match outcome using crowd opinions on social networks: a comparative study of machine learning methods. Malays. J. Comput. Sci. **30**(1), 63–76 (2017)
14. Muthuswamy, S., Lam, S.S.: Bowler performance prediction for one-day international cricket using neural networks. In: IIE Annual Conference. Proceedings, p. 1391. Institute of Industrial and Systems Engineers (IISE) (2008)
15. Passi, K., Pandey, N.: Increased prediction accuracy in the game of cricket using machine learning. arXiv preprint arXiv:1804.04226 (2018)
16. Rushe, S.: Cricsheet. https://cricsheet.org/. Accessed 22 Feb 2020
17. Sankaranarayanan, V.V., Sattar, J., Lakshmanan, L.V.: Auto-play: a data mining approach to ODI cricket simulation and prediction. In: Proceedings of the 2014 SIAM International Conference on Data Mining, pp. 1064–1072. SIAM (2014)
18. Singh, S., Kaur, P.: IPL visualization and prediction using HBase. Procedia Comput. Sci. **122**, 910–915 (2017)
19. Singh, T., Singla, V., Bhatia, P.: Score and winning prediction in cricket through data mining. In: 2015 International Conference on Soft Computing Techniques and Implementations (ICSCTI), pp. 60–66. IEEE (2015)
20. Wikipedia contributors: International cricket in 2004—Wikipedia, the free encyclopedia (2020). https://en.wikipedia.org/w/index.php?title=International_cricket_in_2004&oldid=937461535. Accessed 10 Feb 2020

Clustering of Scientific Activity of Faculty Staff Based on the Results of Publication Activity

O. A. Zyateva[1(✉)], E. A. Pitukhin[1], and M. P. Astafyeva[2]

[1] Petrozavodsk State University, Petrozavodsk, Russia
olga_zyateva@mail.ru, eugene@petrsu.ru
[2] Dubna State University, Protvino, Russia
marines2000@mail.ru
https://petrsu.ru

Abstract. The article is devoted to the study of the publication activity of the faculty staff, which is reflected in the world's leading scientometric systems. The report presents tools for comparative analysis of two main scientometric indicators – the Hirsch index in the RSCI system and the number of publications indexed in Scopus and Web of Science. Significant features that characterize the faculty staff and their publication activity were determined, on the basis of which the clustering of scientific activities was carried out. As a result of cluster analysis, the faculty staff was divided into eight meaningfully interpreted clusters with similar characteristics of employees and their publication activity. A structural model of teacher transitions between selected clusters is presented. The drivers of the University's scientific activity are only two clusters out of eight, in which the publication activity is most pronounced. Employees in the other clusters should also fall into these two groups. From a practical point of view, the presented results can be useful for management to assess the current state of scientific activity and its improvement. This approach is universal and can be carried out in any educational organization of higher education, regardless of its specifics. This will help universities to conduct self-analysis and manage the performance of departments, in particular, to improve the results of publication activity.

Keywords: Data Mining · Clustering · Statistical analysis · Scientific activity · Publication activity · Scientometric indicators · H-index · Scopus and Web of Science

1 Introduction

To date, there are a large number of indicators that allow to assess the effectiveness of the University's scientific activities to some extent [1, 2]. One of the main indicators that characterize scientific activity is the indicator of publication activity [3, 4], in particular, the number of publications of the organization indexed by international databases [5–7]. This is primarily due to the fact that the total indicator for the country depends on the publication activity of each University, which in turn gives an idea of the country's place in the scientific world [8, 9]. The most significant are publications in the scientometric databases Scopus and Web of Science (hereinafter – WofS).

R. Silhavy et al. (Eds.): CoMeSySo 2020, AISC 1295, pp. 771–778, 2020.
https://doi.org/10.1007/978-3-030-63319-6_71

Analysis of the dynamics of publications by Russian authors and determination of the place of Russian science at the world level are presented in [10, 11]. Issues related to the assessment of scientific activity of universities and their employees, as well as ways to improve it both in General [12, 13] and in certain areas [14, 15], remain relevant. Foreign scientists also pay great attention to the issues of evaluating the scientific potential of the University depending on various factors [16–20].

As part of the main work functions, publication activities are not mandatory for faculty staff. It is obvious that all employees work with different productivity in the field of scientific activity, and, as a result, have different indicators of publication activity in Scopus and Web of Science. Accordingly, the contribution of each employee to the overall rating of the University is different. Thus, the University administration faces an important problem – how to most effectively increase the number of publications indexed in Scopus and Web of Science.

In this regard, the task of analyzing the publication activity of the University's faculty staff and identifying features that can be grouped by teachers who have the maximum number of publications in the Scopus and Web of Science databases is relevant. This step is necessary for the University administration to make effective management decisions aimed at increasing the indicators of scientific activity.

2 Materials and Methods

As information base of research were collected and processed data on the main indicators characterizing the scientific activity of the University staff – the parameters of their publication activity and the distribution of the number of publications by year over the past 15 years, existing in the bibliographic database of the Russian science citation index (hereinafter RSCI) and the distribution of articles indexed in the databases Scopus and Web of Science over the same period as 14.04.2020. It is assumed that if the publication was simultaneously indexed in both databases, then in our view the publication was counted twice.

The study used Data Mining methods, the Highlight Exceptions tool, which uses the Microsoft clustering algorithm [21], as well as statistical and comparative analysis. The statistical analysis software used the Excel table processor and the SQL Analysis Services Excel Add-In, which provided clustering algorithms.

3 Analysis of Publication Activity Indicators

We will conduct a study of the indicators of publication activity of the faculty staff on the example of one of the leading universities in the North-West of Russia – Petrozavodsk state University (PetrSU).

The distribution of the total number of PetrSU employees by academic degrees, as well as the results of grouping employees of educational institutions by the growth of the Hirsch index in the RSCI (hereinafter – HI) and the average number of publications in Scopus and Web of Science over the past 2 years is shown in Table 1. The institutions are arranged in descending order of number of publications in the databases:

Scopus/WofS for 2019 from left to right, namely the Institute of mathematics and information technology (IMIT), Institute of physics and technology (FTI), the Medical Institute (MI), Institute of forest, mountain and building Sciences (IFMBS), Institute of biology, ecology and technology (IBEAT), Institute of Philology (IF), Institute of history, political and social Sciences (IHPSS), Institute of foreign languages (IFL), Institute of Economics and law (IEP), the Institute of physical culture, sport and tourism (IPCST) and the Institute of pedagogy and psychology (IPP).

Table 1. Distribution of educational institutions by the value of the H-index and the number of publications in Scopus and Web of Science.

Institute	IMIT	FTI	MI	IFMBS	IBEAT	IF	IHPSS	IFL	IEP	IPCST	IPP	PetrSU
Number of faculty staff:	63	62	107	81	58	66	53	55	46	52	64	707
Candidate of science	32	43	62	58	39	32	37	23	26	18	43	413
Doctor of science	9	7	29	17	15	10	9	2	10	1	4	113
HI growth (ΔH)	0,7	0,4	0,77	0,26	0,46	1,12	0,73	2,2	0,71	0,44	0,78	0,59
Growth in the number of Scopus/WofS publications (ΔSW)	0,53	0,43	0,43	0,79	0,31	2,75	1,2	0,16	0,38	0,75	1	0,50
ΔSW/ΔH	0,75	1,07	0,56	3,09	0,69	2,44	1,64	0,07	0,53	1,69	1,29	0,84
Number of Scopus/WofS publications in 2019	91	54	45	34	13	10	8	5	4	3	1	268
Percentage of doctor of science with Scopus/WoS publications	0,8	0,7	0,6	0,6	0,5	0,1	1	0,5	0,1	0	0	0,51
Percentage of candidates of science with Scopus/WofS publications	0,3	0,5	0,1	0,1	0,1	0,1	0,1	0,1	0	0,1	0,05	0,15
Percentage of non-degree faculty staff with Scopus/WofS publications	0,1	0,1	0,06	0	0	0,04	0,3	0,03	0	0	0	0,04

The largest division is the Medical Institute, which comprises 15.1% of the total number of employees. The largest number with a degree (93%) in the institutions IBEAT and IFMBS, least of all in the IFL (45%) and IPCST (36%). From the presented data, we can see a significant increase in the average value of these H-index for all institutions. The largest growth is observed in IFL (320%) and IF (212%), the smallest – in IFMBS (126%) and FTI (140%). The situation with the publications in Scopus and Web of Science develops in a different way. Despite the fact that all institutions have increased the number of publications over the past two years, the leaders are MI, FTI and IMIT. Their contribution is 76% to the total number of PetrSU publications.

In 2019, Scopus and Web of Science publications have the highest number of publications (34%) from the total number of IMIT, and the lowest (0.4%) from IPP.

The main contribution to the publication activity in terms of degrees is made by doctors of science in technical fields (IMIT, FTI, IFMBS). The Humanities are less active, perhaps due to a relatively small number of magazines that correspond to their interests. A large variation is observed among doctors of science and employees without a degree. Low activity also occurs in candidates of science.

Analysis of growth rates showed that the number of institutions (IF, IHPSS, IPP, IPCST, IFMBS) has set for itself the primary goal to increase the number of publications in Scopus and WofS, FTI – working to increase both indicators. In other institutions, there is a stronger growth of publications in the RSCI, but despite this, it is impossible to conclude that they put this as a priority, given the significant initial data on the number of publications in comparison with other institutions.

4 The Clustering Results

As a result of the analysis of the publication activity of the faculty, a large differentiation was found, both in the context of educational institutions and in the context of academic degrees. In this regard, the problem arises of dividing the entire set of teachers into stable groups with similar characteristics of publication activity in publications indexed in the Scopus and Web of Science databases. As a result of the research, all teachers were divided into eight clusters. Each cluster is characterized by the average values of 7 attributes – age, H-index and length of service, as well as academic degree, academic title, number of publications and citations of articles in the RSCI.

As a result of clustering, the main variable features of clustering were two factors: the employee age factor and a publication activity factor such as the H-index.

At the same time, it is worth noting that another factor of SW publication activity – the average number of Scopus and Web of Science publications per 1 PPP-correlates well with the average value of the H-index for the obtained clusters (see Fig. 1).

Based on the selected clusters, a structural model of transitions between clusters is proposed, which is based on the age factor. It is assumed that employees from more "young" age clusters have a chance, with the appropriate desire and opportunities, to engage in science, and thus strengthen their publication activity, and, consequently, improve their scientometric indicators.

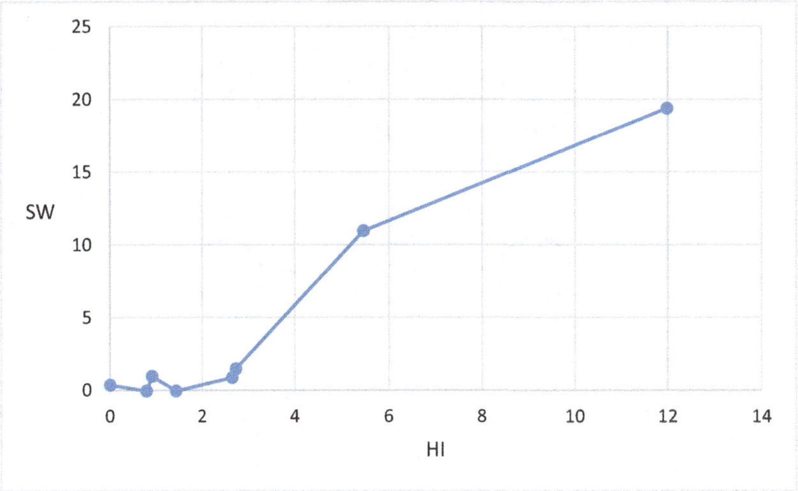

Fig. 1. Dependence of the average number of Scopus and Web of Science publications (SW) on the average value of the Hirsch index (HI)

A structural model of teacher transitions between selected clusters is shown in Fig. 2. Transitions between them are divided into three types: the traditional scenario – the employee does not change his attitude to publication activity (bold arrow), the optimistic scenario – the employee actively publishes and strengthens his position (thin arrow) and the negative scenario – the employee ceases to engage in scientific activities (dotted arrow).

5 Discussions

The first cluster consists of those employees who reduce the average value of the H-index in the whole University (mostly there is no academic degree, the H-index is zero, the average age is about 50 years).

The second cluster includes teachers who stopped actively publishing after defending their PhD thesis (mostly candidates of science with the academic title of associate Professor, the average value of the H-index is 1.43, the average value of age and experience – 50 and 20 years, respectively).

The third cluster includes inactive young scientists who most likely do not plan to get a degree or title in the near future (half are candidates of science, the average value of the H-index is 0.91, the average age is 35 years).

The fourth cluster is the mainstay of the University. Employees are actively engaged in publishing activities that are interesting to others (all have a degree, more than 80% have an academic title, the average value of the H-index is 5.46, the average age is 53 years).

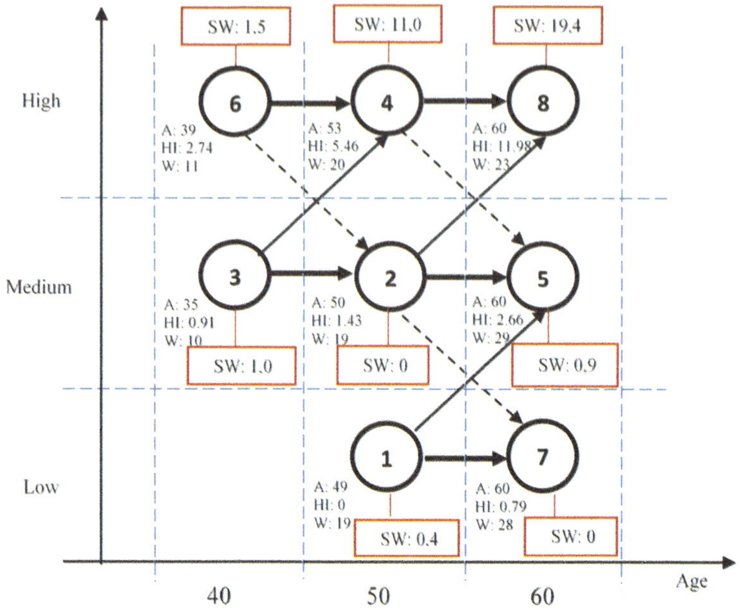

Fig. 2. Structural model of faculty staff transitions between selected clusters (A - average age, HI-average value of the H-index, W-length of service, SW-average number of Scopus and Web of Science publications per 1 faculty staff)

The fifth cluster includes scientists who, having both a degree and a scientific title, continue their publishing activities (the average age is 60 years, experience is more than 30 years, the average Hirsch index is 2.66).

The sixth cluster consists of young scientists who are actively engaged in scientific activities and may be preparing to apply for a scientific title (mainly candidates of science without a scientific title, average age 39 years and experience of 11 years, average H-index 2.74).

In the seventh cluster, there are employees who were active in publishing before receiving the academic title, after which they stopped doing it (the average age is 60 years, experience is more than 28 years, the average Hirsch index is 0.79).

The eighth cluster consists of honorary professors who, despite their age, are still engaged in scientific activities and publish their results (the average age is 61 years, the average value of the Hirsch index is 11.98, mostly doctors of science).

It is interesting to observe which clusters' employees are most actively published in publications indexed in the information and analytical systems of scientific citation Web of Science or Scopus. For this purpose, the average number of Scopus and Web of Science publications per 1 faculty staff in each cluster (SW) was calculated (see Fig. 2). Teachers of 4 and 8 clusters confirm their status – the present and future of the University. They own 79% of the total number of publications. A quarter of the employees from cluster 6 are actively engaged in their activities, thereby strengthening their positions and have the opportunity to become the mainstay of the University in the future (transition to clusters 4 and 8). Teachers from clusters 1, 3, and 5 are inactive. This is also confirmed in publications indexed in the Web of Science or Scopus databases. The data shown in the table shows that there are no representatives of clusters 2 and 7. Perhaps this is due to the fact that after receiving a scientific title, they stopped actively engaged in science, as evidenced by a fairly large average age and a small average value of the Hirsch index, so it is not appropriate to talk about publications in highly rated journals.

6 Conclusion

As a result of the research, seven features were identified, according to which the faculty staff was divided into eight meaningfully interpreted clusters with similar characteristics of publication activity. A structural model of teacher transitions between selected clusters is presented, which allows building individual growth trajectories for those faculty staff who are trying to engage in scientific activities productively. This is especially true for teachers of the sixth cluster.

The proposed approach will help universities conduct self-analysis, adjust the indicators of scientific activity of institutions and employees, taking into account their characteristics. This will undoubtedly have a positive impact on the image of universities and will strengthen their position in the rankings.

References

1. Butvilo, A.I., Korzhov, S.T., Krivonozhenko, A.F., Syunyov, V.S.: Institutes integrated interdisci inary research as a tool of intensification of research activities of the modern University. Univ. Manag. Pract. Anal. **5**(99), 30–38 (2015)
2. Zyateva, O., Pitukhin, E., Peshkova, I.: University performance indicators impact on their ranking position. In: 8th International Conference on Education and New Learning Technologies, pp. 8751–8759. INTED, Spain (2016)
3. Gilyarevskii, R.S.: Publication activity as an indicator of scientific performance. Sci. Tech. Inf. Proc. **41**(3), 170–177 (2014)
4. Klochkov, Y.: Analysis of the publication activity of university researchers. In: Proceedings - 2019 Amity International Conference on Artificial Intelligence. AICAI, pp. 74–79 (2019)
5. Efimova, I.N.: Ratings as comparative assessment of universities for solving management tasks at national level. Manag. Issues **4**(6), 222–231 (2013)
6. Kincharova, A.V.: Methodology of international university rankings: analysis and criticism. Univ. Manag. Pract. Anal. **2**(90), 70–80 (2014)
7. Kim, I.N.: About the measures promoting successful formation of publishing career of the teacher of higher education institution. Eng. Educ. **18**, 64–78 (2015)
8. Smolentseva, A.: Globalization and the research mission of universities in Russia. In: Higher Education in the BRICS Countries, pp. 399–421. Tomsk (2015)
9. Rodionov, D.G., Rudskaia, I.A., Kushneva, O.A.: The importance of the university world rankings in the context of globalization. Life Sci. J. **11**(10), 442–446 (2014)
10. Kotsemir, M.N.: Publication activity of Russian researches in leading international scientific journals. Acta Naturae **4**(2), 15–34 (2012)

11. Moskovkin, V.M., Zdorovtsev, A.D., Munenge, S., Peresypkin, A.P.: Comparative analysis of the publication activity level of the leading Russian Universities conducted in reliance on Web of Science and Scopus databases. Glob. J. Pure Appl. Math. **11**(6), 5121–5133 (2015)
12. Alfred, R., Tanakinjal, G.H., Obit, J.H.: Too few too far: research productivity assessment in Malaysia based on the H-index analysis. In: 6th International Technology, Education and Development Conference, pp. 1925–1934. Spain (2012)
13. Zyateva, O., Pitukhin, E., Peshkova, I.: Research on university key performance indicators: comparative analysis. In: 7th International Conference on Education and New Learning Technologies, pp. 4010–4020. INTED, Spain (2015)
14. Filonov, M.R.: Results of scientific research activity of the national university of science and technology misis (Nust misis) for 2017. Izvestiya Vysshikh Uchebnykh Zavedenij. Chernaya Metallurgiya. **61**, 503–509 (2018)
15. Evdokimov, V.I., Goryachkina, T.G., Leonov, B.I.: Analysis of publication activity in the field of biomedical engineering in 2003–2012. Biomed. Eng. **48**, 1–8 (2014)
16. Bonaccorsi, A., Ancaiani, A., Secondi, L., Setteducati, E.: Exploring the role of third-party research in Italian universities. In: 4th International Conference of Education, Research and Innovation, pp. 1566–1576. Spain (2011)
17. Rodionov, D.G., Fersman, N.G., Kushneva, O.A.: Credibility of the Russian higher education in the world: problems and solutions. Life Sci. J. **11**(11s), 43–47 (2014)
18. Karfaa, M., Haq, F.: The challenges of strategic management in higher education: the case of a Malaysian private university. In: 7th International Technology, Education and Development Conference, pp. 1779–1782. Spain (2013)
19. Berezkina, N.Y., Sikorskaya, O.N., Khrenova, G.S.: Analyzing the publication activities of scientists of the national academy of sciences of Belarus. Sci. Tech. Inf. Process. **39**, 164–168 (2012)
20. Krasovska, O., Andrushchenko, V., Velichko, I.: Scientific cooperation in basic research and higher education. Cult. Psychol. Educ. **7**, 53–58 (2018)
21. LNCS Homepage. https://docs.microsoft.com/ru-ru/sql/analysis-services/highlight-exceptions-table-analysis-tools-for-excel?view=sql-server-2014. Accessed 21 Mar 2020

Machine Learning and Metaheuristics can Collaborate: Image Classification Case Study

Alvaro Valderrama[2], Franklin Johnson[1(✉)], and Carlos Valle[1]

[1] Universidad de Playa Ancha, Valparaiso, Chile
{franklin.johnson,carlos.valle}@upla.cl
[2] Universidad Técnica Federico Santa María, Valparaiso, Chile
alvaro.valderrama.13@sansano.usm.cl

Abstract. Convolutional Neural Networks (CNNs) have provided several real-world solutions to computer vision problems. However, for any given task, their performance heavily depends on the choice of its architecture. In most cases, the structural hyperparameters are optimized manually by researches in a time-consuming trial and error approach. We propose that by using metaheuristics in collaboration with machine learning, we can achieve a good trade-off between computational complexity and learning performance. To this end, we evaluate the performance of a genetic algorithm to optimize the depth, the number of filters, and the kernel size of CNNs for two different tasks of image classification. We achieve a better compromise between computational cost and final accuracy than the state of the art proposals, proving the usefulness of the collaboration of genetic algorithms and convolutional neural networks for image classification tasks.

Keywords: Image classification · Genetic algorithm · Convolutional neural networks · Hyperparameter optimization

1 Introduction

Deep learning has provided state of the art results in a variety of research fields. In particular, for the task of image classification, convolutional neural networks (CNNs) have widely been proved to be an effective and powerful tool [5,7,10]. However, the accuracy attained by each network for a given dataset is strongly dependent on the network's architecture. Even in the simplest setups, the relevant hyperparameters associated with the network's architecture, such as the number of layers, the convolutional filter sizes and the number of filters in each layer need to be tuned. This confronts researchers to a complex and computationally expensive optimization problem, considering that there is no *a priori* direct relationship between hyperparameters and performances, and in some cases even training a single network is already expensive. This leads most researchers without extensive experience with CNN to either use brute force to find a suitable

R. Silhavy et al. (Eds.): CoMeSySo 2020, AISC 1295, pp. 779–787, 2020.
https://doi.org/10.1007/978-3-030-63319-6_72

candidate when possible, or to base their architectures on previously validated networks [11].

Recent contributions to the problem of the optimization of the structural hyperparameters of CNNs for image classification tasks have shown the usefulness of metaheuristics. In [9], Loussaief & Abdelkrim use a genetic algorithm to optimize the kernel number and kernel size of fixed depth networks for the task of image classification of the Caltech-256 dataset, obtaining good results despite the small number of tuned hyperparameters. This approach also allows the authors to achieve the results in comparatively small time. Also, Bhandare & Kaur [1] implement a genetic algorithm to explore the optimization of the structure of CNN (kernel number and size, depth, activation functions), the fully conected section of the network (number of dense layers, number of neurons of that layer), the learning hyperparameters (learn rate, number of epochs) and even the chosen optimizer (stochastic gradient descent, RMSProp), for the MNIST dataset. This approach is slightly more expensive than Loussaief's, but it should be emphasised that the most relevant factors regarding computational complexity in this problem are the number of networks, their complexity and the training process. While Bhandare's choice to optimize the training process gives the algorithm more freedom to explore possible combinations, it also increases the difficulty of balancing the complexity to performance trade off.

A more complex approach is explored by Sun *et al.* [12]. Here, the authors implement a genetic algorithm that searchs for a good configuration of residual convolutional neural networks, using the rather recent advance in CNN of skip connections [4]. This allows the authors to explore much deeper architectures, of which they optimize the sequence and configuration of convolutional residual blocks and max pooling layers. While their approach finds exceptional results for the tailored datasets, this results come at great computational expense and the algorithm doesn't handle consistently different datasets. For this reason, we prefer not to consider skip connections in our research. We find that the problem of the optimization of CNN without skip connections is already a complex and interesting one, however we derive helpful insights from Sun *et al.* proposal, and find in it an interesting benchmark.

We propose that a tailored choice of the relevant hyperparameters to be tuned, based on the CNN differentiating characteristics and on the underlying machine learning optimization problem. Thus, this collaboration between a metaheuristic and the machine learning algorithm can improve the performance in the image classification task without increasing the computational complexity of the algorithms. In particular, we focus only on the hyperparameters of the convolutional section of the network, allowing for depth exploration, in order to achieve networks that are able to find meaningful representations of the involved images regardless of the task complexity. Also we fixed training parameters, to allow for the comparisons between individual structures to be based on the performance achieved by said structure rather than by the quality of its training.

The rest of the article is organized as follow. Firstly in Sect. 2 we briefly introduce the structural hyperparameter optimization problem for CNNs, followed by a description of the implemented algorithm in Sect. 3. Then, experiments on different datasets are presented in Sect. 4. Finally, Sect. 5 presents conclusions and future lines for research based the observed results.

2 Problem Formulation

The problem of hyperparameters optimization in the context of machine learning has a few particularities. Formally, let $\mathcal{A} : \mathcal{S} \longrightarrow \mathcal{H}$ be a learning algorithms, from the space of possible training sets $\mathcal{S}$ over $\mathcal{H}$, a functional space named the hypothesis space. The main goal of this algorithm is, given a training set S_t, find the hypothesis that minimizes the risk. This means to solve:

$$\min_{f:\mathcal{X} \longrightarrow \mathcal{Y}} \mathbb{E}[\ell(f(\mathbf{x}), y)], \tag{1}$$

where ℓ is the loss function, a positive function measuring the difference between the real and predicted value, and the expectation is taken over $(\mathbf{x}, y)$ pairs, drawn from a theoretical joint probability distribution representing the underlying phenomena. We denote by f^* the solution to this optimization problem.

Evidently, as the joint distribution is unknown and we only have access to a finite amount of data this problem cannot be approached directly. Instead, the learning algorithm is tasked with solving:

$$\mathcal{A}(S_t) = \operatorname*{argmin}_{f \in \mathcal{H}} \sum_{(\mathbf{x},y) \in S_t} \ell(f(\mathbf{x}), y). \tag{2}$$

It is important to note that even if the global minimum is found, not only we have not found f^*, we also have not necessarily found the best possible hypothesis in $\mathcal{H}$, since we would need an infinite amount of data pairs to find the best $f_{\mathcal{H}} \in \mathcal{H}$ for Eq. (1). Moreover, in practice, this optimization problem is complex and stochastic methods are used to approximate its solution [8].

However, given the stochastic nature of the phenomena, this is not necessarily concerning. What Eq. (1) conveys, is that the goal of machine learning is not to learn the data, rather infer from it the behaviour of all possible data generated by a phenomenon. The final objective of a learning algorithm is therefore to accurately predict future data pairs, not seen before by the algorithm. This is called generalization. In this regard, Eq. (2) acts as a proxy, an approximation to the real goal of our algorithm. In order to evaluate the actual performance of the algorithm, we use a different set of data examples, the validation set, S_v. Then, while $\mathcal{A}$ chooses an hypothesis f_{S_t} using only the training set S_t, we evaluate the quality of the chosen hypothesis (and by extension of the algorithm) over S_v.

With this in mind, we can formally define the hyperparameter optimization problem. Given a learning algorithm $\mathcal{A}_\lambda$ that depends on the hyperparameter λ, if $f^\lambda_{S_t} = \mathcal{A}_\lambda(S_t)$, the hyperparameter optimization problem is the following:

$$\min_{\lambda} \sum_{(\mathbf{x},y) \in S_v} \ell(f^\lambda_{S_t}(\mathbf{x}), y). \tag{3}$$

Therefore, the problem is to find the best value of the hyperparameter λ, such that given the training set, the learning algorithm is able to find the hypothesis with the greater generalization, or in other words, the smaller error over data pairs not seen before.

While this general problem is relevant for all learning algorithms with hyperparameters, this paper will focus particularly on the structural hyperparameters of convolutional neural networks for image classification problems. We intentionally exclude training hyperparameters, considering these only affect the quality of the approximation of the solution to Eq. (2), which as we see is only a proxy for the real problem. We now present a brief introduction to CNNs and the relevance of the chosen hyperparameters.

CNNs were originally inspired by the organization of the animal visual cortex [6], where different patterns (*e.g.*. vertical lines, horizontal lines) activate different neural pathways. CNNs emulate this behavior using convolutional filters, or kernels. A single kernel corresponds to a moving frame over an image, which outputs the sum of pixels multiplied by the corresponding weight in the kernel. These weights are fixed for the whole image, allowing the filter to identify patterns, since the same pixel pattern will output the same filtered value. Formally, if y is the output image, f is the input image and g is the kernel and σ is a generally non-linear function, we have:

$$y(i,j) = \sigma\left[\sum_{(p,q)} f(p,q)g(i-p,j-q)\right]. \tag{4}$$

A single layer of a CNN is composed of several filters, each with its corresponding kernel and weights. Each filter outputs an image, which is called a feature map. Different feature maps of a layer can be seen as the different channels of an image, and so, the output of a convolutional layer can be feed to another convolutional layer. Thus, a CNN is created by stacking several convolutional layers. Usually other layers, as the max pooling layer (where the output is the pixel with the maximum value over the moving frame), or dense layers (layer with full connections between input and neurons, useful for learning the high-level feature's relationship to the target) are also incorporated for the task of image classification.

However, a successful network needs to have a correct number of layers, filters and size of kernels in order to be able to extract meaningful features. In the sense of the optimization problem presented in Eq. (3), these hyperparameters correspond to λ, while $f^{\lambda}_{S_t}$ corresponds to the trained network given a λ. Each architecture has a high number of trainable parameters, and no direct relationship between the architecture of a network and its final accuracy is known, that is there is no direct functional relationship between λ and our objective function. Therefore, the process of evaluating and exploring network architectures is an expensive and complex one, where the known trade off between computational complexity and learning performance [3] is underscored. An exhaustive search might yield optimal results from the learning performance perspective, however the computational expense is prohibitive. Therefore, an efficient search of the possible configurations must be performed in order to obtain a reasonable balance between performance and computational complexity.

3 Genetic Algorithm

In combinatorial optimization problems, metaheuristics usually show very good results maintaining reasonable computational complexity. Metaheristics are able to efficiently explore large solution spaces, without much *a priori* knowledge of the objective function [2]. This characteristics are compatible for optimizing the hyperparametrs of the CNN, because not only the objective function has no known relationship with the candidate solutions (*i.e.* the chosen hyperparameters), but also the value of the objective function is stochastic since it is itself the result of a stochastic optimization algorithm in each case. This allows for the collaboration of metaheuristics with the machine learning problem of optimizing the structural hyperparameters of CNNs, by exploiting the capacity of metaheuristics to navigate complex search spaces in reasonable computational time.

Genetic algorithms one of the most experimented metaheuristics. They emulate the process of natural evolution observed in living species. To this end, the algorithm maintains a population (*i.e.* a list of possible solutions) that evolves over the generations. In this context, a possible solution to the problem (in this case a particular CNN) is named individual, while the particular values that define the solution (in this case the values of the chosen hyperparameters) are codified in what is known as a chromosome. This chromosome identifies the configuration of the candidate solution. Each individual is evaluated in terms of its performance on the problem at hand. In our case the performance will be evaluated as the validation accuracy of the trained network. For this problem, we represented each chromosome as a list of variable length, containing the depth in number of convolutional layers, the number of filters of each layer and the size of the kernels of each layer for each individual network.

This fitness is latter used by the algorithm to determine which individual is comparatively more successful. Thus, each individual has a probability to reproduce itself to the next generation, by merging the information contained in their chromosome with that of a second individual. This is known as crossover, and in the spirit of natural selection, allows for successful traits to be passed on to latter generations, and so in average improving the whole population. To avoid falling rapidly in local optima, mutation is also implemented. This allows in each iteration for a fraction of individuals to be randomly modified, introducing radically new traits to the population, helping escape local optima.

We implemented a genetic algorithm for this problem, which can be seen in Algorithm 1. Our goal is to evaluate the potential of a simple genetic algorithm tasked with the optimization of the hyperparameters defined in the previous section. In our case, we opted for two simple operators. For the crossover operator, a binary list is randomly created, of length covering both parent's chromosomes. Then, each descendant inherits the values in the chromosome from parent 1 or 2 depending on the value present in the binary list, preserving thus the diversity of depths (since each parent's depth is represented in their descendants) and the general values of their hyperparameters. The mutation operation is also simple, each individual of the population has a probability of 0.3 to have a random value of their chromosome changed to a random value.

Algorithm 1. Genetic Algorithm for collaborating with CNNs

```
Generate an initial population.
Compute the evaluation function of each individual.
while NOT end condition do
    for population size/2 do
        Select two individuals from the previous generation, for the crossover.
        Cross the two chromosomes with some probability obtaining two descendants.
        Insert the two descendants in the new generation
    end for
    for population size do
        Randomly mutate individual with a fixed probability
        Insert mutated individuals in new generation
    end for
    Compute the evaluation function of each individual.
    if Max generation attained then
        Finished := TRUE
    end if
end while
```

Considering the cost of training a single network, our experiments consisted of 12 individuals evolved over 12 generations. In all test the number of filters can take values between 4 and 128, the kernel size between 2 and 9. For the MNIST dataset the depths varied between 2 and 6, while for the CIFAR10 dataset depths between 2 and 7 were explored by the algorithm.

The networks associated to each individual are composed of the convolutional section, with pairs of convolutional layers with hyperparameters defined in the chromosome, ReLu activation function and "same" padding, followed by a max pooling layer. Following the convolutional section, the representation is flattened to a 1-dimensional vector, and passed through 2 dense layers (composed of 64 and 128 units for the MNIST and CIFAR10 respectively) also using ReLu activation, and a final soft-max layer with 10 units that calculates the probabilities associated to each class. Figure 1 gives a representation of a single CNN, where the optimized hyperparameters have influence over the layers depicted in red, while the grey layers have fixed configurations for all individuals.

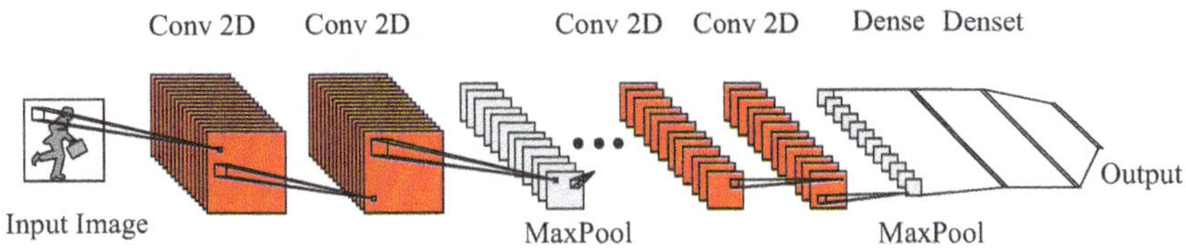

Fig. 1. Structure of an individual CNN. In red the optimized layers, in grey the fixed layers.

4 Experiments

We executed the described algorithm over two datasets, MNIST and CIFAR10, comparing our approaches to those of Loussaief [9], Bhandare [1], and Sun [12]. The MNIST dataset, composed of 70,000 black and white images of resolution 28x28 pixels of handwritten digits (from 0 to 9), of which 45,000 are used as training, 15,000 validating and 10,000 for the test set. On the other hand, CIFAR10 is composed of a total of 60,000 32×32 RGB pixel images labeled in 10 different categories, such as cat, horse, or car. We used 45,000 images for training, with 9,000 for validation and 6,000 for the test set.

Summarized results are depicted in Table 1, for a single run of each algorithm considering time constraints. The last found elite of each algorithm is then trained over the union of the training and validation set and the reported accuracy correspond to the accuracy of the final network over the testing set.

Table 1. Test final test accuracies and total durations in percentages for different algorithms and datasets. The accuracy values correspond to a single run of each algorithm, with the final elite trained over validation and training set and evaluated over the testing set, in percentages. The duration is presented in hour, for a single full run of the algorithm.

	MNIST Acc.	MNIST Time	CIFAR10 Acc.	CIFAR10 Time
Loussaief [9]	99.17%	**1.6**	77.6%	**1.3**
Bhandare [1]	99.02%	6.4	74.43%	6.81
Sun [12]	99.3%	135.42	**82.28%**	257.4
Our Proposal	**99.44%**	4.6	77.82%	12.2

From Table 1, we observe that for the MNIST dataset our simple algorithm yields the best results, in a reasonable time: 4.6 h while the faster algorithm totals 1.6 h. For the CIFAR10 dataset the results are different, the best algorithm in terms of accuracy is by far Sun's proposal with a test accuracy of 82.28%. However, this algorithm is orders of magnitude more time consuming than the rest, totaling 257.4 h, while all others total under 20 h. Of the rest of the algorithms, our proposal achieves the best result in comparable times.

All experiments were performed using GPU acceleration for training (Nvidia GTX 2080), on an Intel i7 Windows machine, with 16 GB of RAM.

5 Conclusion

This paper presented a genetic algorithm for the optimization of a tailored set of hyperparameters, chosen considering both the workings of CNNs and the underlying machine learning problem. From experimental results, we can note that the choice of the hyperparameters to be included in the representation of the individuals of each algorithm has great influence over its trade-off between performance and computational complexity. With this approach, we were able to achieve state of the art results in comparable computational complexity. The only proposal where our algorithm under-performed in terms of final accuracy consumed orders of magnitude more computation time and was not consistently successful. Thus, the proposed collaboration between genetic algorithm and CNNs for image classification defines a better trade off between the learning performance and the computational complexity. We consider that these results can be further improved in the future, without impairing the computational efficiency, by tuning the genetic algorithm, designing specific operators that better capture the underlying nature of CNNs and controlling the diversity in different ways.

Acknowledgments. Franklin Johnson is supported by Grant CONICYT/FONDECYT/INICIACION/11180524.

References

1. Bhandare, A., Kaur, D.: Designing convolutional neural network architecture using genetic algorithms. In: Proceedings on the International Conference on Artificial Intelligence (ICAI), pp. 150–156. The Steering Committee of The World Congress in Computer Science, Computer (2018)
2. Bianchi, L., Dorigo, M., Gambardella, L.M., Gutjahr, W.J.: A survey on metaheuristics for stochastic combinatorial optimization. Nat. Comput. **8**(2), 239–287 (2009)
3. Bottou, L., Bousquet, O.: The tradeoffs of large scale learning. In: Advances in Neural Information Processing Systems, pp. 161–168 (2008)
4. He, K., Zhang, X., Ren, S., Sun, J.: Deep residual learning for image recognition. CoRR abs/1512.03385 (2015)
5. He, K., Zhang, X., Ren, S., Sun, J.: Delving deep into rectifiers: surpassing human-level performance on imagenet classification. In: Proceedings of the IEEE International Conference on Computer Vision, pp. 1026–1034 (2015)
6. Hubel, D.H., Wiesel, T.N.: Receptive fields and functional architecture of monkey striate cortex. J. Physiol. **195**(1), 215–243 (1968)
7. Li, Y.D., Hao, Z.B., Lei, H.: Survey of convolutional neural network. J. Comput. Appl. **36**(9), 2508–2515 (2016)
8. Livieris, I.E., Pintelas, P.: A survey on algorithms for training artificial neural networks. Technical report, Department of Mathematics, University of Patras (2008)

9. Loussaief, S., Abdelkrim, A.: Convolutional neural network hyper-parameters optimization based on genetic algorithms. Int. J. Adv. Comput. Sci. Appl. **9**(10), 252–266 (2018)
10. Rawat, W., Wang, Z.: Deep convolutional neural networks for image classification: a comprehensive review. Neural Comput. **29**(9), 2352–2449 (2017)
11. Smith, L.N., Topin, N.: Deep convolutional neural network design patterns. arXiv preprint arXiv:1611.00847 (2016)
12. Sun, Y., Xue, B., Zhang, M., Yen, G.G.: Automatically designing CNN architectures using genetic algorithm for image classification. arXiv preprint arXiv:1808.03818 (2018)

CLIPS Utilization for Automation of Models' Translation

Maxim Polenov[1(✉)], Artem Kurmaleev[1], Sergey Gushanskiy[1], and Omar Correa Madrigal[2]

[1] Department of Computer Engineering, Southern Federal University, Taganrog, Russia
{mypolenov,kurmaleev,smgushanskiy}@sfedu.ru
[2] Visualization and Virtual Reality Group, Universidad de las Ciencias Informáticas, Havana, Cuba
ocorrea@uci.cu

Abstract. An approach to models' reuse is considered in this paper. It was started with development of the Multitranslator package that allowed translating of the models' source codes into required format. Then it was split into client-server tool called Distributed Storage of Models that has also models storage functions. However, we faced some problems of uncertainty due to lack of source data in translation process. It led into looking for another solution and as a result it was decided to use an expert system based on CLIPS. It was developed an interaction between Distributed Storage of Models, that contains Multitranslator, with CLIPS to automate the translation process in exceptional cases with uncertainty. To utilize CLIPS, we are not just using the expert system itself with knowledge base making user interaction with it, being integrated into the complex of translation, but also allow to edit the rules within the base for the advanced users via interface of intelligent editor. The results of the tests show possibility to reduce time costs of some types of uncertainty cases even at early stages of development. It proves to further increase functionality of the translation complex in terms of automation of models' translation.

Keywords: Models reuse · Translation of models · Multitranslator package · Distributed Storage of Models · Expert system · CLIPS

1 Introduction

The history of the problem starts with the creation of the first modelling tools. One of the first was GPSS and it was used to simulate factories [1]. After the creation of the first and subsequent modeling packages, the problem of transferring existing models for the new modeling tools was already presented. As a result, this entailed time expenditures for each such iteration on both direct transfer and the research of a new modelling tool for the transfer of existing models.

Thus, the task was set to develop a tool for the automated transfer of models and source codes to reduce time costs. As a result of solving this problem the Multi-translator (MT) tool was developed [2]. The MT is a software tool for translating

R. Silhavy et al. (Eds.): CoMeSySo 2020, AISC 1295, pp. 788–796, 2020.
https://doi.org/10.1007/978-3-030-63319-6_73

formalized languages from input to output. However, the greatest effect when using MT is achieved only in the case of previously created translation module [2, 3] or if the number of transferring models is sufficient to justify the creation of a new translation module within estimated time costs. The translation module is a set of grammar instructions in the grammar description language (GDL) [2] and corresponding actions in the action description language (ADL) [2]. To create such modules, you need a specialist in the output and input languages, but in the case of translation of a large amount of source codes, learning the output language will be more profitable in terms of time costs, because one person who has knowledge of ADL and GDL can write a translation module that will allow translation of all models instead of training a number of researchers that would transfer those models manually.

Multitranslator is a standalone application that seems to be difficult for the user to learn, both for creating translation modules and working with the source codes of models, while most users only need the translation function, and when iterating through large volumes of source code, this can also entail some time costs.

When developing MT, we added the ability to connect via the COM interface which opened the possibility of using MT without loading the entire environment with GUI. So, it was decided to create MT in the form of a client-server application [4], where MT itself would be on the server side and process translations instead of the user's PC.

Also, with the transfer of the translation tool to the server, it was decided to create the ability for users to store models on the server side, which simplified sharing access to the source code of the models and translation modules.

Based on these requirements, the Distributed Storage of Models [5] was created, where on the client side a simplified user interface was used to connect to the server, select local or server models, select the translation module and, upon completion of the procedure, access the output source code, thereby utilizing model reuse approach [6].

To continue the development of this approach, it was decided to resolve some exceptional cases arising from the translation of models [3] in case of insufficient input data. Such cases when occurring in teams without a specialist with sufficient qualifications to solve this problem can lead to undefined time costs.

2 Current State of the Problem

First, to continue development of our approach we reviewed how other researchers solve it and what approaches do they use in general.

Models reuse and code reuse are generally solved by how simulation is presented [7]:

1. Multi-formalism, multi-scale modelling. Research states that we have to use new approaches when we develop application and components.
2. Reuse across communities of interest and the implementation spectrum. Next research suggest that higher level of abstraction is required to achieve better results.

3. Exploitation of model and simulation web services. On this point it was offered to use component-based development together with standards and well-made documentation.
4. Quality-centric approaches to component evaluation. Taking it short application requires to obtain such qualities as adaptability, composability, extensibility, interoperability, maintainability, modifiability, openness, performance, reusability, scalability, and usability.

Another big part in the research offers four step reuse that is not compatible with our problem and only relates to model actual reuse, i.e. that you don't modify the source codes with this approach but instead you make a structure of your sources and models so you can properly choose whenever you need a model to create it or take already existing one:

1. Machine learning. It helps monitoring and checking running simulations to sort out models and creation of data bases with labelled models.
2. Context Management. List of context attributes is generally required to help in deciding on reuse.
3. Domain Knowledge Extension for Collaboration and Enhanced Decision Making. It is required to validate reuse in two steps as an assessment and inferring of validity of the results in terms of quality.
4. Reuse Through Model Discovery. In the context it is presented as using sustained metadata that is labelled and marked up.

This overall gives us a picture of non-existing formal languages translation for models in current research with mainly knowledge reuse approach based mostly on the knowledge about models themselves and language that was used to code it. Moreover, none is based on generation of the required models. Even when there is a vast number of models in many languages.

All above leads us into making further improvements of our system for the future works and continuation of development for our research.

3 Decision on Problem Solution

Models and code are always written in a formal language [8] that is constructed with a set of rules with a set of words.

To start with, let us consider the main properties of formal languages that are relevant for our research:

1. Generated by some formal grammar.
2. Described or matched by a regular expression.
3. Accepted by some automaton [9].
4. Accepted by decision procedures [9].

There are few main types of tools to consider:

- Search and Optimization. Where search and evolutionary algorithms will require too much unnecessary computation for this type of problem. The optimization algorithms will also require even more additional time spends for works in terms of formalization and developing.
- Neural networks, classifiers, and statistical learning methods. Neural networks will require precise learning and still will not guarantee the result for all the cases. Also, the requirement of precise learning for each new pair of languages. And classifiers job is already done by MT.
- Probabilistic methods for uncertain reasoning. Probabilistic methods become very unhandy when you use problems with determined rules of formal languages. And Bayesian networks job is already done by MT as well.
- Logic programming and Automated reasoning. Logic appears to fit the predefined rules of the formal languages since it is not needed to determine the correct data when it is already given. Also, logic can process the data in automated way if there is a set of rules by the process of decision making which fits all our criteria at the same time.

In the logic programming tools, there are many different subjects to choose from. To make a proper choice we have to review again the specifics of the task that we are willing to process.

Currently in predetermined tasks like translation, there are always some experts with knowledge of how-to. What if we could gather all those experts so could we cover all the rules that were ever known? It would be their knowledge combined all together. It is called the knowledge base and a tool to operate it will be the expert system that processes all the rules, stored in the knowledge base.

4 Interaction of Distributed Storage of Models with CLIPS

To attach the expert system (ES) to the existing complex of the Distributed Storage of Models [10], which includes a database of models, a user interface and a module for describing models on the client side along with the model manager, the database of models and MT on the server side, it is necessary to develop a new approach [11], that implements:

1. The dialogue between the user and the ES.
2. Automation of work and output ES.
3. The intelligent editor of the knowledge base.
4. Interaction with the Distributed Storage of Models.

As a result of the concept development, the suggested structure of tools is shown in Fig. 1.

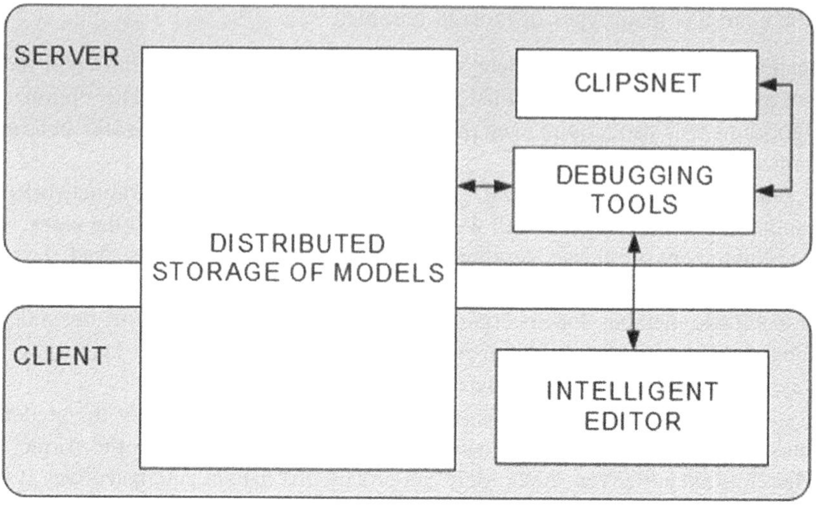

Fig. 1. The structure of the model's translation complex with intelligent support of CLIPS

The composition of the Distributed Storage of Models was described above, the data from it (from MT) is transferred into debugging tools, which are responsible for the automation of the ES and dialogue with a user through an intelligent editor.

Debugging tools are required to work with ES for the ordinary user and get results of the ES work, where data is received using the CLIPS dotNet library [12, 13]. The Distributed Storage of Models is written in C# [14], so the development of an application is conducted in the same language to avoid compatibility issues and time costs.

Such issue occurred during the development when we tried to marshal the CLIPS C++ library. It was not possible for unknown reason to get an "entry point" for some main functions of the library. But decompiling the library to find out the cause could take an indefinite time and would not guarantee any positive result for marshalling. For this reason, a new and modernly supported CLIPS dotNet library was chosen. Its development is not finished yet, as well as our work on a complex of models' translation with intelligent CLIPS support.

5 Developing the Utilization of CLIPS Based on DotNet

When using external dotNet CLIPS libraries, first it is necessary to connect the library "CLIPSNET_Environment.cpp", indicating the corresponding directive in the project:

```
using CLIPSNET;
```

Next, you need to create an environment object for initializing CLIPS, for example:

```
public CLIPSNET.Environment Env = new CLIPSNET.Environment
();
```

The next step is to load the knowledge base and expert system to the CLIPS environment:

```
Env.LoadFromResource (“ES”, “ES.ES_e_v0.1.0.clp”);
Env.LoadFromResource (“ES”, “ES.ES_kb_v0.1.0.clp”);
```

Where the string with the name of the current project is indicated as the first argument, and the expert system or knowledge base file itself, which in our case has already been added to the project earlier, is therefore indicated through the project namespace.

Only after loading the knowledge base and expert system, it is possible to start the launch of the ES itself, which should begin with the preparation of the CLIPS environment by cleaning it:

```
Env.Reset ();
```

After the resetting of the environment, it is necessary to submit facts for ES processing, for example:

```
String commandFactStr = “(assert” + <factStr> + ”)”;
Env.Eval (commandFactStr);
```

Here, as a <factStr> , a line with a fact is passed for processing, often passing lines in a loop from an incoming array of fact lines.

After organizing the transfer of facts, we can start the fact processing the rules of the expert system:

```
ES.Run ();
```

Here we go to the step that could not proceed further earlier due to problem with marshaling. At this step it is necessary to read the result of the output of the rules from the ES memory.

To do this prepare a line for ES which upon execution returns the facts that were processed:

```
String findFactStr = “(find-fact ((? F UI-state)) TRUE)”;
```

Now we have to prepare the address variable to get the values:

```
FactAddressValue factValue = (FactAddressValue) ((Multi-
fieldValue) ES.Eval (findFactStr)) [0];
```

where “FactAddressValue” is a class of the “CLIPSNET” library:

```
public: ref class CLIPSNET:: FactAddressValue
```

And here are the subclasses:

1. FactAddressValue.
2. InstanceAddressValue.
3. LexemeValue.
4. MultifieldValue.
5. NumberValue.
6. VoidValue.

Thus, we get access to the subclasses FactAddressValue, MultifieldValue, and LexemeValue that are necessary for our task.

And so, we can get the output values:

```
String result = (StringValue) factValue.GetFactSlot
(“print-answer”). GetStringValue ();
```

This string variable "result" can already be passed on to complete the work of the translation module of the multilingual translation subsystem of the Distributed Storage of Models, that is based on Multitranslator.

It is worth noting that it is also possible to load individual facts-tokens into the expert system, as well as auxiliary facts as constants for the ES knowledge base, to which the knowledge base will refer to display specific result lines. In this case the classes "MultifieldValue" and "LexemeValue" are required when solving the problem of developing an intelligent editor for the user (see Fig. 1). It makes user interaction in the system to clarify the data transferred to the ES.

To do this, in the expert system it is necessary to write to another field class, from which we obtain the values as follows:

```
MultifieldValue answerValue = (MultifieldValue) ES.GetFact-
Slot ("answer");
```

It is also necessary to get the user-selected answers:

```
String uAnswers = ES.GetFactSlot ("response"). ToString ();
```

For further processing, we require the value of the tokens for the answers:

```
LexemeValue answerLex = (LexemeValue) answerValue;
String answerStr = answerLex. GetLexemeValue ();
```

Where the variable of the string type "answerStr" is the string response of the expert system, obtained from the auxiliary database of tokens with answers that can be sent to the output for the user.

Thus, by using external dotNet CLIPS libraries, it is possible to solve the problem of creating an effective mechanism for the organization of the interaction of the Distributed Storage of Models with an expert system to support of the translation of external models.

6 Developing Interface and Debugging

To improve experience of expert system implementation and make some basic functionality for the intelligent editor before continuing development of the knowledge base and rules of expert system we decided to make an interface. Its main purpose is to output expert system data that we can get, and which considered in the previous section. We chose to use multi-document interface (MDI). Here it is described two main ways via word algorithm how developer can proceed the debug. It is the first one:

1. Choose the 'Manual Input' item in the drop-down 'Data' menu. It will create child window with manual input for your data so you can debug the expert system manually.
2. After entering the data choose the 'Menu' drop-down and pick the 'Expertise' item. It will display debugging information in new child window with 'Result' title and will write the same information into debugging log file.

And the second way involves the translation complex, taking data from Multitranslator:

1. At first the data from the Distributed Storage of Models have to be transferred to our application, so you have to start translation first with sample of code.
2. When the exceptional case occurs during model translation, data will be automatically transferred and displayed within debugging window. In the 'Menu' drop-down another item 'Finish Translation' will become active so you have to choose it.
3. As the result, it will proceed the same way as manual debug has finished on the first example.

MDI class allows us to simply open, close, create the internal windows if it is necessary and even reopen. It makes application flexible and open for the further improvements of the interface of debug and intelligent editor.

7 Conclusion

As a result of the research, an approach to the intellectual support for the multilingual translation of models was proposed and implemented. The approach is based on the use of a Distributed Storage of Models as a repository and models conversion tool. The conversion tool based on the translation modules, created in Multitranslator environment, with the support of an expert system. As part of the approach, the organization of interaction between the Multitranslator and the CLIPS core was implemented using dotNet libraries.

As the result we expand the functionality of the software package of the Distributed Storage of Models, due to the efficient processing of uncertainties that can occur in the process of model translation. But also, we reduce time costs when using external models when modeling complex systems.

Acknowledgments. The reported study was funded by Russian Foundation for Basic Research (RFBR) according to the research project No. 19-07-00936.

References

1. Thesen, A., Schmidt, J.W.: GPSS is a Process Flow Oriented Simulation Language, Computer Methods in Operations Research. Academic Press, Cambridge (2014)
2. Chernukhin, Y., Guzik, V., Polenov, M.: Multilanguage Translation for Virtual Modeling Environments. Publishing house of Southern Scientific Center of Russian Academy of Sciences, Rostov-on-Don (2009). (in Russian)
3. Chernukhin, Y., Guzik, V., Polenov, M.: Multilanguage translation usage in toolkit of modeling systems. WIT Trans. Inform. Commun. Technol. **58**, 397–404 (2014)
4. Tanenbaum, E., Van Sten, M.: Distributed Systems: Principles and Paradigms, 2nd edn. Prentice-Hall, Upper Saddle River (2006)
5. Polenov, M., Guzik, V., Gushanskiy, S., Kurmaleev, A.: Development of the translation tools for distributed storage of models. In: Proceedings of 9th International Conference on Application of Information and Communication Technologies (AICT 2015), pp. 30–34. IEEE Press (2015)
6. Robinson, S., Nance, R.E., Paul, R.J., et al.: Simulation model reuse: definitions, benefits and obstacles. Simul. Model. Pract. Theor. **12**, 479–494 (2004)

7. Fujimoto, R., Bock, C., Chen, W., Page, E., Panchal, J.H.: Research Challenges in Modeling and Simulation for Engineering Complex Systems. Simulation Foundations Methods and Applications Series. Springer, New York (2017)
8. Sander, M.: Developing Interacting Domain Specific Languages, Universiteit Utrecht (2007)
9. Rozenberg, G.: Handbook of Formal Languages. Springer, Heidelberg (1997)
10. Polenov, M., Gushanskiy, S., Kurmaleev, A.: Synthesis of expert system for distributed storage of models. In: Software Engineering Perspectives and Application in Intelligent Systems. Advances in Intelligent Systems and Computing, vol. 575, pp. 220–228. Springer (2017)
11. Polenov, M., Kurmaleev, A., Gorbunov, A., Pereverzev, V.: An approach to the models translation intelligent support for its reuse. In: Advances in Intelligent Systems and Computing, vol. 1046, pp. 477–483. Springer (2019)
12. CLIPS. http://www.clipsrules.net/
13. dotNet. https://docs.microsoft.com/en-us/dotnet/
14. C#. https://docs.microsoft.com/en-us/dotnet/csharp/

Mean-Variance Portfolio Optimization Under Parametric Uncertainty

Anna Andreevna Malakhova[1(✉)], Olga Valeryevna Starova[1], Svetlana Anatolyevna Yarkova[2], Albina Sergeevna Danilova[2], Marina Yuryevna Zdanovich[1,2], Dmitry Ivanovitch Kravtsov[1], and Dmitry Valeryevitch Zyablikov[1]

[1] Siberian Federal University, Krasnoyarsk, Russia
amalakhova@sfu-kras.ru, malahovaanna@mail.ru
[2] Krasnoyarsk Institute of Railway Transport - Branch of the Irkutsk State Transport University, Krasnoyarsk, Russia

Abstract. In this paper portfolio optimization problem is studied under premise that the asset classes in investment portfolio are selected but the values of asset returns and variances are random. The sensitivity of the Markowitz model to market uncertainties is described by 'conservative', 'nominal' and 'optimistic' formulations of the problem. An algorithm for defining the initial parameters to describe the conservative and optimistic formulation of the Markowitz problem is developed. The shift of the efficient frontier of portfolios is evaluated. The applicability of the developed model is demonstrated by considering an illustrative example via the Trading Organiser 'Moscow Exchange'. The nominal, conservative and optimistic portfolio weight distributions are obtained. The difference in the obtained weight distributions and the shift of efficient frontiers confirm that the solution of the Markowitz problem is sensitive to changing market parameters. Mean – standard deviation diagrams for the nominal, optimistic and conservative weight distributions are constructed, the interval values of portfolio means and risk are assessed in real time. It is concluded that portfolios with a conservative weight distribution produce better results than nominal and optimistic portfolios: higher mean return values are obtained at a lower risk. The obtained results are of practical interest under conditions of financial market instability.

Keywords: Efficient frontier · Markowitz model · Investment portfolio · Mean-variance analysis · Optimal solution

1 Introduction

The problem of portfolio optimization remains a widely discussed issue in investment analysis. In the present state of the art there are different formulations and models developed to solve this problem (see e.g. [1–3]). The most commonly used and efficient approach to solve the portfolio optimization problem is the Markowitz model [4]. One of the reasons why the Markowitz model is criticized is the lack of stability ("robustness") of the optimal solution. Small changes in the input parameters can have a

R. Silhavy et al. (Eds.): CoMeSySo 2020, AISC 1295, pp. 797–812, 2020.
https://doi.org/10.1007/978-3-030-63319-6_74

huge impact on the resulting portfolio [1, 2, 5]. From the point of view of financial investment analysis the instability of the optimal solution can be explained by the following reasons.

The premise of a normal distribution of returns on the assets is not always fulfilled. This situation is often observed in emerging markets, which raises the question of the adequacy of the standard deviation as a risk measure when investing in such markets.

The input parameters of the model do not always accurately describe the future development of the market. In fact, they are estimates of the parameters of distributions unknown in the future, thus, the estimation error can be significant [6].

Thus, it is important to consider that in conditions of instability of the emerging markets and volatility of stock quotes, the efficient frontier of the Markowitz model may turn out to be unstable. Therefore, when making an investment decision, it is important to take into account the sensitivity of the optimal solution to changing market parameters.

The aim of this paper is to analyze the sensitivity of the solution of the Markowitz problem by formulating nominal, conservative and optimistic market development scenarios.

The paper is organized as follows. The classical formulation of the Markowitz problem for portfolio optimization is formulated in the second section. An analysis of the sensitivity of the solution of the Markowitz problem is proposed and an algorithm for defining the initial parameters of the investment portfolio optimization problem is developed. A modified Markowitz model is proposed on the basis of the conservative and optimistic formulation of the problem. In the third section an illustrative experimental case-study is performed via 'Moscow Exchange' Trading Organiser. The investment portfolio is optimized on the basis of the modified Markowitz model. Stability ('robustness') analysis of the obtained assets' parameters is held. The portfolios with the nominal, optimistic and conservative weight distribution are tested in real time. The results of model verification are discussed in the forth section. The main findings of the research are summarized in the conclusion.

2 Methods

2.1 The Markowitz Model for Portfolio Optimization

The classical Markowitz problem is formulated as minimizing the risk measured by variance for a fixed portfolio return or maximizing a return for a fixed risk [1, 4, 7].

Initial premises of the classical Markowitz problem:

1. The portfolio consists of risky assets;
2. Short sales are not available to the investor;
3. Variance is an adequate measure to assess portfolio risk.

The mean return of a portfolio of n assets is obtained as:

$$R_p = \sum_{i=1}^{n} w_i \cdot \overline{R}_i, \tag{1}$$

where w_i is the weight of the asset i in the portfolio, $\bar{R}_i$ is the mean return on the asset i.

The variance of a portfolio return is defined as [7]:

$$\sigma_p^2 = \sum_{ij} w_i \cdot w_j \cdot Cov_{ij} \tag{2}$$

where w_i and w_j are the weights of the assets i and j in the portfolio, respectively, Cov_{ij} is the covariance between two assets i and j.

The problem to calculate points of the efficient frontier may be defined as a problem of optimization in two approaches which yield the same result: to find minimum-variance (maximum-mean) portfolio under fixed mean (variance) value at some arbitrary value R (σ^2).

In terms of risk minimization, the Markowitz problem is formulated as follows:

$$J = \sum_{ij} w_i \cdot w_j \cdot Cov_{ij} \to \min \tag{3}$$

subject to

$$\sum_{i} w_i \cdot \overline{R}_i = R \tag{4}$$

$$\sum_{i} w_i = 1$$

$$w_i \geq 0,\ i = \overline{1, n}$$

In terms of mean maximization, the Markowitz problem is formulated as follows:

$$J = \sum_{i} w_i \cdot \overline{R}_i \to \max \tag{5}$$

subject to

$$\sum_{ij} w_i \cdot w_j \cdot Cov_{ij} = \sigma^2,\ i = \overline{1, n} \tag{6}$$

$$\sum_{i} w_i = 1,\ i = \overline{1, n}$$

$$w_i \geq 0,\ i = \overline{1, n}$$

The solution of the problem yields the optimal weight coefficients w_i^* for the assets in the portfolio.

2.2 Sensitivity Analysis of the Markowitz Model: Defining Initial Parameters

The solution of the Markowitz problem is characterized by instability to changing parameters. Estimates of expected returns on assets present the greatest difficulty; small changes in expected returns can have a significant impact on the position of the efficient set. It has been proved that estimates of variances and covariances of assets can be determined more accurately, and their change has less impact on the shift of the efficient set [5, 6]. The interval approach to constructing an efficient set confirms these statements [7].

We propose nominal, optimistic and conservative formulations of the Markowitz model to analyze the location of the efficient set, taking into account possible quote fluctuations in the financial market.

We assume that the real situation corresponds to the nominal formulation of the model, that is, the portfolio parameters are described on the basis of existing data on the return and risk of assets, which corresponds to the statement of problem (1), (2) in terms of risk minimizing. We denote the nominal weight distribution of the assets in the portfolio to be the solution of the nominal formulation of the Markowitz problem.

We propose using the values of maximum risk and minimum assets mean returns observed over a certain time interval to describe the conservative formulation of the model. Such an approach makes it possible to assess the situation in the most unfavorable market scenario.

We propose using the values of the assets maximum mean returns and minimum risk to describe the optimistic formulation of the model. The covariance values between the assets will be left unchanged due to the insignificant effect of their change on the location of the efficient set.

To formulate a conservative and optimistic scenario, we propose introducing modified parameters of the initial formulation of problem (1), (2). To determine the minimum and maximum risk and return on an asset, we will divide the available statistical series of observations of moving returns R_{it} into s intervals (we divide the series of annual observations by months) and estimate the return on the asset and its variance within each interval k, $k = \overline{1, s}$. The minimum and maximum risk and return on the asset will be assessed on the basis of the obtained monthly estimates.

If the distribution differs from the normal, but it is close to a symmetric one, or the errors can be represented as a mixture of normally distributed errors and large but unlikely errors, the best estimate of the mathematical expectation is the median of the sample. To obtain a robust solution, we will use a sample median as an estimate of the mathematical expectation.

Considering the aforesaid, we formulate the following algorithm to assess the input parameters of the model.

Step 1. Divide the period of values of moving stock returns R_{it} into s equal consecutive intervals, the duration of each of which corresponds to the investment period. The number of observations in each interval is equal to n_s. The series of obtained returns within the interval k, $k = \overline{1, s}$, is denoted by R_{ikl}, $l = \overline{1, n_s}$.

Step 2. Inside each interval k, $k = \overline{1,s}$, determine the average returns $\overline{R}_{ik}$ and risk of the asset i σ^2_{ik}, $i = \overline{1,n}$:
Step 3. Determine the values of the maximum and minimum average return on asset i:

$$\overline{R}_{i\min} = \min_k\left(\overline{R}_{ik}\right),\ k = \overline{1,s},\ i = \overline{1,n} \tag{7}$$

$$\overline{R}_{i\max} = \max_k\left(\overline{R}_{ik}\right),\ k = \overline{1,s},\ i = \overline{1,n} \tag{8}$$

Step 4. Assess the maximum and minimum risk of asset i:

$$\sigma^2_{i\min} = \min_k\left(\sigma^2_{ik}\right),\ k = \overline{1,s},\ i = \overline{1,n} \tag{9}$$

$$\sigma^2_{i\max} = \max_k\left(\sigma^2_{ik}\right),\ k = \overline{1,s},\ i = \overline{1,n} \tag{10}$$

Step 5. Determine the covariance matrix $\underline{[Cov_{ij}]}$ for the optimistic scenario using the minimum risk value of asset i:

$$\underline{Cov_{ij}} = \sigma^2_{i\min},\ i = j,\ i = \overline{1,n} \tag{11}$$

$$\underline{Cov_{ij}} = Cov_{ij},\ i \neq j,\ i = \overline{1,n}$$

Determine the covariance matrix for the pessimistic scenario using the maximum risk value of asset i:

$$\overline{Cov_{ij}} = \sigma^2_{i\max},\ i = j,\ i = \overline{1,n} \tag{12}$$

$$\overline{Cov_{ij}} = Cov_{ij},\ i \neq j,\ i = \overline{1,n}$$

2.3 The Modified Markowitz Problem for Nominal, Conservative and Optimistic Formulation

To describe the optimistic scenario, we take the expected return on asset i equal to the maximum mean return determined according to (8) and the modified covariance matrix $\underline{[Cov_{ij}]}$, obtained from the nominal covariance matrix $[Cov_{ij}]$ using (11). The modified Markowitz problem, taking into account the background of the problem, will take the following form:

$$J_{opt} = \sum_{ij} w_i \cdot w_j \cdot \underline{Cov_{ij}} \rightarrow \min \tag{13}$$

subject to

$$\sum_i w_i \cdot \overline{R}_{i\max} = R,\ i = \overline{1,n} \tag{14}$$

$$\sum_i w_i = 1,\ i = \overline{1,n}$$

$$w_i \geq 0,\ i = \overline{1,n}$$

where $\underline{[Cov_{ij}]}$ is the covariance matrix for the optimistic scenario, determined according to (11), w_i is the weight of asset i in the portfolio, $\overline{R}_{i\max}$ is the maximum average return on asset i, determined according to (8), R is the given portfolio return.

We denote the weight distribution w^*_{opt} to be the solution of this problem.

To describe the conservative formulation of the Markowitz problem, we take the expected return on asset i equal to the minimum average return, determined according to (7), and use the covariance matrix $[\overline{Cov_{ij}}]$ obtained from the nominal covariance matrix $[Cov_{ij}]$ and (12). The modified Markowitz problem, taking into account the initial premises (1)–(3) of problem (1), (2), will take the following form:

$$J_{cons} = \sum_{ij} w_i \cdot w_j \cdot \overline{Cov_{ij}} \to \min \tag{15}$$

subject to

$$\sum_i w_i \cdot \overline{R}_{i\min} = R,\ i = \overline{1,n} \tag{16}$$

$$\sum_i w_i = 1,\ i = \overline{1,n}$$

$$w_i \geq 0,\ i = \overline{1,n}$$

where $\overline{[Cov_{ij}]}$ is the covariance matrix for the pessimistic scenario, determined according to (12), w_i is the weight of asset i in the portfolio, $\overline{R}_{i\min}$ is the minimum average return on asset i, determined according to (7), R is the given portfolio return.

The solution to this problem is the weight distribution w^*_{cons}.

Problems (13), (14) and (15), (16) refer to quadratic programming problems, the solution of which can be obtained in the Excel Optimization Toolbox and MatLab software environment.

3 Results

3.1 Portfolio Optimization Under Parametric Uncertainty

As the statistic basis for this part of the experimental study, the Trading Organiser 'Moscow Exchange' – Equity and Bond Market (MOEX) was chosen as the leading trading platform for the securities market in Russia which is of great interest to both domestic and foreign portfolio investors [8].

An important factor in the choice of the 'Moscow Exchange' was the open access to trading information; the prices that are formed on the 'Moscow Exchange' are a generally accepted guideline for investors making transactions with Russian stocks and depositary receipts. Trading information is broadcast to a huge number of consumers in Russia and abroad and is the basis for calculating the RTS and MICEX stock indices.

For the analysis and synthesis of the portfolio, assets were selected representing various sectors of the economy according to the principle of industry diversification: electricity, oil and gas, banking and finance, telecommunication, metallurgy, trading sector. Securities with growth potential and sufficient liquidity were preferred. For further analysis, eight assets were selected that meet the listed characteristics of the following issuers: Public Joint Stock Company Gazprom (ordinary share *GAZP*), Public Joint Stock Company "Mining and Metallurgical Company "NORILSK NICKEL" (ordinary share *GMKN*), Public Joint Stock Company "Aeroflot- Russian Airlines" (ordinary share *AFLT*), Public Limited Liability Company Yandex N.V., shares of a foreign issuer (ordinary share *YNDX*), Rosneft Oil Company, ordinary share (ordinary share *ROSN*), Public Joint Stock Company "Magnit", (ordinary share *MGNT*), Mobile TeleSystems Public Joint Stock Company, ordinary share (ordinary share *MTSS*), Sberbank of Russia, ordinary share (ordinary share *SBER*) [8].

The period of twelve months was chosen as the base period, namely, December 2018–December 2019. The price series of assets were transformed into series of moving returns with a smoothing window of one month. Further analysis was carried out according to the received monthly moving returns for the period January–December 2019.

For the synthesis of the portfolio according to the Markowitz model and modified models of optimistic and pessimistic scenarios, a covariance matrix was obtained (see Table 1). Covariance-correlation analysis revealed that some assets can be negatively, positively or low correlated with other assets at the same time. According to the Markowitz Portfolio Theory including negatively correlated assets into the portfolio is conducive to decreasing its risk [7].

Table 1. Covariance matrix for the selected assets

Asset	Cov_{ij}							
	GMKN	AFLT	GAZP	MGNT	MTSS	ROSN	YNDX	SBER
GMKN	0,000864	0,000525	−0,00082	0,000219	0,001255	0,00085	0,001241	0,000196
AFLT	0,000525	0,000862	0,000655	0,000995	0,000875	0,000183	0,000363	0,000405
GAZP	−0,00082	0,000655	0,004023	0,001683	0,000939	0,000493	−0,00057	0,001351
MGNT	0,000219	0,000995	0,001683	0,001632	0,001478	0,000323	0,002439	0,00165
MTSS	0,001255	0,000875	0,000939	0,001478	0,00082	0,000686	0,001843	0,001375
ROSN	0,00085	0,000183	0,000493	0,000323	0,000686	0,000578	0,000676	0,00105
YNDX	0,001241	0,000363	−0,00057	0,002439	0,001843	0,000676	0,00312	0,001072
SBER	0,000196	0,000405	0,001351	0,00165	0,001375	0,00105	0,001072	0,001735

The assets mean returns and variances correspond to the nominal formulation of the model. On the basis of the developed algorithm proposed in Sect. 2.2. the initial parameters of the model were defined to describe the conservative and optimistic formulations of the Markowitz problem. Maximum average returns, minimum average returns, maximum risk, and minimum risk for assets were obtained using (8), (7), (10) and (9), respectively. The results are presented in Table 2.

Table 2. Statistic parameters of the model

Asset	Asset parameters					
	$\overline{R}_i$	$\overline{R}_{imin}$	$\overline{R}_{imax}$	σ_i^2	σ_{imin}^2	σ_{imax}^2
GMKN	1,030672	0,955652	1,134694	0,000864	0,000347	0,001861
AFLT	1,002309	0,949233	1,080857	0,000862	0,000183	0,002351
GAZP	1,041405	0,97576	1,215828	0,004023	0,000339	0,015351
MGNT	0,995392	0,909809	1,115687	0,001632	0,00019	0,007446
MTSS	1,021866	0,963496	1,099646	0,00082	0,000153	0,001755
ROSN	1,005757	0,965775	1,076357	0,000578	0,000195	0,001174
YNDX	1,029459	0,887676	1,134366	0,00312	0,000311	0,014672
SBER	1,021043	0,945474	1,099559	0,001735	0,00066	0,004095

The portfolio optimization problem (3), (4) was solved using estimates of expected returns $\bar{R}_i$ from Table 2 and covariance matrix from Table 1. The solution of the $(w_i w_j Cov_{ij})$ -minimization problem yields the optimal weight coefficients w_{inom}^* for portfolios from the efficient frontier, represented in Table 3. We consider the formulation above to be the 'nominal' formulation of the problem, in which the solution is obtained on the basis of standard sample characteristics.

Table 3. Portfolio parameters and weight distribution w^*_{inom}

Portfolio return	Standard deviation	Weights of the assets							
		GMKN	AFLT	GAZP	MGNT	MTSS	ROSN	YNDX	SBER
1,0410	0,008535	0,0376	0,0000	0,9624	0,0000	0,0000	0,0000	0,0000	0,0000
1,0400	0,006863	0,1308	0,0000	0,8692	0,0000	0,0000	0,0000	0,0000	0,0000
1,0390	0,005423	0,2240	0,0000	0,7760	0,0000	0,0000	0,0000	0,0000	0,0000
1,0380	0,004214	0,3171	0,0000	0,6829	0,0000	0,0000	0,0000	0,0000	0,0000
1,0370	0,003237	0,4103	0,0000	0,5897	0,0000	0,0000	0,0000	0,0000	0,0000
1,0360	0,002492	0,5035	0,0000	0,4965	0,0000	0,0000	0,0000	0,0000	0,0000
1,0350	0,001978	0,5966	0,0000	0,4034	0,0000	0,0000	0,0000	0,0000	0,0000
1,0340	0,001696	0,6898	0,0000	0,3102	0,0000	0,0000	0,0000	0,0000	0,0000
1,0330	0,001604	0,6916	0,0012	0,2593	0,0000	0,0478	0,0000	0,0000	0,0000
1,0320	0,001540	0,6673	0,0279	0,2485	0,0000	0,0532	0,0032	0,0000	0,0000
1,0310	0,001480	0,6421	0,0407	0,2393	0,0000	0,0532	0,0247	0,0000	0,0000
1,0300	0,001424	0,6169	0,0536	0,2300	0,0000	0,0532	0,0463	0,0000	0,0000
1,0290	0,001371	0,5918	0,0664	0,2208	0,0000	0,0531	0,0679	0,0000	0,0000
1,0280	0,001321	0,5666	0,0792	0,2116	0,0000	0,0531	0,0894	0,0000	0,0000
1,0270	0,001274	0,5414	0,0921	0,2024	0,0000	0,0531	0,1110	0,0000	0,0000
1,0260	0,001230	0,5163	0,1049	0,1932	0,0000	0,0531	0,1326	0,0000	0,0000
1,0250	0,001190	0,4911	0,1177	0,1840	0,0000	0,0530	0,1541	0,0000	0,0000
1,0240	0,001153	0,4659	0,1306	0,1748	0,0000	0,0530	0,1757	0,0000	0,0000
1,0230	0,001119	0,4435	0,1412	0,1658	0,0042	0,0503	0,1949	0,0000	0,0000
1,0220	0,001088	0,4240	0,1494	0,1572	0,0129	0,0449	0,2116	0,0000	0,0000
1,0210	0,001060	0,4046	0,1576	0,1486	0,0216	0,0394	0,2284	0,0000	0,0000
1,0200	0,001034	0,3851	0,1657	0,1399	0,0302	0,0339	0,2451	0,0000	0,0000
1,0190	0,001011	0,3656	0,1739	0,1313	0,0389	0,0284	0,2619	0,0000	0,0000
1,0180	0,000991	0,3461	0,1821	0,1226	0,0476	0,0229	0,2786	0,0000	0,0000
1,0170	0,000973	0,3267	0,1903	0,1140	0,0563	0,0174	0,2954	0,0000	0,0000
1,0160	0,000958	0,3072	0,1985	0,1053	0,0649	0,0119	0,3121	0,0000	0,0000
1,0150	0,000946	0,2877	0,2067	0,0967	0,0736	0,0065	0,3288	0,0000	0,0000
1,0140	0,000937	0,2684	0,2148	0,0881	0,0824	0,0007	0,3457	0,0000	0,0000
1,0130	0,000930	0,2461	0,2228	0,0788	0,0899	0,0000	0,3624	0,0000	0,0000
1,0120	0,000926	0,2235	0,2306	0,0695	0,0973	0,0000	0,3791	0,0000	0,0000
1,0110	0,000925	0,2009	0,2384	0,0602	0,1046	0,0000	0,3959	0,0000	0,0000

To evaluate the sensitivity of the optimal solution we consider 'conservative' and 'optimistic' formulations of the problem. We obtain the 'conservative' optimal solution w^*_{icons}, which provides the investor with the guaranteed values of return and variance, and represents the worst possible result. Similarly, the 'optimistic' optimal solution w^*_{iopt} is determined which yields a value of return impossible to exceed, and the least possible variances as well.

For 'conservative' ('optimistic') formulation of the Markowitz problem we assume $\overline{R}_i = \overline{R}_{i\min}$ ($\overline{R}_i = \overline{R}_{i\max}$) and $\sigma_i^2 = \sigma_{i\max}^2$ ($\sigma_i^2 = \sigma_{i\min}^2$) (see Table 2). Thus, we solve the problem (13), (14) and (15), (16) for these modified parameters. The optimal solutions

w^*_{inom}, w^*_{icons} and w^*_{iopt} have been obtained using the Optimization Toolbox of Excel and the Visual Basic Editor.

The optimal weight coefficients w^*_{icons} and w^*_{iopt} for 'conservative' and 'optimistic' portfolios are given in Table 4 and Table 5, respectively.

Table 4. Portfolio parameters and weight distribution w^*_{icons}

Portfolio return	Standard deviation	Weights of the assets							
		GMKN	AFLT	GAZP	MGNT	MTSS	ROSN	YNDX	SBER
0,9750	0,013178	0,0000	0,0000	0,9238	0,0000	0,0000	0,0762	0,0000	0,0000
0,9740	0,010594	0,0000	0,0000	0,8237	0,0000	0,0000	0,1763	0,0000	0,0000
0,9730	0,008326	0,0000	0,0000	0,7237	0,0000	0,0000	0,2763	0,0000	0,0000
0,9720	0,006363	0,0000	0,0000	0,6234	0,0000	0,0000	0,3766	0,0000	0,0000
0,9710	0,004717	0,0000	0,0000	0,5234	0,0000	0,0000	0,4766	0,0000	0,0000
0,9700	0,003379	0,0000	0,0000	0,4231	0,0000	0,0000	0,5769	0,0000	0,0000
0,9690	0,002356	0,0000	0,0000	0,3231	0,0000	0,0000	0,6769	0,0000	0,0000
0,9680	0,001642	0,0000	0,0000	0,2228	0,0000	0,0000	0,7772	0,0000	0,0000
0,9670	0,001238	0,0000	0,0000	0,1303	0,0000	0,0336	0,8361	0,0000	0,0000
0,9660	0,001046	0,0000	0,0000	0,0669	0,0000	0,1943	0,7389	0,0000	0,0000
0,9650	0,000969	0,0000	0,0493	0,0494	0,0000	0,1983	0,7030	0,0000	0,0000
0,9640	0,000912	0,0000	0,1084	0,0410	0,0000	0,1714	0,6792	0,0000	0,0000
0,9630	0,000873	0,0000	0,1674	0,0326	0,0000	0,1446	0,6554	0,0000	0,0000
0,9620	0,000850	0,0222	0,2164	0,0278	0,0000	0,1083	0,6253	0,0000	0,0000
0,9610	0,000839	0,0683	0,2404	0,0286	0,0044	0,0640	0,5942	0,0000	0,0000
0,9600	0,000831	0,0940	0,2460	0,0292	0,0174	0,0328	0,5806	0,0000	0,0000
0,9590	0,000827	0,1196	0,2516	0,0298	0,0303	0,0016	0,5670	0,0000	0,0000
0,9580	0,000825	0,1309	0,2538	0,0272	0,0451	0,0000	0,5430	0,0000	0,0000

Table 5. Portfolio parameters and weight distribution w^*_{iopt}

Portfolio return	Standard deviation	Weights of the assets							
		GMKN	AFLT	GAZP	MGNT	MTSS	ROSN	YNDX	SBER
1,1480	0,000933	0,7379	0,0590	0,2031	0,0000	0,0000	0,0000	0,0000	0,0000
1,1380	0,000733	0,5996	0,2162	0,1842	0,0000	0,0000	0,0000	0,0000	0,0000
1,1280	0,000533	0,4613	0,3734	0,1653	0,0000	0,0000	0,0000	0,0000	0,0000
1,1230	0,000433	0,3922	0,4520	0,1558	0,0000	0,0000	0,0000	0,0000	0,0000
1,1180	0,000333	0,0025	0,0000	0,2975	0,0000	0,0000	0,7000	0,0000	0,0000
1,1160	0,000328	0,0000	0,0000	0,2842	0,0000	0,0000	0,7158	0,0000	0,0000
1,1140	0,000323	0,0000	0,0000	0,2699	0,0000	0,0000	0,7301	0,0000	0,0000
1,1120	0,000318	0,0000	0,0000	0,2556	0,0000	0,0000	0,7444	0,0000	0,0000
1,1100	0,000312	0,0000	0,0000	0,2412	0,0000	0,0000	0,7588	0,0000	0,0000
1,1080	0,000307	0,0000	0,0000	0,2269	0,0000	0,0000	0,7731	0,0000	0,0000
1,1060	0,000301	0,0000	0,0000	0,2125	0,0000	0,0000	0,7875	0,0000	0,0000
1,1040	0,000295	0,0000	0,0000	0,1982	0,0000	0,0000	0,8018	0,0000	0,0000
1,1020	0,000289	0,0000	0,0000	0,1839	0,0000	0,0000	0,8161	0,0000	0,0000
1,1000	0,000283	0,0000	0,0000	0,1695	0,0000	0,0000	0,8305	0,0000	0,0000

(continued)

Table 5. (*continued*)

Portfolio return	Standard deviation	Weights of the assets							
		GMKN	AFLT	GAZP	MGNT	MTSS	ROSN	YNDX	SBER
1,0980	0,000277	0,0000	0,0000	0,1552	0,0000	0,0000	0,8448	0,0000	0,0000
1,0960	0,000270	0,0000	0,0000	0,1408	0,0000	0,0000	0,8592	0,0000	0,0000
1,0940	0,000263	0,0000	0,0000	0,1265	0,0000	0,0000	0,8735	0,0000	0,0000
1,0920	0,000256	0,0000	0,6223	0,0921	0,0000	0,0000	0,2856	0,0000	0,0000
1,0900	0,000244	0,0000	0,9322	0,0678	0,0000	0,0000	0,0000	0,0000	0,0000
1,0880	0,000231	0,0000	0,9471	0,0529	0,0000	0,0000	0,0000	0,0000	0,0000
1,0860	0,000218	0,0000	0,9619	0,0381	0,0000	0,0000	0,0000	0,0000	0,0000
1,0840	0,000205	0,0000	0,9767	0,0233	0,0000	0,0000	0,0000	0,0000	0,0000
1,0820	0,000191	0,0000	0,9915	0,0085	0,0000	0,0000	0,0000	0,0000	0,0000
1,0800	0,000184	0,0000	0,8096	0,0000	0,0000	0,0000	0,1904	0,0000	0,0000

Tables 3, 4 and 5 demonstrate that nominal, conservative and optimistic solutions to the Markowitz problem produce various weight distributions. Meanwhile, the nominal distribution turned out to be closer to the optimistic distribution than to the conservative one.

The expected values of R_p and σ_p for nominal, conservative and optimistic portfolios vary within the intervals, presented in Table 6.

Table 6. Interval values of R_p and σ_p

Model formulation	Portfolio parameters	
	R_p	σ_p
Conservative	[0,9580; 0,9750]	[0,000825; 0,013178]
Nominal	[1,0110; 1,0410]	[0,000925; 0,008535]
Optimistic	[1,0800; 1,1480]	[0,000184; 0,000933]

The efficient frontiers for the nominal, optimistic and conservative weight distributions are presented in Fig. 1.

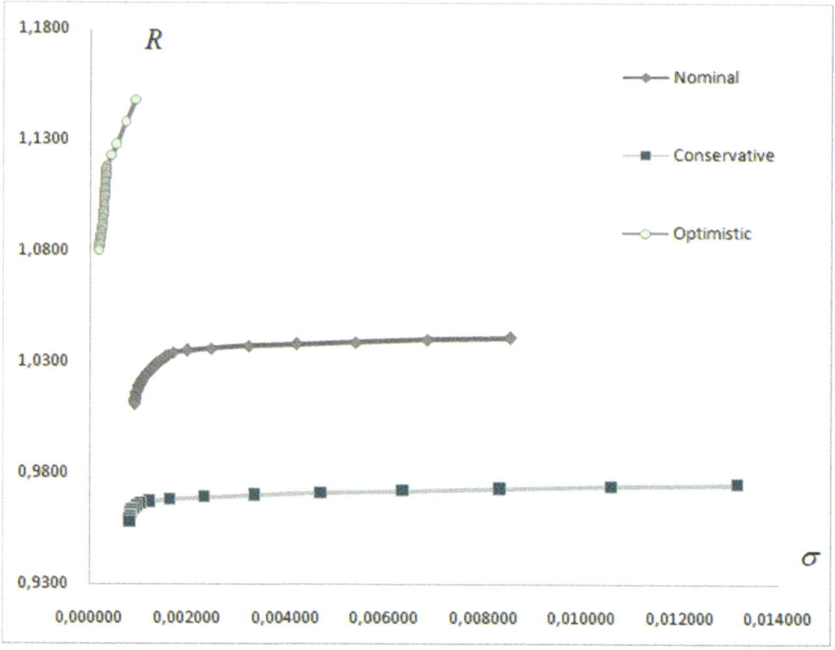

Fig. 1. Efficient frontiers of portfolios for the nominal, optimistic and conservative weight distribution

3.2 Verification of Portfolios with Optimistic, Conservative and Nominal Weight Distribution

In this part of experimental study the constructed portfolios from the nominal, optimistic and conservative efficient frontiers with the weight distributions w^*_{inom}, w^*_{icons} and w^*_{iopt} were verified in real time in a subsequent period, namely January–April 2020.

On the basis of series of the assets' moving returns we calculated portfolio moving returns for the mentioned period. The actual parameters of the portfolios $R_p\left(w^*_{inom}\right)$, $R_p\left(w^*_{icons}\right)$, $R_p\left(w^*_{iopt}\right)$ and $\sigma^2_p\left(w^*_{inom}\right)$, $\sigma^2_p\left(w^*_{icons}\right)$, $\sigma^2_p\left(w^*_{iopt}\right)$ for the named period were evaluated. The values of R_p and σ_p, corresponding to w^*_{inom}, w^*_{icons} and w^*_{iopt}, vary within the intervals, presented in the Table 7.

It has been found that portfolios with w^*_{icons} weight distribution in real time produced better results than portfolios with a nominal and optimistic distribution: higher mean return values were obtained at a lower risk.

Table 7. Interval values of R_p and σ_p for w^*_{inom}, w^*_{icons}, w^*_{iopt}

Model formulation	Portfolio parameters	
	R_p	σ_p
Conservative	[0,934867; 0,992372]	[0,047941; 0,122102]
Nominal	[0,912847; 0,979823]	[0,123421; 0,188832]
Optimistic	[0,913735; 0,924682]	[0,177562; 0,188464]

Note, that the index of the return does not achieve 1,0 in real time, that is none of the model modifications allowed to achieve a positive increase in portfolio return in real time, which is due to financial market decline in 2020.

The mean – standard deviation diagrams for 'conservative', 'optimistic' and 'nominal' cases are presented in Fig. 2.

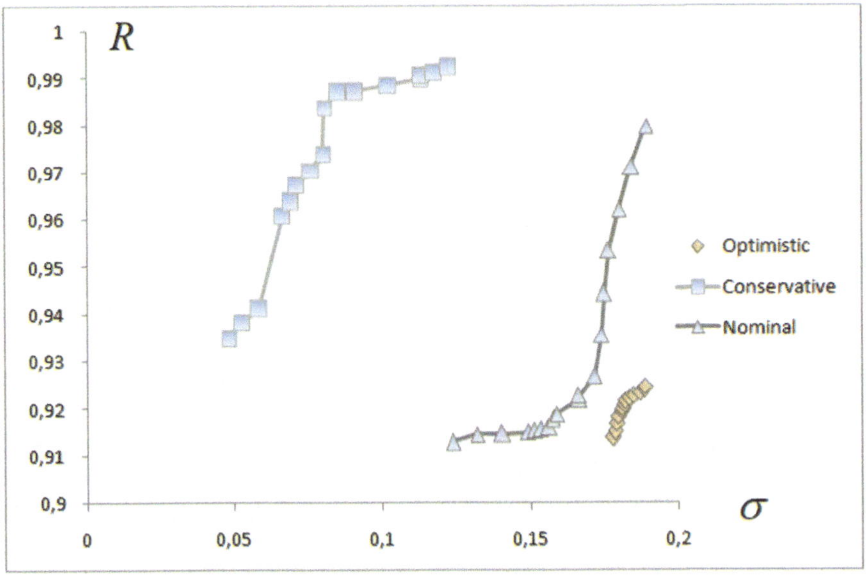

Fig. 2. Mean – standard deviation diagrams for the nominal, optimistic and conservative weight distributions in real time

3.3 Stability ('Robustness') Analysis of Assets' Parameters Estimates

In this part of the experimental study it is verified whether estimates of expected return and variance for the period January–April 2020 fall in the intervals between the minimum and maximum values estimated according to the results of 2019.

In order to assess the stability of estimates of asset parameters, we compare the average return on assets and variances for 2020 with the most pessimistic estimates obtained in 2019 from the Table 2. Data for comparison are given in Table 8.

The results of the Table 8 show that the observed mean returns for *AFLT-*, *GAZP-*, *ROSN-* and *SBER*-assets in 2020 were lower than the estimated minimum average returns estimated for 2019. Financial market decline due to COVID-19 is to blame for the decrease in mean returns of the assets. Actual variances of assets for 2020 for all assets under study turned out to be greater than the worst predicted risk for 2019. Instability of risk estimation is explained by the fact that such a decrease in asset returns had not been observed in the previous period under study. Nevertheless, the use of pessimistic estimates is conducive to reduce the decrease in portfolio parameters in case of sudden unpredicted changes in factors affecting assets prices and other market parameters.

Table 8. Assets' parameters estimates

Asset	Asset parameters					
	$\overline{R}_{i2020}$	$\overline{R}_{imin}$	$\overline{R}_{imax}$	σ^2_{i2020}	σ^2_{imin}	σ^2_{imax}
GMKN	0,993715	0,955652	1,134694	0,014473	0,000347	0,001861
AFLT	0,899621	0,949233	1,080857	0,052201	0,000183	0,002351
GAZP	0,917419	0,975760	1,215828	0,016956	0,000339	0,015351
MGNT	0,979440	0,909809	1,115687	0,027175	0,000190	0,007446
MTSS	0,986559	0,963496	1,099646	0,013769	0,000153	0,001755
ROSN	0,898106	0,965775	1,076357	0,041524	0,000195	0,001174
YNDX	0,992975	0,887676	1,134366	0,013391	0,000311	0,014672
SBER	0,903884	0,945474	1,099559	0,013361	0,000660	0,004095

These findings give us reason to believe that the conservative formulation of the Markowitz problem based on pessimistic estimates of asset parameters, yields better results under market stress. To the best of our knowledge the results of a portfolio with a conservative distribution in real time should not be lower than the results produced by the 'nominal' and 'optimistic' portfolios. The choice of the model formulation depends on the investor's degree of risk preference.

4 Discussions

We evaluated how the optimistic and pessimistic parameters of assets differ from the nominal ones. Nominal, optimistic and pessimistic estimates are presented in Table 2. It was found that the nominal returns on all securities were 3.99%–14.18% higher than the pessimistic ones, and the pessimistic variances were 3.1%–256% higher than the nominal ones. The nominal returns on all securities were lower than optimistic by 7.06%–17.44%, and the nominal variances exceeded the optimistic ones 1.49–10.9 times.

This deviation of the input parameters of the model is conducive to the shift of the efficient frontier as shown in Fig. 1 and led to the following deviations of the portfolio parameters (Table 6): the return on the nominal portfolio is 5.3%–6.6% higher than the return on the conservative portfolio. The maximum return on the optimistic portfolio exceeds the maximum return on the nominal portfolio by 10.7%. The maximum risk of the conservative portfolio exceeds the maximum risk on the nominal portfolio by 54.31%.

The difference in the obtained nominal, conservative and optimistic weight distributions shown in Tables 3, Table 4 and Table 5, respectively, and the deviation of the portfolio parameters confirm that the solution of the Markowitz problem is sensitive to changing market parameters.

Note that portfolios with w^*_{icons} weight distribution in real time produced better results than portfolios with a nominal and optimistic distribution: higher mean return values were obtained at a lower risk. The minimum return on a conservative portfolio exceeds the minimum return on a nominal portfolio by 2.21% and by 2.12% on an optimistic portfolio. The maximum return on the conservative portfolio exceeds by 1.26% the maximum return on the nominal portfolio and by 6.77% on the optimistic portfolio. The minimum risk of a nominal portfolio exceeds 2.57 times the minimum risk of a conservative portfolio. The minimum risk of an optimistic portfolio exceeds 3.7 times the minimum risk of a conservative portfolio. The maximum risk of a nominal portfolio exceeds by 54.65% the risk of a conservative portfolio. The maximum risk of an optimistic portfolio exceeds by 54.35% the maximum risk of a conservative portfolio.

To the best of our knowledge the conservative formulation of the Markowitz problem based on pessimistic estimates of asset parameters yields better results under parametric uncertainty than the nominal and optimistic formulations. The 'optimistic' portfolios have not proved their applicability under financial market conditions.

5 Conclusion

In this paper an approach to analyze the sensitivity of the solution of the Markowitz problem based on the nominal, optimistic and conservative formulation of the model is proposed. An algorithm for defining the initial parameters to describe the conservative and optimistic formulation of the Markowitz problem is developed.

It is proposed to use estimates of maximum risk and minimum asset return observed over a certain time interval to describe the conservative scenario. Such a formulation makes it possible to assess the most unfavorable market situation. It is proposed to evaluate the input parameters on the basis of the maximum average return and minimum risk of assets to describe the optimistic market movement.

The differences of the optimal solutions w^*_{inom}, w^*_{icons} and w^*_{iopt} confirm that the Markowitz model is sensitive to parametric uncertainty. The changes in input parameters produce shift of the efficient frontier of portfolios. The choice of the formulation of the model depends on the degree of the investor's risk aversion. An analysis of the proposed scenarios gives an in-depth understanding of the location of the efficient set under market instability and thus provides the investor with the opportunity to make a more informed choice of the optimal portfolio.

The results of experimental study demonstrate that portfolios with a conservative weight distribution produce better results: higher mean return values were obtained at a lower risk in real time. These findings give us reason to believe that the conservative formulation of the Markowitz problem yields better results under market stress. The use of pessimistic estimates is conducive to reduce the decrease in portfolio parameters in case of sudden unpredicted changes in factors affecting market parameters. The obtained results can be of practical interest especially under conditions of financial market decline that is observed nowadays.

References

1. Sharpe, W.F., Alexander, G.J., Beiley, J.V.: Investments. Prentice Hall International Inc., Upper Saddle River (1995)
2. Bodie, Z., Kane, A., Marcus, A.J.: Essentials of Investments. McGraw-Hill, New York (2001)
3. Fabozzi, F.J.: Quantitative strategies: multi-asset. J. Portfolio Manag. Multi-Asset Special Issue **46**(6), 1–3 (2020)
4. Markowitz, H.M.: Portfolio selection. J. Financ. **7**(1), 77–91 (1952)
5. Chen, Z.-P., Zhao, C.-E.: Is the MV efficient portfolio really that sensitive to estimation errors. Asia-Pac. J. Oper. Res. **19**, 149–168 (2002)
6. Chopra, V.K., Ziemba, W.T.: The effect and errors in means, variances and covariances in optimal portfolio choice. J. Portfolio Manag. Winter **19**(2), 6–11 (1993)
7. Luenberger, D.G.: Investment science. Oxford University Press, New York (1998)
8. Moscow Exchange. https://www.moex.com/. Accessed 21 June 2020

Ontology Based Recommendation System for Predicting Cultivation and Harvesting Timings Using Support Vector Regression

Heba Osman[1(✉)], Nashwa El-Bendary[2], Essam El Fakharany[2], and Mohamed El Emam[1]

[1] College of Computing and Information Technology, Arab Academy for Science, Technology and Maritime Transport, Cairo, Egypt
hebaob51@yahoo.com

[2] College of Computing and Information Technology, Arab Academy for Science, Technology and Maritime Transport, Smart Village, Cairo, Egypt

Abstract. Agriculture is an interesting area of research due to its importance in life. Climate change is one of the main factors that affect farming and it remarkably impacts crops' growth and may lead to several obstacles, like preventing optimum results of the expected harvest. Unfortunately, there are neither reliable indicators nor early alerts for the unexpected weather changes in order to save a crop during its life cycle. Few researchers relate growth crops with the weather condition changes. Whereas most researchers do not give attention to the best cultivation and harvesting timings of any kinds of crop according to the unexpected weather conditions in the future and to the absence of using the ontology with climate forecasting in their proposed models. The aim of the proposed system is to provide useful recommendations for farmers to get more accurate information about the climate change affecting crops and make the best farm resources utilization. The proposed system depends on using support vector regression (SVR) to predict the weather conditions along with using the ontology approach. The crop ontology has been successfully evaluated with common evaluation criteria using a web-based evaluation tool. Besides, the system achieved 94.7% recall and 66.6% rejection ratios, which are considered beneficial results in agriculture field.

Keywords: Climate changes · Machine Learning (ML) · Support Vector Regression (SVR) · Recommendation system · Ontology

1 Introduction

In the agriculture field, accurate and timely forecast of the climate conditions has a significant influence on crop growth and cultivation. Weather prediction is considered to be one of the most important attributes that can affect the agricultural sectors and industries. Decision making in different applications such as monitoring climate conditions is very critical considering the unexpected changes of weather. However, the unstable weather conditions may damage the crops and cause soil erosion [1], which may lead farmers to suffer in both cultivation and harvesting processes.

R. Silhavy et al. (Eds.): CoMeSySo 2020, AISC 1295, pp. 813–824, 2020.
https://doi.org/10.1007/978-3-030-63319-6_75

Moreover, agriculture is considered as a popular field of applying and conducting research on various sides of scientific methods with a distinctive empirical knowledge related to machine learning (ML). The ML approaches can be used to predict the most common natural phenomena that affect crop yield, which are climate and weather changes [2]. Also the ML algorithm can construct a forecast based on large amount of historical agricultural data by enabling machines to learn without strict programming [3].

The farmers need accurate recommendations that could facilitate and enhance the agricultural process based on weather prediction and crop profile to improve the farm utilization and crop quality. Therefore, the recommendation systems were given more attention and have become part of many applications [4].

The recommendation systems can help users through providing guidance on an item based on the techniques used in these systems [5].

The aim of this work is building a recommendation system to provide recommendations about the crops such as the best cultivating timing and the best harvesting timing based on weather condition data and crop ontology to help farmers taking the right decisions. It shall also provide corps as recommendations if the requested crop is not suitable in the context of the given data.

The main goals for the proposed recommendation system are:

- Making the best resources utilization of the farm.
- Enhancing the economic status of farmers and decrease the loss of crops and money using correct recommendations.

The rest of this paper is organized as follows: Sect. 2 presents research work considering ML algorithms for weather condition prediction and ontology usage for recommendations. Section 3 introduces the proposed approach and describes its phases. Section 4 presents the obtained experimental results and system evaluation. Finally, Sect. 5 presents conclusions and discusses future work.

2 Related Work

The complexity of weather conditions is a great challenge to weather prediction approaches. In [1], authors focused on using big data statistical analysis to handle weather data for effective forecasting. The used a supervised ML algorithm, which is Support Vector Machine (SVM) to build their proposed method.

The weather dataset is preprocessed, classified and compared to a single decision tree algorithm. The SVM performance results outperformed the single decision tree.

Authors in [6] proposed a cultivation-time recommender system to forecast the best sowing timings for cereal crops in winter season. In the phase of weather prediction, five algorithms were applied: M5Rules, M5P, Multi-layer Perceptron (MLP) regressor, and IBK to predict the Growing Degree Days (GDD) based on the sowing timing.

The Mean Absolute Error (MAE) was used to evaluate the performance, and the results proved that the two regression algorithms M5P and the IBk gave the best sowing timing prediction.

On the other hand, a semantic web based recommendation system that generates agricultural recommendations was implemented in [7] using spatial data and agricultural knowledge bases. Most agriculture problems can be solved by this project using data mining algorithms like Genetic Algorithm (GA), Association Rule Mapping (ARM) rather than the traditional ways.

Also, the existing independent Agriculture Information System (AIS) platforms give a specific agricultural information service, yet still incomplete. A new AIS system is proposed in [8] based on ontology techniques. The implemented ontology based on integrated AIS is more accurate than the existing one. The proposed system's recall and rejection achieved ratios are 81.6 and 46.7 compared to 53.2 and 78.7 for the current system.

Furthermore, the OntoAgroHidro approach was implemented in [9] as ontology to represent knowledge about the impacts of climatic changes and agricultural activities on water resources. The main steps to develop that system were the analysis of the information sources, core ontology concepts reuse, integration concepts and the visualization of the ontology. The OntoAgroHidro has the best F-Measure of 0.84 by combining good precision 0.83 with good recall 0.86 values.

Also, the grape growers suffer from the problem of total yield due to pests and diseases in India, which causes lots of side effects. Another study proposes a decision support system "PDMGrapes" as a solution based on ontology which represents the IPM knowledge [10], and image processing techniques to detect crop's diseases and forecast pests. Using support vector machine gives most accurate results of detection of powdery mildew disease. The precision values for powdery mildew, downy mildew and anthracnose are 73.33%, 60% and 46.66% respectively.

Based on the previously discussed literature, the use of weather condition prediction in the agricultural field is an interesting area of research. Most of researches focus on prediction of weather conditions and weather phenomena in several domains.

However, few researchers have investigated the relation between the growth of crops and the unexpected weather condition predictions.

Furthermore, the ontology evaluation does not receive much attention as many researches did not utilize it in their proposed models. Besides, the utilization of ontologies in agriculture systems is recommended as it can be used to represent knowledge of a specific domain for sharing and reuse. Also, the integration of ML techniques with ontology can improve the performance of recommendations, as it is considered as one of the best intelligent methodologies in this area.

The proposed model should improve the accuracy of the recommendations, which helps the farmer to know the best sowing dates of the specific crops in the future based on crop profile. This recommendation system relies on two main components: weather condition prediction module to predict the daily average air temperature based on historical weather conditions data for twenty six seasons (1990/91 to 2015/16) by using common ML algorithms, and a crop ontology model, which is built by crops database.

The localization concept is applied to this work by choosing a specific location in "Menoufia" city, Egypt, which is suitable for growing tomato crops that were considered as the case study for the proposed recommender system.

3 The Proposed Approach

The following steps show clearly how the proposed recommendation system works as shown in Fig. 1.

1. Data Preparation and Preprocessing:
   - Preparing the weather condition dataset.
   - Applying the ML algorithms on the weather dataset.
2. Weather Prediction:
   - Investigating and choosing the ML algorithm that showed superior performance.
   - Computing the GDD based on the results of the applied ML algorithm.
3. Ontology construction:
   - Building the crop ontology based on crop database and accumulative GDD then evaluating the crop ontology.
4. Recommendation:
   - Using the ontology reasoner to get the final recommendations then evaluate recommendations as results of the ontology reasoner.

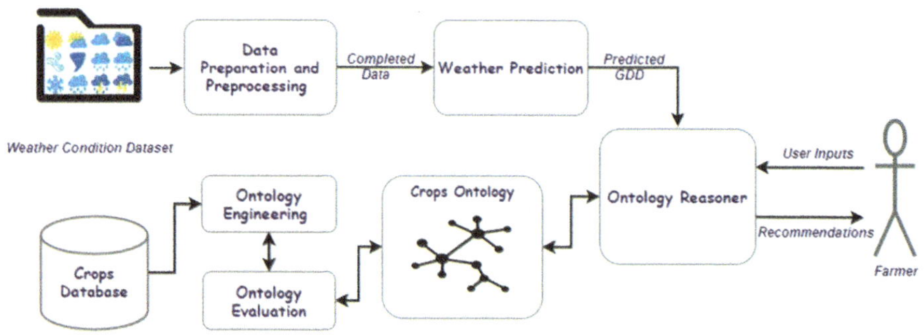

Fig. 1. The proposed model

The proposed model mainly depends on the crop ontology module and the weather condition module for predicting the minimum and maximum temperatures, which are used in computing GDD.

Meteorological data in this work is chosen based on certain locations that should be suitable for the selected crop. There are a set of parameters that must be determined to get the required dataset such as the start and the end year for this data, latitude and longitude.

After obtaining the dataset, an important preparation step should be done, which is data cleaning [13]. The cleaning of data can be made manually or automatically to exponentially remove the high or low values that may affect the entire results. These values may affect the results and accordingly all these extreme negative values would be removed.

Before applying the testing algorithms, the dependent variable must be identified as the first attribute then the other independent variables come after the first one. Additionally, an important attribute was considered, which is season of cultivation.

There are two seasons of cultivation: summer season and winter season, this classification is done based on the months of the year, as a cultivation season division, as summer (June to September) and winter (November to April).

The three selected ML techniques, which were tested and applied on the weather conditions dataset: Support Vector Regression (SVR) [14, 15], Multiple Linear Regression (MLR) [16], and Extreme Learning Machine (ELM) [17].

The weather prediction module is implemented according to the following steps:

1. Applying the selected ML algorithms to the weather condition dataset to get the predicted minimum and maximum air temperature based on historical weather conditions data.
2. Selecting the outperforming ML algorithm based on the comparative results between an input incomplete dataset with virtual gaps and the completed dataset.
3. Computing the accumulative GDD for crops according to the predicted minimum and maximum temperature.

Based on the computed accumulative GDD, user's inputs, and crop ontology, the ontology reasoner gives recommendations to the farmer about the best cultivation date, harvesting date as well as several alternative recommendations.

The ontology-based recommender system with ML techniques is considered as a granted approach to give more accurate recommendations [11]. Ontology has been used in the proposed system as it correlates the concepts that can be used by recommendation rules to discover the associations between the crops.

Moreover, ontology gives reasons behind each of the recommendations. It handles any type of data formats and it can be directly integrated with semantic web [12].

The following steps show how to build the crop ontology based on the crop data and its related computed GDD values in order to apply the reasoner on ontology and get recommendations:

1. Collecting the data from the crops database.
2. Extracting the concepts then defining relationships between concepts.
3. Developing the crop ontology then defining the facts or rules.
4. Evaluating the crop ontology and the results of reasoner.

4 Experimental Results and Discussion

Experiments implemented in this research have been conducted using the Matlab software version R2016a and the Protégé 5.5 ontology development tool.

As previously stated, three ML algorithms have been tested in this research for daily temperature prediction of the following growing seasons (2017/18 to 2020/22) based on the climatic parameters; namely daily maximum and minimum temperatures, for years (1990 to 2016) in the cultivation area in Ashmoun, El Menoufia city, Egypt (near latitude 30.15° North and longitude 30.56° East). The meteorological data that has been used for validating the proposed cultivation-time recommender system was obtained from the NASA project for Prediction of Worldwide Energy Resource (POWER) as Climatology Resource for Agroclimatology [18].

The crop real data is collected from the Agricultural Engineering Research Institute in Egypt. The main attributes of this database are crop's name, crop's type, cultivation season, minimum/maximum temperatures suitable for specific crop, thermal temperature, and crop life cycle.

4.1 Appling ML Algorithms

The results of the tested ML algorithms were compared according to the measures of Mean Absolute Error (MAE), Mean Square Error (MSE), and Root Mean Square Error (RMSE).

The best results were achieved by SVR. In addition the SVR with RBF kernel function was selected as it is more commonly used in the prediction process and it is a superior model than the MLR model in terms of prediction performance, accuracy, and generalization ability [14].

Table 1 summarizes the performance MAE and MSE and RMSE measures of the SVR, MLR, and ELM algorithms for the daily average temperature.

Table 1. Results of SVR, MLR and ELM

Measure	SVR	MLR	ELM
Mean Absolute Error	0.1179	0.1217	0.1265
Mean Square Error	0.1857	0.1872	0.1904
Root Mean Square Error	0.3433	0.3489	0.3557

Based on the obtained experimental results SVR have showed more suitable performance considering the nature of the weather conditions data, which is considered as time series data [19].

Furthermore, SVR is commonly used in the forecasting process and it gives more accurate results compared to other regression algorithms [20, 21]. A set of parameters was used when applying SVR, during the training process in the SVR model. The RBF kernel function is used with an automatic kernel scale in order to standardize the data.

4.2 Computing Crop Growth Indicator

The value of GDD represents the accumulative amount of heat every day in the cultivation season that the crop needs for growth stage [22].

Equation (1) shows how the GDD was computed based on the daily maximum and minimum temperatures.

$$GDD = \sum_{n} [(T_{max} + T_{min})/2] - T_b \quad (1)$$

Where n represents the number of days in each growth stage throughout the growing season, and T_b is the crop's base temperature.

In this experiment, tomato crops were selected to be examined as a case study. It is regarded as a vegetable crop type. The range of needed temperature is between 18C as minimum and 35C as maximum temperature. The cultivation winter season of tomatoes starts from mid-September and its harvesting timing begins in end of December, and the cultivation summer season starts from mid-April and its harvesting is in the end of July. The calculated accumulative GDDs of tomato crops was entered as input into the crop ontology, these GDD values were based on the SVR predicted minimum and maximum temperature of the five years 2018 to 2020 for winter and summer seasons.

4.3 Evaluating the Ontology Results

The Protégé ontology management tool provides a way of graphically representing and visualizing the ontology structure using the onto-graph plug-in. Figure 2 shows the onto-graph provides representation of the ontology classes and relationships using Protégé.

In this research, there were forty tomato profiles that were added as a case study in the crop ontology. They were divided to two categories: The first twenty profiles were in the winter season as the other twenty were in the summer season.

Each profile had its own rules; these rules determine crop type, cultivation season, year of cultivation, predicted accumulative GDD, and number of days. The reasoner was used to check crop ontology consistency and infer the results, which were derived from these reasoning rules that were accordingly constructed during the implementation of crop ontology.

Table 2 summarizes the recommendations of the tomato profile case studies in the winter and the summer seasons from year 2018 to year 2022 that were generated as results from crop ontology.

Evaluating the ontology is a very important step; it can be built from scratch or by modifying existing ontology. There are certain evaluation criteria to make sure ontologies are implemented correctly [23].

The OntOlogy Pitfall Scanner (OOPS) [24] was used to evaluate and validate the crop ontology model against common development mistakes or pitfalls. The OOPS detects some of the most pitfalls appearing when developing ontologies according to the ontology evaluation criteria. Validation of Resource Description Framework (RDF) file of the concerned crop ontology was made after evaluation. The ontology RDF file has been checked against any syntax errors using the RDF validator [25].

The crop ontology model was successfully validated. The evaluation results for Completeness, Consistency, and Consciousness by OOPS web application have shown a successful evaluation of crop ontology in terms of the most common ontology evaluation criteria. The reasoner results have been evaluated according to the GDD

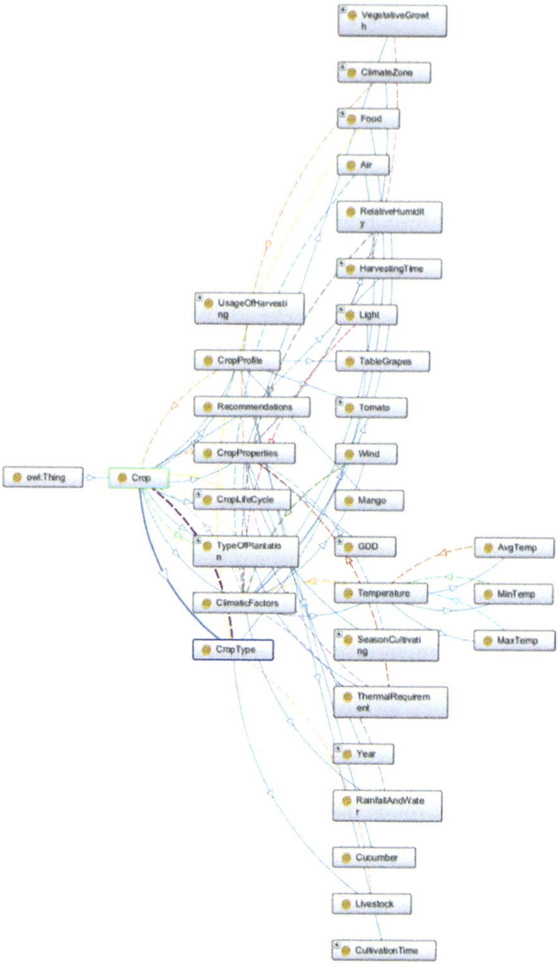

Fig. 2. The main classes and relationships of crop ontology

average range of tomato which is [1367 to 1704]. To demonstrate the precision and efficiency of the system, all recommendation results were submitted into two metrics which are: recall and rejection [8, 26].

Recall is the percentage of recommendation for which the system returns correctly as shown in Eq. (2).

$$\text{Recall} = \text{Number}_{\text{correct}} / \text{Number}_{\text{all}} \tag{2}$$

Such that $\text{Number}_{\text{correct}}$ is the number of recommendations that are in GDD average range given by the system and $\text{Number}_{\text{all}}$ represents the total number of all positive recommendations.

Table 2. Recommendations of winter cultivation and harvesting timings of tomato crop

Crop Profile	Winter Season Results			Summer Season Results		
	GDD	Cultivation Time	Harvesting Time	GDD	Cultivation Time	Harvesting Time
Tomato1-2018	1697.584365	1/9/2018	31/12/2018	1581.999270	8/4/2018	31/7/2018
Tomato2-2018	1596.308330	8/9/2018	30/12/2018	1470.628279	15/4/2018	31/7/2018
Tomato3-2018	1505.191941	11/9/2018	31/12/2018	1408.401453	22/4/2018	31/7/2018
Tomato4-2018	1606.837512	1/9/2018	23/12/2018	1381.987431	15/4/2018	23/7/2018
Tomato1-2019	1692.774294	1/9/2019	31/12/2019	1569.076619	8/4/2019	31/7/2019
Tomato2-2019	1599.522365	8/9/2019	30/12/2019	1482.490272	15/4/2019	31/7/2019
Tomato3-2019	1504.148463	11/9/2019	31/12/2019	1374.390207	22/4/2019	31/7/2019
Tomato4-2019	1583.172425	1/9/2019	23/12/2019	1374.538525	15/4/2019	23/7/2019
Tomato1-2020	1723.764087	1/9/2020	31/12/2020	1570.007549	8/4/2020	31/7/2020
Tomato2-2020	1568.553744	8/9/2020	30/12/2020	1477.725096	15/4/2020	31/7/2020
Tomato3-2020	1503.502635	11/9/2020	31/12/2020	1396.627408	22/4/2020	31/7/2020
Tomato4-2020	1579.412556	1/9/2020	23/12/2020	1364.789844	15/4/2020	23/7/2020
Tomato1-2021	1697.886782	1/9/2021	31/12/2021	1624.877030	8/4/2021	31/7/2021
Tomato2-2021	1595.084687	8/9/2021	30/12/2021	1482.135845	15/4/2021	31/7/2021
Tomato3-2021	1514.095130	11/9/2021	31/12/2021	1400.161948	22/4/2021	31/7/2021
Tomato4-2021	1599.058346	1/9/2021	23/12/2021	1380.829205	15/4/2021	23/7/2021
Tomato1-2022	1707.974932	1/9/2022	31/12/2022	1545.946887	8/4/2022	31/7/2022
Tomato2-2022	1574.528314	8/9/2022	30/12/2022	1484.359922	15/4/2022	31/7/2022
Tomato3-2022	1491.139212	11/9/2022	31/12/2022	1375.658174	22/4/2022	31/7/2022
Tomat04-2022	1598.607266	1/9/2022	23/12/2022	1356.680309	15/4/2022	23/7/2022

Equation (3) shows the rejection, which is the percentage of recommendations for which the system returns incorrectly.

$$\text{Rejection} = \text{Number}_{\text{rejected}} / \text{Number}_{\text{real}} \quad (3)$$

Where $\text{Number}_{\text{rejected}}$ is the number of recommendations that are out of GDD average range given by the system, and $\text{Number}_{\text{real}}$ is the number of recommendations that actually should be out of GDD range.

The proposed system has obtained a recall of 94.7% and rejection of 66.6. In addition, the higher recall and rejection are superior according to the scientific point of view. In addition, the proposed system has recorded 86.9% as accuracy [27] and F-score of 92.2% with recall 94.7% with precision 90%.

The proposed work has been compared to the related work according to the current systems based on ontology such as AIS platform [8], the recall and rejection of AIS are 81.6% and 46.7% compared to 53.2 and 78.7 for the existing system. Also, the OntoAgroHidro [9] returned the F-Measure 84% by combining precision 83% with recall 86%.

5 Conclusion and Future Work

The proposed model generates recommendations for the best cultivation and harvesting timings of tomato crops during the five years starting from year 2018 to year 2022 in correlation with the changes of climate. This recommendation system can help farmers to know the accurate information about the crops and make the best utilization of farm resources.

Each crop has its GDD range, which is the total amount of heat every day that crops need to complete the growth stage. Tomato crops were selected in this research since they are the most suitable crops for the soil at the chosen study location.

Performance measures of MAE, MSE and RMSE were used to evaluate several ML algorithms in this research. The SVR algorithm was selected in the prediction module because it gives the best results in comparison to other tested algorithms according to MAE, MSE and RMSE with values of 0.1179, 0.1857 and 0.3433, respectively. The proposed system has also recorded accuracy of 86.9% and achieved F-score of 92.2%.

For future work, this study can be improved by testing different ML algorithms to predict the weather changes, which have great effects on growth of crops. In addition, any type of crop at different environmental conditions can be added into the proposed crop ontology based on its appropriate agricultural soil location.

References

1. Sankaralingam, B.P., Sarangapani, U., Thangavelu, R.: An efficient agro-meteorological model for evaluating and forecasting weather conditions using SVM. First International Conference on Information and Communication Technology for Intelligent Systems 2, pp. 65–75 (2016)

2. Mohiuddinl, K., Alam, M.M.: A short review on agriculture based on machine learning and image processing. ACTA Sci. Agric. **3**(5), 55–59 (2019)
3. Liakos, K.G., Busato, P., Moshou, D., Pearson, S., Bochtis, D.: Machine learning in agriculture: a review. Sensors **18**(8), 2674 (2018)
4. Sridevi, M., Rao, R.R., Rao, M.V.: A survey on recommender system. Int. J. Comput. Sci. Inform. Secur. **14**(5), 265 (2016)
5. Middleton, S., Roure, D.D., Shadbolt, N.R.: Handbook on ontologies. international handbooks on information systems. In: Staab, S., Studer, R. (eds.) Ontology-Based Recommender Systems, pp. 779–796. Springer, New York (2009)
6. El-Bendary, N., Elhariri, E., Hazman, M., Saleh, S.M., Hassanien, A.E.: Cultivation-time recommender system based on climatic conditions for newly reclaimed lands in Egypt. In: 20th International Conference on Knowledge Based and Intelligent Information and Engineering Systems 96, pp. 110–119 (2016)
7. Sonare, P.M., Boob, P.J., Setiya, P.V., Ganjar, R.R.: Recommendation system for agriculture. Int. J. Res. Comput. Inform. Technol. (IJRCIT) **2**(3), 32–36 (2017)
8. Liao, J., Li, L., Liu, X.: An integrated, ontology-based agricultural information system. SAGE **31**(2), 150–163 (2015)
9. Bonacin, R., Nabuco, O.F., Junior, I.P.: Ontology models of the impacts of agriculture and climate changes on water resources: scenarios on interoperability and information recovery. Future Gener. Comput. Syst. **54**, 423–434 (2016)
10. Chougule, A., Jha, V.K., Mukhopadhyay, D.: Decision support for grape crop protection using ontology. Int. J. Reasoning-based Intell. Syst. **11**(1), 24–37 (2019)
11. Obeid, C., Lahoud, I., El Khoury, H., Champin, P.: Ontology-based recommender system in higher education. In: The 2018 Web Conference Companion, pp. 1031–1034 (2018)
12. Asim, M.N., Wasim, M., Khan, M.U.G., Mahmood, W., Abbasi1, H.M.: A survey of ontology learning techniques and applications. Database. Oxford, **2018**, 1–24 (2018)
13. Kanda, I., Veguillas, J.M.Q.: Data preprocessing and quantile regression for probabilistic load forecasting in the GEFCom2017 final match. Int. J. Forecast. **35**, 1460–1468 (2019)
14. Parveen, N., Zaidi, S., Danish, M.: Support vector regression model for predicting the sorption capacity of lead (II). Perspect. Sci. J. **8**, 629–631 (2018)
15. Mathworks.com. Understanding Support Vector Machine Regression - MATLAB & Simulink - MathWorks United Kingdom (2017). https://www.mathworks.com/help/stats/understanding-support-vector-machine-regression.html. Accessed 15 December 2017
16. Mouhaffel, A.G., Dominguez, C.M., Arcones, B., Redonda, F.M., Martín, R.D.: Using multiple regression analysis lineal to predict occupation market work in occupational hazard prevention services. Int. J. Appl. Eng. Res. **12**(3), 283–288 (2017)
17. Albadr, M.A., Tiun, S.: Extreme learning machine: a review. Int. J. Appl. Eng. Res. **12**(14), 4610–4623 (2017)
18. NASA. Climatology Resource for Agroclimatology (2016). http://power.larc.nasa.gov/cgi-bin/cgiwrap/solar/agro.cgi. Accessed 5 Jan 2016
19. Ju, X., Cheng, M., Xia, Y., Quo, F., Tian, Y.: Support vector regression and time series analysis for the forecasting of bayannur's total water requirement. In: 2nd International Conference on Information Technology and Quantitative Management, ITQM 31, pp. 523–531 (2014)
20. Hasan, N., Rasel, R., Nath, N.: A support vector regression model for forecasting rainfall. In: 2nd International Conference on Electrical Information and Communication Technologies (EICT), pp. 554–559 (2015)
21. Guajardo, J., Weber, R., Miranda, J.: A forecasting methodology using support vector regression and dynamic feature selection. J. Inform. Knowl. Manag. **5**(4), 329–335 (2006)

22. Miller, P., Lanier, W., Brandt, S.: Using growing degree days to predict plant stages. AG/Extension Commun. Coordinator Commun. Serv. Montana State Univ. Extensions Serv. **59717**(406), 994–2721 (2001)
23. Raad, J., Cruz, C.: A survey on ontology evaluation methods. In: Proceedings of the International Conference on Knowledge Engineering and Ontology Development. The 7th International Joint Conference on Knowledge Discovery, Knowledge Engineering and Knowledge Management (2015)
24. O.E. Group, OntOlogy Pitfall Scanner. http://oops.linkeddata.es/response.jsp. Accessed 1 May 2019
25. E. G. Prud'hommeaux, Check and Visualize your RDF documents, W3C, 2 (2006). https://www.w3.org/RDF/Validator/. Accessed 15 March 2019
26. Wang, F., Teng, G., Ren, L., Ma, J.: Research on mechanism of agricultural FAQ retrieval based on ontology. In: International Conference on Software Engineering, Artificial Intelligence, Networking, and Parallel/Distributed Computing, pp. 955–958 (2008)
27. Goutte, C., Gaussier, E.: A probabilistic interpretation of precision, recall and F-score, with implication for evaluation. In: Advances in Information Retrieval, pp. 345–359 (2005)

New Metaheuristic for Priority Guillotine Bin Packing Problem with Incompatible Categories and Sequential Deformation

Voronov Vladimir, Peresunko Pavel, Videnin Sergey, Matyukhin Nikita, and Masich Igor(✉)

Siberian Federal University, 660041 Krasnoyarsk, Russian Federation
{vsvoronov, pvperesunko}@sfu-kras.ru,
sergey.videnin@gmail.com, nmatryukhin@mail.ru,
i-masich@yandex.ru

Abstract. The paper considers the formulation of a new priority packing problem with incompatible categories and dynamically changing bin sizes, which is a variant of the well-known bin packing problem. This is a challenging optimization problem that is often encountered in the context of cutting ingots of non-ferrous and precious metals using a guillotine. We use the concepts of priority and incompatibility of categories combined with the need to split the base bin into components to reflect the specifics of the problem being solved. Incompatibility of categories follows from the grouping of rectangles by thickness, and priorities determine the desired order of production. A problem with independent categories can be represented as a multiple strips packing problem. In our case, however, the bin lengths are limited to the minimum and maximum values and are not known in advance. In this regard, it becomes necessary not only to find the optimal cutting scheme, but also the optimal bin lengths. We propose a sequential metaheuristics (SM) to solve this problem. It is based on splitting a set of rectangles into groups. They are formed according to the priorities and weights of the rectangles. Individual groups are sequentially placed in their own bins. Then bins with rectangles of the same category are combined. The basis is a simple deterministic single-pass priority heuristic (PH). It uses its own priorities to select the most appropriate rectangles for packing. The results presented are based on extensive computational experiments performed on the generated benchmark datasets. The tested instances differ in number of items, categories, priorities and bin capacities. Computational experiments show that the SM algorithm can effectively solve the problem.

Keywords: Bin packing problem · Strip packing problem · Metaheuristic · Incompatible categories · Priority

1 Introduction

Optimization problems are common in many areas such as medicine, metallurgy, logistics and etc. Obtaining optimal parameters allows achieving significant advantages which affects effectively the company in a highly competitive environment. The

R. Silhavy et al. (Eds.): CoMeSySo 2020, AISC 1295, pp. 825–836, 2020.
https://doi.org/10.1007/978-3-030-63319-6_76

problem of optimal use of resources is one of such problems. It allows not only to reduce the amount of waste, but also to increase the volume of production. Metallurgy is one of the areas where production of non-ferrous and precious metals is of particular interest. Inefficient use of such materials leads to irretrievable losses of material and an increase in the cost of waste processing. There is an example of such production. Parts of different shapes are made from one ingot by its deformation. The initial ingot of rectangular shape is rolled to a certain thickness, and then blanks are cut from it. This procedure is repeated as long as the parts can be placed on the sheet. Ingots are cut with a guillotine to save material. One of the problems associated with this method of production is planning. Parts with bigger thickness are located on the sheet first. Thus, the probability of placing thin parts is reduced. However, the priority of the parts may be higher than thick ones. In most cases the worker cannot take this into account during making parts.

This task is one of the varieties of the known bin packing problem (BPP). BPP is an optimization problem that minimizes the amount of waste or resources used. BPP has a great number of varieties. Thus, the most known are versions with guillotine restrictions, fixed or unlimited bin length, and the ability to rotate rectangles [1]. Using additional constraints results in new variants of the problem. This paper considers the problem of priority packing (P) into multiple bins of unknown limited length (M) with incompatible categories (IC) and guillotine restrictions (G). We encode this problem as BPP|P|M|IC|G.

Using priorities in BPP|P|M|IC|G allows you to place priority items first. It regulates the order of production. The incompatibility of categories of parts follows from their thickness which is geometric dimensions. Parts with different thicknesses cannot be placed in the same bin. It is worth noting that the optimal number of bins which it is necessary to split the original bin into is not known in advance, as their length is also unknown. As a result, it is necessary not only to provide a packing scheme but also calculate the optimal partitioning of the original bin into several parts as a solution to the problem.

The paper proposes a new metaheuristic for solving BPP|P|M|IC|G. It is based on the usage of the priority heuristic proposed in [2], with several modifications. The proposed algorithm is tested on sets of rectangles designed specially for this problem.

This article has following structure. Section 2 reviews the existing literature on BPP and its variations. Section 3 describes the priority packing problem with a variable length bin, incompatible categories, and guillotine restrictions. The proposed algorithm and a brief description of the heuristics that the algorithm is based on are given in Sect. 4. In Sect. 5, we estimate the efficiency of the new algorithm on a set of test examples. Finally, Sect. 6 summarizes the results.

2 Literature Review

The bin packing problem (BPP) is well-studied and there are a lot of papers on this topic (see for example [3–5]). There are also a lot of variations of this problem. Consider several variations of the closest problems in the context of our statement. One variant of the packing problem is a priority-based problem. It is often called a dual or priority packing problem (2DBPP|D). The papers pay little attention to the priority of manufacturing parts in industry and is more inclined to the physical and price characteristics of elements. Therefore, it is not surprising that this variation has been developed in the field of medicine for optimal usage of medical equipment. The priority packing problem is considered in [6, 7], where the task of resource allocation is set, taking into account the limited time of the operation. It was further developed in [8], where a mathematical model of the planning problem is described. There is heuristic based on the First Fit Decreasing algorithm was used to solve this problem. In [9], the method of particle swarm optimization (PSO) is used to solve this problem. The specificity of the problem considered by the authors is that only certain types of medical examinations can be done on certain equipment. At the same time, the capabilities of different equipment may coincide.

Other variant of 2DBPP is the multiple bin packing problem (2DMBPP). This problem has two formulations: with identical and variable bin sizes. This problem is applied to optimize the use of computing resources, logistics, and etc. Thus, an evolutionary approach combined with local search is proposed in [10]. In [11], a variety of heuristic methods are proposed, and their application is considered for solving the logistics problem as a generalized bin packing problem with bin-dependent profits (GBPPI). Taking into account the demand and costs of transportation and storage, a similar problem is solved using a two-stage heuristic in [12]. In [13], the authors consider the use of 2DMBPP for resource-allocated batch planning, where heuristics were used to minimize resource consumption. Research of the First Fit algorithm for the problem of packing with a restriction on the number of elements is carried out in [14]. The authors basically consider packing in bins of fixed or different sizes. In their research no restrictions are imposed on dynamically changing sizes or on an indefinite number of bins.

A bin packing problem with conflicts (BPC) considers the incompatibility of elements. For such tasks, a graph is often used to determine compatibility between elements. The vertices of the graph indicate the elements or groups to be packed, and the edges indicate compatibility between them. For example, in [15], a solution based on tree decomposition is considered for this task. The algorithm breaks the task down into subtasks that are solved independently. Then obtained solutions are combined. In [16], the authors provide updated upper and lower bounds for the BPC. The authors also provide an updated deterministic algorithm based on the coverage of the set solved by the branching and price algorithm.

Recently a new bin packing problem with compatible categories (PBCC) had appeared [17]. It is a generalization of the packing problem with conflicts (BPC), where pairwise incompatibility is considered. BPCC assumes the compatibility of groups of packed items. The formulation of the BPCC problem for the first time was given in [17]. The authors provide a modification of the algorithm for finding a variable neighborhood (VNS) and offer a set of test cases for algorithm.

Thus, the packing problems presented here are similar to the problem we are proposing. However, none of the proposed problems considers the case with an unknown number of bins and their dynamically changing sizes. In addition, as far as we know this is the first consideration of the problem with a set of constraints such as guillotine cuts, the presence of priorities and incompatible groups of rectangles.

3 Statement of the Problem

In standard BPP, bins $R = \{r_i, i = 1,\ldots, n\}$ with shapes $r_i = (w_i, l_i)$, where w_i – width, l_i – length of i-th bin, are placed on fixed size sheet with width W_0 and length L_0 without overlapping. The target function in this case is waste area:

$$f(R, I') = W_0 L_0 - \sum_{i \in I'} w_i l_i = \min_{I}. \tag{1}$$

Here $I`$ – set of indexes of placed bins, $I` \subseteq I = \{i = 1,\ldots, n\}$.

To define BPP|P|M|IC|G consider bin as $r_i = (w_i, l_i, h_i, p_i)$, where $h_i \in H = \{h_j, j \in I\}$ – height i-th bin. Bin's height specifies sheet's height, from which bin can be made. For example, on a sheet with 2 mm height can be placed bins with the same height. The sheet is gradually deformed (rolled) till specific height, bins are cut out and remaining of the sheets deformed again. This makes it impossible place bins of different thicknesses in the same time. Height allow to group bins in into $m = |supp\ H|$ groups. In addition

$$n = \sum_{j=1}^{m} |R_j| = \sum_{j=1}^{m} n_j,\ R_j = \{r_i | h_i = \bar{h}_j \forall i \in I, j = \overline{1,m}\} \subseteq R,$$

where $\bar{h}_j$ – height j-th group of bins, $j \in J = \{1,\ldots, m\}$, and multiset of heights of one group consist of n_j same elements $H_j = \{n_j \cdot \bar{h}_j | h_i = \bar{h}_j\ \forall i \in I\}$. These $j \in J$ groups are incompatible.

The priority of i-th bin $p_i \in P = \{p_i, i \in I | p_i \in Z_{>0}\}$ characterizes the degree of importance of than bin for production. The smaller p_i, the more preferable the bin to be placed on the current sheet. In the case of the same priority, the order of placement of the bins is not important. We will take $p_i = 1$ as the maximum priority, i.e., the higher the p_i, the lower the priority. Bins in each group H_j ($j = 1,\ldots, m$) with indexes I_j can be grouped into $q_j = |supp\ P_j|$ subgroups, where $P_j = \{p_i, i \in I_j \subseteq I\} \subseteq P$ is multiset of priorities of j-th groups. In so doing,

$$|R_j| = \sum_{k=1}^{q_j} |R_{jk}| = \sum_{k=1}^{q_j} g_{jk} = n_j,\ R_{jk} \subseteq R_j \subseteq R,$$

where R_{jk} – k-th subgroup of j-й groups of bins with indexes I_{jk}, $k = 1,\ldots, q_j$. Priorities of subgroups of bins R_{jk} are same and equal to p_{jk}, i.e. $P_{jk} = \{g_{jk} \cdot p_i | p_i = \bar{p}_{jk}\ \forall i\ I_{jk} \subseteq I\}$, $P_{jk} \in P_j$, and I_{jk} indexes of bins with the same priority in group R_j. It should be noted that with $p_i = const$, $\forall i \in I$ BPP|P|M|IC|G degenerate into BPP|M|IC|G, i.e. without priorities.

We believe that during sheet deformation (rolling) its shape change according to the volume conversation rule, i.e. $V_0 = V_1$, where $V_0 = W_0 L_0 h_0$ is the volume before deformation, $V_1 = W_1 L_1 h_1$ is the volume after deformation. Neglecting changes in the sizes of one of the sides of the sheet orthogonal to which the deformation occurs (for example, W_0), we obtain

$$L_1 = \gamma \frac{L_0 h_0}{h_1}, \gamma > 0, \tag{2}$$

where γ is a coefficient to account for defects that occur at the ends of the sheet during deformation. Further, for simplicity, $\gamma = 1$ is assumed.

Thus, BPP|P|M|IC|G consists in placing the bins considering the priority and splitting the original bin into components taking into account the dynamically changing length. As mentioned in [9], problem (1), considering priority, can be represented as a multicriteria optimization problem

$$f(R, I'_1, \cdots, I'_m) = (f_1(R, I'_1, \cdots, I'_m), f_2(R, I'_1, \cdots, I'_m)) = \min_{I'_j, j=\overline{1,m}}, \tag{3}$$

$$f_1(R, I'_1, \cdots, I'_m) = \sum_{j=1}^{m} \left(W_j L_j - \sum_{i \in I'_j} w_i l_i \right), \; f_2(R, I'_1, \cdots, I'_m) = \sum_{j=1}^{m} \left(\sum_{i \in I'_j} p_i \right),$$

where $I`_j$ – set of placed bin's indexes of group $R_j$. If $I`_j = \varnothing$, then bins of group R_j were not placed. Shape of a bin (W_j, L_j), allocated for packing the bins of the group R_j, should be determined based on the deformation rule (2) and the configuration of the corresponding group. Moreover, if $I`_j = \varnothing$, then $W_j L_j = 0$.

Imposing additional restrictions on problem (3), one can obtain various variations of the packing problem can be obtained:

1. The sides of the bins must be parallel to the sides of the sheet;
2. The bins are allowed to rotate 90 degrees (R);
3. All cuts must only be done by guillotine (G);
4. The sheet bust be deformed in a fixed direction (FD) (e.g., along L_0);

In this paper, we consider the problem of guillotine priority packing of bins with incompatible categories and a fixed direction of deformation BPP|P|M|IC|G|FD.

4 Priority Packing Algorithm for Rectangles with Incompatible Groups and Dynamically Changing Bin Sizes

4.1 Priority Heuristic

Priority heuristic (PH) is a single-pass deterministic heuristic proposed in [2]. It is a greedy recursive packing procedure to ensure guillotine restrictions. The selection of the rectangle from the set is carried out using the "internal" priorities that characterize one of the five possible placement scenarios. The highest priority is given to rectangles that occupy the entire available area of the bin, then there are rectangles whose length or width coincide with the length or width of the bin. The least priority is given to rectangles whose dimensions are smaller than the dimensions of the bin. The fifth scenario is when the rectangle cannot be placed. Thus, a rectangle with a high priority is selected first. The algorithm has one parameter to configure, namely the option to sort the rectangles: by length or by width. Sorting allows to select the largest rectangles in the first place. In their work, the authors conduct the results of the algorithm using various examples from the literature, including the ones based on the ZDF set proposed in [18, 19]. As a result, it is noted that PH is well suited for large-scale problems.

We have made some changes to PH to apply this algorithm to BPP|P|M|IC|G. So, the sorting of the rectangles is performed in each subgroup R_{jk} separately. A shift parameter y_c along the Y axis has been added to the packing algorithm, which corrects the position of the cutting pattern (the origin point) along the Y axis. This allows to sequentially pack groups of rectangles, preserving their position.

To take into account the priority of the rectangles in the algorithm for packing parts with RG, iterations are performed only for elements of the highest priority p. Moreover, in the recursive packing procedure Recursive Packing(x, y, w, h), the stage of assessing the internal priority is carried out in descending order of production priority (for subgroups R_{jk}) of the entire group of elements R_j with the same thickness. This allows to fill the space with rectangles of lower priority when it is impossible to place the rectangles of high priority.

4.2 New Sequential Metaheuristic Algorithm

Let us take such an arrangement of the initial bin that its horizontal side corresponds to the width W and the vertical side corresponds to the length L (Fig. 1). Let the deformation (rolling) occur along L, and W remains constant.

One of the tasks of the algorithm is to distribute the lengths of bin components for rectangles of the groups $H_j, j \in J$. Obviously the length of each bin component belongs to the range $0 \leq L_j \leq L_0 h_0 / min\ H$, where $h_0 \geq min\ H$ is the thickness of the original bin. It is not necessary for each group to have a bin component. The case $L_j = 0$ means that no rectangle of the j-th group was packed. Thus, one of the results of the algorithm should be the mapping L_{bins}: $supp\ H \to \mathbb{R}^+$, which associates the thickness of the packed group with the length of the bin component. The sequence of all bin component lengths must satisfy the inequality.

$$0 \leq \sum_{h \in \mathrm{supp} H} L_{bins}(h) \leq \frac{L_0 h_0}{\min H}$$

It should be noted that the more the bin size is allocated for a group with one thickness, the less space is left for packing the remaining groups. This fact must be considered during distributing lengths. In other words, the packing of the rectangles of the group R_j leaves less chance to all the rectangles of the first priority of the group R_i $(i \neq j)$ to be packed.

The proposed sequential metaheuristics (SM) considers the subgroups of the rectangles R_{jk} in the order they appear in the sorted sequence $G = \{(\overline{h}_j, \overline{p}_{jk}), k = 1,\ldots, q_j, j = 1,\ldots, m\}$. Sorting is carried out lexicographically by non-decreasing priority and

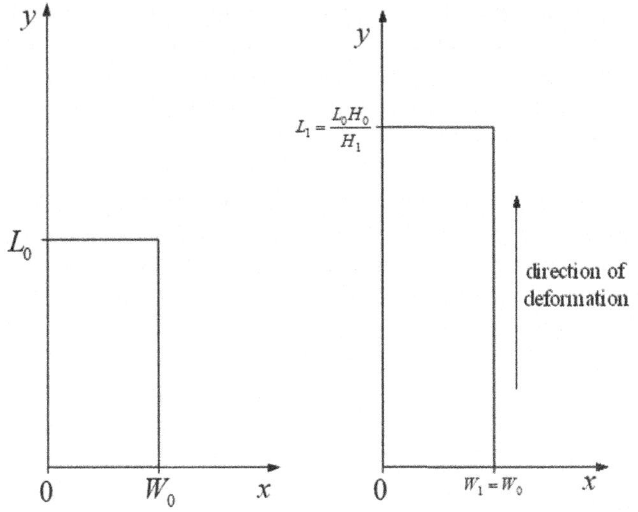

Fig. 1. The position of the original bin and the direction of its deformation.

non-increasing thickness. This allows to pack larger rectangles first. Using $\overline{p}_{jk}$ allows to take into account the priority of packing items. Due to this arrangement, the subgroups R_{jk} with high priority are packed first, then the remaining space is distributed among the subgroups in increasing order of $\overline{p}_{jk}$.

The iterative procedure of Algorithm 1 is constructed as follows. The bin is deformed to the thickness of the group of items R_j according to the rule (2), then the packing is carried out using the PH algorithm. The packing result is the required bin length $L^`$ and a multiset of rectangles $R^`_j$. In the end, the final solution is updated, the length and thickness of the original bin are adjusted. It is worth noting that the elements h_j will not necessarily be ordered in non-increasing order in the sorted sequence G. It is because the values of $\overline{p}_{jk}$ have priority over h_j. Therefore, the bin length can both increase ($h_0 > h_1$ in (2)) and decrease ($h_1 > h_0$ in (2)) in the process of finding a solution. This leads to disordered packing regarding h_j. However, during the manufacturing process, the sheet is rolled in decreasing order of thickness. Therefore, it is necessary to sort the groups of placed rectangles in the correct order at the end of the procedure.

Two cases may occur during packing. The first is that all the rectangles of the subgroup R_{jk} can be placed in the bin of the current length, and the remaining space will be reserved for packing other subgroups. Then it is necessary to use the algorithm for the strip packing problem which returns both the layout and the required length. The procedure PackStrip(W, R_j, y_c) is responsible for this action, where y_c is responsible for the coordinate shift of the layout along the Y axis.

The second case occurs when the bin is not long enough to accommodate all the rectangles of the subgroup R_{jk}. Then it is necessary to explicitly set its length and use the version of the algorithm for the bin packing problem, which is represented by the procedure PackBin(W, $\overline{L}$, R_j, y_c), where $\overline{L}$ is the remaining bin length.

We use the PH algorithm for SPP and BPP as these two procedures. It should be noted that other heuristics can be used to solve these problems. Restrictions are imposed only on the variant of the problem being solved (SPP and BPP). It is not limited to the use of different heuristics in order to obtain an optimal cutting plan. In addition, the properties of the algorithms used are saved, for example, if the heuristic is designed to solve the problem with the fixed orientations of the rectangles (SPP|O, BPP|O) or with the possibility of their rotation (SPP|R, BPP|R), then SM will also solve this version of the problem. This allows to solve a wide range of problems and provide the variability of properties due to the choice of basic heuristics.

Despite the order of priorities in the sequence G, the groups R_j (but not the subgroups R_{jk}) are transferred to the packing procedures. This is to ensure that the empty space is filled with rectangles of lower priority in case when no rectangle of the corresponding priority p can be packed. In this regard, it is necessary to check the multiset R_{jk} for emptiness, since all its elements can be packed earlier.

```
procedure SM(L, W, H, R)
  Lbins ← {0, j ← 1..m}
  R′ ← {}
  G ← {(hj, pjk), k ← 1..qj, j ← 1..m}
  hbins ← H
  sort G descending hj and increasing pjk
  for all (h, p) ∈ G do
    if Rjk = ∅ then
      continue
    end if
    yc ← Lbins[h]
    L̄ ← L·hbins / h
    (L′, Rj′) ← PackStrip(W, Rj, yc)
    if L′ > L̄ then
    (L′, Rj′) ← PackBin(W, L̄, Rj, yc)
   end if
   Lbins[h] ← Lbins[h] + L′
   R′ ← R′ + Rj′
   L ← L̄ - L′
   hbins ← h
   if L == 0 then
     break
   end if
  end for
  return R′, Lbins
end procedure
```

SM can "split" one group R_j into several parts and pack them individually. Despite this, a sequence of cutting patterns for each group is ensured.

A visual demonstration of the algorithm is shown in Fig. 2. It can be seen that the sizes of the unfilled part of the bin are transformed using (2) and act as a bin for packing thinner rectangles.

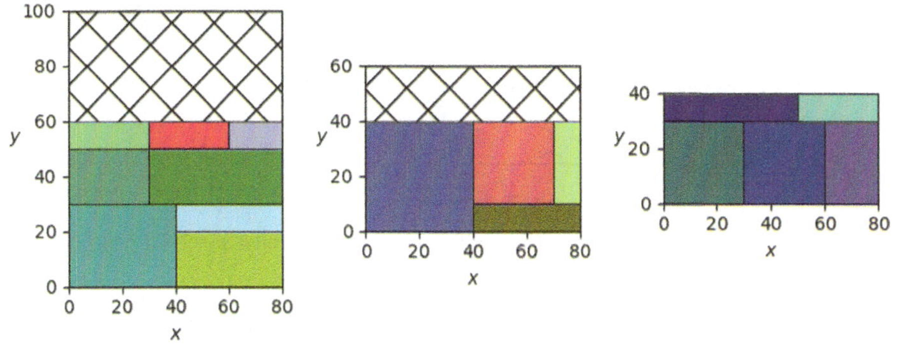

Fig. 2. Packing rectangles with a thickness of 3 mm (left), 2 mm (center), and 1 mm (right) in the original bin with sizes L_0 = 100 mm, W_0 = 80 mm, H_0 = 3 mm.

5 Results

The sequential meta-heuristic (SM) algorithm using priority heuristics (PH) was tested on 10 examples compiled specially for the BPP|P|M|IC|G problem. The examples contain from 3 to 14 incompatible categories and from 3 to 5 priorities. The examples are designed to include problems with the number of elements from a wide range. So, sets 1, 8, and 10 contain relatively few elements ($n < 1{,}000$), while large-scale sets 6 and 7 include more than 10,000 elements.

To allow comparability of the results of the algorithm, the assessment of the packing quality of each group is carried out similarly [2] using the indicator $Gap_j = 100(L_j - L_j^*)/ L_j^*$, where L_j^* is the known optimal bin length or lower bound (LB), and L_j is the bin length obtained using SM, $j = 1, \ldots, m$. The length of the bin is not limited, i.e. SPP is solved. The resulting indicator is defined as the weighted average

$$Gap = \sum_{j=1}^{m} Gap_j S_j \Big/ \sum_{j=1}^{m} S_j, \quad (4)$$

where S_j is the area of all rectangles of the j-th group. Figure 2 reports results of SM for SPP|P|M|IC|G. Figure 3 shows that the influence of the sorting method of the PH algorithm is preserved.

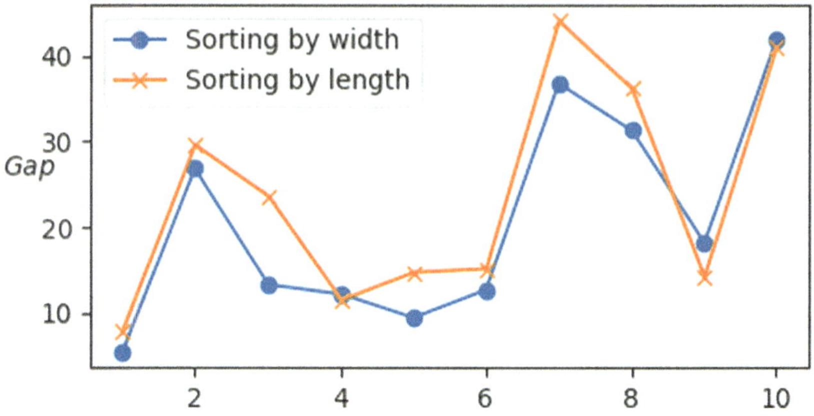

Fig. 3. Results of SM with PH using different sorting methods for SPP|P|M|IC|G.

In the case of BPP|P|M|IC|G, the bin size is limited, which undoubtedly leads to poor performance due to the influence of the priority of the rectangles. The *Gap* index (4) for BPP|P|M|IC|G is presented in Fig. 4 (left). It may be noted that in some cases *Gap* (4) becomes negative. This happens when the bin length allocated by the SM algorithm for some group is less than optimal, so some rectangles cannot be packed. A significant deviation is observed in example 10. It contains several low-priority large rectangles. Despite this fact, the filling density for each bin is high. Figure 4 (right) reports the average relative area for each example

$$\overline{S} = \sum_{j=1}^{m} \overline{S_j} \Big/ m, \ \overline{S_j} = 100 \frac{S'_j}{L_j W}$$

Here $\overline{S}_j$ is the ratio of the area of the placed rectangles to the area of the bin, where $S`_j$ is the area of the placed rectangles, and L_j is the length of the bin. From Fig. 4, it can be seen that the method of sorting by length shows a worse result than sorting by width.

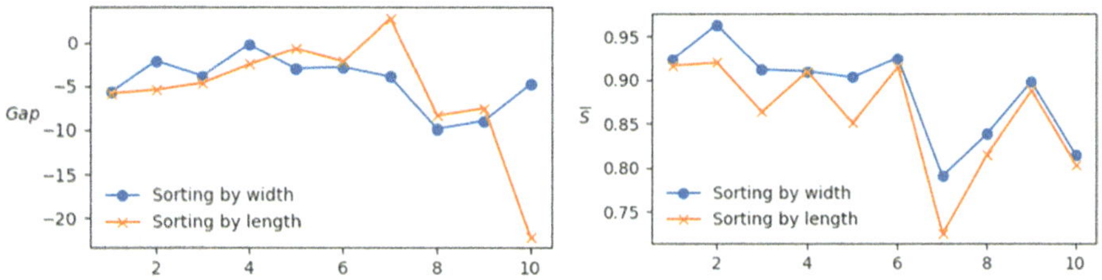

Fig. 4. The results of SM with PH using different sorting methods for BPP|P|M|IC|G.

6 Conclusion

A new problem of priority packing of rectangles with incompatible categories and dynamically changing bin sizes (BPP|P|M|IC|G) is formulated in the work. This problem is characterized by the fact that the sizes of the original bin change during the packing process, and it is necessary to take into account the priority of items, their incompatibility with each other and guillotine restrictions. To solve the presented problem, a new sequential metaheuristic algorithm (SM) is proposed, which in the process of work divides the groups of rectangles according to their priorities and categories. The resulting groups are placed sequentially, but at the same time there remains the possibility to fill the free space with rectangles of a lower level.

As a basic algorithm, a single-pass deterministic PH heuristic algorithm is used, which shows good results among analogues. But SM is not limited to the considered heuristic. As the basic algorithm, any algorithms can be used that allow to solve BPP and SPP. The properties of these algorithms are saved. So, in the examples above it is shown that when using PH, its dependence on the sorting variant is inherited by SM. It is worth noting the flexibility of SM regarding choosing algorithms. They can be combined to achieve the required properties.

The performance of SM was demonstrated on a set of examples. Sets differ both in the number of elements and their distribution by category and priority. When solving BPP|P|M|IC|G and SPP|P|M|IC|G, the algorithm shows good results, and the properties of the solution depend on the properties of the used underlying algorithm.

As we know this is the first study of BPP|P|M|IC|G so there may be many possible options for further development. Thus, the question remains open about the possibility of changing the direction of deformation of the bin. The possibilities of using stochastic and deterministic algorithms for local and global optimization and their combinations are also wide. One option is to distribute bin component lengths using global optimization.

Acknowledgments. The research was carried out within the state assignment of Ministry of Science and Higher Education of the Russian Federation (theme No. FSRZ-2020-0011).

References

1. Wäscher, G., Haußner, H., Schumann, H.: An improved typology of cutting and packing problems. Eur. J. Oper. Res. **183**, 1109–1130 (2007). https://doi.org/10.1016/j.ejor.2005.12.047
2. Zhang, D., Shi, L., Leung, S.C.H., Wu, T.: A priority heuristic for the guillotine rectangular packing problem. Inform. Process. Lett. **116**, 15–21 (2016). https://doi.org/10.1016/j.ipl.2015.08.008
3. Coffman, E.G., Csirik, J., Galambos, G., Martello, S., Vigo, D.: Bin packing approximation algorithms: survey and classification. In: Pardalos, P.M., Du, D.-Z., Graham, R.L. (eds.) Handbook of Combinatorial Optimization, pp. 455–531. Springer, New York (2013)
4. Delorme, M., Iori, M., Martello, S.: Bin packing and cutting stock problems: mathematical models and exact algorithms. Eur. J. Oper. Res. **255**, 1–20 (2016). https://doi.org/10.1016/j.ejor.2016.04.030

5. Christensen, H.I., Khan, A., Pokutta, S., Tetali, P.: Approximation and online algorithms for multidimensional bin packing: a survey. Comput. Sci. Rev. **24**, 63–79 (2017). https://doi.org/10.1016/j.cosrev.2016.12.001
6. Guinet, A., Chaabane, S.: Operating theatre planning. Int. J. Prod. Econ. **85**, 69–81 (2003). https://doi.org/10.1016/S0925-5273(03)00087-2
7. Jebali, A., Hadj Alouane, A.B., Ladet, P.: Operating rooms scheduling. Int. J. Prod. Econ. **99**, 52–62 (2006). https://doi.org/10.1016/j.ijpe.2004.12.006
8. Vijayakumar, B., Parikh, P.J., Scott, R., Barnes, A., Gallimore, J.: A dual bin-packing approach to scheduling surgical cases at a publicly-funded hospital. Eur. J. Oper. Res. **224**, 583–591 (2013). https://doi.org/10.1016/j.ejor.2012.09.010
9. Laurent, A., Klement, N.: Bin packing problem with priorities and incompatibilities using PSO: application in a health care community. IFAC-PapersOnLine **52**, 2596–2601 (2019). https://doi.org/10.1016/j.ifacol.2019.11.598
10. Soak, S.M., Lee, S.W., Yeo, G.T., Jeon, M.G.: Effective evolutionary algorithm for the multiple container packing problem. Prog. Nat. Sci. **18**, 337–344 (2008). https://doi.org/10.1016/j.pnsc.2007.11.007
11. Baldi, M.M., Manerba, D., Perboli, G., Tadei, R.: A generalized bin packing problem for parcel delivery in last-mile logistics. Eur. J. Oper. Res. **274**, 990–999 (2019). https://doi.org/10.1016/j.ejor.2018.10.056
12. Crainic, T.G., Gobbato, L., Perboli, G., Rei, W.: Logistics capacity planning: a stochastic bin packing formulation and a progressive hedging meta-heuristic. Eur. J. Oper. Res. **253**, 404–417 (2016). https://doi.org/10.1016/j.ejor.2016.02.040
13. Liang, X., Zhou, S., Chen, H., Xu, R.: Pseudo transformation mechanism between resource allocation and bin-packing in batching environments. Future Gener. Comput. Syst. **95**, 79–88 (2019). https://doi.org/10.1016/j.future.2019.01.006
14. Dósa, G., Epstein, L.: The tight asymptotic approximation ratio of first fit for bin packing with cardinality constraints. J. Comput. Syst. Sci. **96**, 33–49 (2018). https://doi.org/10.1016/j.jcss.2018.03.004
15. Khanafer, A., Clautiaux, F., Talbi, E.G.: Tree-decomposition based heuristics for the two-dimensional bin packing problem with conflicts. Comput. Oper. Res. **39**, 54–63 (2012). https://doi.org/10.1016/j.cor.2010.07.009
16. Muritiba, A.E.F., Iori, M., Malaguti, E., Toth, P.: Algorithms for the bin packing problem with conflicts. INFORMS J. Comput. **22**, 401–415 (2010). https://doi.org/10.1287/ijoc.1090.0355
17. Santos, L.F.O.M., Iwayama, R.S., Cavalcanti, L.B., Turi, L.M., Morais, F.E. de S., Mormilho, G., Cunha, C.B.: A variable neighborhood search algorithm for the bin packing problem with compatible categories. Expert Syst. Appl. **124**, 209–225 (2019). https://doi.org/10.1016/j.eswa.2019.01.052
18. Leung, S.C.H., Zhang, D.: A fast layer-based heuristic for non-guillotine strip packing. Expert Syst. Appl. **38**, 13032–13042 (2019). https://doi.org/10.1016/j.eswa.2011.04.105
19. Zhang, D., Wei, L., Leung, S.C.H., Chen, Q.: A binary search heuristic algorithm based on randomized local search for the rectangular strip-packing problem. INFORMS J. Comput. **25**, 332–345 (2013). https://doi.org/10.1287/ijoc.1120.0505

Modelling the Water Jet Trajectory of a Robotic Fire Monitor in the SimInTech Dynamic Modelling Environment

Irina Pozharkova[1,2(✉)]

[1] FSBEI HE Siberian Fire and Rescue Academy of the State Fire-Fighting Service of the Ministry of Emergency Situations (Russia), 1, Severnaya Street, Zheleznogorsk 662972, Russian Federation
pozharkova@mail.ru

[2] Siberian Federal University, 79, Svobodny Avenue, Krasnoyarsk 660041, Russian Federation

Abstract. The article describes the algorithm for computing the trajectory of the axisymmetric water jet produced by robotic fire monitors with regard to the condition of the atmospheric friction layer on its lateral surface. This algorithm is designed to be used for finding the optimal pattern of fire robot arrangement at the protected venue. The article presents the model generated in SimInTech, technical system dynamic modelling environment, and the water jet trajectory modelled on its basis. The modelled curves were compared to the experimental data; the model adequacy was assessed with Fisher's test.

Keywords: Robotic fire monitor · Robotic fire fighting units · Water jet trajectory · SimInTech

1 Introduction

At present, robotic fire fighting units [1] are among the most efficient fire safety solutions applied at the sites where the use of any other automatic fire fighting systems may be complicated. In particular, fire fighting robots are used in the large-span buildings and big floorspace rooms, where the establishment of a branch-pipeline system for sprinkler and drencher units would not be appropriate due to the additional load it creates on the engineering structures, and the use of the manual fire monitors is complicated by the fire load, where the fire causes fast smoke spread [2]. Machine houses, hangars, warehouses, expo centres and other facilities are examples of such rooms. The main advantages of the robotic fire fighting units in such kinds of buildings can be listed as follows:

- High fire extinguishing efficiency, cooling of the engineering structures to prevent the collapse;
- Large coverage area;
- No need for branch-pipeline systems and additional load on the engineering structures;
- No need for people to be present in the hazardous fire area to control the robotic fire monitors;

R. Silhavy et al. (Eds.): CoMeSySo 2020, AISC 1295, pp. 837–844, 2020.
https://doi.org/10.1007/978-3-030-63319-6_77

- High precision of jet targeting regardless of the smoke spread level due to the variety of specialized sensors;
- Little delay between the commencement of the extinguishing operation or the cooling of the engineering structures and the beginning of the fire.

One of the essential tasks of the fire fighting robots-based fire safety system design is the optimization of the robot arrangement with due regard to the layout and arrangement specificity of the protected venue. In particular, it is necessary to select the right set up for the minimum number of robotic monitors to cover the required area. The main problem of the task is to determine the robotic monitor jet range. The trajectory of the jet is influenced by several factors, including:

- Location of the fire fighting robot in the space (3D coordinates);
- Fire fighting robot monitor inclination angle with the horizontal axis;
- Fire fighting robot turning angle with the vertical axis;
- Working pressure of the fire fighting robot monitor;
- Water consumption of the fire fighting robot;
- Other conditions of the protected venue that may make an impact on the factor of friction between the jet and the atmospheric friction layer on its surface.

2 Methods

The study [3] presents the method for computation of an axisymmetric jet with regard to the condition of the atmospheric friction layer on its lateral surface. Based on this method, the following computation algorithm for the water jet trajectory of a fire fighting robot monitor was developed:

1. Find the friction coefficient of the atmospheric friction layer:

For the laminar stream:

$$K = \frac{8}{3 \cdot x(2 \cdot \pi \cdot r + x \cdot tg\frac{\phi}{2})} \cdot \sqrt{\frac{v \cdot x \cdot (3 \cdot \pi^2 \cdot r^2 + 3 \cdot \pi \cdot r \cdot x \cdot tg\frac{\phi}{2} + x^2 \cdot tg^2\frac{\phi}{2}}{u_0}} \quad (1)$$

where x is the coordinate of the reference point on the horizontal axis of the jet, m;
r is the minimum jet radius, m;
φ is the sector angle of the net of the jet cone lateral surface, rad;
v is the kinetic viscosity of the environment, m^2/sec;
u_0 is the approach flow speed, m/sec.
For the turbulent stream:

$$K = \frac{2}{x \cdot (2 \cdot \pi \cdot r + x \cdot tg\frac{\phi}{2})} \cdot \sqrt{\frac{v \cdot x \cdot (3 \cdot \pi^2 \cdot r^2 + 3 \cdot \pi \cdot r \cdot x \cdot tg\frac{\phi}{2} + x^2 \cdot tg^2\frac{\phi}{2}}{u_0}} \quad (2)$$

2. Adjust the friction coefficient as follows:

$$k = K \cdot \frac{e^{t^{b_1}}}{sin(\theta)^{b_2}} \tag{3}$$

where K is the coefficient calculated with the formulae (1)–(2);
θ is the fire fighting robot monitor inclination angle with the horizontal axis, rad;
b_1, b_2 are the experimentally determined coefficients.

3. The jet trajectory is computed with the formulae:

$$x = \frac{v_0 \cdot cos(\theta) \cdot t}{1 + k \cdot \frac{v_0^2}{2 \cdot d \cdot g}} \tag{4}$$

$$y = \frac{v_0 \cdot sin(\theta) \cdot t - \frac{g \cdot t^2}{2}}{1 + k \cdot \frac{v_0^2}{2 \cdot d \cdot g}} \tag{5}$$

where x is the horizontal coordinate, m;
y is the vertical coordinate, m;
v_0 is the initial jet speed, m/sec;
t is time, sec;
d is jet diameter, m;
g is the free-fall acceleration, m/sec^2.

To implement the algorithm for computation of the axisymmetric axis trajectory with regard to the condition of the atmospheric friction layer on its lateral surface with the formulae (1)–(5), the SimInTech technical system dynamic modelling environment was used [4]. This software environment functions allow checking the complex technical object control systems, modelling their behaviour and later integrating the developed models in the control systems through the appropriate interfaces. Figure 1–2 present such a model in the SimInTech software environment.

For the convenience of computation in the 3D space, it is necessary to switch to the X, Y, Z coordinates:

$$\begin{aligned} X &= x \cdot cos(\alpha) \\ Y &= x \cdot sin(\alpha) \\ Z &= y \end{aligned} \tag{6}$$

where x, y are the coordinates computed with the formulae (4)–(5);
α is the fire fighting robot monitor inclination angle with the horizontal axis, rad.

The corresponding model in the SimInTech software environment is presented in Fig. 3.

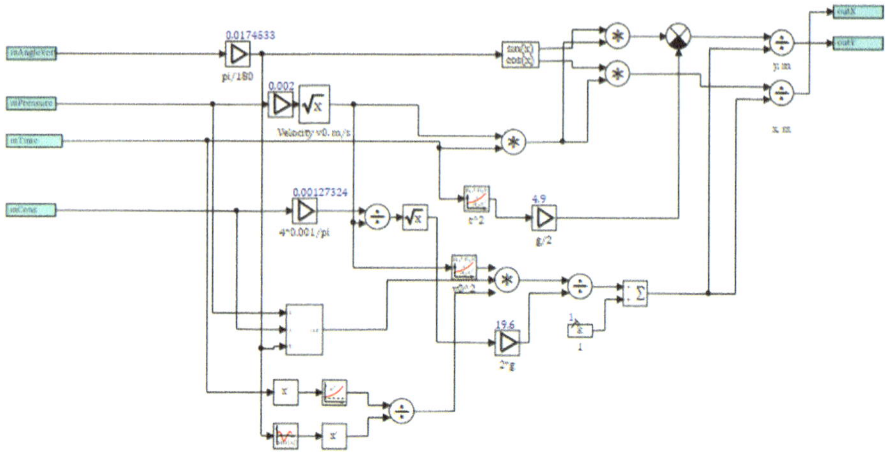

Fig. 1. Fire fighting robot monitor 2D jet stream computation submodel in the SimInTech software environment.

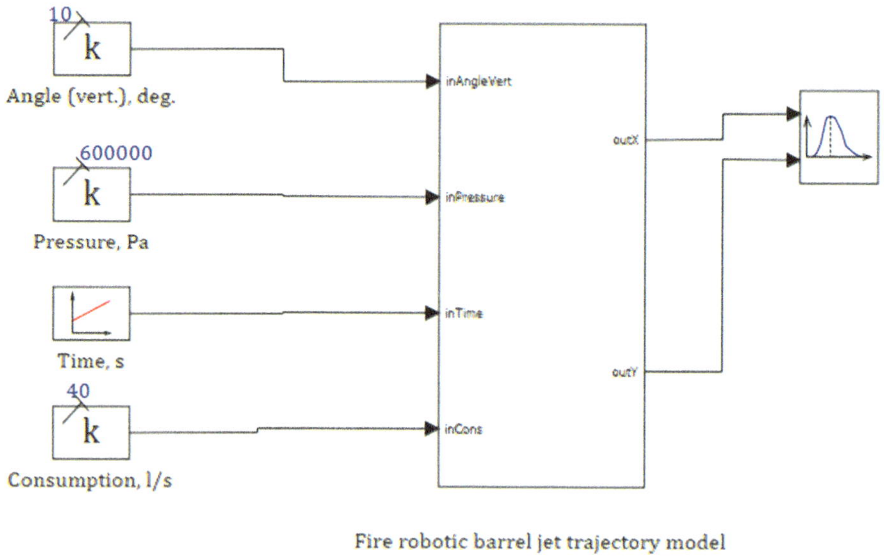

Fig. 2. Fire fighting robot monitor 2D jet stream computation model in the SimInTech software environment.

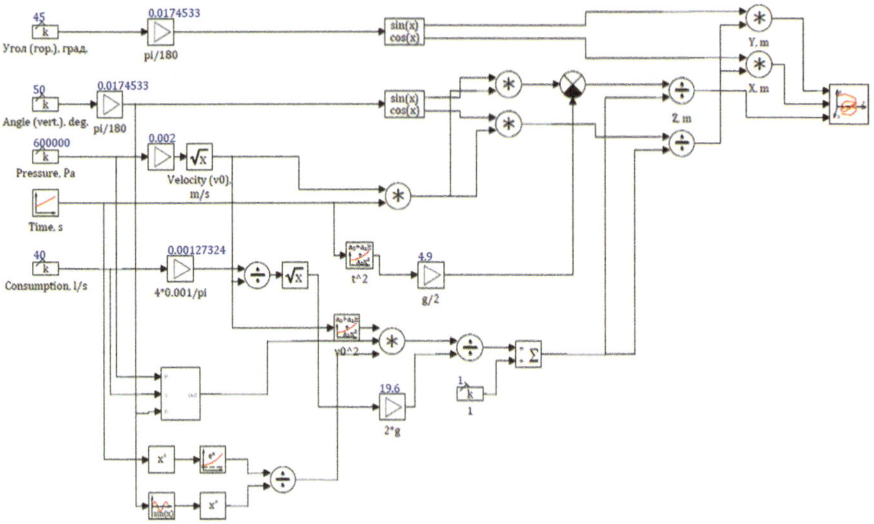

Fig. 3. Fire fighting robot monitor 3D jet stream computation model in the SimInTech software environment.

3 Results

Figure 4 presents the results of modelling the PR-LSD-S40U fire fighting robot jet [5] with the following parameters:

- Working pressure 0.6 MPa;
- Water consumption 40 l/sec;
- Fire monitor inclination angle with the horizontal axis 50°;
- Fire monitor turning angle with the vertical axis 45°.

Based on the model presented in Fig. 1 and 2, the used algorithm adequacy was tested by comparing the estimated trajectories with the experimental outcomes for different values of working pressure, water consumption and fire monitor inclination angle with the horizontal axis. Figure 5 presents the trajectories of the PR-LSD-S40U fire fighting robot jet with the following parameters:

- Working pressure 0.6 MPa;
- Water consumption 40 l/sec;
- Fire monitor inclination angle with the horizontal axis 50°.

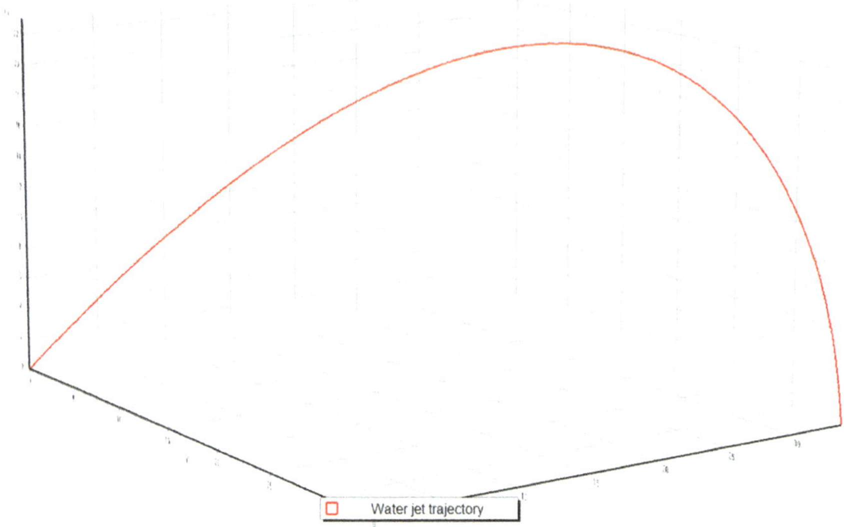

Fig. 4. Results of the fire fighting robot monitor 3D jet stream modelling in the SimInTech software environment.

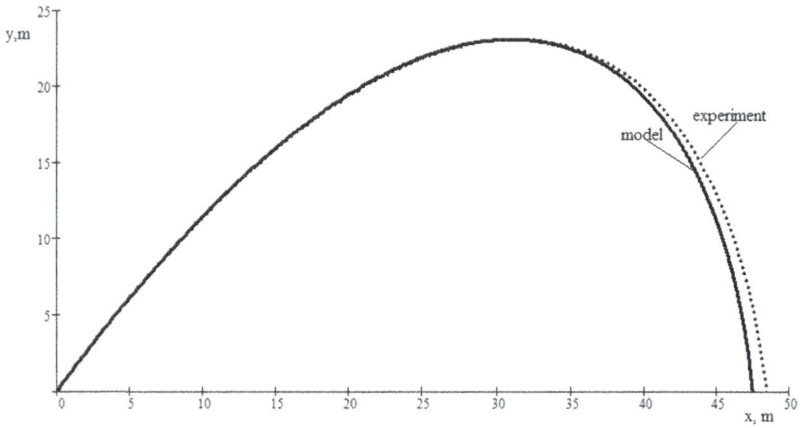

Fig. 5. Comparison of the estimated fire fighting robot monitor water jet trajectory with the experimental values.

4 Discussion

To evaluate the adequacy of the presented model of the fire fighting robot monitor water jet trajectory to the data collected in the full-scale test, Fisher's test was used [6]. The values were synchronized along the horizontal axis based on quantization of the modelled and experimental curves.

Table 1 presents the results of evaluating the adequacy of the estimated model of the water jet trajectory with the experimental data for different values of the fire monitor inclination angle with the horizontal axis θ based on the Fisher's test. The critical value was $F_{cr} = 4.034$, corresponding to the significance value $\alpha = 0.05$ and the freedom degrees $k_1 = 1$, $k_2 = 48$.

Table 1. Results of evaluating the adequacy of the fire fighting robot monitor water jet trajectory model to the experimental data based on the Fisher's test.

Angle value θ	Calculated Fisher's criterion value	Conclusion on the model adequacy
10°	546815	adequate
20°	135425	adequate
30°	60512	adequate
40°	33970	adequate
50°	15038	adequate
60°	3715	adequate
70°	1314	adequate
80°	317	adequate

Based on the performed analysis, the presented model of the fire fighting robot monitor water jet trajectory may be recognized as adequate with Fisher's test and the significance level of 0.05. It should be remarked that as the monitor inclination angle with the horizontal axis increases, the coincidence of the modelled trajectory to the experimental one reduces. At that, the greatest deviations are found in the descending part of the jet. For the optimal arrangement of the fire fighting robots at the protected venue, the deviation values are not critical, as greater θ values are used, as a rule, for the extinguishing or cooling the structures located at the higher elevations than the fire monitor, and, as a rule, exposed to the ascending part of the jet trajectory.

5 Conclusion

The presented model of an axisymmetric jet produced by the robotic fire fighting monitor with regard to the atmospheric friction layer on its lateral surface in the SimInTech dynamic modelling environment can be used for the solution of the following tasks:

- Optimal arrangement of the fire fighting robots;
- Evaluating the fitting of the given reference points of the venue into the water jet coverage area;
- Evaluating the efficiency of the designed robotic fire fighting units etc.

Among the protected venues involved in the mentioned tasks, there may be, for instance, large span buildings and large area facilities, such as machine houses, hangars, warehouses, expo centres etc.

References

1. Gost, R.: Fire Fighting Machinery. Robotic Fire Fighting Units. General Technical Requirements. Test Methods, 53326–2009 (2009)
2. VNPB. Robotic Fire Fighting Unit. Design Norms and Standards, pp. 39–16 (2016)
3. Zdor, G.N., Potekha, A.V.: Determining the Trajectory of an Inclined Hydraulic Jet with Respect to the Friction Force Occurring on Its Surface. Newsletter of Yanka Kupala State University of Grodno. Series 6. Technics vol. 2, pp. 88–97 (2016)
4. Pozharkova, I.N., Chubar, A.V., Grishchenko, I.A., Troyak, E.Y.: Technological Process Modelling in SimInTech Dynamic Modelling Environment. Siberian Firefighting and Rescue Bulletin Scientific and Analytic journal vol. 2, pp. 29–37 (2018). http://vestnik.sibpsa.ru/wp-content/uploads/2018/v2/N9_29-37.pdf. Accessed 18 June 2020
5. TR 4854-003-16820082-2008 Fire-Fighting Monitors (2008)
6. Dennis, B., Ponciano, J.M., Taper, M.L., Lele, S.R.: Errors in statistical inference under model misspecification: evidence, hypothesis testing, and AIC. Front. Ecol. Evol. **7**, 372 (2019)

Methodological Basics of Yoghurt Formula Development for the Far North Population

Marina Nikitina[1], Gennadiy Semenov[2], and Irina Krasnova[2(✉)]

[1] V.M. Gorbatov Federal Research Center for Food Systems of RAS, 26, Str. Talalikhina, Moscow 109316, Russia
nikitinama@yandex.ru

[2] Moscow State University of Food Production, 33, Str. Talalikhina, Moscow 109316, Russia
ira3891@mail.ru

Abstract. Improving the quality of life and worker productivity in the severe climate of the Far North requires creating conditions for adequate and balanced nutrition of the population, the development of food products rich in vitamins, minerals and biologically active substances, beneficial to the human health – first of all - fermented milk products. The most important for providing these regions with fermented milk products is their long-term preservation in unregulated temperatures. This requirement is ensured by the use of vacuum freeze-drying. Freeze-dried foods have a low final water content, a high level of preservation of vitamins, minerals and useful microorganisms for a long time. The article presents one of the food modeling stages - the choice of ingredients for future formula of vacuum freeze-dried fruit yogurts. There has been performed clustering of berries, fruits and vegetables by k-means method according to various criteria (content of vitamins, macro- and microelements, dietary fiber) to enrich yogurt. With particular focus on vitamins, as at low temperatures the body's vitamins requirement increases significantly. Clustering of a database fragment of chemical composition by the "Vitamin C content" feature performed in R software environment has been presented. Visual gradation of products into clusters has been shown as a dendrogram. The proposed methodology can be applied to select any other component or property significant for the product formula.

Keywords: Food product · Freeze-dried foods · Far north population · Database · Clustering · Designing

1 Introduction

Research articles [1–6] of scientists studying the relationship between nutrition and health of the Far North population show that at present their diets are characterized by overconsumption of starchy foods, refined sugars, foods rich in saturates, in contrast to vitamin and some mineral substances deficiency, lack of essential proteins and amino acids. This situation results in development of the most common diet-induced chronic

R. Silhavy et al. (Eds.): CoMeSySo 2020, AISC 1295, pp. 845–850, 2020.
https://doi.org/10.1007/978-3-030-63319-6_78

non-contagious diseases, such as obesity, atherosclerosis, coronary heart disease, arterial hypertension, diabetes mellitus, vitamin deficiency, etc.

Improving the quality of life and worker productivity in the Far North requires creating conditions for adequate and balanced nutrition of the population, the development of food products rich in vitamins, minerals and biologically active substances, beneficial to the human health. Fermented milk products are crucial for the balanced nutrition structure: yoghurts, fermented baked milk, etc. Their fresh delivery to the northern regions is extremely difficult, due to the need to create special temperatures during transportation and short storage periods. Vacuum freeze-drying can provide a high level of preservation of these products in unregulated conditions for a long time.

Development of such products is linked to a system analysis and collected data and knowledge formalization, as well as computer technologies for their processing and making optimal decisions based on complex models of multicriteria optimization and objective estimation of the adequacy of proposed options.

The objective of this work is to classify freeze-dried berries, fruits and vegetables according to various criteria (vitamins, minerals) in order to include them on reasonable grounds in a simulated food product formula for the population working in the Far North regions.

2 Materials and Methods

The authors have chosen a nonhierarchical algorithm - k-means [7, 8] as a clustering algorithm. Its main idea is that at each iteration, the mass center for each cluster obtained in the previous step is recalculated, and then vectors are again divided into clusters according to what new center is closer in the selected metric.

The algorithm is completed when at some iteration there is no change in the intra-cluster distance. This happens in a finite number of iterations, since the number of possible finite collection partitions is finite, and at each step, the total quadratic deviation decreases, and circling becomes impossible.

Software implementation was carried out in R programming language [9, 10].

3 Results

One of the food product modelling stage is a reasonable, well-founded choice of ingredients for the future product formula. The main principle of the balanced nutrition theory one need to remember when modelling a food product is that food nutrients must enter the human body in a certain amount and ratio.

Raw material base of food industry has thousands of ingredients, each having its own features, varying in functional, technological, physicochemical and organoleptic characteristics. Systematization of these parameters makes it possible to simplify the design process to a large extent.

Raw materials used to design food products for the population working in the Far North should meet the following requirements: balance all or individual components in accordance with the balanced nutrition theory; ensure hygienic safety of the resulting product; do not add pronounced flavours, sensations and smells to the products; provide a product with high sensitive characteristics; enrich the product with biologically active substances.

To enrich yogurt, it is required to choose the most preferred fruits and vegetables. Particular attention should be paid to vitamins, as under low temperatures conditions the body's need for vitamins increases significantly, especially in ascorbic acid (vitamin C). The authors have calculated the content of vitamin C in freeze-dried berries, fruits and vegetables.

The authors clustered the database fragment according to one of the features—vitamin C content—in R software environment R (Fig. 1).

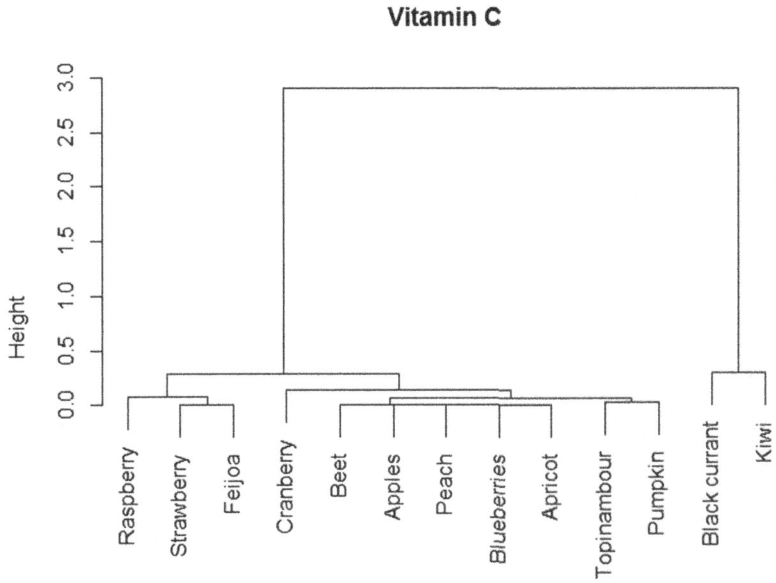

Fig. 1. Dendrogram: general view.

In addition to visual display in a dendrogram form, a table is formed with details for each cluster (Table 1).

Table 1. Clustering results.

	Data			
Cluster	Quantity by the "Vitamin C" field	Average by the "Vitamin C" field. mg.	Maximum by the "Vitamin C" field. mg.	Minimum by the "Vitamin C" field. mg.
1	3.00	17.33	20.00	16.00
2	8.00	7.90	12.00	4.80
3	2.00	152.00	160.00	144.00
Grand total	13.00	32.25	160.00	4.80

Table 1 data analysis shows that the following distribution of products was resulted from clustering using k-means algorithm: products with a low vitamin C content (grade 2), medium (grade 1) and high vitamin C content (grade 3). The largest number of products (8 positions) fell into cluster 2 with an average vitamin C content of 7.90 mg ranging from 4.80 to 12 mg. The fewest number of products (2 positions) fell into cluster 3 with an average vitamin C value of 152 mg ranging from 144 to 160 mg.

The user can see more detailed information when opening each cluster. What product subgroups or products are included in this cluster (Table 2).

Table 2. Clustering results. Detailed information.

	Data			
Cluster	Quantity by the "Vitamin C" field	Average by the "Vitamin C" field. mg.	Maximum by the "Vitamin C" field. mg.	Minimum by the "Vitamin C" field. mg.
1	**3.00**	**17.33**	**20.00**	**16.00**
Feijoa	1.00	16.00	16.00	16.00
Raspberry	1.00	20.00	20.00	20.00
Strawberry	1.00	16.00	16.00	16.00
2	**8.00**	**7.90**	**12.00**	**4.80**
Apples	1.00	8.00	8.00	8.00
Apricot	1.00	8.00	8.00	8.00
Beet	1.00	8.00	8.00	8.00
Blueberries	1.00	8.00	8.00	8.00
Cranberry	1.00	12.00	12.00	12.00
Peach	1.00	8.00	8.00	8.00
Pumpkin	1.00	6.40	6.40	6.40
Topinambour	1.00	4.80	4.80	4.80
3	**2.00**	**152.00**	**160.00**	**144.00**
Black currant	1.00	160.00	160.00	160.00
Kiwi	1.00	144.00	144.00	144.00
Grand total	**13.00**	**32.25**	**160.00**	**4.80**

Thus, if we consider only one indicator (vitamin C), not combined with other features, it is preferable to enrich yogurt with kiwi and black currant.

Computer-aided designing of food products with specified qualitative characteristics is based on optimizing the choice of various types of raw materials and ratios of formula ingredients, which together make it possible to obtain the most suitable composition for the quantitative content and qualitative composition of nutrients according to biomedical requirements.

In design of a food product, it is possible to use criterion of minimum deviation from a given structure of a certain group of indicators of food and biological value [11, 12], for example:

– Criterion of minimum deviation from a given structure of mineral composition

$$P(M) = \sum_{k=1}^{n} \left(M_k^0 - \frac{\sum_{j=1}^{m} \mu_{kj} x_j}{\sum_{j=1}^{m} x_j} \right)^2 \rightarrow min$$

where M_k^0 – reference content of the *k*-th macro- and microelements, %; μ_j – content of *k*-th macro- and microelement in the *j*-th formula component of the simulated product, %; x_j – content of the j-th formula component, %.

– Criterion of minimum deviation from a given structure of vitamin composition

$$P(V) = \sum_{k=1}^{n} \left(V_k^0 - \frac{\sum_{j=1}^{m} v_{kj} x_j}{\sum_{j=1}^{m} x_j} \right)^2 \rightarrow min$$

where V_k^0 – reference content of the *k*-th vitamin, %;- content of the *k*-th vitamin in the *j*-th formula component of the simulated product, %; x_j – content of the *j*-th formula component, %.

It is worth noting that in modern conditions, when creating product formulas of a functional or specialized purpose, multicriteria methods are used as a rule, with a large number of restrictions considering energy content of products, their nutritional value and taste. However, an excessively large list of product initial components, as well as a significant number of physicochemical and technological factors used in their creation, do not allow us to fully solve the task by "manual" methods without using information technologies.

Another important aspect in food production is the use of vacuum freeze-drying. This will increase the shelf life while preserving nutritional value, including biologically active thermolabile components - vitamins, enzymes, amino acids, microorganisms, etc. [13].

4 Conclusion

The example of berries, fruits and vegetables in the article shows the possibility of using cluster analysis to classify the plant-based raw materials for enriching foods with various vitamins, minerals, biologically active compounds beneficial to the human health.

References

1. Eganyan, R.A.: Dietary habits of residents of the Far North of Russia. Prev. Med. **5**, 41–47 (2013)
2. Baturin, A.K., Pogozheva, A.V., Keshabyanc, E.E., Starovojtov, M.L., Kobel'kova, I.V., Kambarov, A.O.: Nutrition, anthropometric indicators and body composition of native and newcomer population of the Russian Arctic. Nutr. Issues **5**(86), 11–16 (2017)

3. Ionova, I.E., Agbalyan, E.V., Buganov, A.A.: Nutrition of residents of the Yamal-Nenets autonomous district and its influence on the formation of some diet-induced risk factors. Healthc. Serv. Yamal **2**, 43–48 (1999)
4. Lobanova, L.P., Agbalyan, E.V., Buganov, A.A.: Adequacy of micronutrient intake among newcomer population of the Far North. Nutr. Issues **5**, 51–54 (2007)
5. Kozlov, A.I., Ateeva, Y.U.A., Vershubskaya, G.G., Ryzhaenkov, V.G., Suvorov, A.V.: D-vitamin status of the population of the perm territory. Komi Udmurtia Nutr. Issues **2**, 31–36 (2013)
6. Beketova, N.A., Kodencova, V.M., Vrzhesinskaya, O.A., Keshabyanc, E.E., Sokol'nikov, A.A., Kosheleva, O.V., Kobel'kova, I.V., Pogozheva, A.V., Kambarov, A.O., Baturin, A.K.: Adequacy of vitamin intake among residents of rural settlements of the Russian Arctic. Nutr. Issues **3**(86), 83–91 (2017)
7. Li, Y., Wu, H.: A clustering method based on K-Means algorithm. Phys. Procedia **25**, 1104–1109 (2012)
8. Bai, L., Liang, J., Cao, F.: A multiple k-means clustering ensemble algorithm to find nonlinearly separable clusters. Inform. Fusion **61**, 36–47 (2020)
9. Wickham, H., Grolemund, G.: R for Data Science: Import, Tidy, Transform, Visualize, and Model Data. O'Reilly Media, Sebastopol (2016)
10. Kabacoff, R.I.: R in Action: Data Analysis and Graphics with R. Manning, Shelter Island (2011)
11. Bruevich, N.A., Ivashkin, Y.A., Nikitina, M.A.: Information technology in the mobile application of analysis and correction of the diet of individual healthy nutrition. In: 2019 Systems of Signals Generating and Processing in the Field of on Board Communications, SOSG 2019, p. 8706763. Institute of Electrical and Electronics Engineers Inc. (2019)
12. Lisitsyn, A., Chernukha, I., Nikitina, M.: Russian methodology for designing multicomponent foods in retrospect. Foods Raw Mater. **1**(8), 2–11 (2020)
13. Semenov, G.V., Krasnova, I.S.: Vacuum Freeze-drying of Food Products. DeLi Plus, Moscow (2018)

Improving MapReduce Process by Mobile Agents

Ahmed Amine Fariz(✉), Jaafar Abouchabka, and Najat Rafalia

LaRIT, Faculty of Sciences, IbnTofail University of Kenitra, Kenitra, Morocco
a.amine.fariz@gmail.com, abouchabaka3@yahoo.fr, arafalia@yahoo.com

Abstract. This article provides information about the processing of big data by MapReduce Framework, Stragglers in MapReduce, their scope, current situation, and influence in the present world and how Stragglers detection can be made more effective by the use of mobile agents. Straggler task detection is one of the primary difficulties in applying MapReduce for parallelizing and distributing enormous data processing. It is characterized as identifying running tasks on weak nodes. Considering two phases in the Map (copy, combine) and three phases of Reduce (shuffle, sort and reduce), the total execution time is the aggregation of the execution time of these five phases. Estimating the correct execution time at each step, which results in correct total execution time, is crucial. The paper proceeds with the details of MapReduce plan for handling Big Data, stragglers, the importance of attenuating stragglers along with their effects. The proposed strategy highlights the use of mobile agent technology and corresponding outcomes.

Keywords: MapReduce · Mobile agent · Hadoop scheduler · Jade

1 Overview

We are living in the modern era where a massive volume of data is created. The data is radically expanding at an unpredicted rate due to the advancement in web technologies, social media, sensing, and mobile devices. MapReduce based systems are frequently being employed for large-scale data analysis operations. As the name shows, MapReduce completes a program in two steps/phases, the first step is data mapping (described by a function-mapper) and the second one is reducing the data (described by a function-reducer).

Stragglers are created due to the variety in the network traffic, I/O discord, or CPU accessibility. Since a task in the MapReduce network is not ended until the execution of all mapping and reducing operations, even a few stragglers or outliers can degrade the reaction period for occupancy. It aims to minimize the remaining time estimation error from the completion of tasks/functions and appropriately identifying straggler functions to expand the Big Data computational infrastructure efficiency. Among other powerful approaches, parallel and distributed processing is outstanding to save and process bulky data. Mantri and LATE are the most productive among several straggler

R. Silhavy et al. (Eds.): CoMeSySo 2020, AISC 1295, pp. 851–863, 2020.
https://doi.org/10.1007/978-3-030-63319-6_79

recognition and mitigation methods. We aim to attain 100% accuracy by reducing and eliminating all stragglers using the mobile agent system.

2 MapReduce

2.1 What is MapReduce

MapReduce is an important programming model that enables us to conduct parallel and distributed processing on big data sets under the distributed environment. To avoid the time consumed parallel architecture can be used for the simultaneous execution of the different tasks. It is also used for big data distribution to avoid processing and storage issues. By using a MapReduce-based methodology. It utilizes parallel processing control using community cluster workers.

This platform is primarily divided into four parts, i.e. MapReduce framework, data load, workers, and analytics or output.

Initially, data is filled into the datastore object, read in mapping variables by stocking and splitting into different portions/pieces for easy processing over the MapReduce model. These pieces of information are preprocessed by the pre-processing technique before applying to MapReduce Phases. This data is pre-processed by averaging theory.

Secondly, parallel processing is performed by phases of the MapReduce framework via data analytics techniques. These analytics techniques are proceeding from the two phases of the MapReduce network.

Thirdly, workstations, or workers play the central role of data processing via internal server-clients communication.

Finally, by combining all reduce phases' output into a common file, the overall output is achieved (see Fig. 1).

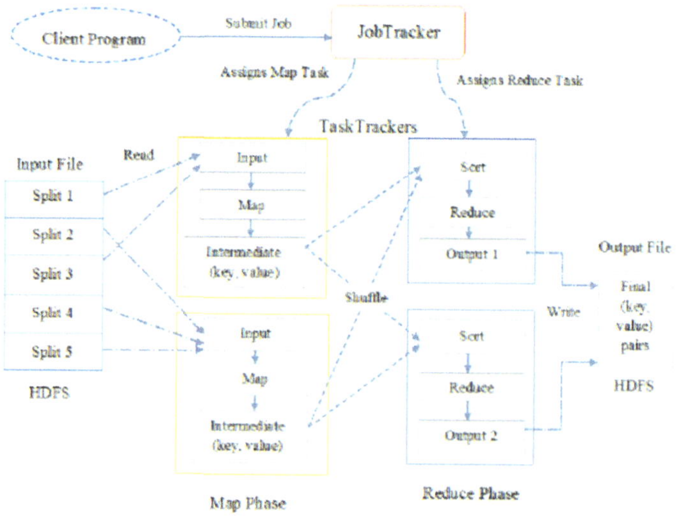

Fig. 1. MapReduce workflow

The MapReduce model consists of two distinct phases i.e. Map and Reduce.

- As proposed by the name MapReduce, a reducer stage comes after mapper stage execution.
- In the map phase, by reading and processing a piece of data, the output is obtained in the form of key-value pairs output.
- The map job output is the input for the Reducer.
- A reducer obtains key-value pair (kvp) from various map jobs.
- A Reducer decreases the number of these intermediate KVPs to give the final output.

2.2 What is MapReduce

MapReduce comprises of the three classes described as:

Mapper Class
This is the first stage of Data Processing using MapReduce. In this phase, RecordReader processes all the Input records and produces the respective KVP. This intermediate data is stored in the local disk by Hadoop's Mapper store.

Input Split: is a logical representation of big data that denotes a chunk of work containing an individual map task in a MapReduce model.

RecordReader: After interaction with Input, split changes the acquired data in Key-Value Pairs form.

Reducer Class
The reducer takes the Intermediate output from the mapper and after processing it, produces the final output. This final output is ultimately saved in HDFS.

Driver Class
It is the major component of a MapReduce job. It sets up the MapReduce task to run by Hadoop.

2.3 MapReduce Scheduling in Heterogeneous Environment

The performance of MapReduce is principally reliant its task scheduler [1], optimizing the general consummation time of a job by properly relegating assignments to the accessible nodes is a shared objective of the MapReduce. In Hadoop cluster if a task is executed in more lengthy way than the other tasks then this condition will be called as straggler. Speculative tasks can induce resource misuse and hinder other job performance in the cluster [2]. Despite the fact that MapReduce schedulers endeavor to dispatch backup tasks for stragglers, the identification of correct straggler tasks is failing because of errors and challenges in estimating remaining execution time of the tasks [3]. This would prompt issues for example, launching a backup task for stragglers that are wrongly identified will not improve the MapReduce performance and the system resources are not used proficiently [4].

2.4 MapReduce Default Scheduling Algorithm

In MapReduce default scheduler [5], the progress score (PS) of a task t is signified by $\mathbf{PS_t}$, which is determined utilizing Eq. 1 for map tasks and Eq. 2 for reduce tasks.

$$PS_t = M/N \tag{1}$$

$$PS_t = (1/3)(K + M/N) \tag{2}$$

Where, the number of (key, value) pairs that have been processed successfully is denoted by, the overall number of (key, value) pairs is N and the stage (shuffle, sort and merge) value in a reduce phase is represented by K.

$\mathbf{PS_{avg}}$ (the average progress score of a job) is calculated using Eq. 3, **PS [i]** is the progress score of a task ti while n is the number of executable tasks in a job.

$$\mathbf{PS_{avg}} = \sum_{i=1}^{n} PS[i]/n \tag{3}$$

The task $\mathbf{t_i}$ is considered a straggler task only if it satisfies Eq. 4, therefore the backup task for that particular straggler task i is launched.

$$PS[i] < PS_{avg} - 20\% \tag{4}$$

The Limitations of the Above Method

In MapReduce default scheduler, the map weights are M1 = 1, M2 = 0 and reduce task weights are (R1 = R2 = R3 = 1/3) but these weights will change when tasks run in a heterogeneous environment.

Default scheduler cannot distinguish the right straggler tasks which need to be re-executed with fast nodes and it happens that in some cases the backup tasks will be mistakenly launched for fast tasks rather than slow tasks.

The average progress score will never go beyond 100% which results in It using difference in progress of 20% as threshold, it signifies that the tasks that has progress score above 80% will no longer speculatively executed.

2.5 Causes of Stragglers

Stragglers are the operations that take more time to complete than other comparable tasks. Stragglers arise due to multiple reasons, for example, faulty hardware, hardware heterogeneity, misconfiguration, flawed machines, I/O discord, background network traffic, resource contention, system obstruction, and conflict for the current assets.

3 Relevant Work

There are many straggler identification and mitigation detection algorithms available like MonTool, Hadoop native scheduler [6], Mantri, LATE, and Dolly [7]. Hadoop scheduler is based on a development score between zero and one. The task execution is principally divided into three stages/phases i.e. Mapper phase, Sorting/shuffling phase, Reducer phase.

In every stage, the score is entirely based on processed data. Hadoop calculates the mean score for all processes to designate a speculative execution limit. If the task advancement score is 0.2 less than the standard of its category, it's recognized as a straggler.

LATE [8] is based on a different scheme that approximates the task completion time to forecast potential stragglers. Progress rate calculation is completed as the advancement score/T (T is the task-running time). Execution time estimation is done as (1-advancement or progress score)/rate of progress. Tasks with the advancement rate below 25% are recognized as stragglers.

MANTRI [9] is based on real-time reports. Therefore, straggler identification by Mantri is done in a shorter time than done by other algorithms. The spot at which a task is unable to make progress at a usual rate is identified by Mantri. Mantri recognizes the straggler by resource perception and cause awareness.

MONTOOL [10] does analysis by tracing system calls. Using this analytic approach, MonTool attempts to find stragglers and their reasons.

DOLLY [11] uses a remarkable mechanism in which it launches task clones and then uses the clone outcome which is completed first.

4 Identification of Problem

There are multiple techniques for the identification and mitigation of stragglers. All the straggler identification and mitigation techniques use different tactics but then ultimately reschedules these approaches to the other place in a nearly similar manner. They hypothetically accomplish them on some other comparatively speedy equipment present in the vicinity of a staggering machine. In the beginning, one may presume that straggler identification will be a simple job, but that is not the case. Various issues which were recognized are as follows:

Firstly, speculative tasks will require other execution resources, e.g. a set-up with other ongoing tasks.

Secondly, the node selection for running a speculative task is not less significant than selecting the task.

Thirdly, identification of nodes that are slower as compared to the stragglers and mean.

So, it is our main concern to overcome these issues related to straggler handling.

5 Offered Work

Software bodies having a certain sort of intelligence are called the Agents. The Agents are those who act, respond, and interconnect with other agents and ultimately give outcome back to users.

Web applications based on agents generate full capabilities of existing network infrastructures and data available in a network. An innovative straggler identification and mitigation strategy using mobile agents, is anticipated in this study.

5.1 Processing of Big Data by MapReduce Framework Based on Mobile Agent

Basic Assumptions: Significant speculations made to specify the suitability of this technique are given below:

- All the machines must have JADE [12]; Environment for agent execution is pre-installed, providing efficient development of mobile agent and proper execution environment to enable parallel computing. In an execution environment, the agent can protect its code, owner identification, and data integrity.
- The file of Big Data must be splitted uniformly so that all agents get an equal load share, leading to an even computed score.
- Each agent acquires a single file split, i.e., the agents' number in the network is equivalent to the splits' number.
- An intermediate site is preset for storing intermediate data from several map agents.
- The file of big data is placed at the chief host itself.

Method: it was found through the literature review that the techniques were employing some sort of heuristics for the detection of stragglers and then speculative execution at two diverse locations in the same period to lessen their effect. The principal of our suggested solution for big data processing is a scheduler agent. This agent supports the creation of mobile agents and rescheduling if the performance of mobile agents is not close to the average performance of all agents.

5.2 Technique or Algorithm

The algorithm utilizes one agent per split, it gains from a user, and then the mobile agent travels from one position to another based on scheduler scheduling, based on the score of performance that the mobile agent has directed towards the scheduler. A scheduler supports two arrays, i.e., one for faster machines and one for slower machines. The scheduler reorganizes the mobile agents moving on slower machines towards faster machines, individually. The scheduler eventually retains all the consequences after each mobile agent has finished its job and then hand over the outcome to the user.

Scheduling Algorithms

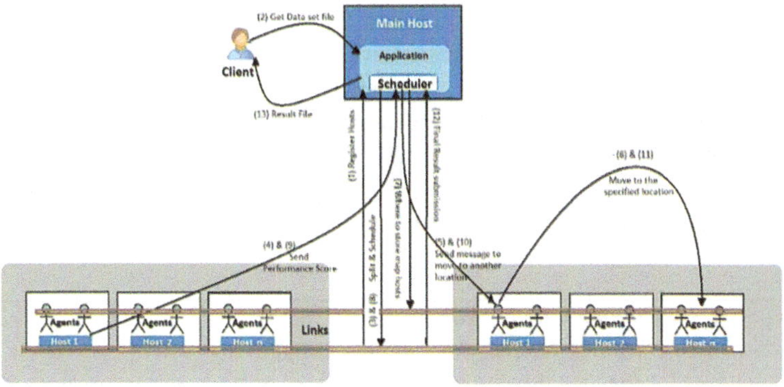

Fig. 2. Scheduling algorithms

1. A host in a network that wants to participate in the processing of big data registers itself with its IP Addresses.
2. These IP addresses of the hosts are used for the creation of the agent containers.
3. By the core application, a Scheduler agent is generated. This agent takes the name of file as evidence and detects all the agent containers moving on the main platform.
4. A scheduler agent produces the mobile agents after the file splitting and allots a split to all of them.
5. After every Ts seconds, each agent directs its score of performance to a scheduler.

   Score agent (*i*) = Total executed or performed/Total number to be performed * Ts

6. In the meantime, when an agent travels from one place to another, it protects its present state, so that it can initiate from the same location where it stopped the task execution.
7. After executing all the processes, the agent delivers the outcome to a scheduler.
8. A scheduler sends this outcome to the original user after the gathering results from every reducer (Fig. 2).

Functions/phases of Agent Life Cycle

Setup: Using this function, an agent is produced. The Agent is loaded with an associated file and the site where it originally has been located. While producing an agent, the scheduler also provides the location in addition to the file-split.

Map-execution: In this phase, the agent performs the map task; defined in terms of word-count algorithm, i.e., the map function counts that how many times a word has appeared in a split allotted to a particular agent.

Ticker Behavior: This is used to send the messages of the score to a scheduler. As implied by the name, after a certain time defined by its time factor, this behavior executes its specific code.

Action: This phase specifies the kind of action (i.e., map or reduce) which is to be performed on a particular split. After that, the specific execution technique is called.

Reduce-execution: This function describes the Reduce job for a specific split designated by a scheduler. The hashing function-based scheduler calculates the specific split to be assigned to a specific reducer and after that, it assigns that specific split to that reducer using this method.

Move/location: This method is present in the agent class. It helps in the movement of the agent class from one position to another. The scheduler directs the parameter of location towards the agent, in turn; the agent calls this procedure on itself through that location parameter.

Destroy: This method is also present in the agent class. It aids in liberating all the means, which an agent had already used (Fig. 3).

```
MainScheduler (file[], containers[])
        ratio = file.length /container.length
        for j=1 to countainer.length
                for i=1 to ratio
                        countainer[j].createAgent(file[i],countainer[j])
                end
        end
        for j=1 to file.length
                score [j] = receive (msg[j]);
                avg+=score/file.length;
                for i=1 to n // number of agents
                        if score[i] < 0.5 * avg
                                slowContainers[k++] = container[i];
                        else if score[i] >= avg
                                fastContainers[k++] = countainer[i];
                        end
                end
                for i=1 to N // N of slowContainers
                        agent[i].move(fastContainer[i])
                end
        end
        If allMessageFinished
return
```

Fig. 3. Pseudo Algorithm

Operational Steps

1. Registration Phase: The node who wants to participate in the computing grid, registers itself with the main node (the application has been running in the main node). The registration procedure is accomplished in a fixed time. For further communication with the nodes, each node directs its IP Address towards the chief application. The IP addresses are kept in an arrangement and for each of the agents, a respective agent container is initiated in the jade runtime environment. This container manages all the tasks to be executed on a specific agent. In the registration

process, each node admits that it's meeting every condition required to satisfy and is providing its services in parallel processing.

2. Fetching File and Creation of Scheduler Agent: After fetching the address of a big data file that is to be processed, the scheduler agent is generated along with the agent containers-array created earlier and the file address. Then the scheduler interacts with all the containers and then they are used for creating the agents. In the meantime, the scheduler splits the file into multiple sections based on the split size taken from a user.
3. Agent Score Computation: An agent determines the score using the following formula:

   Score = (Processed word count/word count to be processed) * unit time

- The Processed word count is calculated by a specific variable.
- The word count to be processed is established by using the JAVA string class.
- Here, the time is constant to 100 ms.

4. Score Submission by the agent: After submitting its calculated score, each agent directs it back towards the scheduler. To deliver the current execution score, the agent utilizes the message format from the class of the agent in the JADE-execution environment. There is an array in the application for the fastest and slowest agents to save the precise values.
5. Relocation of Agent: The arrays saved for the fastest and slowest machines are utilized to deliver the position where a slow agent has to travel where it can complete the actual activity of the agent. The information of location is delivered to an agent in the form of a message specifically to that agent, which has been declared as the straggler by the scheduler.
6. Result Submission by the Agent: The agent conveys the outcome to the application. Then the application combines all the outcomes into the final result which is directed towards the user. The outcome is in the file form, having the overall result.
7. Submission of Final Result: The application delivers the result to the user who has requested the particular task.

5.3 Results

The suggested algorithm is tried on both situations for multiple setups of stragglers as well as for the size of splits. This section explains the results of experiments and compares these results with outcomes of other methods proposed for resolving the issue of straggler identification and mitigation. In the last part of this study, further optimization is provided to make the algorithm execution faster.

Testing

As accuracy is part of the evaluation, we needed to completely control the conduct of the application and to stay away from the startling deviation, which can affect the outcomes. In this manner, we picked WordCount, a straightforward yet representative MapReduce application. The WordCount application counts the occurrences of each word in an enormous amount of input files. Each Map task of WordCount application handles one small part of the total input files. Each word in the input data of this Map task is utilized to make a key/value pair, where the key is the word and the value is 1. The Reduce tasks gather all the key/value pairs within their assigned key range.

Then, they combine all pairs sharing the same key to return the results. In WordCount the output data size is typically small compared to the size of input data which makes it a CPU-intensive application and it has a high input/output ratio. Therefore, by using WordCount application, we can restrict the "real" stragglers to the ones resulted from resource heterogeneity and thus provide an accurate interpretation of the results.

To test the implementation, an Intel Core i7 laptop having 8 GB RAM is utilized for the program execution for both situations with a text file of 500 MB. The operations of the map and reduce are described for word count processes. To create agents on multiple machines, JADE is used. The values of the parameters can be refined according to the properties of the network as they differ with huge amounts. This provides the process, the extensibility to be employed in different kinds of networks. One can achieve different times of execution by differing these parameters. If one comes up with values, which are beyond those restrictions, the results will vary. This method is tested to deliver good outcomes through these values within constraints (Table 1).

Table 1. Parameter values

Parameter	*Value*
Size of Big Data file (Data Set)	*480 MB*
Agent containers per main container	*3 - 4*
Agents per agent container	*4 - 12*
Mapping technique	*Word count*

The difference with HADOOP Scheduler based on Percentage of Straggler

The recommended method having the wide implementation in straggler identification and mitigation process is compared with the HADOOP Scheduler which shows that the proposed algorithm outperforms the HADOOP in time of execution for randomly produced graphs for a variety of split sizes. It does so without a DFS. It saves the additional time taken by the Hadoop to create a DFS and after that recommences them from a specific place, where they left earlier (Table 2).

The proposed algorithm yielded better outcomes than the compared algorithms. However, split sizes can increase the difficulties because we have used the network for transferring the data across the machines. Speed of the network comes into the picture when the split or information is transmitted because it can cause a rise in the overall time of execution (Fig. 4).

Table 2. Execution overview on straggler %

Straggler (%)	*HADOOP native scheduler time (ms)*	*Mobile agent-based MapReduce time (ms)*
10	*36,440*	*38,427*
20	*43,911*	*41,169*
30	*53,217*	*48,432*
40	*65,529*	*55,590*
50	*90,189*	*74,523*

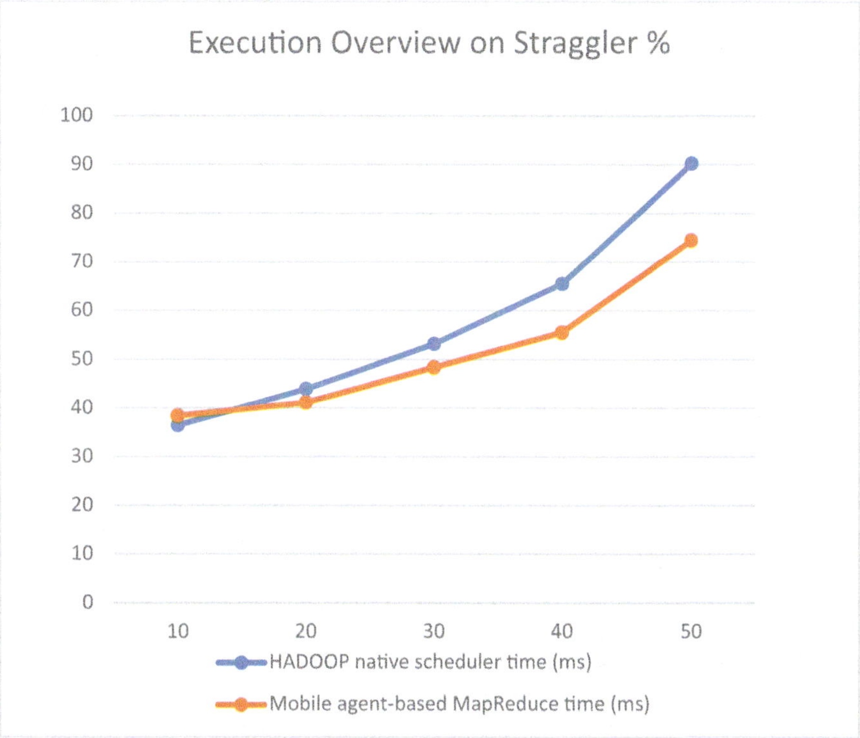

Fig. 4. Performance comparison with HADOOP based on split sizes

Big data processing by MapReduce framework based on mobile agents not only operates well for a single machine but also for a machine network. It improves the time of execution in both situations (Table 3).

Table 3. Execution overview on the basis of various split sizes

*Setup scenario Size * container * agents/container*	*HADOOP native scheduler time (s)*	*Mobile agent-based MapReduce time (ms)*
40 * 3 * 4	42	44
32 * 3 * 5	45	45
16 * 3 * 10	48	47
10 * 4 * 12	52	52

The important points from the outcome of the implementation of the proposed technique are given below:

i. After detection, the stragglers are effectively rescheduled to another machine, by avoiding the wastage of any machine cycle.
ii. The process of rescheduling is delayed to facilitate a node trapped in any other valuable task to continue its job, as it is not considered a straggler. As we are utilizing the mobile agents (popular for their capability to initiate from the location, they left execution), this delay of time does not charge us much.
iii. As compared to the already existing techniques, in this technique there is a remarkable refinement in the time of execution, i.e., approximately 10–12% (without any DFS).

6 Conclusion

In recent years, the speed of data analysis in Big Data processing is of great significance. The purpose of this study is to improve Straggler identification and mitigation methods in order to enhance the efficiency of the computational infrastructure of Big Data to accelerate data processing. Identifying straggler tasks have gained the attention of researchers. Therefore, the number of procedures published for solving this straggler handling issue is rising as it has uses in several domains like parallel computation and big data processing. Speculative execution is a substantial technique used to improve stragglers mitigation and the performance of MapReduce jobs. The speculative execution efficiency needs a profound understanding of the straggler detection mechanism characteristics. Unfortunately, using the existing metrics can be inefficient and sometimes can lead to inaccurate interpretation. The main objective is to offer a more accurate technique than the solutions already existing. The recommended technique performs better than the classical procedures and the present. The results indicate that it is possible to improve the existing straggler detection mechanisms targeting better efficiency. The characteristics, achieved using our metrics, can give useful hints for enhancing the efficiency of speculative execution. As regards of the importance of accurate results in short execution time, the suggested method can be developed further in the future.

References

1. Nanduri, R., Maheshwari, N., Reddyraja, A., Varma, V.: Job awarescheduling algorithm for mapreduce framework. In: Proceedings of the 3rd International Conference on Cloud Computing Technology and Science, pp. 724–729 (2011)
2. Tian, C., Zhou, H., He, Y., Zha, L.: A dynamic mapreduce scheduler for heterogeneous workloads. In: Eighth International Conference on Grid and Cooperative Computing, pp. 218–224, Aug. 2009
3. Rasooli, A., Down, D.G.: Guidelines for selecting hadoop schedulers based on system heterogeneity. J. Grid Comput. **12**(3), pp. 1–26 (2014)
4. Naik, N.S., Negi, A., Sastry, V.N.: A review of adaptive approaches to MapReduce scheduling in heterogeneous environments. In: IEEE International Conference on Advances in Computing, Communications and Informatics, pp. 677–683, Sept 2014
5. Rasooli, A., Down, D.G.: COSHH: a classification and optimization based scheduler for heterogeneous Hadoop systems. J. Future Gener. Comput. Syst. **36**, pp. 1–15 (2014)
6. LNCS Homepage. http://hadoop.apache.org. Accessed 25 June 2020
7. Javadpour, A., Wang, G., Rezaei, S., Li, K.: Detecting straggler MapReduce tasks in big data processing infrastructure by neural network. The Journal of Supercomputing, Springer Science + Business Media, LLC, Part of Springer Nature (2020)
8. Zaharia, M., Konwinski, A., Joseph, A.D., Katz, R., Stoica, I.: Improving MapReduce performance in heterogeneous environment. In: OSDI'08 Proceedings of the 8th USENIX Conference on Operating Systems Design and Implementation, pp. 29–42 (2008)
9. Ananthanarayan, G., Ghodsi, A., Shenker, S., Stoica, I.: Effective straggler mitigation: attack of the clones. In: Proceedings of the 10th USENIX Conference on Networked Systems Design and Implementation, pp. 185–198 (2013)
10. Gandhi, R., Sabne, A.: Finding stragglers in Hadoop. In: European Conference on Computer Systems (2012)
11. Ananthanarayanan, G., Kandula, S., Berg, A.G., Stoica, I., Harris, E., Shaha, B.: Reining in the outliers in MapReduce clusters using Mantri. In: 9th USENIX Symposium on Networked Systems Design and Implementation (2010)
12. Franklin, S., Graesser, A.: Is it an Agent, or Just a Program?: A Taxonomy for Autonomous Agents, vol. 1193, pp. 21–35. Springer, Berlin (1997)

Industry 4.0 Visual Tools for Digital Twin System Design

V. A. Shakhnov, A. E. Kurnosenko, A. A. Demin, and A. I. Vlasov(✉)

Department of Design and Technology of Electronic Devices, Bauman Moscow State Technical University, 2-ya Baumanskaya, 5, Street 1, 105005 Moscow, Russian Federation
shakhnov@iu4.bmstu.ru, vlasov@iu4.ru

Abstract. This work analyses methods of digital model development and uses electronics production as an example. It focuses on the analysis of the visual tools of a digital manufacturing system design. The digital model development process is divided into a sequence of tasks, to solve which a number of models, methods and tools is proposed. These procedures can be used to perform simulation modelling, performance prediction, promising option selection, parameters optimisation of both the device itself and its manufacturing mechanisms along with other lifecycle components, without the expensive sample tests, field tests or introduction of any changes to the already established manufacturing, selling, service and other infrastructures. Digitalisation aspects are covered by creating interrelated digital models of the device and the digital manufacturing process model that take all the technical and economic aspects into account. This approach, called Digital Twin (of the actual production), underlies the modern Industry 4.0 concept that suggests the introduction of different cyber-physical systems in the manufacturing systems. The proposed method of the visual system design uses a complex of software tools to produce digital system models of an electronics manufacturing system. The resulting digital twin of the actual manufacturing is representative of both the produced device and its manufacturing process design features. And it can be used for preparation, analysis and manufacturing improvement purposes.

Keywords: Visual modelling · System design · Digitaltwin · Industry digital transformation · Industry 4.0 · Electronics · Manufacturing preparation system

1 Introduction

Electronics manufacturing is one of the essential branches of the modern industry featuring intense dynamics of changes, severe competition, an abundance of cutting-edge technologies and ever shorter periods of new products market entry. Such an environment requires not only adequate technology, but also managerial, economical, and socio-psychological knowledge and solutions [1]. To be competitive the industry is to consider the technical and economic market analysis, specification formulation, manufacturing flow preparation, actual manufacturing, sales activities, operation,

R. Silhavy et al. (Eds.): CoMeSySo 2020, AISC 1295, pp. 864–875, 2020.
https://doi.org/10.1007/978-3-030-63319-6_80

maintenance and utilisation of a device not as isolated phases, but as constituent parts of the Product Lifecycle Management (PLM). The interaction among these constituent parts shall be bidirectional in the form of data flows, supply chains, etc. Such approach can be implemented in the frame of the modern concept of the digital transformation of industry [2], which suggests digitalisation of the mentioned parts of the lifecycle by producing mathematical, physical and simulation models of the object representatives in terms of parameters, inputs/outputs, control actions, criteria, limitations and optimisation variables.

A decade ago, the opinion prevailed that to solve the tasks mentioned above, a digital model of the device (design/manufacturing object) suffices. The approach was mainly practical for computer-aided design (CAD), computer-aided engineering (CAE) and partly computer-aided manufacturing (CAM) of machine-building operations (such as foundry, stamping, cutting, etc.). The link with manufacturing was unidirectional, allowing only for assessment of how a given device fits within the existing manufacturing structure, while the readjustment, adaptation and flexibility of the manufacturing were strictly limited.

The electronics manufacturing involves multiple assembly operations of PCBs and modules where permanent connections are soldered, welded, glued, etc. It is also a multiproduct, small and medium batch production where the introduction of new products means quick adjustments. There are a wide range of automated assembly lines and cell-type systems [3] and a portion of manual assembly, inspection and repair operations. To be able to take all these specific features into account, not only the device digital model is needed but also the digital model of the manufacturing, representative of all technical and economic aspects: equipment, facilities, personnel, components, technological processes and flowcharts, materials logistics within production site/warehouse, figures of performance, accuracy, fail-safe operation, minimal readjustment time, forced outages, optimisation road maps, number of service personnel, etc.

The purpose of this work is to develop an approach to creating the electronics digital manufacturing and to identify the subsystems responsible for separate tasks of the information content provision, simulation modelling and optimisation of the manufacturing in question.

This approach – Digital Twin of the actual production – underlies the modern Industry 4.0 concept that suggests the introduction of different cyber-physical systems in the manufacturing systems. In this context, one of the critical factors of effective development is economical and social management aspects that should be closely linked with the product and process specifications [4–9].

2 Literature Review

The overall task of electronics manufacturing modelling is complicated and depends on many interdependent parameters. The proposed approach suggests that it shall be broken down to a sequence of individual tasks, each of which shall be solved by a dedicated software module [6]. At the same time, each manufacturing task of the lifecycle is included in the overall information and management manufacturing infrastructure, which ensures manufacturing transparency, repeatability, controllability and reliability of the entire production system [10–14].

The technology of making a digital twin for actual manufacturing includes the creation of digital models for the device, technological operations, equipment, and facilities as well as their breaking down to individual components and electromechanical nodes, for which the links and kinematic implementation patterns will be further introduced [15–17]. Such an approach was implemented in works [11, 18], covering the issue of digital manufacturing modelling of the I level electronics modules for PCBs.

In this context the actions of the personnel are digitally simulated as well using anthropomorphic dummies, allowing for simulating support operations of the technological process (such as above all, order composition, equipment adjustment, in-process relocations and loading/unloading). Such dummies are parts of the digital twin, and they help to simulate manual assembly operations as well (such as solder paste application, component installations, mechanical fixation and attachment, if talking in the context of the example mentioned above). It also facilitates an assessment of operations feasibility, ergonomics and time cost. Works [19, 20] cover initial steps of this approach.

3 Methods

The implementation of the approach requires a unified environment of automated design, modelling and manufacturing preparation, where all the models, as mentioned earlier, can be gathered in one information field.

CAD and PLM systems market analysis results show that the Tecnomatix system by Siemens PLM under Teamcenter PLM system control can be used as such a complex environment [10, 11]. Figure 2 shows the merging of the device, technological process and manufacturing models within a common complex structure. This is a modular system that breaks the task of digital manufacturing creation down to subtasks, which are solved by individual modules [12]. Let us list the subtasks of the electronics assembly manufacturing (see Fig. 2). Opposite each task, there is a relevant Tecnomatix module responsible for its resolution.

Let us focus in detail on the subtasks
Tecnomatix/Teamcenter database information content. Before modelling the manufacture, the database needs to be filled with information of the device to be manufactured and technologies, equipment, accessories, etc. to be used

Three major data groups for a standard I level electronics module on a PCS are as follows (Fig. 3)

- Electronic copy of the design documentation package (including the frozen PCB design in ECAD with routing, components allocation and designation, the PCB work drawing, the device assembly drawing, specification);
- List of components (if several design options are available) and product production plan (Fig. 1);

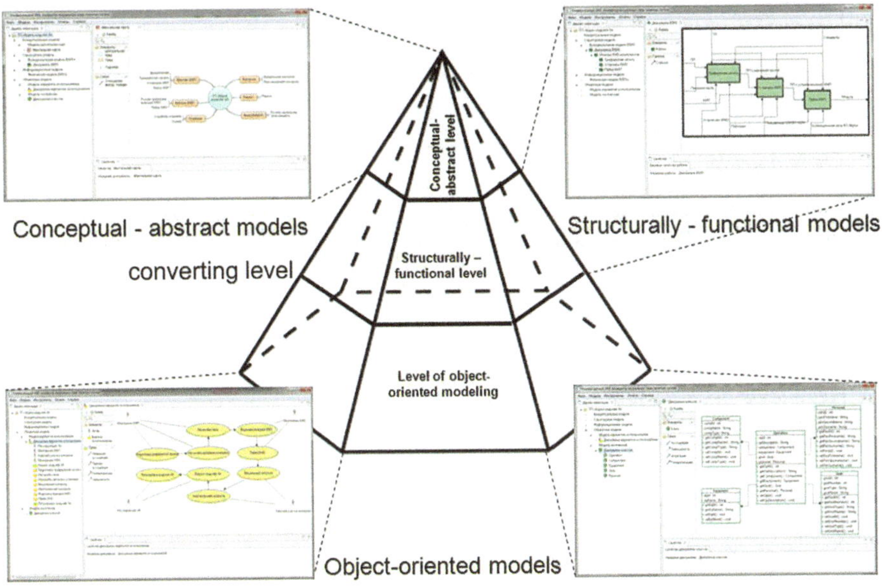

Fig. 1. The pyramid of the relationship between visual methods by the level of expertise.

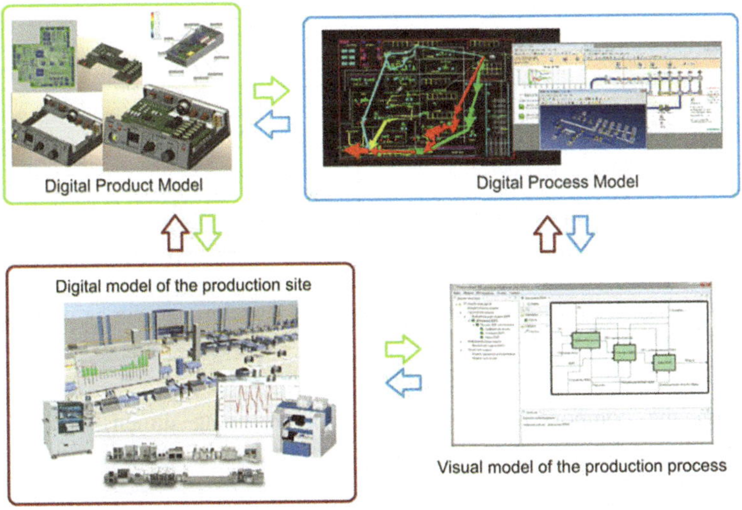

Fig. 2. Interaction of device, technological process and manufacturing models within the common complex structure.

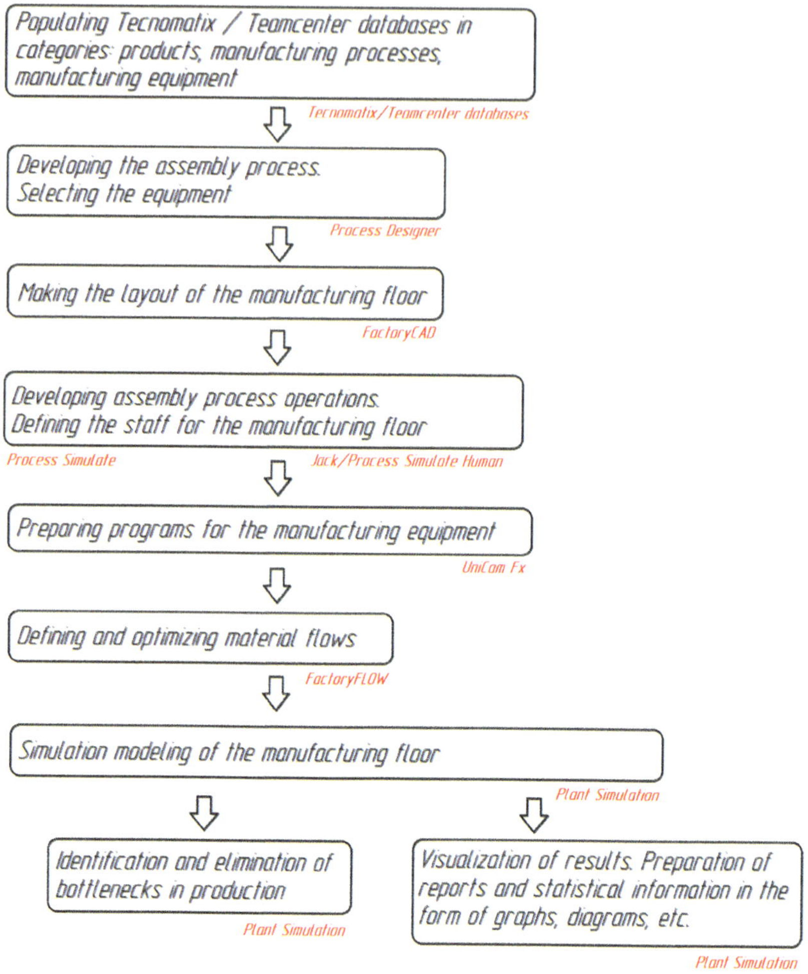

Fig. 3. A sequence of tasks on creating the electronics digital manufacturing model based on Tecnomatix/Teamcenter systems.

- Data on the components (electronic parts), including dimensions, tolerances, packing types for automated assembly, recommended technological materials, solder types etc.;

  Information on the device is stored in the Teamcenter PLM system database. Tecnomatix system database in its turn shall contain the following information:

- Standard and group technological processes for the device class in question;
- Standard allocation schemes of the manufacturing sites with the communication lines;
- Models (3D including) of the technological equipment within the group technological processes with the specified kinematic diagram and a parameter list

(dimensions, capacity, accuracy, reliability, adjustment time etc.) and elements of electrical/mechanical interface for operation within the automated process line;
- Models of standard technological facilities: feed trolleys, feeders (belt, vibration, tray) and vacuum placement heads for pick-and-place machines, card frames and solder paste templates, wave and selective soldering pallets, a bed-of-nails fixture for functional control etc.;
- Models of work stations, manual tools and appliances (for forming, bundling etc.);
- Models of process containers;
- Models of materials in their respective packings – solder alloys and solder pastes (slugs, rollers, cans, one-shot pumps, cartridges etc.), fluxing agents (one-shot pumps, canisters etc.), wash and solder liquids for vapour phase reflow (canisters), moisture-protective materials (one-shot pumps, canisters) etc.;
- Data on manufacturing organisation, production standards and labour safety.

Models of non-automated and automated storage areas of components, end products, materials, fixtures, including dry storage boxes for components and PCBs etc.

Let us limit the standard set of operations for manufacturing of I level electronics module on a PCS:

- Template printing/dosing;
- Components placement;
- Soldering (wave, reflow, vapour-phase, selective etc.);
- Automated optical inspection;
- Washing/drying;
- Application of moisture-protective coatings;
- Visual inspection;
- X-ray inspection;
- Functional check-out;
- Support operations – transportation, accumulation/issue, PCBs turn around etc.;
- Manual/semiautomatic operations – solder paste preparation, lead forming, dosing, mounting, visual inspection, packing.

4 Result

Let us consider the example of a components placement device – Topaz flexible assembly machine by Philips/Assembleon. Figure 4 shows the device modelled in Solid Edge design CAD system.

The digital model is representative of the operation and movement of the machine nodes (conveyor, assembly head and feeders), and PCB and components as per the equipment kinematics diagram specified in Tecnomatix system. The rate and acceleration of each kinematics axis of the machine can be set, so that together with the setting the speed of manual adjustment (feeder replacement, head replacement etc.) the operation cyclogram can be implemented and then a particular device assembly productivity can be defined.

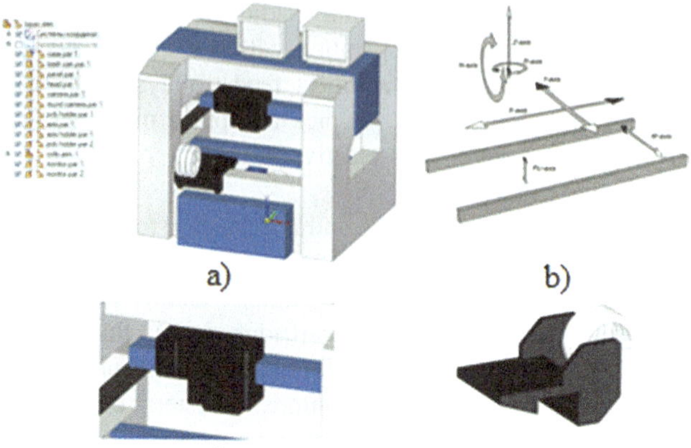

a) b)

c) d)

Fig. 4. Digital model of Topaz flexible assembly machine by Philips/Assembleon: a) Solid Edge CAD system assembly model; b) movements system of the actuators by axes; c) assembly head model; d) model of a feeders with rollers module

Development of an assembly process flowchart. Equipment designation. These tasks are solved in Process Designer module – a 3D environment for manufacturing processes planning. An automated generation of the assembly pattern and process flowchart based on standard/group technological processes from the database is the essence of the task solving. Several alternative structures can be generated and the best option selected based on specific criteria. When the operations are modelled, they are divided into automated and manual. The selected equipment is gathered in one process line.

Manufacturing site arrangement. Then the 3D environment uses Factory CAD module to generate and visualise the manufacturing site layout, to arrange the equipment and other manufacturing environment objects in 3D, to connect communication lines (in our case they are a supply of electric power, compressed air, nitrogen and outlet ventilation).

Operation process development. Definition of the personnel needed on the manufacturing site. 3D-design and modelling of the operations and transitions of the entire process of the device assembly considering the equipment, tools and appliances. For each operation, the appliances and materials are selected, the required number of personnel and group of operations are defined, and then the personnel is placed appropriately with their responsibility areas being defined. These tasks are solved by Process Simulate module. Kinematic 3D-modelling of each operation helps to identify any conflicts (static and dynamic collision of actuators with other objects) and to resolve them.

Manual operations (such as assembly, inspection, in-process relocation and operations with storage area, loading/unloading, packing, feeder readjustment) and repair operations are modelled using the dedicated Jack and Process Simulate Human modules. These modules simulate operator actions in anthropometrically and biometrically representative model of a human, analyse working conditions, productivity, ergonomics, injury risk, comfort, direct visibility areas, energy consumption, load limits considering the angle and field of view and build accessibility areas for express verification of the work station arrangement quality.

Preparation of control programs. An essential phase of any electronics assembly manufacturing is the equipment automation programming (programming of the automated placement machines above all). UniCamFx module can import a project from ECAD system and generate a control program automatically. While for the manual mounting operation, this module generates process flow documentation and user manuals. It can also optimise the sequence of the component placement and perform final optimisation of the feeders and assembly head adjustments.

Factory FLOW module specifies and optimises materials distribution and logistics (of the components, materials, appliances, unfinished manufacturing, end product, gases, energy, etc.). Using the results of the automated analysis, a technology expert may optimise the arrangement of the manufacturing sites based on the distribution, frequency and rate of the materials distribution flows. The module can solve such tasks as materials distribution flows optimisation and identification of problematic spots of the manufacturing, calculation of the transportation equipment requirements, operator movements analyses and routes check, routes generation based on the requirements to the materials relocation, optimisation of the used containers and their storage allocation.

As soon as the above-mentioned procedures are completed, the manufacturing site operation simulation modelling can start. In the Tecnomatix module, this task is solved by the Plant Simulation module – a discrete simulation modelling tool. This module creates structural, hierarchical models of the manufacturing objects, lines and processes and performs experiments and drills scenarios of What If type without any interference with the existing manufacturing systems operation. It also calculates the time (for main, support and in-process service, etc.) automatically and defines one-time, lead, estimate time by operations. This produces a picture of the overall productivity of the process line.

This module has a set of analytic tools aimed at identification and correcting bottlenecks of the manufacturing. For example, it assesses the practicability of dividing the equipment into areas of fast speed chip shooters and flexible multipurpose machine; calculates the feasibility of using dual-lane reflow ovens; and weights the advantages and disadvantages of integration of an automatic optical inspection system in the process line versus using it as stand-alone equipment.

Another critical task is to facilitate the introduction of any changes in the technological process, manufacturing site, material or human resources, etc. with further analysis of the production system response to such changes.

This module also visualises the results, i.e. generates reports, statistical data, plots and diagrams, including data on the equipment utilisation ratio, outage periods, equipment maintenance and repair needed to assess various production scenarios. Such plots and diagrams can be integrated directly into the 3D-environment for better clarity (see Fig. 5).

After all the mentioned modules complete their job, the digital twin of the actual manufacturing will be generated. Figure 6 gives an example of a manufacturing structure 3D-visualisation.

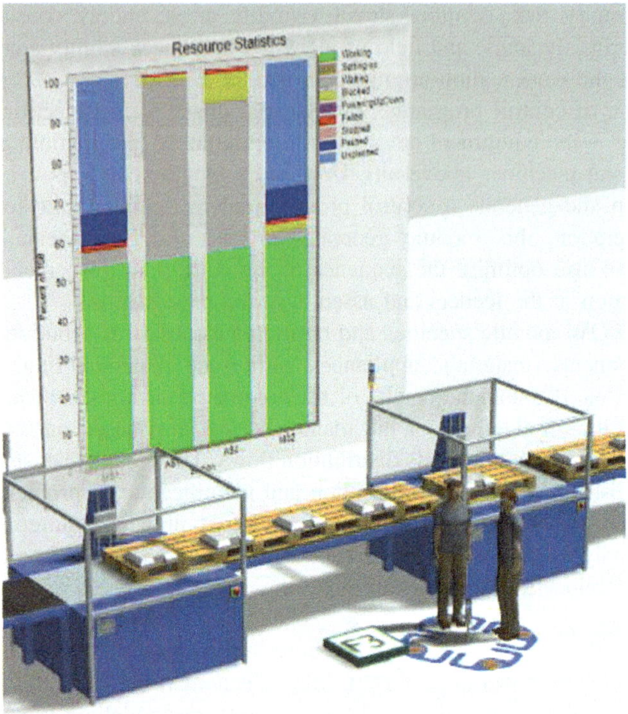

Fig. 5. 3D-diagram of the equipment resource utilisation, integrated into the 3D model of the manufacturing (PlantSimulation) [13]

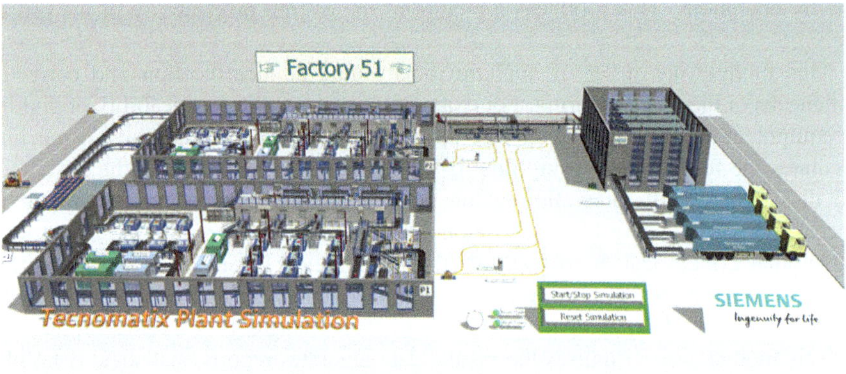

Fig. 6. Example of manufacturing structure 3D-visualisation generated in PlantSimulation module [13].

The model created in such fashion allows considering various scenarios of What If type, to de-bottleneck the manufacturing efficiency and to ensure high effectiveness of the site and to maintain a balance of the technological equipment and work station loadings. The arrangement of the production site can be analysed in digital form only without costly life experiments. It is worth noticing that this model applies to fully- and semi-automated, mechanised and manual operations involving not only technological equipment and accessories but personnel as well.

5 Discussion

Above all, the prospects of usage and further development of the proposed methodology suggest its sophistication and update, making it as much as possible representative of a real-life manufacturing site.

To be able to do so, the uploaded kinematics will need to be elaborated and specified more precisely: not only constant rates but also accelerations and dwell times will need to be uploaded. The flexibility of the model application to various device manufacturing will require to introduce different component compositions. Additional features will have to be introduced for parallel implementation of operations, adjustment of the technological equipment and accessories, orders composition. A full-scale model of storage area for components, semi-finished and end products should be created, and the statistical model of the equipment failures introduced.

After the introduction of the above-mentioned sophistications the model will be able to project more precisely the site productivity and its capabilities to respond to internal and external factors, such as failures, production schedule variations, unscheduled adjustments and maintenance of the equipment and human factor contribution.

6 Conclusion

The authors developed the approach enabling digital modelling of the electronics manufacturing systems based on Tecnomatix solutions by dividing the overall purpose into subtasks that are solved by dedicated software modules. The resulting digital twin of actual manufacturing considers the design information of the produced device and the data of the existing production processes and equipment. It can be used to prepare, analyse and optimise the electronics manufacturing both at the phase of new manufacturing launch and introduction of new products in the existing manufacturing systems.

Acknowledgements. Some results were obtained in the framework of the State task #0705-2020-0041 “Component’s digital transformation methods’ fundamental research for micro- and nanosystems”.

References

1. Shakhnov, V.A., Kurnosenko, A.E.: Modelling of the digital electronics manufacturing in the context of Industry 4.0 concept. In: Akberdina, V.V. (ed.) Proceedings of the 1st International Scientific and Practical Conference Digital Transformation of the Industry: Trends, Management, Strategies, pp. 585–594. Ural State University of Economicis, Ekaterinburg, Russia (2019)
2. Berduygina, O.N., Vlasov, A.I., Kuzmin, E.A.: Investment capacity of the economy during the implementation of projects of public-private partnership. Investment Manag. Financ. Innovations **14**(3), 189–198 (2017)
3. Akberdina, V.V., Kalinina, A.V., Vlasov, A.I.: Transformation stages of the Russian industrial complex in the context of economy digitalization. Probl. Perspect. Manag. **16**(4), 201–211 (2018)
4. Ivanov, Y.V., Kurnosenko, A.E.: Kontur software complex for design of the automated electronics assembly complexes in multiproduct manufacturing. Eng. Bull. **3**, 4 (2013)
5. Kurnosenko, A.E., Arabov, D.I.: Optimization of electronic components mounting sequence for 3D MID assembly process. Breakthrough directions of Scientific Research in NRNU MEPhI: Development Perspectives in the Framework of the Strategic. “KnE-Engineering”, pp. 311–321 (2018)
6. Demin, A.A., Vlasov, A.I.: Visual methods of formalization of knowledge in the conditions of the synchronous technologies of system engineering. In: Proceedings of the 13th Central & Eastern European Software Engineering Conference in Russia (CEE-SECR’17), art. 4, pp. 1–7. Association for Computing Machinery, New York, NY, US (2017). https://doi.org/10.1145/3166094.3166098
7. Yudin, A., Kolesnikov, M., Vlasov, A., Salmina, M.: Project oriented approach in educational robotics: from robotic competition to practical appliance. Adv. Intell. Syst. Comput. **457**, 83–94 (2017)
8. Popova, I.N., Vlasov, A.I., Nikitina, N.I.: Optimization of inventory distribution logistics in industrial enterprises. Espacios **39**(24), 16 (2018)
9. Kurnosenko, A.E., Haritonov, K.P.: Use of teamcenter PLM system to control lifecycle of electronics. Inform. Technol. Des. Manuf. **2**(170), 56–62 (2018)
10. Kurnosenko, A.E., Haritonov, K.P.: Use of teamcenter device data control system for manufacturing technology preparation. Inform. Technol. Des. Manuf. **4**(172), 42–48 (2018)
11. Levin, I.V., Kurnosenko, A.E., Mashina, N.A.: Solutions for design, modelling and optimisation of electronics manufacturing. Inform. Technol. Des. Manuf. **1**(169), 26–32 (2018)
12. PiepenbrockGeorg Tecnomatix Plant Simulation. Validation of Plant Performance and Plant Control. Siemens Industry Software. https://www.plm-europe.org/admin/presentations/2017/2013_PLMEurope_24.10.17-13-30_GEORG-PIEPENBROCK_SPLM_validation_of_plant_performance_and_plant_control_with_tecnomatix_plant_simulati.pdf. Accessed 25 June 2020
13. Adamova, A.A., Adamov, A.P., Vlasov, A.I.: Leading approaches to energy solutions in the 21st century. Adv. Soc. Sci. Educ. Human. Res. **386**, 190–196 (2019). https://doi.org/10.2991/assehr.k.191221.197
14. Berdyugina, O.N., Krivoshein, A.I., Vlasov, A.I.: Technological platform for innovative social infrastructure development on basis of smart machines and principles of internet of things. In: 2018 Global Smart Industry Conference (GloSIC), pp. 1–7. IEEE, Chelyabinsk, Russia (2018)

15. Ragulina, Y.V., Semenova, E.I., Zueva, I.A., Kletskova, E.V., Belkina, E.N.: Perspectives of solving the problems of regional development with the help of new internet technologies. Entrepreneurship Sustain. Issues **5**(4), 890–898 (2018). https://doi.org/10.9770/jesi.2018.5.4(13)
16. Toomsalu, L., Tolmacheva, S., Vlasov, A., Chernova, V.: Determinants of innovations in small and medium enterprises: a european and international experience. Terra Economicus **17**(2), 112–123 (2019)
17. Prudius, A.A., Karpunin, A.A., Vlasov, A.I.: Analysis of machine learning methods to improve efficiency of big data processing in Industry 4.0. J. Phys. Conf. Ser. **1333**, 032065 (2019). https://doi.org/10.1088/1742-6596/1333/3/032065
18. Kurnosenko, A.E., Levin, I.V., Semeniakina, V.O., Zaharov, E.R., Sherstiuk, A.E.: Methods and means of simulation modelling for studying the technical preparation of electronics manufacturing. In: Collected Scientific Works of International Forum Digital Technologies in Engineers' Education: New Trends and Implementation Experience, pp. 310–314. Bauman Moscow State Technical University, Moscow, Russia (2020)
19. Nikolskii, T.V., Lomachenko, A.S., Kurnosenko, A.E.: Use of Jack/Simulate human module for simulation modelling of the electronics assembly operations. Inform. Technol. Des. Manuf. **2**(174), 21–25 (2019)
20. Kurnosenko, A.E., Nikolskii, T.V., Lomachenko, A.S.: Preparation of the equipment model for simulation modelling in a CAD system. Inform. Technol. Des. Manuf. **2**(178), 24–30 (2020)

Ontological Model for Risks Assessment of the Stages of a Smart-Technology for Predicting the "Structure-Property" Dependence of Drug Compounds

Galina Samigulina[1(✉)] and Zarina Samigulina[2(✉)]

[1] Institute of Information and Computing Technologies, Almaty, Kazakhstan
kz.galinasamigulina@gmail.com
[2] Kazakh-British Technical University, Almaty, Kazakhstan
samigulinaresearch@gmail.com

Abstract. Modern researches in the field of medicine and pharmacology is aimed at studying new diseases that threaten humanity and at the search for medicinal compounds for their treatment. A strategically important step is the development of innovative techniques for the computer-aided molecular design of drugs based on the latest achievements in artificial intelligence. The search for new drugs is complicated by the need to analyze a large amount of chemical information and by the increased requirements for computing resources. In this regard, a promising area of research is the artificial immune systems approach and the development of modified algorithms based on it, which allows to process large descriptors databases of drug compounds. A Smart - technology for predicting the "structure-property" dependence of drug compounds is presented in the article. The technology consists of two stages: selection of informative descriptors and solving the prediction problem based on modified algorithms of artificial immune systems. There was developed an ontological model for assessing the risks of the Smart-technology for predicting the "structure-property" dependence of drug compounds using the FMEA (Failure Mode and Effects Analysis) methodology in order to eliminate the potential risks of errors and to improve the quality of the prediction. The ontological model is implemented using Protégé software.

Keywords: Drug design · Prediction of the "structure-property" dependence · Modified immune network algorithm · Smart-technology · Ontological model · FMEA

1 Introduction

In recent years, there has been an increase in researches aimed at the use of artificial intelligence (AI) in the field of pharmacology. Leading international publishers such as Springer Nature, BMC, Elsevier, Scientific American, Apress publish articles on artificial intelligence in medicine and pharmacology. Recent achievements in the field of AI have been successfully used for the computer molecular design of new drugs. A feature of creating new drugs is the need to process a large set of descriptor data

R. Silhavy et al. (Eds.): CoMeSySo 2020, AISC 1295, pp. 876–886, 2020.
https://doi.org/10.1007/978-3-030-63319-6_81

describing the properties of drug compounds. With the help of AI, the tasks of selecting informative descriptors and predicting the "structure-property" dependence of drug compounds (Quantitative Structure Activity Relationship, QSAR) are solved. Since the prediction quality depends on the initial data sample, meta-hereditary swarm intelligence algorithms are widely used to solve the problem of selecting informative descriptors, such as: genetic algorithms (GA) [1], particle swarm (PSO) algorithms [2], ant colony algorithm (ACO) [3], grey wolves optimization method (GWO) [4], flower pollination method (FPA) [5], etc.

When developing new drugs, prediction models must meet the requirements of reliability, high efficiency and a sufficient data processing speed. Currently, neural networks have been successfully used for QSAR modeling. For example, in [6], the application of the Deep Learning model with the Long Short-Term Memory Neural Network is considered to create good quality QSAR models. The article [7] provides an overview of the use of machine learning algorithms to create new drugs. Prediction models based on the naive Bayesian algorithm, decision trees, random forest, and support vector method are considered.

Of particular interest is the approach of artificial immune systems (AIS) and modified algorithms based on it. For example, in [8], there is presented an improved algorithm for clonal selection based on the Biological Forgetting Mechanism. The article [9] considers a modified pattern recognition algorithm by the artificial immune system (Artificial Immune Recognition Systems, AIRS3). Studies [10] are devoted to the development of a hybrid clonal selection algorithm with modified combinatorial recombination and adaptive mutation to solve problems of numerical optimization.

During the development of modified AI algorithms for the creation of new drug compounds, it is necessary to ensure high accuracy of the prediction and to exclude possible modeling errors. In this regard, the development of risk assessment models aimed at improving the efficiency of modified AI algorithms in predicting the structure-property dependence of drug compounds is relevant. Currently, there are many effective risk assessment methodologies, such as: FMEA (Failure Mode and Effects Analysis), PFMEA (Process Failure Mode and Effects Analysis), FTA (Fault Tree Analysis), etc. In the field of pharmacology, the FMEA approach has been successfully applied. For example, in [11], there was developed a modified FMEA approach to assess the risks of the medical software functioning based on the fuzzy logic approach. Article [12] is devoted to the application of the FMEA methodology in the development of new drugs. Using the FMEA approach, the process of pharmaceutical research is visualized in order to identify technological defects and to increase the efficiency of the search for new drug compounds.

Therefore, the development of new innovative methods for searching for new drug compounds candidates based on the latest achievements in the field of AI and the FMEA risk assessment methodology is an urgent task.

The following structure of the article is proposed. Section 2 presents the formulation of the research problem. Section 3 is devoted to research methods. Section 4 describes the modeling results. The final section contains conclusions.

2 Statement of the Problem

The aim of the research is to develop an innovative technology based on bio-inspired AI algorithms to create new drugs that can meet the following requirements: reliability, high accuracy and ease of implementation.

The statement of the research problem is formulated as follows: it is necessary to develop an ontological model for assessing the risks of the functioning stages of the Smart-technology for predicting the "structure-property" dependence of drug compounds based on modified algorithms of artificial immune systems using the FMEA methodology.

3 Methods

The complexity of developing new drug compounds lies in the interdisciplinarity of researches. The following approaches are used in the work: QSAR modeling, ontological approach, FMEA methodology, bio-inspired approach of artificial immune systems.

3.1 QSAR Modeling

For the first time, researches on the quantitative relationship between structure and biological activity based on mathematical methods appeared in the middle of the last century. Then, the progress in the field of IT technologies led to the creation of a computer molecular design of chemical compounds (Computer Aided Molecular Design, CAMD) [13]. Computer molecular drug design (Computer Aided Drug Design, CADD) is one of the areas of CAMD [14]. Using CADD, the following tasks are solved: target identification, optimization, validation, selection of drug candidates, etc. [15]. When creating new drugs, there are widely used the methods on studying the "structure – property/activity" dependence of chemical compounds - QSPR (Quantitative Structure Property Relationships) and QSAR (Quantitative Structure Activity Relationships). The molecular structure of the compounds is described using descriptors of various levels. The selection of informative descriptors is one of the main problems in creating an adequate QSAR model for further prediction of the "structure-property" dependence of drug compounds.

3.2 Ontological Approach

The development of new drugs is a complex field of research, the formal presentation of knowledge in which is an important task. The development of ontological models is a convenient tool for describing the subject area [16], the creation of which uses universal knowledge representation tools. Since the computer molecular drug design is complicated by the interdisciplinarity of researches, ontologies in this area are created by teams of specialists in various subject areas (bioinformatics, pharmacology, chemistry, etc.) and have a modular structure. For example, in [17], an ontological model of a drug-target is considered for classifying and integrating data at searching for new drugs.

Currently, there have been developed a large number of ontologies for computer molecular drug design, such as: BioAssay Ontology, Library of Integrated Network-Based Cellular Signatures (LINCS), Drug Target Ontology [18], as well as software products for their implementation. Application packages Protégé, Ontolingua, Semantic Web, etc., are widely used. Current researches use the Protégé software product to create an ontological model for assessing the risks of the stages of the Smart-technology functioning.

3.3 FMEA Methodology

Currently, a large number of models have been developed for risk assessment. An effective methodology is the FMEA approach, firstly developed in the United States. The FMEA model is widely used in medicine and in the pharmaceutical industry [12] for the timely detection of defects in the development of new drug compounds in pharmaceutical researches. The Fig. 1 shows the general model of FMEA.

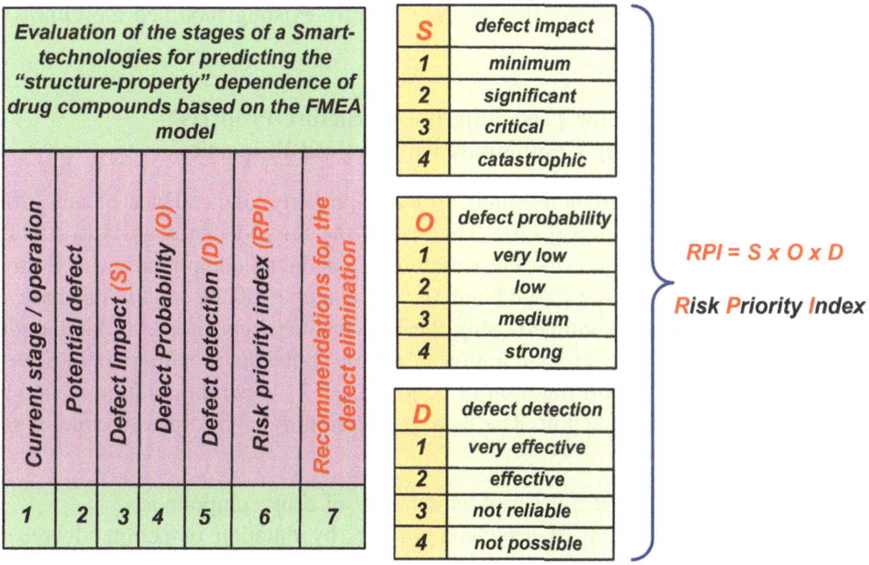

Fig. 1. FMEA risk assessment model

The model consists of the following concepts [19]:

- «Severity» - the degree of significance of the consequences of failure (ranged in the interval $S = \{1, \ldots, n\}$, $n = 4$);
- «Occurrence» - the probability of occurrence of the cause of the defect (ranged in the interval $O = \{1, \ldots, m\}$, $m = 4$);

- «Detection» - the probability of detecting a defect, the causes of their occurrence and the impact of consequences on the entire system (ranged in the interval $D = \{1, \ldots, k\}$, $k = 4$).

The definition of a risk priority index is introduced (RPI).

Remark 1. The risk priority index is a quantitative assessment of complex risk. Calculating this parameter, if for several failures the RPI assumes the same value, the risk is considered as the most serious, for which the criterion "Severity" has the highest value.

3.4 Artificial Immune Systems

Currently, the bio-inspired approach of artificial immune systems and modified algorithms based on it are of great interest for researches in the field of creating new drugs. Artificial immune systems are based on the functioning mechanisms of the human immune system. There are AIS algorithms based on clonal selection, negative selection and the immune network. The advantages of AIS are the ability to make decisions in an unfamiliar situation, the properties of adaptability, autonomy and memory.

Therefore, the development of the general theory of AIS on the basis of an integrated approach and systematization of the currently existing modified algorithms is relevant today.

3.5 Smart-Technology for Predicting the "Structure-Property" Dependence of Drug Compounds with Desired Properties

A Smart-technology for predicting the "structure-property" dependence of new drug compounds has been developed [20]. The aim of the Smart-technology is to analyze descriptor databases that describe the structure of candidate compounds with desired properties. The studies were carried out on the example of the author database of sulfanilamide descriptors compiled using the Mol-Instincts repository. Sulfanilamides are antibacterial drugs of various duration of action. 3 classes of sulfanilamides were distinguished: class 1 - short duration of action, class 2 - medium duration of action, class 3 - long duration of action. The enlarged algorithm of the Smart-technology is presented below:

STAGE 1. Formation of a descriptors database of drug compounds.

STAGE 2. Classification of drug compounds (by duration of action, degree of toxicity, etc.).

STAGE 3. Selection of informative descriptors based on meta-heuristic AI algorithms (genetic algorithm, grey wolf optimization method, flower pollination method, etc.).

STAGE 4. Evaluation of the effectiveness of algorithms for selecting informative descriptors. Formation of an optimal data set based on modeling results.

STAGE 5. Prediction of the properties of sulfanilamide drug compounds based on modified AIS algorithms [20, 21].

STAGE 6. Evaluation of the prediction efficiency based on the estimates: accuracy, classification error, precision, recall, sensitivity, specifity, f-measure,

AUC, ROC-area. Evaluation of the modified AIS algorithm based on the properties of homologous proteins [22].
STAGE 7. Selection of drug compounds candidates with desired properties.

A feature of the developed Smart-technology is the prediction of the "structure-property" dependence of drug compounds based on modified AIS algorithms [20–22].

4 Ontological Model for Risk Assessment of the Stages of a Smart-Technology Based on FMEA

The Fig. 2 shows the developed ontological model for risk assessment of predicting the properties of drug compounds using a Smart-technology in the Protégé ontology editor, and also shows the taxonomy of model classes based on the FMEA approach.

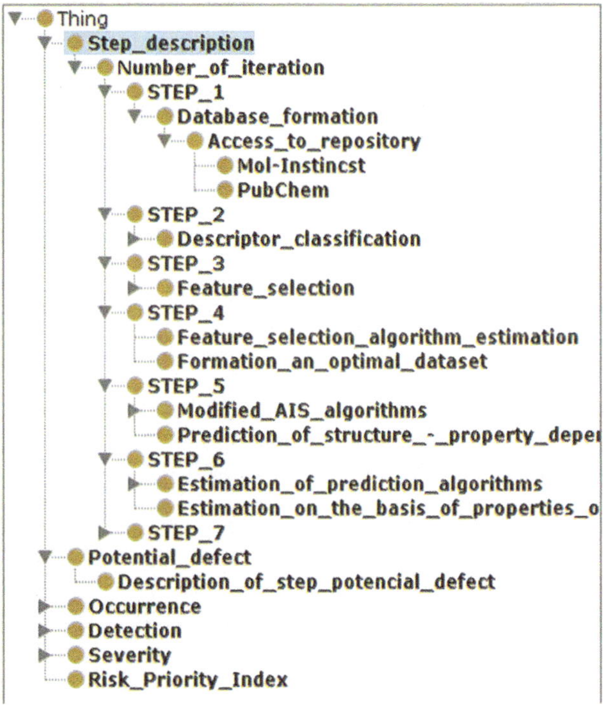

Fig. 2. Taxonomy of classes of an ontological model for risk assessment of the stages of a Smart-technology based on the FMEA approach

The ontological model includes an FMEA-based assessment of each stage of the developed Smart-technology for predicting the "structure-property" dependence of drug compounds and can be used as a knowledge base for software implementation of a Smart-technology based on a multi-agent approach [21, 23]. The Fig. 3 shows a visualization of the ontological model for risk assessment in the Protégé environment using the OntoGraf tool.

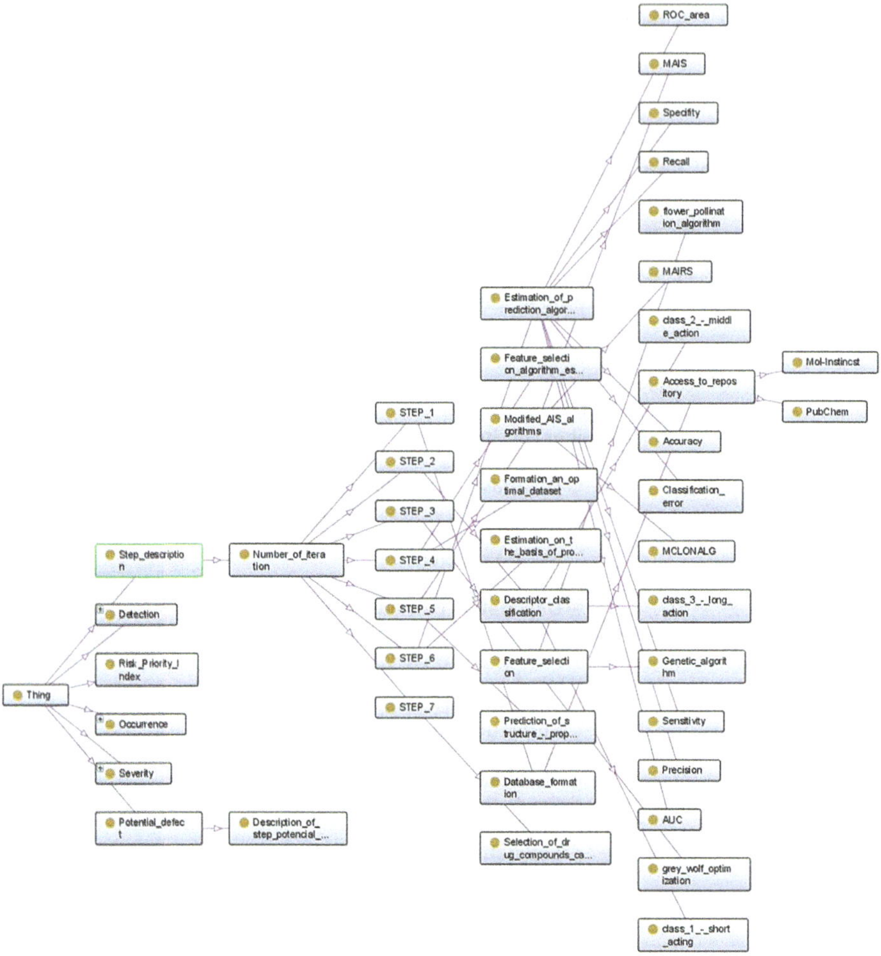

Fig. 3. Visualization of the ontological model for risk assessment of the stages of a Smart-technology

4.1 FMEA-Based Risk Assessment Example

Let consider a risk assessment example based on the FMEA approach. Table 1 presents a fragment of the sulfanilamide descriptor database. The database contains 2005 descriptors of various levels (molecular descriptors, topological indices, quantum chemical descriptors, etc.) and consists of 30,075 data copies. The database fragment under consideration contains the following descriptors: nAT - number of atoms; MW - Molecular weight; AMW - average molecular weight; Sv - sum of atomic van der Waals volumes (scaled on Carbon atom), etc.

Table 1. Fragment of the sulfanilamide database

Drug number	nAT	MW	AMW	Sv	…	Class
Drug 1	27.00	250.275	9.270	17.880	…	Short_acting
Drug 2	33.00	278.328	8.440	21.080	…	Short_acting
Drug 3	31.00	267.301	8.620	19.590	…	Short_acting
Drug 4	27.00	270.324	10.010	17.970	…	Short_acting
Drug 5	46.00	311.441	6.770	26.670	…	Short_acting
Drug 6	33.00	278.328	8.440	21.080	…	Short_acting
Drug 7	46.00	311.441	6.770	26.670	…	Short_acting
…	…	…	…	…	…	…
Drug 2005	24.00	214.238	8.930	15.000	…	Medium_act

The Fig. 4 presents the AUC prediction characteristic based on clonal selection algorithms (CLONALG and CSA), using the Artificial Immune Recognition System (AIRS) algorithm and an algorithm based on immune network modeling (AIS) [21]. According to the modeling results, it is clear that the most effective algorithm is the one that tends to the upper left corner, that is, the value of the AUC characteristic is close to 1.

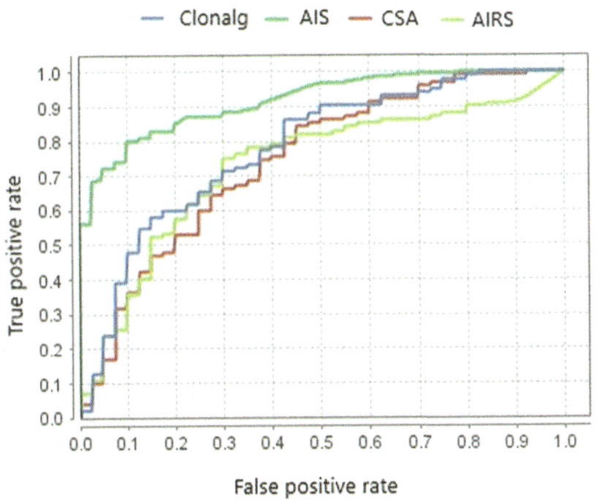

Fig. 4. Prediction of the “structure-property” dependence of drug compounds based on modified AIS algorithms

A modified algorithm based on immune network modeling [20–22] showed the best prognostic result.

A risk assessment example based on the developed ontological model and the FMEA approach is presented on Fig. 4. The table contains information about the stage of the Smart-technology functioning and a potential defect. Then, there is described the name of the failure and the possible cause of its occurrence. According to the evaluation criteria considered in Sect. 3, experts assess the degree of criticality of the defect.

The values for the columns “Severity”, “Occurrence” and “Detection” are filled. The value of the RPI is calculated. The maximum value of the coefficient:

$$RPI = 4 \times 4 \times 4 = 64$$

The lower the value of the coefficient, the less significant the risk for the Smart-technology functioning (Fig. 5).

№ of the stage	Defect description	The cause of a defect	Severity (significance of a defect)	Occurrence (occurrence of a defect)	Detection	RPI index	RISK REDUCTION ACTIONS
STAGE 1	Missing data	parameters beyond detection	4 catastrophic	3 medium	2 effective	24	1. Checking the health of measuring instruments. 2. Filling in the missing data with. 3. Filling with the main component.
STAGE 1	Noisy data	electric and magnetic fields	3 critical	3 medium	2 effective	18	1. Suppression of noise in the source. 2. Noise reduction in the receiver.
STAGE 1	Error in descriptor calculation	human factor	4 catastrophic	4 strong	3 not reliable	48	1. Recalculation of descriptors. 2. Conducting additional experiments. 3. Model validation
STAGE 1	Data redundancy	the result of the data modification operation	2 significant	4 strong	2 effective	16	1. Control over data modification. 2. Check for duplication.
STAGE 2	The assignment not in proper class	human factor	4 catastrophic	3 medium	3 not reliable	36	1. Attracting several experts.
STAGE 3	Selection of uninformative descriptors	model error	4 catastrophic	4 strong	2 effective	32	1. Validation of models. 2. Rebooting the software. 3. Change the model parameters.
...	...	...	...	...	...	...	...

Fig. 5. FMEA risk assessment example for a Smart-technology for predicting the “structure-property” dependence of sulfanilamides

5 Discussion

The developed ontological model for risk assessment of the Smart-technology functioning based on the FMEA methodology has the following advantages:

- The modular structure of the ontological model allows to add additional artificial intelligence algorithms to search for new medicinal compounds.
- The ontological model allows to perform automated knowledge processing at a semantic level.
- The ontological model is available to researchers who do not have programming skills and can be used as a knowledge base in software development.

The practical value of the work lies in the fact that the proposed Smart-technology allows to quickly select the desired modified algorithm of artificial immune systems, which is most suitable for a specific data set and for an effective solution to the problem. Evaluation of the effectiveness of the developed Smart-technology based on the FMEA methodology eliminates the potential risks of errors and improves the prediction quality.

The work was carried out under a grant of the National Academy of Sciences of the Republic of Kazakhstan on the topic: "Development and analysis of databases for the information system for predicting the "structure-property" dependence of drug compounds based on artificial intelligence algorithms" (2018–2020).

References

1. Wutzl, B., Leibnitz, K., Rattay, F., Kronbichler, M., Murata, M., Golaszewski, S.M.: Genetic algorithms for feature selection when classifying severe chronic disorders of consciousness. PLoS ONE **14**(7), 1–16 (2019)
2. Qiu, C.: A novel multi-swarm particle swarm optimization for feature selection. Genet. Program Evolvable Mach. **20**(4), 503–529 (2019). https://doi.org/10.1007/s10710-019-09358-0
3. Dwivedi, R., Kumar, R., Jangam, E., Kumar, V.: An ant colony optimization based feature selection for data classification. Int. J. Recent Technol. Eng. **7**(5S4), 35–40 (2019)
4. Gao, Z., Zhao, J.: An improved grey wolf optimization algorithm with variable weights. Comput. Intell. Neurosci. **1**(3), 1–13 (2019)
5. Abdel-Basset, M., Shawky, L.A.: Flower pollination algorithm: a comprehensive review. Artif. Intell. Rev. **52**(4), 2533–2557 (2018). https://doi.org/10.1007/s10462-018-9624-4
6. Chakravarti, S.K., Alla, S.: Descriptor free QSAR modeling using deep learning with long short-term memory neural networks. Front. Artif. Intell. **2**(1), 1–35 (2019)
7. Priya, N., Shobana, G.: Application of machine learning models in drug discovery: a review. Int. J. Emerging Technol. **10**(3), 268–275 (2019)
8. Yang, C., Chen, L., Jia, L., Wen, H.: Improved clonal selection algorithm based on biological forgetting mechanism. Complexity **10**(1), 10 (2020)
9. Jenhani, I., Elouedi, Z.: Re-visiting the artificial immune recognition system: a survey and an improved version. Artif. Intell. Rev. **42**(4), 821–833 (2012). https://doi.org/10.1007/s10462-012-9360-0
10. Zhang, W., Gao, K., Zhang, W., Wang, X., Zhang, Q., Wang, H.: A hybrid clonal selection algorithm with modified combinatorial recombination and success-history based adaptive mutation for numerical optimization. Appl. Intell. **49**(2), 819–836 (2018). https://doi.org/10.1007/s10489-018-1291-2
11. Batbayar, K., Takács, M., Kozlovszky, M.: Medical device software risk assessment using FMEA and fuzzy linguistic approach: case study. In: Proceeding of 11th IEEE International Symposium on Applied Computational Intelligence and Informatics, pp. 197–202 (2016)
12. Inoue, H., Yamada, S.: Failure mode and effects analysis in pharmaceutical research. Int. J. Qual. Serv. Sci. **2**(3), 369–382 (2010)
13. Batool, M., Ahmad, B., Choi, S.: A structure-based drag discovery paradigm. Int. J. Mol. Sci. **20**(11), 2783 (2019)
14. Surabli, Singh, B.K.: Computer aided drug design: on overview. J. Drug Deliv. Ther. **8**(5), 504–409 (2018)

15. Kore, P.P., Mutha, M.M., Antre, R.V., Oswal, R.J., Kshirsagar, S.S.: Computer-aided drug design: an innovative tool for modeling. Open J. Med. Chem. **2**, 139–148 (2012)
16. Kanza, S., Frey, J.G.: A new wave of innovation in Semantic web tools for drug discovery. Expert Opin. Drug Discov., 1–12 (2019)
17. Lin, Y., Mehta, S., Ginty, H., Turner, J.P., Vidovic, D., Forlin, M., Koleti, A., Nguyen, D., Jensen, L., Guha, R., Mathias, S., Ursu, O., Stathias, V., Nabizadeh, N., Chung, C., Mader, C., Visser, U., Yang, J., Bologa, C., Oprea, T., Schürer, S.: Drug target ontology to classify and integrate drug discovery data. J. Biomed. Simantics **8**, 50 (2017)
18. Küçük-McGinty, H., Lin, Y., Nabizadeh, N., Stathias, V., Vidovic, D., Koleti, A., Mader, C., Duan, J., Visser, U., Schürer, S.: Building concordant ontologies for drug discovery. In: CEUR Workshop Proceedings, vol. 1747, pp. 1–10 (2016)
19. Sharma, K., Srivastava, S.: Failure mode and effect analysis (FMEA) implementation: a literature review. J. Adv. Res. Aeronaut. Space Sci. **5**(1&2), 1–18 (2018)
20. Samigulina, G.A., Samigulina, Z.I.: Modified immune network algorithm based on the Random Forest approach for the complex objects control. Artif. Intell. Rev. **52**(4), 2457–2473 (2018). https://doi.org/10.1007/s10462-018-9621-7
21. Samigulina, G.A., Samigulina, Z.I.: Development of multi-agent technology for prediction of the « structure-property » dependence of drugs on the basis of modified algorithms of artificial immune systems. In: Proceedings of International Work Conference on Bioinformatics and Biomedical Engineering (2018)
22. Samigulina, G.A.: Development of decision support systems based on intellectual technology of artificial immune systems. Autom. Remote Control **73**, 397–403 (2012)
23. Samigulina, G.A., Samigulina, Z.I.: Ontological model of multi-agent Smart-system for predicting drug properties based on modified algorithms of artificial immune systems. Theor. Biol. Med. Modell. **17**(12), 1–22 (2020)

An Iterative Approach for Crowdsourced Semantic Labels Aggregation

Andrew Ponomarev(✉)

St. Petersburg Institute for Informatics and Automation of the Russian Academy of Sciences, St. Petersburg, Russian Federation
ponomarev@iias.spb.su

Abstract. Semantic labeling (describing objects using labels with explicitly defined semantic interpretation, e.g., with a help of ontology) is important for building semantic search systems. However, obtaining semantic labels is a hard and labor-intensive process. One of the approaches to perform semantic labeling is to use crowdsourcing. The use of crowdsourcing, in its turn, usually requires some quality control technique, because labels provided by individual participants may be unreliable. Despite there are many methods for label aggregation, most of them don't take into account semantic relationships that may exist between labels. The paper proposes model, method and algorithm for semantic labels aggregation in crowdsourcing labeling applications. The proposed approach is based on the analysis of semantic relations between ontology classes, and takes into account inaccuracy and incompleteness of the results received from individual participants. It allows one to evaluate both the expected quality of the semantic labels received from individual crowdsourcing participants and the degree of belief of the collected labels. The proposed approach can be employed in semantic labeling crowdsourcing applications, e.g., to structure objects of a certain area in order to ensure semantic search. The only limiting factor in applying the approach is the existence of a pre-designed ontology of semantic relations, written in OWL 2.

Keywords: Crowdsourcing · Crowd computing · Semantic labeling · Label aggregation · Consensus methods

1 Introduction

Semantic information search, taking into consideration not only syntactic constructions describing resources, but also semantics of the resources has great potential and is able to give much more accurate results than simple search based on term analysis. However, to enable semantic search, resources must be described with semantic labels. In recent years, the use of such labels (e.g., with a help of shared vocabularies like schema.org) has become more and more popular. Its main advantage is that it allows you to describe objects (e.g., Internet pages) with tags and attributes that have certain semantics uniformly interpreted by various participants of the information environment. Modern search engines actively use such labels (provided via markup) to obtain more "meaningful" results.

R. Silhavy et al. (Eds.): CoMeSySo 2020, AISC 1295, pp. 887–894, 2020.
https://doi.org/10.1007/978-3-030-63319-6_82

There are two main approaches to perform semantic labeling: automatic and manual. Automatic labeling involves the understanding and analysis of resource information by a computer. With the advances in natural language processing the quality of automatic labeling is constantly improving, but the results are still not perfect. During manual labeling the analysis and structuring of the information content is performed by a human, usually, owner/creator of the resource. This allows one to fully control the set of tags and attributes used in labeling, but the task becomes very difficult in the presence of a large number of objects. Besides, not all resource creators support the necessity of semantic resource labeling.

The complexity of manual labeling (as well as reluctance of some of the resource owners to perform labeling) in some cases can be alleviated with a help of crowdsourcing. Crowdsourcing in general represents the involvement of a large group of people in solving the problems on a voluntary basis with the use of information and communication technologies. Crowdsourcing is now widely used in various fields for tasks that turn out to be relatively difficult for automation, but simple for humans [1–3], and the task of semantic analysis and labeling is one of them.

However, with all the advantages of this technology, its application is associated with a number of difficulties. First of all, the question arises of the competence of participants and the quality of the results obtained with their help [4–6]. In general, quality is understood as the correspondence of the result of a certain task (information collection or processing) obtained via crowdsourcing and the "true" result of this task, which in practice is often the result obtained from a competent "expert" [6]. This problem stems from the general human propensity for errors and incorrect algorithms for evaluating results, and modern crowdsourcing systems usually leverage various mechanisms to ensure the quality of the results (a general overview of such mechanisms is given, for example, in [6]).

In existing quality assurance mechanisms, comparing the results obtained from different participants plays an important role. As a rule, this comparison is binary (the results either match, or not – in the former case they are assumed to also match with the 'true' result, in the latter – at least one is wrong, so further checks are required). This is a universal approach, fully justified in most cases, however, when it comes to semantic labeling, there may be a semantic connection between labels assigned by different participants, that can be taken into account by the quality assurance mechanism. Taking this information into account may allow to compare the results more accurately and obtain more adequate estimates of the individual characteristics of the participants. In particular, the factors that must be considered when reconciling semantic labels are generalization and refinement. For example, the label "Horse" is a refinement w.r.t. the label "Mammal", but does not contradict it. It would be not reasonable to assume that these two labels just mismatch, in fact, they match to the point that the labeled object is a mammal, but one of the labels carries more information.

One of the most common ways of representing semantics in information systems is ontologies, and the most widely used language for describing ontologies is OWL 2 (the specification of which is standardized by the W3C consortium) [7]. OWL 2 language defines many semantic relations that may be leveraged in the process of label aggregation, however, this paper considers only the most important relation serving as a core of any ontology – taxonomy relation (defined in OWL 2 by the object feature *subClassOf*).

The paper proposes an algorithm allowing to 1) aggregate semantic labels of the same object offered by different crowdsourcing participants taking into account possible semantic relationships between these labels defined with a help of OWL 2 ontology, 2) estimate the expected quality of crowdsourcing system participants (e.g., to disable those who provide low quality labelings).

2 Related Work

Quality management of the results obtained using crowdsourcing is an urgent problem that attracts the close attention of researchers in this area. A general perspective of existing quality assurance methods is outlined, for example, in review articles [5] and [6]. So, in the article [6] the following classification of quality assurance methods is given:

- Consensus methods based on the use of redundancy of the collected results to resolve contradictions in them;
- Workflow design methods – an elaboration of consensus methods, which allows, in particular, to consider systems with heterogeneous tasks and a dynamically generated workflow;
- Methods of centralized assignment of work, assuming that the systems is aware of the expected effectiveness (quality of results, task completion time) of each participant and it distributes tasks in such a way as to maximize overall efficiency;
- Game-theoretic methods that allow one to evaluate a particular organization of a crowdsourcing system (validation, rewarding mechanisms) based on the assumption that the participant is a rational agent seeking to maximize his own benefit;
- Methods based on accounting for the specific properties of tasks that allow one to use these properties or their computer processing to increase the final quality of the results of crowd computing;
- Methods based on the analysis of user's actions and the impact of human-machine interface decisions on the user's actions.

According to this classification, the approach proposed in this paper falls into the group of consensus methods in the sense that it does not use additional information other than labels received from various participants. However, most of the existing consensus methods are based on a strict comparison of the results obtained from individual participants, ignoring the relationship between labels. There have been only few attempts to adapt consensus methods to situations where relations between labels can be present. The development of such methods, in particular, is the subject of work [8], the authors of which propose an extension of the DS algorithm and explore different models for representing relations between labels (including a Bayesian network as a compact representation of a joint distribution). In article [9], a probabilistic labeling model for hierarchical classification is proposed, that is, for the situation when the labels that participants assign to objects are classes organized in a hierarchy (classification of books, goods).

However, the results obtained in these papers are not entirely applicable to the problem under consideration: the set of consensus methods proposed in [8] is based on the DS algorithm, which has high computational complexity, therefore, its applicability for labeling using large ontologies is limited. The method proposed in [9] is intended for the scenario of sequential tag refinement by a participant, which is not always reasonable.

There is also a number of publications (for example, LexiTags [10]), where the ontology is used to help the participant in choosing labels. Although this approach to quality assurance can be effective, it belongs to another group in the classification above (methods based on the analysis of user's actions), and can only be used as an auxiliary one, not removing the need for matching labels.

Thus, currently, there are no methods for matching sets of semantic labels for non-sequential labeling, taking into account typical size of the ontologies used in labeling.

3 Model and Method for Label Aggregation

This section describes the proposed model, method and algorithm for semantic label aggregation. The proposed approach allows to obtain a consistent set of labels (ideally close to the 'true' labels) from a set of labels received from crowdsourcing participants. The set of possible labels is defined by an ontology written in OWL 2 language (specifically, the set of classes in such ontology). The proposed approach takes into account the relations between classes (labels) that are represented using the OWL 2 language, when determining a consistent set of object labels.

As a result of the analysis of the various types of relationships described in OWL 2, two relationships were identified that we decided to focus on processing in the first place: *SubClassOf* and *EquivalentClasses*. According to the OWL 2 language specification, the relationship *SubClassOf* between classes C_1 and C_2 reflects the fact that any instance of class C_1 is also an instance of class C_2 (the opposite is not necessarily true). The relation *EquivalentClasses* between classes C_1 and C_2 means that any instance of class C_1 is an instance of class C_2 and any instance of class C_2 is also an instance of class C_1. Thus, the relations *SubClassOf* and *EquivalentClasses* are used to describe the basic taxonomy of concepts, which is the basis (a kind of "skeleton") of any ontology.

When labeling an object i, a participant u offers a set of labels $l_{i,u}$ characterizing, in his opinion, this object. Each element of the set $l_{i,u}$ belongs to C, the set of ontology classes. To reduce the influence of individual characteristics of participants (knowledge, qualifications, motivation) on the labeling result, each object is labeled by several participants. Let c_i be the set of participants who proposed their labels for object i. The result of the label aggregation mechanism M should be the set l_i — a consistent set of labels. Symbolically, this can be written as $l_i = M(l_{i,u}, \xi)$, where ξ denotes all the additional information (except the original labels) that can be used by the matching mechanism. A typical approach to determining a consistent result in crowdsourcing is to obtain some estimate of the possibility that, given the evidence, a certain result is the correct result of the problem. Such an "assessment of the possibility" can be based both on probability-theoretical concepts and on other formal models. When following this approach, the main task of the quality assurance mechanism is to obtain such assessments of opportunity, and, for example, those for which this assessment of opportunity exceeds a certain threshold fall into the final list of agreed labels. Let $Q(i, c| l_{i,u}, \xi)$ be an estimate of the possibility that the label c corresponds to object i with the available evidence $l_{i,u}$, and ξ.

The proposed model and method are an adaptation of the Buchanan model for reasoning under uncertainty. Each participant is assigned a parameter b_u – a degree of belief that takes values in the range [0; 1]. Each labeling received from the participant (elements of the set $l_{i,u}$) is also accompanied with the corresponding degree of belief equal to the degree of belief of the participant from whom this labeling was obtained. The proposed aggregation mechanism:

a) Calculates degree of belief for class labels not reported directly by the user;
b) Adjusts the participants' degrees of belief based on the comparison of the sets of labels.

An assessment of the label possibility for an object is based on the properties of the *SubClassOf* relationship. Since, as mentioned earlier, *SubClassOf*(C_1, C_2) means that if the object c is an instance of the class C_1, then it is also an instance of the class C_2, then the confidence that the object belongs to the class C_1 can also be transferred to the fact of belonging to the class C_2. Thus, if there is evidence that object i belongs to class c with degree of belief μ, this degree of belief is also transferred to the fact that this object belongs to all classes (labels) located above c in the taxonomy formed by the *SubClassOf* relation (see Fig. 1).

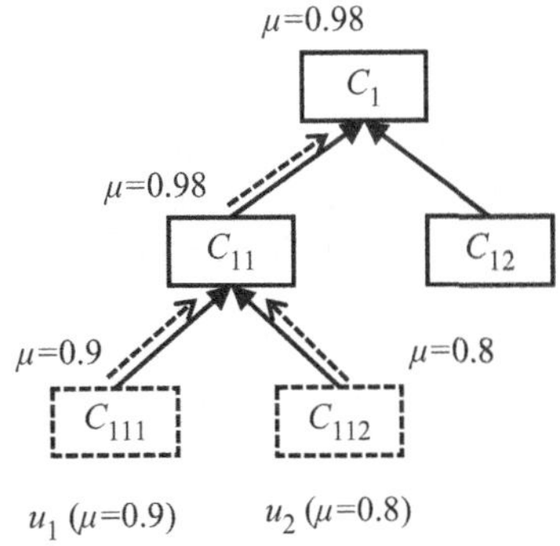

Fig. 1. Propagation of the degree of belief via taxonomy relation.

A special case is the estimation of a degree of belief that a label corresponds to an object in case there are several evidences (possibly received from different participants with similar of different degrees of belief). Let $\mu_1,\ldots, \mu_n$ be the degrees of belief in the statement that the object i belongs to class c obtained from different participants (possibly during the "propagation" of the degree of belief through the class hierarchy described earlier). The function used in combining the results of f_m must satisfy the following restrictions:

1. $f_m(\mu_1, \ldots, \mu_n) \geq \max(\mu_1, \ldots, \mu_n)$. If there are several evidences that some fact is true, then the aggregated degree of belief that this fact is true should not be less that any of the evidences' degrees of belief (evidences reinforce each other);
2. $f_m(\mu_1, \ldots, \mu_n) = fm(\mu_{n1}, \ldots, \mu_{nn})$, where $n1, \ldots, nn$ give some arbitrary permutation of the indices 1, …, *n*. I.e., the value of the function should not depend on the order of the arguments.

In the proposed method, *fm* is defined as $1 - \prod(1 - \mu_i)$ (it is easy to see that this function satisfies the restrictions above).

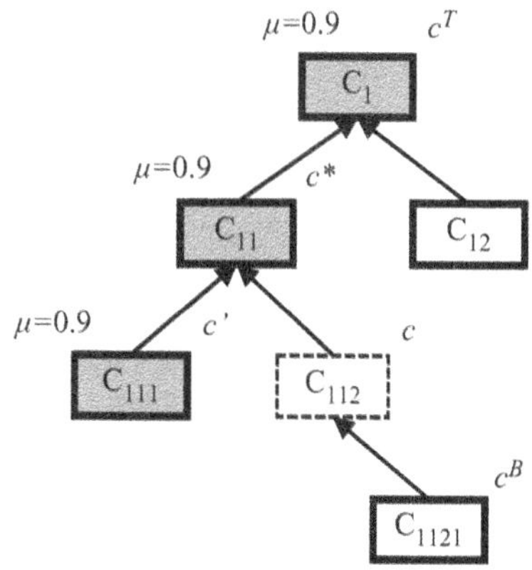

Fig. 2. Adjustment of the participant's degree of belief.

The degree of belief of the labels for the object, in turn, is the basis for adjusting the degree of belief of participants. In particular, for this, various measures of semantic distance between ontology classes are used, which are defined on the oriented graph generated by the taxonomy relation in the ontology. The proposed algorithm for recalculation of the participant's *u* degree of belief is defined as follows:

Step 1. Using the previously described algorithm calculate the consensus degree of belief for all possible labels using data received from all participants except *u* (in particular, this means that if an object was marked up by only one user, then this does not affect his/her degree of belief).

Step 2. For each label $c \in l_{i,u}$ find labels (ontology classes) c^*, *cT*, *cB* и *c'* (see Fig. 2, dashed rectangle is the label provided by the user *u*, filled rectangles – labels with non-zero degree of belief, as estimated in step 1). To find these labels directed acyclic graph *OG* is considered given by the taxonomy relation of the ontology, where nodes are classes and arcs are *SubClassOf* relations. The labels are defined as follows: c^* – is the nearest node of the OG with non-zero degree of belief reachable from *c*, *cT* – is the farthest node of the OG, reachable from *c*, *cB* – is the farthest from *c* node of the *OG*, from which *c* is reachable, and *c'* – is the farthest from c^* node with non-zero degree of belief, from which c^* is reachable.

Step 3. Participant's degree of belief is calculated as an arithmetic average of the degree of correspondence of his/her labels and the labels of other participants for the same objects:

$$\mu_u = \frac{1}{\sum_{i \in n_u} |l_{i,u}|} \left(\sum_{i \in n_u} \sum_{c \in l_{i,u}} (\alpha_1 \phi_1(c, l_{-u}) + \alpha_2 \phi_2(c, l_{-u}) + \alpha_3 \phi_3(c, l_{-u})) \right),$$

where n_u – the set of objects labeled by the participant u, l-u – aggregated labeling without participant u, functions ϕ_i reflect three empirical factors, influencing the evaluation of the labeling performed by the participant u. Coefficients αi give relative importance of these factors (summing up to 1). Particular values should be tuned w.r.t. the size and structure of particular ontology.

Empirical factor evaluation functions use the labels identified in step 2 and are calculated by the following formulae:

$$\phi_1(c, l_{-u}) = \frac{d(c^*, c^T)}{d(c^B, c^T)} \mu(c^*);$$

$$\phi_2(c, l_{-u}) = \frac{\max(d(c', c^*), d(c, c^*))}{d(c^B, c^T)} \mu(c');$$

$$\phi_3(c, l_{-u}) = \frac{d(c, c^T)}{d(c^B, c^T)}.$$

Here, $d(c_1,c_2)$ is distance between the nodes in *OG* (the number of arcs). The factors allow to "transfer" some degree of belief from the aggregated labels to the degree of belief of the participant, depending on how close are his/her labels to the aggregated labels (proposed by other participants). The first factor measures how specific is class $c*$ (the most specific class on which the opinion of the participant agrees with the opinion of other participants) in the hierarchy, the second factor allows to 'horizontally' transfer the degree of belief from the most specific aggregated label (taking into account distance to it in the *OG*), finally, the third factor shows how general the proposed label is (specific labels are generally better).

Recalculation of labels' and participants' degrees of belief occurs iteratively. Based on the initial degrees of belief of participants, degrees of belief of labels are estimated, then participants' degrees of belief are calculated. Using new participants' degrees of belief, again, labels' degrees of belief are estimated and so on until the moment when after a particular iteration the maximum absolute change in the value of degrees of belief will not be less than a specified threshold.

4 Conclusion

The paper proposes model, method and algorithm for semantic labels aggregation in crowdsourcing labeling applications. The proposed approach is based on the analysis of semantic relations between ontology classes, and takes into account inaccuracy and incompleteness of the results received from individual participants. It allows evaluating both the expected quality of the semantic labels received from individual crowdsourcing participants, and the degree of belief of the collected labels.

The proposed approach can be employed in semantic labeling crowdsourcing applications, e.g., to structure objects of a certain area in order to ensure semantic search. This may be a library of scientific articles, a collection of literature, divided by genre or a database for training neural networks. Possible applications are limited by existence of an ontology, written in OWL 2.

Acknowledgement. The reported study was funded by RFBR, project number 19-07-01120. The study was also supported by Russian State Research, project 0073-2019-0005.

References

1. Harvey, D., Kitching, T.D., Noah-Vanhoucke, J., Hamner, B., Salimans, T., Pires, A.M.: Observing dark worlds: a crowdsourcing experiment for dark matter mapping. Astron. Comput. **5**, 35–44 (2014). https://doi.org/10.1016/j.ascom.2014.04.003
2. Shin, D., Aliaga, D., Tunçer, B., Arisona, S.M., Kim, S., Zünd, D., Schmitt, G.: Urban sensing: using smartphones for transportation mode classification. Comput. Environ. Urban Syst. **53**, 76–86 (2014). https://doi.org/10.1016/j.compenvurbsys.2014.07.011
3. Meisel, Z.F., VonHoltz, L.A.H., Merchant, R.M.: Crowdsourcing healthcare costs: opportunities and challenges for patient centered price transparency. In: Healthcare, pp. 1–3 (2015). https://doi.org/10.1016/j.hjdsi.2015.06.004
4. Allahbakhsh, M., Benatallah, B., Ignjatovic, A., Motahari-Nezhad, H.R., Bertino, E., Dustdar, S.: Quality control in crowdsourcing systems: issues and directions. IEEE Internet Comput. **17**, 76–81 (2013). https://doi.org/10.1109/MIC.2013.20
5. Daniel, F., Kucherbaev, P., Cappiello, C., Benatallah, B., Allahbakhsh, M.: Quality control in crowdsourcing: a survey of quality attributes, assessment techniques and assurance actions. ACM Comput. Surv. (CSUR). **51**, 1–40 (2018). https://doi.org/10.1145/3148148
6. Ponomarev, A.V.: Quality control methods in crowd computing: literature review. SPIIRAS Proc. **54**, 152–184 (2017). https://doi.org/10.15622/sp.54.7
7. OWL 2 Web Ontology Language Structural Specification and Functional-Style Syntax. https://www.w3.org/TR/owl2-syntax/
8. Duan, L., Oyama, S., Sato, H., Kurihara, M.: Separate or joint? Estimation of multiple labels from crowdsourced annotations. Expert Syst. Appl. **41**, 5723–5732 (2014). https://doi.org/10.1016/j.eswa.2014.03.048
9. Otani, N., Baba, Y., Kashima, H.: Quality control of crowdsourced classification using hierarchical class structures. Expert Syst. Appl. **58**, 155–163 (2016). https://doi.org/10.1016/j.eswa.2016.04.009
10. Veres, C.: Crowdsourced semantics with semantic tagging: don't just tag it, LexiTag it!. In: CrowdSem'13: Proceedings of the 1st International Conference on Crowdsourcing the Semantic Web vol. 1030, pp. 9:1–15 (2013)

Selection of the Most Informative Genes in the Task of Cancer Tumors Recognition Based on the Gene Expression Profile

Alexey Kruzhalov(✉) and Andrey Philippovich

Moscow Polytechnic University, Moscow 107023, Russian Federation
alexkruzhalov@gmail.com
https://mospolytech.ru

Abstract. The work considers the problem of selecting the most informative genes in the task of cancer tumors recognition based on the gene expression profile. The model of a cancer classifier based on a feedforward neural network is proposed. The most informative genes were selected using the layer-wise relevance propagation (LRP) method. The statistical methods were used to check the selected genes for relations to cancer.

Keywords: Gene expression · Neural network · Layer-Wise relevance propagation

1 Introduction

In recent years, due to the development of computer technology, the intelligent systems based on machine learning methods have become widespread. It allows you to extract hidden patterns from large amounts of data and use it to solve the applied problems. Such systems are indispensable in those areas of human activity in which knowledge of the subject area is incomplete, and it is possible to build an adequate model of the observed processes or phenomena only on the basis of the empirical data. These areas include biology and medicine, as scientists still do not know the mechanisms of development of certain diseases, as well as processes occurring in the human body.

One of the interesting purposes in this area is to study the mechanisms of disease development and its diagnostics based on the genomic data. In particular, many studies are devoted to the study of cancer formation mechanisms [5, 7, 8]. The volume of the accumulated genomic data is growing every year, which is facilitated by modern advances in sequencing technologies, as well as international projects to organize the collection and analysis of this data. The largest project dedicated to the study of the cancer genome is the Cancer Genome Atlas (TCGA) project. One type of the genomic data that was collected in this project is data on gene expression in tumors.

Gene expression is the process by which the hereditary information encoded in DNA is converted into a functional product (RNA or protein). Every cell in the human body can be described using a gene expression profile. It contains information about the entire set of genes expressed in a particular cell or tissue under a particular condition. The cancer formation is caused by mutations in one or more genes from a vast

R. Silhavy et al. (Eds.): CoMeSySo 2020, AISC 1295, pp. 895–909, 2020.
https://doi.org/10.1007/978-3-030-63319-6_83

number that regulate the growth and programmed death of cells. The changes in DNA can affect the expression of mRNA encoded by it and consequently be expressed in changes in the gene expression profile in the tumor [11]. Therefore, the data on gene expression is used to study the cancer formation.

Modern machine learning methods are successfully applied for the analysis of expression data. To study the data structure, the clustering methods are usually used, and the problem of diagnosing diseases can be reduced to solving the classification problem.

The artificial neural networks are widely used today to solve the classification problem. It shows good results, however, its decisions are difficult to interpret. In the task of automated disease diagnostic you should not rely on the response from the «black box». It should be possible to explain the resulting decision. Therefore, this work investigates the possibility of using the layer-wise relevance propagation (LRP) method [1] to interpret the responses of the cancer classifier based on the artificial neural network. The interpretation refers to the determination of the most informative genes, i.e. the genes that had the greatest influence on the decision to assign a sample to a particular type of tumor.

The purpose of this work is to build a cancer classifier and determine on its basis a list of the most informative genes for each type of tumor. These genes can be considered as potential biomarkers, as well as targets for drug design. To test the relation of selected genes with cancer, it is proposed to perform statistical analysis using open cancer genes databases.

2 Methods

2.1 Data

This work uses data from The Cancer Genome Atlas project. The data for analysis was downloaded using the UCSC Xena platform [13]. It contains the information about 32 types of tumors. A complete list of the types of tumors presented in this data can be found in the Appendix. The expression levels were measured for 20530 genes. The tissue samples were taken from 9706 patients. In total, the downloaded data contains 10459 samples, including both tumor tissue samples and healthy tissue samples. The gene expression was measured using the Illumina HiSeq platform.

2.2 Preprocessing

There is a strong class imbalance in the data. The largest number of samples were breast cancer (BRCA) – 1211 examples, while the minority were cholangiocarcinoma (CHOL) – 45 examples and Diffuse Large B-cell Lymphoma (DLBC) – 48 examples.

To smooth out the class imbalance, an attempt was made to generate additional synthetic examples using the Synthetic Minority Oversampling Technique (SMOTE) algorithm, as suggested in some works [5, 8]. The comparison of the recognition accuracy of the neural networks trained on the original data and the data supplemented with synthetic examples showed contradictory results: for some types of tumors the accuracy increased, and for some it fell. The measurements on different test sets showed different results. In addition, the synthetic examples cannot be used in the test set, so SMOTE does not solve the problem of a lack of examples for more accurate estimation of the recognition accuracy. The results of further experiments showed that the neural network can be robust to class imbalance under certain conditions, so it was decided to abandon artificial smoothing of class imbalance and work with data «as is».

To train the neural network, the data was divided into training and test sets in the proportion of 80% to 20%. To maintain the same class distribution between sets, the stratified sampling was used. The genes with a standard deviation less than 0.1 were removed from the data. After that, the data for each gene was normalized: it was reduced to a zero mean and a unit variance.

2.3 Recognition Algorithm

The artificial neural networks have become more widespread in recent years and show amazing results in various fields. However, at the moment, the practical success is poorly supported by theoretical justifications. One of the main results of the artificial neural network theory is the universal approximation theorem, which states that the artificial neural network with two layers is a universal approximator. However, this theorem does not allow us to understand which neural network architecture will be optimal for solving a particular problem.

The recent works [6, 8, 14] present various neural network architectures for solving the problem of the cancer type recognition based on the gene expression profile. Most of it use convolutional neural networks that take into account the spatial data structure. However, as the results of comparisons [14] show, the convolutional networks do not have an advantage over fully-connected networks for this problem. Therefore, in this work, a multilayer perceptron (MLP) was used to construct a cancer classifier.

The built network has the following architecture. The network contains four layers: input, output, and two hidden layers. The hidden layers consist of 100 and 75 neurons respectively. Activation function for hidden layers is ReLU. The choice of the hidden layers activation function was determined by the simplicity of calculation, as well as the absence of saturation problems. The output layer of the network contains 33 neurons and uses the softmax activation function. The cross entropy was used as a loss function for neural network training. The Adam algorithm was used to optimize network parameters.

2.4 Recognition Accuracy Evaluation

The disadvantage of the recognition accuracy evaluation based on a holdout test set is that this method does not allow you to evaluate the classifier reliability. The classification accuracy can vary while the training set is changed. To get more objective estimates, the classic cross-validation method can be used, but it only allows you to get a point estimate, and we would also like to get a confidence interval. Therefore, in this work, the bootstrap method was used to evaluate the accuracy of a trained neural network, which allows us to obtain confidence intervals for the classification accuracy estimates.

The bootstrap method is used for statistical estimation of the distribution parameters of a random variable if there is no a priori information about the distribution law of the studied random variable [12]. The main idea of bootstrap is to repeatedly extract repeated samples, also called pseudo-samples, from an empirical distribution. The value of the estimated parameter is calculated for each pseudo-sample. Based on the distribution of the values of the considered parameter obtained in the simulation process, it is possible to construct confidence intervals for the estimated parameter [12, p. 17].

The calculations were performed as follows. Using a random sampling with replacement, 100 subsamples were formed, the size of each of which was 80% of the size of the source data. A neural network was trained on each of the subsamples. Next, we evaluated the classification accuracy of each network based on data that was not included in the training set. As a result, an empirical distribution of classification accuracy estimates was obtained for each class. Based on this distribution, the boundaries of the 90% confidence interval were calculated as the 5th and 95th percentiles. The median of the received distribution was selected as a point estimate.

2.5 The Most Informative Genes Selection

The problem of selecting the most informative features is key in this work, since our objective is to identify potential oncogenes, which in the future can be considered as biomarkers or goals for drug design. The task of selecting the most informative genes can be reduced to the task of searching for such a subset of genes, using which the selected recognition algorithm allows to get the best results. This task is generally NP-hard, so in practice heuristic algorithms are used.

The feature selection algorithms can be divided into three groups: filtering methods, wrapper methods, and embedded algorithms. The filtering methods are used at the preprocessing stage and do not take into account the special aspects of the recognition algorithm used. In this work, we used a filtering method to discard genes with a small spread of expression values, since these genes are not informative due to the fact that their expression changes slightly in different types of tumors. The wrapper methods for selecting features rely on the quality metric value of the recognition algorithm. The examples of methods in this group are genetic algorithms and algorithms that use a greedy selection strategy. The methods of this group are not suitable for our purpose, since they are computationally expensive. The methods of the third group – embedded algorithms – take into account the special aspects of the recognition algorithm used, which allows them to be more efficient. There is an entire series of algorithms of this type for artificial neural networks.

These algorithms can be divided into two groups: algorithms that use the gradient value of the decision function and algorithms that use the decomposition of the value of the decision function by the sum of «scores» for each attribute. The algorithms of the first group include Grad CAM [8], and the algorithms of the second group include layer-wise relevance propagation (LRP) [1]. One of the main advantages of the methods from the second group is that they are more universal than gradient methods, since they do not require differentiability of nurons' activation functions. It can also be noted that LRP already has successful applications in medicine [2].

In this work, the layer-wise relevance propagation (LRP) method was used to select the most informative genes. It is used to interpret decisions made by neural networks. LRP allows you to determine for each example which input signals of a neural network had the greatest impact on the output signal. In this case, each input signal is associated with a number that shows the degree of impact of this input on the output value of the neural network. If there is more than one neuron in the output layer, each output signal can be considered separately.

The neural network, that is used in this work, contains 33 neurons in the output layer—one for each class. For each example, we took into account only one output signal—the signal of the neuron that corresponds to the correct class for this example. Since only the output signal of the desired neuron is considered, the resulting values of the degree of impact of each of the features on this signal can be interpreted as the informativity (importance) degree of the features for recognizing this class.

Using LRP, we obtained feature informativity (relevance) vectors for each example. Then we summed up the obtained vectors for each class's examples to get informativity vectors for the classes. Since each class contains a different number of examples, the resulting vectors differed in length. To bring them to a comparable form, we divided each vector by its L2-norm.

To see how much the most informative genes selected for different classes differ, a cluster analysis of informativity vectors was performed. The Ward method was used—one of the methods of hierarchical clustering.

2.6 Validation

A number of experiments were conducted to make sure that the genes obtained using LRP are really the most informative. The concept of «informativeness» was considered from two points of view. First, those genes that most significantly affect the functioning of the neural network were considered informative. Secondly, those genes that are related to the cancer formation and can be considered as potential biomarkers of a certain type of tumor were considered informative.

To test informativeness from the first point of view, an experiment was conducted in which the most informative genes were removed from the data and checked how this would affect the accuracy of neural network. The accuracy was expected to decrease significantly.

The procedure for conducting the experiment was as follows. First, the informativity vectors of the genes for each class were summed, then the resulting sum vector was used to sort the genes in descending order of informativity. After that, the genes were gradually «deleted» from the data in this order. The «deletion» consisted of assigning a zero value to the gene being deleted. 50 genes were deleted in one step, and the average recognition accuracy was calculated based on the current data. For comparison, a similar procedure was performed, differing only in that the genes were removed at random order.

After testing the validity of the selected genes in terms of influencing the neural network, it became necessary to determine how many of the most informative genes would be sufficient to recognize a particular type of tumor, and which genes for each type of tumor it makes sense to further analyze for relations to cancer. To determine this, a graph was constructed, which can be used for approximate estimation of the number of genes that account for the majority of information content. To do this, the genes in the informativity vectors of each class were sorted in descending order of informativity and a graph was constructed showing the dependence of the value of the informativity of the gene on its ordinal number in the sorted list. Based on this graph, it was decided how many genes should be selected for further analysis.

To identify the relation with cancer, the selected genes were analyzed using the ToppGene Suite platform [4]. It allows you to perform gene set enrichment analysis, that is, to find associations between a certain list of genes and annotated sets of genes that characterize certain phenotypes (for example, diseases). ToppGene Suite includes annotations from different databases, and these annotations are divided into categories. The analysis of the representation of the analyzed genes in the annotations of each category is performed separately. This work presents the results of the analysis in the following categories: Disease, Pubmed, Coexpression, Coexpression Atlas.

The hypergeometric test was used to check the representation of selected genes in the categories. It is widely used in bioinformatics to check the statistical significance of the representation of genes from a certain set obtained as a research result, among another set of genes associated with a biological trait (for example, a molecular function or biological process). The null hypothesis H_0 of this test is that the genes selected in the study are not related to the genes related to the concerned biological trait, and all matches are nothing more than random.

Since the concerned set of genes is compared with each of the annotations presented in the category, the so-called multiple comparison problem occurs. It consists in increasing the probability of making at least one type I error (that is, a false rejection of the null hypothesis) in comparison with the pre-established probability α. To keep this probability at the same level various corrections for multiple comparisons are used. In this work, the False Discovery Rate (FDR) method was used to control the level of type I errors. The value of FDR is defined as the mathematical expectation of the proportion of type I errors. There are several methods for controlling FDR at the α level. The Benjamini-Hochberg method was used in this work.

3 Results

After preprocessing and filtering, the data contains 19591 genes. The neural network of the previously specified architecture was trained on this data. The following hyperparameters were used for training: the learning rate was set to 0.001, and the batch size was set to 100. These values were obtained empirically. The average accuracy of the trained network was 91.6%.

Then, based on this neural network, the most informative genes were selected using LRP. To verify the validity of the obtained values of informativity, an experiment was conducted to study the effect of removing the most informative genes from the data on the functioning of the neural network. The experiment results are shown in Fig. 1.

The graphs show that when removing genes in descending order of informativity values, the recognition accuracy falls much faster than when removing genes in random order. Moreover, while deleting genes at random, the accuracy is almost unchanged

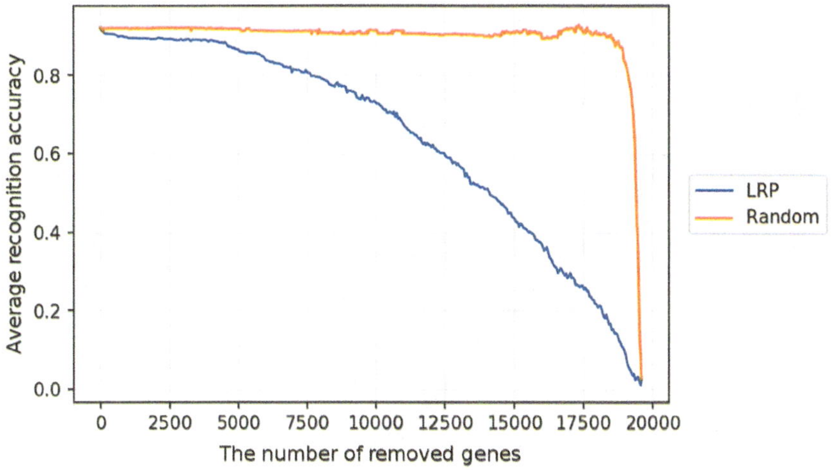

Fig. 1. Graphs of changes in recognition accuracy while removing genes in descending order of informativity (LRP) and in random order (Random)

until the moment when 17000 genes were removed. This means that a small set of genes contains the main information important for classification.

An important question is how similar are the gene informativity vectors obtained for different types of tumors. To find out, a cluster analysis of informativity vectors was performed, the results of which are shown in Fig. 2.

The given distance matrix shows that the greatest similarity is observed in the informativity vectors for rectum adenocarcinoma (READ) and colon adenocarcinoma (COAD). This result is quite expected, since the neural network to which LRP was applied could not separate these two classes. It is known that these two types of tumors are almost indistinguishable at the genomic level [9]. In addition, similarities are also observed between other types of tumors that have a similar location: esophageal carcinoma (ESCA) and stomach adenocarcinoma (STAD), skin cutaneous melanoma (SKCM) and uveal melanoma (UVM), lung adenocarcinoma (LUAD) and lung squamous cell carcinoma (LUSC), kidney renal clear cell carcinoma (KIRC) and kidney renal papillary cell carcinoma (KIRP), cholangiocarcinoma (CHOL) and liver hepatocellular carcinoma (LIHC), ovarian serous cystadenocarcinoma (OV) and uterine corpus endometrial carcinoma (UCEC).

To determine exactly how many genes, carry the bulk of the information, a graph was plotted, shown in Fig. 3. This graph shows the dependence of the value of the informativity of a gene on its ordinal number in the list of genes, sorted in descending order of informativity. The graph shows that for all types of tumors, much of the informativity is related to approximately the first 100 most informative genes.

We combined the genes selected for each class, and got a set of 2732 genes. These genes were further analyzed to determine whether these genes are associated with the cancer formation. For the genes selected for each of the considered types of tumors, an analysis of gene set enrichment was performed using the ToppGene Suite platform.

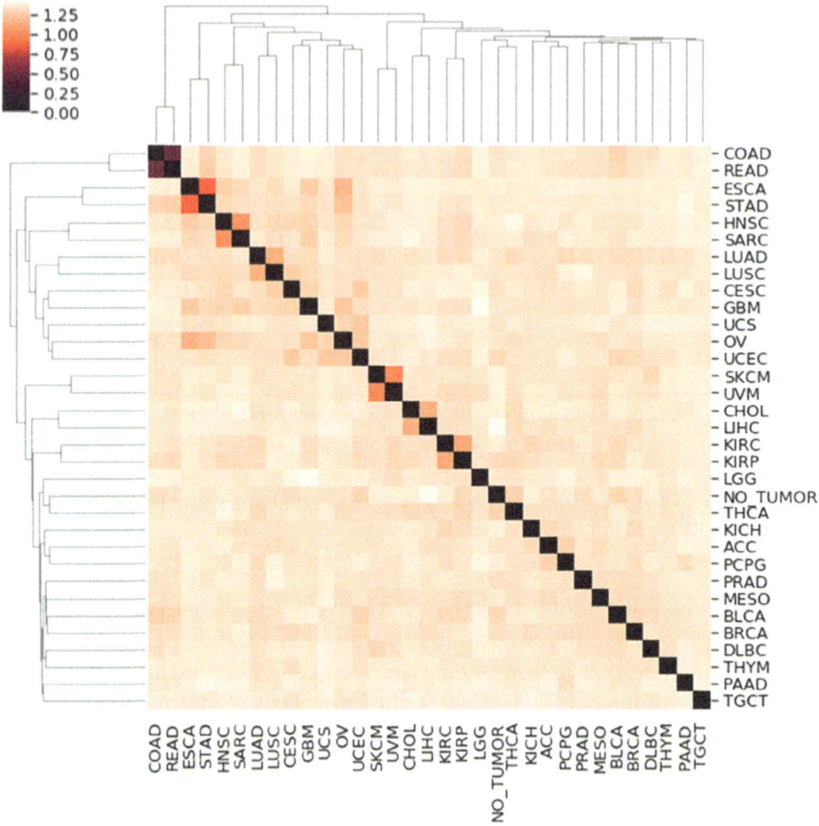

Fig. 2. Distance matrix between gene informativity vectors for different tumor types

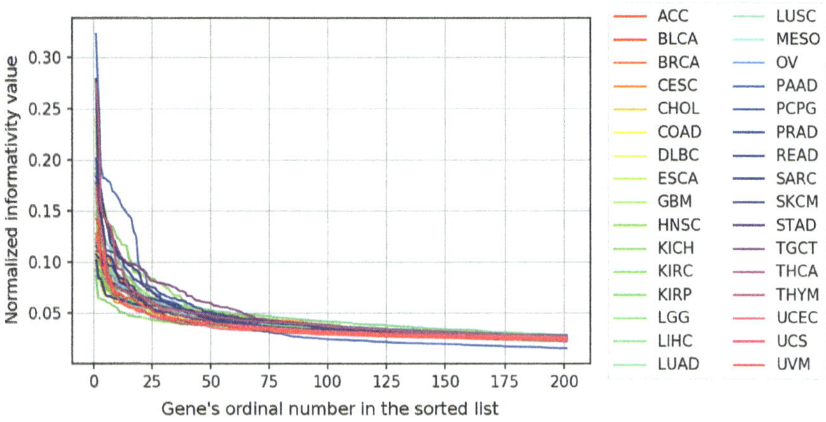

Fig. 3. The dependence of the value of the gene's informativity on its ordinal number in the sorted list

The results of the analysis for some types of tumors are presented in Table 1. The FDR B&H column contains p-values adjusted by the Benjamini-Hochberg method.

Table 1. Verification results obtained using the ToppGene Suite platform

Tumor type	Annotation	FDR B&H
Adrenocortical carcinoma (ACC)	Adrenocortical carcinoma (C0206686)	6,946E−5
Breast invasive carcinoma (BRCA)	Carcinoma breast stage IV (C0278488)	6,525E−3
Cervical squamous cell carcinoma and endocervical adenocarcinoma (CESC)	Squamous cell carcinoma, metastatic (C0334246)	2,617E−2
Colon adenocarcinoma (COAD)	Adenocarcinoma of colon (C0338106)	2,026E−2
Kidney renal clear cell carcinoma (KIRC)	Renal carcinoma (C1378703)	9,768E−4
	Renal Cell Carcinoma (C0007134)	2,803E−3
Liver hepatocellular carcinoma (LIHC)	Liver neoplasms (C0023903)	1,097E−11
Lung adenocarcinoma (LUAD)	Lung Neoplasms (C0024121)	1,663E−5
	Adenocarcinoma of lung (C0152013)	2,207E−5
Mesothelioma (MESO)	Malignant Mesothelioma of Peritoneum (C0346109)	1,399E−4
	Sarcomatoid Mesothelioma (C0334513)	7,325E−3
Pheochromocytoma and Paraganglioma (PCPG)	Pheochromocytoma (C0031511)	1,746E−2
	Paraganglioma (C0030421)	2,330E−2
Prostate adenocarcinoma (PRAD)	Metastatic Prostate Carcinoma (C0936223)	6,921E−4
	Prostatic Neoplasms (C0033578)	1,393E−3
Rectum adenocarcinoma (READ)	Rectal Neoplasms (C0034885)	2,147E−2
Skin cutaneous melanoma (SKCM)	Melanoma, Amelanotic (C0206735)	1,246E−4
	Cutaneous Melanoma (C0151779)	6,296E−3
Thyroid carcinoma (THCA)	Papillary thyroid carcinoma (C0238463)	7,453E-6
	Follicular thyroid carcinoma (C0206682)	1,885E−2

The table above shows that for many types of tumors, it was possible to identify a statistically significant relationship with sets of genes associated with this particular disease. For example, for Adrenocortical carcinoma (ACC), an association with the annotation of Adrenocortical carcinoma (C0206686) was identified, and for Pheochromocytoma and Paraganglioma (PCPG), an association with the annotations of Pheochromocytoma (C0031511) and Paraganglioma (C0030421) was identified.

The neural network was re-trained using selected genes. The bootstrap method was used to calculate confidence intervals of recognition accuracy estimates. For this purpose, 100 pairs of training and test sets were formed, on each of which a neural network of the specified architecture was trained. This allowed us to obtain empirical distributions of recognition accuracy estimates for each class. The point estimates and 90% confidence intervals were calculated from these distributions.

In Table 2 the obtained estimates of recognition accuracy for certain types of tumors are given. The results for the other types of tumors are shown in the Appendix. This table also shows the n – the number of examples of each class in the test set. The results obtained by Li Y. [7] and Lyu B. [8] are presented for comparison.

Table 2. Recognition accuracy evaluation for certain types of tumors

Tumor type	Accuracy Li Y. [7]	Accuracy Lyu B. [8]	Our accuracy	90% conf. interval	n
Esophageal Carcinoma (ESCA)	–	0,77	0,86	[0,70; 0,95]	37
Glioblastoma multiforme (GBM)	0,99	0,94	1,00	[0,97; 1,00]	31
Kidney renal papillary cell carcinoma (KIRP)	0,92	0,93	0,94	[0,90; 0,97]	58
Lung squamous cell carcinoma (LUSC)	0,88	0,91	0,93	[0,89; 0,96]	101
Mesothelioma (MESO)	0,90	0,94	0,95	[0,90; 1,00]	18
Rectum adenocarcinoma (READ)	0,00	0,35	0,43	[0,16; 0,70]	19
Sarcoma (SARC)	0,96	0,97	1,00	[0,97; 1,00]	52
Thymoma (THYM)	0,94	0,99	1,00	[0,96; 1,00]	24

While calculating confidence intervals for each class, we obtained empirical distributions of recognition accuracy estimates. The shape of these distributions was different for different classes, as can be seen in Fig. 4, which shows histograms of distributions of recognition accuracy estimates for COAD, READ, ESCA, and DLBC.

For READ, a symmetric unimodal distribution was obtained, in which a certain most frequently occurring value is present, and deviations from it in both the higher and lower directions occur with approximately the same frequency. For COAD and ESCA, on the contrary, the distributions turned out to be asymmetric, and deviations to the lower side are observed much more often than to the higher. For DLBC, almost all values are close to 1, which means that this class is stably recognized by the neural network when the training set is varied.

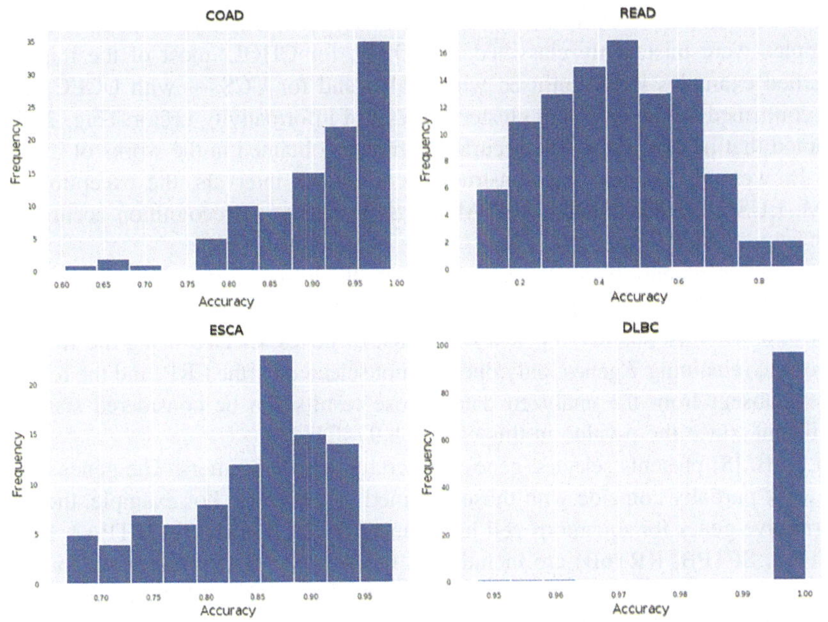

Fig. 4. Histograms of the distributions of recognition accuracy estimates obtained for some tumor types using bootstrap

4 Discussion

Comparing our results with the results of Li Y. [7] and Lyu B. [8], we can say that in general, comparable results were obtained, and for 8 types of tumors, the built neural network showed better results, namely for ESCA, GBM, KIRP, LUSC, MESO, READ, SARC, THYM. The accuracy estimates for these classes are shown in the Table 2. The greatest increase in accuracy was obtained for ESCA (9%), READ (8%), GBM (6%). Among the remaining types of tumors, the same results were obtained for 15 as in the works of Li Y. and Lu B. (for comparison, the best result was chosen). For 9 more types of tumors, lower accuracy values were obtained than in the studies under consideration, but the difference was insignificant — most of all for COAD (5%) and CHOL (4%), and for the rest, the accuracy decreased by 1–3%.

The worst accuracy was obtained for the READ (43%), CHOL (69%), and UCS (81%) classes. The low accuracy of READ recognition can be explained by its similarity at the molecular level to COAD – another type of colorectal cancer. The low accuracy of CHOL recognition can be explained by the insufficient representation of examples of this class in the training set. It should also be noted that a fairly wide confidence interval was obtained for this class's recognition accuracy value [0,39; 0,87], which indicates that the obtained estimate is unreliable. This is the expected result, since only 8 examples were used to evaluate accuracy. In other works, for these classes, too, not very high accuracy was obtained. For example, Lyu B. got 35% for READ and 81% for UCS, and Li Y. received 73% for CHOL. The error analysis confirmed the results obtained by Li Y. and Lyu B., namely, most of the READ examples were mistakenly classified as COAD, for CHOL, most of the mistakenly classified examples were confused with LIHC, and for UCS — with UCEC. This is also confirmed by the results of cluster analysis of informativity vectors (Fig. 2). It can be noted that in most cases, the accuracy estimates obtained in the works of Li Y. and Lyu B. were included in the constructed confidence intervals, the exceptions were GBM, LUSC, READ, SARC, THYM, UCS. The average recognition accuracy was 93.9%.

The selected genes were also compared with the lists of genes given in other works. Li Y. [7] provides a list of the 20 most informative genes obtained using a genetic algorithm. This list shares 13 genes with a list of genes selected using the LRP. Note that of the remaining 7 genes, only one was not selected by the LRP, and the remaining 6 were absent from the analyzed data. These results can be considered statistically significant, since the p-value in this case is 8,974E11.

Lyu B. [8] presents selected genes for certain types of tumors. The genes given in this work partially coincide with those obtained in our work. For example, the 5 most informative genes for squamous cell lung cancer (LUSC) from [8] (SFTPA2, KRT6A, SFTPA1, SFTPB, KRT6B) are included in the list of 100 selected genes for LUSC, with SFTPA2, SFTPA1 and SFTPB at the top of this list. The probability of getting this result randomly is 3,132E12.

Thus, we can conclude that the genes selected for many types of tumors are actually associated with cancer, as shown by the results of statistical tests. This means that the selection of genes using the LRP method takes into account the biological information hidden in the data. In addition, it was found that the selected genes partially coincide with the lists of genes obtained in previous studies. However, not for all types of tumors, cancer genes were statistically significantly represented in the list of selected genes corresponding to this type of tumor.

You can also consider an example of how a neural network works for a specific sample. So, for TCGA-CH-57-63-01 which is a sample of prostate adenocarcinoma, (PRAD), it was found that the PCA3 gene had the greatest influence on the decision to classify this sample in the PRAD class. PCA3 stands for prostate cancer antigen 3. This is a gene that is used for transcribing non-coding RNA. Its expression is highly elevated in prostate cancer, so it is considered a biomarker [10]. The second most informative gene (CDK19) is also associated with prostate cancer. The research shows that during the development of prostate cancer, there is an increase in CDK19 expression, which makes it a potential therapeutic target for advanced prostate cancer [3]. This example suggests that the model can be used by researchers to form new hypotheses about the genes involved in the cancer formation.

5 Conclusion

In this work we proposed a model of the artificial neural network to solve the problem of the cancer type recognition based on the gene expression profile. Comparing the recognition accuracy with the works of other researchers showed that we were able to increase the accuracy for certain types of tumors. In particular, we managed to increase the recognition accuracy for esophageal carcinoma by 9%, rectum adenocarcinoma by 8%, and glioblastoma multiforme by 6%. The confidence intervals we obtained for evaluating recognition accuracy are consistent with the research results by other authors. The reasons for low recognition accuracy for some classes were too few examples of this class in the training set, or the common origin of several types of tumors, which is why the neural network could not separate it.

Based on a trained neural network, the most informative genes were selected using the layer-wise relevance propagation (LRP) method. The experiments have shown that the selection of the most informative genes using LRP is justified. Using the statistical methods, the selected genes were tested for a relation to cancer, and this statistically significant relation was found. Based on this, we can conclude that the proposed method of selecting the most informative genes takes into account the biological patterns hidden in the data.

Of course, the genes selected in this way require further study to determine whether it actually affects the cancer formation. For medical and biological interpretation of the obtained results it is necessary to construct a functional description of the selected set of genes and its further analysis.

Appendix

(See Table 3).

Table 3. Recognition accuracy estimation

Tumor type	Accuracy Li Y. [7]	Accuracy Lyu B. [8]	Our accuracy	90% conf. interval	n
Adrenocortical carcinoma (ACC)	0,97	0,95	0,97	[0,91; 1,00]	16
Bladder urothelial carcinoma (BLCA)	0,91	0,97	0,96	[0,94; 0,99]	82
Breast invasive carcinoma (BRCA)	0,99	0,99	0,99	[0,98; 1,00]	220
Cervical squamous cell carcinoma and endocervical adenocarcinoma (CESC)	0,94	0,93	0,94	[0,91; 0,97]	61
Cholangiocarcinoma (CHOL)	0,73	0,56	0,69	[0,39; 0,87]	8
Colon adenocarcinoma (COAD)	0,99	0,95	0,94	[0,78; 0,99]	58
Esophageal Carcinoma (ESCA)	–	0,77	0,86	[0,70; 0,95]	37
Glioblastoma multiforme (GBM)	0,99	0,94	1,00	[0,97; 1,00]	31
Head and Neck squamous cell carcinoma (HNSC)	0,99	0,98	0,99	[0,97; 1,00]	104
Kidney chromophobe (KICH)	0,96	0,87	0,93	[0,80; 1,00]	14
Kidney renal clear cell carcinoma (KIRC)	0,96	0,95	0,95	[0,92; 0,97]	107
Kidney renal papillary cell carcinoma (KIRP)	0,92	0,93	0,94	[0,90; 0,97]	58
Brain lower grade glioma (LGG)	1,00	0,98	1,00	[0,98; 1,00]	104
Liver hepatocellular carcinoma (LIHC)	0,98	0,97	0,96	[0,92; 0,98]	75
Lung adenocarcinoma (LUAD)	0,96	0,95	0,96	[0,92; 0,98]	103
Lung squamous cell carcinoma (LUSC)	0,88	0,91	0,93	[0,89; 0,96]	101

(*continued*)

Table 3. (*continued*)

Tumor type	Accuracy Li Y. [7]	Accuracy Lyu B. [8]	Our accuracy	90% conf. interval	n
Mesothelioma (MESO)	0,90	0,94	0,95	[0,90; 1,00]	18
Ovarian serous cystadenocarcinoma (OV)	1,00	0,99	1,00	[0,99; 1,00]	61
Pancreatic adenocarcinoma (PAAD)	0,95	0,97	0,97	[0,92; 1,00]	36
Pheochromocytoma and Paraganglioma (PCPG)	1,00	1,00	0,99	[0,98; 1,00]	36
Prostate adenocarcinoma (PRAD)	1,00	1,00	0,99	[0,95; 1,00]	100
Rectum adenocarcinoma (READ)	0,00	0,35	0,43	[0,16; 0,70]	19
Sarcoma (SARC)	0,96	0,97	1,00	[0,97; 1,00]	52
Skin cutaneous melanoma (SKCM)	0,97	0,98	0,98	[0,95; 1,00]	21
Stomach Adenocarcinoma (STAD)	–	0,96	0,96	[0,89; 0,99]	83
Testicular germ cell tumors (TGCT)	1,00	0,99	1,00	[0,98; 1,00]	30
Thyroid carcinoma (THCA)	1,00	1,00	0,99	[0,98; 1,00]	101
Thymoma (THYM)	0,94	0,99	1,00	[0,96; 1,00]	24
Uterine corpus endometrial carcinoma (UCEC)	0,96	0,96	0,96	[0,92; 1,00]	36
Uterine carcinosarcoma (UCS)	0,62	0,81	0,81	[0,64; 0,96]	12

References

1. Bach, S., Binder, A., Montavon, G., Klauschen, F., Müller, K.-R., Samek, W.: On pixel-wise explanations for non-linear classifier decisions by layer-wise relevance propagation. PLoS One **10**(7) (2015). https://doi.org/10.1371/journal.pone.0130140
2. Böhle, M., Eitel, F., Weygandt, M., Ritter, K.: Layer-wise relevance propagation for explaining deep neural network decisions in MRI-based Alzheimer's disease classification. Front. Aging Neurosci. **10**, (2019). https://doi.org/10.3389/fnagi.2019.00194
3. Brägelmann, J., Klümper, N., Offermann, A., von Mässenhausen, A., Böhm, D., Deng, M., Queisser, A., Sanders, C., Syring, I., Merseburger, A.S., Vogel, W., Sievers, E., Vlasic, I., Carlsson, J., Andrén, O., Brossart, P., Duensing, S., Svensson, M.A., Shaikhibrahim, Z., Kirfel, J., Perner, S.: Pan-Cancer analysis of the mediator complex transcriptome identifies CDK19 and CDK8 as therapeutic targets in advanced prostate cancer. Clin. Cancer Res. **23**, 1829–1840 (2017). https://doi.org/10.1158/1078-0432.CCR-16-0094
4. Chen, J., Bardes, E.E., Aronow, B.J., Jegga, A.G.: ToppGene suite for gene list enrichment analysis and candidate gene prioritization. Nucleic Acids Res. **37**, 305–311 (2009). https://doi.org/10.1093/nar/gkp427
5. Danaee, P., Ghaeini, R., Hendrix, D.A.: A deep learning approach for cancer detection and relevant gene identification. Pac. Symp. Biocomput. **22**, 219–229 (2016). https://doi.org/10.1142/9789813207813_0022
6. De Guia, J.M., Devaraj, M., Leung, C.K.: DeepGX: deep learning using gene expression for cancer classification. In: Proceedings of the 2019 IEEE/ACM International Conference on Advances in Social Networks Analysis and Mining, ASONAM 2019, pp. 913–920. Association for Computing Machinery, Inc., New York (2019)
7. Li, Y., Kang, K., Krahn, J.M., Croutwater, N., Lee, K., Umbach, D.M., Li, L.: A comprehensive genomic pan-cancer classification using the cancer genome atlas gene expression data. BMC Genom. **18**, 508 (2017). https://doi.org/10.1186/s12864-017-3906-0

8. Lyu, B., Haque, A.: Deep learning based tumor type classification using gene expression data. In: Proceedings of the 2018 ACM International Conference on Bioinformatics, Computational Biology, and Health Informatics, Washington, DC, USA, pp. 89–96 (2018). https://doi.org/10.1101/364323
9. Muzny, D., Bainbridge, M., Chang, K., et al.: Comprehensive molecular characterization of human colon and rectal cancer. Nature **487**, 330–337 (2012). https://doi.org/10.1038/nature11252
10. Pavlov, K.A., Korchagina, A.A., Abdulina, Y.A., Surenkov, D.N., Zus'man, L.A., Darenkov, S.P., Grigor'ev, M.E., CHekhonin, V.P.: PCA3—perspektivnyj biomarker raka predstatel'noj zhelezy (PCA3 — Promising Prostate Cancer Biomarker). Vestnik Rossijskogo gosudarstvennogo medicinskogo universiteta **3**, 54–58 (2012)
11. Perevodchikova, N.I., Stenina, M.B.: Lekarstvennaya terapiya raka molochnoj zhelezy (Breast Cancer Drug Treatment). Praktika, Moscow (2014)
12. Shitikov, V.K., Rozenberg, G.S.: Randomizaciya i butstrep: statisticheskij analiz v biologii i ekologii s ispol'zovaniem R (Randomization and bootstrap: statistical analysis in biology and ecology using R). Kassandra, Tolyatti (2013)
13. UCSC Xena. The Cancer Genome Atlas (2020). https://tcga.xenahubs.net. Accessed 10 May 2020
14. Yu, H., Samuels, D.C., Zhao, Y., Guo, Y.: Architectures and accuracy of artificial neural network for disease classification from omics data. BMC Genom. **20**, 167 (2019). https://doi.org/10.1186/s12864-019-5546-z

Assessment of the Possibility of Reception Errors of Aperiodic Pseudo-Random Sequences in the Arranged Countermeasures

I. M. Azhmukhamedov(✉) and E. V. Melnikov

Astrakhan State University, Astrakhan, Russian Federation
arhelia@bk.ru

Abstract. The article discusses the obtained estimate of the upper bounds for the probable receiving error of the synchronizing sequence during sensor phasing of aperiodic pseudo-random sequences (CRR) in broadband communication systems in the channels of low quality with strong disturbances of natural and organized structure. The obtained results allow to design synchronization systems of pseudo-random sequences for the worst case, which guarantees their reliable operation in low-quality channels, and, unlike the well-known methods, the estimation of synchronization of aperiodic pseudo-random sequence sensors doesn't depend on the error distribution in the communication channel and the period of sequence. The appointed differences simplify the evaluation of synchronization in the operation of broadband communication systems in low-quality channels.

Keywords: Aperiodic Pseudo-Random sequence · Broadband communication systems · Jamming · Test cut · Synchronization · Bernstein inequality

1 Introduction

In the wideband communication systems, the irresponsiveness to the random and intentionally arranged countermeasures is usually assessed by means of the complicated signals with the structure changing under the aperiodic pseudo-random sequence law (APRS) [1–8]. To decode such signals, the receiving party needs to have the APRS, identical to the one sent [1, 2]. The reliable functioning of wideband communication systems with the intentionally arranged countermeasures is only possible when this condition is satisfied [9]. This creates the task of synchronizing the wideband communication system APRS sensors as an important independent field of study.

The scientific studies of the APRS sensor synchronization in the low-quality channels are a way to improve the efficiency of the designed wideband communication systems with the organized countermeasures [9].

There are many works by different authors dedicated to the problem of synchronizing the pseudo-random sequence (PRS) sensors: by B.I. Bazhenov, L. A. Vaynshteyn, E. B. Brodskaya, B.P. Kalmykov, A. F. Kozlov, R.S. Rachkauskas, V. S. Skvortsov, V. V. Losev, V. I. Korzhik, E. M. Martynov, J. J. Stiffler, E. R. Keyne, V. Lindsay and other authors [10, 11]. These works consider the synchronization process from the point of

R. Silhavy et al. (Eds.): CoMeSySo 2020, AISC 1295, pp. 910–918, 2020.
https://doi.org/10.1007/978-3-030-63319-6_84

view of finding the signal at a given moment of time and determining its delay value with regard to the tact signal of the local generator.

As a rule, the known mathematic evaluations of PRS synchronization are carried out for binomial channels of quite good quality with the moderate amount of countermeasures in the channel and small PRS periods [10, 11].

In the known studies, the PRS synchronization assessment based on the test piece (TP) method for small periods in the low-quality channels suggests that the TP generated by dividing the analysis interval L by the test piece length do not overlap (see Fig. 1).

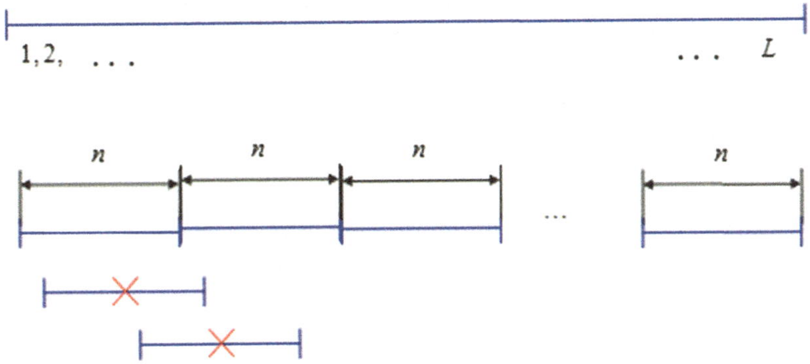

Fig. 1. Method of dividing the analysis interval L into independent TPs with the length n.

However, this method does not account for the test pieces that occur at the junctions of the adjacent TPs, which creates significant errors in assessing the possibility of correct and incorrect reception of the APRS TP in the low-quality channels with the known means of evaluation. Therefore, the task of assessing the synchronization of the APRS sensor with the TP method with account for all the test pieces in the analysis interval L is set. At that, the known accurate assessments create unsolvable transcendent equations in the computation process, becoming unusable for APRS synchronization calculations with an endlessly big period [11]. At the same time, the approximate computation methods have greater error values, therefore becoming unsuitable for the assessment of synchronization parameters with the strong organized countermeasures [10].

For this reason, the best solution for the wideband communication system designers would be a border method of evaluating the APRS sensor synchronization on the low-quality channels that would enable them to design a synchronization system for the worst-case scenario. Let us consider the solution to this problem based on the Bernstein bound evaluation.

Let us find the upper bound of APRS sensor synchronization in low-quality channels with the test piece method with the Bernstein inequality, which, unlike other known methods, does not depend on the distribution of errors in the liaison channel and the APRS period, simplifying the synchronization evaluation in the process of wideband communication system design.

2 Problem Setting

Terms:

Transmitted signal is a test piece, presenting n binary-coded APRS symbols.

Liaison channel is a binary binomial channel with the symbol reception error possibility of P. The channel has tact synchronization; cycle synchronization is absent. For this reason, the number of TP (test piece) symbols corresponding to the received symbols is unknown. However, a priori we know that the receiving party begins processing the signal before the first TP symbol is received.

Processing unit constraints: in each tact, not more than n *of* L binary symbols received from the liaison channel are memorized. At that, there are different reception situation possible: $n = L$, $n < L$.

The task is to receive L symbols from the liaison channel and to make up a judgment on the TP location with respect to the received information, i.e. to find the beginning (or the end) of the test piece.

The processing quality criterion is the possibility of correct TP reception $(P_{cr} = 1 - P_i)$, where P_i is the possibility of incorrect test piece reception.

3 Formal Problem Setting

There are L incoming binary symbols $u_1, u_2, \ldots, u_L$, and the length of TP of n binary symbols $S_1, S_2, \ldots, S_n$ is known. It is required to determine the true TP position with respect to the sequence $u_1, u_2, \ldots, u_L$, if the entire TP is known to be a part of the sequence $u_1, u_2, \ldots, u_L$.

4 Justification of the Solution Method

To solve the problem of determining the correct synchronization possibility, we need to know the function of distribution possibility of the different TP positions. If the given function is uniform or its type is unknown, the optimal method will be the maximum likelihood method, used for every of $L-n$ possible shears to calculate the posterior probability of coincidence of the selected shear to the true one; the solution is the shear with the maximum coincidence probability [40, 42, 52, 74]. Therefore in this case, we calculate the probability of occurrence of the sequence $u_1, u_2, \ldots, u_L$ at every selection of shear, at $L - n$ different distances of the test piece from the reference input signal and select the shear with the maximum probability:

$$P(u_1, u_2, \ldots, u_L / j) = \prod_{k=1}^{j-1} P^{U_k}(1-P)^{1\oplus} \cdot \prod_{k=j}^{j=n-1} P^{U_k S_{k-j+1}} \times (1-P)^{1 \oplus U_k \oplus S_{k-j+1}} \cdot \prod_{K=j+n}^{L} P^{U_k} \cdot (1-u_k)^{1 \oplus U_k}, \tag{1}$$

where $\oplus$ is addition under module two.

This proportion is true in the situation when the liaison channel asymmetry is not big (for example, when amplitude modulation is used).

However, when the liaison channel is significantly asymmetric, the countermeasures (noise) in the discreet channels are not additive, i.e. technically speaking, the proportion cannot be accepted as $u_i = S_i \oplus e_i$, where e_i is a countermeasure in the liaison channel, $P(e_i = 0) = 1-P$, $P(e_i = 1) = P$. Consequently, the proportion (1) does not work any more.

Let us consider the sequence $u_1, u_2, \ldots, u_j \oplus S_1, u_{j+1} \oplus S_2, \ldots, u_{j+n-1} \oplus S_N, u_{j+n}, \ldots, u_l$.

Let us signify the number of units in the given sequence using d_j.

Then proportion (1) gets the following form:

$$P(u_1, u_2, \ldots, u_L / j) = P^{d_j}(1-P)^{L-d_j}. \quad (2)$$

With $P \leq 1/2$, the right-hand side of the Eq. (2) is monotonously decreasing as d_j is growing. Consequently, to apply the maximum likelihood method, it is enough to use the accepted sequence $u_1, u_2, \ldots, u_l$ to form the following sequences:

$u_1, u_2, \ldots, u_{j-1}, u_j \oplus S_1, u_{j+1} \oplus S_2, \ldots, u_{j+n-1} \oplus S_N, u_{j+n}, u_{j+n+1}, \ldots, u_l, j = 0, 1, \ldots, n - L + 1$.

The solution of the true shear computation problem in this case will be the shear value, which corresponds to the sequence of the smallest number of units.

If in the signal transmission process the frequency or phase modulations have been applied, the proportion (1) is not true anymore, as in the absence of a direct code there is an equal chance of having both zeroes and ones, as there is an assumption that in the absence of signal, the white Gaussian noise is coming into the channel. At the frequency or phase modulation, the proportion (1) gets the following form

$$P(u_1, u_2, \ldots, u_L / j) = \prod_{k=1}^{n+j-1} P^{U_k \oplus S_k}(1-P)^{U_k \oplus S_k + 1}. \quad (3)$$

From the proportion (3) we can see that to determine the most likely shear value, only the products need to be compared. In its turn, such comparison can be limited to the selection of such j, for which in the sequence of $u_j \oplus S_1, u_{j+1} \oplus S_2, \ldots, u_{j+n-1} \oplus S_n$ the number of units would be minimal.

Therefore, we will use the rule when the current cross-correlation function is calculated to be compared with the threshold value:

$$\sum_{i=1}^{n} u_{i+j} \cdot S_i \underset{>}{<} u_0.$$

At that, to minimize the possibility of false TP detection (which is critical in case of rare use of sensors), it is rational to use the Neyman–Pearson criterion.

In this situation, the sum of possibilities (2) for different shear values is calculated, and this value is compared with the set threshold value. If the found value is below the threshold value, the absence of TP is concluded.

To accelerate the decision-making process, it is necessary to make comparisons with the threshold values at every step, and if the threshold value was exceeded, the decision on successful TP acceptance shall be immediately made.

In case of a good proportion of the parameters, the excess of two and more threshold values is unlikely, and therefore, this method will be close to optimal.

5 Generation of the Formula for the Upper Bound Possibility of the Synchronizing Sequence Reception Error

Based on the aforesaid, let us find the possibility of failure to receive the TP with the Bernstein bounds, derived for the centered random variable as described below.

The advantage of using the Bernstein bounds is that the possibility of exceeding the cross-correlating function threshold exponentially depends on the number of n components, or, in other words, on the TP length. Moreover, to apply those bounds, there is little a priori information that needs to be used. There is no need to know the distribution laws; it is enough to know the dispersion and the average random value (RV). Let us apply the Bernstein inequality to determine the possibility of TP reception error in the binomial low-quality channels.

Let us have the independent RV ξ_i, $i = 1, 2, 3, \ldots n$ with the finite expectation M $(\xi_i) = a_i$ and dispersion $D(\xi_i) = b_i$.

Let us introduce a new RV as $\eta_i = \xi_i - a_i$ and find the expectation and dispersion of the sum from one to n for the new RV, i.e. we get: $M(\eta_i) = \sum_{i=1}^{n} \eta_i = S_n$; $D(\eta_i) = \sum_{i=1}^{n} b_i = B_n$.

Let there be the inequality of

$$M(\eta_i) \leq \frac{b_i}{2} H^{k-2} \cdot k!,$$

where $k \geq 2, H > 0, 0 < t < \frac{B_n}{2H}$.

Then, for the RV, the following Bernstein inequalities will be true:

$$P\{S_n \geq 2t\sqrt{B_n}\} < e^{-t^2};$$

$$P\{S_n \leq -2t\sqrt{B_n}\} < e^{-t^2};$$

$$P\{|S_n| \geq 2t\sqrt{B_n}\} < 2e^{-t^2};$$

Let us consider an isolated situation of a RV with a zero expectation and equal dispersions of $a_i = 0$, $b_i = \sigma^2$. In this case, the Bernstein inequalities will get the following form

$$P\left\{\sum_{i=1}^{n} \xi_i \geq 2t\sigma\sqrt{n}\right\} < e^{-t^2};$$

$$P\left\{\sum_{i=1}^{n} \xi_i \leq -2t\sigma\sqrt{n}\right\} < e^{-t^2};$$

$$P\left\{\left|\sum_{i=1}^{n} \xi_i\right| \geq 2t\sigma\sqrt{n}\right\} < 2e^{-t^2},$$

where $0 < t < \frac{\sigma\sqrt{n}}{2H}$ $0 < t < \frac{\sigma\sqrt{n}}{2H}$.

At $H > 0$, $M\{\xi_i^k\} \leq \frac{\sigma^2}{2} H^{k-2} \cdot k!, k = 2, 3, \ldots$ will be true.

The Bernstein inequalities are true for any random values ξ_i, if their expectation is $M\{e^{\gamma\xi_i}\} < \infty$, and $|\gamma| < \frac{1}{2H}$:

$$P\left\{\sum_{i=1}^{n} \xi_i \geq 2t\sigma\sqrt{n}\right\} < e^{-t^2};$$

$$P\left\{\sum_{i=1}^{n} \xi_i \leq -2t\sigma\sqrt{n}\right\} < e^{-t^2};$$

$$P\left\{\left|\sum_{i=1}^{n} \xi_i\right| \geq 2t\sigma\sqrt{n}\right\} < 2e^{-t^2}$$

where $0 < t < \frac{\sigma\sqrt{n}}{2H}$; $H > 0$, for which $M\{\xi_i^k\} \leq \frac{\sigma^2}{2} H^{k-2} \cdot k!, k = 2, 3, \ldots$ is true.

From the generated inequalities we see, that if the RV are limited $|\xi_i| < C - const, i = 1, 2, 3, \ldots, n$, the Bernstein inequality can be applied to any t. The majority of physical values are always limited, and even if there are unlimited value models (Gaussian distribution) applied to them, then for the random values themselves, not for the models, the Bernstein inequalities will be also true.

If the expectation of a RV $M\{e^{c\xi_i}\} < \infty$ is a limited value, the Bernstein inequalities can be brought to a more convenient form for practical use. For this purpose, the t parameter can be presented as $\mathrm{t} = \frac{\mathrm{A}\sqrt{\mathrm{n}}}{2\sigma}$, and then the Bernstein inequalities can be recorded as

$$P\left\{\sum_{i=1}^{n} \xi_i \geq An\right\} < e^{\frac{-A^2 n}{4\sigma^2}};$$

$$P\left\{\sum_{i=1}^{n} \xi_i \leq -An\right\} < e^{\frac{-A^2 n}{4\sigma^2}};$$

$$P\left\{\left|\sum_{i=1}^{n} \xi_i\right| \geq An\right\} < 2e^{\frac{-A^2 n}{4\sigma^2}}.$$

The given inequalities can be applied to the sums of mutually independent random values with zero expectation and the same dispersions for any parameter $A > 0$ $A > 0$, if the exponential moments of these random values are limited, which is always true when the RV themselves are limited.

Using the Bernstein inequality, the APRS probability of failure to receive was calculated as follows:

$$\bar{P}_H \leq (L - 2n)\left(\frac{-[2P-1]^2}{e^{4(4P(1-P)+1)}} + \frac{-[(1-2P)(n+R(T))]^2}{e^{16P(1-2)(4d(T)+T)n}}\right), \tag{4}$$

where L is the length of the analyzed PRS segment; n is the TP length; $R(T)$ is the aperiodic mutual correlation function; $d(T)$ is the number of mismatches between the start and reference signals within the equal time periods calculated as

$$d(T) = \begin{cases} 0 \\ \sum_{t=1}^{n-T} S_i \oplus S_{i+T} \end{cases}$$

Then for the possibility of correct TP reception, the following evaluation will be true:

$$P_{\Pi\Pi} \geq 1 - P_H. \tag{5}$$

Unlike the known precise and approximate methods, the developed Bernstein inequality-based mathematical models (4) and (5) make it possible to produce a strict evaluation of the probability of APRS synchronization for any periods of any big size, used in the wideband communication systems for protection from the organized countermeasures.

The produced formula (4) is quite convenient for computation. The results of using the formula for the computation of the incorrect TP reception probability in Matlab medium $\bar{P}_{\mathrm{H}}$ at different values of n are shown in Fig. 2.

For comparison, it also presents similar incorrect TP reception probability curves produced with the known formula suggested by [11].

The systematic analysis of the collected results demonstrates that at the same channel and TP parameters, the developed Bernstein inequality-based mathematical model has an advantage of being countermeasure-resistant compared to the previously known method. For example, at n = 31 and P = 0.1 respectively, in the first situation the incorrect reception probability was $\bar{P}_H = 10^{-1}$, while in the second it was $\bar{P}_H = 3 \cdot 10^{-1}$, and the advantage grows with the increasing TP length, i.e. the developed model is one order more precise than the previously known methods [11].

And as we can see from Fig. 2, with the increase of the test piece to n = 127 and the error probability in the channel to P = 0.1, the mentioned advantage reaches three decimal exponents. The results of the computation shown in Fig. 2 show that, compared with the known approximate APRS synchronization evaluation method [11], the developed mathematical model considering all the test pieces allows to reduce the possibility of failure due to the increased TP length in the low-quality channels. From Fig. 2 we can see, that if at n = 63 and channel error probability P = 0.2, the incorrect reception probability is $\bar{P}_H = 7 \cdot 10^{-2}$, then at the same error probability in the channel the length increase to n = 127 produces the error probability of $\bar{P}_H = 10^{-3}$.

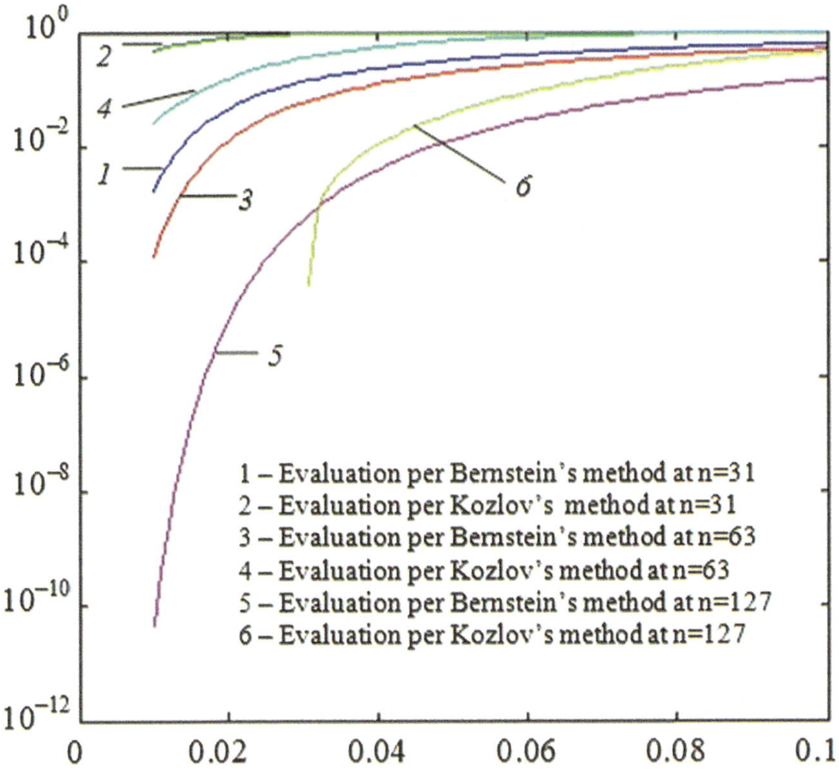

Fig. 2. Probability of incorrect TP reception computed with the Bernstein inequality.

6 Conclusions

This way, the calculations prove that, unlike all the known approximate methods, the Bernstein inequality-based mathematical model allows to produce a strict evaluation of the APRS synchronization probability for any periods of any size, used in the commercial wideband communication systems for protection from the organized countermeasures. For this reason, the developed mathematical model allows improving the operation reliability of the designed commercial wideband communication systems with the efficient methods for evaluation of the low-quality channel APRS synchronization ($P \geq 0.1$), which is especially relevant for improving the reliability of the operation of remotely controlled corporate networks exposed to any organized countermeasures.

References

1. Malygin, I.: Broadband Communication Systems. LAP LAM-BERT Academic Publishing. Moscow, Russia (2012)

2. Dornan, E.: Broadband from space. LAN **2**, 14–25 (2002)
3. Tomasi, W.: Electronic Communications Systems: Fundamentals Through Advanced. Prentice Hall, Englewood Cliffs (1988)
4. Filatov, V.I.: Broad-band high speed radio communication system. Sviaz'. http://trudymai.ru/published.php?ID=57889. Accessed 15 Aug 2019
5. Bychkov, V., Dmitriev, Iu., Khvostov, D.: Broadband communication lines in Russia in the near future through the eyes of cable men. http://lib.broadcasting.ru/articles2/econandmen/shirokopolosn_lin_svyazi_ros_blizh_per-spekt_glaz_kabelsc. Accessed 15 Aug 2019
6. Ericsson Mobility Report Ericsson.com. http://www.ericsson.com/res/docs.2016/ericsson-report-2016. Accessed 29 Aug 2019
7. Odinskii, A.L.: MOTOROLA mobile broadband digital information systems. Evraziia. Vesti. http://www.eav.ru/publs.php?nomber=2008-07. Accessed 29 Aug 2019
8. Yang, L.L., Hanzo, L.: Serial acquisition performance of single-carrier and multi-carrier DS-CDMA over Nakagami-m fading channels. IEEE Trans. Wirel. Commun. **1**(4), 692–702 (2002)
9. Fortinet: forecasting cyber threats in the near future. https://www.itweek.ru/iot/article/detail.php?ID=206001. Accessed 29 Aug 2019
10. Ksenofontov, V., Khisamov, D., Milovanov, M., Zolotuev, A.: Method of analog synchronization of pseudorandom sequences under strong jumming. Inf. Res. Russia **2**, 38–39 (2015)
11. Khisamov, F.G., Podol'tsev, V.V.: Improving accuracy of synchronization when processing short segments of memory bandwidth in MC-CDMA systems. Comput. Sci. Secur. Mod. Soc. **3**(5), 26–35 (2018)

Training and Application of Neural-Network Language Model for Ontology Population

Pavel Lomov, Marina Malozemova(✉), and Maxim Shishaev

Institute for Informatics and Mathematical Modeling, Kola Science Centre of the Russian Academy of Sciences, Apatity, Russia
{lomov,malozemova,shishaev}@iimm.ru

Abstract. This paper considers one of the subtasks of ontology learning - the ontology population, which implies the extension of existing ontology by new instances without changing the ontology structure. A brief overview of existing ontology learning approaches and their software implementations is presented. A highly automated technology for ontology population based on training and application of the neural network language model to identify and extract potential instances of ontology classes from domain texts is proposed. The main stages of its application, as well as the results of its experimental evaluation and the main directions of its further improvement are considered.

Keywords: Ontology population · Neural network · Ontology learning

1 Introduction

The use of ontologies as a component of information systems for domain knowledge representation become a common practice today. Ontology is a formal, explicit specification of a shared conceptualization [1–3]. A conceptualization is a representation of objects that are assumed to exist in some area of interest and the relationships among them. A shared conceptualization means that some group of agents (humans, organizations, information systems, etc.) accept it. A formal character of the specification points out that its definition has been made with using a formal language such as Resource Description Framework (RDF), Web Ontology Language (OWL) or the Simple Knowledge Organization System (SKOS).

However, the development of ontologies is a complex problem requiring the involvement of domain experts as well as ontology engineers. For this reason, attempts are being made to automate it in order to reduce labor costs with obtaining as a result a quality ontology that reflects in the necessary extent knowledge of the subject area. Further support of existing ontologies during their using in information systems oriented on adaptation to changing requirements is also important. The solution to these problems is considered in the context of the

R. Silhavy et al. (Eds.): CoMeSySo 2020, AISC 1295, pp. 919–926, 2020.
https://doi.org/10.1007/978-3-030-63319-6_85

ontology learning, which is defines a set of methods and techniques for creation of new ontologies from scratch or for extension of already existing ontologies in a automatic or semiautomatic way from various resources.

In this paper, we consider the solution to the problem of population of an ontology by a new concepts by using of a neural network language model to the analysis of domain texts. Ontology population is one of the sub-tasks of ontology learning. It consists in addition of a new class instances into existing ontology without any modification of its structure. To solve this problem, we propose a technology for training and subsequent application a neural network language model for detecting and extracting potential instances of ontology classes from domain texts.

The remaining part of the paper is structured as follows: Section 2 gives general observation of existing approaches, techniques and software tools for solution of ontology learning tasks. Section 3 describes the main stages of proposed ontology population technology, Section 4 presents experimental evaluation of the technology and discussion about obtained results. The final section provides conclusions and defines the directions for further research.

2 Overview of Existing Ontology Learning Approaches

Ontology learning can be considered as reconstruction from text the domain model, originally formed in the mind of an expert - text author [4,5]. As a rule, ontology learning involves analysis and extraction of domain terms from texts, their subsequent grouping to determine ontology classes and relations between them. In conclusion, the logical restrictions (axioms) found in the texts can be determined on the generated system of classes. Today, ontology learning approaches differ in techniques that are applied on different stages of ontology formation. These techniques can be divided into three types: linguistic, statistical, and logical [6].

At the initial stage, the text corpus is pre-processed using linguistic techniques such as part-of-speech tagging, parsing, and lemmatization. After text preprocessing, the domain terms are extracted. For this, various natural language processing (NLP) methods are used. Among them the following ones can be distinguished: syntactic parsing, subcategorization frames and seed words search[7]. Also some statistical text analysis methods, such as identifying multi-words terms using C-value/NC-value, contrastive analysis [8], co-occurrence analysis, latent-semantic analysis (LSA), and clustering [9] are applied.

At the next stage, the ontology structure is formed. It involves identifying taxonomic and non-taxonomic relationships between the terms found at the previous stage. Various linguistic and statistical methods for processing text such as dependency analysis, lexico-syntactic analysis, term subsumption, formal concept analysis (FCA), hierarchical clustering (agglomerative [10] and divisive [11]) and association rule mining (ARM) are applied.

At the final stage, inductive logical programming (ILP) methods [12] can be applied to derive new relationships in the form of hypotheses based on the elements and relationships of the obtained ontology.

Among the well-known software tools focused on ontology learning are the following:

- CRCTOL (Concept-Relation-Concept Tuple-based Ontology Learning) [13]. This system parses the text and assigns POS and syntax tags to its lexemes. Further, to form an ontology, the following set of algorithms is applied: a statistical algorithm for extracting key concepts; disambiguation algorithm for the found concepts; an algorithm based on lexical-syntactic patterns for extracting relationships from the text.
- OntoGain [14]. A system for unsupervised ontology acquisition from a text consisting of the following main modules: a preprocessing module that performs text parsing tasks; module for extraction domain concepts; module of taxonomy formation for detected concepts; module for extracting non-taxonomic relations based on associative rules [15].
- OntoLearn [16]. This is a system for ontology generation based on extracted domain concepts. The concepts are extracted with using of the statistical and NLP methods [17]. Then a search for the concepts definitions in thesaurus and domain texts is performed with the subsequent syntactic analysis of the definitions and elimination of irrelevant ones. Finally based on the results of the analysis of obtained definitions, taxonomy of concepts is formed.

The technology proposed in this work is focused on solving one of the subtasks of ontology learning, namely ontology population. Thus, it requires the presence of an existing ontology containing concepts and relations between them. The technology involves the use of linguistic methods (tokenization, lemmatization, syntactic analysis) for text preprocessing and the formation of a training set with subsequent training and application of a neural network language model aimed.

3 Proposed Technology for Ontology Population

The proposed technology involves the use of concepts existing in the ontology to detect and extract from domain texts new concepts which are semantically close to initial ones. Semantic closeness of concepts in this case follows from the their presence in similar contexts, which in turn are represented by text sentences with similar structure. To identify such sentences containing the domain concepts, texts are analyzed using a language model based on a neural network, initially oriented to solving named entity recognition problem (NER). This task implies extraction of named entities belong to some category from texts. A named entity is a word or phrase denoting an object that can be categorized. Examples of named entities are the names of people, organizations and locations. In the proposed technology, the role of a recognized entity is played by a potential instance (individual) of some ontology class. The latter, in turn, is considered as an assigned category.

A key component of the technology is the procedure for training a language model by means of a training set, which is automatically generated by searching in a set of domain texts the sentences containing separate tokens and noun groups corresponding to instances of the ontology. It is assumed that many of such sentences represent contexts that in other texts may contain new terms that are similar in meaning to the initial ones.

Then with the generated set a language model from widespread Python NLP library spaCy is trained. Subsequently, with the help of the trained model, the domain texts are analyzed. The extracted terms are considered by the human expert as candidates for inclusion in the ontology as instances of classes. The application of proposed technology is presented on Fig. 1.

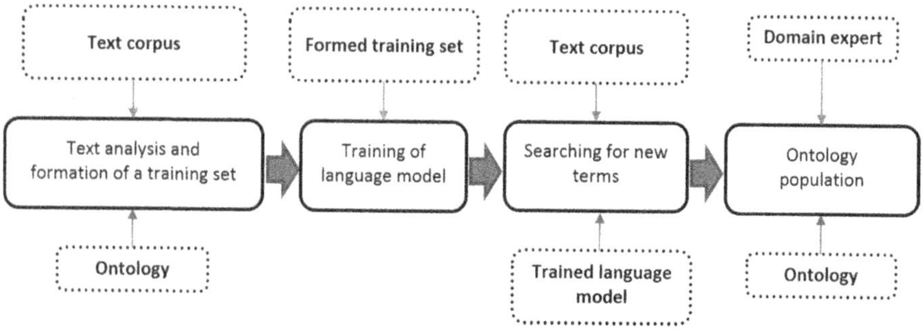

Fig. 1. General scheme of proposed ontology population technology

The effectiveness of the use of technology is largely determined by the quality of the trained model, which in turn depends on the size of a training set, the diversity and truthfulness of the samples presented in it. Consider its formation in more detail.

At the first stage, a set of domain texts is formed, which will be further used as a source for obtaining training samples. To solve this problem, a crawler [18] was developed, which allows you to automatically collect a set of documents (files in PDF, DOC, DOC, DOCX, ODT) of a given topic and form a text corpus. The crawler is based on the widely used python framework Scrapy, which focuses on the search and analysis of web pages.

The developed crawler allows you to automatically download documents, the links to which are published on pages pointed by user, as well as perform searching with helps of the google search engine and download documents from found pages. The latter feature provides the ability to constantly replenish the corpus of domain texts. As a source for the automatic formation of search queries, the conceptual system of initial ontology is used. That ensures the correct thematic orientation of the found texts.

Next, preprocessing of the texts is carried out, during which the obtained texts are divided into sentences and tokens. For this, a set of methods was developed based on the NLTK python library, which allows you to get the necessary data from the text corpus [19]. For example:

("Travel up to Kirovsk at the foot of the Khibiny Mountains and go on a snowmobiling adventure while traveling through rugged mountain regions and beautiful tundra", 13, 19, CITY).

For this purpose, a specialized software module [20] was developed. The module allows to detect tokens that correspond to the names of instances of ontology classes, as well as their boundaries. The main problem in this case is the discovery of instance names consisting of several tokens. In sentences, such complex name can be represented partially and/or its tokens may has other order which.

After the formation of the training set, the language model is trained. As a result of training, the model can be used to detect domain specific concepts in the texts. The concepts found in this way are presented to the expert for final solution about their addition to the ontology as instances of classes.

4 Evaluation

The proposed technology was evaluated in the experiment which implies training the language model and its subsequent using for population of the ontology of Arctic economic activity [21]. Classes and instances of this ontology represent various types of industrial and research activities in the Arctic, as well as objects associated with them. To train the model we collected and analyzed about 400 texts of the Arctic domain, which contain lexemes and noun groups representing names of instance of ontology classes. As a result, a set consisting of 360 000 samples was formed.

The quality assessment of the trained model was carried out on a test set of a similar structure. For its creation the domain texts that were not applied for training set formation were used. Their collecting and analysis was performed manually. Thus, the resulting test set is considered a reference result of the extraction of ontology concepts from texts. Assessment the quality of the model was performed by comparing of the test set and a set which was obtained as a result of applying the language model. For comparison, the multi-language NER model from spaCy library oriented on extraction of person, location and organization names from news text was also evaluated. The following results were obtained:

- Experiment 1. Detection of concepts from the test set:
  - multi-language spaCy model: precision = 0.13, fullness = 0.03;
  - trained model: precision = 0.95, fullness = 0.35.
- Experiment 2. Detection of concepts from the test set that are not represented in the ontology:
  - multi-language spaCy model: precision = 0.013, fullness = 0.012;
  - trained model: precision = 0.25, fullness = 0.006.

Results of the first experiment indicate the presence of some positive effect of training. The model has learned to recognize some contexts of the usage of domain concepts. However, this is not enough to discover new concepts, as shown by results of the second experiment.

This can also be caused by the fact that the initial hypothesis that the names of instances of same class will meet in same contexts is confirmed to a greater extent for classes that potentially have a large variety of instances in the analyzed texts. For example, the classes "Organization" or "Person" can potentially have a large number of instances in news texts. Thus, the level of specificity of the classes of the ontology and the analyzed texts should correspond. In the context of this study, this means that it is necessary not only to increase the size of text corpus, but also to ensure more accurate search and/or additional filtering of the texts according to ontology domain.

On the other hand, the reason for the poor results of new concepts extraction may be a presence in the ontology of a small number of instances of classes, which also does not allow the model to form a sufficient variety of generalized contexts for the discovery of new concepts.

Thus, based on the obtained results it is necessary to increase the size of the training set and the variety of samples included in it. To this end, it is planned to improve the effectiveness of the crawler regarding the collection of text date directly from the web-pages of specialized sites, blogs and forums. Along with this, it is planned to consider the use of the names of relations defined between classes to detect text contexts of the instances of these classes. It also looks promising to use the data augmentation techniques with respect to the formed training set [22] to expand it by generating additional correct training samples based on existing ones.

5 Conclusion

Ontology learning is a multi-aspect problem, for the solution of which it is possible to apply various stacks of technologies, methods and software depending on the task (creating an ontology from scratch, expanding the structure of an existing ontology with new classes and relationships, populating an ontology with instances), extent of initial data structuring, required degree of automation and quality of the ontology obtaining as a result.

The technology involves the automatic formation of a set of thematic texts and their analysis using traditional NLP techniques (tokenization, lemmatization, parsing) to search for sentences that represent the contexts of using instance names from the initial ontology. Based on the selected sentences, a set of training samples is generated for training a language model based on a convolutional neural network that is focused on solving the NER problem. The correspondence of the collected texts and training samples to the topic of the ontology is substantiated by the use of its elements - the names of the instances and classes during the formation of the corresponding sets. As a result of training, the model generalizes the presented contexts, which allows it to find concepts used in similar contexts in other texts of the subject area. It is assumed that some of these concepts will represent new instances for inclusion in the ontology.

An experimental verification of the proposed technology has shown a positive effect in relation to finding the concepts of ontology in arbitrary texts of the subject area. However, for its full-fledged usage it is necessary to increase the number of training samples and their diversity. This determines the further direction of this study.

The proposed technology can be used when reusing ontologies for their actualization and/or bringing their content in line with the tasks of information systems using them. High automation of the technology allows you to reduce the labor costs of domain experts providing ontology support, and thereby reduce the cost of developing and operating ontology-driven information systems.

Acknowledgements. The study was funded by the Russian Foundation for Basic Research (RFBR), project number 20-07-00754.

References

1. Studer, R., Richard Benjamins, V., Fensel, D.: Knowledge engineering: principles and methods. Data Knowl. Eng. **25**(1–2), 161–197 (1998). 25:161–197
2. Gruber, T.R.: A translation approach to portable ontology specifications **5**(2), 199–220 (1993)
3. Borst, W.N., Borst, W.N.: Construction of engineering ontologies for knowledge sharing and reuse (1997)
4. Somodevilla, M., Ayala, D., Pineda, I.: An overview on ontology learning tasks, vol. 22 (2018)
5. Chen, J., Dosyn, D., Lytvyn, V., Sachenko, A.: Smart data integration by goal driven ontology learning, pp. 283–292 (2017)
6. Asim, M.N., Wasim, M., Khan, M.U.G., Mahmood, W., Abbasi, H.M.: A survey of ontology learning techniques and applications (2018)
7. Hwang, C.: Incompletely and Imprecisely Speaking: Using Dynamic Ontologies for Representing and Retrieving Information (1999)
8. Navigli, R., Velardi, P.: Semantic interpretation of terminological strings (2002)
9. Berkhin, P.: Survey Of Clustering Data Mining Techniques, vol. 10 (2002)
10. Zepeda-Mendoza, M.L., Resendis-Antonio, O.: Hierarchical agglomerative clustering. In: Dubitzky, W., Wolkenhauer, O., Cho, K.-H., Yokota, H., (eds.) Encyclopedia of Systems Biology, pp. 886–887. Springer (2013)
11. Dhillon, I.S., Mallela, S., Kumar, R.: A Divisive Information-Theoretic Feature Clustering Algorithm for Text Classification (2013)
12. Lima, R., et al.: An Inductive Logic Programming-Based Approach for Ontology Population from the Web (2013)
13. Jiang, X., Tan, A.-H.: CRCTOL: A semantic-based domain ontology learning system (2010)
14. Drymonas, E., Zervanou, K., Petrakis, E.: Unsupervised Ontology Acquisition from Plain Texts: The OntoGain System, vol. 6177, pp. 277–287 (2010)
15. Srikant, R., Agrawal, R.: Mining generalized association rules **13**(2), 161–180 (1997)
16. Navigli, R., Velardi, P., Gangemi, A.: Ontology Learning and Its Application to Automated Terminology Translation **18**(1), 22–31 (2003)
17. Navigli, R., Velardi, P.: Learning Domain Ontologies from Document Warehouses and Dedicated Web Sites. **30**(2), 151–179 (2004)
18. Lomov, P., Malozemova, M.: Crawler module (2020). https://gitlab.com/iimm-dev/tsg/-/tree/rc/crawler

19. Lomov, P., Malozemova, M.: Text processing module (2020). https://gitlab.com/iimm-dev/tsg/-/tree/rc/text_processing/arctic_text_corpus
20. Lomov, P., Malozemova, M.: Training module (2020). https://gitlab.com/iimm-dev/tsg/-/tree/rc/training
21. Lomov, P.: Ontology of integrated knowledge space (2018). https://github.com/palandlom/ontology-of-integrated-knowledge-space/blob/master/src/poltes.owl
22. Badimala, P., Mishra, C., Venkataramana, R.K.M., Bukhari, S., Dengel, A.: A study of various text augmentation techniques for relation classification in free text, pp. 360–367 (2019)

Skewness in Applied Analysis of Normality

Marek Vaclavik[1(✉)], Zuzana Sikorova[1], and Tomas Barot[2]

[1] Department of Education and Adult Education, Faculty of Education, University of Ostrava, Fr. Sramka 3, 709 00 Ostrava, Czech Republic
{Marek.Vaclavik, Zuzana.Sikorova}@osu.cz

[2] Department of Mathematics with Didactics, Faculty of Education, University of Ostrava, Fr. Sramka 3, 709 00 Ostrava, Czech Republic
Tomas.Barot@osu.cz

Abstract. In the quantitative research, the property of the normality of data is an important assumption. For the particular selection of the concrete statistical methods, e.g. for purposes of the testing the hypotheses, the knowledge of the normality of data has the significant role. There exists a wide spectrum of methods aimed for the purposes of the testing the normality. These tests are based on the consideration of the statistical significance level. In this paper, a descriptive approach to an analysis of the normality is realized with regards to the sample parameter of the skewness. This parameter is explored with a consideration of the changing sample size in the frame of the applied quantitative research with 2671 respondents. The shapes of the histograms are discussed in a context of the obtained results of the parameter of the skewness. The trend of the dependence of the skewness on the Shapiro-Wilk criterion is analyzed using the regression analysis.

Keywords: Normality of data · Skewness · Shapiro-Wilk · Quantitative research

1 Introduction

The quantitative research [1] has been frequently used for a verification of the declared research questions in the social science. This type of research can be complemented with the opposite type of the qualitative research [2], where the individual observations are preferred instead of the using the statistical methods. As the advantage of the quantitative research in a comparison with the qualitative research, the possibility of an induction [1, 3] of achieved results on the whole statistical population is the significant characteristic.

The fulfilling the normality of data [4] is the initial necessary condition before the testing the hypotheses. According to this condition, the final selection of suitable methods is chosen for purposes of the following testing the hypotheses [1]. Due to the resulted property of the normality, either the non-parametrical [5] or the parametrical [5] methods are utilized.

As the generally used methods for the testing the normality [4], the following tests have been applied: the Shapiro-Wilk test [6], the Anderson Darling test [7], the Jarque-Bera test [6] or the Lilliefors test [8]. All of these tests provide results in the form of

R. Silhavy et al. (Eds.): CoMeSySo 2020, AISC 1295, pp. 927–937, 2020.
https://doi.org/10.1007/978-3-030-63319-6_86

p-values [5] as well as in the form of criterions [6, 7], which can be compared with the confidence intervals [5]. The constraints values for confidence intervals can be found in the statistical tables [9].

The testing the normality and the own testing the hypotheses are based on the principle of the mathematical induction [5] with the consideration of an existence of the statistical significance [5]. In the form of the p-value, the results of the testing are being compared with the declared significance level [5]. The consideration of the including the principles based on the statistical significance can be often seen in the research in the field of the social science [10–15] or in the technical applications [16–20].

The normality can be explored also by the descriptive parameters. As the one of these parameters, a skewness [21] can be a suitable characteristic, which is near or equal to zero in case of the appearance of the normal probability distribution of data [4].

In major cases of the practical realizations of the quantitative based researches, the testing normality can be frequently seen, e.g. in [15]. However, there are still many researches, where the testing the normality of data has not been considered. In this paper, a demonstration of the importance of the normality is presented using by the analysis of shapes of histograms [5], by the trend of the parameter of the skewness of the probability distribution and by the trend of the Shapiro-Wilk criterions [6] with regards to the changing sample size. The analysis is extended by the regression and correlation analysis [5] of the relation between the Shapiro-Wilk criterion and the parameter of the skewness. The advantage of this contribution can be seen in the analysis of the stability of behaviour of the normality with dependence on the sample size in case of the applied research [15] with 2671 respondents.

2 Normality of Data in Quantitative Research

The normality [4] of the cardinal random variable X with n values x_i is bound on the distribution of the density probability function [5]. Variable X, which fulfils the normality, has the Gaussian probability distribution [3]. The Gaussian probability distribution can be described by the function $f(x)$ (1) including the exponential part Exp with parameters of the mean value μ and the standard deviation σ [5].

$$f(x) = \frac{1}{\sigma\sqrt{2\pi}} Exp\left(-\frac{(x-\mu)^2}{2\sigma^2}\right) \tag{1}$$

The probability distribution (1) can be also described as (2). For the mean value equal to 0 and for the standard deviation equal to 1, the normalized normal probability distribution is defined by a plot, which can be seen in Fig. 1.

$$X \sim N(\mu, \sigma) \tag{2}$$

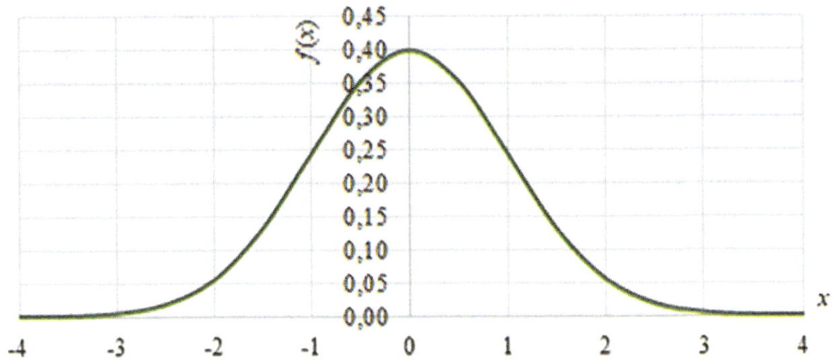

Fig. 1. Probability function of normalized Gaussian probability distribution N (0, 1)

In the frame of the quantitative research [1], the cardinal variable X can be achieved as the respondents' answers regarding the interval scale or can be computed as sequences of an arithmetical averages of more answers of each respondent.

An importance of the testing the normality of variables (further described also as a normality of data) is essential for a consideration, which concrete method should be used in favour of the testing the hypotheses in the quantitative research. With regards to the fulfilling or unfulfilling the property of normality, either the parametrical or non-parametrical tests are then applied on data sets [1, 4].

For situations of a selection of parametrical tests for purposes of the testing the cardinal variable across several items of the nominal variables, the T test (for 2 items) or the ANOVA test (for 3 or more items) are applied. The non-parametrical alternatives are for T-test: the Mann-Whitney test and for the ANOVA test: the Kruskal-Wallis test [1].

In the form of p-value [5] and also as the test criterion [6], the determination of the property of the normality can be realized by methods of: Shapiro-Wilk (with criterion W) [6], Anderson-Darling (with criterion A) [7], Jarque-Bera (with criterion JB) [6] or Lilliefors (with criterion L) [8]. In each test of the normality, the pair (3) of the zero and alternative hypotheses [5] is proven with conclusion considering the significance level α [5]. In the social science, α has been defined as 0.05, e.g. in [23].

$$H_0 : X \sim N(0,1); H_1 : X \nsim N(0,1) \tag{3}$$

In the technical or medical fields, α has been frequently declared as 0.01 or 0.001, as can be seen e.g. in [16].

3 Exploration of Behaviour of Normality by Skewness

In this section, the structure of an observation is proposed. At first, a progress of the statistical criterion of the Shapiro-Wilk test W [6] is measured on the interval of the changing the sample size of the statistical population. On the same principle, the computation of the descriptive parameter of a skewness [21] is being explored.

There exists also the descriptive estimation of the parameter of the skewness (4) [21], which is nearly bound on a shape of the probability distribution of the variables considered in the data sets. The expression (4) is the estimated version for the continuous case with the definition (5) [22] including the probability function *f*(*x*) and the mean value μ. In the rule (5), the improper integral in the numerator is the 3rd central moment [22] and the improper integral in the denominator is the 2nd central moment [22].

$$\zeta = \zeta(n) = \sqrt{n}\frac{\sum_{i=1}^{n}(\overline{x} - x_i)^3}{\sqrt{\left(\sum_{i=1}^{n}(\overline{x} - x_i)^2\right)^3}} \tag{4}$$

$$Z = \frac{\int_{-\infty}^{\infty}(x - \mu)^3 f(x)dx}{\sqrt{\left(\int_{-\infty}^{\infty}(x - \mu)^2 f(x)dx\right)^3}} \tag{5}$$

The normalized probability distribution is characterized by the zero skewness. Other exceptions are projected into the negative or positive values of the skewness [21].

As the further part of the presented analysis, the identification of dependences between the Shapiro-Wilk criterion *W* and the parameter of the skewness ζ is aimed. The regression function (6) [5] is assumed with the determination of the correlation coefficient *R* [5].

$$\zeta(n) = aW(n) + b; a, b \in \mathbb{R} \tag{6}$$

In the final step, the histogram [5] of occurred absolute frequencies of data is complemented by a displaying the arithmetical averages [5] of measured variables, for the purposes of the graphical interpretation of shape of the probability distribution.

4 Results

The behaviour of the Shapiro-Wilk criterion *W* and of the parameter of skewness ζ was analyzed in case of the practically realized quantitative research with 2671 respondents. The educational research of Z. Sikorova and her colleagues [15] was consisted of a wide spectrum of the research questions based on approaches to learning of the university students. Beside the testing the hypotheses, the factor analysis was previously used to improve the validity and reliability of the research. Approaches to learning were divided into three categories: Deep, Strategic and Surface, which were computed as the average values from a high number of the particular answers in the frame of the questionnaire.

For the computations, the possibilities of the statistical software solution PAST Statistics of version 4 [24] and IBM SPSS Statistics were suitably utilized.

In Table 1, resulted p-values (for the testing the normality of data) were obtained with regards to the changing the sample sizes. The Shapiro-Wilk test was applied on the significance level $\alpha = 0.05$, which has been frequently defined in the social science.

Table 1. Achieved results of testing normality using Shapiro-Wilk test for data sets

Data set	Results with regards to sample size	Conclusion about normality
Deep	$\forall n \in \{i+50\}_{i=150}^{2550}: p(n) < 0.05$	Normality was not proven on $\alpha = 0.05$
Strategic	$\forall n \in \{i+50\}_{i=150}^{2550}: p(n) < 0.05$	Normality was not proven on $\alpha = 0.05$
Surface	$\forall n \in \{i+50\}_{i=150}^{2550}: p(n) < 0.05$	Normality was not proven on $\alpha = 0.05$

In the graphical comparison of Fig. 2 and Fig. 3 with inconsideration of the scale of the vertical axes, the shapes of the displayed functions $W(n)$ and $\zeta(n)$ for data sets Deep and Strategy seem to be approximately similar in the trend view. For these data sets, the functions may be caused by the similar mathematical principles, which can be further identified in the following statistical research.

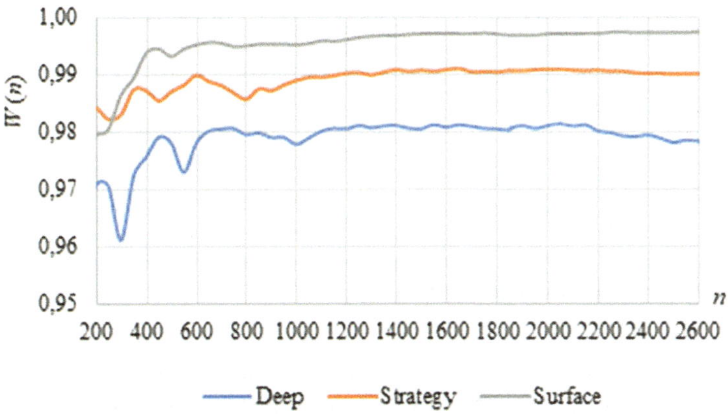

Fig. 2. Shapiro-Wilk testing criterion W in dependence on sample size n for variables of data sets (deep, strategic and surface)

In the following achieved results of the regression and correlation analyses (provided in Fig. 4, Fig. 6 and Fig. 8), the linear relation between the trends of the functions $W(n)$ and $\zeta(n)$ were proven for all data sets. The interval for the realized observation was defined from 1000 to 2600 by step 50 according to the approximately stable parts of the functions in Fig. 2 and Fig. 3. As the significant, the following correlation coefficients can be classified as the square root of achieved R^2: $R = 0.862$ (for data set Deep), $R = 0.888$ (for data set Strategic) and $R = 0.792$ (for data set Surface).

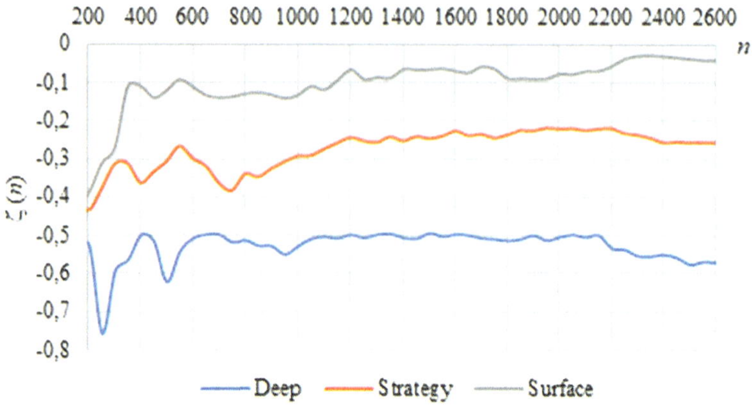

Fig. 3. Skewness coefficient ζ in dependence on sample size *n* for variables of data sets (deep, strategic and surface)

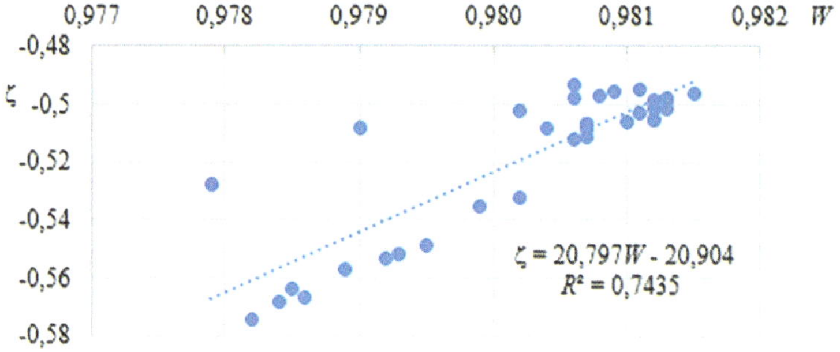

Fig. 4. Dependence of skewness ζ on Shapiro-Wilk criterion *W* with regards to sample size *n* = 1000, 1050, …, 2600 for data set (deep)

On Fig. 5, Fig. 7 and Fig. 9; the influence of the skewness value $\zeta(n)$ can be seen in shapes of the concrete histograms, in which the displaying the absolute frequencies of occurred values is shown. Shapes are more symmetrical for higher values of the sample size *n*. In case of lower values of *n*, the shapes of histograms became more asymmetrical. This described behaviour can be also seen in Fig. 2, where for lower values of the sample size *n*, the skewness is being decreased.

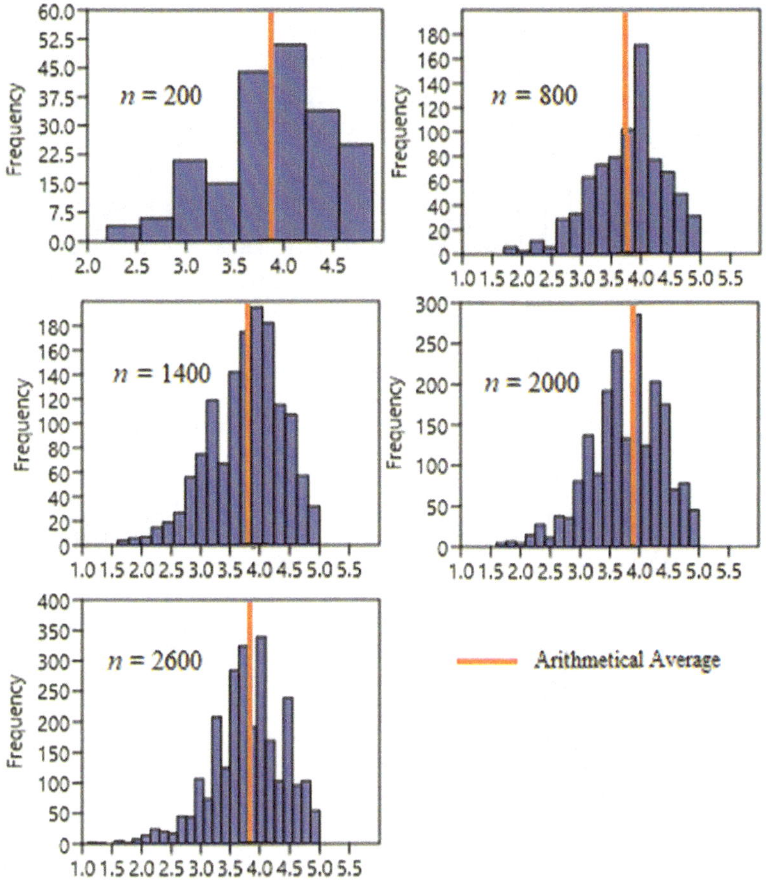

Fig. 5. Occurred frequencies in dependence on sample size for data set (deep)

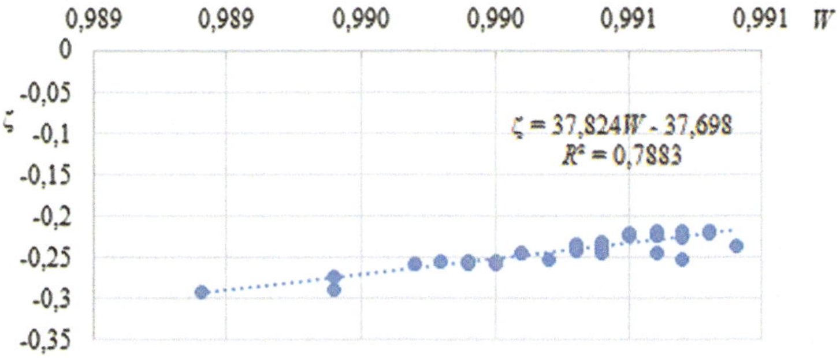

Fig. 6. Dependence of skewness ζ on Shapiro-Wilk criterion W with regards to sample size n = 1000, 1050, …, 2600 for data set (strategic)

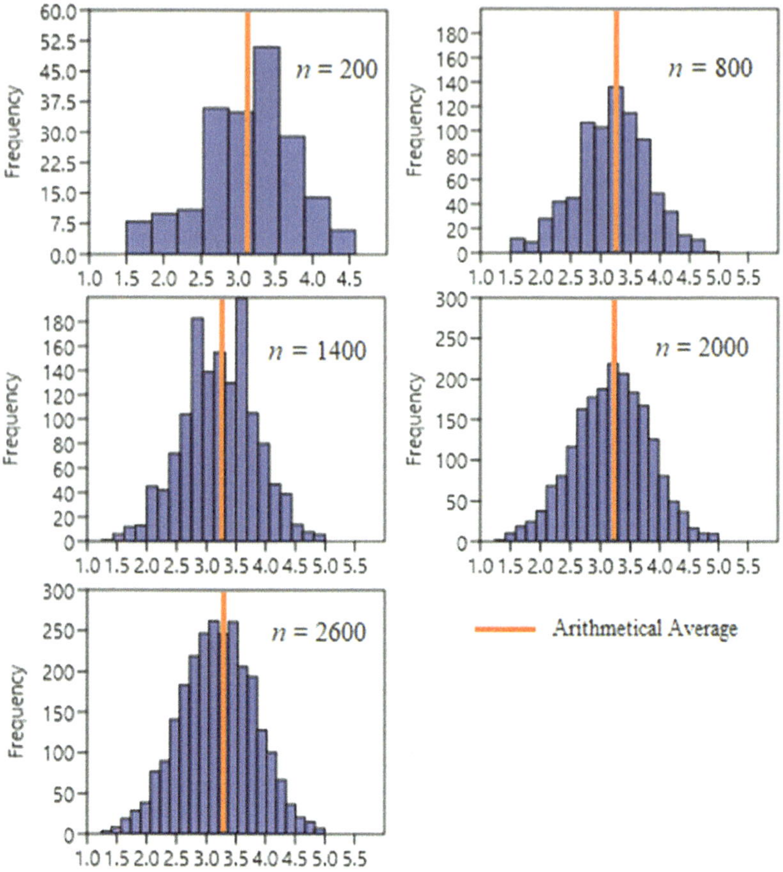

Fig. 7. Occurred frequencies in dependence on sample size for data set (strategic)

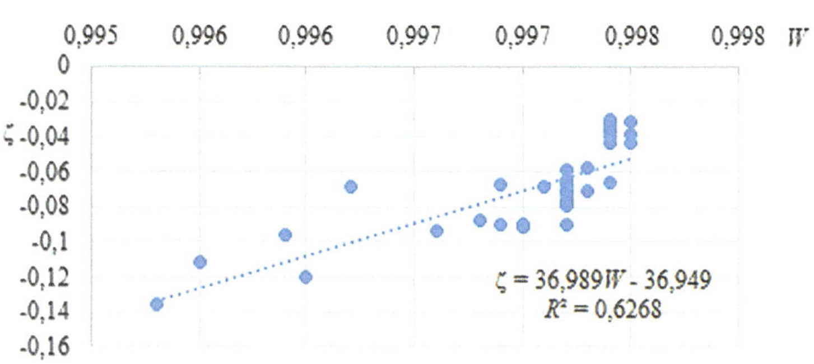

Fig. 8. Dependence of skewness ζ on Shapiro-Wilk criterion *W* with regards to sample size n = 1000, 1050, …, 2600 for data set (surface)

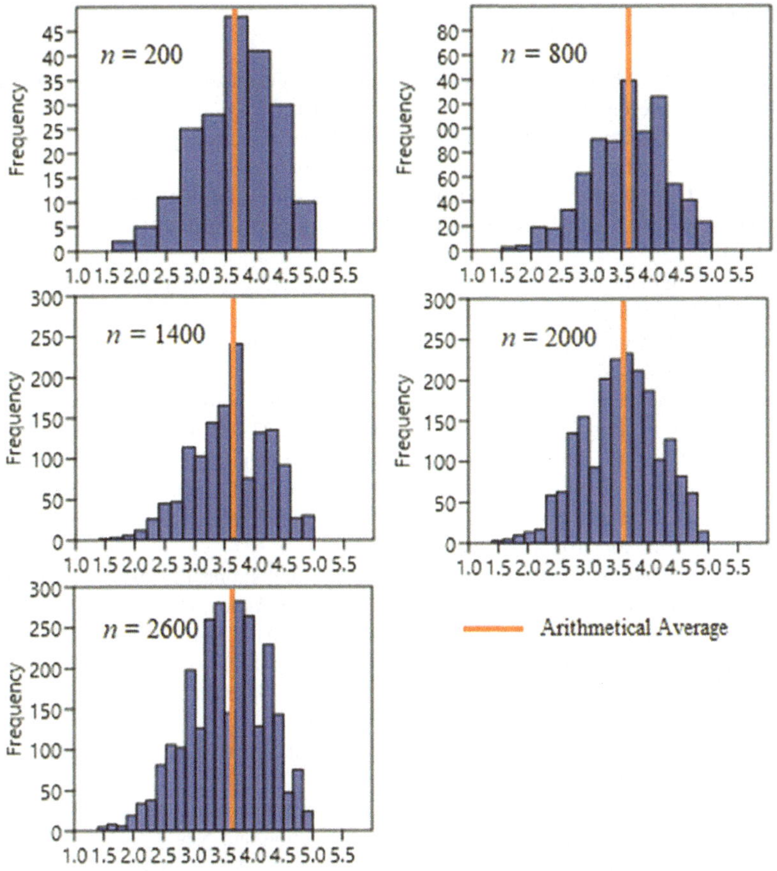

Fig. 9. Occurred frequencies in dependence on sample size for data set (surface)

5 Conclusion

In the contribution, the quantifier of the skewness was explored in favour of the analysis of the property of the normality in case of the applied quantitative research with 2671 respondents. Achieved data was used for purposes of the identification of a behaviour of the normality of data. The skewness was proven in the frame of the regression and correlation analysis in the relation on the Shapiro-Wilk criterion with the achieved linear character. The shapes of histograms of the occurred frequencies of data values expressed the similar trends as the trends of the parameter of the skewness with regards to the sample size. The behaviour of this descriptive parameter of the skewness can guarantee the stable behaviour due to variable sample size in the frame of the considered statistical population from the limit of approximately 1000 of respondents.

References

1. Kitchenham, B., Madeyski, L., Budgen, D., et al.: Robust statistical methods for empirical software engineering. Empir. Softw. Eng., 1–52 (2016)
2. Corrado, M.: Qualitative research methods and evidential reasoning. Philos. Soc. Sci. **49**(5), 385–412 (2019). https://doi.org/10.1177/0048393119862858
3. Gauthier, T.D., Hawley, M.E.: Statistical methods. In: Introduction to Environmental Forensics. 3rd edn., pp. 99–148. Elsevier (2015). https://doi.org/10.1016/b978-0-12-404696-2.00005-9
4. Alizadeh Noughabi, H.: Two powerful tests for normality. Ann. Data Sci. **3**(2), 225–234 (2016). ISSN 2198-5812
5. Stockemer, D.: Quantitative Methods for the Social Sciences. Springer (2019). https://doi.org/10.1007/978-3-319-99118-4
6. Tomsik, R.: Power comparisons of Shapiro-Wilk, Kolmogorov-Smirnov and Jarque-Bera Tests. Scholars J. Res. Math. Comput. Sci. **3**(3), 238–243 (2019). ISSN 2581-3064
7. Muller, C., Kloft, H.: Parameter estimation with the Anderson-Darling test on experiments on glass. Stahlbau **84**(S1), 229–240 (2015). https://doi.org/10.1002/stab.201590081
8. Adhikari, A., Schaffer, J.: Modified lilliefors test. J. Mod. Appl. Stat. Methods **14**(1), 53–69 (2015). https://doi.org/10.22237/jmasm/1430453280
9. Gupta, B.C., Guttman, I., Jayalath, K.P.: Statistics and Probability with Applications for Engineers and Scientists Using MINITAB, R and JMP. Wiley, Hoboken (2020)
10. Kularbphettong, K.: Analysis of students' behavior based on educational data mining. In: 1st Conference on Computational Methods in Systems and Software: Applied Computational Intelligence and Mathematical Methods, pp. 167–172. Springer (2017). https://doi.org/10.1007/978-3-319-67621-0_15, ISBN 978-331967620-3
11. Lackova, L.: Protective factors of university students. New Educ. Rev. **38**(4), 273–286 (2014). ISSN 1732-6729
12. Simbartl, P., Honzikova, J.: Using demonstration to design creative products focused on folk traditions. In: 8th International Conference on Education and New Learning Technologies, pp. 2832–2837. IATED (2016). https://doi.org/10.21125/edulearn.2016.1613
13. Korenova, L.: GeoGebra in teaching of primary school mathematics. Int. J. Technol. Math. Educ. **24**(3), 155–160 (2017)
14. Schoftner, T., Traxler, P., Prieschl, W., Atzwanger, M.: E-learning introduction for students of the first semester in the form of an online seminar. In: Pre-Conference Workshop of the 14th E-Learning Conference for Computer Science, pp. 125–129. CEUR-WS (2016). ISSN 1613-0073
15. Sikorova, Z., Barot, T., Vaclavik, M., Cervenkova, I.: Czech university students' use of study resources in relation to the approaches to learning. New Educ. Rev. **56**(2), 114–123 (2019)
16. Bendickova, K., Tidu, F., De Zuani, M., et al.: Calcineurin inhibitors reduce NFAT-dependent expression of antifungal pentraxin-3 by human monocytes. J. Leukoc. Biol. **107** (3), 497–508 (2020). https://doi.org/10.1002/JLB.4VMA0318-138R
17. Sulovska, K., Belaskova, S., Adamek, M.: Gait patterns for crime fighting: Statistical evaluation. In: Proceedings of SPIE - The International Society for Optical Engineering, vol. 8901. SPIE (2013). ISBN 978-081949770-3, https://doi.org/10.1117/12.2033323
18. Svoboda, Z., Honzikova, L., Janura, M., et al.: Kinematic gait analysis in children with valgus deformity of the hindfoot. Acta Bioeng. Biomech. **16**(3), 89–93 (2014). ISSN 1509-409X

19. Barot, T., Burgsteiner, H., Kolleritsch, W.: Comparison of discrete autocorrelation functions with regards to statistical significance. Advances in Intelligent Systems and Computing. Springer (in print). ISSN 2194-5357
20. Barot, T., Krpec, R., Kubalcik, M.: Applied quadratic programming with principles of statistical paired tests. In: Computational Statistics and Mathematical Modeling Methods in Intelligent Systems, Advances in Intelligent Systems and Computing, vol. 1047, pp. 278–287. Springer (2019). ISBN 978-3-030-31361-6, https://doi.org/10.1007/978-3-030-31362-3_27
21. Mariappan, P.: Skewness, moments, and kurtosis. In: Statistics for Business. Taylor & Francis (2019). https://doi.org/10.1201/9780429443244-6
22. Ruiz Espejo, M., Delgado Pineda, M., Nadarajah, S.: Optimal unbiased estimation of some population central moments. METRON **71**, 39–62 (2013). https://doi.org/10.1007/s40300-013-0006-z
23. Vaclavik, M., Sikorova, Z., Barot, T.: Particular analysis of normality of data in applied quantitative research. In: Computational and Statistical Methods in Intelligent Systems, Advances in Intelligent Systems and Computing, vol. 859, pp. 353–365. Springer (2019). ISBN 978-3-030-00210-7, https://doi.org/10.1007/978-3-030-00211-4_31
24. Hammer, O., Harper, D.A.T., Ryan, P.D.: PAST: paleontological statistics software package for education and data analysis. Palaeontologia Electronica **4**(1) (2001). http://palaeo-electronica.org/2001_1/past/issue1_01.htm

Gradient-Based Algorithm for Parametric Optimization of Variable-Structure PI Controller When Using a Reference Model

V. V. Kulikov, N. N. Kutsyi, and A. A. Podkorytov(✉)

Irkutsk National Research Technical University, Irkutsk, Russia
podkorytovleha@mail.ru

Abstract. For object controlling in an automatic system with a significant delay that reduces the quality of operation of typical controllers (integral, proportional-integral, proportional-integral-differential), a PI controller with variable or switchable parameters is used which is known for its advantages and belongs to the class of variable-structure controllers (hereinafter referred to as VSC) that do not use the sliding mode. Due to the fact that the selected controller contains switchable parameters, filter and an object with delay is considered, it is extremely hard to use the analytical approaches to parametric optimization of the system. This forces one to turn to algorithmic methods. In the present paper, we use the gradient-based algorithm, in which the components of the gradient are calculated using the sensitivity functions with their known advantages. The generated algorithm for automatic parametric optimization (APO) calculated the optimal parameters of the VSC based on the proximity of the output coordinate of the object to the reference process with a minimum of the integral quadratic criterion. The reliability of the vector of parameterization of the controller found by the generated APO algorithm is confirmed by the necessary and sufficient conditions for the extremum of the selected optimization criterion. The positive experience gained in optimizing the PI controller with variable parameters when using the reference model allows it to be applied to other VSCs that do not use the sliding mode, and so further to expand the practice in the application of the gradient-based algorithm through the sensitivity functions for such a VSC class with given indicatives of transition process quality.

Keywords: PI controller · Variable-structure systems · Sensitivity functions · Gradient-based algorithm · Automatic systems with delay · Variable parameters of the controller · Reference model

1 Introduction

The objects of heat energy systems with transport delay (heat exchangers, boiler units) represent an important class [1–5]. With the ratio $\tau_{ob}/T_{obm} > 1$, τ_{ob} — delay value, $T_{obm} = max(T_{ob1}, T_{ob2},\ldots, T_{obn})$ the time constant of the controlled object, serial continuous controllers (integral, proportional-integral, proportional-integral-differential) do not provide the required transition process quality [1, 3, 5]. A common way to control such objects is to use controllers with a delay element in their

R. Silhavy et al. (Eds.): CoMeSySo 2020, AISC 1295, pp. 938–949, 2020.
https://doi.org/10.1007/978-3-030-63319-6_87

structure (for example, a predictive proportional-integral controller) [6–23]. Another way to compensate for the delay of an object is to use variable parameters without sliding mode in the PI controller (hereinafter referred to as VSC) [2]. It is necessary to introduce a deadband in order to avoid switching of VSC structures due to interference in the steady state mode [4, 5]. The presence of switching elements, filter and a deadband in such controllers complicates the application of analytical approaches to solving the parametric optimization problem which consists in calculating the $\boldsymbol{q}$ VSC configuration vector based on the extreme value of the accepted optimality criterion through the reference model.

In this paper, the positive experience [24] of applying a parametric optimization algorithm using a gradient-based procedure following the elements of the theory of sensitivity to calculating configurable parameters of a PI controller with semi-constant integration is applied to the controller under consideration and, thus, the parametric optimization problem for the considered automatic system is solved.

2 Statement of the Problem

The structural diagram of the automatic system under consideration is presented in Fig. 1.

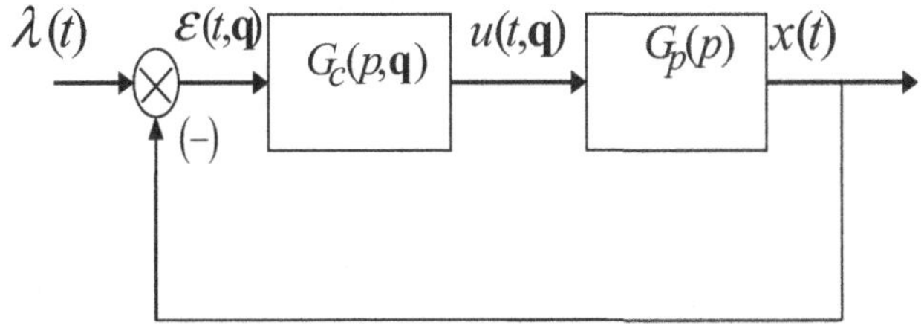

Fig. 1. Structural diagram of the automatic system under research when using the reference model.

Description of processes in the automatic system in Fig. 1 with a specified PI controller with a deadband [4, 5] is presented below:

$$\varepsilon(t,q) = \lambda(t) - x(t)$$

$$u(t,q) = G_c(t,q^i)\varepsilon(t,q), (i = 1,2,3), \tag{1}$$

$$u(t,\boldsymbol{q}) = \begin{cases} u_1(t,\boldsymbol{q}) = \varepsilon(t,\boldsymbol{q})G_c(t,\boldsymbol{q}^1), ((\Psi(t,\varepsilon(t,\boldsymbol{q})) > 0) \vee (0<t<t_{ob})) \wedge (|\varepsilon(t,\boldsymbol{q})| > \alpha\lambda(t)); \\ u_2(t,\boldsymbol{q}) = \varepsilon(t,\boldsymbol{q})G_c(t,\boldsymbol{q}^2), (\Psi(t,\varepsilon(t,\boldsymbol{q}))<0) \wedge (|\varepsilon(t,\boldsymbol{q})| > \alpha\lambda(t)); \\ u_3(t,\boldsymbol{q}) = \varepsilon(t,\boldsymbol{q})G_c(t,\boldsymbol{q}^3), |\varepsilon(t,\boldsymbol{q})| < \alpha\lambda(t); \end{cases}$$

$$x(t) = G_p(p)u(t,q),$$

where $\varepsilon(t,\boldsymbol{q})$ is the error of the control system; $\Psi(t,\varepsilon(t,\boldsymbol{q})) = \varepsilon(t,\boldsymbol{q})\frac{p\varepsilon(t,\boldsymbol{q})}{T_f p+1}$ — switching function; T_f — filter parameter (rationale for choosing a filter parameter is presented below); $\boldsymbol{q} = (\boldsymbol{q}^1(q_1,q_2),\boldsymbol{q}^2(q_3,q_4),\boldsymbol{q}^3(q_5,q_6))$ — vector of the VSC parameters with a specified switching function $\Psi(t,\varepsilon(t,\boldsymbol{q}))q_5 = 0, q_6 = 0$ — non-optimizable parameters; α — deadband of the controller, in the present paper $\alpha = 5\%\lambda(t)$; $\lambda(t)$ — reference-input action; $u(t,\boldsymbol{q})$ — control action; $x(t)$ — output coordinate of the automatic control system (ACS); $G_c(t,\boldsymbol{q}^1) = q_1 + \frac{q_2}{p}, G_c(t,\boldsymbol{q}^2) = q_3 + \frac{q_4}{p}, G_c(t,\boldsymbol{q}^3) = 0 + \frac{0}{p}$ — operator of the typical PI controller with variable parameters; $G_p(p)$ — operator of the controlled object; $p = {}^{d}/_{dt}$ — differentiation operator; $\wedge, \vee$ — logical operations.

The operator of the controlled object $G_p(p)$ is selected in the form with which one can describe the processes of industrial objects of heat energy systems:

$$G_p(p) = \frac{k_{ob}}{(T_{ob1}p+1)(T_{ob2}p+1)}e^{-\tau_{ob}p} \quad (2)$$

where k_{ob} is the static amplifier gain; T_{ob1}, T_{ob2} — time constants; T_{ob} — delay time.

The data from papers devoted to APO algorithms was taken as the basic values of the parameters of the operator of the object (2) [24, 25]:

$$T_{ob1} = 10, T_{ob2} = 40, k_{ob} = 1, \tau_{ob} = 50. \quad (3)$$

Based on the parameters (3), the object (2) has a large delay, since $\tau_{ob}/T_{ob2} > 1$.

The requirements for the transition process (oscillation, response time, maximum deviation, accuracy in the steady state mode) can be implemented by the reference model represented by the oscillating element [27]:

$$G_{et}(p) = \frac{k_{et}}{T_{et}^2p^2 + 2T_{et}\zeta_{et}p + 1}e^{-\tau_{et}p} \quad (4)$$

For a further example of a computational experiment, the following parameters of the reference model (4) were taken: $\tau_{et} = \tau_{ob}, T_{et} = 40, k_{et} = 1, \zeta_{et} = 1,25$.

A graphic illustration of filter parameter selection is presented below T_f.

According to the obtained dependence (based on the weight function of the aperiodic element) of the filter parameter (see Fig. 2), T_f is selected equal to the maximum value $\frac{\tau_{ob}}{5\div6}$, which allows to minimize the phase lag in the switching principle with a

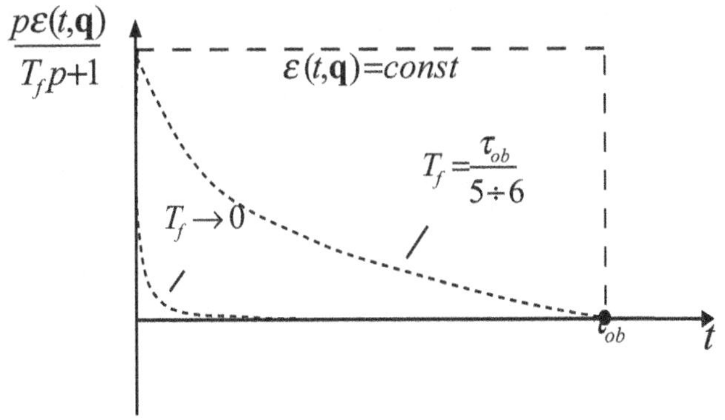

Fig. 2. Dependence of the parameter T_f on the delay time of the object τ_{ob}.

filter in comparison with the switching principle without a filter and not to "jump over" the first switching point t_0.

An optimized integral criterion that determines the degree of proximity of the output coordinate $x(t)$ to the reference process $x_{et}(t)$ is presented below:

$$I = \int_0^\infty (x_{et}(t) - x(t, \boldsymbol{q}))^2 \tag{5}$$

where $x_{et}(t) = G_{et}(p)\lambda(t)$.

3 Optimization Algorithm

The basis of the generated APO algorithm is a gradient-based procedure, and it is necessary to calculate the gradient components of the selected optimization criterion which are obtained using the sensitivity functions in this paper. The sensitivity equations for system (1) with criterion (5) [26] are presented below:

$$\xi_j(t) = G_p(p)\frac{\partial u(t, \boldsymbol{q})}{\partial q_j} - \sum_k \Delta u_{t_k} \frac{\partial t_k}{\partial q_j} G_p(p)\delta(t - t_k) \tag{6}$$

$$(j = 1, 2 \ldots, 4; k = 0, 1, \ldots)$$

where Δu_{t_k} is the magnitude of the control action jump at the time of its rupture $t_k t_k$; $\delta(t - t_k)$ — delta function offset by time t_k.

Configurable parameters $\boldsymbol{q} = (q_1, \ldots, q_4)$ do not depend on the deadband of the VSC, since it is determined by the specified parameter α. It is shown below that $\boldsymbol{q}$ also does not depend on the selected switching function $\Psi(t, \varepsilon(t, \boldsymbol{q}))$ when using the criterion (5). For the accepted condition for switching the controller structures (1), the first moment of rupture t_0 depends on τ_{ob}, which allows to write the following expression:

$$\frac{\partial t_0}{\partial q_j} = 0. \tag{7}$$

The expression for the derivative $\frac{\partial t_k}{\partial q_j}(k) = 1, 2, \ldots$ is determined from the switching condition in the VSC:

$$\Psi(t_k(\boldsymbol{q}), \varepsilon(t_k, \boldsymbol{q}), \boldsymbol{q}) = 0. \tag{8}$$

If we differentiate the condition (8) by q_j, we obtain:

$$\Psi'_{\varepsilon}\left[\dot{\varepsilon}(t_k)\frac{dt_k}{dq_j} + \frac{\partial\varepsilon(t_k)}{\partial q_j}\right] + \Psi'_{t_k}\frac{dt_k}{dq_j} + \Psi'_{q_j} = 0, (j = 1, 2, \ldots, 4). \tag{9}$$

Let us express $\frac{dt_k}{dq_j}$ from Eq. (10) with the replacement of $\frac{\partial\varepsilon(t_k)}{\partial q_j}$ by $\xi_j(t_k)$:

$$\frac{dt_k}{dq_j} = -\frac{\Psi'_{\varepsilon}\xi_j(t_k) + \Psi'_{q_j}}{\Psi'_{\varepsilon}\dot{\varepsilon}(t_k) + \Psi'_{t_k}} \tag{10}$$

Let us present expressions for the calculation of $\Psi'_{\varepsilon}, \Psi'_{t_k}, \Psi'_{q_j}$ in the moments $t_k(k = 1, 2, \ldots)$:

$$\Psi'_{\varepsilon} = (\varepsilon(t_k, \boldsymbol{q})\dot{\varepsilon}(t_k, \boldsymbol{q}))'_{\varepsilon} = \dot{\varepsilon}(t_k, \boldsymbol{q}) + \frac{\partial\dot{\varepsilon}(t_k, \boldsymbol{q})}{\partial\varepsilon}. \tag{11}$$

$$\Psi'_{t_k} = (\varepsilon(t_k, \boldsymbol{q})\dot{\varepsilon}(t_k, \boldsymbol{q}))'_{t_k} = \dot{\varepsilon}(t_k, \boldsymbol{q})\dot{\varepsilon}(t_k, \boldsymbol{q}) + \dot{\varepsilon}(t_k, \boldsymbol{q})\ddot{\varepsilon}(t_k, \boldsymbol{q}). \tag{12}$$

$$\Psi'_{q_j} = (\varepsilon(t_k, \boldsymbol{q})\dot{\varepsilon}(t_k, \boldsymbol{q}))'_{q_j} = \xi_j(t_k)\dot{\varepsilon}(t_k, \boldsymbol{q}) + \varepsilon(t_k, \boldsymbol{q})\frac{\partial\dot{\varepsilon}(t_k, \boldsymbol{q})}{\partial q_j}. \tag{13}$$

According to (1), for the accepted switching condition $\Psi(t_k, \varepsilon(t_k, \boldsymbol{q}))$ in the moments $t_k(k = 1, 2, \ldots)$, the derivative $\dot{\varepsilon}(t_k, \boldsymbol{q}) = 0$, which allows us to rewrite (11)–(13) as follows:

$$\Psi'_{\varepsilon} = 0; \Psi'_{t_k} = 0; \Psi'_{q_j} = 0; \frac{\partial t_k}{\partial q_j} = 0(k = 1, 2, \ldots). \tag{14}$$

Based on the above expressions, we rewrite the Eqs. (5):

$$\xi_j(t) = G_p(p)\frac{\partial u(t,\boldsymbol{q})}{\partial q_j}, (j = 1, 2, \ldots, 4). \tag{15}$$

Let us set the following designations in this paper to compactly represent expressions that define $\frac{\partial u(t,\boldsymbol{q})}{\partial q_j}$:

$$\begin{aligned} \Psi^+ &= ((\Psi(t, \varepsilon(t,q))0) \vee (0tt_{ob})) \wedge (|\varepsilon(t,q)|\alpha\lambda(t)); \\ \Psi^- &= (\Psi(t, \varepsilon(t,q))0) \wedge (|\varepsilon(t,q)|\alpha\lambda(t)); \\ \Psi^0 &= |\varepsilon(t,q)|\alpha\lambda(t). \end{aligned} \tag{16}$$

Due to the limited volume of the article, we present only the expression for the calculation of

$$\frac{\partial u(t,\boldsymbol{q})}{\partial q_1} = \begin{cases} \frac{\partial u_1(t,\boldsymbol{q})}{\partial q_1} = -\xi_1(t)\left(q_1 + \frac{q_2}{p}\right) + \varepsilon(t,\boldsymbol{q}), \Psi^+ ; \\ \frac{\partial u_2(t,\boldsymbol{q})}{\partial q_1} = -\xi_1(t)\left(q_3 + \frac{q_4}{p}\right), \Psi^- ; \\ \frac{\partial u_3(t,\boldsymbol{q})}{\partial q_1} = -\xi_1(t)\left(0 + \frac{0}{p}\right), \Psi^0 . \end{cases}$$

$$\frac{\partial u(t,\boldsymbol{q})}{\partial q_2} \begin{cases} \frac{\partial u_1(t,\boldsymbol{q})}{\partial q_2} = -\xi_2(t)\left(q_1 + \frac{q_2}{p}\right) + \varepsilon(t,\boldsymbol{q}), \Psi^+ ; \\ \frac{\partial u_2(t,\boldsymbol{q})}{\partial q_2} = -\xi_2(t)\left(q_3 + \frac{q_4}{p}\right), \Psi^- ; \\ \frac{\partial u_3(t,\boldsymbol{q})}{\partial q_2} = -\xi_2(t)\left(0 + \frac{0}{p}\right), \Psi^0 . \end{cases} \tag{17}$$

According to the gradient-based algorithm in the optimization process, the vector $\boldsymbol{q}$ of configurable parameters changes in accordance with the expression [27]:

$$\boldsymbol{q}[l] = \boldsymbol{q}[l-1] - \Gamma\nabla_{\boldsymbol{q}} I(x_{et}(t) - x(t, \boldsymbol{q}[l-1])), (l = 1, 2, \ldots), \tag{18}$$

where $\Gamma = \{\gamma_i\}$ is the vector of weighting factors obtained in the process of preliminary research $\Gamma = \{0.1, 0.001, 0.1, 0.001\}$; l — algorithm step number; $\nabla_{\boldsymbol{q}} I(x_{et}(t) - x(t,\boldsymbol{q}))$ — gradient vector (4). Let us present an expression for determining the vector gradient $\nabla_{\boldsymbol{q}} I(x_{et}(t) - x(t,\boldsymbol{q}))$ associated with the calculation of the vector of sensitivity functions (14):

$$\frac{\partial I(x_{et}(t) - x(t,\boldsymbol{q}))}{\partial q_i} = -2\int_0^{\infty} (x_{et}(t) - x(t,\boldsymbol{q}))\xi_i(t)dt, (i = 1, \ldots 4). \tag{19}$$

The problem to be solved belongs to the field of nonlinear programming and implies an infinite number of steps to achieve the result, which requires the use of the process stop condition for the optimization process in order to be implemented on a computer. In this paper, the stop condition is formed according to the following rule

[28] selected based on the simplicity of implementation and the minimum number of parameters: every n steps of the algorithm (18) the minimum of the selected optimization criterion (5) is checked up to two digits in comparison with the previous value of the criterion. Algorithm (18) is satisfied while the condition is true:

$$I_{min}(l \in [1;n]) > I_{min}(l \in [i \bullet n+1;(i+1)n]) > \ldots, (i = 1,2,\ldots), \quad (20)$$

where I_{min} is the minimum value of the criterion (5) for each n steps of the algorithm (18). During the algorithm testing, the reference value $n = 10$ was selected.

4 Checking the Performance of the Optimization Algorithm

For the generated APO algorithm, it is necessary to conduct a research of its performance, which consists in verifying the reliability of the calculated values of the configurable parameters $\boldsymbol{q}^*$ from the point of view of finding the local minimum of the criterion (5).

When starting the APO algorithm from various initial values of the vector of configurable parameters $\boldsymbol{q}_k^0 = \left(q_{1k}^0, q_{2k}^0, q_{3k}^0, q_{4k}^0\right)(k = 1,2\ldots)$, the corresponding total values at the optimal point $\boldsymbol{q}_k^* = \left(q_{1k}^*, q_{2k}^*, q_{3k}^*, q_{4k}^*\right)$ calculated by the APO algorithm must ensure that the necessary condition for the extremum at these points is satisfied:

$$\frac{\partial I(x_{et}(t) - x(t,\boldsymbol{q}^*))}{\partial \boldsymbol{q}} = (0 \pm \Delta), \quad (21)$$

where Δ is a calculation error.

Also at the optimal point, from the position of the minimum of the criterion (4), a sufficient optimality condition must be fulfilled:

$$\frac{\partial^2 I(x_{et}(t) - x(t,\boldsymbol{q}^*))}{\partial q_i \partial q_j} > 0, (i,j = 1,\ldots,4). \quad (22)$$

The basis of the condition (22) is the square Hessian matrix, which consists of the second derivatives of the optimality criterion (4) with respect to the configurable parameters q_i:

$$\frac{\partial^2 I(x_{et}(t) - x(t,\boldsymbol{q}^*))}{\partial q_i \partial q_j} = \begin{bmatrix} \frac{\partial^2 I}{\partial q_1^2} & \frac{\partial^2 I}{\partial q_1 \partial q_2} & \frac{\partial^2 I}{\partial q_1 \partial q_3} & \frac{\partial^2 I}{\partial q_1 \partial q_4} \\ \frac{\partial^2 I}{\partial q_2 \partial q_1} & \frac{\partial^2 I}{\partial q_2^2} & \frac{\partial^2 I}{\partial q_2 \partial q_3} & \frac{\partial^2 I}{\partial q_2 \partial q_4} \\ \frac{\partial^2 I}{\partial q_3 \partial q_1} & \frac{\partial^2 I}{\partial q_3 \partial q_2} & \frac{\partial^2 I}{\partial q_3^2} & \frac{\partial^2 I}{\partial q_3 \partial q_4} \\ \frac{\partial^2 I}{\partial q_4 \partial q_1} & \frac{\partial^2 I}{\partial q_4 \partial q_2} & \frac{\partial^2 I}{\partial q_4 \partial q_3} & \frac{\partial^2 I}{\partial q_4^2} \end{bmatrix} \quad (23)$$

that must be positively defined or its own values must be strictly positive. In order to calculate the elements of the matrix (23), it is necessary to refer to the second-order sensitivity functions; hereinafter, we present the second-order sensitivity equations obtained in this paper for the system (1):

$$\xi_{ij}(t) = G_p(p)\frac{\partial^2 u(t,\boldsymbol{q})}{\partial q_i \partial q_j}, (i,j = 1,\ldots,4), \tag{24}$$

$$\frac{\partial^2 u(t,\boldsymbol{q})}{\partial q_i \partial q_j} = \begin{cases} \frac{\partial^2 u_1(t,\boldsymbol{q})}{\partial q_i \partial q_j} = -\xi_{ij}(t)G_c(p,\boldsymbol{q}^1) - \xi_i(t)\frac{\partial G_c(p,\boldsymbol{q}^1)}{\partial q_j} \\ -\xi_i(t)\frac{\partial G_c(p,\boldsymbol{q}^1)}{\partial q_i} + \varepsilon(t,\boldsymbol{q})\frac{\partial^2 G_c(p,\boldsymbol{q}^1)}{\partial q_i \partial q_j}, \Psi^+; \\ \frac{\partial^2 u_2(t,\boldsymbol{q})}{\partial q_i \partial q_j} = -\xi_{ij}(t)G_c(p,\boldsymbol{q}^2) - \xi_i(t)\frac{\partial G_c(p,\boldsymbol{q}^2)}{\partial q_j} \\ -\xi_i(t)\frac{\partial G_c(p,\boldsymbol{q}^2)}{\partial q_i} + \varepsilon(t,\boldsymbol{q})\frac{\partial^2 G_c(p,\boldsymbol{q}^2)}{\partial q_i \partial q_j}, \Psi^-; \\ \frac{\partial^2 u_3(t,\boldsymbol{q})}{\partial q_i \partial q_j} = 0, \Psi^0 \end{cases}$$

The differentiation $\frac{\partial^2 u(t,\boldsymbol{q})}{\partial q_i \partial q_j}$ occurs similarly to (17) according to the rules of conventional differentiation.

An expression for determining the matrix $\frac{\partial^2 I(\boldsymbol{q})}{\partial q_i \partial q_j}$ is presented below:

$$\frac{\partial I(\varepsilon(t,\boldsymbol{q}^*))}{\partial q_{ij}} = 2\int_0^\infty (\xi_i(t)\xi_j(t) - \xi_{ij}(t)(x_{et}(t) - x(t,\boldsymbol{q})))dt, (i,j = 1,\ldots,4). \tag{25}$$

Having regard to the above in this paper, the performance indicator of the generated APO algorithm is the fulfillment of criteria (21), (22) for $\boldsymbol{q}^*$.

5 Research Results

Let us present the above-described methods for verifying the performance of the APO algorithm on practical examples with a reference-input action $\lambda(t) = 1 \cdot 0.5(t)$. Figure 3 illustrates that the generated APO algorithm of the system with the controller (1) provides the calculation of $\boldsymbol{q}_k^*(k = 1,2,3)$ for different types of starting transition processes in the automatic system with different $\boldsymbol{q}_k^0(k = 1,2,3)$. It is shown in this figure that the resulting transition processes coincide with the transition process of the reference model with a reasonable degree of accuracy.

Figure 4 shows that when starting the APO algorithm from various initial values of the vector of configurable parameters $\boldsymbol{q}_k^0(k = 1,2,3)$, the corresponding total values at the optimal point $\boldsymbol{q}_k^*(k = 1,2,3)$ calculated by the APO algorithm ensure that the necessary condition for the extremum (21) of the criterion (5) at these points is satisfied (Fig. 5).

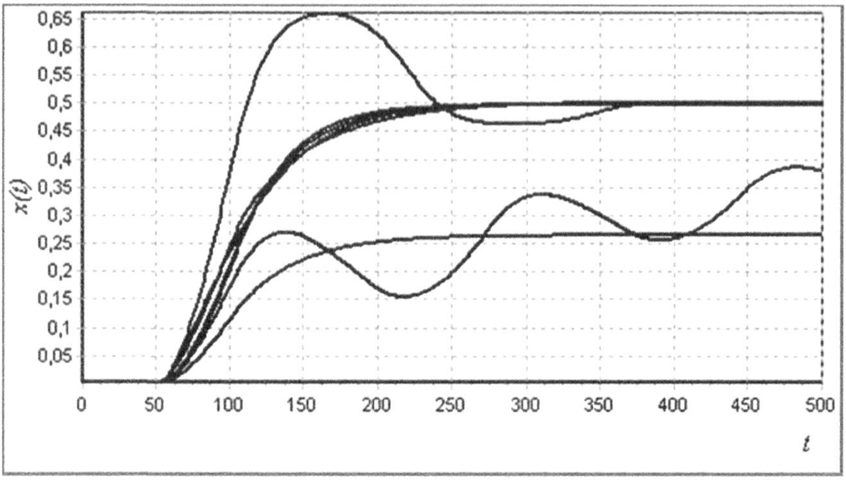

Fig. 3. Graphs of transition processes at the initial (1) and final (2) points of the performance of the APO algorithm and the reference model (3).

Table 1 shows the results of the APO algorithm performance with various $q_k^0(k = 1, 2, 3)$ and presents the Hessian matrix.

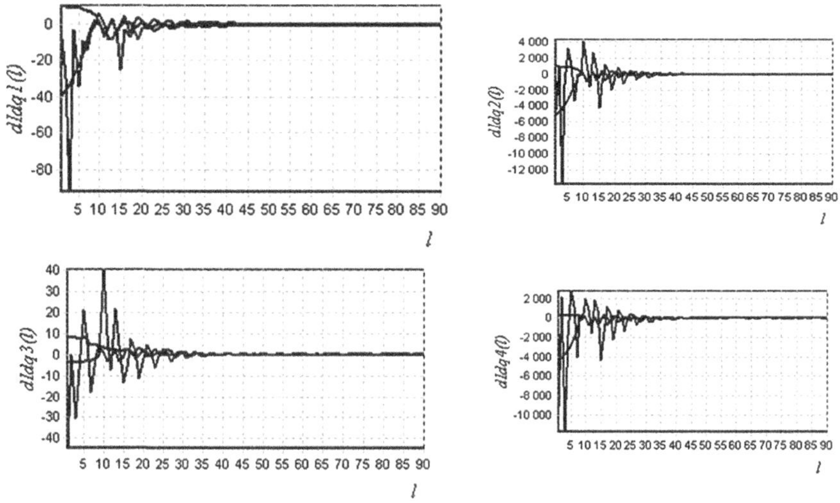

Fig. 4. Values of the gradient components $dI/dq_1(l)$, $dI/dq_2(l)$, $dI/dq_3(l)$, $dI/dq_4(l)$ of the optimization criterion (5) during the APO algorithm performance with object parameters (3).

For the parameters of the object (3), according to Table 1, the Hessian matrix is positively determined, which confirms that the optimal parameters of the controller (1) are determined by the generated APO algorithm based on the minimum of the criterion (5).

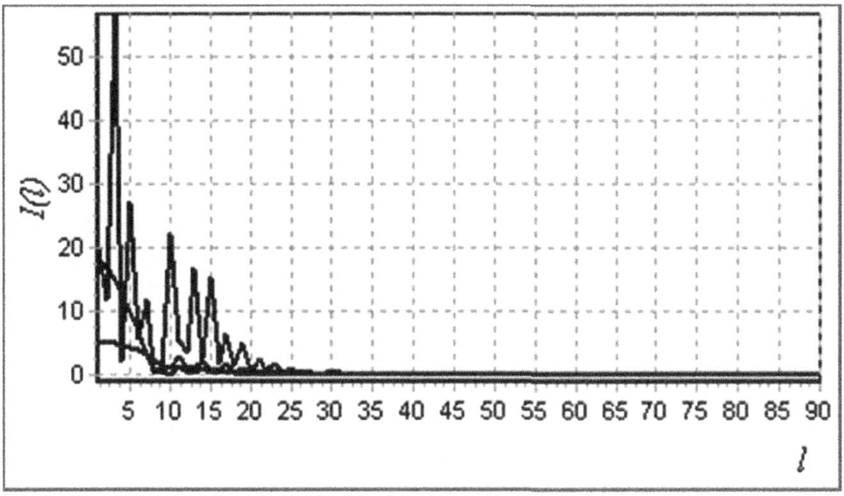

Fig. 5. Values of the optimization criterion I (5) during the APO algorithm performance with object parameters (3).

Table 1. Results of the APO algorithm performance with object parameters (3).

k	q_1^0	q_2^0	q_3^0	q_4^0	q_1^*	q_2^*	q_3^*	q_4^*	$I(\mathbf{q}^0)$	$I(\mathbf{q}*)$	Hessian matrix for the criterion (5)
1	0.1	0.01	0	0	0.1013	0.0204	0.0003	−0.0014	20.1	0.08	12; $9.4 \cdot 10^2$; 6.9; $3.9 \cdot 10^2$; $9.4 \cdot 10^2$; $2 \cdot 10^2$; $1.5 \cdot 10^3$; $1.4 \cdot 10^5$; 6.9; $1,5 \cdot 10^3$; 13; $9.2 \cdot 10^2$; $3.9 \cdot 10^2$; $1.4 \cdot 10^5$; $9.2 \cdot 10^5$; $1,1 \cdot 10^5$;
2	0.15	0.015	−0.01	-0.01	0.1504	0.0196	−0.0102	-0.0006	18.0	0.05	12; $8 \cdot 10^2$; 6; $3 \cdot 10^2$; $8 \cdot 10^2$; $1.9 \cdot 10^5$; $1.4 \cdot 10^3$; $1.4 \cdot 10^5$; 6; $1,4 \cdot 10^3$; 12; $8.8 \cdot 10^2$; $3 \cdot 10^2$; $1.4 \cdot 10^5$; $8.8 \cdot 10^2$; $1.2 \cdot 10^5$;
3	0.3	0.034	−0.015	−0.0125	0.2995	0.0179	−0.0154	0.0011	5.2	0.02	11; $5.7 \cdot 10^2$; 4.9; 73; $5.7 \cdot 10^2$; $1.7 \cdot 10^5$; $1.2 \cdot 10^3$; $1.2 \cdot 10^5$; 4.9; $1.2 \cdot 10^3$; 11; $7.4 \cdot 10^2$; 73; $1.2 \cdot 10^5$; $7.4 \cdot 10^2$; $1.1 \cdot 10^5$;

6 Conclusion

In this paper, using the generated gradient-based algorithm, we solved the problem of parametric optimization of a PI controller with variable parameters for an object with a large delay for given requirements for the transition process using error filtering in the switching principle. The performance of the generated APO algorithm is confirmed by calculating the Hessian matrix for the criterion (5).

The reported study was funded by RFBR, project number 19–31-90083\19

References

1. Guretskiy, H.: Analysis and Synthesis of Control Systems with Delay. Mashinostroenie, Moscow (1974)
2. Kulakov, G.T.: Theory of Automatic Control of Heat Energy Processes. Vysheishaya Shkola, Minsk, Belarus (2017)
3. Denisenko, V.V.: Computer control of the technological process, experiment, equipment. Goryachaya liniya – Telecom, Moscow, Russia (2009)
4. Govorov, A.A.: Methods and means of controller formation with advanced functionality for continuous technological processes: dis. of Doctor of Engineering: 05.13.06: defended on 15.11.2002, Moscow, Russia, p. 499 (2002)
5. Shygin, E.K.: Automatic control of an object with a pure delay by a controller with switchable parameters II. Autom. Telemech. **6**, 72–81 (1966)
6. Åström, K.J., Hägglund, T.: The future of PID control. Control Eng. Pract. **9**(11), 1163–1175 (2001)
7. Ramírez, A., Mondié, S., Garrido, R.: Proportional integral retarded control of second order linear systems. In: IEEE 52nd Annual Conference on Decision and Control (CDC), pp. 2239–2244 (2013)
8. Ramírez, A., Garrido, R., Mondié, S.: Integral retarded control velocity control of DC servomotors. In: IFAC TDS Workshop, Grenoble, France, pp. 558–563 (2013)
9. Kharitonov, V.L., Niculescu, S.I., Moreno, J., Michiels, W.: Static output feedback stabilization: necessary conditions for multiple delay controllers. IEEE Trans. Autom. Control **50**, 82–86 (2005)
10. Niculescu, S.I., Michiels, W.: Stabilizing a chain of integrators using multiple delays. IEEE Trans. Autom. Control **49**, 802–807 (2004)
11. Abdallah, C., Dorato, P., Benites-Read, J.: Delayed positive feedback can stabilize oscillatory systems. In: Proceedings American Control Conference, San Francisco, USA, pp. 3106–3107 (1993)
12. Suh, I.H., Bien, Z.: Proportional minus delay controller. IEEE Trans. Autom. Control **24**, 370–372 (1979)
13. Villafuerte, R., Mondié, S., Garrido, R.: Tuning of proportional retarded controllers: theory and experiments. IEEE Trans. Control Syst. Technol. **21**(3), 983–990 (2013)
14. Ramírez, A., Mondié, S., Garrido, R.: Integral retarded velocity control of dc servomotors. In: 11th IFAC Workshop on Time-Delay Systems, vol. 46, no. 3, pp. 558–563 (2013)
15. Ramírez, A., Mondié, S., Garrido, R.: Velocity control of servo systems using an integral retarded algorithm. ISA Trans. **58**, 357–366 (2015)
16. Arousi, F., Schmitz, U., Bars, R., Haber, R.: PI controller based on first-order dead time model. In: Proceedings of the 17th World Congress, Seoul, Korea, pp. 5808–5813 (2008)
17. Airikka, P.: Extended predictive proportional-integral controller for typical industrial processes. In: 18th IFAC World Cogress, Milano, Italy, pp. 7571–7576 (2011).
18. Larsson, P., Hagglund, T.: Comparison between robust PID and predictive PI controllers with constrained control signal noise sensitivity. In: 2nd IFAC Control Conference on Advances on PID Control, Brescia, Italy, pp. 175–180 (2012).
19. Airikka, P.: Another novel modification of predictive PI controller for processes with long dead time. In: IFAC Conference on Advances in PID Control, Brescia, Italy, p. 5 (2012)

20. Airikka, P.: Stability analysis of a predictive PI controller. In: 21st Mediterranean Conference on Control and Automation, Chania, Crete, Greece, pp. 1380–1385 (2013)
21. Shinde, D., Hamde, S., Waghmare, L.: Nonlinear predictive proportional integral controller for multiple time constants. Int. J. Innov. Res. Adv. Eng. **1**(7), 53–58 (2014)
22. Airikka, P.: Robust predictive PI controller tuning. In: 19th World Congress, IFAC, Cape Town, South Africa, pp. 9301–9306 (2014)
23. Prakash, G., Alamelumangai, V.: Design of predictive fractional order PI controller for the quadruple tank. Process. WSEAS Trans. Syst. **6**(11), 258–265 (2017)
24. Kulikov, V.V., Kutsyi, N.N.: Searchless algorithm for parametric optimization of a PI controller with semi-constant integration. Bull. Nat. Res. Irkutsk State Tech. Univ. **22**(6), 98–108 (2018)
25. Osipova, E.A.: Automatic parametric optimization of control systems with integrated pulse-width modulation: dis. of Ph.D. in Engineering Science: 05.13.06: defended on 21.02.2013. Irkutsk, p. 170 (2013)
26. Gorodetskiy, V.I., Zakharin, F.M., Rozenwasser, E.N., Yusupov, R.M.: Methods of the Sensitivity Theory in Automatic Control. Energoizdat, Moscow (1971)
27. Kostyuk, V.I., Shyrokov, L.A.: Automatic Parametric Optimization of Control Systems. Energoizdat, Moscow (1981)
28. Boyarinov, A.I., Kafarov, V.V.: Optimization Methods in Chemical Technology. Khimiya, Moscow (1969)

Author Index

R. Silhavy et al. (Eds.): CoMeSySo 2020, AISC 1295, pp. 951–954, 2020.
https://doi.org/10.1007/978-3-030-63319-6

The manufacturer's authorised representative in the EU is Springer Nature Customer Service Centre GmbH, Europaplatz 3, 69115 Heidelberg, Germany. If you have any concerns regarding our products, please contact ProductSafety@springernature.com

Printed and bound by CPI Group (UK) Ltd, Croydon, CR0 4YY

30/09/2026

02225441-0002